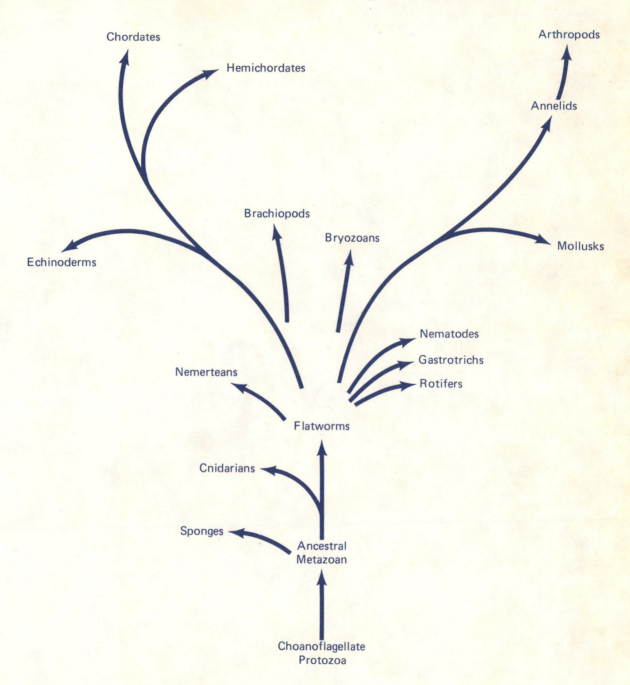

Invertebrate Zoology

FIFTH EDITION

ROBERT D. BARNES

Gettysburg College, Pennsylvania

SAUNDERS COLLEGE PUBLISHING

Philadelphia   New York   Chicago   San Francisco   Montreal   Toronto
London   Sydney   Tokyo   Mexico City   Rio de Janeiro   Madrid

Address orders to:
383 Madison Avenue
New York, NY 10017

Address editorial correspondence to:
210 West Washington Square
Philadelphia, PA 19105

QL 362
B27
1987

Text Typeface: Trump Medieval
Compositor: University Graphics, Inc.
Acquisitions Editor: Edward F. Murphy
Project Editor: Martha Hicks-Courant
Copy Editor: Sally Atwater
Art Director: Carol C. Bleistine
Text Design: Adrianne Onderdonk Dudden
Cover Design: Lawrence R. Didona
Layout Artist: Dorothy Chattin
New Text Artwork: J&R Technical Services
Chapter Opening Illustrations: Curtis Ether
Production Manager: Tim Frelick
Assistant Production Manager: JoAnn Melody

Cover credit: Blue bottle jellyfish © Animals Animals/Keith Gilletti

Library of Congress Cataloging-in-Publication Data

Barnes, Robert D.
 Invertebrate zoology.

 Includes bibliographies and index.

 1. Invertebrates. I. Title.
QL362.B27 1986   592   86-10023
ISBN 0-03-008914-X

INVERTEBRATE ZOOLOGY   ISBN 0-03-008914-X

Library of Congress catalog card number 86-10023

7890   032   98765432

CBS COLLEGE PUBLISHING
Saunders College Publishing
Holt, Rinehart and Winston
The Dryden Press

*To Cazlyn G. Bookhout, who introduced me
to the wonderful world of invertebrate animals.*

Preface

An enormous amount of new information on invertebrate animals appears in the research literature each year, and with some justification each edition of *Invertebrate Zoology* has grown in size. But the last edition, the fourth, was felt by many to be undesirably large. This edition has therefore been reduced by approximately 170 pages. The reduction was made by removing some detail from the text and eliminating redundant and less useful illustrations. However, the contraction was not made at the expense of new material, which was added before the final trimming, nor at the expense of the level of depth that has characterized this textbook since its inception.

A new feature of this edition is short, boxed essays on topics such as bioluminescence, invertebrate eyes, and trochophore larvae that are applicable to more than one group of invertebrates. These essays appear in the first place in the text in which they are relevant, and they are referred to elsewhere in the text.

Various invertebrate groups have been moved about in this edition to reflect new ideas on phylogeny. Thus annelids now precede the mollusks, onychophorans have been returned to the chapter on lesser protostomes, and priapulids have rejoined the aschelminths. I have always attempted to use a system of classification that is up to date but generally accepted. In some cases I have let pedagogy be the determining factor. For example, it is simpler to discuss the cubomedusae with the scyphozoans than as a separate class, and protobranchs, lamellibranchs, and septibranchs are still very useful reference groups for bivalves. Nevertheless, the student is made aware of current thinking on the classification of these animals.

I continue to believe that this text should provide the student entry to the vast literature on invertebrates. The references at the end of each chapter therefore include most books and review articles on invertebrates, as well as many research papers. Some of these would be appropriate for further student reading if the instructor wished to so designate them. The listings for the larger chapters have been subdivided to make them easier to use. The glossary, which continues to be restricted to recurring terms, has been expanded and combined with the index.

I am very grateful to the many persons who over the past seven years have provided me with corrections, information, and suggestions. They have contributed much to the improvements of this edition. I wish especially to acknowledge those persons who reviewed chapters or sections on particular invertebrate groups or topics. They are not responsible for any errors that may remain. These reviewers were Lawrence G. Abele, Florida State University (decapod crustaceans); Thomas W. Cronin, University of Maryland, Baltimore County (invertebrate eyes); Virginia R. Ferris, Purdue University (nematodes); Bernard Fried, Lafayette College (parasites); Elizabeth H. Gladfelter, West Indies Laboratory of Fairleigh Dickinson University (cnidarians); William G. Gladfelter, West Indies Laboratory of Fairleigh Dickinson University (cnidarians); Richard L. Hoffman, Radford College (myriapods); Saburo Ishii, Fukushima Medical College of Japan (protonephridia); Clarence D. Johnson, Northern Arizona University (insects); Mimi Koehl, University of California at Berkeley (crustacean suspension feeding); Reinhardt M. Kristensen, Copenhagen University (aschelminths other than nematodes and rotifers); Michael F. Land, University of Sussex (invertebrate eyes); Herbert W. Levi, Harvard University (arachnids); David R. Lindberg, University of California at Berkeley (gastropod mollusks); Brian M. Marcotte, McGill University (trilobites and nonmalacostracan crustaceans); David L. Meyer, University of Cincinnati (echinoderms); Brian Morton, University of Hong Kong (bivalve mollusks); Clyde F. E. Roper, Smithsonian Institution (cephalopod mollusks); Julian Smith, University of Maine (turbellarian flatworms); Richard D. Stemberger, Dartmouth College (rotifers); and Seth Tyler, University of Maine (turbellarian flatworms).

General reviews of this edition as a textbook were provided by Robert C. Bullock, University of Rhode Island; Robert W. Mead, University of Nevada, Reno; Edward K. Mercer, California State Polytechnic University; and Naida Zucker, New Mexico State University.

The burden of the various revision tasks was lightened by the help of the following persons, for which I am most appreciative: David Barnes, Michael Locher, Grace Myers, Timothy Stapleford, Thomas Stapleford, Denise Telford.

I wish to thank especially Sally Atwater and Martha Hicks-Courant of Saunders College Publishing for their help in bringing the manuscript to its final form.

Contents

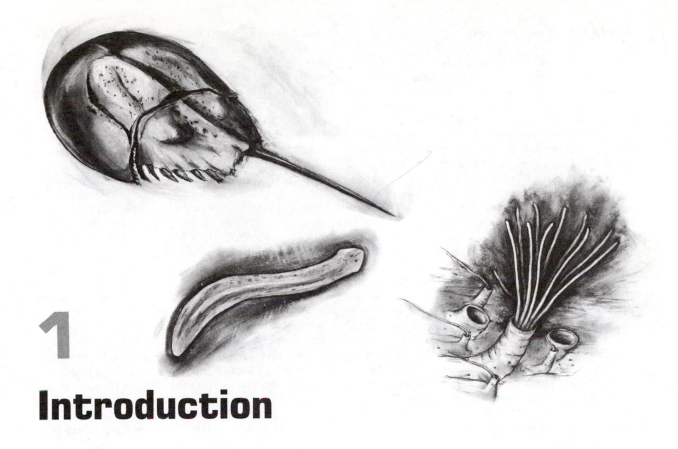

1

Introduction

There are over a million described species of animals. Of this number about 5 per cent possess a backbone and are known as vertebrates (figure inside front cover). All others, constituting the greater part of the Animal Kingdom, are invertebrates. These animals are the subject of this book.

Division of the Animal Kingdom into vertebrates and invertebrates is artificial and reflects a historical human bias in favor of man's own relatives. One characteristic of a single subphylum of animals is used as the basis for separation of the entire Animal Kingdom into two groups. One could just as logically divide animals into mollusks and nonmollusks or arthropods and nonarthropods. The latter classification could be supported at least from the standpoint of numbers, since approximately 85 per cent of all animals are arthropods (figure inside front cover).

The artificiality of the invertebrate concept is especially apparent when one considers the vast and heterogeneous assemblage of groups that are lumped together in this category. There is not a single positive characteristic that invertebrates hold in common, aside from general animal features also shared with vertebrates. The range in size, in structural diversity, and in adaptations to different modes of existence is enormous. Some invertebrates have common phylogenetic origins; others are only remotely related. Some are much more closely related to the vertebrates than to other invertebrate groups.

Quite obviously, invertebrate zoology cannot be considered a special field of zoology, certainly not in the same sense as protozoology or entomology. A field that embraces all biological aspects—morphology, physiology, embryology, and ecology—of 95 per cent of the Animal Kingdom represents no distinct area of zoology itself. For the same reason, no zoologist can truly be called an invertebrate zoologist. He or she is a protozoologist, a malacologist, or an acarologist, or studies some aspect of physiology, embryology, or ecology of one or more animal groups. Beyond such limited areas the number and diversity of invertebrates are too great to permit much more than a good general knowledge of the major groups.

The Marine Environment

The Animal Kingdom is generally believed to have originated in Archeozoic oceans long before the first fossil record. Every major phylum of animals has at least some marine representatives; some groups, such as cnidarians and echinoderms, are largely or entirely marine. From the ancestral marine environment, different groups of animals have invaded fresh water; some have moved onto land.

Compared with fresh water and with land, the marine environment is relatively uniform. Oxygen is generally available, and the salinity of the open ocean is relatively constant, ranging from 34 to 36 parts per thousand (3.4–3.6 per cent), depending upon the latitude. Light and temperature vary greatly, however, largely as a consequence of depth. Thus, life is not uniformly distributed throughout the depth and breadth of the world's oceans, which cover approximately 71 per cent of the earth's surface. The margins of the continents extend seaward in the form of underwater shelves to depths of 150 to 200 meters and then slope more steeply to depths of 3000 meters or more. Before reaching the ocean floor, the continental slope is interrupted by a terrace or more gradual incline, formed by the continental rise (Fig. 1–1). The floor of the ocean basins, called the abyssal plain, ranges from 3000 to 5000 meters in depth and may be marked by such features as sea mounts, ridges, and trenches. Widths of the different continental shelves vary considerably. The edge of the western Atlantic shelf is some 75 miles from the shore, but along the Pacific coast of North America the continental shelf is very narrow.

Waters over the continental shelves constitute the *neritic zone*, and those beyond the shelf make up the *oceanic zone* (Fig. 1–1). The edge of the sea, which rises and falls with the tide, is the *intertidal (littoral) zone*. The region above is the *supratidal (supralittoral)*, and that below is the *subtidal (sublittoral)*. The continental slopes form the *bathyal zone*, the abyssal plains form the *abyssal zone*, and the trenches form the *hadal zone*.

Vertical distribution of marine organisms is largely controlled by the depths of light penetration. Light sufficient for photosynthesis to exceed respiration penetrates to only a short distance below the surface or to depths as great as 200 meters, depending upon the turbidity of the water. Below this upper *euphotic zone* is a transition zone, where some photosynthesis can occur but the production rate is less than the loss through respiration. From the transition zone down to the ocean floor, total darkness prevails. This region constitutes the *aphotic zone*. The animals that are per-

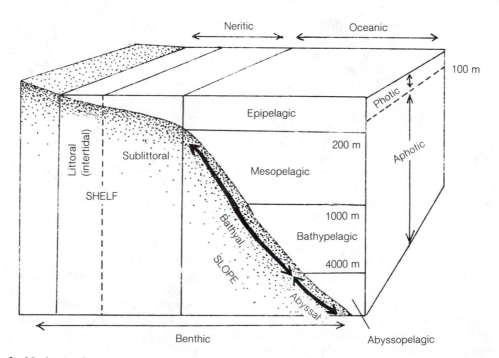

Figure 1–1 Marine environments.

manent inhabitants of the aphotic and transition zones are carnivorous, suspension, or detritus feeders and depend ultimately on the photosynthetic activity of the microscopic algae in the upper, lighted regions.

The suspended or swimming animals of ocean waters constitute the *pelagic fauna,* and those that live on the bottom compose the *benthic fauna.* Bottom dwellers may live on the surface *(epifauna)* or beneath the surface *(infauna)* of the ocean floor and usually strikingly reflect the character of the substratum, that is, whether it is a hard bottom of coral or rock, or a soft bottom of sand or mud. Many animals are adapted for living in the spaces between sand grains and compose what is commonly referred to as the *interstitial fauna* or *meiofauna.* This group includes representatives of virtually every major phylum of animals, and a number of previously unknown groups of animals have been discovered here in recent years. Pelagic and benthic animals are found in all of the horizontal zones. For example, one can refer to neritic pelagic animals or to the infauna of the abyssal zone.

Many intertidal and shallow-water habitats contain a rich diversity of species. They also contain the best-studied invertebrates, since these habitats are the most accessible to investigators. Surf beaches and rocky shorelines are certainly the most familiar shallow-water marine habitats, but there are others that are rich areas for collecting and study. In protected bays and sounds, there are usually shallow expanses of sand and mud that are exposed at low tide but covered at high water. Many marine animals live on such sand or mud flats, usually buried but often with evidence of occupancy, such as burrow openings, on the surface above.

Sea grass beds of subtidal shallow water are also habitats for many invertebrates. Sea grasses, such as eel grass, turtle grass, manatee grass, and others of both temperate and tropical seas, are flowering plants adapted for a life of total submersion in sea water. The long leaves are flat or stringlike, and the plants are anchored to the bottom by roots.

Freshwater and Estuarine Environments

The lakes of the world also exhibit a horizontal and vertical zonation, but their smaller size, shallower depth, and freshwater content make them ecologically different in many ways from oceans.

Temperature is a primary factor controlling the environment of lakes. In contrast to salt water, which becomes increasingly dense at decreasing temperatures, fresh water reaches its greatest density at 4°C; thus, when lakes in temperate parts of the world are warmed during spring and summer, the warm water stays at the surface while the heavier, colder water remains at the bottom. Little circulation occurs between the upper and lower levels, so that not only is the bottom zone dark, but it is also relatively stagnant from lack of oxygen and supports only a limited fauna. A general circulation occurs in fall and spring with temperature changes.

Tropical lakes either have a single winter turnover or exhibit a highly stable condition, with little vertical circulation.

The junction of freshwater rivers and streams with the sea is not abrupt. Rather, the two environments grade into one another, creating the estuarine environment, characterized by brackish water, i.e., salinities considerably below the 3.5 per cent typical of the open sea. The estuarine environment embraces river mouths and surrounding deltas, coastal marshes, small embayments, and the finger-like extensions of the sea that probe the coast or margins of sounds. It is usually affected by tides, from which the word *estuary (aestus,* tide) is derived. The majority of marine animals are osmoconformers and stenohaline and cannot survive greatly reduced salinities. The lower and fluctuating salinities of estuaries thus restrict the estuarine fauna to those euryhaline marine invaders and few freshwater species that can tolerate these conditions. The fauna also contains some animals that have become especially adapted for estuarine conditions and are found nowhere else.

In temperate regions, a characteristic estuarine community is the salt marsh, composed principally of various grasses and sedges. Salt marshes differ from sea grass beds in being intertidal and emergent. Only the lower half of the plant is covered at high tide. Along the east coast of the United States, cord grass *(Spartina)* forms great expanses of salt marsh where the salinity is not too low.

In the tropics, the ecological counterpart of salt marshes is mangroves. Mangroves are species of small trees that can tolerate saline conditions. They occupy the intertidal zone and commonly possess prop roots or special aerial roots (pneumatophores) that project above the water's surface. The most highly developed mangrove communities are found in the Indo-Pacific, where numerous species form a number of zones extending seaward. Such mangroves may occupy vast coastal areas and are virtually impenetrable. Red mangrove, *Rhizophora mangle,* which possesses long prop roots extending straight downward from the limbs, is the common

Figure 1–2 A mangrove at low tide. This is red mangrove, *Rhizophora mangle.* Note the bolsters of algae surrounding the prop roots. (By Betty M. Barnes.)

mangrove of tropical America (Fig. 1–2). Mangroves trap sediment and contribute to land building. They create a habitat that is occupied by many animals and other plants.

Plankton, Primary Production, and Food Chains

Both oceans and freshwater lakes contain a large assemblage of microscopic organisms that are free swimming or suspended in the water. These organisms constitute the plankton and include both plants (phytoplankton) and animals (zooplankton). Although many planktonic organisms are capable of locomotion, they are too small to move independently of currents. Phytoplankton is composed of enormous numbers of diatoms and other microscopic algae.

Marine zooplankton includes representatives from virtually every group of animals, either as adults or as developmental stages. Some species (holoplankton) spend their entire lives in the plankton; the larvae of others (meroplankton) enter and leave the plankton at different points in the course of their development. The animal constituents of freshwater plankton are more limited in number. Plankton, especially marine plankton, is of primary importance in the aquatic food chain. The photosynthetic phytoplankton—chiefly diatoms, dinoflagellates, and minute flagellates—

form the primary trophic level and serve as food for larger animals. As would be expected, plankton attains its greatest density in the upper, lighted zone of waters with high nutrient levels (nitrates, phosphates, and so on). The inorganic nutrients are necessary in the synthesis of organic compounds by phytoplankton. In general, higher nutrient levels are found in shallow coastal waters, in areas of upwelling, and in the surface waters of cold and temperate seas, where mixing with deeper levels is not impeded.

Tropical and subtropical oceanic surface waters are generally impoverished because mixing with nutrient-rich deeper water is minimal. The surface water, which is warm and thus less dense, rides on top of the colder and therefore heavier water of deeper levels. Oceanic waters that have a low productivity, such as the Gulf Stream and the Sargasso Sea, are clear and blue. The low concentration of plankton allows light to penetrate to a considerable depth, and the blue wavelengths are reflected from the water molecules. Sea water that is rich in plankton is green or gray. Plankton and organic detritus reflect yellow wavelengths which, combined with the blue wavelengths reflected by the water molecules, produce a green or gray color.

Fecal particles and fragments of dead plant and animal remains, derived from plankton or benthic plants and animals, settle to the bottom and become mixed with mineral particles (sand). This deposited material is a food source for many animals.

Some of these *deposit feeders* digest the organic matter itself; others feed on bacteria and other microorganisms that the deposit material supports.

Animal Diversity

For the student who is making the first attempt to study invertebrates in some depth, the task may seem overwhelming. Each group has certain structural peculiarities, a special anatomical terminology, and a distinct classification. All of these factors tend to magnify the differences between groups and to obscure functional and structural similarities that result from similar modes of existence and similar environmental conditions, as well as mask the homologies arising from close evolutionary relationships. It will be helpful to keep in mind a few basic biological principles. All animals must solve the same problems of existence—procurement of food and oxygen, maintenance of water and salt balance, removal of metabolic wastes, and perpetuation of the species. The body structure necessary to meet these problems is, in large part, correlated with three factors: the type of environment in which the animal lives, the size of the animal, and the mode of existence of the animal.

Of the three major environments—salt water, fresh water, and land—the marine environment is generally the most stable. Wave action, tides, and vertical and horizontal ocean currents produce a continual mixing of sea water and ensure a medium in which the concentration of dissolved gases and salts fluctuates relatively little. The buoyancy of sea water reduces the problem of support. It is therefore not surprising that the largest invertebrates have always been marine. Since sea water is more or less isotonic to the body tissue fluids of most invertebrate marine animals, there is little dif-

ficulty in maintaining water and salt balance, and most marine animals are osmoconformers.

The buoyancy and uniformity of sea water provide an ideal medium for animal reproduction. Eggs in sea water can be shed and fertilized and can undergo development as floating embryos, with little danger of desiccation and salt imbalance or of being swept away by rapid currents into less favorable environments. Larvae are characteristic of many marine animals. Larval stages provide a means of wide dispersal of the species, and feeding larvae can obtain food material for completion of their development without the necessity for large amounts of yolk material within the egg (Fig. 1–3). Also, some selection of substratum is possible at the time of settling.

The principal disadvantage of early planktonic development and larval stages is the high mortality rate from predation and other factors. Thus in most marine animals the eggs are deposited within coverings of various sorts and are attached to the bottom. Here early development occurs, and hatching takes place at the larval stage (Fig. 1–3). If the larva feeds and has a relatively long larval life, it is said to be planktotrophic. If hatching is further delayed and the larva has a brief life with yolk providing for nutrition, it is said to be lecithotrophic. In many marine animals, larval stages are completely suppressed and development is direct; i.e., the young have the adult form on hatching (Fig. 1–3). In shallow tropical waters, about 75 per cent of the benthic marine fauna have a larval stage. The percentage declines toward the poles, and in polar regions, where the brief period of primary production would perhaps restrict larval feeding, development is generally direct.

Fresh water is a much less constant medium than sea water. Streams vary greatly in turbidity, velocity, and volume, not only along their course but also from time to time as a result of droughts

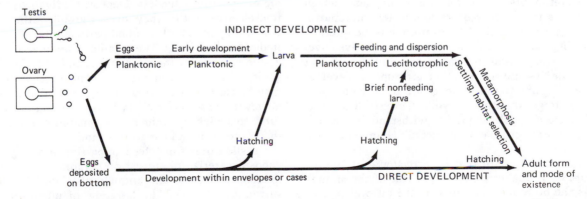

Figure 1–3 Patterns of development in animals.

or heavy rains. Small ponds and lakes fluctuate in oxygen content, turbidity, and water volume. In large lakes the environment changes radically with increasing depth.

Like salt water, fresh water is buoyant and aids in support. The low salt concentration, however, creates some difficulty in maintaining water and salt balance. Since the body of the animal contains a higher salt concentration than that of the external environment, there is a tendency for water to diffuse inward. The animal thus has the problem of getting rid of excess water. As a consequence, freshwater animals usually have some mechanism for pumping water out of the body while holding onto the salts; i.e., they osmoregulate. In general, the eggs of freshwater animals are either retained by the parent or attached to the bottom of the stream or lake, rather than being free floating, as is often true of marine animals. Larval stages are usually absent. Floating eggs and free-swimming larvae are too easily swept away by currents. Since development is usually direct, the eggs typically contain considerable amounts of yolk.

Nitrogenous waste of aquatic animals, both marine and freshwater, is usually excreted as ammonia. Ammonia is very soluble and toxic and requires considerable water for its removal, but since there is no danger of water loss in aquatic animals, the excretion of ammonia presents no difficulty.

Terrestrial animals live in the harshest environment. The supporting buoyancy of water is absent. Most critical, however, is the problem of water loss by evaporation. The solution of this problem has been a primary factor in the evolution of many adaptations for life on land. The integument of terrestrial animals presents a better barrier between the inner and the outer environment than that of aquatic animals. Respiratory surfaces, which must be moist, have developed in the interior of the body, reducing desiccation. Nitrogenous wastes are commonly excreted as urea or uric acid, which are less toxic and require less water for removal than does ammonia. Fertilization must be internal; the eggs are usually enclosed in a protective envelope or deposited in a moist environment. Except for insects and a few other arthropods, development is direct, and the eggs are usually endowed with large amounts of yolk. Terrestrial animals that are not well adapted to withstand desiccation are either nocturnal or restricted to humid or moist habitats.

A second factor that is correlated with the nature of animal structure is the size of the animal. As the body increases in size, the ratio of surface area to volume decreases, for volume increases by the cube, and surface area by the square, of body length. In small animals the surface area is sufficiently great in comparison with the body volume that exchange of gases and waste can be carried out efficiently by diffusion through the general body surface. Also, internal transport can take place by diffusion alone.

However, as the body increases in size, distances become too great for internal transport to take place by diffusion alone, and more efficient transportation mechanisms are necessary. In larger animals this has led to the development of coelomic and blood-vascular circulatory systems. Also, through folding and coiling, the surface area of internal organs and portions of the external surface may be increased to facilitate secretion, absorption, gas exchange, and other processes.

A third factor that is related to the nature of animal structure is the mode of existence. Free-moving animals are generally bilaterally symmetrical. The nervous system and sense organs are then concentrated at the anterior end of the body, since this is the part that first comes in contact with the environment. Such an animal is said to be cephalized.

Attached, or sessile, animals are radially symmetrical or tend toward a radial symmetry, in which the entire body (as in sea anemones) or some part of the body (as in barnacles) consists of a central axis around which similar parts are symmetrically arranged. Radial symmetry is an advantage for sessile existence, since it allows the animal to meet the challenges of its environment from all directions. Skeletons, envelopes, or tubes are commonly present for support or for protection against motile predators.

The feeding mechanisms of animals are usually correlated with their modes of existence. Animals that can actively swim or crawl are frequently raptorial. The slower movers may be scavengers or herbivores. Those that inhabit bottom sediments are often deposit feeders. Deposit feeders may be nonselective, in that they ingest the substratum in which they live, both mineral particles and deposited organic material. The organic component is digested away from the sand grains and other inorganic substances, and the latter are egested and may be ejected as castings. Many deposit feeders, however, are selective. They possess various structures that pick up light organic material, leaving the heavier mineral particles behind.

Sessile animals may feed on passing prey, ingest deposit material, or subsist on organic detritus or microscopic plants and animals suspended in the surrounding water. The last type of nutrition is called suspension feeding. Some suspension-feeding animals utilize suspended material that simply adheres to the general body surface. The collected particles are then transported by ciliary currents to the mouth. More often the animal is adapted for filtering a water current for suspended particles. The water current may be produced by the animal or may be a natural current. The term *suspension feeding* will generally be used in the ensuing chapters only when filtering is not employed; otherwise the animal will be called a filter feeder. It should be kept in mind, however, that the distinction is not always clear.

Phylogeny

In the subsequent chapters, the evolutionary histories of the various phyla are explored. Their evolution is frequently used as a basis for understanding the adaptive diversity within the phylum or class. In discussing phylogenetic (evolutionary) relationships, it is convenient to use such terms as *primitive, lower, higher,* and *specialized.* Unfortunately, these terms tend to create the erroneous impression that evolution has proceeded from one group to another toward some ideal state or goal of perfection. Such terms as *primitive* are relative and are significant primarily in discussing evolution within a particular group of animals. For example, *primitive species* are those that possess many, or conspicuous, characteristics believed to have been possessed by the ancestral stock from which the living members of the group arose. *Specialized* usually refers to characteristics of species that are especially adapted to a particular ecological niche. The term *specialized,* however, should not be thought of as meaning *superior* or *better,* because the changes usually result from invasions of new environmental conditions or assumption of different habits and life styles. Moreover, although certain species possess some primitive characteristics, these same species frequently are specialized in other respects.

The term *primitive* can be particularly misleading when comparing different phyla because usually only one or a few characteristics are being referred to. For example, since multicellular animals evolved from single-celled forms, the protozoa are, in respect to this characteristic, primitive compared with multicellular phyla. However, in other ways protozoa are not necessarily primitive when compared with metazoans, for they have undergone at the unicellular level a great evolutionary development, leading to an intracellular specialization unequaled by the cells of metazoan animals.

The terms *higher* and *lower* usually refer to the levels at which species or groups have stemmed from certain main lines of evolution. Thus, sponges and cnidarians are often referred to as *lower* phyla, since they are believed to have originated near the base of the Animal Kingdom phylogenetic tree. This does not imply that sponges and cnidarians are primitive in all respects, for they, like all other groups of animals, have followed certain independent lines of specialization. Furthermore, this does not necessarily imply that higher groups have evolved *directly* through sponges and cnidarians.

The terms *primary* and *secondary* are frequently used to distinguish a primitive condition from one that is similar but specialized. For example, flatworms probably evolved from an ancestor that lacked an anus. Thus, the lack of an anus in flatworms is primitive, or primary. Brittle stars also lack an anus, but the absence represents a loss of this opening from ancestors that possessed it and is therefore specialized or secondary.

No treatment of invertebrates is complete, nor can there be a proper understanding of invertebrate phylogeny, without some consideration of fossil forms and extinct groups. As much invertebrate paleontology has been included as space permits. From time to time, reference will be made to geological eras and periods. For the student who has only a slight background in geology, the geological timetable inside the back cover may be of some value.

References

The references at the end of each chapter are not intended to be a selection of titles recommended for further reading. The literature on invertebrates is enormous, as one would expect, considering the vast area of biology it covers. Most of this literature consists of research papers scattered through a great number of biological journals published throughout the world over the last 100 years. The references in this text may be placed in two categories. One category comprises the literature cited in the text, largely more recent papers, and the other category consists of larger reference works on general biology or the systematics of the group of invertebrates with which the chapter deals. Those reference citations whose content is not clearly indicated by their titles have been provided brief annotations. The papers, review articles, and books listed at the end of each chapter and the references that they provide in turn will lead the student to much of the literature available for any group of invertebrates.

REFERENCES

These references are composed of general works on invertebrates. Specific volumes especially relevant to certain groups are cited again in later chapters.

Multivolume Works Covering Invertebrate Groups

Bronn, H. G. (Ed.), 1866–   : Klassen und Ordnungen des Tierreichs. C. F. Winter, Leipzig and Heidelberg. (Many volumes; series still incomplete.)

Grassé, P. (Ed.), 1948–   : Traité de Zoologie. Masson et Cie, Paris. (Covers the entire Animal Kingdom; still incomplete.)

Hyman, L. H., 1940–1967: The Invertebrates. Six volumes. McGraw-Hill Book Co., N.Y. (Volumes on annelids and arthropods were never completed.)

Kaestner, A., 1967–1970. Invertebrate Zoology. Three volumes. Interscience Publishers, N.Y. (Completed, although lophophorates and echinoderms are not included.)

Moore, R. C. (Ed.), 1952–   : Treatise on Invertebrate Paleontology. Geological Society of America and University of Kansas Press. (A detailed treatment of fossil invertebrates in many volumes.)

Parker, S. P. (Ed.), 1982: Synopsis and Classification of Living Organisms. McGraw-Hill Book Co., N.Y. Vol. 1, 1166 pp. Vol. 2, 1236 pp. (A brief description of the families and all higher taxa of living organisms, including some information on the biology of the group. A very useful reference.)

Works on Morphology, Physiology, or Ecology of Invertebrates

Alexander, R. M., 1982: Locomotion of Animals. Chapman and Hall, London. 192 pp.

Autrum, H. (Ed.), 1980: Handbook of Sensory Physiology. Eight volumes. Springer-Verlag, Berlin.

Beklemishev, V. N., 1969: Principles of Comparative Anatomy of Invertebrates. Two volumes. University of Chicago Press, Chicago.

Bereiter-Hahn, J., Matoltsy, A. G., and Richards, K. S. (Eds.), 1984: Biology of the Integument. Springer-Verlag, Berlin. Vol. 1, Invertebrates. 841 pp. (The integument of each phylum of invertebrates is covered separately.)

Boardman, R. S., Cheetham, A. H., and Rowell, A. J., 1986: Fossil Invertebrates. Blackwell Scientific Publ., Boston.

Brusca, G. J., 1975: General Patterns of Invertebrate Development. Mad River Press, Eureka, Ca. 134 pp.

Bücherl, W., and Buckley, E. E. (Eds.), 1971. Venomous Animals and Their Venoms. Academic Press, N.Y. Vol. 3, Venomous Invertebrates. 560 pp.

Bullock, T. H., and Horridge, G. A., 1965: Structure and Function of the Nervous System of Invertebrates. Two volumes. W. H. Freeman, San Francisco.

Chia, F., and Rice, M. E. (Eds.), 1978: Settlement and Metamorphosis of Marine Invertebrate Larvae. Elsevier North Holland. 290 pp.

Conway Morris, S., George, J. D., Gibson, R., and Platt, H. M. (Eds.), 1985: The Origins and Relationships of Lower Invertebrates. Systematics Association Spec. Vol. No. 28. Clarendon Press, Oxford. 394 pp.

Corning, W. C., Dyal, J. A., and Willows, A. O. D. (Eds.), 1973: Invertebrate Learning. Plenum Press, N.Y. Vol. 1, 296 pp. Vol. 2, 284 pp.

Crawford, C. C., 1981: Biology of Desert Invertebrates. Springer-Verlag, Berlin. 314 pp.

Daiber, F. C., 1982: Animals of the Tidal Marsh. Van Nostrand Reinhold, N.Y. 432 pp.

Elder, H. Y., and Trueman, E. R. (Eds.), 1980: Aspects of Animal Movement. Cambridge University Press, N.Y. 250 pp.

Eltringham, S. K., 1971: Life in Mud and Sand. Crane, Russak, and Co., N.Y. 218 pp. (An ecology of marine mud and sand habitats.)

Florkin, M., and Scheer, B. T. (Eds.), 1967–1978: Chemical Zoology. Ten volumes. Academic Press, N.Y. (A compilation of articles covering various aspects of biochemistry and physiology of animal groups.)

Fretter, V., and Graham, A., 1976: A Functional Anatomy of Invertebrates. Academic Press, London. 600 pp.

Giese, A. C., and Pearse, J. S., 1974–1979: Reproduction of Marine Invertebrates. Five volumes. Academic Press, N.Y.

Habermehl, G. G., 1981: Venomous Animals and Their Toxins. Springer-Verlag, Berlin. 195 pp.

Halstead, B. W., 1978: Poisonous and Venomous Marine Animals of the World. Revised Edition. Darwin Press, Princeton, N.J. 1043 pp.

Harrison, F. W., and Cowden, R. R. (Eds.), 1982: Developmental Biology of Freshwater Invertebrates. Alan R. Liss, N.Y.

Highnam, K. C., and Hill, L., 1977: The Comparative Endocrinology of the Invertebrates. 2nd Edition. University Park Press, Baltimore. 357 pp.

House, M. R. (Ed.), 1979: The Origin of Major Invertebrate Groups. Systematics Association Spec. Vol. No. 12. Academic Press, London. 515 pp.

Kennedy, G. Y., 1979: Pigments of marine invertebrates. Adv. Mar. Biol., 16: 309–381.

Kerfoot, W. C. (Ed.), 1980: Evolution and Ecology of Zooplankton Communities. University Press of New England, Hanover, N.H. 794 pp.

Kumé, M., and Dan, K., 1968: Invertebrate Embryology. Clearinghouse for Federal Scientific and Technical Information, Springfield, Va.

Little, C., 1984: The Colonization of Land: Origins and Adaptations of Terrestrial Animals. Cambridge University Press, N.Y. 480 pp.

MacGinitie, G. E., and MacGinitie, N., 1968: Natural History of Marine Animals. 2nd Edition. McGraw-Hill Book Co., N.Y. 523 pp.

Maloiy, G. M. O. (Ed.), 1979: Comparative Physiology of Osmoregulation in Animals. Vols. 1 and 2. Academic Press, London.

Mann, T., 1984: Spermatophores: Development, Structure, Biochemical Attributes and Role in the Transfer of Spermatozoa. Springer-Verlag, Berlin.

Mill, P. J., 1972: Respiration in the Invertebrates. St. Martin's Press, London. 212 pp.

Newell, R. C., 1979: Biology of Intertidal Animals. 3rd Edition. Marine Ecological Surveys Ltd., Faversham, Kent, U.K. 560 pp.

Nicol, J. A. C., 1969: Biology of Marine Animals. 2nd Edition. John Wiley and Sons, N.Y. 707 pp.

Prosser, C. L. (Ed.), 1973: Comparative Animal Physiology. 3rd Edition. W. B. Saunders Co., Philadelphia. 966 pp.

Rankin, J. C., and Davenport, J. A., 1981: Animal Osmoregulation. John Wiley and Sons, N.Y. 202 pp.

Russell, F. E., 1984: Marine toxins and venomous and poisonous marine plants and animals. Adv. Mar. Biol., *21*:60–233.

Schaller, F., 1968: Soil Animals. University of Michigan Press, Ann Arbor. 114 pp.

Schmidt-Nielsen, K., 1979: Animal Physiology: Adaptation and Environment. 2nd Edition. Cambridge University Press, Cambridge. 560 pp.

Smith, D. C., and Tiffon, Y. (Eds.), 1980: Nutrition in the Lower Metazoa. Pergamon Press, Elmsford, N.Y. 188 pp. (Proceedings of a symposium.)

Stephenson, T. A., and Stephenson, A., 1972: Life Between Tidemarks on Rocky Shores. W. H. Freeman, San Francisco. 425 pp. (Ecology of the intertidal zone of rocky shores. Systematic coverage of specific regions of the world.)

Trueman, E. R., 1975: The Locomotion of Soft-Bodied Animals. American Elsevier Publishing Co., N.Y. 200 pp.

Vernberg, F. J., and Vernberg, W. B. (Eds.), 1981: Functional Adapations of Marine Organisms. Academic Press, N.Y. 347 pp.

Vernberg, W. B., and Vernberg, F. J., 1972: Environmental Physiology of Marine Animals. Springer-Verlag, N.Y. 346 pp.

Welsch, U., and Storch, V., 1976: Comparative Animal Cytology and Histology. University of Washington Press, Seattle. 343 pp.

General References for Collecting and Identification

Brusca, G. J., and Brusca, R. C., 1978: A Naturalist's Seashore Guide: Common Marine Life of the Northern California Coast and Adjacent Shores. Mad River Press, Eureka, Ca. 215 pp.

Brusca, R. C., 1980: Common Intertidal Invertebrates of the Gulf of California. 2nd Edition. University of Arizona Press, Tucson. 513 pp.

Campbell, A. C., 1976: The Hamlyn Guide to the Seashore and Shallow Seas of Britain and Europe. The Hamlyn Publishing Group, London. 320 pp.

Colin, P. L., 1978: Caribbean Reef Invertebrates and Plants. T.F.H. Publishers, Neptune City, N.J. 478 pp.

Fotheringham, N., and Brunenmeister, S. L., 1975: Common Marine Invertebrates of the Northwestern Gulf Coast. Gulf Publishing Co., Houston.

Gosner, K. L., 1979: A Field Guide to the Atlantic Seashore. The Peterson Field Guide Series. Houghton Mifflin Co., Boston. 329 pp. (Covers an area from the Bay of Fundy to Cape Hatteras.)

Hurlbert, S. H., and Villalobos-Figueroa, A. (Eds.), 1982: Aquatic Biota of Mexico, Central America and the West Indies. Aquatic Biota–SDSU Foundation, San Diego State University, San Diego. 529 pp.

Kaplan, E. H., 1982: A Field Guide to Coral Reefs of the Caribbean and Florida. Peterson Field Guide Series. Houghton Mifflin Co., Boston. 289 pp.

Kozloff, E. N., 1983: Seashore Life of the Northern Pacific Coast: An Illustrated Guide to Northern California, Oregon, Washington and British Columbia. Revised Edition. University of Washington Press, Seattle. 370 pp.

Lincoln, R. J., and Sheals, J. G., 1979: Invertebrate Animals: Collection and Preservation. British Museum, Cambridge University Press, Cambridge. 150 pp.

Meinkoth, N. A., 1981: The Audubon Society Field Guide to North American Seashore Creatures. Alfred Knopf, N.Y. 799 pp.

Morris, R. H., Abbott, D. P., and Haderlie, E. C., 1980: Intertidal Invertebrates of California. Stanford University Press, Palo Alto, Ca. 690 pp. (This impressive work not only provides a guide to the intertidal invertebrates of California but also summarizes the information known about them and gives references to the literature.)

Morton, J., and Miller, M., 1973: The New Zealand Sea Shore. 2nd Edition. Collins, London. 653 pp.

Newell, G. E., and Newell, R. C., 1973: Marine Plankton, a Practical Guide. Hutchinson Educational Ltd., London. 244 pp.

Pennak, R. W., 1978: Freshwater Invertebrates of the United States. 2nd Edition. John Wiley and Sons, N.Y. 803 pp.

Smith, D. L., 1977: A Guide to Marine Coastal Plankton and Marine Invertebrate Larvae. Kendall/Hunt Publishing Co., Dubuque, Iowa. 161 pp.

Smith, R. I. (Ed.), 1964: Keys to Marine Invertebrates of the Woods Hole Region. Contribution No. 11, Systematics-Ecology Program, Mar. Biol. Lab., Woods Hole, Mass.

Smith, R. I., and Carlton, J. T., 1975: Light's Manual: Intertidal Invertebrates of the Central California Coast. 3rd Edition. University of California Press, Berkeley. 716 pp.

Sterrer, W. E. (Ed.), 1986: Marine Fauna and Flora of Bermuda. Wiley-Interscience, N.Y. (This work is also a valuable reference for Florida and the Caribbean.)

Voss, G. L., 1976: Seashore Life of Florida and the Caribbean. E. A. Seemann Publishing Co., Miami. (A guide to common marine invertebrates other than mollusks.)

Wickstead, J. H., 1965: An Introduction to the Study of Tropical Plankton. Hutchinson and Co., London. 160 pp.

Zinn, D. J., 1985: Handbook for Beach Strollers from Maine to Cape Hatteras. Globe Pequot Press, Chester, Conn.

Laboratory Guides

Dales, R. P., 1981: Practical Invertebrate Zoology. 2nd Edition. John Wiley and Sons, N.Y. 356 pp.

Freeman, W. H., and Bracegirdle, B., 1971: An Atlas of Invertebrate Structure. Heinemann Educational Books, London, 129 pp.

Sherman, I. W., and Sherman, V. G., 1976: The Invertebrates: Function and Form. 2nd Edition. Macmillan Co., N.Y. 334 pp.

Welsh, J. H., Smith, R. I., and Kammer, A. E., 1968: Laboratory Exercises in Invertebrate Physiology. 3rd Edition. Burgess Publishing Co., Minneapolis. 219 pp.

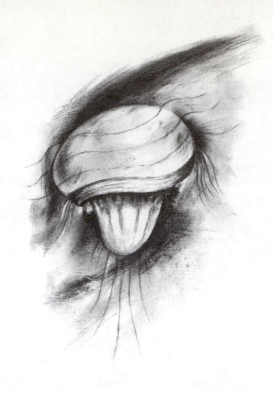

2

The Protozoa

The protozoa are a heterogeneous assemblage of some 50,000 single-cell organisms possessing typical (eukaryote) membrane-bound cellular organelles. Because most are motile and many are heterotrophic, this assemblage was treated in the past as a single phylum within the Animal Kingdom—the phylum Protozoa. They are now known to consist of a number of different unicellular phyla, which together with most algal phyla are placed in the Kingdom Protista. Some of these protozoan groups are related to each other, some probably evolved independently from remote eukaryote ancestors, and some are members of various algal groups.

The unicellular level of organization is the only characteristic by which the protozoa as a whole can be described; in all other respects they display extreme diversity. Protozoa exhibit all types of symmetry, a great range of structural complexity, and adaptations for all types of environmental conditions. As organisms, the protozoa have remained at the unicellular level but have evolved along numerous lines through the specialization of parts of the protoplasm (organelles) or of the skeletal structure. Thus, simplicity and complexity in protozoa are reflected in the number and nature of their organelles and skeletons in the same way that simplicity and complexity in multicellular animals can be reflected in the development of tissues and organ systems. A protozoan cell may be far more complex than a metazoan cell, but a protozoan cell is an entire organism, not part of an organism, as is a metazoan cell.

Protozoa occur wherever moisture is present—in the sea, in all types of fresh water, and in the soil. There are commensal, mutualistic, and many parasitic species.

Although most protozoa occur as solitary individuals, there are numerous colonial forms. Some colonial forms, such as species of *Volvox*, attain such a degree of cellular interdependence that they approach a true multicellular level of structure (Fig. 2–6). Both solitary and colonial species may be either free moving or sessile.

This chapter is included to accommodate those invertebrate zoology courses that survey the protozoa. The coverage, however, is not as detailed as that provided for the free-living metazoan groups, and references to the literature have been largely restricted to secondary sources.

Protozoan Organelles and General Physiology

The protozoan body is usually bounded only by the cell membrane, which possesses the typical bilayered lipid ultrastructure of cells in general. The rigidity or flexibility of the protozoan body is largely dependent on the nature of the underlying cortical cytoplasm, called ectoplasm, which is rather gelatinous, in contrast to the more fluid, internal cytoplasm called endoplasm. Nonliving external coverings or shells occur in many different groups. Such coverings may be simple gelatinous or cellulose envelopes, or they may be distinct shells, composed of various inorganic and organic materials, or sometimes foreign particles cemented together.

Depending on the species, there are one to many nuclei. The locomotor organelles may be flagella, cilia, or flowing extensions of the body called pseudopodia. Since the type of locomotor organelle is important in the classification of the phylum, discussion of the structure of these organelles is deferred until later.

All types of nutrition occur in protozoa. Some are autotrophic or saprozoic; many ingest food particles or prey and digest this food intracellularly within food vacuoles. Food reaches the vacuole by engulfment, or phagocytosis, often through a mouth, or cytostome. Soluble food may enter by pinocytosis. Intracellular digestion has been most studied in amebas and ciliates. The food vacuoles undergo definite changes in hydrogen ion concentration (pH) and in size during the course of digestion. Following ingestion, the vacuole contents become increasingly acid and smaller, as excess water is removed. Lysosomes deliver hydrolytic enzymes (Fig. 2-1), and the vacuole increases in size and becomes alkaline. The enzymes digest the vacuole contents, and products of digestion then pass into the cytoplasm by pinocytosis. The undigestible remnants are egested.

Protozoa that live in water where there is active decomposition of organic matter or in the digestive tract of other animals can exist with little or no oxygen present. Some protozoa are facultative anaerobes, utilizing oxygen when present but also capable of anaerobic respiration. Changing availability of food supply and of oxygen associated with decay typically results in a distinct succession of populations and protozoan species (see Bick, 1973).

Metabolic wastes diffuse to the outside of the organism. Ammonia is the principal nitrogenous waste, and the amount eliminated varies directly with the amount of protein consumed.

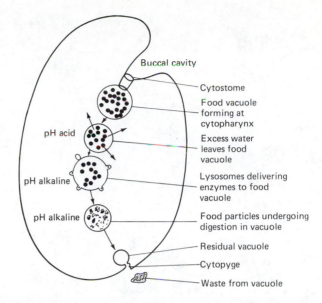

Figure 2-1 Formation of food vacuoles and digestion in a ciliated protozoon, such as *Tetrahymena*.

Characteristic of many protozoa is an organelle system called the contractile vacuole complex (Fig. 2-2). The complex is composed of a spherical vesicle—the contractile vacuole proper—and a surrounding system of small vesicles or tubules termed the spongiome. The complex functions primarily in water balance (osmoregulation), pumping excess water out of the organism. The spongiome provides for the collection of water, which is delivered to the contractile vacuole. The latter expels the fluid to the outside of the organism through a temporary or permanent pore. In some protozoa (some amebas and flagellates) the vacuole completely disappears following contraction and is reformed by fusion of small vesicles. In others (many ciliates) the vacuole collapses at discharge and is refilled by fluid from the surrounding tubules of the spongiome.

Reproduction and Life Cycles

The protozoan reproductive processes and life cycles are varied. Only a few of the more common terms are described here.

Asexual reproduction occurs in most protozoa and is the only known mode of reproduction in some species. Division of the animal into two or more daughter cells is called fission. When this process results in two similar daughter cells, it is termed binary fission; when one daughter cell is

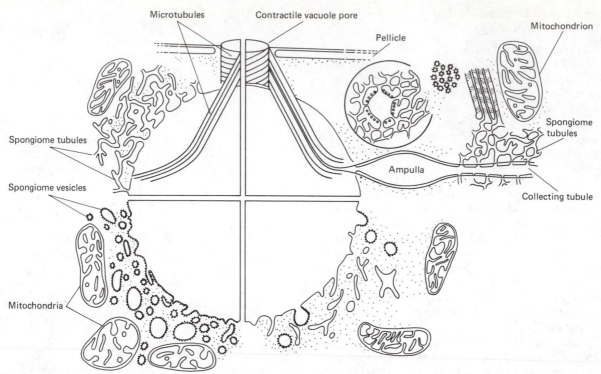

Microtubules Contractile vacuole pore Pellicle Mitochondrion

Spongiome tubules Spongiome tubules

Spongiome vesicles Ampulla

Collecting tubule

Mitochondria

Figure 2–2 Diagram of the four types of contractile vacuoles found in protozoa. Types A and B are both characteristic of ciliates, where the spongiome is composed of irregular tubules and bands of microtubules support the pore and extend over the vacuole surface. In A, the network of spongiome tubules empties directly into the vacuole; in B, the network of irregular tubules of the spongiome first empties into straight collecting tubules, which may dilate as ampullae before discharging into the vacuole. In type C, typical of amebas, the spongiome contains small vesicles, which empty into the vacuole. In type D, typical of many flagellates and small amebas, the spongiome contains both vesicles and tubules. (After Patterson, D. J., 1980: Contractile vacuoles and associated structures: their organization and function. Biol. Rev., 55:1–46. Copyright Cambridge University Press; reprinted by permission.)

much smaller than the other, the process is called budding. In some protozoa, multiple fission, or schizogony, is the rule. In schizogony, after a varying number of nuclear divisions, the cell divides into a number of daughter cells. With few exceptions, asexual reproduction involves some replication of missing organelles following fission.

Sexual reproduction may involve fusion (syngamy) of identical gametes (called isogametes) or gametes that differ in size and structure. The latter, called anisogametes, range from types that differ only slightly in size to well-differentiated sperm and eggs. Meiosis commonly occurs in the formation of gametes, but in many flagellate protozoa and sporozoans meiosis is postzygotic, that is, it occurs following the formation of the zygote as in most algae (Fig. 2–3). In ciliate protozoa there is no formation of distinct gametes; instead, two animals adhere together in a process called conjuga-

tion, and they exchange nuclei. Each migrating nucleus fuses with a stationary nucleus in the opposite conjugant to form a zygote nucleus (synkaryon). Less common is a process called autogamy, in which two nuclei, each representing a gamete, fuse to form a zygote, all within a single individual.

Encystment is characteristic of the life cycle of many protozoa, including the majority of freshwater species. In forming a cyst, the protozoon secretes a thickened envelope about itself and becomes inactive. Depending on the species, the protective cyst is resistant to desiccation or low temperatures, and encystment enables the animal to pass through unfavorable environmental conditions. The simplest life cycle includes only two phases: an active phase and a protective, encysted phase. However, the more complex life cycles are often characterized by encysted zygotes or by for-

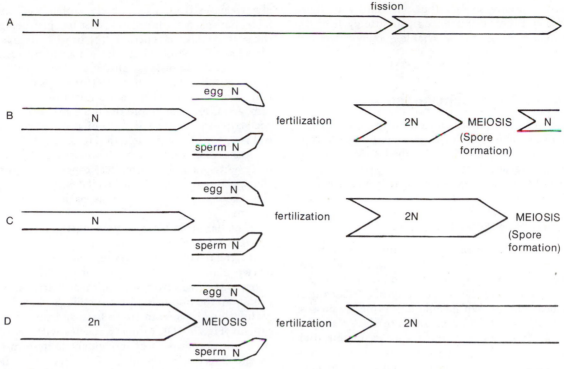

Figure 2–3 Types of life cycles in protozoa, showing the occurrence of meiosis and fertilization. *A*, Individuals are haploid, and reproduction is entirely by fission. This is illustrated by the flagellate *Trypanosoma*. *B*, Individuals are haploid for most of their life cycle. Meiosis occurs shortly after fertilization in the formation of spores. From each spore a new haploid individual arises. This is illustrated by some flagellates and algae and by sporozoans. *C*, This stage is the same as in *B*, but the diploid phase is extended so that there are haploid and diploid generations and individuals. This is illustrated by foraminiferans, some algae, and most higher plants. *D*, Individuals are diploid, and meiosis occurs in the formation of gametes; the diploid number is restored with fertilization. This stage is characteristic of animals and various protozoa, such as ciliates. (Based on a figure by Sleigh.)

mation of special reproductive cysts, in which fission, gametogenesis, or other reproductive processes take place.

Protozoa may be dispersed long distances in either the motile or encysted stages. Water currents, wind, and mud and debris on the bodies of water birds and other animals are common agents of dispersion.

SUMMARY

1 Protozoa are unicellular organisms that are animal-like in being heterotrophic and motile. In all other respects, protozoa are a very diverse assemblage, and the major groups are now commonly treated as separate phyla of eukaryote protistans. The old phylum name *Protozoa* can be used as a convenient term of reference for any member of these phyla.

2 Most protozoa inhabit the sea or fresh water, but there are many parasitic, commensal, and mutualistic species.

3 In contrast to metazoans, complexity in protozoa has proceeded through development and specialization of organelles or skeletal structures. Although a protozoon is a single cell, it is also a complete organism.

4 Digestion occurs intracellularly within a food vacuole, and food reaches the vacuole through a cell mouth or by engulfment.

5 Excess water is usually eliminated by a contractile vacuole.

6 Most of the members of the protozoan phyla are distinguished, in part, by their type of locomotor organelles: flagella, pseudopodia, or cilia.

7 Reproduction by fission occurs at some time in the life history of almost all protozoa. Meiosis, gamete formation, and fertilization have been observed in many species, but the nature of these events and their occurrence in the life cycle of the organism is highly variable. Encystment is common.

Phylum Sarcomastigophora

These protozoa possess flagella or pseudopodia as locomotor or feeding organelles and a single type of nucleus. The 18,000 described species constitute the largest phylum of protozoa, which is composed of two principal groups (subphyla), the flagellates and the sarcodines.

SUBPHYLUM MASTIGOPHORA

The flagellates, or mastigophorans, include those protozoa that possess flagella as adult organelles. They can be conveniently divided into phytoflagellates and zooflagellates. The phytoflagellates usually bear one or two flagella and typically possess chloroplasts. These organisms are thus plantlike, and phycologists treat most species in this division as algae. The phytoflagellate division contains most of the free-living members of the class and includes such common forms as *Euglena*, *Chlamydomonas*, *Volvox*, and *Peranema*. The zooflagellates possess one to many flagella, lack chloroplasts, and are either holozoic or saprozoic. Some are free living, but the majority are commensal, symbiotic, or parasitic in other animals, particularly arthropods and vertebrates. Many groups have become highly specialized. It is generally agreed that this division does not represent a closely related phylogenetic unit.

Locomotion

The presence of flagella is the distinguishing feature of flagellates. The phytoflagellates usually have one or two flagella, the zooflagellates one to many. When two or many flagella occur, they may be of equal or unequal length, and one may be leading and one trailing, as in *Peranema* (Fig. 2–5*B*) and the dinoflagellates.

The ultrastructures of flagella and cilia are fundamentally similar in all eukaryote organisms. A single flagellum (or cilium) is constructed very much like a cable. Two central microtubules form a core that is in turn encircled by nine double outer microtubules (Fig. 2–4*A*). One microtubule of each doublet bears two rows of projections (arms) directed toward the adjacent doublet (Fig. 2–4*B*). The entire bundle is enclosed within a sheath that is continuous with the cell membrane. The flagellum always arises from a basal body, sometimes called a blepharoplast in flagellates, that lies just beneath the surface. Like a centriole, the basal body has an ultrastructure somewhat like a flagellum, except that the central fibrils are absent and the nine fibrils in the outer circle are in triplets, two of the three being continuous with the doublets of the flagellum. Arms are absent and the inner microtubule of each triplet is connected by a radial strand to a central ring for part of its length (Fig. 2–4*B*).

A fibrillar root system connecting the basal body with various organelles, especially the nucleus, characterizes many flagellates. In some species the basal body functions as a centriole in mitosis.

Flagellar propulsion in most mastigophorans essentially follows the same principle as that of a propeller, the flagellum undergoing undulations in one or two planes that either push or pull. The undulatory waves pass from base to tip and drive the organism in the opposite direction (Fig. 2–4*C*), or the undulations pass from tip to base and pull the organism (Fig. 2–4*D*). In many species with two flagella, the actual path of movement is determined by the flagellar orientation. Other types of beat have been described for flagella besides undulatory.

The relationship of flagellar (or ciliary) ultrastructure to movement has received much attention, and the sliding tubule model is now widely accepted. According to this theory, the microtubules do not change length but adjacent doublets slide past each other, causing the entire organelle to bend. Sliding involves the establishment of cross bridges and utilization of adenosine triphosphate (ATP), as in muscle contraction (see Sleigh, 1974).

Mastigophorans that have thin, flexible pellicles are often capable of ameboid movement; some forms, such as chrysomonads, may cast off their flagella and assume an ameboid type of locomotion entirely.

Nutrition

Phytoflagellates are primarily autotrophic and contain chlorophyll. When the chlorophyll is not masked by other pigments, a flagellate appears green in color, like the phytomonads and euglenids. If the xanthophylls dominate, the color is red, orange, yellow, or brown.

Strict heterotrophic nutrition occurs in zooflagellates as well as some other groups, and there are many parasitic species. The mechanisms of food capture and ingestion vary greatly, and the methods employed by some of the better known groups will be described in the following sections.

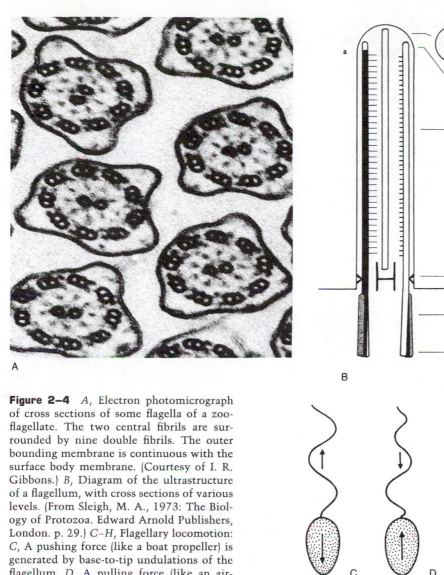

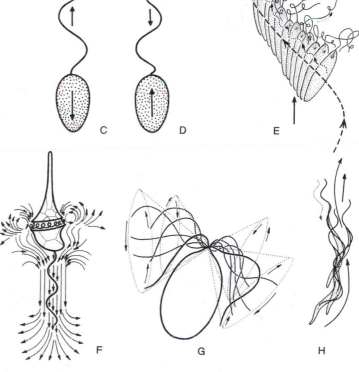

Figure 2–4 *A*, Electron photomicrograph of cross sections of some flagella of a zooflagellate. The two central fibrils are surrounded by nine double fibrils. The outer bounding membrane is continuous with the surface body membrane. (Courtesy of I. R. Gibbons.) *B*, Diagram of the ultrastructure of a flagellum, with cross sections of various levels. (From Sleigh, M. A., 1973: The Biology of Protozoa. Edward Arnold Publishers, London. p. 29.) *C–H*, Flagellary locomotion: *C*, A pushing force (like a boat propeller) is generated by base-to-tip undulations of the flagellum. *D*, A pulling force (like an airplane propeller) is generated by tip-to-base undulations of the flagellum. *E*, Movement in *Euglena viridis*. The actual path is indicated by dashed arrows. *F*, Locomotion in the dinoflagellate *Ceratium*. Arrows indicate the water currents generated by the transverse and posterior flagella. *G*, Locomotion in the phytoflagellate *Polytomella*. Arrows indicate the direction of the flagellar beat. *H*, Locomotion of the blood parasite *Trypanosoma*. The dotted arrow indicates movement of the undulating membrane; the fine solid arrow, the actual path of movement. (From Jahn, T. L., and Bovee, E. C., 1967: Motile behavior of Protozoa. *In* Chen, T. (Ed.): Research in Protozoology. Pergamon Press, N.Y. pp. 41–200.)

Phytoflagellates store reserve foods, such as oils or fats, or they may store carbohydrates as typical plant starch or in other forms. In zooflagellates, glycogen is the usual reserve food product.

Biology of Some Flagellate Groups

Flagellates vary so greatly in structure that a description of the assemblage as a whole is difficult. Most possess distinct anterior and posterior ends, although almost any plan of symmetry occurs. Most are free swimming, but there are some sessile forms. There are also many colonial species. Space permits description of only a few of the many different groups.

EUGLENIDA

The members of this marine and freshwater order contain chlorophyll b and are classified by botanists with other green algae in the Chlorophyta.

The genera *Peranema* and *Euglena* contain perhaps the most familiar flagellates. The body is elongate with an invagination (reservoir) at the anterior end. The contractile vacuole discharges into the pocket, and two flagella arise from its wall. In *Euglena* one flagellum is very short and terminates at the base of the long flagellum (Fig. 2–5*A*). A pigment spot, or stigma, shades a swollen basal area of the long flagellum, which is thought to have a photoreceptive function. In the colorless *Peranema* both flagella are long but one trails backward (Fig. 2–5*B*). The body covering, or pellicle, is commonly flexible, and helical striations or sculpturing can be seen with electron microscopy.

Green species, such as *Euglena*, are of course autotrophic, and carbohydrate is stored as granules of paramylon. Colorless heterotrophic species may depend on organic compounds absorbed from the surrounding water or may be holozoic, like *Pera-*

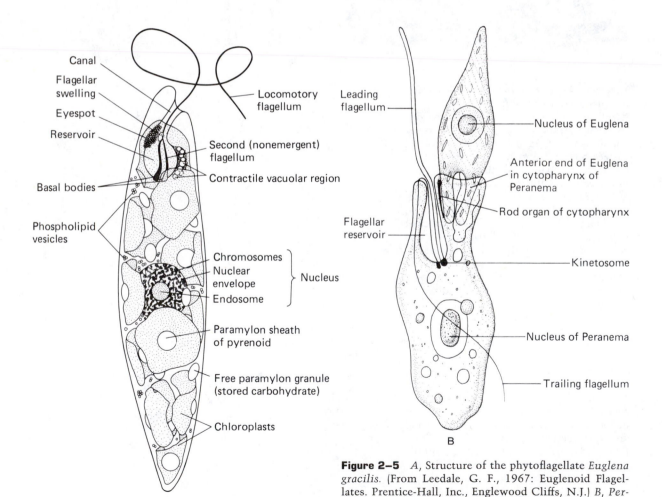

Figure 2–5   *A*, Structure of the phytoflagellate *Euglena gracilis*. (From Leedale, G. F., 1967: Euglenoid Flagellates. Prentice-Hall, Inc., Englewood Cliffs, N.J.) *B*, *Peranema* swallowing a *Euglena*. (Modified after Chen.)

nema. The mode of ingestion in *Peranema* has been studied in some detail. The anterior end of the body contains two parallel rods, making up a so-called rod organ, located adjacent to the reservoir (Fig. 2–5*B*). The anterior of the rod organ terminates at the cytostome, which is just below the outer opening of the canal leading from the reservoir. *Peranema* feeds on a wide variety of living organisms, including *Euglena*, and the cytostome can be greatly distended to permit engulfment of large prey. In feeding, the rod organ is protruded and used as an anchor to pull in prey, which is swallowed whole (Fig. 2–5*B*), or the rod organ can cut into the victim, whose contents are then sucked out. Following ingestion, the prey is digested within the food vacuoles.

VOLVOCIDA

These flagellates are placed by botanists in the large group of marine and freshwater green algae, which includes nonmotile filamentous and thalloid forms, as well as some motile species with two or four apical flagella. The cells are bounded by a cellulose wall and are similar in organization to those of multicellular green plants. Among the flagellate species, some are solitary, such as *Chlamydomonas*, and others are colonial. The colonies have the form of flattened curved plates or hollow spheres. In *Gonium, Pandorina,* and *Eudorina* the cells composing the colony number from 4 to 64 and are held together by a mucoid material (Fig. 2–6). The largest colonies are the hollow spheres of *Volvox* and *Pleodorina,* which contain hundreds of cells interconnected by protoplasmic strands in addition to mucoid material (Fig. 2–6). All of these colonies possess an oriented flagellar beat and swim with a definite anterior pole forward. In *Pleodorina* and *Volvox,* a degree of cellular specialization exists. Some of the cells are strictly somatic; others are reproductive. The reproductive cells are larger and located away from the anterior pole.

DINOFLAGELLATES

The Dinoflagellida are common marine and freshwater flagellates, allied with the brown algae and diatoms by the presence of xanthophyll pigments that give them a brown or golden brown color and by the absence of chlorophyll b (chlorophylls a and c are present). The primitive dinoflagellate body is roughly ovoid but asymmetrical (Fig. 2–7*A*). Typical dinoflagellates possess two flagella. One is attached a short distance behind the middle of the body; it is directed posteriorly and lies in a longitudinal groove called the sulcus. The other flagellum is transverse and is located in a groove that

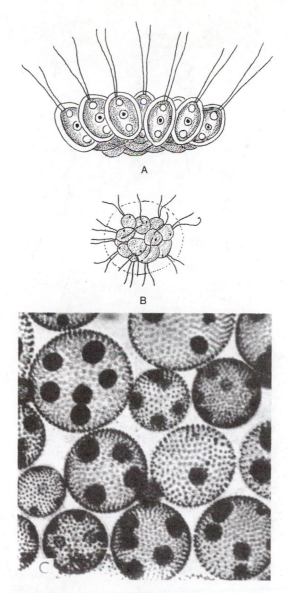

Figure 2–6 Colonial volvocids. *A,* Side view of *Gonium pectorale.* (After Stein from Pavillard.) *B, Pandorina morum.* (After Smith from Hall.) *C, Volvox.* Note daughter colonies within parent colonies. (Courtesy of General Biological Supply House, Inc.)

either rings the body or forms a spiral of several turns. The transverse groove is called the girdle, if it is a simple ring, or the annulus, if it is spiraled. The transverse flagellum causes both rotation and forward movement; the longitudinal one drives water posteriorly.

Dinoflagellates possess a relatively complex thickened pellicle, or theca, which contains deposits of cellulose within flattened membranous sacs

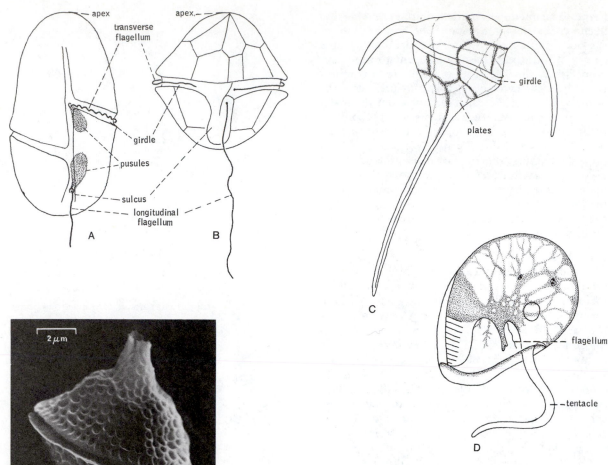

Figure 2–7 Dinoflagellates. *A*, A naked dinoflagellate, *Gymnodinium.* (After Kofoid and Swezy from Hall.) *B*, A freshwater armored dinoflagellate, *Glenodinium cinctum.* (After Pennak R. W., 1978: Freshwater Invertebrates of the United States. 2nd Edition. John Wiley and Sons, N.Y.) *C*, *Ceratium.* (After Jörgenson from Hyman.) *D*, *Noctiluca.* An aberrant bioluminescent dinoflagellate. Only one of the small flagella is visible within the "oral" depression. (After Robin from Kudo.) *E*, Scanning electron micrograph of *Gonyaulax digitale*, a marine species that causes red tides. (By J. D. Dodge.)

(alveoli). Where the theca is thin and flexible, as in the common freshwater and marine genus *Gymnodinium*, the dinoflagellate is said to be unarmored (Fig. 2–7*A*). Armored dinoflagellates have a highly developed theca composed of two valves or of plates (Fig. 2–7*B*). Frequently the armor is sculptured, and often long projections or winglike extensions protrude from the body, creating very bizarre shapes. For example, some species of *Ceratium*, a common marine and freshwater genus, possess three large horns, one at one end of the body and two at the other, which give the body an anchor-like appearance (Fig. 2–7*C*). The large and aberrant *Noctiluca* (Fig. 2–7*D*) and smaller species

of some common genera are luminescent and are the principal contributors to planktonic bioluminescence. When present in large numbers they may produce a striking effect in a quiet sea at night.

Many dinoflagellates are autotrophic, but the colorless forms are heterotrophic; some pigmented species, such as the common marine *Ceratium*, exhibit both modes of nutrition. The prey is captured with pseudopodia and ingested through an oral opening associated with the longitudinal flagellar groove. There are parasitic and mutualistic dinoflagellates. The latter, symbionts of corals and other animals, are of great significance in reef ecology and are discussed on page 133.

Dinoflagellates occur in countless numbers in marine plankton, and a number of forms are abundant in fresh water. Marine species of the genera *Gymnodinium* and *Gonyaulax* are responsible for outbreaks of the so-called red tides off the coasts of New England, Florida, California, Europe, and elsewhere (Fig. 2–7E). Under ideal environmental conditions, and perhaps with the presence of a growth-promoting substance, populations of certain species increase to enormous numbers. Red tides are not always red. The water may be yellow or brown. Concentrations of certain toxic metabolic substances reach such high levels that other marine life may be killed. The 1972 red tides off the coasts of new England and Florida killed thousands of birds, fish, and other animals and wreaked havoc with the shellfish industry because of the danger of eating clams and oysters that had fed upon the dinoflagellates.

CHOANOFLAGELLIDA
This group of marine and freshwater zooflagellates that contain a number of colonial forms is peculiar in having a cylindrical collar around the base of the single flagellum. Many zoologists now believe that the choanoflagellates are most closely related to the metazoan animals of any group of protozoans. They are described in Box 4–1.

KINETOPLASTIDA
This order of zooflagellates contains a few common free-living species as well as the most important parasites. All possess an organelle, called a kinetoplast, that contains DNA and is located within a large, elongated mitochondrion. The one or two flagella arise from a pit, and their basal bodies are located on or near the kinetoplast. Species of the free-living biflagellate *Bodo* are commonly found in the sea and in fresh water, where they feed on bacteria.

The trypanosomids are gut parasites of insects and blood parasites of vertebrates. Only the anterior flagellum is present, the second flagellum being represented by a basal body (Fig. 2–8C). Commonly, the flagellum trails and is connected along the sides of the body by an undulating membrane.

Species of the trypanosomid genera *Leishmania* and *Trypanosoma* are agents of numerous diseases of humans and domesticated animals in subtropical and tropical regions of the world. Part of the life cycle is passed within or attached to gut cells of blood-sucking insects, mostly various kinds of flies, and part in the blood or in white blood cells and lymphoid cells of the vertebrate host, although other tissues may be invaded. Intracellular stages are aflagellate, but during the life cycle there are motile, extracellular flagellate stages in the bloodstream or in the invertebrate host (Fig. 2–8B and C).

Leishmania is the agent of the widespread kala azar and related diseases of Eurasia, Africa, and America, causing skin lesions (Fig. 2–8A) and interference with immune responses, among other effects. Sand flies are the blood-sucking insect host.

Chagas's disease of tropical America is caused by *Trypanosoma cruzi* and is transmitted by blood-sucking hemipteran bugs. Extensive damage may be caused in the human host when the parasite leaves the circulatory system and invades the liver, spleen, and heart muscles. *Trypanosoma brucei rhodesiense* and *Trypanosoma brucei gambiense* are the causal agents of African sleeping sickness and are transmitted by the tsetse fly (Fig. 2–8B and C). The parasite invades the cerebrospinal fluid and brain, producing the lethargy, drowsiness, and mental deterioration that mark the terminal phase of the disease.

There are also various trypanosome diseases of horses, cattle, and sheep that are of considerable economic importance.

METAMONAD FLAGELLATES
Most zooflagellates other than the kinetoplastids belong to a number of small orders* collectively called metamonads. There are a few free-living species (*Hexamita*), and some are parasitic in the gut or genital tract of vertebrates (*Trichomonas*, Fig. 2–9A), but most are mutualistic symbionts in the guts of insects, especially termites and wood roaches. All of the metamonads possess complex organelles associated with the flagella, of which there may be from four to many. One organelle, the axostyle, is composed of microtubules that run from one of the anterior flagellar basal bodies posteriorly through the flagellate, partially enclosing the nucleus. Two other organelles extending posteriorly from the basal bodies are the costa, a contractile fiber, and the parabasal fiber, which is associated with Golgi vesicles. All of these organelles and their associated flagella make up what is called the karyomastigont system. In the evolution of the metamonads great complexity has resulted from multiplication of the karyomastigonts, leading to species with many flagella. The hypermastigote flagellates, which are gut symbionts of termites and wood roaches, are extremely complex. They are nearly all multiflagellate, with a saclike or elongated body, usually bearing an anterior rostrum

*Retortamonadida, Diplomonadida, Oxymonadida, Trichomonadida, and Hypermastigida.

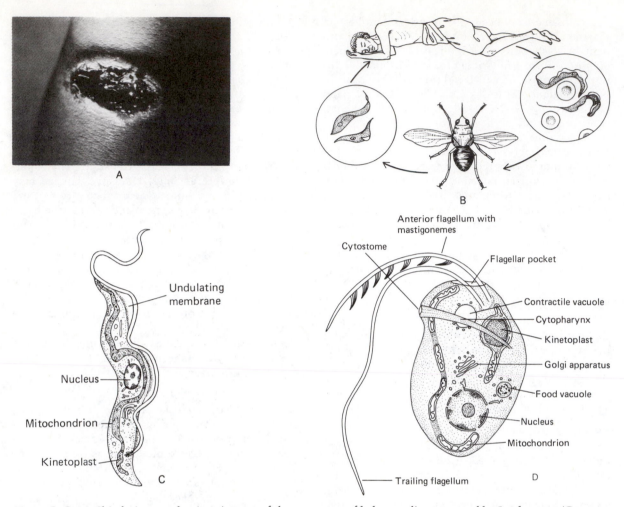

Figure 2–8   *A*, Skin lesion on a boy's wrist, one of the symptoms of kala azar disease caused by *Leishmania*. (Courtesy of S. S. Hendrix.) *B*, Life cycle of *Trypanosoma brucei*. *C*, Structure of *Trypanosoma brucei*. (From Sleigh, M. A., 1973: The Biology of Protozoa. Edward Arnold, London. p. 141.) *D*, *Bodo saltans*, a free-living member of the Kinetoplastida. (Modified after Brooker from Farmer, J. N., 1980: The Protozoa: Introduction to Protozoology. C. V. Mosby Co., St. Louis. p. 214.)

and cap. The ultrastructure of *Trichonympha*, which possesses thousands of flagella, virtually defies description (Fig. 2–9*B*). However, there is also a hypermastigote *(Mixotricha)* in termites that has few flagella and is moved by attached bacteria (spirochetes).

In many termites and wood-eating cockroaches, the host is dependent on its flagellate fauna for the digestion of wood. The wood consumed by the roach or termite is ingested by the flagellates, and the products of digestion are also utilized by the insect. The ingestion of the wood particles occurs at the posterior end of the flagellate by pseudopodial engulfment. The termite host loses its fauna with each molt, but by licking other individuals, by rec-

tal feeding, or by eating cysts passed in feces (in the case of roaches), a new fauna is obtained. In wood-eating cockroaches, the life cycles of the flagellates are closely correlated with the production of molting hormones by the late nymphal insect.

Reproduction and Life Cycles

In the majority of flagellates, asexual reproduction occurs by binary fission, and most commonly the animal divides longitudinally. Division is thus said to be symmetrogenic, that is, producing mirror-image daughter cells (Fig. 2–10). In multiflagellate species the flagella are divided between the daugh-

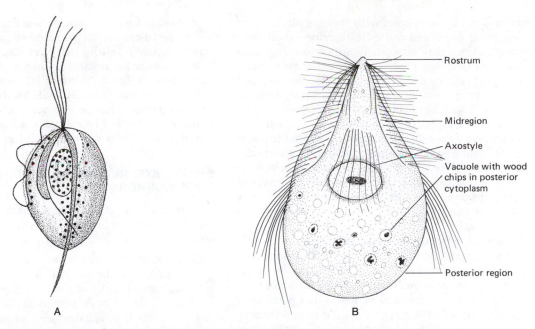

Figure 2–9 *A, Trichomonas vaginalis,* a trichomonad, parasitic in the human vagina and male reproductive tract. In addition to the four anterior flagella, there is a trailing flagellum bordering an undulating membrane. A supporting rodlike structure, the axostyle, extends posteriorly from the blepharoplast. (After Wenrich from Cheng.) *B, Trichonympha campanula,* a hypermastigote flagellate that lives in the gut of termites. (From Farmer, J. N., 1980: The Protozoa: Introduction to Protozoology. C. V. Mosby Co., St. Louis. p. 266.)

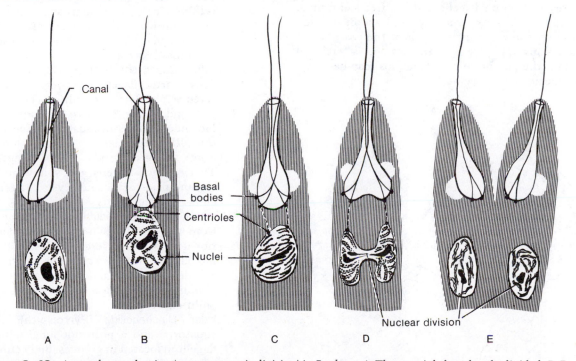

Figure 2–10 Asexual reproduction (symmetrogenic division) in *Euglena. A,* The centriole has already divided. *B,* Each centriole produces a new basal body and flagellum. The nucleus is in prophase and the contractile vacuole is double. *C,* The old pair of flagellar roots separate and fuse with the new roots. *D,* Mitosis proceeds and the gullet begins to divide. *E,* Anterior end dividing following duplication of organelles. (Redrawn from Ratcliffe, 1927.)

ter cells. In those species with few flagella, the one or several flagella may duplicate prior to cell division, or they may be equally apportioned to each daughter cell, resorbed, and formed anew in each daughter cell, or even unequally apportioned. The same may apply to other organelles. Thus, division in many flagellates is not perfectly symmetrogenic.

In the armored dinoflagellates the two fission products regenerate the missing plates, but in a few species, such as *Glenodinium*, naked daughter cells form new plates. In *Volvox* any one of a certain few cells at the posterior of the colony may undergo fission to form a daughter colony. The daughter colonies usually escape by rupturing the parent wall.

Sexual reproduction in flagellates is poorly known except in the Volvocida and the metamonads. The Volvocida display all gradations of gamete differentiation, from isogamy to highly developed heterogamy, and meiosis takes place after the formation of the zygote, i.e., it is postzygotic. In some species of *Chlamydomonas*, the cells act as gametes (isogametes). Other species show the beginnings of sex differentiation by having gametes that differ just slightly in size. In *Platydorina*, heterogamy is well developed, but the large macrogametes still retain flagella and are free swimming. Finally, in *Volvox*, true eggs and sperm develop from special reproductive cells at the posterior of the colony. The egg is stationary and is fertilized within the parent colony. Colonies may be either monoecious or dioecious.

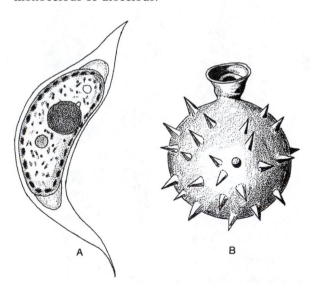

Figure 2–11 *A*, Palmella stage of the dinoflagellate, *Cystodinium steinii*. *B*, Siliceous cyst of the chrysomonad, *Ochromonas fragilis*. (After Conrad from Hollande.)

Cysts are formed in many flagellate groups (Fig. 2–11). Nonflagellated (palmella) stages are characteristic of most well-known phytoflagellates. In the palmella stage the organism loses its flagellum and becomes a ball-like, nonmotile, usually floating, relatively undifferentiated cell, located inside the original parental envelope when such an envelope is present (Fig. 2–11*A*). Fission often follows, so that the palmella consists of a cluster of cells.

SYSTEMATIC RÉSUMÉ OF SUBPHYLUM MASTIGOPHORA*

One to many flagella present. Asexual reproduction by binary, more or less symmetrogenic, fission. Autotrophic or heterotrophic or both.

Class Phytomastigophora. Mostly free-living, plantlike flagellates with or without chromoplasts and usually one or two flagella.

*Order Chrysomonadida (Chrysophyta). Small flagellates with yellow or brown chromoplasts and two unequal flagella. Siliceous scales commonly cover the body. Marine and freshwater. *Chromulina, Ochromonas, Synura.*

*Order Silicoflagellida (Chrysophyta). Flagellum single or absent and chromoplasts brown. Internal siliceous skeleton. Marine. *Dictyocha.* Known mostly from fossil forms.

*Order Coccolithophorida (Haptophyta). Tiny marine flagellates covered by calcareous platelets—coccoliths. Two flagella and yellow to brown chromoplasts. No endogenous siliceous cysts. *Coccolithus, Rhabdosphaera.*

*Order Heterochlorida (Xanthophyta). Two unequal flagella and yellow-green chromoplasts. Siliceous cysts. *Heterochloris, Myxochloris.*

*Order Cryptomonadida (Cryptophyta). Compressed, biflagellate, with an anterior depression or reservoir. Two chromoplastids, usually yellow to brown or colorless. Marine and freshwater. *Chilomonas* is a common colorless genus in polluted water.

*Order Dinoflagellida (Pyrrophyta). An equatorial and a posterior longitudinal flagellum located in grooves. Body either

---

*Those groups treated as algae are preceded by asterisks, and the name of the algal phylum to which each belongs is given in parentheses.

naked or covered by cellulose plates or valves or by a cellulose membrane. Brown or yellow chromoplasts and stigma usually present, but there are many colorless species. Largely marine; some parasites. Includes the marine genera *Gonyaulax, Noctiluca, Histiophysis,* and *Ornithocercus,* and the marine and freshwater genera *Glenodinium, Gymnodinium, Ceratium, Oödinium,* and *Symbiodinium.*

Order Ebriida. Biflagellate, with no chromoplasts; internal siliceous skeleton. Mainly fossil. *Ebria.*

*Order Euglenida (Euglenophyta). Elongated green or colorless flagellates with two flagella arising from an anterior recess. Stigma present in colored forms. Primarily freshwater. *Euglena, Phacus, Peranema, Rhabdomonas.*

*Order Chloromonadida (Chloromonadophyta or Rhaphidiophyta). Small, dorsoventrally flattened flagellates with numerous green chromoplasts. Two flagella, one trailing. *Gonyostomum.*

*Order Volvocida (Chlorophyta; order Volvocales). Body with green, usually single, cup-shaped chromoplast, stigma, and often two to four apical flagella per cell. Some colorless forms. Many colonial species. Largely freshwater forms. *Chlamydomonas, Polytomella, Haematococcus, Gonium, Pandorina, Platydorina, Eudorina, Pleodorina, Volvox.*

Class Zoomastigophora. Flagellates with neither chromoplasts nor leucoplasts. One to many flagella, in most cases with basal granule complex. Many commensals, symbionts, and parasites.

Order Choanoflagellida. Freshwater flagellates, with a single flagellum surrounded by a collar. Sessile, sometimes stalked, sometimes with lorica; solitary or colonial. *Codosiga, Proterospongia, Salpingoeca.*

Order Rhizomastigida. Ameboid forms, with one to many flagella. Chiefly freshwater. *Mastigamoeba, Dimorpha.*

Order Kinetoplastida. One or two flagella emerging from a pit. Mostly parasitic. *Bodo, Leishmania, Trypanosoma.*

Order Retortamonadida. Gut parasites of insects or vertebrates, with two or four flagella. One flagellum associated with ventrally located cytostome. *Chilomastix.*

Order Diplomonadida. Bilaterally symmetrical flagellates, with one or two nuclei, each nucleus associated with one to four flagella. Mostly parasites. *Hexamita, Giardia.*

Order Oxymonadida. Commensal or mutualistic flagellates in the guts of insects; a few in vertebrates. One to many nuclei, each nucleus associated with four flagella, some of which are turned posteriorly and adhere to body surface. *Oxymonas, Pyrsonympha.*

Order Trichomonadida. Parasitic flagellates. Four to six flagella, one of which is trailing. *Trichomonas* (Fig. 2–9A).

Order Hypermastigida. Many flagella, with kinetosomes arranged in a circle, plate, or longitudinal or spiral rows. Symbionts in guts of termites, cockroaches, and wood roaches. *Lophomonas, Trichonympha, Barbulanympha.*

Superclass Opalinata. Body covered by longitudinal, oblique rows of cilia rising from anterior subterminal rows. Two or many monomorphic nuclei. Binary fission generally symmetrogenic. Sexual reproduction involves syngamy with flagellated gametes. Gut commensals of anurans; less commonly of fishes, salamanders, and reptiles. *Opalina, Zelleriella.*

SUMMARY

1 Flagellates are distinguished by the presence of one or more flagella.

2 Classically, the group has included many autotrophic groups (phytoflagellates), such as chrysomonads, euglenids, and the volvocids. They possess chlorophyll plus other pigments and store such food materials as oils, fats, and starches (other than glycogen). These groups are more properly assigned to the various algal phyla.

3 The remaining heterotrophs (zooflagellates) are a small, heterogeneous assemblage. A few are free living, but most are parasitic, commensal, or mutualistic in other animals.

4 Flagella (and cilia) are composed of microtubules surrounded by the plasma membrane. The arrangement of microtubules in which nine pairs (doublets) surround two central microtubules is with few exceptions characteristic of flagella and cilia in all eukaryote organisms. Movement of the organelle is thought to result from the sliding of the microtubules relative to each other. Each flagellum (or cilium) arises from a basal body, or kinetosome.

5 Flagella commonly beat by undulation in two planes. The beat pushes or pulls the flagellate, and the path of movement depends on the point of flagellum attachment and the combined action, when there is more than one flagellum.

SUBPHYLUM SARCODINA

The subphylum Sarcodina contains those protozoa in which adults possess flowing extensions of the body called pseudopodia. Pseudopodia are used for capturing prey in all Sarcodina, and in benthic groups, pseudopodia are also used as locomotor organelles. The subphylum includes the familiar amebas as well as many other marine, freshwater, and terrestrial forms. The slime molds are sometimes included in the Sarcodina, but in the following discussion, the slime molds will be considered to be fungi and left to the mycologists.

The Sarcodina either are asymmetrical or have a spherical symmetry. They possess relatively few organelles and in this respect are perhaps the simplest protozoa. However, skeletal structures, which are found in the majority of species, reach a complexity and beauty that is surpassed by few other organisms.

The presence of flagellated gametes among many Sarcodina and the tendency of many flagellates to lose their flagella during some phase of the life cycle, often becoming ameboid, are important reasons for uniting the mastigophorans and sarcodines within a single phylum. These facts would also seem to indicate that the Mastigophora are the ancestral group.

Form and Structure

The subphylum Sarcodina contains four distinct groups: the amebas, the foraminiferans, the heliozoans, and the radiolarians.

AMEBAS

The amebas may be naked or enclosed in a shell. The naked amebas, which include the genera *Amoeba* (Fig. 2–12*B*) and *Pelomyxa*, live in the sea, in fresh water, and in the water films around soil particles. They are asymmetrical and have a constantly changing body shape. Some giant species reach several millimeters in length. The cytoplasm in amebas is divided into a stiff external ectoplasm and a more fluid internal endoplasm. The pseudopodia are one of two types: lobopodia, which are

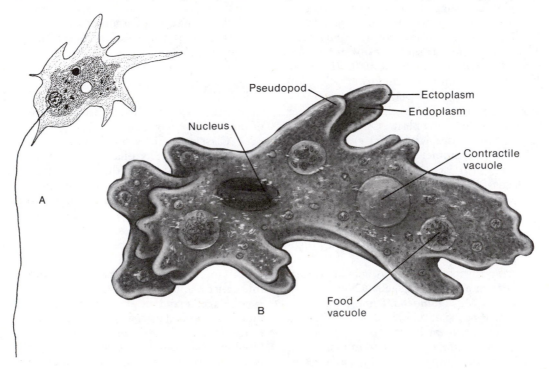

Figure 2–12 *A, Mastigamoeba,* an amoeboid protozoa with a long flagellum. (After Calkins, modified from Hyman.) *B, Amoeba.*

typical of most amebas, are rather wide with rounded or blunt tips, are commonly tubular, and are composed of both ectoplasm and endoplasm; filopodia, which occur in some small amebas, tend to have pointed ends, are composed of ectoplasm only, and are sometimes branched.

In shelled amebas, which are largely inhabitants of fresh water, damp soil, and mosses, either the radially or bilaterally symmetrical shell is secreted by the cytoplasm, in which case it is organic, siliceous, or both, or it is composed of foreign materials embedded in a cementing matrix. The ameba is attached by protoplasmic strands to the inner wall of the shell, and there is a large opening through which the pseudopodia or body can be protruded. In *Arcella* (Fig. 2–13*A* and *B*), one of the most common freshwater amebas, the brown or straw-colored protein shell has the shape of a flattened dome with the aperture in the middle of the

underside. In *Euglypha* the secreted shell is constructed of overlapping siliceous scales (Fig. 2–13*C*). *Difflugia* has a shell composed of mineral particles that are ingested by the animal and embedded in a secreted matrix (Fig. 2–13*D*).

FORAMINIFERANS

The order Foraminiferida is primarily marine. The pseudopodia are threadlike, branched, and interconnected and are called reticulopodia (Fig. 2–14*B*). Foraminiferans construct a shell of secreted organic material or of cemented foreign mineral particles or of secreted calcium carbonate. Calcareous shells are most common, and there are many fossil species.

Some species of forams live within a single-chambered shell and are said to be unilocular, but most forms are multilocular, having many-chambered shells. Multilocular forams begin life in a

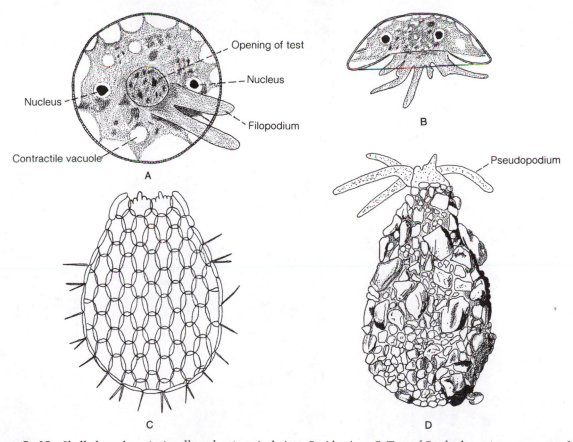

Figure 2–13 Shelled amebas. *A*, *Arcella vulgaris*, apical view; *B*, side view. *C*, Test of *Euglypha strigosa*, composed of siliceous scales and spines. (After Wailes from Deflandre.) *D*, *Difflugia oblonga* with test of mineral particles. (*A*, *B*, and *D* after Deflandre, G., 1953: *In* P. Grassé's Traité de Zoologie, Masson and Co., Paris. Vol. I, pt. II.)

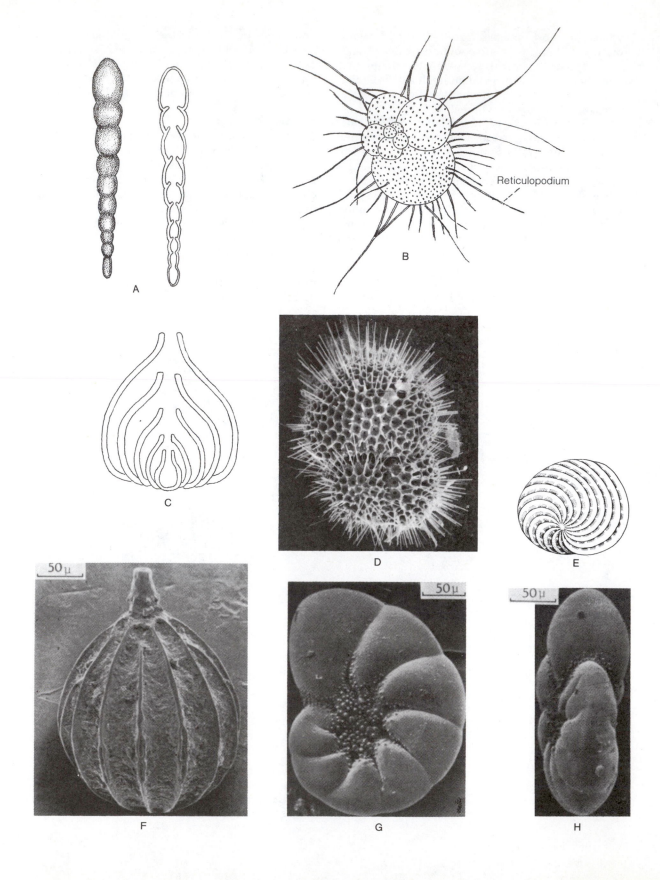

A

B

Reticulopodium

C

D

E

50μ

F

50μ

G

50μ

H

single chamber, the proloculum, but as the animal increases in size, the protoplasm overflows through a large opening in the first chamber and secretes another compartment. This process is continuous throughout the life of the animal and results in the formation of a series of many chambers, each of which may be larger than the preceding one. Since the addition of new chambers follows a symmetrical pattern, the shells of multilocular forams have a distinct shape and arrangement of chambers (Fig. 2–14).

Each chamber in multilocular species opens successively into another chamber; the entire shell is filled with protoplasm that is continuous from one chamber to the next. The outside of the foram shell is covered with a thin layer of cytoplasm that overflows from the large aperture. Since the shell is actually internal, it should be called a test. Pseudopodia may be restricted to the cytoplasm of the aperture, or they may arise from the layer over the shell. They sometimes appear to be emerging through the shell pores, which are actually sealed.

Multilocular forams are not colonies but represent single individuals. Many are visible to the naked eye and a few, such as the so-called mermaid's pennies of Australia, attain a size of several centimeters in diameter.

Most forams are benthic, but species of *Globigerina* and related genera are common planktonic forms. The chambers of these multilocular species are spherical but arranged in a somewhat spiral manner (Fig. 2–14*B* and *D*). Planktonic forams have more delicate shells than do benthic species, and the shells commonly bear spines. The spines are so long in some species that the foram is visible to the naked eye and can be collected undamaged with a jar by a scuba diver.

A few forams are sessile. *Homotrema* forms large, red, calcareous tubercles about the size of a wart on the underside of coral heads. The pink sands of the beaches of Bermuda result from the large number of pieces of *Homotrema* tests.

HELIOZOANS

Members of this group of spherical Sarcodina occur primarily in fresh water and may be floating or benthic. Some benthic forms are stalked. The fine, needle-like pseudopodia, called axopodia, radiate from the surface of the body (Fig. 2–15). Each axopod contains a central axial rod, which is covered with a moving, adhesive cytoplasm. Although the axial rod has a supporting function, it is not a permanent skeleton but a bundle of microtubules that can shorten or even "melt."

The body of a heliozoan consists of two parts (Figs. 2–15*A* and 2–16*B*). There is an outer ectoplasmic sphere, called the cortex, which is often greatly vacuolated. The inner part of the body, or medulla, is composed of dense endoplasm, containing one to many nuclei and the bases of the axial rods.

Although heliozoans may be naked, skeletons are not uncommon and may be composed of organic or siliceous pieces secreted by the animal and embedded in an outer gelatinous covering. The siliceous pieces assume a great variety of forms, such as scales (Fig. 2–16*A*), tubes, spatulas, or needles. These siliceous pieces either may be arranged tangentially to the body or, when the skeleton is composed of long needles (Fig. 2–16*B*), may radiate like the axopods. In a few cases the skeleton is a beautiful organic lattice sphere. Regardless of the nature and arrangement of the skeleton, openings are present through which the axopods project.

RADIOLARIANS

Among the most beautiful of the protozoa are members of three classes, collectively called radiolarians. They are entirely marine and primarily planktonic. Radiolarians are relatively large protozoa; some species are several millimeters in diameter, and some colonial forms attain a length of up to 20 cm (*Collozoum*). Like heliozoans, the bodies of radiolarians are usually spherical and divided into inner and outer parts (Fig. 2–17). The inner region, which contains one to many nuclei, is bounded by a central capsule with a membranous wall. The capsule membrane is perforated by openings, which may be evenly distributed (Fig. 2–18*A*) or restricted to three pores located at specific sites in the membrane. The perforations allow the cytoplasm of the central capsule (or intracapsular cytoplasm) to be continuous with the cytoplasm of the outer division of the body. This extracapsular cytoplasm forms a broad cortex, called the calymma, that surrounds the central capsule.

◄**Figure 2–14** Foraminiferans. *A*, Shell of *Rheophax nodulosa*, entire, and in section. (After Brady from Calvez.) *B*, Living *Globigerina. C*, Shell of an ellipsoidinid foraminiferan, in section. (*B* and *C* after Hyman, L. H., 1940: The Invertebrates. Vol. 1. McGraw-Hill Book Co., N.Y.) *D*, Cleaned test of *Globigerinoides sacculifer*, a tropical planktonic foram with spines. (By A. W. H. Bé, 1968: Science, *161*:881–884. Copyright 1968 by the American Association for the Advancement of Science.) *E*, *Archaias* sp., a common benthic foram of shallow tropical waters. *F–H*, Scanning electron micrographs of foram tests: *F*, *Lagena sulcata. G* and *H*, Surface and edge views of *Nonion depressulus.* (From Murray, J. W., 1971: An Atlas of British Recent Foraminiferida. Heinemann Educational Books, London.)

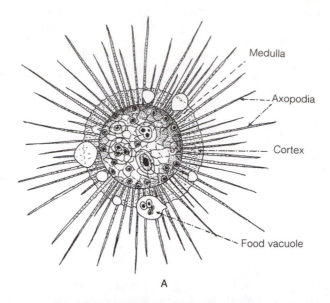

Medulla

Axopodia

Cortex

Food vacuole

A

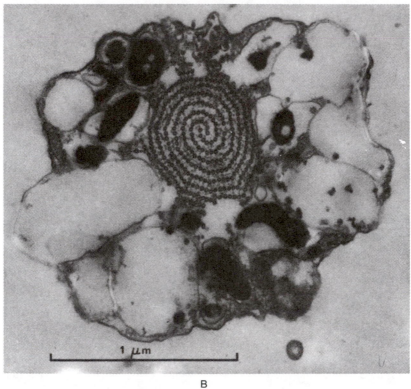

1 μm

B

Figure 2–15   An multinucleate heliozoan, *Actinosphaerium eichorni. A*, Entire animal. (After Doflein from Trégouboff.) *B*, Electron micrograph of a section through an axopod of *Actinosphaerium.* The axial rod is composed of a double spiral of microtubules. (Photomicrograph by A. C. Macdonald. *In* Sleigh, M. A., 1973: The Biology of Protozoa. Edward Arnold Publishers, London. p. 162.)

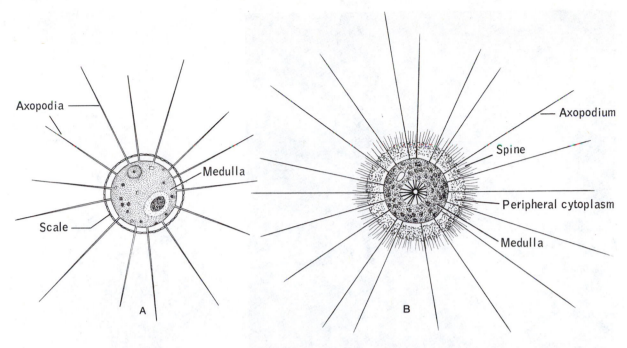

Figure 2–16 Heliozoans. *A, Pinaciophora fluviatilis* with skeleton of scales. *B, Heterophrys myriopoda* with skeleton in form of spines. (Both after Penard from Hall.)

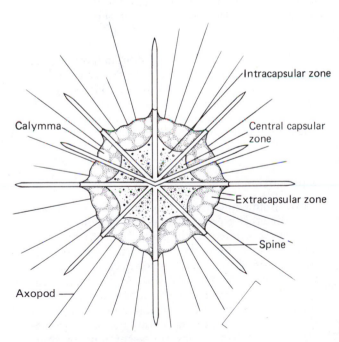

Figure 2–17 *Acanthometra*, a radiolarian with a skeleton of radiating strontium sulfate rods. (From Farmer, J. N., 1980: The Protozoa. C. V. Mosby Co., St. Louis. p. 353.)

In many species the calymma contains large numbers of symbiotic dinoflagellates (zooxanthellae) in a nonflagellate stage as well as many vacuoles, which give the cytoplasm a frothy appearance (Fig. 2–18). As in corals and some other marine animals (p. 134), the excess organic photosynthate produced by the zooxanthellae is utilized by the radiolarian host as an accessory food source.

The pseudopodia are axopods and radiate from the surface of the body. Their axes of microtubules originate within the central capsule and extend through the calymma.

A skeleton is almost always present in radiolarians and is usually siliceous, but in the class Acantharia it is composed of strontium sulfate. Several types of skeletal arrangements occur. One type has a radiating structure, in which the skeleton is composed of long spines or needles that radiate from the center of the central capsule and extend beyond the outer surface of the body (Fig. 2–17). The points where the skeletal rods leave the body surface are surrounded by contractile fibrils (as in Acantharia). The action of these fibrils can cause the calymma to be expanded. A second type of skeleton is constructed in the form of a lattice sphere, which is often ornamented with barbs and spines (Fig. 2–18*B*).

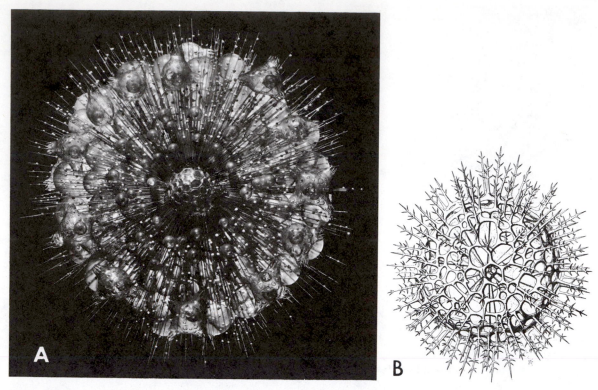

Figure 2–18   *A*, Glass model of a colonial radiolarian, *Trypanosphaera transformata.* (Neg./Trans. No. 318863. Courtesy Department Library Services, American Museum of Natural History.) *B*, The internal, siliceous skeleton of a radiolarian. (After Haeckel.)

The planktonic radiolarians display a distinct vertical stratification from the ocean surface down to 4600-meter depths. The great numbers in which planktonic foraminiferans and radiolarians occur are indicated by the fact that their shells, sinking to the bottom at death, form a primary constituent of many ocean bottom sediments. Where they compose 30 per cent or more of the sediment, it is called a foram or radiolarian ooze. However, at depths below 4000 meters the great pressure tends to dissolve foram shells. The strontium sulfate skeletons of acantharians do not seem to fossilize but dissolve rapidly with the death of the cell.

The Sarcodina are the only class of protozoa that has an extensive fossil record. Fossil forms are, of course, restricted to those with skeletons or shells—the shelled amebas, the foraminiferans, and the radiolarians. The fossil record of the shelled amebas is relatively brief and recent. The group appears as fossils only in the Cenozoic era, and the fossils consist of forms that are virtually identical to those living today. However, the foraminiferans and radiolarians have a long and abundant fossil record. In fact, the radiolarians are among the oldest known fossils. The foraminiferans first appeared in the Cambrian period, and from the late Paleozoic era on, there is an abundant fossil record. Extensive accumulations of foram shells occurred during the Mesozoic and early Cenozoic eras and contributed to the formation of great limestone and chalk deposits in different parts of the world.

Locomotion

Flowing ameboid movement is limited to those Sarcodina that possess lobopods or filopods and has been most studied in the naked amebas. In these animals locomotion may involve a single large, tubular lobopod or several small ones with caps of hyaline protoplasm at the tips. At points, a pseudopodium is anchored to the substratum. Electron microscopy has revealed a filamentous fringe on the outer mobile plasma membrane of some species. Such a fringe may represent mucoproteins that facilitate adhesion to the substratum.

The theory of ameboid movement accepted by most zoologists at present assumes that cytoplasmic flow results from contraction of protoplasmic elements in the change from the fluid endoplasm to the gelatinous ectoplasm. As a result of some initial stimulus, ectoplasm becomes endoplasm at some point on the body surface and internal pressure causes the cytoplasm to flow out at this point, forming a pseudopodium. In the interior of the pseudopodium, the fluid endoplasm flows forward along the line of progression. Around the periphery, endoplasm is converted to ectoplasm, thus building up and extending the sides of the pseudopodium like a sleeve. At the posterior of the body, ectoplasm is assumed to be undergoing conversion to endoplasm.

Protein chains in the endoplasm are believed to undergo contraction at the anterior end. Here the endoplasm is converted to ectoplasm, and in the gelatinous ectoplasm the protein chains are in the contracted state. The protein chains become extended at the posterior end where the ectoplasm is liquefied during its conversion to endoplasm. Thus, according to this front-contraction theory, the body of the animal is pulled forward by contraction at the anterior end.

A different type of ameboid flow has been demonstrated in the ectoplasmic reticulopods of *Allogromia*, a foraminiferan. *Allogromia* extends a rigid reticulopod net for some distance from the body. Particles adhering to the surface can readily be followed with the light microscope and have been found to move up one side and down the other side of the reticulopod. Thus, in each pseudopodium there exist two opposing cytoplasmic streams. A kind of shearing or sliding effect of opposing cytoplasmic units or bundles has been suggested.

Although some foraminiferans are pelagic or cling to a surface film, the majority are bottom dwellers and creep slowly over the substratum. The body is pulled or dragged along by the reticulopods, which may project several millimeters into the sand.

The pseudopodia of most heliozoans and radiolarians are food-capturing rather than locomotor organelles. However, radiolarians are able to move vertically in the water by extending or contracting the calymma and axopods, by increasing or decreasing the vacuolated condition of the calymma, and by the presence of endoplasmic oil droplets.

Nutrition

The Sarcodina are entirely heterotrophic. Their food consists of all types of small organisms: bacteria, algae, diatoms, protozoans, and even small multicellular animals such as rotifers, copepod larvae, and nematodes. The prey is captured and engulfed by means of the pseudopodia.

In the amebas, pseudopodia extend around the prey, eventually enveloping it completely with cytoplasm, or the body surface invaginates to form a food cup. The enclosing of the captured organism by cytoplasm results in the formation of a food vacuole within the ameba (Fig. 2–12*B*).

In foraminiferans, heliozoans, and radiolarians the numerous radiating pseudopodia (chiefly reticulopods or axopods) act primarily as traps in the capture of prey. Any organism that comes in contact with the pseudopodia becomes fastened to the granular, adhesive surface of these organelles. A granular mucoid film is especially evident on the surface of foraminiferan reticulopods and quickly coats the surface of captured prey. This film contains lysosomes, and their proteolytic secretions aid in paralyzing the prey and initiate digestion even during capture. The long spines of many planktonic forams are also covered with the mucoid film and are important as a food-trapping surface rather than as a flotation mechanism. They are able to capture fairly large prey, such as small crustaceans. In all three groups food particles are enclosed in food vacuoles and drawn toward the interior of the body. The axial rods of heliozoans may contract, drawing the prey into the ectoplasmic cortex, or the axopods may liquefy and surround the food, forming a vacuole at the site of capture. The vacuole will then be moved inward.

Digestion occurs in the cortex of heliozoans and the calymma of radiolarians. In acantharian radiolarians, digestion of at least small particles takes place largely inside the central capsule. Where an enveloping skeletal sphere is present, the food passes through the openings for the pseudopodia. In foraminiferans food is initially digested outside the shell, and then digestion is completed within small food vacuoles within the shell.

Egestion can take place at any point on the surface of the body, and in the actively moving amebas, wastes are usually emitted at the posterior, as the animal crawls about.

Some naked amebas are parasitic. The majority are endoparasites in the digestive tracts of annelids, insects, and vertebrates. Several species occur in the human digestive tract, but only *Entamoeba histolytica*, which is responsible for amebic dysentery, is ordinarily pathogenic. The life cycle of these intestinal amebas is direct, and the parasites are usually transmitted from the digestive tract of one host to that of another by means of cysts that are passed in feces. Certain free-living soil and water amebas (*Naegleria* and *Acanthamoeba*) are known to invade humans and cause a fatal meningoencephalitis.

Water Balance

Contractile vacuoles (Fig. 2–2) occur in freshwater species but are absent in marine Sarcodina. There may be one to many contractile vacuoles, and the expulsion of fluid can occur anywhere on the body surface. In amebas the single or few contractile vacuoles are carried about in the flowing endoplasm.

Reproduction and Life Cycle

Asexual reproduction in most amebas, heliozoans, and radiolarians is by binary fission (Fig. 2–19A). In amebas with a soft shell, the shell divides into two parts, and each daughter cell forms a new half.

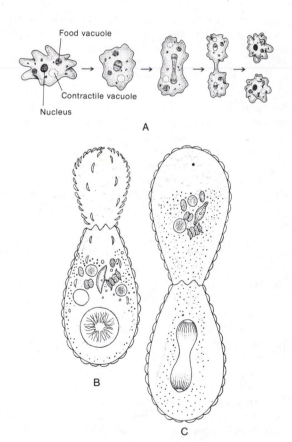

Figure 2–19 *A*, Fission in a naked ameba: *B* and *C*, two stages in the division of *Euglypha*, a shelled ameba: *B*, Formation of skeletal plates on protoplasmic mass protruding from aperture. *C*, Division of nucleus. (*B* and *C* after Sevajakov from Dogiel.)

When the shell is dense and continuous, as in *Arcella*, a mass of protoplasm extrudes from the opening prior to division; this extruded mass secretes a new shell (Fig. 2–19B). The double-shelled animal now divides.

Division in the radiolarians is somewhat similar to that in the shelled amebas. Either the skeleton itself divides and each daughter cell forms the lacking half, or one offspring receives the skeleton and the other secretes a new one.

Multiple fission is common in multinucleated amebas and heliozoans. In certain shelled amebas the parent animal sporulates a large number of little, naked amebas, each of which produces a new shell in the process of growth.

Sexual reproduction has been observed infrequently among the amebas, but hologamy—the fusion of two individuals, each acting as a gamete— has been reported in some species. Among the heliozoans, sexual reproduction is known in some genera, such as *Actinosphaerium* and *Actinophrys*, and a number of species produce flagellated stages. Flagellated stages have also been reported in some species of radiolarians.

Reproduction in the foraminiferans is complex but relatively uniform in the species observed and involves a definite alteration of asexual and sexual generations. Each species of foraminiferans is dimorphic, and in most multilocular species the two types of individuals differ in the size of the proloculum. One type of individual, known as a schizont, reproduces asexually and has a shell with a small proloculum, called a microspheric shell. The other type, known as a gamont, has a megalospheric shell and reproduces sexually with fusion of flagellated gametes. An example of a multilocular foraminiferan life cycle is illustrated in Figure 2–20.

SYSTEMATIC RÉSUMÉ OF SUBPHYLUM SARCODINA

Protozoa with pseudopodia as feeding and locomotor organelles; flagella, when present, only in developmental stages. Little development of cortical organelles. Skeletons of various forms and composition characteristic of some groups.

Superclass Rhizopoda. Lobopodia, filopodia, or reticulopodia used for locomotion and feeding.
　Class Lobosa. Pseudopodia, usually lobopods.
　　Subclass Gymnamoeba. Amebas that lack shells.
　　　Order Amoebida. Naked amebas that lack flagellated stages. Largely freshwater,

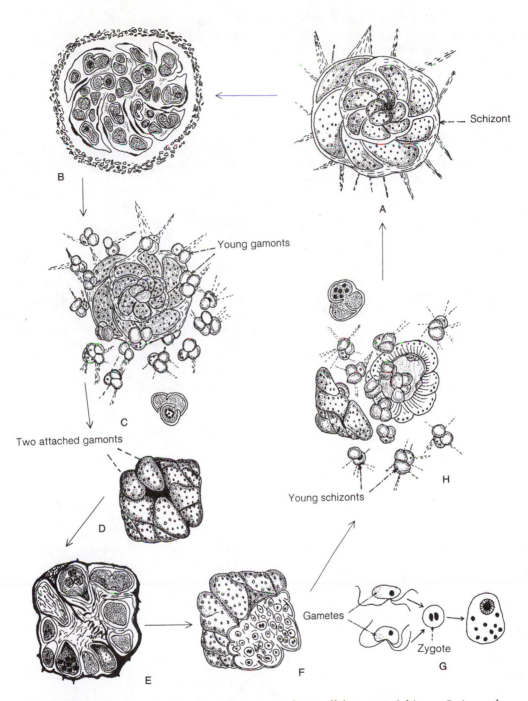

Figure 2–20 Life cycle of the foraminiferan, *Discorbis patelliformis. A*, Schizont. *B*, Asexual development of gamonts within chambers of schizont. *C*, Liberation of young gamonts from parent schizont. *D–G*, Formation and fusion of gametes within two attached gamonts. *H*, Separation of attached gamonts accompanied by liberation of young schizonts. (After Myers from Calvez.)

some marine; many parasites. *Chaos, Amoeba, Entameoba, Hydramoeba.*

Order Schizopyrenida. Naked amebas with flagellated stages. Marine, freshwater, and soil species. *Naegleria, Acanthamoeba.*

Order Pelobiontida. Naked, multinucleated amebas with one pseudopod and no flagellated stages. *Pelomyxa.*

Subclass Testacealobosa. Amebas with shells.

Order Arcellinida, or Testacida. Body enclosed in a shell or test with an aperture through which the pseudopodia protrude. Free living, largely in fresh water. *Arcella, Difflugia, Centropyxis.*

Class Filosa. Amebas with filopods.

Order Aconchulinida. Naked amebas. Freshwater and parasites of algae. *Vampyrella.*

Order Testaceafilosida. Shelled amebas. Mostly in fresh water and soil. *Gromia, Euglypha.*

Class Granuloreticulosa. Sarcodina with delicate granular reticulopodia.

Order Foraminiferida. Chiefly marine species with mostly multichambered shells. Shells may be organic, but most commonly are calcareous. *Globigerina, Orbulina, Discorbis, Spirillina, Nummulites.*

Superclass Actinopoda. Primarily floating or sessile Sarcodina with actinopodia radiating from a spherical body.

Class Acantharia. Radiolarians with a radiating skeleton of strontium sulfate; axopodia. Most without a central capsule separating endoplasm and ectoplasm. Marine. *Acanthometra.*

Class Polycystina. Radiolarians with a siliceous skeleton and a perforated capsular membrane. *Thalassicola, Collozoum, Sphaerozoum.*

Class Phaeodaria. Radiolarians with a siliceous skeleton but a capsular membrane containing only three pores. *Aulacantha.*

Class Heliozoa. Without central capsule. Naked, or if skeleton present, of siliceous scales and spines. Primarily in fresh water. *Actinophrys, Actinosphaerium, Camptonema.*

SUMMARY

1 Members of the phylum Sarcodina are distinguished by the presence of flowing extensions of cytoplasm called pseudopodia, which are used in feeding and, in some, for locomotion. The pseudopodia are given different names, depending on their shape and structure.

2 Although organelles have remained relatively simple, many Sarcodina have evolved complex skeletons. The various classes of Sarcodina are distinguished by the nature of their skeletons and their pseudopodia.

3 The marine, freshwater, and parasitic naked amebas have no special skeletal structures and possess large, commonly tubular lobopodia or small, straplike filopodia, which are used for both feeding and locomotion.

4 Shelled amebas, which are largely restricted to fresh water, are covered by a shell composed of secreted organic material or of foreign mineral material cemented together. A large aperture permits the protrusion of lobopodia or filopodia.

5 Foraminiferans, which are largely benthic marine Sarcodina, possess a calcareous test that is usually multichambered. A single large opening permits the protrusion of cytoplasm, which may cover the exterior of the test. Long, delicate, and often anastomosing reticulopodia extend from the protruded cytoplasm and are used in food trapping and locomotion.

6 Heliozoans are spherical, radially arranged, floating, and benthic Sarcodina that are largely restricted to fresh water. Long, radiating, needle-like pseudopodia (axopods) are used in trapping food. The axopods arise from the interior (medulla) and extend through an outer ectoplasmic cortex, which is commonly vacuolated. The cortex often contains a siliceous skeleton of plates, tubes, and needles.

7 Radiolarians are marine planktonic Sarcodina with spherical bodies and radiating axopods. An organic capsule wall separates a central cortex from extracapsular cytoplasm. Radiolarians have complex skeletons of silicon dioxide or strontium sulfate within the extracapsular cytoplasm, organized in the form of lattice spheres or radiating spines or both.

The Sporozoans: Phyla Apicomplexa and Microspora

Sporozoans are parasitic protozoa, living within or between cells of their invertebrate or vertebrate hosts. They belong to two phyla, the Apicomplexa and the Microspora, both formerly composing an old protozoan grouping, the Sporozoa. *Sporozoan,* which refers to the presence of sporelike stages, continues to be used as a common name.

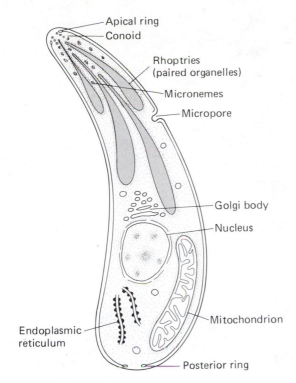

Apical ring
Conoid
Rhoptries
(paired organelles)
Micronemes
Micropore
Golgi body
Nucleus
Endoplasmic
reticulum
Mitochondrion
Posterior ring

Figure 2–21 Lateral view of a generalized apicomplexan sporozoan. (From Farmer, J. N., 1980: The Protozoa. C. V. Mosby Co., St. Louis. p. 360.)

Most known sporozoans and all those of known economic and medical importance belong to the phylum Apicomplexa, so named because of a complex of ringlike, tubular, filamentous organelles at the apical end, visible only with the electron microscope (Fig. 2–21). The function of the apical complex is uncertain but may include entry into the host cell. One or more feeding pores are located on the side of the body.

The life cycle of apicomplexans typically involves an asexual and a sexual phase (Fig. 2–22). An infective stage, called a sporozoite, invades the host and undergoes asexual multiplication by fission, producing individuals called merozoites. Merozoites can continue schizogamy but eventually form gametes (gamogony) that fuse to form a zygote. The zygote undergoes meiosis to form sporozoites.

The nature and life cycle of apicomplexan sporozoans can be illustrated by the coccidians, which include the parasites that cause malaria in humans. Malaria continues to be one of the worst scourges of mankind. About 300 million people are believed to be infected each year. The untreated disease can be long lasting and terribly debilitating. Malaria has played a major and often unrecognized role in human history. The name means literally "bad air" because the disease was originally thought to be caused by the air of swamps and marshes. Although malaria had been recognized since ancient times, the causative agent was not discovered until 1880, when a physician with the French army in North Africa identified the coccidian parasite *Plasmodium* in the blood cells of a malarial patient. In 1887 the mosquito was recognized to be the vector.

The introduction of the parasite into a human host is brought about by the bite of certain species of mosquitoes, which inject the sporozoites along with their salivary secretions into the capillaries of the skin (Fig. 2–23). The parasite is carried by the bloodstream to the liver, where it invades a liver cell. Here further development results in asexual reproduction through multiple fission. These daughter cells invade other liver cells and continue to reproduce. After a week or so there is an invasion of red blood cells by parasites produced in the liver. Within the red cell the parasite increases in size and undergoes multiple fission. The individuals (merozoites) produced by fission within the red cells escape and invade other red cells. The liberation and reinvasion are not continuous but occur simultaneously from all infected red blood cells. The timing of the event depends on the period of time required to complete the developmental cycle within the host's cells. The release causes chills and fever, the typical symptoms of malaria.

Eventually, some of the parasites invading red cells do not undergo fission but become transformed into gametocytes. The gametocyte remains within the red blood cell. If such a cell is ingested by a mosquito, the gametocyte is liberated within the new host's gut. After some further development, a male gametocyte (microgametocyte) fuses with a female gametocyte (macrogametocyte) to form a zygote. The zygote penetrates the stomach wall and gives rise to a large number of spore stages (sporozoites). It is these stages, which migrate to the salivary glands, that are introduced into the human host by the bite of the mosquito.

The asexual stage of other coccidians occurs in blood cells or in gut cells. A number of diseases of domesticated animals are caused by coccidians. The genus *Eimeria*, for example, affects chickens, turkeys, pigs, sheep, and cattle (Fig. 2–22).

Another common group of apicomplexans contains the gregarines, which attain the largest size among the sporozoans. They are parasites of invertebrates, especially annelids and insects, and there-

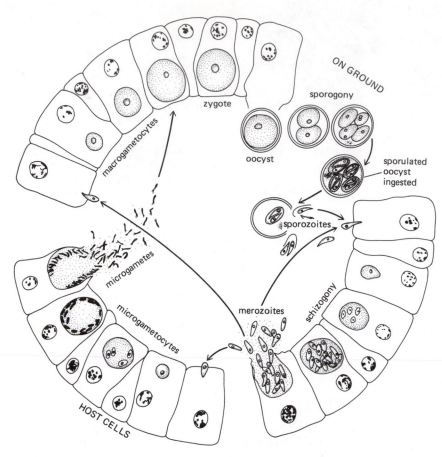

Figure 2–22 Life cycle of an eimeriid coccidian, a destructive intracellular parasite of the gut epithelium of many vertebrates, including domesticated birds and mammals. (From Noble, E. R., and Noble, G. A., 1982: Parasitology. 5th Edition. Lea and Febiger, Philadelphia.)

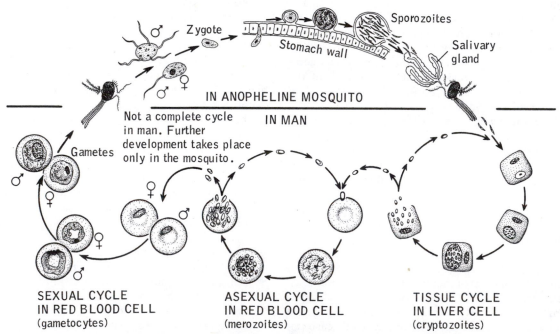

Figure 2–23 The life cycles of *Plasmodium* in a mosquito and in man. Reinvasion of liver cells in the tissue cycle does not occur in *Plasmodium falciparum*. (Redrawn and modified from Blacklock and Southwell.)

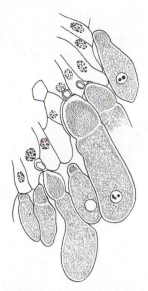

Figure 2–24 Trophozoites of the gregarine *Gregarina garnhami* attacking the midgut epithelium of a locust. (After Canning from Noble and Noble.)

fore not of economic importance. Intracellular parasitic species are only a few microns long, but those that inhabit the body or gut cavities of the host may reach 10 mm in length. The body of a gregarine trophozoite is elongate (Fig. 2–24), and the anterior part sometimes possesses hooks, a sucker or suckers, or a simple filament or knob for anchoring the parasite into the host's cells. The host becomes infected through ingesting spores containing sporozoites of the parasites (Fig. 2–25). Depending on the species, the liberated gregarine sporozoites either remain in the gut of the host or penetrate the gut wall to reach other areas of the body. The life cycle commonly lacks schizogony.

Members of the class Piroplasmea are another small group of sporozoans that also attack the red blood cells of vertebrates. Spores and gametes are not produced, and the parasites are transmitted by ticks. Pathogenic infections in cattle and other domesticated animals are of considerable economic importance.

The phylum Microspora contains a smaller number of intracellular parasites, but they are found in most animal groups, especially arthropods. They lack the apical complex of other sporozoans, and the sporelike stage is characterized by a polar filament that is extruded when this stage is taken into the host. The filament appears to be involved in some way with the invasion of the host's cell. As parasites of the honeybee and silkworm, the microsporidians are of economic importance. One of the early studies of sporozoans was that of Pasteur in 1870, on *Nosema bombycis* in silkworms.

SYSTEMATIC RÉSUMÉ OF THE SPOROZOANS

PHYLUM APICOMPLEXA
With apical complex at some stage. Spores usually present but lacking polar filaments. All species parasitic.
Class Sporozoa. Reproduction sexual and asexual.
  Subclass Gregarinia. Mature trophozoites large and occur in host's gut and body cavities. Parasites of annelids and arthropods. *Gregarina, Monocystis* (common parasite of earthworm's seminal receptacles).
  Subclass Coccidia. Mature trophozoites small and intracellular. *Eimeria, Isospora, Aggregata, Plasmodium, Toxoplasma.*
Class Piroplasmea. Parasites of vertebrate red blood cells transmitted by ticks. No spores. *Theileria, Babesia.*

PHYLUM MICROSPORA
Spores with polar filament present. All species parasitic. *Nosema.*

SUMMARY

1 Sporozoans are parasitic protozoa belonging to two phyla, the Sporozoa and the Microspora. Some species possess sporelike infective stages, from which the name *sporozoan* is derived.

2 The phylum Apicomplexa contains the gregarines, which are parasites of insects and anelids, and the coccidians, which are intracellular parasites of gut and blood cells of vertebrates and invertebrates. *Plasmodium*, the causal agent of malaria, is the best known and most familiar coccidian.

3 The complex life cycles usually involve fission (schizogony), sexual reproduction (gamogony), and spore formation (sporogony).

4 The phylum Microspora contains intracellular parasites, especially of insects. The name *microspora* is derived from the spore, which contains filaments that can be everted.

Phylum Ciliophora

The phylum Ciliophora is the largest and the most homogeneous of the principal protozoan groups, and all evidence indicates that they share a common evolutionary ancestry (Fig. 2–43). Some 7200 species have been described, and many groups are still not well known.

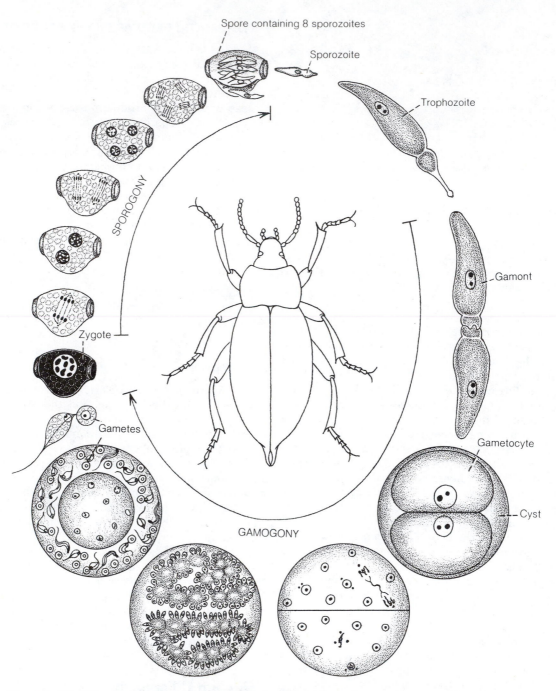

Spore containing 8 sporozoites

Sporozoite

Trophozoite

SPOROGONY

Gamont

Zygote

Gametes

Gametocyte

Cyst

GAMOGONY

Figure 2–25   Life cycle of a gregarine, *Stylocephalus longicollis*, an intestinal parasite of a beetle. There is no schizogony in this species. (After Grell from Noble and Noble.)

All possess cilia or compound ciliary structures as locomotor or food-acquiring organelles at some time in the life cycle. Also present is an infraciliary system, composed of ciliary basal bodies, or kinetosomes, below the level of the cell surface and associated with fibrils that run in various directions. Such an infraciliary system may be present at all stages in the life cycle even with marked reduction in surface ciliation. Most ciliates possess a cell mouth, or cytostome. In contrast to the other protozoan classes, ciliates are characterized by the presence of two types of nuclei: one vegetative (the macronucleus, concerned with the synthesis of RNA as well as DNA) and the other reproductive (the micronucleus, concerned only with the synthesis of DNA). Fission is transverse, and sexual reproduction never involves the formation of free gametes.

Ciliates are widely distributed in both fresh and marine waters and in the water films of soil. About one third of ciliate species are ecto- and endocommensals or parasites.

Form and Structure

The body shape is usually constant and in general is asymmetrical; however, radial symmetry with an anterior mouth is probably the primitive condition (Fig. 2–26). Although the majority of ciliates are

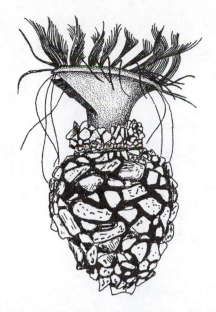

Figure 2–27 *Tintinnopsis*, a marine ciliate with lorica, or test, composed of foreign particles. Note conspicuous membranelles and tentacle-like organelles interspersed between them. (After Fauré-Fremiet from Corliss.)

solitary and free swimming, there are both sessile and colonial forms. The bodies of tintinnids and some heterotrichs, peritrichs, and suctorians are housed within a lorica, a girdle-like encasement, which is either secreted or composed of foreign material cemented together. In the peritrichs the lorica is attached to the substratum, but in many others the lorica is carried about by the organism (Fig. 2–27).

The ciliate body is typically covered by a complex, living pellicle, usually containing a number of different organelles. The pellicular system has been studied in detail in numerous genera, including *Paramecium*. There is an outer limiting plasma membrane, which is continuous with the membrane, surrounding the cilia. Beneath the outer membrane are closely packed vesicles, or alveoli, which are moderately to greatly flattened (Figs. 2–28 and 2–29). The outer and inner membranes bounding a flattened alveolus would thus form a middle and inner membrane of the ciliate pellicle. Between adjacent alveoli emerge the cilia and mucigenic or other bodies (Fig. 2–29). Beneath the alveoli are located the infraciliary system, the kinetosomes, and fibrils. The alveoli contribute to the stability of the pellicle and perhaps limit the permeability of the cell surface (Pitelka, 1970).

The pellicle of the familiar *Paramecium* has inflated kidney-shaped alveoli (Fig. 2–29). The in-

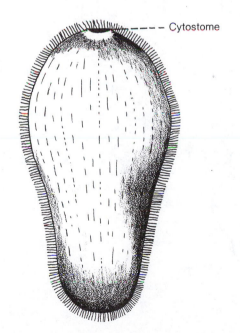

— Cytostome

Figure 2–26 *Prorodon*, a primitive ciliate. (After Fauré-Fremiet from Corliss.)

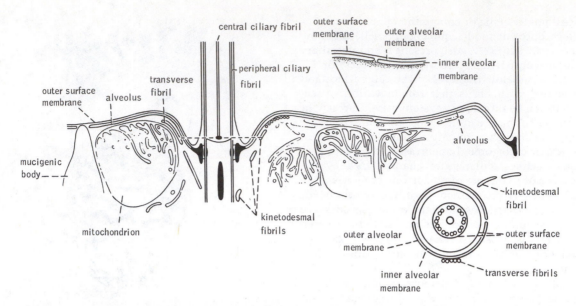

Figure 2–28 Section through cilium and pellicle of *Colpidium*. Note that alveoli are greatly flattened and their inner and outer membranes fused at base of cilium. At top right is an enlarged view of surface and alveolar membranes. At lower right is a cross section of a cilium and surrounding pellicle taken at the level indicated by the dashed line. Note the circle of nine doubled peripheral ciliary fibrils. (After Pitelka.)

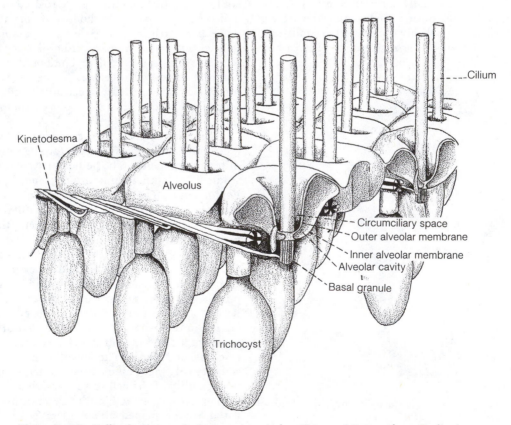

Figure 2–29 Pellicular system in *Paramecium*. (After Ehret and Powers from Corliss.)

flated condition and the shape of the alveoli produce a polygonal space about the one or two cilia that arise between them. Alternating with the alveoli are bottle-shaped organelles, the trichocysts, which form a second, deeper, compact layer of the pellicular system.

The trichocyst is a peculiar rodlike organelle characteristic of many ciliates that can be explosively discharged as a filament. In the undischarged state they are oriented at right angles to the body surface. The discharged trichocyst consists of a long, striated, threadlike shaft surmounted by a barb, which is shaped somewhat like a golf tee (Fig. 2–30). The shaft is not evident in the undischarged state and is probably polymerized in the process of discharge. The function of trichocysts is uncertain, but they may be used in anchoring the ciliate when feeding.

Toxicysts are vesicular organelles found in the pellicle of gymnostomes (*Dileptus* and *Didinium*), which on discharge consist of bulbous bases that taper into long threads (Fig. 2–31*A* and *B*). Toxicysts are used for defense or for capturing prey by paralysis and cytolysis. They are commonly restricted to the parts of the ciliate body that contact

prey, such as around the smooth region in *Didinium*.

Mucigenic bodies (mucocysts) are another group of pellicular organelles found in some ciliates. They are arranged in rows like trichocysts

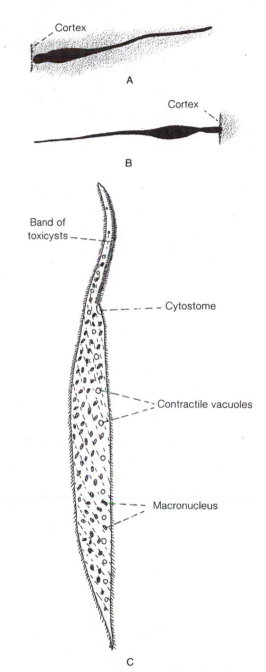

Figure 2–31 *A* and *B*, Toxicyst of *Dileptus anser*: *A*, Before discharge. *B*, After discharge. (After Hayes from Hall.) *C*, *Dileptus gigas*. (After Hyman, L. H., 1940: The Invertebrates, McGraw-Hill Book Co., N.Y. Vol. I.)

Figure 2–30 Electron micrograph of discharged trichocysts of *Paramecium*. Note golf-tee–shaped barb and long, striated shaft. (By Jacus and Hall, 1946: Biol. Bull., *91*:141.)

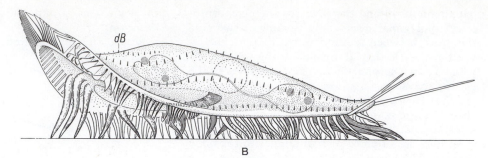

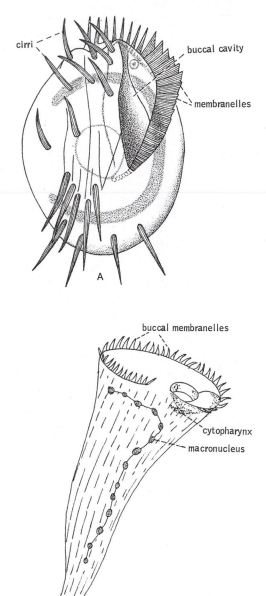

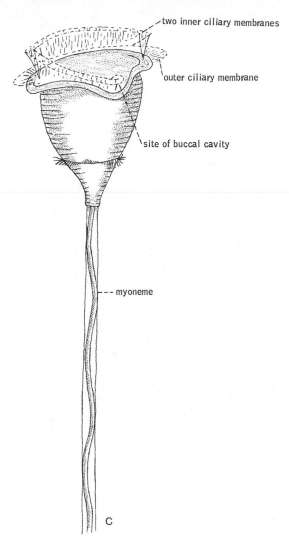

Figure 2–32 *A,* Ventral view of the hypotrich *Euplotes.* (After Pierson from Kudo.) *B,* Lateral view of the hypotrich *Stylonychia mytilus.* The ventral view would be rather similar to *Euplotes.* (By H. Machemer, from Grell, K. G., 1973: Protozoology. Springer-Verlag, N.Y. p. 304.) *C,* The peritrich *Vorticella. D,* The heterotrich *Stentor.* (After Tartar from Manwell.)

and discharge a mucoid material that may function in the formation of cysts or protective coverings (Fig. 2–28).

Cilia have the same structure as flagella; they differ from flagella chiefly in that they are generally more numerous and shorter. Compound ciliary organelles, evolved from the adhesion of varying numbers of individual cilia, are of common occurrence and will be described later.

The ciliature can be conveniently divided into the body (or somatic) ciliature, which occurs over the general body surface, and the oral ciliature, which is associated with the mouth region. Distribution of body cilia is quite variable. In the primitive groups, cilia cover the entire animal and are arranged in longitudinal rows (Fig. 2–26), but in many of the more specialized groups they have be-

come limited to certain regions of the body (Figs. 2–32*A* and *C*).

As mentioned earlier, each cilium arises from a basal body or kinetosome, located in the alveolar layer (Figs. 2–29 and 2–33). The kinetosomes that form a particular longitudinal row are connected by means of fine, striated fibrils, called kinetodesma. The cilia, kinetosomes, and fibrils of a row make up a kinety. The longitudinal bundle of fibrils runs to the right side of the row of kinetosomes, and each kinetosome gives rise to one kinetodesmos (fibril), which joins the longitudinal bundle and extends anteriorly. Single kinetodesmata are tapered and extend for varying distances as parts of the bundle. At the kinetosome, the kinetodesmata are connected to certain of the kinetosome triplets.

Figure 2–33 Reconstruction of section of the pellicle of *Tetrahymena*. Right side is on viewer's left. Abbreviations: kinetodesmos (k); transverse microtubules (tm); postciliary microtubules (pm); longitudinal microtubules (lm); basal microtubules (bm); alveolus (a); cilium (c); epiplasm (e); mitochondrion (m); mucigenic body (mb). (From Allen, R. D., 1967: Fine structure, reconstruction and possible function of components of the cortex of *Tetrahymena pyriformis*. J. Protozool., *14*:553–565.)

In addition to kinetodesmata, there are also ribbons of microtubules that extend posteriorly and transversally from the kinetosome and apparently function as part of the cilium anchorage system.

A kinety system is characteristic of all ciliates, although there are variations in details of the pattern. Even groups such as the Suctorida, which possess cilia only during developmental stages, retain part of the kinety system in the adult.

Locomotion

The ciliates are the fastest moving of the protozoa. In its beat each cilium performs an effective and a recovery stroke. During the effective stroke the cilium is outstretched and moves from a forward to a backward position (Fig. 2–34*A* and *B*). In the recovery stroke the cilium is bent over to the right against the body (when viewed from above and looking anteriorly) and is brought back to the forward position in a counterclockwise movement. The recovery position offers less water resistance and is somewhat analogous to feathering an oar. A cilium moves in three different planes in the course of a complete cycle of beat, and the positions have been captured and recorded in scanning electron micrographs of freeze-dried *Paramecium* (Tamm, 1972).

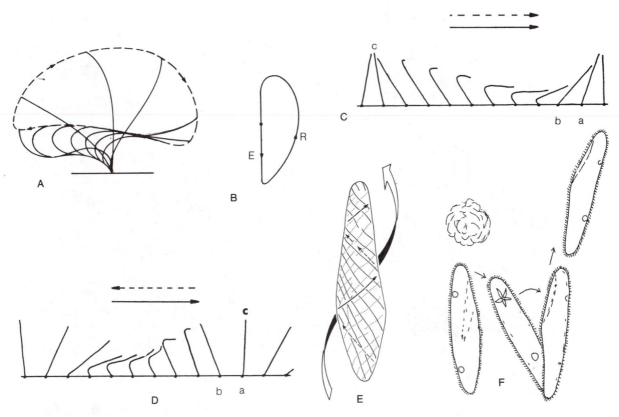

Figure 2–34 Ciliary beating and locomotion. *A*, Cycle of a ciliary beat seen from the side. In the effective stroke the cilium is outstretched and moves from left to right. *B*, Path described by tip of cilium during beat cycle as seen from surface. E is effective stroke and R is the recovery stroke. (From Sleigh, M. A., 1973: The Biology of Protozoa. Edward Arnold Publishers, London. p. 38.) *C*, Series of adjacent cilia in a metachronal wave in various stages of the beat cycle. Direction of effective stroke (solid arrow) is same as metachronal wave (dashed arrow). Letter *c* indicates wave crest. *D*, As in *B* except metachronal wave is moving in opposite direction to effective stroke. (*C* and *D* from Jones, A. R., 1974: The Ciliates. St. Martin's Press, Hutchinson in London. p. 70.) *E*, Metachronal waves in *Paramecium* during forward swimming. Wave crests are shown by lines and their direction by arrows (dotted on opposite side). Movement of ciliate is indicated by large arrow. (From Machemer, H., 1974: Ciliary activity and metachronism in Protozoa. *In* Sleigh, M. A. (Ed.): Cilia and Flagella. Academic Press, London. p. 224.) *F*, The avoiding reaction of *Paramecium*. (After Hyman, L. H., 1940: The Invertebrates, McGraw-Hill Book Co., N.Y. Vol. I.)

The movements of adjacent cilia are coupled as a result of interference effects of the surrounding water layers. Thus, hydrodynamic forces impose a coordination on the cilia. The beat of individual cilia, rather than being random or synchronous, is part of the metachronal waves that sweep along the length of the body (Fig. 2–34C). Most commonly, the metachronal waves pass at right angles to the beat stroke, but there are variations in the pattern (see Sleigh, 1973). There is no evidence that the infraciliature of fibrils functions as a conducting system in coordination. They may serve primarily in anchorage.

In forms such as *Paramecium* the direction of the effective stroke is oblique to the long axis of the body (Fig. 2–34E). This causes the ciliate to swim in a spiral course and at the same time to rotate on its longitudinal axis. The ciliary beat can be reversed, and the animal can move backward. This backward movement is associated with the so-called avoiding reaction. In *Paramecium*, for example, when the animal comes in contact with some undesirable substance or object, the ciliary beat is reversed (Fig. 2–34F). The animal moves backward a short distance, turns slightly clockwise or counterclockwise, and moves forward again. If unfavorable conditions are still encountered, the avoiding reaction is repeated. External stimuli are probably detected through certain long, stiff cilia that play no role in movement and are probably entirely sensory. The direction and intensity of the beat are controlled by levels of Ca^{++} and K^{+} ions (Eckert, 1972).

The highly specialized hypotrichs, such as *Urostyla*, *Stylonychia*, and *Euplotes* (Fig. 2–32A), have greatly modified body cilia. The body has become differentiated into distinct dorsal and ventral surfaces, and cilia have largely disappeared except on certain areas of the ventral surface. Here the cilia occur as a number of tufts, called cirri. The cilia of a cirrus beat together, and coordination here is believed to result from some sort of structural coupling as a result of the close proximity of their bases.

Some ciliates, especially sessile forms, can undergo contractile movements, either shortening the stalk by which the body is attached, as in *Vorticella* (Fig. 2–32C), or shortening the entire body, as in *Stentor* (Fig. 2–32D). Contraction is brought about by bundles of contractile filaments, or myonemes, that lie in the pellicle. In *Vorticella* and the colonial *Carchesium*, both of which have bell-shaped bodies attached by a long slender stalk, the myonemes extend into the stalk as a single, large, spiral fiber. The contractions of this spiral myoneme, which functions very much like a coiled spring, produce the familiar popping movements that are so characteristic of *Vorticella* and related genera.

Nutrition

Feeding in ciliates parallels, on a microscopic level, feeding in multicellular animals. Typically a distinct mouth, or cytostome, is present, although it has been secondarily lost in some groups. In primitive groups the mouth is located anteriorly (Fig. 2–26), but in most ciliates it has been displaced posteriorly to varying degrees. The mouth opens into a canal or passageway called the cytopharynx, which is separated from the endoplasm by a membrane. It is this membrane that enlarges and pinches off as a food vacuole. The wall of the cytopharynx is strengthened with rods (nemadesmata) arranged like the staves of a barrel. Primitively, the ingestive organelles consist only of the cytostome and cytopharynx (Figs. 2–26 and 2–35A), but in the majority of ciliates the cytostome is preceded by a preoral chamber. The preoral chamber may take the form of a vestibule, which varies from a slight depression to a deep funnel, with the cytostome at its base (Fig. 2–35B). The vestibule is clothed with simple cilia derived from the somatic ciliature.

In the higher ciliates the preoral chamber is typically a buccal cavity, which differs from a vestibule by containing compound ciliary organelles instead of simple cilia (Fig. 2–35C to F). There are two basic types of such ciliary organelles: the undulating membrane and the membranelle. An undulating membrane is a row of adhering cilia forming a sheet (Fig. 2–36A and B). A membranelle is derived from two or three short rows of cilia, all of which adhere to form a more or less triangular or fan-shaped plate and typically occur in a series (Figs. 2–27, 2–32D, and 2–36B). Although there is no actual fusion of adjacent cilia in these compound organelles, their kinetosomes and bases are sufficiently close together to produce some sort of structural coupling that causes all of the cilia of a membranelle to beat together.

The term *peristome*, which is commonly encountered in the literature, is synonymous with *buccal cavity*. In members of a number of orders the buccal organelles project from the buccal cavity, or, as in the Hypotrichida (Fig. 2–32A), the buccal cavity is somewhat shallow so that the organelles occupy a flattened area around the oral re-

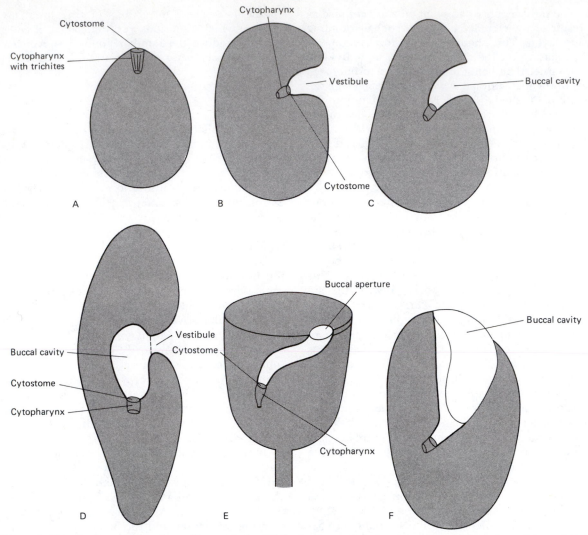

Figure 2–35 Oral areas of various ciliates. *A*, In rhabdophorine gymnostomes, such as *Coleps, Prorodon,* and *Didinium. B*, In a trichostome such as *Colpoda,* with a vestibule that is displaced from anterior end. *C*, In a tetrahymenine hymenostome, such as *Tetrahymena. D*, In a peniculine hymenostome, such as *Paramecium. E*, In a peritrich, such as *Vorticella. F*, In a hypotrich, such as *Euplotes.* (Modified after Corliss, J. O., 1961: The Ciliated Protozoa. Pergamon Press, N.Y.)

gion. Such an area is called the peristomial field. In forms like *Paramecium* there is a vestibule in front of the buccal cavity (Figs. 2–35*D* and 2–36*D*).

The free-swimming holozoic species display several types of feeding habits. Some are raptorial, and attack and devour rotifers, gastrotrichs, protozoans, and other ciliates. A smaller number, including *Nassula,* are herbivorous on algae and diatoms. Many have become specialized for suspension feeding. The oral apparatus of raptorial ciliates is typically limited to the cytostome and cytopharynx.

Didinium has perhaps been most studied of all the raptorial feeders. This little barrel-shaped ciliate feeds on other ciliates, particularly *Paramecium* (Fig. 2–37*A*). When *Didinium* attacks a *Paramecium,* it discharges toxicysts into the *Paramecium* and the proboscis-like anterior end attaches to the prey through the terminal mouth, which can open almost as wide as the diameter of the body.

An interesting group of raptorial ciliates is the aberrant subclass Suctoria, formerly considered a separate class. Free-living suctorians are all sessile

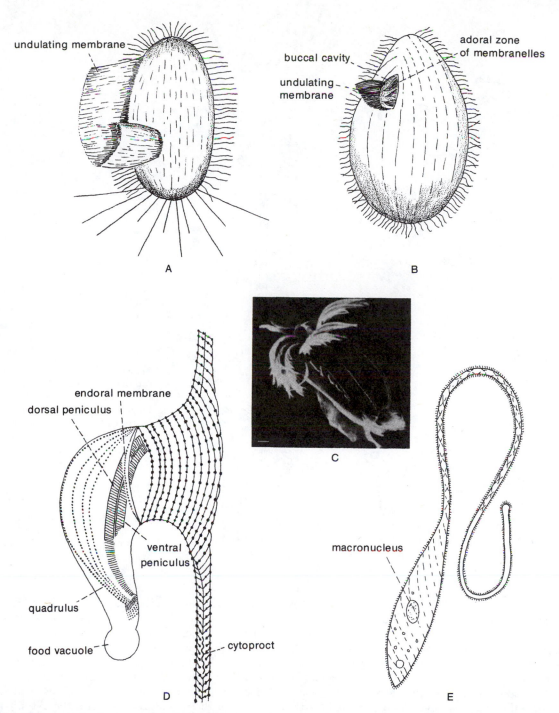

Figure 2–36   *A, Pleuronema.* (After Noland from Corliss.) *B, Tetrahymena.* (After Corliss, J. O., 1961: The Ciliated Protozoa. Pergamon Press, N.Y.) *C,* Scanning electron photomicrograph of *Uronychia,* a marine ciliate, showing the highly developed membranelles. (By Small, E. B., and Marszalek, D. S., 1969: Science, 163: 1064–1065. Copyright 1969, American Association for the Advancement of Science.) *D,* Buccal organelles of *Paramecium.* (After Yusa from Manwell.) *E, Lacrymaria.* (After Conn from Hyman.)

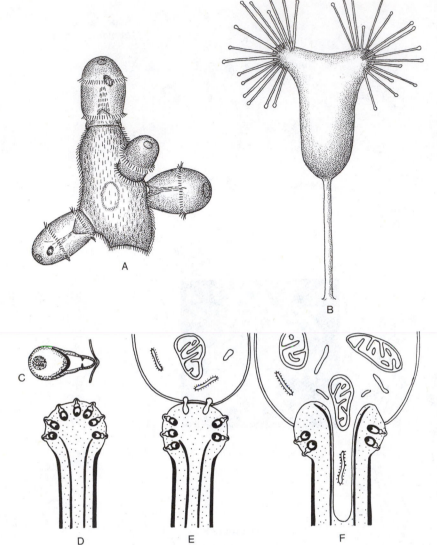

Figure 2–37 *A*, Four *Didinium*, raptorial ciliates, attacking one *Paramecium*. (After Mast from Dogiel.) *B*, *Acineta*, a suctorian. (After Calkins from Hyman.) *C–F*, Suctorian haptocysts and prey capture. Haptocyst isolated (*C*) and within tentacle tip (*D*). Attachment of tentacle to prey (*E*) and engulfment through tentacle (*F*). (From Sleigh, M. A., 1973: The Biology of Protozoa, Edward Arnold Publishers, London. p. 64. Based on micrographs of Rudzinska, Bardele, and Grell.)

and are attached to the substratum directly or by means of a stalk (Fig. 2–37). Cilia are present only in the immature stages. The body bears tentacles, which may be knobbed at the tip or shaped like long spines (Fig. 2–41*B*). Each tentacle is supported by a cylinder of microtubules and carries special organelles, called haptocysts (Fig. 2–37*D*). Suctorians feed on other ciliates, and when prey strikes the tentacles, the haptocysts are discharged into the prey body, anchoring it to the tentacles (Figs. 2–37*D* to *F* and 2–38). The contents of the prey are then sucked through the tubular tentacle into the suctorian, where they are collected into food vacuoles.

Typically characteristic of suspension feeders is the buccal cavity. Food for suspension feeders consists of any small organic particles, dead or living, particularly bacteria that are suspended in water. Food is brought to the body and into the buccal cavity by the compound ciliary organelles. From the buccal cavity the food particles are driven through the cytostome and into the cytopharynx. When the particles reach the cytopharynx, they collect within a food vacuole.

The order Hymenostomatida—"membrane-mouthed"—contains some of the most primitive suspension feeders. *Tetrahymena* is a good example of such a primitive type (Fig. 2–36*B*). The cy-

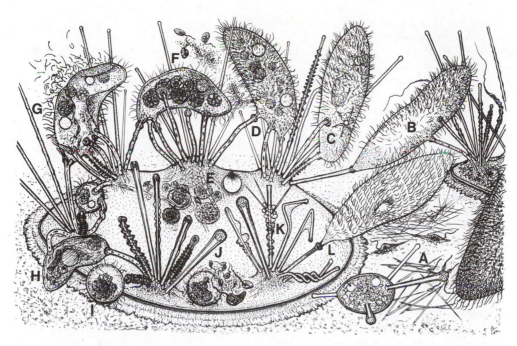

Figure 2–38 A colony of the suctorian *Heliophrya* feeding on *Paramecium*. Some individuals of *Paramecium* have just been captured. Others have been ingested to various degrees. (From Spoon et al., 1976: Observations on the behavior and feeding mechanisms of the suctorian *Heliophrya erhardi* preying on *Paramecium*. Trans. Am. Micros. Soc., 95:443–462.)

tostome is located a little behind the leading edge of the body. Just within the broad opening to the buccal cavity are four ciliary organelles—an undulating membrane on the right side of the chamber and three membranelles on the left. The three membranelles constitute an adoral zone of membranelles, which in many higher groups of ciliates is much more developed and extensive.

In *Paramecium*, the most familiar genus of the order, an oral groove along the side of the body leads posteriorly to a vestibule, located about midway back from the anterior end. The vestibule, buccal cavity, and cytopharynx together form a large, curved funnel (Figs. 2–35*D* and 2–36*D*). The undulating membrane, here called the endoral membrane, runs transversely along the right wall and marks the junction of the vestibule and buccal cavity. The three membranelles are also modified. Two, called peniculi, are greatly lengthened and thus tend to be more similar to an undulating membrane in function than to the more typical membranelle.

In feeding, the cilia of the oral groove produce a current of water that sweeps in an arclike manner down the side of the body and over the oral region. The ciliature of the vestibule and buccal cavity pull in food particles and drive them into the forming food vacuole.

In the subclass Peritricha, whose members possess little or no somatic cilia, the buccal ciliary organelles are highly developed and form a large, disclike, peristomial field at the apical end of the animal. In the much-studied peritrich genus *Vorticella*, a peripheral shelflike projection can close over the disc when the animal is retracted (Figs. 2–32*C* and 2–35*E*). The ciliary organelles lie in a peristomial groove between the edge of the disc and the peripheral shelf and consist of two ciliary bands, which wind in a counterclockwise direction around the margin of the disc and then turn downward into the funnel-shaped buccal cavity. The inner ciliary band produces the water current and the outer band acts as the filter. Suspended particles, mostly bacteria, are transported along with a stream of water between the two bands into the buccal cavity.

Ciliates of the subclass Spirotricha, which includes such familiar forms as *Stentor*, *Halteria*, *Spirostomum*, and *Euplotes*, are typically suspension feeders. They usually possess a highly developed adoral zone of many membranelles (Fig. 2–32*A* and *D*).

Within the cytopharynx of all ciliates, food particles enter the food vacuole. When the food vacuole reaches a certain size, it breaks free from the cytopharynx, and a new vacuole forms in its place.

Detached vacuoles then begin a more or less circulatory movement through the endoplasm.

Digestion follows the usual pattern, and a pH as low as 1.4 has been reported during the acid phase in some species *(Paramecium)*. Following digestion, the waste-laden food vacuole moves to a definite anal opening, or cytopyge (Fig. 2–1), at the body surface and expels its contents. The cytopyge varies in position. In *Tetrahymena* it is located on the side of the animal, near the posterior end (Fig. 2–1), whereas in the peritrichs it opens into the buccal cavity.

There are relatively few parasitic ciliates, although there are many ecto- and endocommensals. Many suctorians are commensals, and a few are parasites. Hosts include fishes, mammals, various invertebrates, and other ciliates. *Sphaerophyra*, for example, lives within the endoplasm of *Stentor*, and *Endosphaera* is parasitic within the body of the peritrich *Telotrochidium*.

Other interesting commensal ciliates include *Kerona*, a little crawling hypotrich, and *Trichodina*, a mobile peritrich, which are ectocommensals on the surface of hydras. There are also some free-swimming peritrichs that occur on the body surfaces of freshwater planarians, tadpoles, sponges, and other animals.

The genus *Balantidium* includes many species that are endocommensals or endoparasites in the guts of insects and many different vertebrates. *Balantidium coli* is an endocommensal in the intestines of pigs and is passed by means of cysts in the feces. This ciliate has occasionally been found in humans, where in conjunction with bacteria it erodes pits in the intestinal mucosa and produces pathogenic symptoms. The related highly specialized Entodiniomorphida (Fig. 2–39) live as harmless commensals in the digestive tracts of many different hoofed mammals. Like the flagellate symbionts of termites and roaches, some of them ingest and break down the cellulose of the vegetation eaten by their hosts. The products of digestion are utilized by the host.

A few ciliates display symbiotic relationships with algae. The most notable of these is *Paramecium bursaria*, in which the endoplasm is filled with green zoochlorellae.

Water Balance

Contractile vacuoles are found in both marine and freshwater species, but especially in the latter, in which they may discharge as rapidly as every few

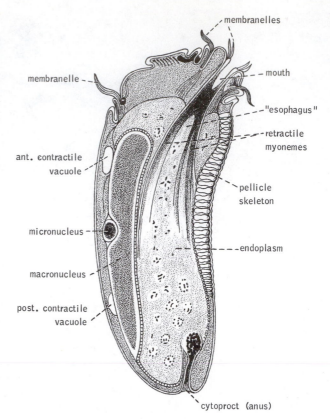

Figure 2–39 The entodiniomorph, *Diplodinium ecaudatum*, a commensal in the rumen of cattle. (After Manwell, R. D., 1961: Introduction to Protozoology. St. Martin's Press, N.Y.)

seconds. In some species a single vacuole is located near the posterior, but many species possess more than one vacuole (Fig. 2–31C). In *Paramecium* one vacuole is located at both the posterior and the anterior of the body (Fig. 2–34F). The vacuoles are always associated with the innermost region of the ectoplasm and empty through one or two permanent pores that penetrate the pellicle. The spongiome contains a network of irregular tubules, which may empty into the vacuole directly or by way of collecting tubules (Fig. 2–2).

When there is more than one vacuole present, they pulsate at different rates depending on their positions. For example, in *Paramecium* the posterior vacuole pulsates faster than the anterior vacuole because of the large amount of water being delivered into the posterior region by the cytopharynx. Although contractile vacuoles may be present in marine species, the rate of pulsation is considerably slower than that in freshwater species; they are probably removing ingested water.

Reproduction

Ciliates differ from almost all other organisms in possessing two distinct types of nuclei—a usually large macronucleus and one or more small micronuclei. The micronuclei are small, rounded bodies and vary in number from 1 to as many as 20, depending on the species. They are diploid, with little RNA. The micronucleus is a store of genetic material, is responsible for genetic exchange and nuclear reorganization, and also gives rise to the macronuclei. The macronucleus is sometimes called the vegetative nucleus, since it is not critical in sexual reproduction. However, the macronucleus is essential for normal metabolism, for mitotic division, and for the control of cellular differentiation, and it is responsible for the genic control of the phenotype through protein synthesis.

One or more macronuclei are present, and they may assume a variety of shapes (Fig. 2–40). The large macronucleus of *Paramecium* is somewhat oval or bean shaped and is located just anterior to the middle of the body. In *Stentor* and *Spirostomum* the macronuclei are long and arranged like a string of beads. Not infrequently the macronucleus is in the form of a long rod bent in different configurations, such as a C in *Euplotes* or a horseshoe in *Vorticella*. The macronucleus is highly polyploid, the chromosomes having undergone repeated duplication following the micronuclear origin of the macronucleus. The macronuclei include numerous nucleoli with RNA.

ASEXUAL REPRODUCTION

Asexual reproduction is always by means of binary fission, which is typically transverse. More accurately, fission is described as being homothetogenic, with the division plane cutting across the kineties—the longitudinal rows of cilia or basal bodies (Fig. 2–41A). This is in contrast to the symmetrogenic fission of flagellates, in which the plane of division (longitudinal) cuts between the rows of basal granules. Mitotic spindles are formed only in the division of the micronuclei. Division of the macronuclei is usually accomplished by constriction. When a number of macronuclei are present, they may first combine as a single body before dividing. The same is true of some forms with beaded or elongated macronuclei.

Modified fission in the form of budding occurs in some ciliate groups, notably the Suctoria. In most members of this subclass the parent body buds off a varying number of daughter cells from the outer surface (Fig. 2–41B); or there is an internal cavity or brood chamber, and the buds form internally from the chamber wall. In contrast to the sessile adults, which lack cilia, the daughter cells, or buds, are provided with several circlets of cilia and are free swimming (Fig. 2–41C). Following a few hours of free existence, the "larva" attaches and assumes the characteristics of the sessile adults.

Although there are no centrioles, the kinetosomes of many ciliates, like the basal granules of flagellates, divide at the time of fission. Furthermore, the kinetosomes play a primary role in the re-formation of organelles. It has been found that all of the organelles can be re-formed providing the cell contains a piece of macronucleus and some kinetosomes. In the more primitive ciliates, in which the cilia have a general distribution over the body surface, the kinetosomes have equal potentials in the re-formation of organelles. However, in the

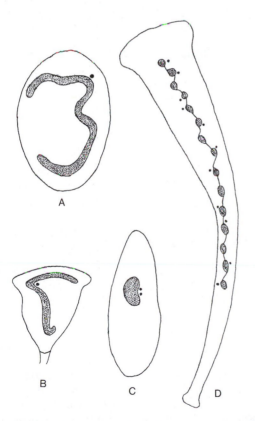

Figure 2–40 Macronuclei (in gray) of various ciliates (micronuclei, in black). *A, Euplotes. B, Vorticella. C, Paramecium. D, Stentor.* (After Corliss, J. O., 1961: The Ciliated Protozoa. Pergamon Press, N.Y.)

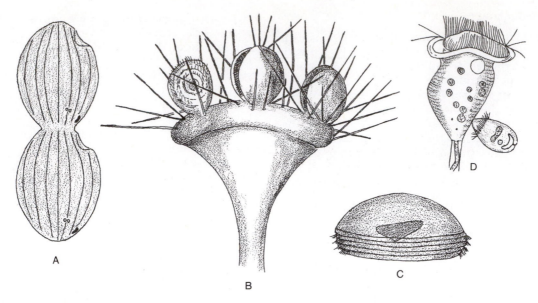

A

B

C

D

Figure 2–41 *A*, Homothetogenic type of fission, in which the plane of division cuts across the kineties. (After Corliss.) *B*, Suctorian *Ephelota* with external buds. (After Noble from Hyman.) *C*, Detached bud of *Dendrocometes*. (After Pestel from Hyman.) *D*, Conjugation in *Vorticella*. Note the small nonsessile microconjugant. (After Kent from Hyman.)

specialized ciliates there is a corresponding specialization of the kinetosomes; only certain ones are involved in the re-formation of new cellular structures during fission. For example, in hypotrichs such as *Euplotes*, all of the organelles are resorbed at the time of fission, and certain of the kinetosomes on the ventral side of the animal divide to form a special group that is organized in a definite field or pattern. These special "germinal" kinetosomes then migrate to different parts of the body, where they form all of the surface organelles—cirri, peristome, cytopharynx, and other structures.

SEXUAL REPRODUCTION

An exchange of nuclear material by conjugation is involved in sexual reproduction. By apparently random contact in the course of swimming, two sexually compatible members of a particular species adhere in the oral or buccal region of the body. Following the initial attachment, there is degeneration of trichocysts and cilia (but not kinetosomes) and a fusion of protoplasm in the region of contact. Two such fused ciliates are called conjugants; attachment lasts for several hours. During this period a reorganization and exchange of nuclear material occurs (Fig. 2–42*A* to *F*). Only the micronuclei are involved in conjugation; the macronu-

cleus breaks up and disappears either in the course of or following micronuclear exchange.

The steps leading to the exchange of micronuclear material between the two conjugants are fairly constant in all species. After two meiotic divisions of the micronuclei, all but one of them degenerate. This one then divides, producing two gametic micronuclei that are genetically identical. One is stationary; the other will migrate into the opposite conjugant. The migrating, or "male," nucleus in each conjugant moves through the region of fused protoplasm into the opposite member of the conjugating pair. There the "male" and "female" nuclei fuse with one another to form a "zygote" nucleus, or synkaryon.

Shortly after nuclear fusion the two animals separate; each is now called an exconjugant. After separation, there follow in each exconjugant a varying number of nuclear divisions, leading to the reconstitution of the normal nuclear condition characteristic of the species. This reconstitution usually, but not always, involves a certain number of cytosomal divisions. For example, in some forms where there is but a single macronucleus and a single micronucleus in the adult, the synkaryon divides once. One of the daughter nuclei forms a micronucleus; the other forms the macronucleus. Thus, the normal nuclear condition is restored without any cytosomal divisions.

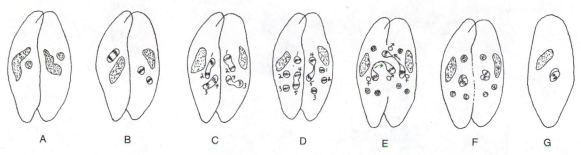

Figure 2–42 Sexual reproduction in *Paramecium caudatum*. *A* to *F*, Conjugation. *B* to *D*, Micronuclei undergo three divisions, the first two of which are meiotic. *E*, "Male" micronuclei are exchanged. *F*, They fuse with the stationary micronucleus of the opposite conjugant. *G*, Exconjugant with macronucleus and synkaryon micronucleus; other micronuclei have been resorbed. (Modified after Calkins from Wichterman.)

However, in *Paramecium caudatum*, which also possesses a single nucleus of each type, the synkaryon divides three times, producing eight nuclei. Four become micronuclei and four become macronuclei. The animal now undergoes two cytosomal divisions, during the course of which each of the four resulting daughter cells receives one macronucleus and one micronucleus. In those species that have numerous nuclei of both types, there is no cytosomal division; the synkaryon merely divides a sufficient number of times to produce the requisite number of macronuclei and micronuclei.

In some of the more specialized ciliates, the conjugants are a little smaller than nonconjugating individuals, or the two members of a conjugating pair are of strikingly different sizes. Such dioecious macro- and microconjugants occur in *Vorticella* (Fig. 2–41*D*) and represent an adaptation for conjugation in sessile species. The macroconjugant, or "female," remains attached, while the small bell of the microconjugant, or "male," breaks free from its stalk and swims about. On contact with an attached macroconjugant the two bells adhere. A synkaryon forms only in the macroconjugant from one gametic nucleus contributed by each conjugant. However, there is no separation after conjugation, and the little "male" conjugant degenerates. In the Suctoria conjugation takes place between two attached individuals that happen to be located side by side.

The frequency of conjugation is extremely variable. Some species have rarely been observed to undergo the sexual phenomenon of conjugation; others conjugate every few days or weeks. In some species a period of "immaturity," in which only fission occurs, precedes a period in which individuals are capable of conjugation. Numerous factors, such as temperature, light, and food supply, are known to induce or influence conjugation.

In some ciliates the nuclear reorganization following conjugation seems to have a rejuvenating effect and is necessary for continued asexual fission. For example, it has been shown that some species of *Paramecium* can pass through only approximately 350 continuous asexual generations. If nuclear reorganization does not occur, the asexual line (or clone) will die out. Another type of nuclear reorganization called autogamy may take place and has the same effect on fission as does conjugation. Autogamy involves the same nuclear behavior as does conjugation, but there is no conjugation and no exchange of micronuclear material between two individuals. The macronucleus degenerates and the micronucleus divides a number of times to form eight or more nuclei. Two of these nuclei fuse to form a synkaryon; the others degenerate and disappear. The synkaryon then divides to form a new micronucleus and macronuclei, as occurs in conjugation.

Definite mating types have been shown to exist in species of *Paramecium*, *Tetrahymena*, *Euplotes*, *Stylonychia*, and some other ciliates. For example, there are a number of varieties, or syngens, of *Paramecium caudatum* and *Paramecium aurelia*, each with two or more mating types. Conjugation is always restricted to a member of the opposite mating type within the same syngen and does not occur between members of the same type, apparently owing to a failure of the surfaces to adhere. The mating types are hereditary.

Most ciliates are capable of forming resistant cysts in response to unfavorable conditions, such as lack of food or desiccation. Encystment is important in carrying the species through winter or

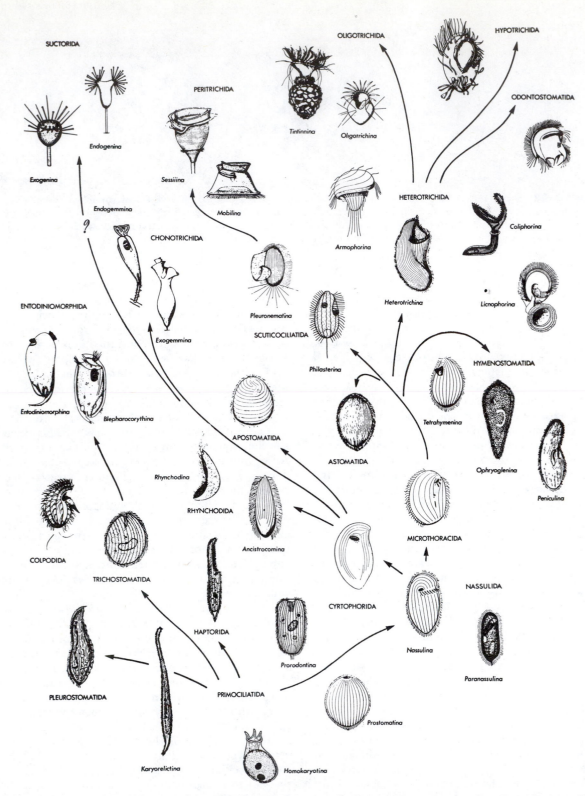

Figure 2–43 Possible phylogeny of the ciliated protozoans. See the Systematic Résumé for characters of main subdivisions. (From Corliss, J. O., 1974: The changing world of ciliate systematics: historical analysis of past efforts and a newly proposed phylogenetic scheme of classification for the protistan phylum Ciliophora. Syst. Zool., 23(1):91–138.)

dry periods and providing a condition in which the organism can be transported by wind or in mud on the feet of birds or other animals. In some forms, such as *Colpoda*, reproductive processes occur only during the encysted state. There are also some ciliates, including *Paramecium*, that are believed never to encyst.

SYSTEMATIC RÉSUMÉ OF PHYLUM CILIOPHORA

The following system of classification follows that of Corliss (1979).

Class Kinetofragminophora. Isolated kineties in oral region of body bearing cilia but not compound ciliary organelles.
  Subclass Gymnostomata. Cytostome at or near surface of body and located at the anterior end or laterally. Somatic ciliation generally uniform. *Stephanopogon, Loxodes, Coleps, Prorodon, Actinobolina, Didinium, Dileptus, Lacrymaria, Litonotus,* and *Loxophyllum.*
  Subclass Vestibulifera. Cytostome within a vestibulum (vestibule) bearing distinct ciliature. Free-living and symbiotic species. *Balantidium, Colpoda, Blepharocorys, Entodinium.*
  Subclass Hypostomata. Body cylindrical or dorsoventrally flattened, with the mouth on the ventral side in either case. Somatic ciliature often reduced. Free-living and many symbiotic species. *Synhymenia, Nasula, Microthorax, Hypocoma, Trochilioides, Chilodochona, Lobochona, Spirochona, Stylochona, Ancistrocoma, Foettingeria, Chromidina, Ascophrys.*
  Subclass Suctoria. Sessile, generally stalked, with tentacles at the free end. Cilia lacking in the adult but present in the free-swimming larval stage. Most are ectosymbionts on aquatic invertebrates. *Ephelota, Podophrya, Acineta.*
Class Oligohymenophora. Oral apparatus usually well developed and containing compound ciliary organelles.
  Subclass Hymenostomata. Body ciliation commonly uniform and oral structures not conspicuous. *Colpidium, Glaucoma, Tetrahymena, Paramecium, Pleuronema.*
  Subclass Peritricha. Mostly sessile forms with reduced body ciliation. Oral ciliary band usually conspicuous. *Carchesium, Epistylis, Lagenophrys, Vorticella, Zoothamnium, Trichodina.*
Class Polyhymenophora. Oral region with conspicuous adoral zone of buccal membranelles. Some species with uniform body ciliation; others with compound organelles, such as cirri.
  Subclass Spirotricha. With characteristics of class.
    Order Heterotrichida. Mostly large ciliates with uniform body ciliation. *Blepharisma, Bursaria, Spirostomum, Stentor, Folliculina.*
    Order Odontostomatida. Laterally compressed, wedge-shaped ciliates with reduced body ciliation. *Saprodinium.*
    Order Oligotrichida. Ciliates with reduced somatic ciliature but with extensive projecting buccal ciliary organelles. *Halteria.* The suborder Tintinnina contains loricate species—*Codonella, Favella, Tintinnopsis, Tintinnus.*
    Order Hypotrichida. Dorsoventrally flattened ciliates with cirri on the ventral side. *Urostyla, Euplotes, Uronychia, Stylonychia.*

SUMMARY

1 The ciliates constitute the largest phylum of protozoa (phylum Ciliophora). They are also the most animal-like and exhibit a very high level of organelle development.

2 They possess cilia for locomotion and in many species for suspension feeding. Ciliary beat is more or less planar, with an effective and recovery stroke. Associated with the kinetosomes is a complex anchorage system of fibrils, all of which make up the infraciliature.

3 The body wall of ciliates is a complex living pellicle, containing alveoli, trichocysts, and other organelles, in addition to the infraciliature.

4 Primitively, the body surface is covered with uniform cilia, which function in locomotion. There has been a tendency, however, for this somatic ciliature to become reduced or in some groups to become specialized as well (cirri).

5 The cilia around the mouth region (buccal ciliature) have become specialized as compound ciliary organelles called membranelles and undulating membranes in many ciliates that employ suspension feeding.

6 In addition to suspension feeding on bacteria and other particles, some ciliates feed on algae, and many are carnivorous on other protozoa and microscopic animals. The cytostome and cytopharynx open into a food vacuole. The undigestible remains are discharged through a fixed cell anus (cytoproct).

The discharge position of the contractile vacuoles is also constant.

7 Ciliates reproduce asexually by transverse fission and sexually by conjugation. Conjugation involves an exchange of micronuclei, each of which fuses with a nonmigratory micronucleus to form a zygote nucleus. Conjugation is preceded by meiotic divisions of one micronucleus and is followed by reconstitution of the normal nuclear condition, which may involve fission.

REFERENCES

The general parasitology texts listed at the beginning of the references for the chapter on flatworms (p. 204) contain additional information on parasitic protozoa.

Allen, R. D., Francis, D., and Zek, R., 1971: Direct test of the positive pressure gradient theory of pseudopod extension and retraction in amoebae. Science, 174:1237–1240. (See also Science [1972], 177:636–638 for additional contributions to the debate on the nature of ameboid movement.)

Anderson, O. R., 1983: Radiolaria. Springer-Verlag, Berlin. 355 pp.

Anderson, O. R., 1983: The radiolarian symbiosis. In Goff, L. J. (Ed.): Algal Symbiosis. Cambridge University Press, p. 69.

Bannister, L. H., 1972: The structure of trichocyst in Paramecium caudatum. J. Cell Sci., 11:899–929.

Bé, A. W. H., 1982: Biology of planktonic Foraminifera. Univ. Tenn., Studies in Geology, 6:51–92.

Bick, H., 1972: Ciliate Protozoa. An illustrated guide to the species as biological indicators in freshwater biology. Am. Public Health Assoc., Washington, D.C., 198 pp.

Bick, H., 1973: Population dynamics of Protozoa associated with the decay of organic materials in fresh water. Am. Zool., 13:149–160.

Bold, H. C., and Wynne, M. J., 1985: Introduction to the Algae: Structure and Reproduction. Prentice-Hall, Englewood Cliffs, N.J. 848 pp.

Borror, A. C., 1973: Protozoa: Ciliophora. Marine Flora and·Fauna of the Northeastern United States. NOAA Tech. Report NMFS Circular 378. U. S. Printing Office, 62 pp.

Boynton, J. E., and Small, E. B., 1984: Ciliates by the slice. Science Teacher (Feb.), pp. 35–38.

Corliss, J. O., 1972: The ciliate Protozoa and other organisms; some unresolved questions of major phylogenetic significance. Am. Zool., 12(4):739–753.

Corliss, J. O., 1973: Protozoa. In Gray, P. (Ed.): Encyclopedia of Microscopy and Microtechnique. Van Nostrand Reinhold Co., N.Y

Corliss, J. O., 1974: The changing world of ciliate systematics: historical analysis of past efforts and a newly proposed phylogenetic scheme of classification for the phylum Ciliophora. Syst. Zool., 23(1):91–138.

Corliss, J. O., 1975: Taxonomic characterization of the superfamilial groups in a revision of recently proposed schemes of classification of the phylum Ciliophora. Trans. Am. Micros. Soc., 94(2):224–267.

Corliss, J. O., 1977: Annotated assignment of families and genera to the orders and classes currently comprising the Corlissian scheme of higher classification for the phylum Ciliophora. Trans. Am. Micros. Soc., 96(1):104–140.

Corliss, J. O., 1979: The Ciliated Protozoa: Characterization, Classification, and Guide to the Literature. 2nd Edition. Pergamon Press, N.Y. 455 pp.

Corliss, J. O., and Esser, S. C., 1974: Comments on the role of the cyst in the life cycle and survival of free-living Protozoa. Trans. Am. Micros. Soc., 93(4):579–593.

Dodson, E. O., 1971: The kingdoms of organisms. Syst. Zool., 20:265–281.

Eckert, R., 1972: Bioelectric control of ciliary activity. Science, 176:473–481.

Elliott, A. M. (Ed.), 1973: Biology of Tetrahymena. Dowden, Hutchinson, and Ross, Stroudsburg, Pa. 508 pp.

Farmer, J. N., 1980: The Protozoa: Introduction to Protozoology. The C. V. Mosby Co., St. Louis. 732 pp.

Gibbons, I. R., and Grimstone, A. V., 1960: On flagellar structure in certain flagellates. J. Biophys. Biochem. Cytol., 7:697–716.

Giese, A. C., 1973: Blepharisma: The Biology of a Light-Sensitive Protozoan. Stanford University Press, Stanford, Ca. 366 pp.

Grell, K. G., 1973: Protozoology. Springer-Verlag, N.Y. 554 pp.

Hammond, D. M., and Long, P. L. (Eds.), 1973: The Coccidia. University Park Press, Baltimore, Md. 482 pp.

Haynes, J. R., 1981: Foraminifera. John Wiley and Sons, N.Y. 434 pp.

Jahn, T. L., and Bovee, E. C., 1967: Motile behavior of Protozoa. In Chen, T. (Ed.): Research in Protozoology. Pergamon Press, N.Y. pp. 41–200.

Jeon, J. W. (Ed.), 1973: The Biology of Amoeba. Academic Press, N.Y.

Jones, A. R., 1974: The Ciliates. St. Martin's Press, N.Y. 207 pp. (A general biology of the ciliates.)

Kidder, G. W. (Ed.), 1967: Protozoa. In Florkin, M., and Scheer, B. J. (Eds.): Chemical Zoology. Vol. 1. Academic Press, N.Y.

Kudo, R. R., 1966: Protozoology. 5th Edition. Charles C. Thomas, Springfield, Ill.

Laybourn-Parry, J., 1985: A Functional Biology of Free-Living Protozoa. University of California Press, Berkeley, Ca.

Lynn, D. H., 1981: The organization and evolution of microtubular organelles in ciliated protozoa. Biol. Rev., 56:243–292.

Margulis, L. 1974: Five-kingdom classification and the origin and evolution of cells. Evol. Biol., 7:45–78.

Moore, R. C. (Ed.), 1964 and 1954: Treatise on Invertebrate Paleontology. Protista. Vols. C and D. Geological Society of America and University of Kansas Press, Lawrence, Kans.

Murray, J. W., 1973: Distribution and Ecology of Living Benthic Foraminiferids. Crane, Russak and Co., N.Y. 274 pp.

Murray J. W., 1979: British Nearshore Foraminiferids. Synopses of the British Fauna No. 16. Academic Press, London. 68 pp.

Ogden, C. G., and Hedley, R. H., 1980: An Atlas of Freshwater Testate Amebae. British Museum, Oxford University Press, Oxford. 222 pp.

Patterson, D. J., 1980: Contractile vacuoles and associated structures: their organization and function. Biol. Rev., *55*:1–46.

Pitelka, D. R., 1970: Ciliate ultrastructure; some problems in cell biology. J. Protozool., *17*(1):1–10.

Read, C. P., 1970: Parasitism and Symbiology. Ronald Press, N.Y. 316 pp.

Sarjeant, W. A. S., 1974: Fossil and Living Dinoflagellates. Academic Press, London. 1002 pp.

Sleigh, M. A., 1973: The Biology of Protozoa. American Elsevier Publishing Co., N.Y. 315 pp. (An excellent general account of the Protozoa.)

Sleigh, M. A. (Ed.), 1974: Cilia and Flagella. Academic Press, London. 500 pp. (Papers reviewing the structure and function of cilia and flagella.)

Spoon, D. M., Chapman, G. B., Cheng, R. S., and Zane, S. F., 1976: Observations on the behavior and feeding mechanisms of the suctorian *Heliophyra erhardi* (Reider) Matthes preying on *Paramecium*. Trans. Am. Micros. Soc., *95*:443–462.

Steidinger, K. A., and Haddad, K., 1981: Biologic and hydrographic aspects of red tides. BioScience, *31*(11):814–819.

Tamm, S. L., 1972: A scanning electron microscope study. J. Cell Biol., *55*:250–255.

Tartar, V., 1961: The Biology of *Stentor*. Pergamon Press, N.Y.

Trainor, F. R., 1978: Introductory Phycology. John Wiley and Sons, N.Y. 525 pp.

Whittaker, R. H., 1969: New concepts of kingdoms of organisms. Science, *163*:150–159.

3

Introduction to the Metazoa

Metazoans are multicellular, motile, heterotrophic organisms that develop from embryos. The gametes never form within unicellular structures but rather are produced within multicellular sex organs, or at least within surrounding somatic cells. Correlated with the multicellular condition has been the evolution of tight junctions between the epithelial cells that separate the external environment from the internal fluids of the animal. Collagen fibers are a distinctive product of metazoans, and metazoan sperm, which are typically monoflagellated, are relatively uniform compared with those of other eukaryotic organisms. Metazoans comprise almost all of what are generally considered to be animals. The diversity is enormous. There are 29 phyla, according to the most widely accepted groupings, and only 1, the Chordata, contains some animals that are not invertebrates.

The diagram depicted in Figure 3–1 illustrates a phylogeny of the Animal Kingdom that, in the main, is supported by most zoologists today. The arrangement of phyla reflects a number of ways of grouping animals that not only have evolutionary significance but also are useful for purposes of reference.

Symmetry

Three phyla of animals—sponges, cnidarians, and ctenophores—exhibit a radial symmetry (Fig. 3–2), which all evidence indicates to be primary; i.e., the radial symmetry appears to be primitive and not secondarily derived from bilaterality. These animals are sometimes grouped together as the Radiata. Many species of these groups have become irregular, such as sponges, or have become biradial, such as sea anemones and ctenophores, but the primitive symmetry for each group is radial.

All of the remaining animal phyla are bilateral, or if radial, like starfishes, the radial symmetry has clearly been secondarily derived from bilateral ancestors.

Body Architecture

Internal architecture is another criterion used to group animals. The radiate phyla (sponges, cnidarians, and ctenophores) have few or no organs, and their cellular differentiation is not as great as in other animals, or the cell types are different (sponges).

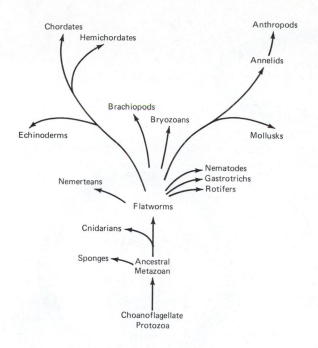

Figure 3–1 A possible phylogeny of the Animal Kingdom.

The bilateral animals possess internal organs, but they differ in the relationship of the organs to the internal body cavity. Although *body cavity* can mean any internal space, the term generally refers to a large, fluid-filled cavity lying between the body wall and the internal organs. The fluid within the body cavity may serve a variety of functions. It often functions as a hydrostatic skeleton, and it may serve as a circulatory medium. The space may be utilized as a temporary site for the accumulation of excess fluids and waste and may be the site for maturation of eggs and sperm. It provides an area for the enlargement (and increase in surface area) of internal organs.

The flatworms, which lack a body cavity, have a solid type of body construction and are said to be acoelomate. The space between the gut and the body wall is filled with a network of cells called parenchyma (Figs. 3–3 and 3–4).

Two types of body cavities are found among the remaining bilateral metazoans. One group of phyla possesses what is called a pseudocoel—an unfortunate name, since it implies that there is something false about the body cavity. The pseudocoel is derived from the blastocoel of the embryo; that

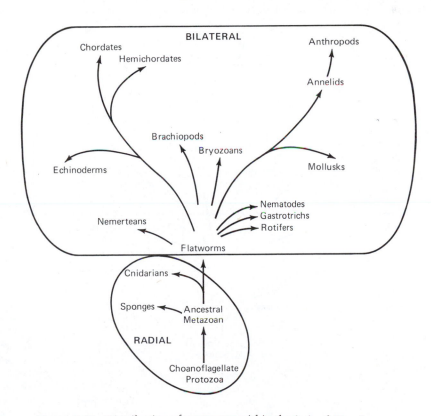

Figure 3–2 Distribution of symmetry within the Animal Kingdom.

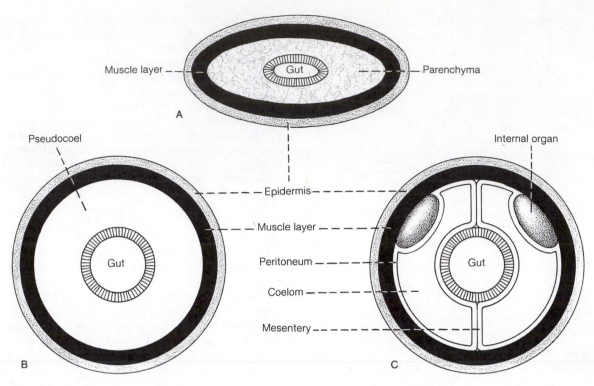

Figure 3–3   Metazoan body plans. *A*, Acoelomate. *B*, Pseudocoelomate. *C*, Coelomate.

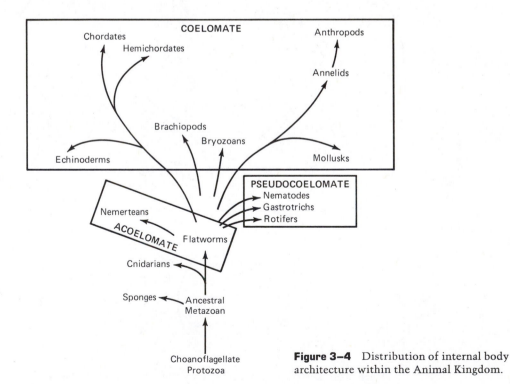

Figure 3–4   Distribution of internal body architecture within the Animal Kingdom.

is, it represents a persistent blastocoel. The internal organs are actually free within the space, for there is no peritoneum bounding the cavity (Fig. 3–3*B*). The pseudocoelomate Bilateria include such familiar animals as rotifers and roundworms.

All other metazoans possess a body cavity called a coelom, or if a body cavity is absent, it clearly represents a secondary loss. A coelom has a different embryogeny than does a pseudocoel. A coelom arises as a cavity within the embryonic mesoderm. The mesoderm provides the cellular lining to the cavity. The lining is called the peritoneum. None of the internal organs are actually free within the coelom. They may bulge into it, but all are bounded by peritoneum. The internal organs are thus all located behind peritoneum, i.e., they are retroperitoneal (Fig. 3–3*C*).

We are not certain whether the coelom evolved once or several times. Its mode of embryonic origin varies, however, and is an important basis for recognizing two evolutionary lines of coelomates, which are described in the following section.

Metazoan Embryogeny

The embryonic development of bilateral metazoans provides the principal basis for the division of these phyla into two rather well defined groups. One division embraces flatworms, annelids, mollusks, and arthropods, as well as a number of smaller allied phyla. These groups constitute the Protostomia (Fig. 3–5). The other division includes echinoderms, chordates, and several smaller phyla; the members of this division are known as the Deuterostomia.

The protostomes and deuterostomes each display a basic plan of development that is characteristic and distinct. This is not to say that all members of each group have identical patterns of development. There are many examples of modification and deviation in every phylum, largely through changes in the distribution and the amount of yolk material. But each line does display certain characteristic features.

TYPES OF CLEAVAGE AND EMBRYONIC DEVELOPMENT

Determinate and Indeterminate Cleavage

The embryonic fate of blastomeres is not determined until relatively late in the embryogeny of deuterostomes. This late establishment of the embryonic fate of cells allows the phenomenon of

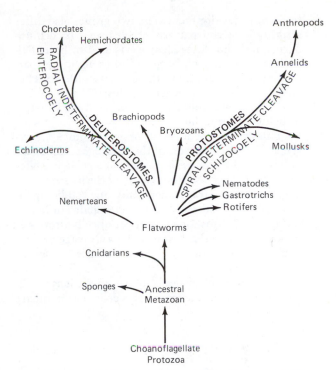

Figure 3–5 A phylogeny of the Animal Kingdom showing the protostome and deuterostome lines and their embryonic characteristics.

twinning to occur. If a starfish egg is allowed to cleave to the four-cell stage and the cells (blastomeres) are then separated, each cell is capable of forming a complete gastrula and then a larva. This formation of blastomeres in the embryo with unfixed fates, known as indeterminate cleavage, is characteristic of deuterostomes.

The flatworm–annelid–mollusk division (the protostomes) displays a very early fixation of the fate of embryonic cells. If a marine annelid egg is allowed to undergo two cleavage divisions and the resulting four blastomeres are separated, each blastomere will develop into only a fixed quarter of the gastrula and larva. Thus, each cell has a predetermined and fixed fate that cannot be altered, even if the cell is moved from its original position in the embryo. The formation of blastomeres with fixed embryonic contributions, known as determinate cleavage, is characteristic of development in protostomes.

Spiral Cleavage

Determinate versus indeterminate cleavage is not the only difference between protostome and deuterostome development. The two groups also differ strikingly in the pattern of cleavage. In echinoderms and chordates cleavage is said to be radial. The axes of early cleavage spindles are either parallel or at right angles to the polar axis (the axis between animal and vegetal poles). The resulting blastomeres are thus always situated directly above or below one another (Fig. 3–6).

This arrangement is rare in protostomes. Cleavage is total, but the axes of the cleavage spindles are oblique to the polar axis, rather than at right angles or parallel. This position results in the blastomeres' having a spiral arrangement, any one cell being located between the two blastomeres above or below it (Fig. 3–7). Thus, one tier or set of cells alternates in position with the next tier. This cleavage pattern is characteristic of protostomes and is known as spiral cleavage.

Since spiral cleavage is determinate, it is possible to trace the lineage of the cells and compare

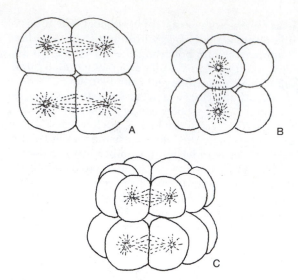

Figure 3–6 Radial cleavage in the sea cucumber *Synapta*: *A*, Polar view. *B* and *C*, Lateral views. (After Selenka from Balinski.)

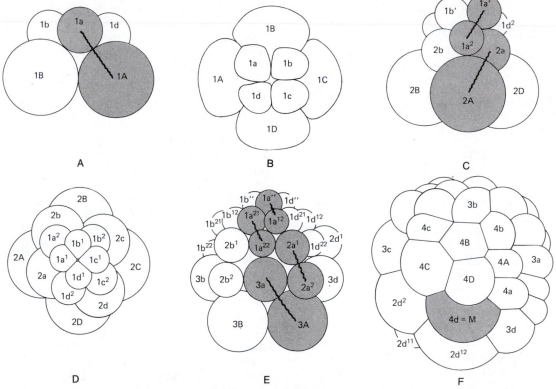

Figure 3–7 *A*, *C*, and *E*, Spiral cleavage (lateral views). All cells derived from the original A blastomere are shaded. (After Villee, Walker and Smith.) *B*, *D*, and *F*, Same stages viewed from animal pole. (After Hyman, L. H., 1951: The Invertebrates, Vol. II. McGraw-Hill Book Co., N.Y.) The numbers and letters indicate the lineage of particular blastomeres (Wilson system). Each of the four blastomeres produced by the second-cleavage division takes the letter A, B, C, or D, and all their descendants take the same letter. Micromeres are designated by lowercase letters, and macromeres by capitals. The numerical prefix indicates at which spiral cleavage division the micromeres were separated from the macromeres.

them in different groups of animals. The fates of the various blastomeres are actually quite similar throughout the protostome phyla. By the time the blastula stage has been reached, all the germ layers have been laid down. For example, the neural ectoderm is derived from one blastomere (designated the 2d cell) that is formed at the fourth cleavage in the animal hemisphere.

At gastrulation, the blastopore of protostomes divides to give rise to both mouth and anus; in deuterostomes the mouth arises a considerable distance from the site of the blastopore, which forms the anus, or the anus forms near the site of the closed blastopore. This difference is the basis for the terms *protostome* (first mouth) and *deuterostome* (second mouth).

Mesoderm and Coelom Origin

In addition to the patterns of cleavage and the site of the origin of the mouth, two further embryogenic differences between protostomes and deuterostomes are evident. In protostomes all the entomesoderm arises from a single cell, the mesentoblast cell, formed at the sixth cleavage. This cell, also called the 4d cell from its cleavage lineage, gives rise to two teloblasts, or primordial mesoderm cells. Located originally at the posterior end of the animal in front of the site of the future anus, each cell proliferates to form two masses of mesodermal cells, one on either side of the body (Fig. 3–8). In metameric (segmented) protostomes the mesodermal masses form a linear series of segmented blocks of cells. In any event, a split forms within each mesodermal mass, and the resulting cavity enlarges to form the coelom. This mode of coelom formation is termed *schizocoely*, and coelomate protostomes are therefore often referred to as schizocoelous coelomates, or schizocoelomates.

Deuterostomes, on the other hand, have a different method of mesoderm and coelom formation. The mesoderm primitively arises by a process called enterocoelic pouching, in which the wall of the archenteron evaginates to form pouches. The pouches separate from the archenteron, either as a pair or, in the case of metameric deuterostomes, as a series of lateral pairs (Fig. 3–8). The cavity of the evagination and later pouch becomes the coelom, and the wall becomes the mesoderm. Because of this mode of coelom formation, deuterostomes are commonly referred to as enterocoelous coelomates, or enterocoelomates.

Although embryonic evidence is the principal basis for differentiating the protostomes and deuterostomes, these two main lines of metazoan evolution are also reflected in a number of structural and biochemical features. The evolution of surface ciliation (p. 222) seems to have been different in the two lines. Chitin is a common skeletal compound among protostomes but is rarely encountered among deuterostomes. Arginine phosphate is the typical phosphate store of protostomes, but creatine phosphate is typical of deuterostomes.

Origin of Metazoa

Most zoologists agree that metazoans have a common ancestry from some unicellular organism, but there have been differing views as to the particular group of unicellular forms involved and the mode of origin.

Hadži (1953) and Hanson (1977) have been the chief proponents of a ciliate origin for metazoans. Their theory, which may be called the *Syncytial Theory*,* holds that multicellular animals arose from a primitive group of multinucleate ciliates. The ancestral metazoan was at first syncytial in structure but later became compartmented or cellularized when it acquired cell membranes, which produced a typical multicellular condition. Since many ciliates tend toward bilateral symmetry, proponents of the Syncytial Theory maintain that the ancestral metazoan was bilaterally symmetrical and gave rise to the acoel flatworms, which are therefore held to be the most primitive living metazoans. That the acoels are in the same size range as the ciliates, are bilaterally symmetrical, are ciliated, and tend toward a syncytial condition is considered evidence supporting the primitive position of this group. The ciliate macronucleus, which is absent in acoels, is assumed to have been absent in the multinucleate protociliate stock from which the metazoans arose; according to this theory, it developed later in the evolutionary line leading to the higher ciliates.

There are a number of objections to the Syncytial Theory. Nothing comparable to cellularization occurs in the ontogeny of any of these groups.† Furthermore, a ciliate ancestry does not explain the general occurrence of flagellated sperm in metazoans. No comparable cells are produced in ciliates, and it is necessary to assume a *de novo* origin of

---

*Syncytial refers to the histological condition in which cell membranes are absent between adjacent nuclei.

†Cellularization does occur in the superficial cleavage of arthropod eggs, but this is a highly specialized condition associated with abundant yolk material.

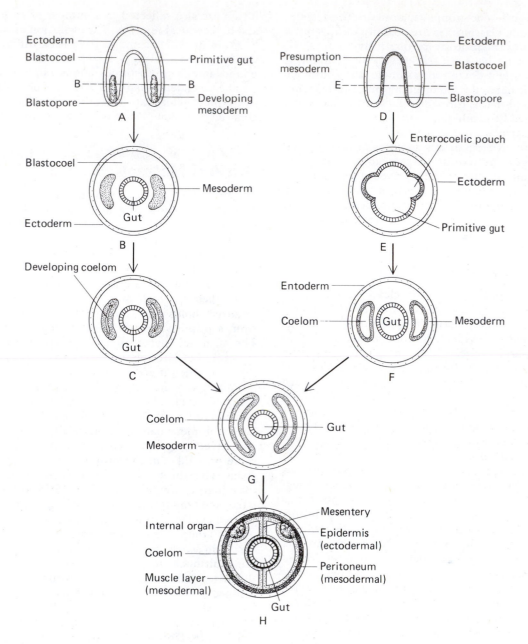

Figure 3–8   Comparison of the two principal modes of mesoderm and coelom formation in animals: *A–C,* Coelom formation by schizocoely as it occurs in mollusks and annelids. *A,* Frontal section through gastrula, showing developing mesoderm derived from cells set aside very early in cleavage. *B,* Cross section of *A* taken at level *B–B. C,* Mesodermal mass splits to form coelomic cavity. *D–F:* Coelom formation by enterocoely as it occurs in echinoderms and primitive chordates. *D,* Frontal section through gastrula showing presumptive (future) mesoderm as part of primitive gut wall. *E,* Cross section of *D* taken at level *E–E.* Presumptive mesoderm forms wall of outpocketing of primitive gut, the enterocoelic pouches. *F,* Pouches separate as coelomic vesicles, the walls forming the mesoderm and the cavity of the coelom. *G–H:* The fate of the coelomic vesicles is more or less the same regardless of their mode of origin. The outer wall of the vesicle becomes associated with ectoderm to form body wall muscles; the inner wall of the vesicle becomes associated with the gut to form the muscles and blood vessels of the gut wall.

motile sperm in the metazoan ancestor. The most serious objection to the Syncytial Theory is the necessity for making the acoels the most primitive living metazoans. Bilateral symmetry then becomes the primitive symmetry for metazoans, and the radially symmetrical cnidarians must be derived secondarily from the flatworms. Many specialists now doubt that acoels are even the most primitive flatworms (p. 185).

The *Colonial Theory*, in which the metazoans are derived by way of a colonial flagellate, is the classic and most frequently encountered theory of the origin of multicellular animals. There is increasing evidence in its support, and it is the most widely held view among contemporary zoologists. This idea was first conceived by Haeckel (1874), later modified by Metschnikoff (1887), and revived by Hyman (1940). The Colonial Theory maintains that the flagellates are the ancestors of the metazoans, and in support of such an ancestry the following facts are cited as evidence. Flagellated sperm cells occur throughout the Metazoa. Flagellated body cells commonly occur among lower metazoans, particularly among sponges and cnidarians. True sperm and eggs have evolved in the phytoflagellates. The phytoflagellates display a tendency toward a type of colonial organization that could have led to a multicellular construction; in fact, a differentiation between somatic and reproductive cells has been attained in *Volvox.*

Although *Volvox* is frequently used as a model for the design of the flagellate colonial ancestor, these autotrophic organisms with plantlike cells are not likely ancestors of metazoans. Ultrastructural evidence points to the choanoflagellates, a small group of animal-like, monoflagellated protozoa, as the best candidates. Some are solitary and some are colonial (see Box 4–1).

The Colonial Theory holds that the ancestral metazoan probably arose from a spherical, hollow, colonial flagellate. Like *Volvox*, the cells were flagellated on the outer surface; the colony possessed a distinct anterior-posterior axis and swam with the anterior pole forward; and there was a differentiation of somatic and reproductive cells. This stage was called the blastaea in Haeckel's original theory, and the hollow blastula, or coeloblastula, was considered a recapitulation of this stage in the embryogeny of living metazoans (Fig. 3–9). According to Haeckel, the blastaea invaginated to produce a double-walled, saclike organism, the gastraea (Fig. 3–9). This gastraea was the hypothetical metazoan ancestor, equivalent to the gastrula stage in the embryonic development of living metazoans. In addition

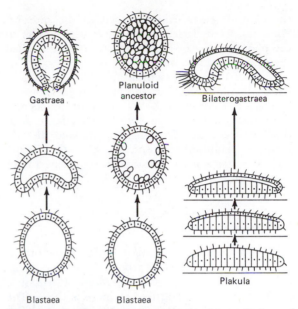

Figure 3–9 Hypothetical stages in the evolution of early metazoans according to Haeckel (left), Metschnikoff (middle), and Grell-Butschli (right). (Greatly modified from Grell, 1981.)

tion to embryological evidence, Haeckel noted the close structural similarity between the gastraea and some lower metazoans, such as the hydrozoan cnidarians and certain sponges. Both of these latter organisms are double walled, with a single opening into a saclike cavity.

Haeckel's blastaea and gastraea stages are still widely held as starting points in metazoan evolution and have been recently elaborated in the phylogenetic scheme of Nielsen (1985).

A popular revision of Haeckel's theory still encountered today was initiated by Metschnikoff (1887), who noted that the primitive mode of gastrulation in cnidarians is by ingression, in which cells are proliferated from the blastula wall into the interior blastocoel. This produces a solid gastrula. Invagination may have been a secondary embryonic shortcut. Metschnikoff therefore argued that through the migration of cells into the interior, the originally hollow sphere (blastaea) became transformed into an organism having a solid structure (gastraea) (Fig. 3–9). The body of this hypothetical ancestral metazoan is believed to have been ovoid and radially symmetrical. The exterior cells were flagellated and, as such, assumed a locomotor sensory function. The solid mass of interior cells functioned in nutrition and reproduction. There was no mouth, and food could be engulfed anywhere on the exterior surface and passed to the interior. Since this hypothetical organism is very similar to the planula larva of cnidarians, it has been called the planuloid ancestor.

From such a free-swimming, radially symmetrical, planuloid ancestor the lower metazoans are believed to have arisen. On the basis of this theory, the primary radial symmetry of the cnidarians can thus be accounted for as being derived directly from the planuloid ancestor. The bilateral symmetry of the flatworms would then represent a later modification in symmetry.

Phylum Placozoa: The Most Primitive Metazoans?

In 1883 a minute, animal-like, multicellular organism was discovered in a European seawater aquarium and named *Trichoplax adhaerens*. It has since been observed and cultured numerous times. The flattened body, which reaches 2 to 3 mm in diameter, is composed of two outer layers of monociliated epitheloid cells, which house between them an inner layer of loose, contractile, stellate cells rather like mesenchyme. The lower surface of *Trichoplax* is somewhat concave, and the upper

surface is flat. The margins of the body are irregular and change shape like an ameba (Fig. 3–10). It creeps over the substratum by means of its cilia and feeds upon protoza and algae, which are digested extracellularly. *Trichoplax* undergoes asexual reproduction by fission and by budding. Eggs have been observed within the internal mesenchyme-like layer, but they may arise from the ventral epithelium. The DNA content is smaller than that determined for any other animal. Grell (1971) has proposed the phylum Placozoa for *Trichoplax* and places it with sponges in an early branch of the Animal Kingdom.

Trichoplax is certainly the simplest known metazoan and may represent the most primitive. It shares a number of features with the proposed planuloid ancestor described earlier. The flattened, asymmetrical body, however, does not coincide with the radial symmetry postulated for the earliest metazoans. Is the symmetry of *Trichoplax* primary or secondary? Grell (1981) believes the *Trichoplax* form is primary and argues that a bilateral gastraea evolved from such an ancestral metazoan (Fig. 3–9). Nielsen (1985) believes that *Trichoplax* is simply a flattened blastaea, and that its symmetry is a secondary condition and not primitive.

Figure 3–10 *A*, Dorsal view of *Trichoplax adhaerens*. The organism is greatly flattened, but note the irregular outline. (Drawn from a photograph by K. G. Grell.) *B*, Diagrammatic section through *Trichoplax* showing part of the ventral and dorsal layers and the cells that lie in the space between them. (After Grell, K. G., 1981: *Trichoplax adhaerens* and the origin of Metazoa. International congress on the origin of the large phyla of metazoans. Accademia Nazionale dei Lincei. Atti dei Convegni Lincei 49:113.)

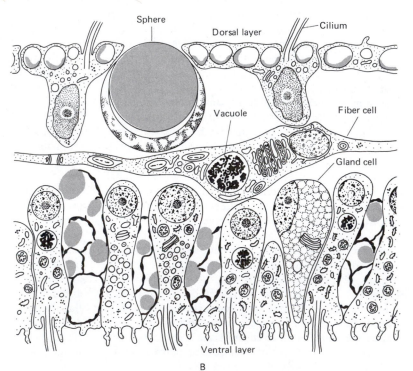

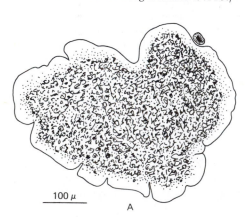

100 μ

A

B

SYNOPSIS OF THE METAZOAN PHYLA

The following diagnoses are limited to distinguishing characters. Only existing classes of the larger phyla are listed. The approximate number of described species is indicated in parentheses.

Parazoa. Metazoans with poorly differentiated tissues and no organs.

Phylum Placozoa (1). *Trichoplax adhaerens.* Microscopic, flattened, marine animal composed of ventral and dorsal epitheloid layers enclosing loose, mesenchyme-like cells.

Phylum Porifera (5000). Sponges. Sessile; no anterior end; primitively radial, but most are irregular. Mouth and digestive cavity absent; body organized about a system of water canals and chambers. Marine, a few in fresh water.

Eumetazoa. Metazoans with organs, mouth, and digestive cavity.

Radiata. Tentaculate, radiate animals with few organs. Digestive cavity, with mouth the principal opening to the exterior.

Phylum Cnidaria, or Coelenterata (9000). Free swimming or sessile, with tentacles surrounding mouth. Specialized cells bearing stinging organoids called nematocysts. Solitary or colonial. Marine, few in fresh water.

Class Hydrozoa. Hydras, hydroids.

Class Scyphozoa. Jellyfish.

Class Anthozoa. Sea anemones, corals.

Phylum Ctenophora (50). Comb jellies. Free swimming; biradiate, with two tentacles and eight longitudinal rows of ciliary combs (membranelles). Marine.

Bilateria. Bilateral animals.

Protostomes. Cleavage spiral and determinate; mouth arising from or near blastopore.

Acoelomates. Area between body wall and internal organs filled with parenchyma.

Phylum Platyhelminthes (12,700). Body dorsoventrally flattened; digestive cavity, when present, with a single opening, the mouth. Free living and parasitic. Marine, freshwater, a few terrestrial.

Class Turbellaria. Free-living flatworms.

Class Monogenea. Flukes.

Class Trematoda. Flukes.

Class Cestoda. Tapeworms.

Phylum Mesozoa (50). An enigmatic group of minute parasites of marine invertebrates. No organs; body with few cells.

Phylum Rhynchocoela, or Nemertea (900). Nemerteans. Long, dorsoventrally flattened body with complex proboscis apparatus. Marine, few terrestrial and freshwater.

Phylum Gnathostomulida (80). Minute, ciliated acoelomate worms. Buccal region bearing cuticular pieces; digestive tract with anus.

Pseudocoelomates. A persistent blastocoel forming a body cavity. Digestive tract with mouth and anus. Body covered with a collagen cuticle.

Phylum Gastrotricha (460). Gastrotrichs. Slightly elongated body with flattened, ciliated ventral surface. Few to many adhesive tubes present; cuticle commonly ornamented. Microscopic. Marine and freshwater.

Phylum Nematoda (12,000). Roundworms. Slender, cylindrical worms with tapered anterior and posterior ends. Cuticle thick and complex. Free living and parasitic. Free-living species usually only a few millimeters or less in length. Inhabit salt water, fresh water, and soil.

Phylum Nematomorpha (230). Hairworms. Extremely long, threadlike bodies. Adults free living in damp soil and fresh water; a few marine. Juveniles parasitic.

Phylum Rotifera (1500). Rotifers. Anterior end bearing a ciliated crown; posterior tapering to a foot. Pharynx containing movable cuticular pieces. Microscopic. Largely freshwater, some marine, some inhabitants of mosses.

Phylum Acanthocephala (1150). Small, wormlike endoparasites of arthropods. Anterior retractile proboscis bearing recurved spines.

Phylum Kinorhyncha (100). Slightly elongated body. Cuticle segmented and bearing recurved spines. Spiny anterior end retractile. Less than 1 mm in length. Marine.

Phylum Loricifera (1). Body composed of a spiny, anterior introvert and a trunk encased within a cuticular lorica. Microscopic in marine shelly gravel.

Phylum Priapulida (9). Cucumber-shaped or wormlike marine animals, with a retractile anterior introvert. Body surface covered with spines and tubercles.

Schizocoelous Coelomates. Body cavity a coelom that develops as a schizocoel, or if a body cavity is absent, the coelom has been lost. Digestive tract with mouth and anus.

Phylum Sipuncula (320). Cylindrical marine worms. Retractable anterior end, bearing lobes or tentacles around mouth.

Phylum Mollusca (50,000). Ventral surface modified in the form of a muscular foot, having various shapes; dorsal and lateral surfaces of body modified as a shell-secreting mantle, although shell may be reduced or absent. Marine, freshwater, and terrestrial.

    Class Monoplacophora.

    Class Polyplacophora. Chitons.

    Class Aplacophora.

    Class Gastropoda. Snails, whelks, conchs, slugs.

    Class Bivalvia, or Pelecypoda. Bivalve mollusks.

    Class Scaphopoda. Tusk, or tooth, shells.

    Class Cephalopoda. Squids, cuttlefish, octopods.

Phylum Echiura (140). Cylindrical marine worms with flattened, anterior, nonretractile proboscis. Trunk with a large pair of ventral setae.

Phylum Annelida (8700). Segmented worms. Body wormlike and metameric. A large, longitudinal, ventral nerve cord. Marine, freshwater, and terrestrial.

    Class Polychaeta. Marine annelids.

    Class Oligochaeta. Freshwater annelids and earthworms.

    Class Hirudinea. Leeches.

Phylum Pogonophora (80). Marine, deepwater animals with a long body housed within a chitinous tube. Anterior end of body bearing one to many long tentacles. Digestive tract absent.

Phylum Tardigrada (400). Water bears. Microscopic, segmented animals. Short, cylindrical body bearing four pairs of stubby legs terminating in claws. Freshwater and terrestrial in lichens and mosses; few marine.

Phylum Onychophora (70). Terrestrial, segmented, wormlike animals, with an anterior pair of antennae and many pairs of short, conical legs terminating in claws. Body covered by a thin, chitinous cuticle.

Phylum Arthropoda (923,000+). Body metameric with jointed appendages and encased within a chitinous exoskeleton. Coelom vestigial. Marine, freshwater, and terrestrial.

    Subphylum Chelicerata (68,400+). No antennae; one pair of chelicerae. Body composed of a cephalothorax and abdomen.

        Class Merostomata. Horseshoe crabs.

        Class Arachnida. Scorpions, spiders, mites.

        Class Pycnogonida. Sea spiders.

    Subphylum Crustacea (42,000). With two pairs of antennae and one pair of mandibles.

        Class Cephalocarida.

        Class Branchiopoda. Fairy shrimps, water fleas.

        Class Ostracoda.

        Class Mystacocarida.

        Class Copepoda.

        Class Branchiura.

        Class Remipedia.

        Class Cirripedia. Barnacles.

        Class Malacostraca. Amphipods, isopods, shrimps, crabs.

    Subphylum Uniramia. With one pair of antennae and one pair of mandibles.

        Class Insecta. (800,000+)

        Class Diplopoda. Millipedes. (7500)

        Class Pauropoda. (380)

        Class Symphyla. (120)

        Class Chilopoda. Centipedes. (2500)

Phylum Pentastomida (90). Wormlike endoparasites of vertebrates. Anterior end of body within two pairs of leglike projections terminating in claws and a median, snoutlike projection bearing the mouth. Phylum status very questionable.

Lophophorate Coelomates. With a crown of hollow tentacles (a lophophore) surrounding or partially surrounding mouth.

Phylum Phoronida (10). Marine, wormlike animals with the body housed within a chitinous tube.

Phylum Bryozoa (4000). Colonial, sessile; the body housed within a gelatinous or more commonly a chitinous or chitinous and calcareous exoskeleton. Mostly marine, few freshwater.

Phylum Entoprocta (150). Body attached by a stalk. Mouth and anus surrounded by the tentacular crown. Mostly marine.

Phylum Brachiopoda (325). Lamp shells. Body attached by a stalk and enclosed within two unequal dorsoventrally oriented, calcareous shells.

Deuterostomes, or Enterocoelous Coelomates. Cleavage radial and indeterminate; mouth arising some distance away (anteriorly) from blastopore. Mesoderm and coelom develop primitively by enterocoelic pouching of the primitive gut.

Phylum Chaetognatha (70). Marine planktonic animals with dart-shaped bodies bearing fins. Anterior end with grasping spines flanking a ventral preoral chamber.

Phylum Echinodermata (6000). Secondarily pentamerous radial symmetry. Most existing forms free moving. Body wall containing calcareous ossicles usually bearing projecting spines. A part of the coelom modified into a system of water canals with external projections used in feeding and locomotion. All marine.

Class Crinoidea. Sea lilies and feather stars.

Class Stelleroidea.

Subclass Asteroidea. Sea stars.

Subclass Ophiuroidea. Brittle stars and basket stars.

Class Echinoidea. Sea urchins, sand dollars, and heart urchins.

Class Holothuroidea. Sea cucumbers.

Phylum Hemichordata (85). Acorn worms. Body divided into proboscis, collar, and trunk. Anterior part of trunk perforated with varying number of pairs of pharyngeal clefts. Marine.

Class Enteropneusta. Acorn worms.

Class Pterobranchia.

Phylum Chordata (39,000). Pharyngeal clefts, notochord, and dorsal hollow nerve cord present at some time in life history. Marine, freshwater, and terrestrial.

Subphylum Urochordata (1250). Sessile, nonmetameric invertebrate chordates enclosed within a cellulose tunic. Notochord and nerve cord present only in larva. Solitary and colonial. Marine.

Class Ascidiacea. Sea squirts, or sessile tunicates.

Class Thaliacea. Planktonic urochordates.

Class Larvacea. Planktonic urochordates.

Subphylum Cephalochordata (45). Amphioxus. Metameric, fishlike invertebrate chordates.

Subphylum Vertebrata (37,790). The vertebrates. Metameric. Trunk supported by a series of cartilaginous or bony skeletal pieces (vertebrae) surrounding or replacing notochord in adult.

REFERENCES

Anderson, D. T., 1982: Origins and relationships among the animal phyla. Proc. Linn. Soc. New South Wales, *106*(2):151–166.

Clark, R. B., 1964: Dynamics in Metazoan Evolution. Clarendon Press, Oxford. (This volume contains a concise review of past and contemporary theories of the evolution of the coelom and metamerism.)

Conway Morris, S., George, J. D., Gibson, R., and Platt, H. M. (Eds.), 1985: The Origins and Relationships of Lower Invertebrates. Systematics Association Spec. Vol. No. 28. Clarendon Press, Oxford. 394 pp.

Grell, K. G., 1971: *Trichoplax adhaerens*, F. E. Schulze und die Entstehung der Metazoen. Naturwissenschaftliche Rundschau, *24*(4):160–161.

Grell, K. G., 1981: *Trichoplax adhaerens* and the origin of Metazoa. International congress on the origin of the large phyla of metazoans. Accademia Nazionale dei Lincei. Atti dei Convegni Lincei *49*:107–121.

Grell, K. G., and Benwitz, G., 1971: Die ultrastruktur von *Trichoplax adhaerens* F. E. Schulze. Cytobiologie, *4*(2):216–240.

Grell, K. G., 1972: Formation of eggs and cleavage in *Trichoplax adhaerens.* Z. Morphol. Tiere, *73*(4):297–314.

Hadži, J., 1953: An attempt to reconstruct the system of animal classification. Syst. Zool., *2*:145–154.

Hadži. J., 1963: The Evolution of the Metazoa. Macmillan, N.Y. (An exhaustive elaboration and defense of Hadži's views.)

Haeckel, E., 1874: The gastraea-theory, the phylogenetic classification of the Animal Kingdom and the homology of the germ-lamellae. Q. J. Micr. Sci., *14*:142–165; 223–247.

Hanson, E. D., 1977: The Origin and Early Evolution of Animals. Wesleyan University Press, Middletown, Conn. 670 pp.

Hyman, L., 1940: The Invertebrates: Protozoa Through Ctenophora, Vol. 1. McGraw-Hill, N.Y. (This volume contains a discussion of the Colonial Theory of the origin of the metazoans.)

Hyman, L., 1951: The Invertebrates: Platyhelminthes and Rhynchocoela, Vol. 2. McGraw-Hill, N.Y. (The introductory section of this volume discusses the origin of bilaterality, the origin of body cavities, and spiral cleavage.)

Lankester, E. R., 1875: On the invaginate planula, or diploblastic phase of *Paludina vivipara.* Q. J. Micr. Sci., *15*:159–166.

Metschnikoff, E., 1887: Embryologische Studien an Medusen, mit Atlas. A. Holder, Vienna.

Nielsen, C., 1985: Animal phylogeny in the light of the trochaea theory. Biol. Jour. Linn. Soc., *25*:243–299.

Salvini-Plawen, L. V., 1978: On the origin and evolution of the lower Metazoa. Z. Zool. Syst. Evolutionsforsch., *16*:40–88.

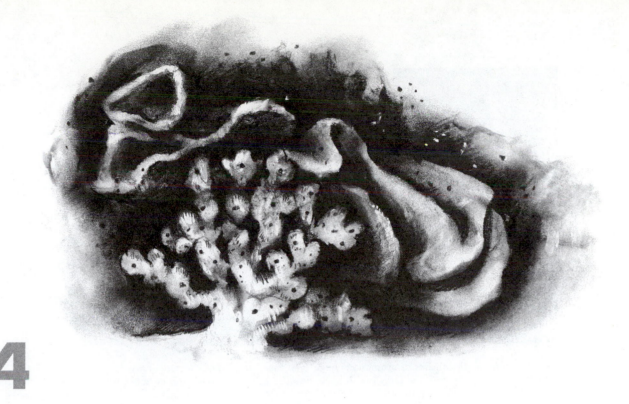

4

The Sponges

Sponges, which constitute the phylum Porifera, are the most primitive of the multicellular animals. Neither true tissues nor organs are present, and the cells display a considerable degree of independence. All members of the phylum are sessile and exhibit little detectable movement. This combination of characteristics convinced Aristotle, Pliny, and other ancient naturalists that sponges were plants. In fact, it was not until 1765, when internal water currents were first observed, that the animal nature of sponges became clearly established.

Except for some 150 freshwater species, sponges are marine animals. They abound in all seas, wherever rocks, shells, submerged timbers, or coral provide a suitable substratum. Some species even live on soft sand or mud bottoms. Most sponges prefer relatively shallow water, but some groups, including most glass sponges, live in deep water.

Sponge Structure

Sponges vary greatly in size. Certain calcareous sponges are about the size of a grain of rice, but a large loggerhead sponge may exceed a meter in height and diameter. Some are radially symmetrical, but the majority are irregular and exhibit massive, erect, encrusting, or branching growth patterns (Figs. 4–1, 4–2, and 4–8). Drabness is more the exception than the rule among the Porifera; most of the common species are brightly colored. Green, yellow, orange, red, and purple sponges are frequently encountered. The significance of the coloration is uncertain. Protection from solar radiation and warning coloration have been suggested.

Sponge architecture is unique in being constructed around a system of water canals, an arrangement that is correlated with sponge sessility. The basic structure and histology of sponges can be most easily understood by beginning with primitive radial forms. Sponges with the simplest, most primitive type of structure are called asconoid sponges, a structural term rather than a taxonomic one. The asconoid sponge is shaped like a tube and is always small (Fig. 4–3). *Leucosolenia*, which is one of the few living genera of asconoid sponges, rarely exceeds 10 cm in height. Asconoid sponges are not usually solitary but are composed of clusters of tubes fused together along their long axes or at their bases.

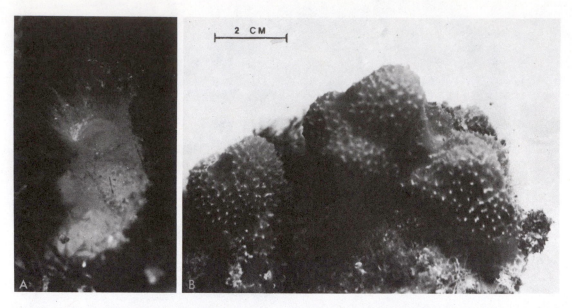

Figure 4–1 *A*, A Small, calcareous syconoid sponge. The vase-shaped body of this individual is no more than 5 mm in length. Long spicules fringe the large osculum. *B*, *Dysidea etheria*, a West Indian leuconoid sponge, which is pale blue. One osculum is visible at left. (Both by Betty M. Barnes.)

The surface of an asconoid sponge is perforated by many small openings, called incurrent pores, from which the name *Porifera* (pore-bearer) is derived. These pores open into the interior cavity, the atrium (spongocoel). The latter in turn opens to the outside through the osculum, a large opening at the top of the tube. A constant stream of water passes through the incurrent pores into the spongocoel and out the osculum.

The body wall is relatively simple. The outer surface is covered by flattened cells, the pinacocytes, which together make up the pinacoderm. Unlike the epithelium of most other animals, there is no basement membrane, and the margins of pinacocytes can be contracted or withdrawn so that the entire animal may decrease slightly in size. The basal pinacocytes secrete material that fixes the sponge to the substratum. Each pore is formed by a porocyte, a cell that is shaped like a short tube and extends from the external surface to the spongocoel. The bore, or lumen, of the tube forms the incurrent pore, or ostium, and can be closed or opened by contraction. A porocyte is derived from a pinacocyte through the formation of an intracellular perforation or perhaps by an infolding.

Beneath the pinacoderm lies a layer called the mesohyl (sometimes referred to as mesenchyme), which consists of a gelatinous protein matrix containing skeletal material and ameboid cells.

The skeleton is relatively complex and provides a supporting framework for the living cells of the animal. (To avoid repetition, the discussion presented here on the sponge skeleton applies to the phylum in general, not just to the asconoid sponges.) The skeleton may be composed of calcareous spicules, siliceous spicules, protein spongin fibers, or a combination of the last two (Figs. 4–4 and 4–5). The spicules exist in a variety of forms and are important in the identification and classification of species. An extensive nomenclature has developed through the use of these structures in sponge taxonomy.

Monaxon spicules are shaped like needles or rods and may be curved or straight. The ends are pointed, knobbed, or hooked. Triaxons have three rays, tetraxons four, and hexaxons six. Polyaxons are composed of a number of short rods radiating from a common center. They may be grouped so they appear like a bur, like a star, or something like a child's jack. Two additional terms are commonly encountered and can apply to any of the spicule types just described: *megascleres*, which are larger spicules forming the chief supporting elements in the skeleton, and *microscleres*, which are considerably smaller.

The skeleton is located primarily in the mesohyl, but spicules frequently project through the pinacoderm. The arrangement of spicules is orga-

Figure 4–2 Relationship of sponge form to utilization of substratum. The two massive sponges at the right on top of the rock require an exposed surface, but their elevated form enables them to utilize water well above the substratum, and their attachment area is a relatively small part of the total body surface area. The encrusting sponges below the rock utilize much of their surface area for attachment, but their low encrusting form enables them to exploit the space of crevices and other confined areas. The sponge on the vertical surface at left utilizes space *within* the substratum. Small arrows indicate the movement of water into the sponge; large arrows indicate the exit of water from oscula.

nized, with various types often combined in distinct groupings (Fig. 4–5F and G). They may interlock or be fused together. The organization in one part of the body may differ from that in another. Microscleres, for example, support the pinacoderm lining of the water canals. The presence of spicules increases the overall stiffness of sponge tissue by restricting the ability of organic components (polymers) to undergo molecular rearrangements in response to mechanical loads. Thus, for example, the spicule-filled sponge body resists deformation in strong water currents (Koehl, 1982).

The mesohyl of all sponges contains dispersed collagen fibrils, but some sponges also possess a skeleton of coarse fibers, called spongin. A spongin skeleton is organized as interconnecting fibers of varying thickness. The substance composing spongin is a fibrous protein similar to collagen. Some sponges contain so much spongin that they are tough and rubbery, and in many species siliceous spicules are embedded partially or completely in the fibers (Fig. 4–4C and D).

Ameboid cells in the mesohyl include a number of types. Large cells with large nuclei are called archeocytes. They are phagocytic and play a role in digestion. Archeocytes are capable of transforming themselves into other types of cells that are needed within an animal; they are said to be totipotent.

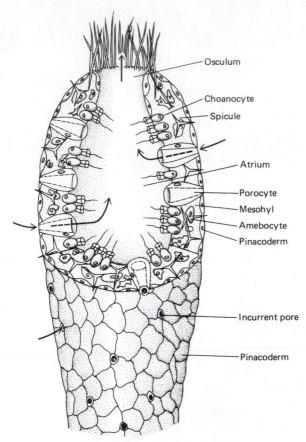

Osculum

Choanocyte

Spicule

Atrium

Porocyte

Mesohyl

Amebocyte

Pinacoderm

Incurrent pore

Pinacoderm

Figure 4–3 Diagram of a partially sectioned asconoid sponge. (Based on a figure by Buchsbaum.)

There are also fixed cells, called collencytes, that are anchored by long, cytoplasmic stands. They secrete the dispersed collagen fibers. Many sponges possess mobile lophocytes, which also secrete these fibers.

The spicule or spongin skeleton is secreted by ameboid sclerocytes or spongocytes. One to several sclerocytes are usually involved in the secretion of a single spicule in the calcareous sponges, and the process is relatively complex. A three-pronged spicule, for example, has its beginnings in three sclerocytes derived from an amebocyte, called a scleroblast. The three sclerocytes partially fuse to form a trio of cells (Fig. 4–5*B* to *E*). Each member of the trio then divides, and between each pair of daughter cells one prong, or ray, of the spicule is secreted. The three prongs fuse at the base. Each of the three pairs of sclerocytes now moves outward along a ray, one cell secreting the end and one thickening the base of the spicule (Fig. 4–5*E*).

On the inner side of the mesohyl, and lining the

atrium, is a layer of cells, the choanocytes, which are very similar in structure to choanoflagellate protozoa (Box 4–1). The choanocyte is ovoid, with one end adjacent to the mesohyl. The opposite end of the choanocyte projects into the atrium and bears a flagellum surrounded by a contractile collar of microvilli. The choanocytes are responsible for moving water through the sponge and for obtaining food. Both of these processes are described in detail later.

The primitive asconoid structure imposes very definite size limitations. An increase in the volume of the atrium is not accompanied by a sufficient increase in surface area of the choanocyte layer to provide for water movement. Thus, asconoid sponges are always small.

The problems of water flow and surface area have been overcome during the evolution of sponges by the folding of the body wall and the reduction of the atrium. The folding increases the surface area of the choanocyte layer, and the reduction of the atrium lessens the volume of water that must be circulated. The net result of these changes is a greatly increased and more efficient water flow through the body. A much greater size now becomes possible, although the primitive radial symmetry is commonly lost.

Living sponges display various stages in the changes just described. Sponges that exhibit the first stages of body wall folding are called syconoid sponges. Syconoid sponges include the well-known genera *Grantia* and *Sycon* (=*Scypha*). In syconoid structure, the body wall has become "folded" horizontally, forming finger-like processes (Fig. 4–6*B*). This development produces external pockets, extending inward from the outside, and evaginations, extending outward from the atrium. The two pockets produced by a fold do not meet but bypass each other and are blind.

In this more advanced type of sponge, the choanocytes no longer line the spongocoel but are now confined to the evaginations, which are called flagellated or radial canals. The corresponding invaginations from the epidermal side are known as incurrent canals and are lined by pinacocytes. The two canals are connected by openings called prosopyles, which are equivalent to the pores of asconoid sponges. Water now flows through the incurrent canals, the prosopyles, the flagellated canals, and the atrium and flows out the osculum.

A slightly more specialized stage of the syconoid structure develops when pinacocytes and mesohyle plug the open ends of the incurrent canals (Fig. 4–6*C*). Openings remain to permit entrance of

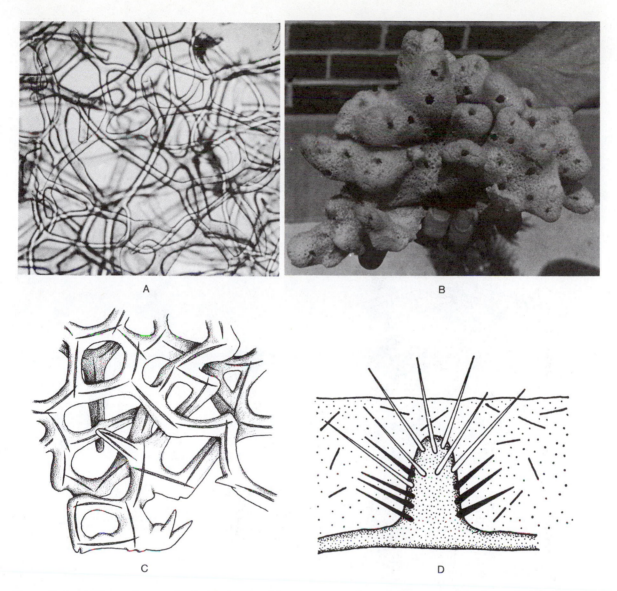

A

B

C

D

Figure 4–4 *A*, Photomicrograph of spongin fibers (they appear translucent). (Courtesy of the General Biological Supply House, Inc.) *B*, The spongin skeleton of a commercial sponge *(Spongia)* from the Mediterranean. The large openings are oscula. (By Betty M. Barnes.) *C*, Spongin network of *Endectyon*, showing embedded spicules. *D*, Sponge with one type of spicule partially embedded within spongin. (From Berquist, P. R., 1978: Sponges. Hutchinson, London. p. 46.)

water into incurrent canals. Despite the folding of the body wall, syconoid sponges still retain a radial symmetry.

The highest degree of folding takes place in the leuconoid sponge (Fig. 4–6*D*). The flagellated canals have undergone folding, or evagination, to form small, rounded, flagellated chambers, and the atrium has commonly disappeared except for water channels leading to the osculum. Water enters the sponge through the dermal pores and passes into

subdermal spaces, which in some species may form an extensive vestibule. The spaces lead into branching incurrent channels, which eventually open into the flagellated chambers through prosopyles. Water leaves the chamber through an apopyle and courses through excurrent channels, which become progressively larger as they are joined by other excurrent channels. A single large channel eventually opens to the outside through the osculum. The canals are lined by pinacocytes, called endopinaco-

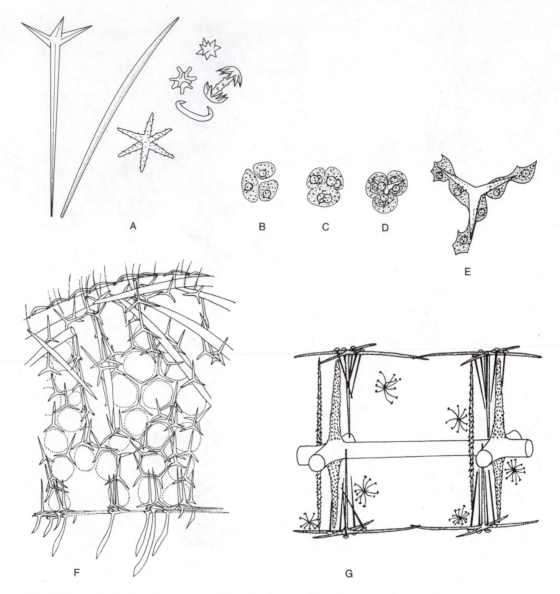

Figure 4–5   *A*, Sponge spicules, showing variations in shape and in size (megascleres and microscleres). (Drawn from various sources.) *B* to *E*, Secretion of a calcareous triradiate spicule. (After Minchin from Jones.) *F*, Section through a part of a leuconoid calcareous sponge, showing spicules in their natural position. *G*, Spicules of the hexactinellid sponge *Farrea sollasii* in their natural position. (*F* after Borojevic, *G* after Schulze, both from Bergquist, P. R., 1978: Sponges. Hutchinson, London. pp. 147 and 151.)

cytes to differentiate them from the exopinaco-cytes that cover the exterior of the sponge. The number of flagellated chambers may be enormous. For example, *Microciona prolifera* contains 10,000 chambers per mm^3, each 20 to 39 microns in diameter and containing about 57 choanocytes (Reiswig, 1975a). The mesohyl is usually considerably thicker than in asconoid sponges. The presence of porocytes in leuconoid sponges was long debated,

but Weissenfels's studies (1980) on the structure and function of the freshwater sponge *Ephydatia fluviatilis*, certainly the best known of any sponge species, have clearly demonstrated that porocytes form the dermal pores, the apopyles, and proso-pyles in the endopinacoderm (Fig. 4–7). However, there are also some ostia which are formed by the margins of several adjacent pinacocytes (see Simpson, 1984, for review of subject).

BOX 4–1 CHOANOFLAGELLATES

Choanoflagellates are a small order of marine and freshwater zooflagellate protozoans that have a single flagellum surrounded by a collar of microvilli. Their striking similarity to the choanocytes of sponges, not only in structure but in their function as well, has provoked continual speculation about the evolutionary relationship of the two groups. Choanoflagellates may be solitary or colonial, attached or free swimming. Sessile species are attached by a stalk, part of vaselike theca (A and B) that is sometimes present. The individuals of colonial planktonic forms, such as species of *Proterospongia*, are united by a jelly-like matrix (C) or by their collars (B). In the latter case the colony somewhat resembles a plate, with all of the collars and flagella located on the same side. Recently, the marine *Proterospongia choanojuncta* (B) was found to include both a colonial planktonic stage and a solitary, aflagellate, attached stage, which had been previously assigned to a different genus (*Choanoecia*).

A number of zoologists have suggested that sponges evolved independently from the choanoflagellates, but there is growing ultrastructural evidence that choanoflagellates are not only related to sponges but are perhaps the best candidates to be ancestral to all the metazoa. Choanoflagellates possess flat mitochondrial cristae, as do metazoans; other protozoans have tubular cristae. Moreover, the ultrastructural details of the flagellar kinetosomes and associated structures are essentially like those of monociliated metazoan cells (Box 9–1) (Nielsen, 1985). (A, Farmer, J. N., 1980: The Protozoa. C. V. Mosby Co., St. Louis; B, Leadbeater, B. S. C., 1983: Life-history and ultrastructure of a new marine species of *Proterospongia*. Jour. Mar. Biol. Assoc. U.K., *63*:135–160. C, after Kent from Leadbeater.)

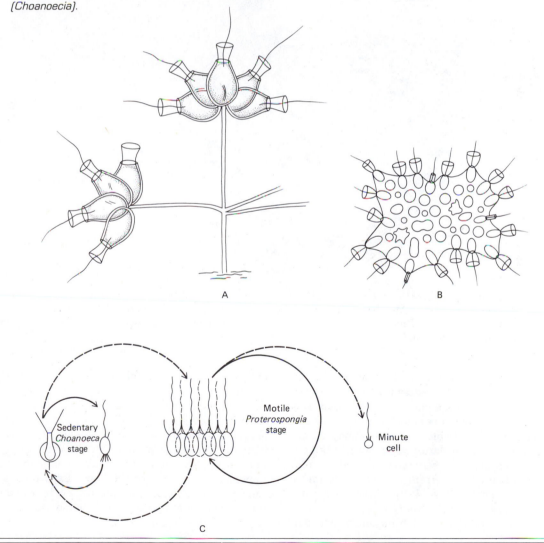

A

B

Sedentary *Choanoeca* stage

Motile *Proterospongia* stage

Minute cell

C

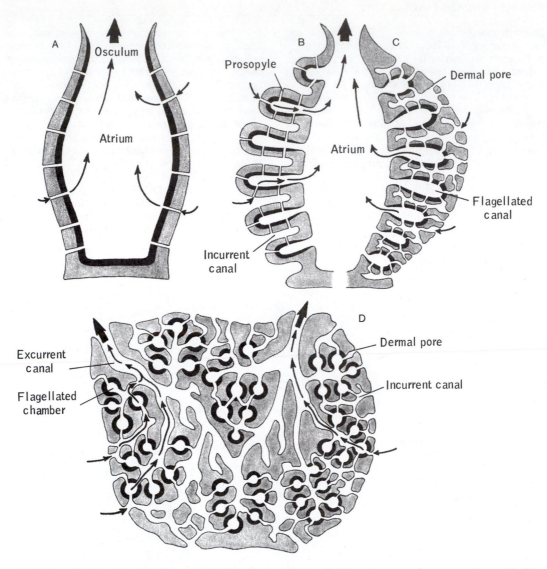

Figure 4-6 Morphological types of sponges (pinacoderm and mesohyl in pale gray; choanocyte layer, black): *A*, Asconoid type. *B*, Syconoid type. *C*, More specialized syconoid type, in which entrance to incurrent canals has been partially filled with pinacoderm and mesohyl. *D*, Leuconoid type. (All modified from Hyman, L. H., 1940: The Invertebrates. Vol. I. McGraw-Hill Book Co., N.Y.)

Most sponges are built on the leuconoid plan, which is evidence of the efficiency of this type of structure. Leuconoid sponges are composed of a mass of flagellated chambers and water canals and may attain a considerable size. They may be encrusting or erect with flattened or branching bodies. Many species are vase-shaped and tubular forms in which the excurrent canals empty into a large, central chamber (Fig. 4–8). Rather than a single osculum, there may be many.

Most leuconoid sponges exhibit one of two general types of internal architecture, regardless of the external form. In one type the body is solid, and adjacent incurrent and excurrent canals run parallel, conducting water inward to flagellated chambers and outward to oscula, which are scattered over the surface (Fig. 4–9). In the other type the body is hollow, with the oscula confined to the upper or distal parts of the body. Excurrent water canals do not run back to the surface but open into the interior cavity, which leads to a distal osculum (Figs. 4–9 and 4–10).

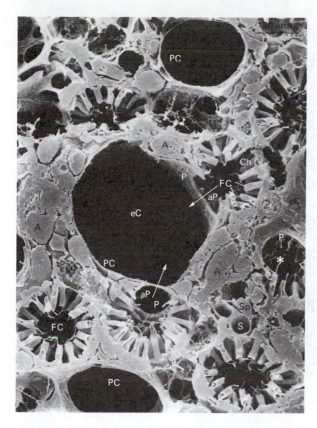

Figure 4–7 SEM of a section taken from the freshwater sponge *Ephydatia fluviatilis.* In this section a number of flagellated chambers (FC) with choanocytes (Ch) surround a large excurrent canal (eC) with a wall formed by pinacocytes (PC). The apopyles (aP) from the flagellated chambers pass through porocytes (P). Within the mesohyl (M) can be seen archeocytes (A), spicules (S), and spongin (Sp). (From Weissenfels, N., 1982: Bau und Funktion des Süsswasserschwamms *Ephydatia fluviatilis.* IX. Rasterelektronenmikroskopische Histologie und Cytologie. Zoomorphology, *100*:75–87.)

The leuconoid plan clearly evolved more than one time within sponges and in some instances may have involved a preceding syconoid stage.

Physiology

The physiology of a sponge largely depends on the current of water flowing through the body. The water brings in oxygen and food and removes waste. Even sperm and eggs are moved in and out by the water currents. The volume of water pumped by a sponge is remarkable. A specimen of *Leuconia (Leucandra),* a leuconoid sponge

10 cm in height and 1 cm in diameter, has roughly 2,250,000 flagellated chambers and pumps 22.5 liters of water per day through its body. The rate of flow is fastest through the osculum and slowest through the flagellated chambers, for these two regions have, respectively, the smallest and the largest total cross-sectional area of the various passageways (Fig. 4–12). By regulating the size of the osculum and closing the ostia, the animal can control the rate of flow and even stop it altogether. In some Demospongiae control of the osculum is facilitated by a special type of mesohyl cell called a myocyte, which displays some similarities to a smooth muscle cell in shape and contractility. However, the myocytes surrounding an osculum do not touch each other.

The current is produced by the beating of the choanocyte flagella, but there is neither coordination nor synchrony of the choanocytes in a particular chamber. The choanocytes are turned toward the apopyle, and each flagellum beats in a spiral manner from its base to its tip (Fig. 4–13). As a result, water is sucked into the flagellated chamber through the small prosopyles located between the bases of the choanocytes. It is then driven to the center of the chamber and out the larger apopyle into an excurrent canal.

The flagellated chambers of many leuconoid sponges possess a central cell, which can regulate water flow (Fig. 4–13C). By shifting position, the central cell can effect closure of the apopyle, and grooves on its surface can immobilize the flagella of the surrounding choanocytes.

Vogel (1974) has demonstrated that when exposed to an external current, water will flow passively through the body of a sponge, given certain structural conditions, such as elevated oscula. It is uncertain to what extent such passive flow is important in the biology of sponges. Many species live in strong to moderate currents, but others reach their greatest development in quiet water or inhabit very confined spaces out of direct exposure to strong currents.

Reiswig's studies on tropical Demospongiae (1971) demonstrated that some exhibit a diurnal rhythm in the propulsion of water through their bodies; others exhibit an erratic, endogenous water flow. External conditions, such as turbulent water caused by storms, may halt water flow regardless of internal conditions. The amount of water filtered by a large population of sponges can be enormous.

Sponges feed on extremely fine particulate material. Studies on three species of Jamaican sponges (Reiswig, 1971) have demonstrated that 80 per cent

(*Text continued on p. 82*)

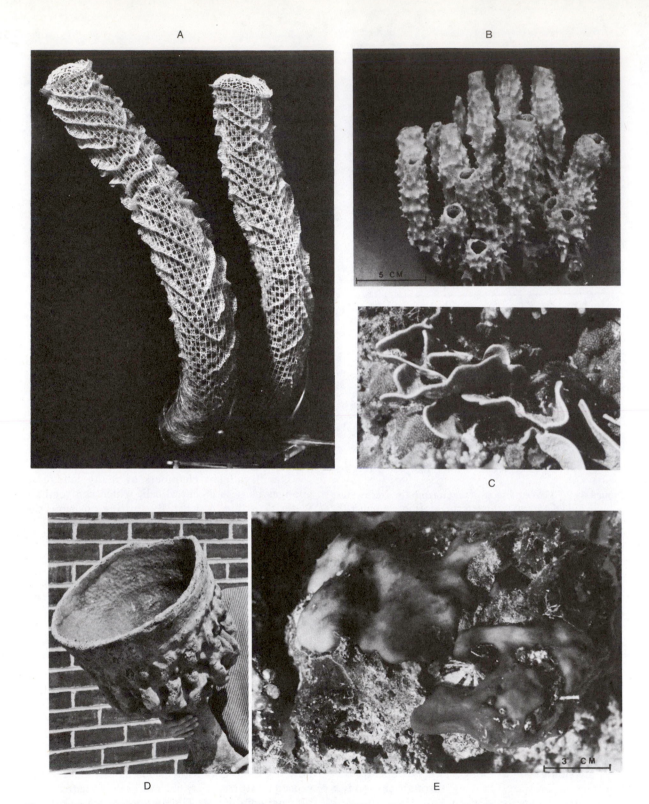

Figure 4–8 *A*, Venus's-flower-basket, *Euplectella*, a hexactinellid sponge in which the spicules are fused to form a lattice. (Courtesy of the American Museum of Natural History.) *B*, *Callyspongia*, a tropical leuconoid sponge (Demospongiae) with a tubular body form. *C*, *Phyllospongia*, a leaflike sponge on a reef flat in Fiji. *D*, *Poterion*, a large, goblet-shaped leuconoid sponge (Demospongiae). *E*, "Chicken-liver," *Chondrilla*, a very common West Indian encrusting sponge, with a tough, almost cartilage-like spongin skeleton (Demospongiae). (*B–E* by Betty M. Barnes.)

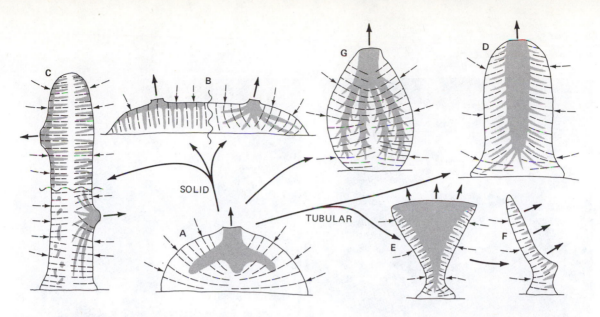

Figure 4–9 Diagram of the two types of sponge architecture, solid and tubular, and their relationship to sponge form. The incurrent system is shown with dotted arrows and lines, and the excurrent system with heavy black arrows and lines. See text for further details. (From Reiswig, H. M., 1975: The aquiferous systems of three marine Demospongiae. J. Morphol., *145*(4):493–502.)

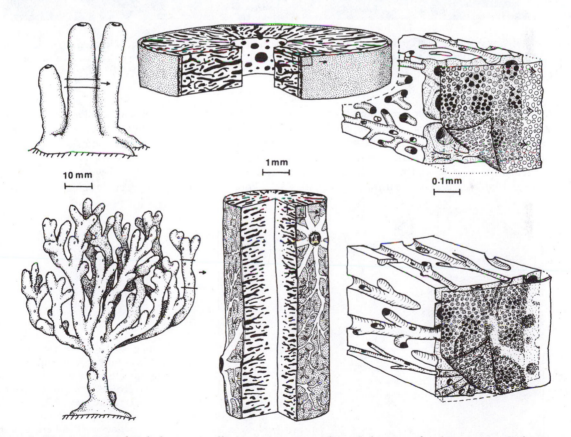

Figure 4–10 Anatomy of *Haliclona permollis* (top), a sponge with a tubular type of architecture, and of *Microciona prolifera* (bottom), a sponge with a solid type of architecture. Three successive levels of magnification are shown for each. (From Reiswig, H. M., 1975: The aquiferous systems of three marine Demospongiae. J. Morphol., *145*(4):493–502.)

Figure 4–11 *A*, A vaselike *Xestospongia muta* from the West Indies. A scleractinian coral is located to the left of the sponge. *B*, *Verongia*, a tubular West Indian sponge, which commonly reaches 50 cm or more in length. The surrounding plantlike animals are gorgonian corals. (*A* and *B* by David Barnes.)

of the filterable organic matter consumed by these sponges is too small to be resolved with ordinary microscopy. The other 20 per cent consists of bacteria, dinoflagellates, and other fine plankton.

In tropical waters, at least, there is about seven times more available carbon in the unresolved frac-

tion than at the planktonic level. The sponges' ability to utilize this food source, directly or indirectly by way of symbiotic cyanobacteria (p. 86), undoubtedly accounts for their long success as sessile animals, especially in tropical waters.

Food particles are apparently selected largely on

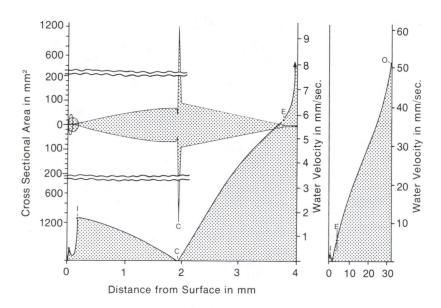

Figure 4–12 Velocity of water passing through different parts of the sponge canal system *(Microciona prolifera)* as related to the cross-sectional area of the passageway. *I* is the inhalant surface; *C* is the choanocyte chambers, *E* is the apertures of the exhalant canals, *O* is the osculum. (From Reiswig, H. M., 1975: The aquiferous systems of three marine Demospongiae. J. Morphol., *145*(4):493–502.)

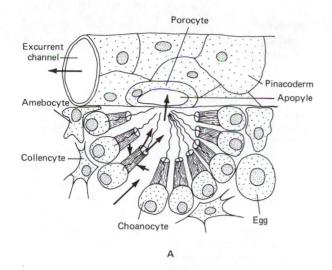

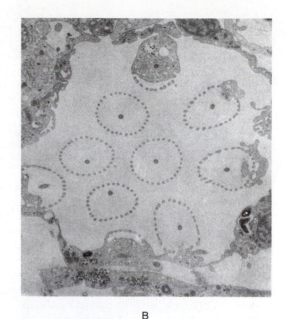

B

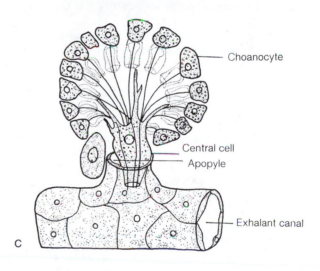

C

Figure 4–13 *A*, Section through flagellated chamber of freshwater sponge, *Ephydatia*. Arrows indicate direction of water currents. *B*, Cross section through a flagellated chamber of the boring sponge *Cliona lampa*. The rings are cross sections of the collars of choanocytes, showing the circles of microvilli and the central flagellum. (From Rutzler, K., and Rieger, G., 1973: Sponge burrowing: fine structure of *Cliona lampa* penetrating calcareous substrata. Mar. Biol., 21(2):144–162.) *C*, Choanocyte chamber of *Suberites massa* containing a central cell, which contributes to the regulation of water flow. *D*, Choanocyte chambers within the body wall of the hexactinellid *Euplectella* (Venus's flower-basket). (*C* after Connes, and *D* after Schulze; both from Bergquist, P. R., 1978: Sponges. Hutchinson, London. pp. 59 and 26.)

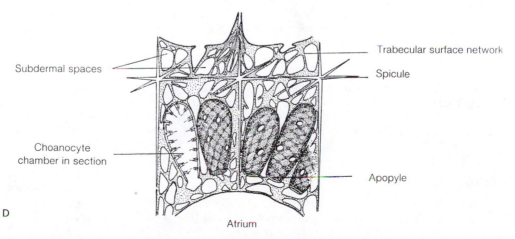

D

the basis of size and are screened in the course of their passage into the flagellated chambers. Only particles smaller than a certain size can enter the dermal pores or pass through the prosopyles. Screening is also provided by protoplasmic strands stretched across the incurrent canals. Food particles are finally filtered by the choanocytes. The trapping of particles by these cells occurs when water passes through the microvillar mesh composing the collar (Fig. 4–13).

All sponge cells can phagocytize particles. Large particles (5–50 μ) are phagocytized by pinacocytes lining the inhalant pathways. Particles of bacterial size and below (less than 1 μ) are removed and engulfed by the choanocytes. Both choanocytes and amebocytes can transfer their engulfed particles to another cell, and amebocytes rather than choanocytes appear to be the principal sites of digestion (Weissenfels, 1976, 1983; Willenz, 1980). Regardless of the cell in which the particle ends up, digestion is intracellular within a food vacuole. The amebocytes probably also act as storage centers for food reserves.

The great volume of water passing through the extensive canal system of a sponge means that most cells, even those deep within the body, are interacting directly with the external medium. Egested waste and nitrogenous waste (largely ammonia) leave the body in the water currents. Gas exchange occurs by simple diffusion between the flowing water and the cells in the sponge along the course of water flow. Most of the cells throughout the body of freshwater sponges possess one or more contractile vacuoles, which constitute the principal osmoregulatory system of these animals (Brauer, 1975).

There is no nervous system in sponges, and reactions are largely local. Coordination depends upon transmission of messenger substances by diffusion within the mesohyl, by wandering ameboid cells and along fixed cells in contact with each other. Electrical conduction appears to take place by the latter route (Lawn et al, 1981).

Many sponges produce metabolites that may prevent other organisms from settling on their surfaces or may deter some potential grazing predators. Some species have very distinctive odors, such as that of garlic, and a few, like the red Caribbean fire sponge *Tedania ignis*, can cause a rash when handled.

The Classes of Sponges

Approximately 5000 species of sponges have been described and are placed within four classes (see Bergquist, 1978, or McGraw-Hill, 1982, *Synopsis and Classification of Living Organisms*, for characterization of orders and families).

Class Calcarea, or Calcispongiae

Members of this class, known as calcareous sponges, are distinctive in having spicules composed of calcium carbonate. All the spicules are of the same general size and are monaxons or three- or four-pronged types; they are usually separate. Spongin fibers are absent. All three grades of structure—asconoid, syconoid, and leuconoid—are encountered. Many Calcarea are drab, although brilliant yellow, red, and lavender species are known. They are not as large as species of other classes; most are less than 10 cm in height. Species of calcareous sponges exist throughout the oceans of the world, but most are restricted to relatively shallow coastal waters. Genera such as *Leucosolenia* and *Sycon* are commonly studied examples of asconoid and syconoid sponges.

The subclass Sphinctozoa contains a single recently discovered representative *(Neocoelia)* from shaded recesses on Indo-Pacific reefs. The Sphinctozoa were abundant from the late Paleozoic through the Mesozoic. There are no spicules, but a calcareous skeleton forms an outer perforated wall and also the walls of interior chambers.

Class Hexactinellida, or Hyalospongiae

Representatives of this class are commonly known as glass sponges. The name *Hexactinellida* is derived from the fact that the spicules include a hexaxon, or six-pointed type (Fig. 4–5G). Furthermore, some of the spicules often are fused to form a skeleton that may be lattice-like and built of long, siliceous fibers. Thus, they are called glass sponges. The glass sponges, as a whole, are the most symmetrical and the most individualized of the sponges—that is, they show less tendency to form interconnecting clusters or large masses with many oscula. The shape is usually cup-, vase-, or urnlike, and they average 10 to 30 cm in height. The coloring in most of these sponges is pale. There is a well-developed atrium, and the single osculum is sometimes covered by a sieve plate—a gratelike covering formed from fused spicules. Lattice-like skeletons composed of fused spicules in species such as Venus's-flower-basket *(Euplectella)* retain the general body structure and symmetry of the living sponge and are very beautiful; the white, filmy skeleton looks as if it were fashioned from rock wool (Fig. 4–8A). Basal tufts of spicule fibers implanted in sand or sediments adapt many species for living on soft bottoms.

The histology of hexactinellids is very different from that of other sponges. All surfaces exposed to water are covered not by pinacoderm but by a syncytial layer (trabecular syncytium), through which long spicules may project. Another syncytium, containing flagella with collars, lines the flagellated chambers. Archeocytes are one of the few discrete cell types. The flagellated chambers are commonly thimble shaped and oriented at right angles in parallel planes to the body wall and central antrium (Fig. 4–13*D*). Hexactinellids are thus somewhat syconoid in structure.

In contrast to the Calcarea, the Hexactinellida are chiefly deepwater sponges. Most live between depths of 200 and 1000 meters, but some have been dredged from the abyssal zone. Although found throughout the world, hexactinellids are the dominant sponges in the Antarctic.

Species of *Euplectella*, Venus's-flower-basket, display an interesting commensal relation with certain species of shrimp *(Spongicola)*. A young male and a young female shrimp enter the atrium and, after growth, are unable to escape through the sieve plate covering the osculum. Their entire life is spent in the sponge prison, where they feed on plankton brought in by the sponge's water currents. A spider crab *(Chorilla)* and an isopod *(Aega)* are also found as commensals with some species of *Euplectella*.

Class Demospongiae

This large class contains 90 per cent of sponge species and includes most of the common and familiar forms. These sponges range in distribution from shallow water to great depths.

Coloration is frequently brilliant because of pigment granules located in the amebocytes. Different species are characterized by different colors, and a complete array of hues is encountered.

The skeleton of this class is variable. It may consist of siliceous spicules or spongin fibers or a combination of both. The genus *Oscarella* is unique in lacking both a spongin and a spicule skeleton. These Demospongiae with siliceous skeletons differ from the Hexactinellida in that their larger spicules are monoaxons or tetraxons, never hexaxons. When both spongin fibers and spicules are present, the spicules are usually connected to, or completely embedded in, the spongin fibers.

All Demospongiae are leuconoid, and the majority are irregular, but all types of growth patterns are displayed. Some are encrusting (Fig. 4–8*E*);

some have an upright branching habit or form irregular mounds; others are stringlike or foliaceous (Fig. 4–8*C*). There are also species, such as *Poterion* (Fig. 4–8*D*), that are goblet or urn shaped, and others, such as *Callyspongia* (Fig. 4–8*B*), that are tubular. The great variation in the shapes of the Demospongiae reflects, in part, adaptations to limitations of space, inclination of substrate, and current velocity. Large upright forms can exploit vertical space and use only a small part of their surface area for attachment. Encrusting forms, although they require more surface area for attachment, can utilize vertical surfaces and very confined habitats, such as crevices and spaces beneath stones (Fig. 4–2). The largest sponges are members of the Demospongiae; some of the tropical loggerhead sponges *(Spheciospongia)* form masses over a meter in height and diameter.

Several families of Demospongiae deserve mention. The boring sponges, composing the family Clionidae, are able to bore into calcareous structures, such as coral and mollusk shells (Fig. 4–2), forming channels that the body of the sponge then fills. At the surface the sponge body projects from the channel opening as small papillae. These papillae represent either clusters of ostia opening into an incurrent canal or an osculum. Excavation, which is begun by the larva, occurs when special amebocytes remove chips of calcium carbonate. The amebocyte begins the process, etching the margins of the chip by digesting the organic framework material and dissolving the calcium carbonate (Pomponi, 1979) (Fig. 4–14*B*). The chip is then undercut in the same manner, the amebocyte enveloping the chip in the process. Eventually, the chip is freed and is eliminated through the excurrent water canals. *Cliona celata,* a common boring sponge that lives in shallow water along the Atlantic coast, inhabits old mollusk shells. The bright sulfur yellow of the sponge is visible where the bored channels reach the surface of the shell. *Cliona lampa* of the Caribbean is red, and it commonly overgrows the surface of the coral or coralline rock that it has penetrated as a thin encrusting sheet. Boring sponges are important agents in the decomposition of shell and coral (Fig. 4–14).

Members of two families of sponges occur in fresh water, but the family Spongillidae contains the majority of freshwater species. The Spongillidae are worldwide in distribution and live in lakes, streams, and ponds where the water is not turbid. They have an encrusting growth pattern, and some are green because of the presence of symbiotic zoochlorellae in the amebocytes. The algae are brought in by water currents and are transferred

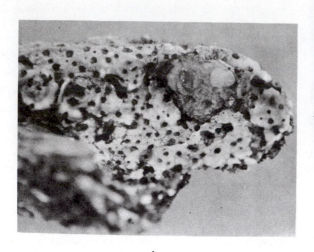

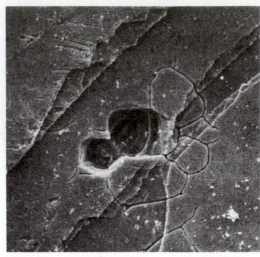

A   B

Figure 4–14 *A*, Remains of a clam shell that has been riddled with boring sponge. *B*, Calcareous surface from which two chips have been removed; four more are partially etched. (From Rützler, K., and Rieger, G., 1973: Sponge burrowing: fine structure of *Cliona lampa* penetrating calcareous substrata. Mar. Biol., *21*:144–162.)

from the choanocytes to the amebocytes. The growth rate of sponges deprived of zoochlorellae is less than half the normal rate (Frost and Williamson, 1980).

Many marine sponges, both Demospongiae and Calcarea, are now known to harbor symbiotic organisms. A few species contain nonmotile dinoflagellates (zooxanthellae), but the most common symbionts are cyanobacteria (blue-green algae), which live within the mesohyl or within specialized amebocytes. The cyanobacterial symbionts of some keratose sponges, including *Verongia*, may make up more than 33 per cent of the sponge. Such sponges live in shallow, well-lighted habitats. Excess photosynthate in the form of glycerol and a phosphorylated compound are utilized by the sponge host. Although bacteria filtered from the water currents are an important part of the sponge diet, there is no evidence that the symbiotic bacteria are digested (Vacelet, 1979; Wilkinson, 1978, 1979, and 1983).

The family Spongiidae contains the common bath sponges. The skeleton is composed only of spongin fibers. *Spongia* and *Hippospongia*, the two genera of commercial value, are gathered from important sponge-fishing grounds in the Gulf of Mexico, the Caribbean, and the Mediterranean. (There is no longer any large, commercial sponge fishing in the United States.) The sponges are gathered by divers, and the living tissue is allowed to decom-

pose in water. The remaining undecomposed skeleton of anastomosing spongin fibers is then washed (Fig. 4–4*B*). The colored block "sponges" seen on store counters are a synthetic product.

Class Sclerospongiae

A fourth class of sponges, the Sclerospongiae, contains a small number of species found in grottoes and tunnels associated with coral reefs in various parts of the world (Jackson et al, 1971). These leuconoid sponges differ from other sponges in having not only siliceous spicules and spongin fibers but also a solid basal skeleton of calcium carbonate. The numerous oscula are raised on the calcareous skeletal mass and have a starlike configuration from the converging excurrent canals (Hartman and Goreau, 1970).

Reproduction

Asexual reproduction by the formation of buds that are liberated from the parent is not common in sponges, although it does occur in some species. Somewhat different from budding is the formation and release of packets of essential cells. Spongillid sponges, as well as some marine forms, have aggregates called gemmules (Fig. 4–15). In freshwater sponges a mass of food-filled archeocytes becomes

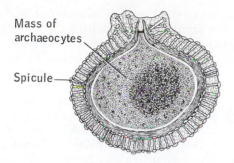

Mass of
archaeocytes

Spicule

Figure 4–15 Section through gemmule of freshwater sponge. (After Evans from Hyman.)

surrounded by other amebocytes (spongocytes) that deposit a hard covering composed of a material similar to spongin. Spicules are also added, so that a thick resistant shell is formed. Gemmule formation takes place primarily in the fall; a large number of these bodies are formed by each sponge. With the onset of winter, the parent sponge disintegrates. The gemmules are able to withstand freezing and drying and thus are able to carry the species through the winter. In spring the interior cells undergo some initial development, and the primordium eventually emerges through an opening (micropyle) in the shell. The primordium continues development into an adult sponge and may attain a large biomass by the end of the summer (Höhr, 1977; Frost et al, 1982).

Considering the relative independence of sponge cells, it is not surprising that many of these animals should have remarkable powers of regeneration. Regeneration was employed in the propagation of commercial sponges in overfished areas off the Florida coast. Pieces of sponge called cuttings can be attached to cement blocks and dumped into the water. Regeneration and several years' growth produce a sponge of marketable size.

The classic experiment demonstrating the regenerative ability of sponges begins with the forcing of living sponge tissue through a silk mesh. The separated cells reorganize after a short time through progressive association of similar cells bounded by pinacocytes, and they may form themselves into several new sponges. Archeocytes are essential for reaggregation. Also essential is a minimum number of cells, about the same number required to form a gemmule. Calcium and magnesium ions plus some cell surface macromolecule are necessary for reaggregation. Whether successful reaggregation will occur only with dissociated

cells from the same species is still being debated. If an individual of certain sponges—*Tethya* for example—is sliced and a piece from another sponge of the same species is inserted into the wound, host and graft will grow together in a short time. The host will reject a graft from a different species. There are also certain sponges, such as species of *Halichondria* and the freshwater *Spongilla*, in which developing individuals, following dense larval settlement, will fuse and form sponges that are genetic mosaics (Fell and Jacob, 1979). On the other hand, grafting experiments with the tropical *Calyspongia diffusa* indicate immunocompetence in some sponges, i.e., the recognition of self and nonself: an individual of this species will accept grafts that come from itself but will reject those from other members of the species. (Hildemann et al, 1979). (See Simpson, 1984, for a review of aggregation studies.)

The sexual reproduction and the embryogeny of sponges display a number of peculiar features. Both hermaphroditic and dioecious sponge species exist, although most are hermaphroditic, but usually producing eggs and sperm at different times. The sperm arise from choanocytes. For example, the choanocytes of an entire flagellated chamber will lose their collars and flagella and form spermatogonia. The cluster becomes surrounded by a cellular wall and is then a spermatic cyst, or a spermatic cyst may be derived from the division of a single sperm mother cell.

Eggs are reported to arise from choanocytes or archeocytes, but there is still no real proof of either origin (see Simpson, 1984). Eggs generally accumulate their food reserves by engulfing adjacent nurse cells and are usually located within a follicle of surrounding cells. Gamete production appears to be initiated by water temperature, photoperiod, or cellular regression (p. 88), depending on the species (Van de Vyver and Willenz, 1975; Elvin, 1976).

Sperm leave the sponge by means of water currents and enter other sponges in the same manner. Certain tropical sponges have been observed to release their sperm suddenly in great milky clouds (Fig. 4–16). Reiswig (1970) reported that a sperm cloud extended 2 to 3 meters above the bottom and induced other sponges to release their sperm. Sudden sperm release may be characteristic of most sponges.

After a sperm has reached a flagellated chamber, it enters a choanocyte, which transports the sperm to the egg. Both cells lose their flagella. After the carrier with its sperm has reached an egg (which would be close by in the surrounding mesohyl), the

Figure 4–16 Sperm release from a specimen of the tubular West Indian sponge *Verongia archeri*. Sponge is about 1.5 meters long. (By Reiswig, H. M., 1970: Science, *170*:538–539.)

carrier fuses with the egg and transfers the sperm to it. Fertilization thus occurs *in situ*.

In the majority of sponges, development to the larval stage takes place within the body of the parent (viviparous). Among the Demospongiae, however, there are some species that liberate fertilized eggs (oviparous). Development in the latter occurs in the sea water.

Cleavage is complete but may be equal or unequal depending on the species. The pattern also varies. Development leads to a larval stage, which displays varying degrees of differentiation. The majority of sponges possess a parenchymula larva, in which flagellated cells cover all of the outer surface, except often the posterior pole (Fig. 4–17*B* and *C*). Spicules are often present, and the interior of the larva commonly contains most of the adult cells, with the exception of choanocytes. The parenchymula larva breaks out of mesohyl of the parent and has a brief free-swimming existence (no more than two days).

Some calcareous sponges, such as *Grantia*, *Sycon*, and *Leucosolenia*, and among the Demospongiae, *Oscarella*, have an amphiblastula larva

(Fig. 4–17*A*). This larva is hollow, and one hemisphere is composed of small flagellated cells and the other of large nonflagellated macromeres.

Following settling and attachment by the anterior end, the sponge larva undergoes an internal reorganization that is comparable to gastrulation in other animals. In the parenchymula, the external flagellated cells lose their flagella and move to the interior, where they form choanocytes, and interior cells move to the periphery to form pinacocytes. The parenchymula larva of freshwater sponges and some marine species develops choanocytes before leaving the parent sponge. In these species, the external flagellated cells are sloughed off or move to the interior but are then phagocytized by amebocytes. In the hollow amphiblastula larva, reorganization following settling occurs either by epiboly or by invagination, or by both, but the macromeres overgrow the micromeres (Fig. 4–17*D*); in other metazoans the macromeres typically become internal. The macromeres in these sponges give rise to the pinacoderm and the micromeres to the choanocytes; both layers produce the amebocytes of the mesohyl.

In most animals development proceeds from the establishment of gross form (morphogenesis) to the addition of more and more histological detail. Sponge development differs in that cell differentiation precedes morphogenesis. Moreover, there is a great deal of cell mobility and reversal of cell differentiation.

In many of the calcareous sponges with a leuconoid structure, the final stages of development after attachment of the larva recapitulate the more primitive asconoid and syconoid structures. For example, a leuconoid sponge undergoes developmental stages resembling the asconoid and syconoid structures before finally reaching the adult leuconoid condition. In other leuconoid sponges, especially the Demospongiae, the leuconoid condition is attained more directly. The first stage is known as a rhagon (Fig. 4–17*E*). It resembles either the asconoid or the syconoid structure except that the walls are quite thick. The leuconoid plan develops directly from the rhagon stage by means of the formation of channels and flagellated chambers in the thick body wall.

Some marine sponges live only one year; others live many years. Those in temperate regions are usually dormant in the winter. *Microciona* in Long Island Sound, for example, passes the water in a reduced state, lacking flagellated chambers and other parts of the water canal system (Simpson, 1968). With an increase in water temperature, the sponge redevelops the adult functional condition. Fresh-

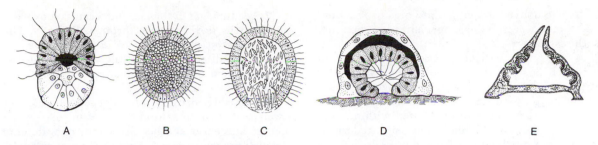

Figure 4–17   Sponge larvae and postlarval development: *A*, An amphiblastula larva. *B* and *C*, Parenchymula larvae. *D*, Gastrulation of an amphiblastula larva following settling. (*C* after Wilson; *D* after Hammer.) *E*, Postlarval rhagon stage. (After Sollas.)

water sponges also winter in a regressed state or else die, releasing gemmules.

The growth rates of encrusting species are very slow, some species requiring ten years to reach diameters of 1 to 5 cm. Large encrusting sponges may be over 75 years old. Growth at the margin is very uneven, accounting for the irregular shape of encrusting species (Ayling, 1983).

The Phylogenetic Position of Sponges

Sponges certainly arose prior to the Paleozoic era, and there have been a number of claims of pre-Cambrian fossils, but none of these claims has been clearly established. However, beginning with the Cambrian period and extending to the present, the fossil record of sponges is abundant. Early Paleozoic reefs were composed of calcareous blue-green algae (stromatolites) and two now-extinct groups of calcareous spongelike animals (archaeocyathids and stromatoporoids*), There are also some flints composed entirely of fused sponge spicules.

The first Calcarea known are from the Devonian period, and during this same period there was an especially great development of glass sponges (Hexactinellida). This class underwent another period of development during the Cretaceous period, when the phylum as a whole attained its greatest diversity and abundance.

The evolutionary origin of sponges poses a number of interesting problems. The absence of organs and the low level of cellular differentiation and interdependence in sponges certainly seem to be primitive characteristics. But a specialized body

structure built around a water canal system and lacking distinct anterior and posterior ends is found in no other groups of animals. Moreover, the cellular differentiation is unlike that of other metazoans. All these features suggest that sponges are phylogenetically remote from other metazoans.

Some zoologists have suggested that the multicellular condition of sponges evolved independently from that leading to the rest of the metazoan animals. However, most now believe that sponges had a common origin with other metazoans but diverged early in metazoan history (Bergquist, 1985). There is a growing belief that some colonial choanoflagellate (Box 4–1) probably gave rise to the metazoans and that the sponge choanocyte retains the flagellated collar cell of such an ancestor (Nielsen, 1985).

Many of the specialized sponge features, such as the water canal system, certainly developed in conjunction with a sessile existence and the flagellated, exterior cells probably moved into the interior to become choanocytes. Such movement occurs in the embryogeny of living sponges.

There can be little doubt that sponges diverged early from the main line of metazoan evolution and have given rise to no other members of the Animal Kingdom. Because of their isolated phylogenetic position, the sponges have often been separated from the other multicellular animals (Eumetazoa) and placed in a separate subkingdom, the Parazoa.

Summary

1 Sponges are sessile aquatic animals, mostly marine and largely inhabitants of hard substrates.

2 They are primitive in their lack of organs, including mouth and gut. There are different kinds of cells, but cellular differentiation has not followed the common designs of other animals.

*The stromatoporoids are now considered a subphylum of sponges similar to the sclerosponges (see Stearn, 1975).

3 The bodies of sponges are organized around a system of water canals, a specialization correlated with sessility.

4 The small, vase-shaped asconoid body form, in which flagellated choanocytes line an interior atrial chamber, is the primitive sponge form. The evolution of the common leuconoid form, in which the flagellated cells are distributed within a vast number of minute chambers, has permitted the attainment of much larger size and great diversity of shape, since each addition to the sponge body brings with it all of the units necessary to provide the required additional water flow.

5 The growth form of sponges is, in part, an adaptive response to the availability of space, the inclination of the substrate, and the current velocity.

6 Support is provided by a skeleton of organic spongin fibers or siliceous or calcareous spicules, or a combination of spongin fibers and siliceous spicules.

7 Feeding, gas exchange, and waste removal depend on the flow of water through the body. The ability of the choanocyte collar to remove extremely small particles from the water stream has probably been an important factor in the long, successful history of sponges.

8 Probably because of their sessility, most sponges are hermaphroditic. Sperm leave one sponge and enter another in the currents flowing through the water canals. Eggs in the mesophyl are fertilized *in situ*. They may then be released by way of the water canals or brooded up to the larval stage. In most sponges the flagellated larva is a blastula, and reorganization equivalent to gastrulation occurs following settling.

9 Sponges are probably an early evolutionary side branch that gave rise to no other groups of animals.

REFERENCES

The literature included here is restricted to books and papers on sponges alone. The introductory references on page 8 list many general works and field guides that contain sections on sponges.

Ayling, A. L., 1983: Growth and regeneration rates in thinly encrusting Demospongiae from temperate waters. Biol. Bull., 165:343–352.

Bergquist, P. R., 1978: Sponges. Hutchinson and Co., London. 268 pp. (An excellent general account of sponges.)

Bergquist, P. R., 1985: Poriferan relationships. *In* Conway Morris, S., George, J. D., Gibson, R., and Platt, H. M. (Eds.), 1985: The Origins and Relationships of Lower Invertebrates. Systematics Association Spec. Vol. No. 28. Clarendon Press, Oxford. 344 pp.

Brauer, E. B., 1975: Osmoregulation in the freshwater sponge, *Spongilla lacustris*. J. Exper. Zool., 192(2):181–192.

Brien, P., Lévi, C., Sara, M., Tuzet, O., and Vacelet, J., 1973: Spongaires. Traite de Zoologie. Vol. III. Pt. 1. Masson et Cie. 716 pp.

Brill, B., 1973: Untersuchungen zur Ultrastruktur der Choanocyte von *Ephydatia fluviatilis*. L. Z. Zellforsch., 144:231–245.

Cobb, W. R., 1969: Penetration of calcium carbonate substrates by the boring sponge, *Cliona*. Am. Zool., 9:783–790.

Elvin, D. W., 1976: Seasonal growth and reproduction of an intertidal sponge, *Haliclona permollis* (Bowerbank). Biol. Bull., 151:108–125.

Fell, P. E., 1974: Porifera. *In* Giese, A. C., and Pearse, J. S. (Eds.): Reproduction of Marine Invertebrates. Vol. I. Academic Press, N.Y. pp. 51–132.

Fell, P. E., and Jacob, W. F., 1979: Reproduction and development of *Halicondria* sp. in the Mystic estuary, Connecticut. Biol. Bull., 156:62–75.

Frost, T. M., Nagy, G. S., and Gilbert, J. J., 1982: Population dynamics and standing biomass of the freshwater sponge *Spongilla lacustris*. Ecology, 63(5):1203–1210.

Frost, T. M., and Williamson, C. E., 1980: *In situ* determination of the effect of symbiotic algae on the growth of the freshwater sponge *Spongilla lacustris*. Ecology, 61(6):1361–1370.

Fry, W. G. (Ed.), 1970: The Biology of Porifera. Academic Press, N.Y. 512 pp. (A collection of papers presented at a symposium of the Zoological Society of London.)

Goreau, T. F., and Hartman, W. D., 1963: Control of coral reefs by boring sponges. *In* Sognnaes, R. F. (Ed.): Mechanisms of Hard Tissue Destruction. American Association for the Advancement of Science, Washington, D.C. Vol. 75, pp. 25–54.

Harrison, F. W., and Cowden, R. R., 1975: Cytochemical observations of gemmule development in *Eunapius fragilis* (Leidy): Porifera; Spongillidae. Differentiation, 4:99–109.

Harrison, F. W., and Cowden, R. R., 1975: Cytochemical observations of larval development of *Eunapius fragilis* (Leidy): Porifera; Spongillidae. J. Morph., 145:125–142.

Harrison, F. W., and Cowden, R. R., 1976: Aspects of Sponge Biology. Academic Press, N.Y. 354 pp. (Papers presented at a symposium held in Albany, N.Y., in 1975.)

Hartman, W. D., and Goreau, T. F., 1970: Jamaican coralline sponges: their morphology, ecology, and fossil relatives. *In* Fry, W. G. (Ed.): The Biology of Porifera. Academic Press, N.Y. pp. 205–240.

Hildemann, W. H., Johnson, I. S., and Jokiel, P. L., 1979: Immunocompetence in the lowest metazoan phylum: transplantation immunity in sponges. Science, *204*:420–422.

Höhr, D., 1977: Differenzierungsvorgänge in der keimenden Gemmula von *Ephydatia fluviatilis*. Wilhelm Roux's Archives, *182*:329–346.

Jackson, J. B. C., Goreau, T. F., and Hartman, W. D., 1971: Recent brachiopod-coralline sponge communities and their paleoecological significance. Science, *173*:623–625.

Koehl, M. A. R., 1982: Mechanical design of spicule-reinforced connective tissue: stiffness. Jour. Exp. Biol., *98*:239–267.

Lawn, I. D., Mackie, G. O., and Silver, G., 1981: Conduction system in a sponge. Science, *211*:1169–1171.

Lévi, C., and Boury-Esnault, N. (Eds.), 1979: Biologie des Spongiaires. Colloques Internat. du Centre National de la Recherche Scientfique No. 291. (A collection of papers presented at a symposium.)

Moore, R. D. (Ed.), 1955: Treatise on Invertebrate Paleontology. Archaeocyatha, Porifera. Vol. E. Geological Society of America and University of Kansas Press, Lawrence, Kans.

Nielsen, C., 1985: Animal phylogeny in the light of the trochaea theory. Biol. Jour. Linn. Soc., *25*:243–299.

Pavans de Ceccatty, M., 1974: Coordination in sponges. The foundations of integration. Am. Zool., *14*:895–903.

Pomponi, S. A., 1979: Ultrastructure and cytochemistry of the etching area of boring sponges. See Lévi and Boury-Esnault, pp. 317–323.

Reiswig, H. M., 1970: Porifera: sudden sperm release by tropical Demospongiae. Science, *170*:538–539.

Reiswig, H. M., 1971: *In situ* pumping activities of tropical Demospongiae. Mar. Biol., *9*(1):38–50.

Reiswig, H. M., 1971: Particle feeding in natural populations of three marine demosponges. Biol. Bull., *141*(3):568–591.

Reiswig, H. M., 1975a: Bacteria as food for temperate-water marine sponges. Can. J. Zool., *53*(5):582–589.

Reiswig, H. M., 1975b: The aquiferous systems of three marine Demospongiae. J. Morph., *145*(4):493–502.

Rützler, K., and Rieger, G., 1973: Sponge burrowing: fine structure of *Cliona lampa* penetrating calcareous substrata. Mar. Biol., *21*:144–162.

Simpson, T. L., 1968: The biology of the marine sponge *Microciona prolifera*. Temperature related, annual changes in functional and reproductive elements with a description of larval metamorphosis. J. Exp. Mar. Biol. Ecol., *2*:252–277.

Simpson, T. L., 1984: The Cell Biology of Sponges. Springer-Verlag, N.Y. 662 pp. (A detailed review of most aspects of sponge biology.)

Simpson, T. L., and Gilbert, J. J., 1973: Gemmulation, gemmule hatching, and sexual reproduction in freshwater sponges: the life cycle of *Spongilla lacustris* and *Tubella pennsylvanica*. Trans. Am. Microsc. Soc., *92*(3):422–433.

Stearn, C. W., 1975: The stromatoporoid animal. Lethaia, *8*:89–100.

Vacelet, J., 1979: La place des spongiaires dans les systems trophicques marins. See Lévi and Boury-Esnault, pp. 259–270.

Van de Vyver, G., and Willenz, P., 1975: An experimental study of the life-cycle of the fresh-water sponge *Ephydatia fluviatilis* in its natural surroundings. Wilhelm Roux' Archiv., *177*:41–52.

Vogel, S., 1974: Current induced flow through the sponge, *Halichondria*. Biol. Bull., *147*(2):443–456.

Weissenfels, N., 1976: Bau und Funktion des Süsswasserschwamms *Ephydatia fluviatilis*. III. Nahrungsaufnahme, Verdauung und Defäkation. Zoomorph., *85*:73–88.

Weissenfels, N., and Landschoff, H. W., 1977: Bau und Funktion des Süsswasserschwamms *Ephydatia fluviatilis*. IV. Die Entwicklung der monaxialen SiO_2-Nadeln in Sandwich-Kulturen. Zool. Jb. Anat., *98*:355–371.

Weissenfels, N., 1980: Bau und Funktion des Süsswasserschwamms *Ephydatia fluviatilis*. VII. Die Porocyten. Zoomorph., *95*:27–40.

Weissenfels, N., 1983: Bau und Funktion des Süsswasserschwamms *Ephydatia fluviatilis*. X. Der Nachweis des offenen Mesenchyms durch Verfütterung von Bäckerhefe. Zoomorph. *103*:15–23.

Wiedenmayer, F., 1977: Shallow-Water Sponges of the Western Bahamas. Birkhauser Verlag. Basel, Switzerland. 287 pp.

Wilkinson, C. R., 1978: Microbial associations in sponges: III. Ultrastructure in the *in situ* associations in coral reef sponges. Mar. Biol., *49*(2):177–185.

Wilkinson, C. R., 1979: Nutrient translocation from symbiotic cyanobacteria to coral reef sponges. See Lévi and Boury-Esnault, pp. 373–380.

Wilkinson, C. R., 1983: Net primary productivity in coral reef sponges. Science, *219*:410–411.

Willenz, P., 1980: Kinetic and morphological aspects of particle ingestion by the freshwater sponge *Ephydatia fluviatilis*. In Smith, D. C., and Tiffon, Y. (Eds.): Nutrition in the Lower Metazoa. Pergamon Press, Oxford, pp. 163–178.

5

The Cnidarians and Ctenophores

The phylum Cnidaria, or Coelenterata, includes the familiar hydras, jellyfish, sea anemones, and corals. The bright coloring of many species, combined with a radial symmetry, creates a beauty that is surpassed by few other animals. The radial symmetry is commonly considered justification for uniting the cnidarians and the ctenophores within a division of phyla of the Animal Kingdom called the Radiata.

The cnidarians possess two basic metazoan structural features. There is an internal space for digestion, called in cnidarians a gastrovascular cavity (Fig. 5–1). This cavity lies along the polar axis of the animal and opens to the outside at one end to form a mouth. The presence of a mouth and digestive cavity permits the use of a much greater range of food sizes than is possible in the protozoa and sponges. In cnidarians a circle of tentacles, representing extensions of the body wall, surrounds the mouth to aid in the capture and ingestion of food.

The cnidarian body wall consists of three basic layers (Fig. 5–1): an outer layer of epidermis, an inner layer of cells lining the gastrovascular cavity, and between these a layer called mesoglea. The mesoglea ranges from a thin, noncellular membrane to a thick, fibrous, jelly-like, mucoid material with or without wandering cells. The mesoglea probably evolved from a basement membrane, and in forms like *Hydra* it has hardly progressed beyond that level. In others it is more like connective tissue, but the cells appear to be derived from the epidermis. Thus, cnidarians are diploblastic; i.e., the body is constructed from only two germ layers, ectoderm and endoderm.

Histologically, the cnidarians have remained rather primitive, although they anticipate some of the specializations that are found in higher metazoans. A considerable number of cell types compose the epidermis and gastrodermis, but there is only a limited degree of organ development.

Although all cnidarians are basically tentaculate and radially symmetrical, two structural types are encountered within the phylum. One type, which is sessile, is known as a polyp. The other form is free swimming and is called a medusa. Typically, the body of a polyp is a tube or cylinder, in which the oral end, bearing the mouth and tentacles, is di-

Figure 5–1 *A*, Polypoid body form. *B*, Medusoid body form.

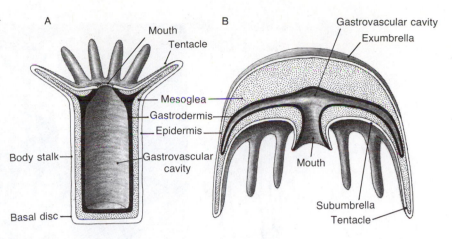

rected upward, and the opposite, or aboral, end is attached (Fig. 5–1*A*).

The medusoid body resembles a bell or an umbrella, with the convex side upward and the mouth located in the center of the concave undersurface (Fig. 5–1*B*). The tentacles hang down from the margin of the bell. In contrast to the polypoid mesoglea (middle layer), which is more or less thin, the medusoid mesoglea is extremely thick and constitutes the bulk of the animal. Because of this mass of jelly-like mesogleal material, these cnidarian forms are commonly known as jellyfish. Some cnidarians exhibit only the polypoid form, some only the medusoid form, and others pass through both in their life cycle. Colonial organization has evolved numerous times within the phylum, especially in polypoid forms.

Except for the hydras and a few other freshwater hydrozoans, cnidarians are marine. Most are inhabitants of shallow water; sessile forms abound on rocky coasts or on coral formations in tropical waters. The phylum is composed of approximately 9000 living species, and a rich fossil record dates from the Cambrian period.

Cnidarian Histology and Physiology: Hydras

For the sake of simplicity, the following introduction to the histology and physiology of cnidarians is based primarily on the familiar hydras, but the greater part of this discussion also applies to the other cnidarians; the more important exceptions are included here, as well as being described in the survey of the three classes later in the chapter.

Hydras are cylindrical, solitary polyps that range from a few millimeters to 1 cm or more in length (Fig. 5–2). However, the diameter seldom exceeds 1 mm. The aboral end of the cylindrical body stalk forms a basal disc, by which the animal attaches to the substratum. The oral end contains a mound, or cone, called the hypostome, with the mouth at the top. Around the base of the cone is a circle of about six tentacles.

The Epidermis

The epidermis is composed of five principal types of cells.

EPITHELIOMUSCULAR CELLS

The most important type of cell, in terms of body covering, is the epitheliomuscular cell (Figs. 5–3 and 5–4*A* and *E*). These cells are somewhat columnar in shape, with the base resting against the mesoglea and the slightly expanded distal end forming most of the epidermal surface. Unlike true columnar epithelium, however, epitheliomuscular cells possess two, three or more basal extensions, each containing a contractile myofibril. The basal extensions are oriented parallel to the axis of the body stalk and tentacles and interwoven with those of other epitheliomuscular cells. The ends of successive extensions are connected to each other (West, 1978). Collectively, they form a cylindrical, longitudinal, contractile layer. The epitheliomuscular cell as just described is widespread among cnidarians, but there are also many modifications resulting from suppression of either the epithelial or the contractile part of the cell. Moreover, the cell may be flattened or squamous, rather than columnar (Fig. 5–5).

INTERSTITIAL CELLS

Located beneath the epidermal surface and wedged between the epitheliomuscular cells are small,

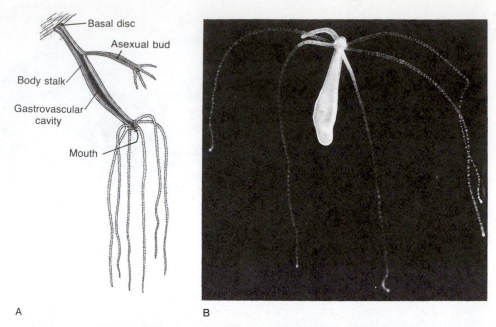

Figure 5–2 *A,* Section through a hydra with an asexual bud. (After Hyman, L. H. 1940: The Invertebrates: Protozoa through Ctenophora. (Vol.1) McGraw-Hill Book Co., N.Y.) *B,* Photograph of a living hydra. Note extended tentacles and the wartlike batteries of nematocysts that appear along the tentacles. (By Charles Walcott.)

rounded cells with relatively large nuclei. These are known as interstitial cells (Fig. 5–3). The interstitial cells, acting as the germinal or formative cells of the animal, give rise to the sperm and eggs as well as to any other type of cells.

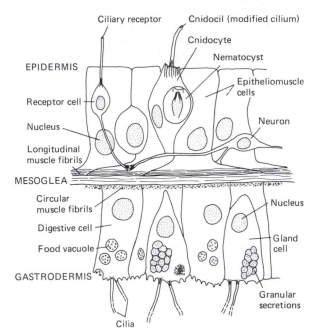

Figure 5–3 Body wall of a hydra (longitudinal section).

CNIDOCYTES

Cnidocytes, a third type of cell, are located throughout the epidermis and are lodged between or invaginated within epitheliomuscular cells. They are especially abundant on the tentacles. These specialized cells, which are unique to and characteristic of all cnidarians, contain organelles capable of eversion, known as cnidae; these include the stinging structures called nematocysts. A cnidocyte is a rounded or an ovoid cell with a basal nucleus (Fig. 5–4*B*). One end of the cell contains a short, stiff, bristle-like process called a cnidocil in hydrozoans and scyphozoans. It has an ultrastructure similar to that of a flagellum and is exposed to the surface. Although the cnidocil is not present in anthozoans, a ciliary cone complex of presumably similar function is associated with at least some of the types of cnidocytes in this group (Fig. 5–6*C* and *D*). The interior of the cell is filled by a capsule containing a coiled, usually pleated tube (Fig. 5–4*B*), and the end of the capsule that is directed toward the outside is covered by an operculum or by lid-like flaps. The base is anchored to the lateral extensions of one or more epitheliomuscular cells and may also be associated with a neuron terminal.

The discharge mechanism apparently involves a change in the permeability of the capsule wall. Under the combined influence of mechanical and chemical stimuli, which are initially received and

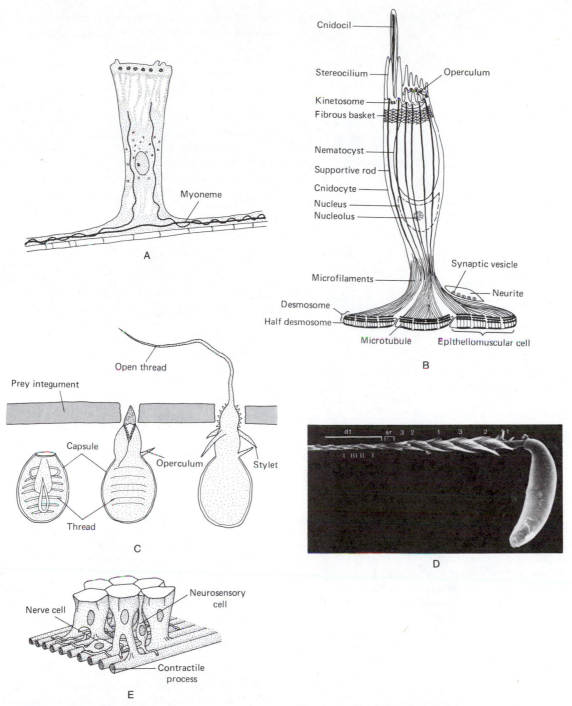

Figure 5–4 *A,* Epitheliomuscle cell of *Hydra.* (After Gelei from Hyman.) *B,* Diagram of a cnidocyte from the hydrozoan jellyfish *Gonionemus.* (From Westfall, J., 1970: Z. Zellforsch., *110*:457–470.) *C,* Puncturing of prey integument by a barbed nematocyst of *Hydra.* See text for description. (Based on figures from Holstein, T., and Tardent, P., 1984: An ultra-high-speed analysis of exocytosis: nematocyst discharge. Science, *223*:830–833.) *D,* Open-tubed nematocyst of the hydroid *Laomedea.* (From Ostman, C., 1982: Nematocysts and taxonomy in *Laomedea, Gonothyraea* and *Obelia.* Zoologica Scripta, *11*(4):227–241.) *E,* Diagram of the cnidarian epidermis, showing epitheliomuscle cells, sensory cell, and nerve net. (From Mackie, G. O., and Passano, L. M., 1968: Epithelial conduction in hydromedusae. J. Gen. Physiol. *52*:600.)

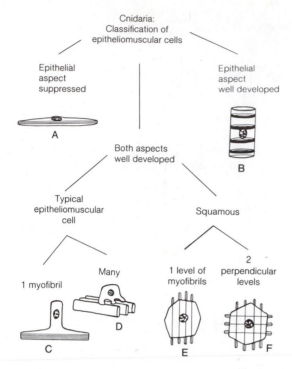

Figure 5–5   Types of cnidarian epitheliomuscular cells. (From Chapman, D. M., 1974: Cnidarian histology. *In* Muscatine, L., and Lenhoff, H. M. [Eds.]: Coelenterate Biology. Academic Press, N.Y. p. 45.)

conducted by the cnidocyte surface structure, the operculum or apical flaps of the cnida open. Hydrostatic pressure within the capsule everts the tube (turns it inside out). Although cnidocytes usually fire as independent effectors, discharge can apparently be effected by nerve impulses from an associated neuron terminal, and neuronal connections may serve to bring about coordinated firing by a large number of nematocysts.

A discharged nematocyst consists of a capsule and a threadlike tube of varying length (Fig. 5–4*C* and *D*). The tube or thread is commonly armed with spines, particularly around the base, and may be open or closed at the tip. Variations in the arrangement and length of the spines and the diameter of the tube give rise to some 30 kinds of cnidae, which are constant for the species and of great value to cnidarian taxonomists.

Nematocysts are the most important type of cnidae and are found in all cnidarians. The other two types, spirocysts and ptychocysts, are found only in anthozoans and will be described later. Nematocysts function in prey capture, and many are penetrating and capable of delivering a toxin (Fig. 5–4*C*). The thread is generally open at the tip and frequently armed with spines. Upon discharge, the

thread bores its way into the tissues of a prey and injects a protein toxin that has a paralyzing action. At least in some types, the spines aid in puncturing the integument of the prey. Under high-speed microcinematography the eversion of a barbed nematocyst (stenotele) of *Hydra* is seen to occur in two phases. In the very rapid first phase, the capsule lid opens and the everting barbed region with three stylets directed forward punches a hole in the prey's integument (Fig. 5–4*C*). In the second phase the stylets flip back and the thread everts into the body of the prey through the opening created by the stylets (Holstein and Tardent, 1984).

The toxic effect of the nematocysts of most cnidarians is not perceptible to humans. However, some marine forms have nematocysts that can produce a very painful burning sensation and irritation or even death. Other nematocysts, such as the desmonemes, do not possess any known toxic properties; instead, they function by adhesion or wrapping and entangling prey. The thread is closed at the end and may be unarmed and coiled or have a long spiny shaft. The spines appear to be an adaptation for adhesion to the prey surface (Purcell, 1984).

Different species of cnidarians possess one to seven structural types of nematocysts, the number and types depending on the nature of the prey. *Hydra*, for example, possesses four types, which are arranged on the tentacles in groups of batteries (Fig. 5–2*B*). Each battery represents a large number of nematocysts that have become invaginated within one epitheliomuscular cell (Fig. 5–6). The batteries appear as bumps or warts when the tentacles are extended.

Nematocysts and other cnidae are used but once; new cnidocytes are formed from nearby interstitial cells. Studies on *Hydra littoralis* (Kline, 1961) indicate that about 25 per cent of the nematocysts in the tentacles are lost in the process of eating a brine shrimp. The discharged nematocysts are replaced within 48 hours.

MUCUS-SECRETING CELLS

Mucus-secreting gland cells are a fourth type of cell found in the epidermis. They are particularly abundant in the basal disc of *Hydra* and possess contractile extensions similar to the epitheliomuscular cells.

RECEPTOR AND NERVE CELLS

The remaining two types of cells are the sensory cells and the nerve cells. Receptor cells are elongated cells oriented at right angles to the epidermal surface (Figs. 5–3 and 5–4*E*). The base of each cell gives rise to a number of neuron processes, and the

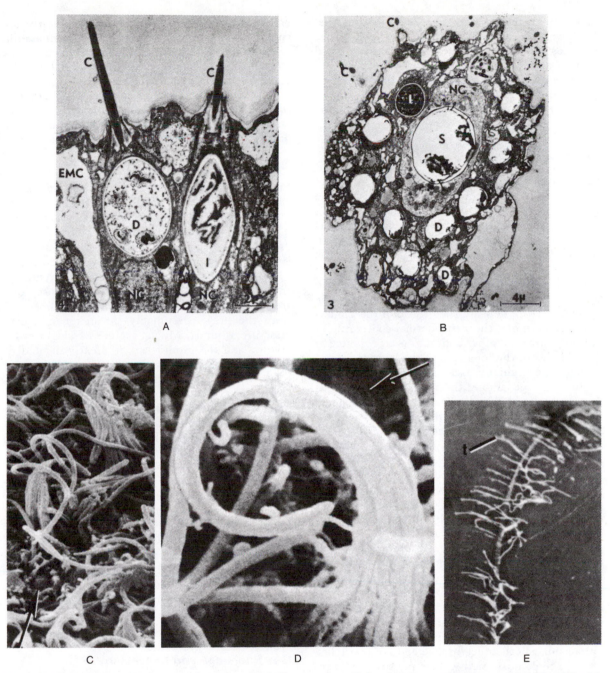

Figure 5-6 *A*, Longitudinal section through an epitheliomuscular cell showing two nematocyst capsules (D and I) with their projecting cnidocils (C). *B*, Cross section of an epitheliomuscular cell containing a central stenotele nematocyst (S) surrounded by smaller nematocysts (D and I). Cnidocils of adjacent cnidocytes are labeled C, and NC is the central cnidocyte. (By Westfall, J., Yamataka, S., and Enos, P., 1971: 20th Ann. Proc. Electron Micro. Soc. Am.) *C*, Tentacle surface of the coral *Balanophyllia*, showing a number of ciliary cone complexes. Arrow indicates a probable spirocyst. *D*, A ciliary cone complex on the tentacle of the sea anemone *Corynactis californica*. A single central cilium is surrounded by shorter stereocilia. (From Mariscal, R. N., 1974: Scanning electron microscopy of the sensory surface of the tentacles of sea anemones and corals. Z. Zellforsch., *147*:149–156.) *E*, Spirocyst thread from the coral *Paracyathus*, showing large numbers of unsolubilized tubules. (From Mariscal, R. N., McLean, R. B., and Hand, C., 1977: The form and function of cnidarian spirocysts: 3. Ultrastructure of the thread and the function of spirocysts. Cell Tiss. Res., *178*:427–433.)

distal end terminates in a sphere or a sensory bristle. Receptor cells are particularly abundant on the tentacles and, like cnidocytes, may be invaginated within epitheliomuscular cells.

The nerve cells, which are more or less similar to the multipolar neurons of other animals, are located at the base of the epidermis next to the mesoglea (Fig. 5–3).

The Gastrodermis

The histology of the gastrodermis, or inner layer of the body wall, is somewhat similar to that of the epidermis (Fig. 5–3). However, the cells corresponding to epitheliomuscular cells in the epidermis are called nutritive-muscle cells in the gastrodermis. The two types are similar in shape, but the nutritive-muscle cells are usually flagellated, and the basal contractile extensions, which develop to the highest degree in the hypostome and tentacle bases, are more delicate than those of the epidermis. Furthermore, the gastrodermal contractile fibers in hydras and other hydrozoans are oriented at right angles to the long axis of the body stalk and thus form a circular muscle layer.

Interspersed among the nutritive-muscle cells are enzymatic-gland cells. These are wedge-shaped, flagellated cells with their tapered ends directed toward the mesoglea. Enzymatic-gland cells do not exhibit the basal contractile processes.

Mucus-secreting gland cells are abundant around the mouth. Nerve cells also exist, but in far fewer numbers. Nematocysts are lacking in the gastrodermis of hydras and other hydrozoans but are present in this layer in the other classes of cnidarians, although restricted in distribution.

In many cnidarians the gastrodermal cells contain symbiotic algae. Some species of *Hydra* harbor the green zoochlorellae, as do certain freshwater sponges. However, the symbiotic algae of most marine cnidarians are nonmotile dinoflagellates, called zooxanthellae. The green or yellow-brown color of these algae gives a similar color to the cnidarian host. Both zoochlorellae and zooxanthellae provide their hosts with excess photosynthate (p. 134) (Muscatine, 1974; Pardy and White, 1977).

Locomotion

The body stalk and tentacles of hydras can extend, contract, or bend to one side or the other. In *Hydra* the gastrodermal fibers in most parts of the body are so poorly developed that movement is due almost entirely to the contractions of the longitudi-

nal, epidermal fibers. Fluid within the gastrovascular cavity plays an important role as a hydraulic skeleton. By taking in water through the mouth as a result of the beating of the gastrodermal flagella, a relaxed hydra may stretch out to a length of 20 mm. Contraction of the epidermal fibers can shorten the same animal to 0.5 mm. Hydras can detach and shift locations by somersaulting or floating.

Nutrition

Most cnidarians, including hydras, are carnivorous and feed mainly on small crustaceans. Contact with the tentacles brings about a discharge of nematocysts, which entangle and then paralyze the prey. The tentacles then pull the captured organism toward the mouth, which opens to receive it. The retraction and bending inward of the tentacles and the opening of the mouth are a response initiated by various amino acids and peptides liberated from the prey, presumably through nematocyst puncture wounds. Mucus secretions aid in swallowing, and the mouth can be greatly distended. Eventually the prey arrives at the gastrovascular cavity. Enzymatic-gland cells then discharge proteolytic enzymes that begin the digestion of proteins, and the tissues of the prey are gradually reduced to a soupy broth. The beating of the flagella of the gastrodermal cells ensures mixing. In colonial cnidarians such flagellar action distributes food products among members of the colony.

After this initial extracellular phase, digestion continues intracellularly. The nutritive-muscle cells engulf small fragments of tissue (see McNeil, 1981). Continued digestion of proteins and the digestion of fats occur within food vacuoles in the nutritive-muscle cells, and the food vacuoles undergo the acid and alkaline phases characteristic of protozoa. Products of digestion are circulated by cellular diffusion. Undigestible materials are ejected from the mouth when the body contracts.

Gas Exchange and Excretion

Gas exchange occurs across the general body surface. Nitrogenous wastes (ammonia) also diffuse through the general body surface. As in many other freshwater animals, there is a continual influx of fresh water into the bodies of hydras through the external surface. Excess water is removed during periodic oral elimination of fluid from the gastrovascular cavity, which is hypo-osmotic to the tissue fluids. The gastrovascular cavity thus acts like a giant contractile vacuole (Benos and Prusch, 1972).

The Nervous System

The nerve cells are arranged in an irregular nerve net, or plexus, in the base of the epidermis and are particularly concentrated around the mouth. Hydras have only an epidermal nerve net, but many cnidarians possess a gastrodermal nerve net as well. Some synapses are symmetrical; that is, neuron terminals on both sides of the synapse secrete a transmitter substance, and an impulse can be initiated in either direction across the synapse. Neurons serving symmetrical synapses obviously transmit impulses in both directions and are thus directionally nonpolarized, in contrast to the neurons of higher animals.

The association of nerve cells to form conducting chains between receptor and effector shows all degrees of complexity. The neurons contain two, three, or more processes. These processes may terminate in muscle fibers, in sensory cells, or with the processes of other ganglion cells. Some neurons have even been shown to possess two branches, each serving a different kind of effector (cnidocyte and muscle fiber). The length of the processes varies, and they may give rise to motor and sensory branches. Further, some neurons have only motor processes; some have only sensory processes; and some act as interneurons.

A double nerve net system in the same body layer is very common in cnidarians other than hydras. One nerve net acts as a diffuse, slow-conducting system of multipolar neurons; the other is a rapid, through-conducting system of bipolar neurons.

Reproduction

Hydras reproduce asexually by budding; in fact, this is the usual means of reproduction during the warmer months of the year. A bud develops as a simple evagination of the body wall and contains an extension of the gastrovascular cavity. The mouth and tentacles form at the distal end. Eventually, the bud detaches from the parent and becomes an independent hydra.

Considering their facility for asexual reproduction, it is not surprising that hydras, like many cnidarians, have considerable powers of regeneration. A classic experiment is that of Trembley, performed first in 1744. By inserting a knotted thread through the basal disc and drawing it out through the mouth, Trembley turned a hydra wrong side out. After a short period the gastrodermal cells reoriented themselves on the inner side of the mesoglea, and the epidermal cells migrated to the outside. Although interstitial cells can contribute to regeneration, the process is not dependent upon these cells; rather, dedifferentiated epidermal and gastrodermal cells are the principal source of regenerate material. Entire animals have been produced from gastrodermis and from epidermis in other cnidarians.

A polarity or gradient of dominance exists from oral to aboral end. If the body stalk of a hydra is severed into several sections, each will regenerate into a new individual. Furthermore, the original polarity is retained, so the tentacles always form on the end that was closest to the oral end of the intact animal, and a basal disc forms at the other end. A piece of oral end grafted onto the middle of the body stalk will induce the formation of another oral end. The oral-aboral gradient is also reflected in the rate of regeneration, an aboral piece regenerating more slowly than a more oral piece (Rose, 1970; Campbell, 1974; Gierer, 1974).

Mitotic activity occurs throughout the stalk, but the rate of production of new cells is especially high in the region just below the hypostome. Since the tips of the tentacles, basal disc, and buds are sites of cell death or cell loss (bud), there is a gradual migration of cells down the stalk and out along the tentacles (Fig. 5–7). Within a period of several weeks, all the cells in the body of an individual are replaced. Hydras never grow old.

Sexual reproduction in hydras occurs chiefly in the fall, because the eggs are a means by which the species survives the winter. The majority of hydras are dioecious. As in all cnidarians, the germ cells originate for the most part from interstitial cells, which aggregate to form ovaries or testes (Fig. 5–20C). In general, the testes are located in the epidermis of the upper half of the stalk and the ovaries in the lower half.

A single egg is produced in each ovary; the other interstitial cells of the ovarian aggregate merely serve as food in the egg's formation. As the egg enlarges, a rupture occurs in the overlying epidermis, exposing the egg. The testis is a conical swelling with a nipple through which the sperm escape. Sperm liberated from the testes into the surrounding water penetrate the exposed surface of the egg, which is thus fertilized *in situ*.

The egg then undergoes cleavage and simultaneously becomes covered by a chitinous shell. When shell formation is complete, the encapsu-

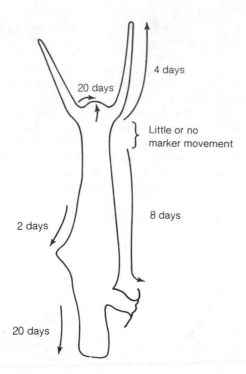

Figure 5–7 Tissue movement in *Hydra* as a result of mitotic activity. Arrows indicate direction of movement, and numbers indicate days required for cells to move along path of arrow. (From Campbell, R. D., 1967: J. Morphol., *121*(1):19–28.)

lated embryo drops off the parent and remains in its protective casing through the winter. With the advent of spring the shell softens, and a young hydra emerges. Since each individual may bear several ovaries or testes, a number of eggs may be produced each season. The reproductive pattern described for *hydra* is not typical of most marine cnidarians. Most cnidarians possess a ciliated planula larva which will be described in the next section.

SUMMARY

1 Cnidarians are aquatic, radially symmetrical animals with tentacles encircling the mouth at one end of the body. The mouth is the only opening into the gut cavity.

2 Cnidarians exhibit two body forms: the medusa, which is adapted for a pelagic existence, and the polyp, which is adapted for an attached, benthic existence. Colonial organization has evolved in many polypoid groups.

3 The body wall consists of an outer epidermis, an inner gastrodermis, and an intervening meso-

glea. The latter may be thin or thick, acellular or cellular.

4 Cnidarians are primitive in their lack of organs, their lack of fully differentiated epithelial and muscle cells, and the diploblastic origin of the adult body.

5 Most feed on zooplankton, although some utilize larger animals and some are suspension feeders on fine particulate matter. Prey is caught with the tentacles and immobilized by explosive cells, called cnidocytes, which are unique to the phylum. Digestion is initially extracellular, then intracellular.

6 The neurons are usually arranged as a nerve net at the base of the epidermal and gastrodermal layers, and impulse transmission tends to be radiating. Synaptic junctions are commonly nonpolarized.

7 A ciliated, free-swimming stereogastrula, called the planula larva, occurs in the life cycle of most cnidarians.

Class Hydrozoa

The class Hydrozoa contains about 2700 species of common cnidarians, but because of their small size and plantlike appearance, the layman is largely unaware of their existence. A considerable part of the marine growth attached to rocks, shells, and wharf pilings, usually dismissed as "seaweed," is frequently composed of hydrozoan cnidarians.

The few known freshwater cnidarians belong to the class Hydrozoa. They include the hydras and some small, freshwater jellyfish.

Hydrozoans display either the polypoid or the medusoid structure, and some species pass through both forms in their life cycle. Three characteristics unite the members of this class. The mesoglea is never cellular; the gastrodermis lacks cnidocytes; and the gonads are epidermal, or if gastrodermal, the eggs and sperm are shed directly to the outside and not into the gastrovascular cavity.

Hydroid Structure

Although some hydrozoans display only the medusoid form, most species possess a polypoid stage in their life cycle. Some forms, such as the hydras, exist as solitary polyps, but the vast majority are colonial. In hydras, buds form on the stalk as simple evaginations of the body wall. The distal end of the bud forms a mouth and a circle of tentacles (Fig. 5–2*A*); then the whole bud drops off to form a new individual. In the development of colonial

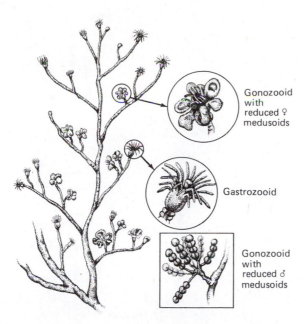

Figure 5–8 *Eudendrium*, a genus of hydroids having an arborescent growth form. The medusoids, called gonophores, are reduced and never liberated from parent colony.

forms, the buds remain attached; these in turn produce buds, so each polyp is connected to the others. Such a collection of polyps is known as a hydroid colony (Figs. 5–8 and 5–11). The three body layers in a hydroid colony—epidermis, mesoglea, and gastrodermis—and the gastrovascular cavities are all continuous.

In describing hydrozoan individuals or colonies, it is convenient to use the term *hydranth*, which refers to the oral end of the polyp bearing the mouth and tentacles, and the term *hydrocaulus*, which refers to the stalk of the polyp.

In most species of hydrozoans the colony is anchored by a horizontal, rootlike stolon, called the hydrorhiza, which creeps over the substratum (Fig. 5–11). From the stolon arise single upright polyps or branches of polyps. The branches develop through different growth and budding patterns and vary in form. They may be arborescent, resembling a tree (Fig. 5–8), or pinnate (feather-like) (Fig. 5–9), or single polyps may arise from basal stolons (Fig. 5–10B).

Most hydroid colonies are 5 to 15 cm high; commonly, individual polyps are the size of the oral end of hydras. The coloration is usually not very striking because of the small size. The largest hydrozoan polyps are solitary species belonging to the genera *Branchiocerianthus*, a deep-sea giant that may reach a length of over 2 meters, and the

shallow-water *Corymorpha. Corymorpha palmata*, which is found on intertidal mud flats along the California coast, may reach a height of 14 cm when submerged and erect.

Probably because of the increased size resulting from a colonial organization, most hydroids are, at least in part, surrounded by a supporting, nonliving, chitinous envelope secreted by the epidermis. Such a cylinder is known as the perisarc (Fig. 5–11), and the living tissue that it surrounds is called the coenosarc. The perisarc may be confined to the hydrocaulus (Fig. 5–10A), but often it continues up-

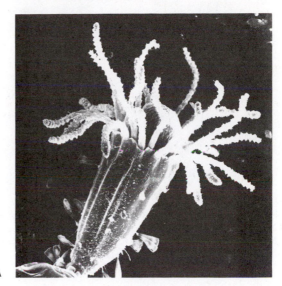

A

B

Figure 5–9 *A*, SEM of a gastrozooid of the thecate hydroid *Gonothyrae loveren*. Vaselike structure is the skeleton around the hydranth. Diatoms are growing on it. (Courtesy of C. Östman.) *B, Pennaria*, a hydroid with a pinnate growth form. (By Betty M. Barnes.)

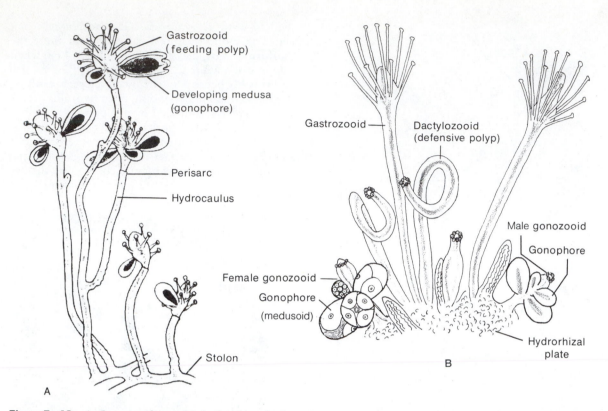

Figure 5–10   *A, Coryne,* a dimorphic hydroid in which medusoids are formed directly on gastrozooids. (Modified after Naumov.) *B, Hydractinia,* a tetramorphic hydroid. Polyps arise separately from a mat of stolons, or hydrorhizae. Skeleton limited to covering of stolons, and fusion of adjacent skeletons forms a plate. (After Hyman, L. H., 1940: The Invertebrates, Vol. I. McGraw-Hill Book Co., N.Y.)

ward to enclose the hydranth itself in a casing known as the hydrotheca, as in *Obelia* and *Campanularia.* The hydrotheca may be bell-like and open (Figs. 5–9*A* and 5–11), or its opening may be covered by a lid of one to several pieces. The lid opens when the polyp is extended and feeding and closes when the polyp contracts. Hydroids with a hydrotheca surrounding the polyp proper are said to be thecate; those without the hydrotheca, athecate.

The hydras and the solitary polyps of other hydrozoans have no such external skeleton, nor is a skeleton very extensive in some colonial species in which all polyps arise from hydrorhizae. For example, in *Hydractinia echinata,* which lives on snail shells inhabited by hermit crabs, the closely placed, naked polyps arise directly from hydrorhizae. Only the hydrorhizae are provided with skeletons, which are fused to form a more or less continuous plate anchoring the colony to the shell (Fig. 5–10*B*).

Polymorphism is another characteristic of hydroids associated with their colonial organization.

Most hydroid colonies are at least dimorphic; that is, the colony consists of at least two structurally and functionally different types of individuals. The most numerous and conspicuous type of individual is the nutritive, or feeding, polyp, called a gastrozooid or trophozooid (Figs. 5–10 and 5–11). The feeding polyp looks like a short hydra with a distal mouth, a hypostome, and tentacles.

The gastrozooids capture and ingest prey, and thus provide nutrition for the colony. Most hydrozoans feed on any zooplankton that is small enough to be handled by the gastrozooids. Extracellular digestion takes place in the gastrozooid itself; the partially digested broth then passes into the common gastrovascular cavity of the colony, where intracellular digestion occurs. Circulation is probably facilitated by rhythmic pulsations and contraction waves, which have been observed in many hydroids. Some hydroids contain zooxanthellae within the gastrodermal cells (see p. 134 and also Muscatine, 1974).

In most species the gastrozooids also fulfill the defensive functions of the colony, but in some hy-

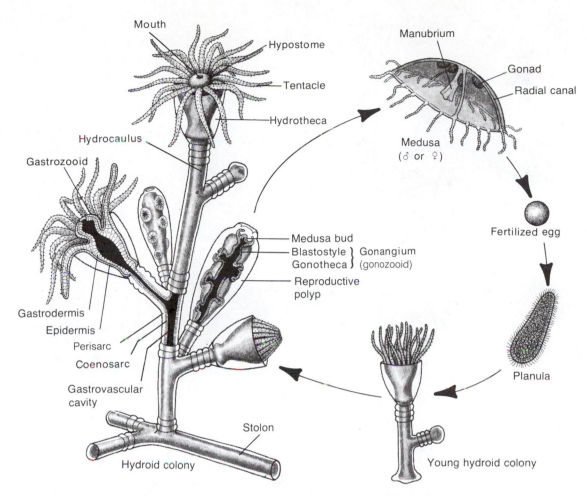

Figure 5–11   Life cycle of *Obelia*, showing structure of hydroid colony. (Adapted from various sources.)

droids there are special defensive polyps (dactylozooids). The defensive polyps assume a variety of forms but are frequently club-shaped structures, well supplied with cnidocytes and adhesive cells (Fig. 5–10*B*). Defensive polyps commonly are located around the gastrozooid and probably contribute to prey capture (Fig. 5–12).

All hydroids possess reproductive individuals as part of the colony. The sexually reproductive individuals are produced as asexual medusoid buds from some part of the hydroid colony (Fig. 5–11). The medusoids may either develop into free medusae or be retained on the colony; in either case, they produce the gametes to complete the sexual phase of the life cycle.

Medusoids assume a variety of shapes and locations. They may arise from the hydrocaulus, from the hydrorhiza, from the gastrozooid stalk, or frequently from the body of the gastrozooid itself (Figs. 5–10*A* and 5–13). Medusoids may be pro-

duced only on certain polyps, called gonozooids, which may be greatly reduced, lacking mouth and tentacles. Such a reduced gonozooid is called a blastostyle. In thecate hydroids, such as *Obelia* and *Campanularia*, the medusoids are restricted to a type of gonozooid called gonangium, which consists of a central blastostyle surrounded by an extension of the perisarc, called the gonotheca (Fig. 5–11). Although free medusae are produced in some species, such as the well-known *Obelia* (Fig. 5–11), the majority of hydroids do not release their medusae.

An unusual group of hydroids consists of members of the genera *Porpita* and *Velella*. These are pelagic species, floating on or near the surface, which range from 2 to 10 cm in diameter. They have been interpreted as a colony or as a single large polyp. Each is suspended from a rather flattened, chambered, chitinous float (Fig. 5–14). At the center of the colony is a large gastrozooid. Gon-

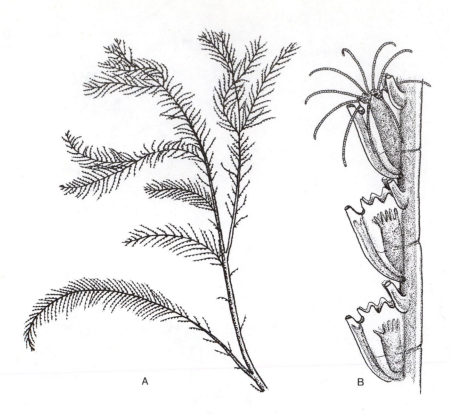

Figure 5–12 The hydroid *Aglaophenia: A,* Branch of colony. *B,* Three successive gastrozooids, each of which is surrounded by small defensive polyps.

ozooids bearing gonophores hang down between the central mouth of the gastrozooid and the marginal, tentacle-like defensive polyps. Found throughout the world's oceans, these floating colonies are often washed onto beaches. They are sometimes placed in a separate order, the Chondrophora.

Medusoid Structure

Unlike the medusae of the Scyphozoa, hydroid medusae are usually small, ranging from 0.5 cm to 6 cm in diameter (Fig. 5–15). The upper surface of the bell is called the exumbrella, and the lower surface, the subumbrella. The margin of the bell projects inward to form a shelf called a velum (characteristic of most hydromedusae). The tentacles that hang down from the margin of the bell are richly supplied with cnidocytes.

The mouth opens at the end of a tubelike extension called the manubrium, which hangs down from the center of the subumbrella. The manubrium also possesses cnidocytes and is often lobed or frilled. The mouth leads into a central stomach, from which typically extend four radial canals lined with gastrodermis. The radial canals join with a ring canal running around the margin of the umbrella.

As in all medusoid forms, the mesoglea is thick and gelatinous and constitutes the bulk of the animal. The mesoglea of hydromedusae is devoid of cells but does contain fibers.

The muscular system of hydromedusae is best developed around the bell margin and subumbrella surface, where epidermal contractile extensions are organized as circular and radial sheets. Their contractions produce rhythmic pulsations of the bell, driving water out from beneath the subumbrella (Fig. 5–16). The compressed mesoglea with its elastic fibers provides the antagonistic force to restore the bell shape between contractions. Joints of soft mesoglea are commonly present (Fig. 5–16*A* and *B*) (see Gladfelter, 1973). The velum, which is most effective in bell-shaped medusae, reduces the subumbrella aperture and thus increases the force of the water jet. Although the medusa often turns when swimming, the general direction is more or less vertical, and following a series of pulsations that drive the animal upward, it slowly sinks. Horizontal movement largely depends on water currents. There are a few species of hydromedusae, such as *Gonionemus,* that crawl about over the

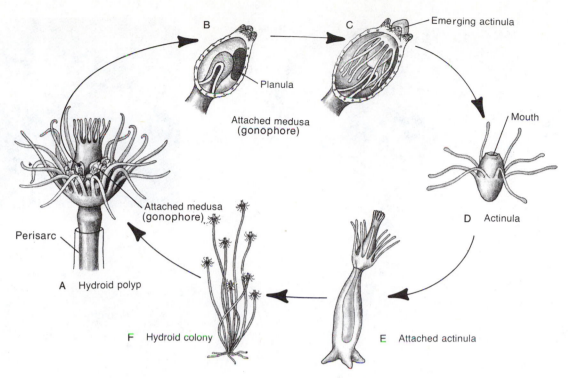

Figure 5–13 Life cycle of *Tubularia:* *A,* Skeleton is restricted to stalk of polyp; medusae are formed on gastrozooid and remain attached. *B,* Egg develops into planula within attached parent medusa. *C,* Actinula larva is released from medusa (*D*) and eventually settles to the bottom and develops into a new hydroid colony (*E* and *F*). (After Allman from Bayer and Owre.)

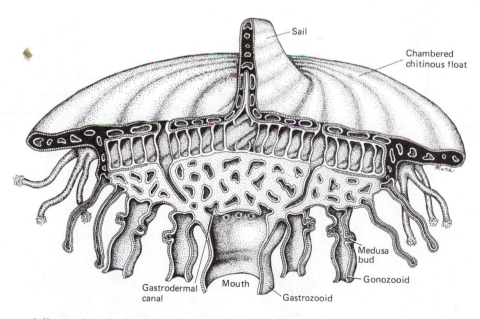

Figure 5–14 *Velella,* a pelagic hydroid. These pelagic hydroids float near the surface, and the aboral surface bears a sail. (Adapted from Fowler and other sources.)

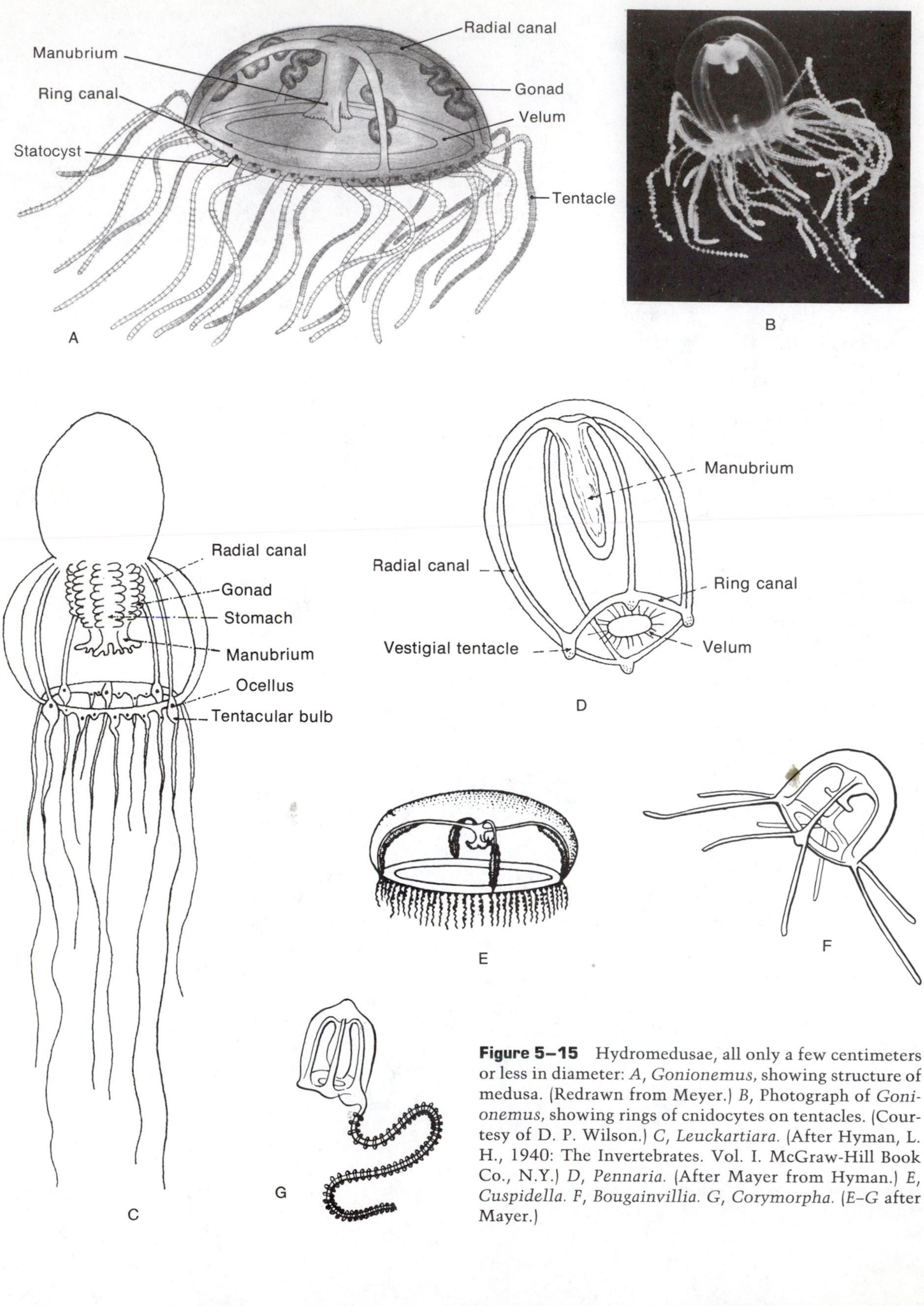

Manubrium
Ring canal
Statocyst
Radial canal
Gonad
Velum
Tentacle

A

B

Radial canal
Gonad
Stomach
Manubrium
Ocellus
Tentacular bulb

C

Manubrium
Radial canal
Ring canal
Vestigial tentacle
Velum

D

E

F

G

Figure 5–15 Hydromedusae, all only a few centimeters or less in diameter: *A, Gonionemus,* showing structure of medusa. (Redrawn from Meyer.) *B,* Photograph of *Gonionemus,* showing rings of cnidocytes on tentacles. (Courtesy of D. P. Wilson.) *C, Leuckartiara.* (After Hyman, L. H., 1940: The Invertebrates. Vol. I. McGraw-Hill Book Co., N.Y.) *D, Pennaria.* (After Mayer from Hyman.) *E, Cuspidella. F, Bougainvillia. G, Corymorpha.* (*E–G* after Mayer.)

 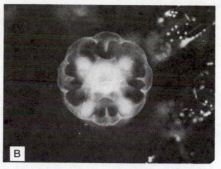

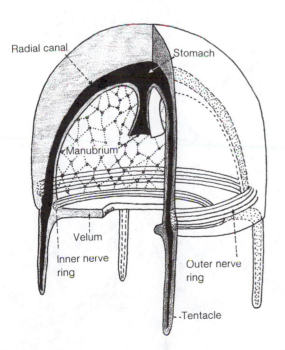

Figure 5–16 Swimming pulsations of hydromedusae: *A* and *B*, Aboral views of the relaxed and contracted phases of *Euphysa flammea*, showing action of the joints. *C*, Lateral views of *Bougainvillia multitentaculata* during swimming. (From Gladfelter, W. G., 1973: A comparative analysis of the locomotory systems of medusoid Cnidaria. Helgol. wiss. Meeresunters, 25:228–272.)

bottom, attaching to vegetation with their tentacles.

The nervous system of the medusa (Fig. 5–17) is also more highly specialized than that of the polyp. In the margin of the bell, the epidermal nerve cells are usually organized and concentrated into an inner and an outer nerve ring. These nerve rings, which represent one of the highest levels of nervous organization in cnidarians, connect with fibers innervating the tentacles, the musculature,

and the sense organs. The inner ring contains large motor neurons to the swimming muscles and is the center of rhythmic pulsation; i.e., it contains the pacemakers (Satterlie and Spencer, 1983).

In addition to dispersed receptor cells, the bell margin contains two types of true sense organs— light-sensitive ocelli and statocysts. The ocelli consist of patches of pigment and photoreceptor cells organized within either a flat disc or a pit on the outer side of the tentacular bulbs. Statocysts are located between the tentacles or associated with the tentacular bulb at the tentacle base. They may be in the form of either pits or closed vesicles in the velum or within pendent clubs below the exumbrellar nerve ring. In the vesicular type a calcium sulfate concretion lies next to a sensory cilium, and tilting of the bell margin probably moves the concretion in such a way as to stimulate the cilium (Fig. 5–18). Statocyst stimulation appears to inhibit muscular contractions on that side of the bell, and the opposite side throws water beneath the tilted margin to bring about righting.

Like polyps, medusae are carnivorous and feed on other planktonic animals and even small fish that come in contact with the tentacles.

Reproduction and Life Cycle

Some hydromedusae can bud off daughter medusae from the manubrium or tentacular bulbs, and some can undergo fission. All medusae reproduce sexually, and most are dioecious. The eggs and sperm arise from epidermal or gastrodermal interstitial cells that have migrated to specific locations in the epidermis. Here these cells cluster to form a gonad. However, as in hydras, the gonads are only aggre-

Figure 5–17 Nervous system of a hydromedusa. (After Bütschli from Kaestner.)

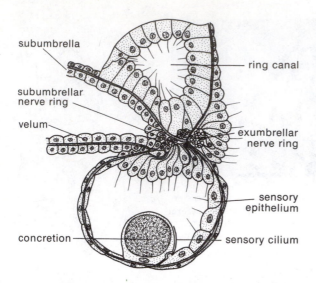

subumbrella

ring canal

subumbrellar
nerve ring

velum

exumbrellar
nerve ring

sensory
epithelium

concretion

sensory cilium

Figure 5–18   Closed vesicular statocyst of a hydromedusa in horizontal position. (From Singla, C. L., 1975: Statocysts of hydromedusae. Cell Tiss. Res., *158*:391–407.)

gations of developing gametes. The gonads are most commonly located beneath the radial canals in the epidermis of the subumbrella (Fig. 5–15*A*).

Fertilization may be external in the sea water, on the surface of the manubrium, or internal, with the eggs beginning development in the gonad. Cleavage is complete and leads to a hollow blastula and then a stereogastrula (solid). The endoderm forms the future gastrodermis and the ectoderm the epidermis. The stereogastrula rapidly elongates to become a ciliated, free-swimming planula larva. The planula is elongated and radially symmetrical but with distinct anterior and posterior ends. After a free-swimming existence lasting from several hours to several days, the planula larva attaches to an object by the anterior end and develops into a hydroid colony.

Such a life cycle, with a free-swimming medusoid generation as well as a hydroid stage, is displayed by *Obelia* (Fig. 5–11), *Pennaria, Syncoryne,* and other genera of hydrozoans.

The majority of hydroids, such as *Tubularia, Sertularia,* and *Plumularia,* do not produce a free-swimming medusa. Instead, the medusa remains attached to the parent hydroid (Fig. 5–13) and displays various degrees of arrested development. Despite the attachment and incomplete development of the medusa, it remains a sexually reproducing individual. In some hydroids the attached medusa is represented by only the gonadal tissue. Such an incomplete medusa, often called a gonophore, or,

when greatly reduced, a sporosac, represents nothing more than a gamete-producing structure (Figs. 5–10*B* and 5–13). Finally, in the hydras no vestige remains of a medusoid generation. Gonads form directly in the epidermis of the polyp stalk (Fig. 5–20*C*).

As is the case in free-swimming medusae, the eggs of attached medusoids may be retained and pass through their early embryonic stages while in the gonad. In some species, such as *Orthopyxis* (Fig. 5–19), the egg may develop through gastrulation within a sporasac, and the embryo escapes as a planula larva. In *Tubularia* even the planula stage is passed in the gonophore, and an actinula larva is eventually released. An actinula looks like a stubby hydra and creeps about on its tentacles (Fig. 5–13). The lack of a larval stage in freshwater hydras is probably correlated with their existence, in fresh water, where larval stages are generally absent.

From the discussion thus far it might appear that all hydrozoans are hydroids, but this is by no means true. The medusoid generation is the dominant form in some hydrozoans. In *Liriope* and *Aglaura,* there is no polypoid stage. The planula forms a planktonic actinula larva that transforms directly into a medusa (Fig. 5–20*A*). *Gonionemus*

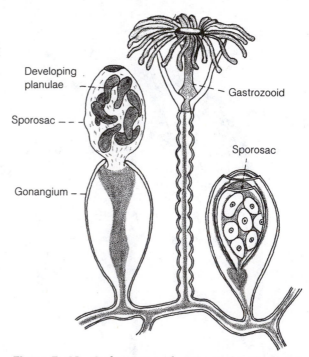

Developing
planulae

Gastrozooid

Sporosac

Sporosac

Gonangium

Figure 5–19   *Orthopyxis* with uneverted and everted sporosacs. (After Nutting from Hyman.)

Figure 5–20 Some hydrozoan life cycles: *A*, *Aglaura*, a hydrozoan that has no polypoid stage. Planula larva develops into an actinula, which develops directly into a medusa. (From Bayer and Owre.) *B*, *Craspedacusta*. (Life cycle of *Gonionemus* is similar.) Polyp is solitary. *C*, *Hydra*, a freshwater hydrozoan in which medusoid stage has disappeared and planula larva is suppressed. (From Bayer, F., and Owre, H. B., 1968: The Free-Living Lower Invertebrates. Macmillan Co., N.Y.)

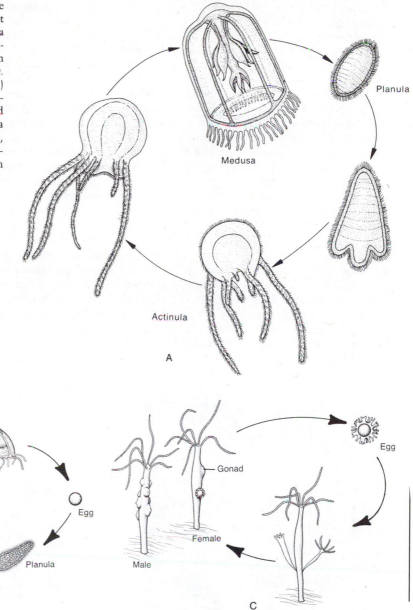

and *Craspedacusta* also belong to this group. Although a polypoid stage is present, the polyp is tiny and solitary, and medusae bud off from the sides (Fig. 5–20*B*).

The appearance of hydroid colonies and hydromedusae is usually highly seasonal and, in temperate regions at least, is closely correlated with water temperatures. Hydrozoans that have free medusae may liberate enormous numbers of them in a short time. A 7-cm colony of *Bougainvillia* produced 4450 medusae over a three-day period; as a result of such production, hydromedusae may often make a dramatic appearance in the plankton. The life span of a medusa ranges from a few days to many months.

Hydroids have been the subject of numerous developmental and regeneration studies, and many of the results are similar to those described on page 99 for hydras (see review of Rose, 1970, and Campbell, 1974).

Hydrozoan Evolution

The evolutionary significance of the life cycle of hydrozoans is a fascinating problem. Which came first, the polyp or the medusa? In 1886, W. K. Brooks worked out a theory of cnidarian evolution that is still supported by many zoologists today. According to Brooks's theory, the ancestral cnidarian form was medusoid. The tendency among hydrozoans has been to suppress the medusoid stage so that, in such forms as *Hydra,* the medusa has completely disappeared. The polyp, on the other hand, represents an evolutionary retention and development of the polyp-like actinula larva.

The evolutionary sequence may have occurred in the following manner. The ancestral cnidarians were medusoid, and development led through a planktonic planula larva and a later planktonic actinula larva. The actinula, in which the cnidarian characters first make their appearance, developed into the adult medusa. In some groups of such ancestral medusoid cnidarians, the actinula took up an attached benthic existence to give rise to a polyp. Such an attached condition could have been an adaptation to exploit a new food supply, to extend larval life, or to facilitate asexual reproduction of additional polyps by budding. Hydromedusae may have first developed by direct transformation of the attached actinula-polyp and later by budding, the latter process making possible the formation of numerous medusae from one polyp. The life cycle of such a species would have been: medusa→ planula→ polyp→ medusa.

The medusa, which primitively was free, became retained and then gradually suppressed until it disappeared completely. Living hydrozoans display life cycles that illustrate different stages in this evolutionary sequence.

The hydroid colony evolved through the retention of budded polyps. Correlated with the evolution of colonial organization was the development of skeletons and polymorphism, but these events probably occurred independently in different evolutionary lines. Medusoid suppression was probably also repeated numerous times, and carried to various degrees, throughout the Hydrozoa. For example, it is unlikely that the hydras evolved from a colonial hydroid; they are probably derived from

A

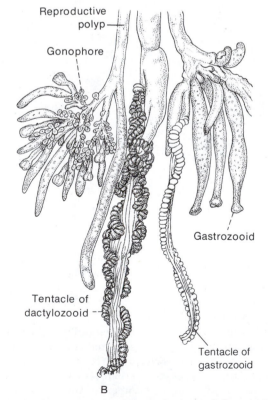

Reproductive polyp

Gonophore

Gastrozooid

Tentacle of dactylozooid

Tentacle of gastrozooid

B

Figure 5–21 *A, Physalia,* the Portuguese man-of-war, a siphonophoran which has a large, horizontal float and no swimming bell. (Courtesy of the New York Zoological Society.) *B,* Part of Portuguese man-of-war colony. (After Lane.)

a line of hydrozoans, similar to *Gonionemus* and *Craspedacusta*, in which polyps were always solitary but originally budded off free-swimming medusae.

Specialized Hydrozoan Orders

Before the conclusion of the discussion on Hydrozoa, a few specialized orders must be described briefly.

SIPHONOPHORA

Members of the order Siphonophora, which includes the familiar *Physalia* (the Portuguese man-of-war, Fig. 5–21), exist as large pelagic colonies composed of modified polypoid and medusoid individuals, which display a remarkable degree of polymorphism. The individuals of the colony are commonly attached to a long stem (Fig. 5–22). A conspicuous feature of many species is a gas-filled sac that acts as a float for the colony. The large float

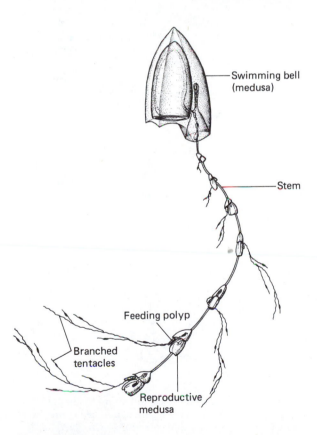

Figure 5–22 *Muggiaea*, a submergent, more typical siphonophore than *Physalia*. Clusters of feeding polyps and reproductive medusae are connected by a long stem hanging beneath a swimming bell, which is about 1 cm long. (Drawn from several sources.)

of the Portuguese man-of-war may attain a length of 30 cm and keeps the colony at the surface. Most siphonophores, however, live below the surface and may migrate great distances vertically. During the migration, gas (which is over 90 per cent carbon monoxide) is secreted into or released from the floats. The Pacific *Nanomia bijuga* has been reported to migrate 300 meters in less than an hour. Locomotion is also brought about by swimming bells, or nectophores, which are mouthless, pulsating medusae.

Feeding is carried on by gastrozooids with terminal mouths and a single, long, fishing tentacle, which is contractile and armed with batteries of cnidocytes (Figs. 5–21 and 5–22). There are also defensive polyps with long tentacles. The 4 to 150 polyps and tentacles form an effective fishing net for catching various planktonic invertebrates and small fish. Many siphonophores interrupt their fishing with a short interval of swimming, after which their "nets" are let out in a new location (Biggs, 1977). In the Gulf of California colonies of *Rhizophysa eysenhardti*, which reaches a length of somewhat less than a meter and may contain up to 28 gastrozooids, was found to consume about nine fish larvae (5–15 mm long) per day. This siphonophore fed only during the day and only on fish larvae, each of which took about 8 minutes to ingest and 3 to 7 hours to digest (Purcell, 1981).

Siphonophores are largely tropical and semitropical, but numbers of Portuguese men-of-war (*Physalia*) are often seen on the North Atlantic coast, especially following storms that have blown them in from the Gulf Stream. An accidental encounter with the tentacles of this siphonophore, which may hang several meters below the float (Fig. 5–21), can be a painful and even a dangerous experience for a swimmer.*

HYDROCORALLINA

The two small groups of hydrocorals, sometimes placed in separate orders Milleporina (*Millepora*) and Stylasterina (*Stylaster*), secrete a calcareous skeleton (Fig. 5–23). Both are colonial polypoid hydrozoans with either an encrusting or an upright growth form. Both may attain considerable size. Gastrozooids and defensive polyps emerge from pores in the skeleton, which are sometimes arranged in rings (Fig. 5–23). The defensive polyps are numerous, and the Milleporina are sometimes called fire corals because of their sting. The polyps

---

*Meat tenderizer, commonly carried in the first-aid box of snorklers and scuba divers, can be effective if quickly rubbed on the sting of this or other cnidarians. The protease breaks down the protein toxin.

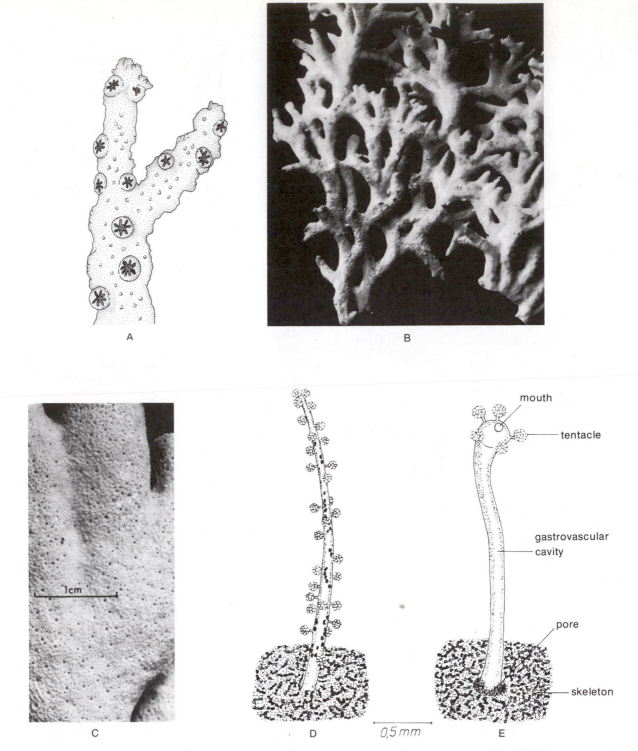

Figure 5–23 Hydrocorals: *A*, *Allopora* of the order Stylasterina. These hydrocorals are usually more finely branched than the Milleporina. The gastrozooids and surrounding defensive polyps are located in the large, notched cups. (After Moseley from Bayer, F. M., and Owre, H. B., 1968: The Free-Living Lower Invertebrates. Macmillan Co., N.Y.) *Millepora* of the order Milleporina: *B*, Part of a colony. (Courtesy of the Encyclopedia Britannica.) *C*, Pores in skeleton through which polyps emerge. Gastrozooids emerge through large pores, defensive polyps from surrounding small pores. *D*, Defensive polyp (dactylozooid). *E*, Gastrozooid. (*D* and *E* from de Kruijf, H. A. M., 1975: General morphology and behaviour of gastrozooids and dactylozooids in two species of *Millepora*. Mar. Behav. Physiol., 3:181–192. © Gordon and Breach Science Publishers, Inc.)

are connected by canals beneath the surface of the skeleton. *Millepora* is a common component of coral reefs and is yellow brown like stony corals because of the symbiotic zooxanthellae. The order Stylasterina contains many species, which are found in both temperate and tropical seas.

SYSTEMATIC RÉSUMÉ OF CLASS HYDROZOA

Order Trachylina. Medusoid hydrozoans lacking a polypoid stage. Medusa develops directly from an actinula. This order contains perhaps the most primitive members of the class. *Liriope, Aglaura.*

Order Hydroida. Hydrozoans with a well-developed polypoid generation. Medusoid stage present or absent. The majority of hydrozoans belong to this order.

> **Suborder Limnomedusae.** Mostly freshwater hydrozoans possessing small solitary polyps and free medusae. The marine *Gonionemus*; the freshwater *Craspedacusta.*

> **Suborder Anthomedusae.** Skeletal covering, when present, does not surround hydranth (athecate). Free medusae, which are tall and bell-shaped, are commonly present. *Tubularia, Pennaria, Syncoryne, Eudendrium, Hydractinia, Polyorchis, Branchiocerianthus,* the freshwater hydras.

> **Suborder Leptomedusae.** Hydranth surrounded by a skeleton (thecate). Free medusae are commonly absent, but when present, they are more or less flattened. *Obelia, Campanularia, Abietinaria, Sertularia, Plumularia, Aglaophenia.*

> **Suborder Chondrophora.** Pelagic, polymorphic, polypoid colonies. (These cnidarians can also be interpreted as large, single, inverted polyps.) *Velella, Porpita.*

Order Actinulida. Tiny, solitary hydrozoans resembling actinula larvae. No medusoid stage present. Interstitial inhabitants. *Halammohydra, Otohydra.*

Order Siphonophora. Pelagic hydrozoan colonies of polypoid and medusoid individuals. Colonies with float or large swimming bells. Largely in warm seas. *Physalia* (Portuguese man-of-war), *Stephalia, Nectalia.*

Order Hydrocorallina. Colonial polypoid hydrozoans that secrete a calcium carbonate skeleton.

> **Suborder Milleporina.** Stinging, or fire, coral. Skeleton covered by only a thin epidermal layer. Defensive polyps arising from separate pores encircling a central gastrozooid. *Millepora* is the only genus.

> **Suborder Stylasterina.** A thick layer of tissues overlying skeleton. Defensive and feeding polyps located within star-shaped openings on the skeleton. *Stylaster, Allopora.*

SUMMARY

1 Members of the class Hydrozoa are medusoid or polypoid or exhibit both forms in their life cycle. The mesoglea is acellular, cnidocytes are restricted to the epidermis, and gametes develop in the epidermis. Hydrozoans may be the most primitive of the three classes of cnidarians.

2 Hydromedusae are usually small and planktonic.

3 The most primitive hydrozoans are probably medusoid species, in which the pelagic actinula develops directly into an adult medusa. Such a life cycle may also be primitive for the phylum.

4 The polypoid form may have arisen in some medusoid species in which the actinula passed through a period of attachment prior to development into a pelagic adult; i.e., the attached actinula was the first polyp.

5 Early polypoid stages, including the attached actinula, probably reproduced asexually by budding. Persistent attachment of the buds led to colonial polypoid species, called hydroids, which now compose the majority of hydrozoans.

6 Associated with colonial organization has been the evolution of a skeleton (support) and polymorphism (division of labor).

7 Naked solitary species, such as hydras and the *Gonionemus* polyp, probably stem from early polypoid forms that were not colonial.

8 Suppression of the medusa through attachment to the polyp and subsequent reduction has evolved independently in different hydrozoan lines, and living species exhibit all degrees of reduction in the medusoid form.

Class Scyphozoa

Scyphozoans are the cnidarians most frequently referred to as jellyfish. In this class the medusa (Figs. 5–24*A* and 5–25) is the dominant and conspicuous individual in the life cycle; the polypoid form is restricted to a small larval stage. In addition, scyphozoan medusae are generally larger than hydromedusae. The majority of scyphozoan medusae have a bell diameter ranging from 2 to 40 cm; some species are even larger. The bell of *Cyanea capillata* may reach 2 meters in diameter. Coloration is often striking; the gonads and other internal struc-

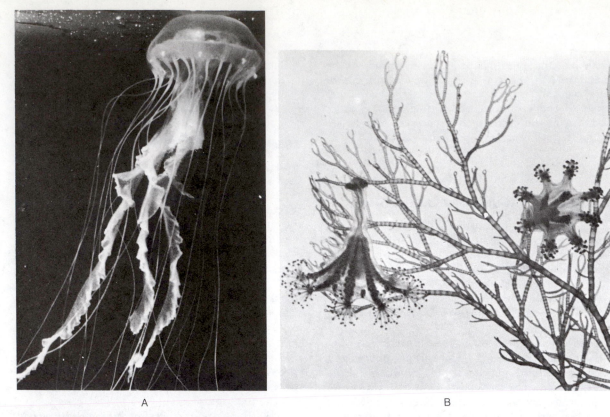

A   B

Figure 5–24 *A*, The sea nettle, *Chrysaora quinquecirrha*, a common scyphozoan along the Atlantic coast. (Courtesy of William H. Amos.) *B*, Two species of sessile scyphomedusae (stauromedusae) attached to an alga. The specimen on the left, *Haliclystus auricula*, has a diameter of less than 1 inch. The specimen on the right is *Craterolophus convolvulus*. (Courtesy of D. P. Wilson, *In* Buchsbaum, R. M., and Milne, L. J., 1961: The Lower Animals. Chanticleer Press, N.Y.)

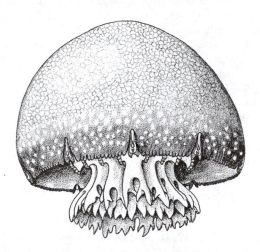

Figure 5–25 *Stomolophus meleagris*, sometimes called the cannonball jellyfish, is found along the southeastern coast of the United States and in the Caribbean. This species is a rapid swimmer, and its hemispherical, relatively rigid bell may reach the size of a football. *Stomolophus* is a rhizostome medusa. The members of this order lack tentacles around the bell margin and have multiple secondary mouths in the oral arm area. (After Mayer.)

tures, which may be deep orange, pink, or other colors, are visible through the colorless or more delicately tinted bell.

The some 200 described species of scyphozoans live in all seas from the Arctic to tropical oceans. Although there are deep-sea and oceanic forms, many species inhabit coastal waters and are often a nuisance on bathing beaches. Their large size and their nematocysts make them unpleasant and sometimes dangerous swimming companions. The so-called sea nettles, which inhabit the North Atlantic coast in large numbers during late summer, are members of this class. The sea wasps, including such species as *Chironex fleckeri* (Fig. 5–26) of the tropical Cubomedusae, produce such virulent stings that they are extremely dangerous, more so than the notorious hydrozoan Portuguese man-of-war. Numerous fatalities have been recorded from Australian coasts. Death, if it occurs, takes place 3 to 20 minutes after stinging. Lesions from nonfatal stings may be very severe and slow to heal.

Although these cnidarians are typically free-swimming animals, one order, the Stauromedusae,

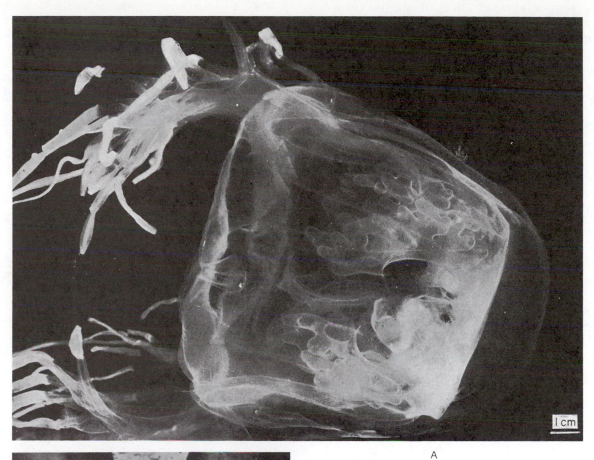

A

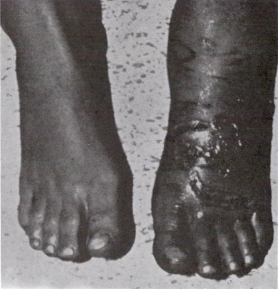

B

Figure 5–26 *A*, The Indo-Pacific sea wasp, *Chironex fleckeri*, a dangerous scyphozoan of the order Cubomedusae. The tentacles have been cut to permit handling of the specimen. *B*, Five-day-old lesions produced by a small sting of this sea wasp. (By J. H. Barnes. *In* Rees, W. J. (Ed.), 1966: The Cnidaria and Their Evolution. Zoological Society of London.)

is sessile (Fig. 5–24B). In this group, the exumbrellar surface is drawn out into a stalk by which the animal attaches to algae and other objects. Members of this group are thus polypoid in general structure.

In general, scyphozoan medusae are similar to hydromedusae. The bell varies in shape from a shallow saucer to a deep helmet, and the margin is typically scalloped to form lobes called lappets (Fig.

5–25). A velum is absent except in the Cubomedusae. The manubrium of many common coastal species (Semaeostomeae) is drawn out into four or eight often frilly, oral arms, which bear cnidocytes and aid in the capture and ingestion of prey. The tentacles around the bell margin vary from four to many but are absent in rhizostomes (Fig. 5–25). In *Aurelia*, the type most often studied in introductory courses, the tentacles are small and form a

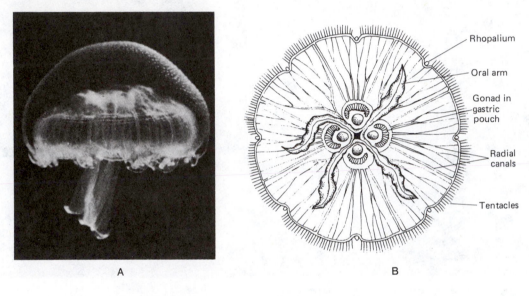

A

B

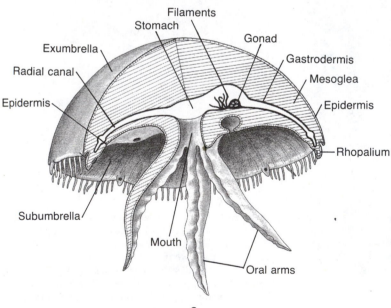

C

Figure 5–27 *Aurelia*, a scyphozoan medusa: *A*, Young specimen with bell contracted. *B*, Ventral view; *C*, side view in section. (*A* courtesy of D. P. Wilson.)

short fringe around the margin (Fig. 5–27); but in other species the tentacles are much longer (Fig. 5–24*A*).

The mesoglea of the scyphozoan is thick, gelatinous, and fibrous, but unlike that of the hydrozoans, it contains ameboid cells, which originate from the epidermis.

Locomotion is brought about by a band of powerful circular fibers (the coronal muscle) on the subumbrella; the fibers comprise contractile cells located within the mesoglea (epithelial portion reduced). Swimming pulsations are similar to those of hydromedusae. As in hydromedusae, the mesoglea provides some elastic recoil, and some species have areas around the bell that "fold" during contraction. Although scyphomedusae drift with currents or waves, most can swim horizontally, and some tropical Cubomedusae and rhizostome me-

dusae (Figs. 5–25 and 5–26) are very rapid swimmers. The velum-like flap (velarium) of a cubomedusa greatly increases the force of the water jet leaving the subumbrellar cavity.

The ground plan of the scyphozoan gastrovascular system is somewhat different from that seen in the hydromedusae. The mouth opens through the manubrium into a central stomach, from which extend four gastric pouches (Fig. 5–29). Between the pouches are septa, each of which contains an opening to help circulate water. Thus, all four pouches are in lateral communication with each other. The margin of the septum, which faces the central portion of the stomach, bears a large number of filaments containing cnidocytes and gland cells. Canals typically run from the stomach pouches to the bell margin. A ring canal may be present or absent.

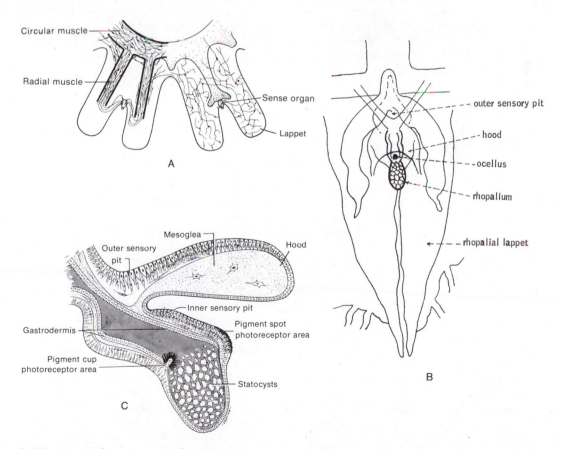

Figure 5–28 Marginal sense organs of scyphozoans: *A*, Diagram of a section of bell margin showing nerve net, lappets, and position of sense organ, or rhopalium. (After Horridge.) *B*, Rhopalium of *Aurelia* (aboral view). (After Hyman, L. H., 1940: The Invertebrates, Vol. I. McGraw-Hill Book Co., N.Y.) *C*, Section through a rhopalium showing the hood and the various sensory areas. (Modified from Hyman after Schewiakoff.)

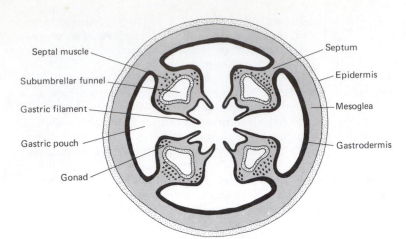

Septal muscle

Subumbrellar funnel

Gastric filament

Gastric pouch

Gonad

Septum

Epidermis

Mesoglea

Gastrodermis

Figure 5–29 Section through a scyphozoan with gastric pouches. (Greatly modified after Hyman.)

There is considerable modification of this plan in different groups. Some, such as *Aurelia*, lack the gastric pouches. Nevertheless, filaments are present and are attached interradially to the periphery of the stomach floor. *Aurelia* has an extensive system of branched and unbranched canals extending from the stomach to a ring canal in the bell margin (Fig. 5–27).

Adult scyphozoans feed on all types of small animals, especially crustaceans. Some scyphozoans feed upon fish; however, larval fish of a number of species swim with certain species of scyphozoans for protection. As the medusa gently swims or slowly sinks, prey is captured on contact with the tentacles or oral arms of the manubrium. The tentacles may contract to bring the prey to the vicinity of the manubrium. Some species, including *Aurelia*, are actually suspension feeders, trapping plankton in mucus on the flagellated subumbrellar surface. Flagella then sweep the food to the bell margin, where it is scraped off by the oral arms. Flagellated grooves on the oral arms carry the food to the mouth and stomach. *Cassiopea*, a common jellyfish of Florida and the West Indies, rests upside down on the bottom in the quiet shallow water of mangrove embayments (Fig. 5–30). This genus, like other members of the order Rhizostomeae, possesses many small secondary mouths that open into the stomach by way of canals in the oral arms. Small animals trapped on the surface of the frilly oral arms are carried into the mouths within mucous cords. However, *Cassiopea* possesses symbiotic algae (zooxanthellae) in its mesoglea, and in adequate light it can survive and grow entirely on the products of the algal photosynthesis.

Digestion is essentially as described in hydras. The gastric filaments are the source of extracellular enzymes, and the gastrodermal cnidocytes are probably used to quell prey that is still active.

Nerve rings occur only in the orders Coronatae and Cubomedusae. In other scyphozoans the pulsation control is centered in marginal concentrations of neurons. Each such concentration is situated in a little club-shaped structure called a rhopalium; these are located around the bell margin between lappets and number four or multiples of four. Each rhopalium is flanked by a pair of small specialized lappets called rhopalial lappets and is covered by a hood (Fig. 5–28).

Two sensory pits, a statocyst, and sometimes an ocellus are borne by each rhopalium (Fig. 5–28*B*). Cubomedusans have complex eyes containing a lens and a retina-like arrangement of sensory cells and can orient to small points of light. Many other scyphozoans display distinct phototaxis. They come to the surface of the water during cloudy weather and at twilight but move downward in bright sunlight and at night.

With few exceptions, scyphozoans are dioecious, and the gonads are located in the gastrodermis, in contrast to the usual epidermal gonad in hydrozoans. In septate groups with gastric pouches, the eight gonads are located on both sides of the four septa (Fig. 5–29). In semaeostome medusae, which lack septa, four horseshoe-shaped gonads lie on the floor of the stomach periphery (Fig. 5–27). When mature, the eggs or sperm break into the gastrovascular cavity and pass out through the mouth. In some semaeostomes, including *Aurelia*, the eggs become lodged in pits on the oral arms. This temporary brood chamber is the site of fertilization and early development through the planula stage.

Cleavage leads to a typical planula larva (Fig. 5–31), which, after a brief free-swimming exis-

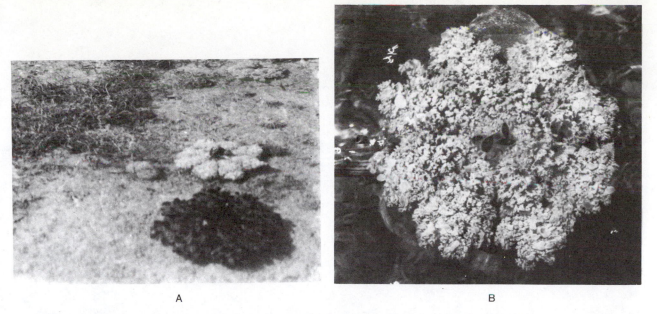

Figure 5–30 *Cassiopea*, a common tropical genus of rhizostome scyphozoans, which live upside down on the bottom in quiet shallow water: *A*, Specimens resting on bottom, viewed through water. *B*, A single specimen about 21 cm in diameter, which has been lifted to the surface. (Both by Betty M. Barnes.)

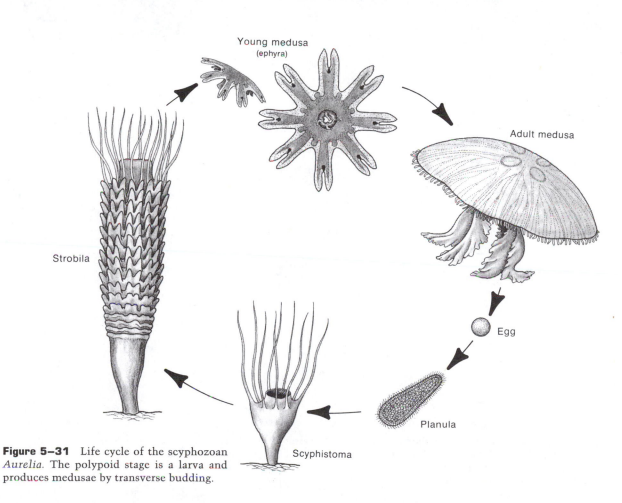

Figure 5–31 Life cycle of the scyphozoan *Aurelia*. The polypoid stage is a larva and produces medusae by transverse budding.

tence, settles to the bottom and becomes attached by its anterior end. The attached planula then develops into a little polypoid larva called a scyphistoma.

The scyphistoma looks very much like a hydra (Fig. 5–31), and it feeds and produces new scyphistomae by asexual budding, either directly from the midcolumn wall or from stolons (in *Aurelia*). At certain periods of the year young medusae are formed. In the one species of Cubomedusae for which the life cycle is known, the scyphistoma transforms directly into a small medusa (Fig. 5–32).

In other scyphozoans, medusa formation is accomplished by transverse fission of the oral end of the scyphistoma, a process called strobilation. Medusae may form one at a time (monostrobilation) or many simultaneously (polystrobilation) so that the immature medusae, called ephyrae, are stacked up like saucers at the oral end of the body stalk (Fig. 5–31). As formation of the ephyrae is completed, they break away from the oral end of the scyphistoma one by one.

After strobilation, the scyphistoma may resume its polypoid existence until the following year, when formation of ephyrae is repeated. A scyphistoma may live for one or several years.

The ephyra is almost microscopic and has a deeply incised bell margin (Fig. 5–31). Some ephyrae take two years to grow into sexually repro-

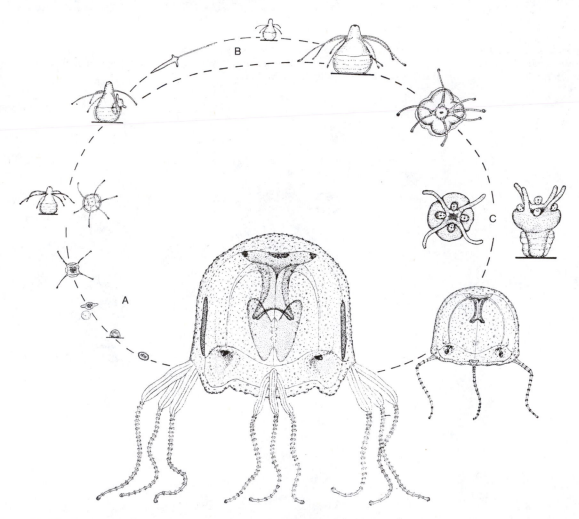

Figure 5–32   Life cycle of the cubomedusa *Tripedalia cystophora*. The planula (A) develops into a solitary attached larval polyp (B), which eventually metamorphoses directly (C) into the medusa. Additional polyps may be derived by budding. (From Werner, B., 1973: New investigations on systematics and evolution of the class Scyphozoa and the phylum Cnidaria. Proc. 2nd Internat. Symp. Cnidaria, Publ. Seto Mar. Biol. Lab., *20*:35–61.)

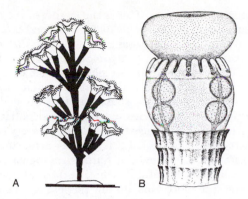

Figure 5–33 *Stephanoscyphus racemosus* (order Coronatae): *A,* Colonial scyphistome. *B,* Medusa, just released from strobila, already possesses ripe eggs. (From Werner, B., 1973: New investigations on systematics and evolution of the class Scyphozoa and the phylum Cnidaria. Proc. 2nd Internat. Symp. Cnidaria, Publ. Seto Mar. Biol. Lab., 20:35–61.)

ducing adult medusae; others are relatively short-lived. The ephyrae of *Aurelia aurita* on the west coast of the United States are produced in March and reach sexual maturity by June.

Modifications of the life cycle that avoid the necessity of a substratum are known. In some species of *Chrysaora* and *Cyanea* the larva is retained on the parent in cysts.

At the other extreme, species of *Stephanoscyphus* and *Nausithoe* (order Coronatae) have branching colonial polyps with a supportive skeletal tube (Fig. 5–33). Some even have reduced medusoid stages, with precocious gamete production occurring during or immediately after strobilation (Werner, 1973).

The larval nature of the scyphozoan polyp and its direct transformation into a medusa in Cubomedusae seems to support the belief that the medusoid form is primitive in cnidarians and that the polypoid form evolved as a larval adaptation. There are zoologists, such as Werner (1971 and 1973), who take the opposite view and consider the scyphistoma to recapitulate the ancestral cnidarian condition.

SYSTEMATIC RÉSUMÉ OF CLASS SCYPHOZOA

Order Stauromedusae, or Lucernariida. Sessile polypoid scyphozoans attached by a stalk on the aboral side of the trumpet-shaped body. Chiefly in cold littoral wa-

ters. *Haliclystus, Craterolophus, Lucernaria.*

Order Coronatae. Bell of medusa with a deep encircling groove or constriction, the coronal groove, extending around the exumbrella. Many deep-sea species. *Periphylla, Stephanoscyphus, Nausithoe, Linuche, Atolla.*

Order Semaeostomeae. Scyphomedusae with bowl-shaped or saucer-shaped bells having scalloped margins. Manubrium divided into four oral arms. Gastrovascular cavity with radial canals or channels extending from central stomach to bell margin. Occur throughout the oceans of the world, especially along coasts. *Cyanea, Pelagia, Aurelia, Chrysaora, Stygiomedusa.*

Order Rhizostomeae. Bell of medusa lacking tentacles. Oral arms of manubrium branched and bearing deep folds into which food is passed. Folds, or "secondary mouths," lead into arm canals of manubrium, which pass into stomach. Original mouth lost through fusion of oral arms, except in *Stomolophus.* Mostly tropical and subtropical shallow water scyphozoans. *Cassiopea, Rhizostoma, Mastigias, Stomolophus.*

Although the Cubomedusae have, in this edition, been discussed with the Scyphozoa, the nature of their nematocysts, the possession of a velum, and their life cycle are considered evidence that they are not closely related to the other scyphozoans and should be placed within a separate class, the Cubozoa (Werner, 1975; Calder and Peters, 1975).

Class Cubomedusae. Medusoid cnidarians with bells having four flattened sides. Bell margin simple and bearing four tentacles or tentacle clusters. An attached polypoid larva follows planula. Tropical and subtropical oceans. *Carybdea, Chiropsalmus, Chironex.*

SUMMARY

1 Members of the class Scyphozoa are pelagic cnidarians in which the medusa is the dominant and conspicuous form. A polypoid larva, equivalent to an actinula, follows the planula. Assuming the primitive nature of the medusoid form, scyphozoans are primitive in their life cycle and perhaps evolved early from the ancestral hydrozoans.

2 Within the Scyphozoa, specialization has led to complexities in medusoid structure, as evidenced by such features as the following: larger size than that of most hydromedusae, more highly developed manubrium, cellular mesoglea, septate gut or at least a gut with gastric filaments, gastrodermal cnidocytes, and some development of sense organs.

3 The gonads are gastrodermal, and the eggs, which are shed through the mouth, develop into planula larvae. Following settling, the planulae develop into polypoid larvae, which feed and may reproduce asexually.

4 In some species the polypoid larva transforms directly into a young medusa, which can be taken as additional evidence that the polypoid form was derived from a larval stage in the evolution of the cnidarians. In most species of scyphozoans, young medusae are budded off transversely from the oral end of the polypoid larva.

Class Anthozoa

Anthozoans are either solitary or colonial polypoid cnidarians in which the medusoid stage is completely absent. Many familiar forms, such as sea anemones, corals, sea fans, and sea pansies, are members of this class. This is the largest of the cnidarian classes and contains over 6000 species.

Although the anthozoans are polypoid, they differ considerably from hydrozoan polyps. The mouth leads into a tubular pharynx that extends more than halfway into the gastrovascular cavity (Fig. 5–34). The gastrovascular cavity is divided by longitudinal mesenteries, or septa, into radiating compartments, and the edges of the mesentery bear nematocysts. The gonads, as in the scyphozoans, are gastrodermal, and the fibrous mesoglea contains cells. The nematocysts, unlike those of hydrozoans and scyphozoans, do not possess an operculum, or lid. Some anthozoan nematocysts have a three-part tip that folds back on expulsion; in others, the thread appears to rupture directly through the end of the capsule.

In order to simplify the survey of this class, we will deal with the sea anemones, the stony corals, and the octocorallian corals separately.

Sea Anemones

Sea anemones are solitary polyps and are considerably larger and heavier than the polyps of hydrozoans (Fig. 5–35). Most sea anemones range from 1.5 to 10 cm in length and from 1 to 5 cm in diameter, but specimens of *Tealia columbiana* on the

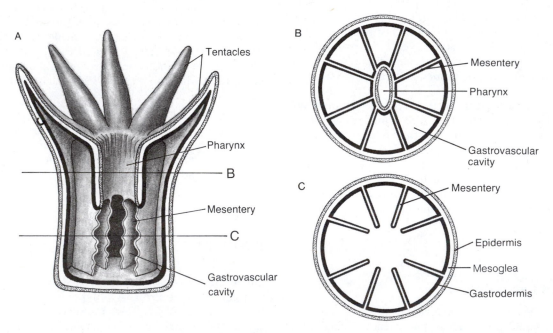

Figure 5–34 Structure of an anthozoan polyp: *A*, Longitudinal section. *B*, Cross section at the level of the pharynx. *C*, Cross section below the pharynx.

North Pacific coast of the United States and *Sticho-dactyla* on the Great Barrier Reef of Australia may have a diameter of more than a meter at the oral end. Sea anemones are often brightly colored. They may be white, green, blue, orange, or red, or a combination of colors.

Sea anemones inhabit deep and coastal waters throughout the world but are particularly diverse in tropical oceans. They commonly live attached to rocks, shells, and submerged timbers, and some forms burrow in mud or sand. Some are commensal on other animals, such as the shells of hermit

Figure 5–35 Sea anemones: *A* and *B*, *Anthopleura*, a genus of common sea anemones found in shallow water along the Pacific coast of the United States. The column of the partially closed specimen in *B* is covered with shell fragments, which adhere to epidermal tubercles. (*A* courtesy of Turtox News.) *C*, A West Indian species of *Actinia*. (*B* and *C* by Betty M. Barnes.) *D*, The venomous *Alicia mirabilis* from the Caribbean. The column is covered with berry-like evaginations containing nematocyst batteries. (By Charles Arneson.)

Figure 5–36 Relationship of body form to habitat in Caribbean reef-dwelling sea anemones: *A, Rhodactis sanctith-omae,* a surface-dwelling form with short column and tentacles. *B, Phymanthus crucifer,* and *C, Bartholomea annulata,* sand pocket forms. *D, Heteractis lucida,* a hole dweller with long column and tentacles. (From Sebens, K., 1976: The ecology of Caribbean sea anemones in Panama: utilization of space on a coral reef. *In* Mackie, G. O. (Ed.): Coelenterate Ecology and Behavior. Plenum Press, N.Y. p. 69.)

crabs (see p. 608). The shape of the body is often related to the habitat in which the sea anemone lives (Fig. 5–36).

The major part of the sea anemone body is formed by a heavy column (Fig. 5–37), which may be smooth or bear tubercles or even tentacle-like outgrowths. At the aboral end of the column there is a flattened pedal disc for attachment. At the oral end, the column flares slightly to form the oral disc, which bears eight to several hundred hollow tentacles, and in some species is drawn out into lobes. In the center of the oral disc is a slit-shaped mouth, bearing at one or both ends a ciliated groove called a siphonoglyph. The groove provides for the circulation of water into the gastrovascular cavity. The current of water functions to maintain an internal fluid, or hydrostatic, skeleton against which the muscular system can act, and also provides for the exchange of gases through the gastrodermal surface.

When a sea anemone contracts, the upper surface of the column of most species is pulled over and covers the oral disc. In many sea anemones, including the familiar *Metridium* of the Atlantic coast, the column bears a circular fold at its junction with the oral disc. This fold, known as a collar, covers the oral surface when the animal contracts (Fig. 5–37A).

The mouth leads into a flattened pharynx, which extends approximately two thirds of the way into the column (Fig. 5–37). The wall of the pharynx, being derived from an infolding, contains the same layers as the body wall. The inner side is covered by a ciliated epidermis, and the outer by gastrodermis. Between the two is a layer of mesoglea. The pharynx is kept closed and flat by the water

pressure in the gastrovascular cavity. The siphonoglyph is kept open by an especially thick mesoglea and very heavy epidermal cells.

As in all anthozoans, the gastrovascular cavity of the sea anemone is partitioned by longitudinal, radiating mesenteries. In the sea anemones there are usually two types of mesenteries, called complete and incomplete. Complete mesenteries are connected to the body wall on one side and to the wall of the pharynx on the other (Fig. 5–37B). Incomplete mesenteries are connected only to the body wall and extend only partway into the gastrovascular cavity. The mesenteries, both complete and incomplete, are arranged in adjacent pairs. The pairs at each end of the tapered pharynx are called directives. The mesenteries usually occur in multiples of 12. When only 12 mesenteries are present, as in the primitive *Halcampoides*, they are complete and are called the primary cycle. The addition of a cycle of 12 secondary mesenteries between the primary ones brings the total to 24. A tertiary cycle brings the number to 48. Only the first cycle may be complete, with subsequent cycles incomplete and successively smaller. Many exceptions to the numerical symmetry just described exist. Moreover, asexual reproduction produces considerable irregularity, particularly in *Metridium.*

In the upper part of the pharyngeal region, the mesenteries are pierced by openings that facilitate water circulation (Fig. 5–37A). Below the pharynx, the complete mesenteries have free margins and recurve toward the body wall.

Histologically, each mesentery consists of two layers of gastrodermis separated by a layer of mesoglea. Both the gastrodermis and the mesoglea are continuous with their corresponding layers in

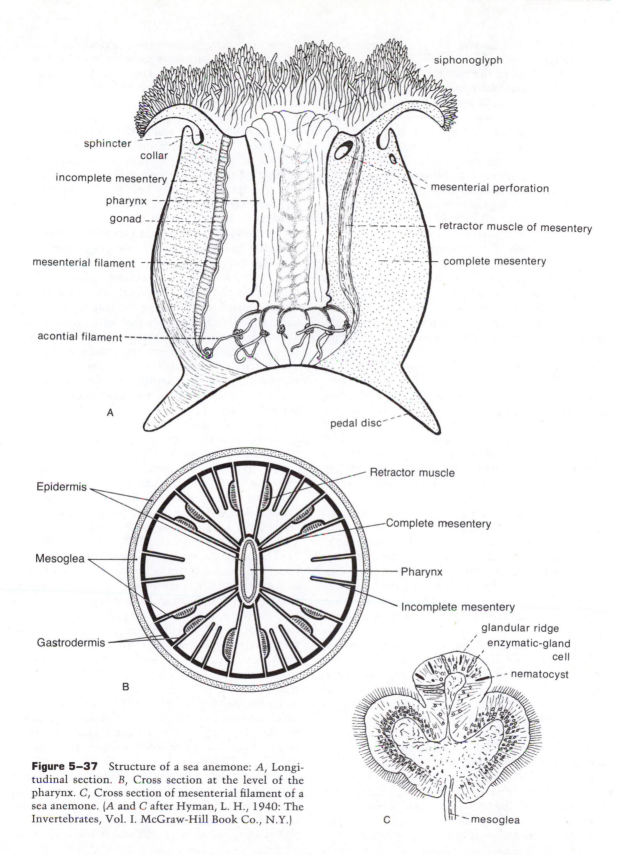

siphonoglyph

sphincter

collar

incomplete mesentery

pharynx

gonad

mesenterial filament

acontial filament

mesenterial perforation

retractor muscle of mesentery

complete mesentery

pedal disc

A

Epidermis

Mesoglea

Gastrodermis

B

Retractor muscle

Complete mesentery

Pharynx

Incomplete mesentery

glandular ridge

enzymatic-gland cell

nematocyst

mesoglea

C

Figure 5–37 Structure of a sea anemone: *A*, Longitudinal section. *B*, Cross section at the level of the pharynx. *C*, Cross section of mesenterial filament of a sea anemone. (*A* and *C* after Hyman, L. H., 1940: The Invertebrates, Vol. I. McGraw-Hill Book Co., N.Y.)

the body wall and also in the pharynx when the mesenteries are complete.

The free edge of the mesentery is trilobed and is called a mesenterial filament (Fig. 5–37C). The mesenterial filament is longer than the mesentery and thus tends to be somewhat convoluted. The filament's lateral lobes are composed of ciliated cells and aid in water circulation. The middle lobe contains cnidocytes and enzymatic-gland cells. The middle lobe is present on all mesenteries below the pharynx. In some sea anemones, including *Metridium*, the middle lobe continues at the base of the mesentery as a thread called an acontium, which projects into the gastrovascular cavity.

The epidermis of sea anemones may be ciliated and in some species is covered with a cuticle. The mesoglea is much thicker than that of hydrozoan polyps and contains a large number of fibers, as well as amebocytes. Associated with the tip of at least some anthozoan nematocysts is a cnidocil-like cilium and microvilli complex that is thought to be involved in the reception of stimuli for discharge. In addition to usually two types of ordinary nematocysts, sea anemones also possess spirocysts, which have a capsule with a single wall and a long adhesive thread. The spirocysts function in prey capture and in attachment to the substrate (Fig. 5–6E). The common intertidal *Anthopleura elegantissima* along the west coast of the United States uses spirocysts to attach shells and pebbles to the tubercles on its column, providing itself with a protective cover when the tide is out.

There are a few sea anemones in European and American waters whose nematocysts can produce a severe toxic reaction in humans. They include the berried sea anemones *(Alicia mirabilis)* (Fig. 5–35D), whose column is covered with berry-like clusters of nematocyst batteries, and the Caribbean sea anemone *Lebrunia danae.* The West Australian *Dofleina armata* is believed to be the most toxic sea anemone.

Sea anemones feed on various invertebrates, and large species can capture fish. Species with delicate tentacles lodge their bodies in protective crevices. Many large intertidal species feed on crabs and bivalves washed down from higher intertidal levels by waves or predators. The prey is caught by the tentacles, paralyzed by nematocysts, and carried to the mouth. The mouth is opened by radial muscles in the mesenteries, and the prey is swallowed.

When the prey passes from the pharynx into the lower part of the gastrovascular cavity, it is brought into contact with the free edges of many mesenteries and also the acontia, when present.

These structures produce the enzymes for extracellular digestion. Not only proteins but also fats are digested extracellularly. These filaments also appear to be the principal site of intracellular digestion and absorption.

Some large sea anemones with short tentacles, such as *Stichodactyla* and *Radianthus*, feed on fine particles. Planktonic organisms are trapped on the surface of the column and tentacles. Cilia on the surface of the column beat toward the oral disc, and cilia on the tentacles beat toward the tentacle tips. The tentacles then bend down and deposit the food in the mouth.

In the Indo-Pacific, little fish of the genus *Amphiprion* live symbiotically among the tentacles of large sea anemones. Following behavioral acclimation to the anemone, the altered mucous coat on the surface of the fish apparently raises the threshold of nematocyst discharge, making it possible for the fish to live in an otherwise lethal habitat. The anemone provides protection and some food scraps; the fish in turn protects the anemone from some predators, removes necrotic tissue, and by its swimming and ventilating movements reduces fouling of the anemone by sediment of various sorts (Mariscal, 1970; Lubbock, 1980). Other commensals of sea anemones include cleaning shrimps (p. 601), snapping shrimps, arrow crabs, brittle stars, and other types of fish.

Symbiotic zooxanthellae or zoochlorellae or even both are found in the gastrodermal cells of many sea anemones, especially in the tentacles and the oral disc. The relationship is similar to that described for corals on page 134 (Muscatine et al, 1975; Steele and Goreau, 1977).

The muscular system in sea anemones is much more specialized than that in the other two classes of cnidarians. The longitudinal, epidermal fibers of the column and pharynx have disappeared except in primitive species. They are present, however, in the tentacles and oral disc. Thus, the muscular system is largely gastrodermal. Bundles of longitudinal fibers in the mesenteries form retractor muscles for shortening the column (Fig. 5–37). Circular muscle fibers in the columnar gastrodermis are well developed. The presence of complete mesenteries may facilitate the function of the internal hydraulic skeleton by limiting the maximum extent of the diameter of the column when the retractor muscles contract.

Although sea anemones are essentially sessile animals, many species are able to change locations by slow gliding on the pedal disc, by crawling on the side, or by walking on the tentacles. A few species can detach the pedal disc and swim briefly with lashing motions of the column or tentacles. Burrowing sea anemones slowly plant the body column into sand or mud by peristaltic contractions, which change the column diameter. Members of the genus *Minyas* are pelagic and hang upside down from a chitinous float secreted by the pedal disc.

The nervous system exhibits the typical cnidarian pattern, and no specialized sense organs are present.

Some species of sea anemones exhibit aggressive behavior toward members of other clones or toward other species. Cnidocytes on specialized searching tentacles or on column projections (acrorhagi) are fired on contact with the "foreign" sea anemone. There is some withdrawal between combatants, and one or both parties may suffer tissue damage. Such aggressive behavior apparently provides for some spacial separation between species or clones (Francis, 1973; Purcell, 1977; Williams, 1978; Bigger, 1982).

Asexual reproduction is common in sea anemones. One method is by pedal laceration, in which parts of the pedal disc are left behind as the animal moves. In some instances the disc puts out lobes that pinch off. These detached portions then regenerate into small sea anemones. Many sea anemones reproduce asexually by longitudinal fission, and a few species do so by transverse fission.

Most sea anemones are hermaphroditic but produce only one type of gamete during any one reproductive period. The gonads are located in the gastrodermis on all or certain of the mesenteries in the form of longitudinal, bandlike cushions behind the mesenterial filament (Fig. 5–37A).

The eggs may be fertilized in the gastrovascular cavity, with development taking place in the mesenteric chambers, or fertilization may occur outside the body in the sea water. The planula larva may be planktotrophic (feeding) or lecithotrophic (getting nutrition from yolk) and has a variable larval life span. Mesenteries develop from the column wall and grow toward the pharynx. There are still no tentacles, and the young sea anemone lives as a ciliated ball, unattached and free swimming. With further development, the "larva" settles, attaches, and forms tentacles. Studies indicate that a New Zealand intertidal sea anemone (*Actinia tenebrosa*) requires 8 to 66 years to reach a column diameter of 40 mm and has an average longevity of 50 years (Ottaway, 1980).

CERIANTHARIANS AND ZOANTHIDEANS

Two small orders of anemone-like anthozoans contain a number of commonly encountered species.

Figure 5–38 *Cerianthus,* a large, burrowing anthozoan. The animal secretes a tube into which the body can be retracted. The members of this order (Ceriantharia) are similar to sea anemones in size and general appearance. (Courtesy of Buchsbaum, R. M., and Milne, L. J., 1961: The Lower Animals. Chanticleer Press, N.Y.)

The order Ceriantharia includes the "tube-dwelling" sea anemones. These large, solitary anthozoans are adapted for life in soft bottoms. The body is lodged in a heavy, secreted mucous tube, which is buried within the substratum. This tube is formed of mucus and fired threads and capsules of ptychocysts, a nematocyst-like organelle (Mariscal et al, 1977.) When feeding, the animal projects the tentacles and oral disc from the surface or elevated opening of the tube (Fig. 5–38).

Members of the order Zoanthidea are largely tropical, and some are common reef inhabitants. Most are 1 to 2 cm in diameter, and the majority are colonial, being connected by a stolon or a common mat. The body may be columnar, but many species are rather short and button-like (Fig. 5–39). A short fringe of tentacles surrounds the broad oral disc. The column and mat are covered by a thick cuticle, and many species have sand or other debris embedded in the surface layer. Some reef species pave rocks or form large encrusting masses and harbor zooxanthellae (Fig. 5–39). Some are commensal.

Stony, or Scleractinian, Corals

Closely related to sea anemones are the stony, or scleractinian, corals (also called madreporarian corals), which constitute the largest order of anthozoans. In contrast to sea anemones, stony corals produce a calcium carbonate skeleton. Some corals, such as the Indo-Pacific reef-inhabiting *Fungia* and some deep-sea species, are solitary, with polyps as large as 25 cm in diameter (Fig. 5–42E), but the ma-

A   B

Figure 5–39   Order Zoanthidea, colonial and semicolonial anemone-like anthozoans: *A,* Contracted individuals of *Palythoa.* In the tropics species of *Palythoa* commonly carpet rocks in shallow water. (By Betty M. Barnes.) *B,* Expanded specimens of *Palythoa psammophilia.* The letters indicate mouth, oral disc, peristome, marginal area, and tentacles. (From Reimer, A. A., 1971: Specificity of feeding chemoreceptors in *Palythoa psammophilia.* Comp. Gen. Pharm., Pergamon Press *2* (8):383–396.)

jority are colonial with very small polyps averaging 1 to 3 mm in diameter (Fig. 5–40). However, the entire colony can become very large. Coral polyps are very similar in structure to sea anemones but do not possess siphonoglyphs (Figs. 5–40 and 5–44*D*).

The skeleton is composed of calcium carbonate crystals and is secreted by the epidermis of the lower half of the column as well as by the basal disc. Their secreting process produces a skeletal cup, within which the polyp is fixed. The cup is termed the calyx; the surrounding walls of the cup, the theca; and the floor of the cup, the basal plate. The floor contains thin, radiating, calcareous septa (Figs. 5–41*A* and 5–42*C* and *F*). Each scleroseptum projects upward into the base of the polyp, folding the basal layers and inserting them between a pair of mesenteries. As long as a colony is alive, calcium carbonate is deposited beneath the living tissues.

In addition to providing a uniform substratum on which the living colony can attach, the skeleton (and especially the sclerosepta) also serves as protection. When contracted, the polyps project little above the skeletal platform and are difficult for most fish and other predators to remove.

The polyps of colonial corals are all interconnected, but the attachment is lateral rather than

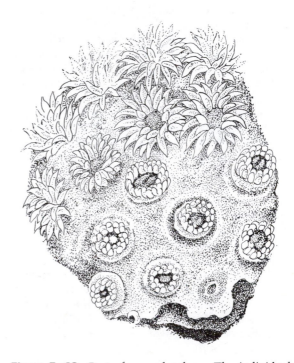

Figure 5–40   Part of a coral colony. The individuals, some contracted and some expanded, are connected by a lateral sheet of tissue.

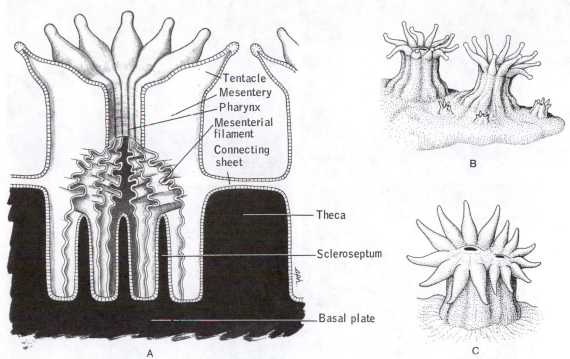

Tentacle
Mesentery
Pharynx
Mesenterial filament
Connecting sheet

Theca

Scleroseptum

Basal plate

A

B

C

Figure 5–41 *A*, A coral polyp in its theca (longitudinal section). (Modified after Hyman, L. H., 1940: The Invertebrates. Vol. I. McGraw-Hill Book Co., N.Y.) *B*, Extratentacular budding. *C*, Intratentacular budding.

aboral, as in hydroids. The column wall folds outward above the skeletal cup and connects with similar folds of adjacent polyps. Thus, all the members of the colony are connected by a horizontal sheet of tissue (Figs. 5–40 and 5–41). Since this sheet represents a fold of the body wall, it contains an extension of the gastrovascular cavity as well as an upper and lower layer of gastrodermis and epidermis. The lower epidermal layer secretes the part of the skeleton that is located between the cups in which the polyps lie. The living coral colony, therefore, lies entirely above the skeleton and completely covers it.

The skeletal configurations of various species of corals are due in part to the growth pattern of the colony and in part to the arrangement of polyps in the colony (Figs. 5–42 and 5–44). Some species form flat or rounded skeletal masses; others have an upright and branching growth form. Some are large and heavy; others are small and delicate (Fig. 5–43). When the polyps are well separated, the coral skeleton has a pitted appearance, as in *Oculina*, the eyed coral, in *Astrangia*, one of the corals living along the North Atlantic coast, and in the reef coral *Montastrea* (Fig. 5–42*C*). The polyps of brain coral are arranged in rows (Figs. 5–42*A* and

B, 5–44*A*). The rows are well separated, but the polyps in each row are fused together so that their cups are confluent. As a result, the skeleton of the colony has the appearance of a human brain, containing troughs or valleys separated by skeletal ridges.

The coral colony expands by the budding of new polyps from the bases of old polyps (Fig. 5–41*B*) or from the oral discs of old polyps. In the latter case, the oral disc of the parent legnthens in one direction (Fig. 5–41*C*). Gradually, the oral disc constricts and the separation extends down the length of the column to form two new polyps. The budding process is accompanied by simultaneous changes in the deposition of the underlying sclerosepta. As might be expected, brain corals arise by intratentacular budding, in which the oral discs and columns never constrict after new mouths are formed. Thus the polyps in a row of brain coral share a common oral disc bearing many mouths (Fig. 5–42*A*).

Calcium carbonate is continually deposited by the basal epidermis of the living colony that rests upon it. In many corals the polyps periodically lift their bases and secrete a new floor to their cup. This closes off a minute chamber in the skeleton

(Text continued on p. 132)

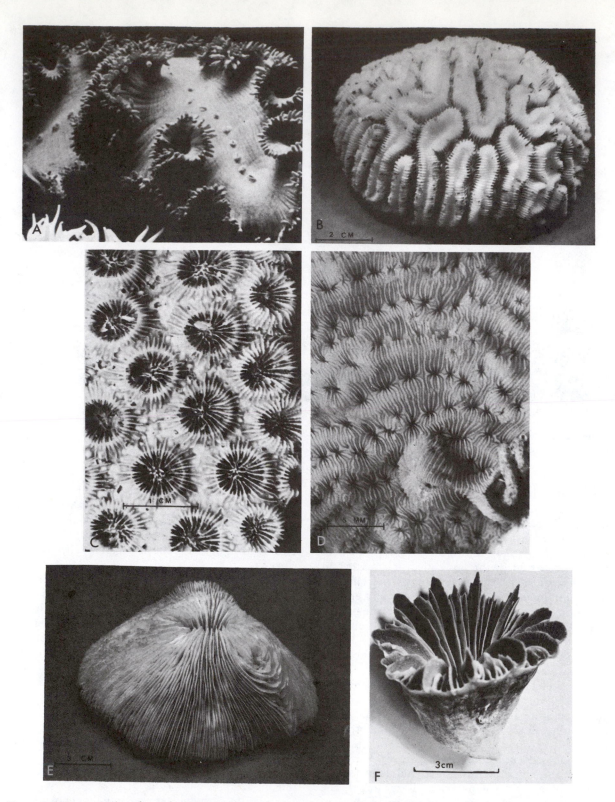

Figure 5–42 *A*, Oral surface of a living brain coral. Note the row of mouths. (By Catala, R. L., 1964: Carnival under the Sea. R. Sicard, Paris.© R. L. Catala.) *B*, Skeleton of the brain coral *Diploria*. Note the arrangement of sclerosepta. *C*, Skeleton of *Montastrea*. In life the polyps are in the large, distinct cups. *D*, Skeleton of lettuce coral, *Agaricia*, in which the polyps are arranged in rows. *E*, Skeleton of *Fungia*, a solitary coral of the Indo-Pacific. The skeleton of this very large polyp is limited to sclerosepta projecting from the basal plate. There is no wall, hence no cup. (*B–E* by Betty M. Barnes.) *F*, Solitary deep-water coral. (By Katherine E. Barnes.)

Figure 5–43 Part of a colony of a species of *Seriatopora*, an Indo-Pacific genus of delicate branching corals. Specimen is 6 cm across. (By Katherine E. Barnes.)

Figure 5–44 Scleractinian, or stony, corals: *A*, Brain coral on the Great Barrier Reef. (Courtesy of Fritz Goro.) *B*, Staghorn coral from the Great Barrier Reef. (By Allen Keast, courtesy of Buchsbaum, R. M., and Milne, L. J., 1961: The Lower Animals. Chanticleer Press, N.Y.) *C*, Knobbed and lettuce coral on a Bahamian reef. (By John Storr.) *D*, Cup corals. (By D. P. Wilson.)

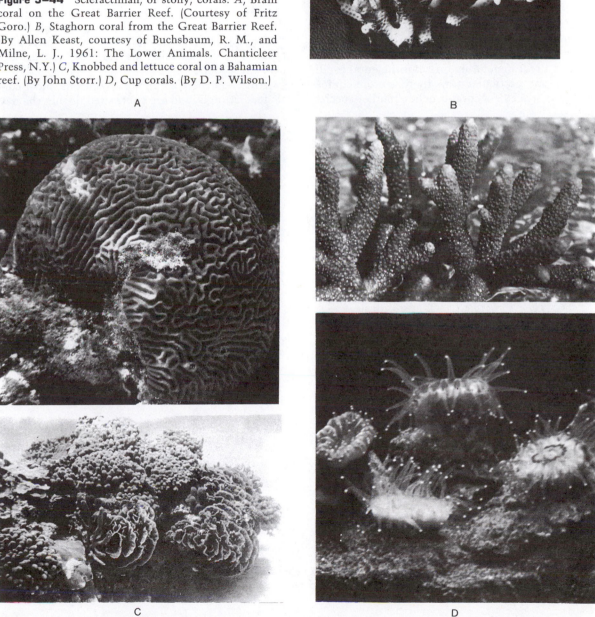

A

B

C

D

(Fig. 5–45A). The growth rate varies greatly, depending upon the species and water temperature. Many dome and plate corals grow only 0.3 to 2 cm a year through vertical or linear deposition of calcium carbonate. Some branched corals, on the other hand, grow rapidly, increasing in the linear direction as much as 10 cm per year (Gladfelter et al, 1978).

The density of the secreted $CaCO_3$ is not the same throughout the year, the change being governed by seasonal shifts of temperature and light. Thus many coral skeletons exhibit seasonal growth bands like tree rings (Knutson et al, 1972; Isdale, 1977; Wellington and Glynn, 1983). They can be revealed with x-radiographs to determine the age and growth rate of the coral (Fig. 5–45B).

Although coral species are restricted to various parts of a reef depending on environmental conditions, at any particular zone there is often competition for space. If a coral touches another species,

it may extrude its mesenterial filaments and digest the intruder's tissues. Or, like certain sea anemones, it may use specialized tentacles. Aggressive responses depend on the species, some being very aggressive, others only slightly so (Lang, 1973; Richardson et al, 1979).

Corals feed like sea anemones, and the prey ranges from small fish down to small zooplankton, depending on the size of the polyps. When expanded, the outstretched tentacles of adjacent polyps present a broad, continuous mesh that prey might touch. In addition to capturing zooplankton, many corals also collect fine particles in mucous films or strands, which are then driven by cilia to the mouth (Lewis and Price, 1976). Some, such as the foliaceous agaricids (Fig. 5–42D), which have reduced or no tentacles, are entirely mucous suspension feeders.

Although there are many exceptions, most corals feed at night and are contracted during the day

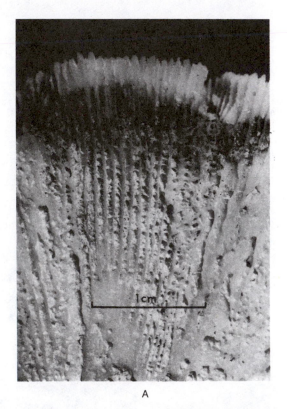

A

B

Figure 5–45 *A,* Minute chambers in coral skeleton that result from the lifting of the polyp and base and subsequent secretion of a new underlying calcareous plate. (By Katherine E. Barnes.) *B,* X-radiograph of a radial section through a 15-year-old specimen of *Porites lobata* from the Great Barrier Reef, Australia. One year of growth is represented by a dark band (low-density skeleton, formed in winter) and a light band (high-density skeleton, formed in summer). (From Isdale, P., 1977: Variation in growth rate of hermatypic corals in a uniform environment. Proceedings of the 3rd International Coral Reef Symposium, University of Miami. *2* (Geology):406.)

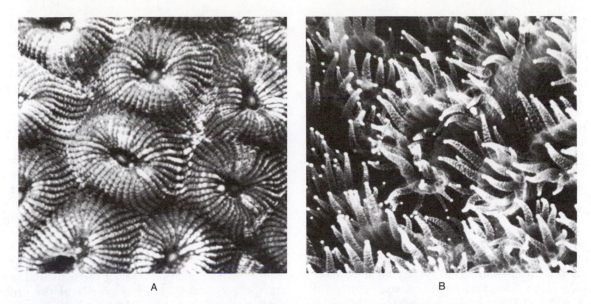

A   B

Figure 5-46 *Favia favus* from the Great Barrier Reef during the day (*A*) and at night (*B*). (From Veron, J. E. N., Pichon, M., and Wijsman-Best, M., 1977: Scleractinia of Eastern Australia. Part II. Australian Institute of Marine Science Monograph. Series 3, p. 29.)

(Fig. 5–46), and some species display a persistent rhythm of expansion, even when kept in constant darkness or light (Fig. 5–47).

Over 60 genera of corals contain symbiotic zooxanthellae within the gastrodermal cells. Deepwater and some cold-water corals lack zooxanthellae, but virtually all reef-dwelling (hermatypic) corals possess them. The algae may reach such concentrations as to account for 50 per cent of the protein nitrogen of the coral (Muscatine, 1974; Muscatine et al, 1975). The algal symbionts give most of their coral host a yellow-brown to dark brown color.

Our knowledge of the physiological relationship between hermatypic corals and their symbiotic algae has grown considerably in recent years.

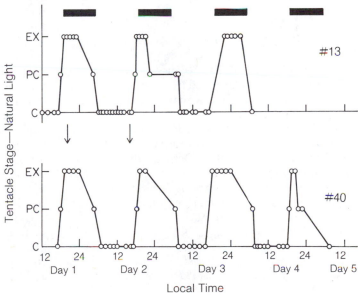

Figure 5-47 Rhythm of expansion and contraction of two species of fungiid corals in natural light: *Fungia repanda* (#13), *Fungia fungites* (#40). Dark periods are indicated by black horizontal bars; arrows, the introduction of zooplankton into the aquarium. Ex, PC, and C refer to expansion, partial contraction, and contraction. (From Sweeney, B. M., 1976: Circadian rhythms in corals, particularly Fungiidae. Biol. Bull., *151*:236–246.)

The nutritive needs of the coral are supplied in part by the planktonic animals upon which it feeds and in part by its algal symbionts. A large portion of the carbon fixed by the algae in photosynthesis is passed to the coral, largely in the form of glycerol but including glucose and alanine. The food caught by the coral probably supplies both coral and algae with nitrogen, which is then cycled back and forth between the two. Thus, the coral requires zooplankton prey even though the photosynthate supplies a major part of its energy needs. The symbiosis also facilitates the deposition of the coral skeleton, for corals that are deprived of their algae or kept in the dark deposit calcium carbonate at a much slower rate than under normal conditions. The degree to which the coral depends on the algae varies by species and even within species populations (Lewis and Smith, 1971; Clayton and Lasker, 1982; Jokiel and York, 1982; Cook, 1983; Falkowski et al, 1984; Rinkevich and Loya, 1984).

Sexual reproduction is similar to that in the sea anemones, and there are both dioecious and hermaphroditic species (see review by Fadlallah, 1983). The planula, which is produced by sexual reproduction, attaches and the subsequent first polyp (which develops by asexual budding) becomes the parent of all other members of the colony.

Corals are subject to injury or death from storms, extremely low tides, predation, and disease. The living colony can regenerate about 1 cm of destroyed tissue but not much more (Bak and Steward-Van Es, 1980). However, in some branching corals, such as the Caribbean *Acropora palmata*, whole colonies may regenerate from broken fragments.

Octocorals

Sea anemones and corals, because of their structural similarities, are grouped in the subclass Zoantharia. The remaining anthozoans, including such common marine forms as sea pens, sea pansies, sea fans, whip corals, and pipe corals, form the subclass Octocorallia. The Octocorallia possess a number of distinctive features. Octocorallians always have eight tentacles, and these are pinnate—that is, they possess side branches, as does a feather (Fig. 5–48). There are always eight complete mesenteries on either side of a tentacle base. Only one siphonoglyph is present.

The octocorallians are colonial cnidarians, and the polyps are usually rather small, similar to those of stony corals. The polyps of an octocorallian colony are connected by a mass of tissue called coenenchyme (Fig. 5–48). This consists of a thick mass of mesoglea, which is perforated by gastrodermal tubes that are continuous with the gastrovascular cavities of the polyps. The surface of the entire fleshy mass is covered by a layer of epidermis, which joins the epidermis of the polyp column. Only the upper portion of the polyp projects above the coenenchyme (Fig. 5–48).

The amebocytes of the mesoglea secrete calcareous skeletal material that supports the colony. Thus, the skeleton of the Octocorallia is internal and is an integral part of the tissue. This arrangement is in sharp contrast to that of the stony corals, whose skeletons are entirely external. The octocorallian skeleton may be composed of separate or fused calcareous spicules or of a horny material.

Among the most familiar of the octocorallians are the gorgonian, or horny, corals of the order Gorgonacea (Fig. 5–49), which include the whip corals, sea feathers, sea fans, and precious red coral *(Corallium)*. Gorgonians are common and conspicuous members of reef faunas, especially in the West Indies (see Fig. 6–11A). The body of most gorgonian corals contains a central axial rod composed of an organic substance called gorgonin (proteins plus mucopolysaccharides). The axial rod is commonly impregnated with calcium carbonate, and in some species a calcified section alternates with a noncalcified section, so the rod is jointed. Around the axis is a cylinder of coenenchyme and polyps (Fig. 5–48). The coenenchyme contains embedded calcareous ossicles or spicules of different shapes and colors (Fig. 5–50). It is the color of the calcareous skeletal components that accounts for the yellow, orange, or lavender color of some species. The yellow-brown color of many reef species results from the presence of symbiotic zooxanthellae. The colonies of most gorgonian corals are erect branching rods and are thus rather plantlike (Fig. 5–49). Whip corals consist of slightly branched, long, cylindrical filaments about the diameter of drinking straws. Many gorgonians are branched only in one plane, and in sea fans the branches may be connected by cross bars to form a lattice (Fig. 5–49). Sea fans are usually oriented at right angles to the water current. In general, the design of gorgonian corals requires a small amount of surface area for attachment but provides a large surface area for feeding, as a consequence of the branching vertical development. The flexible skeleton permits bending in currents (see Wainwright and Koehl, 1976, for review). Depending upon the

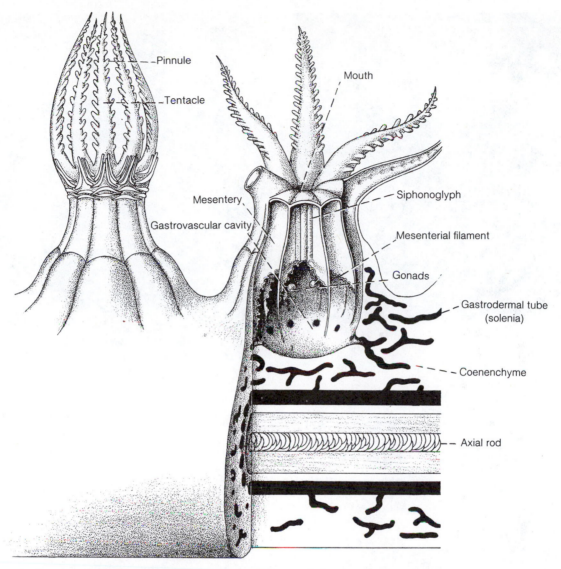

Figure 5–48 Structure of a gorgonian coral of the anthozoan subclass Octocorallia. (Redrawn from Bayer, F. M., 1956: Octocorallia. *In* Moore, R. C., Treatise on Invertebrate Paleontology. Courtesy of Geological Society of America and University of Kansas, Lawrence, Kans. p. F169.)

species, gorgonians reach the maximum size their architecture will permit in 10 to 30 years (Velimirov, 1975; Wineberg and Wineberg, 1979).

Gorgonians are more commonly expanded during the day than during the night. They have been assumed to feed on zooplankton like most scleractinian corals, but recent studies on a number of gorgonians and other octocorallians indicate that they have small numbers of cnidocytes and feed on smaller particles than zooplankton (Lasker, 1981).

Gorgonians harbor many symbiotic animals that either are attached to or crawl over the gorgonian surface—colonial tunicates, barnacles, bivalves, snails, and gobies. Some take on the colors of their gorgonian host.

The tropical Indo-Pacific organ-pipe coral, *Tubipora*, which belongs to another group of octocorallians, is differently organized (Fig. 5–51). Its long, parallel polyps are encased in calcareous tubes of fused spicules. The tubes are connected at

A

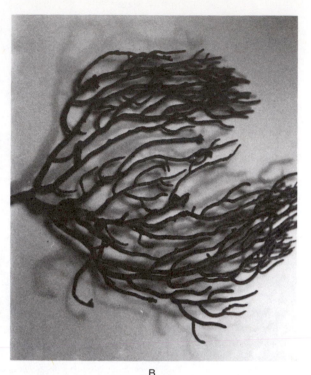

B

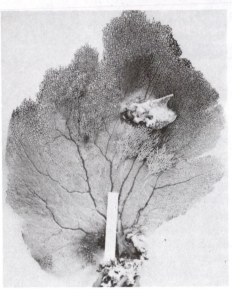

C

Figure 5–49 Gorgonian corals (order Gorgonacea): *A*, Sea rods, the most common gorgonian growth form. See also Figure 6–15. (Courtesy of T. Parkinson.) *B*, A species of sea fan in which there is extensive branching in one plane but branches do not interconnect. *C*, *Gorgonia ventilina*, the common West Indian sea fan, in which the branches anastomose to form a lattice. Rule is 15 cm. A winged oyster *(Pteria)* is attached to the fan. *(B* and *C* by Katherine E. Barnes.)

intervals by transverse, calcareous plates, or platforms.

The sea pens and sea pansies (order Pennatulacea) are inhabitants of soft bottoms and are quite different from coral-like members of the Octocorallia. There is a large, primary polyp with a stem-like base, which is anchored in the sand. The body is fleshy, although the coenenchyme contains cal-

careous spicules, or ossicles. The upper part of this primary polyp gives rise to secondary dimorphic polyps. The more typical and conspicuous of these secondary polyps are termed autozooids. Highly modified polyps, called siphonozooids, pump water into the interconnected gastrovascular cavities and thereby keep the colony turgid and erect. In the sea pens, the primary polyp is elongate and

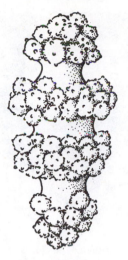

Figure 5–50 Spicule of a gorgonian coral.

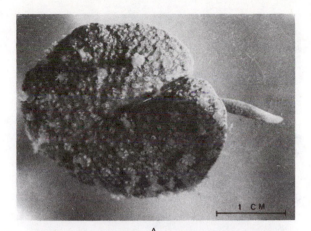

A

cylindrical. In the sea pansy, *Renilla* (Fig. 5–52*A*), which is common along the Atlantic, Gulf, and southern California coasts of the United States, the primary polyp is leaflike, with the secondary polyps limited to the upper surface. The flattened horizontal surface reduces resistance to the turbulent near-shore waters in which these colonies live. Moreover, they are able to uncover themselves when buried by shifting sediment and reanchor themselves when dislodged (Kastendiek, 1976). There are deep-sea species. *Umbellula*, which lives on the Atlantic abyssal plain, looks like a pinwheel (secondary polyps) mounted at the end of a long stick (primary polyp) embedded in the bottom (Fig. 5–52*B*).

B

Figure 5–51 Skeleton of organ-pipe coral, a member of the small tropical octocoral order Stolonifera. (By Betty M. Barnes.)

Figure 5–52 Order Pennatulacea. *A*, Sea pansy, *Renilla*. (By Betty M. Barnes.) *B*, *Umbellula*, a strange deep-sea octocoral related to sea pansies. Photograph was taken at a depth of over 5000 meters some 350 miles off the west coast of Africa. Stalk is estimated to be about a meter in length. (By Walter Jahn, U.S. Naval Oceanographic Service.)

A

B

Figure 5–53 Soft corals of the order Alcyonacea: *A*, Soft corals (pale) and scleractinian corals (dark) from Fiji. Mound of corals is about 2 meters across and about 2 meters below the water surface. *B*, Part of a colony of *Capnella gaboensis* from Sidney Harbour, Australia, showing polyps. (Both courtesy of Penny Farrant.)

Members of the order Alcyonacea are known as soft corals or leather corals. These octocorallians possess soft fleshy or leathery colonies that may reach a large size, 1 meter in diameter in the case of *Sarcophyton*. The colonies are irregular in shape, some encrusting and some massive, often with lobate or finger-like projections (Fig. 5–53). Separate calcareous spicules are embedded in the coenenchyme. Alcyonacean corals are often conspicuous on Indo-Pacific reefs. In some respects they resemble scleractinian corals, with the fleshy coenenchymal mass acting as the substrate for the colony instead of calcium carbonate.

SYSTEMATIC RÉSUMÉ OF CLASS ANTHOZOA

Subclass Octocorallia, or Alcyonaria. Polyp with eight pinnate tentacles and eight mesenteries. Almost entirely colonial.

Order Stolonifera. No coenenchymal mass; polyps arising from a creeping mat or stolon. Skeleton of calcareous tubes of separate calcareous spicules, or horny external cuticle. Tropical and temperate oceans in shallow water. *Tubipora* (organ-pipe coral), *Clavularia*.

Order Telestacea. Lateral polyps on simple or branched stems. Skeleton of calcareous spicules. *Telesto*.

Order Alcyonacea. Soft corals. Coenenchyme forming a rubbery mass. Colony may have a massive mushroom shape or an encrusting growth form. Skeleton of separate calcareous spicules. Largely tropical. *Alcyonium*, *Gersemia*, *Sarcophyton*.

Order Helioporacea. Contains only the Indo-Pacific blue coral, *Heliopora*, having a massive, blue, calcareous skeleton.

Order Gorgonacea. Horny corals or gorgonian corals. Common tropical and subtropical octocorallian cnidarians having a largely upright, plantlike growth form and an axial skeleton of a horny organic material. Separate or fused calcareous spicules may also be present. *Gorgonia* (sea fan), *Leptogorgia* (sea whip), *Corallium* (precious red coral), *Muricea* (sea rod).

Order Pennatulacea. Sea pens. Colony having a fleshy, flattened or elongate body, or rachis. Skeleton of calcareous spicules. *Stylatula* (sea pen), *Veretillum*, *Renilla* (sea pansy), *Umbellula*.

Subclass Zoantharia. Polyps with more than eight tentacles and mesenteries, the latter typically in cycles of 12. Tentacles rarely pinnate. Solitary or colonial.

Order Zoanthidea. Small, anemone-like anthozoans having one siphonoglyph and no skeleton. Solitary or colonial. *Palythoa* and *Zoanthus*. Some, such as *Epizoanthus* and *Parazoanthus*, epizoic on other invertebrates.

Order Actiniaria. Sea anemones. Solitary anthozoans with no skeleton, with mesenteries in hexamerous cycles, and usually with two siphonoglyphs. *Halcampoides*, *Edwardsia*, *Metridium*, *Epiactis*, *Stichodactyla*.

Order Scleractinia, or Madreporaria. Stony corals. Mostly colonial anthozoans secreting a heavy, external, calcareous skeleton. Sclerosepta arranged in hexamerous cycles. *Fungia, Acropora, Porites, Astrangia, Oculina.* Many fossil species.

Order Corallimorpharia. Solitary species with radially arranged tentacles. Resemble true corals but lack skeletons. *Corynactis, Rhodactis, Ricordea.*

Order Ceriantharia. Anemone-like anthozoans with greatly elongate bodies adapted for living with secreted tubes buried in sand or mud. One siphonoglyph; mesenteries all complete. *Cerianthus, Ceriantheopsis.*

Order Antipatharia. Black or thorny corals. Gorgonian-like species with upright, plant-like colonies. Polyps arranged around an axial skeleton composed of a black horny material and bearing thorns. Largely in deep water in tropics. *Antipathes.*

SUMMARY

1 Members of the class Anthozoa are polypoid cnidarians; the medusoid stage is entirely lacking.

2 The anthozoan polyp is more specialized than that of hydrozoans, and its cellular mesoglea, septate gastrovascular cavity, cnidocytes in gastric filaments, and gastrodermal gonads indicate a closer phylogenetic relationship with the Scyphozoa than with the Hydrozoa.

3 The difference in the body form of the Scyphozoa and the Anthozoa (medusa versus polyp) may be reconciled if the anthozoans are derived through the polypoid larva of scyphozoans.

4 The two subclasses, the Zoantharia and the Octocorallia, reflect different levels of structural evolution within the Anthozoa. The Octocorallia have retained an arrangement of eight complete mesenteries and eight tentacles, which may be the primitive anthozoan condition. Colonial organization is characteristic of almost all octocorallians, and the polyps are interconnected through a complex mass of mesoglea and gastrodermal tubes. The zoantharia display a more complex system of mesenteries, which always exceed eight in number. There are many solitary forms, and colonial species are connected by more or less simple outfoldings of the body wall.

5 Sea anemones are the principal group of solitary anthozoans, and perhaps because of their solitary condition, many species have evolved a larger size than most other anthozoan polyps. The number and complexity of their mesenteries, providing a large surface area of gastric filaments, may be related to the utilization of larger prey.

6 The majority of anthozoans are colonial, and this type of organization has evolved independently a number of times within the class. Although colonies may reach a large size, the individual polyps are generally small. There are some groups with polymorphic colonies, but this condition is not as widespread as in the hydrozoans.

7 Scleractinian corals, although similar to sea anemones, are largely colonial and are unique in their secretion of an external calcareous skeleton. The skeleton provides the colony with a uniform substrate on which the living colony rests. The sclerosepta may contribute to the adherence of the polyps within the thecal cups and provide some protection against grazing predators when the polyps are withdrawn.

8 The majority of scleractinian corals are tropical reef inhabitants (hermatypic) and harbor zooxanthellae. Zooxanthellae are found in many other anthozoans as well as some scyphomedusae and some hydrozoans.

9 The colonial alcyonaceans, or soft corals, which are most abundant on Indo-Pacific reefs, in many ways parallel the scleractinian corals, for the massive coenenchymal mass forms the substrate from which the individual polyps arise.

10 The branching, rodlike colonies or gorgonian corals are adapted for exploiting the vertical water column while using only a small area of the substrate for attachment. Flexible support is provided by a central, organic skeletal rod and separate calcareous spicules embedded in the coenenchyme.

11 The pennatulaceans, which include sea pens, sea feathers, and sea pansies, are adapted for life on soft bottoms. A large, primary polyp, which determines the form of the colony, not only provides anchorage in the sand but also acts as the substrate from which the small, secondary polyps arise.

12 A planula larva is characteristic of most anthozoans and develops into the polyp. Colonial forms are derived by budding from the first polyp.

The Ctenophores

The Ctenophora constitute a small phylum of marine animals that are commonly known as sea walnuts or comb jellies. The phylum contains approximately 50 known species, some of which are abundant in coastal waters; many others are oceanic. The group is still poorly known even though a great deal has been learned about them over the last ten years. Ctenophores are so delicate that they are difficult to collect with plankton nets. Much of our recent knowledge of oceanic ctenophores is based on specimens collected by scuba divers with hand-held jars (Harbison et al, 1978).

Ctenophores have been thought to be an offshoot from some medusoid cnidarian because the general body plan is somewhat similar to that of a medusa. The gastrovascular cavity has a canal system, and the thick, middle body layer is comparable to the cnidarian mesoglea. However, the similarities may represent more convergence than common ancestry (Harbison, 1985). In any event, ctenophores have undergone considerable specialization and display a number of innovations that indicate a sharp divergence from the cnidarian line.

The more primitive ctenophores, such as the common coastal *Pleurobrachia* (Fig. 5–54), are spherical or ovoid in shape and range in size from that of a pea to that of a golf ball. Most ctenophores are transparent, but some are brightly colored or possess spots of bright pigment.

The body wall is composed of an outer epidermis containing sensory cells and often mucous gland cells. Beneath the epidermis lies a thick mesoglea, composed of a jelly-like material strewn with fibers and amebocytes derived from ectoderm (Siewing, 1977). The mesoglea of ctenophores also contains smooth muscle cells, which are arranged as an anastomosing network.

The biradial body can be divided into two hemispheres (Figs. 5–54 and 5–55). The mouth, on one side, forms the oral pole; the diametrically opposite point on the body bears an apical organ and marks the aboral pole. The body is further divided into equal sections by eight ciliated bands. These bands, called comb rows, are characteristic of ctenophores and are the structures from which the name of the phylum is derived (*ktenes* in Greek means *combs*). Each band extends about four-fifths of the distance from the aboral pole to the oral end of the body and is made up of short transverse plates of long, fused cilia called combs. The combs are arranged in succession one behind the other, to form a comb row.

The combs provide the locomotor power in ctenophores, although lobate forms can also swim by contractions of the lobes. The ciliary beat functions in waves beginning at the aboral end of the row. The effective sweep of each comb is toward the aboral pole, so the animal is driven with the mouth, or oral end, forward.

From each side of the aboral hemisphere is suspended a long, branched, contractile tentacle, which emerges from the bottom of a deep, ciliated, epidermal canal called the tentacular sheath, or pouch. There are two pouch openings, located between comb rows on opposite sides of the body, each at an approximately 45-degree angle from the aboral pole.

The tentacular epidermis possesses peculiar adhesive cells called colloblasts. A colloblast cell is somewhat pear-shaped, with the narrowed end anchored in the tentacular mesenchyme and having a synaptic junction with a nerve cell. An intracellular helical thread winds around the long axis of the cell and at its distal end gives rise to a large number of smaller radiating fibers (Fig. 5–56). Each radiating fiber terminates in a granule filled with an adhesive mucoid material, which is liberated on contact with prey (Franc, 1978).

The digestive system is composed of an elaborate, biradial system of canals that arise from a central stomach (or infundibulum) (Fig. 5–55A).

The mouth leads into a long, tubular pharynx, which extends along the polar axis to the stomach. The digestive system, excluding the pharynx, is lined by gastrodermal epithelium.

Ctenophores are carnivorous, feeding on other planktonic animals. *Pleurobrachia* fishes with its branched tentacles, which form a very large net when fully expanded. Prey, especially copepods, are caught on the adhesive colloblasts, hauled in by the tentacle retraction, and wiped into the mouth and pharynx. The body is rotated to bring the mouth near the tentacles. The lobate ctenophores, such as *Mnemiopsis* and *Leucothea*, use both the tentacles and the mucus-covered oral surfaces of the lobes to capture prey, especially small crustaceans. The cylindrical *Beroe*, which lacks tentacles, feeds on other ctenophores. Contact of the large mouth with the prey causes an inward gulp, and the prey is swallowed. Digestion is both extracellular and intracellular, and indigestible wastes are passed out through the anal pores and mouth.

Ctenophores are noted for their luminescence, which is characteristic of many species. Light production takes place in the walls of the meridional canals, so externally the light appears to emanate from the comb rows (see Box 5–1).

The nervous system of ctenophores is a subepidermal nerve network that is particularly well developed beneath the comb rows. The only sense organ in an apical organ containing a statolith, which rests on four tufts of balancer cilia in a deep pit (Fig. 5–55B). When the animal is tilted, the pressure exerted by the statolith on the respective balancer cilia changes the rate of the beat of these cilia and then the comb rows. The change is transmitted by way of the ciliated grooves to the corresponding comb rows, and the animal turns.

All members of the phylum are hermaphroditic. The gonads are in the form of two bands located in the thickened wall of each meridional canal. One band is an ovary and the other a testis. The eggs and sperm are usually shed to the exterior through the mouth, and fertilization takes place in the sea water, except in the few species that brood their eggs.

Cleavage is total and determinate. The gastrula develops into a free-swimming, cydippid larva that closely resembles the adult of ctenophores with the more ovoid or spherical body structure described previously. The flattened species of ctenophores also possess a spherical cydippid larva that under-

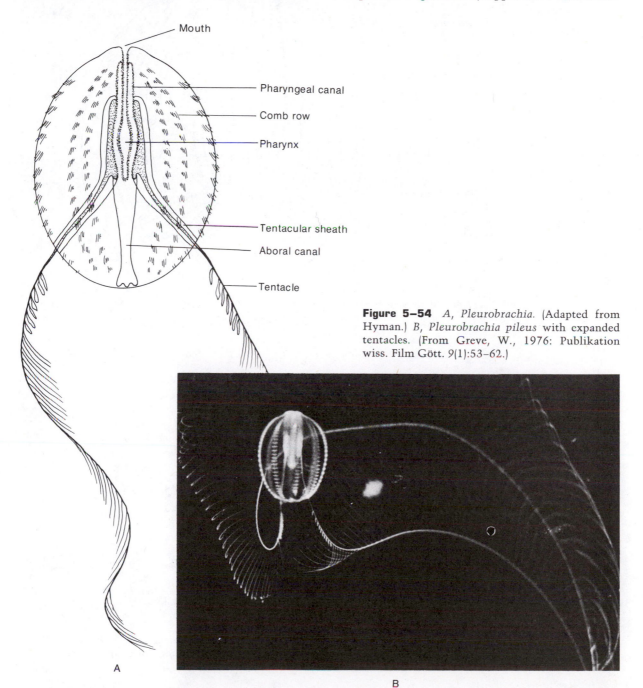

Mouth

Pharyngeal canal

Comb row

Pharynx

Tentacular sheath

Aboral canal

Tentacle

A

B

Figure 5–54 *A, Pleurobrachia.* (Adapted from Hyman.) *B, Pleurobrachia pileus* with expanded tentacles. (From Greve, W., 1976: Publikation wiss. Film Gött. *9*(1):53–62.)

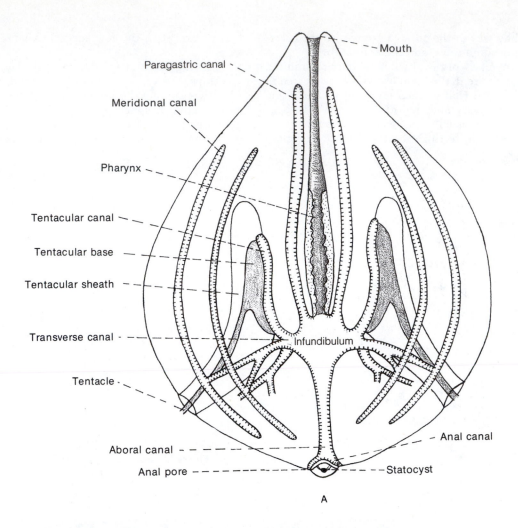

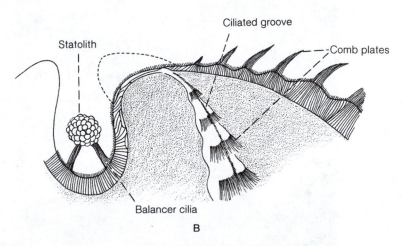

Figure 5–55 *A*, Digestive system of a cydippid ctenophore. (After Hyman, L. H., 1940: The Invertebrates, Vol. I. McGraw-Hill Book Co., N.Y.) *B*, Part of anterior end of a ctenophore showing statocyst, ciliated groove, and comb plates. (After Horridge.)

Figure 5–56 Colloblast. (From Franc, J., 1978: Organization and function of ctenophore colloblasts: an ultrastructural study. Biol. Bull., *155*:527–541.)

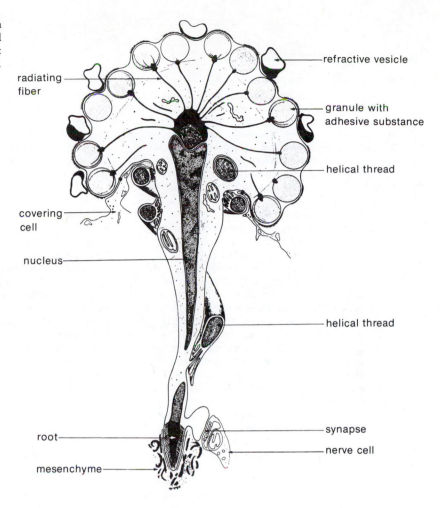

radiating fiber

refractive vesicle

granule with adhesive substance

helical thread

covering cell

nucleus

helical thread

synapse

root

nerve cell

mesenchyme

goes a more extensive transformation to attain the adult structure. This general existence of a spherical larva in ctenophores seems to substantiate the belief that the primitive shape was spherical or ovoid.

The spherical shape is restricted to some species (*Pleurobrachia*) of one order. Both *Mertensia* and *Mnemiopsis* are moderately flattened laterally, and in *Mnemiopsis* and *Leucothea* the middle of the body has become constructed along the tentacular plane, leaving the expanded outer portions in the form of large lobes (Fig. 5–57*A*). The resulting shape is somewhat similar to that of a clam. In the lobate ctenophores, tentacles are short, lack sheaths, and have moved to a position near the mouth.

Velamen (Fig. 5–57*B*) and *Cestum*, a genus known as Venus's girdle, have become so expanded and flattened that they look like transparent belts.

One species of *Cestum* reaches a length of over 1 meter. These animals swim not only by means of the comb rows but also by muscular undulations.

The Beroida *(Beroe)* are conical but somewhat flattened along the oral-aboral axis (Fig. 5–58).

SYSTEMATIC RÉSUMÉ OF PHYLUM CTENOPHORA

Class Tentaculata. Ctenophores with tentacles.

Order Cydippida. Body rounded or oval; tentacles branched and retractile into pouches. *Mertensia, Pleurobrachia*.

Order Lobata. Body moderately compressed, with two large oral lobes to either side of tentacular plane. Small tentacles not in pouches. *Mnemiopsis, Bolinopsis, Leucothea*.

BOX 5–1 BIOLUMINESCENCE

The terrestrial fireflies are perhaps the most familiar example of animals capable of producing light, but bioluminescence is far more widespread among marine organisms. On a still night light may appear to trail behind a swimmer's arms or slow-moving boat. In shallow coastal waters bacteria, dinoflagellates, jellyfish, ctenophores, crustaceans, brittle stars and fishes include the majority of luminescent species, although they are but a small part of the total shallow-water fauna. On the other hand, in the mesopelagic zone of the open ocean (200–1000 meters) 70 to 80 per cent of the jellyfishes, ctenophores, shrimps, squids and fishes are luminescent.

Bioluminescence involves the oxidation of a substrate substance, called luciferin, involving a molecule of oxygen and an enzyme, called luciferase. The luminescent product molecule is sufficiently excited to give off a photon. Luciferin and luciferase are different compounds in different luminescent groups, and the reaction pathways also vary. Given the variation in luminescent biochemistry and the taxonomic distribution of luminescence, the ability to produce light has clearly evolved independently many times.

The light source of luminescent animals may be symbiotic bacteria (in thaliaceans, some fishes, and squids), cellular secretions that are mixed extracellularly (in ostracods), or specialized cells called photocytes, within which light is produced. The photocytes may be distributed in various parts of the body (e.g., under the comb rows in ctenophores) or within complex organs called photophores, which are provided with various accessory structures, such as shutters, reflectors, and lenses (in shrimps, squids and fishes). Most fishes and some squids and shrimps have complex light organs, called glandular light organs, in which the light-producing system is provided by symbiotic bacteria or by glandular secretion.

The light produced by luminescent animals may be emitted as a flash or as a sustained glow, depending on the species and functions. Some species can flash and glow, and one species can have several kinds of photophores depending on functional requirements. The light may serve one or several functions, but the most common and widespread function appears to be predator avoidance. The predator may be startled and repelled by the light, which is usually a flash on contact. Or the light may save the animal from a predator by disguising its form, by creating a confusing substitute shape (luminescent cloud), or by making it invisible. Invisibility is achieved in oceanic animals such as squids and fishes by counterillumination. Prey above a deeper-swimming predator will stand out as a silhouette against light coming down from the surface. Glowing light organs located over the ventral surface can provide counterillumination to reduce or obliterate the silhouette.

Intraspecific communication is the second most common function of bioluminescence. Flashes or patterns of glowing light over the body act as signals to other indivduals and may serve to bring together members of the opposite sex.

The functional significance of light production has been well established for relatively few marine animals, and in gelatinous zooplankton it is especially poorly understood. Much is surmised. Excellent reviews of luminescence in marine organisms are presented by Morin (1983), Hastings (1983), and Young (1983).

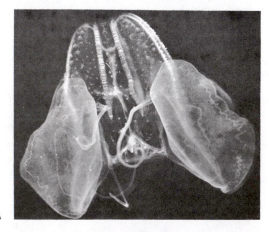

A

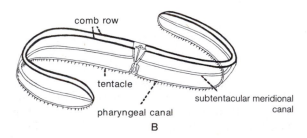

comb row

tentacle

pharyngeal canal

subtentacular meridional canal

B

Figure 5–57 *A, Leucothea multicornis,* an oceanic lobate ctenophore. (From Harbison, G. R., Madin, L. P., and Swanberg, N. R., 1978: On the natural history and distribution of oceanic ctenophores. Deep-Sea Research, Pergamon Press, 25:233–256.) *B, Velamen.* (After Mayer from Hyman.)

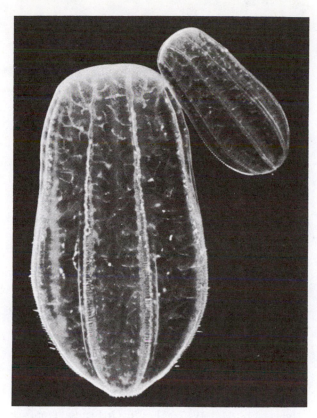

Figure 5–58 Lateral view of *Beroe cucumis*. (From Greve, W., 1976: Publikation wiss. Film Gött., 9(1):53–62.)

Order Cestida. Body ribbon shaped and greatly compressed along tentacular plane. The two principal tentacles and two of the four comb rows reduced. *Cestum, Velamen.*

Order Platyctenida. Body greatly flattened as a result of a reduction in the oral-aboral axis. Adapted for creeping. Comb rows reduced or absent in adult. *Ctenoplana, Coeloplana.*

Class Nuda. Ctenophores without tentacles.

Order Beroida. Body conical or cylindrical, somewhat flattened along the tentacular plane. Mouth muscular. *Beroe.*

SUMMARY

1 Members of the phylum Ctenophora, or comb jellies, are pelagic marine animals that are somewhat similar to medusae in their globose biradial form, their jelly-like mesoglea, and their transparency. However, the similarities to medusae may represent convergence and not evidence of a close evolutionary relationship to cnidarians.

2 Ctenophores are distinguished by eight meridional ciliary comb rows that propel the body in swimming, oral end forward.

3 Ctenophores are predatory on other pelagic invertebrates. Two branched retractile tentacles bear special adhesive structures (colloblasts) that are used in prey capture in many species. Some ctenophores have reduced tentacles and capture prey directly with the mouth.

4 Primitive ctenophores have spherical bodies, but many species have become flattened to varying degrees, compressed or depressed.

5 Ctenophores are hermaphroditic. In the development of flattened species, the body is initially spherical (cydippid larva).

REFERENCES

The literature included here is restricted in large part to the cnidarians and ctenophores. The introductory references on page 8 include many general works and field guides that contain sections on these animals.

Bak, R. P. M., and Steward-Van Es, Y., 1980: Regeneration of superficial damage in the scleractinian corals *Agaricia agaricites*, f. *purpurea* and *Porites asteroides*. Bull. Mar. Sci., 30(4):883–887.

Bayer, F. M., 1961: The shallow-water Octocorallia of the West Indian Region. Martinus Nijhoff, Hague.

Benos, D. J., and Prusch, R. D., 1972: Osmoregulation in fresh-water hydra. Comp. Biochem. Physiol. [A], 43A:165–171.

Benson, A. A., and Lee, R. F., 1975: The role of wax in oceanic food chains. Sci. Am., 232(3):76–86.

Bigger, C. H., 1982: The cellular basis of the aggressive acrorhagial response of sea anemones. J. Morph., 173(3):259–278.

Biggs, D. C., 1977: Field studies of fishing, feeding and digestion in siphonophores. Mar. Behav. Physiol., 4:261–274.

Boardman, R. S., Cheetham, A. H., and Oliver, W. A. (Eds.), 1973: Animal Colonies. Dowden, Hutchinson and Ross, Stroudsburg, Pa. 603 pp.

Burnett, A. L. (Ed.), 1973: Biology of Hydra. Academic Press, N.Y. 466 pp.

Cairns, S., 1976: Guide to the commoner shallow-water gorgonians of Florida, the Gulf of Mexico, and the Caribbean region. Sea Grant Field Guide Series #6, University of Miami.

Calder, D. R., 1975: Biotic census of Cape Cod Bay: Hydroids. Biol. Bull., *149*:287–315.

Calder, D. R., and Peters, E. C., 1975: Nematocysts of *Chiropsalmus quadrummanus* with comments on the systematic status of the Cubomedusae. Helgoländer wiss. Meeresunters., *27*(3):364–369.

Campbell, R. D., 1974a: Cnidaria. *In* Giese, A. C., and Pearse, J. S. (Eds.): Reproduction of Marine Invertebrates I. Academic Press, N.Y. pp. 133–200.

Campbell, R. D., 1974b: Development. *In* Muscatine, L., and Lenhoff, H. N. (Eds.): Coelenterate Biology. Academic Press, N.Y. pp. 179–210.

Carlgren, O., 1949: A survey of the Ptychodactiaria, Corallimorpharia and Actiniaria. Kungl. Svenska Vetensk. Handl., Ser. 4, Vol. 1, No. 1. (A monograph on the sea anemones of the world; includes a list of all known species and keys to genera.)

Chapman, D. M., 1966: Evolution of the Scyphistoma. (See Rees, The Cnidaria and Their Evolution.)

Cheng, L., 1975: Marine pleuston—animals at the sea-air interface. Oceanogr. Mar. Biol. Ann. Rev., *13*:181–212.

Chia, F. S., 1976: Sea anemone reproduction: patterns and adaptive radiation. *In* Mackie, G. O. (Ed.): Coelenterate Ecology and Behavior. Plenum Press, N.Y. pp. 261–270.

Christensen, H. E., 1967: Ecology of *Hydractinia echinata*. I. Feeding biology. Ophelia, *4*:245–275.

Clayton, W. S., Jr., and Lasker, H. R., 1982: Effects of light and dark treatment of feeding by the reef coral *Pocillopora damicornis*. J. Exp. Mar. Biol. Ecol. *63*(3):269–280.

Cook, C. B., 1983: Metabolic interchange in algae-invertebrate symbiosis. Internat. Rev. Cytology, Suppl. *14*:177–209.

Fadlallah, Y. H., 1983: Sexual reproduction, development and larval biology in scleractinian corals. A review. Coral Reefs, *2*(3):129–150.

Falkowski, P. G., Dubinsky, Z., Muscatine, L., and Porter, J. W., 1984: Light and the bioenergetics of a symbiotic coral. BioSci., *34*(11):705–709.

Franc, J.-M., 1978: Organization and function of ctenophore colloblasts: an ultrastructural study. Biol. Bull., *155*:527–541.

Francis, L., 1973: Intraspecific aggression and its effect on the distribution of *Anthopleura elegantissima* and some related sea anemones. Biol. Bull., *144*:73–92.

Franzisket, L., 1970: The atrophy of hermatypic reef corals maintained in darkness and their subsequent regeneration in light. Inter. Rev. Gesamten Hydrobiol., *55*(1):1–12.

Fraser, C., 1954: Hydroids of the Atlantic Coast of North America. University of Toronto Press, Toronto.

Gierer, A., 1974: Hydra as a model for the development of biological form. Sci. Am. *231*(6):44–54.

Gladfelter, E. H., Monohan, R. K., and Gladfelter, W. G., 1978: Growth rates of five species of reef-building corals in the northeastern Caribbean. Bull. Mar. Sci., *28*(4):728–734.

Gladfelter, W. G., 1973: A comparative analysis of the locomotory system of medusoid Cnidaria. Helgoländer wiss. Meeresunters., *25*:228–272.

Gosner, K. K., 1971: Guide to identification of marine and estuarine invertebrates: Cape Hatteras to the Bay of Fundy. Wiley-Interscience, N.Y. pp. 66–165.

Greve, W., 1975: Ctenophora. Fich. Ident. Zooplancton 146. 6 pp.

Greve, W., 1976: Die Rippenquallen der sudlichen Nordsee und ihre interspezifischen Relationen. (The comb jellies of the southern North Sea and their interspecific relations). Publ. wiss. Film Gott. (1976) Bd., *9*(1):53–62.

Hand, C., 1959: On the origin and phylogeny of the coelenterates. Syst. Zool., *8*(4):191–202.

Hand, C., 1966: On the evolution of the Actiniaria. (See Rees, The Cnidaria and Their Evolution.)

Harbison, G. R., 1985: On the classification and evaluation of the Ctenophora. *In* Conway Morris, S., George, J. D., Gibson, R., and Platt, H. M. (Eds.): The Origins and Relationships of Lower Invertebrates. Systematics Association Spec. Vol. No. 28. Clarendon Press, Oxford. 394 pp.

Harbison, G. R., Madin, L. P., and Swanberg, N. R., 1978: On the natural history and distribution of oceanic ctenophores. Deep-Sea Research, *25*:233–256.

Hastings, J. W., 1983: Chemistry and control of luminescence in marine organisms. Bull. Mar. Sci., *33*(4):818–828.

Hayes, R. L., and Goreau, N. I., 1977: Intracellular crystal-bearing vesicles in the epidermis of scleractinian corals, *Astrangia danae* (Agassiz) and *Porites porites* (Pallas). Biol. Bull., *152*:26–40.

Holstein, T., and Tardent, P., 1984: An ultrahigh-speed analysis of exocytosis: nematocyst discharge. Science, *233*:830–833.

Horridge, G. A., 1974: Recent studies on the Ctenophora. *In* Muscatine, L., and Lenhoff, H. M. (Eds.): Coelenterate Biology. Academic Press, N.Y., pp. 439–458.

Hyman, L. H., 1940: The Invertebrates: Protozoa Through Ctenophora. Vol. 1. McGraw-Hill, N.Y. pp. 365–696. (An old but still useful account.)

Isdale, P., 1977: Variation in growth rate of hermatypic corals in a uniform environment. Proceedings 3rd International Coral Reef Symposium, University of Miami, *2*:403–408.

Jokiel, P. L., and York, R. H., Jr., 1982: Solar ultraviolet photobiology of the reef coral *Pocillopora damicornis* and symbiotic zooxanthellae. Bull. Mar. Sci., *32*(1):301–315.

Kastendiek, J., 1976: Behavior of the sea pansy *Renilla kollikeri* Pfeffer and its influence on the distribution and biological interactions of the species. Biol. Bull., *151*:518–537.

Knutson, D. W., Buddemeier, R. W., and Smith, S. V., 1972: Coral chronometers: seasonal growth bands in reef corals. Science, *177*:270–272.

Kramp, P. L., 1961: Synopsis of the medusae of the world. Mar. Biol. Assoc. U.K., *40*:1–469.

Lang, J., 1973: Interspecific aggression by scleractinian corals: 2. Why the race is not only to the swift. Bull. Mar. Sci., *23*(2):260–279.

Larson, R. J., 1976: Cnidaria: Scyphozoa. Marine Flora and Fauna of the Northeastern United States. NOAA Tech. Rep. NMFS circular 397. U.S. Gov. Printing Office. 18 pp.

Larson, R. J., 1976: Cubomedusae: feeding-functional morphology, behavior and phylogenetic position. *In* Mackie, G. O. (Ed.): Coelenterate Ecology and Behavior. Plenum Press, N.Y. pp. 237–245.

Lasker, H. R., 1981: A comparison of particulate feeding abilities of three species of gorgonian soft corals. Marine Ecol. Prog. Ser., 5:61–67.

Lenhoff, H. M., and Loomis, W. F. (Eds.), 1961: The Biology of Hydra and Some Other Coelenterates. University of Miami Press, Coral Gables, Fla.

Lenhoff, H. M., and Muscatine, L. (Eds.), 1971: Experimental Coelenterate Biology. University Press of Hawaii, Honolulu. 288 pp.

Lewis, D. H., and Smith, D. C., 1971: The autotrophic nutrition of symbiotic marine coelenterates with special reference to hermatypic corals. Proc. R. Soc. Lond. [Biol.], 178:111–129.

Lewis, J. B., and Price, W. S., 1975: Feeding mechanisms and feeding strategies of Atlantic reef corals. J. Zool. [London], 176:527–544.

Lewis, J. B., and Price, W. S., 1976: Patterns of ciliary currents in Atlantic reef corals and their functional significance. J. Zool. [London], 178:77–89.

Lubbock, R., 1980: Why are clown fishes (*Amphiprion clarkii*) not stung by sea anemones (*Stichodactyla haddoni*)? Proc. Roy. Soc. Lond. B. Biol. Sci., 207(1166):35–62.

Mackie, G. O. (Ed.), 1976: Coelenterate Ecology and Behavior. Plenum Press, N.Y. (Papers presented at the 3rd International Symposium on Coelenterate Biology at the University of Victoria, B.C.)

Manuel, R. L., 1981: British Anthozoa. Synopses of the British Fauna No. 18. Academic Press, London. 250 pp.

Mariscal, R. N., 1970: Nature of symbiosis between Indo-Pacific anemone fishes and sea anemones. Mar. Biol., 6:58.

Mariscal, R. N., 1974: Nematocysts. *In* Muscatine, L., and Lenhoff, H. M. (Eds.): Coelenterate Biology. Academic Press, N.Y. pp. 129–178.

Mariscal, R. N., Conklin, E. J., and Bigger, C. H., 1977: The ptychocyst, a major new category of cnida used in tube construction by a cerianthid anemone. Biol. Bull., 152:392–405.

Mariscal, R. N., McLean, R. B., and Hand, C., 1977: The form and function of cnidarian spirocysts. 3. Ultrastructure of the thread and the function of spirocysts. Cell Tiss. Res., 178:427–433.

Mayer, A. G., 1910 (Republished 1977): The Medusae of the World. I-II. Hydromedusae. III. Scyphomedusae. A. Asher, Amsterdam.

McNeil, P. L., 1981: Mechanisms of nutritive endocytosis. I. Phagocytic versatility and cellular recognition in *Chlorohydra* digestive cells, a scanning electron microscope study. Jour. Cell Science, 49:311–339.

Morin, J. G., 1983: Coastal bioluminescence: patterns and functions. Bull. Mar. Sci., 33(4):787–817.

Muscatine, L., 1974: Endosymbiosis of cnidarians and algae. *In* Muscatine, L., and Lenhoff, H. M. (Eds.): Coelenterate Biology. Academic Press, N.Y. pp. 359–396.

Muscatine, L., and Lenhoff, H. M. (Eds.), 1974: Coelenterate Biology. Academic Press, N.Y. 501 pp.

Muscatine, L., Pool, R. R., and Trench, R. K., 1975: Symbiosis of algae and invertebrates: aspects of the symbiont surface and the host-symbiont interface. Trans. Am. Micros. Soc., 94(4):450–469.

Newell, N. D., 1972: The evolution of reefs. Sci. Am., 226(6):54–65.

Ottaway, J. R., 1980: Population ecology of the intertidal anemone, *Actinia tenebrosa*: 4. Growth rates and longevities. Aust. J. Mar. Freshw. Res., 31(3):385–396.

Pardy, R. L., and White, B. N., 1977: Metabolic relationships between green hydra and its symbiotic algae. Biol. Bull., 153:228–236.

Porter, J. W., 1976: Autotrophy, heterotrophy, and resource partitioning in Caribbean reef-building corals. Am. Nat., 110:731–742.

Purcell, J. E., 1977: Aggressive function and induced development of catch tentacles in the sea anemone *Metridium sinile*. Biol. Bull., 153:355–368.

Purcell, J. E., 1981: Feeding ecology of *Rhizophysa eysenhardti*, a siphonophore predator of fish larvae. Limnol. Oceanogr., 26(3):421–432.

Purcell, J. E., 1984: The functions of nematocysts in prey capture by epipelagic siphonophores. Biol. Bull., 166:310–327.

Rees, W. J. (Ed.), 1966: The Cnidaria and Their Evolution. Academic Press, N.Y.

Rees, W. J., 1966: The evolution of the Hydrozoa. (See Rees, The Cnidaria and Their Evolution.)

Reeve, M. R., and Walter, M. A., 1978: Nutritional ecology of ctenophores—a review of recent research. Adv. Mar. Biol., 15:249–287.

Richardson, C. A., Dustan, P., and Lang, J. C., 1979: Maintenance of living space by sweeper tentacles of *Montastrea cavernosa*, a Caribbean reef coral. Mar. Biol., 55(3):181–186.

Rinkevich, B., and Loya, Y., 1984: Does light enhance calcification in hermatypic corals? Mar. Biol., 80:1–6.

Rose, S. M., 1970: Regeneration. Appleton-Century-Crofts, N.Y. 264 pp.

Ross, D. M., 1974: Behavior patterns in associations and interactions with other animals. *In* Muscatine, L., and Lenhoff, H. M. (Eds.): Coelenterate Biology. Academic Press, N.Y. pp. 281–312.

Russell, F. S., 1954–1970: Medusae of the British Isles. Vol. I (1954), Vol. II (1970). Cambridge University Press, London.

Satterlie, R. A., and Spencer, A. D., 1983: Neuronal control of locomotion in hydrozoan medusae: a comparative study. Jour. Comp. Physiol. A Sens Neural Behav. Physiol., 150(2):195–206.

Sebens, K. P., 1976: The ecology of Caribbean sea anemones in Panama: utilization of space on a coral reef. *In* Mackie, G. O. (Ed.): Coelenterate Ecology and Behavior. Plenum Press, N.Y. pp. 67–77.

Shih, C. T., 1977: A guide to the Jellyfish of Canadian Atlantic Waters. Nat. Mus. Canada. University of Chicago Press, Chicago. 90 pp.

Siewing, R., 1977: Mesoderm in ctenophores. Z. Zool. Syst. Evolutions forsch., 15(1):1–8.

Singla, C. L., 1975: Statocysts of Hydromedusae. Cell Tiss. Res., 158:391–407.

Smith, F. G. W., 1971: Atlantic Reef Corals. University of Miami Press, Coral Gables, Florida. 164 pp.

Steele, R. D., and Goreau, N. I., 1977: The breakdown of symbiotic zooxanthellae in the sea anemone *Phyllactis flosculifera*. J. Zool. [London], 181:421–437.

Swedmark, B., and Teissier, G., 1966: The Actinulida and their evolutionary significance. (See Rees, The Cnidaria and Their Evolution.)

Taylor, D. L., 1973: The cellular interactions of algal-invertebrate symbiosis. Adv. Mar. Biol., 11:1–56.

Thiel, H., 1966: The evolution of the Scyphozoa. (See Rees, The Cnidaria and Their Evolution.)

Tokioka, T., and Nishimura, S., 1973: The Proceedings of the Second International Symposium on Cnidaria. Seto Mar. Biol. Lab., Japan. (53 papers presented at a 1972 symposium held in Japan.)

Vaughn, T. W., and Wells, J. W., 1943: Revision of the suborders, families, and genera of the Scleractinia. Special Paper, Geological Society of America, 44:1–363.

Velimirov, B., 1975: Wachstum und Altersbestimmung der Gorgoni *Eunicella carolinii.* Oecologia, 19:259–272.

Veron, J. E. N., Pichon, M., and Wijsman-Best, M., 1976–1984: Scleractinia of Eastern Australia. Pts. I–IV. Australian Inst. Mar. Sci., Monog. Ser. (Almost complete, this superb treatise on the scleractinian corals of the Australian Great Barrier Reef will be of great value in the study of corals of other parts of the tropical Pacific.)

Wainwright, S. A., and Koehl, M. A. R., 1976: The nature of flow and the reaction of benthic Cnidaria to it. *In* Mackie, G. O. (Ed.): Coelenterate Ecology and Behavior. Plenum Press, N.Y. pp. 5–21.

Wellington, G. M., and Glynn, P. W., 1983: Environmental influences on skeleton banding in eastern Pacific corals. Coral Reefs, 1:215–222.

Werner, B., 1971: Neue Beitrage zur Evolution der Scyphozoa und Cnidaria. *In* First International Symposium on Zoophylogeny, Salamanca, 1969. Acta Salmant. I. Ciencas, 36:223–244.

Werner, B., 1973: New investigations on systematics and evolution of the class Scyphozoa and the phylum Cnidaria. Proc. 2nd. Internat. Symp. on Cnidaria, Publ. Seto Mar. Biol. Lab., 20:35–61.

Werner, B., 1975: Structure and life history of the polyp of *Tripedalia cystophora* (Cubozoa, class. nov., Car-

ybdeidae) and its importance for the evolution of the Cnidaria. Helgoländer wiss. Meeresunters., 27(4):461–504.

West, D. L., 1978: The epitheliomuscular cell of *Hydra:* its fine structure, three-dimensional architecture and relation to morphogenesis. Tissue Cell, 10(4):629–646.

Westfall, J. A., 1970: The nematocyte complex in a hydromedusan, *Gonionemus vertens.* Z. Zellforsch., 110:457–470.

Westfall, J. A., 1973: Ultrastructural evidence for neuromuscular systems in coelenterates. Am. Zool., 13:237–246.

Westfall, J. A., Yamataka, S., and Enos, P. D., 1971: Scanning and transmission microscopy of nematocyst batteries in epitheliomuscular cells of Hydra. 29th Annual Proceedings of the Electron Microscopy Society of America. Boston, edited by C. J. Arceneaux.

Williams, R. B., 1978: Some recent observations on the acrorhagi of sea anemones. Jour. Mar. Biol. Assoc. U.K., 58(3):787–788.

Wineberg, S., and Wineberg, F., 1979: The life cycle of a gorgonian: *Eunicella singularis.* Bijdragen tot de Dierjunde, 48(2):127–140.

Wood, E. M., 1983: Corals of the World. T. F. H. Publications, Neptune City, N.J. 256 pp.

Wyttenbach, C. R. (Ed.), 1974: The developmental biology of the Cnidaria. Am. Zool., 14(2):540–866. (Papers presented at a 1972 symposium held in Washington, D.C.)

Yonge, C. M., 1963: The biology of coral reefs. Adv. Mar. Biol., 1:209–261.

Yonge, C. M., 1968: Living corals. Proc. R. Soc. London [Biol.], 169:329–344.

Young, R. E., 1983: Oceanic bioluminescence: an overview of general functions. Bull. Mar. Sci., 33(4):829–847.

6

Coral Reefs

Coral reefs are tropical, shallow-water, calcareous structures supporting a diverse association of marine plants and animals. A unique characteristic of coral reefs is that they are formed by certain of the plants and animals that inhabit them.

Distribution of Reefs

Of all of the calcium carbonate–secreting organisms that contribute to modern reef formation, the scleractinian corals are the most important. Not only are they a major source of calcium carbonate, but the environmental demands of these animals also describe the limits of reef distribution.

Coral reefs occur in shallow water, ranging to depths of 60 meters. Reef-building, or hermatypic, corals contain gastrodermal symbiotic algae (zooxanthellae), which require light for photosynthesis. Thus, the vertical distribution of living reef corals is restricted to the depth of light penetration. That relationship is dramatically reflected in the graph in Figure 6–1, in which the 150 species of corals on Bikini Atoll are plotted against depth. The number

of species declines very rapidly in deeper water, the curve closely following that for light extinction.

Many reef-inhabiting octocorallians, zoanthids, sea anemones, and hydrozoan corals also contain zooxanthellae and are similarly restricted in their vertical distribution.

Because of their dependence on light, reef corals require clear water. Thus, coral reefs are found only where the surrounding water contains relatively small amounts of suspended material, i.e., water of low turbidity and low productivity.

Coral reefs are further restricted by water temperature and occur only in tropical and semitropical seas, where the average minimum water temperature is not less than 20°C. Existing reefs are restricted to the Caribbean, the Indian Ocean, and the tropical Pacific (Fig. 6–2). The diversity of reef corals, i.e., the number of species, decreases in higher latitudes to about 30° north and south, beyond which reef corals are usually not found. Bermuda, at latitude 32° north and lying in the path of northward-moving warm water, is an exception to this rule. Staghorn and elkhorn corals (*Acropora*), which are so conspicuous on West Indian

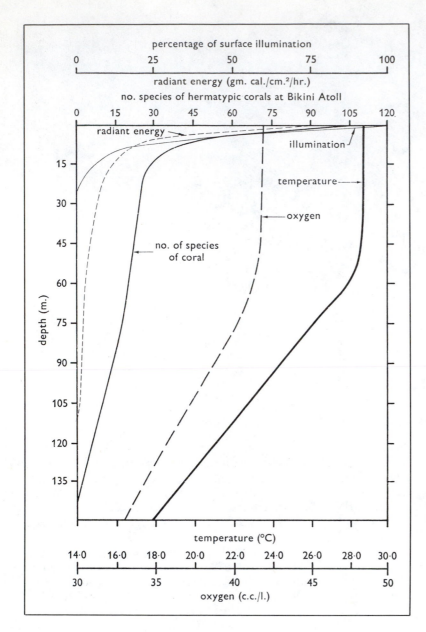

Figure 6–1 Relationship between the vertical distribution of hermatypic corals at Bikini Atoll and environmental factors. (After Wells and Motoda from Stoddard, D. R., 1969: Biol. Rev. *44*:433–498; copyrighted and reprinted by permission of Cambridge University Press.)

reefs, are absent in Bermuda, however, because of lower water temperatures in winter. The greatest concentration occurs near the equator, where water temperatures are highest.

Coral reefs are conspicuously absent from much of the Atlantic. Water turbidity is a limiting factor on the east coast of South America, where the Amazon and other rivers disgorge great quantities of sediment. Rivers have a similar inhibiting effect on the west coast of Africa, where there is also a more restricted range of warm water because of cold currents moving up the coast from the south.

Thus, the two important physical parameters of coral reef ecosystems are warm water (20° to 28°C) and light (clear water).

Reef Structure

Three general types of reefs can be recognized on the basis of their structure and the underlying substratum.

Fringing reefs project seaward directly from the shore (Figs. 6–3*A* and *B* and 6–4*A*). They surround islands as well as border continental land masses

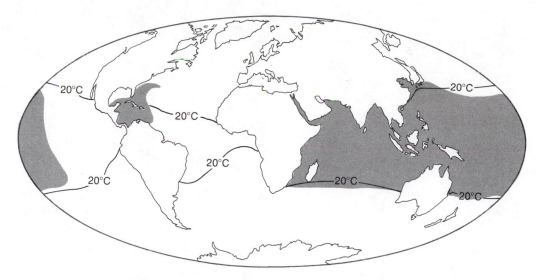

Figure 6–2 Distribution of coral reefs today (heavy shading).

and are the most common type of reef. *Barrier reef* platforms are separated from the adjacent land mass by a lagoon (Fig. 6–3*C* and *D*). The Great Barrier Reef of Australia is the longest in the world, stretching over 1000 miles along Australia's northeast coast. Another long barrier reef is located in the Caribbean off the coast of Belize between Mexico and Guatemala.

Atolls, the third type of reef, rest on the summits of submerged volcanos (Fig. 6–3*E* and *F*). They are usually circular or oval with a central lagoon, and parts of the reef platform may emerge as one or more islands. Breaks in the reef provide passes into the central lagoon. Over 300 atolls are present in the Indo-Pacific but only 10 are found in the western Atlantic. Bermuda is one of the latter, and the islands of Bermuda are located on the south side of a 16- by 32-mile atoll ring. *Table reefs* are small, open-ocean reefs that have no central islands or lagoons.

The term *patch reef* refers to small circular or irregular reefs that rise from the floor of lagoons behind barrier reefs or within atolls. They are rather typical lagoon features and often are very numerous (Fig. 6–3*E*).

All three reef types—fringing, barrier, and atoll—show certain similarities in profile (Fig. 6–5*A*). The seaward side of the reef, the reef front or fore-reef slope, rises from lower depths to a level just below or just at the water surface. The inclination of the reef front varies from gentle to steep and very often is interrupted by a terrace. In some reefs, the lower part of the front is a vertical wall, referred to as the drop-off.

The reef front does not present a flat, wall-like barrier to incoming wave energy. Instead, there are finger-like, seaward projections of the reef that alternate with deeper, sand-filled pockets from the sea (Figs. 6–5*B* and 6–6). This so-called spur-and-groove formation aids in dispersing wave energy.

Behind the reef crest is a reef flat, or back-reef, which is highly variable in character. The flat may be very short or it may extend back several 100 meters, and it may be dissected by channels several meters deep. Further back the flat usually becomes shallower and may even include intertidal areas. The flat may be paved with rock or sand or strewn with coralline boulders (Fig. 6–4*B*). The sand often supports beds of sea grasses. Extensive reef flats are rarely uniform in character and exhibit zones (Fig. 6–5*B*). The reef flat ends at the shore on fringing reefs and descends into the lagoon of atolls and barrier reefs.

The profile just described clearly indicates differing environments that support different species of corals and other animals (Fig. 6–7). On the reef front coral populations extend to depths between 10 and 60 meters. On the intermediate slopes there is a rich zone of massive dome and columnar corals. Below this coral head zone, platelike species predominate. Higher up, where wave stress is greater, still other species are found. In the Caribbean this is commonly the zone of *Acropora palmata,* or elkhorn coral (Fig. 6–7*A*), whose heavy, spreading branches may form a seaward-projecting thicket. Wave stress is an important factor determining the species of corals or other organisms that occupy the reef crest (Geister, 1977).

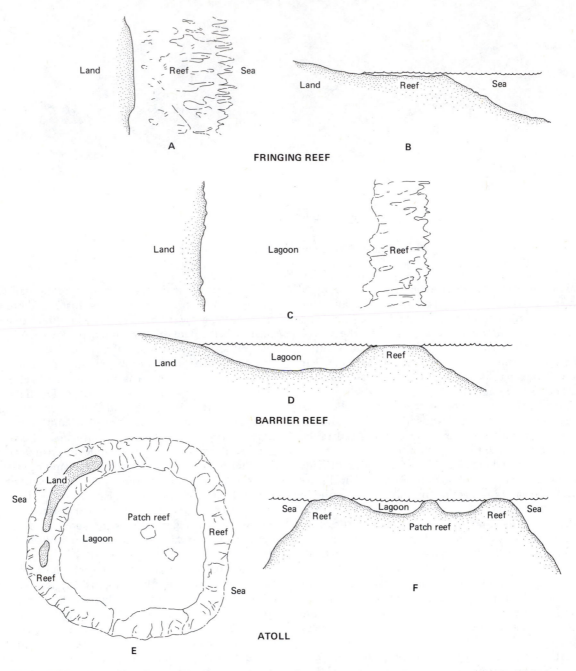

Figure 6–3 Types of reefs: *A* and *B*, Fringing reef, surface view and in section. *C* and *D*, Barrier reef. *E* and *F*, Atoll.

A

Figure 6–4 *A*, Fringing reef around a small island in the Celebes. *B*, Reef flat on Glovers Reef, Belize.

B

Behind the protected reef front, pools, channels, and other areas of deeper, less turbulent water support more delicate corals. In the Caribbean, such a species can be *Acropora cervicornis*, or staghorn coral, although it is not limited to this zone, and in the Indo-Pacific there are many species of small, fragile, branching corals (Fig. 5–43). Still other species are characteristic of quiet, shallow water farther from the reef front. Some corals occupy several zones and gradually appear and disappear along the reef profile (Fig. 6–8).

The diversity of scleractinian corals on Caribbean reefs is much less than that of corals on reefs of the Indo-Pacific. Six genera and ten species of scleractinians plus the hydrozoan *Millepora* (stinging coral) make up most of the coral biomass of Caribbean reefs. These species are found throughout most of the Caribbean, but on a particular reef only five or six are especially conspicuous.

The Indo-Pacific contains a very rich and diverse scleractinian fauna. In contrast to 20 known Caribbean genera, there are 80 in the Indo-Pacific. There are only three species of *Acropora* in the Ca-

ribbean but 150 in the Indo-Pacific and over 200 species in the Great Barrier Reef of Australia. The areas being compared are not equivalent, however. The Indo-Pacific is a much vaster geographical region than the Caribbean, and the Great Barrier Reef stretches over 1000 miles.

The reef platform supports a great array of other animals and plants. Algae are an important component and are found in every zone. The rapidly growing calcareous reds and greens, such as *Halimeda*, are a major contributor to calcareous reef sands. The coralline red algae, some of which lay down very resistant sheets of calcium carbonate on exposed surfaces of old coral skeletons, contribute to the processes of calcium carbonate deposition and cementation. On parts of many reefs, especially on the crests, coralline red algae are at least as important as scleractinian corals in reef formation. In the Indo-Pacific calcareous algae generally dominate the crest of the reef front (Fig. 6–5*B*).

Large sponges are found on the lower slopes of the reef front, and other sponge species live behind the crest. Gorgonian corals—the branching sea

(Text continued on p. 156)

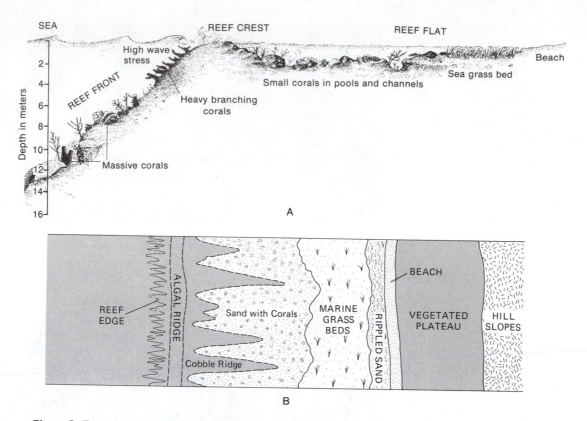

A

B

Figure 6–5 *A*, Generalized profile of a fringing reef. *B*, Profile of a reef on the Seychelles in the Indian Ocean. (*B* after Lewis and Taylor from Stoddard, D. R., 1973: Coral reefs of the Indian Ocean. *In* Jones, O. A. and Endean, R. (Eds.): Biology and Geology of Coral Reefs, *1* (Geology 1):51–92. Academic Press, N.Y.)

Figure 6–6 Reef front of a Barbados fringing reef looking landward and showing spur and groove formation. (From Stearn, C. W., Scoffin, T.P., and Martindale, W., 1977: Calcium carbonate budget of a fringing reef on the west coast of Barbados. Bull. Mar. Sci., 27(3):479–510.)

Figure 6–7 *A, Acropora palmata*, elkhorn coral, on a reef off the coast of Belize in the Caribbean. One "horn" may cover an arc of over a meter. (Courtesy of K. Ruetzler and I. G. Macintyre.) *B*, Coral diversity and cover on Great Barrier Reef off the coast of northeast Australia. Coral is growing in shallow, protected water of reef flat. (Courtesy of the Australian News and Information Bureau.)

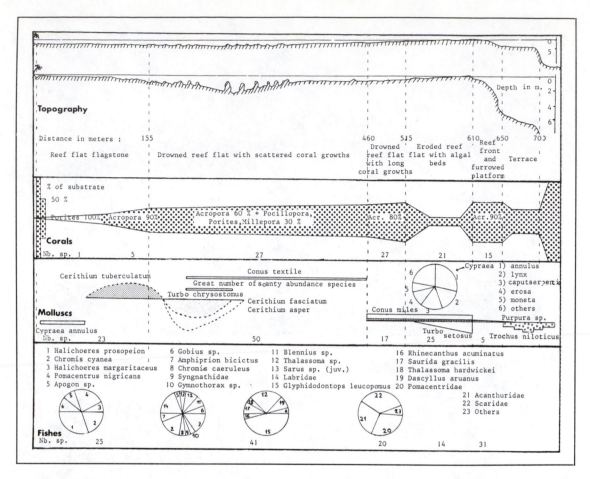

Figure 6–8 Zonation of corals, gastropod mollusks, and fish across a fringing reef of Lakeba Island, Fiji. Reef profile is shown at top. (From Salvat, B., Ricard, M., Richard, G., Galzin, R., and Toffart, J. L., 1977: Reef lagoon complex of Lakeba Island (Lau Group, Fiji): geomorphology, biotic associations, and socio-ecology. Proceedings 3rd International Coral Reef Symposium, University of Miami. *2* (Geology):297–303.)

rods and sea fans—are a very conspicuous component of Caribbean reefs, particularly on the shallow fore-reef; in the Indo-Pacific gorgonians are usually less conspicuous, but large massive or encrusting alcyonacean, or soft, corals are often abundant.

Besides sponges and cnidarians many other attached animals live on reefs, including snails, clams, tunicates, and bryozoans. There are also certain sponges, clams, and worms that bore into exposed coral surfaces. Reef topography is highly irregular (much more so than the diagram in Fig. 6–5A suggests) and contains innumerable holes and passages of all sizes, in which dwell a great cryptic fauna of shrimp, crabs, snails, worms, fish, and other animals.

Over 270 cryptic animal and algal species have been reported from a Madagascan reef (Vasseur,

1977). Sponges and tunicates were the most abundant.

Reef Ecosystem

Much of the blue wavelength of light passing through the photic zone of the sea is scattered by water molecules. Suspended material in productive seas—those with large amounts of plankton and detritus—tends to reflect yellow wavelengths. This accounts for the typically green color of productive seas. Light transmission in productive seas is reduced, and light extinction occurs rapidly below the surface.

Clear water—water of low productivity—is blue. Indeed, blue has been called the color of ocean deserts. It is such impoverished water that bathes coral reefs, yet paradoxically coral reefs are among the most productive marine environments. Reef organisms are largely benthic and include not only many attached algae but also an enormous population of symbiotic zooxanthellae. These photosynthetic organisms are at the base of the food chain and constitute such an important energy source that reefs are generally considered to be autotrophic; i.e., resident primary production is the source of most energy flow through reef food webs.

Not all reef energy has a resident origin, however. The plankton in the surrounding water, although scarce compared with most nontropical seas, is still an important food source for the many suspension-feeding animals, including corals. Some, or most, of this plankton develops off the reef and is washed to the platform by currents.

To varying degrees the reef platform constitutes an energy and nutrient trap. Rather than being lost to deep-water sediments, some organic compounds and nutrients (e.g., nitrogen and phosphorus compounds) are retained on the platform and recycled (see Lewis, 1977).

Reef Formation

The building of the reef platform is not simply a matter of secretion of new calcium carbonate on top of old. The building involves constructive and destructive phases. Scleractinian corals and the larger skeletons of some other organisms form the framework material, or the "bricks," of the reef platform; finer skeletal material forms the "mortar" (Figs. 6–9 and 6–10).

The destructive phase may begin long before the death of a living coral colony. Any exposed surface of the coral is quickly attacked by boring organisms, particularly boring sponges and bivalves (Fig. 6–9). Living corals are attacked from the underside, which sometimes leads to the toppling of a large dome. When a living colony dies, regardless of the cause of death, the underlying skeleton also becomes riddled with boring organisms. Eventually, a large coral is fragmented or reduced to chunks. Deeper surfaces exposed by borers may be colonized by encrusting bryozoans and coralline algae, both depositing additional calcium carbonate.

The duration of the destructive phase of reef formation is determined by the deposition of fine, excavated material produced by the borers. This material accumulates; eventually the borers are

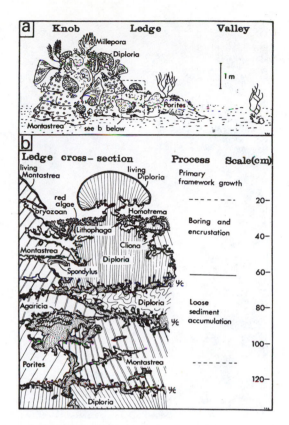

Figure 6–9 Section (b) through a Bermuda patch reef (a) showing phases of reef formation. (From Scoffin, T. P., 1972: Fossilization of Bermuda patch reefs. Science, 178:1280–1282. Copyright 1972 by the American Association for the Advancement of Science.)

smothered in their own debris and boring stops. The more rapidly the boring is halted, the larger will be the chunks of coral left intact. The debris produced by boring, as well as fine skeletons and shells of gorgonian corals, calcareous algae, sponges, mollusks, and other organisms, settles into crevices and holes, slowly filling the spaces between the larger framework pieces with a calcareous mortar. The fusion, or lithification, of this material actually occurs within 10 to 15 cm of the surface. In agitated water, magnesium calcite particles $(4–30\mu)$ precipitate and act as a cement (Macintyre, 1977). A core taken from a living reef will reveal all phases of platform formation (Figs. 6–9, 6–10 and 6–11).

Reefs may grow seaward, but vertical growth is limited by light and water depth. Yet core sampling on most modern reefs reveals a platform thickness of 6 to more than 1000 meters. Moreover, the lower parts of many platforms are located *below*

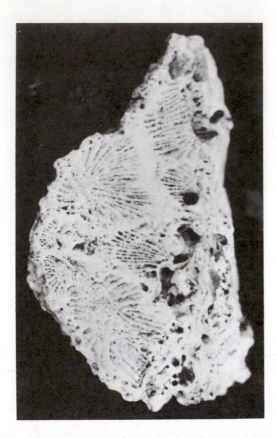

Figure 6–10 Section of a piece of reef rock composed of chunks of scleractinian coral cemented with bits of the skeletal remains of other organisms. Worm shells are conspicuous. Rock was taken from the lowest level shown in Figure 6–9.

the photic zone (Fig. 6–14). How can this be explained?

Extensive vertical growth of reef platforms is a result of changes in sea level or subsidence of the substratum. Virtually all modern reefs reflect some growth associated with the most recent sea level rise. During the last Pleistocene glacial period, when enormous quantities of sea water were locked in ice over much of the present north temperate regions, the sea level dropped 120 meters below the present level (Fig. 6–12). Coastal areas now covered by water were exposed, islands were larger or connected to mainlands, and submerged platforms, such as the Great Bahamas Bank, were out of water. Fringing and other reef types developed in favorable areas, as is true today.

Glacial melting and sea level rise toward modern levels, called the Holocene transgression, began

about 18,000 years before the present (B.P.), during the Holocene (Recent) epoch. From 18,000 to 7000 B.P. sea level rise was rapid, averaging about 1 cm per year, or 10 meters per 1000 years. Sea level rise slowed abruptly about 7000 B.P., and from that date to the present the rate of rise has been 0.1 to 0.2 cm per year, or 1 to 2 meters per 1000 years.

Some modern reefs followed this rise throughout much of the Holocene transgression, but most made their appearance later. Upward growth rates of modern reefs vary from 3 to 15 meters per 1000 years, with some of the most rapid rates recorded for sites in the Caribbean. One of the thickest recorded Holocene reefs is Alacran Reef off Yucatan, Mexico, which has been cored to 33.5 meters.

The rate of upward reef growth is affected by other factors besides sea level rise. Reefs or zones subjected to high wave energy grow more slowly and more compactly. They are typically dominated by coralline algae and, in the Caribbean, by the hydrocoral *Millepora* as well. Reefs subjected to moderate wave energy grow more rapidly. Reefs dominated by agroporids, e.g., *Acropora palmata* in the Caribbean, develop thick, porous accumulations. Such growth rates demand clear water and occur at depths of less than 4 to 6 meters.

Carbon 14 dating of core samples taken from Hanauma Reef, a fringing reef on Oahu, Hawaii, illustrates the generalizations that have just been made. Hanauma Reef began growing about 7000 B.P., with most vertical growth from 5800 to 3500 B.P. (Easton and Olson, 1976) (Fig. 6–13). During this period vertical growth was 1 meter per 300 years, and the seaward advance was 1 meter per 45 years. Vertical growth has been about 1 meter per 2900 years since 3500 B.P., reflecting the even slower rise in sea level since that time.

Reef platforms of great thickness are the result of substrate subsidence. Such movement has been especially important in the formation of atolls and certain barrier reefs, such as the Great Barrier Reef of Australia. Atolls are typically located on top of volcanic seamounts. Slow subsidence of the seamount has been matched by compensating upward growth of the reef platform.

Core drillings taken on Eniwetok Atoll provide evidence for reconstructing the geological history of the reef. The drillings go through 1283 meters (almost a mile) of coral rock before striking a basaltic base. This base marks the top of a volcanic cone that rises 3.2 km (almost 2 miles) from the ocean bottom (Fig. 6–14). Analysis of the core samples reveals strata containing recrystallized sediments and fossils of land plants. The oldest and

A   B

Figure 6–11 *A*, Coring of the reef structure on the St. Croix shelf edge. Diver is holding the drilling rig. The bit is ring shaped and cuts out the cores shown in *B*. Note the growth of gorgonians on the sea floor. (From Adey, W. H., Macintyre, I. G., Stuckenrath, R., and Dill, R. F., 1977: Relict barrier reef system off St. Croix: its implications with respect to late Cenozoic coral reef development in the western Atlantic. Proceedings 3rd International Coral Reef Symposium, University of Miami. Vol. II, p. 18.) *B*, Sections of cores taken from a geological well drilled at the Bermuda Biological Station. The white core is part of the coralline cap of the Bermuda atoll. The black core was taken from the underlying basaltic seamount on which the coralline cap rests.

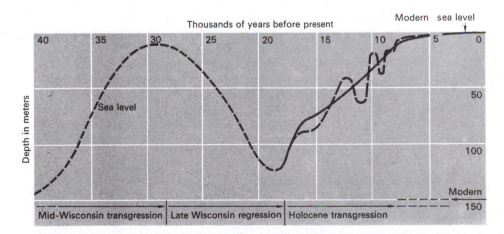

Figure 6–12 Sea level changes during the last Pleistocene interglacial and glacial periods. Transgression refers to encroachment by the sea as a result of glacial melting and sea level rise; regression refers to the reverse conditions. (From Curray, J. H., 1965: Late Quaternary history, continental shelves of the United States. *In* The Quaternary of the United States. Princeton University Press, p. 725.)

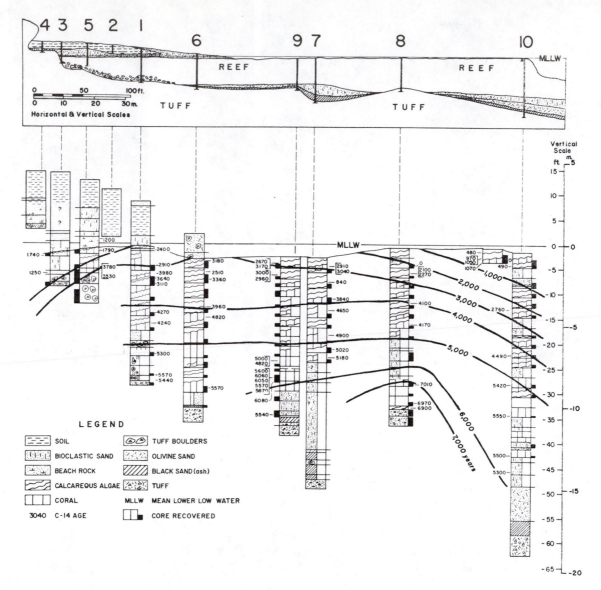

Figure 6–13 Cross section of Hanauma Reef, Oahu. Large numbers at top and columns below represent core samples. Numbers beside columns are dates. Horizontal black lines connect levels of cores having same age and give profile of reef at 1000-year intervals. Depth below and height above mean lower low water (MLLW) are provided by vertical scale on right. (From Easton, W. H., and Olson, E. A., 1976: Radiocarbon profile of Hanauma Reef, Oahu, Hawaii. Geol. Soc. Am. Bull., 87:711–719.)

deepest sediments date from the beginning of the Cenozoic era.

At the dawn of the Cenozoic era, Eniwetok was a volcanic seamount whose peak emerged above the ocean as an island. A fringing reef developed around the island. During the Cenozoic era the emergent volcanic cone gradually subsided. The fringing reef continued to grow upward, keeping pace with subsidence and forming a vertical wall resting on the volcanic basaltic base. Within the coral walls the present lagoon of the atoll filled the area left behind by the submerged cone. The extent of submergence can be measured by the depth of the reef revealed by drilling—1283 meters. But submergence was not steady, as a result of fluctuations in sea level caused in part by periods of glaciation. When the sea level fell, the reef was an emergent island ring on which vegetation devel-

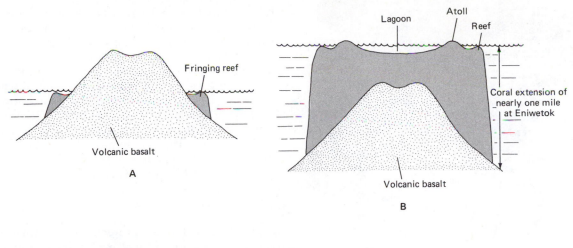

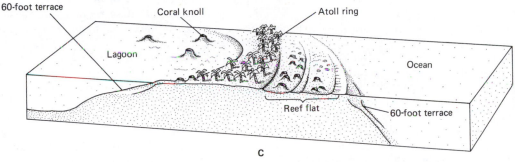

Figure 6–14 Formation of an atoll: *A*, Fringing reef around an emergent volcano. *B*, Continuous deposition of coral as volcanic cone subsides leads to the formation of a great coralline cap; emergent part of cap is atoll. *C*, Section through part of atoll. (After Ladd.)

oped, erosion occurred, and the level of most active coral growth shifted. These changes are recorded in the pollen and fossil vegetation embedded in the coral walls of the existing atoll and the submergent erosion terraces that ring its outer perimeter. With rise in sea level above the present level, which occurred during certain of the Pleistocene interglacial periods, the ring was washed by waves. The rise in sea level since the last glacial low has resulted in the present extent of emergence of Eniwetok. Thus, the formation of this atoll, like many others, reflects both subsidence and sea level changes.

Reefs of the Past

Reefs are ancient marine structures and appear in the geological record from the beginning of the Paleozoic. Many paleontologists believe that the same environmental conditions that characterize reefs today, i.e., warm, clear water, also characterized reefs of the past (Newell, 1972). The organisms responsible for reef construction have continually changed, however.

The earliest reefs were formed by blue-green algae (stromatolites) and spongelike archeocyathids. During the 350 million years of the Paleozoic, certain now-extinct calcareous sponges and the tabulate and rugose corals (Figs. 6–15 and 19–61*B*) were important reef builders. Other contributors were algae, bryozoans, brachiopods, and crinoids.

Reefs were widespread during much of the Mesozoic and exhibited a marvelous diversity of inhabitants, some familiar and some now extinct. Scleractinian and hydrozoan corals had made their appearance, replacing the tabulate and rugose corals, which died out at the end of the Paleozoic. By the Cretaceous, the number of scleractinian corals far exceeded the number existing today. The association of zooxanthellae with reef species was probably an early event.

Rudists, large, hornlike bivalve mollusks, were also important reef-forming animals (p. 423). At

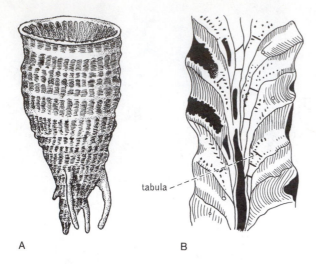

tabula

Figure 6–15 *A*, A rugose coral. (After Orbigny.) *B*, Longitudinal section of a tabulate coral. (After Rominger.)

end of the Mesozoic there was a great and abrupt extinction of many reef organisms, including the rudists and two thirds of the Mesozoic genera of scleractinian corals.

The last great flowering of reef builders occurred in the Eocene of the early Cenozoic. Eocene reefs were much like those today in composition but were more widespread, for warm seas and moderate climates characterized much of the earth at that time. A cooling trend followed the Eocene, associated in part with spreading of the sea floor, deepening of the ocean basins, and elevation of continental crusts. Tropical seas became steadily more restricted, eventually resulting in our present distribution of reefs.

The Great Barrier Reef appeared in the Miocene with subsidence of the eastern continental shelf of Australia. The Isthmus of Panama became emergent during the Pliocene, separating for the first time the Pacific reef fauna from that of the Caribbean.

Coral reefs today experience natural disasters just as they have in the past. Hurricanes and unusually low tides can cause the deaths of many corals and other organisms, but in time the reef recovers (Loya, 1976). Disasters produced by man are much more serious and recovery may be prolonged beyond the lifetime of this text's readers. Reefs in many parts of the world are in jeopardy from a variety of conditions. Construction causes silting and changes patterns of water flow. The patch reefs in Castle Harbor, Bermuda, have never recovered from the construction of the airbase there during the Second World War. Silt and fertilizer in runoff from farmland have destroyed or damaged fringing reefs in many places. Oil pollution from tank washing and spills is taking a toll in certain areas. Removal of larger corals for construction or for making cement is a common practice in undeveloped countries. The damage is compounded when quarriers drag large coral heads over the bottom, destroying smaller corals in the process. Fishermen sometimes dynamite reefs in order to collect the stunned fish. Although some of these damaging practices, such as dynamiting, are outlawed, enforcement is difficult in many countries.

Summary

1 Coral reefs are calcareous rock formations supporting a great array of marine plants and animals, and certain of these reef organisms secrete the calcium carbonate that forms the underlying reef formation.

2 The distribution of coral reefs is restricted to clear tropical seas, permitting the penetration of light required by symbiotic zooxanthellae. Reefs of past geological ages were probably limited by the same conditions.

3 The principal types of reefs are fringing reefs, barrier reefs, and atolls. Most exhibit a somewhat similar cross-sectional profile, consisting of a seaward-facing and sloping reef front, which receives most of the wave energy, a shallow or just-emergent reef crest, or ridge, and a shallow reef flat behind the ridge.

4 Since environmental conditions differ across the reef, there is a corresponding zonation of reef organisms.

5 The reef structure grows upward and seaward by the accumulation of calcium carbonate. The principal framework builders are scleractinian corals and coralline algae. Following death, scleractinian corals are attacked by boring organisms that reduce the framework skeletons to smaller chunks. Intervening spaces are filled with the excavated debris and the fine skeletal material of other organisms. This "brick and mortar" is cemented, or lithified, by the precipitation of magnesium calcite from the surrounding sea water.

6 The great thickness of many reefs resulted from sea level rise or subsidence of the platform on which the reef rests, or from both.

7 Reefs are known from all geological ages, but there has been a progression of different framework builders. Reef-building corals did not become important until the Mesozoic, and the greatest diversity of reef organisms was probably attained in the Cretaceous.

REFERENCES

Adey, W. H., 1977: Shallow water Holocene biotherms of the Caribbean Sea and West Indies. Proceedings Third International Coral Reef Symposium, University of Miami. Vol. 2 (Geology):xxi–xxiv.

Adey, W. H., 1978: Coral reef morphogenesis: a multidimensional model. Science, *202*:831–837.

Easton, W. H., and Olson, E. A., 1976: Radiocarbon profile of Hanauma Reef, Oahu, Hawaii. Geol. Soc. Am. Bull., *87*:711–719.

Fifth International Coral Reef Congress, Tahiti, 27 May to 1 June 1985, Proceedings. Antenne Museum—Ephe, Moorea, French Polynesia, 1985. (These volumes contain many papers on reef ecology and geology and the biology of reef organisms.)

Geister, J., 1977. The influence of wave exposure on the ecological zonation of Caribbean coral reefs. Proceedings Third International Coral Reef Symposium, University of Miami. Vol. 1 (Biology):23–29.

Goreau, T. F., Goreau, N. I., and Goreau, T. J., 1979: Corals and Coral Reefs. Sci. Amer., *241*(2):124–136.

Jones, O. A., and Endean, R., 1973–1977: Biology and Geology of Coral Reefs. Vols. I–IV. Academic Press, N.Y.

Kaplan, E. H., 1982: A Field Guide to Coral Reefs of the Caribbean and Florida. Peterson Field Guide Series. Houghton Mifflin Co., Boston. 289 pp.

Kühlmann, D., 1985: Living Coral Reefs of the World. Arco. Publ., N.Y.

Lewis, J. B., 1977: Processes of organic production on coral reefs. Biol. Rev., *52*:305–347.

Loya, Y., 1976: Recolonization of Red Sea coral affected by natural catastrophes and man-made perturbations. Ecology, *57*(2):278–289.

Macintyre, I. G., 1977: Distribution of submarine cements in a modern Caribbean fringing reef, Galeta Point, Bahamas. J. Sedimentary Petrology, *47*(2):503–516.

Newell, N. D., 1972: The evolution of reefs. Sci. Am. *226*(6):54–65. (A good account of the changing nature of the world's reefs since the Precambrian.)

Salvat, B., Ricard, M., Richard, G., Galzin, R., and Toffart, J. L., 1977: Reef lagoon complex of Lakeba Island (Lau Group, Fiji): Geomorphology, biotic associations, and socio-ecology. Proceedings Third International Coral Reef Symposium, University of Miami. Vol. 2 (Geology):297–303.

Scoffin, T. P., 1972. Fossilization of Bermuda patch reefs. Science, *178*:1280–1282.

Smith, S. V., and Kinsey, D. W., 1976: Calcium carbonate production, coral reef growth, and sea level change. Science, *194*:937–939.

Stoddard, D. R., 1969: Ecology and morphology of recent coral reefs. Biol. Rev., *44*:433–498.

Vasseur, P., 1977. Cryptic sessile communities in various coral formations on reef flats in Tulear vicinity (Madagascar). Proceedings Third International Coral Reef Symposium, University of Miami. Vol. 1 (Biology):95–100.

Yonge, C. M., 1963: The biology of coral reefs. Adv. Mar. Biol., *1*:209–261.

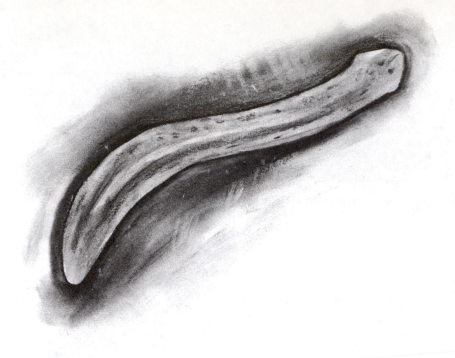

7

The Flatworms

The remaining phyla of the Animal Kingdom are bilaterally symmetrical and are collectively called the Bilateria. Of all the bilateral phyla, the free-living flatworms of the phylum Platyhelminthes have long been considered the most primitive. However, many zoologists are beginning to question this view. The flatworms' position here, first in our survey of bilateral invertebrates, is therefore one of convenience and may not have phylogenetic significance.

The phylum Platyhelminthes embraces four classes of worms. Three are entirely parasitic. The Monogenea and Trematoda comprise flukes, and the Cestoda, tapeworms. The fourth class, the Turbellaria, are free living and are certainly the ancestors of the three parasitic classes. Members of the phylum are dorsoventrally flattened—accounting for the name *flatworm*—and display a solid, or acoelomate, grade of structure, in which there is no body cavity; a loose tissue (parenchyma) of mesenchymal origin fills the space between the internal organs and body wall. The mouth is the only opening to the digestive tract, when a digestive tract is

present. Protonephridia are present, and the reproductive system is hermaphroditic.

Class Turbellaria

The Turbellaria vary in shape from oval to elongate and, like the other flatworm classes, are usually dorsoventrally flattened. In general, the larger the worm, the more pronounced the flattened condition. Head projections are present in some species. These may be in the form of short marginal or dorsal tentacles (Fig. 7–1B) or lateral projections of the head called auricles (Fig. 7–1E). The auricles are frequently found in freshwater planarians. Coloration is mostly in shades of black, brown, and gray, although some groups display brightly colored patterns. Turbellarians range in size from microscopic to more than 60 cm long, but most are small, less than 10 mm in length.

Turbellarians are primarily aquatic, and the great majority are marine. Although there are a few pelagic species, most are bottom dwellers that live

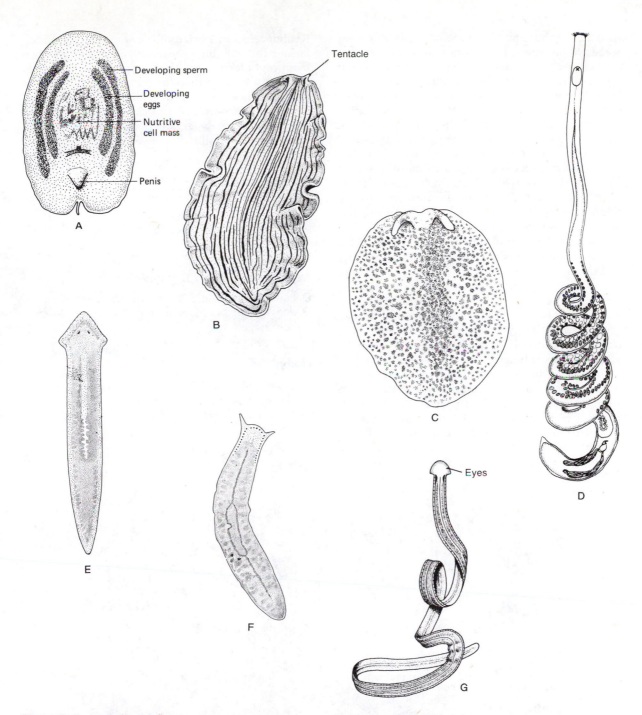

Figure 7–1 Turbellarian flatworms. *A, Polychoerus*, an acoel. *B, Prostheceraeus*, a polyclad. *C, Eurylepta leoparda*, a polyclad that preys upon ascidians. Spots are brick red, and the length is about 1.5 cm. (Based on a photograph by Ching, H. L., 1977: Redescription of *Eurylepta leoparda* Freeman, 1933, predator of the ascidian *Corella willmeriana*. Can. J. Zool., 55:338–342.) *D, Nematoplana*, an interstitial member of the order Proseriata. (From Rieger, R., and Ott, J., 1971: Vie et Milieu, Supplement 22.) *E, Dugesia*, a common freshwater planarian. *F, Polycelis*, a freshwater planarian. (*E* and *F* after Steinmann and Bresslau.) *G, Bipalium kewense*, a cosmopolitan land planarian found in Florida, Louisiana, and California and in greenhouses throughout the United States. (After Hyman, L. H., 1951: The Invertebrates. Vol. II. McGraw-Hill Book Co., N.Y.)

in sand or mud, under stones and shells, or on sea-weed. Many species are common constituents of the interstitial fauna. Freshwater forms live in lakes, ponds, streams, and springs, where they occupy bottom habitats. Some species have become terrestrial (Fig. 7–1*G*), but these are confined to very humid areas and usually hide beneath logs and leaf mold during the day, emerging only at night to feed. The land planarians are the giants of the Turbellaria, some reaching 60 cm or more in length. They are largely tropical, but a few, such as the North American *Bipalium adventitium*, live in temperate regions. Some 3000 species of turbellarians have been described.

Despite their similarity in appearance, turbellarians exhibit considerable internal complexity, and the class is composed of a relatively large number of diverse groups. Those orders that must be frequently referred to in the following discussion can perhaps be more easily kept in mind if they are grouped and designated at the outset. The members of the orders Acoela, Catenulida, Macrostomida, and Polycladida exhibit a more primitive level of organization (often referred to as the archoophoran level). All four are marine, but the Macrostomida and Catenulida are also represented in freshwater habitats. The Acoela, Catenulida and Macrostomida are small to microscopic in size; members of the Polycladida are large, some species reaching 5 cm or more in length (Fig. 7–1*C*). The

Rhabdocela and Tricladida are among those turbellarians that exhibit a more advanced level of organization (referred to as the neoophoran level). They are found both in the sea and in fresh water. The rhabdocoels are generally small; triclads, which include the freshwater species called planarians, may attain a moderately large size. These groups may be summarized as follows:

*Archoophoran Level of Organization**

Order Acoela. Marine. Small.
Order Catenulida. Mostly freshwater. Body size small.
Order Macrostomida. Marine and freshwater. Small.
Order Polycladida. Marine. Large.

Neoophoran Level of Organization

Order Rhabdocoela. Marine and freshwater. Small.
Order Tricladida. Marine and freshwater. Moderately large.

Body Wall

The body of a turbellarian is covered by a ciliated epidermis bearing surface microvilli (Figs. 7–2 and 7–3). In some triclads and polyclads cilia are con-

*This synopsis is abbreviated. See the systematic résumé on page 186 for a complete classification.

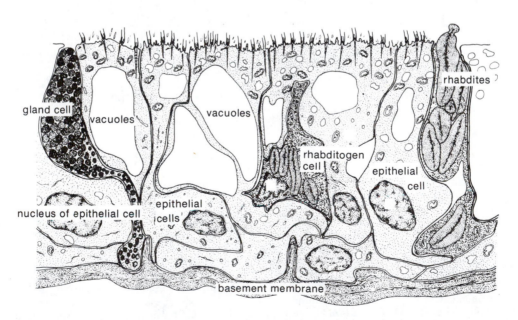

Figure 7–2 Reconstruction of the epidermis of the polyclad *Thysanozoon brocchii*. Only the bases of the cilia are shown. (From Bedini, C., and Papi, F., 1974: Fine structure of the turbellarian epidermis. *In* Riser, N. W., and Morse, M. P. (Eds.): Biology of the Turbellaria. McGraw-Hill Book Co., N.Y. p. 112.)

fined to the ventral surface (Fig. 7–4). Beneath the epidermis are a basement membrane (except in the order Acoela and Catenulida) and several layers of muscle—an outer circular and an inner longitudinal layer, with diagonal fibers lying between them. In addition, dorsoventral muscles extend through the interior of the body (Fig. 7–4).

A characteristic feature of turbellarians is the presence of numerous gland cells (Figs. 7–2, 7–3, and 7–4). The gland cells may be entirely within the epidermis but are commonly sunk in the muscle layers, with only the neck of the gland penetrating the epidermis. The glands may provide for adhesion and for mucus secretion. Temporary adhesion to the substratum is made possible by adhesive glands, adhesive cilia, or muscular suckers. Many interstitial marine species adhere to the sand grains of their environment with glandular adhe-

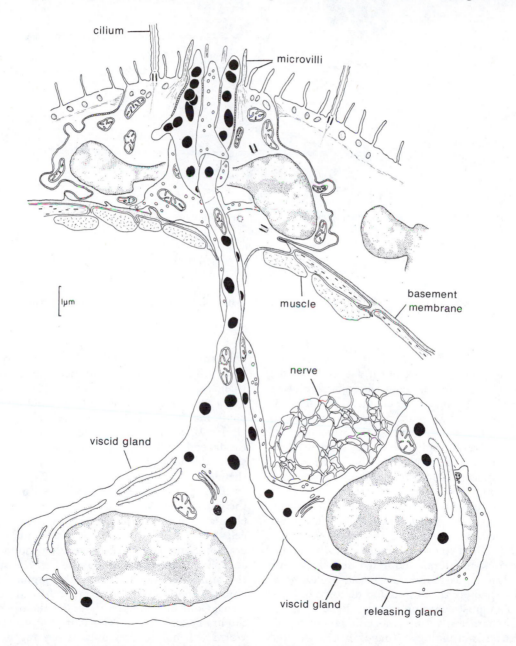

cilium

microvilli

1μm

muscle

basement membrane

nerve

viscid gland

viscid gland   releasing gland

Figure 7–3   Section through the integument of *Haplopharynx* (Macrostomida) showing a duogland adhesive organ. (From Tyler, S., 1976: Comparative ultrastructure of adhesive systems in the Turbellaria. Zoomorphologie, *84*:1–76.)

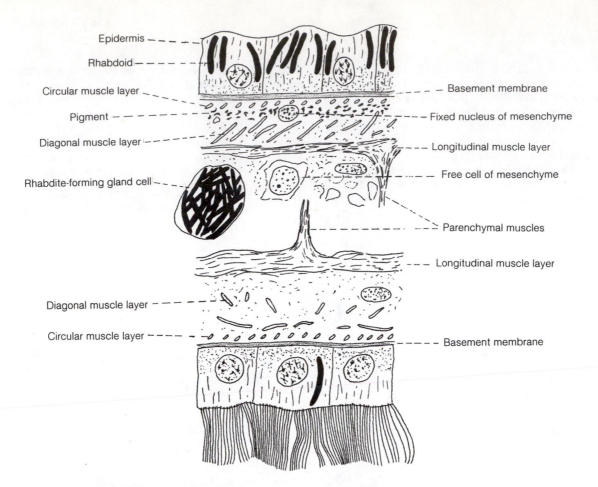

Epidermis

Rhabdoid

Circular muscle layer

Pigment

Diagonal muscle layer

Rhabdite-forming gland cell

Basement membrane

Fixed nucleus of mesenchyme

Longitudinal muscle layer

Free cell of mesenchyme

Parenchymal muscles

Longitudinal muscle layer

Diagonal muscle layer

Circular muscle layer

Basement membrane

Figure 7–4 Dorsal and ventral body walls of a freshwater planarian (longitudinal section). (After Hyman, L. H., 1951: The Invertebrates. Vol. II. McGraw-Hill Book Co., N.Y.)

sive organs known as duogland organs. Each organ, which may project from the surface as a papilla, contains a viscid gland, a releasing gland, and an anchor cell (Fig. 7–3).The viscid gland provides adhesive material. The releasing gland is believed to produce a substance that breaks the attachment of the viscid material and enables the animal to release from the substratum. The anchor cell, through which the glands pass and open, functions as the attachment surface (Tyler, 1976).

An anterior aggregation, called a frontal gland, is characteristic of many turbellarians and is believed to be a primitive turbellarian feature (Figs. 7–5 and 7–7B). Typical of almost all turbellarians are numerous membrane-bounded, rod-shaped secretions known as rhabdoids that are released to the surface, where they swell and dissolve to form mucus. The most common kind of rhabdoid is the rhabdite, characterized by a specific lamellate, ultrastructural morphology. Rhabdites are secreted by epidermal gland cells, usually sunk below the epidermis (Fig. 7–4), and are found largely in macrostomids, polyclads, and neoophorans. Other rhabdoids with different ultrastructural morphologies are found in gland cells in acoels and in the epidermal (epithelial) cells of planarians (Smith et al, 1982). Some turbellarians have glands at the caudal end. In *Bdelloura*, which lives as a commensal on the book gill of the Atlantic horseshoe crab, the glands form an adhesive plate (Fig. 7–6).

The space between the body wall muscles and the gut is filled with epithelial glands, with parenchyma cells of various sorts, and with neoblasts, cells that provide a source of new tissue in regeneration. Some planarians have been found to possess parenchyma chromatophores, which can cause the animal to lighten when the pigment in the chromatophore is concentrated toward the nucleus. Control is apparently under the direction of the brain, for the posterior half of a bisected dark worm will not lighten until it has regenerated a brain (Palladini et al, 1979).

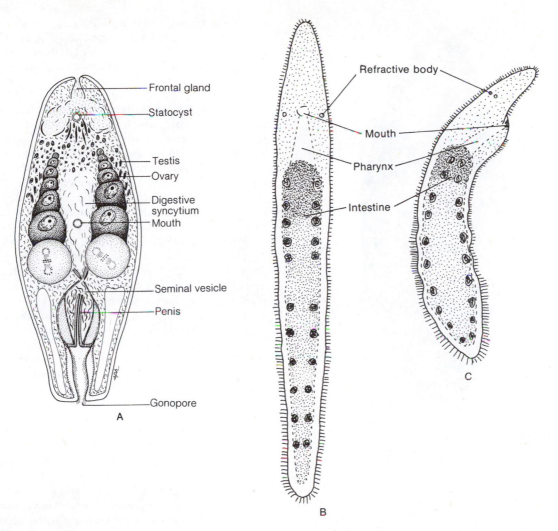

Figure 7–5 *A, Pseudaphanostoma brevicaudatum,* an acoel flatworm. (After Dörjes.) *B* and *C,* The catenulid *Stenostomum,* a common, microscopic, freshwater turbellarian: *B,* Dorsal view. *C,* Contracted specimen with anterior end turned in lateral view. A nonmuscular, ciliated pharynx connects the anterior, ventral mouth and the intestine, which is a simple, elongated sac. (From Villee, C. A., Walker, W. F., and Barnes, R. D., 1978: General Zoology. 5th Edition. W. B. Saunders Co., Philadelphia.)

Locomotion

Minute turbellarians swim about within bottom debris. Larger species creep over the bottom. Cilia provide the force for propulsion, and the flattened body of creeping species gives greater contact with the substratum. Aquatic planarians utilize cilia, but muscular undulations are important in the locomotion of large polyclads and terrestrial planarians. All flatworms, minute and large, depend on muscle contractions to turn and twist the body.

Like some snails, turbellarians that glide over surfaces on cilia depend on an adhesive, mucous blanket to gain traction. The mucus is produced by rhabdites or other glands, and the tips of the cilia dig into the mucus during the effective stroke. It is postulated that the cilia lift above the mucous film during recovery (Martin, 1978).

The interstitial fauna contains many minute turbellarians. Those that live in the intertidal or swash zone, where the spaces between sand grains

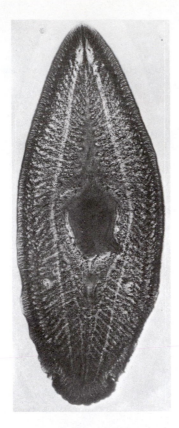

Figure 7–6 *Bdelloura*, a marine triclad commensal that lives on gills of horseshoe crabs. The large, central, dark mass is the pharynx, in front of which can be seen the anterior branch of the intestine. The two longitudinal streaks passing to either side of the pharynx are the ventral nerve cords. The pair of ganglia from which they arise, lying beneath the eyes, can be seen at the anterior end. (Courtesy of Ward's Natural Science Establishment.)

are not continuously filled with water, are among the most specialized. Adhesive glands and elongate bodies are common adaptations for gaining traction to move or maintain position among sand grains (Fig. 7–1D).

Nutrition

The digestive cavity, or intestine, of turbellarians is a blind sac, with the mouth serving for both ingestion and egestion. The wall of the intestine is single layered and is composed of phagocytic and gland cells. The primitive intestine, as in the Macrostomida, Catenulida, and some polyclads, is ciliated (Fig. 7–7A).

The form of the intestine is in part related to the size of the worm. Small turbellarians, such as

the macrostomids, catenulids, and rhabdocoels, have intestines that are simple sacs. The members of the order Acoela lack a permanent intestinal cavity, and the intestinal cells form a solid, often syncytial mass (Fig. 7–7B). It is to the lack of a gut cavity that the name *Acoela* refers.

The larger turbellarians have intestines with lateral diverticula, which increase the surface area for digestion and absorption and compensate for the absence of an internal transport system. In polyclads the intestine consists of a central tube, from which a great many lateral branches arise (Figs. 7–8A and 7–9A). These in turn are subdivided and may anastomose with other branches. The members of the Tricladida, to which the planarians belong, have an intestine composed of three principal lateral branches—one anterior and two posterior (Figs. 7–10 and 7–11C). Each of these principal branches, in turn, has many lateral diverticula. The three branches join in the middle of the body, anterior to the mouth and pharynx. The names *Polycladida* and *Tricladida* refer to the branching intestine of these groups of turbellarians.

The mouth is primitively located on the midventral surface but may be situated anywhere along the midventral line. The connection between the mouth and the intestine shows increasing complexity within the class. This region, called the pharynx, is a simple, ciliated tube (simple pharynx) in the small, primitive Macrostomida and Catenulida (*Stenostomum*) (Figs. 7–7A and 7–12A). In higher turbellarians (polyclads and all neophorans) the pharynx has become a more complex ingestive organ as a consequence of folding and development of the muscle layers. A folded, or plicate, pharynx is characteristic of the polyclads and triclads, which are large turbellarians with branched intestines. The folded condition is believed to have evolved from a simple pharynx and has resulted in a muscular pharyngeal tube lying within a pharyngeal cavity (Figs. 7–10A–C, 7–11C, and 7–12B). The free end of the tube projects from the mouth during feeding (Fig. 7–11D). The pharynx may project backward (Figs. 7–10A–C and 7–11), as in the common freshwater planarians, or the pharynx may be attached posteriorly and extend forward (Fig. 7–8A). In many polyclads it hangs down from the roof of the pharyngeal cavity like a circular, ruffled curtain (Figs. 7–8B and 7–9A).

A bulbous pharynx, encountered in such orders as the Rhabdocoela, is believed to be derived from the plicate condition through reduction of the outer pharyngeal cavity and separation by a special muscular septum from the parenchyma of the body

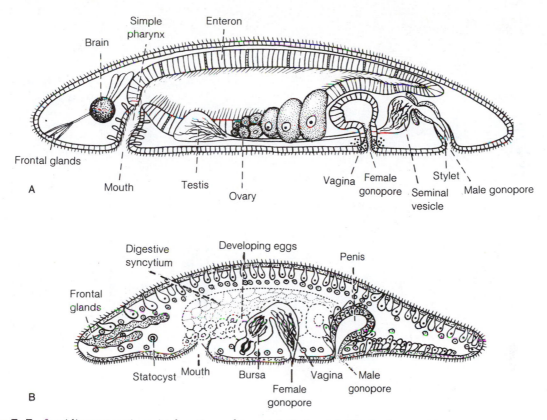

Figure 7–7 Semidiagrammatic sagittal sections of two turbellarians: *A*, The freshwater *Macrostomum*, a member of the order Macrostomida. (After Ax.) *B*, The marine acoel *Convoluta*. (After Westblad.)

(Fig. 7–12). The muscular bulb, which characterizes this type of pharynx, can be protruded from the mouth in many species (Fig. 7–11*B*).

Turbellarians are largely carnivorous and prey on various invertebrates that are small enough to be captured, as well as on the dead bodies of animals that sink to the bottom. Feeding behavior is elicited, at least in some species (planarians), by substances emitted from the potential food source. Protozoa, rotifers, insect larvae, small crustaceans (water fleas, copepods, amphipods, isopods), snails, and small annelid worms (see Reynoldson and Sefton, 1976) are common prey, but there are marine species that feed on sessile animals, such as bryozoans and little tunicates. The polyclad *Stylochus frontalis* feeds on living oysters, and *Stylochus tripartitus* preys on barnacles. The ectocommensal triclad *Bdelloura*, which lives on the book gill of horseshoe crabs, shares in the food of its host.

Not all turbellarians are predaceous. Some feed on algae, especially diatoms, and some predaceous species feed on diatoms as juveniles (Kozloff, 1972; Bush, 1975).

Many turbellarians capture living prey by wrapping themselves around it, entangling it in slime, and pinning it to the substratum by means of the adhesive organs. Species of *Mesostoma* have been found to paralyze their prey with mucus (Case and Washino, 1979). A few species are known to stab prey with the penis, which terminates in a hardened stylet and projects from the mouth; the interstitial kalyptorhynch rhabdocoels possess an anterior, raptorial proboscis that may have hooks (Fig. 7–13*A*). The proboscis of these species is not connected with the mouth and bulbous pharynx (Fig. 7–13*B*).

Prey is swallowed whole by those turbellarians with a simple pharynx, by those with a protrusible bulbous pharynx, and even by the polyclads, which have a plicate pharynx. In the triclads, the pharyngeal tube is extended from the mouth and inserted into the body of the prey or the carrion. The exoskeleton of crustaceans is penetrated at thin points, such as the articulations between segments. Penetration by the pharynx and ingestion of prey contents are aided by proteolytic enzymes (endopepti-

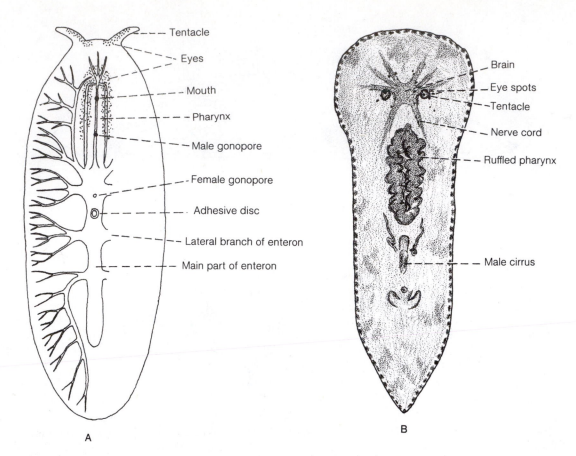

Figure 7–8 *A*, Digestive system of a polyclad with a tubular pharynx. (After Hyman, L. H., 1951: The Invertebrates. Vol. II. McGraw-Hill Book Co., N.Y.) *B*, Polyclad with dorsal tentacles and ruffled pharynx.

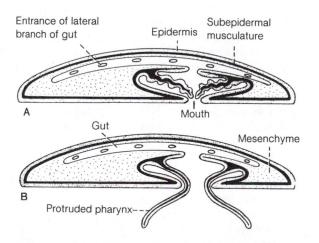

Figure 7–9 The polyclad *Leptoplana*, showing retracted and protruded, ruffled, plicate pharynx (sagittal section). (After Jennings, J. B., 1957: Studies on feeding, digestion, and food storage in free-living flatworms. Biol. Bull., *112*:63–80.)

dases) produced by pharyngeal glands that open onto the tip of the pharynx. The contents are then pumped into the enteron by peristaltic action (Figs. 7–10*D* and 7–14).

In studies on the acoel *Convoluta paradoxa*, Jennings (1957) found that small prey is captured and engulfed by the internal mass of digestive cells, which are partially everted through the mouth. Larger prey is pressed into the mouth and swallowed. After ingestion the prey passes into the internal mass of nutritive cells.

Digestion is first extracellular. Disintegration of the ingested food is initiated by pharyngeal enzymes, and additional endopeptidase is supplied by gland cells of the intestine. The resulting food fragments are then engulfed by the phagocytic cells, where digestion by endopeptidases continues in an acid medium. About 8 to 12 hours following ingestion the vacuole becomes alkaline, which marks

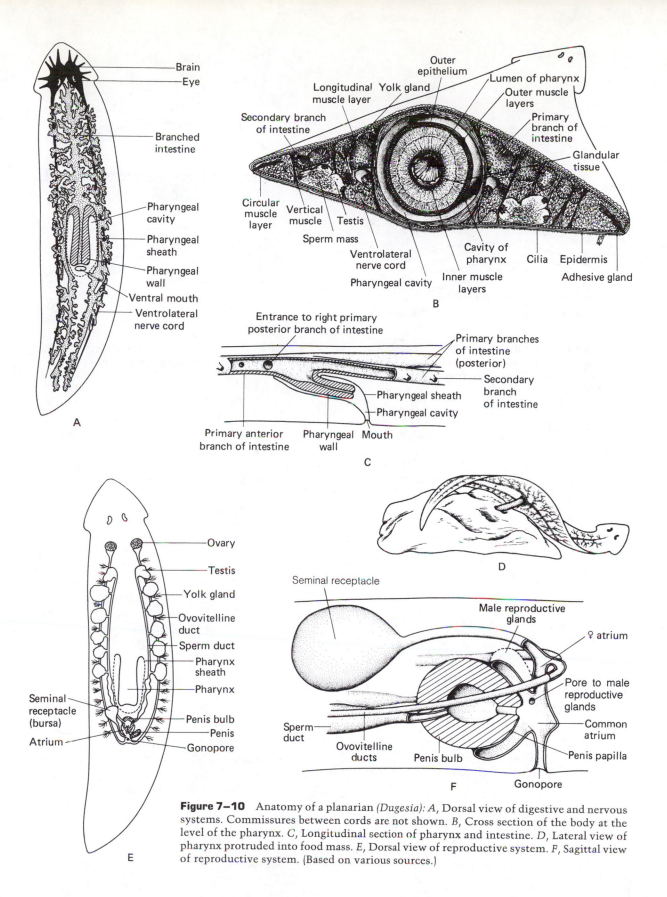

Figure 7–10 Anatomy of a planarian *(Dugesia): A,* Dorsal view of digestive and nervous systems. Commissures between cords are not shown. *B,* Cross section of the body at the level of the pharynx. *C,* Longitudinal section of pharynx and intestine. *D,* Lateral view of pharynx protruded into food mass. *E,* Dorsal view of reproductive system. *F,* Sagittal view of reproductive system. (Based on various sources.)

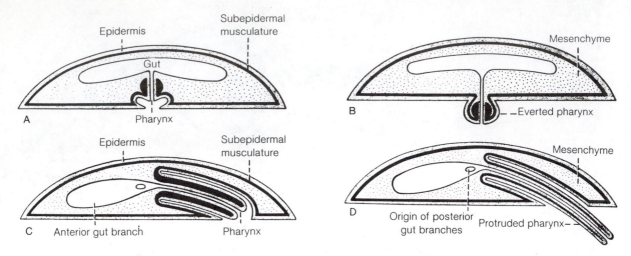

Figure 7–11 *A* and *B*, The rhabdocoel, *Mesostoma*, showing retracted and bulbous pharynx (sagittal section). *C* and *D*, The freshwater triclad *Polycelis*, showing retracted and protruded tubular, plicate pharynx (sagittal section). (All after Jennings, J. B., 1957: Studies on feeding, digestion, and food storage in free-living flatworms. Biol. Bull., *112*:63–80.)

the appearance of exopeptidases, lipases, and carbohydrases, and digestion is completed. During the course of intracellular digestion, the vacuole sinks lower and lower into the phagocytic cell and eventually disappears (see Jennings review, 1974).

Freshwater planarians are able to withstand prolonged experimental starvation. In extreme cases they utilize part of the enteron and all of the parenchyma and reproductive system. In fact, the body volume may be reduced to as little as 1/300 of the original.

Some species of *Convoluta* harbor green zoochlorellae or golden zooxanthellae or diatoms (without shells) within their parenchyma. These worms may occur in such enormous numbers in certain regions on the European coast that they form green blotches in shallow water.

Although parasitism in flatworms is usually associated with flukes and cestodes, there are a number of commensal and parasitic turbellarians. They are largely rhabdocoels (suborders Dalyellioida and Temnocephalida) and are found in both the sea and

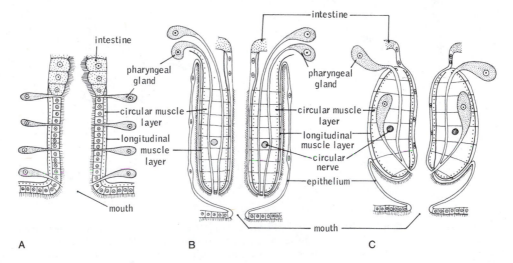

Figure 7–12 The three types of turbellarian pharynx: *A*, Simple pharynx, found in a few acoels, the Macrostomida, and the Catenulida. *B*, Plicate pharynx, found in the polyclads and planarians. *C*, Bulbous pharynx, found in the rhabdocoels. (Modified after Ax.)

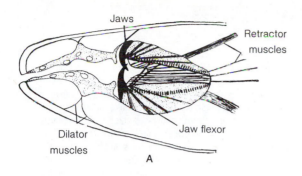

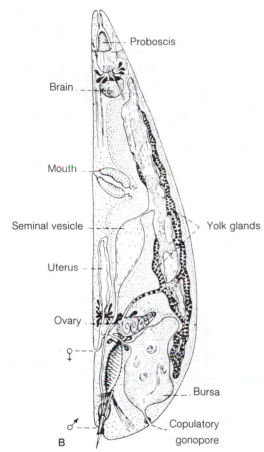

Figure 7–13 Kalyptorhynchs, rhabdocoel turbellarians that possess an anterior proboscis apparatus with jaws: *A*, Proboscis apparatus showing jaws. *B*, Lateral view of *Gyratrix hermaphroditus*. Note that proboscis apparatus is anterior and not part of gut. (Both after Meixner. *A* from Hyman, L. H., 1951: The Invertebrates: Platyhelminthes and Rhynchocoela. Vol. 2. McGraw-Hill Book Co., N.Y. p. 146. *B* from de Beauchamps, P., et al, 1961: Platyhelminthes et Mesozoaires. *In* Grassé, P. (Ed.): Traité de Zoologie, Vol. 4, pt. 1. Masson et Cie, Paris. p. 169.)

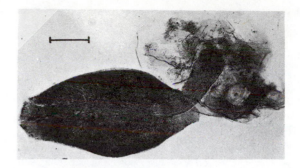

Figure 7–14 The triclad *Polycelis*, feeding on large *Daphnia*. (Courtesy of J. B. Jennings.)

fresh water. Commensals include species that live within the mantle cavities of mollusks and on the gills of crustaceans. Parasitic forms inhabit the guts and body cavities of mollusks, crustaceans, and echinoderms, and the skin of fish.

Excretion and Osmoregulation

Turbellarians eliminate nitrogenous wastes (ammonia) across the general body surface, but they osmoregulate with protonephridia. The turbellarian protonephridium is of the flame-cell type, each consisting of a branched tubule terminating in a number of blind capillaries (Box 7–1). The number of protonephridia varies, but they are typically paired. As many as four pairs occur in the triclads, and they often form an anastomosing network with many nephridiopores (Fig. 7–15). The position of the nephridiopores is also variable.

Nervous System and Sense Organs

Although there are some acoels with only an epidermal nerve-net type of nervous system, the primitive turbellarian plan is perhaps a subepithelial nerve net plus three or four pairs of longitudinal cords—dorsal or dorsolateral, lateral or marginal, ventrolateral, and ventral (Fig. 7–16*A*). The cords are interconnected along their length by commissures and anteriorly by a brain. Within the brain is a statocyst. This primitive plan is best developed in the lower groups, such as acoels and catenulids (Reisinger, 1972; Moraczewski et al, 1977). In most other turbellarians the statocyst has disappeared and there have been tendencies toward fewer pairs of cords and increased prominence of the ventral pair. Reduction to a dominant pair of ventral cords has been more or less attained in freshwater triclads and most rhabdocoels (Figs. 7–10*A* and

BOX 7–1   PROTONEPHRIDIA

Protonephridia are branching, osmoregulatory tubules found in a number of animal groups. Although often described as excretory organs, excretion is not usually a primary function, since most of the animals in which protonephridia occur are aquatic and too small to require special organs for the removal of nitrogenous waste. Protonephridia open to the exterior through a nephridiopore, but the inner end is blind (A). There are two types of protonephridia, those whose blind ends terminate in flame cells (A, B) and those that terminate in solenocytes. Solenocytes, which are best known from annelids (p. 279), are long, cylindrical cells enclosing a single cilium (flagellum).

Flame cells, also called cyrtocytes, are characteristic of the protonephridia of flatworms and other groups. Each encloses one cilium or a cluster of cilia, whose beating reminded early zoologists of a candle flame (B and C). *Flame bulb* is a term often applied to flame cells in which the nucleus of the terminal cell is displaced some distance from the apical end containing the cilia. It will not be used here.

In all protonephridia the barrel-like wall around the terminal ciliary flame cell is partially perforated by slits, usually running parallel to the long axis of the tubule (B and C). It is through these slits, or fenestrations, that filtration occurs. The beating of the cilia drives fluid down the tubule so that the fluid pressure inside the tubule at the terminal end is lower than that of the surrounding tissues. Fluid thus filters inward through the fenestrations. Distal to the flame cell the protonephridial tubule is ciliated, at least in part, and fluid is driven out to the nephridiopore. In some animals, such as planarians, there are cytological indications that selective reabsorption occurs in the tubule (Prusch, 1976; Ishii, 1980).

The basket-like wall of flatworm cyrtocytes is usually formed by a circle of rods that extend from the apical end of the cell bearing the ciliary tuft (C). The slits between the rods are fenestrations, but they are not actually open. A membranous material of uncertain origin closes the slits. Often an inner cir-

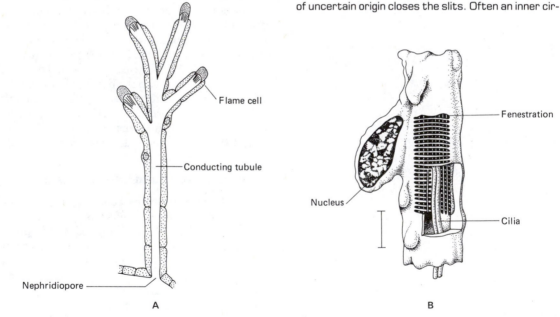

Flame cell

Conducting tubule

Nephridiopore

A

Fenestration

Nucleus

Cilia

B

7–16*C*). The longitudinal nerve cords of polyclads undergo continuous branching, so the nervous system takes the form of a complex network (Fig. 7–16*B*). In all turbellarians the nervous system is relatively primitive in lacking ganglia, except in the brain, but typical types of neurons are present. The structure and physiology of the nervous system has been best studied in the polyclad *Notoplana acticola* from the coast of California (Faisst et al, 1980; Keenan et al, 1981).

Eyes are common in most turbellarians and are of the pigment-cup type (Fig. 7–17*B*) (Box 7–2).

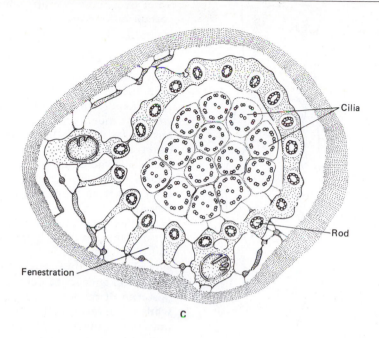

C

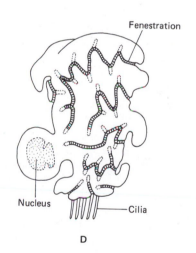

D

cle of rods alternates with the outer circle. Wilson and Webster (1974) have suggested that one set of rods from the apical cell (cyrtocyte) interdigitates with those of the first cell, forming the tubule wall, but such a relationship has been demonstrated only in parasitic flatworms. The cyrtocytes of planarians have highly irregular processes that interdigitate like the processes of a vertebrate podocyte. The fenestration is thus a very convoluted slit, and the entire cyrtocyte is somewhat star shaped (D).

A, Diagrammatic representation of an entire protonephridium. *B*, Cyrtocyte of *Stenostomum*, which is unusual in having only two cilia (flagella). (*B* from Kummel, G., 1962: Z. Zellforsch, *57*:172–201.) *C*, Section through a cyrtocyte of *Temnocephala novaezealandiae*. (From Williams, J. B., 1981: The protonephridial system of *Temnocephala novaezealandiae*: Structure of the flame cells and main vessels. Aust. J. Zool., *29*:131–146.) *D*, Cyrtocyte of the freshwater planarian *Bdellocephala brunnea*. (From Ishii, S., 1980: The ultrastructure of the protonephridial flame cell of the freshwater planarian *Bdellocephala brunnea*. Cell Tiss. Res., *206*:441–449.)

Two is the usual number (Figs. 7–1*E*, 7–17*A*, and 7–18), but two or three pairs are not uncommon; in the polyclads and the land planarians, there may be a great many eyes (Fig. 7–8*A*). The eyes function largely in orienting to light, and most turbellarians are negatively phototactic.

Dispersed ciliary receptors are distributed over the entire body but are particularly concentrated on the tentacles, the auricles, and the body margins. Ciliated patches are sunken in pits or grooves in the head region and probably contain chemoreceptors. The cilia maintain a continual current of

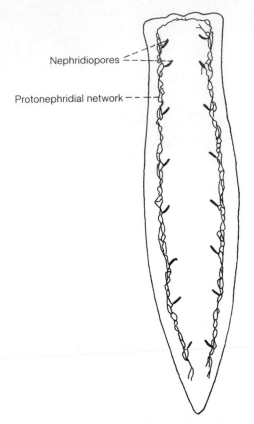

Nephridiopores

Protonephridial network

Figure 7–15   Excretory system of triclad *Dendrocoelum lacteum.* (After Wilhelmi from Hyman.)

water over the receptors, which appear to be important in food location.

Asexual Reproduction and Regeneration

Many turbellarians, especially freshwater species, reproduce asexually by means of fission. In the genera *Catenula, Stenostomum,* and *Microstomum,* whose members are all small, fission is transverse, but the individuals may remain attached, so that chains are formed (Fig. 7–19B). The individuals in such chains are known as zooids. When a zooid attains a fairly complete degree of development, it detaches from the chain as an independent individual.

Freshwater planarians also reproduce by transverse fission, but no chains of zooids are formed, and regeneration occurs after separation. The fission plane usually forms behind the pharynx, and separation appears to depend on locomotion. The posterior of the worm adheres to the substratum, while the anterior half continues to move forward until the worm snaps in two. Each half then regenerates the missing parts to form a new small worm.

A few species of freshwater planarians, such as members of the genus *Phagocata,* and some land planarians fragment rather than undergo transverse fission. In *Phagocata* each piece forms a cyst in which regeneration takes place and from which a small worm emerges.

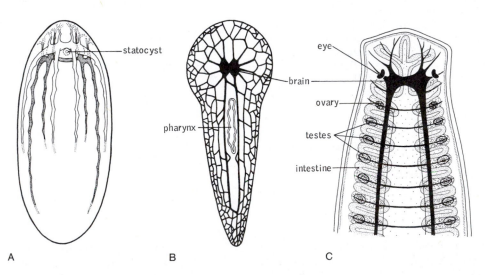

A   B   C

Figure 7–16   Turbellarian nervous systems: *A,* Radial arrangement of nerves in the acoel *Anaperus.* (After Westblad.) *B,* Ventral submuscular nerve net of polyclad *Gnesioceros. C,* Anterior end of the nervous system of the planarian *Procerodes.* (Modified after Lang.)

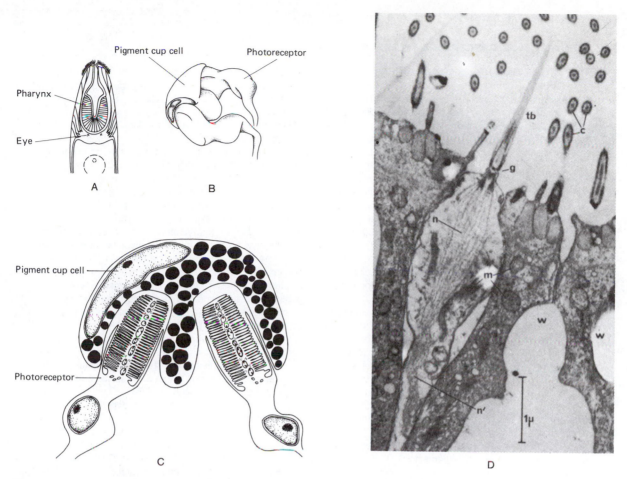

Figure 7–17 *A*, Anterior end of the kalyptorhynch *Polycystis naegelii*, showing the pharynx and the position of the two eyes, which lie beneath the surface within the parenchyma. *B*, Enlarged view of one eye. The pigment cup is formed by one cell and encloses two photoreceptor cells. The organization of the eyes of other turbellarians, including planarians, is similar, but there may be more photoreceptors and more cells forming the pigment cup. *C*, Diagrammatic section of the eye. The pigment cup is not always subdivided by median partition in other turbellarians. (Modified from Lanfranchi, A., and Bedini, C., 1982: The ultrastructure of the sense organs of some Turbellaria Rhabdocoela. I. The eyes of *Polycystis naegelii* Kolliker. Zoomorph., *101*:95–102.) *D*, Electron photomicrograph of a sensory cilium (tb) of the marine *Bothrioplana* (order Seriata). The letter *c* designates the ordinary cilia. (From Reisinger, E., 1968: Z. Zool. Syst. Evolutionsforsch., 6:1–55.)

Reproduction may be controlled by day length and temperature. Freshwater planarians, for example, which are almost all inhabitants of temperate regions, reproduce asexually by fission during the summer and sexually during the fall under the stimulus of shorter day lengths and lower temperatures. Asexually reproducing laboratory cultures of *Catenula* have been maintained for as long as six years without occurrence of sexual reproduction (Moraczewski, 1977).

The common planarian *Dugesia dorotocephala*, which is easily maintained in laboratory culture

and has been studied extensively, undergoes fission only at night. During the day the brain produces some substance that inhibits fission, and the production of the inhibitor appears to be under photoperiod control (Morita and Best, 1984).

Most flatworms have considerable powers of regeneration, and their regenerative ability has been investigated by numerous workers (see Rose, 1970, for review; Moraczewski, 1977). Only a few aspects can be mentioned here. A distinct physiological gradient exists in flatworms, so the body is polarized, the anterior representing one pole and the

---

BOX 7–2 PHOTORECEPTORS AND EYES I

The photoreceptor cells of most animals have evaginations in some part of the surface, providing greater surface area for photochemical reactions. There are two principal types of photoreceptors—ciliary and rhabdomeric. In ciliary photoreceptors the photosensitive surface is derived from the membrane of a cilium (A), and in rhabdomeric photoreceptors it is derived from the microvilli of the cell surface (B; see also Fig. 7–17). Both types certainly evolved numerous times in the Animal Kingdom, and epidermal cells were probably the usual precursors; it is epidermal cells that typically bear cilia and microvilli, and outer epidermal cells would be the first cells through which light would pass on entering an animal body. A third, less common type, called an epigenous photoreceptor, is derived from the dendritic surface of a sensory neuron and is best known in rotifers and nematodes (p. 247).

Both rhabdomeric and ciliary photoreceptors have now been found in larvae or adults of many of the animal groups. However, the rhabdomeric type predominates in the protostome line of evolution (flatworms, mollusks, annelids, arthropods) and is the only type found among arthropods. Certain polyclad flatworms, which have rhabdomeric photoreceptors in the adult eye, have ciliary eyes in addition to rhabdomeric eyes in the larval stage (Eakin and Brandenberger, 1981).

Although vertebrates possess only the ciliary type, echinoderms are rhabdomeric. The ciliary type, then, does not predominate in deuterostomes, as was once thought.

The photoreceptors of some animals, such as earthworms, are dispersed over the integument. Thus, earthworms can respond to light even though they have no eyes. However, in most animals the photoreceptors are concentrated. Such a concentration, called an eye, enables the animal to utilize other information provided by light besides general light intensity.

The simplest eye, called an ocellus, is organized as a pigment spot or pigment cup, and is most commonly associated with the integument. A pigment spot is a patch of photoreceptors interspersed with pigment cells (C). In a pigment-cup ocellus the pigment cells form a cup into which the photoreceptor elements project (E). When the photoreceptors project between the pigment cells into the lumen of the cup, the ocellus is said to be everted (D); this arrangement is typical of integumental ocelli. In turbellarians the pigment cup is in the parenchyma, and the photoreceptors project into the cup opening, not between pigment cells (E). An ocellus of this type is said to be inverted. The pigment shades the photoreceptors, enabling the photoreceptors to detect the direction of the light source. Directionality is increased if the ocellus takes the form of a cup (E). An excellent review of the origin of photoreceptors and eyes is provided by Burr (1984). Box 11–1 discusses further specialization of invertebrate eyes. *A* and *B* greatly modified from Eakin, R. M., 1968: Evolution of photoreceptors. *In* Dobzhansky, T., et al (Eds.): Evolutionary Biology. Vol. 2. Appleton-Century-Crofts, N.Y. p. 206. *C* and *D* adapted from Singla, C. L., 1974: Ocelli of hydromedusae. Cell Tiss. Res., *149*:413–429.)

---

posterior representing the other. Regeneration is correlated with this polarity. For example, an excised piece retains its polarity, with the cut surface toward the anterior producing a new head and the posterior surface producing a new tail. One region suppresses the regeneration of the same region at another level of the body. For example, head extract added to the culture medium of a headless worm inhibits regeneration of a new head. The rapidity of the process and the size of pieces necessary for regeneration are related to the original gradient. The process is fastest anteriorly; for normal regeneration, pieces cut more posteriorly must be larger. Neurosecretion appears to be involved in asexual reproduction and regeneration, but its precise role is uncertain (see Highnam and Hill, 1977).

Sexual Reproduction

Except in acoels, the gonads are distinct from the surrounding parenchyma, although the germ cells apparently originate in the parenchyma and migrate into the gonads.

The male and female systems are complicated and variable, but *Macrostomum* will serve to illustrate the basic and perhaps primitive plan. A sperm duct leads from the single pair of testes to the seminal vesicle (Fig. 7–7A). The latter passes into the

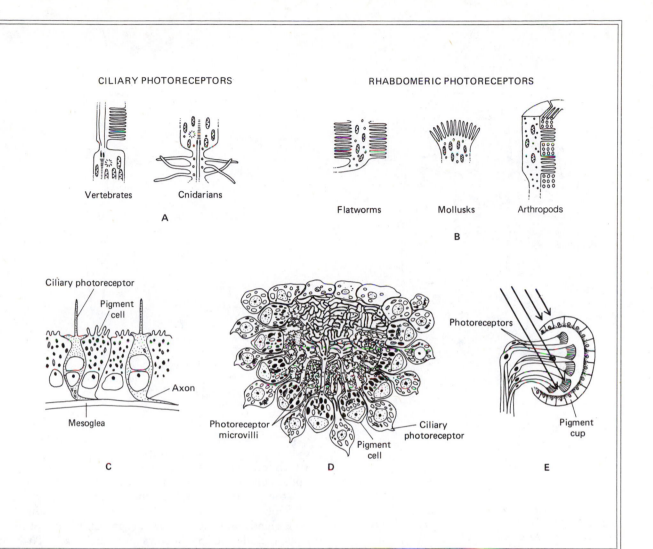

CILIARY PHOTORECEPTORS

Vertebrates   Cnidarians

A

RHABDOMERIC PHOTORECEPTORS

Flatworms   Mollusks   Arthropods

B

Ciliary photoreceptor
Pigment cell
Axon
Mesoglea

C

Photoreceptor microvilli
Pigment cell
Ciliary photoreceptor

D

Photoreceptors
Pigment cup

E

penis bulb, which bears duct openings of the prostatic glands and is armed with a hollow stylet (Fig. 7–20*A*). The penis lies within the male genital canal, which opens through the male gonopore onto the posterior ventral surface. An oviduct leads from the pair of ovaries to the bursa, a sperm storage center, and the latter communicates with the vagina and female gonopore located in front of the male opening. Cement glands surround the vagina.

In other turbellarians more than one pair of testes may be present. The penis is generally muscular but does not always bear a stylet. There are also a number of turbellarians, including some polyclads, that display the peculiar condition of hav-

ing multiple male parts, such as prostatic glands, seminal vesicles, and penises (Fig. 7–20*B*). However, in at least some of these worms the multiple penis bulbs and stylets function in defense rather than in reproduction. Most turbellarians have biflagellate sperm with a 9-1 axoneme (or 9-0 in some acoels), a condition that does not seem primitive. Perhaps the original condition is reflected in *Nemertoderma*, which has conventional uniflagellate sperm (Tyler and Rieger, 1975).

Similarly, the female system may have one to numerous pairs of ovaries but only one pair of oviducts. In some turbellarians the ovaries are like those of other animals and produce eggs in which

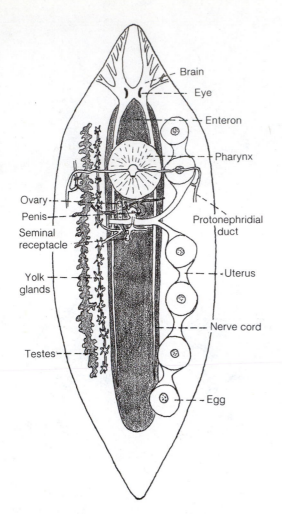

Figure 7–18   The rhabdocoel *Mesostoma ehrenbergii* (ventral view). (After Ruebush.)

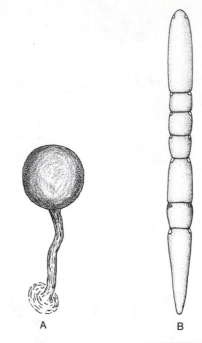

Figure 7–19   *A*, Stalked cocoon of a freshwater planarian. (Modified from Pennak.) *B*, Chain of zooids in *Stenostomum*. (After Child from Hyman.)

yolk material is an integral part of the cytoplasm (Fig. 7–7). In many turbellarians, however, the ovary is of a more specialized form. A division of labor has evolved, and part of the ovary, called the vitellarium (or sometimes yolk gland), has become specialized for the production of yolk cells (modified eggs), while part of the ovary, called the germarium, has become specialized for the production of yolkless eggs. The germarium and vitellarium may be united, or they may be completely separated, with a special duct bringing the yolk cells to the oviduct (Fig. 7–10). In either case, the egg, after being released from the germinarium, becomes surrounded by a number of yolk cells, and the entire mass is deposited.

The two types of ovaries are the basis for distinguishing the two levels of organization in the Tur-

bellaria, namely the Archoophora and the Neoophora. The more primitive Archoophora, in which there are no yolk glands and the eggs contain yolk as in other animals, include the acoels, the macrostomids, the catenulids, and the polyclads. The Neoophora, in which the female system contains yolk glands and the eggs are accompanied by yolk cells, include the rhabdocoels and the triclads. The parasitic classes of flatworms, containing the flukes and tapeworms, also possess specialized ovaries with vitellaria.

In addition to the bursa, there may be still another storage center, a seminal receptacle (Fig. 7–18). The sperm are stored in the seminal receptacle until fertilization. The seminal receptacle may receive sperm from the copulatory bursa or, if a bursa is absent, directly after copulation.

Another modification sometimes present in the female system is a temporary storage center, or uterus, for ripe eggs. The uterus may be a blind sac (Fig. 7–18), as in some rhabdocoels, or it may be merely a dilated part of the oviduct, as in the polyclads. However, most turbellarians lack uteri because only a few eggs are laid at a time.

In acoels and catenulids the female system is less well developed than in other turbellarians.

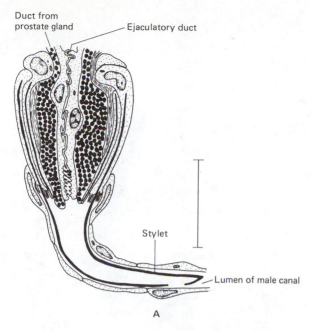

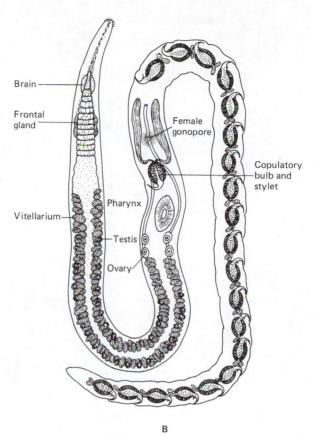

Figure 7–20 *A*, Male copulatory organ of *Macrostomum*. The ejaculatory duct from the seminal vesicle opens into the base of the hollow stylet, which carries an opening just behind the tip. The male canal leads to the gonopore (see Fig. 7–7*A*). (Redrawn from Doe, D. A., 1984: Ultrastructure of copulatory organs in Turbellaria. 1. *Macrostomum* sp. and *Microstomum* sp. Zoomorph., *101*:39–59.) *B*, *Polystyliphora* (suborder Proseriata), a turbellarian with multiple copulatory bulbs and stylets. Since only the first receives sperm, the others may serve a defensive function. (From Ax, P., 1958: Zool. Anz. Suppl., *21*:227–249.)

Some possess no female ducts at all, not even a gonopore. In others, there are no oviducts, but a short blind vagina for receiving the penis leads from a female gonopore (Fig. 7–7*B*). Some zoologists believe that this condition in acoels is a reduction from the more developed condition described for *Macrostomum*. Others consider it primitive. Separate male and female gonopores are characteristic of most macrostomids, the acoels, and the polyclads, and this is probably the primitive turbellarian condition. Many turbellarians, however, including the common planarian, possess a single gonopore and genital atrium into which both male and female systems open.

Sperm transfer in turbellarians involves copulation and is usually reciprocal. In most turbellarians the penis is inserted into the female gonopore or common gonopore of the partner (Fig. 7–21*C*). During copulation the worms orient themselves in a variety of ways with the ventral surfaces around

the genital region pressed together and elevated (Fig. 7–21*B*). Hypodermic impregnation occurs in some acoels, macrostomids, rhabdocoels, and polyclads. The penis, which bears stylets, is rammed through the body wall of the copulating partner, depositing sperm into the parenchyma (Figs. 7–20*A* and 7–21*A*). The sperm then migrate to the ovaries. Self-fertilization probably occurs in some species but is the exception and not the rule.

Turbellarians that possess oviducts but lack yolk glands release only a small number of eggs. The marine polyclads lay their eggs in gelatinous strings, and an individual may lay a number of egg masses. Acoels that have no gonoducts release their eggs through the mouth or by temporary rupture of the body wall.

Rhabdocoels and triclads possess yolk glands, and as a result, egg production is somewhat modified. As the fertilized eggs pass through the oviduct, they are accompanied by yolk cells produced

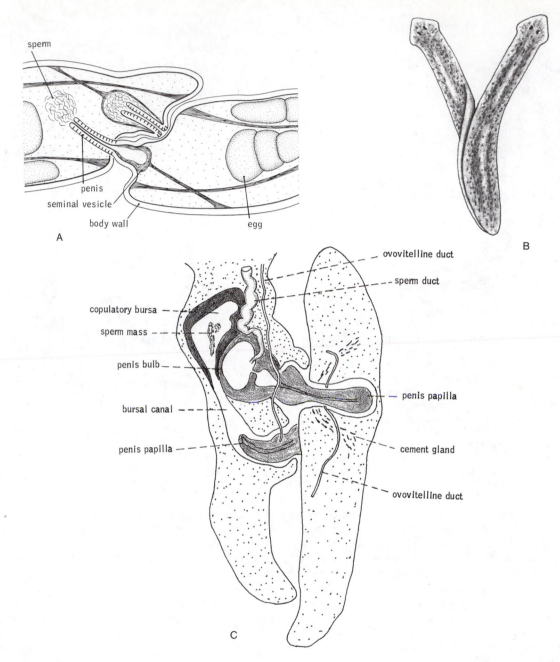

Figure 7–21   Copulation in turbellarians: *A*, Detailed section of hypodermic impregnation between two individuals of the acoel *Archaphanostoma agile.* (After Apelt.) *B*, Copulating planarians. *C*, Section through a pair of copulating *Dugesia.* (After Burr from Hyman.)

by the yolk glands. On reaching the atrium, one or several eggs, along with many yolk cells, are enclosed by a hard capsule that is cemented to the substratum. In many forms, including some freshwater triclads, such as the common *Dugesia*, the

brownish capsules are attached to stalks (Fig. 7–19*A*) and resemble little balloons stuck to rocks and other debris. One worm can produce a number of capsules, in each of which several embryos may develop (Fig. 7–23*E*).

Figure 7–22 *Mesostoma lingua* with resting eggs. (From Mac-Fira, V., 1974: The turbellarian fauna of the Romanian littoral waters of the Black Sea and its annexes. *In* Riser, N. W., and Morse, M. P. (Eds.): Biology of Turbellaria. McGraw-Hill Book Co., N.Y. p. 263.)

Some freshwater turbellarians often produce two types of eggs—summer eggs (subitaneous eggs), which are enclosed in a thin capsule and hatch in a relatively short period, and autumn eggs (resting eggs), which have a thicker and more resistant capsule (Fig. 7–22). Resting eggs remain dormant during the winter, resisting freezing and drying, and hatch in the spring with the rise in water temperature. In *Mesostoma ehrenbergii* the generation time is between 16 and 75 days, depending on water temperatures, and one individual produces about 15 summer eggs or 45 resting eggs. The life span is between 65 and 140 days (Heitkamp, 1977). Parthenogenesis is characteristic of some turbellarians, and there are some parthenogenic species in which males are unknown.

As we have seen, the eggs of archoophoran turbellarians—acoels, polyclads, and macrostomids—are entolecithal; i.e., the yolk is an integral part of the egg cytoplasm. Early development in these groups involves spiral cleavage like that in other protostome phyla (see p. 61). Gastrulation is by epiboly and produces a stereogastrula. The mouth and pharynx form a stomodeal invagination near the original site of the blastopore. This invagination connects with the enteron, which has formed from an entodermal mass and become hollow.

In most polyclads there are no larvae, but in some there is a free-swimming stage called a Müller's larva (Fig. 7–23*A*). Eight arms or lobes, which are directed posteriorly and bear long cilia, form extensions of the body. The ciliary tuft of the frontal organ projects forward. The larva swims about for a few days and then settles to the bottom as a young worm (Fig. 7–23*B,C,* and *D*) (Ruppert, 1978).

The eggs of neoophorans—triclads and many rhabdocoels—are ectolecithal, i.e., there are yolk glands and the yolk is separate and outside the egg. These groups have undoubtedly evolved from forms that had spiral cleavage, but in most species the presence of external yolk cells has so altered the pattern of development that it bears little resemblance to the ancestral type. There is no free-swimming larval stage in the orders of Turbellaria that have ectolecithal eggs; the young worms emerge from the capsule in a few weeks.

Turbellarian Origins

Although most zoologists agree that the neoophorans represent an advanced level of organization and that acoels, macrostomids, and catenulids are primitive groups, there are different viewpoints as to the nature of the ancestral turbellarians. A widely held view considers the Macrostomida to be closest to the base of the turbellarian phylogenetic tree (Fig. 7–24). As elaborated by Ax (1963), such a "macrostomid" ancestor possessed a simple pharynx, a ciliated intestine without diverticula, a radial arrangement of longitudinal nerve cords, a statocyst, a frontal gland, a pair of protonephridia, an hermaphroditic system like that described for *Macrostomum*, and entolecithal eggs with spiral cleavage. The Acoela and Catenulida, although displaying some primitive features, differ significantly from the Macrostomida, Polycladida, and Neoophora. Most specialists now consider the acoels too remote to hold the ancestral position formerly accorded them. Indeed, Smith et al (1985) believe that the Acoela-Nemertodermatida, the Catenulida, and the remaining orders constitute three different turbellarian assemblages, each difficult to relate to the other (Fig. 7–24).

Whichever living group of turbellarians is considered the most primitive, there is still the question of how turbellarians are related to other meta-

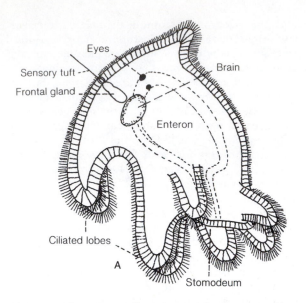

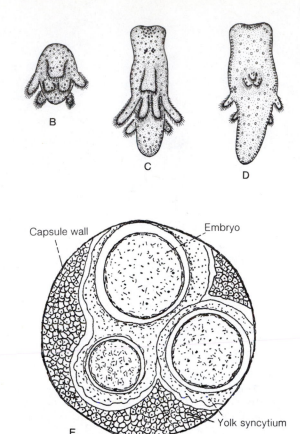

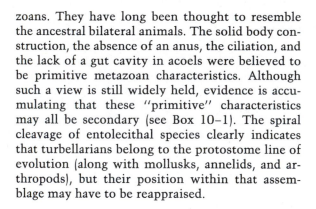

Figure 7–23 *A*, A Müller's larva (lateral view). (After Kato from Hyman.) *B–D*, Metamorphosis of the Müller's larva of *Yungia aurantiaca*. (After Lang from Ruppert, E. E., 1978: A review of metamorphosis of turbellarian larvae. *In* Chia, F., and Rice, M. E. (Eds.): Settlement and Metamorphosis of Marine Invertebrate Larvae. Elsevier-North Holland Biomedical Press, Amsterdam. p. 73.) *E*, Section through a triclad capsule, showing three embryos embedded in yolk syncytium. (After Metschnikoff from Hyman.)

zoans. They have long been thought to resemble the ancestral bilateral animals. The solid body construction, the absence of an anus, the ciliation, and the lack of a gut cavity in acoels were believed to be primitive metazoan characteristics. Although such a view is still widely held, evidence is accumulating that these "primitive" characteristics may all be secondary (see Box 10–1). The spiral cleavage of entolecithal species clearly indicates that turbellarians belong to the protostome line of evolution (along with mollusks, annelids, and arthropods), but their position within that assemblage may have to be reappraised.

SYSTEMATIC RÉSUMÉ OF CLASS TURBELLARIA

Archoophoran turbellarians. Orders that reflect a more primitive level of organization. Yolk glands absent; eggs entolecithal; cleavage spiral.

Order Acoela. Small marine flatworms, usually less than 2 mm in length. Mouth and sometimes a simple pharynx present, but no digestive cavity. Protonephridia absent. Gonads often not bounded by a cellular wall. Oviducts absent. *Amphiscolops, Anaperus, Afronta, Polychoerus, Convoluta, Archaphanostoma, Pseudaphanostoma.* A few species are commensal within the intestine of various echinoderms.

Order Nemertodermatida. Small marine species similar to acoels but possessing uniflagellate sperm and an epithelial digestive tract. *Nemertoderma.*

Order Macrostomida. Small marine and freshwater species having a simple saclike ciliated intestine, a simple pharynx, and one pair of ventrolateral nerve cords. *Macrostomum, Microstomum.*

Order Haplopharyngida. Small marine species similar to macrostomids but possessing a proboscis and a temporary anus. *Haplopharynx.*

Order Catenulida. Mostly small, freshwater species having a simple pharynx and ciliated, saclike intestine, and unpaired gonads, with the male gonopore dorsal above the pharynx. No female gonoducts. *Stenostomum, Catenula.*

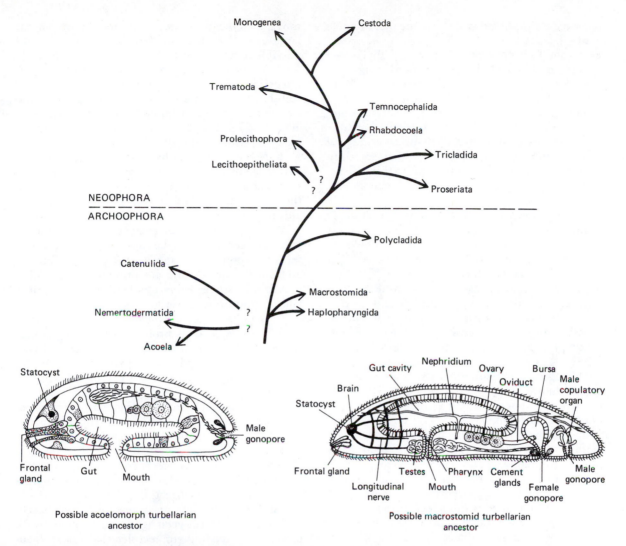

Figure 7–24 Possible phylogenetic tree of the Platyhelminthes based in part on the ideas of Ehlers (1984), Smith et al (1985), and Llewellyn (1965). A turbellarian ancestor similar to living macrostomids (Ax, 1963) is more widely accepted at the present time than one similar to the acoels, although the latter view was long held in the past. (Figure of acoelomorph ancestor is after Smith et al, 1985.)

Order Polycladida. Marine flatworms of moderate size, averaging 3 to 20 mm in length, with a greatly flattened and more or less oval shape. A pair of anterior marginal or dorsal tentacles may be present. Many are brightly colored. Intestine elongate and centrally located, with many highly branched diverticula. Plicate pharynx either an anteriorly directed tube or pendant from the roof of the pharyngeal cavity. Eyes numerous. *Gnesioceros, Leptoplana, Notoplana, Stylochus, Prostheceraeus.*

Neoophoran turbellarians. Orders that reflect an advanced level of organization. Yolk glands present, eggs ectolecithal, and development greatly modified from the spiral pattern.

Order Prolecithophora. Usually small, marine and freshwater species having a plicate or bulbous pharynx and a simple intestine. Ovary produces eggs and follicle-like yolk cells. *Plagiostomum.*

Order Lecithoepitheliata. Marine and freshwater species in which ovary produces eggs surrounded by follicle-like yolk cells. Mouth and complex pharynx at anterior end of body, intestine simple. *Prorhynchus.*

Order Rhabdocoela. A large group of small marine and freshwater turbellarians, having a bulbous pharynx, a simple intestine, and one pair of nerve cords.

Suborder Typhloplanoida. Mouth usually located in the middle and pharynx oriented at right angles to long axis. Contains marine and freshwater, free-living species. *Mesostoma.*

Suborder Dalyellioida. Mouth typically at anterior end of body and pharynx oriented parallel to long axis of body. Contains marine and freshwater species, some of which are commensal and parasitic on and within snails, clams, sea urchins, and sea cucumbers. *Anoplodiera.*

Suborder Kalyptorhynchia. Mouth anterior to middle of body. Anterior protrusible; muscular proboscis often bears cuticularized hooks or teeth. Contains mostly marine interstitial species. *Gyratrix, Gnathorhynchus.*

Suborder Temnocephalida. Commensal and parasitic on crustaceans, mollusks, and turtles. Posterior ventral surface provided with an adhesive disc, and anterior margin bears fingerlike projections, by which worm moves leechlike on its host.

Order Proseriata. Small, mostly marine turbellarians, including many interstitial forms. Pharynx is plicate and tubular, but gut is not branched. *Otoplana, Nemertoplana.*

Order Tricladida. Relatively large, marine, freshwater, and terrestrial turbellarians. Pharynx is plicate, tubular, and posteriorly directed; gut has three branches. Among marine species, *Bdelloura* is commensal on the book gills of horseshoe crabs. The freshwater species are known as planarians and include *Planaria, Dendrocoelum, Procerodes, Dugesia, Phagocata, Polycelis, Procotyla.* Land planarians include *Bipalium, Orthodemus, Geoplana.*

SUMMARY

1 The free-living Platyhelminthes, members of the class Turbellaria, are small, bilaterally symmetrical animals with a low level of cephalization and an acoelomate type of body construction.

2 The majority of turbellarians are marine, but there are freshwater species and a few terrestrial forms in humid environments. Turbellarians are benthic animals, living on or beneath stones, algae, and other objects. They are common members of the interstitial fauna.

3 Most turbellarians move entirely by cilia; large species (polyclads) are markedly flattened and move by cilia plus muscular undulation on the large ventral surface area. Adhesive organs make possible temporary attachment in many species.

4 Turbellarians are predators and scavengers. Digestion is initially extracellular and then intracellular. Small species have a simple saclike intestine with a simple or bulbous pharynx. Large species have a branched intestine and a plicate pharynx, usually tubular.

5 Mucus produced by rhabdites and other epidermal glands plays an important role in the life of turbellarians, coating the substratum over which the animal crawls and swathing prey. Mucus also aids in prey trapping and swallowing.

6 The small size, flattened shape, and branched gut (in larger forms) make unnecessary special systems for internal transport, gas exchange, and excretion. Protonephridia are present in many flatworms and are probably involved in internal fluid balance and osmoregulation.

7 A radial arrangement of four pairs of longitudinal nerve cords is probably primitive, and arrangements with lesser numbers have likely evolved through reduction. Pigment cup ocelli, which may be numerous, are the principal sense organs.

8 Turbellarians are simultaneous hermaphrodites with the reproductive system adapted for internal fertilization and egg deposition.

9 At the primitive, archoophoran level the eggs are entolecithal, cleavage is spiral, and there is a free-swimming larva. In most archoophoran species development is direct, however.

10 Many turbellarians have evolved an ovarian division of labor between egg production and yolk production leading to ectolecithal eggs (neoophoran level). Spiral cleavage has been lost, and development is always direct.

Flukes

The classes Monogenea and Trematoda, to which flukes belong, contain over 8000 species of ectoparasites and endoparasites. The majority are parasites of vertebrates, especially fish, but immature stages are harbored by invertebrates. Many species are of great economic and medical importance.*

*The following accounts of the flukes and tapeworms are designed to accommodate those courses that cover the parasites in their survey of the invertebrates. They are of necessity much briefer than the account of the turbellarians.

Structure and Physiology

The body of flukes* is oval to elongate, usually not more than a few cm long, and the mouth is typically at the anterior end. Adhesive suckers are usually present around the mouth and may also be present midventrally. The monogenetic flukes possess large posterior attachment organs, called opisthaptors, provided with various structures, such as suckers and hooks (Fig. 7–25).

In contrast to the ciliated epidermis of the turbellarian, the body of a trematode is covered by a

*The older system of classification, still widely encountered, makes the Monogenea an order of the class Trematoda. In such a system the terms *fluke* and *trematode* are synonymous.

nonciliated cytoplasmic syncytium, the tegument, overlying consecutive layers of circular, longitudinal, and diagonal muscle. The syncytium represents extensions of cells that are located in the parenchyma (Fig. 7–26).

The mouth leads into a muscular pharynx that pumps into the digestive tract the cells and cell fragments, mucus, tissue fluids, or blood of the host on which the parasite feeds. The pharynx passes into a short esophagus and one or, more commonly, two blind intestinal ceca that extend posteriorly along the length of the body (Fig. 7–27A). The physiology of nutrition is still incompletely understood, but secretive and absorptive cells have been reported, so digestion is apparently extracellular in part.

The tegument plays a vital role in the physiology of flukes. It provides protection, especially against the host's enzymes in gut-inhabiting species. Nitrogenous wastes are passed to the exterior through the tegument, and it is the site of gas exchange. In endoparasites the tegument absorbs some amino acids. The protein synthesis involved in fluke egg production and in larval reproduction places especially heavy demands on the amino acid supply.

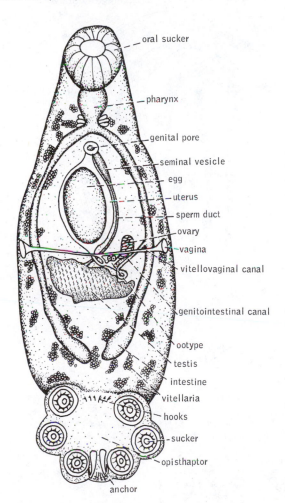

oral sucker
pharynx
genital pore
seminal vesicle
egg
uterus
sperm duct
ovary
vagina
vitellovaginal canal
genitointestinal canal
ootype
testis
intestine
vitellaria
hooks
sucker
opisthaptor
anchor

Figure 7–25 *Polystomoidella oblongum,* a monogenetic fluke parasitic in the urinary bladder of turtles. (After Cable.)

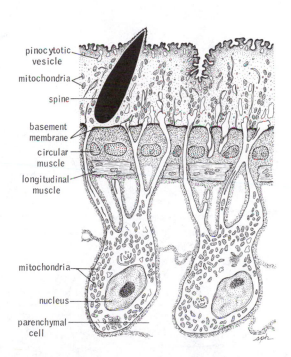

pinocytotic vesicle
mitochondria
spine
basement membrane
circular muscle
longitudinal muscle
mitochondria
nucleus
parenchymal cell

Figure 7–26 Section through the integument of the sheep liver fluke, *Fasciola hepatica.* (After Threadgold.)

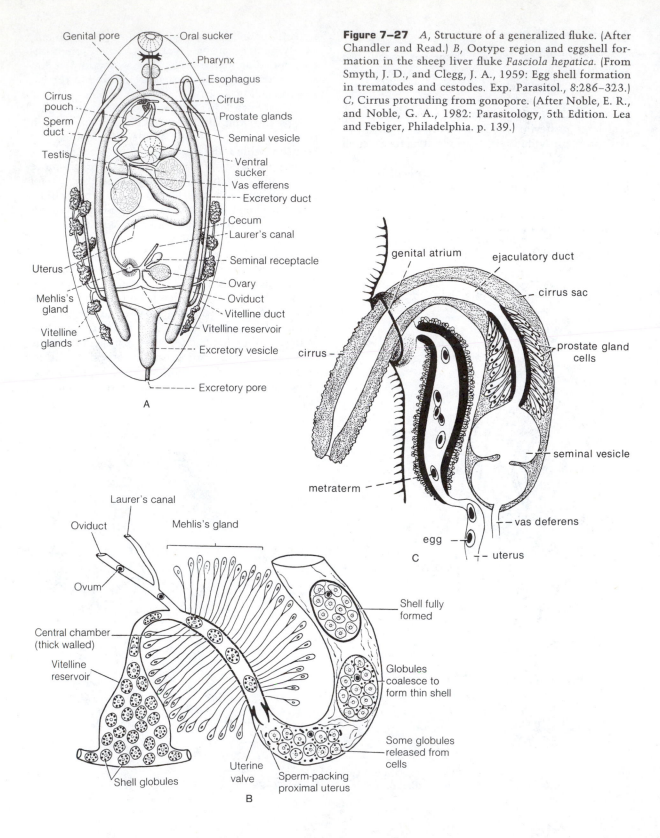

Figure 7–27 *A*, Structure of a generalized fluke. (After Chandler and Read.) *B*, Ootype region and eggshell formation in the sheep liver fluke *Fasciola hepatica*. (From Smyth, J. D., and Clegg, J. A., 1959: Egg shell formation in trematodes and cestodes. Exp. Parasitol., 8:286–323.) *C*, Cirrus protruding from gonopore. (After Noble, E. R., and Noble, G. A., 1982: Parasitology, 5th Edition. Lea and Febiger, Philadelphia. p. 139.)

A labels:
Genital pore, Oral sucker, Pharynx, Esophagus, Cirrus pouch, Cirrus, Sperm duct, Prostate glands, Testis, Seminal vesicle, Ventral sucker, Vas efferens, Excretory duct, Cecum, Laurer's canal, Seminal receptacle, Uterus, Ovary, Mehlis's gland, Oviduct, Vitelline duct, Vitelline glands, Vitelline reservoir, Excretory vesicle, Excretory pore

C labels:
genital atrium, ejaculatory duct, cirrus sac, prostate gland cells, cirrus, seminal vesicle, metraterm, vas deferens, egg, uterus

B labels:
Laurer's canal, Oviduct, Mehlis's gland, Ovum, Shell fully formed, Central chamber (thick walled), Globules coalesce to form thin shell, Vitelline reservoir, Some globules released from cells, Shell globules, Uterine valve, Sperm-packing proximal uterus

The ectoparasitic flukes are aerobic, but the endoparasites are facultative anaerobes. The amount of oxygen utilized in respiration depends on the location within the host and also on the developmental stage of the parasite (see Smyth and Halton, 1983, for physiology of trematodes).

Flukes, like other flatworms, have protonephridia, and there is typically a pair of longitudinal collecting ducts. There may be two anterior, dorsolateral nephridiopores (in Monogenea) or a single posterior bladder and nephridiopore (in Trematoda) (Fig. 7–32*A*). In the ectoparasites, the protonephridia are probably only osmoregulatory in function. The function of the protonephridia in endoparasites is still uncertain.

The fluke nervous system is essentially like that of turbellarians. There is a pair of anterior cerebral ganglia from which usually three pairs of longitudinal nerve cords extend posteriorly. The ventral pair is most highly developed, and the dorsal pair is absent in the Trematoda. The fluke body surface has a variety of sensory papillae, and in many ectoparasites there are one or two pairs of ocelli.

Reproduction

The reproductive system is relatively uniform throughout the two classes (Fig. 7–27). There are usually two testes, which is probably the primitive number, and the position of the testes is of taxonomic importance. Sperm ducts, one from each testis, unite anteriorly and then enter a copulatory organ or cirrus sac. The latter (mostly in the Trematoda) contains the seminal vesicle, prostate glands, and copulatory apparatus. The copulatory apparatus is at the terminal end of the male system and is called a cirrus if it is eversible and a penis if not. The copulatory apparatus opens into a genital atrium shared with the female system. The gonopore is usually located on the midventral surface in the anterior half of the worm. There are many variations of the general plan just described.

The central structure of the female system is a small chamber called the ootype, which receives eggs, sperm, and yolk cells via a short ovovitelline duct (Fig. 7–27). The eggs are produced by the usually single ovary, and the oviduct is joined by a duct from the seminal receptacle and a common duct from the right and left yolk glands. The ootype is surrounded by unicellular gland cells, called collectively the Mehlis's gland. Leaving the ootype is the uterus, which runs anteriorly to the genital atrium. One or two vaginas (Monogenea) open separately to the exterior on the dorsal, lateral, or ventral surface. A part of the length may also be modified as a seminal receptacle. In many trematodes there is a vestigial copulatory canal (Laurer's canal) that extends from the duct of the seminal receptacle to the dorsal surface of the worm.

Sperm, on leaving the testes, are stored in the seminal vesicle. Copulation is mutual and cross fertilization is the general rule, although self-fertilization does occur. During copulation the cirrus or penis of the male system of one worm is inserted into the uterine or vaginal opening of the other worm and sperm are ejaculated. The prostate gland provides semen for sperm survival. Sperm travel down the uterus or the vaginal canal to be stored in the seminal receptacle.

Once released from the ovary, the eggs are fertilized either en route to the ootype or within the ootype. Fluke eggs are ectolecithal, as are those of neoophoran turbellarians. The vitelline glands supply yolk material for the egg and also a material that hardens around the egg to form the shell.

The encapsulated eggs pass through the uterus to be expelled. The function of the Mehlis's gland (around the ootype) is uncertain, but its secretions may provide lubrication for the passage of eggs through the uterus. Compared with the free-living turbellarians, the number of eggs produced by flukes is enormous. Cheng (1973) cites a figure 10,000 to 100,000 times that for turbellarians.

The fluke life cycle involves one to several hosts. The primary, or definitive, host, that of the adult parasite, is almost always a vertebrate. Fish are by far the principal victims of fluke parasitism. Mammals, which by no means escape infection, are hosts for a relatively small number of flukes.

LIFE CYCLES: CLASS MONOGENEA

There is but a single host in the life cycle, and the monogenetic flukes are so named because there is but one generation in the life cycle; i.e., one egg produces one adult. Monogenea are largely parasitic on marine and freshwater fish, but amphibians, reptiles, and cephalopod mollusks also serve as hosts. The majority are ectoparasites, but some invade body chambers with external openings, such as the mouth, gill chamber, and urogenital tract.

The shell of the elongate egg is provided with a lid and usually one or two threads that attach the egg to the host or serve as flotation devices (Fig. 7–28*A*). The egg, on hatching, releases a free-swimming, ciliated larva called an onchomiracidium, which enables the parasite to reach a new

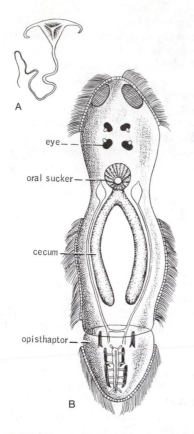

eye

oral sucker

cecum

opisthaptor

A

B

Figure 7—28 Egg *(A)* and onchomiracidium *(B)* of *Benedenia melleni,* a monogenetic fluke parasitic on fish. (Both after Jahn and Kuhn.)

host (Fig. 7–28*B*). The development of the adhesive organ enables the larva to attach. Space permits only a few specific life cycles to be described, and it should be realized that there are many variations.

Polystoma integerrimum is found in the bladders of frogs and toads and is an example of remarkable synchronization of the life cycle with that of the amphibian host (Fig. 7–29). The eggs are shed when the frog or toad returns to the water to breed. The onchomiracidium attaches to the gills of the tadpoles. When the tadpoles metamorphose, the parasite leaves the gill chamber, crawls over the host's belly, and enters the bladder. When the tadpole is very young, some of the larvae may attain a precocious sexual maturity and produce eggs. This ectoparasitic generation dies when metamorphosis occurs.

Neobenedenia melleni is parasitic on the epidermis and eyes of a variety of marine fish. The parasite can cause blindness and serious damage to the host's integument.

Species of the genus *Gyrodactylus* are common ectoparasites on the gills and body surface of marine and freshwater fish and tadpoles (Fig. 7–30) and may be a serious pest in fish hatcheries. A complex and precocious larval development takes place within the uterus, and a single worm can give rise to about 140 descendants in the short space of three weeks.

LIFE CYCLES: CLASS TREMATODA

The class Trematoda (digenetic flukes) is the largest group of parasitic flatworms. Over 6000 species have been described, and new descriptions are continually being published. There are many species that cause parasitic diseases in man and domesticated animals.

In contrast to the monogenetic trematodes, the life cycles of the digenetic trematodes involve two to four hosts. The host for the adult is the definitive host, and the one to three hosts for the numerous developmental stages are termed intermediate hosts. The adhesive organs are typically two large suckers (Fig. 7–32*A*). One sucker, called the oral sucker, is located around the mouth. The other sucker, the acetabulum, is located ventrally in the middle or posterior end of the body.

Most digenetic trematodes are endoparasitic. The definitive hosts include all groups of vertebrates, and virtually any organ system may be infected. The intermediate hosts are largely invertebrates, commonly snails.

The life cycle is complex and will be introduced by a generalized scheme followed by more specific examples. The egg is enclosed within an oval shell with a lid, deposited in the gut, and passed to the outside with the definitive host's feces. A snail may ingest an egg containing a miracidium or the ciliated, free-swimming miracidium hatched from the egg (Fig. 7–31*A*), or the larval stage may penetrate the snail's epidermis. It thus comes to inhabit the hemocoel.

Inside the snail the miracidium, which loses its cilia when it enters the host, begins a second developmental stage, called a sporocyst (Fig. 7–31*B*). Inside the hollow sporocyst, germinal cells give rise to a number of embryonic masses. Each mass develops into another developmental stage, called a redia or daughter sporocyst, which is also a chambered form (Fig. 7–31*C*). Germinal cells within the redia again develop into a number of larvae called cercariae (Fig. 7–31*D* and *E*). The term *digenetic*

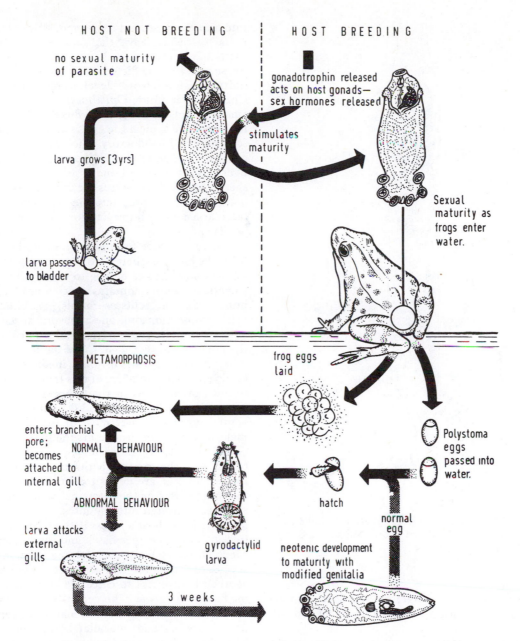

HOST NOT BREEDING | HOST BREEDING

no sexual maturity
of parasite

gonadotrophin released
acts on host gonads—
sex hormones released

stimulates
maturity

larva grows [3yrs]

Sexual
maturity as
frogs enter
water.

larva passes
to bladder

METAMORPHOSIS

frog eggs
laid

Polystoma
eggs
passed into
water.

enters branchial
pore;
becomes
attached to
internal gill

NORMAL BEHAVIOUR

ABNORMAL BEHAVIOUR

hatch

normal
egg

larva attacks
external
gills

gyrodactylid
larva

neotenic development
to maturity with
modified genitalia

3 weeks

Figure 7–29 Life cycle of *Polystoma integerrimum*, a monogenetic fluke found in the bladder of frogs. Diagram also shows a neotenic population parasitic on the gills of the tadpole. (From Smyth, J. D., 1976: Introduction to Animal Parasitology. 2nd Edition, Hodden and Stoughton Educational, Kent. p. 139.)

refers to this second generation of individuals, produced asexually.

The cercaria, a fourth developmental stage, possesses a digestive tract, suckers, and a tail. The cercaria leaves the host and is free swimming. Its sec-ond intermediate host may be an invertebrate (commonly an arthropod) or a vertebrate, in which it encysts. The encysted stage is called a metacercaria (Fig. 7–31F). If the host of the metacercaria is eaten by the final vertebrate host, the metacercaria

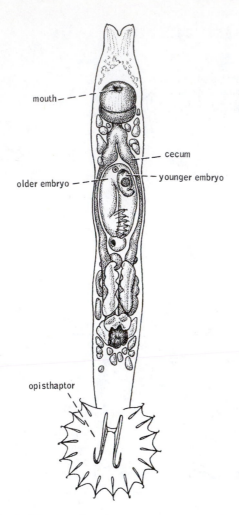

mouth

cecum

older embryo — younger embryo

opisthaptor

Figure 7–30 *Gyrodactylus*, a monogenetic fluke ectoparasitic on fish. (Modified from Mueller and Van Cleave.)

escapes from its cyst, migrates, and develops into the adult form within a characteristic location in the host.

A great many trematodes infect the gut or gut derivatives of their definitive host. Lungs, bile ducts, pancreatic ducts, and intestines are common sites. The life cycle of *Echinostoma ilocanum*, which infects the intestines of a variety of Philippine mammals, including cats, dogs, and man, is a good illustration of the generalized life cycle just described. The eggs pass out with the host's feces and hatch in water. The free-swimming miracidium can penetrate and give rise to a sporocyst in certain species of snails. The sporocyst develops

into a first and second redial generation. From the redia, cercariae leave the snail and enter other snails and clams, where they form metacercariae. If the second intermediate host is ingested by the definitive host, the worm develops into an adult. The Chinese liver fluke, *Opisthorchis* (= *Clonorchis*) *sinensis*, and the sheep liver fluke, *Fasciola hepatica*, also have typical life cycles and are usually described in introductory biology and zoology courses. The intermediate hosts for the Chinese liver fluke are a snail and a fish (Fig. 7–32). The sheep liver fluke possesses the typical larval stages, but a snail is the only intermediate host. The metacercaria encysts on vegetation along the edges of ponds and streams.

There are three families of trematodes that inhabit the blood of their hosts, but certainly the best known blood flukes belong to the family Schistosomatidae, which contains species producing the human disease schistosomiasis, or bilharziasis. *Schistosoma mansoni*, one of several species parasitic in man, occurs in Africa and tropical areas of the New World (Fig. 7–33). The adult, like that of other species of schistosomes, inhabits the intestinal veins. The members of this family, in contrast to most other flatworms, are dioecious. The male is 6 to 10 mm in length and 0.5 mm in width. A ventral groove extends most of the length of the male, and into this groove fits the longer but more slender female. When laying eggs, the female extends from the groove in the male or leaves the male. The eggs are deposited in small venules of the mesenteries or the intestinal wall and eventually break through into the intestinal lumen. They leave with the host's feces. If the feces are deposited in water, the eggs hatch and the miracidia escape. The miracidia penetrate snails belonging to species of the pulmonate genus *Biomphalaria*, and sporocysts give rise to cercariae without an intermediate redia stage. The cercariae leave the snail hosts and, on contact with human skin, penetrate, using enzymes and muscular boring movements. The now tailless larvae are carried by the bloodstream first to the lungs, then to the liver, and finally to the mesenteric veins. During this period the larvae gradually transform into adults.

The two other species that attack man are the oriental *Schistostoma japonicum*, which has a life cycle similar to that of *Schistosoma mansoni*, and *Schistostoma haematobium* of North Africa. The eggs of *Schistosoma haematobium* leave the primary host via the bladder and urine.

Schistosomiasis is seriously debilitating and can

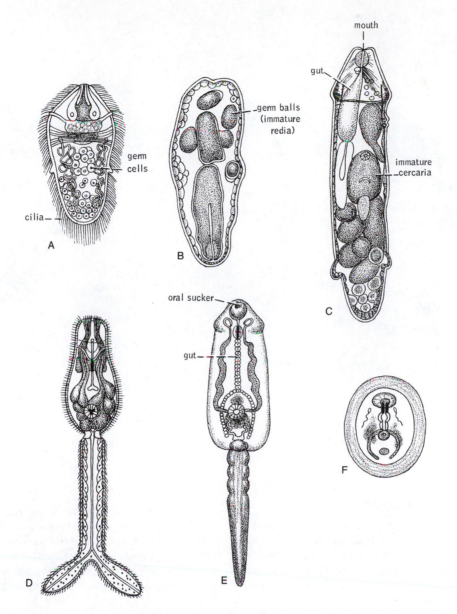

Figure 7–31 Larval types of digenetic flukes: *A*, Miracidium. *B*, Sporocyst. *C*, Redia. *D* and *E*, Cercariae. *F*, Metacercaria. (From the U.S. Naval Medical School Laboratory Manual.)

be lethal. Egg penetration through the intestinal wall and bladder, aberrant lodging of eggs in various organs, and the developmental stages of the larvae in the lung and liver can result in inflammation, necrosis, or fibrosis, depending on the degree of infection. Pathogenic response to the eggs is generally more serious than that to the larvae or adults. With malaria and hookworm, schistosomiasis is one of the three greatest parasitic scourges of mankind. The percentage of the population infected in endemic areas is enormous, and about 300 million humans are estimated to be infected by one of the three species.

Other members of the Schistosomatidae infect various birds and mammals, including domestic species. "Swimmer's itch" is an irritation produced by the incomplete penetration into human skin by cercariae of blood flukes of birds.

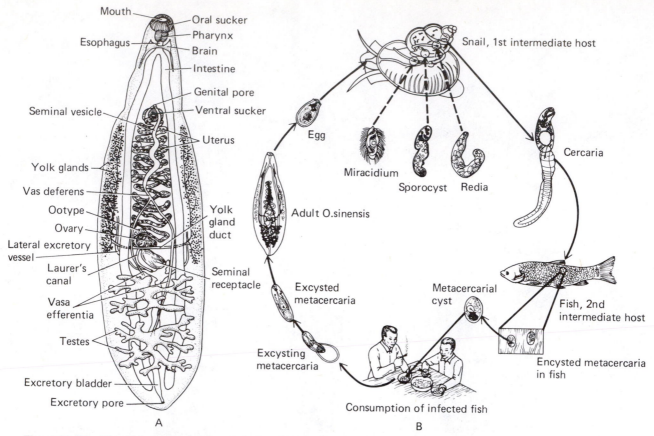

Figure 7–32 The Chinese liver fluke, *Opisthorchis sinensis: A*, Dorsal view of adult worm. *B*, Life cycle. (*A* after Brown from Noble and Noble; *B* after Yoshimura from Noble, E. R., and Noble, G. A., 1982: Parasitology, 5th Edition. Lea and Febiger, Philadelphia.)

SUBCLASS ASPIDOGASTREA

The subclass Aspidogastrea includes a small group of flukes that show similarities to both classes of flukes but appear to be more closely related to the Trematoda. The distinguishing feature is the adhesive organ, which is either a single, lobulate sucker covering the entire ventral surface or a longitudinal row of suckers (Fig. 7–34). The digestive tract contains only one intestinal cecum. The reproductive system is essentially like that of the digenetic trematodes, but there is typically only one testis.

The aspidogastreans are mostly endoparasites in the gut of fish and reptiles and in the pericardial and renal cavities of mollusks. Their life cycles involve one or two hosts.

Evolution of the Flukes

Among the free-living turbellarian flatworms, the rhabdocoels may well have been the ancestors of the trematodes (Fig. 7–24). Some rhabdocoels are ectocommensal on echinoderms and mollusks, the latter being an important host for trematodes. The reproductive systems of both groups are similar. It is quite possible that the monogenetic and digenetic trematodes had independent origins.

Class Cestoda (Cestoidea)

The Cestoda are the most highly specialized of the flatworm classes. All are endoparasites, and the body is covered by a tegument, as in trematodes. The cestodes differ, however, from the members of the other two classes in the complete absence of a digestive tract. The class is divided into two subclasses, the Cestodaria and the Eucestoda. The subclass Cestodaria is a small group showing certain similarities to the trematodes and will be considered briefly after the subclass Eucestoda.

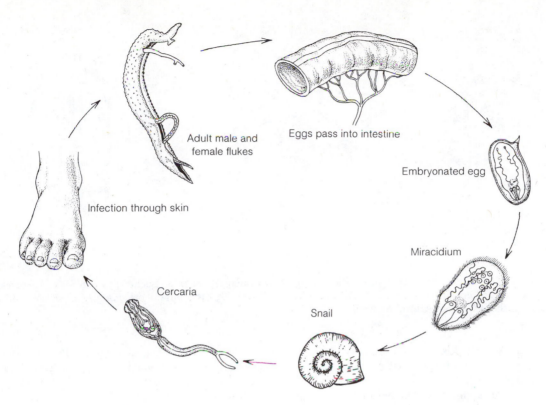

Figure 7–33 Life cycle of *Schistosoma mansoni*. (Modified from several authors.)

SUBCLASS EUCESTODA

Structure and Physiology

The great majority of cestodes belong to the sub-class Eucestoda and are known as tapeworms. The body of the adult tapeworm is unlike that of the other flatworms. There is an anterior region, called the scolex, which is adapted for adhering to the host (Fig. 7–35). Behind the scolex is a narrow neck region that gives rise to the body, sometimes called the strobila. The strobila consists of linearly arranged individual segments, called proglottids, and constitutes the greater part of the worm. Tapeworms are generally long, and some species reach lengths of 40 feet. The scolex, neck, and body are regarded as a single individual and not a colony.

Compared with the mature proglottids, the scolex is relatively small. It generally has the form of a more or less four-sided knob and is provided with suckers or hooks for adhering to the host. Although there are generally four large suckers arranged around the sides of the scolex, the scolex is often a more complicated structure than that of the familiar *Taenia* (Fig. 7–35*A*). The main suckers

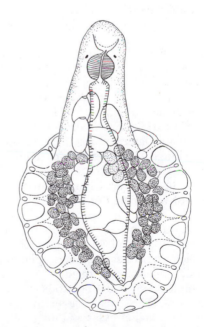

Figure 7–34 Dorsal view of *Cotylaspis insignis*, an aspidogastrean fluke parasitic in freshwater mussels. Note the large ventral sucker with subdivisions. (After Hendrix and Short.)

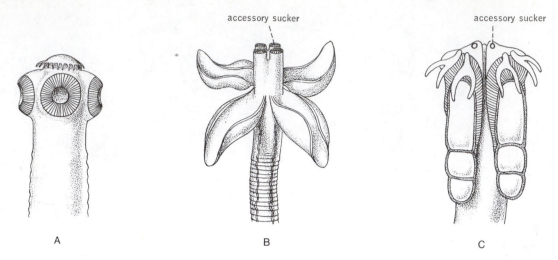

accessory sucker   accessory sucker

A   B   C

Figure 7–35   Scoleces of three tapeworms, showing the four principal suckers or sucker-like adhesive structures, the small accessory suckers, and the hooks: *A, Taenia.* (After Southwell.) *B, Myzophyllobothrium.* (After Shipley and Hornell.) *C, Acanthobothrium.* (After Southwell.)

may be leaflike or ruffled, and there may be terminal accessory suckers in place of or in addition to hooks (Fig. 7–35C).

The neck is a short region behind the scolex, which produces the proglottids by means of transverse constrictions (Fig. 7–35B). The youngest proglottids are thus at the anterior end and increase in size and maturity toward the posterior end of the strobila.

Extending through the chain of proglottids are the nervous system and protonephridial system. An anterior nerve mass lies in the scolex, and two lateral, longitudinal cords extend posteriorly through the strobila (Fig. 7–36). There may also be a dorsal and ventral pair of cords and, quite commonly, accessory lateral cords. Ring commissures connect the longitudinal cords in each proglottid.

Flame cells and tubules in the mesenchyme drain into four peripheral longitudinal collecting canals, two of which are dorsolateral and two ventrolateral (Fig. 7–36). The ventral canals are usually connected by a transverse canal in the posterior end of each proglottid. After proglottids have begun to shed, the collecting ducts open to the exterior through the last proglottid.

The complex body wall of cestodes is illustrated in Figure 7–37. The muscle layer of tapeworms consists of the usual circular and longitudinal layers, but in addition there is a secondary parenchymal musculature of longitudinal, transverse, and dorsoventral fibers, which encloses the interior parenchyma.

The tapeworm tegument plays a vital role in the absorption of food, since tapeworms have no digestive system. The surface of the outer syncytial cytoplasm is thrown into folds, the microtriches (singular, microthrix), which increase the surface area (Fig. 7–37). Anaerobic metabolism apparently predominates in tapeworms but is not their exclusive mode of metabolism. (See Arne, 1983, for review of tapeworm physiology.)

Reproduction

A complete reproductive system occurs within each proglottid and makes up a major part of each of these body sections. As shown in Figure 7–36, the tapeworm reproductive system is basically like that of trematodes. There are usually a common male and female atrium and gonopore. The tapeworm differs from many trematodes, however, in that a vaginal canal extends between the atrium and the ootype. The canal is enlarged as a seminal receptacle. The uterus is usually a blind sac from the ootype or from the atrium and functions solely in egg storage.

Cross fertilization is probably the rule where there are adjacent individuals in the host's gut, but self-fertilization between two proglottids in the same strobila or even within the same proglottid is known to occur. The tendency for the male system to develop before the female system would be an obstacle to self-fertilization within the same proglottid in many species, however.

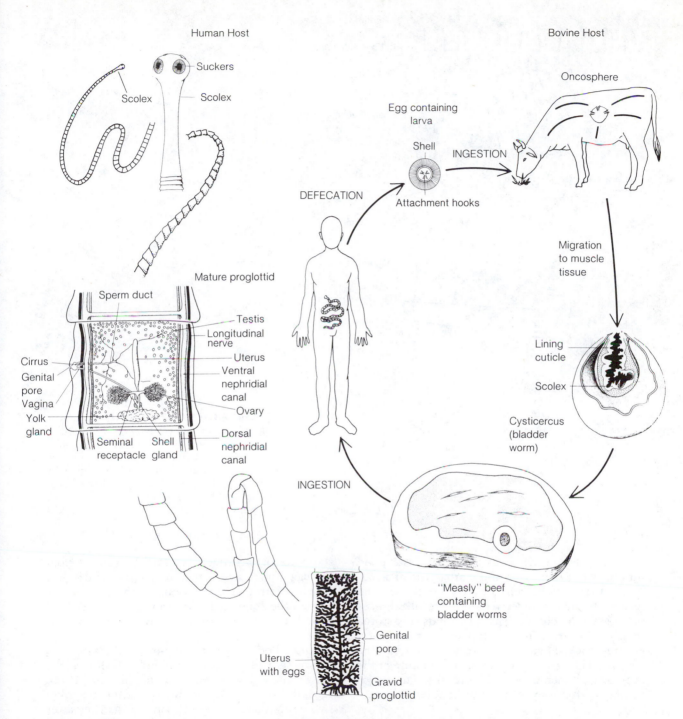

Human Host

Suckers

Scolex

Scolex

Bovine Host

Oncosphere

Egg containing larva

Shell

INGESTION

DEFECATION

Attachment hooks

Migration to muscle tissue

Mature proglottid

Sperm duct

Testis

Longitudinal nerve

Uterus

Cirrus

Genital pore

Vagina

Yolk gland

Ventral nephridial canal

Ovary

Dorsal nephridial canal

Seminal receptacle

Shell gland

Lining cuticle

Scolex

Cysticercus (bladder worm)

INGESTION

"Measly" beef containing bladder worms

Genital pore

Uterus with eggs

Gravid proglottid

Figure 7–36 Structure and life cycle of the beef tapeworm, *Taeniarhynchus saginatus*. (Adapted from various sources.)

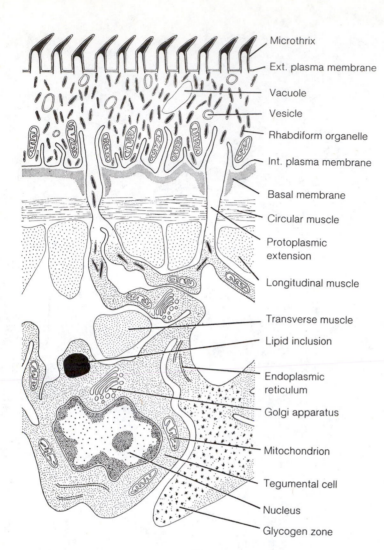

Microthrix
Ext. plasma membrane
Vacuole
Vesicle
Rhabdiform organelle
Int. plasma membrane
Basal membrane
Circular muscle
Protoplasmic extension
Longitudinal muscle
Transverse muscle
Lipid inclusion
Endoplasmic reticulum
Golgi apparatus
Mitochondrion
Tegumental cell
Nucleus
Glycogen zone

Figure 7–37 Section through the tegument of the tapeworm *Caryophyllaeus*. Other than the folding of the surface membrane, note the similarity to the fluke tegument shown in Figure 7–26. (After Beguin from Smyth, J. D., 1969: The Physiology of Cestodes, W. H. Freeman and Co., San Francisco.)

At copulation the cirrus of the male is everted into the vaginal opening of the proglottid of an adjacent worm. Sperm cells are stored in the seminal receptacle and then liberated for the fertilization of the eggs in the ootype. The eggs are usually stored in a blind uterus. In the latter case terminal proglottids packed with eggs break away from the strobila. The eggs are freed with the rupture of the proglottid, which may occur within the host's intestine or after they leave with the feces.

Life Cycles

Tapeworms are endoparasites in the guts of vertebrates. Their life cycles require one, two, or sometimes more intermediate hosts, which are arthropods and vertebrates. The basic developmental stages are an oncosphere larva, which hatches from the egg, and a cysticercus or plerocercoid stage, which is terminal and develops into an adult. Although the following few examples illustrate the basic life cycle patterns of tapeworms, variations exist.

Diphyllobothrium latum, one of the fish tapeworms, is widely distributed and parasitic in the gut of many carnivores, including humans. If the egg is deposited with feces in water, a ciliated, free-swimming oncosphere (coracidium) hatches after an approximately ten-day development (Fig. 7–38A). The larva is ingested by certain copepod crustaceans. It penetrates the intestinal wall and develops within the hemocoel into a six-hooked stage called a procercoid (Fig. 7–38B). When the copepod is ingested by a variety of freshwater fish,

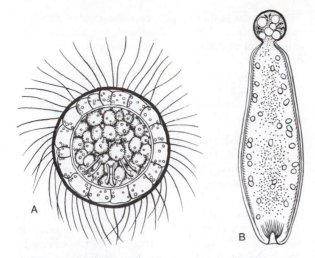

Figure 7–38 Some stages in the life cycles of the fish tapeworm, *Diphyllobothrium latum: A,* Ciliated oncosphere. (After Vergeer.) *B,* Mature procercoid. (After Brumpt.)

the procercoid, like the oncosphere, penetrates the fish's gut and eventually reaches the striated muscles of the fish to develop into a plerocercoid stage. The plerocercoid, which looks like an unsegmented tapeworm, develops into an adult tapeworm when ingested by a definitive host.

Species of the family Taeniidae are among the best known tapeworms. *Taeniarhynchus saginata,* the beef tapeworm, is one of the most common species in humans, where it lives in the intestine and frequently reaches a length of over 3 meters (Fig. 7–36). Proglottids containing embryonated eggs are eliminated through the anus, usually with feces. If an infected person defecates in a pasture, the eggs may be eaten by grazing cattle, sheep, or goats. On hatching in the intermediate host, an oncosphere larva, bearing three pairs of hooks, bores into the intestinal wall, where it is picked up by the circulatory system and transported to striated muscle. Here the larva develops into a cysticercus stage (Fig. 7–36). The cysticercus, sometimes called a bladder worm, is an oval worm about 10 mm in length, with the scolex invaginated. If raw or insufficiently cooked beef is ingested by humans, the cysticercus is freed, the scolex evaginates, and the larva develops into an adult worm in the gut.

Taenia solium, the pork tapeworm, is also a parasite of humans, but the intermediate host is the pig and the cysticercus is obtained from pork. *Taenia pisiformis* occurs in cats and dogs, with rabbits as the intermediate hosts. This order (Cyclo-

phyllidea) contains tapeworms that are largely parasitic in birds and mammals. Vertebrates, insects, mites, annelids, and mollusks serve as intermediate hosts.

A severe infection of adult tapeworms may cause diarrhea, weight loss, and reactions to the toxic wastes of the worm. The worms may be eliminated with drugs. Much more serious is cysticercus infection. Fortunately, the cysticercus stage of the beef tapeworm will not develop within humans, but this is not the case for the pork tapeworm, *Taenia solium,* and for the dog tapeworm, *Echinococcus granulosus.* The adult *Echinococcus,* which lives in the intestine of a dog, is minute, with only a few proglottids present at any one time. Many different mammals, including humans, can act as intermediate hosts, although herbivores are the most important in completing the life cycle. The cysticerci of the pork tapeworm develop in subcutaneous connective tissue and in the eye, brain, heart, and other organs. The bladder worm, or hydatid, of *Echinococcus* develops mostly in the lung or liver but can develop in many other sites as well. The bladder worms of both species can be very dangerous when growing in such places as the brain and can do much damage elsewhere. Hydatid cysts can reach a large size and contain a great volume of fluid (up to many liters), which if released into the host can cause severe reactions. Bladder worm cysts can be removed only by surgery.

SUBCLASS CESTODARIA

The subclass Cestodaria is a small group of cestodes that show some similarities to trematodes. A member of this subclass lacks a scolex and strobila, and the body contains only one hermaphroditic reproductive system. Trematode-like suckers are sometimes present (Fig. 7–39). However, the absence of digestive systems and the similarity of the larvae to those of many tapeworms would place them with the cestodes.

The cestodarians are intestinal and coelomic parasites of sharks, rays, and primitive bony fish. The intermediate hosts are invertebrates.

ORIGIN OF THE CESTODES

Some parasitologists believe that cestodes evolved independently of the flukes from a rhabdocoel turbellarian ancestor; others believe that the cestodes may have evolved from the Monogenea (Fig. 7–24). The position of the cestodarians is an enigma, and their relationship to the eucestodes may be very remote.

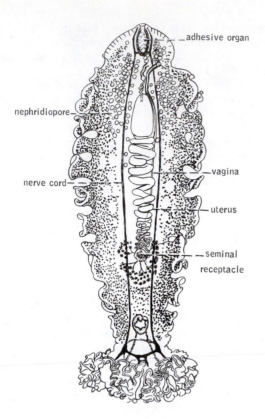

Figure 7–39 *Gyrocotyle fimbriata,* a cestodarian cestode in fish. (After Lynch.)

HOST-PARASITE RELATIONSHIPS

An extensive discussion of host-parasite relationships is not possible in a book of this scope, and the student who is especially interested in the subject should refer to one of the parasitology texts cited in the bibliography of this chapter. Here, however, are a few general principles.

The trematodes and cestodes illustrate most of the adaptations of animals for a parasitic existence:

1 Adhesive organs for attachment to the exterior of the host or to the walls of interior cavities are usually present.

2 Sense organs are reduced, especially in endoparasites.

3 There are various modifications in nutrition, including development of storage areas for ingested food (in the case of ectoparasites), adaptations of the ingestive organs, and direct absorption of food through the body wall.

4 There is an increase in the reproductive capabilities of the parasite through greater egg production, and often asexual reproductive stages.

5 Larval stages that permit passage of the parasite from one host to another are common. The utilization of a single host may be a primitive condition, from which forms having multiple hosts evolved, or a single host may have resulted from the dropping out of intermediate hosts. Undoubtedly, both conditions have evolved in the history of parasitism.

The parasite may have no noticeable effect on the host, or there may be marked effects from a variety of factors. The parasite utilizes food materials that would normally be available to the host. This deprivation usually has little effect on the host unless the infestation is very great. More serious is the damage to the walls of the organ to which the parasite is attached or obstruction of the cavity in which the parasite is living. For example, in heavy infestations flukes and tapeworms may completely occlude the gut or associated ducts. Egg and larval penetration and encystment of developmental stages in various organs are perhaps the causes of the most serious effects of parasitism in the host. The liberation of toxins by the adult or developing parasite can produce deleterious effects on the host and may also be a cause of pathogenicity.

Phylum Mesozoa

The Mesozoa are a small group of some 50 species of minute parasitic animals that have very simple structure but complex life cycles. Marine invertebrates are the hosts. Members of the class Orthonectida have been found in flatworms, nemerteans, polychaetes, bivalve mollusks, brittle stars, and other groups. The dioecious or hermaphroditic adults are unattached within the host, and the microscopic, wormlike body, which lacks organs, consists of an outer layer of ciliated cells enclosing an internal mass of either sperm or egg cells. Adult males and females are released from the host simultaneously, and sperm from the males penetrate the bodies of the females and fertilize the eggs (Fig. 7–40A and B). Cleavage leads to a ciliated larva that is released from the female parent and host and infects a new host, entering through the reproductive openings (Fig. 7–40C). Within host cells the larva loses its cilia and gives rise to a plasmodium that reproduces asexually, forming cell masses that dissociate and are carried to other parts of the host's body (Fig. 7–40D). These asexually produced stages develop into sexually reproducing individuals.

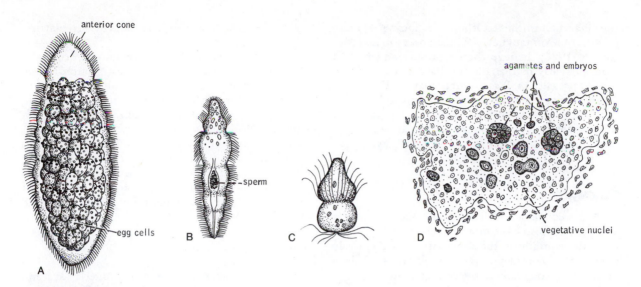

Figure 7–40 Stages in the life cycle of orthonectid mesozoans: *A* and *B*, Mature female and male of the clam parasite, *Rhopalura granosa.* (Both after Atkins.) *C*, Ciliated larva of same species. (After Atkins.) *D*, Male plasmodium of another species of *Rhopalura.* (After Caullery and Mesnil.)

The mesozoans in class Rhombozoa live attached within the nephridial cavities of squids, cuttlefish, and octopods. Although this class was known before the orthonectids, knowledge of its life cycle is still incomplete. The adult rhombozoan is called a nematogen and is 0.5 to 7 mm long. The body, composed of only 20 to 30 cells, consists of a long, central axial cell surrounded by a single layer of ciliated cells. The anterior cells are used for attachment. The population of individuals within the hosts, especially young cephalopods, increases through the production of young having the same form as the parent. Such young are called vermiform embryos and are formed *within* the axial cell of the parent (Fig. 7–41).

In mature cephalopods the nematogens become rhombogens, which have a similar structure to nematogens except that the axial cell of the parasite gives rise to another type of larva, called an infusoriform larva. The axial cell first forms a structure that is interpreted as being a hermaphroditic gonad (infusorigen). An egg produced by this structure and fertilized by sperm produced in the same "gonad" gives rise to the infusoriform larva. All of this takes place within the axial cell of the parent. The infusoriform larvae are small with posteriorly directed cilia (Fig. 7–41). They are passed out with the host's urine, and, although their fate is unknown, it is now believed that they are picked up by their cephalopod host without an intervening intermediate host (Lapan and Morowitz, 1972).

The phylogenetic position of the Mesozoa is an enigma. One view held by many zoologists is that the mesozoans are degenerate flatworms; in fact, Stunkard (1954, 1972) would make them a class of

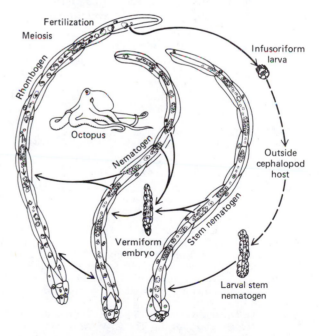

Figure 7–41 Life cycle of a rhombozoan mesozoan, *Dicyemennea*, which lives in the kidney of *Octopus.* (From Hochberg, F. G., 1982: The "kidneys" of cephalopods: a unique habitat for parasites. Malacologia, 23:121–134.)

the Platyhelminthes. Others, including Hyman (1940), McConnaughey (1963), and Lapan and Morowitz (1972), believe they represent an offshoot from the early metazoans and have had a long history of parasitism. It is possible that the Orthonectida are related to flatworms but not the Rhombozoa. The only significance to their being placed in the present chapter is to enable the student to better appreciate the view regarding their phylogeny.

SUMMARY

1 Three classes of flatworms—the Trematoda and the Monogenea, containing the flukes, and the Cestoda, containing the tapeworms—are entirely parasitic. In contrast to the turbellarians, all have nonciliated body coverings, or teguments.

2 Adult flukes are external or internal parasites of vertebrates. They are less modified from the ancestral turbellarian condition than the tapeworms. The oval to elongate body is flattened and provided with a ventral sucker or other attachment organs. A gut is present, and the anterior mouth in some species (Trematoda) is associated with a second sucker.

3 Most flukes are hermaphroditic, and the reproductive systems are adapted for copulation, internal fertilization, ectolecithal development, and the formation of eggshells.

4 The monogenetic flukes are mostly ectoparasites of fish. The life cycle involves a host only for the adult, and a new host is reached by a free-swimming onchomiracidium larva.

5 The digenetic flukes (Trematoda) are endoparasites and constitute the largest group of parasitic flatworms. The life cycle involves two to four hosts and a number of developmental stages, including two types of larvae (miracidium and cercaria). The definitive host is always a vertebrate, and snails are common intermediate hosts. Species of blood flukes *(Schistosoma)* are among the most widespread and serious groups of human parasites.

6 Cestodes, or tapeworms, are gut parasites of vertebrates. They are structurally more specialized than flukes, having a scolex, bearing attachment organs, a neck region, and a strobila, which consists of a chain of segments (proglottids) budded off from the neck region.

7 A gut is absent, and the reproductive system, which is somewhat similar to that of flukes, is repeated in each segment.

8 The life cycle involves an oncosphere larva and an intermediate host.

9 The phylum Mesozoa contains two small groups of minute endoparasites of invertebrates. The group inhabiting the excretory organs of squids, cuttlefish, and octopods is the best known.

10 The simple, elongated, ciliated body lacks the usual cell types and typical organs. Mesozoans are thought to be either derived from flatworms by regression or representative of an offshoot from some early, free-living metazoan. The two groups of mesozoans may have had different origins.

REFERENCES

General Parasitology Textbooks

The following general parasitology texts will provide information on trematodes and cestodes.

Chappell, L. H., 1980: Physiology of Parasites. John Wiley and Sons, N.Y. 230 pp.

Noble, E. R., and Noble, G. A., 1982: Parasitology. 5th Edition. Lea & Febiger, Philadelphia. 522 pp.

Read, C. P., 1970: Parasitism and Symbiology. Ronald Press, N.Y. 316 pp. (A text that examines host-parasite relationships rather than surveying parasitic groups.)

Schmidt, G. D., and Roberts, L. S., 1981: Foundations of Parasitology. 2nd Edition. The C. V. Mosby Co., St. Louis. 795 pp.

Smyth, J. D., 1977: Introduction to Animal Parasitology. 2nd Edition. John Wiley and Sons, N.Y. 466 pp.

Whitfield, P. J., 1979: The Biology of Parasitism: An Introduction to the Study of Associating Organisms. University Park Press, Baltimore. 277 pp.

Flatworms

The literature included here is restricted in large part to flatworms. The introductory references on page 8 include many general works and field guides that contain sections on flatworms.

Apelt, G., 1969: Fortpflanzungsbiologie, Entwicklungszyklen und vergleichende Fruehenwicklung acoeler Turbellarien. Mar. Biol., 4(4):267–325. (Reproductive biology, developmental cycles, and comparative early development of acoel turbellarians.)

Arne, C., 1983: Aspects of tapeworm physiology. J. Biol. Education, 17(4):352–357. (An excellent brief review of tapeworm physiology.)

Arne, C., and Pappas, P. W. (Eds.), 1983: Biology of the Eucestoda. Vols. I and II. Academic Press, London.

Ax, P., 1963: Relationships and phylogeny of the Turbellaria. *In* Dougherty, E. D. (Ed.): The Lower Metazoa. University of California Press, Berkeley. pp. 191–224.

Ax, P., and Apelt, G., 1965: Die "Zooxanthellen" von *Convoluta convoluta* (Turbellaria, Acoela) entstehen aus Diatomen. Naturwissenschaften, *52*(15):444–446.

Ax, P., and Borkett, H., 1968: Organisation und Fortpflanzung von *Macrostomum romanicum* (Turbellaria, Macrostomida). Zool. Anz. Suppl., *32*:344–347.

Bedini, C., and Papi, F., 1974: Fine structure of the turbellarian epidermis. *In* Riser, N. W., and Morse, M. P. (Eds.): Biology of the Turbellaria. McGraw-Hill Book Co., N.Y. pp. 108–147.

Bronsted, H. V., 1969: Planarian Regeneration. Pergamon Press, N.Y.

Burr, A. H., 1984: Evolution of eyes and photoreceptor organelles in the lower phyla. *In* Ali, M. A.: Photoreception and Vision in Invertebrates, Plenum Press, N.Y. pp. 131–178.

Bush, L., 1975: Biology of *Neochildia fusca* n. gen., n. sp. from the northeastern coast of the United States. Biol. Bull., *148*(1):35–48.

Case, T. J., and Washino, R. K., 1979: Flatworm control of mosquito larvae in rice fields. Sci., *206*:1412–1414.

de Beauchamp, P., Caullery, M., Euzet, L., Grasse, P., and Joyeux, C., 1961: Plathelminthes et Mesozaires. *In* Grasse, P. (Ed.): Traite de Zoologie. Vol. 4 pt. 1. Masson et Cie, Paris. pp. 1–729.

Doe, D. A., 1981: Comparative ultrastructure of the pharynx simplex in Turbellaria. Zoomorph. *97*:133–193.

Doe, D. A., 1982: Ultrastructure of copulatory organs in Turbellaria. I. *Macrostomum* sp. and *Microstomum* sp. Zoomorph. *101*:39–59.

Eakin, R. M., and Brandenberger, J. L., 1981: Unique eye of probable evolutionary significance. Sci., *211*:1189–1190.

Ehlers, U., 1984: Phylogenetisches System der Plathelminthes. Verh. Naturwiss. Ver. Hamburg, *27*:291–294.

Erasmus, D. A., 1972: The Biology of Trematodes. Crane, Russak & Co., N.Y. 312 pp.

Faisst, J., Keenan, C. L., and Koopowitz, H., 1980: Neuronal repair and avoidance behavior in the flatworm, *Notoplana acticola*. Jour. Neurobiology, *11*(5):483–496.

Fournier, A., 1984: Photoreceptors and photosensitivity in Platyhelminthes. *In* Ali, M. A. (Ed.): Photoreception and Vision in Invertebrates, Plenum Press, N.Y. pp. 217–240.

Heitkamp, U., 1977: The reproductive biology of *Mesostoma ehrenbergii*. Hydrobiologia, *55*(1):21–32.

Henley, C., 1974: Platyhelminthes (Turbellaria). *In* Giese, A. C., and Pearse, J. S. (Eds.): Reproduction of Marine Invertebrates. Vol. I. Acoelomate and Pseudocoelomate Metazoans. Academic Press, N.Y. pp. 267–343.

Highnam, K. C., and Hill, L., 1977: The Comparative Endocrinology of the Invertebrates. 2nd Edition. University Park Press, Baltimore.

Hyman, L. H., 1951: The Invertebrates: Platyhelminthes and Rhynchocoela. Vol. 2. McGraw-Hill, N.Y. pp. 52–219.

Ishii, S., 1980: The ultrastructure of the protonephridial flame cell of the freshwater planarian *Bdellocephala brunnea*. Cell Tiss. Res., *206*:441–449.

Jennings, J. B., 1957: Studies on feeding, digestion, and food storage in free-living flatworms. Biol. Bull., *112*:63–80.

Jennings, J. B., 1968: Nutrition and digestion in Platyhelminthes. *In* Florkin, M., and Scheer, B. T. (Eds.) Chemical Zoology, Vol. 2. Academic Press, N.Y. pp. 305–327.

Jennings, J. B., 1974: Digestive physiology of the Turbellaria. *In* Riser, N. W., and Morse, M. P. (Eds.): Biology of the Turbellaria. McGraw-Hill Book Co., N.Y. pp. 173–197.

Karling, T. G., 1974: On the anatomy and affinities of the turbellarian orders. *In* Riser, N. W., and Morse, M. P. (Eds.): Biology of the Turbellaria. McGraw-Hill Book Co., N.Y. pp. 1–16.

Keenan, C. L., Coss, R., and Koopowitz, H., 1981: Cytoarchitecture of primitive brains: Golgi studies in flatworms. Jour. Comp. Neurobiology, *195*:697–716.

Kenk, R., 1972: Freshwater planarians (Turbellaria) of North America. Biota of Freshwater Ecosystems. Identification Manual No. 1. Environmental Protection Agency, Washington, D.C. 81 pp.

Koopowitz, H., 1974: Some aspects of the physiology and organization of the nerve plexus in polyclad flatworms. *In* Riser, N. W., and Morse, M. P. (Eds.): Biology of the Turbellaria. McGraw-Hill Book Co., N.Y. pp. 198–212.

Kozloff, E. N., 1972: Selection of food, feeding and physical aspects of digestion in the acoel turbellarian *Otocelis luteola*. Trans. Am. Microsc. Soc., *91*(4):556–565.

Kummel, G., 1962: Zwei neue Formen von Cyrtocyten, Vergleich der bisher bekannten Cyrtocyten und Eröterung des Begriffes "Zelltype." Z. Zellforsch., *57*:172–201. (Two new forms of cyrtocytes, comparison to previously known cyrtocytes, and discussion of concepts of "cell type.")

Lapan, E. A., and Morowitz, H., 1972: The Mesozoa. Sci. Am., *227*(6):94–101.

Lauer, D. M., and Fried, B., 1977: Observations on nutrition of *Bdelloura candida*, an ectocommensal of *Limulus polyphemus*. Am. Midl. Nat., *97*(1):240–247.

Llewellyn, J., 1965: The evolution of parasitic platyhelminths. *In* Taylor, A. E. R. (Ed.): Evolution of Parasites. 3rd Symposium for the British Soc. for Parasitology. Blackwell, London.

Martin, G. G., 1978: Ciliary gliding in lower invertebrates. Zoomorphologie, *91*:249–261.

Martin, G. G., 1978: A new function of rhabdites: mucus production for ciliary gliding. Zoomorphologie, *91*:235–248.

McConnaughey, B. H., 1963: The Mesozoa. *In* Dougherty, E. C. (Ed.): The Lower Metazoa. University of California Press, Berkeley. pp. 151–168.

McKanna, J. A., 1968: Fine structure of the protonephridial system in planaria. I. Flame cells. Z. Zellforsch. Mikrosk. Anat., *92*:509–523.

Moraczewski, J., 1977: Asexual reproduction and regeneration of *Catenula*. Zoomorphologie, *88*(1):65–80.

Moraczewski, J., Czubaj, A., and Bakowska, J., 1977: Organization and ultrastructure of the nervous system in Catenulida. Zoomorphologie, *87*(1):87–95.

Morita, M., and Best, J. B., 1984: Effects of photoperiods and melatonin on planarian asexual reproduction. Jour. Exp. Zool., *231*:273–282.

Muscatine, L., Boyle, J. E., and Smith, D. C., 1974: Symbiosis of the acoel flatworm *Convoluta roscoffensis* with the alga *Platymonas convolutae*. Proc. R. Soc. Lond. [Biol.], *187*(1087):221–234.

Nentwig, M. R., 1978: Comparative morphological studies after decapitation and after fission in the planarian *Dugesia dorotocephala.* Trans. Am. Micros. Soc., 97(3):297–310.

Palladini, G., Medolago-Albani, L., Margotta, V., Conforti, A., and Carolei, A., 1979: The pigmentary system of planaria: 2. Physiology and functional morphology. Cell Tiss. Res., 199(2):203–212.

Prusch, R. D., 1976: Osmotic and ionic relationships in the freshwater flatworm *Dugesia dorotocephala.* Comp. Biochem. Physiol., 54A:287–290.

Reisinger, E., 1968: *Xenoprorhynchus* ein Modellfall für progressiven Funktionswechsel. Z. Zool. Syst. Evolutionsforsch., 6:1–55.

Reisinger, E., 1972: Die Evolution des Orthogons der Spiralier und der Archicoelomatenproblem. Z. Zool. Syst. Evolutionsforsch., 10(1):1–43.

Reynoldson, T. B., and Sefton, A. D., 1976: The food of *Planaria torva,* a laboratory and field study. Freshwater Biol. 6(4):383–393.

Rieger, R. M., 1981: Morphology of the Turbellaria at the ultrastructural level. Hydrobiologia, 84:213–229.

Riser, N. W., and Morse, M. P. (Eds.), 1974: Biology of the Turbellaria. McGraw-Hill Book Co., N.Y. 530 pp. (Papers from a 1970 symposium in Chicago sponsored by the American Society of Zoologists.)

Rose, S. M., 1970: Regeneration. Appleton-Century-Crofts, N.Y. (The chapter on regeneration in worms provides a good introduction to studies on flatworm regeneration.)

Ruppert, E. E., 1978: A review of metamorphosis of turbellarian larvae. *In* Chia, F. S. and Rice, M. (Eds.): Settlement and Metamorphosis of Marine Invertebrate Larvae. Elsevier and North Holland Biomedical Press, pp. 65–81.

Schell, S. C., 1970: How to Know the Trematodes. W. C. Brown Co., Dubuque, Iowa. 355 pp.

Schockaert, E. R., and Ball, I. R. (Eds.), 1981: The Biology of the Turbellaria. Dr. W. Junk Publ., The Hague. (Papers from a symposium on the Turbellaria.)

Smith, J. P. S. III, Tyler, S., Thomas, M. B., Rieger, R. M., 1982: The morphology of turbellarian rhabdites: phylogenetic implications. Trans. Amer. Microsc. Soc., 101(3):209–228.

Smith, J. P. S. III, Tyler, S., Rieger, R. M., 1985: Is the Platyhelminthes polyphyletic? *In* Research in Turbellarian Biology. 4th Internat. Symp., 1984.

Smyth, J. D., and Halton, D. W., 1983: The Physiology of Trematodes, 2nd Edition. Cambridge University Press, Cambridge. 446 pp.

Stunkard, H. W., 1954: The life history and systematic relations of the Mesozoa. Q. Rev. Biol., 29:230–244.

Stunkard, H. W., 1972: Clarification of taxonomy in Mesozoa. Syst. Zool., 21(2):210–214.

Tyler, S., 1976: Comparative ultrastructure of adhesive systems in the Turbellaria. Zoomorphologie, 84:1–76.

Tyler, S., and Rieger, R. M., 1975: Uniflagellate spermatozoa in Nemertoderma and their phylogenetic significance. Science, 188:730–732.

Wilson, R. A., and Webster, L. A., 1974: Protonephridia. Biol. Rev., 49:127–160.

Wright, C. A., 1971: Flukes and snails. *In* Carthy, J. D., and Sutcliffe, J. F. (Eds.): The Science of Biology. Series 4. Allen and Unwin, Publishers, London.

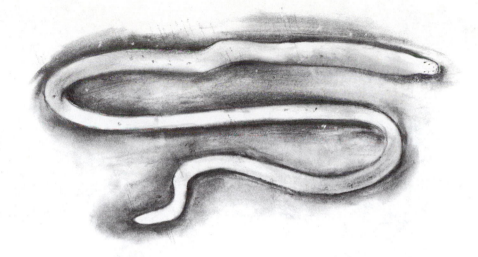

8

The Nemerteans

The phylum Rhynchocoela, or Nemertea, comprises about 900 species of elongated and often flattened worms. They are sometimes known as proboscis worms because of the presence of a remarkable proboscis apparatus used in capturing food. Most nemerteans, or nemertines, are marine bottom dwellers, but there are some deepwater pelagic species. The many common shallow-water nemerteans live beneath shells and stones or in algae, or burrow in mud and sand. Some species form semipermanent burrows lined with mucus or even distinct tubes. There is a single genus of freshwater species, and one genus containing terrestrial forms is confined to the tropics and subtropics. A few species live as parasites or commensals in the gill chambers of crabs, in the mantle cavities of bivalve mollusks, and in the atria of tunicates.

Nemerteans appear to be related to the free-living flatworms. Parenchyma fills the space between the body wall and the intestine, as in flatworms. Moreover, both flatworms and nemerteans possess protonephridia, ciliated epidermis, and somewhat similar nervous systems.

External Structure

In appearance nemerteans resemble flatworms, but most are larger and more elongated. The anterior end is commonly pointed or shaped like a spatula (Fig. 8–1B). Although most species are less than 20 cm in length and a number of forms are only a few millimeters long, those of some genera, such as the burrowing, shallow-water *Cerebratulus* and *Lineus* (Fig. 8–1A), are ribbon shaped and reach lengths of several meters or more. The greatest reported length is about 30 meters, for some specimens of the European *Lineus longissimus*. Most nemerteans are pale, but some have bright hues or patterns of yellow, orange, red, and green. Strangely, many bathypelagic species are bright red, orange, or yellow.

Body Wall and Locomotion

The body wall is composed of an outer layer of ciliated epithelium, which rests on a dermis of connective tissue (Figs. 8–2 and 8–4A). Unicellular

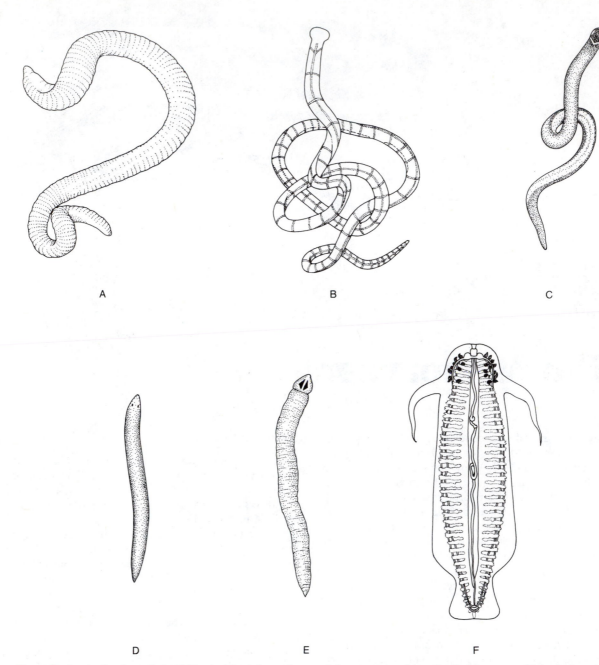

A   B   C

D   E   F

Figure 8–1 *A, Cerebratulus californiensis,* A large, burrowing nemertean from the Pacific coast of North America. (Drawn from a photograph by Robert Morris: *In* Intertidal Invertebrates of California.) *B,* A large, banded species of *Tubulanus* from the Pacific coast of North America. (After Coe.) *C, Paranemertes peregrina,* a small nemertean common on rocky shorelines of the California coast. *D,* A species of *Carcinonemertes,* which lives in the gill chambers of crabs. *E,* A species of *Amphiporus,* a genus of mostly small, burrowing, shallow-water nemerteans. *F, Nectonemertes mirabilis,* a common, deep-water pelagic nemertean. (*C–F* from Gibson, R. *In* The Synopsis and Clasification of Living Organisms. Vol. 1. McGraw-Hill Book Co., N.Y.)

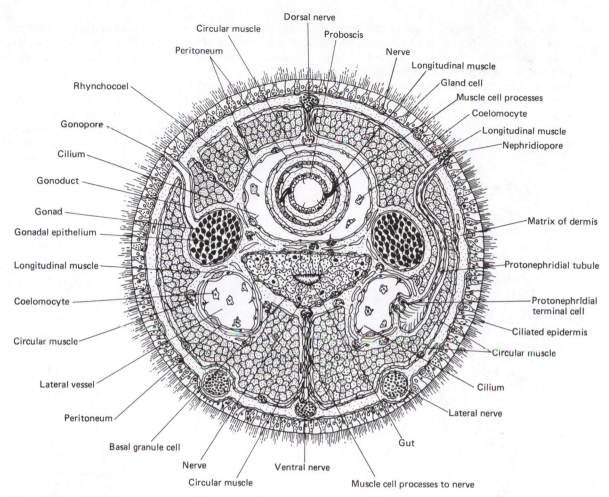

Circular muscle
Peritoneum
Rhynchocoel
Gonopore
Cilium
Gonoduct
Gonad
Gonadal epithelium
Longitudinal muscle
Coelomocyte
Circular muscle
Lateral vessel
Peritoneum
Basal granule cell
Nerve
Circular muscle

Dorsal nerve
Proboscis
Nerve
Longitudinal muscle
Gland cell
Muscle cell processes
Coelomocyte
Longitudinal muscle
Nephridiopore
Matrix of dermis
Protonephridial tubule
Protonephridial terminal cell
Ciliated epidermis
Circular muscle
Cilium
Lateral nerve
Gut
Ventral nerve
Muscle cell processes to nerve

Figure 8–2   Diagrammatic cross section of a nemertean at the level of the rhynchocoel. The diagram interprets the rhynchocoel, blood vessels, and gonadal cavities as derivatives of the coelom. The presence of rudimentary cilia in the lining of these cavities is part of the evidence for such an interpretation. (See p. 264). (From Turbeville, J. M., and Ruppert, E. E., 1985: Ultrastructure and evolution of the nemertines. Amer. Zool., 25(1):53–72.)

mucus-secreting glands are scattered throughout the epidermis. Beneath the dermis is a body wall musculature, composed of circular and longitudinal smooth muscle in two or three layers, of which the longitudinal is most highly developed (Fig. 8–2). Parenchymal tissue occupies the space between the digestive tract and the muscle layers; bands of muscle extend through the parenchyma, connecting the dorsal and ventral body walls.

Most nemerteans move, like flatworms, by gliding over the substratum on a trail of slime. Depending on the size of the worm, cilia and peristaltic muscular contractions account for locomotion. Peristaltic contractions, which are particularly important in burrowing, involve the fluid-filled rhynchocoel of the proboscis apparatus as a hydrostatic

skeleton and probably account for the rather large proportion of muscle tissue in nemerteans. Indeed, some nemerteans are unique among metazoans in having extensions of muscle cells into the epidermal layer. (See Turbeville and Ruppert, 1983, for comparison of nemertean and annelid peristaltic burrowing.) Some species, particularly certain pelagic forms, are able to swim and use dorsoventral muscular undulations exclusively.

Nutrition and Digestive System

The most characteristic feature of the phylum is the proboscis apparatus. Although a proboscis, when present, is usually associated with the digestive tract in invertebrates, the proboscis of nemerteans is not connected to the digestive tract except secondarily in some species. The opening of the proboscis apparatus is through a pore at the anterior tip of the worm (Figs. 8–3 and 8–4*A*). This part leads into a short canal known as the rhynchodeum, which extends to the level of the brain. The histology of the rhynchodeum wall, as well as the remainder of the proboscis apparatus, is similar to that of the body wall, because the entire structure develops as an ectodermal invagination. Rhabdites are present in the proboscis wall of some species, providing further evidence of a flatworm affinity for the phylum.

The lumen of the rhynchodeum is continuous with that of the proboscis proper (Fig. 8–4*A*). The latter consists of a long tube, often coiled, lying free in a fluid-filled cavity called the rhynchocoel, which is very similar to a coelom in both structure

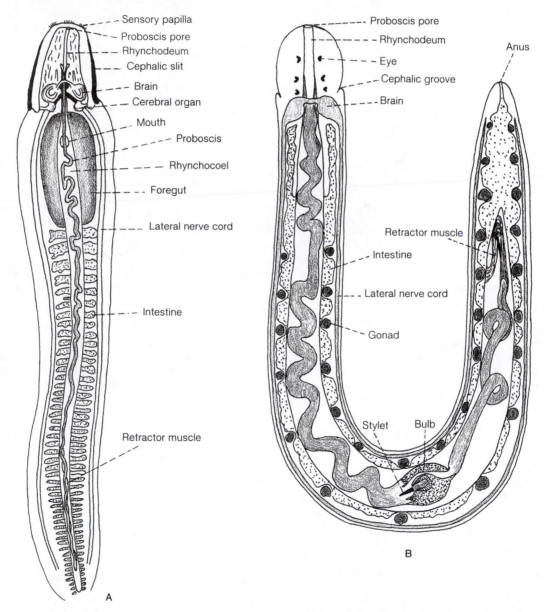

Figure 8–3 *A, Cerebratulus,* a heteronemertean (dorsal view). (After Bürger from Hyman.) *B, Prostoma rubrum,* a freshwater hoplonemertean. (Redrawn from Pennak, R. W., 1978: Fresh-Water Invertebrates of the United States, 2nd Edition. John Wiley and Sons, N.Y.)

and origin (p. 61). The posterior of the proboscis is blind and is attached to the back of the rhynchocoel by a retractor muscle.

In some nemerteans (class Anopla) the proboscis is a simple tube (Fig. 8–3*A*), but in others (class Enopla) the proboscis has become more specialized and is armed with a heavy, calcareous barb called a stylet, which is set in the proboscis wall (Figs. 8–3*B* and 8–4*B*). Where the stylet is attached, about two thirds of the way from the anterior of the animal, the proboscis wall is greatly swollen, particularly on one side, virtually occluding the lumen of the tube. Accessory or reserve stylets are present to replace the main stylet as the animal increases in size or when the main stylet is lost during feeding (Stricker and Cloney, 1982).

The mouth is ventral and located anteriorly at the level of the brain (Figs. 8–3*A*, 8–4*A*, and 8–5). It opens into a foregut consisting of a buccal cavity, an esophagus, and a glandular stomach (Figs. 8–3*A* and 8–5). The foregut opens into a long intestine, which has lateral diverticula (Fig. 8–5), and in some species it extends anteriorly beyond the junction with the foregut as a cecum. In contrast to the flat-

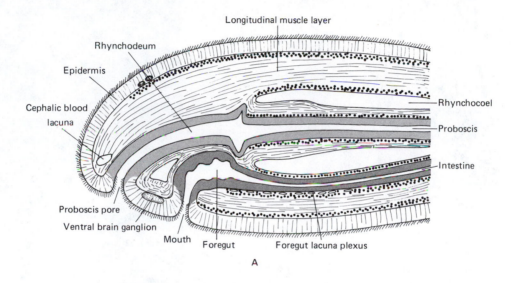

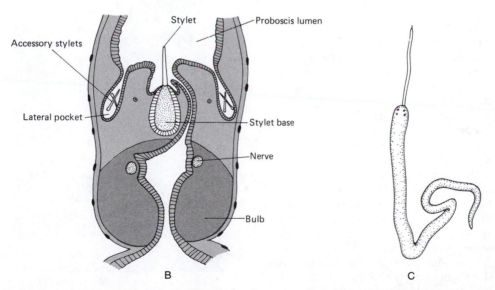

Figure 8–4 *A*, Anterior end of the paleonemertean *Procarina* (sagittal section). (Modified after Nawitzki from Hyman.) *B*, Stylet apparatus of the freshwater nemertean *Prostoma*. (Modified after Böhmig from Hyman.) *C*, Armed nemertean with protruded proboscis.

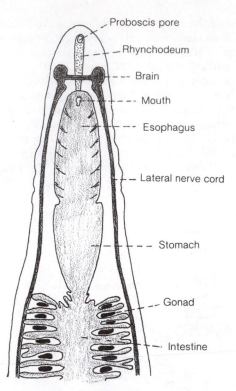

Figure 8–5 Digestive tract of *Carinoma*, a paleonemertean. (After Coe.)

worm, whose wastes are eliminated through the mouth, a nemertean egests wastes through an anus at the posterior end of the body.

In many armed nemerteans, the mouth has disappeared and the esophagus opens into the rhynchodeum; in the commensal bdellonemerteans, the rhynchodeum has disappeared and the proboscis opens into the anterior part of the gut. In all other nemerteans the digestive system is completely separate from the proboscis apparatus.

Nemerteans are entirely carniverous and feed primarily on annelids and crustaceans. The proboscis is used to capture prey. It is shot out of the body. Because the posterior of the proboscis is attached to the back of the rhynchocoel by the retractor muscle, the proboscis can never be completely everted. In the armed nemerteans the proboscis is projected only as far as the stylet, which then occupies its tip (Fig. 8–4C). The force for eversion is provided by muscular pressure on the fluid in the rhynchocoel, and retraction is effected by the posterior retractor muscle. The proboscis coils around the prey, and glandular secretions from the proboscis wall aid in holding it. The

stylet of armed nemerteans stabs the prey repeatedly, and toxic secretions produced by the posterior part of the proboscis are pumped into the wound by the muscular stylet bulb (Fig. 8–6). The prey, after capture by the proboscis, is swallowed whole, or its contents are sucked into the mouth (see review by McDermott and Roe, 1985).

The large unarmed *Cerebratulus lacteus* of the east coast of the United States enters the burrows

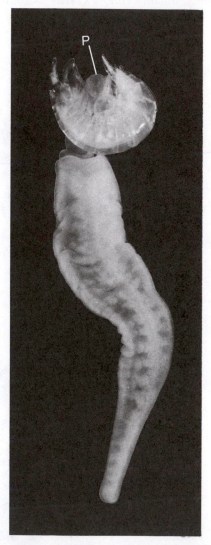

Figure 8–6 The armed nemertean *Nipponnemertes pulcher* attacking an amphipod. The proboscis (P) has wrapped around the amphipod and struck the ventral surface. (From McDermott, J. J., 1984: The feeding biology of *Nipponnemertes pulcher*, with some ecological implications. Ophelia, *23*(1):1–21.)

of the razor clam *Ensis directus* from below and swallows the anterior end of the clam (McDermott, 1976). *Paranemertes peregrina,* an armed, intertidal nemertean found on the Pacific coast of the United States, feeds on polychaete annelids. The nemertean leaves its burrow to feed and can follow the mucous trails of prey, but it must touch the polychaete for the feeding response to be initiated. Then the proboscis is everted and wraps around and stabs the prey several times with the stylet. The nemertean returns to its burrow by following its own mucous trail (Roe, 1970 and 1976). Other armed nemerteans feed on small crustaceans, such as amphipods. They kill the prey with a stylet strike on the ventral exoskeleton and then force the head through the opening (Fig. 8–6). The esophagus is everted, and the contents of the prey are sucked out and digested.

McDermott (1976, 1984) has estimated that in Danish waters, where the tube-dwelling amphipod *Haploops* is the dominant prey, there are about 146 individuals of *Nipponnemertes pulcher* and 1956 amphipods per square meter, and each worm consumes about 146 amphipods per year.

Carcinonemertes feeds on eggs brooded by female crabs and can be very destructive.

Digestion, which takes place in the intestine, is initially extracellular but is concluded intracellularly in phagocytic cells (for details, see Gibson, 1972).

Nervous System and Sense Organs

The nervous system, which is somewhat similar to that of higher flatworms, consists of an anterior brain of four ganglia surrounding the rhynchocoel and a pair of larger, lateral ganglionated nerve cords (Figs. 8–3*A* and 8–5). A number of other minor longitudinal nerves are frequently present

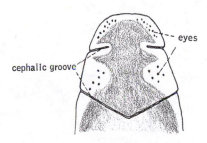

Figure 8–7 Head of the hoplonemertean *Amphiporus angulatus.* (After Hyman, L. H., 1951: The Invertebrates, Vol. II. McGraw-Hill Book Co., N.Y.)

(Fig. 8–2) and may represent vestiges of a primitive, radial arrangement of cords, as is found in flatworms.

Sense organs consist of sensory epidermal pits, pigment-cup eyes; ciliated cephalic slits, or grooves; and cerebral organs (Fig. 8–7). The last two are probably chemoreceptors. The cerebral organs are a pair of ciliated canals associated with the brain. The external openings of the canals are in the cephalic slits or in a pair of pits over the brain area. Water currents in the canals appear to be activated in the presence of food. The cerebral organs also appear to have a neuroendocrine function.

Internal Transport and Excretion

Nemerteans possess a blood-vascular system that is closed—that is, the blood is confined to vessels or direct channels. Moreover, the nemertean vessels are lined by epithelium, an unusual condition in invertebrates. In the simplest form (Fig. 8–8*A*) there are only two vessels, one on each side of the gut but connected anteriorly and posteriorly. In many nemerteans considerable elaboration on this basic plan has occurred with the development of additional longitudinal and transverse vessels.

The large vessels are contractile, but blood flow depends on contraction of the blood vessels and body wall musculature and is very irregular. It does not follow a definite circuit. The blood may flow forward and backward in the longitudinal vessels.

Nemertean blood is usually colorless, but in many species it contains corpuscles bearing yellow, red, orange, or green pigments of uncertain function. In addition to the corpuscles, the blood also contains amebocytes.

The primitive excretory system consists of one pair of protonephridia. A nephridiopore is located on each side of the foregut, and a tubule extends anteriorly from the opening of each nephridiopore. The terminal cells in most nemerteans project into the wall of the lateral blood vessel (Fig. 8–8*B*), which in some cases has even disappeared, so the terminal cells are directly bathed in blood. The role of the protonephridia in excretion, if any, is still unknown.

Regeneration and Reproduction

Nemerteans, especially the larger species, display a marked tendency to fragment when irritated. Collecting large, intact specimens is therefore usually difficult. Very frequently the proboscis becomes detached when everted. The proboscis soon regen-

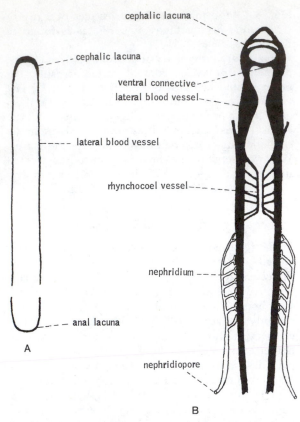

cephalic lacuna

cephalic lacuna

ventral connective

lateral blood vessel

lateral blood vessel

rhynchocoel vessel

nephridium

anal lacuna

A

nephridiopore

B

Figure 8–8 Circulatory system of various nemerteans: *A, Cephalothrix.* (After Oudemans from Hyman.) *B, Tubulanus.* (After Bürger from Hyman.)

erates, but when the body has fragmented, the ability of the fragments to regenerate varies greatly, depending on the species. Some species, including certain members of the genus *Lineus*, reproduce asexually by fragmentation, and even posterior sections of the body are capable of regeneration, which takes place within a mucous cyst.

The majority of nemerteans are dioecious, and the reproductive system is very simple. The gametes develop from parenchymal cells that aggregate and become surrounded by a thin-walled sac to form a gonad. A considerable number of such gonads usually form a regular row on each side of the body in the parenchyma (Fig. 8–5).

After maturation of the gametes, a short duct grows from the gonad to the outside, allowing the gametes to escape. Each gonad produces 1 to 50 eggs, depending on the species. The shedding of eggs or sperm does not necessarily require contact between two worms, although some species aggregate at the time of spawning or a pair of worms may occupy a common burrow. Fertilization is ex-

ternal in most nemerteans, and the eggs are either shed and dispersed into the sea water or deposited within the burrow or in gelatinous strings.

Embryogeny

Cleavage in nemerteans is spiral. All orders except heteronemerteans have direct development. Heteronemerteans typically pass through a free-swimming and feeding larval stage, called a pilidium (Fig. 8–9), which possesses an apical tuft of cilia and is somewhat helmet shaped. After a free-swimming existence, the pilidium metamorphoses into a young worm. *Paranemertes peregrina* has a life span of about a year and a half (Roe, 1976). Spawning occurs in spring and summer, and adults die in the winter. Juveniles resulting from the spring and summer spawn attain sexual maturity the following spring and summer.

Phylogenetic Position of the Nemerteans

Nemerteans have long been thought to have been derived from turbellarian flatworms, and there are many zoologists who still hold to such a relationship. However, recent work by Turbeville and Ruppert (1985) on the ultrastructure of nemerteans suggests that both nemerteans and flatworms

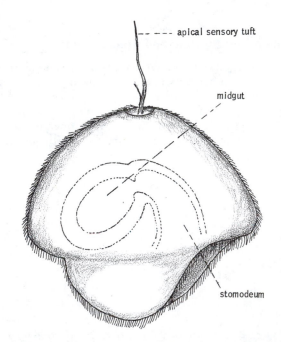

apical sensory tuft

midgut

stomodeum

Figure 8–9 Pilidium larva of the nemertean *Cerebratulus marginatus.* (After Coe.)

evolved from annelids and that nemerteans are intermediate in structure between flatworms (which are entirely acoelomate) and annelids (which are coelomate). The nemerteans' rhynchocoel, blood vessels, and gonadal cavities are all interpreted as being coelomic derivatives (see also Box 10–1).

Systematic Résumé of Phylum Rhynchocoela

Class Anopla. Unarmed nemerteans. Mouth located below or posterior to brain.

> **Order Paleonemertea.** Primitive neritic nemerteans. Body wall musculature is two layered or three layered, with the longitudinal layer located between two circular layers. Nerve cords outside muscle layers or within longitudinal muscle layers. *Tubulanus, Carinina, Carinoma, Cephalothrix.*

> **Order Heteronemertea.** Body wall musculature is three layered, with a circular layer between two longitudinal layers. Nerve cords within outer, longitudinal muscle layer. *Cerebratulus, Lineus.*

Class Enopla. Armed nemerteans. Mouth located anterior to brain. Nerve cords are located inside body wall musculature, composed of outer circular and inner longitudinal layers.

> **Order Hoplonemertea.** All of the armed nemerteans belong to this order, which contains several interesting ecological groups, such as the floating or swimming pelagic nemerteans, which live in the open ocean, many at depths below 1000 meters; the freshwater nemerteans (*Prostoma*); the tropical and subtropical terrestrial nemerteans (*Geonemertes*); and many marine, shallow-water forms (*Amphiporus, Paranemertes*).

> **Order Bdellonemertea.** This small order contains but one genus, *Malacobdella*, with four species; three are commensal in the mantle cavity of marine clams, and one is commensal in the mantle cavity of a freshwater snail. Although the proboscis is unarmed, it has probably been derived from the armed type. The proboscis and esophagus open into a common chamber. Bdellonemerteans feed on bacteria, algae, and protozoa removed from the ventilating current of the host.

Summary

1 Nemerteans, members of the phylum Rhynchocoela, are similar to flatworms in being dorsoventrally flattened and acoelomate. Like flatworms, the epidermis is ciliated, and some nemerteans even have rhabdites.

2 In general, nemerteans are larger than flatworms. They live in rocky crevices, beneath stones, and within mats formed by algae and sessile animals, and many burrow in soft bottoms. They move by cilia or muscular undulations, or by a combination of the two.

3 Nemerteans are predatory on other invertebrates, which are captured with a unique proboscis apparatus. The proboscis is explosively everted by hydrostatic pressure, and the prey is enrolled or is stabbed by a proboscis stylet. The proboscis apparatus is not connected to the gut except secondarily in some groups. As in flatworms, digestion is initially extracellular and then intracellular.

4 There are no gas exchange organs, but there is a blood-vascular system with which protonephridia are intimately associated.

5 The nervous system, composed of longitudinal cords, is somewhat similar to that of flatworms.

6 The sexes are separate in most nemerteans, and the gonoducts are simple and transitory. Fertilization is external, although individuals may be in contact or close together at the time of spawning. Eggs are shed into the sea water or encased in gelatinous masses, and development may be indirect with a pilidium larva or direct.

REFERENCES

The literature included here is restricted in large part to nemerteans. The introductory references on page 8 include many general works and field guides that contain sections on nemerteans.

Cantell, C. E., 1969: Morphology, development, and biology of the pilidium larvae from the Swedish west coast. Zool. Bidrag. Fran Uppsala, 38:61–111.

Gibson, R., 1972: Nemerteans. Hutchinson University Library, London. 224 pp. (An excellent general account of the nemerteans.)

Gibson, R., 1974: Histochemical observations on the localization of some enzymes associated with digestion in four species of Brazilian nemerteans. Biol. Bull., 147(2):352–368.

Gibson, R., and Moore, J., 1976: Freshwater nemerteans. Zool. Linn. Soc., *58*(3):117–218.

Hyman, L. H., 1951: The Invertebrates: Platyhelminthes and Rhynchocoela. Vol. 2. McGraw-Hill, N.Y. pp. 459–531. (An old but still valuable general treatment of the nemerteans.)

McDermott, J. J., 1976: Predation of the razor clam *Ensis directus* by the nemertean worm *Cerebratulus lacteus*. Chesapeake Sci., *17*(4):299–301.

McDermott, J. J., 1976: Observations on the food and feeding behavior of estuarine nemertean worms belonging to the order Hoplonemertea. Biol. Bull., *150*:57–68.

McDermott, J. J., 1984: The feeding biology of *Nipponnemertes pulcher*, with some ecological implications. Ophelia, *23*(1):1–21.

McDermott, J. J., and Roe, P., 1985: Food, feeding behavior and feeding ecology of nemerteans. Amer. Zool., *25*(1):113–126.

Riser, N. W., 1974: Nemertinea. *In* Giese, A. C., and Pearse, J. S. (Eds.): Reproduction of Marine Invertebrates. Vol. 1. Academic Press, N.Y. pp. 359–389.

Roe, P., 1970: The nutrition of *Paranemertes peregrina*. I. Studies on food and feeding behavior. Biol. Bull., *139*:80–91.

Roe, P., 1976: Life history and predator-prey interactions of the nemertean *Paranemertes peregrina*. Biol. Bull., *150*:80–106.

Stricker, S. A., and Cloney, R. A., 1982: Stylet formation in nemerteans. Biol. Bull., *162*:387–403.

Turbeville, J. M., and Ruppert, E. E., 1983: Epidermal muscles and peristaltic burrowing in *Carinoma tremaphoros:* correlates of effective burrowing without segmentation. Zoomorph., *103*:103–120.

Turbeville, J. M., and Ruppert, E. E., 1985: Comparative ultrastructure and the evolution of the nemertines. Amer. Zool., *25*(1):53–72.

9

The Aschelminths

The aschelminths are a heterogeneous assemblage of common marine and freshwater animals, such as gastrotrichs, rotifers, and roundworms (nematodes). They are sometimes placed within a phylum called the Aschelminthes, but each of the component groups is much more distinctive than the combination. The common practice, and the one used here, is to consider each of the groups a separate phylum and to use the name *aschelminths* as a convenient term of reference for the entire assemblage.

The majority of free-living aschelminths are minute animals, ranging from microscopic size to a centimeter in length. Although the anterior of the body bears the mouth and sense organs, there is no well-formed head. Surface ciliation is reduced or absent, and the body is covered with a cuticle. Adhesive glands in varying numbers are characteristic of many species and often open to the outside of the body by projecting cuticular tubes. The digestive tract is usually a complete tube with a mouth and anus, and a specialized pharyngeal region is almost always present. Since most species are aquatic and very small, no gas exchange organs or circula-

tory system is present. However, protonephridia are characteristic of a number of aschelminth phyla. A peculiar feature of many members (nematodes and rotifers) is the numerical constancy of the cells, or the nuclei, that compose the various organs, a condition known as eutely. The numbers are characteristic for each species but vary from one species to another. For example, in the rotifer *Epiphanes senta* there are 958 nuclei, of which 35 form the stomach, 120 the musculature, etc. The condition is perhaps associated with the extremely small size characteristic of many members of the group. Commonly, mitosis ceases following embryonic development, and growth continues through increase in the size of the cells. As would be expected, regeneration in eutelic animals is very limited. Most aschelminths are dioecious, and cleavage is strongly determinate.

The aschelminths are commonly said to exhibit a pseudocoelomate grade of construction. As explained in Chapter 3, a pseudocoel represents a persistent embryonic blastocoel. The mesodermal derivatives are, for the most part, located at the outer side of the cavity, and the internal organs are actually free within the pseudocoel; i.e., they do not lie behind the peritoneum (Fig. 3–3*C*). The increasing numbers of electron microscopy studies on various aschelminth groups, however, reveal only slit-like spaces that hardly justify the term *body cavity*. Only in some nematodes and a few other groups is there a large functional pseudocoel.

Phylum Gastrotricha

The gastrotrichs are a small phylum of some 460 marine and freshwater aschelminths that inhabit the interstitial spaces of bottom sediments and superficial detritus, the surfaces of submerged plants and animals, and the water films of soil particles. The phylum is considered to contain a single class, divided into two orders. The order Macrodasyida is composed of marine and brackish water species, and the order Chaetonotida contains all the freshwater species as well as some marine forms. Many gastrotrichs are common animals of ponds, streams, and lakes. They are also common marine interstitial animals. Hummon (1971) reported 42 species of gastrotrichs from the intertidal zone of New England beaches.

External Structure

Like many other pseudocoelomates, most gastrotrichs are microscopic in size. They range in length from 50 to 1000 μm, though members of some species may approach 4 mm. The bottle-shaped or strap-shaped body is usually flattened ventrally, and the posterior end is sometimes forked (Figs. 9–1, 9–2, 9–3, 9–4, and 9–5).

The locomotor cilia are always restricted to the ventral surface of the trunk and head. The entire ventral surface may be ciliated, or the cilia may be arranged in longitudinal bands, transverse rows, or patches (Fig. 9–6). Locomotion takes place by a gliding motion over the substratum.

The ventral ciliation of gastrotrichs suggests that these animals may be the most primitive of the aschelminth pseudocoelomates. Moreover, the primitive groups of gastrotrichs have monociliated epidermal cells, a feature shared only with the gnathostomulids (p. 256) among the lower bilateral animals (see Box 9–1).

Adhesive tubes provide for temporary adhesion to the substratum (Figs. 9–1 and 9–3) and contain duogland systems, as in turbellarians, i.e., a viscid gland coupled with a releasing gland (Tyler and Rieger, 1980). The adhesive tubes may be located

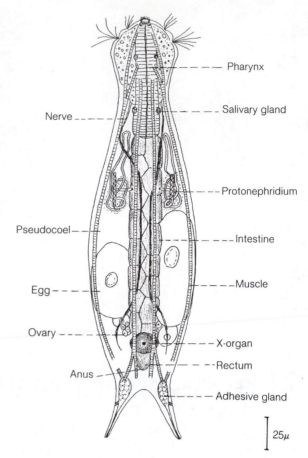

Figure 9–1   A chaetonotid gastrotrich (ventral view). (Modified after Zelinka from Pennak.)

along the sides of the body and numerous (macrodasyids), or they may be restricted to the forked, posterior end of the body (chaetonotids) (Figs. 9–1 and 9–5).

Internal Structure

The body wall is composed of an external cuticle, an epidermis, and underlying bands of longitudinal and circular muscle fibers; there is no pseudocoel. The cuticle in its fine structure consists of a fibrous basal layer and an outer lamellar layer composed of bilayered sheaths (Fig. 9–7). In some gastrotrichs the basal layer is locally thickened and specialized to form scales, spines, and hooks (Figs. 9–5 and 9–7) (Rieger and Rieger, 1977).

The terminal mouth opens directly into the pharynx or, in some chaetonotids, into a small, protrusible, buccal capsule. The pharynx is an elongated tube, which may contain one to four bulbous

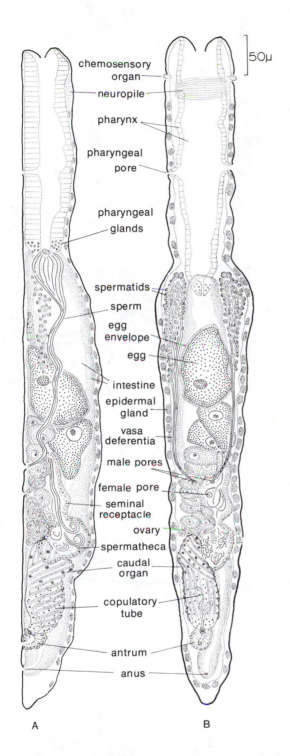

chemosensory
organ

neuropile

pharynx

pharyngeal
pore

pharyngeal
glands

spermatids

sperm

egg
envelope

egg

intestine

epidermal
gland

vasa
deferentia

male pores

female pore

seminal
receptacle

ovary

spermatheca

caudal
organ

copulatory
tube

antrum

anus

50μ

A

B

Figure 9–2 *A* and *B,* Lateral and ventral views of *Macrodasys* sp. (From Ruppert, E. E., 1978: The reproductive system of gastrotrichs. II. Insemination in *Macrodasys:* A unique mode of sperm transfer in Metazoa. Zoomorphologie, 89:207–228.)

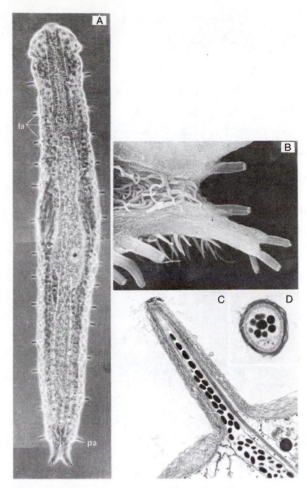

Figure 9–3 Adhesive tubes of *Turbanella: A,* Photograph of living animal, showing lateral and posterior adhesive tubes. Each lateral adhesive tube contains a sensory cilium. *B,* Enlarged view of posterior adhesive tubes. *C* and *D,* Longitudinal and cross sections of an adhesive tube. (pa, posterior adhesive tube; la, lateral adhesive tube; f, fiber; vg, viscid gland; rg, releasing gland) (From Tyler, S., and Rieger, G. E., 1980: Adhesive organs of the Gastrotricha. I. Duo-gland organs. Zoomorphologie, 95:1–15.)

swellings (Figs. 9–1 and 9–2). In macrodasyids, the pharynx opens to the exterior through a pair of pores. The pharyngeal wall is composed largely of myoepithelial cells, i.e., cells that not only have a lining function but also possess radial myofibrils. The lumen is triangular and lined with cuticle (Ruppert, 1982). Posteriorly, the pharynx opens into a cylindrical, tapered stomach-intestinal region, the walls of which are composed of a single layer of cuboidal or columnar epithelium. The anus is located near the terminal end of the trunk.

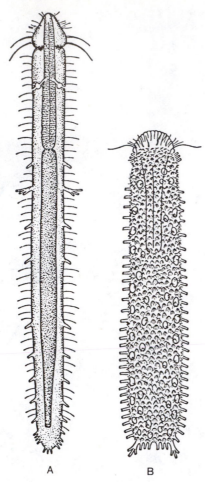

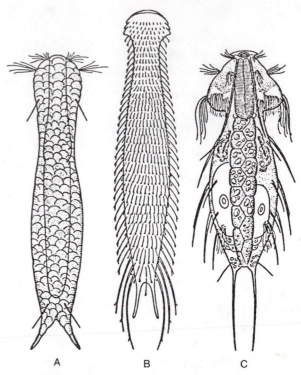

Figure 9-5 Gastrotrich diversity: *A, Lepidodermella*, a chaetonotid. *B, Chaetonotus. C, Dasydytes*, a chaetonotid. (*A* after Zelinka, *B* after Grunspan, *C* after Remane, all from Grassé, P., 1965: Traité de Zoologie, Vol. 4, p. 3.)

Figure 9-4 Gastrotrich diversity: *A, Pleurodasys*, a marine macrodasyid. *B, Tetranchyroderma*, a marine macrodasyid. (After Remane from Grassé, P., 1965: Traité de Zoologie, Vol. 4, p. 3.)

lenocyte, of the protonephridium is composed of cytoplasmic rods enclosing one or two flagella (Fig. 9-8). The fenestrations are closed by a net of fine fibrils. The rods are enclosed in turn by a cyto-

Gastrotrichs feed on small, dead or living organic particles, such as bacteria, diatoms, and small protozoa, all of which are sucked into the mouth by the pumping action of the pharynx or driven in by buccal cilia. In the marine macrodasyids the paired pharyngeal pores permit the release of excess water ingested with the prey. Digestion takes place in the stomach-intestinal region and is very rapid.

Protonephridia, typically a single pair, are best developed in freshwater gastrotrichs and probably function as osmoregulatory organs (see Box 7-1). The long, coiled, protonephridial tubules open through separate nephridiopores on the ventrolateral surface near the middle of the trunk (Fig. 9-1). In the Macrodasyida each terminal cell, or so-

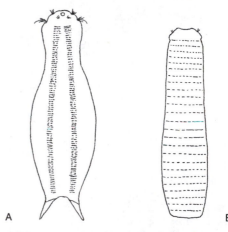

Figure 9-6 *A*, Ventral ciliation in *Chaetonotus. B*, Ventral ciliation in *Thaumastoderma*. (After Hyman, L. H., 1951: The Invertebrates, Vol. 3. McGraw-Hill Book Co., N.Y.)

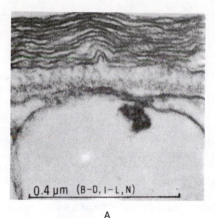

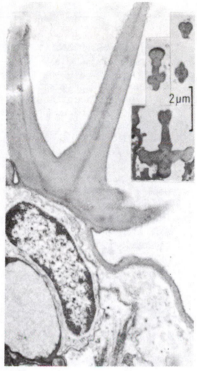

Figure 9–7 *A*, Cuticle of *Turbanella ocellata*, showing basal layer overlying a vesicle and the outer lamellar layer composed of numerous bilayers. *B*, Cuticular spines of *Tetranchyroderma* sp. Inset shows cross sections of spines at various levels. (From Rieger, G. E., and Rieger, R. M., 1977: Comparative fine structure study of the gastrotrich cuticle and aspects of cuticle evolution within the Aschelminthes. Z. Zool. Syst. Evolutionsforsch., *15*(2):81–124.)

plasmic skirt perforated by pores and slits. The skirt is also a part of the terminal cell. The solenocytes join a collecting cell, which is provided with a flagellated outlet cell that conducts fluid to the exterior (Fig. 9–8).

The brain is composed of a ganglionic mass (Fig. 9–2) on each side of the anterior of the pharynx; the two masses are connected dorsally by a commissure. Each brain mass gives rise to a ganglionated cord that extends the length of the body. The sense organs are represented by ciliated tufts and bristles and ciliated pits on the head, as well as by sensory bristles located over the general body surface.

Reproduction

In contrast to other aschelminths, gastrotrichs are hermaphroditic. In the marine *Macrodasys*, which probably represents the primitive plan, there is a pair of hermaphroditic gonads, testis anteriorly and ovary posteriorly (Fig. 9–2). Sperm ducts carry sperm, which may be formed into spermatophores (sperm packets), to a pair of ventral male gonopores located about two thirds of the body length from the anterior end. Sperm transfer is indirect, and at copulation each individual so flexes its body

that it is able to pick up its own sperm released from the male pores with a more posteriorly located caudal organ. The caudal organ then functions as a penis to transfer the sperm to the opposite individual (Fig. 9–9), where they are stored within a seminal receptacle. The internally fertilized eggs are released from the body by rupture. From an indirect mode, other gastrotrichs have evolved direct sperm transfer. In all forms the accessory organs are complex (see reviews by Hummon and Hummon, 1983).

The male system of the freshwater chaetonotids has become so degenerate that functionally all individuals are females and reproduce parthenogenetically. However, recently sperm were observed for the first time in the common and widespread *Lepidodermella squamata* and six other species (Weiss and Levy, 1979; Kisielewska, 1980; Hummon, 1984).

Laboratory studies on *Lepidodermella* indicate a maximum life of about 40 days, during the first 10 of which a female lays four or five eggs. However, the life expectancy is probably far shorter under natural conditions.

In the freshwater parthenogenetic chaetonotids, two types of eggs are produced and attached to the substratum. One type is the dormant, or rest-

BOX 9–1 MONOCILIATED CELLS OF METAZOANS

There is increasing evidence that the primitive ciliated epidermal cell of animals carries only one cilium, which is surrounded by a ring of microvilli. The basal body from which the cilium arises is diplosomal (i.e., it contains nine microtubule pairs) and is anchored by two rootlets, which extend in opposite directions. An accessory centriole lies at right angles to the basal body. Such a monociliated condition is characteristic of most lower animals (like placozoans, sponges, cnidarians, gnathostomulids, and some gastrotrichs) and is important evidence that metazoans derive from a colonial, flagellated protozoan. A flagellum and a cilium do not differ in ultrastructure. The choanoflagellates, whose ultrastructure is very similar to that of the metazoans, are the most probable flagellate ancestors (see Box 4–1). The monociliated condition is fairly common within the deuterostome line of evolution (echinoderms, chordates) but rare in the protostome line (Spiralia or flatworms, annelids, mollusks).

Along the right margin in the figure below is shown the derivation of the multiciliated cells of some gastrotrichs from the monociliated condition of primitive species (bottom). The possible evolution of the ciliary rootlet system of flatworms, nemerteans, and annelids is depicted at the top. [From Rieger, R. M., 1976: Monociliated epidermal cells in Gastrotricha: significance for concepts of early metazoan evolution. Z. Zool. Syst. Evolutionsforsch., *14*(3):198–226.]

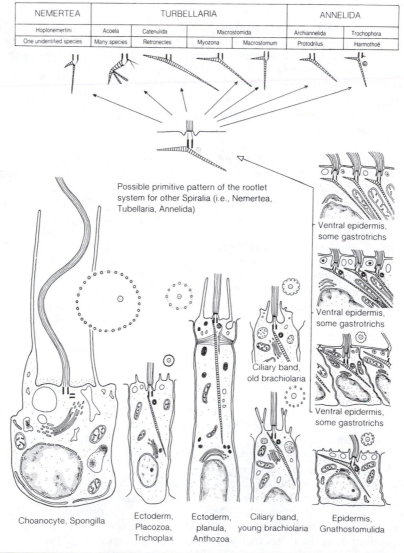

NEMERTEA	TURBELLARIA				ANNELIDA	
Hoplonemertini	Acoela	Catenulida	Macrostomida		Archiannelida	Trochophora
One unidentified species	Many species	Retronectes	Myozona	Macrostomum	Protodrilus	Harmothoë

Possible primitive pattern of the rootlet system for other Spiralia (i.e., Nemertea, Tubellaria, Annelida)

Ventral epidermis, some gastrotrichs

Ventral epidermis, some gastrotrichs

Ciliary band, old brachiolaria

Ventral epidermis, some gastrotrichs

Choanocyte, Spongilla

Ectoderm, Placozoa, Trichoplax

Ectoderm, planula, Anthozoa

Ciliary band, young brachiolaria

Epidermis, Gnathostomulida

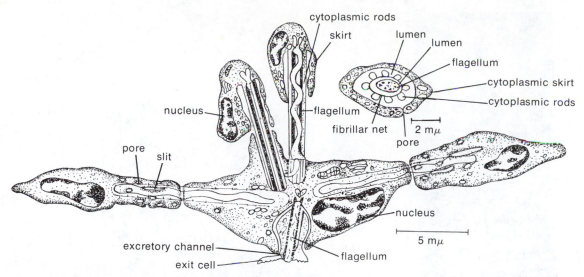

Figure 9–8 Protonephridium of the gastrotrich *Turbanella cornuta*. Solenocytes I and IV are depicted in external view; II and III are shown in section. Between III and IV a solenocyte is shown in cross section. The four solenocytes join a collecting cell, from which fluid departs by a flagellated exit cell. (From Teuchert, G., 1973: Die Feinstruktur des Protonephridialsystems von *Turbanella cornuta* Remane, einem marinen Gastrotrich der Ordnung Macrodasyoidea. Z. Zellforsch., *136*:277–289.)

ing, egg, which like those of freshwater flatworms, can withstand desiccation and low temperatures; the other type hatches in one to four days. The eggs are enormous and are laid one at a time. Cleavage is bilateral but determinate. Young gastrotrichs have most of the adult structures on hatching and reach sexual maturity in about three days. The growth stages of *Turbanella* are illustrated in Figure 9–10.

SYSTEMATIC RÉSUMÉ OF PHYLUM GASTROTRICHA

Order Macrodasyida. Body usually strap shaped. Adhesive tubes located on anterior, posterior, and sides of body. Pharyngeal pores present. Hermaphroditic. Marine and estuarine species. *Macrodasys, Urodasys, Turbanella, Dactylopodola.*

Order Chaetonotida. Freshwater and marine gastrotrichs. Body usually bottle shaped. Adhesive tubes usually restricted to the posterior end. Pharyngeal pores absent. Most freshwater species are parthenogenetic females. *Chaetonotus, Lepidodermella.*

Phylum Nematoda

The Nematoda, or Nemata, called roundworms, form the largest aschelminth phylum (12,000 described species) and include some of the most widespread and numerous of all multicellular animals. Free-living nematodes are found in the sea, in fresh water, and in the soil. They occur from the polar regions to the tropics in all types of environments, including deserts, hot springs, high mountain elevations, and great ocean depths. Nematodes are benthic animals and live in interstitial spaces of algal mats and especially aquatic sediments and soil (Table 9–1). They are often present in enormous numbers. One square meter of bottom mud off the Dutch coast has been reported to contain as many as 4,420,000 nematodes. An acre of good farm soil has been estimated to contain several hundred million to billions of terrestrial nematodes. A single decomposing apple on the ground of an orchard has yielded 90,000 roundworms belonging to a number of species.

In addition to free-living species, there are many parasitic nematodes. The parasitic forms display all degrees of parasitism and attack virtually all groups of plants and animals. The numerous species that infest food crops, domesticated animals, and man himself make this phylum one of the most important of the parasitic animal groups. The phylum also contains the best known animal, *Caenorhabditis elegans*, whose every cell has been traced throughout the course of development (Sulston et al, 1984).

The initial discussion of this phylum will emphasize the free-living members, and the parasitic forms will be dealt with at the end of the section.

Figure 9–9 Copulating pair of *Macrodasys* sp. during process of sperm transfer. (From Ruppert, E. E., 1978: The reproductive system of gastrotrichs. II. Insemination in *Macrodasys:* a unique mode of sperm transfer in Metazoa. Zoomorphologie, 89:207–228.)

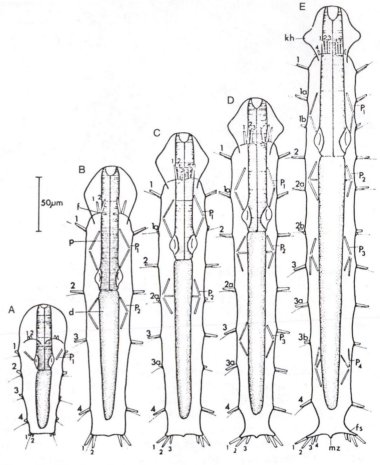

Figure 9–10 Growth in the marine gastrotrich *Turbanella cornuta: A,* Embryo prior to hatching. *B–D,* Juvenile states. *E,* Individual at sexual maturity. p, pharynx; p_1–p_4, protonephridia; d, intestine; 1–4, adhesive tubes. (From Teuchert, G., 1968; Z. Morph. Tiere, 63:343–418.)

External Structure

The size and form of nematodes are important adaptations for living in interstitial spaces. They have slender, elongated bodies with both ends gradually tapered in most species (Fig. 9–11). The majority of free-living nematodes are less than 2.5 mm in length and are often microscopic. However, some soil nematodes are as long as 7 mm, and some marine species attain a length of 5 cm.

The body of a nematode is perfectly cylindrical; hence the name *roundworm*. The mouth is located at the somewhat rounded anterior end and is surrounded by lips and sensory papillae or bristles. In many marine nematodes (Fig. 9–12), which include the most primitive members of the phylum, six lip-like lobes border the mouth, three on each side, but as a result of fusion, there are often only three lips in terrestrial and parasitic species. Primitively, the lips and the anterior surface outside the lips bear 16 sensory papillae or setae and may also carry a variety of cuticular projections (Figs. 9–12 and 9–13*A* and *D*). As in other aschelminths, the cuticle of the general body surface is often sculptured or ornamented in different ways (Fig. 9–13*C* and *D*). Members of the marine interstitial family Stilbonematidae have body surfaces clothed with a symbiotic blue-green alga and appear to be hairy (Fig. 9–13*F*).

A caudal gland, also called a spinneret, is typical of many free-living nematodes, including most marine species. The gland opens at the posterior tip of the body, which is sometimes drawn out to resem-

TABLE 9–1 **The Diversity and Biomass of Marine Nematodes from Various Habitats**

Habitat	Locality	Maximum Depth	Number of Species	Maximum Number Reported per m² × 10⁶
Intertidal sand beach	North Britain	12 cm in summer	104	2.44
Sublittoral silty sand	North Sea, Britain	Most 7 cm	38	0.82
Sublittoral silty sand	New England	80% 3 cm.	72	1.86
Sublittoral silty sand	Florida	Most 1–2 cm	100	
Various	N. American continental shelf and slope	95% 1–2 cm	212	0.59
Algae, rocky shore	Britain		70	
Sublittoral coral heads	North Carolina		41	

(Adapted from Nicholas, W. L., 1975: The Biology of Free-Living Nematodes. Clarendon Press, Oxford. pp. 150–151.)

ble a tubelike tail. The caudal gland of nematodes is similar to the adhesive tubes of other aschelminths, and in at least some nematodes it is a duogland system (Adams and Tyler, 1980).

Body Wall

The nematode cuticle is considerably more complex than that of other aschelminths. It contains collagen, as well as other compounds, and it is organized within three main layers (Fig. 9–14). The outer cortical layer is bounded externally by a thin epicuticle, which may exhibit quinone tanning. It is typically annulated (ringed). The median layer varies from a uniform granular structure in some species to the occurrence of struts, skeletal rods, fibrils, or canals in others. The basal layer may be striated or laminated or contain spiral fibers.

Growth in nematodes is accompanied by four molts of the cuticle. Beginning at the anterior end,

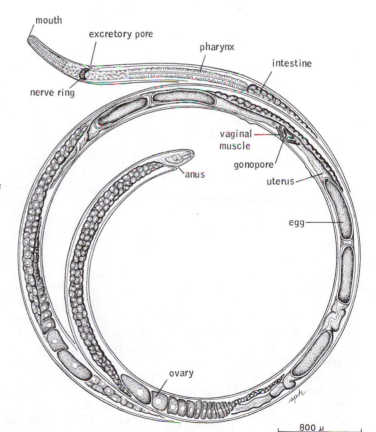

Figure 9–11 Female of the marine nematode *Pseudocella.* (After Hope.)

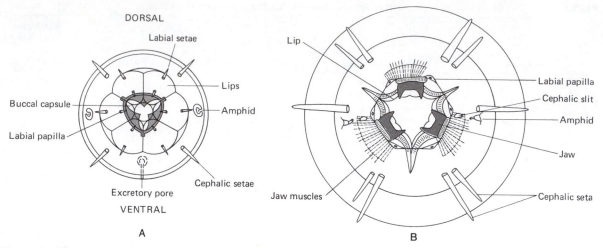

Figure 9–12 *A,* Anterior end of a generalized nematode. *B,* Anterior end of the marine nematode *Enoplus.* (*A* and *B* after de Coninck from Hyman.)

the old cuticle separates from the underlying epidermis and a new cuticle is secreted, at least in part. The old cuticle is shed in fragments or intact. Molting does not occur after the worm becomes adult, but the cuticle continues to grow.

The epidermis, also called hypodermis, is usually cellular but may be syncytial in some species. A striking feature of the nematode epidermis is the expansion of the cytoplasm into the pseudocoel along the middorsal, midventral, and midlateral lines of the body (Fig. 9–15). The bulging epidermis thus forms longitudinal cords that extend the length of the body. The epidermal nuclei are commonly restricted to these cords and are typically arranged in rows.

The muscle layer of the body wall is composed entirely of longitudinal, obliquely striated fibers arranged in bands, each strip occupying the space between two longitudinal cords. The fibers may be relatively broad and flat, with the contractile filaments limited to the base of the fiber, or they may be relatively tall and narrow, with filaments at the base and sides. In both types the base of the cell containing the contractile fiber is located against the hypodermis, and the side of the cell with the nucleus is directed toward the pseudocoel (Fig. 9–16). Each nematode muscle fiber possesses a slender arm that extends from the fiber to either the dorsal or the ventral longitudinal nerve cord, where innervation occurs (Fig. 9–15). In most animals a nerve process extends from the nerve cord to the main body of the muscle.

The nematode pseudocoel is spacious and filled with fluid. No free cells are present, but fixed cells, located either against the inner side of the muscle layers or against the wall of the gut and the internal organs, are characteristic of many nematodes.

Locomotion

Most nematodes move by undulatory waves of muscular contraction passing along the longitudinal muscle fibers of the body wall. The hydrostatic skeleton of the pseudocoel fluid and the elasticity of the cuticle are antagonistic to the bending of the body produced by muscle contractions and are thus important contributors to the nematode undulatory ability. When nematodes are removed from sediment, as in laboratory observations, their movement becomes undirected and of a whipping and thrashing nature. It should be kept in mind that free-living nematodes are largely interstitial inhabitants, and their undulatory movements are effective for progression only when applied against substratum particles or the surface tension of water films. Nematodes move through soil pores of 15 to 45 μm in diameter, and the pore size for optimum movement is about 1.5 times the worm's diameter. (See review in Nicholas, 1975.)

Many nematodes may swim intermittently for short distances. This is true, for example, of moss-inhabiting species when the plant is flooded. A few species can crawl. The cuticle is sculptured, which aids in gripping the surface. In one species, which possesses a ringed cuticle, crawling is similar to that of earthworms; others crawl like caterpillars; and still others move like inchworms (Fig. 9–13E). The caudal gland, or spinneret, which is present in

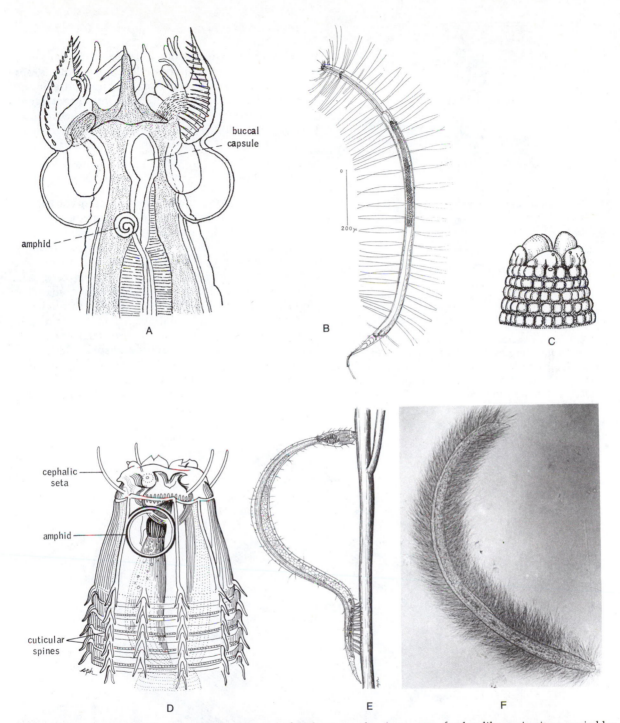

Figure 9–13 Nematode diversity: *A*, Anterior end of *Wilsonema*, showing ornate, feather-like projections carried by the lips. (After Steiner from Hyman.) *B*, *Trichotheristus*, possessing long body setae. *C*, Anterior end of *Placodira*, in which cuticle is ornamented with plaques. (*B* after Schuurmans-Stekhoven and de Coninck; *C* after Thorne; both from de Coninck, L., 1965: *In* Grassé, P. (Ed.): Traité de Zoologie. Vol. IV, 2nd Fascicle. Masson et Cie, Paris.) *D*, Anterior end of *Monopisthia*, in which the cuticle is annulated and bears spines. (After de Coninck.) *E*, Marine draconematid, with preanal ambulatory setae. (Adapted from several sources.) *F*, A stilbonematid nematode *(Eubostrichus dianae)*, with the body clothed with algae. (Courtesy of Bruce E. Hopper.)

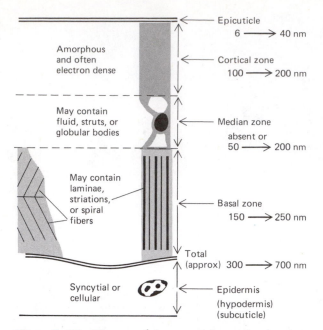

Figure 9–14 Diagram of the nematode cuticle, showing ultrastructural features that may be present in the different layers. (From Bird, A. F., 1984: Nematoda. *In* Bereiter-Hahn, J., Matoltsy, A. G., and Richards, K. S. (Eds.): Biology of the Integument. 1. Invertebrates. Springer-Verlag, N.Y. pp. 212–233.)

most marine nematodes, provides for temporary attachment.

A great many freshwater and terrestrial nematodes have a cosmopolitan distribution. Birds, animals, and floating debris to which small amounts of mud adhere are undoubtedly important agents in the spread of nematodes. Many of the saprophagous nematodes that inhabit dung utilize dung insects to move from one habitat to another.

Nutrition

Many free-living nematodes are carnivorous and feed on small metazoan animals, including other nematodes. Other species are phytophagous. Many marine and freshwater species feed on diatoms, algae, and fungi (Table 9–2). Algae and fungi are also important food sources for many terrestrial species, but there are fungi that trap nematodes. The worms are caught when they pass through special hyphal (threadlike) loops, which close on stimulation. A large number of terrestrial nematodes pierce the cells of plant roots and suck out the contents. Such nematodes can be responsible for serious damage to commercial plants. There are also many deposit-feeding marine, freshwater, and ter-

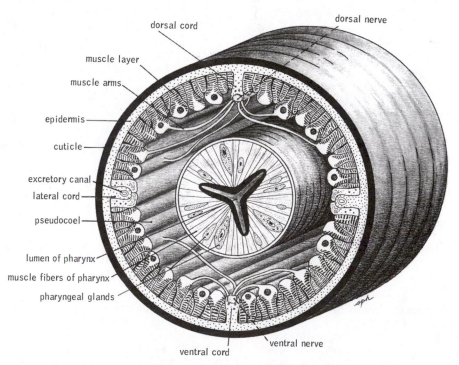

Figure 9–15 Diagrammatic cross section through the body of a nematode at the level of the pharynx. Note that the muscles send extensions, or arms, to the nerve cord, instead of the more usual reverse arrangement. Only a few of the many muscle arms are shown. (Adapted from several sources.)

TABLE 9–2 Classification of the Feeding Habits of Freshwater and Terrestrial Nematodes

Food	Method of Feeding	Taxonomic Groups Well Represented
Bacteria	Ingest as suspension, or on detritus	Rhabditida Araeolaimida, Monhysterida
Fungi, yeasts	Ingest whole	Rhabditida
Fungi, mycelial	Pierce and suck	Tylenchida
Algae	Ingest whole	Rhabditida
Algae	Pierce and suck	Dorylaimida, Tylenchida
Higher plants		
External root browsers	Pierce and suck	Dorylaimida, Tylenchida
Sedentary root parasites	Pierce and suck	Dorylaimida, Tylenchida
Sedentary foliar parasites	Pierce and suck	Tylenchida
Migratory tissue feeders	Pierce and suck	Tylenchida
Predators	Ingest whole	Dorylaimida, Rhabditida, Enoplida
Predators	Pierce and suck	Dorylaimida, Tylenchida

(From Nicholas, W. L., 1975: The Biology of Free-Living Nematodes. Clarendon Press, Oxford. p. 5. Based largely on Yeates, 1971.)

restrial species, which ingest substratum particles. Deposit feeders and the many nematodes that live on dead organic matter, such as dung, or on the decomposing bodies of plants and animals, feed only on associated bacteria and fungi, however. This is true of the common vinegar eel, *Turbatrix aceti,* which lives in the sediment of nonpasteurized vinegar. Nematodes are the largest and most ubiquitous group of organisms feeding on fungi and bacteria and are of great importance in the food chains leading from decomposers.

The mouth of the nematode opens into a buccal cavity, or stoma (Figs. 9–11, 9–12, and 9–17), which is somewhat tubular and lined with cuticle. The cuticular surface is often strengthened with ridges, rods, or plates, or it may bear a large number of teeth (Fig. 9–18). The structural details of the buccal cavity are correlated with feeding habits and are of primary importance in the identification of nematodes. Teeth are especially typical of carnivorous nematodes; they may be small and numerous or limited to a few, large, jawlike processes

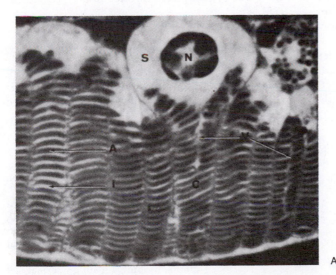
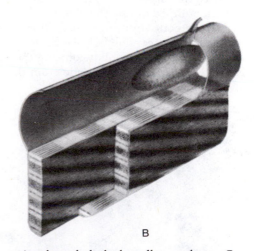

Figure 9–16 Nematode musculature: *A,* Photomicrograph of a cross section through the body wall musculature. Contractile part of cell (C) is directed toward hypodermis, which appears black at bottom of photograph. Nucleus (N) and noncontractile part of cell (S) are located on pseudocoel side of body wall. *B,* Diagram of a section of nematode muscle cell. The oval body is the nucleus, and the noncontractile part of the cell appears hollow. Note the obliquely striated fibrils. (Both from Hope, D., 1969: Proc. Helminthological Soc. Washington, 36(1):10–29.)

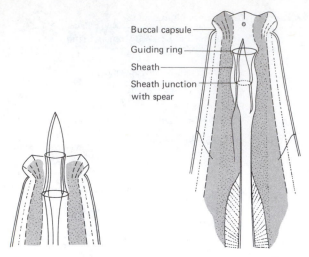

Buccal capsule
Guiding ring
Sheath
Sheath junction
with spear

Figure 9–17 Anterior end of *Aparcelaimus regius*, showing extruded and retracted spear. This dorylaimid soil nematode preys on small earthworms. The spear is hollow, and prey contents pass through its lumen into the pharynx. (From Thorne, G., 1974: Nematodes of the northern Great Plains. Part II. Dorylaimoidea in part. Agric. Exp. Stat., S. Dakota Univ., Tech. Bull. *41*:7.)

(Fig. 9–12*B*). The feeding habits of *Mononchus papillatus*, which is a toothed nematode, have been described by Steiner and Heinly (1922). This terrestrial carnivore, which has a large dorsal tooth opposed by a buccal ridge, consumes as many as 1000 other nematodes during its life-span of approximately 18 weeks. In feeding, this nematode attaches its lips to the prey and makes an incision in it with the large tooth. The contents of the prey are then pumped out by the pharynx.

In some carnivores, as well as in many species that feed on the contents of plant cells, the buccal capsule carries a long hollow or solid spear (stylet) (Fig. 9–17), which can protrude from the mouth.

Both kinds of stylet are used to puncture prey, and the hollow stylet may act as a tube through which the contents of the victim are pumped out. In a stylet-bearing herbivore, it is used to penetrate the root cell walls, being thrust rapidly forward and backward (Fig. 9–18*B*). Both groups secrete pharyngeal enzymes that initiate digestion of the prey or the plant cell contents and may even aid in the penetration of the plant cell wall.

The buccal cavity leads into a tubular pharynx, referred to as the esophagus by nematologists (Fig. 9–11). The pharyngeal lumen is triradiate in cross section and lined with cuticle (Fig. 9–15). The wall is composed of myoepithelium (as in gastrotrichs) and gland cells (Ruppert, 1982). Frequently, the pharynx contains more than one muscular swelling or bulb. The pharynx or pharyngeal bulbs act as pumps and bring food from the mouth into the intestine. Valves are frequently present.

From the pharynx a long tubular intestine composed of a single layer of epithelial cells extends the length of the body. A valve located at each end of the intestine prevents food from being forced out of the intestine by the fluid pressure of the pseudocoel. A short, cuticle-lined rectum (cloaca in the male) connects the intestine with the anus, which is on the midventral line just in front of the posterior tip of the body (Fig. 9–11).

Digestive enzymes are produced by the pharyngeal glands and the intestinal epithelium. Digestion begins extracellularly within the intestinal lumen but is completed intracellularly (Deutsch, 1978).

Excretion and Osmoregulation

Protonephridia are absent in all nematodes and apparently disappeared with the ancestral members of the class. Some nematodes have no specialized

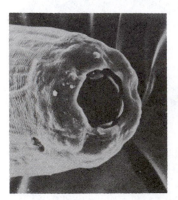

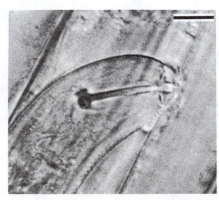

A   B

Figure 9–18 *A*, Scanning electron micrograph of the mouth of the predatory nematode *Prionchulus*. Edges of jawlike plates are visible. (From Croll, N. A., and Matthews, B. E., 1977: Biology of Nematodes. John Wiley and Sons, N.Y. 201 pp.) *B*, *Heterodera cruciferae* probing a plant epidermal cell with its stylet. (From Doncaster, C. C., and Seymour, M. K., 1973: Exploration and selection of penetration site by Tylenchida. Nematologica, 19(2):137–145.)

excretory system, but many do possess a peculiar system of gland cells, with or without tubules, that has some excretory function. In the class Adenophorea, which includes most marine and freshwater nematodes, there is usually one large gland cell, called a renette gland (Fig. 9–19*A*), located ventrally in the pseudocoel near the pharynx. The gland cell is provided with a necklike duct that opens ventrally on the midline as an excretory pore.

All members of the class Secernentea, which includes many terrestrial species, have a more specialized tubular system, still composed of only a few cells. Three long canals are arranged to form an H (Fig. 9–19*B*). Two are lateral and extend inside the lateral longitudinal cords. The two lateral canals are connected by a single transverse canal, from which a short, common, excretory canal leads to the excretory pore, located ventrally on the midline. In many nematodes, that part of each lateral canal anterior to the transverse canal has disappeared, so the system is shaped like a horseshoe; in others the tubules on one side have been lost, so the system is asymmetrical.

The excretory gland cell or tubules are known to eliminate foreign substances, but may have other functions as well. Ammonia is the principal nitrogenous waste of nematodes and is removed through the body wall and eliminated from the digestive system along with the indigestible residues.

Maintenance of fluid pressure within the pseudocoel is important. Water passes readily through the cuticle and body wall, and this is probably the principal pathway for water and ion regulation. Freshwater and terrestrial nematodes must maintain the pseudocoelomic fluid hypertonic to the surrounding medium. Some unusual aquatic habitats of nematodes include hot springs, in which the water temperature may reach 53°C and the water in tropical, epiphytic bromeliads. In large lakes, there is often a distinct zonation of the nematode species from the shoreline to deeper water.

Terrestrial species live in the film of water that surrounds each soil particle, and they are therefore actually aquatic. In fact, there are species that are found in both soil and fresh water. Although nematodes exist in enormous numbers in the upper soil, the population decreases rapidly at greater depths. Moreover, the numbers are greater in the vicinity of plant roots. In addition to the more typical terrestrial habitats, nematodes have also been reported from accumulations of detritus in leaf axils and in the angles of tree branches. Mosses and lichens maintain a characteristic nematode fauna, and these and many soil species are able to withstand periodic desiccation. During such times the worm passes into a state of anabiosis or cryptobiosis (i.e., inactivity accompanied by some water loss and a very low metabolic rate). In such a state they can survive extremely dry conditions and very low temperatures. Recovery after several years in cryptobiosis has been recorded (see Nicholas, 1984). Rotifers and tardigrades (p. 725), which live in the same habitat, show similar adaptations.

Nervous System

The brain is represented by a circumenteric nerve ring with ganglia attached dorsally, laterally, and ventrally (Fig. 9–11). From the brain, ring nerves extend anteriorly, innervating the labial and cephalic papillae or bristles and the amphids (sense organs that are described later). Dorsal, lateral, and ventral nerves extend posteriorly from the brain and run within the longitudinal cords. To these nerves the muscle cell tails make contact (Fig.

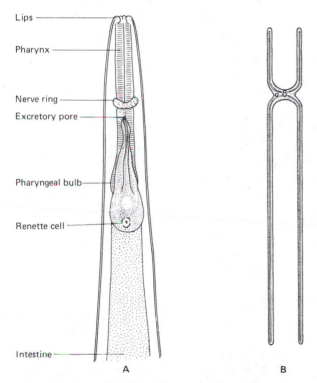

Figure 9–19 *A*, Renette gland excretory system of an adenophoran. (Greatly modified from Chitwood.) *B*, Tubular excretory system of a secernentean. Central circle is excretory pore, and solid black circles are the nuclei of the three cells forming the sytem. (Modified after Chitwood from Maggenti, A., 1981: General Nematology. Springer-Verlag, N.Y. pp. 76, 77.)

Lips
Pharynx
Nerve ring
Excretory pore
Pharyngeal bulb
Renette cell
Intestine

A   B

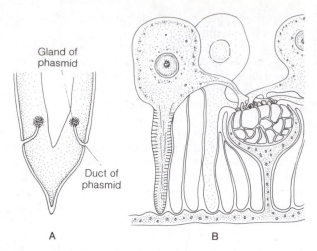

Gland of
phasmid

Duct of
phasmid

A   B

Figure 9–20 *A*, Phasmids at posterior of *Rhabditis*. (After Chitwood from Hyman.) *B*, Junction of muscle cell tails with nerve cord, which is embedded in a hypodermal cord cell. (From Rosenbluth, J., 1965: Ultrastructure of somatic cells in *Ascaris lumbricoides*. II. Jour. Cell Biol., 26:580.)

9–20). The largest of these is the ventral nerve, which arises as a double cord from the ventral side of the nerve ring. In the region of the excretory pore, the double cord fuses and continues posteriorly as a ganglionated chain.

The principal sense organs in nematodes are papillae, setae, and amphids. The labial and cephalic papillae are projections of the cuticle (Fig. 9–12), and each contains a nerve fiber from the papillary nerves. Sensory setae may be especially prevalent on the head, but they are also present elsewhere on the body surface. The setae appear to function as mechanoreceptors, and the sensory element is a modified cilium (Croll and Smith, 1974). The amphids, which reach their highest development in the aquatic nematodes, especially in the marine species, are blind, pouchlike or tubelike invaginations of the cuticle. One amphid is situated on each side of the head; they are commonly located just posterior to the cephalic setae (Figs. 9–12 and 9–13). The amphids contain chemoreceptors but probably have other sensory functions as well. Electron microscopy of certain nematodes has revealed that the sensory processes of the amphids are also modified cilia.

A pair of simple ocelli are located one on each side of the pharynx of some marine and freshwater nematodes, but their function is uncertain. Stretch receptors have been found in the epidermal cords of nematodes and probably regulate locomotor movements (Lorenzen, 1978).

In the tail region of some groups of nematodes (Secernentea) is a pair of unicellular glands, called phasmids (Fig. 9–20A), which open separately on either side of the tail. The phasmids are perhaps glandulosensory structures that function in chemoreception. They reach their best development in parasitic nematodes.

Reproduction

Most nematodes are dioecious. Males are often smaller than females, and the posterior of the male is curled like a hook (Fig. 9–21A). The gonads are tubular and may be especially long and coiled in parasitic species. In most nematode orders, the germ cells arise from a large terminal cell located at the distal end of the tube; the germ cells gradually pass through the length of the gonad, during which time growth and maturation take place.

In males there may be one or two tubular testes, which pass more or less imperceptibly into a long sperm duct. The sperm duct eventually widens to form a long seminal vesicle. A muscular ejaculatory duct, containing a varying number of prostatic glands, connects the seminal vesicle with the rectum. The prostatic secretions are adhesive and supposedly aid in copulation. The wall of the rectum, or cloaca (since it functions as a part of the reproductive system), is evaginated to form two pouches, which join before they open into the cloacal chamber. Each pouch contains a spicule, which is usually short and shaped somewhat like a pointed, curved blade (Fig. 9–21A). Special muscles cause the spicules to protrude through the cloaca and out of the anus or vent. In many nematodes the dorsal walls (and sometimes even the ventral and lateral walls) of the pouch bear special, cuticular pieces (the gubernaculum) that guide the spicules through the cloacal chamber (Fig. 9–21).

There may be one or, more typically, two ovaries, which are usually oriented in opposite directions (Fig. 9–11). The ovary gradually extends into a tubular oviduct and then into a much-widened, elongated uterus. Each of the two uteri opens into a short, common, muscular tube called the vagina, which leads to the outside through the gonopore. The female gonopore is located on the midventral line, usually in the midregion of the body. The upper end of the uterus usually functions as a seminal receptacle.

The females of some nematodes are known to produce a pheromone that attracts males (Somers et al, 1977; Jensen, 1982) and the mechanism is probably widespread. In copulation the curved pos-

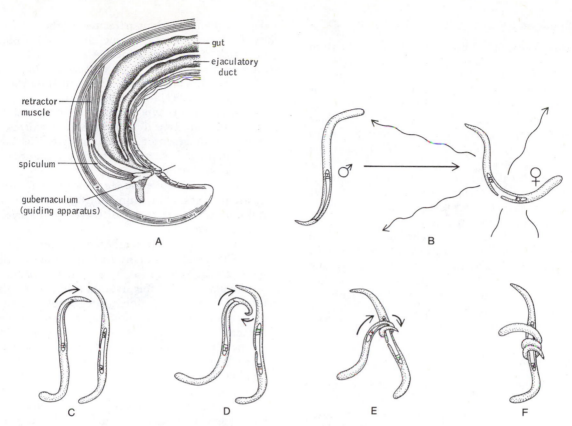

Figure 9–21 *A*, Posterior end of the male of *Pseudocella*. (Modified from Hope.) *B–F*, Diagrammatic representation of sexual behavior in the marine nematode *Chromadorita tenuis*. The pointed end of the male is the posterior, bearing the male cloacal opening and penial spicule. Arrows indicate movements of male. *B*, Male attracted by female phero-mone. *C* and *D*, male approaching female and beginning to coil posterior end. *E* and *F*, Posterior end of male coiled about female until male penial setae contact female gonopore (middle of body). Sensory seta in front of the male's cloacal opening probably aids in the orientation process. (From Jensen, P., 1982: Reproductive behaviour of the free-living marine nematode *Chromadorita tenuis*. Mar. Ecol. Prog. Ser., 10:89–95.)

terior of the male nematode is usually coiled around the body of the female in the region of the genital pores (Fig. 9–21B to F). The copulatory spicules of the male are extended through the cloaca and anus and are used to hold open the female gonopore during the transmission of the sperm into the vagina. Nematode sperm are peculiar in that they lack flagella, and some may move in a more or less ameboid manner.

After copulation the sperm migrate to the upper end of the uterus, where fertilization takes place. The fertilized egg secretes a thick fertiliza-tion membrane, which hardens to form the inner part of the shell. To this inner shell is added an outer layer, which is secreted by the uterine wall. The outer surface of nematode eggs is sculptured in various ways and is often important in the de-tection of parasitic infections from fecal smears. Nematode eggs may be stored in the uterus prior

to deposition, and not infrequently embryonic de-velopment begins while the eggs are still in the female.

Some terrestrial nematodes, particularly the rhabditoids, are hermaphroditic. Sperm develop before the eggs within the same gonad (ovotestis) and are stored. Self-fertilization takes place after the formation of the eggs. Parthenogenesis also oc-curs in some nematodes, especially terrestrial spe-cies, and there are some parthenogenetic nema-todes for which males are unknown.

The deposition of the eggs of free-living nema-todes is still not well known. Marine species rarely produce more than 50 eggs, which are often depos-ited in clusters. Terrestrial species may produce up to several hundred eggs, which are deposited in the soil. Many parasitic nematodes and some freeliving forms, such as the vinegar eel, are viviparous.

Embryogeny

Cleavage is not spiral but does follow a fixed asymmetrical pattern. Cleavage is also determinate, and there is an early separation of future germ cells from somatic cells. As in some other aschelminths, the various organs of the body contain a relatively fixed number of cells, and these cells have been largely attained by the time hatching takes place. For example, in certain *Rhabditis* adults there are 200 nerve cells and 120 epidermal cells, and the digestive tract is composed of 172 cells. Although there is a limited increase in the number of cells during juvenile stages, most growth in the size of nematodes results from increases in cell size.

The young, sometimes called larvae, have almost all of the adult structures when they hatch except for certain parts of the reproductive system. Growth is accompanied by four molts of the cuticle, the first two of which may occur within the shell before hatching (as in Secernentea). Adults do not molt but many continue to increase in size.

Parasitism

The great numbers of parasitic species of nematodes attack virtually all groups of animals and plants and, except for the absence of ectoparasitic forms on animals, display all degrees and types of parasitism. Moreover, some of the major groups of nematodes contain both free-living and parasitic species. All of these facts suggest that parasitism has evolved many times within the phylum. Schaefer (1971) suggests that some nematode radiation was coupled with the evolution of flowering plants, insects, and higher vertebrates, all of which contain important nematode host species. The types of nematode host-parasite relationships have been outlined by Hyman (1951). A modification of that outline is followed here:

1 Completely free living. Life cycle is direct, and all stages are free living.

2 Ectoparasites of plants. Worms feed on the external cells of plants by puncturing the cell wall with stylets and sucking out the cell contents. (These nematodes could just as readily be called herbivores.)

3 Endoparasites of plants. Worms in juvenile stage enter the plant body and feed on the living cells, producing gall-like structures or causing tissue death. Reproduction takes place within the host and the new generation of juveniles migrates to other plants. *Heterodera* contains many such species.

Most of the plant parasites, both ectoparasites and endoparasites, are members of the orders Tylenchida and Aphelenchida. All parts of a plant are attacked, depending upon the species. Many of these nematodes are of great economic importance (see Maggenti, 1981, for a brief review).

4 Saprophagous type of zooparasitism. Adults and juveniles are free living in soil, but worms in late juvenile stage enter an invertebrate. The host is not injured, and the worms feed on the dead tissues when the host dies.

5 Zooparasitic juvenile stages only. The juveniles are parasitic in an animal host, usually an invertebrate, for at least part of the early life cycle. Adults are free living.

6 Phytoparasitic juveniles and zooparasitic adults. The female worm produces juveniles within a plant-feeding insect host. When the insect punctures the plant tissue, the juvenile worms enter the plant and remain as endoparasites. When mature and after copulation, the female enters the larva of the insect host, which lives on the same plant. The larva metamorphoses into an adult and deposits a new generation of juvenile worms. *Heterotylenchus aberrans*, a parasite of onion flies, illustrates this type of life cycle.

7 Zooparasitic juveniles and phytoparasitic adults. Early stages of development take place within an invertebrate host. Later, worms in juvenile stages leave the host and enter the plant on which the host feeds. Worms complete development and reproduce as phytoparasites. The new generation of young then enter the host.

8 Zooparasitism in adult females only. The young become adults in the soil. After copulation, the male dies, and the female enters an invertebrate host to produce the next generation.

9 Adult zooparasites with one host. Adult worms of both sexes are parasitic within a vertebrate or invertebrate host. Transmission from one host to another is by eggs or newly hatched young, which may be free living for a part of their development. Many economically or medically important nematode parasites possess this type of life history; a few of the more interesting or important examples follow.

The ascaroid nematodes, which feed on the intestinal contents of humans, dogs, cats, pigs, cattle, horses, chickens, and other vertebrates, include the largest species of nematodes. They are entirely parasitic within a single host, and the life cycle typically involves transmission by the ingestion of eggs or juveniles passed in the feces of another host. The juvenile stages usually penetrate the intestinal wall

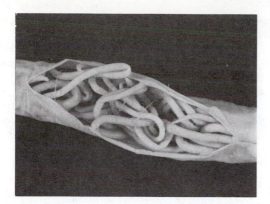

Figure 9–22 Adult specimens of *Ascaris suum* within the intestine of a pig. *Ascaris lumbricoides*, which affects humans, is very similar. (From Schmidt, G. D., and Roberts, L. S., 1985: Foundations of Parasitology. 3rd Edition. Times Mirror/Mosby College Publ., St. Louis. p. 491.)

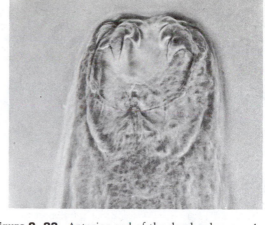

Figure 9–23 Anterior end of the dog hookworm *Ancylostoma caninum*, showing buccal region and teeth.

to enter the circulatory system, where they are carried to the lungs. Here they break into the alveoli and migrate back to the intestine via the trachea and esophagus.

The human ascaroid, *Ascaris lumbricoides*, reaches a length of 49 cm and is one of the best known parasitic nematodes (Fig. 9–22). The species is widely distributed throughout the world, including the southeastern United States, particularly in children. The embryonated eggs are notoriously resistant to adverse environmental conditions and may remain viable in soil for 10 years. The very closely related species in pigs probably had a common evolutionary origin with the human species, the common ancestor being confined to one host or the other prior to human domestication of the pig.

Toxocara canis and *cati* are two small ascaroid species common in dogs and cats. It is these species for which puppies and kittens are usually wormed.

Physiological studies suggest that *Ascaris* produces enzyme inhibitors that protect the worm from the host's digestive enzymes. If this is true, such a mechanism is probably utilized by most other gut-inhabiting nematodes. The ascarids feed on the host's intestinal contents.

The hookworms are another group of parasites of the digestive tract of vertebrates. Most members of this group feed on the host's blood. The mouth region is usually provided with cutting plates, hooks, teeth, or combinations of these structures for attaching to and lacerating the gut wall (Fig. 9–23).

An infection of more than about 25 worms will produce symptoms of hookworm disease, and a heavy infection can produce serious danger to the host through loss of blood and tissue damage. An adult worm may live as long as two years in the intestine. Hookworm infection is widespread in humans. It is estimated that over 380 million people are infected with *Necator americanus*, the most important species throughout the tropical regions of the world (despite the species' name).

The life cycle of hookworms involves an indirect migratory pathway by the juveniles, as in ascaroids. The fertilized eggs leave the host in its feces and hatch outside the host's body on the ground. The juvenile gains reentry by penetrating the host's skin (feet in humans) and is carried in the blood to the lungs. From the lungs the juvenile migrates to the pharynx, where it is swallowed and passes to the intestine.

Oxyurid nematodes, known as pinworms, have a simpler life cycle. These small nematodes are parasitic in the gut of vertebrates and invertebrates. Infection usually occurs through the ingestion of eggs passed in feces. The eggs hatch, and juveniles develop within the gut of the new host. The human pinworm, *Enterobius vermicularis*, affects children throughout the world. The female worm deposits eggs at night in the perianal region. Itching is caused by the migration of the female depositing her eggs, and scratching by the child contaminates the fingernails and hands with eggs. The eggs thus easily spread to other children or reinfect the same child.

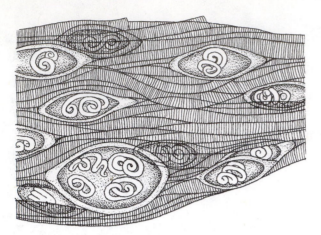

Figure 9–24 Larvae of *Trichinella spiralis* within calcareous cyst in striated muscle tissue of host. (After Chandler, A. C., and Read, C. P., 1961: Introduction to Parasitology. John Wiley and Sons, N.Y.)

Trichinelloids are also parasites of the alimentary tracts of vertebrates, especially birds and mammals. The whipworms, *Trichuris*, which infect human beings, dogs, cats, cattle, and other mammals, are relatively small (the human whipworm, *Trichuris trichiura*, is about 4 cm) and have a life cycle similar to that of pinworms. Certainly the most familiar of the trichinelloids is *Trichinella spiralis* of mammals, the cause of the disease trichinosis. The minute worm, which lives in the intestinal wall, is viviparous, and its juveniles are carried in the blood to the striated muscles. There the juveniles form calcified cysts and, if infection is high, can produce pain and stiffness (Fig. 9–24). Transmission to another host can occur only if flesh containing encysted juveniles is ingested. Thus, in some animals, such as the rat, this can be a one-host parasite; in others, such as man and the pig, it would normally require two hosts.

10 Zooparasites with one intermediate host. Varying degrees of juvenile development take place within an intermediate host, after which there is reinfection of the definitive host, where reproduction occurs.

This type of life cycle is illustrated by a number of nematode parasites, including the familiar filarioids and dracunculoids. The filarioids are threadlike worms that inhabit the lymphatic glands and some other tissue sites of the vertebrate host, especially birds and mammals. The female is viviparous and the larvae are called microfilariae. Bloodsucking insects, such as fleas, certain flies, and especially mosquitoes, are the intermediate hosts.

A number of species parasitize man, producing filariasis.

The chiefly African and Asian *Wuchereria bancrofti* illustrates the life cycle. (The male is 40 mm × 0.1 mm, and the female is about 90 mm × 0.24 mm.) The adults live in the ducts adjacent to the lymph glands. The microfilariae are found in the blood and are present in the peripheral bloodstream. A distinct periodicity of larval migration to the peripheral circulation, coinciding with the activity of the mosquito, has been demonstrated by a number of investigators. When certain species of mosquitoes bite the host, the microfilariae enter with the host's blood. Development within the intermediate host involves a migration through the gut to the thoracic muscles and after a certain period into the proboscis. From the proboscis the microfilariae are introduced back into the definitive host when the mosquito feeds. In severe filariasis the blocking of the lymph vessels by large numbers of worms results in serious short-term lymphatic inflammation marked by pain and fever. Over a long period, increase of connective tissue in affected areas may result in terrible enlargement of the legs, breast, and scrotum. Such enlargement is called elephantiasis (Fig. 9–25), but extreme cases are no longer common.

Dirofilaria immitis, heartworms, which live as adults in the hearts or pulmonary arteries of dogs, wolves, and foxes, are also transmitted by mosquitoes.

Loa loa, the African eye worm, lives in the subcutaneous tissues of humans and baboons. The worm migrates about in the tissue and sometimes passes across the eyeball, accounting for the name *eye worm* (Fig. 9–26).

The dracunculoids are also rather threadlike worms. They live in the connective tissues and body cavities of vertebrate hosts. The most notable example is the guinea worm, *Dracunculus medinensis*, which parasitizes humans and many other mammals, especially in Asia and Africa. The female is about 1 mm in diameter and up to 120 cm in length. After a period of development in the body cavity and connective tissue of the host, the gravid female migrates to the subcutaneous tissue and produces an ulcerated opening to the exterior. If the ulcerated area of the host comes in contact with water, larvae are released. After a short free-living stage, the larvae are ingested by species of copepod crustaceans *(Cyclops)* and continue their development in the host's hemocoel. When the definitive host swallows copepods in drinking water,

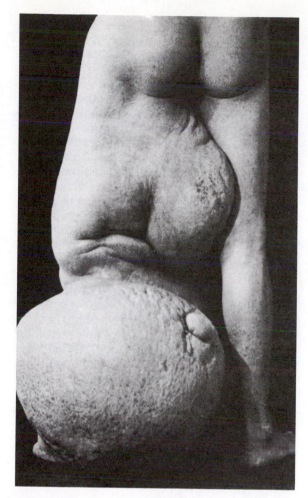

Figure 9–26 The eye worm *Loa loa* in the cornea. (From Chandler, A. C., and Read, C. P., 1961: Introduction to Parasitology. John Wiley and Sons, N.Y.)

and almost all of the marine forms. To this class belong the orders Enoplida, Isolaimida, Mononchida, Dorylaimida, Trichocephalida, Mermithida, Muspiceida, Chromadorida, Desmodorida, Desmoscolecida, Monhysterida, and Araeolaimida.

The class Secernentea, or Phasmida, contains nematodes that usually possess porelike amphids in the lateral lips. Many parasitic forms are mem-

Figure 9–25 A victim of elephantiasis, which results from severe filariasis. (Courtesy of Mayo Clinic.)

the nematode larvae are released and penetrate the intestinal wall to reach the coelom or subcutaneous tissue. The worms can be removed surgically, but the ancient method, still practiced, is to slowly wind them out on a small stick (Fig. 9–27). Breaking of the worm will cause severe inflammation.

Classification

The phylum Nematoda is divided into 2 classes and 14 orders. The class Adenophorea, or Aphasmida, contains nematodes that have variably shaped amphids behind the lips. There are both free living and parasitic members. The free-living species include terrestrial and freshwater forms

Figure 9–27 A guinea worm, *Dracunculus medinensis*, being removed from ulcerated opening of arm by slow winding on a match stick. (Courtesy, Institute of Public Health Research, Teheran University School of Public Health.)

bers of this class, and the free-living species are largely soil inhabitants. To this class belong the orders Rhabditida, Strongylida, Ascaridida, Spirurida, Camallanida, Diplogasterida, Tylenchida, and Aphelenchida.

Phylum Nematomorpha

Closely related to the nematodes is a small group (about 230 species) of long worms, known as horsehair worms or hairworms, which constitute the phylum Nematomorpha. The adults are free living, but the juveniles are all parasitic in arthropods. In most hairworms, which form the class Gordioida, the adults live in fresh water and damp soil. The single genus *Nectonema*, which makes up the class Nectonematoida, is pelagic in marine waters. The body of nematomorphs (Fig. 9–28) is extremely long and threadlike, without a distinct head. Lengths of 36 cm or more are typical. The diameter, however, is usually not much more than 1 mm. The hairlike nature of these worms is so striking that they were formerly thought to arise spontaneously from the hairs of a horse's tail.

Like that of nematodes, the nematomorph body wall is composed of a thick, outer, brown to black cuticle; there are a cellular epidermis with longi-tudinal cords and a muscle layer of longitudinal fibers. The pseudocoel is usually reduced. The digestive tract is vestigial, and the adults do not feed. The nervous system is composed of an anterior nerve ring and a ventral cord. The sexes are separate, and two long, cylindrical gonads extend the length of the body. As in nematodes, the sperm ducts empty into the rectum, or cloaca, but there are no copulatory spicules. The oviducts also empty into the cloaca.

Hairworms live in all types of freshwater habitats in temperate and tropical regions of the world. The females are inactive, but the males commonly swim or crawl about by whiplike motions. There is copulation, and an egg string is deposited in the water. On hatching, the nematomorph larva has a protrusible proboscis armed with spines (Fig. 9–29).

After hatching, the larvae enter an arthropod host living in water or along the water's edge. Common hosts are beetles, cockroaches, crickets, grasshoppers, centipedes, and millipedes. Recently, leeches have been reported as hosts. The larvae of *Nectonema*, the only marine hairworms, parasitize hermit crabs and true crabs. The larvae either penetrate the host or are ingested as cysts. The young then enter the hemocoel, where development is completed. Their nutrition as parasites

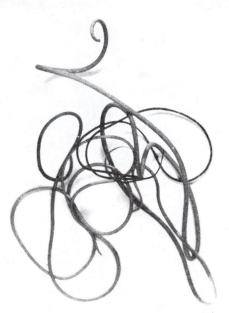

Figure 9–28 Three female gordioid nematomorphs, or horsehair worms. (From Pennak, R. W., 1978: Freshwater Invertebrates of the United States. 2nd Edition. John Wiley and Sons, N.Y.)

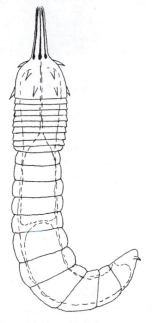

Figure 9–29 Gordioid worm larva. Proboscis protrudes. (After Pennak, R. W., 1978: Freshwater Invertebrates of the United States. 2nd Edition. John Wiley and Sons, N.Y.)

is apparently accomplished by direct absorption of food materials through the body wall. After several weeks to several months of development, during which a number of molts occur, the worms leave the host as almost completely formed adults. They emerge only when the host is near water. Sexual maturity is shortly attained during the free-living adult phase of the life cycle.

Phylum Rotifera

The phylum Rotifera, or Rotatoria, contains the very common freshwater animals known as rotifers. Although some marine species exist and some species live in mosses, the majority inhabit fresh water. Over 1500 species have been described, and most have a widespread distribution.

Most rotifers are 0.1 to 1 mm long, only a little longer than ciliated protozoa. The body is composed of about 1000 cells, and the organ systems are very eutelic. They are solitary, free-swimming animals, but there are some sessile species (Fig. 9–33) as well as some colonial forms (Fig. 9–36A). The body is usually transparent, although some rotifers appear green, orange, red, or brown, owing to coloration of the digestive tract.

External Structure

The elongated or saccate body, which is relatively cylindrical, can be divided into a short anterior region, a large trunk composing the major part of the body, and a terminal foot (Figs. 9–30 and 9–34B). The body is always covered by a distinct cuticle, which may be ringed, sculptured, or ornamented in various ways.

The broad or narrowed anterior end forms the head region and bears a ciliated organ called the corona, which is characteristic of all members of the phylum. The primitive corona is believed to have consisted of a large, ventral, ciliated area called the buccal field (Fig. 9–31), which surrounded the mouth. If rotifers evolved from a small, ciliated, creeping ancestor, as is generally believed, then perhaps the buccal field represented a vestige of the ancestral ventral ciliation. From the buccal field, cilia extended around the anterior margins of the head to form a crownlike ring called the circumapical band. The area inside the ring, which is devoid of cilia, is called the apical field.

The different types of coronas characteristic of different groups of rotifers are believed to have evolved from this basic structural plan. Various parts of the buccal field and the circumapical band have either been lost or become more highly developed. The corona of *Euchlanis* is a good illustration of the modifications that can be found (Fig. 9–32). Not infrequently, certain cilia have become modified to form cirri, membranelles, or bristles. A few of the more common types of coronal ciliation are described here.

The buccal field in *Collotheca* and related forms is modified into a funnel, with the mouth at the bottom, and ciliation is reduced (Fig. 9–33). The edges of the buccal field have become expanded, forming a varying number of lobes bearing bristles, or setae, which may be arranged in bundles or tufts.

In *Polyarthra* and related forms the corona is derived entirely from the circumapical band, which has been transformed into two circlets of modified cilia—an anterior band called the trochus and a posterior band called the cingulum which passes below the mouth. The corona of bdelloid rotifers, which include many common species, also possess these two ciliated bands, but the anterior circlet of cilia (the trochus) is raised on a pedestal and divided into two discs, called trochal discs (Fig. 9–30A). The posterior circlet passes around the base of the pedestals and runs beneath the mouth. In living species, the beating membranelles of the trochal discs resemble two rotating wheels at the anterior of the body, and it is from this type of corona that the names *rotifer* and *wheel animalcule* are derived. The trochal discs are used both in swimming and in feeding, and the pedestals can be retracted when the discs are not functioning.

The head in bdelloid and notommatid rotifers carries a middorsal projection called the rostrum (Fig. 9–30B). This little projection bears cilia and sensory bristles at its tip. Other anterior structures in rotifers include the eyes (which vary in number and location), one or two short antennae, and the mucus-secreting retrocerebral organ (Figs. 9–31 and 9–32B).

The trunk forms the major part of the body. The cuticle is frequently much thickened to form a conspicuous encasement, called a lorica (Fig. 9–34). The lorica may be divided into distinct plates or ringlike sections and is usually ornamented with ridges, spines, or articulated appendages. The spines may be long, and in some rotifers they are movable. In *Polyarthra* the appendages are long, flat, skipping blades that are grouped in four clusters of three each (Fig. 9–35).

The terminal portion of the body, or foot, is considerably narrower than the trunk region (Figs. 9–30 and 9–34B). The cuticle is frequently ringed,

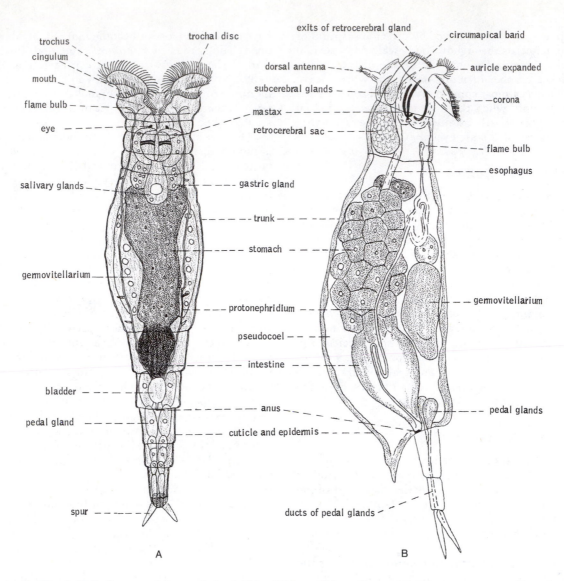

Figure 9–30 *A, Philodina roseola* (ventral view). (After Hickernell from Hyman.) *B, Notommata copeus* (lateral view). (After Hyman, L. H., 1951: The Invertebrates, Vol. III. McGraw-Hill Book Co., N.Y.)

and in many bdelloids (Fig. 9–30A) the resulting segments or joints of the foot are able to telescope into similar larger joints of the trunk. Even the head may be retracted in this manner. The end of the foot usually bears one to four projections, called toes. In both the crawling and the sessile rotifers, the foot is used as an attachment organ; in these groups the foot contains pedal glands that open by ducts at the tips of the toes or other parts of the foot. These pedal glands produce an adhesive substance for temporary attachment. Duogland organs have not yet been found in rotifers.

In a number of common bdelloids, such as *Philodina* (Fig. 9–30A) and *Rotaria*, the pedal glands open onto the ends of two long and diverging conical spurs located near the end of the foot. Functionally, the spurs replace the toes, which are very small in these genera. In planktonic rotifers, the foot is usually reduced or has disappeared altogether.

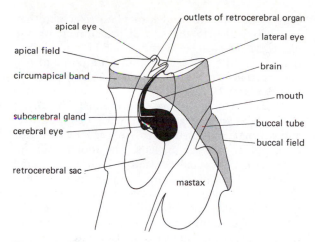

Figure 9–31 Primitive corona (lateral view). (After Beauchamp from Hyman.)

Body Wall and Pseudocoel

The cuticle of rotifers, unlike that of gastrotrichs and nematodes, is intracellular rather than external. It is part of the epidermis, which is thin and syncytial and always possesses a constant number of nuclei. Beneath the epidermis are the body muscles. Although some of the muscle fibers are circular (ring muscles) and some are longitudinal (retractor muscles), the body wall musculature is not organized into distinct circular and longitudinal sheaths, as in flatworms. The small pseudocoel lies beneath the body wall and surrounds the gut and other internal organs. The pseudocoel is filled with fluid and a syncytial network of branching ameboid cells.

Locomotion

Rotifers move by creeping or swimming. Propulsion for swimming is provided by the beating cilia

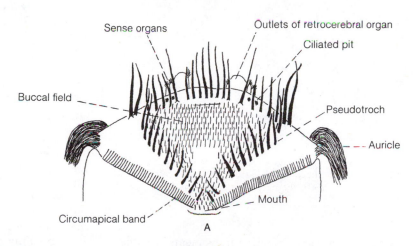

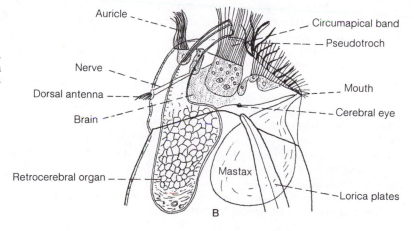

Figure 9–32 Corona of *Euchlanis: A,* Ventral view. (After Stoszberg from Hyman.) *B,* Lateral view. (After Beauchamp from Hyman.)

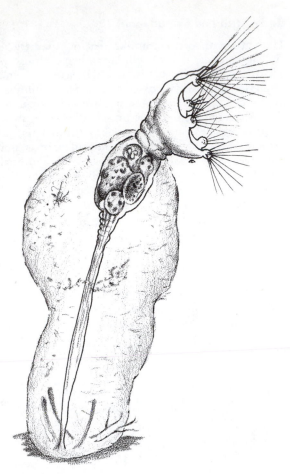

Figure 9–33 *Collotheca ornata,* A sessile rotifer with funnel-shaped buccal field. Trunk and stalk surrounded by a mucous sheath. Female. (After Hudson.)

of the corona, and crawling movements are aided by the foot. Most species of freshwater rotifers inhabit the substratum or live on submerged vegetation and other objects. Some of these benthic species never swim, but many both creep and swim. In the common bdelloids, for example, the corona is retracted when the animal creeps, and the foot adheres to the substratum using the adhesive secretion produced by the pedal glands. The animal then extends the body, attaches the rostrum, and detaches the foot to move forward and again grip the substratum. During swimming, which is only for short distances, the corona is extended, and the foot is retracted.

Many species of sessile rotifers attach to vegetation and display a remarkable restrictiveness, not only to the species of alga or plant to which they attach but also to the site of attachment on the plant. Many species of the order Flosculariaceae live in vaselike tubes, commonly composed of foreign particles embedded within a secreted material (see Wallace, 1980).

Pelagic rotifers swim continuously. Usually, the body is somewhat saccate with a thin cuticle, and various flotation devices, such as long spines or oil droplets, may be present. The foot has disappeared or has turned ventrally. Among the many strictly pelagic species are a few colonial forms, such as *Conochilus,* whose members resemble trumpets radiating from a common center (Fig. 9–36A). The combined ciliary action of the coronae propels the colony through the water.

Many pelagic species undergo seasonal changes in body shape or proportions, a phenomenon known as cyclomorphosis and one that also takes place in small crustaceans. For example, certain individuals during one season of the year have spines that are longer or shorter than those during another season. In *Brachionus calyciflorus* spines can be induced by starvation, low temperature, and by some substance produced by the predator rotifer *Asplanchna.* Elongated spines protect *Brachionus* from being eaten by this predator.

Nutrition

The mouth of the rotifer is typically ventral and is usually surrounded by some part of the corona. The mouth may open directly into the pharynx, or a ciliated buccal tube may be situated between the mouth and the pharynx, as in suspension feeders. The pharynx, or mastax (Figs. 9–30 and 9–34B), is characteristic of all rotifers, and its structure is a distinguishing feature of the phylum. The mastax is usually oval or elongated and highly muscular. The inner walls carry seven large, interconnected, projecting pieces, or trophi, composed of an acid mucopolysaccharide material (Fig. 9–37). The mastax is used both in capturing and in triturating food, and its structure therefore varies considerably, depending on the type of feeding behavior involved.

Most rotifers are either suspension or raptorial feeders, although the latter group is rather omnivorous. The suspension feeders, of which the bdelloids are the most notable examples, feed on minute organic particles that are brought to the mouth in the water current produced by the coronal cilia. In the bdelloids, which have trochal

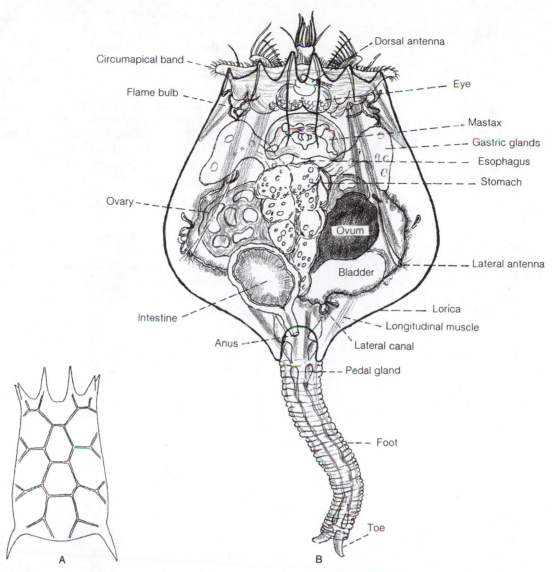

Figure 9–34 *A.* Lorica of *Keratella quadrata* (dorsal view). (Modified from Ahlstrom from Pennak.) *B,* A brachionid rotifer. (After Hudson.)

discs, the larger, preoral cilia (trochus) produce the principal water current, which is directed backward, and function in both swimming and feeding. Food particles brought in by the water current are swept by both preoral and postoral cilia into a food groove that lies between them. Food groove cilia then carry the particles around to the mouth (Strathmann et al, 1972). The mastax of suspension feeders (Figs. 9–37*A* and 9–38) is adapted to grinding; two of the pieces are extremely large, platelike,

and ridged. The two plates oppose each other, and the ridges form a surface for grinding. The mastax of suspension feeders probably also acts as a pump, sucking in particles that have collected at the mouth. Food intake can be regulated in various ways. In *Brachionus*, for example, the ciliated buccal field can be screened or exposed by certain large coronal cirri, or the buccal field's ciliary beat can be reversed, or the mastax can reject particles (Gilbert and Starkweather, 1977).

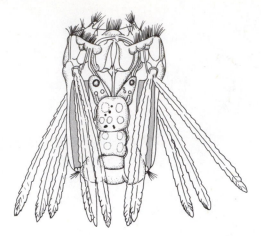

Figure 9–35 *Polyarthra dolichoptera*, a planktonic rotifer with skipping blades. (From Koste, W., 1978: Rotatoria, Vol. II. Gebrüder Borntraeger, Berlin. p. 155.)

The carnivorous species, which feed on protozoa, rotifers, and other small metazoan animals, capture their prey by trapping or suction. The forceps-like trophi of suction feeders are used to hold or manipulate prey once it is in the mastax cavity (Fig. 9–37) (Gilbert and Stemberger, 1985). After the prey is broken up, the indigestible parts of its body are commonly discarded.

Those rotifers that capture prey by trapping include *Collotheca* and other forms that possess funnel-like buccal fields (Fig. 9–33). When small protozoa accidentally swim into the funnel, the setae-bearing lobes of the funnel fold inward, preventing escape. The captured organisms then pass into the mouth and pharynx. The mastax of a trapping rotifer is often very much reduced.

Some rotifers, especially notommatids, feed on the cell contents of filamentous algae.

There are number of epizoic and parasitic rotifers that live primarily on small crustaceans, particularly on the gills. Endoparasitic species inhabit snail eggs, heliozoans, the interior of *Volvox*, and the intestine and coelom of earthworms, freshwater oligochaetes, and slugs. One genus, *Proales*, is parasitic within the filaments of the freshwater alga *Vaucheria* and produces gall-like swellings. In parasitic rotifers either the foot or the mastax becomes modified as an attachment organ, and the corona is reduced.

Located in the mastax walls of most rotifers are glandular masses called salivary glands (Fig. 9–30*A*), which open through ducts just in front of the mastax proper. A tubular esophagus connects the pharynx with the stomach. At the junction of the esophagus and stomach are a pair of gastric glands (Fig. 9–38*B*), each of which opens by a pore

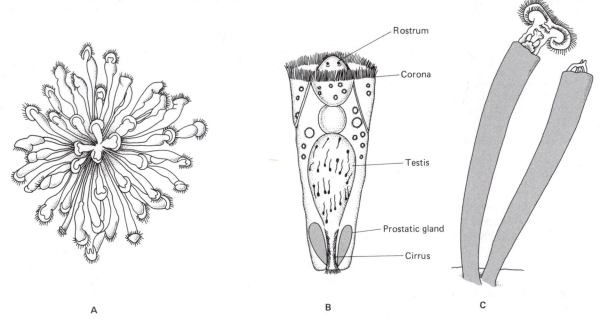

Figure 9–36 *A*, *Conochilus hippocrepis*, a colonial pelagic rotifer. *B*, Motile, solitary male of *Conochilus*. (After Wesenberg-Lund from Hyman.) *C*, *Limnias ceratophylli*, a tubicolous rotifer in which the tube is secreted. (*A* and *C* after Hudson.)

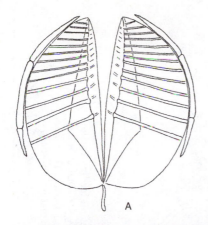

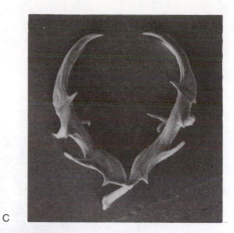

Figure 9–37 Types of mastax trophi. *A*, Mastax with trophi adapted for grinding. Note the two large ridged plates that provide grinding surfaces. (After Beauchamp from Hyman.) *B*, Scanning electron micrograph of the open mouth of the carnivorous rotifer *Asplanchna brightwelli*, showing protruding mastix pieces (arrows). (From Amsellem, J., and Clement, P., 1980: A simplified way to prepare rotifers for transmission and scanning electron microscopy. Hydrobiologia, 73:119–122.) *C*, Scanning electron micrograph of the cleaned trophi of *Asplanchna brightwelli*. Compare with *A* and *B*. (From Salt, G. W., Sabbadini, G. F., and Commins, M. L., 1978: Trophi morphology relative to food habits in six species of rotifers (Asplanchnidae). Trans. Amer. Micros. Soc., 97(4):469–485.)

into each side of the digestive tract. The stomach is a large sac or tube that passes into a short intestine. The excretory organs and the oviduct also open into the terminal end of the intestine, which is sometimes called the cloaca. The anus opens to the dorsal surface near the posterior of the trunk.

The gastric glands, and perhaps the salivary glands, are believed to produce enzymes, and digestion and absorption occur in the stomach.

Water Balance

Typically, two protonephridia are present in the pseudocoel, one on either side of the body. The protonephridia are of the flame-cell type, but the nucleus is some distance from the terminal, fenestrated ends of the branched tubules (Fig. 9–39 and Box 7–1). Two collecting tubules empty into a bladder (Fig. 9–34B), which opens into the ventral side of the cloaca; in the bdelloids, there is a constriction between the stomach and the intestine, and the somewhat bulbous cloaca acts as a bladder (Fig. 9–30A). The contents of the bladder or cloaca are emptied through the anus by contraction. The pulsation rate is often between one and four times per minute.

The protonephridia of rotifers function in osmoregulation. The excreted fluid is hypoosmotic to the fluid of the pseudocoel, and the rate of bladder discharge is determined by the ionic content of the

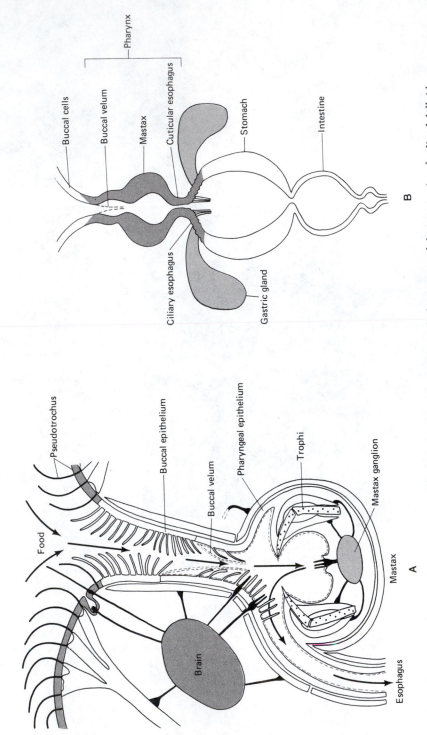

Figure 9–38 *A,* Diagram showing the ingestion of food and its passage through the mastax of the suspension-feeding bdelloid rotifer *Philodina.* Food is driven into the mouth by the trochal and buccal cilia, ground by mastix pieces, screened by mastix cilia, and passed into the esophagus. The brain and mastax ganglion control muscles of the trochal discs and mastax. The buccal velum prevents food backflow. *B,* Digestive system of a rotifer. (From Clement, P., et al, 1980: An ultrastructural approach to feeding behavior in *Philodina roseola* and *Brachionus calyciflorus.* Hydrobiologia, 73:133–141.)

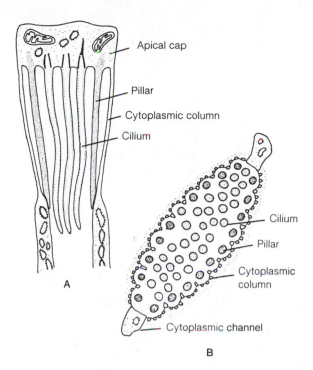

Figure 9–39 Flame bulb of a protonephridium of the rotifer *Asplanchna. A,* Longitudinal section. *B,* Cross section. See text for explanation. (After Warner from Wilson, R. A., and Webster, L. A., 1974: Protonephridia. Biol. Rev., *49:*127–160; copyrighted and reprinted by permission of Cambridge University Press.)

environmental medium (see review by Wilson and Webster, 1974). The high rate of discharge from the bladder suggests that some fluid may be entering the body through the mouth in feeding.

Most terrestrial rotifers are associated with soil, leaf mold, mosses, and lichens and are active only during the short periods when these surfaces are filled with water. During this time these terrestrial rotifers swim about in the water films. They are capable of undergoing desiccation, usually without the formation of cysts, and can remain in a dormant, highly resistant state (cryptobiosis) for as long as three to four years (see review on soil rotifers by Pourriot, 1979).

Nervous System

The brain consists of a dorsal ganglionic mass lying over the mastax (Fig. 9–31*A*). It gives rise to a varying number of nerves that extend to the anterior sense organs and to other parts of the body. The sense organs consist of sensory bristles in various parts of the ciliated crown, two ciliated pits (Fig. 9–32*A*), and one or two cerebral eyes or a pair of anterior eyes or both (Fig. 9–31). The eyes are simple pigment-cup ocelli composed of one to two photoreceptor cells plus an accessory, red pigmented cell. The eyes provide for locomotor orientation to light and for photic regulation of reproduction (see Clément and Wurdak, 1984, for eye structure and light responses).

Reproduction

The great ecological success of rotifers is certainly correlated in large part with their reproductive adaptations. Reproduction is entirely sexual, and like most other aschelminths, rotifers are dioecious. The males are always smaller than the females, and certain structures, such as the cloaca, are degenerate or absent in the male. Parthenogenesis is characteristic of most groups, and in most species males are present in the population only at certain times; in the bdelloids no males have ever been reported. Thus, females make up all or most of the population of rotifers usually encountered.

The female reproductive system in the majority of species consists of one or two ovaries combined with a yolk-producing vitellarium located anteriorly in the pseudocoel (Figs. 9–30 and 9–34*B*). The vitellarium supplies yolk to the eggs by direct flow through cytoplasmic bridges. The eggs then pass through an oviduct into the cloaca, or to a genital pore if there is no intestine.

The male is short lived, and the gut is vestigial or absent. A single, saclike testis and a ciliated sperm duct (Fig. 9–36*B*) are present. Because a cloaca is usually absent in the male, the sperm duct runs directly to a gonopore that is homologous to the anus in the female and has the same position. Two or more glandular masses called prostate glands are associated with the sperm duct, and the end of the sperm duct is usually modified to form a copulatory organ.

Copulation is by hypodermic impregnation or by insertion through the cloaca. In the planktonic *Asplanchna* the penis adheres to the female cuticle, and the male may be pulled about by the female (Aloia and Moretti, 1973). The mechanism by which penetration takes place is still uncertain. In *Brachionus* penetration occurs only in the softer coronal region. In some species copulation occurs within a few hours of hatching, when the female's cuticle is still soft.

Each ovarian nucleus forms one egg, and since only some 8 to 20 such nuclei exist in most species, there is a corresponding limit to the number of eggs produced in the lifetime of a particular female. Each egg is surrounded by a shell and a number of egg membranes, all of which are secreted by the egg itself. The eggs may be free floating, attached to objects on the substratum, or attached to the body of the female. A few rotifers, such as *Asplanchna* and *Rotaria*, brood their eggs internally.

Bdelloids, in which males are unknown and development is parthenogenetic, produce eggs mitotically; the eggs hatch into females only. In other rotifers (the monogonontans) several types of eggs are produced.

One type, called an amictic egg, is thin shelled, cannot be fertilized, and develops by parthenogenesis into amictic females. Typical meiosis does not take place in maturation, and the eggs are diploid (Fig. 9–40). A second type of egg, called a mictic egg, is also thin shelled but is haploid. If these eggs are not fertilized, they produce males parthenogenetically. If they are fertilized, they secrete heavy, resistant shells. Such fertilized eggs are called dormant eggs; in contrast to the thin-shelled, unfertilized amictic and mictic eggs, which hatch in several days, these dormant eggs are capable of withstanding desiccation and other adverse conditions and may not hatch for several months or even years. Dormant eggs hatch into females. As females may produce amictic or mictic eggs, but not both, the type of egg produced appears to be determined at the time of oocyte development. The life cycle may be further complicated, as in *Asplanchna*, by the presence of different morphological types of females (see Fig. 9–41).

Parthenogenesis and the production of two kinds of eggs are probably adaptations for life in fresh water, especially temporary ponds and streams. Dormant eggs are able to withstand adverse environmental conditions—desiccation and temperature extremes—and parthenogenesis facilitates a rapid increase in population. One or both of these adaptations are found in other animals that live in temporary bodies of fresh water—turbellarians, gastrotrichs, and water fleas.

The reproductive pattern of such rotifers tends to be cyclic. After spring rains, with the advent of warmer temperatures, dormant eggs that have passed through the winter hatch into amictic females. These females produce a number of generations of parthenogenetic females, each having a life-span of one to two weeks. Some species can double their population every two days. In the late

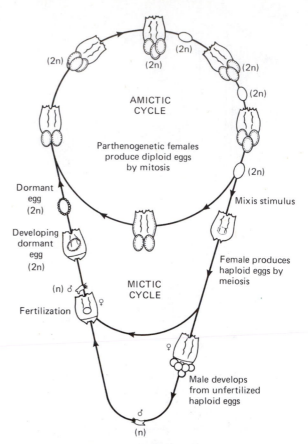

Figure 9–40 Life cycle of a monogonont rotifer. This particular species, *Branchionus leydigi*, lives in temporary ponds. Dormant eggs hatch with melting snows and spring rains to begin a first amictic cycle. Stagnating water stimulates the production of mictic eggs, and dormant eggs carry the species through summer, when the pond dries up. With autumn rains, there is a second amictic cycle. Frost stimulates the production of mictic eggs, and dormant eggs carry the species over the winter. (Modified from Koste, W., 1978: Rotatoria. I. Textband, p. 34. Gebrüder Borntraeger, Berlin.)

spring or early summer, when this population reaches a peak, mictic eggs are produced and males appear (Fig. 9–40). Dormant eggs carry the species through until the next season and, if the pond or stream dries up, can be dispersed by birds or blown with dust. Rotifers inhabiting large, permanent bodies of fresh water may display a number of cycles or population peaks during the warmer months or may be present during the whole year. The production of mictic eggs is induced by specific environmental factors, such as high population density, changes in the amount or kind of

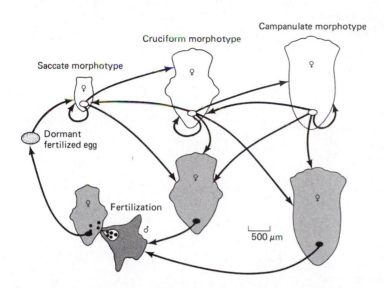

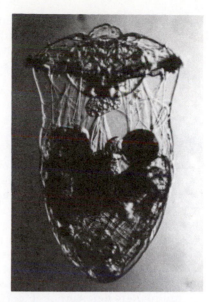

Figure 9–41 Life cycle of *Asplanchna sieboldi*. This planktonic rotifer, common in lakes and ponds, has the same life cycle illustrated for other monogonont rotifers in Fig. 9–40. Here, however, there are three female forms, differing in size and shape (morphotypes). Morphotype development is controlled by the amount of tocopherol (vitamin E) in the diet of the mother, higher levels causing the formation of the cruciform and campanulate types, and by the type of prey eaten by the mother. Eggs develop within the female. White eggs are amictic (diploid); black eggs are mictic (haploid). Photograph illustrates a campanulate female containing developing eggs and advanced embryo. (Modified after Gilbert, J. J., 1980: Developmental polymorphism in the rotifer *Asplanchna sieboldi*. Amer. Sci., 68:636–646.)

food, photoperiod, and temperature, but the importance of different factors varies from species to species (see Thane, 1974; Gilbert, 1974 and 1980).

Embryogeny

Cleavage in rotifers is spiral and determinate but modified. Nuclear division is completed early in development and never occurs again. In free-moving species no larval development takes place. When the females hatch, they have all the adult features and attain sexual maturity after a growth period of a few days. For example, the common bdelloid rotifer *Philodina roseola* has an average life-span of 48 days, and the adult is 28 times heavier than a newly hatched individual. A female lays 45 eggs, and the generation time is 4 days (Lebedeva and Gerasimova, 1981). The smaller male rotifers do not undergo a growth period but are sexually mature when they leave the egg. The sessile rotifers hatch as free-swimming "larvae" that are structurally very similar to free-swimming species. After a short period they settle down, attach, and assume the characteristics of the sessile adults.

SYSTEMATIC RÉSUMÉ OF PHYLUM ROTIFERA

Class Seisonidea. A single genus of marine rotifers commensal on certain crustaceans. Large, elongate body with reduced corona. *Seison.*

Class Bdelloidea (Digononta). Anterior end retractile and usually bearing two trochal discs. Mastax adapted for grinding, with one pair of flattened trophi. Ovaries paired. Telescopic cylindrical body. Swimming and creeping species. Males absent. *Philodina, Embata, Rotaria, Adineta.*

Class Monogononta. Rotifers with one ovary. Mastax, if adapted for grinding, not designed as in the bdelloids.

> **Order Flosculariaceae.** Sessile, many tubicolous; or free-swimming, toeless rotifers. Corona with a double wreath of cilia. *Conochilus, Floscularia, Hexarthra, Testudinella.*

> **Order Collothecaceae.** Mostly sessile rotifers. Mouth at the bottom of a shallow concavity. Anterior end often surrounded by arms or bundles of setae. *Stephanoceros, Collotheca.*

Order Ploima. Swimming rotifers. Body with or without a lorica, often short, sometimes saclike. This order contains the majority of rotifers. *Notommata, Proales, Polyarthra, Synchaeta, Chromogaster, Gastropus, Asplanchna, Brachionus, Euchlanis, Keratella.*

Phylum Acanthocephala

The acanthocephalans are a phylum of some 1150 species of parasitic, wormlike pseudocoelomates. All are endoparasites requiring two hosts to complete the life cycle. The juveniles are parasitic within crustaceans and insects; the adults live in the digestive tracts of vertebrates, especially fish. The body of the adult is elongated and composed of a trunk and a short, anterior proboscis and neck region (Fig. 9–42). Most acanthocephalans are only a few millimeters long, although one species may attain a length of 80 cm. The proboscis is covered with recurved spines (Fig. 9–42*B*), hence the name *Acanthocephala*—spiny head.

The acanthocephalan proboscis and neck can be retracted into a muscular proboscis sac in the anterior of the trunk. The trunk is frequently covered with spines and often is divided into superficial segments. The retractable proboscis and anchoring spines provide the means of attachment in the host's gut and also enable the acanthors (larvae) to move within the host. To either side of the neck an evagination of the body wall projects posteriorly into the trunk pseudocoel. These two evaginations, called lemnisci, are filled with fluid and function as part of the hydraulic system in proboscis protrusion.

The body wall of acanthocephalans is covered by a living syncytial integument. As in rotifers, there is a thick, cuticle-like lamina that lies within the outer plasma membrane. The lamina is penetrated by numerous branches of pore canals. The syncytial cytoplasm beneath the lamina and pore canals surround the large basal nuclei and contain bundles of hollow fibers (Whitfield, 1984). Lipid-filled spaces between the fibers represent what has been called a lacunar canal system.

These worms possess no digestive system, and food is absorbed directly through the body wall from the host. Two protonephridia, which are associated with the reproductive system, are present in one family. The nervous system is composed of a ventral, anterior ganglionic mass, from which arise varying numbers of single and paired nerves for different body structures.

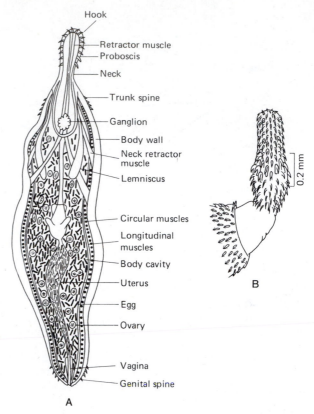

Figure 9–42 *A*, Diagrammatic sagittal section through a female acanthocephalan. (From Conway Morris, S., and Crompton, D. W. T., 1982: The origins and evolution of the Acanthocephala. Biol. Rev., 57:85–115; copyrighted and reprinted by permission of Cambridge University Press.) *B*, The spiny proboscis of *Andracantha phalacrocoracis*. (From Schmidt, G. D., 1975: *Andracantha*, a new genus of Acanthocephala from fish-eating birds, with descriptions of three species. Jour. Parasit., 61:615–620.)

The sexes are separate, and the gonopore is located at the end of the trunk. The male system includes a protrusible penis, and fertilization is internal following copulation. Development of the egg takes place within the female pseudocoel. Eventually, a larval stage having an anterior crown or rostellum with hooks and enclosed within a shell is passed out of the vertebrate host with the feces. If the egg is eaten by certain insects, such as roaches or grubs (Fig. 9–43), or by aquatic crustaceans, such as amphipods, isopods, or ostracods, the larva emerges from the egg, bores through the gut wall of the host, and becomes lodged in the hemocoel. The larva, called an acanthor, possesses a rostellum with hooks that are used in penetrating the host's tissues. When the intermediate host is

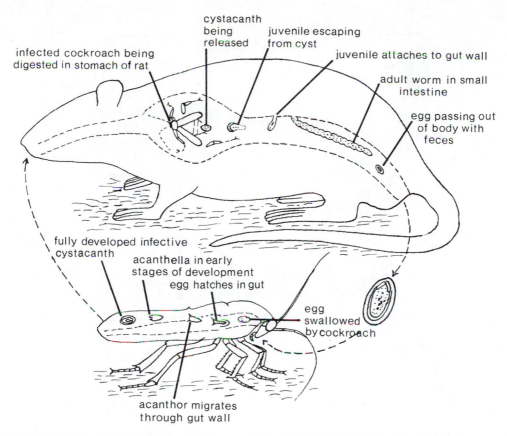

cystacanth being released

juvenile escaping from cyst

juvenile attaches to gut wall

infected cockroach being digested in stomach of rat

adult worm in small intestine

egg passing out of body with feces

fully developed infective cystacanth

acanthella in early stages of development egg hatches in gut

egg swallowed by cockroach

acanthor migrates through gut wall

Figure 9–43 Life cycle of *Moniliformis dubius*, an acanthocephalan that inhabits the gut of mice, rats, dogs, and cats. Acanthella and cystacanths are stages following the acanthor, in which adult features are developing. (From Olsen, O. W., 1967: Animal Parasites. Burgess Publishing Co. Minneapolis.)

eaten by a fish, bird, or mammal, the worm attaches to the intestinal wall of the vertebrate host by using the spiny proboscis. Acanthocephalans often exist in great numbers within the vertebrate host and can do considerable damage to the intestinal wall. As many as 1000 acanthocephalans have been reported in the intestine of a duck, and 1154 in the intestine of a seal.

Phylum Kinorhyncha

The phylum Kinorhyncha, also known as Echinoderida, consists of a small group of some 100 described marine pseudocoelomates that burrow in the surface layer of marine mud or live in the interstitial spaces of marine sand. They have been found from the intertidal zone to depths of several thousand meters. The members of this phylum are somewhat larger than rotifers and gastrotrichs but are usually less than 1 mm in length. The general body shape is similar to that of the gastrotrichs, with a head and trunk that are not sharply demarcated, (Fig. 9–44). However, kinorhynchs differ from both rotifers and gastrotrichs in their lack of external free cilia.

A distinguishing feature of the phylum is the division of the chitinous cuticle into clearly defined segments (zonites). The head composes the first segment, the neck makes up the second segment, and the trunk in all adults contains 11 segments (Fig. 9–44). The dorsal cuticular plate of some trunk segments bears large median and lateral, recurved movable spines, from which the name *Echinoderida*—spiny neck—is derived. Usually a single pair of adhesive tubes is located on the anterior, ventral surface of the trunk, but there are species that possess numerous lateral adhesive tubes.

The anteriorly located mouth is situated at the end of a protrusible cone, which is surrounded at

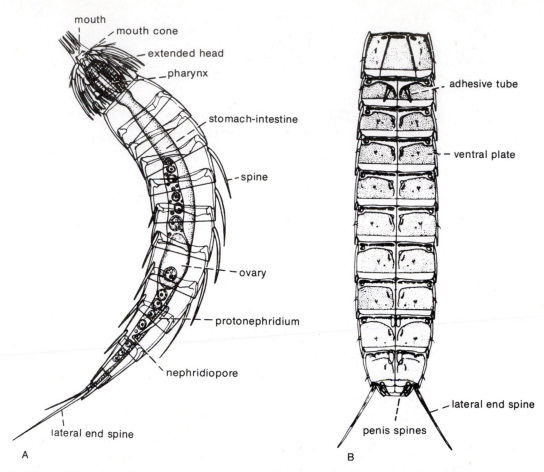

Figure 9–44 *A, Echinoderes* (lateral view). *B, Pycnophyes frequens* (ventral view with head retracted). (Both courtesy of R. P. Higgins.)

the tip and base by circlets of spines (scalids) (Fig. 9–44*A*). The entire head can be withdrawn either into the neck or into the first trunk segments, hence the name *Kinorhyncha*—movable snout. In the former case the cuticular plates of the neck (placids) are adapted for closing over the retracted head. A similar closing apparatus is present on the first trunk segments in those species in which both head and neck retracted into the trunk.

The epidermis of the body wall is thickened along the middorsal line and along each lateral angle to form longitudinal epidermal cords (Fig. 9–45). The pseudocoel is well developed and contains ameboid cells.

A kinorhynch burrows by anchoring the extended head with its recurved spines into the mud or interstitial spaces. The body is then drawn forward until the head is retracted into the neck or trunk. The head is again pushed forward into the mud, and the process is repeated.

Most kinorhynchs feed on diatoms. A buccal cavity in the mouth cone leads into a barrel-shaped pharynx, esophagus, and stomach-intestine (Fig. 9–44*A*). A short hindgut connects the stomach-intestine with the anus, which opens to the outside at the end of the trunk. The details of digestion are unknown.

Two protonephridia with solenocytes open to the dorsolateral surface of the 11th segments. The nervous system, which is closely associated with the epidermis, consists of a nerve ring around the anterior of the pharynx, and from this ring arises a single, midventral nerve cord containing one cluster of ganglion cells in each segment. The sense organs consist of the scalids, which contain ciliary receptors, and sensory bristles located over the general body surface, especially the trunk. A few species bear anterior pigment cup ocelli.

Kinorhynchs are dioecious, with a pair of ovaries or testes and a single gonopore on the last seg-

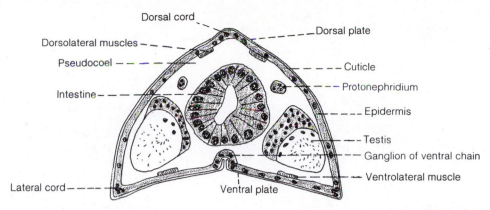

Figure 9-45 The kinorhynch, *Pycnophyes* (transverse section at level of intestine). (After Zelinka from Hyman.)

ment. The end of the sperm duct usually carries two to three penial spines or spicules. Copulation has never been observed, but species in two genera have been found to extrude spermatophores, which are directed toward a female with the penial spines (Brown, 1983). The young have less distinct segmentation and may possess spine arrangements unlike those of the adult. Periodic molts of the cuticle occur in the attainment of adulthood, after which molting does not occur.

Phylum Loricifera

The Loricifera, described in 1983 by the Danish zoologist R. M. Kristensen, are the most recently described phylum of animals and bring to three the number of phyla discovered in this century—Pogonophora (1914), Gnathostomulida (1956), and Loricifera (1983). The loriciferan *Nanaloricus mysticus* is a minute animal, about a quarter of a millimeter in length, that lives in the interstitial spaces of shelly, marine gravel. Although specimens have been found off the coasts of North Carolina and Greenland and in the Coral Sea, the phylum description is based on juvenile and mature specimens of *Nanaloricus* collected from bottom samples taken off the coast of Roscoff, France, at about 25 meters deep.

The animals adhere so tightly to the gravel substrate that they cannot be extracted by the usual magnesium chloride method, which probably accounts for their late discovery. Most loriciferans have been collected by submerging the bottom sample briefly in fresh water. This shocks the animals osmotically and they release.

Nanaloricus mysticus looks like a cross between a rotifer and a kinorhynch (Fig. 9–46). The

major part of the body, called the abdomen, is encased within a cuticular lorica composed of a dorsal, a ventral, and two lateral plates. It is to the lorica that the name of the phylum refers (lorica bearer). The cone-shaped anterior end, or introvert, of the animal bears many recurved spines (scalids) on its lateral surface. The spines are continuous with similar spines on the neck region, called the thorax, which connects the introvert with the abdomen. Both introvert and thorax can be retracted into the anterior end of the lorica.

Eight stylets surround the mouth cone at the end of the introvert, but their tips open to the side of the mouth and not within it. A buccal tube leads into a pharyngeal bulb, and a long midgut region forms most of the gut tube. A terminal rectum opens to the outside through a posterior anal cone. A large brain lies within the introvert. The sexes are separate, each with a pair of gonads. At present little is known about the habits and physiology of loriciferans.

A larval stage, called a Higgins larva, is somewhat similar to the adult except that the buccal cone lacks stylets and the thorax lacks spines. The thorax can enclose the introvert but cannot itself be retracted into the lorica. The posterior end of the larval trunk carries a pair of toes, which are used in swimming.

Possible phylogenetic relationships of the loriciferans are discussed at the end of the chapter (p. 258).

Phylum Priapulida

The phylum Priapulida consists of only 13 known species of cucumber-shaped or wormlike animals, which live buried in the bottom sand and mud in

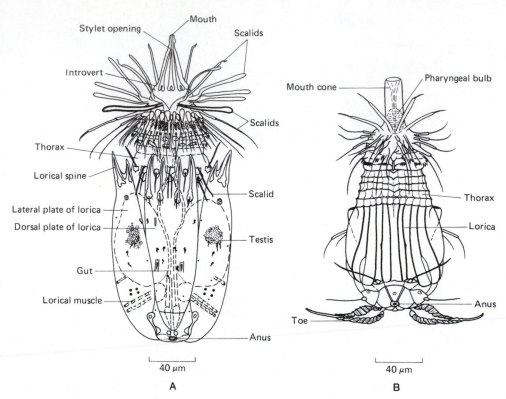

Figure 9–46   A loriciferan, *Nanaloricus mysticus: A*, Dorsal view of adult male. *B*, Dorsal view of Higgins larva. (From Kristensen, R. M., 1983: Loricifera, a new phylum with Aschelminthes characters from the meiobenthos. Z. f. zool. Systematik u. Evolutionsforschung, *21*(3):163–180.)

shallow and deep water. Off the North American coast they range northward from Massachusetts and California. They are also found in the Baltic Sea and off Siberia; some species have been taken from Antarctic waters. The tiny *Tubiluchus* lives in coralline sediments. Another minute species is tubicolous.

The group has had an unsettled history, being moved back and forth between the pseudocoelomates and coelomates because of uncertainty about the nature of the body cavity.

The cylindrical body of priapulids ranges in length from 0.5 mm to 20 cm and is divided into an anterior proboscis region, or introvert, and a posterior trunk region, or abdomen (Fig. 9–47). The introvert, which constitutes the anterior third of the animal, is somewhat barrel shaped and ornamented with longitudinal, riblike, conical projections, called scalids. In one species the mouth is surrounded by a crown of branched tentacles in addition to the scalids (Fig. 9–48). The introvert invaginates into the anterior of the abdomen, which is covered with small, variable projections. In the little *Tubiluchus corallicola*, the abdomen bears a

long, terminal tail (Fig. 9–49), and in the genus *Priapulus*, the posterior end of the abdomen carries one or two caudal appendages, each consisting of a hollow stalk to which are attached many spherical vesicles. A gas exchange or chemoreceptive function has been suggested for these structures, but neither function has yet been demonstrated.

The body wall is composed of one layer of epithelial cells lying beneath the cuticle and relatively well developed circular and longitudinal muscle layers. The cuticle contains chitin and is periodically molted.

The body cavity is bounded by a delicate membrane, which was claimed to be cellular, and thus a coelomic peritoneum. Other zoologists have been dubious, and more recent studies by Malakhov (1980) and McLean (1984) deny the presence of any peritoneal lining. Without such a lining the body cavity would be a pseudocoel. Unfortunately, the embryonic origin of the priapulid body cavity is unknown.

The fluid within the body cavity functions as a hydrostatic skeleton and as a circulatory system. It contains amebocytes and, at least in *Priapulus cau-*

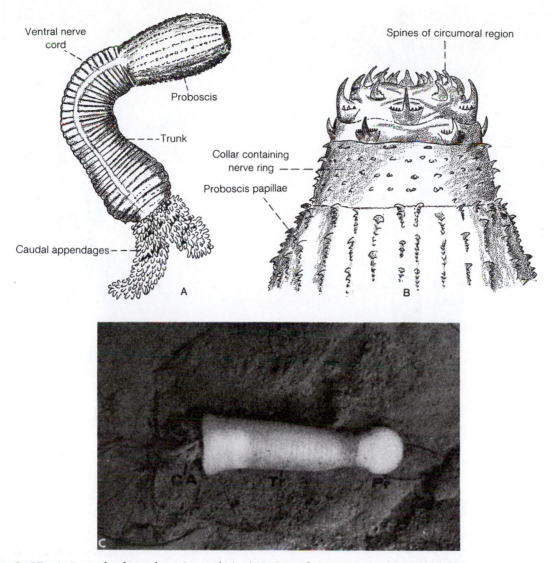

Figure 9–47 *A, Priapulus bicaudatus* (ventral view). *B, Priapulus;* anterior end with mouth region everted. (Both after Theel from Hyman.) *C,* A recently molted and contracted specimen of *Priapulus caudatus.* The ball-like swelling represents the retracted proboscis. The retracted caudal appendage partially projects from the posterior end of the trunk. (From Shapeero, W. L., 1962: The American Midland Naturalist, 68:237.)

datus, a large number of corpuscles bearing hem-erythrin.

Priapulids are capable of burrowing through the substratum by contractions of the body, along with protrusions and retractions of the introvert.

Some priapulids are reported to be predacious and feed on soft-bodied, slow-moving inverte-brates, particularly polychaete worms. The mouth and pharyngeal regions can be everted during feed-ing (Fig. 9–47B); the spines are used to seize prey. Food is passed to a muscular pharynx, which is lined with cuticle and has teeth. A straight, tubular intestine connects the pharynx with the short, ter-minal rectum. The anus is located at the posterior end of the trunk.

The priapulid nervous system is closely associ-ated with the epidermis of the body wall and con-sists of a nerve ring surrounding the anterior end of the pharynx, and a single, midventral, ganglion-ated cord. The papillae on the proboscis and abdo-men probably have a sensory function.

On each side of the intestine is a long, proto-nephridial tubule with solenocytes. A gonad is also associated with the tubule, which thus functions as

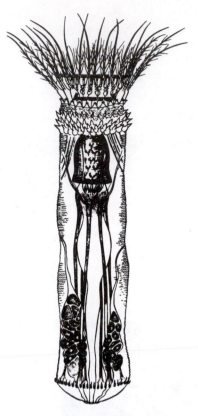

Figure 9–48 *Maccabeus tentaculatus*, a tiny, tubicolous priapulid from the Mediterranean. Adult. Tentacles are probably homologous to the anterior spines of other priapulids. (From Por, F. D., 1972: Priapulida from deep bottoms near Cyprus. Isr. J. Zool., 21(3/4):525–528.)

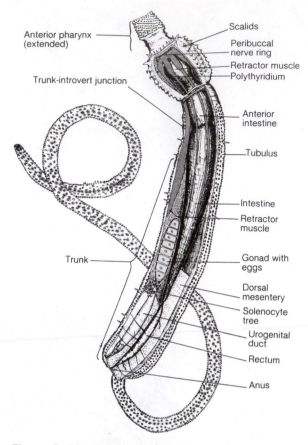

Figure 9–49 Structure of *Tubiluchus corallicola*, a minute priapulid that inhabits coralline sediments. (By Brian Marcotte.)

an excretory canal and as a genital duct. Each protonephridial tubule opens separately through a nephridiopore at the end of the abdomen.

The sexes are separate, and fertilization is external. The egg undergoes radial cleavage and develops into a stereoblastula. The intervening period of development before the stereogastrula is the hatching stage. The priapulid larva, which inhabits bottom mud, as does the adult, is surrounded by a lorica formed from the cuticle of the abdomen. The anterior part of the body can be withdrawn into the lorica, giving the larva a superficial resemblance to certain rotifers. Larval growth is accompanied by molting.

Phylum Gnathostomulida

The Gnathostomulida are a small phylum of minute, acoelomate worms that live in the interstitial spaces of marine gravels and muds, especially in

black, anaerobic sediments. This group of worms represents one of the latest phyla of animals to be discovered. Although observed earlier, the first description of a gnathostomulid was not published until 1956 by Ax. Since that time over 80 species and 18 genera have been described from many parts of the world, especially from the east coast of North America. They are widespread and may occur in large numbers, but their ability to tolerate anaerobic conditions results in their being among the last interstitial animals to come to the surface of a sample of stagnating bottom material.

Gnathostomulids are mostly between 0.5 and 1 mm in length. All are elongate, and some are even threadlike. The more or less cylindrical, transparent body consists of an anterior head separated from a trunk by a slightly constricted neck region (Fig. 9–50). The body tapers posteriorly into a tail.

Like flatworms, the gnathostomulids are ciliated, but each epithelial cell bears but a single cil-

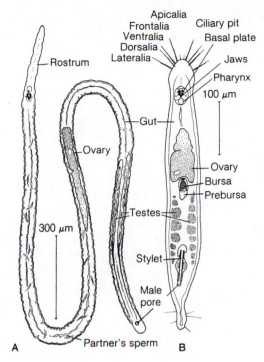

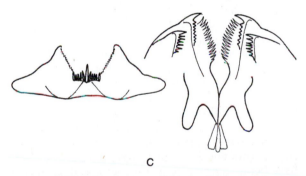

Figure 9–50 Gnathostomulida: *A, Haplognathia simplex. B, Gnathostomula jenneri. C*, Basal plate and jaws of *Gnathostomula mediterranea.* (All from Sterrer, W., 1972: Syst. Zool. 21(2):151–173.)

ium, a condition shared with primitive gastrotrichs (Box 9–1). Gnathostomulids move by a slow, gliding type of swimming, and the body can be contracted and twisted by the three or four paired groups of longitudinal fibers that constitute the body wall musculature. Parenchyma is poorly developed. The nervous system is intraepidermal, and the sense organs are ciliary pits and sensory cilia, which are especially well developed at the anterior end. There are two to five pairs of uncon-

nected, monociliated protonephridia (Lammert, 1985).

Gnathostomulids were thought to lack an anus, but indications of an anal pore have been recently reported (Knauss, 1979). The mouth and pharynx, which are located ventrally behind the head, bear a comblike plate on the ventral lip and a pair of toothed lateral jaws within the pharyngeal bulb cavity (Fig. 9–50B and C). The remainder of the gut is a tubular intestinal sac. The principal food appears to be bacteria and fungi, which are scraped up by the comb on the ventral lip and passed into the intestinal sac by snapping movements of the jaws.

Gnathostomulids are hermaphroditic, although individuals with only male or female systems occur. The female reproductive system usually consists of a single ovary and associated bursa, or sperm storage sac. A vagina is present in some species. The male system consists of one or a pair of posterior testes and usually a copulatory organ, which in some species bears a stylet. The male gonopore is at the posterior end. Copulation has not yet been observed, but sperm transfer probably occurs by penetration of the body wall through hypodermic impregnation or attachment of the male gonopore to the integument of the partner, or by injection into the vagina (Sterrer, 1974). A single large egg is laid at each oviposition. The egg ruptures through the body wall and briefly adheres to the bottom. The worm regenerates very quickly. Cleavage is spiral, and development is direct. It appears that at least some gnathostomulids may exhibit nonsexual feeding stages that alternate with sexual nonfeeding stages, the reproductive system degenerating and regenerating from one stage to another.

Phylogenetic Relationships

The aschelminths are clearly a diverse assemblage. Despite the common occurrence of a cuticle, and adhesive glands, there are no distinctive features that unite the various aschelminth groups. For this reason most zoologists are opposed to placing them within a single phylum.

There appear to be two evolutionary lines among the aschelminths, one embracing the gastrotrichs, nematodes, and nematomorphs, and one embracing the rotifers, kinorhynchs, loriciferans, priapulids, and acanthocephalans (Fig. 9–51). The ventral locomotor cilia of gastrotrichs are probably derived from those ancestors, which like tubellarians and gnathostomulids were completely ciliated.

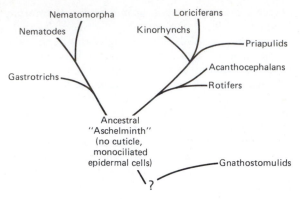

Figure 9–51 Possible phylogenetic relationships of the aschelminth phyla. The cuticle of the gastrotrich, nematode, and nematomorph line is extracellular; the cuticle of rotifers and acanthocephalans is intracellular but extracellular in the kinorhynchs and priapulids. (Based on Conway Morris, S., 1982; Kristensen, R. M., 1983; and others.)

Moreover, some gastrotrichs exhibit the primitive monociliated condition. Gastrotrichs seem to be most closely related to nematodes. Both groups have an external cuticle with lamellar components, and both have a pharynx with myoepithelial walls. These same features are found in tardigrades, or water bears, a little group of minute animals that live in the water films of mosses and other habitats (p. 725). The tardigrades have been considered allied to arthropods, but they may have to be reassigned to the aschelminths.

The nematomorphs are certainly related to the nematodes, which they closely resemble. The nematomorph larva is thought by some (Riemann, 1972) to show striking similarities to a primitive, free-living, marine nematode *(Kinonchulus)* that has a spiny, eversible pharynx.

Gastrotrichs and rotifers appear to be remote from each other. The epithelium of rotifers is composed of multiciliated cells, and the cuticle of both rotifers and acanthocephalans is intracellular. The rotifer pharynx, unlike that of gastrotrichs, is composed of separate epithelial and muscle cells.

The acanthocephalans, kinorhynchs, loriciferans, and priapulids all have a spiny anterior that can be retracted into more posterior trunks, and all are probably related (Fig. 9–51) (Whitfield, 1971). Loriciferans appear to be most closely related to kinorhynchs. Both groups have very similar introverts, but some of the loriciferan scalids are strikingly like those of priapulids.

The Gnathostomulida are placed at the end of this chapter only for the benefit of comparison. They are like flatworms in their hermaphroditic reproductive system and the absence of an anus. But they lack the flatworms 1-9 flagellar ultrastructure; the parenchyma is greatly reduced; the epidermis is monociliated like that of some gastrotrichs, not multiciliated; and the ciliary beat can be reversed. Moreover, they display a number of features, such as head, jaws, and sensory, bristle-like cilia, that are found in rotifers and gastrotrichs. The monociliated epidermis suggests that both gnathostomulids and gastrotrichs are close to some ancestral monociliated form. Although the gastrotrichs are clearly a part of the aschelminth assemblage, the relationship of the gnathostomulids to the other lower bilateral animals is uncertain. Kristensen and Nørrevang (1977) believe that they are unrelated to turbellarians and that the gnathostomulid pharyngeal bulb is most like that of certain annelids.

Summary

1 Members of the phyla Gastrotricha, Rotifera, Nematoda, and several others share such features as a collagen cuticle, adhesive tubes, muscular pharynx, protonephridia (except nematodes), minute size, and eutely. When a body cavity is present, as in the nematodes, it is derived from the embryonic blastocoel and called a pseudocoel. These common features are the basis for placing most of these groups within an assemblage, perhaps a superphylum, called the aschelminths.

2 Gastrotrichs are inhabitants of fresh water and the sea, the marine species being largely interstitial. They glide over the bottom on ventral cilia and temporarily anchor by means of adhesive tubes that contain duogland systems. The ventral ciliation is thought to be primitive, gastrotrichs perhaps being derived from some ancestral form that was more uniformly ciliated.

3 Free-living nematodes (roundworms) occur in large numbers in the sea, in fresh water, and in soil. Their long, tapered, cylindrical bodies are adapted for living in minute spaces, especially interstitial spaces of aquatic sediments and algal mats. Terrestrial species live in the water films around soil particles and are actually aquatic.

4 The cylinder of longitudinal body wall muscles, the complex elastic cuticle, and the hydrostatic pressure of the fluid within the pseudocoel make possible the undulatory movements that drive nematodes through interstitial spaces.

5 Nematodes exhibit a wide range of feeding habits, most of which involve a muscular pharynx for ingestion. Teeth or stylets are commonly present in carnivorous and plant-feeding nematodes.

The large number of parasitic nematodes includes many species of economic and medical importance, such as hookworms, ascarid worms, pinworms, *Trichinella*, and filarial worms. Parasitic species vary greatly in their life cycles.

6 Members of the phylum Nematomorpha, called horsehair worms, have threadlike bodies as long as 36 cm. The free-living adults of most species live in fresh water or in damp soil. The juvenile stages are parasitic in arthropods.

7 Rotifers are distinguished by a ciliated crown, foot, and mastax. The cuticle is not external.

8 Rotifers are largely inhabitants of fresh water. Benthic species swim intermittently by using their ciliated crowns and temporarily attach by using the adhesive glands that open at the ends of the toes on the feet. Planktonic rotifers swim continuously and have reduced feet. The cuticular pieces of the pharynx forming the mastax are adapted for raptorial or suspension feeding.

9 Nematodes and rotifers that inhabit the high-stress environments of water films around soil particles and mosses are capable of remarkable degrees of anabiosis.

10 Parthenogenesis and two types of eggs—rapid-hatching and dormant—are common adaptations of animals living in temporary bodies of fresh water. They permit rapid population growth and survival during periods of adverse conditions. One or both adaptations are encountered among some species of flatworms, gastrotrichs, and rotifers.

11 Members of the phylum Acanthocephala are gut parasites of aquatic and terrestrial vertebrates. The elongate parasite hooks to the gut wall with an anterior, retractile proboscis bearing recurved spines. There is no digestive tract, and food is absorbed through the complex body wall. The life cycle requires a crustacean or insect intermediate host.

12 The phylum Kinorhyncha is composed of a small number of marine interstitial species. The cuticle is segmented, and the mouth is at the end of a spiny, anterior, retractile head cone.

13 The Loricifera is the most recently discovered phylum of animals. They are minute, interstitial species whose spiny, cone-shaped anterior region (introvert) can be retracted into a lorica-encased posterior region (abdomen).

14 The phylum Priapulida contains nine species of wormlike marine animals that live in mud and sand. Some are microscopic; others reach 20 cm in length. The cylindrical body is composed of an anterior retractile introvert bearing the mouth, a trunk, and in some species a terminal tail region. Introvert and trunk bear papillae or spinelike projections.

15 The Gnathostomulida are a small phylum of elongate, marine, interstitial animals. The body resembles the flatworm in having a ciliated external surface and a gut with a temporary anus. The epithelium is monociliated, however, and the mouth region bears a pair of jaws and a comblike ventral plate.

16 The aschelminth assemblage appears to represent two evolutionary lines. One includes the gastrotrichs, nematodes, and nematomorphs; the other includes the rotifers, acanthocephalans, kinorhynchs, loriciferans, and priapulids.

REFERENCES

The literature included here is restricted to aschelminths. The introductory references on page 8 include many general works and field guides that contain sections on the aschelminths. General parasitology texts, which contain much additional information on parasitic nematodes, are listed at the beginning of the references for flatworms, Chapter 7, page 204.

Adams, P. J. M., and Tyler, S., 1980: Hopping locomotion in a nematode: functional anatomy of the caudal gland apparatus of *Theristus caudasaliens* sp. n. Journ. Morphol., *164*:265–285.

Aloia, R. C., and Moretti, R. L., 1973: Mating behavior and ultrastructural aspects of copulation in the rotifer *Asplanchna brightwelli*. Trans. Am. Microsc. Soc., *92*(3):371–380.

Baer, J. C., 1961: Acanthocéphales. *In* Grassé, P. (Ed.): Traité de Zoologie, Vol. 4, pt. 1. Masson et Cie, Paris. pp. 733–782.

Bird, A. F., 1971: The structure of Nematodes. Academic Press, N.Y. 318 pp.

Brown, R., 1983: Spermatophore transfer and subsequent sperm development in homalorhagid kinorhynchs. Zool. Scripta, *12*(3):257–266.

Chitwood, B. G., and Chitwood, M. B., 1974: Introduction to Nematology. University Park Press, Baltimore. 334 pp.

Clément, P., and Wurdak, E., 1984: Photoreceptors and photoreception in rotifers. *In* Ali, M. A. (Ed.): Photoreception and Vision in Invertebrates. Plenum Press, N.Y.

Conway Morris, S., and Crompton, D. W. T., 1982: The origins and evolution of the Acanthocephala. Biol. Rev., *57*:85–115.

Croll, N. A., 1970: The Behavior of Nematodes. St. Martin's Press, N.Y. 117 pp.

Croll, N. A., and Matthews, B. E., 1977: Biology of Nematodes. John Wiley and Sons, N.Y. 201 pp. (A general account of the phylum.)

Croll, N. A., and Smith, J. M., 1974: Nematode setae as mechanoreceptors. Nematologica, *20*(3):291–296.

Croll, N. A., and Viglierchio, D. R., 1969: Osmoregulation and uptake of ions in a marine nematode. Proc. Helminthol. Soc. Wash., *36*(1):1–9.

Crompton, D. W. T., 1970: An Ecological Approach to Acanthocephalan Physiology. Cambridge University Press, N.Y. 136 pp.

Crompton, D. W. T., and Nichol, B. B. (Eds.), 1985: Biology of Acanthocephala. Cambridge University Press, N.Y. 500 pp.

Crowe, N. H., and Maden, K. A., 1974: Anhydrobiosis in tardigrades and nematodes. Trans. Am. Micros. Soc., *93*:513–524.

Deutsch, A., 1978: Gut ultrastructure and digestive physiology of two marine nematodes, *Chromadorina germanica* (Butschli, 1874) and *Diplolaimella sp.* Biol. Bull., *155*:317–355.

D'Hondt, J. L., 1971: Gastrotricha. Oceanogr. Mar. Biol. Ann. Rev., *9*:141–192.

Doncaster, C. C., and Seymour, M. K., 1973: Exploration and selection of penetration site by Tylenchida. Nematologica, *19*(2):137–145.

Dropkin, V. H., 1980: Introduction to Plant Nematology. Wiley-Interscience, N.Y. 293 pp.

Dumont, H. J., and Green, J., 1980: Rotatoria. Proceedings of 2nd International Rotifer Symposium. Hydrobiologia, *73*. W. Junk, The Hague. 263 pp.

Ferris, V. R., 1971: Taxonomy of the Dorylaimida. *In* Zuckerman, B. M., Mai, W. F., and Rohde, R. A. (Eds.): Plant Parasitic Nematodes. Vol. 1. Academic Press, N.Y. pp. 163–189.

Ferris, V. R., Ferris, J. M., and Tjepkema, J. P., 1973: Genera of freshwater nematodes (Nematoda of eastern North America). Biota of Freshwater Ecosystems. Ident. Manual 10. EPA. U.S. Government Printing Office.

Gilbert, J. J., 1974: Dormancy in rotifers. Trans. Amer. Micros. Soc., *93*:490–513.

Gilbert, J. J., 1980: Female polymorphism and sexual reproduction in the rotifer, *Asplanchna:* evolution of their relationship and control by dietary tocopherol. Amer. Nat., *116*(3):409–431.

Gilbert, J. J., and Starkweather, P. L., 1977: Feeding in the rotifer *Brachionis calyciflorus:* I. Regulatory mechanisms. Oecologica, *28*(2):125–132.

Gilbert, J. J., and Stemberger, R. S., 1985: Prey capture in the rotifer *Asplanchna girodi.* Verh. Internat. Verein. Limnol., *22*:2997–3000.

Golden, A. M., 1971: Classification of the genera and higher categories of the order Tylenchida (Nematoda). *In* Zuckerman, B. M., Mai, W. F., and Rohde, R. A. (Eds.): Plant Parasitic Nematodes. Vol. 1. Academic Press, N.Y. pp. 191–232.

Goodey, J. B., 1951: Soil and Freshwater Nematodes. John Wiley and Sons, N.Y.

Grassé, P. (Ed.), 1965: Némathelminthes, Rotifères, Gastrotriches, et Kinorhynques. *In* Traité de Zoologie, Vol. 4, pts. 2 and 3. Masson et Cie, Paris. (A detailed general account of the pseudocoelomate invertebrates.)

Higgins, R. P., 1971: A historical overview of kinorhynch research. *In* Hullings, N. C. (Ed.): Proceedings of the First International Conference on Meiofauna. Smithsonian Contributions to Zoology, *76*:25–31.

Hope, W. D., and Murphy, D. G., 1972: A Taxonomic Hierarchy and Checklist of the Genera and Higher Taxa of Marine Nematodes. Smithsonian Contribution to Zoology, *137*:1–101.

Hummon, M. R., 1984: Reproduction and sexual development in a freshwater gastrotrich. 3. Postparthenogenic development of primary oocytes and the x-body. Cell Tiss. Res., *236*:629–636.

Hummon, W. D., 1971: Biogeography of sand beach Gastrotricha from the northeastern United States. Biol. Bull., *141*(2):390.

Hummon, W. D., and Hummon, M. R., 1983: Gastrotricha. *In* Adiyodi, K. G., and Adiyodi, R. G. (Eds.): Reproductive Biology of Invertebrates. Vol. I:211–221. Vol. II: 195–205.

Hyman, L. H., 1951: The Invertebrates: Acanthocephala, Aschelminthes, and Entoprocta. Vol. 3. McGraw-Hill, N.Y. (Out of date in many areas, but still useful.)

Jennings, J. B., and Colam, J. B., 1970: Gut structure, digestive physiology and food storage in *Pontonemi vulgaris* (Nematoda: Enoplida). J. Zool. London, *161*:211–221.

Jensen, P., 1982: Reproductive behaviour of the free-living marine nematode *Chromadorita tenuis.* Marine Ecology, Prog. Ser., *10*:89–95.

Knauss, E. B., 1979: Indication of an anal pore in Gnathostomulida. Zool. Scripta, *8*:181–186.

Koste, W., 1978: Rotatoria. Die Rädertiere Mitteleuropas. I. Text, II. Illustrations. Gebrüder Borntraeger, Berlin. (The wide distribution of most species of rotifers makes this work a very important reference for identification, but the section on bdelloids is not yet completed.)

Kozloff, E. N., 1972: Some aspects of development in *Echinoderes* (Kinorhyncha). Trans. Amer. Microsc. Soc., *91*:119–130.

Kristensen, R. M., 1983: Loricifera, a new phylum with Aschelminthes characters from the meiobenthos. Z. f. zool. Systematik u. Evolutionsforschung, *21*(3):163–180. (See also the short report in Science, Oct. 14, 1983, p. 149.)

Kristensen, R. M., and Nørrevang, A., 1977: On the fine structure of *Rastrognathia macrostoma* gen. et sp. n. Zool. Scripta, *6*:27–41.

Lammert, V., 1984: The fine structure of spiral ciliary receptors in Gnathostomulida. Zoomorph., *104*:360–364.

Lammert, V., 1985: The fine structure of protonephridia in Gnathostomulida and their comparison within some Bilateria. Zoomorph., *105*:308–316.

Lapan, E. A., and Morowitz, H., 1972: The Mesozoa. Scientific American, *227*(6):94–100.

Lebedeva, L. I., and Gerasimova, T. N., 1981: Growth, breeding and production of *Philodina roseola* upon individual cultivation. Zool. Zh., *60*(11):1614–1620.

Lee, D. L., and Atkinson, H. J., 1977: Physiology of Nematodes. 2nd Edition. Columbia University Press, N.Y. 215 pp.

Levine, N. D., 1968: Nematode Parasites of Domestic Animals and of Man. Burgess Publishing Co., Minneapolis. 600 pp.

Lorenzen, S., 1978: Discovery of stretch receptor organs in nematodes—structure, arrangement and functional analysis. Zool. Scripta, *7*:175–178.

Maggenti, A., 1981: General Nematology. Springer-Verlag, N.Y. 372. pp. (Emphasis on parasitic nematodes.)

Mai, W. F., and Lyon, H. H., 1975: Pictorial Key to the Plant-Parasitic Nematodes. 4th Edition. Comstock Publishing Associates, Cornell University Press, Ithaca, N.Y.

Malakhov, V. V., 1980: Cephalorhyncha, a new type of animal kingdom uniting Priapulida, Kinorhyncha, Gordiacea, and a system of Aschelminthes worms. Zoologicheski Zhurnal, *59*:485–499. (In English.)

McLean, N., 1984: Amoebocytes in the lining of body cavity and mesenteries of *Priapulus caudatus.* Acta Zool., *65*(2):75–78.

Nicholas, W. L., 1967: The Biology of the Acanthocephala. Adv. Parasitol., *5*:205–246.

Nicholas, W. L., 1973: The Biology of the Acanthocephala. Adv. Parasitol., *11*:671–706.

Nicholas, W. L., 1984: The Biology of Free-Living Nematodes, 2nd Edition. Oxford University Press, London. (An excellent introductory account of free-living nematodes, including sections on methods of collection and preparation of material.)

Platt, H. M., and Warwick, R. M., 1983: Freeliving Marine Nematodes. Part I: British Enoplids—pictorial key to world genera and notes for identification of British species. Synopses of the British Fauna, No. 28. Cambridge University Press, Cambridge. 307 pp.

Poinar, G. O., 1983: The Natural History of Nematodes. Prentice-Hall, Inc., Englewood Cliffs, N.J. 323 pp.

Por, F. D., 1983: Class Seticoronaria and phylogeny of the phylum Priapulida. Zool. Scripta, *12*:267–272.

Pourriot, R., 1979: Soil rotifers. Rev. Ecol. Biol. Sol., *16*(2):279–312.

Pratt, I., 1969: The biology of the Acanthocephala. *In* Florkin, M., and Scheer, B. T. (Eds.): Chemical Zoology. Vol. 3. Academic Press, N.Y. pp. 245–252.

Riedl, R. J., 1969: Gnathostomulida from America. Science, *163*:445–452.

Rieger, G. E., and Rieger, R. M., 1977: Comparative fine structure study of the gastrotrich cuticle and aspects of cuticle evolution within the Aschelminthes. Z. f. zool. Systematik u. Evolutionsforschung, *15*(2):81–124.

Rieger, R. M., 1976: Monociliated epidermal cells in Gastrotricha: Significance for concepts of early metazoan evolution. Z. f. zool. Systematik u. Evolutionsforschung, *14*(3):198–226.

Rieger, R. M., and Mainitz, M., 1977: Comparative fine structure study of the body wall in Gnathostomulida and their phylogenetic position between Platyhelminthes and Aschelminthes. Z. f. zool. Systematik u. Evolutionsforschung, *15*(1):9–35.

Riemann, F., 1972: *Kinonchulus sattleri* n.g.n. sp., an aberrant free-living nematode from the lower Amazonas. Veröff. Inst. Meersforsch. Bremerh. *13*:317–326.

Roggen, D. R., Raski, D. J., and Jones, N. O., 1966: Cilia in nematode sensory organs. Science, *152*:515–516.

Ruppert, E. E., 1978: The reproductive system of gastrotrichs. II. Insemination in *Macrodasys:* A unique mode of sperm transfer in Metazoa. Zoomorphologie, *89*:207–228.

Ruppert, E. E., 1978: The reproductive system of gastrotrichs. III. Genital organs of Thaumastodermatinae subfam. n. and Diplodasyinae subfam. n. with discussion of reproduction in Macrodasyida. Zool. Scripta, *7*:93–114.

Ruppert, E. E., 1982: Comparative ultrastructure of the gastrotrich pharynx and the evolution of myoepithelial foreguts in Aschelminthes. Zoomorph., *99*:181–220.

Salt, G. W., Sabbadini, G. F., and Commins, M. L., 1978: Trophi morphology relative to food habits in six species of rotifers (Asplanchnidae). Trans. Amer. Microsc. Soc., *97*(4):469–485.

Schaefer, C., 1971: Nematode radiation. Syst. Zool., *20*:(1)77–78.

Siddiqui, I. A., and Viglierchio, D. R., 1977: Ultrastructure of the anterior body region of marine nematode *Deontostoma californicum.* J. Nematology, *9*(1):56–82.

Somers, J. A., Shorey, H. H., and Gastor, L. K., 1977: Sex pheromone communication in the nematode, *Rhabditis pellio.* J. Chem. Ecol., *3*(4):467–474.

Steiner, G., and Heinly, H., 1922: Possibility of control of *Heterodera radicola* by means of predatory nemas. J. Washington Acad. Sci., *12*:367–396.

Sterrer, W., 1972: Systematics and evolution within the Gnathostomulida. Syst. Zool., *21*(2):151–173.

Sterrer, W., 1974: Gnathostomulida. *In* Giese, A. C., and Pearse, J. S. (Eds.): Reproduction of Marine Invertebrates. Vol. I. Acoelomate and Pseudocoelomate Metazoans. Academic Press, N.Y. pp. 345–357.

Storch, V., 1984: Minor Pseudocoelomates. *In* Bereiter-Hahn, J., Matoltsy, A. G., and Richards, K. S.: Biology of the Integument, Vol. 1, Springer-Verlag, N.Y. pp. 242–268.

Strathmann, R. R., John, T. L., and Fonseca, J. R. C., 1972: Suspension feeding by marine invertebrate larvae: clearance of particles by ciliated bands of a rotifer, pluteus, and trochophore. Biol. Bull., *142*:505–519.

Sulston, J. E., Schieringberg, E., and White, J. G. The work of these investigators was reviewed in Science, *225*(6 July 1984):40–42 (*Caenorhabditis elagans:* getting to know you). (See also Science 22 June and 13 July 1984 for additional reviews.)

Tarjan, A. C., Esser, R. P., and Chang, S. L., 1977: An illustrated key to nematodes found in freshwater. J. Water Pollut. Control Fed., *49*(11):2318–2337.

Teuchert, G., 1973: Die Feinstruktur des Protonephridialsystems von *Turbanella cornuta* Remane, einem marinen Gastrotrich der Ordnung Macrodasyoidea. Z. Zellforsch., *136*:277–289.

Thane, A., 1974: Rotifera. *In* Giese, A. C., and Pearse, J. S. (Eds.): Reproduction of Marine Invertebrates. Vol. I. Acoelomate and Pseudocoelomate Metazoans. Academic Press, N.Y. pp. 471–484. (This review covers freshwater as well as marine forms.)

Tietjen, J. H., 1969: The ecology of shallow water meiofauna in two New England estuaries. Oecologia, *2*:251–291.

Travis, P. B., 1983: Ultrastructural study of body wall organization and Y-cell composition in the Gastrotricha. Z. f. zool. Systematik u. Evolutionsforschung, *21*:52–68.

Tyler, S., and Rieger, G. E., 1980: Adhesive organs of the Gastrotricha. I. Duo-gland organs. Zoomorphologie, *95*:1–15.

Wallace, H. R., 1970: The movement of nematodes. *In* Fallis, A. M. (Ed.): Ecology and Physiology of Parasites. University of Toronto Press, Toronto. pp. 201–212.

Wallace, R. L., 1980: Ecology of sessile rotifers. Hydrobiologia, 73:181–193.

Weiss, M. J., and Levy, D. P., 1979: Sperm in "parthenogenetic" freshwater gastrotrichs. Science, 205:302–303.

Whitfield, P. J., 1971: Phylogenetic affinities of Acanthocephala: An assessment of ultrastructural evidence. Parasitology, 63(1):49–58.

Whitfield, P. J., 1984: Acanthocephala. In Bereiter-Hahn, J., Matoltsy, A. G., and Richards, K. S.: Biology of the Integument, Vol. I. Invertebrates. Springer-Verlag, N.Y. pp. 234–241.

Wilson, R. A., and Webster, L. A., 1974: Protonephridia. Biol. Rev., 49:127–160.

Yeats, G. W., 1971: Feeding types and feeding groups in plant and soil nematodes. Pedobiologia, 11(2):173–179.

Zuckerman, B. M., Mai, W. F., and Rohde, R. A., 1971: Plant Parasitic Nematodes: Cytogenetics, Host-Parasite Interactions and Physiology. Vol. 2. Academic Press, N.Y. 347 pp.

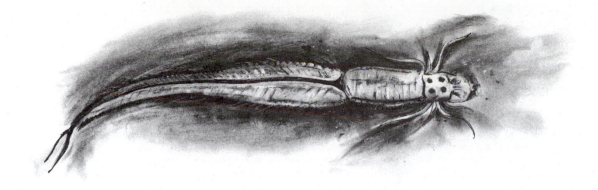

10

The Annelids

The phylum Annelida comprises the segmented worms and includes the familiar earthworms and leeches, plus a number of marine and freshwater species of which most people are completely unaware. A shovelful of muddy sand taken from the shore along a coastal sound at low tide usually brings to light a much richer and far more spectacular collection of "worms" than could be found in a backyard garden.

In general, the annelids attain the largest size of any of the wormlike invertebrates and display the greatest structural differentiation. The most distinguishing characteristic of the phylum is metamerism, the division of the body into similar parts, or segments, which are arranged in a linear series along the anteroposterior axis. The segmented part of the body is always limited to the trunk. The head, or acron, represented by the prostomium and containing the brain, is not a segment, nor is the pygidium, the terminal part of the body that carries the anus. However, in all metameric animals there has been a tendency for anterior trunk segments to fuse, in varying degrees, with the unsegmented prostomium or acron. This fusion gives rise to a

secondary, compound "head." The formation of new segments in a metameric animal always takes place just in front of the pygidium. The oldest body segments are therefore anterior, and the youngest are posterior.

Metamerism appears to have evolved twice within the Animal Kingdom, in the chordates as an adaptation for undulatory swimming and in the annelids as an adaptation for burrowing. In chordates the primary metameric structures are the body wall muscles. In annelids the primary metameric structures are the coelomic compartments created by segmentation of the coelom with transverse septa. Each septum is composed of two layers of peritoneum, one derived from the segment in front and one from the segment behind. As an accommodation to serve the primary segmentation of the coelom, the lateral nerves to the body wall musculature, blood vessels, and excretory organs are also segmentally arranged.

Coelomic fluid functions as a hydraulic skeleton against which the muscles act to change the body shape. Contraction of the longitudinal muscles causes the coelomic fluid to exert a laterally di-

---

BOX 10–1   ORIGIN OF COELOMATES AND ACOELOMATES

Annelids, like all the remaining phyla to be described in this text, possess a type of body cavity called a coelom. Although the coelom may be absent in the adult, as in arthropods, it is still present in the embryo. The coelom arises during embryonic development as a cavity within mesoderm, either by the splitting of a mass of mesodermal cells (schizocoely) or by outpocketing of the archenteron in the initial separation of mesoderm from endoderm (enterocoely) (p. 63 and Fig. 3–8). The inner mesodermal cells differentiate to form a mesothelial lining of the coelom, called peritoneum. Typically, the coelom is composed of two or more paired chambers (right and left), depending on the animal design. Annelids, for example, being metameric, possess numerous pairs separated by transverse septa.

The fluid that fills the coelom serves a variety of functions in different coelomate groups. It may transport gases and nutritive materials, provide fluid for processing excretory wastes, be a hydrostatic skeleton, and function as a site for gamete maturation and brooding of embryos. Most or all of these functions are performed by the coelomic fluid of annelids. The internal transport function is important for most coelomates and may have been a primary factor in the evolution of the coelom.

The evolution of the coelom is uncertain. One theory postulates that coelomates evolved from an ancestral acoelomate, such as some flatworm, by a hollowing out of mesenchymal parenchyma cells, some of which would form peritoneum. This schizocoel theory of coelom origin would claim as supporting evidence the embryonic development of annelids and mollusks, whose coeloms form in this way.

The enterocoel theory, on the other hand, argues that the coelom evolved from the gastric pouches of some cnidarian ancestor, such as schyphozoans or anthozoans, and would claim as supporting evidence the enterocoelous mode of mesoderm and coelom formation in the embryogeny of echinoderms, hemi-chordates, and chordates (see Clark, 1964, for review of these and other theories).

Depending on the theory, then, the acoelomate condition is either ancestral to or derived from the coelomate condition. According to the schizocoel theory, the acoelomate body plan is primary and ancestral to the coelomate plan. The acoelomate flatworms would be the stem group in the evolution of bilateral animals. The enterocoel theory proposes that all bilateral animals are basically coelomate and that acoelomate forms, such as the flatworms, are secondarily derived from coelomate ancestors by loss of the cavity.

Although both theories date from the end of the 19th century, the enterocoel theory has never gained much acceptance, at least in part because it is difficult to postulate functional steps that would have led to a change in both design (from coelomate to acoelomate) and symmetry (from bilateral to radial). In contrast, the ancestral position of flatworms among the Bilateria and the primitive nature of the acoelomate body plan has been widely accepted in the literature, including all of the previous editions of this text. However, recent ultrastructural research on flatworms and annelids suggests that the acoelomate condition may indeed be secondary. Studies on the parenchyma cells of turbellarian flatworms clearly indicate that this "filling" is not a uniform, mesenchyme-like tissue and its composition differs among turbellarian groups (p. 168 and Fig. 7–4). This lack of uniformity suggests an independent origin of the acoelomate condition either by invasion of cells or by retention of larval features in the adult (Rieger, 1985). Studies on the peritoneal lining of the coelom also raise questions about the coelom's evolving within an acoelomate. Myoepithelium, cells that both line surfaces and contract, was once thought limited to the cnidarians, e.g., the epitheliomuscular and nutritive muscular cells. However, myoepithelium is now known to occur in many animals, including poly-

---

rected force, and the body widens. Contraction of the circular muscles causes the coelomic fluid to exert an anteriorly and posteriorly directed force, and the body elongates (Fig. 10–1). The significance of the compartmentation of the coelom by transverse septa now becomes apparent. The coelomic fluid skeleton is localized, and the widening and elongation can be restricted to certain segments. Waves of peristaltic contraction pass down the length of the body, bringing about elongation and then shortening in each segment in sequence. Those segments in which the longitudinal muscles are contracted have an increased diameter and are anchored against the burrow wall. Segments in which the circular muscles are contracted are elongated and move forward (Fig. 10–1). Powerful force can be generated, enabling the animal to thrust the anterior end through the substratum or chaetes, where it may contribute to the coelomic lining (Fransen, 1980). This is not the sort of cell that would be expected to line the coelom if the peritoneum differentiated from the mesenchymal parenchyma of an acoelomate ancestor, but it would be expected if the coelom were derived from gastric pouches of some cnidarian ancestor, where the myoepithelial gut lining (nutritive muscle cells) became the lining of the coelom.

Still other clues about the origin of the acoelomate condition come from some strange little worms that have been recently discovered in the marine interstitial fauna. If flatworms were derived from some annelidan, coelomate ancestor, these worms appear to be somewhat intermediate (Rieger, 1980). Called *Lobatocerebrum*, the members of this genus are completely ciliated, like turbellarians. But like an annelid, *Lobatocerebrum* has a cuticle, a pair of ganglionated ventral nerve cords, and metamerically arranged protonephridia that open separately to the exterior. The digestive tract exits by way of a posterior anus. On the other hand, *Lobatocerebrum* has no setae, septa, or coelom; it is acoelomate, like flatworms. However, the parenchyma is composed of baglike, fluid-filled cells with very little intercellular matrix. On the whole, *Lobatocerebrum* displays more annelidan than turbellarian features and is therefore tentatively assigned to the Annelida and, within that phylum, to the class Oligochaeta because it is hermaphroditic and has a glandular region a little like a clitellum. Besides *Lobatocerebrum*, there are some small interstitial polychaetes (*Dinophilus*, protodrilids) that are ciliated and display coelom reduction, the cavity being partially filled by muscle cells or large, baglike cells (Fransen, 1980).

Many of the peculiarities of these interstitial worms may be related to the assumption of a life among sand grains, but Rieger (1985) believes that just such changes may have been the route by which turbellarians evolved from the coelomate annelids.

As our knowledge of comparative ultrastructure and the interstitial fauna grows, evidence may enable us to trace the major pathways that led to the evolution of lower invertebrates—or at least to know which were likely and which were improbable.

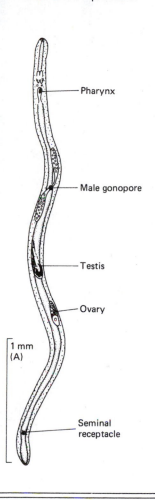

Pharynx

Male gonopore

Testis

Ovary

1 mm
(A)

Seminal
receptacle

to move rapidly through a previously excavated burrow. Chitinous, paired, lateral bristles, or setae, on each segment increase traction with the substratum.

Annelids possess a more or less straight digestive tract running from the anterior mouth to the posterior anus. The gut is suspended within the coelom by longitudinal mesenteries and by the septa, through which the gut penetrates (Fig. 10–

2). Digestion is extracellular. Excretion takes place by means of nephridial tubules, and characteristically there is one pair per segment. There is usually a well-developed blood-vascular system, in which the blood is usually confined to vessels—that is, the system is closed. The nervous system consists of an anterior, dorsal ganglionic mass, or brain; a pair of anterior connectives surrounding the gut; and a long double or single ventral nerve cord with

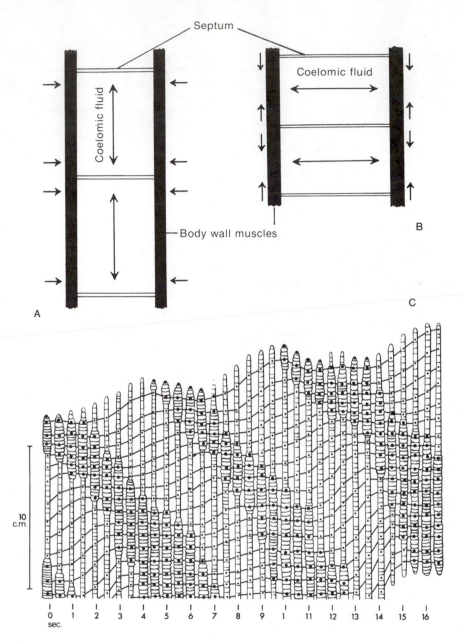

Figure 10–1 *A* and *B*, Diagrammatic frontal section through two annelid segments. External arrows indicate direction of force exerted by body wall muscles. Internal arrows indicate direction of force exerted by coelomic fluid pressure: (*A*) during circular muscle contraction and (*B*) during longitudinal muscle contraction. *C*, Diagram showing mode of locomotion of an earthworm. Segments undergoing longitudinal muscle contraction are marked with larger dot and drawn twice as wide as those undergoing circular muscle contraction. The forward progression of a segment during the course of several waves of circular muscle contraction is indicated by the horizontal lines connecting the same segments. (After Gray and Lissmann.)

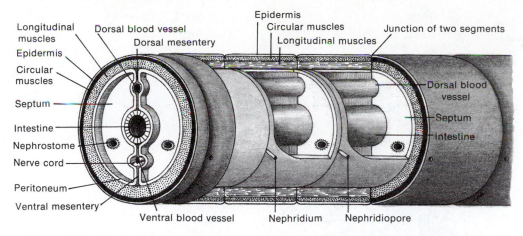

Figure 10–2   Annelid segments. (After Kaestner, A., 1967: Invertebrate Zoology, Vol. 1. Interscience Publ., N.Y.)

ganglionic swellings and lateral nerves in each segment.

The phylum contains over 8700 described species, which are placed into three classes: Polychaeta, Oligochaeta, and Hirudinea. The ancestral annelids were probably marine animals burrowing in the bottom sand and mud of shallow coastal waters. The class Polychaeta contains most of the living marine species. The class Oligochaeta, which includes the freshwater annelids and the terrestrial earthworms, may have stemmed from some early polychaetes, but more likely the oligochaetes evolved independently from the ancestral annelids. The class Hirudinea, the leeches, clearly arose from some stock of freshwater oligochaetes.

SUMMARY

1 Members of the phylum Annelida are vermiform (wormlike) metameric animals. Metamerism, which is a distinguishing feature, probably evolved as an adaptation for peristaltic burrowing in soft substrata.

2 The primary metameric structures are the coelomic compartments, which permit localization of shape changes along the length of the body.

3 Segmental setae increase traction with the substratum.

4 The nervous, circulatory, and excretory systems are metameric, for as maintenance systems, they accommodate the primary metamerism of the coelomic compartments.

5 The gut is typically a straight tube extending through the body between the anterior mouth and posterior anus.

Class Polychaeta

Polychaete worms are very common marine animals, but their secretive habits result in their being overlooked by casual observers. Over 5300 species have been described. The majority are less than 10 cm long with a diameter ranging from 2 to 10 mm, but some interstitial forms are less than 1 mm, whereas one species of *Eunice* may attain a length of 3 meters. Many polychaetes are strikingly beautiful, colored red, pink, or green or a combination of colors.

The generalized polychaete is perfectly metameric, with identical cylindrical body segments, each bearing a pair of lateral, fleshy, paddle-like appendages called parapodia (Fig. 10–3). At the anterior of the worm is a well-developed prostomium, which bears sense organs. The mouth is located on the ventral side of the body between the prostomium and a postoral region called the peristomium. The terminal, unsegmented region, the pygidium, carries the anus. Few polychaetes possess such a typical structure, however. The different life-styles of the worms of this class have led to varying degrees of modification in the basic plan.

Polychaetes can be errant (free moving) or sedentary, but the distinction is not always sharp. The errant polychaetes include some species that are strictly pelagic, some that crawl about beneath rocks and shells, and some that are active burrowers in sand and mud. Many sedentary species construct and live in stabilized burrows or tubes of varying degrees of complexity. The completely tubicolous forms usually cannot leave the tubes and can only project their heads from the tube openings.

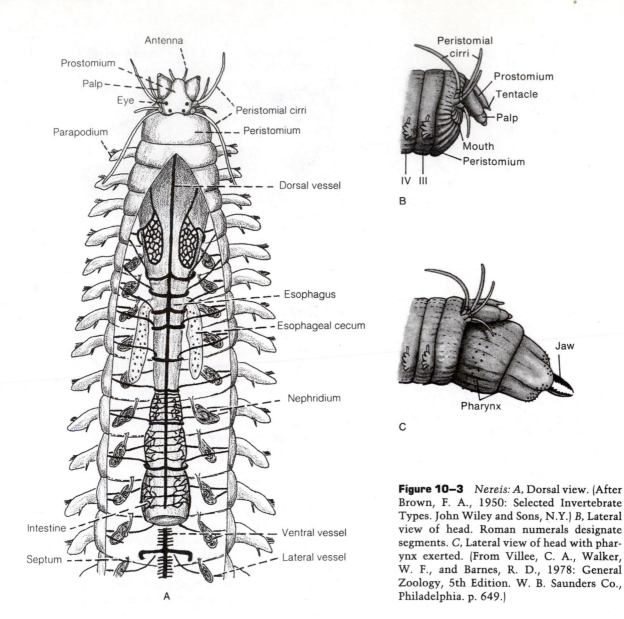

Figure 10–3 *Nereis: A*, Dorsal view. (After Brown, F. A., 1950: Selected Invertebrate Types. John Wiley and Sons, N.Y.) *B*, Lateral view of head. Roman numerals designate segments. *C*, Lateral view of head with pharynx exerted. (From Villee, C. A., Walker, W. F., and Barnes, R. D., 1978: General Zoology, 5th Edition. W. B. Saunders Co., Philadelphia. p. 649.)

Polychaete Structure

The generalized polychaete body plan is most nearly attained in the errant families. The dorsal, preoral prostomium is well developed and bears numerous sensory structures. The prostomial sense organs usually consist of eyes, antennae, and ventrolateral palps (Figs. 10–3 and 10–4). The prostomium projects like a shelf over the mouth. Behind it is the peristomium, which forms the lateral and ventral margins of the mouth.

The most distinguishing feature of polychaetes is the presence of parapodia, the paired, lateral appendages extending from the body segments. A typical parapodium is a fleshy projection extending from the body wall and is more or less laterally compressed (Fig. 10–5). The parapodium is basically biramous, consisting of an upper division, the notopodium, and a ventral division, the neuropodium. Each division is supported internally by one or more chitinous rods, or acicula. A tentacle-like process (cirrus) projects from the dorsal base of the notopodium and from the ventral base of the neuropodium. The notopodia and neuropodia assume various shapes in different families and may be subdivided into several lobes or even greatly reduced (Figs. 10–5, 10–6*B*, and 10–7*A*).

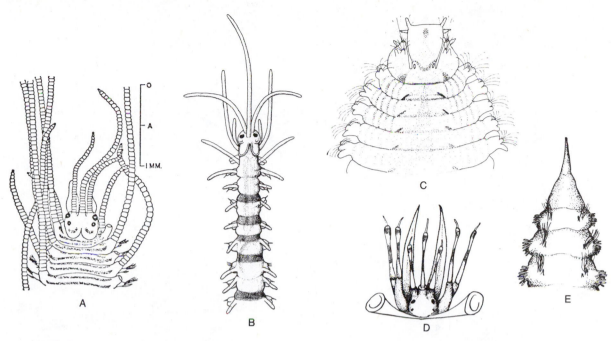

Figure 10–4 Dorsal views of the anterior ends of four polychaetes. *A*, A syllid, *Typosillis corallicola*. (From Jones, M. L., 1962: On some polychaetous annelids from Jamaica, the West Indies. Bull. Am. Mus. Nat. Hist., *124*(5):181.) *B*, Anterior end of *Proceraea fasciata*, a common syllid on rocks, shells, and pilings in shallow water along the coasts of the United States. *C*, *Nephtys australiensis*. (From Paxton, H., 1974: Contribution to the study of Australian Nephtyidae. Records of the Australian Museum, *29*(7):205.) *D*, The scale worm *Lepidonotus variabilis*. *E*, *Scoloplos rubra*. (*D* and *E* from Renaud, J. C., 1956: A report on some polychaetous annelids from the Miami-Bimini area. Am. Mus. Novit., *1812*:1–40.)

The parapodial rami contain pockets or setal sacs from which many chitinous bristles, or setae, project. Each simple seta is secreted by a single cell at the base of the setal sac and usually projects a considerable distance beyond the end of the parapodial lobe. New setae are continually produced by the setal sac as older setae are lost (see review by Schroeder, 1984). They assume a great variety of shapes, and the setal bundles of a particular species may be composed of more than one type of seta. (Fig. 10–5C).

The members of the largely tropical family Amphinomidae, which live in coral and beneath stones, have brittle, tubular, calcareous setae containing poison (Fig. 10–7C). The setae are used for defense. The amphinomids are commonly known as fireworms because of the pain produced by poison liberated when the setae break after easily penetrating the skin. These worms are avoided by fish.

The body segments of polychaetes are generally similar; however, in some burrowers and tube dwellers there has been a tendency for the trunk to become differentiated into distinct regions as a result of variations in the parapodia or the presence or absence of gills (Fig. 10–8A).

The polychaete epidermis, or integument, is composed of a single layer of cuboidal or columnar epithelium, which is covered by a thin collagen cuticle (see Fig. 10–9 for evolution of cuticle). Mucus-secreting gland cells are a common component of the epithelium.

Beneath the epithelium lie, in order, a layer of circular muscle fibers, a much thicker layer of longitudinal muscle fibers, and a thin layer of peritoneum (Fig. 10–2). Although the muscles of the body wall essentially comprise two sheaths, the longitudinal fibers typically are broken up into four bundles—two dorsolateral and two ventrolateral (Fig. 10–5A).

Within the spacious coelom, the gut is suspended by septa and mesenteries (Fig. 10–2). Thus, each coelomic compartment is divided into right and left halves, at least primitively. However, the septa have partially or completely disappeared in many polychaetes.

Adaptive Diversity in Polychaetes

If the first annelids were burrowers in sand and mud of shallow water, then the polychaetes may

(Text continued on p. 272)

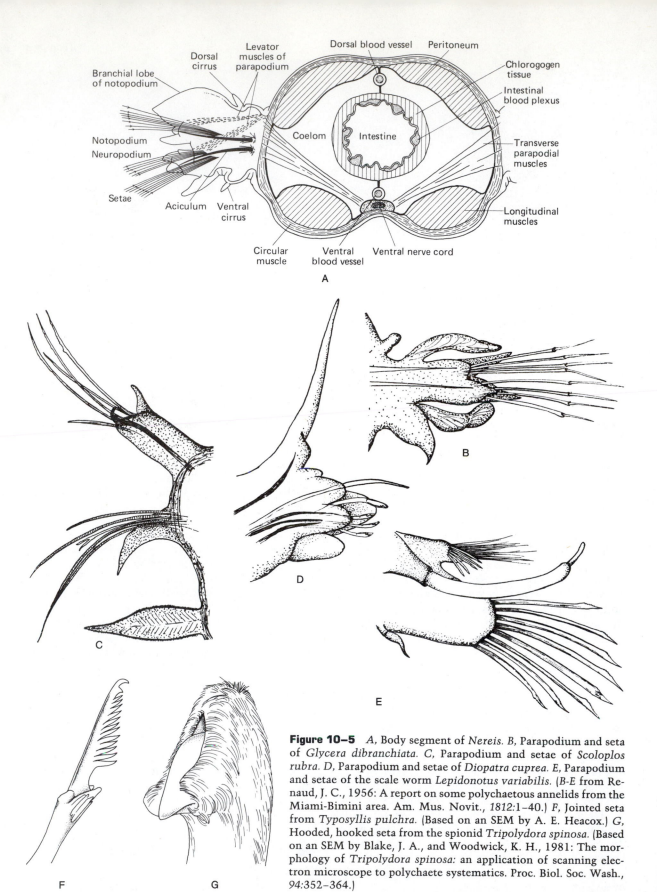

Figure 10–5 *A*, Body segment of *Nereis. B*, Parapodium and seta of *Glycera dibranchiata. C*, Parapodium and setae of *Scoloplos rubra. D*, Parapodium and setae of *Diopatra cuprea. E*, Parapodium and setae of the scale worm *Lepidonotus variabilis.* (*B-E* from Renaud, J. C., 1956: A report on some polychaetous annelids from the Miami-Bimini area. Am. Mus. Novit., *1812:1–40.) F*, Jointed seta from *Typosyllis pulchra.* (Based on an SEM by A. E. Heacox.) *G*, Hooded, hooked seta from the spionid *Tripolydora spinosa.* (Based on an SEM by Blake, J. A., and Woodwick, K. H., 1981: The morphology of *Tripolydora spinosa:* an application of scanning electron microscope to polychaete systematics. Proc. Biol. Soc. Wash., *94:352–364.)*

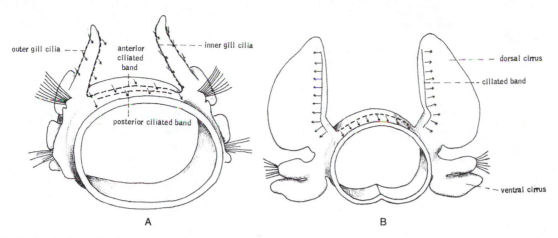

Figure 10–6 Surface ciliation in two polychaetes. Arrows indicate direction of water currents. *A, Scolelepis squamata. B, Phyllodoce laminosa.* (After Segrove.)

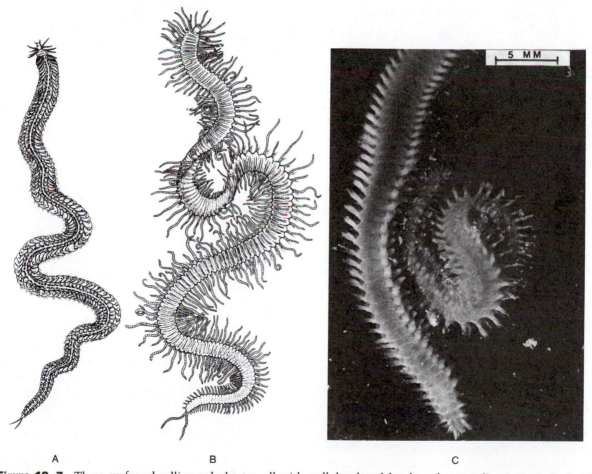

Figure 10–7 Three surface-dwelling polychaetes, all with well-developed heads and parapodia: *A,* A phyllodocid, *Phyllodoce maculata,* with large, flattened, dorsal cirri. *B,* A syllid, *Trypanosyllis zebra,* with long, tentacular, dorsal cirri. (*A* and *B* after McIntosh.) *C,* An amphinomid, a fireworm. Note the bundles of calcareous setae, which contain an irritant. This West Indian species is very common beneath stones. (By Betty M. Barnes.)

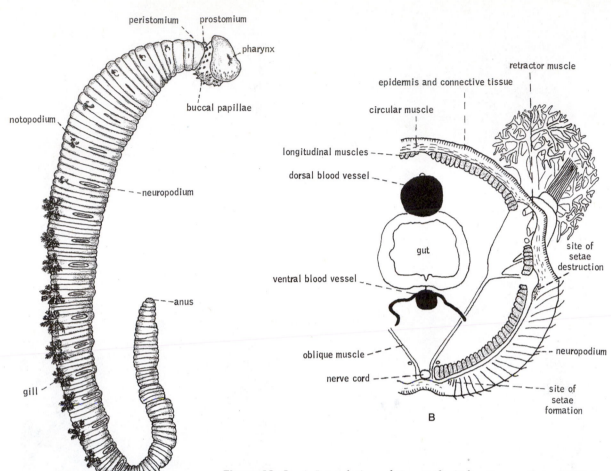

Figure 10–8 *A*, Lateral view of *Arenicola*. (After Brown, F. A., 1950: Selected Invertebrate Types. John Wiley and Sons, N.Y.) *B*, Transverse section of a setigerous segment of *Arenicola marina*. (Modified from Wells.)

constitute a line that originally diverged in the exploitation of a surface existence, for which parapodia and a more developed head would be adaptive. From perhaps several different stocks of ancestral polychaetes inhabiting soft bottoms, there may have been radiation into other habitats and the assumption of other modes of existence—hard bottoms, tube dwelling, burrowing, and so on.

SURFACE DWELLERS

The surface-dwelling, crawling polychaetes, which live beneath stones and shells, in coral crevices, in algae, and among hydroids and other sessile organisms, most nearly approach the generalized body form just described. In these worms the prostomium is equipped with eyes and other sensory organs, the parapodia are well developed, and the body segments are generally similar (Fig. 10–4*B* and 10–7).

Among the many groups of surface-dwelling polychaetes are certain species called scale worms, so named because of the peculiar platelike scales, or elytra, carried by short stalks on the dorsal side of the body (Figs. 10–10*B* and 10–11). The scales may provide a protective channel for the ventilating current when the worms are wedged beneath stones or (like the sigalionids) burrowing in sand and mud (p. 293). In the scale worm *Aphrodita* (called a sea mouse) the entire dorsal surface, including the elytra, is covered by hairlike "felt," composed of setae that arise from the notopodia and trail back over the dorsal surface of the animal (Fig. 10–11*A*).

Movement in polychaetes is brought about by the combined action of the parapodia, the body wall musculature, and to some extent the coelomic fluid. The longitudinal muscle layer is better developed than the circular layer, and the septa tend

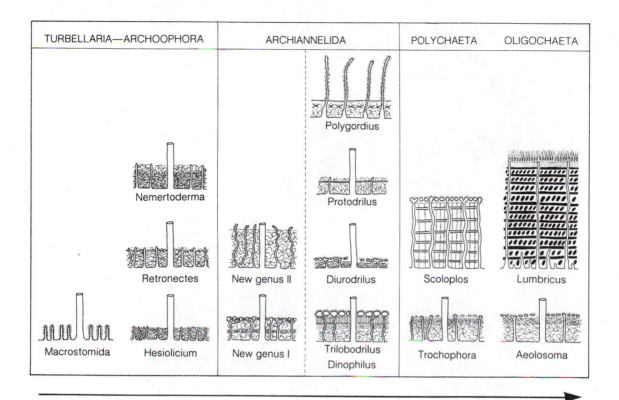

TURBELLARIA—ARCHOOPHORA	ARCHIANNELIDA	POLYCHAETA	OLIGOCHAETA

Polygordius

Nemertoderma   Protodrilus

Retronectes   New genus II   Diurodrilus   Scoloplos   Lumbricus

Macrostomida   Hesiolicium   New genus I   Trilobodrilus   Trochophora   Aeolosoma
Dinophilus

Figure 10–9 Evolution of the cuticle in flatworms, nemerteans, and annelids. The cuticle of annelids is composed of mucopolysaccharide fibers deposited between microvilli. The tips of the microvilli are commonly enlarged as vesicles. Primitive flatworms, such as the macrostomids, have microvilli but no cuticle. Between the primitive condition of macrostomids and the complex cuticle of such annelids as *Scoloplos* and *Lumbricus*, all degrees of cuticle development exist as a result of the deposition of fibers between microvilli. Rieger believes that the cuticle originally was an adaptation for the absorption of dissolved organic compounds and secondarily assumed a protective function. The cuticle of aschelminths (Fig. 9–7) is believed to have had a different origin. (From Rieger, R. M., and Rieger, G. E., 1976: Fine structure of the archiannelid cuticle and remarks on the evolution of the cuticle within the Spiralia. Acta Zool. (Stockh.), 57:53–68.)

to be incomplete. The parapodia and setae push against the substratum like legs. Parapodial movement involves an effective backward stroke in which the acicula and setae are extended and the parapodium is in contact with the substratum. After the backward stroke of the parapodium, the acicula and setae are retracted, and the parapodium lifts off the substratum and moves forward. The combined sweeps of the numerous parapodia propel the worm forward.

As soon as one parapodium begins its effective stroke, the preceding parapodium sweeps forward and starts its effective stroke. The waves of activation thus move forward rather than backward. Activation takes place on both sides of the body, but the waves alternate with each other; when a particular parapodiun on one side is executing an

effective stroke, the parapodium on the opposite side is executing the recovery stroke (Fig. 10–12).

When worms such as nereids crawl rapidly or swim, lateral body undulations are produced by waves of contraction in the longitudinal muscles of the body wall. These waves of contraction coincide with the alternating waves of parapodial activity just described, and the backward power thrust of the parapodia occurs when they are at the crest of an undulatory wave.

PELAGIC POLYCHAETES

Six families of polychaetes contain exclusively planktonic or pelagic species (Dales and Peter, 1972). They are rather like crawling forms but tend to be transparent, as are many other planktonic animals. Their swimming movements are similar to

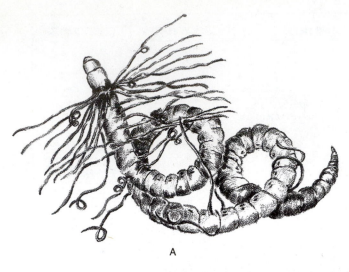

A

Figure 10–10 *A, Cirratulus cirratus*, a polychaete with long, threadlike, dorsal cirri. (After McIntosh from Fauvel.) *B, Lagisca flocculosa*, a scale worm. (After McIntosh.)

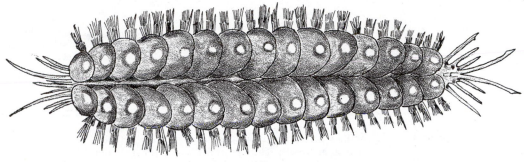

B

those just described for crawling species. The Alciopidae, which have enormous eyes (Fig. 10–13*B*), and the Tomopteridae, which have lost the setae and possess membranous parapodial pinnules, are the most highly specialized of the pelagic families (Fig. 10–13*A*).

BURROWERS

Many polychaetes have become adapted for burrowing. The more active burrowers construct a system of mucus-lined galleries, within which they move about (Fig. 10–15). The adaptations of many of these species remarkably parallel those of the earthworms among the oligochaetes. The prostomium is small and pointed, and eyes, palps, and antennae are usually absent (Figs. 10–14*A* and *C* and 10–15). Parapodia tend to be smaller than those of crawling surface dwellers.

Like earthworms, many active burrowers, such as the gallery-dwelling lumbrinerids and capitellids, move through the substratum by peristaltic contractions. The circular muscle layer of the body wall is well developed, and the septa effectively compartmentalize the coelomic fluid and localize its skeletal function. However, there are some burrowers, such as the nephtyids and some species of nereids, that have retained the surface-dwelling design with well-developed parapodia and prostomial sense organs.

Many polychaetes occupy more or less fixed, simple vertical or U-shaped burrows excavated in the substratum. These sedentary burrowers include members of such families as the Arenicolidae (the lugworms, Figs. 10–8 and 10–16) and the Terebellidae (Fig. 10–17). The prostomial sensory appendages are generally absent, but the head region may carry specialized feeding structures (Fig. 10–17). Movement through the burrow is usually by peristaltic contractions; the parapodia are greatly reduced and are in part represented by transverse ridges provided with setae modified into hooks, called uncini (Fig. 10–5*G*).

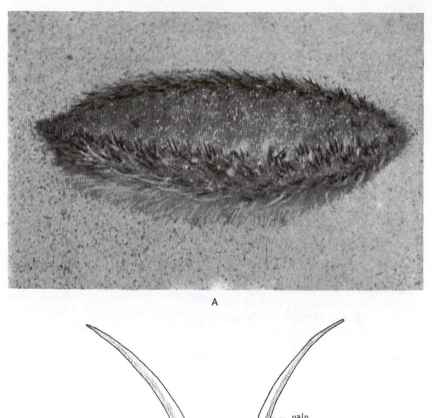

A

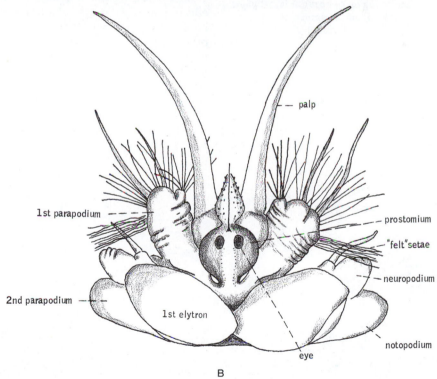

B

Figure 10–11 The sea mouse, *Aphrodita aculeata: A,* Dorsal view. The wide, uniform dorsal strip is covered by felt setae. (Courtesy of D. P. Wilson.) *B,* Anterior end, including the first pair of dorsal scales, or elytra. (After Fordham.)

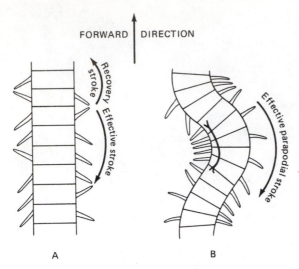

Figure 10–12 Parapodial crawling in *Nereis: A,* Slow crawling, in which only parapodia are involved. Effective and recovery strokes alternate on opposite sides of body. *B,* Fast crawling, in which body is thrown into undulating waves. Effective parapodial stroke is on the crest of the wave. Parapodia are recovering on opposite side, which is also the side on which longitudinal muscles are contracted (heavy arrow). (Based on Mettam.)

TUBICOLOUS POLYCHAETES

A tube-dwelling habit has evolved in many families of polychaetes. The tube may serve the worm as a protective retreat or as a lair for catching passing prey. It may provide access to clean, oxygenated water above a muddy or sandy bottom. It may permit the worm to inhabit hard, bare surfaces such as rock, shell, or coral. Polychaete tubes may be composed of secreted material or sand grains cemented together or a combination of secreted and foreign material. Tube secretions are commonly produced by glands on the ventral surfaces of the segments.

Some tubicolous worms, including members of the families Eunicidae and Onuphidae, are commonly carnivorous and extend from the opening of the tube to seize passing prey. They are not greatly different from the surface polychaetes. Prostomial sensory appendages are well developed, and the parapodia, which are used in crawling through the tube, are not markedly reduced (Figs. 10–18 and 10–19). Species of the onuphid genus *Hyalinoecia* construct secreted, somewhat transparent, quill-like tubes, which lie horizontally on the bottom and may be moved about by the worm (Fig. 10–18). The onuphids *Diopatra* and *Onuphis* build heavy, conspicuous, membranous tubes that may occur in great numbers in intertidal areas. The projecting chimney of the tube is bent over and flares at the end like a ship's funnel (Fig. 10–19B). The chimneys are covered with bits of shell, seaweed, and other debris that the worm collects and places in position with its jaws. The ornamentation probably provides a cryptic refuge and aids in the detection of possible predators or prey by more readily transmitting disturbances in the surrounding water or from contact with the tube (Brenchley, 1976).

The majority of tubicolous polychaetes are highly modified for a tube-dwelling existence, with

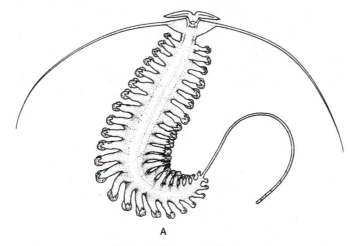

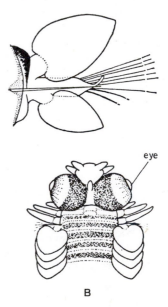

Figure 10–13 Pelagic polychaetes: *A, Tomopteris renata,* a tomopterid. *B,* Dorsal view of head of *Rhynchonerella angelina,* an alciopid. *C,* Parapodium of *Rhynchonerella.* (From Day, J., 1967: Polychaeta of South Africa. British Museum, London.)

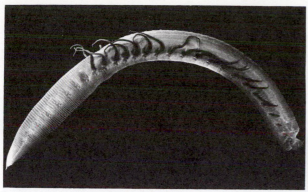
A

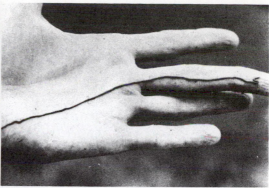
B

Figure 10–14 Burrowing polychaetes: *A*, Lateral view of *Ophelia denticulata*. Note the small, pointed prostomium. The long projections are gills. (Courtesy of C. R. Gilmore.) *B*, The threadworm *Drilonereis*, a gallery dweller. (By Betty M. Barnes.) *C*, Dorsal view of anterior end of *Drilonereis*.

C

adaptations similar to the more sedentary burrowers. Prostomial sensory appendages are reduced or absent, and special anterior feeding structures are common. The worms usually move within the tube by peristaltic contractions, and the parapodia are reduced and provided with uncinate setae for gripping the tube wall. The body segments are commonly differentiated into regions.

The sand-grain tubes of some members of the Maldanidae, called bamboo worms, are common in the intertidal zone. These worms, which live anterior end downward in their tubes, have truncate heads and parapodia that are reduced to ridges having the appearance of cane joints (Fig. 10–20*A*).

Both the Sabellariidae and the Pectinariidae construct sand-grain tubes and have highly modified heads bearing heavy, conspicuous setae. In the Sabellariidae two fused segments have grown forward and dorsally to form an operculum for blocking the tube entrance (Fig. 10–21*A* and *B*). Where the water is turbulent, species of *Sabellaria* and related genera build their tubes on top of each other, creating honeycomb-like aggregations (Fig. 10–21*C* and *D*). Such colonies may be composed of millions of individuals and assume reeflike proportions.

The sand-grain tube of a pectinariid is conical, with the smaller end opening at the surface (Fig. 10–20*E*). The head of this worm bears rows of large, conspicuous setae that are used in digging in soft sand or mud.

The tube of *Owenia* is composed of an inner, membranous, secreted lining and an outer layer of sand grains. The animal gives it flexibility by using flat sand grains, attached at one edge and overlapping adjacent grains (Figs. 10–20*D* and 10–29). Sand grains are collected during feeding, and those grains suitable for tube construction are stored in a ventral pouch located beneath the mouth. During construction of the tube, the pouch projects outward and downward and fastens a grain to the margin of the tube. The membranous lining is secreted by paired glands in each of the first seven trunk segments. The secretion is applied by parapodial setae as the worm revolves.

Among the most beautiful of the sedentary polychaetes are the fan worms, or feather dusters or Christmas tree worms, of the families Sabellidae and Serpulidae. In both groups the prostomial palps have developed to form a funnel-shaped or spiral crown consisting of a few to many pinnate processes called radioles (Figs. 10–22). The radioles are rolled up or closed together when the worm withdraws its anterior into the free end of the tube. Sabellids build membranous or sand-grain tubes.

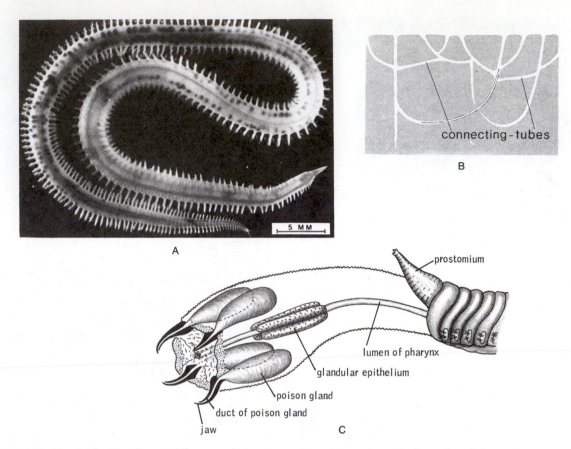

connecting-tubes

B

prostomium

lumen of pharynx

glandular epithelium

poison gland

duct of poison gland

jaw

C

Figure 10–15 *A,* The bloodworm, *Glycera americana,* a common burrowing polychaete found along the east coast of the United States. Note the pointed prostomium. (Courtesy of G. M. Moore.) *B,* Burrow system of *Glycera alba,* showing worm lying in wait for prey. (After Ockelmann and Vahl.) *C,* Structure of the proboscis of *Glycera.* (After Michel.)

Serpulids secrete calcareous tubes that are attached to rocks, shells, or algae and thus can live on an otherwise inhospitable hard substratum. The most dorsal radiole on one or both sides of a serpulid is modified into a long, stalked knob called an operculum (Fig. 10–22*B*), which acts as a protective plug at the end of the tube when the crown is withdrawn.

In both sabellids and serpulids the peristomium is folded back to form a very distinct collar, which fits over the tube opening and is the principal structure used to mold additions on the end of the tube. In a serpulid two large glands secreting calcium carbonate open beneath the collar folds. Crystals of $CaCO_3$ are added to an organic matrix material secreted by the ventral surface (shields) of the anterior segments, and when additions are made to the tube, secretions flow out between the collar and the body wall. This space then acts as a mold in which

the secretion hardens and is simultaneously fused as a new ring on the end of the tube.

The fan worm *Sabella* constructs a tube of sand grains embedded in mucus. The worm sorts detritus collected by the ciliated radioles, and sand grains of suitable size for the tube construction are stored in a pair of opposing ventral sacs located below the mouth. The walls of the sacs produce mucus, which is mixed with the sand grains. To make additions at the end of the tube (Fig. 10–31), the ventral sacs deliver a ropelike string of mucus and sand grains to the collar folds below, which are divided midventrally, like the front of a shirt collar. The string of building materials is received at the collar folds. The worm rotates slowly in the tube, and the collar folds act like a pair of hands, molding and attaching the rope to the end of the tube. The operation is quite similar to an Indian method of making pottery.

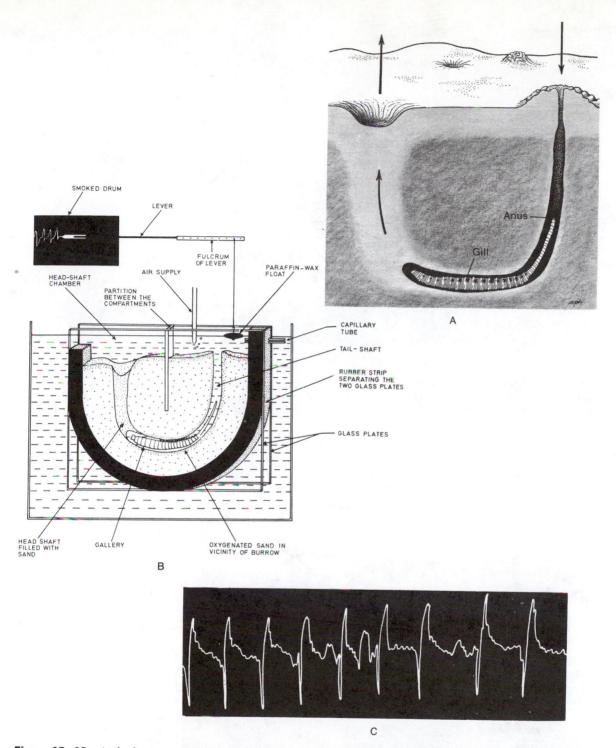

Figure 10–16 *A,* the lugworm, *Arenicola,* in burrow. Arrows indicate the direction of water flow produced by the worm. The worm ingests the column of sand on the left, through which water is filtered. Pile of sand at burrow opening is defecated castings. *B,* Apparatus devised for recording activity cycles of *Arenicola marina. C,* Kymograph tracing of activity cycles of *Arenicola* over a period of six hours. The downstroke reflects the worm backing up to the burrow opening to defecate; the sharp upstroke reflects the worm moving back down to the head of the burrow and vigorously resuming ventilation contractions and deposit feeding. Intervals between defecations are about 40 minutes. (*B* and *C* after Wells, 1959, from Newell, R. C., 1970: Biology of Intertidal Animals. American Elsevier Co., N.Y.)

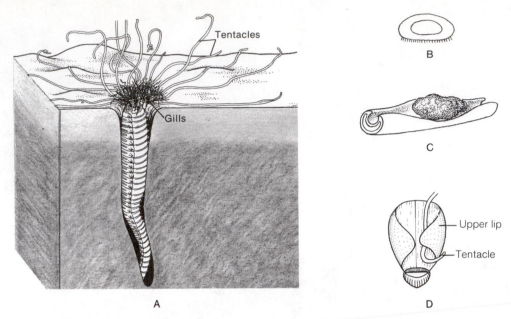

Figure 10–17 *A, Amphitrite* at aperture of burrow with tentacles outstretched over substratum. *B,* Cross section through tentacle of *Terebella lapidaria,* creeping over substratum. *C,* Section of tentacle of *Terebella lapidaria* rolled up to form a ciliary gutter, transporting deposit material. *D,* A tentacle being wiped by one of the lips. (*B, C,* and *D* after Dales.)

In both sabellids and serpulids, the ventral surface of each segment bears a pair of large, mucus-secreting pads, or glandular shields. When the worm rotates, these glands lay down a mucous coating on the inner surface of the tube.

BORING POLYCHAETES

Representatives of a number of different polychaete families excavate protective retreats within dead or living calcareous shells and coral. On coral reefs the brightly colored, spiral radioles of Christ-

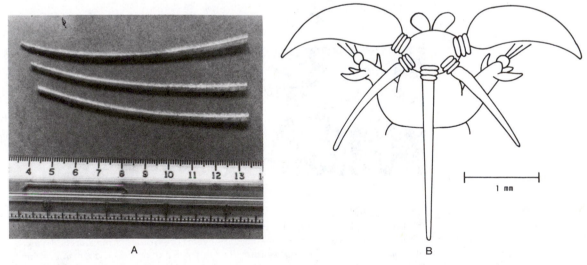

Figure 10–18 A quill worm, the tubicolous eunicid polychaete *Hyalinoecia artifex,* which inhabits the Atlantic continental slope of North America. The tubes lie free on the surface. *A,* Tube. Head of the worm is located at the large end of the tube. *B,* Anterior end of the worm. (*B* from Mangum, C. P., and Rhodes, W. R., 1970: The taxonomic status of quill worms, genus *Hyalinoecia* from the North American Atlantic continental slope. Postilla [Peabody Museum of Natural History], *144*:1–13.)

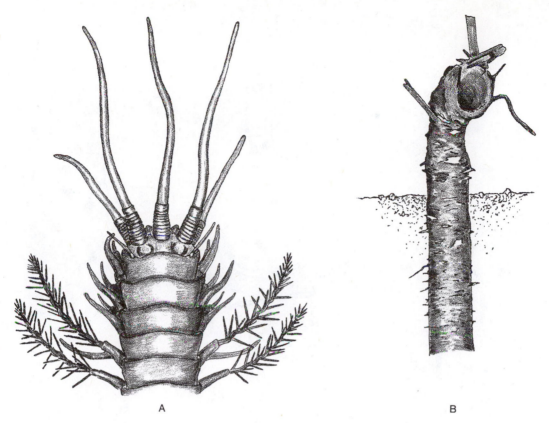

Figure 10–19 *A*, Head and first two gill-bearing segments of *Diopatra* (dorsal view). *B*, Funnel-shaped parchment tube of *Diopatra*.

mas tree worms (the sabellid *Spirobranchus*) can often be seen on the surface of living coral heads. Boring is begun by a newly settled young worm, but the mechanisms are still largely unknown. Species of the spionid *Polydora* excavate chemically in live oyster shells (Zottoli and Carriker, 1974). The worm partially fills the excavation with debris, leaving room for a U-shaped burrow (Fig. 10–23). When boring takes place between the mantle edge and shell, the oyster attempts to wall out the worm with new shell, creating unsightly "mud blisters" and reducing its market value.

INTERSTITIAL POLYCHAETES

The interstitial fauna includes polychaetes of many families. All are minute and some are very aberrant (Figs. 10–24*B* and *C*).

COMMENSALISM

In their commensal relationship with other animals, polychaetes may be hosts or guests. As might be expected, the role of host is played primarily by the tube dwellers and the burrowing polychaetes.

Their guests include scale worms and crustaceans, particularly species of little crabs. Commensal polychaetes, some facultative, are distributed in numerous families throughout the class, but the scale worms include the largest number. They live in tubes and burrows of other polychaetes and crustaceans, with hermit crabs, on echiuroid worms, on corals, in the ambulacral grooves of sea stars, and on sea urchins and sea cucumbers and other animals. Many display colors similar to those of the host and have setae modified for clinging.

PARASITISM

Parasitism is not common among polychaetes. *Labrorostratus* and other arabellids live in the coelom of other polychaetes and may be almost as big as their host. Polychaete ectoparasites include the bloodsucking Ichthyotomidae, which attach to the fins of marine eels.

The myzostomes, sometimes placed in a separate class, the Myzostomida, are a strange group of commensal and parasitic polychaetes. These little

(Text continued on p. 285)

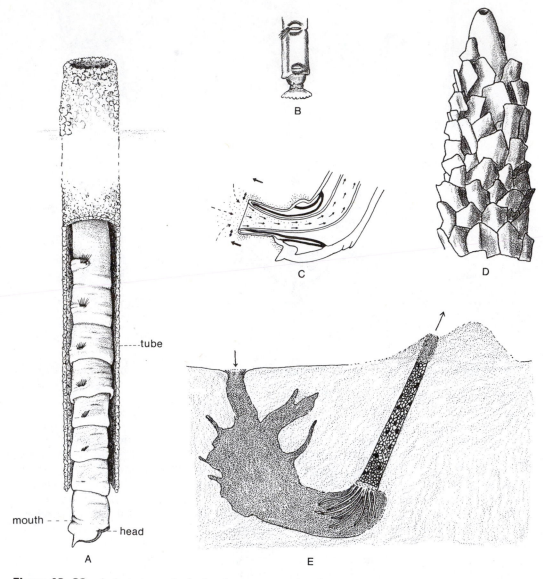

Figure 10–20 *A*, Anterior end of a bamboo worm within sand-grain tube. Head of worm is directed downward. *B*, Posterior end of bamboo worm. *C*, Lateral view of the everted pharynx of the bamboo worm *Axiothella rubrocincta.* Heavy arrows indicate movement of pharynx into substratum; dashed arrows show passage of particles into gut. (From Kudenov, J. D., 1977: The functional morphology of feeding in three species of maldanid polychaetes. Zool. J. Linn. Soc., *60*:95–109.) *D*, Anterior of *Owenia* sand-grain tube, with overlapping sand grains. The membranous, secreted part of the tube projects at the tip. (After Watson.) *E*, *Pectinaria* and sand-grain tube in normal, buried position. Arrows indicate path of water current. (After Wilcke from Kaestner.)

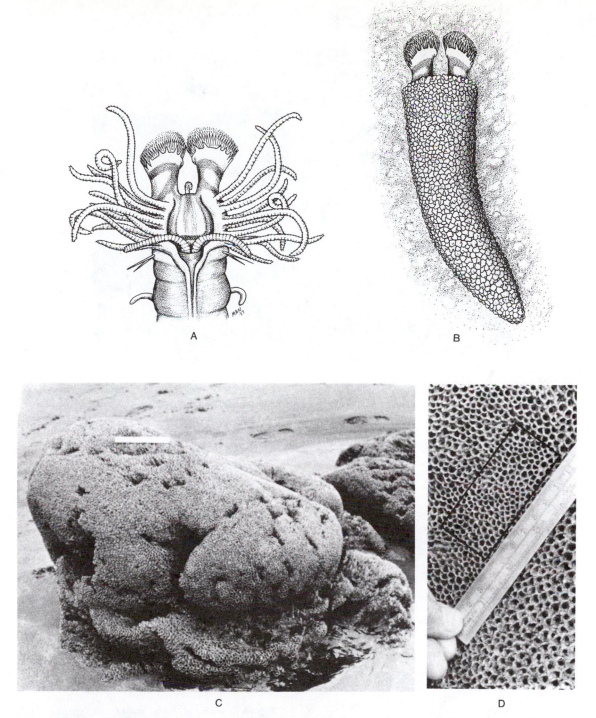

Figure 10–21 *Sabellaria: A*, Anterior end of *Sabellaria vulgaris*, a common shallow-water species along the Atlantic coast of North America. Worm builds sand-grain tubes attached horizontally to shells, stones, pilings, and other objects. The two projections of the first segment carry bundles of golden setae, which form an operculum for closing end of tube. *B*, Opercular bundles of worm projecting from end of tube. *C*, An intertidal rock on the Cornish coast of England encrusted with the tubes of *Sabellaria alveolata*, a species that forms colonies. Tubes are oriented at right angles to substratum and are attached to each other like a honeycomb. Rule is 15 cm long. *D*, Enlarged view of a small area of crust surface showing the tube openings. (*C* and *D* from Wilson, D. P., 1974: *Sabellaria* colonies at Duckpool, North Cornwall, 1971–1972, with a note for May 1973. J. Mar. Biol. Assoc. U.K., 54:393–436.)

A

C

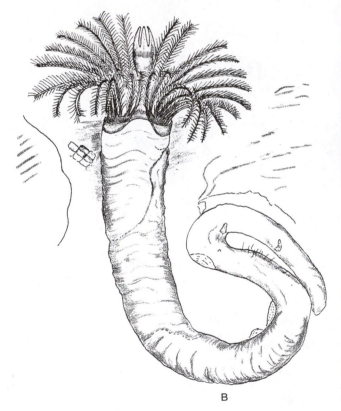

B

Figure 10–22   *A,* Fan, or peacock, worms. The sabellid *Sabella pavonina,* showing the expanded radioles projecting from the apertures of the tubes. (Courtesy of D. P. Wilson.) *B,* The serpulid *Hydroides* with radioles and operculum extended from end of calcareous tube attached to a rock. *C, Spirorbis,* a common serpulid with a snail-like tube found attached to a variety of substrates, including algae. Cutaway of shell shows eggs being brooded within tube.

Figure 10–23 Burrow of the spionid *Polydora websteri* excavated in an oyster shell. Burrow is about 17 mm long. (From Zottoli, R. A., and Carriker, M. R., 1974: Burrow morphology, tube formation, and microarchitecture of shell dissolution by the spionid polychaete *Polydora websteri.* Mar. Biol., 27:307–316.)

worms are rarely more than 5 mm long and resemble flatworms (Fig. 10–24*A*). The body is oval and greatly flattened, and the five pairs of parapodia are carried on the undersurface. They are found only on echinoderms, especially crinoids.

Nutrition

The feeding methods of polychaetes are closely correlated with the various life-styles of the class (see review by Fauchald and Jumars, 1979).

RAPTORIAL FEEDERS

Raptorial feeders include members of many families of surface-dwelling species, many pelagic groups, tubicolous eunicids and onuphids, and gallery dwellers like the glycerids and nephtyids. The prey consists of various small invertebrates, including other polychaetes, which are usually captured by means of an eversible pharynx (proboscis). The pharynx commonly bears two or more horny jaws composed of tanned protein (Fig. 10–25). The pharynx is rapidly everted; this places the jaws at the anterior of the body and causes them to open. The food is seized by the jaws and the pharynx is retracted. Although protractor muscles may be present, an increase in coelomic pressure resulting from the contraction of body wall muscles is an important factor in the eversion of the pharynx. When pressure on the coelomic fluid is reduced, the pharynx is withdrawn by the retractor muscles, which extend from the body wall to the pharynx (Fig. 10–3*B* and *C*).

Raptorial tube dwellers may leave the tube partially or completely when feeding, depending on the species. *Diopatra* uses its hood-shaped tube as a lair (Fig. 10–19*B*). Chemoreceptors monitor the ventilating current of water passing into the tube, and when approaching prey is detected, the worm partially emerges from the tube opening and seizes the victim with a complex pharyngeal armature of teeth. During feeding the prey may be clasped with the enlarged anterior parapodia. Species of *Diopatra* may also feed on dead animals, algae, organic debris, and small organisms, such as forams, that are in the vicinity of the tube or become attached to it.

Some raptorial feeders, such as syllids and glycerids, have long, tubular proboscises. Species of *Glycera* live within a gallery system constructed in muddy bottoms. The system contains numerous loops that open to the surface (Fig. 10–15*B*). Lying in wait at the bottom of a loop, the worm can detect the surface movements of prey such as small crustaceans and other invertebrates, by changes in water pressure. It slowly moves to the burrow opening and then seizes the prey with the proboscis.

When the proboscis is retracted, it occupies approximately the first 20 body segments. At the back of the proboscis are four jaws arranged equidistantly around the wall. The proboscis is attached to an S-shaped esophagus. No septa are present in these anterior segments, and the proboscis apparatus lies free in the coelom. Just prior to eversion of the proboscis the longitudinal muscles contract violently, sliding the proboscis forward and straightening out the esophagus. The proboscis is then everted with explosive force, and the four jaws emerge open at the tip (Fig. 10–15*C*). Each jaw contains a canal that delivers poison from a gland at the jaw base.

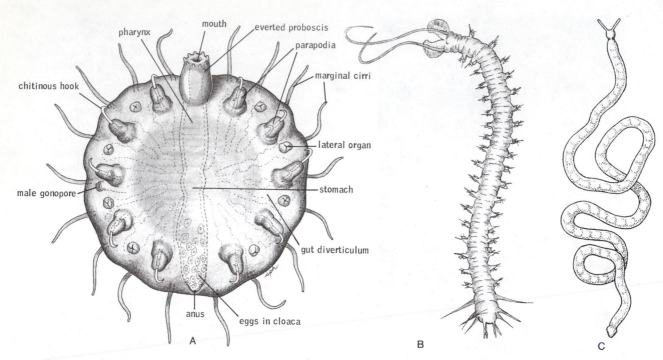

Figure 10–24   *A, Myzostoma,* a commensal on crinoids. *B, Hesionides arenaria,* a minute polychaete, no more than 2 mm in length, that inhabits interstitial spaces in the intertidal zones of surf beaches of north Europe. The ventrally directed parapodia are adapted for crawling between sand grains. (From Ax, P., 1966: Veroeffentlungen des Institute für Meeresforschung Bremerhaven. Suppl. II, pp. 15–66.) *C, Polygordius neapolitanus,* an interstitial polychaete. Segmentation is poorly indicated externally. (After Fraipont.)

HERBIVORES, OMNIVORES, SCAVENGERS, AND BROWSERS

Not all errant polychaetes that possess jaws are carnivores. Scavenging and omnivorous habits have evolved in many polychaetes. The jaws may, for example, be used to tear off pieces of algae. These polychaetes generally belong to the same families as do the carnivores, and they are similarly adapted.

Studies on *Nereis* by Goerke (1971) demonstrate the diversity of feeding habits that may exist even within a genus. Species of *Nereis* possess a muscular, eversible pharynx with a pair of heavy jaws (Fig. 10–3C). Some, such as *Nereis pelagica,* *Nereis virens,* and *Nereis diversicolor,* are omnivorous and feed on algae, other invertebrates, and even detritus. *Nereis succinea* and *Nereis longissima* feed primarily on detritus material in the substratum. However, *Nereis fucata,* a commensal in hermit crab shells, is carnivorous.

NONSELECTIVE DEPOSIT FEEDERS

Some deposit-feeding polychaetes consume sand or mud directly when the mouth is applied against the substratum and are not selective. Ingestion is generally facilitated by means of a simple, nonmuscular proboscis, which is everted by elevated coelomic fluid pressure (Figs. 10–26, 10–27, and 10–28). Among polychaetes, such deposit feeders include burrowers and tube dwellers. The less stationary burrowers include some capitellids and opheliids, both of which ingest the substratum through which they burrow.

McConnaughey and Fox (1949) took population counts of the little burrowing opheliid *Euzonus (Thoracophelia) mucronatus,* which is about 25 mm long and not more than 2 mm in diameter. These worms inhabit the intertidal zone on the Pacific coast of the United States and form colonies that occupy extensive stretches of protected (low-energy) beaches. In such colonies the number of worms averages 2500 to 3000 per square foot. Worms occupying a typical strip of beach 1 mile long, 10 feet wide, and 1 foot thick would ingest approximately 14,600 tons of sand each year.

The lugworms, members of the family Arenicolidae, are common deposit feeders. *Arenicola*

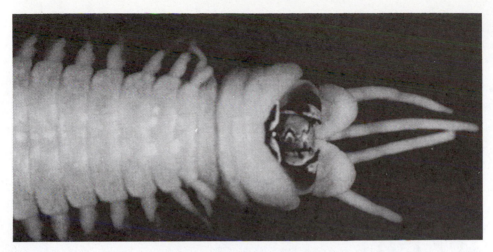

Figure 10–25 Ventral view of anterior end of *Eunice,* showing protruded jaws. The complex buccal armature is composed of a number of pieces, of which two are especially large and opposing. (From Bennett, I., 1966: Fringe of the Sea. Rigby Ltd., Adelaide, Australia.)

lives in an L-shaped burrow whose vertical part opens to the surface (Fig. 10–16*A*). The head of the worm is directed toward the blind, horizontal part of the burrow, where sand is continually ingested by means of a simple proboscis. The worm irrigates the burrow by peristaltic contractions that drive water into the burrow opening. Water leaves the burrow by percolating up through the sand. At cyclic intervals the worm backs up to the surface to defecate mineral material (castings) (Fig. 10–16).

Bamboo worms, the Maldanidae, are examples of deposit-feeding tube dwellers. The worm lives upside down and ingests the substratum at the bottom of the sand-grain tube. Cilia within the everted pharynx drive loose particles into the gut, and there is actually some particle selection (Fig. 10–20) (Kudenov, 1977). Following a distinct rhythm, the feeding halts and the worm backs up to the top of the tube to defecate the mineral particles passed through the gut.

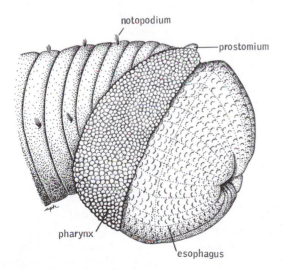

Figure 10–26 Lateral view of the anterior end of the capitellid *Notomastus,* a burrowing, nonselective deposit feeder. The everted pharynx obscures the greatly reduced prostomium. (After Michel.)

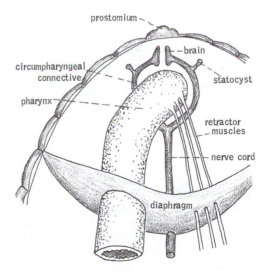

Figure 10–27 Anterior of *Arenicola* (doral dissection). (After Ashworth from Brown.)

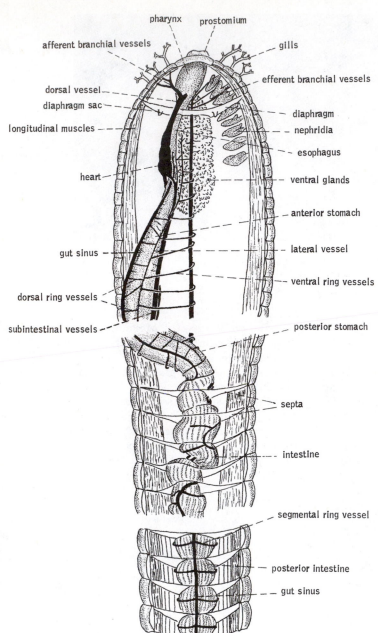

Figure 10–28 *Amphitrite* (dorsal dissection). (After Brown, F. A., 1950: Selected Invertebrate Types. John Wiley and Sons, N.Y.)

SELECTIVE DEPOSIT FEEDERS

Selective deposit feeders lack proboscises. Special head structures extend over or into the substratum. Deposit material adheres to mucous secretions on the surface of these feeding structures and is then conveyed to the mouth along ciliated tracts or grooves. These polychaetes thus select organic deposit material from between sand particles.

The prostomial ingestive organs of the terebellids, for example, *Amphitrite* and *Terebella*, are formed of large clusters of contractile tentacles, which stretch over the surface of the substratum by ciliary creeping (Fig. 10–17). Surface detritus adheres to the mucus secreted by the tentacular epithelium. Particles are moved down a ciliated gutter formed by the rolled tentacle, and food accumu-

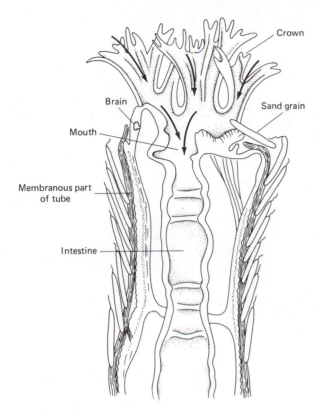

Figure 10–29 *Owenia* adding a sand grain to its tube (longitudinal section). (Modified from Watson.)

Crown

Brain

Mouth

Sand grain

Membranous part of tube

Intestine

lates at the base of the tentacles, each of which is wiped over the upper lip bordering the mouth. Cilia on the lip then drive the food into the mouth. The tubicolous *Owenia* feeds in a somewhat similar manner but utilizes a prostomial crown of flattened, branched, ribbon-like filaments, each of which ends in a bifid lobe (Fig 10–29).

The tentaculate, deposit-feeding cone worm, *Pectinaria*, which lives buried in the sand head down (Fig. 10–20E), selects organic aggregates and mineral particles encrusted with organic material (Whitlatch, 1974), and by selection the worm concentrates the organic matter from about 32 per cent (in sediment) to 42 per cent (in gut). Of the ingested organic fraction, about 30 per cent is utilized and the remainder passes through the gut. By differentially staining the organic matter, it is possible to determine something of the relative amounts of proteins and carbohydrates present and utilized.

The tubicolous spionids are suspension feeders that probably depend on particles stirred up from the bottom. They possess two long, tentacular palps that lash the water or project from the tube opening and extend over the bottom (Fig. 10–30D and E). Particles that adhere to the surface are propelled toward the mouth down the length of the palps in a ciliated channel.

FILTER FEEDERS

Many of the sedentary burrowers and tubicolous polychaetes are filter feeders. The head is usually equipped with special feeding processes that collect detritus and plankton from the surrounding water. The particles adhere to the surface of the feeding structures and are then conveyed to the mouth along ciliated tracts.

The crownlike, bipinnate radioles of serpulid and sabellid fan worms form a funnel or one or two spirals when expanded outside the end of the tube (Fig. 10–31). Beating of the cilia located on the pinnules produces a current of water that flows through the radioles into the funnel and then flows upward and out. Particles are trapped on the pinnules and are driven by cilia into a groove running the length of each radiole. The particles are carried along the groove down to the base of the radiole, where a rather complex sorting process takes place (Fig. 10–32). The largest particles are rejected, and fine material is carried by ciliated tracts into the mouth. Many sabellids sort particles into three grades and store the medium grade for use in tube construction.

The feeding mechanism of the chaetopterids differs from that of the other filter feeders (Figs. 10–30 and 10–33). *Chaetopterus*, which lives in U-shaped parchment tubes, has a highly modified body structure. The notopodia on the 12th segment are extremely long and aliform (winglike), and the epithelium is ciliated and richly supplied with mucus glands. The notopodia on segments 14 to 16 are modified and fused, forming semicircular fans that project like piston rings against the cylindrical wall of the tube (Fig. 10–33A). The beating of the fans produces a current of water that enters the chimney of the U-shaped tube near the anterior of the worm, flows through the tube, and then flows out of the opposite chimney.

The paired, aliform notopodia are stretched out around the walls of the tube, and a sheet of mucus is secreted between them. The mucous film is continuously secreted from each notopodium, and so the sheet assumes the shape of a bag. The posterior of the bag is grasped by a ciliated cupule on the middorsal side of the worm a short distance behind the aliform notopodia. Water brought into the tube by the rhythmic beating of the fan parapodia passes

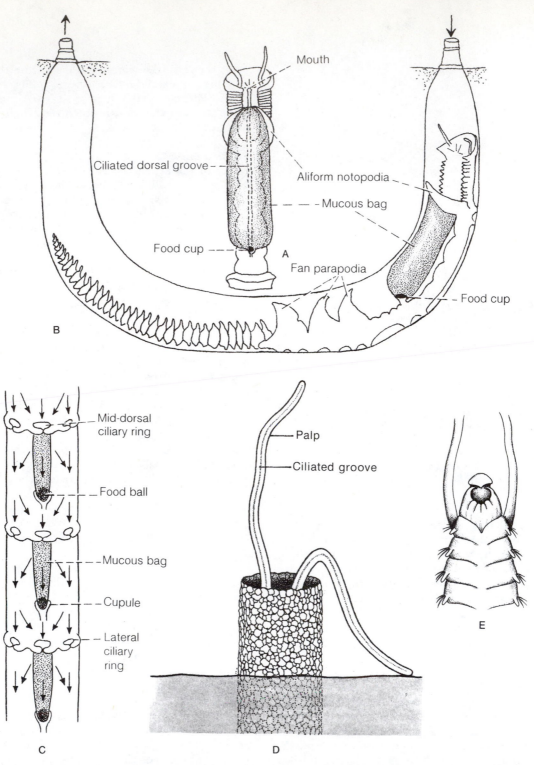

Figure 10–30 *Chaetopterus* during feeding: *A*, Anterior part of body (dorsal view). *B*, Worm in tube (lateral view). Arrows indicate direction of water current through the tube. (After MacGinitie.) *C*, Dorsal view of three segments of middle body region of *Spiochaetopterus*, showing the position of mucous bags and the formation of food balls. Arrows indicate the direction of water currents. (After Barnes.) *D*, Palps of a spionid polychaete projecting from its sand-grain tube. Ciliated groove of palp conveys to mouth detritus material picked up from substratum or suspended in water. *E*, Ventral view of anterior end of *Spio pettiboneae* showing base of palps and mouth.

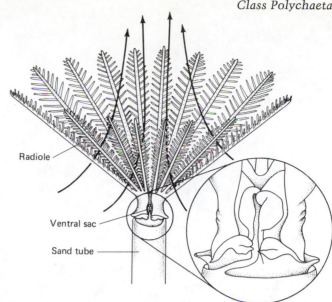

Figure 10–31 Anterior end of the fan worm *Sabella*, showing the filter-feeding currents and tube building. (Modified after Nicol from Newell, R. C., 1970: Biology of Intertidal Animals. American Elsevier Co., N.Y.)

through the mucous bag, which strains suspended detritus and plankton (Flood and Fiala-Médioni, 1982).

Large objects brought into the tube by the water current are detected by peristomial cilia; the aliform notopodia then are pulled back to let the large objects pass by. The food-laden mucous bag is continuously being rolled up into a ball by the dorsal

cupule. When the ball reaches a certain size, the bag is cut loose from the notopodia and rolled up with the ball. The cupule then projects forward and deposits the mucous food ball onto a ciliated mid-dorsal groove, which extends to the anterior of the worm, and the ball is carried to the mouth. An 18 to 24 cm long specimen of *Chaetopterus* may produce mucous film for the bag at the rate of approx-

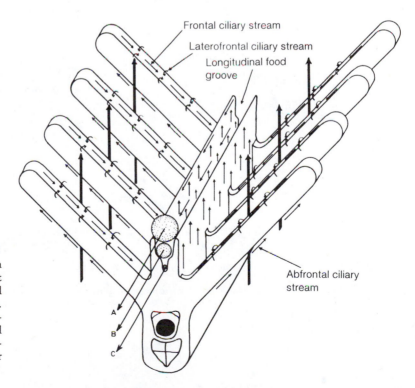

Figure 10–32 Filter feeding in the fan worm *Sabella*, showing water current (large arrows) and ciliary tracts (small arrows) over a section of one radiole. The letters A, B, and C indicate different sized particles sorted. (After Nicol from Newell, R. C., 1970: Biology of Intertidal Animals. American Elsevier Co., N.Y.)

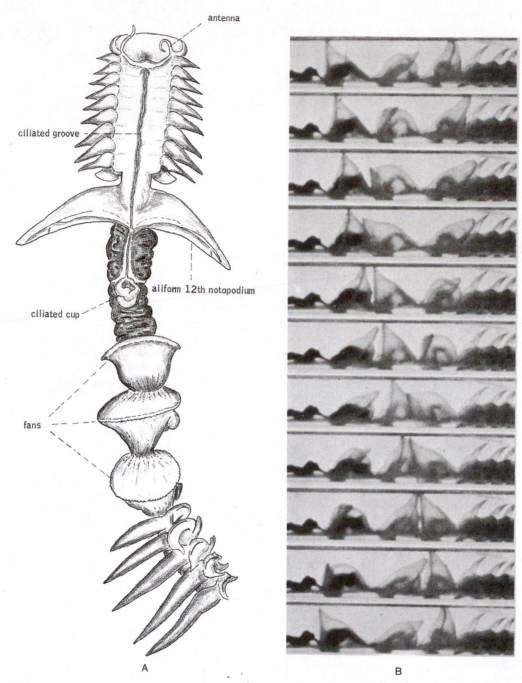

Figure 10–33   *A, Chaetopterus variopedatus.* (From life.) *B,* Side views of the parapodial fans of *Chaetopterus* during one pumping cycle. Worm is in a glass tube. (From Brown, S. C., 1975: Biomechanics of water-pumping by *Chaetopterus variopedatus* Renier. Skeletomusculature and kinematics. Biol. Bull.,.*148*(4):136–150.)

imately 1 mm per second, with food balls averaging 3 mm in diameter.

The other members of the Chaetopteridae build straight, vertical tubes but utilize mucous bags for filter feeding. The number of mucous bags and the site of their formation vary—as many as 13 are formed at one time in *Spiochaetopterus*. In a number of genera the water current is activated by cilia rather than by pumping (Fig. 10–30C).

THE ALIMENTARY CANAL

Typically, the alimentary canal of polychaetes is a straight tube extending from the mouth at the anterior of the worm to the anus located in the pygidium. The canal is commonly differentiated into a pharynx (or buccal cavity if the pharynx is absent), short esophagus, stomach (in sedentary species), intestine, and rectum (Fig. 10–3). However, in many species these regions can be detected only histologically, and the gross appearance of the digestive tract behind the pharynx is that of a simple, uniform tube. The stomach or anterior intestine elaborates enzymes for extracellular digestion. The intestine is the site of absorption, and not infrequently the walls are folded, increasing the intestinal surface area. In *Nereis* two large, glandular ceca open into the esophagus (Fig. 10–3). They, along with the anterior end of the intestine, secrete digestive enzymes.

The egested wastes from a worm living in a tube with double openings, such as *Chaetopterus*, are readily removed by water currents. Such flushing, however, is less efficient when the tube is deeply buried in mud and sand. Many polychaetes turn around in the tube or burrow and thrust the pygidium out of the opening during defecation. Some species produce fecal pellets or strings, which reduce the risks of fouling. A fan worm has a ciliated groove, which carries fecal pellets from the anus anteriorly out of the tube.

Gas Exchange

Gills are common among the polychaetes, but they vary greatly in both structure and location, indicating that they have arisen independently within the class a number of times. They are never enclosed within protective chambers; many species that possess gills are already protected, since they live in tubes and burrows. Gills are lacking in polychaetes that are very small or that possess long, threadlike bodies, as in many burrowing Lumbrineridae, Arabellidae, and Capitellidae.

In the scale worms gas exchange is largely restricted to the dorsal body surface, which is roofed over by the elytra. Cilia on the dorsal surface create a current of water flowing posteriorly beneath the elytra. The felt-covered sea mice (*Aphrodita*) lack cilia (Fig. 10–11), but a similar dorsal water current is produced by the animal, which tilts the elytra upward and then rapidly brings them down in sequence.

Most commonly the gills are associated with the parapodia and in many cases are modified parts of the parapodium. The notopodium may possess a flattened branchial lobe, which acts as a gill, as in nereids (Fig. 10–5A). Commonly, the dorsal cirrus of the parapodium is modified to serve as a gill (Fig. 10–6A and B), or the gills arise from the base of the dorsal cirrus (Fig. 10–19A). Cirratulids have long, contractile, threadlike gills, each attached to the base of the notopodium (Fig. 10–10A).

The gills are not always associated with the parapodia. Many sedentary species have gills at the anterior ends near the opening of the tubes or burrows. For example, the gills of some terebellids, such as *Amphitrite*, are arborescent and are located on the dorsal surface of the anterior segments (Fig. 10–17A). The bipinnate radioles composing the fans serve as sites of gas exchange in the sabellid and serpulid fan worms.

Ventilation may be provided by gill cilia or by gill contractions, but many burrowing and tube-dwelling polychaetes drive water through their burrows or tubes by undulating or peristaltic contractions of the body. Worms that ventilate by muscular activity typically exhibit a spontaneous ventilating rhythm in which a period of ventilation alternates with a period of rest (Fig. 10–34). Although ventilation activity increases the worm's

Figure 10–34 Trace of the rhythmic irrigation of the tube of *Hyalinoecia tubicola* (see Fig. 10–18). The downward deflections are bursts of irrigation activity, followed by a gradual return to the resting interval. Recording covers a period of 2 hours and 5 minutes. (From Dales, R. P., Mangum, C. P., and Tichy, J. C., 1970: Effects of changes in oxygen and carbon dioxide concentrations on ventilation rhythms in onuphid polychaetes. J. Mar. Biol. Assoc. U.K., 50:365–380; copyrighted and reprinted by permission of Cambridge University Press.)

oxygen requirement as much as 15-fold, there is still a net gain of about 50 per cent over ventilating costs (Mangum, 1976).

Internal Transport

In most polychaetes there exists a well-developed blood-vascular system, in which the blood is enclosed within vessels. The basic plan of circulation is relatively simple. Blood flows anteriorly in a dorsal vessel situated over the digestive tract; at the anterior of the body, the dorsal vessel is connected to a ventral vessel by one to several vessels or by a network of vessels passing around the gut. The ventral vessel carries blood posteriorly beneath the alimentary tract (Fig. 10–35*A*).

In each segment the ventral vessel gives rise to one pair of ventral, parapodial vessels, which supply the parapodia, the body wall, and the nephridia, and to several ventral, intestinal vessels that supply the gut (Fig. 10–35*A*). The dorsal vessel, in turn, receives a corresponding pair of dorsal parapodial vessels and a dorsal intestinal vessel. The dorsal and ventral parapodial vessels and the dorsal and ventral intestinal vessels are interconnected by a network of smaller vessels.

There are many variations of this basic circulatory pattern, and the circulatory mechanisms in polychaetes are not nearly so uniform as this description might suggest. All polychaetes rely to varying degrees on the transporting capacity of the coelomic fluid, and some have lost the blood-vascular system completely.

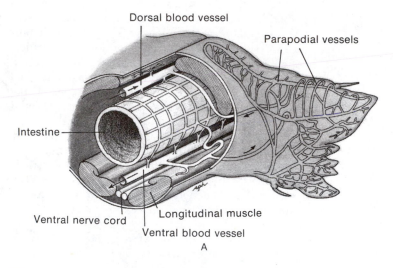

A

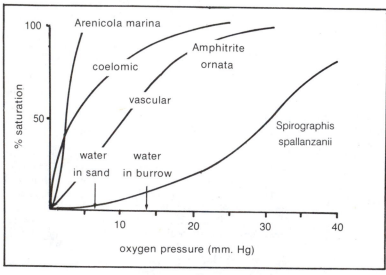

B

Figure 10–35 *A*, Vascular system within a segment of *Nereis virens.* Arrows indicate direction of blood flow. (Modified from Nicol.) *B*, Oxygen dissociation curves of the respiratory pigments of three species of polychaete. *Arenicola* possesses vascular hemoglobin. *Amphitrite ornata* possesses two hemoglobins, one vascular and one coelomic. The respiratory pigment of *Spirographis spallanzani* is chlorocruorin. (See p. 295 of text for the significance of the positions of the curves.) (In part after Jones from Dales.)

Although the polychaete system is usually a closed system, in that blood is restricted to vessels, all the vessels tend to be relatively thin walled and not always lined with endothelium (Nakao, 1974). Moreover, the endothelium, when present, is myoendothelium, and the basal membrane is directed toward the lumen instead of away from it, as in vertebrates (Fig. 10–36A) (Ruppert and Carle, 1983.)

The gills are usually provided with afferent and efferent vascular loops permitting a two-way flow. This is true, for example, of the gills of lugworms and the branchial, notopodial lobes of nereids (Fig. 10–36D). On the other hand, the radioles of fan worms, which function in both food gathering and gas exchange, contain only a single vessel, within which blood flows tidelike, in and out. In many polychaetes, such as glycerids, the gills are irrigated with coelomic fluid and not blood.

In general, blood is driven by peristaltic waves of contraction that sweep over the blood vessels, particularly the dorsal vessels. The vessel wall in *Magelona*, for example, consists of a single layer of myoepithelial cells that contain striated myofibrils arranged in a circular direction or in both circular and longitudinal directions (Boilly and Wissocq, 1977). Many polychaetes have accessory, heartlike pumps located in various places within the blood-vascular system.

The blood contains few cells compared with coelomic fluid. In small polychaetes it is usually colorless, but in larger species and those that burrow in soft bottoms, the blood contains respiratory pigments dissolved in the plasma. In fact, within the Polychaeta are found three of the four respiratory pigments of animals. Hemoglobin is the most common of these pigments, but chlorocruorin is characteristic of the blood of the serpulid and sabellid fan worms and also of the Flabelligeridae and Ampharetidae. Chlorocruorin is an iron porphyrin like hemoglobin, but the slight difference in side chains gives it a green rather than a red color. There is little reason for separating chlorocruorin from plasma hemoglobin, which it resembles more than either resembles intracellular hemoglobin. *Magelona* has a blood-vascular system with anucleated corpuscles containing a third iron-bearing protein pigment (not a porphyrin) called hemerythrin. Hemerythrin is a protein similar to hemocyanin rather than a porphyrin, and the molecule of O_2 is carried between two iron atoms.

Plasma, or extracellular, hemoglobin and chlorocruorin molecules are always very large. The plasma hemoglobin of *Arenicola*, for example, contains 96 heme units. The entire molecule attains a molecular weight of 3,000,000. This compares with a molecular weight of 60,000 in mammalian hemoglobin, in which there are four hemes, each attached to a 15,000 molecular weight unit. There are numerous polychaetes, including *Glycera, Capitella*, and some terebellids, in which the blood-vascular system is reduced or absent and the coelomic fluid functions in internal transport. The coelom of these worms contains hemoglobin located in coelomic corpuscles. Such coelomic hemoglobin, like the corpuscular hemoglobin of vertebrates, is always a small molecule.

Mangum (1985) believes that Hb packed in red blood cells may be the primitive condition in animals and has been retained in various animal groups, including some polychaetes. The more specialized extracellular condition appears to be related to the lack of capillary beds in most polychaetes. In the larger vessels through which polychaete blood is pumped, the blood is less viscous with its Hb in solution than it would be with red blood cells.

There are a number of interesting exceptions to the usual disposition of respiratory pigments just described. The blood of *Serpula* contains both hemoglobin and chlorocruorin. Most terebellids and ophelids possess not only coelomic red corpuscles but also a blood-vascular system with a different hemoglobin. The two hemoglobins are not alike. The coelomic hemoglobin (Hb) of *Amphitrite* has a greater affinity of O_2 at low oxygen tensions (dissociation curve to the left) than does the blood-vascular Hb. This difference facilitates the passage of oxygen from the blood-vascular system to the coelomic fluid, which is the principal source of oxygen for internal tissues (Fig. 10–35B).

In the majority of polychaetes the respiratory pigments function in oxygen transport, although for only a part of the oxygen consumed. When the blood from the gills does not become mixed with unoxygenated blood before delivering its oxygen load to the target tissues, the oxygen affinity of the hemoglobin is relatively low (oxygen dissociation curve to the right). This is the situation for polychaetes like *Amphitrite* and the fan worms (e.g., *Sabella*), in which the gills are at the anterior end. In worms with segmental gills, in which the blood from the gills is mixed with unoxygenated blood en route to the target tissues, the oxygen affinity of the hemoglobin is high; i.e., the hemoglobin holds on to its oxygen at relatively low oxygen tensions (Fig. 10–35B).

In some polychaetes, such as the bloodworm *Glycera*, the Hb may also store oxygen during the resting periods between ventilation or at low tide,

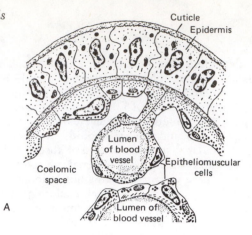

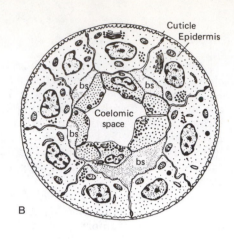

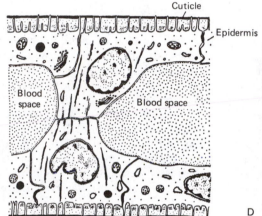

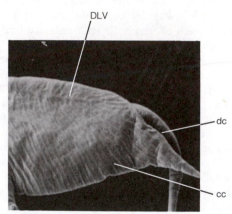

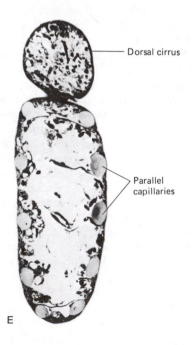

Figure 10–36 *A–C*, Cross sections of various polychaete gills, showing the relationship of the blood channels to the gill epithelium. Note that only in *Scalibregma* (a) is the vessel bounded by its own epithelium. (a) *Malacoceros fuliginosus*; (b) *Scalibregma inflatum*, bs: blood space; (c) *Pectinaria koreni*. (From Storch, V., and Alberti, G., 1978: Ultrastructural observations on the gills of polychaetes. Helgol. wiss. Meeresunters, *31*:169–179.) *D*, SEM view of the lateral surface of the branchial lobe of the parapodium of *Nereis succinea*. Parallel capillaries (cc) can be seen beneath the surface connecting a dorsal (DLV) and a ventral lateral vessel. *E*, Section through part of branchial lobe and dorsal cirrus (dc) shown in *D*. Walls of the capillaries (cc) are composed of cells (exothelium) in which the basal lamina lines the lumen. In vertebrates the basal lamina of the capillary wall cells is on the opposite side from the lumen. (Courtesy of J. M. Johnson.)

BOX 10–2   METANEPHRIDIA

The most common type of excretory organ among coelomate animals is a metanephridium. In contrast to the blind protonephridial tubule, a metanephridial tubule opens internally into the coelom. The opening is often funnel-like and clothed with ciliated peritoneum, in which case it is called a nephrostome. In nonsegmented coelomates there may be one nephridium or one to several pairs of metanephridia; in segmented groups, such as the annelids, the metanephridia are serially repeated, one pair per segment.

In general, a metanephridium processes coelomic fluid. Blood filtrate passes into the coelom at various sites of filtration, depending on the species. For example, in a mollusk part of the heart wall is the major site of filtration and is composed of podocytes, cells with finger-like processes that interdigitate (Fig. 11–90). The slits between processes are the sites of filtration. Podocytes are found at the filtration sites of many animals, e.g., the glomeruli of the vertebrate kidney.

Coelomic fluid, derived from blood filtrate, passes through the nephrostome into the ciliated nephridial tubule. Here it becomes modified by selective reabsorption and secretion, and the product is finally expelled through the nephridiopore as urine. The extent of tubular secretion and reabsorption depends in part on the environment in which the animal lives,

i.e., whether it is an osmoconformer or osmoregulator. The tubule wall is correspondingly specialized and provided with a vascular backing.

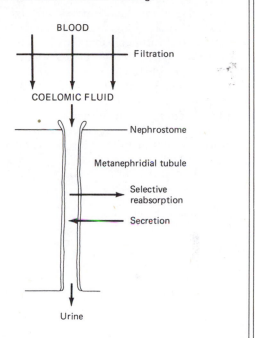

---

when the oxygen tension of the water in the burrow and surrounding sand is considerably decreased. The amount of hemoglobin present would seem to permit an oxygen reserve lasting only a few minutes. However, like many invertebrates, polychaetes are oxyconformers; i.e., oxygen consumption is regulated in part by the amount of oxygen available in the surrounding environment. Mangum (1970) has suggested that the decreasing metabolic demands for oxygen during stagnation enable the worm to stretch out over a significant period of time what would otherwise be an inadequate store of oxygen provided by the hemoglobin. The intertidal burrower *Euzonus mucronatus* (p. 286) uses its oxyhemoglobin storage to carry it through regular 2- to 4-hour periods of anoxia during low tides. If subjected to longer periods of anoxia, the worm switches over to anaerobic respiration, on which it can survive as long as 20 days (Ruby and Fox, 1976).

Excretion

Polychaete excretory organs are either protonephridia or metanephridia (Box 10–2). In primitive polychaetes there is one pair of nephridia per segment, but reduction to few or even one pair for the entire worm has occurred in some families. The anterior end of the nephridial tubule is located in the coelom of the segment immediately anterior to that from which the nephridiopore opens (Fig. 10–2). The tubule penetrates the posterior septum of the segment, extends into the next segment, where it may be coiled, and then opens to the exterior in the region of the neuropodium. Both the preseptal portion of the nephridium and the postseptal tubule are covered by a reflected layer of peritoneum from the septum.

Protonephridia of a type called solenocytes are found in phyllodocids, alciopids, tomopterids, glycerids, nephtyids, and a few others. The solenocytes are always located at the short preseptal end of the nephridium and are bathed by coelomic fluid. The solenocyte tubules are very slender and delicate and arise from the nephridial wall in bunches (Fig. 10–37) Each tubule contains a single flagellum, and the wall is composed of parallel rods connected by the thin lamellae (Fig. 10–37C). The latter represent the fenestrations through which fluid passes; this arrangement is characteristic of other types of protonephridia (Box 7–1, p. 176).

All other polychaetes possess metanephridia (Box 10–2), in which the preseptal end of the nephridium possesses an open, ciliated funnel, the nephrostome, instead of solenocytes. Typical metanephridia are found in the nereids (Fig. 10–38C), where the nephrostome possesses an outer investment of peritoneum and the interior is heavily cil-

iated. The postseptal canal, which extends into the next successive segment, becomes greatly coiled to form a mass of tubules, which are enclosed in a thin, saclike covering of peritoneal cells. Coiling is probably an adaptation that increases the surface area for tubular secretion or reabsorption. The nephridiopore opens at the base of the neuropodium on the ventral side. The entire lining of the tubules is ciliated.

The metanephridia of most other polychaetes differ only in minor details (Fig. 10–38) but may display various degrees of regional restriction in the more specialized families. In the fan worms, where only one pair of functional nephridia remain, the two nephridia join at the midline to form a single median canal, which extends forward to open through a single nephridiopore on the head

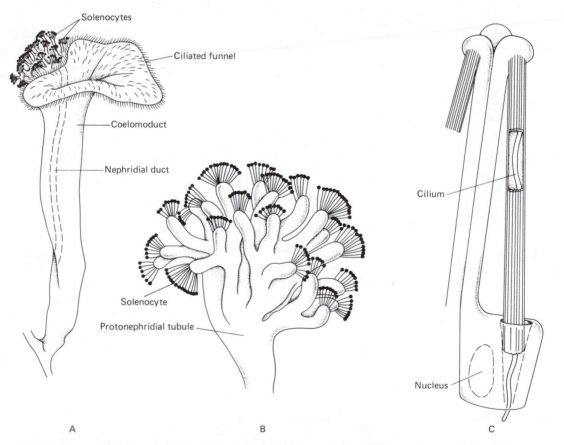

A   B   C

Figure 10–37   Protonephridium and coelomoduct of *Phyllodoce paretti: A,* Relation of protonephridium and coelomoduct. *B,* Branched end of protonephridium. (*A* and *B* after Goodrich.) *C,* Ulstrastructure of a solenocyte of *Nephtys.* Each solenocyte, which is one member of a triplet, bears a long projecting arm. The filtration cylinder, composed of rods and membrane and surrounded by a single internal cilium, extends back to the base of the cell, where it is inserted but free within an intracellular passage to the nephridial tubule. (After Kümmel from Welsch, U., and Storch, V. 1976: Comparative Animal Cytology and Histology. University of Washington Press, Seattle. p. 298.)

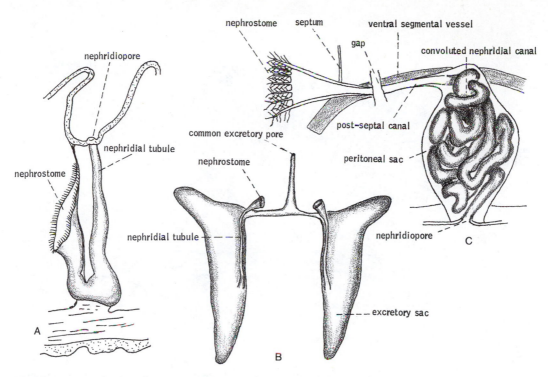

Figure 10–38 Metanephridia of three polychaetes: *A, Scolelepis.* (After Goodrich.) *B,* The serpulid fan worm, *Pomatoceros.* (After Thomas.) *C, Nereis vexillosa.* (After Jones.)

(Fig. 10–38*B*). Excretory waste is deposited directly outside, and fouling of the tube is avoided.

In polychaetes the association of the blood vessels with the nephridia is variable. The fan worms and the arenicolids lack a well-developed nephridial blood supply, and the coelomic fluid must be the principal route for waste removal. In other polychaetes the nephridia are surrounded by a network of vessels. In the nereids the nephridial blood supply is greater in those species that live in brackish water.

Many polychaetes, particularly nereids, can tolerate low salinities and have become adapted to life in brackish sounds and estuaries. The gill (notopodial lobe) of *Nereis succinea* contains cells specialized for absorbing ions. A small number of species live in fresh water. The sabellid *Manayunkia speciosa,* for example, occurs in enormous numbers in certain regions of the Great Lakes, such as around the mouth of the Detroit River. There are a few terrestrial polychaetes, all tropical Indo-Pacific nereids, which burrow in soil or live in moist litter.

Chloragogen tissue, coelomocytes, and the intestinal wall may play accessory roles in excretion. Chloragogen tissue is composed of brown or greenish cells located on the wall of the intestine or on various blood vessels. Chloragogen tissue, which has been studied much more extensively in earthworms (see p. 316), is an important center of intermediary metabolism and hemoglobin synthesis.

Nervous System

The polychaete brain, usually bilobed, lies in the prostomium beneath the dorsal epithelium (Figs. 10–39 and 10–40*A*). Depending on the degree of development of sense organs, the brain supplies nerves to the palps, antennae, eyes, and nuchal organs. Typically, a pair of circumpharyngeal or circumesophageal connectives surround the anterior gut and interconnect the brain and the ventral nerve cord.

The primitive ventral nerve cord is completely double throughout with transverse connectives between the separate ganglia as in fan worms, but in most polychaetes the two cords are fused in varying degrees. There is typically one ganglionic swelling per segment, and from each ganglion emerge usually three or four pairs of lateral nerves that in-

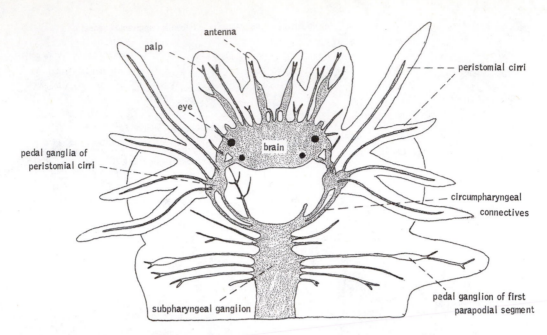

antenna

palp

peristomial cirri

eye

brain

pedal ganglia of
peristomial cirri

circumpharyngeal
connectives

subpharyngeal gangiion

pedal ganglion of first
parapodial segment

Figure 10–39 Anterior part of *Nereis* nervous system. (After Henry from Kaestner.)

nervate the body wall of that segment (see Box 10–4, p. 321, for ganglionic organization).

An important defense of most polychaetes against their many predators is the ability to contract very rapidly. The rapid end-to-end contraction reflex is particularly well developed in the tube dwellers, which project from their protective housing to feed. Correlated with this ability to contract rapidly is the presence of giant axons in the ventral nerve cords. The enlarged diameter of the axon increases the rapidity of conduction and therefore makes possible simultaneous contractions of the segmental muscles (Fig. 10–40*B* and *C*). For example, the single giant fiber of *Myxicola* can be fired at any level along the length of the body and will conduct an impulse in either direction. Conduction along the giant fiber is 12 meters per second, compared with about 0.5 meter per sec-

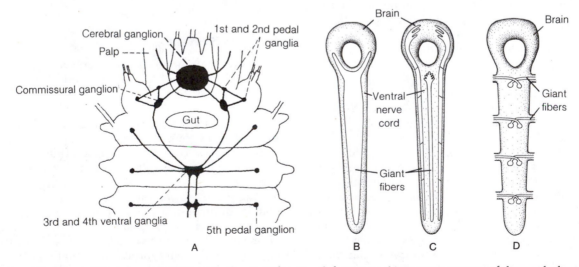

Cerebral ganglion

Palp

1st and 2nd pedal
ganglia

Commissural ganglion

Gut

3rd and 4th ventral ganglia

5th pedal ganglion

A

Brain

Ventral
nerve
cord

Giant
fibers

B

Brain

C

Brain

Giant
fibers

D

Figure 10–40 *A*, Nervous system of *Lepidasthenia*, (After Storch from Fauvel.) Nervous systems of three polychaetes, showing types of giant axons: *B*, *Eunice*, with a single giant axon. *C*, *Nereis*, with medial and lateral giant fibers. *D*, *Thalanessa*, with intrasegmental giant axons. (*B* and *C* after Nicol. *D* after Rhodes from Nicol.)

ond along ordinary longitudinal tracts. Efferent branches from the giant axon run to the longitudinal muscles. The giant fiber is not involved in the conduction of ordinary locomotor impulses, for if the fiber is severed, locomotion is not inhibited; only rapid contraction is blocked.

Sense Organs

The principal specialized sense organs of polychaetes are eyes, nuchal organs, and statocysts.

The eyes, which are best developed in errant polychaetes, are found on the surface of the prostom-

ium in two, three, or four pairs (Figs. 10–3 and 10–11). In general, the polychaete eye is of the retinal-cup variety, the wall of which is composed of rodlike photoreceptors, pigment, and supporting cells (Fig. 10–41).

The eyes of most polychaetes can probably determine only light intensity and light source, but the huge bulging eyes of the pelagic, raptorial Alciopidae are capable of image formation (Fig. 10–41*C*) (see Box 11–1, p. 458).

The radioles of some sabellids bear eye spots, and those of serpulids have dispersed photoreceptors. Fan worms are very sensitive to sudden light

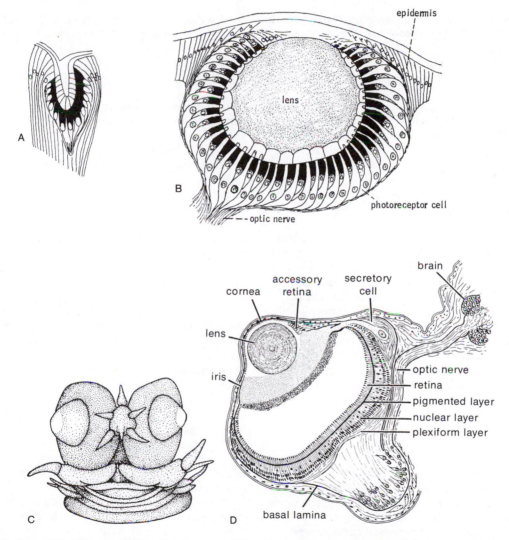

Figure 10–41 *A*, Simple eye of *Mesochaetopterus*. (After Hesse from Fauvel.) *B*, Eye of *Nereis*. (After Hesse from Fauvel.) *C*, Ventral view of the head of the pelagic alciopid *Vanadis tagensis*. Each of the paired eyes consists of the large and small bulges to either side of the central, starlike prostomium. *D*, Section through the eye of *Vanadis formosa*. (*C* after Tebble and *D* after Hesse; both from Hermans, C. O., and Eakin, R. M., 1974: Fine structure of the eyes of an alciopid polychaete *Vanadis tagensis*. Z. Morph. Tiere, 79:245–267.)

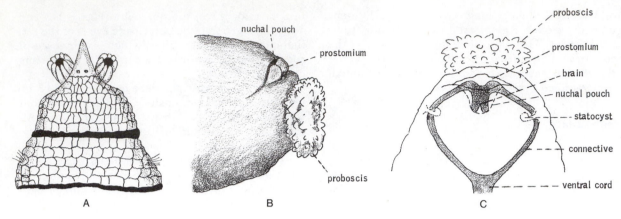

Figure 10–42 *A*, Anterior of *Notomastus latericeus* with everted nuchal organs. (After Rullier from Fauvel.) *B*, Head of *Arenicola* (side view). (After Wells.) *C*, Statocysts and anterior part of nervous system of *Arenicola*. (After Wells.)

reduction and will immediately withdraw into their tubes. This behavior probably represents a protective adaptation against passing predators.

Nuchal organs consist of a pair of ciliated sensory pits, or slits, often eversible, located in the head region of most polychaetes (Fig. 10–42*A* and *B*). These sense organs are important for detecting food and attain their greatest development in the predatory species.

Statocysts are found in many sedentary burrowers or tube dwellers. The statocysts of *Arenicola* are located within the body wall of the head with a canal that opens to the outer, lateral body surface (Fig. 10–42*C*). Wells (1950) reports that the statocysts of *Arenicola* contain spicules, diatom shells, and quartz grains, all covered with a chitinoid material. *Arenicola* always burrows head downward, and if an aquarium containing a worm is tilted 90 degrees, the worm makes a compensating 90-degree turn in burrowing. If the statocysts are destroyed, this compensating ability is lost.

Regeneration

Polychaetes have relatively great powers of regeneration. Tentacles, palps, and even heads ripped off by predators are soon replaced. Such replacement is a common occurrence in burrowers and tube dwellers. In general, the potential for regeneration is somewhat greater in worms with undifferentiated trunks than in those with thoracic and abdominal regions. Cells for regeneration are supplied by the remains of whatever tissues have been lost. Experimental studies indicate that the nervous system plays an important inductive role in regeneration and that the neuroendocrine system is involved in some way. If the nerve cord alone is severed, a new head will form where the cut is made. A lateral secondary head will form if the severed end of a cord, cut just behind the subesophageal ganglion, is pulled through a hole in the lateral body wall.

A good review of some of the studies on polychaete regeneration is presented by Rose (1970) and Highnam and Hill (1977).

Reproduction

Asexual reproduction is known in some polychaetes, including cirratulids, syllids, sabellid fan worms, and spionids; it takes place by budding or division of the body into two parts or into a number of fragments.

As far as we know, most polychaetes reproduce only sexually, and the majority of species are dioecious. Polychaete gonads are not distinct organs but are masses of developing gametes, which develop as projections or swellings of the peritoneum in different parts of some segments. For example, in *Nereis* the ventral septal peritoneum is the site of gamete formation.

In the primitive state most of the segments produce gametes; this is true of many polychaetes. When there are distinct thoracic and abdominal regions, the gonads are usually limited to the abdomen. Among the few hermaphroditic polychaetes, some fan worms have anterior abdominal segments that produce eggs and posterior ones that produce sperm.

The gametes are usually shed into the coelom as gametogonia or primary gametocytes, and maturation takes place in the coelomic fluid. When the worm is mature, the coelom is packed with eggs or sperm; in species in which the body wall is thin or not heavily pigmented, the gravid condition is easily apparent. For example, the abdomen of a ripe male *Pomatoceros* appears white, and that of a female, bright pink or orange, because of the color of the sperm and eggs.

There is considerable diversity in the ways gametes reach the exterior. Many polychaetes, such as the capitellids, have separate coelomoducts, or gonoducts. They develop at the time of sexual maturity, one pair per segment, somewhat like nephridia, and possess a ciliated funnel that receives the eggs or sperm. In many species the coelomoduct joins the nephridium so that the gametes leave the body through the nephridiopores (Fig. 10–37A). The nephridia alone serve as gonoducts in such groups as syllids and spionids, the organs enlarging with gamete development. In many male nereids the sperm may exit through special anal apertures.

The escape of gametes through a rupture in the body wall is perhaps a specialized condition. Rupturing is found among nereids, syllids, and eunicids, all of which become pelagic at sexual maturity, and in nephtyids, which do not. After rupture of the body wall, the adults die.

EPITOKY

Epitoky is a reproductive phenomenon characteristic of many polychaetes and is especially well known in nereids, syllids, and eunicids. Epitoky is the formation of a pelagic reproductive individual, or epitoke, that is adapted for leaving bottom burrows, tubes, and other habitations. Epitokal modifications include changes in the formation of the head, the structure of the parapodia and setae, the size of the segments, and the segmental musculature.

Often the gamete-bearing segments are the most strikingly modified, and the body of the worm appears to be divided into two markedly different regions. For example, the epitoke of *Nereis irrorata* has large eyes and reduced prostomial palps and tentacles (Fig. 10–43). The anterior 15 to 20 trunk segments are not greatly modified, but the remaining segments, forming the epitokal region and packed with gametes, are much enlarged; their parapodia contain fans of long, spatulate swimming setae. In *Palolo* (=*Eunice*) *viridis*, the Samoan palolo worm, the anterior of the worm is unmodified, and the epitokal region consists of a chain of egg-filled segments (Fig. 10–44A).

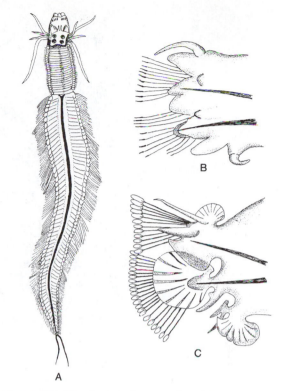

Figure 10–43 *A*, Epitokous male of *Nereis irrorata*. (After Rullier from Fauvel.) *B* and *C*, Parapodia of atoke (*B*) and epitoke (*C*) of *Nereis irrorata* male. (After Fauvel.)

Epitokous individuals arise from an atokous form either by direct transformation of the entire individual, as in nereids, or by transformation and separation of the posterior end from the atoke, as in eunicids and syllids. Syllid epitokes usually arise as buds at the caudal end of the epitoke (Fig. 10–44B).

SWARMING

Usually, epitokous polychaetes swim to the surface during the shedding of the eggs and sperm. This synchronous behavior, known as swarming, congregates sexually mature individuals in a relatively short time and increases the likelihood of fertilization. Experimental evidence indicates that the female produces a pheromone that attracts the male and stimulates shedding of the sperm. The sperm in turn stimulate the shedding of the eggs. The male syllid *Autolytus*, for example, swims in circles around the female, touching her with the antennae and releasing sperm (Fig. 10–44C).

Swarming is induced in *Autolytus edwardsi* by changes in light intensity, and the worms leave the bottom and swim to the surface at dawn and dusk. Swarming often coincides with lunar periods.

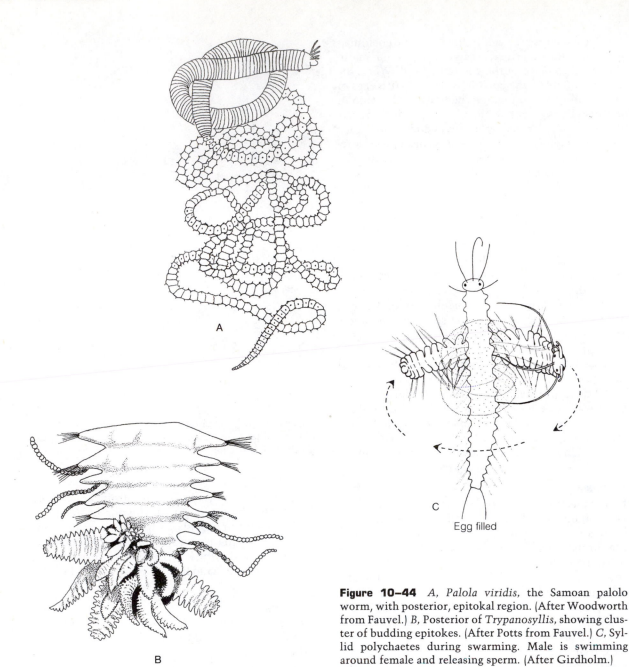

Figure 10–44 *A, Palola viridis*, the Samoan palolo worm, with posterior, epitokal region. (After Woodworth from Fauvel.) *B*, Posterior of *Trypanosyllis*, showing cluster of budding epitokes. (After Potts from Fauvel.) *C*, Syllid polychaetes during swarming. Male is swimming around female and releasing sperm. (After Girdholm.)

Odontosyllis enopla in the West Indies and Bermuda swarms in the summer about 50 to 60 minutes after sunset up to 12 days following a full moon. The worms luminesce when they reach the surface, and when males and females swim around each other, releasing gametes, they create small circles of light. Striking examples of swarming lunar periodicity are displayed by the so-called palolo

worms. The name palolo originally referred to the Samoan species of eunicid *Palolo*, but it is now applied to other species as well. The Samoan palolo worm occupies rock and coral crevices below the low-tide mark and releases epitokes in October or November at the beginning of the last lunar quarter (Caspers, 1984). The natives, who consider the epitokes a great delicacy, eagerly await the pre-

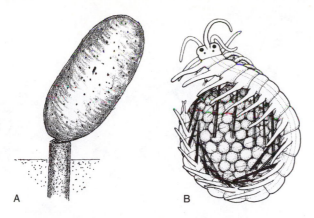

Figure 10–45 *A, Clymenella mucosa* egg mass attached to end of tube. *B, Autolytus* carrying egg mass beneath body. (After Thorson.)

dicted night of swarming and scoop up great numbers of the worms from the ocean surface.

The West Indian palolo, *Eunice schemacephala*, lives in habitats similar to those of the Samoan palolo. The worm is negatively phototactic and emerges from the burrow to feed only at night. Swarming takes place in July near the last quarter of a lunar cycle. At three or four o'clock in the morning during such a period, the worm backs out of its burrow, and the caudal, sexual epitokal region breaks free. The epitoke swims to the surface, where it makes spiral motions. By dawn the ocean surface is covered with sexual bodies, and at the rising of the sun the epitokes rupture. Fertilization immediately follows dehiscence, and acres of eggs may cover the sea. A ciliated larval stage is attained by the next day, and in three days the larvae sink to the bottom.

CONTROL OF REPRODUCTION

Polychaete reproductive events are regulated by hormones. The hormones are neurosecretions produced by the brain or, in the case of syllids, by nervous elements of the proventriculus. In worms such as nereids and syllids, which reproduce only once and then die, the hormone regulates the entire reproductive state, i.e., both the production of gametes and the development of epitokal features. In worms that breed more than once, a hormone is required for gamete development, especially for eggs, and the hormone effect is largely limited to that process.

The precise mechanisms that control swarming and the relationship between swarming and normal control of reproduction are still poorly understood.

The relation of lunar phases to swarming periods differs among species, and swarming even occurs on cloudy nights. This makes any hypothesis based merely on light intensity difficult to prove. For reviews of these problems as well as experimental studies on normal control of reproduction, see Schroeder and Hermans (1975), Baskin (1976), Highnam and Hill (1977) and Olive and Clark (1978).

EGG DEPOSITION

Many polychaetes shed their eggs freely into the sea water, where they become planktonic. Some polychaetes, however, retain the eggs within the tubes or burrows or lay them in mucous masses that are attached to tubes or to other objects. For example, *Clymenella* (Fig. 10–45A), a bamboo worm, produces a small ovoid egg mass that is attached to the chimney of the tube.

Many polychaetes brood their eggs. There are tubicolous species, such as some spionids and serpulids, which brood their eggs within the tubes. Some species of *Spirorbis* brood their eggs in the cavity of the operculum, and *Autolytus* broods its eggs within a secreted sac attached to the ventral surface of the body (Fig. 10–45B). A few species, such as *Nereis limnicola*, brood their eggs within the coelom.

Embryogeny

The polychaete egg is telolecithal with a variable amount of yolk, and cleavage is spiral. A displaced blastocoel is usually present, but a stereoblastula develops in *Nereis, Capitella*, and others. Gastrulation takes place by invagination, epiboly, or both.

THE TROCHOPHORE

After gastrulation, the embryo rapidly develops into a top-shaped trochophore larva (Figs. 10–46 and 10–47; Box 10–3).

The greatest development of larval structures is attained in planktotrophic trochophores, those that feed on plankton (e.g., *Owenia, Polygordius*, phyllodocids, serpulid fan worms). The trochophores of many species, however, are lecithotrophic, i.e., yolky and nonfeeding (e.g., nereids and eunicids), and their short larval existence is spent near the bottom.

METAMORPHOSIS

Additional development takes place as the trochophore metamorphoses into the adult body form (Fig. 10–47). The most conspicuous feature of metamorphosis is the gradual lengthening of the

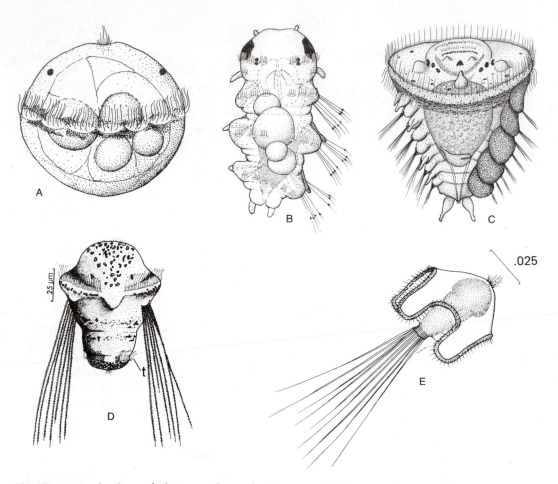

Figure 10–46 *A*, Trochophore of *Platynereis bicanaliculata. B*, Later three-setiger stage of *Platynereis. C*, Late meta-trochophore of the scale worm *Halosydna brevisetosa*. (*A–C* from Blake, J. D., 1975: The larval development of Poly-chaeta from the northern California coast. III. Eighteen species of Errantia. Ophelia, *14*:23–84.) *D*, Larva of the sabel-lariid *Phragmatopoma* (t-telotroch). (From Eckelbarger, K. J., 1976: Larval development and population aspects of the reef-building polychaete *Phragmatopoma lapidosa* from the east coast of Florida. Bull. Mar. Sci., 26(2):117–132.) *E*, Larva, called a mitraria, of *Owenia*. (From Smith, D. L., 1977: A Guide to Marine Coastal Plankton and Marine Inver-tebrate Larvae. Kendall/Hunt Publishing Co., Dubuque, Iowa.)

growth zone—the region between the mouth and the telotroch—as trunk segments form and de-velop (Fig. 10–47). The segments develop from an-terior to posterior, and the germinal region re-mains just in front of the terminal pygidium. Thus, in adult polychaetes the oldest segments are those closest to the head of the worm. In the prototrochal region, which originally formed the major part of the body of the trochophore, the cells of the apical plate form the prostomium and the brain.

Metamorphosis may result in the immediate termination of a planktonic existence, but more often the elongated, metamorphosing larvae re-main planktonic for varying lengths of time. The

metamorphosing stages of spionids, sabellariids, and oweniids even possess greatly enlarged, erec-tile, anterior setae that serve as flotation or protec-tive devices.

In many polychaetes the trochophore stage is passed in the egg prior to hatching, which occurs at various times during advanced development. In such species metamorphosis is more direct, since larval structures are never greatly developed to begin with. There may still be a free-swimming, posttrochophoral, larval stage. For example, in *Au-tolytus* an elongated larva breaks free from the brood sac of the mother. On the other hand, *Cly-menella mucosa* and *Scoloplos armiger* have no

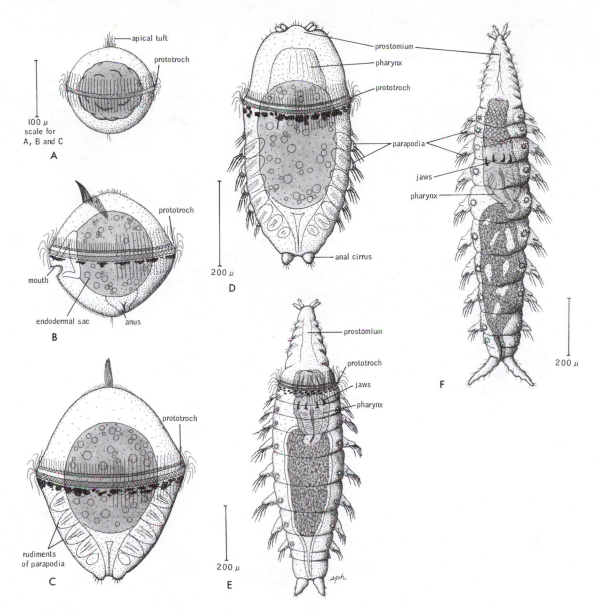

Figure 10–47 Larval stages of *Glycera convoluta*. *A*, Early trochophore (15 hours). *B*, Later trochophore (ten days). *C*, Young metatrochophore (four weeks). *D*, Metatrochophore at seven weeks. Although still a swimming stage, it frequently comes to rest on the bottom. *E*, Postlarva at eight weeks. Metamorphosis follows metatrochophore illustrated in *D*. Larva becomes benthic and raptorial. *F*, Young worm at two months. Compare with Figure 10–15. (All after Cazaux, C., 1967: Vie et Milieu, *18*:559–571.)

planktonic stages and assume the adult mode of existence on emerging from the jelly egg case.

In general, the developmental patterns of the relatively small number of polychaetes that have been studied appear to fall into three categories (Fauchald, 1983). Annual species, those that live only one or two years and spawn only once, pro-

duce a large number of relatively small eggs. They have well-developed feeding larvae that are planktonic for a week or more. Perennial species, which live and breed more than one year, produce a small number of large, yolky eggs and nonfeeding, benthic larvae. Multiannual species, those with such short life spans that several generations can be

BOX 10–3 TROCHOPHORE LARVAE

The characteristic larva of polychaetes, mollusks, and many other marine protostomes—animals with spiral cleavage—is a trochophore. A typical trochophore larva is top shaped with a tuft of cilia at the apical end (*A* and *B*). The distinguishing feature is a conspicuous girdle of cilia, called the prototroch, which rings the body about one third to one half of the distance from the apical tuft. The gut is a complete tube, and the mouth opens just below the prototroch. In the trochophore of many polychaetes and some other groups, a second girdle of cilia, called the metatroch (*A*), develops below the mouth, and a third, the telotroch, forms just before the anus at the posterior end.

Internally, the old blastocoel remains as a large cavity lying between the gut and the outer ectoderm (*B*). Larval muscle bands cross the cavity, and within are a pair of protonephridia, one on either side of the gut. There is a pair of ventrolateral, mesodermal bands from which adult mesodermal structures are derived. The cells in the vicinity of the apical tuft contain the primordium of the cerebral ganglia.

The fully developed trochophore larva can be divided into three regions: the prototrochal region, consisting of the apical plate, the prototroch and the mouth region; the pygidium, consisting of the telotroch and the anal region behind it; and the growth zone, which includes all of the larva between the mouth and the telotroch. In polychaetes the growth region eventually forms all of the trunk segments of the body.

Trochophores may feed (planktotrophic) and have long planktonic lives; or they may be yolk laden and nonfeeding (lecithotrophic) with a short planktonic existence. The prototroch is the swimming organ and, in trochophores that feed, also collects suspended food particles. In feeding trochophores a food groove lies between the preoral prototroch and the postoral metatroch. The longer cilia of the preoral band drive water and suspended food particles backward. The shorter cilia of the postoral band beat in

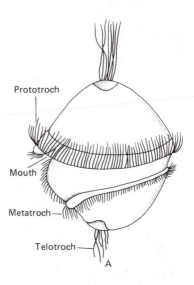

produced in one year, produce numerous small batches of large, yolky eggs that hatch nonfeeding, benthic larvae.

Population Biology

Burrowing and tubicolous polychaetes commonly occur in enormous numbers in the ocean floor and compose a major part of the soft bottom infauna. A study in Tampa Bay, Florida, reported the average density of polychaetes to be 13,425 individuals per square meter, i.e., living in the sediment beneath that area. They belonged to some 37 species (Santos and Simon, 1974). On the upper continental slope and the deep ocean floor polychaetes compose 40 to 80 per cent of the infauna.

In general, such populations do not appear to be limited by the resources available, at least not in shallow water. Predation and other pressures usually prevent annelidan, molluscan, and other infaunal populations from ever reaching the carrying capacity of the habitat. When areas in the York River estuary of the Chesapeake Bay were protected from fish and crabs by means of wire cages, over half of the species in the polychaete population increased from two to many times their numbers in unprotected conditions (Virnstein, 1977).

SYSTEMATIC RÉSUMÉ OF CLASS POLYCHAETA

Polychaete families are in many ways more distinctive and useful taxa than are the orders. The following list includes the more common families to which references have been made in the preceding the opposite direction, and suspended particles are driven into the food groove between the two bands (*C*). The short cilia of the food groove transport the particles to the mouth. (*A*, Trochophore of *Polygordius*, after Dawydoff. *B*, An annelid trochophore, after Shearer from Hyman. *C*, Two-band suspension feeding system from Strathmann, R. R., Jahn, T. L., and Fonseca, J. R., 1972: Suspension feeding by marine invertebrate larvae: clearance of particles by ciliated bands of rotifer, pluteus and trochophore. Biol. Bull., *142*:505–519.)

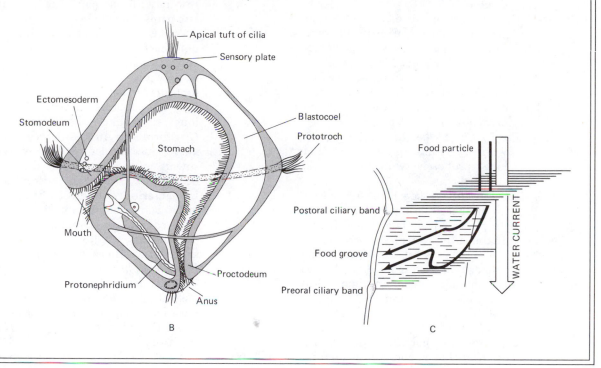

discussion. Lack of space permits mention of only the distinguishing characteristics. See Fauchald (1977) or Parker (1980) for complete definitions of orders and families.

Subclass Errantia. Polychaetes with most of the following characteristics: segments numerous and similar; parapodia well developed, with acicula and setae; head with sensory structures; pharynx with jaws or teeth. Swimming, crawling, burrowing, and tube-dwelling worms.

> Family Aphroditidae. Sea mice. Long setae forming a feltlike covering of the dorsal surface. *Aphrodita, Laetmonice.*

> Families Polynoidae and Sigalionidae. Scale worms. Dorsal surface bears elytra. Chiefly crawling polychaetes. *Lepidonotus, Harmothoe, Polynoe, Lepidasthenia, Lagisca, Iphinoe, Arctonoe, Euthalenessa, Sigalion, Thalanessa.*

> Family Phyllodocidae. Uniramous parapodia with flattened, leaflike cirri. Crawling polychaetes. *Notophyllum, Phyllodoce.*

> Family Amphinomidae. Crawling polychaetes with brittle, poisonous setae. Fireworms. *Hermodice, Amphinome.*

> Family Alciopidae. Planktonic worms with transparent bodies and two large eyes. *Alciopa, Vanadis.*

> Family Tomopteridae. Planktonic worms with membranous pinnules in place of setae. *Tomopteris.*

Family Hesionidae. Crawling polychaetes with well-developed prostomial sense organs and long, dorsal parapodial cirri. *Podarke.*

Family Syllidae. Small crawling worms with long, delicate bodies and uniramous parapodia. *Syllis, Odontosyllis, Pionosyllis, Autolytus, Procerastea, Trypanosyllis, Brania (= Grubia).*

Family Nereidae. Four eyes and four pairs of peristomal cirri. Pharynx contains a pair of jaws. Large crawling species. *Nereis, Platynereis.*

Family Nephtyidae. Rapid-crawling worms with well-developed prostomial sense organs located between parapodial rami. *Nephtys.*

Family Glyceridae. Errant burrowing worms. Conical prostomium and long proboscis with four jaws. *Glycera.*

Superfamily Eunicacea. Elongated worms with proboscis armature of at least two pieces. This large superfamily contains a number of ecologically diverse but closely related families. The Eunicidae and Onuphidae, with eyes and antennae, contain many tubicolous species. *Eunice, Marphysa, Palola, Lysidice; Onuphis, Diopatra, Hyalinoecia, Nothria.* The Lysaretidae are crawling polychaetes. *Oenone.* The Arabellidae and Lumbrineridae are threadlike, errant burrowers with reduced prostomial sensory structures. *Arabella, Drilonereis, Labrorostratus; Lumbrinereis.*

Family Histriobdellidae. Ectoparasitic. *Histriobdella.*

Family Ichthyotomidae. Ectoparasitic. *Ichthyotomus.*

Family Myzostomidae. Greatly flattened commensals and parasites of echinoderms, particularly crinoids. *Myzostoma.*

Subclass Sedentaria. Polychaetes with most of the following characteristics: body regions commonly differentiated, parapodia reduced, without acicula or compound setae; prostomium without sensory appendages but head commonly provided with palps, tentacles, and other structures for feeding; no teeth or jaws.

Family Orbiniidae. Sedentary burrowers with conical or globular prostomium, without appendages. *Orbinia, Scoloplos.*

Family Spionidae. Tubicolous polychaetes with two long prostomial palps. *Spio, Scolelepis (= Nerine), Polydora, Mala.*

Family Magelonidae. Sedentary burrowers with one pair of long papillate palps. *Magelona.*

Family Chaetopteridae. Tubicolous polychaetes with one pair of long palps. *Chaetopterus, Phyllochaetopterus, Spiochaetopterus, Mesochaetopterus (= Ranzania).*

Family Cirratulidae. Segments bear long, threadlike gills. *Cirratulus, Cirriformia (= Audouinia).*

Family Flabelligeridae. Sedentary burrowers with cephalic gills that can be retracted with the prostomium into anterior segments. *Flabelligera.*

Family Opheliidae. Errant burrowers with conical prostomium. Number of segments, usually relatively low, are fixed within each species. *Ophelia, Polyophthalmus, Euzonus (= Thoracophelia).*

Family Capitellidae. Errant burrowers with conical prostomium and long body. *Dasybranchus, Capitella, Notomastus.*

Family Arenicolidae. Sedentary burrowers without head appendages. Lugworms. *Arenicola.*

Family Maldanidae. Tubicolous polychaetes with small prostomium fused to peristomium, without head appendages. The bamboo worms. *Clymenella, Maldane, Axiothella.*

Family Oweniidae. Tubicolous species without prostomial appendages, or with foliaceous prostomial crown. *Owenia.*

Family Sabellariidae. Tubicolous polychaetes. Setae of anteriorly projecting segments are modified to form an operculum for closing tube. *Sabellaria.*

Family Pectinariidae, or Amphictenidae. Tubicolous polychaetes constructing conical tubes. Head with heavy, forward-directed, golden setae used as an operculum and for digging. *Pectinaria* (=*Cistenides*).

Family Ampharetidae. Tubicolous polychaetes with retractile buccal tentacles. *Amphicteis, Amparete.*

Family Terebellidae. Prostomium bears numerous long, nonretractile tentacles. Tubicolous or sedentary burrowers. *Terebella, Amphitrite, Polymnia.*

Family Sabellidae. The fan worms or feather-duster worms. Noncalcareous tubes. *Branchiomma* (=*Dasychone*), *Sabella, Myxicola, Potamilla, Manayunkia, Megalomma, Spirographis.*

Family Serpulidae. The fan worms or feather-duster worms. Tubes calcareous. The coiled spirorbids are placed within a separate family (Spirorbidae) by some authorities. *Serpula, Apomatus, Spirorbis, Pomatoceros, Filograna* (=*Salmacina*), *Hydroides* (=*Eupomatus*).

Archiannelida. Most modern authorities on the annelids agree that the old phylum or class Archiannelida represents an assortment of modified annelids, such as *Polygordius* (Fig. 10–24C), *Protodrilus, Nerilla,* and *Dinophilus,* that properly belong to the Polychaeta. Most are members of the interstitial fauna (Fauchald, 1975).

SUMMARY

1 The evolution of the class Polychaeta from the ancestral burrowing annelids is perhaps correlated with a shift to a crawling, surface existence. Various lines then diverged to invade other habitats and to assume other modes of existence, including burrowing in soft bottoms.

2 Surface-dwelling, errant polychaetes possess well-developed parapodia and heads (prostomium) with sense organs. They crawl with the parapodia, using them as leglike appendages. Most are predacious, but some are herbivorous or scavengers. They typically possess an eversible pharynx equipped with jaws.

3 Gallery-dwelling polychaetes show some convergence with earthworms. The prostomium is usually small and more or less conical with poorly developed sense organs. The parapodia are smaller than those of surface dwellers and provide anchor-

age in peristaltic movement through the galleries. Gallery dwellers are predators or deposit feeders.

4 The more sedentary burrowers, which live in simple vertical or U-shaped excavations, move by peristaltic contractions and possess parapodial ridges with hooked setae for gripping the mucus-lined burrow walls. Those species that are nonselective deposit feeders have a small prostomium without conspicuous sense organs; those that are selective deposit feeders usually possess head appendages, such as tentacles or long palps, which are used in feeding.

5 Tubicolous polychaetes live in tubes composed of secreted materials or of sand grains or shell fragments cemented together. The majority are selective or nonselective deposit feeders or filter feeders and are adapted much like sedentary burrowers. The few that are predatory are similar to errant surface dwellers.

6 Most large polychaetes possess gills (thin-walled evaginations with an interior vascular supply), or some part of the parapodium is especially modified as a gas exchange surface.

7 Internal transport is provided by a more or less closed blood-vascular system, by coelomic fluid, or by a combination. Gas transport frequently involves respiratory pigments, of which hemoglobin is the most common, but some polychaetes possess chlorocruorin or hemerythrin. Blood-vascular hemoglobin and chlorocruorin are always extracellular and large molecules; blood-vascular hemerythrin and coelomic hemoglobin are always intracellular and small molecules.

8 Most polychaetes possess paired, segmental metanephridia, in which the nephrostome opens into the coelomic compartment that is anterior to the one housing the tubule. Some polychaetes possess protonephridia, which may be primitive for the phylum.

9 The sexes are separate in most polychaetes. The gametes are produced by the peritoneum. After maturing in the coelom, they exit by coelomoducts, coelomoducts joined to nephridia, nephridia alone, or rupture of the body wall. In primitive polychaetes the gametes are produced by the peritoneum of most segments.

10 Copulation is rare, and synchronous emission of sperm and eggs is important. Epitoky and swarming bring a dispersed benthic population together for a brief pelagic existence, when gametes are shed and the likelihood of fertilization is increased.

11 A trochophore larva is the basic larval stage of polychaetes.

Class Oligochaeta

The class Oligochaeta contains some 3100 species of annelids, including the familiar earthworms and many species that live in fresh water. Some freshwater forms burrow in bottom mud and silt; others live among submerged vegetation. About 200 marine species have been described, even some that inhabit sediments of the deep sea. Oligochaetes approximate the polychaetes in size; however, the giant earthworms of Australia and other parts of the world may exceed 3 meters in length (Fig. 10–48). In general, aquatic species are smaller than earthworms.

Oligochaetes are believed to have evolved directly from the burrowing ancestral marine annelids, independently of the polychaetes (Brinkhurst, 1982). The first oligochaetes were probably burrowers in freshwater sediment. They may have given rise in one direction to the strictly freshwater species that invaded loose bottom debris and in another direction to the earthworms that invaded successively drier sediments.

External Anatomy

Metamerism is well developed, parapodia are absent, and the prostomium is usually a small, rounded lobe or a small cone without sensory appendages (Figs. 10–49 and 10–52). In a few genera, such as *Stylaria* (Fig. 10–49C), the prostomium is drawn out into a tentacle.

Oligochaetes have no parapodia, but with few exceptions, they have setae. Although not as diverse as those of polychaetes, oligochaete setae commonly have tips that are simple, bifid, pectinate, or in other ways different from the shaft (Fig. 10–49B). Genital setae may be more complex.

In general, the longer setae are characteristic of aquatic species (Fig. 10–49C); the setae of earthworms project only a short distance beyond the integument and are commonly sigmoid. On each side of a segment there are setal sacs, in which the setae are secreted and from which they emerge as groups or bundles. Two of the groups are ventral, and two are ventrolateral or dorsolateral. The number of setae per bundle varies from 1 to 25 (Fig. 10–49C). In any case, they are generally less numerous in these worms than in polychaetes, hence the origin of the name *Oligochaeta*—"few setae."

In most earthworms, such as *Lumbricus*, and in some aquatic families, the setae are limited to eight, with two setae forming each group (Fig. 10–49A). Attached to the base of each seta are protractor and retractor muscles that allow the seta to be extended or withdrawn (Fig. 10–49B).

In mature oligochaetes certain adjacent segments in the anterior half of the body are thickened and swollen by glands that secrete mucus for copulation and also secrete the cocoon. The glandular area of these segments, collectively called the clitellum, partially or completely covers the segments and often forms a conspicuous girdle around the

Figure 10–48 An Australian giant earthworm. (Courtesy of Globe Photos.)

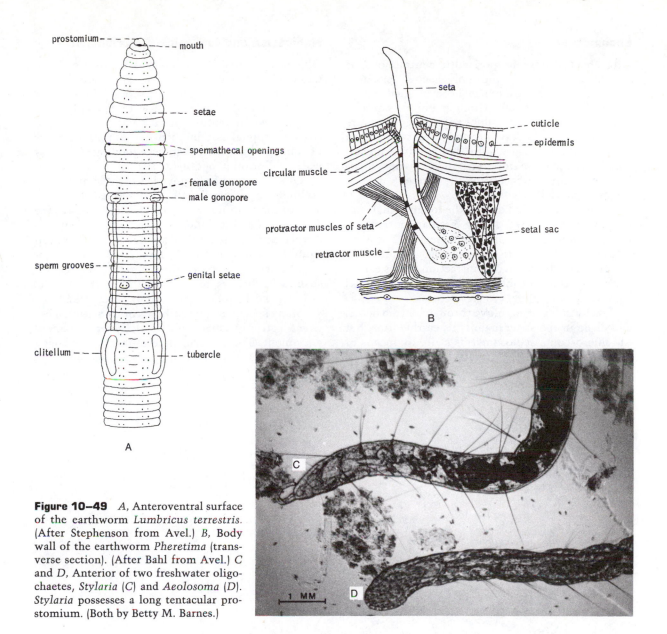

Figure 10–49 *A*, Anteroventral surface of the earthworm *Lumbricus terrestris*. (After Stephenson from Avel.) *B*, Body wall of the earthworm *Pheretima* (transverse section). (After Bahl from Avel.) *C* and *D*, Anterior of two freshwater oligochaetes, *Stylaria* (*C*) and *Aeolosoma* (*D*). *Stylaria* possesses a long tentacular prostomium. (Both by Betty M. Barnes.)

body (Fig. 10–49*A*). The clitellum is discussed in more detail in the section on reproduction.

Body Wall and Coelom

The structure and histology of the oligochaete body wall, especially in terrestrial species, is essentially like that of burrowing polychaetes. A thin cuticle overlies an epidermal layer, which contains mucus-secreting gland cells. Circular muscles are well developed, and the septa partitioning the coelom are relatively complete. Earthworms, which have the best developed septa, may possess sphincters around septal perforations to control the flow of coelomic fluid from one segment to another.

In most earthworms each coelomic compartment, except at the extremities, is connected to the outside by a middorsal pore located in the intersegmental furrows and provided with a sphincter. These pores exude coelomic fluid, which aids in keeping the integument moist. When disturbed, some giant earthworms squirt fluid several centimeters.

Locomotion

Oligochaetes move by peristaltic contractions, as described for burrowing polychaetes (p. 263). Earthworm locomotion has been studied extensively (Seymour, 1969). Circular muscle contraction and the consequent elongation of segments are most important in crawling and always generate a coelomic fluid pressure pulse. Longitudinal muscle contraction is more important in burrowing, dilating the burrow, or anchoring the segments against the burrow wall.

Setae are extended during the longitudinal contraction and retracted during circular contraction. Each segment moves forward in steps of 2 to 3 cm at the rate of seven to ten steps per minute (Fig. 10–1C). The direction of contraction waves can be reversed, thus enabling the worm to crawl backward.

Freshwater species move through bottom debris and algae in the same manner as earthworms, but the microscopic aeolosomatids swim by means of ciliated prostomia (Fig. 10–50).

Habitation and Ecological Distribution

Oligochaetes live in all types of freshwater habitats, where they usually burrow in bottom debris. Only a small number construct tubes. Oligochaetes are most abundant where the water is shallow, although several families have benthic representatives in deep lakes. Abundance of different species of aquatic oligochaetes can be a good indication of water pollution (Goodnight, 1973).

Oligochaetes have reinvaded the sea, and some 200 marine species have been described. Most belong to the families Enchytraeidae and Tubificidae and are chiefly inhabitants of the supratidal and intertidal zones, but subtidal species are known even from abyssal and hadal depths. Marine oligochaetes are members of the interstitial fauna, are shallow burrowers, or live beneath intertidal rocks or in algal drift (Giere and Pfannkuche, 1982).

Many species are amphibious or transitional between a strictly aquatic and a strictly terrestrial environment. These worms live in marshy or boggy land and on the margins of ponds and streams.

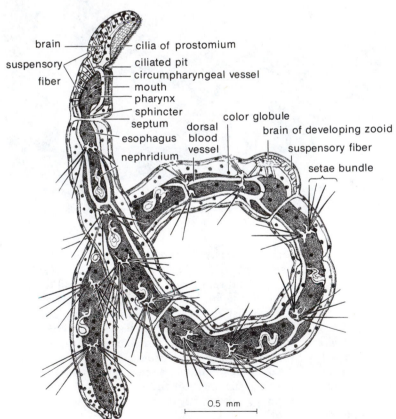

Figure 10–50 The freshwater oligochaete *Aeolosoma*. (Courtesy of R. Singer.)

There are ten families of earthworms, of which four contain large numbers of species—Glossoscolecidae, Lumbricidae, Megascolecidae, Moniligastridae. All are burrowers and are found everywhere except in deserts. They often occur in tremendous numbers. As many as 8000 enchytraeids and 700 lumbricids have been reported from 1 square meter of meadow soil.

Soils containing considerable organic matter, or at least a layer of humus on the surface, maintain the largest fauna of worms, but other soil factors are important to the distribution of terrestrial species. Acid soils are favorable habitats for most earthworms.

A cross section through soil reveals a distinct vertical stratification of worms. The tunnels of larger species, such as *Lumbricus terrestris*, range from the surface to several meters deep, depending on the nature of the soil. Young worms and small species are restricted to the few inches of upper humus; others have a wider distribution but are still limited to the upper level of the soil that contains some organic matter (Martin, 1982).

The terrestrial oligochaete constructs its burrow by forcing its anterior end through crevices and by swallowing soil. The egested material and mucus are plastered against the burrow wall, forming a distinct lining. Some egested material is removed from the burrow as castings. The gut turnover time in the lumbricid *Allolobophora rosea* is 1 to 2.5 hours (Bolton and Phillipson, 1976). The burrows may be complex, with two openings and horizontal and vertical ramifications. *Lumbricus terrestris* always plugs its burrow with debris pulled into the hole. It is not yet clear why large numbers of *Lumbricus terrestris* leave their burrows after heavy rains, for high mortality certainly results. Moreover, many species of earthworms, including *Lumbricus terrestris*, can survive in submerged ground for months.

The activities of earthworms have a beneficial effect on the soil. The extensive burrows increase soil drainage and aeration, but more important are the mixing and churning of the soil. Deeper soil is brought to the surface as castings, and organic material is moved to lower levels. Some tropical species produce enormous castings. For example, the tower-like castings of *Hyperoidrilus africanus* may reach 8 cm in height and 2 cm in diameter.

Earthworms' ability to churn soil can be demonstrated in a container filled halfway with sand and halfway with potting soil. Five worms will thoroughly mix 500 cc of sand with 500 cc of soil in several months.

In the tropics a number of arboreal species have been reported. Such terrestrial forms live in accumulated humus and detritus in leaf axils and branches of trees. Aquatic species inhabit the water reservoirs of bromeliad epiphytes, living on the trunks and branches of tropical trees.

Members of the major aquatic families occur throughout the world wherever suitable habitats exist. Many terrestrial species, however, have very limited geographical distribution. Humans have certainly been an important agent in the spread of earthworms. For example, the earthworm fauna of the larger Chilian cities consists solely of European species, which have displaced the endemic forms.

Nutrition

The majority of oligochaete species, both aquatic and terrestrial, are scavengers and feed on dead organic matter, particularly vegetation. Earthworms feed on decomposing matter at the surface and may pull leaves into the burrow. They also utilize organic material obtained from mud or soil that is ingested in the course of burrowing. The food source and feeding habits of earthworms are related to the species zonation described in the previous section.

Fine detritus, algae, and other microorganisms are important food sources for many tiny, freshwater species. The common, minute *Aeolosoma* collects detritus with its prostomium (Fig. 10–50). The ciliated ventral surface of the prostomium is placed against the substratum, and the center is elevated by muscular contraction. The partial vacuum dislodges particles, which are then swept into the mouth by cilia. Members of the genus *Chaetogaster*, little oligochaetes that are commensals on freshwater snails, are raptorial and catch amebas, ciliates, rotifers, and trematode larvae by a sucking action of the pharynx.

The digestive tract is straight and relatively simple (Fig. 10–52). The mouth, located beneath the prostomium, opens into a small buccal cavity, which in turn opens to a more spacious pharynx. The dorsal wall of the pharyngeal chamber is muscular and glandular and forms a bulb or pad, which is the principal ingestive organ. In aquatic forms the pharynx is everted and the mucus-covered muscular disc collects particles on an adhesive pad (Fig. 10–51). In earthworms the pharynx acts as a pump. Pharyngeal glands produce a salivary secretion containing mucus and enzymes.

The pharynx opens into a narrow, tubular esophagus, which may be modified at different levels to form a gizzard or, in lumbricid earthworms,

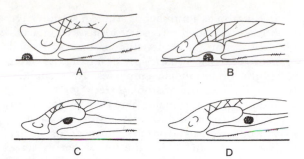

Figure 10–51 Mechanism of ingestion in *Aulophorus carteri*, which possesses a dorsal, padlike pharynx. (After Marcus and Avel.)

a crop. In some forms there are two to ten gizzards, each occupying a separate segment. The gizzard, which is used for grinding food particles, is lined with cuticle and is very muscular. The crop is thin walled and acts as a storage chamber.

A characteristic feature of the oligochaete gut is the presence of calciferous glands in certain parts of the esophageal wall. When highly developed, the glandular region becomes completely separated from the esophageal lumen and may appear externally as lateral or dorsal swellings (Fig. 10–52). The calciferous glands are involved in ionic regulation rather than digestion. They function in ridding the body of excess calcium taken up from food. The calcium is excreted into the esophagus as calcite, which is not absorbed in transit through the intestine.

The intestine forms the remainder of the digestive tract and extends as a straight tube through all but the anterior quarter of the body. The anterior half of the intestine is the principal site of secretion and digestion, and the posterior half is primarily absorptive. In addition to the usual classes of digestive enzymes, the intestinal epithelium of earthworms, at least, secretes cellulase and chitinase. The absorbed food materials are passed to blood sinuses that lie between the mucosal epithelium and the intestinal muscles. The surface area of the intestine is increased in many earthworms by a ridge or fold, called a typhlosole, which projects internally from the middorsal walls.

Surrounding the intestine and investing the dorsal vessel of oligochaetes is a layer of yellowish cells, called chloragogen cells, which play a vital role in intermediary metabolism, similar to the role of the liver in vertebrates. Chloragogen tissue is the chief center of glycogen and fat synthesis and storage. Deamination of proteins, the formation of ammonia, and the synthesis of urea also take place in these cells. In terrestrial species silicates obtained from food material and the soil are removed from the body and deposited in the chloragogen cells as waste concretions.

Gas Exchange

Gas exchange in amost all oligochaetes, both aquatic and terrestrial, takes place by the diffusion of gases through the general body integument, which in the larger species contains a capillary network within the outer epidermal layer.

True gills occur in only a few oligochaetes. Species of the aquatic genera *Dero* (Fig. 10–53*B*) and *Aulophorus* have a circle of finger-like gills at the posterior of the body. A tubificid, *Branchiura* (Fig. 10–53*A*), has filamentous gills located dorsally and ventrally in the posterior quarter of the body.

The larger oligochaetes usually have hemoglobin dissolved in the plasma. The hemoglobin of *Lumbricus* transports 15 to 20 percent of the oxygen utilized under ordinary burrow conditions, where the partial pressure of oxygen is about the same as that in the atmosphere above ground. When the partial pressure drops, the hemoglobin compensates by increasing its carrying capacity (Weber, 1978).

Many aquatic oligochaetes tolerate relatively low oxygen levels and, for a short period, even a complete lack of oxygen. Members of the family Tubificidae, which live in stagnant mud and lake bottoms, are notable examples. There are members of this family, such as *Tubifex tubifex*, that die from long exposure to ordinary oxygen tensions. *Tubifex* ventilates in stagnant water by exposing its posterior end out of the mud and waving it about (Fig. 10–53*C*).

Internal Transport

The circulatory system of oligochaetes (Fig. 10–52) is basically like that of polychaetes. Branches from the segmental vessels send blood into capillaries in the integument (best developed in earthworms) and supply the various segmental organs. The vessels are lined with a reversed myoendothelium (i.e., the basal membrane is directed toward the lumen, as in polychaetes (Jamieson, 1981). The dorsal vessel is contractile and is the principal means by which the blood is propelled. The vessels in oligochaetes commonly refered to as hearts are certain

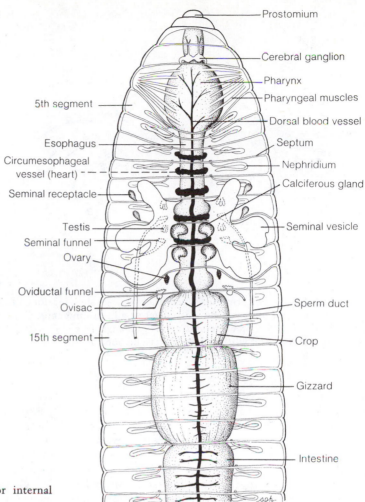

Prostomium

Cerebral ganglion

Pharynx

Pharyngeal muscles

Dorsal blood vessel

Septum

Nephridium

Calciferous gland

Seminal vesicle

Sperm duct

Crop

Gizzard

Intestine

5th segment

Esophagus

Circumesophageal vessel (heart)

Seminal receptacle

Testis

Seminal funnel

Ovary

Oviductal funnel

Ovisac

15th segment

Figure 10–52 Dorsal view of the anterior internal structures of the earthworm *Lumbricus*.

anterior, commissural vessels that are conspicuously contractile and function as accessory organs for blood propulsion. The number of such hearts varies. Five are present in *Lumbricus* and surround the esophagus (Fig. 10–52). Only one pair of hearts is present in *Tubifex*, and this pair is circumintestinal. Valves in the form of endothelial folds are present in the hearts and may also be found in the dorsal vessel at junctions containing segmental vessels.

Excretion, Water Balance, and Quiescence

The adult oligochaete excretory organs are metanephridia, and typically, there is one pair per segment except at the extreme anterior and posterior ends. In the segment following the nephrostome, the tubule is greatly coiled, and in some species, such as *Lumbricus*, there are several separate groups of loops or coils (Fig. 10–54). Before the nephridial tubule opens to the outside, it is sometimes dilated to form a bladder. The nephridiopores are usually located on the ventrolateral surfaces of each segment.

In contrast to the majority of oligochaetes, which possess in each segment a single, typical pair of nephridia called holonephridia, many earthworms of the families Megascolecidae and Glossoscolecidae are peculiar in possessing additional nephridia, which are multiple or branched. Either typical or modified nephridia may open to the outside through nephridiopores, or they may open

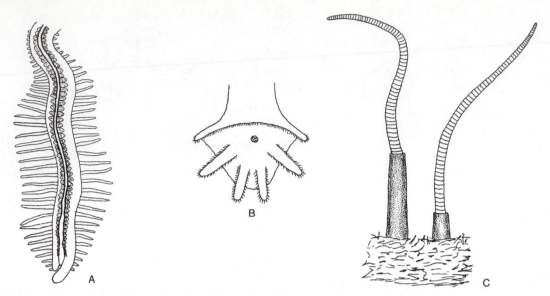

Figure 10–53 *A,* Posterior of *Branchiura sowerbyi,* showing dorsal and ventral filamentous gills. (After Beddard from Avel.) *B,* Posterior of *Dero,* showing circlet of gills around anus. *C,* Posterior of body of *Tubifex* projected from tube and waved about in water, facilitating gas exchange. (*B* and *C* after Pennak, R. W., 1978: Freshwater Invertebrates of the United States, 2nd Edition. John Wiley and Sons, N.Y.)

into various parts of the digestive tract, in which case they are termed enteronephric. A single worm may possess a number of different types of these nephridia, each being restricted to certain parts of the body.

Earthworms excrete urea, but they are less perfectly ureotelic than are other terrestrial animals. Although urea is present in the urine of *Lumbricus* and other earthworms and although the level of urea depends on the condition of the worm and the environmental situation, ammonia remains an important excretory product.

Salt and water balance, which is of particular importance in freshwater and terrestrial environments, is regulated in part by the nephridia (Fig. 10–54B). The urine of both terrestrial and freshwater species is hypoosmotic, and considerable reabsorption of salts must take place as fluid passes through the nephridial tubule. Some salts are also actively picked up by the skin.

In the terrestrial earthworms water absorption and loss occur largely through the skin. Under normal conditions of adequate water supply, the nephridia excrete a copious hypoosmotic urine. It is not certain whether reabsorption by the ordinary nephridia is of importance in water conservation, but the enteronephric nephridia do appear to represent an adaptation for the retention of water. By passing the urine into the digestive tract, much of the remaining water can be reabsorbed as it goes through the intestine. Worms with enteronephric systems can tolerate much drier soils or do not have to burrow so deeply during dry periods.

A few aquatic oligochaete species are capable of encystment during unfavorable environmental conditions. The worm secretes a tough, mucous covering that forms the cyst wall. Some species form summer cysts for protection against desiccation; others form winter cysts when the water temperature becomes low.

During dry seasons or during the winter, earthworms migrate to deeper levels of the soil, down to 3 meters in the case of certain Indian species. After moving to deeper levels, an earthworm often undergoes a period of quiescence and in dry periods may lose as much as 70 per cent of its water. Balance is restored and activity resumed as soon as water is again available.

Nervous System

In most oligochaetes the two ventral nerve cords are fused and located inside the muscle layers of the body wall. The oligochaete brain has shifted

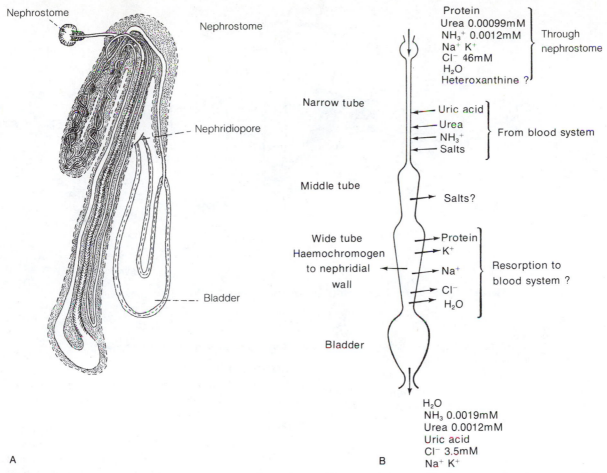

Figure 10–54 *A*, Nephridium of *Lumbricus*. (After Maziarski from Avel.) *B*, Possible functions of various parts of the *Lumbricus* nephridium. (Modified slightly from Laverack, M. S., 1963: The Physiology of Earthworms. Pergamon Press, Oxford. p. 67.)

posteriorly and in lumbricids lies in the third segment, above the anterior margin of the pharynx (Fig. 10–55).

Like polychaetes, oligochaetes have giant nerve fibers. The earthworms possess five giant nerve fibers. Three are quite large and are grouped at the middorsal side of the ventral nerve cord. The other two are less conspicuous; they are situated midventrally and are rather widely separated. The middorsal fiber is fired by anterior stimulation, and the two dorsolateral fibers are fired by posterior stimulation. They make connection in each ganglion with giant motor neurons that innervate the longitudinal muscles. Neuronal organization within the ganglia of annelids and other protostomes is described in Box 10–4.

The basic peristaltic locomotor rhythm of body wall contraction appears to arise from the ventral nerve cord, but contraction can be initiated indirectly by way of reflex arcs involving sensory neurons in the body wall. A wave of longitudinal muscular contraction exerts a pull on the following segments. This pull apparently stimulates the sensory neurons of those following segments, and a reflex action is initiated that causes the contraction of the circular muscle layer and the elongation of the segments. This traction-stimulated reflex was illustrated by Friedlander (1888), who severed a worm and loosely connected the two parts by a thread. Although the nerve cord was cut, a peristaltic wave continued from one part of the worm to the other, resulting from the pull of the thread

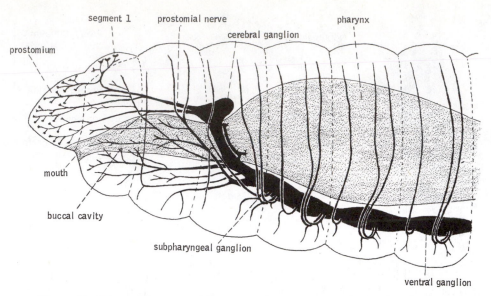

Figure 10–55 *Lumbricus* nervous system (lateral view). (After Hess from Avel.)

on the severed posterior half and from the initiation of the traction reflex.

The subpharyngeal ganglion is the principal center of motor control and vital reflexes and dominates the succeeding ganglia in the chain. All movement ceases when the subpharyngeal ganglion is destroyed. Motor control continues normally following removal of the brain, but the worm loses its ability to correlate movement with external environmental conditions. The relation of the subpharyngeal ganglion to the brain is thus somewhat analogous to the relation of the medulla to the higher brain centers in vertebrates.

Sense Organs

Oligochaetes lack eyes, except for a few aquatic forms that have simple pigment-cup ocelli. However, the integument is well supplied with dispersed photoreceptors located in the inner part of the epidermis, especially at the anterior end. The other receptors have a more or less general distribution in the integument. Clusters of sensory cells that form a projecting tubercle, with sensory processes extending above the cuticle, appear to be chemoreceptors. The tubercles form three rings around each segment and are particularly numerous on the more anterior segments and especially on the prostomium, where (in *Lumbricus*) there may be as many as 700 per square millimeter.

Reproduction and Development

ASEXUAL REPRODUCTION

Asexual reproduction is very common among many species of aquatic oligochaetes, particularly the aeolosomatids and the naidids. In fact, there are many asexually reproducing naidids in which sexual individuals are rare or have never been observed; a clone of *Aulophorus furcatus* was traced through 150 generations for a three-year period with no appearance of sexual individuals and with no diminishing of the asexual fission rate. Other oligochaetes reproduce asexually in the summer and sexually in the fall. Asexual reproduction always involves a transverse division of the parent worm into two or more new individuals. Regeneration commonly precedes the separation of the daughter individuals, and not infrequently, as in species of *Nais*, a new fission zone forms before division has occurred in an old one. Such delayed divisions produce chains of individuals or zooids similar to those in certain turbellarian flatworms.

THE REPRODUCTIVE SYSTEM

The reproductive system of oligochaetes differs from that of polychaetes in a number of striking respects. Oligochaetes are all hermaphroditic, they possess distinct gonads, and the number of reproductive segments present is very limited. The oligochaete arrangement is undoubtedly a specialized one, and the polychaete condition is probably

---

BOX 10–4 NEURONAL ORGANIZATION WITHIN THE GANGLIA OF PROTOSTOME INVERTEBRATES

The ganglia of annelids, mollusks, and arthropods are organized differently from those of vertebrates. The periphery of a ganglion is occupied by cell bodies of various motor neurons or interneurons, forming a rind, or cortical layer. The interior of the ganglion, the neuropile, is filled with processes. A considerable number of the processes arise from the cell bodies of the cortical layer, but there are also present the processes of interneurons whose cell bodies are in other ganglia, and processes of sensory neurons whose cell bodies lie near receptor terminals. Parallel bundles of fibers run between ganglia and compose nerve tracts or connectives.

It is within the neuropile that synaptic connections occur. The branching processes make possible multiple connections and contribute to the complexity of the circuitry. A single interneuron may have connections in the neuropiles of several ganglia, and it may receive and distribute impulses at many syn-

aptic junctions. Some single interneurons, called command neurons, have such an array of connections that impulses from them generate relatively complex motor responses.

The figure above is a diagrammatic cross section through a ventral nerve cord ganglion of the earthworm *Lumbricus*. The cortex, located between the inner and outer cord sheaths, contains cell bodies of motor and interneurons, of which there are about 800 per ganglion. The cortex of this worm contains no cell bodies on the dorsal side of the cord. The neuropile, located within the inner sheath, contains giant interneurons (clear circles), interneuron axon bundles (stippled areas), and sensory axon bundles (shaded areas); it is the site of synaptic junctions. (Based on figures from Gunther, J., 1971: Mikroanatomie des Bauchmarks von *Lumbricus terrestris*. Z. für Morphologie der Tiere, *70*:141–182.)

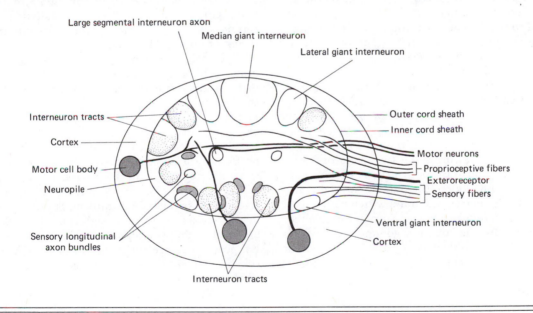

---

primitive. Brinkhurst and Jamieson (1972) have suggested that the ancestral annelids might have looked like oligochaetes but had polychaete reproductive systems.

In most aquatic groups there is usually one ovarian segment followed by one testicular segment; in terrestrial families, two male segments may be present. The gonad-containing segments

are located in the anterior half of the worm, and the female segment or segments are located behind the male segments (Fig. 10–56). The exact location of fertile segments along the trunk varies in different families.

The ovaries and testes, both of which are typically paired, are situated in the fertile segments on the lower part of the anterior septum and project

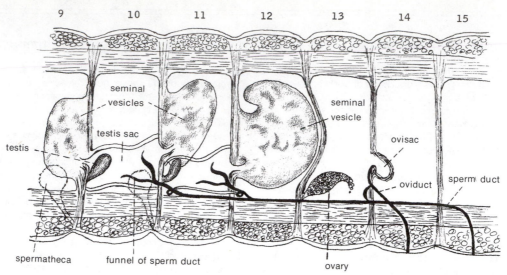

Figure 10–56 Reproductive segments of the earthworm *Lumbricus* (lateral view). (After Hess from Avel.)

into the coelom. Although maturation of the gametes is completed in the coelom, it is typically restricted to special coelomic pouches called seminal vesicles and ovisacs. Both arise as outpocketings of the septa of the reproductive segments, but the number, size, and position vary.

The reproductive segments, whether male or female, are each provided with a pair of sperm ducts or oviducts for the exit of sperm or eggs. The ducts extend backward and pass through one or more segments before opening on the ventral surface of the body. In earthworms the two pairs of sperm ducts on each side of the body usually become confluent before opening to the outside through a single, common, male genital pore, which has a raised border or lips. In many aquatic species the chamber (atrium) within the common gonopore contains a penis or an eversible area of the body wall or atrium tip (pseudopenis) (Fig. 10–57C). Glandular tissues, called prostate glands, are commonly associated with the male gonoducts. In some megascolecid earthworms, the prostates are not connected to the vas deferens, and they open separately onto the ventral surface of segments adjacent to those bearing the male gonopore. Prostates are absent from most lumbricids.

Forming a part of the female reproductive system, but completely separated from the female gonoducts, are the spermathecae (seminal receptacles). These storage chambers are simple pairs of sacs, usually opening into the ventral intersegmental groove adjacent to the segment containing them. The number of spermathecae ranges from one to many pairs, each pair commonly in a separate segment. Although they are usually located in certain segments anterior to the ovarian segment, the exact position is variable.

The position of reproductive structures in *Lumbricus* is illustrated in Figure 10–56. The male segment in this genus is partitioned so that the testes, the sperm duct funnel, and the opening to the seminal vesicles are enclosed in a special ventral compartment called a testis sac. The testis sac is completely separated from the larger, remaining portion of the coelomic cavity. For this reason, the testes are not visible in the usual dorsal dissection.

The general plan of the oligochaete reproductive system is relatively uniform, but the numbers of various structures, the segments in which they are located, and the segments onto which the genital pores open are extremely variable. This variation is of considerable importance in the taxonomy of oligochaetes.

The only exception to the oligochaete pattern appears in the Aeolosomatidae, which, although hermaphroditic, are similar to polychaetes in that they lack distinct gonads and have a large number of segments capable of producing gametes. Also, there is no gonoduct, and the sperm, at least, utilize the nephridia to exit from the body. However, some zoologists question whether these little worms are really oligochaetes.

THE CLITELLUM

The clitellum is a reproductive structure characteristic of oligochaetes. It consists of certain adjacent

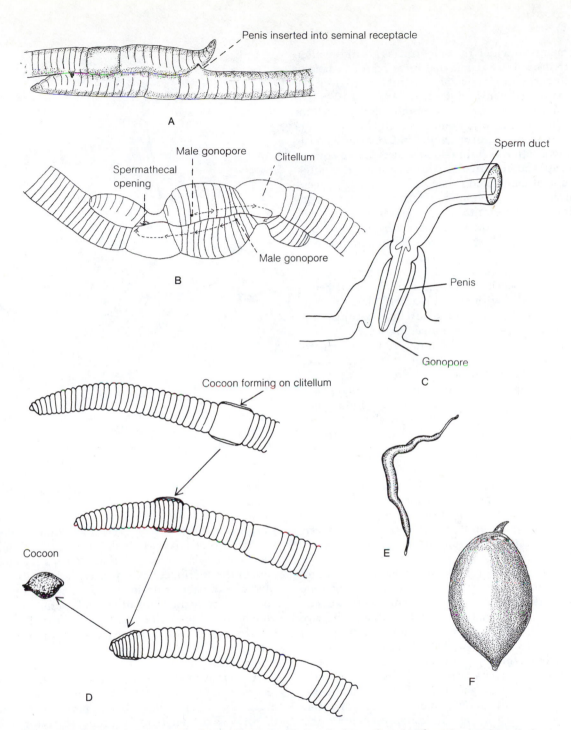

Penis inserted into seminal receptacle

A

Spermathecal opening

Male gonopore

Clitellum

Male gonopore

B

Sperm duct

Penis

Gonopore

C

Cocoon forming on clitellum

Cocoon

D

E

F

Figure 10–57   *A*, Copulation by direct sperm transmission in the earthworm *Pheretima communissima*. (After Oishi from Avel.) *B*, Copulation by indirect sperm transmission in the lumbricid *Eisenia foetida*. Arrows indicate path of sperm from male gonopores to openings of spermathecae. (After Grove and Cowley from Avel.) *C*, Penis of the lumbriculid *Rhynchelmis*. *D*, Cocoon formation in a lumbricid earthworm. (Modified from Tembe and Dubash. *In* Edwards, C. A., and Lofty, J. R., 1972: Biology of Earthworms. Chapman and Hall, London. p. 64.) *E*, Cocoon of the glossocolecid *Alma nilotica*. (*C* and *E* from Brinkhurst, R., and Jamieson, B. G. M., 1972: Aquatic Oligochaeta of the World. Toronto University Press, Toronto.) *F*, Cocoon of the lumbricid *Allobophora terrestris*. (After Avel.)

segments in which the epidermis is greatly swollen with unicellular glands that form a girdle, partially or almost completely encircling the body from the dorsal side downward (Figs. 10–49*A* and 10–57*D*). The number of segments composing the clitellum varies considerably; there are 2 clitellar segments in many aquatic forms, 6 or 7 in *Lumbricus*, and as many as 60 in certain Glossoscolecidae. In aquatic species and megascolecid earthworms, the clitellum is often located in the same region as the genital pores; in the lumbricids, the clitellum is considerably posterior to the gonopores. The degree of development of the clitellum varies from group to group. In aquatic species the clitellum may be only one cell thick, whereas in many earthworms it forms a heavy girdle. The development of the clitellum also varies from season to season. It generally coincides with sexual maturity, but there are some worms in which the clitellum becomes conspicuous only during the breeding season.

The glands of the clitellum produce mucus for copulation, secrete the wall of the cocoon, and secrete the albumin in which the eggs are deposited within the cocoon. In the earthworms the glands performing each of these three functions form three distinct layers with the large, albumin-secreting glands forming the deepest and thickest (Fig. 10–58).

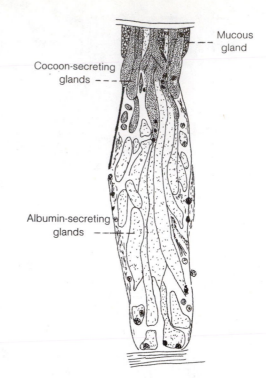

Figure 10–58 Section through clitellum of *Lumbricus terrestris.* (After Grove from Avel.)

COPULATION

Most oligochaetes breed semicontinually in contrast to the noncontinuous breeding of most polychaetes. Copulation is the rule, and mutual sperm transfer occurs. The ventral anterior surfaces of a pair of copulating worms are in contact, with the anterior of one worm directed toward the posterior of the other. In most oligochaetes except the lumbricids, the male genital pores of one worm directly appose the spermathecae of the other. The two worms are held in position by an enveloping mucous coat secreted by the clitella; they may also be hooked together by genital setae. The genital setae are modified ventral setae generally located in the region of the male gonopore of the spermathecae. Transmission of sperm into one pair of spermathecae in the earthworm *Pheretima communissima* takes more than 1.5 hours and then is repeated for each of the other two pairs of receptacles (Fig. 10–57*A*).

In the lumbricids the male genital pores do not appose the spermathecae during copulation, so sperm must travel a considerable distance from one opening to the other (Fig. 10–57*B*). During copulation the posteriorly placed clitellum of one worm attaches to the segments containing the spermathecae of the opposite worm. Attachment is accomplished by means of an adhesive slime tube and by the genital setae. The intervening region between the two clitella is less rigidly attached, although each worm is covered by a slime tube. At the emission of the sperm, certain muscles in the body wall of the segments posterior to the male gonopores contract and form a pair of ventral sperm grooves, which extend posteriorly to the clitellum. Because the grooves are roofed over by the enveloping slime tube, the sperm are actually passed through an enclosed channel.

The movement of sperm down the sperm groove is effected by a greater contraction of the muscles producing the groove. A pit is thus formed that travels the length of the groove, carrying small amounts of semen. When the semen reaches the region of the clitellum, it passes over to the other worm and enters the spermathecae. The mucous tube obviously must be incomplete in this region. The emission of semen may or may not be accomplished simultaneously in both members of the copulating pair. Copulation in *Lumbricus* continues for 2 to 3 hours.

THE COCOON

A few days after copulation, a cocoon is secreted for the deposition of the eggs. First a mucous tube is secreted around the anterior segments, including the clitellum. Then the clitellum secretes a tough, encircling, chitin-like material; this material forms the cocoon. The deeper layer of clitellar glands secretes albumin into the space between the wall of the cocoon and the clitellum.

When the cocoon is completely formed, it slips forward over the anterior end as the worm pulls backward (Fig. 10–57D). Eggs are discharged into the cocoon from the female gonopores as it passes forward from the clitellum over these openings. Sperm are deposited in the cocoon as it passes over the spermathecae. As the cocoon slips over the head of the worm and is freed from the body, the mucous tube quickly disintegrates, and the ends of the cocoon constrict and seal themselves. The cocoons of terrestrial species are left in the soil, and the cocoons of aquatic species are left in the bottom debris or mud or are attached to vegetation.

The cocoons are yellowish in color and ovoid in shape (Fig. 10–57D to F). Cocoons of *Tubifex* are 1.60 mm × 0.85 mm; those of *Lumbricus terrestris* are approximately 7 mm × 5 mm. The largest cocoons, 75 mm × 22 mm, are produced by the giant Australian earthworms, *Megascolides australis.* A cocoon contains anywhere from 1 to 20 eggs, depending on the species. A succession of cocoons may be produced; under favorable conditions, lumbricids may mate continually during the spring, and cocoons are formed every three or four days. Tubificids and lumbriculids generally reproduce only once a year. Their reproductive systems are then resorbed and are reformed the following year.

There is increasing evidence that growth and reproduction in oligochaetes are regulated by neurosecretions of the brain, as is true of polychaetes. However, in contrast to polychaetes, the hormone produced by the brain appears to stimulate rather than inhibit reproduction (see Highnam and Hill, 1977).

EMBRYOGENY

In general, the eggs of aquatic groups, particularly the primitive families, contain relatively large amounts of yolk. On the other hand, terrestrial species have much smaller eggs with much less yolk; the abundant albumin in the cocoon supplies most of the nutritive needs of the embryo. In both aquatic and terrestrial groups, development is direct, with no larval stages, and all development takes place within the cocoon.

The cleavage pattern, although retaining some traces of the spiral character of the cleavage pattern of the polychaetes, is considerably modified in oligochaetes, especially in earthworms.

The oligochaete young emerge from the end of the cocoon after eight days to several months of development. *Lumbricus* hatches in 12 to 13 weeks. Usually, only some of the eggs deposited in the cocoon hatch; in *Lumbricus terrestris* only one egg develops.

Many oligochaetes live several years. Earthworms in aquaria have been known to live for six years, but their life span is probably much shorter in natural, unprotected conditions. Lumbricid earthworms reach sexual maturity in six months to a year, depending on environmental conditions and the species. *Lumbricus terrestris* requires at least 200 days. Some aquatic oligochaetes have shorter generation times. At elevated temperatures enchytraeids inhabiting sewage percolating filters reach sexual maturity in 13 to 28 days, depending on the species, and the generation time between cocoons ranges from one to two months (Learner, 1972). On the other hand, some tubificids have a two-year life span, breeding once and then dying.

SYSTEMATIC RÉSUMÉ OF CLASS OLIGOCHAETA

The following classification is based on that of Brinkhurst and Jamieson (1972), modified by Jamieson (1978).

Order Lumbriculida. Four pairs of setae per segment. At least one pair of male funnels in same segment as male gonopores. Clitellum one cell thick and including male and female gonopores. A single family of freshwater oligochaetes, the Lumbriculidae. *Lumbriculus.*

Order Tubificida. Setal bundles usually with two or more setae, rarely absent. Setae often hairlike or otherwise modified. One pair of testes followed by one pair of ovaries, in adjacent segments. Male gonopores in segment immediately in front of or behind testicular segment. Clitellum one cell thick and including male and female gonopores.

Suborder Tubificina. Spermathecal pores in the segment immediately in front of or behind that bearing the male pores, rarely in the same segment. Setae diverse.

Family Tubificidae. Marine and freshwater. Some species widely distributed in poorly oxygenated and polluted waters.

(*Tubifex, Branchiura, Limnodrilus*). Many genera.

Family Naididae. Very small, aquatic, predominantly freshwater oligochaetes. Some with elongate proboscises. Reproduction often asexual. *Nais, Ripistes, Dero* (including *Aulophorus*), *Slavina, Stylaria*, the carnivorous and parasitic *Chaetogaster*.

Family Phreodrilidae. Small family of freshwater, marine, and commensal species.

Family Opistocystidae and Family Dorydrilidae.

Suborder Enchytraeina. Spermathecae anterior to the testes, separated by a gap of five segments. Setae simple batons, rarely forked or absent.

Family Enchytraeidae. Marine, freshwater, and terrestrial. *Enchytraeus*.

Order Haplotaxida. Basically having two segments with testes followed by two segments with ovaries. One pair of testes or ovaries or both often absent. If one pair of testes, these are separated from the ovaries by one or two segments. Male pores one or more segments behind the corresponding funnels.

Suborder Haplotaxina. Four simple or forked setae per segment. Testes in segments X and XI. Ovaries in segments XII and XIII or sometimes absent from XIII. Male gonopores in the segment immediately behind the two testicular segments. Clitellum one cell thick. A single family of freshwater and semiterrestrial species, the Haplotaxidae.

Suborder Alluroidina. Four pairs of simple setae per segment. Testes in segments X or X and XI. Ovaries in XIII. Clitellum one cell thick. Families Alluroididae and Syngenodrilidae in Africa and South America. Aquatic to semiterrestrial. *Alluroides, Syngenodrilus*.

Suborder Moniligastrina. Four pairs of simple setae per segment. One or two pairs of testes in testis sacs, each of which is suspended in the posterior septum of its segment. One or two pairs of male gonopores at the posterior border of the segment behind the corresponding testis sac. The anterior pair of ovaries lost. Clitellum one cell thick. A single widespread family of sometimes giant earthworms, mostly in India or Burma. *Moniligaster, Drawida*.

Suborder Lumbricina. Setae simple, eight or sometimes multiplied in a ring, in each segment. One or two pairs of testes, usually in seg-

ments X and XI. One pair (exceptionally two pairs) of male pores two or more segments behind the posterior testes. Ovaries in segment XIII, rarely XII also. Clitellum more than one cell thick, eggs therefore with little yolk. Mostly earthworms, some swamp worms and aquatic species.

The suborder contains a number of small aquatic or semiaquatic families: Biwadrilidae (Japan), Spargonophilidae (North America), Almidae (Africa, South America), Criodrilidae (Europe), Lutodrilidae (North America). It also comprises small families of earthworms: Komarekionidae (North America), Kynotidae (Madagascar), Hormogastridae (Europe and North Africa), and Microchaetidae (South Africa, containing giant forms). The Lumbricina also include the following five large families of earthworms.

Family Glossoscolecidae. With esophageal gizzard. Male gonopores on or (rarely) behind the clitellum. *Glossoscolex* South America. *Pontoscolex* throughout warmer regions.

Family Lumbricidae. Male gonopores on segment XV, anterior to the clitellum. With intestinal but never esophageal gizzard. Temperate Old and New Worlds. *Lumbricus, Eisenia, Allolobophora*.

Family Ocnerodrilidae. Male gonopores usually in segment XVII with a single pair of tubular prostate glands, less commonly in XVIII with prostates in XVII and XIX. Usually with esophageal diverticula in segment IX. Circumtropical. *Ocnerodrilus, Eukerria*.

Family Megascolecidae. Male gonopores usually in segment XVIII, fused with or near a pair of prostate glands or with prostates in XVII and XIX. Esophageal diverticula, if present, not restricted to IX. Worldwide excepting western Europe. *Megascolides* includes the giant Australian worm. *Megascolex* and *Pheretima* with a ring of setae.

Family Eudrilidae. Near the Ocnerodrilidae but with distinctive prostates and, often, internal fertilization. Africa. *Eudrilus*, with one species widespread elsewhere.

Members of the family Aeolosomatidae, which are minute freshwater species, have in the past been considered primitive oligochaetes. On the basis of their reproductive systems (see p. 322), and other features, Brinkhurst and Jamieson (1972) believe that they are unrelated to other oligochaetes and should be removed from the class.

SUMMARY

1 Members of the class Oligochacta are believed to have evolved independently from the ancestral annelids. Most have retained a peristaltic mode of locomotion and have well-developed metamerism and a simple prostomium.

2 The first oligochaetes were probably inhabitants of freshwater sediments. Some lines then successively invaded drier substrates to give rise to the earthworms. Others remained aquatic but became adapted for living in loose debris and algae.

3 The digestive tract is adapted for a diet of decomposing organic matter, largely plant material.

4 There are no gills, but cutaneous vascular networks are well developed in larger forms, especially earthworms.

5 The excretory organs are metanephridia, segmentally arranged as in polychaetes.

6 Most of the adaptations of earthworms for life on land are behavioral.

7 Oligochaetes are hermaphroditic, with well-developed reproductive systems limited to a few segments. There is copulation and reciprocal sperm transfer. Fertilization and direct development occur within a cocoon secreted by the clitellum.

Class Hirudinea

The Class Hirudinea contains over 500 species of marine, freshwater, and terrestrial worms, commonly known as leeches. Although they are all popularly considered to be bloodsuckers, a large number of leeches are not ectoparasitic.

As a group, the leeches are certainly the most specialized annelids, and most of the distinguishing characteristics of the class have no counterpart in the other two annelid groups. But they do display many oligochaete features. Both leeches and oligochaetes lack parapodia and head appendages; both are hermaphroditic, with gonads and gonoducts restricted to a few segments; and both possess direct development within cocoons secreted by a clitellum. These similarities clearly suggest a common ancestry, and the two groups are commonly considered subclasses or classes within the class or subphylum Clitellata. Like oligochaetes, leeches are basically a freshwater group, with some invasions of land and a secondary invasion of the sea.

Leeches are never as small as many polychaetes and oligochaetes. The smallest leeches are 1 cm in length, and most species are 2 to 5 cm long. Some species, including the medicinal leech *(Hirudo medicinalis)*, may attain a length of 12 cm, but the giant of the class is the Amazonian *Haementeria ghiliani*, which reaches 30 cm. Black, brown, olive green, and red are common colors, and striped and spotted patterns are not unusual.

External Anatomy

The anatomy of leeches is remarkably uniform. The body is typically dorsoventrally flattened and frequently tapered at the anterior (Fig. 10–59). The segments at both extremities have been modified to form suckers. The anterior sucker is usually smaller than the posterior one and frequently surrounds the mouth. The posterior sucker is disc shaped and turned ventrally. Metamerism is very much reduced. Unlike other annelids, leeches have a fixed number of segments, 34, but secondary external annulation has obscured the original segmentation. There are no setae.

The head, or cephalic region, contains the reduced prostomium plus a number of segments (Fig. 10–60). Dorsally, the head bears a number of eyes; ventrally, it bears the anterior sucker surrounding the mouth (Figs. 10–61 and 10–62*B*). The clitellum, covering three segments, is never conspicuous, except during reproductive periods. The large, ventral posterior sucker is derived from seven segments.

The number of annuli per segment varies not only in different regions of the body but also in different species. The best means of determining the primary segmentation of leeches is by study of the nervous system and the innervation of the annuli by segmental nerves. However, the occurrence of a ring of sensory papillae around the first annulus of each segment, the serial repetition of color patterns, and the placement of nephridiopores give good indications of the segmentation.

Body Wall, Coelom, and Locomotion

The body wall contains a more distinct connective tissue dermis than is present in other annelids, and some of the unicellular gland cells of the integument are very large and sunken into the connective tissue layer (Fig. 10–63). The longitudinal muscle layer of the body wall is powerfully developed, but there are also circular, oblique, and dorsoventral muscle fibers.

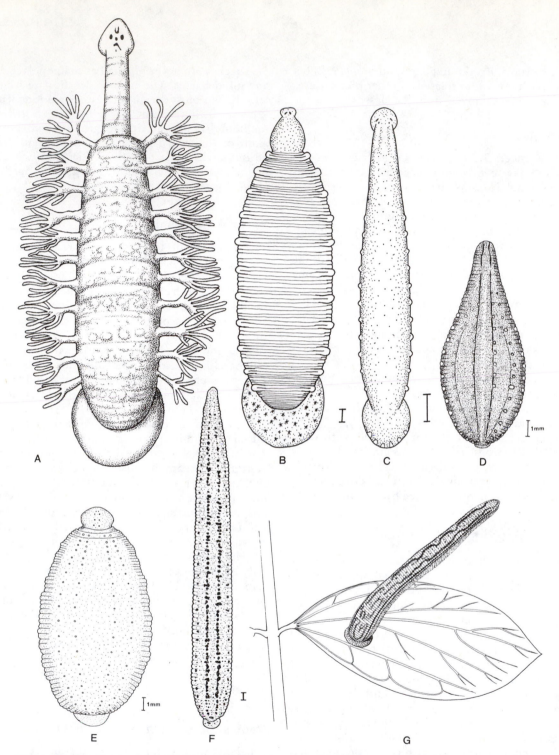

Figure 10–59 External dorsal views of different species of leeches: *A* to *C*, Fish leeches (Piscicolidae). *A, Ozobranchus,* showing lateral gills. (After Oka from Mann.) *B, Cystobranchus. C, Piscicola. D* and *E,* Glossiphoniid leeches. *D, Glossiphonia complanata,* a very common European and North American leech that feeds on snails. *E, Theromyzon,* a cosmopolitan genus of leeches that attack birds. *F, Erpobdella punctata* (Erpobdellidae), a common North American scavenger and predatory leech. *G, Haemadipsa* (Haemadipsidae), a bloodsucking terrestrial leech of south Asia poised on a leaf. (*B–F* from Sawyer, R. J., 1972: North American Freshwater Leeches, University of Illinois Press, Champaign, Ill. *G* adapted from Keegan et al.)

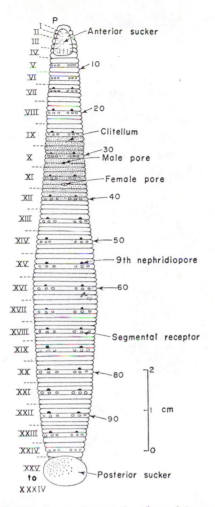

Figure 10–60 External, ventral surface of the medicinal leech, *Hirudo medicinalis.* (From Mann, K. H., 1962: Leeches. Pergamon Press, Elmsford, N.Y.)

A striking difference between leeches and other annelids is the loss of a distinct coelom in the leeches. Only in the five anterior segments of the primitive *Acanthobdella* are there coelomic compartments and separating septa. Interestingly, this same leech has no anterior sucker, and the anterior segments with coelomic compartments bear setae, the only known exception to the absence of setae in leeches. *Acanthobdella* provides additional evidence linking leeches and oligochaetes. In all other leeches septa have disappeared, and the coelomic cavity has been invaded by tissue (connective, chlorogogen, and botryoidal). This tissue invasion has reduced the coelom to a system of interconnected coelomic sinuses and channels (Figs. 10–63 and 10–64).

Many of the peculiarities of leeches—the presence of suckers, loss of setae, loss of septa, reduction of the coelom—are related to a change from peristaltic burrowing to a new mode of locomotion. Leeches crawl, but only the anterior and posterior suckers anchor to the substratum. When the posterior sucker is attached, a wave of circular contraction sweeps over the animal, and the body is lengthened and extended forward. Contraction of the diagonal muscles maintains hydrostatic pressure and keeps the animal rigid while it is raised and attached posteriorly. The anterior sucker then attaches, and the posterior sucker releases. A wave of longitudinal contraction then occurs, shortening the animal and moving the posterior sucker forward. When a leech swims, the body is flattened, and waves of contraction along the longitudinal muscles produce vertical undulations.

Ecological Distribution

Although some leeches are marine, most aquatic species live in fresh water. Relatively few species tolerate rapid currents; most prefer the shallow, vegetated water bordering ponds, lakes, and sluggish streams. In favorable environments, often high in organic pollutants, overturned rocks may reveal an amazing number of individuals; more than 10,000 individuals per square meter have been reported from Illinois (Richardson, 1925). Some species estivate during periods of drought by burrowing into mud at the bottom of a pond or stream and can survive a loss of as much as 90 per cent of their body weight.

There is a tendency toward amphibious habits in the hirudinid and erpobdellid leeches. For example, there is a terrestrial hirudinid, *Haemopis terrestris,* which is occasionally plowed up in fields in midwestern United States. Complete terrestriality has been attained in the Haemadipsidae, which inhabit humid jungles of southern Asian and Australian regions.

Although leeches are found through the world, they are most abundant in north temperate lakes and ponds. Much of the North American leech fauna is shared with Europe.

Internal Transport and Gas Exchange

The glossiphoniids and piscicolids (rhynchobdellids) have retained the blood-vascular system of oligochaetes (Fig. 10–64), but the coelomic sinuses act as a supplemental circulatory system. In the other leech orders the ancestral circulatory system has

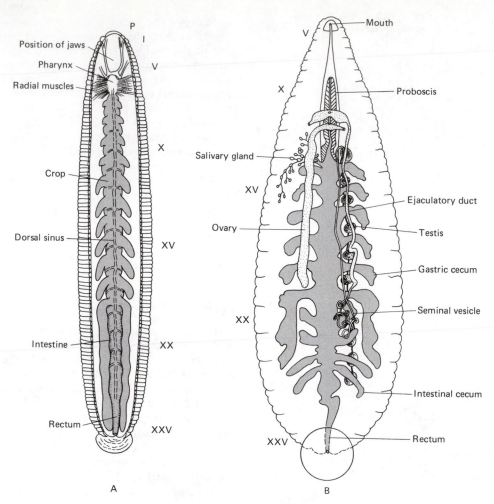

Figure 10–61   *A*, Internal structure of *Hirudo medicinalis* (dorsal view). (From Mann, K. H., 1962: Leeches. Pergamon Press, Elmsford, N.Y.) *B*, Internal structure of *Glossiphonia complanata* (ventral view). (After Harding and Moore from Pennak.)

disappeared, and the coelomic sinuses and fluid have been converted to a blood-vascular system. The hemocoelomic fluid is propelled by the contractions of the lateral longitudinal channels.

Gills are found only in the Piscicolidae, the general body surface providing for gas exchange in other leeches. The piscicolid gills are lateral leaflike or branching outgrowths of the body wall (Fig. 10–59*A*).

Respiratory pigment (extracellular hemoglobin) is found only in the gnathobdellid and pharyngobdellid leeches and is responsible for about half of the oxygen transport.

Nutrition

Leeches possess either a proboscis or a sucking pharynx and jaws. The proboscis (order Rhynchob-

dellida) is an unattached tube lying within a proboscis cavity, which is connected to the ventral mouth by a short, narrow canal (Fig. 10–62*A*). The proboscis is highly muscular, has a triangular lumen, and is lined internally and externally with cuticle. Ducts from large, unicellular salivary glands open into the proboscis. When feeding, the animal extends the proboscis out of the mouth, forcing it into the tissue of the host.

In jawed leeches (order Gnathobdellida), which lack a proboscis, the mouth is located in the anterior sucker (Fig. 10–62*B* and *C*). Just within the mouth cavity are three large, oval, bladelike jaws, each bearing a large number of small teeth along the edge. The three jaws are arranged in a triangle, one dorsally and two laterally. When the animal feeds, the anterior sucker is attached to the surface of the prey or host, and the edges of the jaws slice

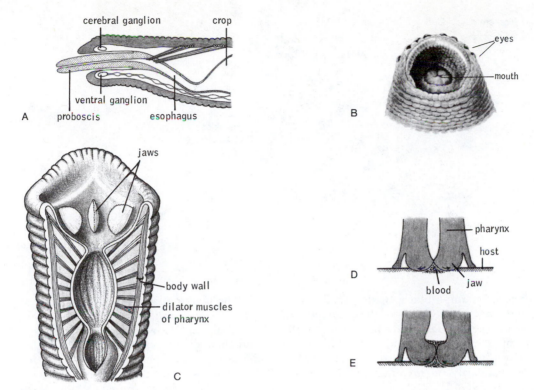

Figure 10–62 *A*, Sagittal section through anterior end of *Glossiphonia*, showing protruded tubular pharynx. (Modified after Scribin.) (Figure 10–61 shows nonprotruded state.) *B*, Ventral view of oral region of a terrestrial, bloodsucking, haemadipsid leech. Jaws are not exposed. (From Keegan, H. L., et al, 1968: 406th Medical Laboratory Special Report. U.S. Army Med. Command, Japan.) *C*, Ventral dissection of anterior end of *Hirudo*, showing three jaws. (Modified after Pfurtscheller.) *D* and *E*, Ingestion by *Hirudo*. Outward movement of teeth (*D*), followed by medial movement of teeth and dilation of pharynx (*E*). (After Herter.)

through the integument (Fig. 10–62*D* and *E*). The jaws swing toward and away from each other, activated by muscles attached to their bases. Salivary glands secrete an anticoagulant called hirudin.

Immediately behind the teeth, the buccal cavity opens into a muscular, pumping pharynx. The erpobdellids also have a pumping pharynx, but the jaws are replaced by muscular folds.

The remainder of the digestive tract is relatively uniform throughout the class. A short esophagus opens into a relatively long stomach, or crop. The stomach may be a straight tube, as in the erpobdellids, but more commonly it is provided with 1 to 11 pairs of lateral ceca (Fig. 10–61). Following the stomach is an intestine, which may be a simple tube or, as in the rhynchobdellids, may have four pairs of slender lateral ceca. The intestine opens into a short rectum, which empties to the outside through the dorsal anus, located in front of the posterior sucker.

Many leeches are predacious, but about three fourths of the known species are bloodsucking ec-

toparasites. However, in many cases the difference lies only in the size of the host. The Hirudinidae especially demonstrate a gradation from predation to parasitism. The Erpobdellidae contain the greatest number of predacious leeches, but this type of feeding habit is found in other families as well. Predatory leeches always feed on invertebrates. Prey includes worms, snails, and insect larvae. Feeding is relatively frequent, and the prey is usually swallowed whole. Many glossiphoniids suck all the soft parts from their hosts and are best regarded as specialized predators. In laboratory studies *Erpobdella punctata* consumed 1.78 tubificids (oligochaete worms) per day and *Helobdella stagnalis* 0.57 per day (Cross, 1976).

The bloodsucking leeches attack a variety of hosts. Some, primarily species of *Glossiphonia* and *Helobdella*, feed only on invertebrates, such as snails, oligochaetes, crustaceans, and insects, but vertebrates are hosts for most species. Piscicolidae are parasites of both freshwater and marine fish, sharks, and rays (Figs. 10–59*A* and *B*). The glossi-

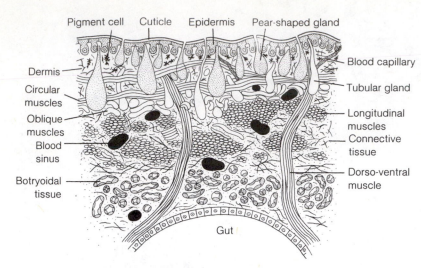

Pigment cell   Cuticle   Epidermis   Pear-shaped gland

Dermis

Circular
muscles

Oblique
muscles

Blood
sinus

Botryoidal
tissue

Blood capillary

Tubular gland

Longitudinal
muscles

Connective
tissue

Dorso-ventral
muscle

Gut

Figure 10–63   Section through the body wall of *Hirudo*. (After Mann, K. H., 1962: Leeches. Pergamon Press, Elmsford, N.Y.)

phoniids feed on amphibians, turtles, snakes, alligators, and crocodiles. Species of the cosmopolitan glossiphoniid genus *Theromyzon* attach to the nasal membranes of shore and water birds. The aquatic Hirudinidae and the terrestrial Haemadipsidae feed primarily on mammals, including humans (Fig. 10–59G).

Parasitic leeches are rarely restricted to one host, but they are usually confined to one class of vertebrates. For example, *Placobdella* will feed on almost any species of turtles and even alligators, but they rarely attack amphibians or mammals. On the other hand, mammals are the preferred hosts of *Hirudo*. Furthermore, some species of leeches that are exclusively bloodsuckers as adults are predacious during juvenile stages.

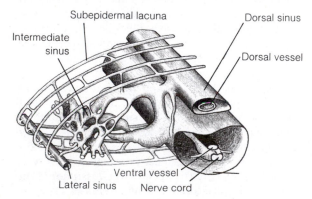

Subepidermal lacuna

Intermediate
sinus

Dorsal sinus

Dorsal vessel

Ventral vessel

Lateral sinus   Nerve cord

Figure 10–64   Section of blood-vascular and coelomic sinus systems of the glossiphoniid *Placobdella costata.* (After Oka from Harant and Grassé.)

The mammalian bloodsuckers, such as *Hirudo*, on contacting a thin area of the host's integument, attach the anterior sucker very tightly to this area, and then slit the skin. The jaws of *Hirudo* make about two slices per second. The incision is anesthetized by a substance of unknown origin. The pharynx provides continual suction, and the secretion of hirudin prevents coagulation of the blood. Penetration of the host's tissues is not well understood in the many jawless, proboscis-bearing species that are bloodsuckers. The proboscis becomes rigid when extended, and it is possible that penetration is aided by enzymatic action.

Leech digestion is peculiar in a number of respects. The gut secretes no amylases, lipases, or endopeptidases. The presence of only exopeptidases perhaps explains the fact that digestion in bloodsucking leeches is so slow. Also characteristic of the leech gut is a symbiotic bacterial flora that is important in nutrition. In both the bloodsucking medicinal leech *Hirudo medicinalis* and the predacious *Erpobdella octoculata*, the gut bacteria are responsible for a considerable part of digestion; they may be significant in the digestion of all leeches. The bacterium *Pseudomonas hirudinicola* of *Hirudo medicinalis* breaks down high-molecular-weight proteins, fats, and carbohydrates, and the bacterial population increases significantly following the ingestion of blood by the leech (Wilde, 1975). The bacteria may also produce vitamins and other compounds that are used by the leech host.

Bloodsuckers feed infrequently, but when they do, they can consume an enormous quantity of blood. *Haemadipsa* may ingest ten times its own weight, and *Hirudo* two to five times its own weight. Following ingestion, water is removed from the blood and excreted through the nephridia. The digestion of the remaining blood cells then takes place very slowly. These leeches can then tolerate long periods of fasting. Medicinal leeches have been reported to have gone without food for one and half years, and since they may require 200 days to digest a meal, they need not feed more than twice a year in order to grow.

Excretion

Leeches contain 10 to 17 pairs of metanephridia, located in the middle third of the body, one pair per segment (Fig. 10–60). As a result of the coelom reduction and the loss of septa in the leech body, the nephridial tubules are embedded in connective tissue, and the nephrostomes project into the coelomic channels. Each nephrostome opens into a nonciliated capsule (Fig. 10–65).

In most leeches the cavities of the capsule and the nephridial canal do not connect, and the two parts of the nephridium may even have lost all structural connection. Many branching, intracellular canals drain into the nephridial canal, which opens to the outside through the ventrolateral nephridiopore. Secretion into the intracellular canaliculi is the initial source of nephridial fluid, but the urine is very hypoosmotic to the blood, indicating reabsorption of salts. The nephridia are thus important organs of osmoregulation (Haupt, 1974).

The function of the nephridial capsules is believed to be the production of coelomocytes. The

coelomocytes are phagocytic and engulf particulate matter, but the eventual fate of the waste-laden cells is not certain. They may migrate to the epidermis or to the epithelium of the digestive tract. Particulate waste is also picked up by botryoidal and vasofibrous tissue of the hirudinid leeches and by pigmented and coelomic epithelial cells of glossiphoniids and piscicolids.

Nervous System and Sense Organs

The nervous system of leeches reflects their specializations of body structure. In segments X and XII a large ganglionic nerve ring surrounds the pharynx or proboscis. This collar represents the brain, the circumpharyngeal connectives, and the subpharyngeal ganglion of other annelids and the ganglia of the first three or four segments that have migrated posteriorly. There are two ventral nerve cords, although each pair of segmental ganglia is fused (Fig. 10–66).

The relatively small number and large size of the neurons has made leeches, along with sea hares and crayfish, favorite invertebrate subjects of neuroanatomists and neurophysiologists. Each of the 21 segmental ganglia in *Hirudo* contains 175 pairs of neuron cell bodies arranged bilaterally around a central neuropil where synaptic junctions are made (see Box 10–4 on p. 321). The cell bodies are large enough to be probed with electrodes and mapped (Fig. 10–66) (Nicholls and Van Essen, 1974; Stent et al. 1978; Stent and Weisblat, 1982).

Some rhynchobdellid leeches can dramatically change color as a result of pigment movements in large chromatophores, which are under neurohumoral control. The significance of the color change is not certain, for these leeches do not adapt to background coloration.

The specialized sense organs in leeches are two to ten pigment-cup eyes and sensory papillae (Fig. 10–67). The sensory papillae are small, projecting discs arranged in a dorsal row or in a complete ring around one annulus of each segment (Fig. 10–60). Each papilla consists of a cluster of many sensory cells and supporting epithelium.

Depite the lack of highly organized, concentrated sense organs, leeches can detect low levels of many types of stimuli, and the sensitivity is often an adaptation for finding prey or a host. Fish leeches respond to moving shadows and water pressure vibrations. Both predatory and bloodsucking leeches will attempt to attach to an object smeared with various host or prey substances—fish scales, tissue juices, oil gland secretions, sweat,

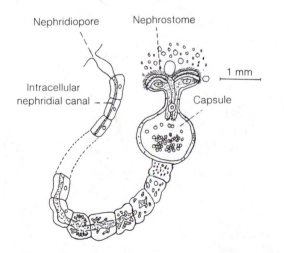

Figure 10–65 Nephridium of a glossiphoniid. (After Graf from Harant and Grassé.)

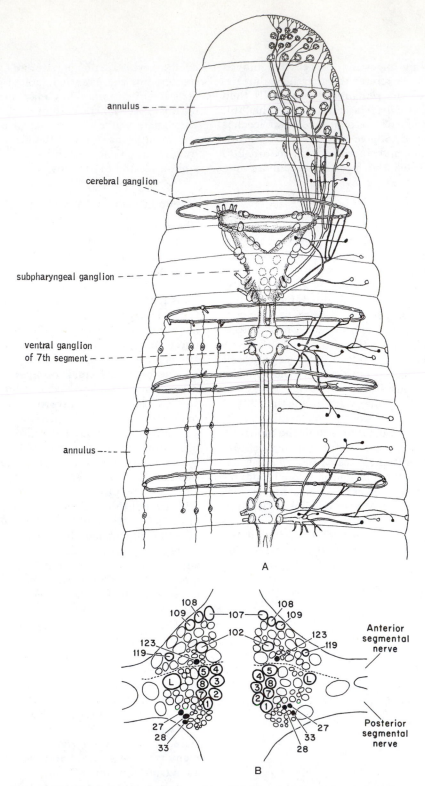

annulus

cerebral ganglion

subpharyngeal ganglion

ventral ganglion
of 7th segment

annulus

A

108
109
107
123
119
102

108
109
123
119

Anterior
segmental
nerve

L
5 4
8 3
7 2
1

4 5
3 8
7
1
L

27
28
33

27
33
28

Posterior
segmental
nerve

B

Figure 10–66 *A*, Nervous system of *Erpobdella punctata*. (After Bristol from Harant and Grassé.) *B*, Dorsal view of a segmental ganglion of a leech. Circles represent cell bodies; solid black circles are interneurons, and those enclosed with heavy black lines are motor neurons. (From Stent, G. S., et al, 1978: Neuronal generation of the leech swimming movement. Science, *200*:1348–1356. Copyright 1978 by the American Association for the Advancement of Science.)

Figure 10–67 Section through eye of *Erpobdella*. (After Hesse from Harant and Grassé.)

and other substances. *Hirudo*, which is a bloodsucker of warm-blooded animals, will swim into waves, which may be generated by a possible host (Young et al, 1981; Dickinson and Lent, 1984). The same leech is also attracted by body secretions and elevated temperatures and has been reported to swim toward a man standing in water. Supposedly, the terrestrial bloodsucking Haemadipsidae of the tropics, which are attracted by passing warm-blooded mammals, will move over vegetation and converge on a man standing in one place.

Reproduction

Unlike many other annelids, leeches do not reproduce asexually, nor can they regenerate lost parts. Like oligochaetes, all leeches are hermaphroditic, but they are protandric, not simultaneous hermaphrodites. The reproductive system is similar to that of oligochaetes; however, there are no separate seminal receptacles (spermathecae), and fertilization is internal (Fig. 10–68).

Sperm transfer in the hirudinids, most of which possess a penis, is similar to direct sperm transmission in earthworms (Fig. 10–69A). The ventral surfaces of the clitellar regions of a copulating pair come together, with the anterior of one worm directed toward the posterior of the other. Thus, the male gonopore of one worm apposes the female gonopore of the other. The penis is everted into the female gonopore, and sperm are introduced into the vagina, which probably also acts as a storage center.

Sperm transfer in most glossiphoniids, piscicolids, and erpobdellids, all of which lack a penis, is by hypodermic impregnation. The two copulating worms commonly intertwine and grasp each other with their anterior suckers. The ventral clitellar regions are in apposition, and by muscular contraction of the atrium, a spermatophore is expelled from one worm and penetrates the integument of the other. The site of penetration is usually in the clitellar region, but spermatophores may be inserted some distance away. As soon as the head of the spermatophore has penetrated the integument, perhaps by a combination of expulsion pressure and a cytolytic action by the spermatophore itself, the sperm are discharged into the tissues (Fig. 10–69F). Following liberation from the spermatophore, the sperm are carried to the ovisacs in the coelomic channels or by a special tissue pathway where there is a restricted region of integument for the reception of the spermatophore.

Eggs are laid from two days to many months after copulation. At this time, the clitellum becomes conspicuous and in most families secretes a cocoon, as in the oligochaetes (Fig. 10–69B and C). The cocoon is filled with a nutritive albumin produced by certain of the clitellar glands. The cocoon then receives the one to many fertilized eggs as it passes over the female gonopore. The eggs are secondarily small and relatively yolkless. Beginning in May, *Erpobdella punctata* in Michigan, for example, produces some ten cocoons, each with five eggs, which hatch three to four weeks later. The cocoons are affixed to submerged objects or vegetation. Some piscicolids attach their cocoons to their hosts. Terrestrial species place them in damp soil beneath stones and other objects, and the hirudinids, such as *Hirudo* and *Haemopis*, leave the water to deposit their cocoons in damp soil.

The glossiphoniids brood their eggs. In some species the cocoons are attached to the bottom and covered and ventilated by the ventral surface of the worm. In others, the atypical cocoons are membranous and transparent and are attached to the ventral surface of the parent (Fig. 10–69E). During the course of development the embryonic leeches break free of the cocoon and attach themselves directly to the ventral surface of the parent.

Most leeches have an annual or two-year cycle, breeding in the spring or summer and maturing by the following year. Life cycles are correlated in part with feeding habits. Some, such as *Hirudo*, are associated with the host only during actual feeding; others, such as the marine fish leech *Hemibdella*, never leave the host. But most leeches leave the host at least to breed. Less than three months is required to complete the entire life cycle in *Hemibdella*.

SYSTEMATIC RÉSUMÉ OF CLASS HIRUDINEA

Order Acanthobdellida. A primitive order, contains a single North European species parasitic on sal-

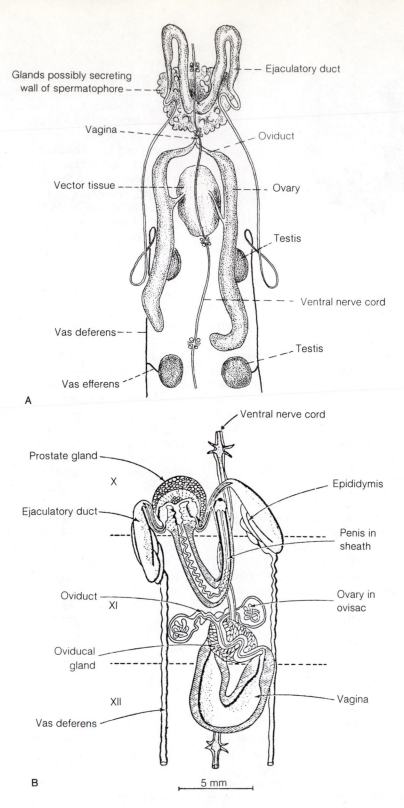

Figure 10–68 *A*, Reproductive system of *Piscicola geometra*. (After Brumpt from Harant and Grassé.) *B*, Reproductive organs of *Hirudo*. Testes are associated with vas deferens in more posterior segments. (After Leuckart and Brandes from Mann.)

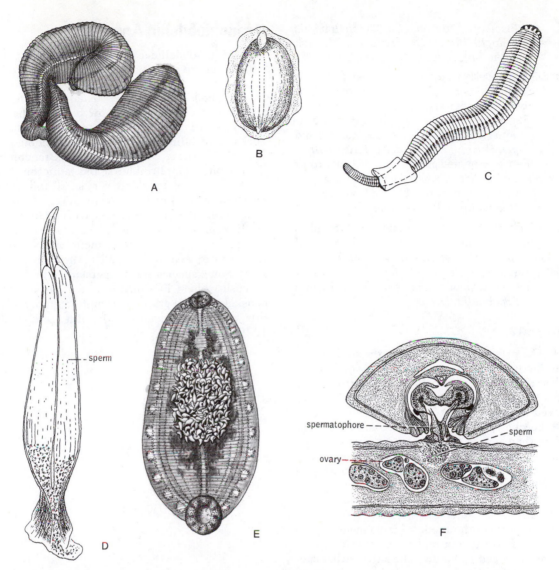

Figure 10–69 *A, Hirudinaria* copulating. (After a photograph by Keegan et al.) *B,* Cocoon of *Erpobdella octoculata.* (After Pavlovsky from Harant and Grassé.) *C, Erpobdella* withdrawing body from cocoon. (From Nagao, Z., 1957: J. Fac. Sci. Hokkaido Univ. Ser. VI, Zool., *13:*192–196.) *D,* Spermatophore of the rhynchobdellid *Haementeria.* (After Pavlovsky from Harant and Grassé.) *E,* Glossiphoniid brooding young. *F,* Section through two copulating *Erpobdella* individuals. Upper leech is injecting spermatophores into lower individual. (After Brumpt.)

monid fish. Setae and a compartmented coelom are present in the five anterior segments.

Order Rhynchobdellida. Strictly aquatic leeches with a proboscis and a circulatory system that is separate from the coelomic sinuses.

Family Glossiphoniidae. Flattened leeches, typically with three annuli per segment in the midregion of the body. Includes many ectoparasites of both invertebrates and vertebrates. *Marsupiobdella, Glossiphonia, Helobdella, Placobdella, Theromyzon, Haementeria, Hemiclepsis.*

Family Piscicolidae. The fish leeches. Subcylindrical body that often bears lateral gills. Usually more than three annuli per segment. Most marine leeches belong to this family. Parasites of marine and freshwater fish and, rarely, crustaceans.

Piscicola, Pontobdella, Trachelobdella, Branchellion, Ozobranchus, Illinobdella, Myzobdella, Cystobranchus.

Order Gnathobdellida. Aquatic or terrestrial leeches having a noneversible pharynx and three pairs of jaws. Five annuli per segment.

Family Hirudinidae. Chiefly amphibious or aquatic bloodsucking leeches. *Haemopis, Hirudo, Macrobdella, Philobdella.*

Family Haemadipsidae. Terrestrial, tropical leeches of Australasian region attacking chiefly warm-blooded vertebrates. *Haemadipsa, Phytobdella.*

Order Pharyngobdellida. Contains five families. Nonprotrusible pharynx; teeth are lacking, although one or two stylets may be present. Primarily aquatic, with some semiterrestrial forms.

Family Erpobdellidae. Predacious leeches. *Erpobdella, Dina.*

SUMMARY

1 The presence of a clitellum, the similarity of reproduction, and the existence of a single genus with anterior setae and coelomic compartments are evidence that members of the class Hirudinea evolved from oligochaetes.

2 Leeches possess an anterior and a posterior sucker; they lack setae, external segmentation, and septa; the coelom is reduced. All of these features are probably correlated with a change from peristaltic locomotion to movement involving extension and contraction of the entire body and attachment with the suckers.

3 The shift in the mode of movement is probably related to the assumption of predacious and ectoparasitic feeding habits. The latter is characteristic of about three quarters of the species of leeches.

4 A proboscis or jaw with a pumping pharynx is utilized in both predacious and ectoparasitic feeding.

5 Vertebrates are the principal hosts of ectoparasitic, or bloodsucking, leeches. Digestion of blood, which is slow, depends on exopeptidases produced by the leech and a symbiotic bacterial flora.

6 Metamerism is reflected internally in the nephridia and the ganglia of the ventral nerve cord. Gills are absent, and the circulatory system is like that of other annelids or has been replaced by coelomic channels.

7 Reproduction is similar to that of oligochaetes, but some species transfer sperm by hypodermic impregnation with spermatophores.

Branchiobdellid Annelids

The Branchiobdellidae is an enigmatic group of annelids that are parasitic or commensal on freshwater crayfish. They show similarities to both oligochaetes and leeches. They are usually included with the oligochaetes, but it is now believed that they diverged early from the base of the common oligochaete-hirudinean stock and justify being placed within a separate class or subclass (Branchiobdellida) (Holt, 1968; Brinkhurst and Jamieson, 1972).

All branchiobdellids are very small and are composed of only 14 or 15 segments (Fig. 10–70). The head is modified into a sucker with a circle of finger-like projections. The buccal cavity contains two teeth. The posterior segments are also modified to form a sucker, and all of the segments lack setae. Some species are ectoparasitic on the gills of crayfish; others live on the outer surface of the exoskeleton and graze on accumulated organic debris and microorganisms (Jennings and Gelder, 1979).

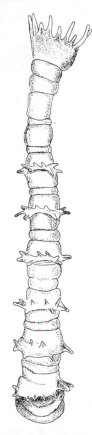

Figure 10–70 *Stephanodrilus*, a member of the leech-like family Branchiobdellidae; these worms are parasitic and commensal with freshwater crayfish. (After Yamaguchi from Avel.)

REFERENCES

The literature included here is devoted largely to annelids. The introductory references on page 8 include many general works and field guides that contain sections on annelids.

Anderson, D. T., 1973: Embryology and Phylogeny in Annelids and Arthropods. Pergamon Press, Oxford. 495 pp.

Baskin, D. G., 1976: Neurosecretion and the endocrinology of nereid polychaetes. Am. Zool., *16*:107–124.

Boilly, B., and Wissocq, J. C., 1977: Occurrence of striated muscle fibers in a contractile vessel of a polychaete: the dorsal heart of *Magelona papillicornis.* Biol. Cell, *28*(2):131–136.

Bolton, P. J., and Phillipson, J., 1976: Burrowing, feeding, egestion and energy budgets of *Allolobophora rosea.* Oecologia, *23*(3):225–245.

Brenchley, G. A., 1976: Predator detection and avoidance: ornamentation of tube-caps of *Diopatra* spp. Mar. Biol., *38*:179–188.

Brinkhurst, R. O., 1982: Evolution in the Annelida. Can. J. Zool., *60*(5):1043–1059.

Brinkhurst, R. O., and Cook, D. G., 1979: Aquatic Oligochaete Biology. Plenum Press, N.Y. 529 pp. (A collection of papers from a symposium, largely on ecological and taxonomic aspects of oligochaete biology.)

Brinkhurst, R. O., and Jamieson, B. G. M., 1972: Aquatic Oligochaeta of the World. Toronto University Press, Toronto. 860 pp.

Caspers, H., 1984: Spawning periodicity and habitat of the palolo worm *Eunice viridis* in the Samoan Islands. Mar. Biol., *79*:229–236.

Cather, J., 1971: Cellular interactions in the regulation of development in annelids and molluscs. Adv. Morphogenesis, *9*:67–124.

Clark, L. B., and Hess, W. N., 1940: Swarming of the Atlantic palolo worm, *Leodice fucato.* Tortugas Lab. Papers, *33*(2):21–70.

Clark, R. B., 1964: Dynamics in Metazoan Evolution: The Origin of the Coelom and Segments. Clarendon Press, Oxford. 313 pp.

Cross, W. H., 1976: A study of predation rates of leeches on tubeficid worms under laboratory conditions. Ohio J. Sci., *76*(4):164–166.

Dales, R. P., 1957: The feeding mechanism and structure of the gut of *Owenia fusiformis.* J. Mar. Biol. Assoc. U.K., *36*(1):81–89.

Dales, R. P., 1963: Annelids. Hutchinson University Library, London.

Dales, R. P., and Peter, G., 1972: A synopsis of the pelagic polychaeta. J. Nat. Hist., *6*(1):55–92.

Day, J., 1967: Polychaeta of Southern Africa. Pt. I, Errantia; Pt. II, Sedentaria. British Museum of Natural History.

Dickinson, M. H., and Lent, C. M., 1984: Feeding behavior of the medicinal leech, *Hirudo medicinalis.* Jour. Comp. Physiol. Sens. Neural Behav. Physiol., *154*(4):449–456.

Edwards, C. A., and Lofty, J. R., 1977: Biology of Earthworms. 2nd Edition. Chapman and Hall, London. (A general account of the earthworms, with special emphasis on ecology. Includes a key to common genera and methods for study.)

Fauchald, K., 1975: Polychaete phylogeny: a problem in protostome evolution. Syst. Zool., *23*:493–506.

Fauchald, K., 1977: The Polychaete worms. Definitions and keys to the orders, families and genera. Nat. Hist. Mus. of Los Angeles Co., Sci. Ser. *28*:1–190.

Fauchald, K. 1983: Life diagram patterns in benthic polychaetes. Proc. Biol. Soc. Wash., *96*(1):160–177.

Fauchald, K., and Jumars, P. A., 1979: The diet of worms: a study of polychaete feeding guilds. Oceanogr. Mar. Biol., Ann. Rev., *17*:193–284.

Fauvel, P., Avel, M., Harant, H., Grassé, P., and Dawydoff, C., 1959: Embranchement des Annélides. *In* Grassé, P. (Ed.): Traité de Zoologie, Vol. 5, pt. 1. Masson et Cie, Paris. pp. 3–686.

Flood, P. R., and Fiala-Médioni, A. 1982: Structure of the mucous feeding filter of *Chaetopterus variopedatus.* Mar. Biol., *72*(1):27–34.

Fransen, M. E., 1980: Ultrastructure of coelomic organization in annelids. I. Archiannelids and other small polychaetes. Zoomorphologie, *95*:235–249.

Friedlander, B., 1888: Über das Kriechen der Regenwürmer. Biol. Zbl., *8*:363–366.

Gardener, S. L., 1975: Errant polychaete annelids from North Carolina. Jour. Elisha Mitchell Sci. Soc., *91*(3):77–220. (Keys, descriptions, and figures for the North Carolina fauna.)

Giere, O., and Pfannkuche, O., 1982: Biology and ecology of marine Oligochaeta, a review. Oceanog. Mar. Biol., Ann. Rev., *20*:173–308.

Goerke, H., 1971: Die Ernahrungweise der Nereis-Arten der deutschen Kusten. Veroff. Inst. Meeresforsch. Bremerh, *13*:1–50.

Goodnight, C. J., 1973: The use of aquatic macroinvertebrates as indicators of stream pollution. Trans. Am. Micr. Soc., *92*(1):1–13.

Gray, J., 1939: Studies in animal locomotion, VIII. The kinetics of locomotion of *Nereis diversicolor.* J. Exp. Biol., *16*:9–17.

Gray, J., and Lissmann, H. W., 1938: Studies in animal locomotion, VII. Locomotory reflexes in the earthworm. J. Exp. Biol., *15*:506–517.

Gray, J., Lissmann, H. W., and Pumphrey, R. J., 1938: The mechanism of locomotion in the leech. J. Exp. Biol., *15*:408–430.

Hartman, O., 1959 and 1965: Catalog of the polychaetous annelids of the world. Allan Hancock Foundation Occas. Papers. Vol. 23. 628 pp. Supplement and Index (1965). 197 pp.

Haupt, J., 1974: Function and ultrastructure of the nephridium of *Hirudo medicinalis* L. II. Fine structure of the central canal and the urinary bladder. Cell Tiss. Res., *152*:385–401.

Hermans, C. O., 1969: The systematic position of the Archiannelida. Syst. Zool., *18*:85–102.

Highnam, K. C., and Hill, L., 1977: The Comparative Endocrinology of the Invertebrates. 2nd Edition. University Park Press, Baltimore. 357 pp.

Holt, T. C., 1968: The Branchiobdellida: Epizootic annelids. The Biologist. Vol. L, Nos. 3–4. pp. 79–94.

Jamieson, B. G. M., 1978: Phylogenetic and phenetic systematics of the opisthoporous Oligochaeta. Evolutionary Theory, *3*:195–233.

Jamieson, B. G. M., 1981: The Ultrastructure of the Oligochaeta. Academic Press, N.Y. 462 pp.

Jennings, J. B., and Gelder, S. R., 1979: Gut structure, feeding and digestion in the branchiobdellid oligochaete *Cambarinicola macrodonta* Ellis, 1912, an ectosymbiote of the freshwater crayfish *Procambarus clarkii*. Biol. Bull., *156*:300–314.

Kay, D. G., 1974: The distribution of the digestive enzymes in the gut of the polychaete *Neanthes virens*. Comp. Biochem. Physiol., *47*(A):573–582.

Klemm, D. J., 1972: Freshwater Leeches of North America. Biota of Freshwater Ecosystems Identification Manual No. 8. Environmental Protection Agency, U.S. Government Printing Office.

Klemm, D. J., 1985: A Guide to the Freshwater Annelida (Polychaeta, Naidid and Tubificid Oligochaeta, and Hirudinea) of North America. Kendall/Hunt Publishing Co., Dubuque, Iowa. 226 pp.

Kudenov, J. D., 1977: The functional morphology of feeding in three species of maldanid polychaetes. Zool. J. Linn. Soc., *60*:95–109.

Lasserre, P., 1975: Clitellata. *In* Giese, A. C., and Pearse, J. S. (Eds.): Reproduction of Marine Invertebrates. Vol. III. Academic Press, N.Y. pp. 215–275.

Learner, M. A., 1972: Laboratory studies on the life histories of four enchytraeid worms which inhabit sewage percolating filters. Ann. Appl. Biol., *70*(3):251–266.

MacGinitie, G. E., 1939: The method of feeding of *Chaetopterus*. Biol. Bull., *77*:115–118.

Mangum, C. P., 1970: Respiratory physiology in annelids. Amer. Sci., *58*(6):641–647.

Mangum, C. P., 1976: Primitive respiratory adaptations. *In* Newell, R. C. (Ed.): Adaptation to Environment. Butterworth Group Publishing, Mass. pp. 191–278. (Excellent review of gas exchange and internal transport in annelids, especially polychaetes.)

Mangum, C. P., 1976: The oxygenation of hemoglobin in lugworms. Physiol. Zool., *49*(1):85–99.

Mangum, C. P., 1977: Annelid hemoglobins: a dichotomy in structure and function. *In* Reish, D. J., and Fauchald, K. (Eds.): Essays in memory of Dr. Olga Hartman. Allan Hancock Foundation, University of Southern California, Los Angeles.

Mangum, C. P., 1982. The function of gills in several groups of invertebrate animals. In Houlihan, D. F., Rankin, J. C., and Shuttleworth, T. J. (Eds.): Gills. Soc. Exp. Biol. Sem. Ser. No. 16, Cambridge University Press, Cambridge.

Mangum, C. P., 1985: Oxygen transport in invertebrates. Am. J. Physiol., *248*:505–514.

Mangum, C. P., Woodin, B. R., Bonaventura, C., Sullivan, B., and Bonaventura, J., 1975: The role of coelomic and vascular hemoglobin in the annelid family Terebellidae. Comp. Biochem. Physiol., *51A*:281–294.

Mann, K. H., 1962: Leeches (Hirudinea), Their Structure, Physiology, Ecology, and Embryology. Pergamon Press, N.Y.

Martin, N. A., 1982: The interaction between organic matter in soil and the burrowing activity of three species of earthworms. Pedobiologia, *24*(4):185–190.

McConnaughey, B., and Fox, D. L., 1949: The anatomy and biology of the marine polychaete *Thoracophelia mucronata*. Univ. Calif. Publ. Zool., *47*(12):319–339.

Meinhardt, U., 1974: Comparative observations on the laboratory biology of endemic earthworm species: II. Biology of bred species. Z. Angew. Zool., *61*(2):137–182.

Mileikovskii, S. A., 1968: Morphology of larvae systematics of Polychaeta. Zool. Zh., *47*:49–50.

Mill, P. J. (Ed.), 1978: Physiology of Annelids. Academic Press, London, 684 pp. (Chapters on various topics by different contributors.)

Morin, J. G., 1983: Coastal bioluminescence: patterns and functions. Bull. Marine Sci., *33*(4):787–817.

Nakao, T., 1974: An electron microscopic study of the circulatory system in *Nereis japonica*. J. Morphol., *144*(2):217–236.

Nicholls, J. G., and Van Essen, D., 1974: The nervous system of the leech. Sci. Amer., *230*(1):38–48.

Nicol, E. A. T., 1931: The feeding mechanism formation of the tube, and physiology of digestion in *Sabella pavonia*. Trans. R. Soc. Edinburgh, *56*(3):537–598.

Nott, J. A., and Parkes, K. R., 1975: Calcium accumulation and secretion in the serpulid polychaete *Spirorbis spirorbis* at settlement. J. Mar. Biol. Assoc. U.K., *55*:911–923.

Olive, P. J. W., and Clark, R. B., 1978: Physiology of reproduction. *In* Mill, P. J. (Ed.): Physiology of Annelids. Academic Press, London. pp. 271–368.

Parker, S. P. (Ed.), 1982: Synopsis and Classification of Living Organisms. Vol. 2, McGraw-Hill Book Co., N.Y. 1236 pp.

Richardson, R. E., 1925: Illinois River bottom fauna in 1923. Illinois Natural History Survey Bulletin, *15*:391–423.

Rieger, R. M., 1980: A new group of interstitial worms, Lobatocerebridae nov. fam. (Annelida) and its significance for metazoan phylogeny. Zoomorphologie, *95*:41–84.

Rieger, R. M., 1985: The phylogenetic status of the acoelomate organization within the Bilateria: a histological perspective. In Conway Morris, S., et al (Eds.): The Origins and Relationships of the Lower Invertebrates. Systematics Association, Spec. Vol. No. 28. Clarendon Press, Oxford. pp. 101–122.

Roe, P., 1975: Aspects of life history and of territorial behavior in young individuals of *Platynereis bicanaliculata* and *Nereis vexillosa*. Pac. Sci., *29*(4):341–348.

Rose, S. M., 1970: Regeneration: Key to Understanding Normal and Abnormal Growth and Development. Appleton-Century-Crofts, N.Y.

Ruby, E. G., and Fox, D. L., 1976: Anerobic respiration in the polychaete *Euzonus (Thoracophelia) mucronata*. Mar. Biol., *35*(2):149–153.

Ruppert, E. E., and Carle, K. J., 1983: Morphology of metazoan circulatory systems. Zoomorph., *103*:193–208.

Santos, S. L., and Simon, J. L., 1974: Distribution and abundance of the polychaetous annelids in a south Florida estuary. Bull. Mar. Sci., *24*(3):669–689.

Satchell, J. E. (Ed.), 1983: Earthworm Ecology. Chapman and Hall, N.Y. 512 pp. (A collection of papers.)

Sawyer, R. T., 1972: North American freshwater leeches, exclusive of the Piscicolidae, with a key to all species.

Sawyer, R. T., 1984: Leech Biology and Behavior. 3 Vols. Oxford Univ. Press, N.Y. 500 pp. (An excellent account of all aspects of the biology of leeches.)

Schroeder, P. C., 1984: Chaetae. *In* Bereiter-Hahn, J., Matoltsy, A. G., Richards, K. D. (Eds.): Biology of the Integument. I. Invertebrates. Springer-Verlag, Berlin. pp. 297–309.

Schroeder, P. C., and Hermans, C. O., 1975: Annelida: Polychaeta. *In* Giese, A. C., and Pearse, J. S. (Eds.): Reproduction of Marine Invertebrates. Vol. III. Academic Press, N.Y. pp. 1–205.

Seymour, M. K., 1969: Locomotion and coelomic pressure in *Lumbricus.* J. Exp. Biol., *51:*47.

Singer, R., 1978: Suction-feeding in *Aeolosoma.* Trans. Am. Micros. Soc., *97*(1):105–111.

Stent, G. S., Kristan, W. B., Friesen, W. O., Ort, C. A., Poon, M., and Calabrese, R. L., 1978: Neuronal generation of the leech swimming movement. Science, *200:*1348–1356.

Stent, G. S., and Weisblat, D. A., 1982: The development of a simple nervous system. Sci. Amer., *246*(1):136–146.

Storch, U., and Alberti, G., 1978: Ultrastructural observations on the gills of polychaetes. Helgol. wiss. Meeresunters, *31:*169–179.

Uebelackei, J. M., and Johnson, P. G. (Eds.), 1984: Taxonomic guide to the polychaetes of the Northern Gulf of Mexico. 7 vols. NOAA Tech. Report NMFS CIRC-375.

Virnstein, R. W., 1977: The importance of predation by crabs and fishes on benthic infauna in Chesapeake Bay. Ecology, *58:*1199–1217.

Warren, L. M., 1976: A population study of the polychaete *Capitella capitata* at Plymouth. Mar. Biol., *38*(3):209–216.

Waxman, L., 1971: The hemoglobin of *Arenicola cristata.* J. Biol. Chem., *246*(23):7318–7327.

Weber, R. E., 1978: Respiration. *In* Mill, P. J. (Ed.): Physiology of Annelids. Academic Press, London. pp. 369–446.

Wells, G. P., 1950: Spontaneous activity cycles in polychaeta worms. Symp. Soc. Exp. Biol., *4:*127–142.

Wells, G. P., 1959: Worm autobiographies. Sci. Amer., *200*(6):132–141.

Whitlatch, R. B., 1974: Food-resource partitioning in the deposit feeding polychaete *Pectinaria gouldii.* Biol. Bull., *147*(1):227–235.

Wilde, V., 1975: Investigations on the symbiotic relationship between *Hirudo officinalis* and bacteria. Zool. Anz., *195*(5/6):289–306.

Wilson, D. P., 1974: *Sabellaria* colonies at Duckpool, North Cornwall, 1971–1972, with a note for May 1973. J. Mar. Biol. Assoc. U.K., *54:*393–436.

Wilson, R. A., and Webster, L. A., 1974: Protonephridia. Biol. Rev., *49:*127–160.

Young, S. R., Dedwylder, R. D., and Friesen, W. O., 1981: Responses of the medicinal leech *(Hirudo medicinalis)* to water waves. J. Comp. Physiol. and Sens. Neural Behav. Physiol., *144*(1):111–116.

Zottoli, R. A., and Carriker, M. R., 1974: Burrow morphology, tube formation, and microarchitecture of shell dissolution by the spionid polychaete *Polydora websteri.* Mar. Biol., *27*(4):307–316.

11

The Mollusks

Members of the phylum Mollusca are among the most conspicuous and familiar invertebrate animals and include such forms as clams, oysters, squids, octopods, and snails. In abundance of species, mollusks constitute the largest invertebrate phylum aside from the arthropods. Over 50,000 living species have been described. In addition, some 35,000 fossil species are known, for the phylum has had a long geological history, and the animals' mineral shells, which increase the chances of preservation, have resulted in a rich fossil record that dates back to the Cambrian.

Despite the striking differences among snails, clams, and squids, mollusks are built on the same fundamental plan. To understand the basic design, we will begin by examining a generalized mollusk. Many features of this hypothetical animal, such as the structure and function of the digestive tract and gills, are encountered among primitive living species of different classes.

The molluscan archeotype would be bilaterally symmetrical, probably not much over 1 cm in length, and have a somewhat ovoid shape (Fig. 11–1A). The ventral surface is flattened and muscular

to form a creeping sole, or foot. The dorsal surface is covered by an oval, convex, shieldlike shell that protects the underlying internal organs or visceral mass. The underlying epidermis, called the mantle (or pallium), secretes the animals' shell, and the most active secretion occurs around the edge of the mantle, although some new material is added to the older portions of the shell. Thus the shell increases in diameter and thickness at the same time.

A series of pairs of retractor muscles enable the animal to pull its shield-shaped shell down against the bottom on which it lives. Each retractor muscle is attached to the inner surface of the shell and is inserted into each side of the foot.

The periphery of the shell, as well as its underlying mantle, overhangs the body only slightly, except toward the posterior, where the overhang is so great that it creates a chamber called the mantle cavity. Within this protective chamber are a number of pairs of gills, as well as openings from a pair of nephridia.

Each gill (ctenidium) consists of a long, flattened axis projecting from the anterior wall of the mantle cavity and contains blood vessels, muscles,

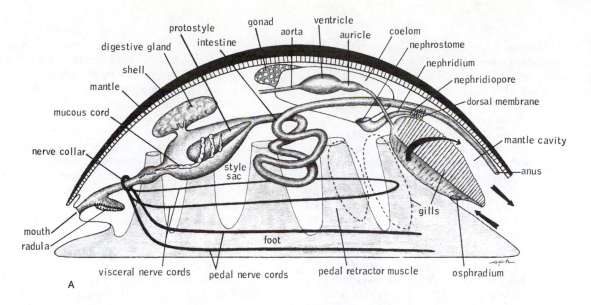

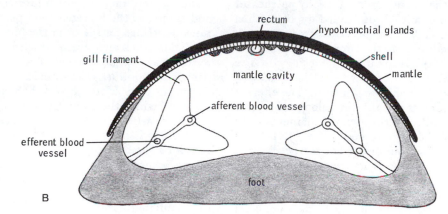

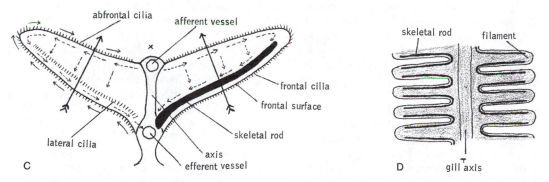

Figure 11–1 *A,* Generalized mollusk (lateral view). Arrows indicate path of water current through mantle cavity. (Adapted from various authors.) *B,* Transverse section through body of ancestral mollusk at level of mantle cavity. *C,* Transverse section through gill of the primitive gastropod *Haliotis.* Large arrows indicate direction of water current over gill filaments; small solid arrows indicate direction of cleansing ciliary currents; small broken arrows indicate direction of blood flow within gill filaments. *D,* Frontal section through primitive gill, showing alternating filaments and supporting chitinous rods. (*B–D* after Yonge, C. M., 1947: The pallial organs of the aspidobranchiate Mollusca, a functional interpretation of their structure and evolution. Phil. Trans. Roy. Soc. London B, 443–518.)

and nerves (Fig. 11–1*A*). To each side of the broad surface of the axis are attached flattened, triangular filaments, which alternate in position with those filaments on the opposite side of the axis (Fig. 11–1*C* and *D*). The gills are located on opposite sides of the mantle cavity and are held in position by a ventral and dorsal membrane. Water enters the lower part of the mantle cavity from the posterior, passes upward through the gills, and then moves posteriorly back out of the cavity.

Propulsion of water through the mantle cavity is largely effected by the beating of a powerful band of lateral cilia located on the gills just behind the frontal margin (which has first contact with the inhalent water stream). Sediment brought in by water currents and trapped on the gills is carried upward first by frontal cilia and then by abfrontal cilia toward the axis, where it is swept out by the exhalant current. On the mantle roof are two patches of mucus-secreting epithelium, called hypobranchial glands (Fig. 11–1*B*). They lie downstream to each gill and trap sediment in the efferent water current.

Two blood vessels run through the gill axis. The afferent vessel, which carries blood into the gill, runs just within the abfrontal margin. The efferent vessel, which drains the gill, runs along the frontal margin. Blood diffuses through the filaments from the afferent to the efferent vessel (Fig. 11–1*C*) and thus constitutes a countercurrent to the external water stream flowing from the frontal to the abfrontal margin.

In most living mollusks not only the mantle epidermis but also the epidermis of the remainder of the exposed body parts, including the foot, are covered by cilia and contain mucous gland cells. Mucous glands are especially prevalent on the foot, where they lubricate the substratum for locomotion.

We will assume that our generalized mollusk, like many living mollusks, is a grazer of fine algae and other organisms growing on rocks. The anterior mouth opens into a chitin-lined buccal cavity (Fig. 11–2*A*). The posterior wall of the buccal cavity is evaginated to form the pocket-like radula sac, which contains on its floor the uniquely molluscan feeding organ, called the radula. The radula apparatus consists of an elongated, cartilaginous base, the odontophore. Over the odontophore and around its anterior is stretched a membranous belt, the radula proper, which bears a number of longitudinal rows of chitinous teeth (Fig. 11–2*D*).

Not only can the odontophore be projected out of the mouth, but also the radula can move to some extent over the odontophore. Within the sac the lateral margins of the radula tend to roll up, but as the odontophore is projected out of the mouth over the substratum, the changing tension causes the radula belt to flatten over the odontophore tip. The flattening in turn brings about the erection of the teeth. In living forms that are grazers, the radula functions as a scraper and collector (Fig. 11–2*B* to *D*). Since the radula teeth recurve posteriorly, the effective scraping stroke is forward and upward when the odontophore is retracted (like licking).

To compensate for the hard wear caused by scraping, there is a gradual loss of membrane and radular teeth at the anterior end of the ribbon, while new ones are continuously secreted at the posterior end. The radula slowly grows forward over the odontophore at a rate of one to five rows of teeth per day.

At least one pair of salivary glands opens onto the anterior dorsal wall of the buccal cavity. These glands secrete mucus, which lubricates the radula and entangles the ingested food particles. Food in mucous strings passes from the buccal cavity into a tubular esophagus, from which it is moved posteriorly toward the stomach (Fig. 11–3*A*). In primitive living mollusks the stomach is shaped like an ice cream cone with a broad, hemispherical anterior, into which the esophagus opens, and a tapered posterior, which leads into the intestine. The anterior region of the stomach is lined with chitin except for a ciliated, ridged sorting region and the entrance point for two ducts from a pair of lateral digestive glands (liver), or diverticula. The posterior conical region of the stomach, called the style sac, is ciliated.

The contents of the stomach are rotated by the style sac cilia. The rotation winds up the mucous food strings, drawing them along the esophagus and into the stomach (Fig. 11–1*A*). The rotating mucus mass is called a protostyle. The size and consistency of particles within the string vary greatly, and the chitinous lining of the anterior part of the stomach protects the wall from damage by sharp surfaces (Fig. 11–3*A*). The acidity of the stomach fluid (pH of 5 to 6 in living mollusks) decreases the viscosity of the mucus and aids in freeing the contained particles. Such particles are eventually swept against the sorting region, in which they are graded by size. Lighter and finer particles are driven by the cilia of the ridges to the duct openings of the digestive diverticula. Heavier and larger particles are carried in the grooves between the ridges to a large groove running along the floor of the stomach to the intestine.

Particles utilized as food pass into the ducts of the digestive glands, and digestion occurs intracellularly within the cells of the distal tubules. Al-

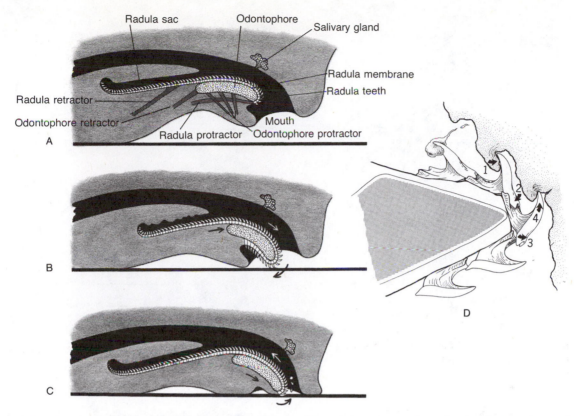

Figure 11-2 *A-C,* Molluscan radula: *A,* Mouth cavity, showing radula apparatus (lateral view). *B,* Protraction of the radula against the substratum. *C,* Forward retracting movement, during which substratum is scraped by radula teeth. *D,* The cutting action of radula teeth when they are erected over the end of the odontophore. Numbers indicate various forces impinging on teeth. (*D* from Solem, A., 1974: The Shell Makers: Introducing Mollusks. Reprinted by permission of John Wiley and Sons, N.Y. pp. 135 and 150.)

though intracellular digestion appears to be primitive in mollusks, at least some extracellular digestion occurs within the stomach cavity of most living species.

The long, coiled intestine functions largely in the formation of fecal pellets. The anus opens middorsally at the posterior margin of the mantle cavity, and wastes are swept away by the exhalant current (Fig. 11–1*A*).

The relatively small coelomic cavity is located in the middorsal region of the body (Fig. 11–1*A*); it surrounds the heart dorsally and a portion of the intestine ventrally. The heart consists of a pair of posterior auricles and a single anterior ventricle. The auricles drain blood from each gill and then pass it into the muscular ventricle, which pumps it anteriorly through a single aorta. The aorta branches into smaller blood vessels that deliver the blood into tissue spaces. This, then, is an open cir-

culatory system. Return drainage through the siback to the auricles. The blood contains amebocytes as well as the respiratory pigment called hemocyanin (see p. 382).

The excretory organs are a pair of tubular metanephridia, commonly called kidneys in living species (Fig. 11–1*A*). Although the typical metanephridium has one end connected to the pericardial cavity and the other opening to the outside through a nephridiopore, in most mollusks the connection with the pericardial cavity (renopericardial canal) and the nephridiopore are at the same end of the nephridium (Fig. 11–44*C*). The nephridium is thus a blind sac. The nephridiopore opens at the back of the mantle cavity. The pericardial coelom receives waste from two sources. The heart wall delivers a filtrate from the blood, and glands in the pericardium secrete waste into the coelom. The pericardial fluid then passes through the ne-

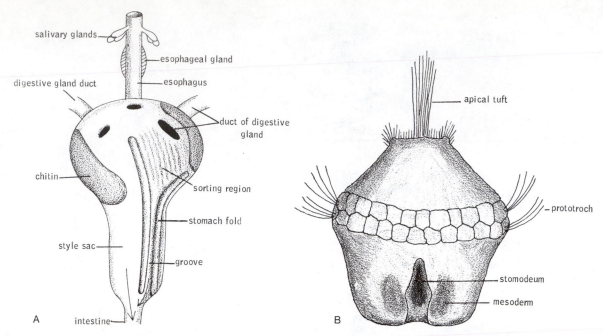

Figure 11–3 *A,* Diagram of a primitive molluscan stomach. (Modified from Owen.) *B,* Trochophore larva of the gastropod *Patella.* (After Patten.)

phrostome into the nephridium. Here further secretion of wastes, as well as some selective reabsorption, occurs through the tubule wall, and the final urine is emptied into the mantle cavity.

The ground plan of the molluscan nervous system consists of a nerve ring around the esophagus, from the underside of which two pairs of nerve cords extend posteriorly. The ventral pair, called the pedal cords, innervates the muscles of the foot; the dorsal pair, called the visceral cords, innervates the mantle and visceral organs.

The sense organs of many living mollusks include tentacles, a pair of eyes, a pair of statocysts in the foot, and osphradia. The osphradia are patches of sensory epithelium located on the posterior margin of each afferent gill membrane (Fig. 11–1*A*); they function as chemoreceptors and also determine the amount of sediment in the inhalant current.

In primitive mollusks a pair of anterior, dorsolateral gonads are present in each side of the coelom. When ripe, the eggs or sperm break into the coelomic cavity and are transported to the outside through the nephridia. Fertilization takes place externally in the surrounding sea water. The gastrula develops into a free-swimming trochophore larva (Fig. 11–3*B*) (see Box 10–3, p. 308, for description).

In most of the molluscan classes the trochophore passes into a more highly developed veliger larva, in which the foot, shell, and other structures make their appearance (Fig. 11–47). Metamorphosis occurs at the end of larval life, and the larva sinks to the bottom to assume the benthic habit of the adult.

Of the seven classes of living mollusks, the monoplacophorans, the chitons, and the gastropods each display some primitive features. We will begin with the gastropods, which are more familiar and better known.

SUMMARY

1 Members of the phylum Mollusca constitute one of the largest phyla of animals and are found in the sea, in fresh water, and on land. They are distinguished by a muscular foot, a calcareous shell secreted by the underlying integument, called the mantle, and a feeding organ, the radula.

2 A generalized mollusk would possess a flat, creeping foot, a dorsal, shield-shaped shell, and a poorly developed head.

3 Several pairs of gills are housed within a mantle cavity created by the overhanging mantle and shell. The gills are composed of flattened filaments projecting to either side of a supporting axis (bipectinate). Each filament bears lateral cilia, which create the ventilating current, and frontal cilia, which remove particulate matter.

4 The molluscan radula, a belt of recurved chitinous teeth stretched over a cartilage base, functions as a scraper in feeding, although in many mollusks it has been secondarily modified for other modes of feeding.

5 The primitive stomach is adapted for processing fine particles of food (especially algae) scraped from hard surfaces by the radula. A rotating, mucous mass in the style sac acts as a windlass to pull in a food-laden mucous string from the esophagus. Particles are separated over a sorting region, and fine particles are sent up the ducts of the surrounding digestive gland, where intracellular digestion occurs.

6 The blood-vascular system is open, and blood drains from the gills into one or more pairs of auricles. From each auricle blood passes into the central ventricle, which pumps it out through the aorta for distribution to the sinuses.

7 The heart is surrounded by a coelomic cavity (pericardial cavity). The excretory organs are metanephridia, which drain the pericardial cavity and empty into the mantle cavity.

8 The ground plan of the nervous system consists of a nerve ring around the esophagus, from which extend a pair of pedal nerve cords innervating the foot and a pair of visceral cords innervating the mantle and visceral mass. The most frequently encountered sense organs are tentacles, eyes, statocysts, and one or two osphradia in the mantle cavity.

9 The primitive mollusk is dioecious, with a pair of gonads in the visceral mass adjacent to the coelom. Maturation occurs in the coelomic cavity, and the metanephridia function as gonoducts. In such primitive mollusks fertilization is external and development is planktonic.

10 Cleavage is spiral and a trochophore is the first larval stage. A later larva, called a veliger, is typical of many groups, and the trochophore may be suppressed.

Class Gastropoda

The class Gastropoda is the largest class of mollusks. About 35,000 existing species have been described, and to this total should be added some 15,000 fossil forms. The class has had an unbroken fossil record beginning with the early Cambrian period and has undergone the most extensive adaptive radiation of all the major molluscan groups. Considering the wide variety of habitats the gastropods have invaded, they are certainly the most successful of the molluscan classes. Marine species have become adapted to life on all types of bottoms as well as to a pelagic existence. They have invaded fresh water, and the pulmonate snails and several other groups have conquered land by eliminating the gills and converting the mantle cavity into a lung.

Origin and Evolution

The evolution of gastropods involved three major changes: (1) the development of a head, (2) the conversion of the shell from a shield to a protective retreat, and (3) torsion. Although gastropods retain the flat, creeping ancestral foot, most are relatively active, mobile animals and are more highly cephalized than such mollusks as chitons and bivalves. Primitively, the head bears a pair of tentacles with an eye at each tentacle base.

The shell of a gastropod is an asymmetrical spiral that functions as a portable retreat instead of a shield. The animal therefore does not depend on clamping against a hard substratum for protection. The change in shell design involved an increase in height and a decrease in aperture, thus changing the shape from a shield to a cone (Fig. 11–4C). However, a cone not only would be unwieldy to carry but would also make it difficult for the animal to exploit crevices and holes for food and shelter. The problem was avoided by the spiraling of the shell over the head as it became higher and more conical (Fig. 11–4C).

The early shell was a planospiral; i.e., the shell was bilaterally symmetrical, with each spiral located completely outside of the one preceding it and in the same plane, like a hose coiled flat on the ground. Reduction of the shell aperture and resulting limitation of space within the mantle cavity perhaps account for the reduction of gills, retractor muscles, and nephridia to a single pair, which is the maximum number in any gastropod. The evolution of an asymmetrical shell is a later event to which we will return shortly.

The most significant modification of gastropods is the twisting, or torsion, that the body has undergone. Torsion is not the coiling of the shell; all evidence indicates that a planospiral shell evolved *before* torsion. Torsion and the spiraling of the shell were therefore separate evolutionary events.

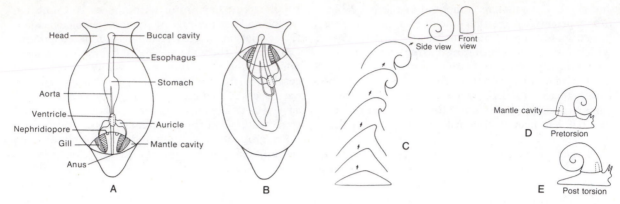

Figure 11–4 Dorsal views of ancestral gastropod: *A*, Prior to torsion. *B*, After torsion. *C*, Evolution of a planospiral shell. Height of the shieldlike shell of hypothetical ancestral mollusk increases and peak forms. Peak is pulled forward and coiled under. Aperture is reduced and animal can withdraw into spiral shell, which is more compact and less awkward to carry than a straight conical shell. Note that shell is bilaterally symmetrical. *D*, Hypothetical pretorsion gastropod with a planospiral shell. *E*, Posttorsion gastropod. Torsion does not affect planospiral shell except to place coils of shell posteriorly. Mantle cavity is now anterior. (*A* and *B* modified from Graham.)

Torsion was a much more drastic change than the spiraling of the shell. When viewed dorsally most of the body behind the head, including the visceral mass, mantle, and mantle cavity, was twisted 180 degrees counterclockwise (Fig. 11–4*A* and *B*). The mantle cavity, gills, anus, and two nephridiopores were now located in the anterior part of the body behind the head. Internally, the digestive tract was looped and the nervous system was twisted into a figure eight. The shell remained a symmetrical spiral.

Torsion is not merely an evolutionary hypothesis, for it appears in the ontogeny of living gastropods. The larva is at first bilaterally symmetrical and then quite suddenly undergoes twisting as a result of differential growth.

No widely accepted explanation of the evolutionary significance of torsion has yet been advanced, despite many contributions. A number of authors have postulated that torsion represents a larval adaptation for protection of the head (Garstang, 1928; Ghiselin, 1966), but this idea has recently been tested and refuted (Pennington and Chia, 1985). Others believe that torsion was an adult adaptation, with head protection and utilization of the anterior water stream as principal advantages. Stasek (1972) postulates that torsion permitted withdrawal of the larger and more developed head that probably evolved in the line of mollusks (monoplacophorans) leading to the gastropods.

Up to this point we have been considering a gastropod with a symmetrical, planospiral shell. Such a form is not entirely hypothetical, for there are early fossil species (the Bellerophontacea, Fig. 11–5*A* and *B*) with a symmetrical, planospiral shell bearing a cleft along the anterior, middorsal edge. The cleft indicates the midline of the mantle cavity and a corresponding cleft in the mantle (Fig. 11–5*C*). Such a cleft shell and mantle, which is also found in some living gastropods, reflects a modified ventilating current.

The inhalant current continued to come in over the head and pass over the gills, but now, instead of making a U-turn and passing out in the same direction, the water current flowed up and out through the cleft in the shell, removing wastes from the nephridia and anus (which was withdrawn to a position beneath the inner margin of the cleft).

Eventually, a final change took place that resulted in the typical gastropod structure. This change involved the shell. Although there are fossil species with planospiral shells, all existing gastropods possess asymmetrical shells, or if the shells are symmetrical, this symmetry has been secondarily derived. The planospiral shell had the disadvantage of not being very compact; since each coil lay completely outside of the preceding one, the diameter of the shell could become relatively great (Fig. 11–5*C*). The problem was solved with the evolution of asymmetrical coiling, in which the coils

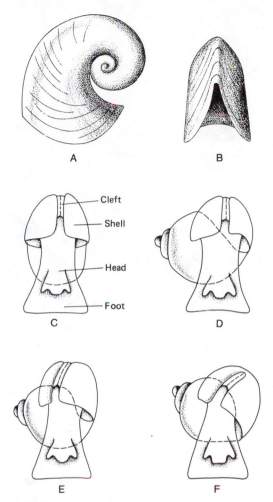

Figure 11–5 Side view (A) and front view (B) of the planospiral shell of *Strepsodiscus*, a genus of the fossil Bellerophontacea, possibly the earliest known archaeogastropod. C to F, Evolution of the asymmetrical gastropod shell. Slot in shell for exhalant water current marks location of mantle cavity. C, Ancestral posttorsion gastropod with planospiral shell. D, Apex of spiral is drawn out, producing a more compact shell. E, Position of shell over body shifted, providing more equal distribution of weight. F, Final position of shell over body, typical of most living gastropods. Axis of shell is oblique to long axis of body, and mantle cavity is on left side. Right side is compressed by shell. (C–F after Yonge.)

The new asymmetrical cone shell obviously could not be carried like the old planospiral shell, because all the weight would hang on one side of the body (Fig. 11–5D). To obtain the proper distribution of weight, the shell position shifted so that the axis of the spiral slanted upward and somewhat posteriorly. The shell was eventually carried obliquely to the long axis of the body, as in living gastropods. The changes in shell symmetry and carriage would have occurred simultaneously.

The new position of the shell restricts the mantle cavity to the left side of the body, for the mantle cavity on the right side is occluded by the bulging whorl of the visceral mass (Fig. 11–5F). This occlusion has had profound effects; it has resulted in the decrease in size, or the complete loss, of the gill, auricle, and nephridium on the right side of the body.

For other ideas about the evolution of gastropods, see Solem (1974) and Linsley (1978).

With this background of possible gastropod evolutionary origins, we must now consider the manner in which existing gastropods are classified. Gastropods are divided into three subclasses. The first, known as the Prosobranchia, includes all gastropods that respire by gills and in which the mantle cavity, gills, and anus are located at the anterior of the body—in other words, those gastropods in which torsion is evident. The majority of gastropods are prosobranchs, about 18,000 species.

From the Prosobranchia evolved the two other subclasses, the Pulmonata and Opisthobranchia. In the subclass Pulmonata, which includes the land snails, the gills have disappeared, and the mantle cavity has been modified into a lung. The Opisthobranchia display detorsion. The shell and mantle cavity are usually either reduced in size or absent, and many species have become secondarily bilaterally symmetrical. The sea hares and the sea slugs (nudibranchs) are perhaps the most familiar members of this subclass.

Shell and Mantle

The typical gastropod shell is a conical spire composed of tubular whorls and containing the visceral mass of the animal (Fig. 11–6A). Starting at the apex, which contains the smallest and oldest whorls, successively larger whorls are coiled about a central axis, called the columella; the last and largest whorl, called the body whorl, eventually terminates at the opening, or aperture, from which the head and foot of the living animal protrude. The whorls above the body whorl constitute the

are laid down around a central axis called the columella and each coil lies beneath the preceding coil (Fig. 11–5D). Such a shell is relatively compact and may even have a globular shape despite a long whorl length.

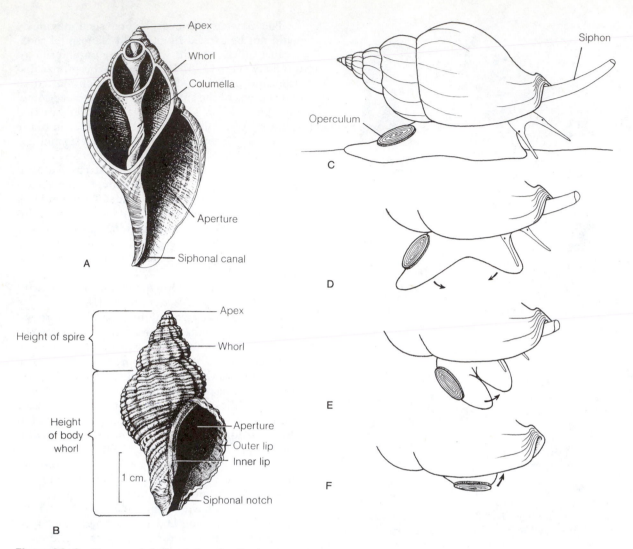

Figure 11–6 Gastropod shells: *A*, Longitudinal section through a shell. *B*, Shell of the oyster drill *Urosalpinx cinerea*, showing commonly designated features. (After Turner.) *C*, Gastropod with an operculum. *D–F*, Withdrawal into shell and closure by operculum.

spire. A shell may be spiraled clockwise or counterclockwise or, as it is more frequently stated, displays a right-handed (dextral) or left-handed (sinistral) spiral. A spiral is right-handed when the aperture opens to the right of the columella (if the shell is held with spire up and aperture facing observer) and left-handed when it opens to the left. Most gastropods are right-handed, a few are left-handed, and some species have both right-handed and left-handed individuals.

A gastropod shell typically consists of four layers. The outer periostracum is composed of a qui-

none-tanned, horny protein material called conchiolin or conchin. Although usually thin, the periostracum may be absent, as in the cowries, or thick and hairy, as in some welks. The inner shell layers consist of calcium carbonate. The outermost calcareous layer is generally prismatic; i.e., the mineral is deposited as vertical crystals, each surrounded by a thin protein matrix. The inner calcareous layers, usually two but sometimes more, are laid down as sheets, or lamellae, over a thin organic matrix. The calcium carbonate may be in the form of aragonite or calcite, or there may be both calcite

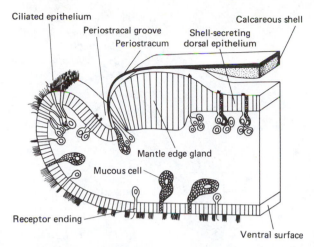

Ciliated epithelium
Periostracal groove
Periostracum
Calcareous shell
Shell-secreting dorsal epithelium
Mantle edge gland
Mucous cell
Receptor ending
Ventral surface

Figure 11–7 Diagrammatic section (not to scale) through mantle edge of an aquatic pulmonate snail *(Heliosoma)*. Periostracum is secreted initially at bottom of periostracal groove and is thickened as it passes over the mantle edge gland. Secretion of the calcareous portion of the shell begins to the inner side of the mantle edge gland. (Modified slightly from Jones, G. M., and Saleuddin, A. S. M., 1978: Cellular mechanisms of periostracum formation in *Physa spp.* Can. Jour. Zool., 56:2299–2311.)

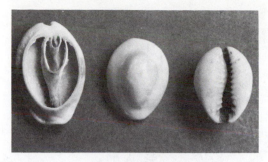

Figure 11–8 Cowrie shells. The shells of these tropical gastropods are superficially bilateral because each whorl completely encloses the previous whorls. The specimen on the right is viewed from the aperture side. The specimens on the left and in the middle are viewed from the side opposite the aperture, and the one on the left has been cut away to show the younger whorls. (By Betty M. Barnes.)

and aragonite layers present. Organic material accounts for about one third of the dry weight of the shell, depending on the species.

The color of the shell results from pigments in the periostracum or in the calcareous layers. The shell is laid down by the outer of two folds at the edge of the mantle (Fig. 11–7). Growth is usually not continuous, and the intervals can often be determined by interval growth lines, as in bivalves (p. 405), and by the sculpturing of the shell surface.

The gastropod head and foot are withdrawn into the shell by a retractor muscle. This muscle, called the columella muscle, arises in the foot and is inserted onto the columella of the shell. Primitively, two retractor muscles are present (Fig. 11–9B); this paired arrangement is found in a few living species, although the left muscle is usually very small. In most gastropods the left muscle has disappeared, and only the right one remains.

The foot of most prosobranch gastropods bears a horny disc, called the operculum, on its posterior dorsal surface (Fig. 11–6C). The operculum neatly fills the shell aperture and thus acts as a protective door or lid (Fig. 11–6D and E).

Gastropod shells display an infinite variety of colors, patterns, shapes, and sculpturing, but at this point only two of the more radical modifications in shell form will be mentioned. In a considerable number of gastropods, the shell is conspicuously spiraled only in the juvenile stages. The coiled nature disappears with growth, and the adult shell represents a single, large, expanded body whorl. In the abalone, *Haliotis* (Fig. 11–9A), and in the slipper shells, *Crepidula*, the shell remains asymmetrical, but in the limpets, of which there are a number of unrelated groups, the shell has become secondarily symmetrical and looks like a Chinese hat (Fig. 11–9D). Bilaterality has been derived in a very different way in the beautiful cowrie shells. Here the last whorl has completely overgrown the previous whorls and the aperture is greatly narrowed (Fig. 11–8).

The second modification is shell reduction and shell loss, a condition that has occurred many times in the history of gastropods. When the shell is greatly reduced, it often becomes buried within the mantle tissues.

Other shell modifications will be described later in connection with ventilation, movement, and habitation.

Water Circulation and Gas Exchange— Evolution of the Gastropod Groups

The great diversity of gastropods reflects adaptive radiation at various points in their evolutionary history. The main lines of this evolution are perhaps best reflected in the modifications for water circulation and gas exchange. We will therefore use these two problems as a way of gaining an initial

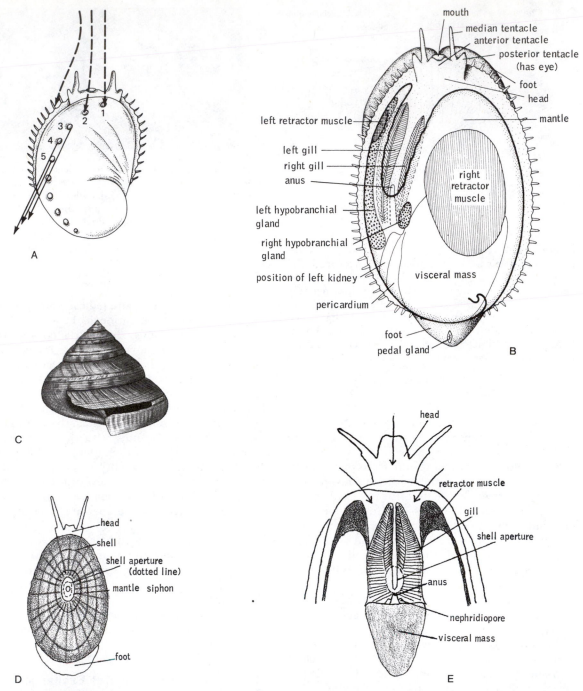

Figure 11–9 *A*, Dorsal view of the abalone *Haliotis kamtschatkana* showing the path of the ventilating current. Part of the inhalant water stream enters the mantle cavity beneath the shell to the left of the head, and part enters the anterior through shell perforations. The exhalant water stream leaves the mantle cavity through perforations 3, 4, and 5. Old, posterior, sealed perforations appear as a line of short tubercles on the shell. (Based on the work of Voltzow, J. 1983: Flow through and around the abalone *Haliotis kamtschatkana.* Veliger, 26(1):18–21.) *B*, Dorsal view of *Haliotis* with shell removed and mantle cavity exposed. (After Bullough.) *C, Pleurotomaria*, a gastropod with a slotted shell for exhalant current. (Drawn from a photograph by Abbott.) *D*, The keyhole limpet, *Diodora*, dorsal view. *E*, Dorsal view of exposed mantle cavity of *Diodora*, showing paired, bipectinate gills. (*D* and *E* after Yonge.)

overview of the class. Such an overview has been diagrammatically depicted in Figure 11–12, which can be used in following the discussion below.

PROSOBRANCHS

The most primitive type of gill structure and water circulation occurs in those prosobranchs with cleft or perforated shells, which may have been the primitive gastropod solution to possible sanitation problems caused by torsion, i.e., having the exhalant water stream located over the head. These prosobranchs belong to the primitive order Archaeogastropoda and include slit shells, abalones, and the keyhole limpets. In all three groups two bipectinate gills are present.* The rectum and anus are removed from the edge of the mantle cavity and open beneath the shell perforation or cleft. The ventilating current produced by the action of the lateral cilia of the gills enters the mantle cavity at the anterior of the body. It passes between the gill filaments and then continues upward and out through the shell cleft.

The slit shells (*Scissurella* and the deep-water *Pleurotomaria*) have typical spiral shells, but the anterior margin of the shell and the underlying mantle are deeply cleft (Fig. 11–9C). The abalones (*Haliotis*) and the keyhole limpets are intertidal and shallow-water inhabitants of wave-swept rocks. The broad shells of both groups are designed for minimum water resistance and as protective shields when the animal is clamped against rock. They are perforated instead of slotted, avoiding the structural weakness of a deep notch.

The low, shieldlike shell of the abalone is asymmetrical and constitutes in large part a single expanded whorl (Fig. 11–9A). As in most other prosobranch gastropods, the mantle cavity is displaced to the left side of the body. In *Haliotis tuberculata* the shell above the cavity contains a line of five holes (Fig. 11–9A). The mantle is cleft along the line of shell perforations, and the edges of the mantle fit together and project into the shell openings to form a lining for each hole. The ventilating current enters the mantle cavity through the anterior two holes and beneath the shell to the left of the head. It leaves through the last three holes (Fig. 11–9A). The anus and nephridial openings lie beneath one of the posterior perforations. A succession of holes develop as the shell grows. Each hole arises as a notch at the front margin of the shell and eventually becomes sealed at the rear.

---

*The Archaeogastropoda is also sometimes called the Aspidobranchia because of the bipectinate gills or the Diotocardia because there are two auricles (although one may be reduced).

The keyhole limpet has a conical, secondarily symmetrical shell, which has either a cleft at the anterior margin or a hole at the apex (Fig. 11–9D and E). The opening arises as a notch along the shell margin during early stages of development. The notch then becomes enclosed, and through differential growth it gradually assumes a position at the apex of the shell. Water enters the mantle cavity anteriorly, flows over the gills, and issues as a powerful stream from the opening at the shell apex. The anus and urogenital openings are located just beneath the posterior margin of the shell opening.

The Patellacea, another group of archaeogastropods, contain the largest assemblage of limpets. They are found from shallow water to the deep sea, and many are common inhabitants of the rocky intertidal zone. Their shells evolved independently from those of the keyhole limpets, and shell openings or clefts are lacking. In *Acmaea*, *Notoacmaea*, and *Collisella*, which include the common limpets on the west coast of North America, there is only a left gill, which projects to the right side of the body. As in all limpets, the mantle and shell overhang produces a distinct groove on each side between the foot and the mantle edge (Fig. 11–10A). The inhalant ventilating current enters the mantle cavity anteriorly on the left side. Part of the current flows posteriorly in the left lateral mantle groove; the rest of the current flows over the gill and then down the right mantle groove. The two exhalant streams converge and exit posteriorly. *Patella*, another genus of widespread intertidal limpets, lacks a gill in the mantle cavity. Instead, mantle folds form secondary gills, which project into the pallial groove along each side of the body (Fig. 11–10B).

The remaining archaeogastropods—the Trochacea (top shells and turban shells) and the Neritacea—possess only left gills, and their ventilating current enters the mantle cavity on the left side of the head and exits on the right (Figs. 11–11 and 11–12). The anus opens at the right edge of the mantle cavity, and wastes are carried away in the exhalant water stream. Such an oblique water current is apparently an efficient solution to ventilation, for it is found in most prosobranchs.

The neritaceans include many common, rocky, intertidal species, such as the semitropical and tropical species of *Nerita* (Fig. 11–11C). The gill, if present, is secondary. Members of this group have also invaded fresh water (*Theodoxus*), and perhaps from some freshwater stock evolved a family of tropical land snails, the Helicinidae (Fig. 11–12).

Of the some 18,000 species of prosobranchs, the majority are not archaeogastropods but mesogastropods and neogastropods. The archaeogastro-

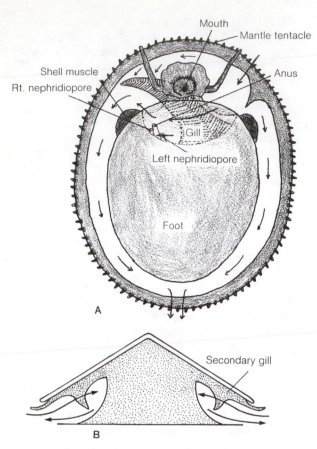

Figure 11–10 Patellacean limpets: *A*, *Acmaea* (ventral view). (After Yonge.) *B*, Cross section through a patellid limpet, showing secondary gills and ventilating currents (arrows). (Redrawn and modified from Yonge and Thompson, 1977.)

pods are largely restricted to the surfaces of rock and kelps. The great adaptive diversity of the mesogastropods and neogastropods is in part correlated with their ability to exploit other types of habitats, especially soft bottoms. This ability may be related to a major change in the structure of the gills. The dorsal and ventral membranes that suspend the bipectinate gills of the archaeogastropods present considerable surface areas that could be fouled by sediment carried within the ventilating current, and this perhaps accounts for the restriction of these primitive prosobranchs to the cleaner water over rocky bottoms. In the mesogastropods and neogastropods, the membranous suspension of the gills has disappeared, and the gill axis is attached directly to the mantle wall (Fig. 11–12). The filaments on the side of the attachment have dis-

appeared; those on the opposite side project into the mantle cavity. The gill of mesogastropods and neogastropods is thus monopectinate.

A further modification associated with the ventilating current of many species of these two higher orders of prosobranch, especially the neogastropods, is the development of an inhalant siphon by the extension and inward rolling of the mantle margin (Fig. 11–13*B*). In many species the anterior margin of the shell aperture is notched (Fig. 11–6*B*) or drawn out as a siphonal canal to house the siphon (Fig. 11–33*B*). The siphon may provide access to surface water in some species that burrow, or by making possible the selection of restricted areas of water, the mobile siphon may function as a sense organ, especially in carnivores. There has also been a tendency in neogastropods to direct the exhalant current toward the rear.

The ventilating current and unipectinate gill are relatively uniform among mesogastropods and neogastropods,* and the diversity of these higher prosobranchs results from variations in adaptations for locomotion, habitation, feeding, and other functions. Only their immigrations from the sea need be mentioned here. Mesogastropods are well represented in fresh water as a result of several independent invasions (Fig. 11–12). The majority are tropical, but there are many genera, such as *Goniobasis*, *Pleurocera*, *Viviparus*, *Campeloma*, and *Valvata*, that contain temperate species.

There are two large families of mesogastropod land snails, the Cyclophoridae and the Pomatiasidae (Fig. 11–12). Like the archaeogastropod Helicinidae, they are largely tropical and operculate, they have no gill, and gas exchange occurs across a vascularized mantle wall within the mantle cavity (lung). A notch or a breathing tube in the shell aperture of some species permits the entrance of air when the operculum is closed (Fig. 11–13*A*).

OPISTHOBRANCHS
The remaining two subclasses of gastropods, the Opisthobranchia and the Pulmonata, are probably both derived from prosobranchs that possessed only the left gill. The some 2000 highly diverse species of opisthobranchs are largely marine and are characterized by detorsion, in which the mantle cavity and the structures it contains have shifted to the right side (Fig. 11–12). The reasons for detorsion are unknown. Complicating the study of the

---

*The two groups are often treated as suborders of a single order, Caenogastropoda or Pectinibranchia or Monotocardia (referring to the single monopectinate gill or the single auricle of the heart).

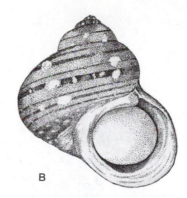

B

C

Figure 11–11 Archaeogastropods with a single gill (left) and an oblique ventilating current. *A*, Top shells (Trochacea). The large specimen is *Trochus niloticus*, a common species around South Pacific islands. *B*, Turban shell (Trochacea), *Turbo*, a genus of common tropical Pacific and Indian Ocean species. The heavy, calcareous operculum of the South Pacific species is often washed up on beaches in large numbers and is called a cat's eye. *C*, *Nerites* (Neritacea). These specimens belong to Pacific species, but the group is found in the intertidal zone throughout the tropics and subtropics. (By Katherine E. Barnes.)

origin of opisthobranchs is the fact that the gill is plicate, or folded, rather than filamentous and may not be homologous with the prosobranch gill (Fig. 11–14*C*).

A primitive opisthobranch is asymmetrical and possesses the more or less typical, coiled, gastropod shell (Fig. 11–12), although an operculum is usually lacking. But throughout the subclass, apparently correlated with detorsion, there has been a tendency toward shell reduction and loss, reduction of the mantle cavity and associated loss of the original gill, and attainment of a secondary bilateral symmetry. Characteristic of most opisthobranchs is a second pair of tentacles, called rhinophores, that are located behind the first pair, and are commonly surrounded at the base by a collarlike fold (Figs. 11–12 and 11–14*L*).

The Cephalaspidea, containing the bubble shells (*Acteon, Scaphander, Hydatina, Bulla*), are the largest order of opisthobranchs, and the one to which the more primitive members of the subclass belong (Fig. 11–14*A*). In *Acteon*, the most primitive known opisthobranch, the nervous system is still twisted and the shell is closed by an operculum. Many bubble shells burrow or crawl on the surface of soft bottoms, and the lateral, skirtlike folds of the foot reduce fouling of the mantle cav-

ity and other parts of the body. Perhaps detorsion was originally an antifouling adaptation, shifting the mantle cavity to the side of the body, out of the line of forward movement.

The order Anaspidea contains the sea hares, which reach the largest size (40 cm) of any opisthobranch. The reduced shell is buried in the mantle or completely lost, and the body is bilaterally symmetrical. The mantle cavity and gills are still present, and the posterior edge of the mantle can be rolled to form an exhalant siphon (Fig. 11–14*B*). When disturbed, many sea hares release a defensive purple ink derived from the pigments of the red algae on which they feed.

The pteropods comprise two orders of small, swimming, pelagic opisthobranchs. The shelled pteropods (order Thecosomata) possess a shell, often with an operculum (Fig. 11–14*D*). The naked pteropods (order Gymnosomata) lack a shell (Fig. 11–22*D*). A gill is absent from most pteropods, and the naked forms have no mantle cavity. Gas exchange occurs across the general body surface.

The sea slugs, members of the order Nudibranchia, certainly rank among the most spectacular and beautiful mollusks. Shell, mantle cavity, and original gill have disappeared, and the body is secondarily bilaterally symmetrical, with the anus at

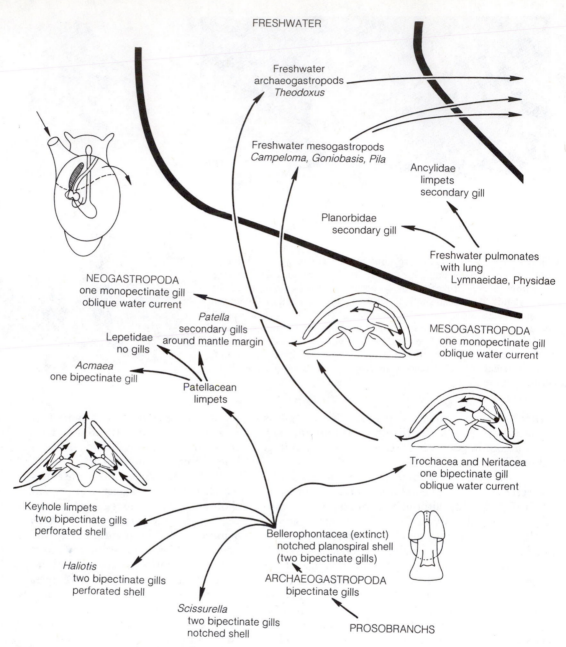

FRESHWATER

Freshwater
archaeogastropods
Theodoxus

Freshwater mesogastropods
Campeloma, Goniobasis, Pila

Ancylidae
limpets
secondary gill

Planorbidae
secondary gill

Freshwater pulmonates
with lung
Lymnaeidae, Physidae

NEOGASTROPODA
one monopectinate gill
oblique water current

Patella
secondary gills
around mantle margin

MESOGASTROPODA
one monopectinate gill
oblique water current

Lepetidae
no gills

Acmaea
one bipectinate gill

Patellacean
limpets

Keyhole limpets
two bipectinate gills
perforated shell

Trochacea and Neritacea
one bipectinate gill
oblique water current

Haliotis
two bipectinate gills
perforated shell

Bellerophontacea (extinct)
notched planospiral shell
(two bipectinate gills)

ARCHAEOGASTROPODA
bipectinate gills

Scissurella
two bipectinate gills
notched shell

PROSOBRANCHS

Figure 11–12 Evolution of water circulation and gas exchange in gastropods. Diagram reflects phylogenetic relationships of the subclasses and orders. Families and genera listed represent only examples. (Figures adapted from Graham, Hyman, Morton, and Yonge.)

the rear. The dorsal body surface is greatly increased in many nudibranchs by numerous projections called cerata (Figs. 11–14*I* to *L*). The cerata may be club shaped, as in *Eolidia*, branched, as in *Dendronotus*, or look like a cluster of grapes, as in *Doto*. Cerata are lacking in some nudibranchs, such as *Doris*, but these sea slugs have secondary gills arranged in a circle around the posterior anus (Fig.

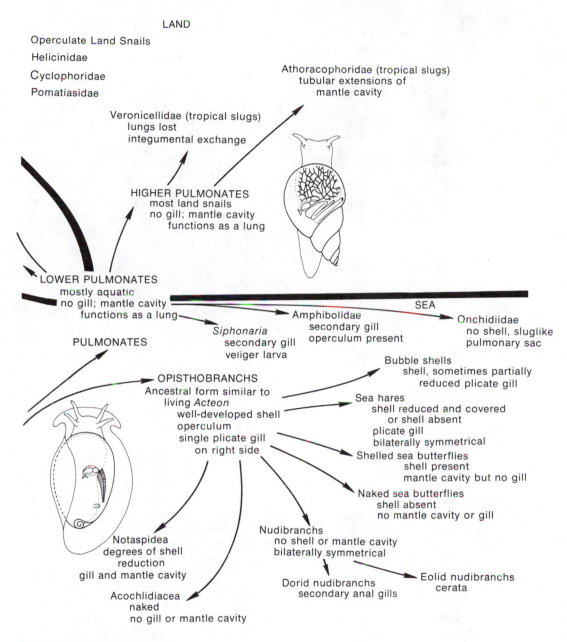

LAND

Operculate Land Snails
Helicinidae
Cyclophoridae
Pomatiasidae

Athoracophoridae (tropical slugs)
tubular extensions of
mantle cavity

Veronicellidae (tropical slugs)
lungs lost
integumental exchange

HIGHER PULMONATES
most land snails
no gill; mantle cavity
functions as a lung

LOWER PULMONATES
mostly aquatic
no gill; mantle cavity
functions as a lung

SEA

PULMONATES

Siphonaria
secondary gill
veliger larva

Amphibolidae
secondary gill
operculum present

Onchidiidae
no shell, sluglike
pulmonary sac

OPISTHOBRANCHS
Ancestral form similar to
living *Acteon*
well-developed shell
operculum
single plicate gill
on right side

Bubble shells
shell, sometimes partially
reduced plicate gill

Sea hares
shell reduced and covered
or shell absent
plicate gill
bilaterally symmetrical

Shelled sea butterflies
shell present
mantle cavity but no gill

Naked sea butterflies
shell absent
no mantle cavity or gill

Notaspidea
degrees of shell
reduction
gill and mantle cavity

Nudibranchs
no shell or mantle cavity
bilaterally symmetrical

Eolid nudibranchs
cerata

Acochlidiacea
naked
no gill or mantle cavity

Dorid nudibranchs
secondary anal gills

Figure 11–12 *(continued)*

11–14*J*). The cerata, as well as other parts of the nudibranch body, are usually brilliantly colored and commonly are red, yellow, orange, blue, green, or a combination of colors.

The sea slugs, as well as some other opisthobranchs, have evolved other defenses in the absence of well-developed shells. Escape swimming is

a common ability. Many have skin glands that produce sulfuric acid or a nonacidic noxious substance that repels potential predators, especially fish. Some utilize nematocysts from the prey on which they feed (p. 373). Some have spicules embedded in the mantle. The flamboyant coloration of some species of sea slugs probably represents warning

A

B

Figure 11–13 Prosobranchs: *A*, Shell of a cyclophorid land snail, showing the breathing tube, which permits gas exchange when the animal is withdrawn and the aperture is sealed by the operculum. *B*, Anterior end of the neogastropod *Mitra*, showing the folded origin of the siphon. (*A* after Rees from Purchon, R. D., 1968: Biology of the Mollusca. Pergamon Press, N.Y. *B* based on a photograph by Paul Zahl.)

coloration; for others it may be camouflage (see review by Todd, 1981).

Several other small orders—Acochlidioidea, Notaspidea, and Sacoglossa—contain sluglike forms, some with cerata (Fig. 11–14*G* and *H*).

PULMONATES

The subclass Pulmonata contains the highly successful land snails as well as many freshwater forms, and the more than 16,000 described species are widely distributed in both tropical and temperate regions throughout the world. The group takes its name from the conversion of the mantle cavity into a lung. The edges of the mantle cavity have become sealed to the back of the animal except for a small opening on the right side called the pneumostome (Figs. 11–12 and 11–15*A*). The gill has disappeared, and the roof of the mantle cavity has become highly vascularized. Ventilation is facilitated by the arching and flattening of the mantle cavity floor (actually the back of the animal). The pneumostome generally remains open at all times, or opens and closes with the ventilating cycle. Gas exchange by diffusion through the pneumostome is probably important in most pulmonates and predominant in small species.

The first pulmonate land snails appeared in the Carboniferous, but their origin is obscure. They probably evolved from some group of operculate prosobranchs that had a single gill. These ancestral forms perhaps inhabited estuarine marshes and mud flats, and the pulmonate condition could have evolved as a means of gas exchange when the animals were confined to small, stagnant puddles or to

wet but exposed surfaces. These conditions are similar to those postulated for the origin of amphibians.

There are a few primitive, marine species, all of which live at the edge of the sea on intertidal rocks or in estuarine habitats. Tropical limpets of the genus *Siphonaria* and the temperate *Melampus* of salt marshes and drift are among the few pulmonates that possess a veliger larva, indicating that the marine habit does not represent a secondary return to the sea.

Amphibola, another marine pulmonate, has a typical shell but is unusual in possessing an operculum. In all other pulmonates the operculum is lost during the course of development. This is a distinguishing characteristic. All freshwater and terrestrial prosobranchs are operculate; pulmonates, which are found in the same habitats, are not.

The lower pulmonates (order Basommatophora), with one pair of tentacles and with eyes at the tentacle base, include the few marine and all the freshwater forms. Many freshwater species, such as the cosmopolitan *Lymnaea* and *Physa*, come to the surface to obtain air for gas exchange. In *Lymnaea* the edges of the mantle cavity can be extended as a long tube for this purpose. The pneumostome is closed when the animal is submerged, and submergence may last from 15 minutes to more than an hour, depending on the time of year and other conditions. However, some deep-lake lymnaeids have abandoned air breathing and fill the mantle cavity with water. A secondary gill (pseudobranch) has evolved in other aquatic pulmonates as folds of the mantle near the pneumostome. Such a secondary gill is found in many planorbids and in the ancylids. The latter are limpets adapted for life in fast-running streams.

Various species of freshwater pulmonate snails are important hosts for certain human parasites. The African genus *Bulinus*, for example, is the principal host for trematodes causing schistosomiasis.

The higher pulmonates (order Stylommatophora) include the terrestrial species and are a considerably larger group than that containing the aquatic forms. There are two pairs of tentacles (the anterior lower pair may be inconspicuous), and the eyes are mounted at the top of the upper pair (Fig. 11–15B). The usually calcareous shell of these terrestrial pulmonates is not as heavy as those of many marine gastropods, although some variation may result from the availability of calcium in the soil. The periostracum protects the calcareous layers from humic acid and may also function as a water repellent. In many small species the aperture of the shell is partially occluded by teeth or ridges, which keep out such predators as insects but allow the soft body of the snail to protrude. The largest shells, 23 cm in height, belong to members of the African species *Achatina fulica*, but the South American strophocheilids are also large, reaching 15 cm in height. Many species of land snails have shells that measure less than 1 cm. Although such species are found throughout the world in leaf mold and beneath bark and stones, they are especially abundant on oceanic islands, as in the Pacific.

Shell reduction or loss has occurred independently a number of times within the higher pulmonates, and such naked species are called slugs *(Arion, Philomycus, Limax)*. The shell is generally absent or reduced and buried within the mantle, but in *Testacella* a little shell is perched on the back (Fig. 11–15C). The pneumostome of slugs is usually a conspicuous opening on the right side of the body (Fig. 11–15A). The evolution of the slug form is perhaps an adaptive response to low availability of calcium, for their original centers of distribution are restricted to areas of high humidity and low soil calcium. Slugs have now been introduced into many parts of the world from which they were originally absent.

It should be remembered that not all land snails are pulmonates. The operculate land snails, although a smaller group (4000 species of terrestrial prosobranchs, compared with some 20,000 species of terrestrial pulmonates), are very common in the tropics, and their adaptation for life on land parallels that of the pulmonates in many ways.

Locomotion and Habitation

The typical gastropod foot is a flat, creeping sole, but it has become adapted for locomotion over a variety of substrata. Typically, the sole is ciliated and provided with numerous gland cells or, in the pulmonates, with a large pedal gland. The glands of the foot elaborate a mucous trail over which the animal moves. Some very small snails, as well as species that live on sand and mud bottoms, move by ciliary propulsion. Most hard-bottom gastropods, terrestrial pulmonates, and even large soft-bottom species, when moving rapidly, are propelled by waves of fine muscular contraction that sweep along the foot. The sole of the foot is firmly anchored to the substratum by gelatinous mucus except in the region of a wave, where the foot slides forward over liquefied mucus. In species in which the waves move anteriorly (see below), the mucus changes very rapidly from gel to sol as a consequence of the shearing force produced by muscle contraction (Fig. 11–16) (Denny, 1981; review by Trueman, 1983). Thus, each wave performs a small step. In some species a wave extends across the entire width of the foot (monotaxic) (Fig. 11–17A and B), but in many forms a wave involves only half of the width of the foot, and the waves on the right side move alternately to those on the left (ditaxic) (Fig. 11–17C and D).

The waves may be direct, progressing in the same direction as the movement of the animal, i.e., from back to front (Fig. 11–16A). Or the waves may be retrograde, passing from front to back in the opposite direction of the animal's movement (Fig. 11–16B). Direct waves involve contraction of longitudinal and dorsoventral musculature beginning at the posterior end of the foot. Successive sections of the foot are in effect pushed forward (Fig. 11–16A). Retrograde waves involve contraction of transverse muscles, which along with blood pressure extends the front of the foot forward. This backward moving wave of elongation is followed by contraction of longitudinal muscles, and successive areas of the foot are pulled forward (Fig. 11–16B). Direct and indirect waves may be associated with either the monotaxic or the ditaxic condition; i.e., the pattern may be direct monotaxic, direct ditaxic, and so on. The most common pat-

(Text continued on p. 362)

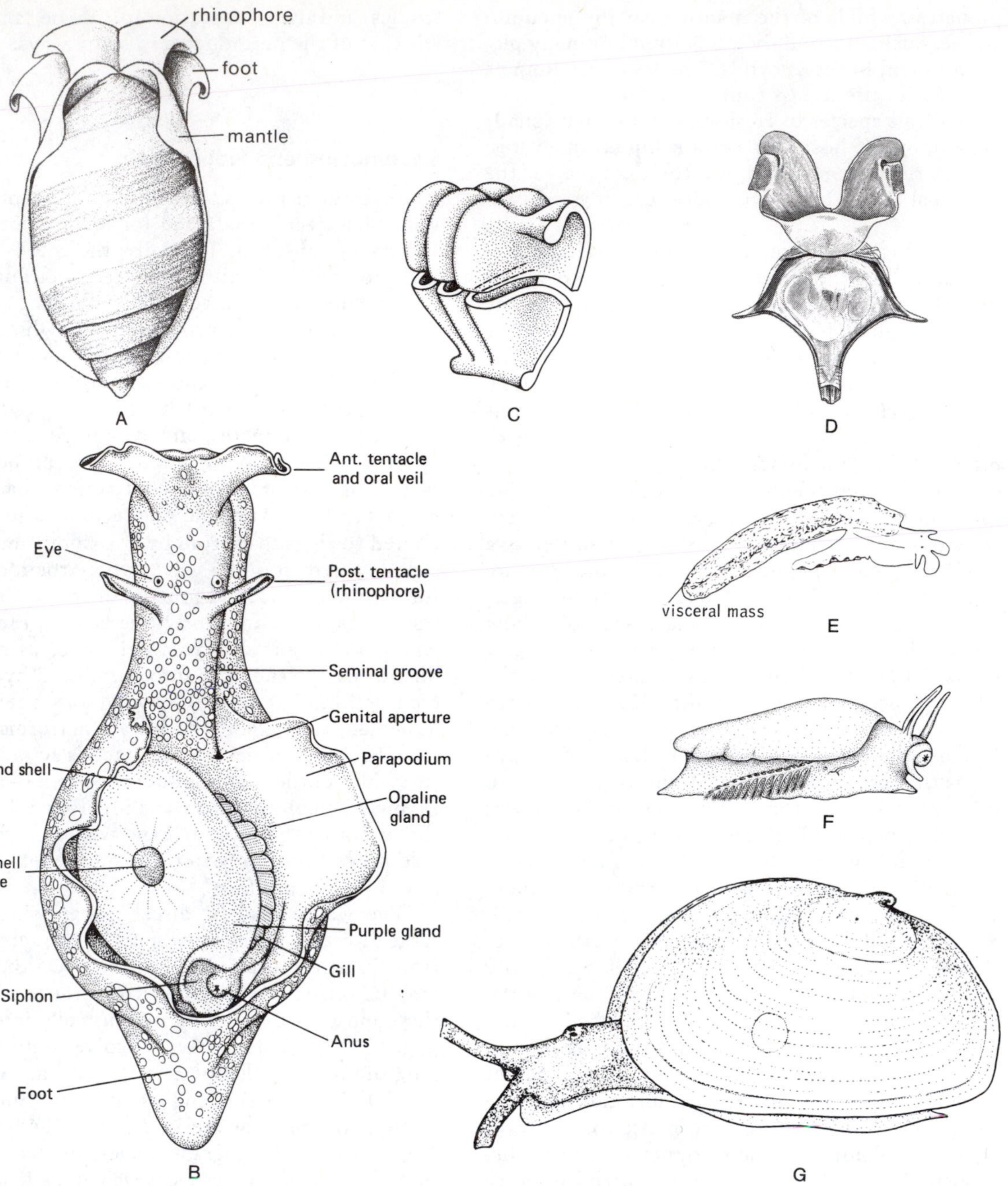

Figure 11–14 Opisthobranchs: *A*, The bubble shell *Hydatina*, a cephalaspidean. (Modified from several authors.) *B*, Diagrammatic dorsal view of a sea hare, an anaspidean. (After Guiart from Kandel, E. R., 1979: Behavioral Biology of *Aplysia*. W. H. Freeman and Co., San Francisco) *C*, Part of gill of *Aplysia* cut horizontally to show folded condition. (Modified after Carew et al.) *D*, *Cavolina*, a shelled sea butterfly (pteropod; Thecosomata). (After a photograph by Abbott.) *E*, *Microhedyle* (Acochlidioidea). (After Odhner from Hyman.) *F*, Sluglike *Pleurobranchus* (Notaspidea). (After Vayassière and Hyman.) *G*, *Berthelinia*, an opisthobranch with a bivalve shell (Sacoglossa). (After Kawaguti from Hyman.)

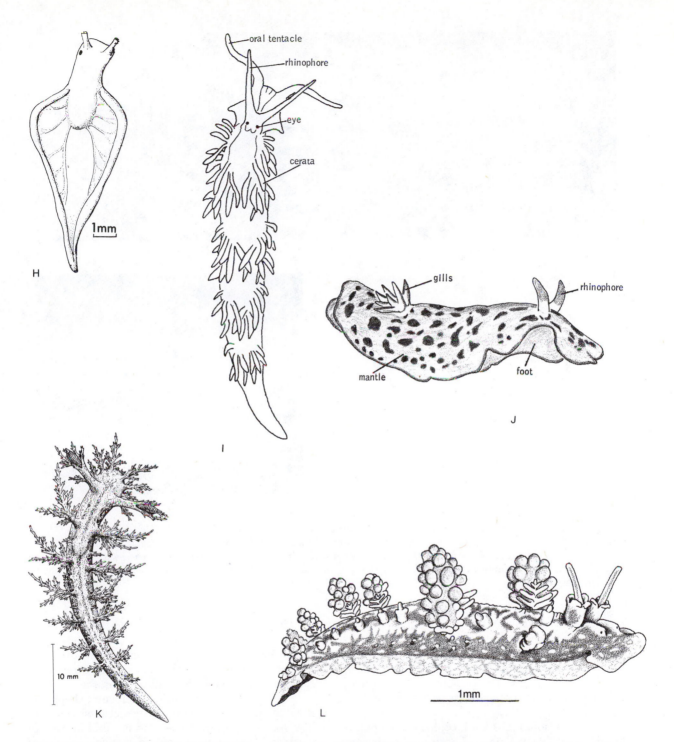

Figure 11–14 *(continued)* H, *Elysia viridis*, a sluglike sacoglossan. (After Gascoigne, T., 1975: Methods of mounting sacoglossan radulae. Microscopy, *32:*513.) I, *Eolidia papillosa*, a nudibranch with cerata (dorsal view). (After Pierce.) J, *Glossodoris*, a nudibranch with secondary anal gills. (After a photograph by P. Zahl.) (F, G, and H from Hyman, L. H., 1967: The Invertebrates. Vol. 6, Mollusca, Pt. I. McGraw-Hill Book Co., N.Y.) K, *Dendronotus frondosus*, a nudibranch with branched cerata. (From Thompson, T. E., and Brown, G. H., 1976: British Opisthobranch Molluscs. Academic Press, London. p. 67.) L, *Doto chica*, a nudibranch with cerata that look like clusters of grapes.

A

B

C

D

Figure 11–15 Pulmonates: *A*, A terrestrial slug. Opening into lung seen at lower edge of saddle-like mantle. *B*, A land pulmonate (Stylommatophora), showing the eyes mounted at the ends of the tentacles. *C*, *Testacella*, a land slug with a reduced shell. *D*, Shell of specimen of the large South American pulmonate *Strophocheilus*, compared with two common species of temperate American land snails, *Polygyra* and *Retinella* (smallest on ruler). (*C* after Baker; photographs by Betty M. Barnes.)

tern among prosobranchs is retrograde ditaxic. Most pulmonates exhibit direct monotaxic waves. In general, pedal waves, whether direct or indirect, occupy about one third of the foot length and only one or two waves are present simultaneously. Commonly, retrograde ditaxic waves are oblique (Fig. 11–17*D*).

Some prosobranchs and cephalaspid opisthobranchs that live on soft sand bottoms have become adapted for burrowing. In the moon shells *Natica* and *Polinices*, the front of the foot, called the propodium, acts like a plough and anchor, and a dorsal flaplike fold of the foot covers the head as a protective shield (Fig. 11–18*A*).

The conch *Strombus* crawls over sand in a very different fashion from other gastropods (Fig. 11–18*B*). The large, clawlike operculum digs into the sand, and the animal then "poles" forward by rapid contraction of the columella muscle.

There are correlations between shell shape and movement and habitation. In general, shells with low spires are more stable and better adapted for carriage upside down or on the vertical surfaces of rocks and vegetation. Shells with long spires are carried horizontally or even dragged over soft bottoms. Spines and other shell projections and sculpturing may contribute to shell strengthening, to protection, to stabilization in soft bottoms, to bur-

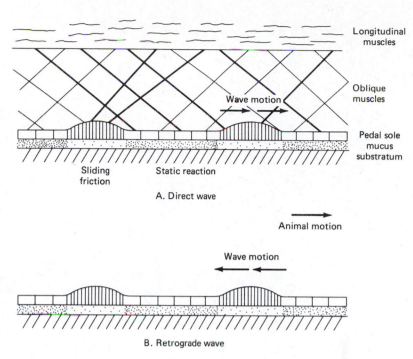

Figure 11–16 Pedal muscular waves in gastropods: *A*, Direct waves. Waves of pedal contraction sweep over foot in same direction as animal is moving. In region of wave (close vertical lines) sole slides forward over liquefied mucus. *B*, Retrograde waves. Waves sweep from front to back of foot opposite to direction of animal movement. (Modified after Denny, M. W., 1980: A quantitative model for the adhesive locomotion of the terrestrial slug, *Ariolimax columbianus*, J. Exp. Biol., 91:195–218; and after Trueman, E. R., and Jones, H. D., 1977: Crawling and burrowing, *In* Alexander, R. M., and Goldspink, G. (Eds.): Mechanics and Energetics of Animal Locomotion, Chapman and Hall, London.)

rowing, or even to landing right side up if the animal is knocked off a rock.

Limpets, abalones, and slipper shells are especially adapted for clinging to rocks and shells. All have low, broad shells that can be pulled down tightly and offer less resistance to waves and currents. The large, mucus-covered foot functions as an adhesive organ as well as for movement, and the surrounding shirtlike mantle margin, which may bear tentacles, serves as an important sense organ (Fig. 11–19B).

The homing ability of many intertidal limpets—*Collisella, Patella, Fissurella, Siphonaria*—has been the subject of numerous studies. A slight depression in the rock surface, over which the edges of the shell have come to fit very tightly, constitutes the limpet's "home" (Fig. 11–19A). At high tide the animal may wander 10 to 150 cm away in feeding, depending on the species, but then returns to its home (Fig. 11–20). The experimental studies to date indicate that homing ability depends primarily on chemical cues in the mucus laid down by the limpet as it makes its feeding excursions (Croll, 1983). Homing may reduce intraspecific competition by establishing a grazing territory and may reduce desiccation and predation, as a result of the tight fit into the home site. Especially remarkable is the discovery that the mucus of at least some homing species stimulates algal growth

in their territory, but this is not true of the mucus of nonterritorial, migratory species (Connor and Quinn, 1984).

Most intertidal limpets fall within two general adaptive groups, migratory and nonmigratory. Migratory limpets are "generalists." They settle low down in the intertidal zone and during the course of growth migrate to higher positions. They are thus found over a relatively wide band of the intertidal zone. They are rather unselective in their grazing habits and do not have territories or homes. Nonmigratory limpets are confined to relatively narrow zones within the intertidal zone. They are selective in their grazing and may be territorial (Branch, 1981).

Many terrestrial snails and slugs are able to return to shelters beneath logs and stones, especially when the shelter is occupied by other individuals of the same species. The homing cue appears to be a pheromone in mucous trails or air-borne pheromones from fecal pellets (Cook, 1979; Rollo and Wellington, 1982; Croll, 1983).

A small number of gastropods are adapted for a sessile existence. The worm shells, members of three unrelated mesogastropod families (Vermetidae, Turritellidae, and Siliquariidae), have typical larval and juvenile shells, but as the animal grows older, the whorls become completely separated, and the adult shell looks like a corkscrew or is com-

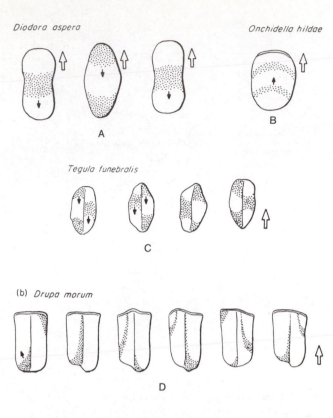

Diodora aspera

Onchidella hildae

A

B

Tegula funebralis

C

(b) *Drupa morum*

D

Figure 11–17 Patterns of pedal waves in gastropods: *A*, Retrograde monotaxic waves in the keyhole limpet *Diodora aspera*. *B*, Direct monotaxic waves in the intertidal pulmonate slug *Onchidella hildae*. *C*, Retrograde ditaxic waves in the archaeogastropod *Tegula funebralis*. *D*, Direct ditaxic waves in the neogastropod *Drupa morum*. Wave sequence is from left to right in each case. Large white arrows indicate direction of animal's movement, small black arrows the direction of the waves. (From Miller, S. L., 1974: The classification, taxonomic distribution, and evolution of locomotor types among prosobranch gastropods. Proc. Malac. Soc. London, *41*:233–272.)

pletely irregular (Fig. 11–21). Worm shells live attached to sponges, other shells, or rocks, and the separated whorls provide greater surface area for attachment. The foot is reduced, but an operculum is present (Fig. 11–21*B* and *C*).

A pelagic existence has been adopted by some mesogastropods (heteropods) and by some opisthobranchs, notably the pteropods, or sea butterflies. In most of these groups the foot has become modified as an effective finlike swimming organ. The Heteropoda are laterally compressed, and the foot is transformed into a ventral fin, even though these animals swim upside down (Fig. 11–22*G*).

The swimming foot of opisthobranchs is modified differently. Two fins, called parapodia, arise as lateral projections from the side of the foot. In the sea hares, which swim intermittently, the fins arise from the middle of the body and are very broad (Fig. 11–23). The pteropods, or sea butterflies, have anteriorly located parapodia, which function as oars (Fig. 11–22*A* to *F*), and the animals swim upside down.

Of still other modes of existence that might be described, space permits only brief mention of the common but minute prosobranch species of *Caecum*, which have short, tusklike shells; the prosobranch violet shells *(Janthina)*, which float beneath a raft of bubbles secreted by the foot (Fig. 11–24*A*); planktonic sea slugs *(Glaucus* and *Glaucilla)*, which stay afloat by means of a bubble of air held in the stomach; the Coralliophilidae, whose often strangely shaped shells bore into coral (Fig. 11–24*B*); the carrier shell, *Xenophora*, which attaches foreign objects, including other gastropod and bivalve shells, to its own shell with a foot secretion (Fig. 11–24*C*); a few minute, naked, interstitial opisthobranchs; and a group of algae-inhabiting, sacoglossan opisthobranchs that have secondarily derived bivalve shells (Fig. 11–14*G*).

Nutrition

Virtually every type of feeding habit is exhibited by gastropods. There are herbivores, carnivores, scavengers, deposit feeders, suspension feeders, and parasites. Despite great differences in feeding habits, it is possible to make a few generalizations. (1) A radula is usually employed in feeding. (2) Digestion is always at least partly extracellular. (3) With few exceptions, the enzymes for extracellular digestion are produced by the salivary glands, esophageal pouches, the digestive diverticula, or a

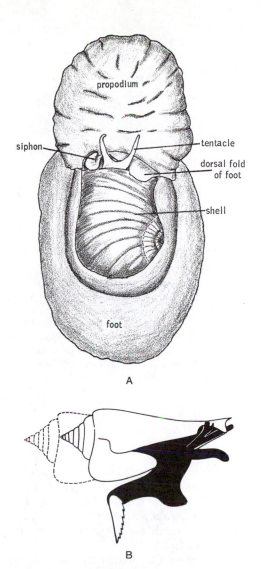

A

A

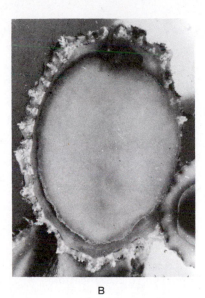

B

Figure 11–18 *A, Polinices,* a burrowing gastropod (dorsal view). (Drawn from life.) *B,* Lateral view of *Strombus.* Black is foot, to which the toothed, bladelike operculum is attached. Dotted outline indicates length of one "leap." (From Morton, J. E., 1964: *In* Wilbur, K. M., and Yonge, C. M. (Eds.): Physiology of Mollusca. Vol. 1. Academic Press, N.Y.)

Figure 11–19 *A,* Photograph of a small area of intertidal rock on the west coast of Scotland. Three specimens of the neogastropod *Thais* are feeding on barnacles. Four patellacean limpets belonging to the genus *Patella* can be seen to the right and above the snails. Numerous barnacles have settled on the limpets' shells. *B,* Ventral view of the keyhole limpet *Fissurella,* showing the sensory tentacles on the mantle margin. (By Betty M. Barnes.)

combination of these structures. (4) The stomach is the site of extracellular digestion, and the digestive diverticula are the sites of absorption and of intracellular digestion, if such digestion takes place. (5) As a result of torsion, the stomach has been rotated 180 degrees, so the esophagus enters the stomach posteriorly and the intestine leaves anteriorly (Fig. 11–4*B*). In the higher gastropods there has been a tendency for the esophageal opening to

migrate forward again toward the more usual anterior position.

In most gastropods the radula has become a highly developed feeding organ, acting as a grater, rasp, brush, cutter, grasper, or conveyor. The teeth vary in number from 16 to thousands and are al-

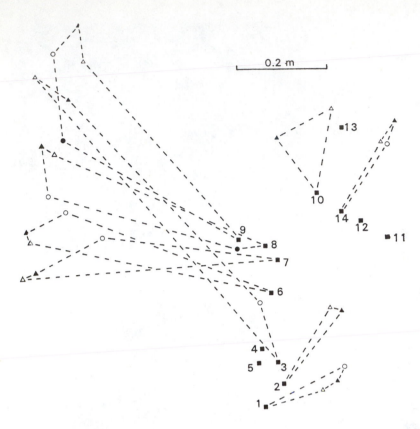

Figure 11–20 Movements of 14 homing limpets *(Patella vulgata).* Square indicates home site; open triangle, position 1.5 hours after immersion; black triangle, position 2.5 hours after immersion; open circle, 3.5 hours after immersion; black circle, 4.5 hours after immersion. (From Hartnoll, R. G., and Wright, J. R., 1977: Foraging movements and homing in the limpet *Patella vulgata.* Anim. Behav., 25:808.)

ways arranged in rows. Usually there is a median, longitudinal row of teeth, on each side of which are rows of lateral and outer fine marginal teeth (Fig. 11–25B). The median, lateral, and marginal teeth usually differ from one another in shape and structure. The character and form of the radula teeth are relatively constant at the family level and above, and are important in classification.

PRIMITIVE GASTROPODS

The most primitive feeding habit and digestive tract are found in the archaeogastropods. The marine archaeogastropods are largely microphagous, grazing on fine algae, sponges, or other organisms growing on rocks or kelps. The radula usually bears many teeth, at least 12 in each transverse row (Fig. 11–25B and C). When the radula is retracted, the numerous, fanlike, marginal teeth direct the ingested particles into the center of the gutter produced by the lateral folding of the radula ribbon. The food-laden mucous string formed within the gutter is pulled into the esophagus and then the stomach by the rotating mucous mass within the style sac.

The grazing activity of large populations of intertidal limpets and chitons may greatly limit the growth of algae (Fig. 11–26). Patellacean limpets have a rasping radula with 6 to 20 teeth in a transverse row; the teeth are stout and impregnated with iron and silicon.

With the exception of the patellacean limpets, the stomach (Fig. 11–25A) is like that described for the generalized mollusk (p. 344). The large surface area of the sorting region is accommodated within a cecum (Fig. 11–25A). Digestion is partly extracellular, by enzymes that are elaborated by glands in the esophageal region, and partly intracellular within the digestive diverticula.

HIGHER GASTROPODS

Perhaps because of the invasion of soft bottoms and other habitats, the diets and feeding habits of higher gastropods became extremely diverse, especially among the mesogastropods and opisthobranchs (Fig. 11–27). Most higher gastropods are macrophagous. Digestion has become entirely extracellular and takes place in the stomach, which has lost most of its primitive features—chitinous

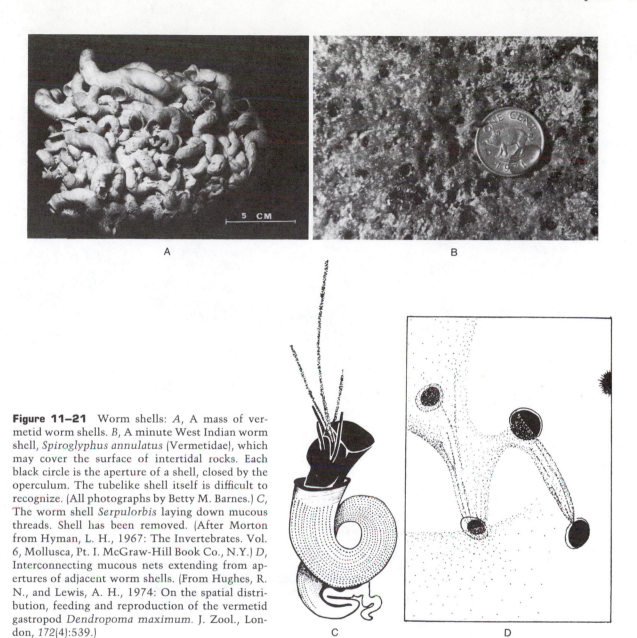

Figure 11–21 Worm shells: *A*, A mass of vermetid worm shells. *B*, A minute West Indian worm shell, *Spiroglyphus annulatus* (Vermetidae), which may cover the surface of intertidal rocks. Each black circle is the aperture of a shell, closed by the operculum. The tubelike shell itself is difficult to recognize. (All photographs by Betty M. Barnes.) *C*, The worm shell *Serpulorbis* laying down mucous threads. Shell has been removed. (After Morton from Hyman, L. H., 1967: The Invertebrates. Vol. 6, Mollusca, Pt. I. McGraw-Hill Book Co., N.Y.) *D*, Interconnecting mucous nets extending from apertures of adjacent worm shells. (From Hughes, R. N., and Lewis, A. H., 1974: On the spatial distribution, feeding and reproduction of the vermetid gastropod *Dendropoma maximum*. J. Zool., London, *172*(4):539.)

lining, sorting area, style sac—and become more or less a simple sac (Fig. 11–28*C*). Enzymes are supplied by the digestive diverticula or by glands associated with the esophagus or buccal region.

The highly adaptable mesogastropod radula bears seven teeth in a transverse row (Fig. 11–28*A*); the marginal teeth are hook shaped. The neogastropods, which are mostly carnivores, have radulas with only three teeth (sometimes only one tooth)

per transverse row, but the teeth are heavy and usually bear several cusps (Fig. 11–28*B*). The outer, hooked-shaped teeth collect torn or detached particles and bring them into the center when the radula is retracted. The largely herbivorous pulmonates have radulae with the largest number of teeth of any gastropod: up to 750 small teeth per transverse row (Fig. 11–29). The opisthobranch radula is highly variable. The efficiency of the radula results

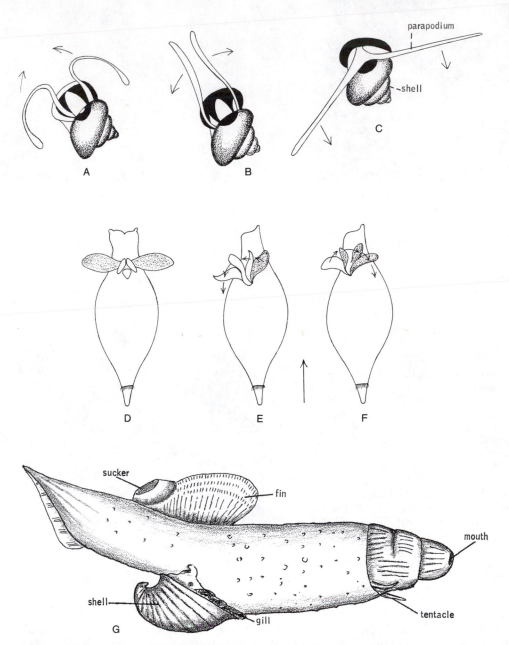

Figure 11–22 *A–C,* Swimming in the shelled sea butterfly, *Limacina.* Arrows indicate the direction of movement of the parapodia. *A* shows the recovery stroke; *B,* the beginning of the effective stroke; and *C,* the middle of the effective stroke. *D–F,* Parapodial movement in the swimming of the naked sea butterfly, *Clione limacina: D,* ventral view; *E* and *F,* side views. (All after Morton, J. E., 1967: Molluscs, 4th Ed. Hutchinson University Library, London.) *G, Carinaria,* a pelagic prosobranch (Heteropoda). (After Abbott.)

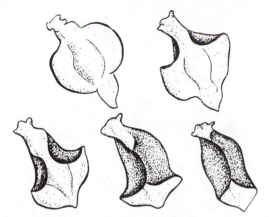

Figure 11–23 Swimming in the sea hare *Aplysia*. The lateral swimming fins are foot folds, or parapodia. (After Pruvot-Fol from Farmer, W. M., 1970: Swimming gastropods. Veliger, *13*(1):73.)

not only from the adaptive design of particular teeth but also from the complex ways in which the teeth interact with each other. Radula function is described at length by Solem (1974) with beautiful scanning electron photomicrographs.

Feeding is also facilitated in many gastropods, including most opisthobranchs and pulmonates, by jaws, which are thickened cuticular pieces in the front of the buccal cavity.

HERBIVORES

The many herbivorous gastropods include some marine prosobranchs, the freshwater prosobranchs, the operculate land snails, a variety of opisthobranchs, and the majority of the pulmonates. Most marine species feed on fine algae that can be rasped from a rock or other surfaces, or on large algae, such as kelps, that can carry the weight of the snail (Steneck and Watling, 1982). Freshwater and land forms also consume the tender parts of aquatic and terrestrial vascular plants, decaying vegetation, or fungi. A few terrestrial snails and slugs are serious agricultural pests. The giant African snail *Achatina fulica*, introduced into Hawaii and the United States, can be very destructive, and considerable effort has been expended to prevent its spread. Jaws are an adaptation of many herbivorous gastropods, including pulmonates.

Members of the mesogastropod family Littorinidae, called periwinkles, are found on rocky shores,

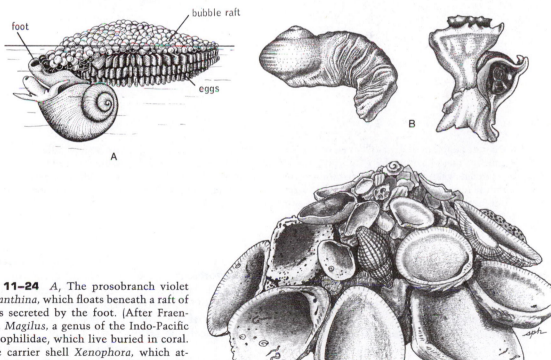

Figure 11–24 *A*, The prosobranch violet shell, *Janthina*, which floats beneath a raft of bubbles secreted by the foot. (After Fraenkel.) *B*, *Magilus*, a genus of the Indo-Pacific Coralliophilidae, which live buried in coral. *C*, The carrier shell *Xenophora*, which attaches shells and other foreign objects to its own shell. (*B* based on photograph by T. Abbott.)

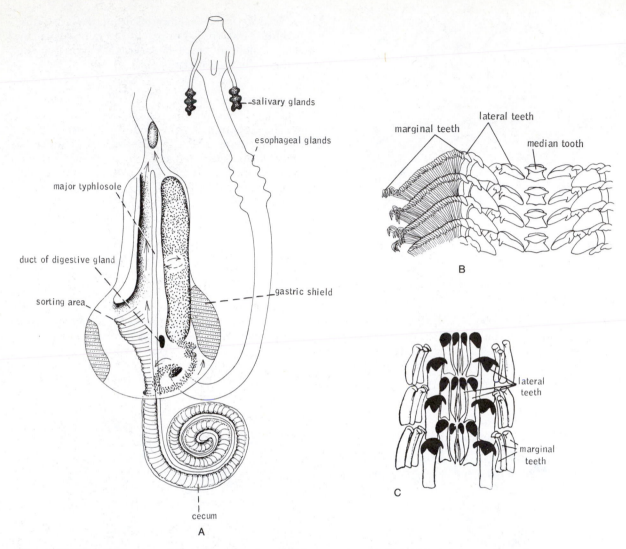

Figure 11–25 *A*, Diagram of digestive tract of a primitive prosobranch (Trochidae). Arrows show ciliary currents and rotation of mucous mass (protostyle) within style sac. (After Owen.) *B*, Part of radula (rhipidoglossan type) of the abalone *Haliotis* (archaeogastropod), with many transverse rows of teeth. Marginal teeth on one side have been omitted. *C*, Part of the radula (docoglossate type) of the archaeogastropod *Patella*, showing three transverse rows of 12 teeth each. (*B* and *C* from Fretter, V., and Graham, A., 1962; British Prosobranch Molluscs. Ray Society, London. p. 171.)

mangroves, and even marsh grasses throughout the world. These common and often abundant snails live in the intertidal zone, each species occupying a characteristic level between the low tide mark and the high splash region. At low tide the animal withdraws into its shell behind the operculum, attaching the lip of the shell to the substratum with mucus. At high tide it emerges to graze on fine algae, including endolithic forms that penetrate just below the surface. The gut contents include large quantities of mineral material.

The sacoglossan opisthobranchs are specialized herbivores. In these tiny, sluglike gastropods the radula is reduced to a single, longitudinal row of teeth, used to slit open algal cells (Fig. 11–30). The contents of the cells are then sucked out. Sacoglossans tend to be rather restrictive in the species of algae used for food. Some species, including the European *Elysia viridis*, are remarkable in incorporating the chloroplast of their food into their own digestive gland cells, where photosynthesis then occurs (Jensen, 1980; Clark et al, 1981;

Figure 11–26 Effect of grazing by limpets on the Isle of Man, Great Britain. The dark zone is a strip through the intertidal zone from which all limpets were removed, permitting an algal cover to form on the rocks. (By Norman Jones.)

Hinde, 1983; Brandley, 1984). Some nudibranch sea slugs are now known to harbor zooxanthellae, but they obtain them from the octocorallians on which they feed (Rudman, 1981).

In many herbivorous species the esophagus or anterior part of the stomach is modified as a crop and gizzard. The gizzard may be lined with cuticle (as in the sea hare) or contain sand grains (as in many freshwater snails). Amylases and cellulases are produced by the esophageal glands or the digestive glands. Terrestrial pulmonates possess a powerful array of digestive enzymes, and digestion is initiated in the large crop that to a great extent replaces the stomach (Fig. 11–31). There is no gizzard.

CARNIVORES

Although a few of the many carnivorous gastropods are pulmonates, feeding on earthworms or other snails and slugs, the majority of carnivorous families are prosobranchs and opisthobranchs. The radula of these marine families is variously modified for cutting, grasping, tearing, scraping, or conveying. Jaws are sometimes present. The most common adaptation of carnivorous prosobranchs is a highly extensible proboscis, which enables the animal to reach and penetrate vulnerable areas of the prey. The proboscis (which contains the esophagus, buccal cavity, radula, and at the tip, the true mouth) lies within a proboscis sac, or sheath, which has a mouthlike opening to the outside (Fig. 11–32). In feeding, the proboscis is projected out of the opening of the proboscis sac by blood pressure (Fig. 10–32*A* and 11–33). Specific proteins liberated by prey or carrion, once detected in the ventilating current, aid in locating the food source and will elicit protrusion and search with the proboscis (Smith, 1977; Croll, 1983).

The larger bottom-dwelling carnivores, the neogastropods and some mesogastropods, commonly feed on bivalve mollusks, other gastropods, sea urchins, sea stars, polychaetes, and even fish. Many burrow into the sand to reach their prey. Bonnets (*Phalium*), helmets (*Cassis*), tuns (*Tonna*), and tritons (*Cymatium*) narcotize their prey with salivary secretions containing sulfuric acid or another toxin. Some olives and volutes smother victims with their feet. A whelk (*Buccinum, Busycon, Fasciolaria, Murex*) may grip the bivalve with its foot, pulling or wedging the two valves apart with the edge of its shell or siphonal canal (Fig. 11-34*A*). To accomplish the wedging, the gastropod may first erode the valve margin with the lip of its own shell.

A number of prosobranchs are adapted for drilling holes in the shells of prey, such as limpets, barnacles, and especially bivalves (Fig. 11–34*C* and *D*). The two best known such families are the neogastropod Muricidae (*Urosalpinx, Murex, Thais, Nucella, Eupleura*) and the mesogastropod Naticidae (moon shells, *Natica, Polinices*). The mechanism has been most extensively studied in the Muricidae, and particularly in *Urosalpinx*, which causes great damage in oyster beds. Both the American drill, *Urosalpinx*, and the Japanese drill, *Rapana*, have been introduced into other parts of the world with shipments of oysters. The anterior sole of the foot contains an eversible gland, which is applied to the area to be drilled. The acidic secretion produced by this gland reduces the organic framework and demineralizes the shell. Penetration is primarily a result of glandular activity rather than the radula. The animal drills with the radula for about a minute and then applies the eversible gland for about 30 to 40 minutes, repeating the cycle until the bivalve shell is penetrated. Approximately 8

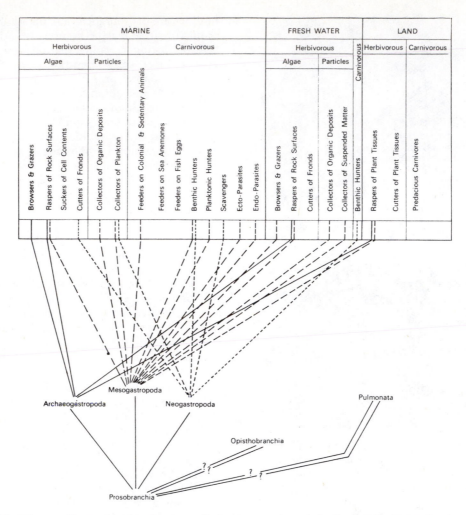

Figure 11–27 Diagram illustrating the adaptive radiation in prosobranch feeding habits. (From Purchon, R. D., 1968: Biology of the Mollusca. Pergamon Press, London. p. 73.)

hours would be required to penetrate a shell 2 mm thick, and penetration to a depth of 5 mm has been recorded. The beveled sides readily identify the hole as having been drilled by a gastropod. When drilling is completed, the proboscis is extended through the hole, and the soft tissues of the prey are torn by the radula and ingested (see Carriker, 1981, for review). In the naticids the shell-softening gland is located at the proboscis tip rather than on the foot.

One of the most remarkable groups of carnivores is the neogastropod genus *Conus*. Cone shells are tropical and subtropical and are found mainly in western Atlantic and Indo-Pacific oceans. They feed primarily on polychaete worms, other gastropods, or fish, which they stab and poison with the radular teeth. The odontophore has disappeared, and the radula is greatly modified (Fig. 11–35A). The teeth, which function singly, are long, grooved, and barbed at the end (Fig. 11–35B), and they are attached to the radula membrane by a slender cord of tissue. A large muscular bulb functions as an injector and is connected to the buccal cavity by a duct, which secretes the poison.

The cone shell has a long, highly maneuverable proboscis. When the proboscis is projected, the barbed end of a single radula tooth slips out of the radula sac into the buccal cavity and is then thrust into the prey. In those species that feed on snails, the tooth is freed from the proboscis, but in forms that feed on polychaetes and fish (Fig. 11–35C), the cone lies buried in the sand and does not strike

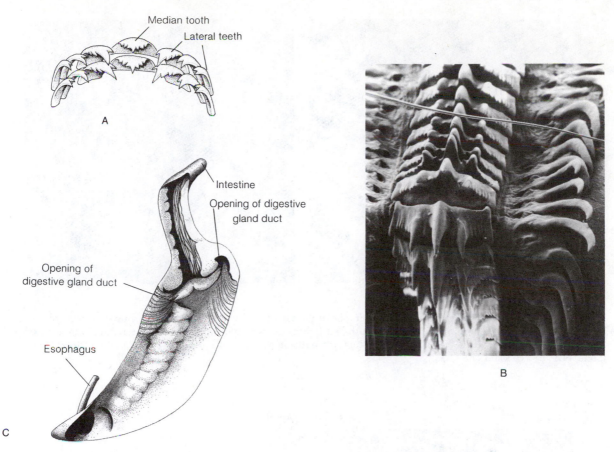

Median tooth
Lateral teeth

A

Intestine
Opening of digestive
gland duct

Opening of
digestive gland duct

Esophagus

C

B

Figure 11–28 *A*, Two transverse rows of radula teeth of the mesogastropod *Lanistes* (taenioglossate nodula). (Modified from Turner.) *B*, Scanning electron micrograph of the radula of *Urosalpinx*, a carnivorous drilling neogastropod. There are only three teeth to a transverse row, but the middle tooth bears several cusps (stenoglossate nodula). (From Carriker, M. R., 1969: Excavation of boreholes by the gastropod *Urosalpinx*. Am. Zool., 9:917–933.) *C*, Saclike stomach of a carnivorous prosobranch *(Natica)*. Stomach has been opened dorsally. (From Fretter, V., and Graham, A., 1962: British Prosobranch Molluscs. Ray Society, London. p. 226.)

until the fish pauses over the bottom. Then the harpoon is thrust in the soft underbelly of the victim, and the cone retains a grip on the end of the tooth. The victim is very quickly immobilized by the neurotoxic peptide poison, which enters the wound through the hollow cavity of the tooth (see brief review by Kohn, 1983).

The bite, or sting, of a number of South Pacific species is highly toxic to humans; a few deaths have been reported, in one case within 4 hours.

The possible adaptive evolution of the carnivorous prosobranchs is illustrated in Figure 11–36.

There are many carnivorous opisthobranchs. The principal raptorial groups are the naked sea butterflies and the bubble shells. The former prey on shelled sea butterflies, and the latter upon bi-

valves and gastropods, which are seized with the hooked teeth of the radula and swallowed whole.

The nudibranchs are grazing carnivores that feed on sessile animals, such as hydroids, sea anemones, soft corals, bryozoans, sponges, ascidians, barnacles, and fish eggs, and each family is generally restricted to one type of prey. There usually is no proboscis, but jaws are commonly present. In the Aeolidiidae, most of which feed on hydroids and sea anemones, the pair of bladelike jaws are used to cut small pieces of tissue from the prey.

The most remarkable feature of some nudibranchs with cerata is their utilization of the prey's nematocysts. Ciliary tracts in the stomach and in the ducts from the digestive diverticula carry the undischarged, even immature nematocysts to the

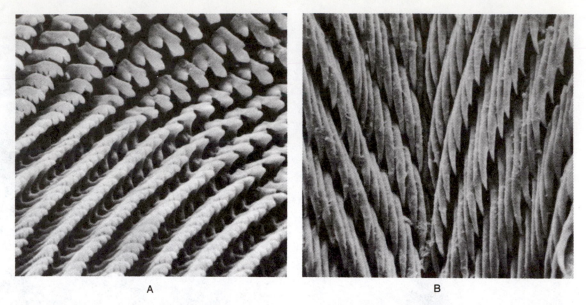

A   B

Figure 11–29 Scanning electron photomicrographs of pulmonate radulas: *A*, The herbivorous snail *Diastole conula*. *B*, The carnivorous *Euglandina rosea*. Each tooth is a long cone, which is erected when the radula is in a functional position. (From Solem, A., 1974: The Shell Makers: Introducing Mollusks. Reprinted by permission of John Wiley and Sons, N.Y. pp. 163 and 168.)

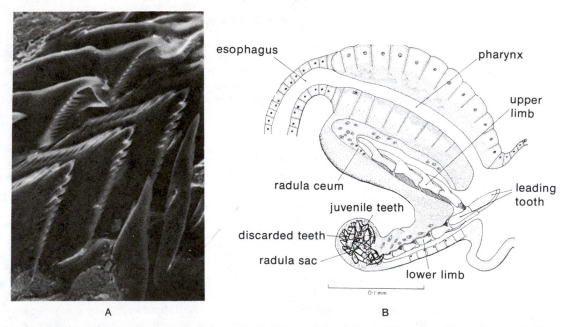

A   B

Figure 11–30 *A*, Scanning electron photomicrograph of the radula of the aeolidacean nudibranch *Coryphella*. (From Cowen, R. K., and Laur, D. R., 1977: A new species of *Coryphella* from Santa Barbara, California. Veliger, *20*(3):292–294.) *B*, Lateral view of the radula and buccal mass of the bivalve sacoglossan *Limapontia*. The single row of radula teeth is adapted for puncturing algal cells. (From Gascoigne, T., and Sartory, P. K., 1974: The teeth of three bivalved gastropods and three other species of the order Sacoglossa. Proc. Malac. Soc. London, *41*:11.)

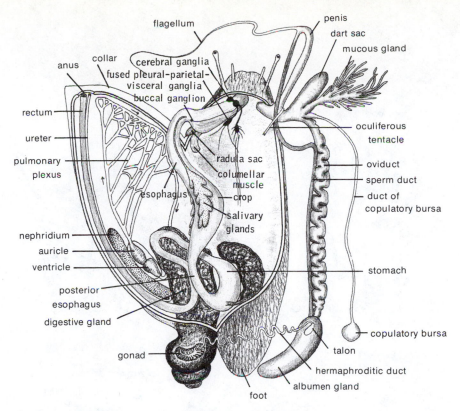

Figure 11–31 Dissection of the land snail *Helix*. (Adapted from several sources.)

cerata, where they are engulfed but not digested. The undischarged nematocysts are moved to the distal tips of the cerata, called cnidosacs, which open to the exterior. There the nematocysts are used by the nudibranchs for defense. The discharge mechanism is not thoroughly understood, but may result from pressure exerted either by circular muscle fibers around the cnidosac or else by the enemy. Nematocysts are replaced in 3 to 12 days, and most nudibranchs utilize only certain types of the various kinds of nematocysts present in their cnidarian prey (Kaelker and Schmekel, 1976; Conklin and Mariscal, 1977; Day and Harris, 1978; Greenwood and Mariscal, 1984).

The prosobranch counterparts of the predacious sea butterflies are the pelagic heteropods, which feed on other small, pelagic invertebrates; the cowries (mesogastropods), like the nudibranchs, are grazers that feed on sessile invertebrates (Fig. 11–34B).

SCAVENGERS AND DEPOSIT FEEDERS

A scavenging habit has been adopted by numerous gastropods, of which *Nassarius* and allies are notable examples. The feeding habits of these little neogastropods range all the way from a carnivorous habit to deposit feeding. On quiet, protected beaches along the Atlantic coast of the southeastern United States, *Ilyanassa obsoleta* may occur in enormous numbers at low tide, feeding on organic material deposited in the intertidal zone (Fig. 11–37A). This same species, however, is also a facultative carrion feeder and will consume the flesh

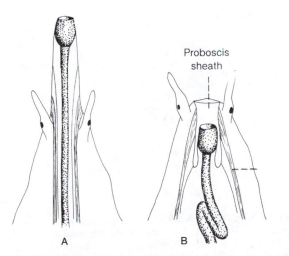

Figure 11–32 Diagram showing the proboscis of a prosobranch protracted (*A*) and retracted (*B*). (After Fretter, V., and Graham, A., 1962: British Prosobranch Molluscs. Ray Society, London.)

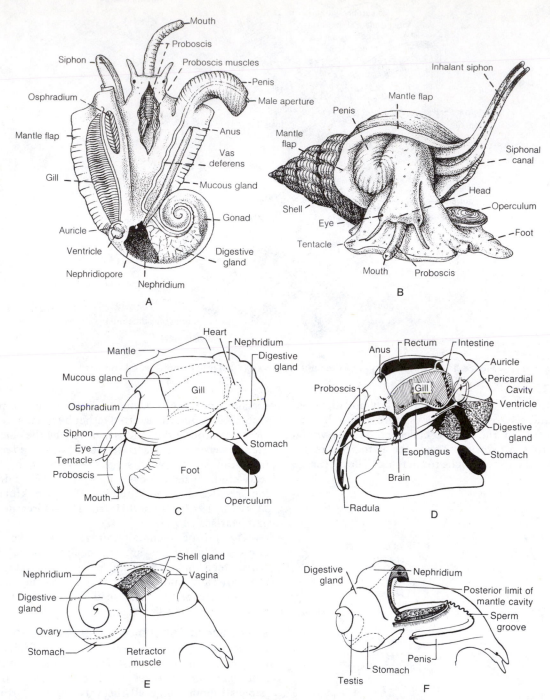

Figure 11–33 Anatomy of two marine prosobranchs (neogastropods): *A* and *B*, *Buccinum undatum*. *A*, Dorsal view, with shell removed and wall of mantle cavity cut and reflected. *B*, Animal crawling with proboscis protuded. (After Cox, L. R., 1960, Gastropoda. *In* Moore, R. C. (Ed.): Treatise on Invertebrate Paleontology, Vol. I (pt. 1). Geol. Soc. America and University of Kansas Press. Lawrence, Kansas. pp. I89 and I91.) *C–F*, *Busycon canaliculatum*, with shell removed. Shell is similar to that of *Buccinum* and is carried in the same position. *C*, Left side, showing external organs and internal organs visible through the integument. *D*, Same view with digestive, respiratory, circulatory, and nervous systems indicated. *E*, Female, showing portion of the right side. *F*, Male, portion of right side with mantle and retractor muscle cut short. In *E* and *F* the proboscis is withdrawn.

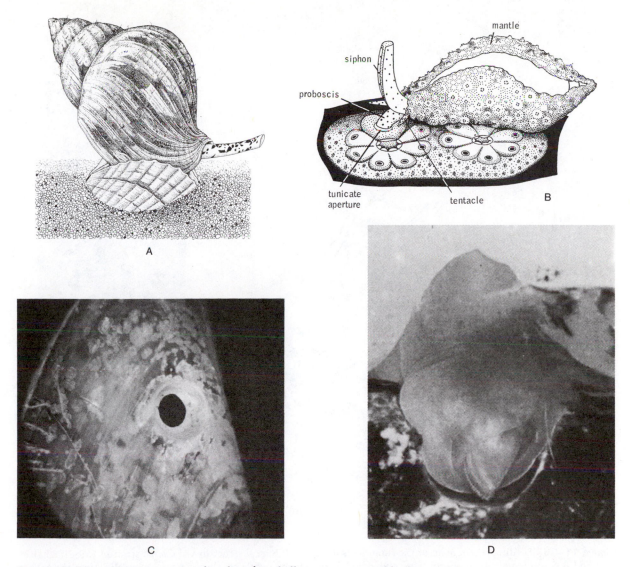

Figure 11–34   *A, Buccinum* using the edge of its shell to pry open a cockle. (From Nielsen, C., 1975: Observations on *Buccinum undatum* attacking bivalves and on prey responses, with a short review on attack methods of other prosobranchs. Ophelia, *13*:87–108.) *B,* The cowrie *Erato voluta* feeding on a colonial tunicate. The proboscis of the snail is thrust into the buccal opening of the tunicate. The shell of the cowrie is partially covered by the reflexed mantle; the erect structure is the siphon. (From Fretter, V., 1951: Proc. Malac. Soc. London, *29*:15.) *C,* Photograph of a perforation through a bivalve shell produced by the radula of a drilling gastropod. Note the beveled edge. *D,* Radula of the oyster drill *Urosalpinx* rasping across bottom of a partially excavated borehole. (From Carriker, M. R., 1969: Amer. Zool., *9*:920.)

of fresh fish. Along the coasts of Britain and Europe the tiny deposit-feeding mesogastropod *Hydrobia* also occurs in enormous numbers (as many as 30,000 per square meter) on muddy, intertidal flats.

Other deposit feeders include many common species of horn shells (*Cerithium, Cerithidea, Batillaria*). The conch *Strombus* and the related burrower *Aporrhais* are deposit feeders or grazers on fine algae, and the large, mobile proboscis sweeps across the bottom like a vacuum cleaner but under the protective canopy of the flaring shell aperture.

SUSPENSION FEEDERS

Suspension feeding has evolved a number of times within the Gastropoda. In the slipper shell (*Crepidula*), a filtering suspension feeder, the gill filaments have been tremendously lengthened to pro-

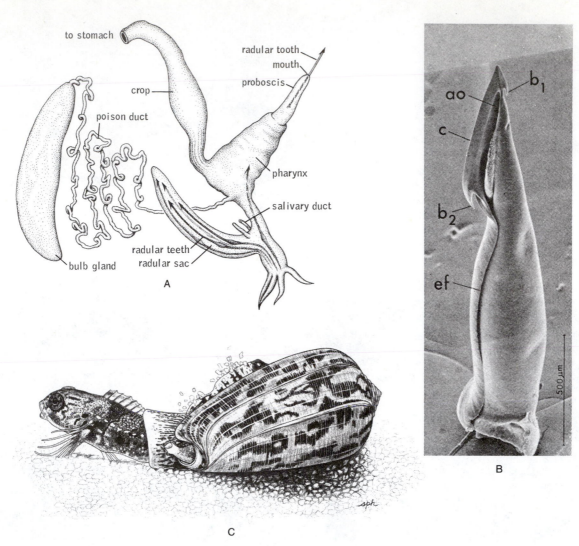

Figure 11–35 Feeding mechanism of the cone shells *(Conus). A*, Buccal structures of *Conus striatus*. (Modified after Clench.) *B*, Scanning electron photomicrograph of the harpoon-like radula tooth of *Conus imperialis*. Note the folded structure and barbed end. (From Kohn, A. J., et al, 1972; Science, 176:49–51. Copyright 1972 by the American Association for the Advancement of Science.) *C*, A cone shell swallowing a fish. (Based on a photograph by Robert F. Sisson and Paul Zahl.)

vide an increased surface area for trapping plankton on a mucous sheet (Fig. 11–38*A* and *B*). Many of the sessile worm shells use the gills as a food-trapping surface. Others secrete a net or veil of mucous threads in the vicinity of the shell opening. The net is produced by the pedal gland, laid down by pedal tentacles, and spread by wave action (Fig. 11–21*C* and *D*), covering as much as 50 cm² in *Serpulorbis squamigerous* of the American Pacific coast. The Red Sea *Dendropoma maxima* takes 2 minutes to haul in the net with the radula, which it does every 13 minutes (Hughes and Lewis, 1974).

The shell-bearing sea butterflies are suspension feeders, trapping food particles in the mucus covering the parapodia on their mantle cavities. Some sea butterflies, such as *Gleba* and *Corolla*, secrete enormous, floating mucous nets as big as 2 meters in diameter (Fig. 11–37*B*). The animal hangs beneath the net by its extended proboscis. There is no radula, and food particles trapped in the net are pulled in by proboscis cilia.

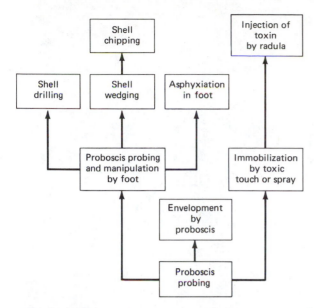

Figure 11–36 Possible evolutionary relationships of carnivorous feeding modes in prosobranchs. See text for descriptions. (From Taylor, J. D., Morris, N. J., and Taylor, C. N., 1980: Food specialization and the evolution of predatory prosobranch gastropods. Palaeontology, 23(2):375–409.)

Crystalline styles are found in many suspension and deposit feeders—*Crepidula*, *Struthiolaria*, many worms shells, *Strombus*, species of *Nassarius*, *Cerithium*, and some sea butterflies among the opisthobranchs. This structure is typical of bivalve mollusks and is described on page 414. The presence of a crystalline style in gastropods is associated with the more or less continuous feeding of the animal on a diet of phytoplankton or organic detritus.

PARASITES

Parasitism has evolved in a number of gastropods, which along with certain free-living forms present an interesting adaptive series leading from an epizoic to an ectoparasitic existence and thence to endoparasitism. The ectoparasites are modified chiefly with respect to the nature of the buccal region and digestive system. Members of the opisthobranch family Pyramidellidae possess chitinous jaws, stylets, and a pumping pharynx for sucking blood from bivalve mollusks and polychaetes (Fig. 11–38*D*).

The remaining commensals and parasites belong to a related group of small families (Eulimacea) that live on or within echinoderms (Fig. 11–38*C*). The most modified are members of the wormlike Entoconchidae, which live within sea cucumbers (Fig. 11–38*E*).

Excretion and Water Balance

Many archaeogastropods possess two nephridia* but in all other gastropods the right nephridium has disappeared, except for a small section that contributes to the reproductive duct. Also, as a result of torsion, the nephridium is located anteriorly in the visceral mass (Fig. 11–33). The structure of the nephridium is that of a blind sac, with the walls

---

*The left nephridium is reduced in patellacean limpets.

A

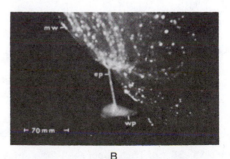

B

Figure 11–37 *A*, *Ilyanassa obsoleta* feeding at low tide. This intertidal prosobranch scavenger and deposit feeder occurs in enormous numbers on protected beaches on the east coast of the United States. (By Betty M. Barnes.) *B*, The sea butterfly *Gleba cordata* feeding from its large, delicate, mucous web. The long process labeled *ep* is the proboscis. The label *wp* indicates one of the winglike extensions of the foot. (From Gilmer, R. W., 1972: Science, 176:1240. Copyright 1972 by the American Association for the Advancement of Science.)

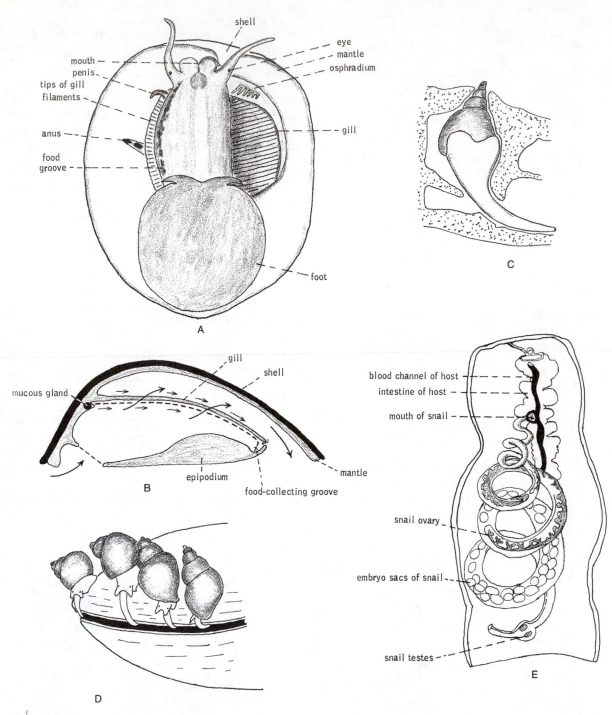

Figure 11–38   *A*, The slipper shell *Crepidula fornicata*, a ciliary feeder (ventral view). (Drawn from life.) *B*, *Crepidula*, showing direction of water current (large arrows) through mantle cavity. Ciliary currents (small arrows) carry food particles. Dashed lines indicate filtering sheet of mucus. Coarse particles are removed by the mucous web at the entrance to the mantle cavity; fine particles are trapped by the mucous sheet on the gill. The latter is driven by frontal gill cilia to a food-collecting groove along the side of the body, where it is rolled up into a string and periodically pulled into the mouth by the radula. (Modified from Morton.) *C*, The parasite *Stilifer* embedded in body wall of a sea star. *D*, *Brachystomia*, an ectoparasite, shown feeding on body fluids of a clam. (*C* and *D* after Abbott.) *E*, The endoparasite *Entoconcha* within body cavity of a sea cucumber. (After Baur from Hyman.)

greatly folded to increase the surface area for secretion. At the end of the sac, near the nephridiopore, the nephridium connects with the pericardial cavity via a renopericardial canal (Fig. 11–44A). The nephridiopore of both the prosobranchs and the lower opisthobranchs opens at the back of the mantle cavity, and wastes are removed by the circulating water current (Fig. 11–12). Such an arrangement is not possible in most pulmonates because the mantle cavity functions as a lung. As a result, the pulmonate ureter has lengthened along the right wall of the mantle and opens to the front of the mantle cavity or to the outside near the anus and the pneumostome.

Freshwater gastropods maintain a rather low level of blood salts, and the nephridia are capable of excreting a hyposmotic urine by reabsorption of salts. Freshwater species expel large amounts of water through the nephridia. Terrestrial pulmonates and operculate land snails conserve water by converting ammonia to relatively insoluble uric acid. Perhaps as an adaptation for water conservation, the renopericardial canal of pulmonates has a small orifice opening into the pericardial cavity, and there are no pericardial glands. Wastes are thus largely a result of nephridial secretion. This is also true of some intertidal prosobranchs, such as the littorinids.

Terrestrial pulmonates have not been very successful in controlling desiccation through the body surface, and considerable water is lost in the production of the slime trail for crawling. However, many pulmonates can survive excessive desiccation: *Helix* can survive a water loss equal to 50 per cent of the body weight, and the slug *Limax* can survive an 80 per cent loss.

As would be expected, the majority of pulmonates require a humid environment. They either are nocturnal or live in damp places, such as beneath logs or in leaf mold on forest floors. There are pulmonates that inhabit dry, rocky areas, dunes, and even deserts. Such species are active only at night or following rains. During dry periods in the tropics and in deserts and during the winter in temperate regions, snails become inactive, either estivating or hibernating. They may first burrow into humus or soil, or climb into vegetation and attach the aperture edge of the shell with dried mucus (Fig. 11–39). Estivation in elevated positions is probably an adaptation to avoid ground-dwelling predators, such as mice, reptiles, and beetles. Terrestrial prosobranchs close the aperture with the operculum, but in pulmonates the edges of the mantle are drawn together in front of the shell aperture, and a protective mucous membrane

Figure 11–39 Estivating snails attached to vegetation in Israel. (By David T. Barnes.)

that hardens when dry (or a thin calcareous membrane) is secreted over the opening. Several such epiphragms may be secreted as the snail withdraws further into the shell. Freshwater snails also estivate when ponds dry up and hibernate when the water is frozen. Estivation may last a number of months and there are records of snails estivating for several years (Boss, 1974). Reactivation commonly results from temperature changes, rise in humidity, or jarring, as from rain drops beating on the shell. Handling can cause reactivation.

Internal Transport

As a result of torsion, the heart of a gastropod is located anteriorly in the visceral mass (Fig. 11–4B). The primitive archaeogastropods have retained two auricles. However, in all other gastropods the right auricle either has become vestigial or, in most instances, has disappeared as a result of the loss of the right gill, which supplied it with blood (Figs. 11–12 and 11–33). The ventricle gives rise to a pos-

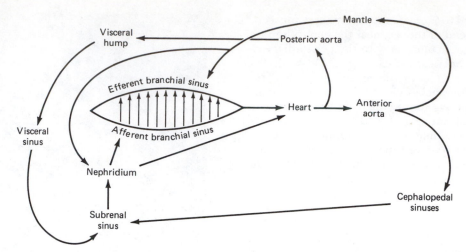

Figure 11–40 Circulatory pathways in the whelk *Busycon canaliculatum*. (From Mangum, C. P., and Polites, G., 1980: Oxygen uptake and transport in the prosobranch mollusc *Busycon canaliculatum*. I. Gas exchange and the response to hypoxia. Biol. Bull., *158*:76–90.)

terior aorta supplying the visceral mass and an anterior aorta supplying the head and foot, or there may be a single short aorta, which then divides into an anterior and a posterior artery.

From the arterial sinuses, blood eventually collects in a large, venous, cephalopedal sinus (Fig. 11–40). Much of this blood passes through the nephridium before entering the branchial circulation, but some may return directly to the heart. In the return flow to the heart in pulmonates, all blood from the venous sinuses passes through the pulmonary capillary network in the roof of the lung (Fig. 11–31).

Prosobranchs and pulmonates possess the respiratory pigment hemocyanin, which is dissolved in the plasma. Hemocyanin is a protein in which a molecule of oxygen is carried between two copper atoms. The functional subunits, which have a molecular weight of 50,000, each carry eight pairs of copper atoms. Ten to 20 subunits are combined to form large molecules. (See Mangum, 1985, for review.) Oxyhemocyanin is pale blue; deoxyhemocyanin is colorless. The freshwater Planorbidae (pulmonates) possess hemoglobin instead of hemocyanin in the plasma. The opisthobranch gas transport provisions are poorly known, but some species of sea hares *(Aplysia)*, which have been extensively studied, possess hemocyanin and some lack it.

Nervous System

The nervous system of gastropods may be more easily understood if the ground plan is first de-

scribed as if torsion had not taken place (Fig. 11–41*A*). A pair of adjacent cerebral ganglia lie over the posterior of the esophagus and give rise to nerves that connect anteriorly to the eyes, tentacles, statocysts, and a pair of buccal ganglia, which are located in the back wall of the buccal cavity. The buccal ganglia innervate the muscles of the radula and other structures in this vicinity. A nerve cord issues ventrally from each cerebral ganglion on each side of the esophagus. These are the two pedal connectives, which extend ventrally to a pair of ganglia located in the midline of the foot. The pedal ganglia innervate the foot muscles.

A pair of connectives arise from the cerebral ganglia and extend back to a pair of pleural ganglia, which supply the mantle and the columella muscle. One pair of connectives join the pleural and pedal ganglia. A second pair of cords leave the pleural ganglion and extend posteriorly until they terminate in a pair of visceral ganglia that are located in the visceral mass and that supply organs in this region. A pair of parietal, or intestinal, ganglia innervate the gills and osphradium; they are located along the length of the visceral nerves.

As a result of torsion, the visceral cords are twisted into a figure eight (Fig. 11–41*B*), and one intestinal ganglion (supraintestinal ganglion) is now located higher in the visceral mass than the other.

The twisted condition is one primitive feature of the gastropod nervous system. Another is the separation of the ganglia by nerve cords, as previously described. Such a primitive nervous system is found, with some modifications, in many proso-

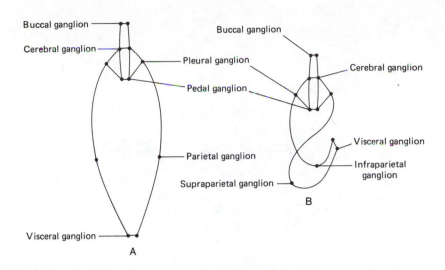

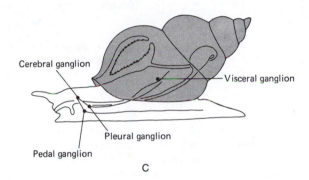

Figure 11–41 *A,* Hypothetical pretorsion nervous system. *B,* Posttorsion nervous system. *C,* Lateral view of gastropod showing position of nervous system. (After Naef from Kandel, E. R., 1979: Behavioral Biology of *Aplysia.* W. H. Freeman and Co., San Francisco.)

branchs. However, in most gastropods the original arrangement is obscured because of two evolutionary tendencies. First, there has been a tendency toward both concentration and fusion of ganglia with a consequent shortening of the connectives between ganglia. Second, there has been a tendency for the ganglia and cords to adopt a secondary bilateral symmetry (Fig. 11–42*A* and *B*). The ganglia of mollusks are organized in the same way as those of annelids (see Box 10–4, p. 321). Sea hares and some other opisthobranchs have been the principal molluscan subjects of neurophysiological investigations of specific neuronal circuitry and its control of behavioral reflexes. Interested students should consult Kennedy and colleagues (1969), Willows (1971), and Kandel (1979). Knowledge of endocrine control, especially for reproduction, has recently increased (see review by Joosse and Geraerts, 1983).

SENSE ORGANS

The sense organs of gastropods include eyes, tentacles, osphradia, and statocysts. The primitive eye, as in *Patella,* is a simple pit containing photoreceptor and pigment cells (Fig. 11–43*A*), but in most higher gastropods the pit has become closed over and differentiated into a single, chambered, vesicular eye with a lens (Fig. 11–43*B*), although the number of photoreceptors varies from few (in nudibranchs) to many. The eyes of most gastropods appear to detect only changes in general light intensity (see Box 11–1, p. 458).

A pair of closed statocysts are located in the foot near the pedal ganglia of gastropods, although they may be innervated by the cerebral ganglia. Statocysts are absent from sessile forms.

The evolution of the osphradium of gastropods closely parallels that of the gills. In the primitive Archaeogastropoda an osphradium is present for each gill. In the other prosobranchs, which possess but one gill, there is also only one osphradium, which is located on the mantle cavity wall anterior and superior to the attachment of the gill (Figs. 11–33 and 11–38*A*). In most cases the osphradium has become either filamentous or folded, thereby increasing the surface area (see Haszprunar, 1985).

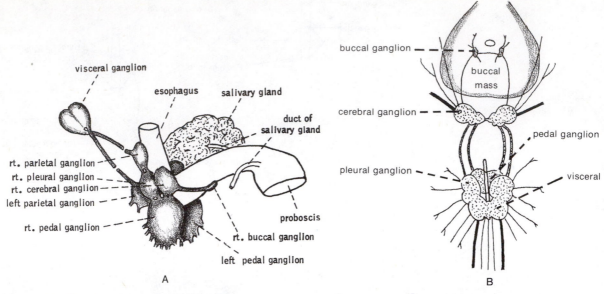

Figure 11–42 *A*, Concentrated nervous system in *Busycon*. (After Pierce, M. *in* F. A. Braun, 1950: Selected Invertebrate Types. John Wiley and Sons, N.Y.) *B*, Secondarily symmetrical nervous system of the pulmonate *Helix*. (After Bullough.)

This organ is highly developed in prosobranch carnivores and scavengers, which can locate carrion, animal juices, or prey from a considerable distance, as much as 2 meters in the case of scavenger *Cominella*. Many gastropods continually wave their siphons about as they move over the bottom, selecting and monitoring water from various parts of the environment.

Reproductive System

Most prosobranchs are dioecious. The single gonad, either ovary or testis, is located in the spirals of the visceral mass buried in the digestive gland (Fig. 11–33). The gonoduct ranges from a very simple to a highly complex structure, but in all cases it has developed in close association with the right nephridium. In the primitive Archaeogastropoda both nephridia are functional, and the right nephridium provides outlet for either the sperm or the eggs (Fig. 11–44*A*). The gametes pass through a short duct that extends from the gonad and opens into the distal part of the nephridium; the gametes are then conducted by the nephridium into the mantle cavity through the nephridiopore. The genital duct is thus formed from two elements—the gonoduct proper and the right nephridium. In this type of reproductive system, the eggs are provided, at most, with gelatinous envelopes produced by the ovary or the terminal part of the nephridium. There is no copulation, and fertilization takes place in the sea water after the eggs are swept out of the mantle cavity.

In all the other gastropods there is copulation and internal fertilization and a correspondingly complex reproductive system. The right nephridium has degenerated except for a very small portion that functions solely as part of the genital duct. Furthermore, the genital duct has become considerably lengthened by a third addition, derived from the mantle; as a result, the genital pore is located at the opening of the mantle cavity. This third section, which might be called the pallial duct, probably first arose as a ciliated groove extending from the nephridiopore, but in at least the females of existing gastropods the groove has become closed over to form a distinct tube (Fig. 11–44*B*). It is the pallial portion of the genital duct that has undergone elaboration or differentiation to provide for sperm storage and for egg membrane formation. Certainly the freeing of the right nephridium from excretory functions and the subsequent development of a complex reproductive system has contributed to the success of the mesogastropods and neogastropods.

In the male of these gastropods, the entire gonoduct consists of a coiled tube (vas deferens) leading from the testis. It functions in sperm stor-

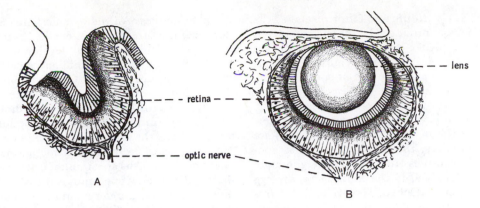

Figure 11–43 Eyes of two marine prosobranchs. *A, Patella,* a limpet. *B, Murex.* (Both after Helger from Parker and Haswell.)

age and transport. The pallial vas deferens, containing a prostate, runs in the floor of the mantle cavity and out to a tentacle-like penis located behind the right cephalic tentacle. In some prosobranchs (neritids, littorinids, cerithiids) this pallial portion is represented by a ciliated groove instead of a duct.

In the female the pallial section of the oviduct is modified to form both an albumin gland and a large jelly gland or capsule gland (Fig. 11–44*B*).

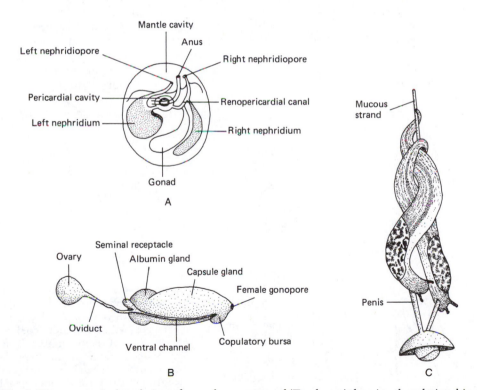

Figure 11–44 *A,* Diagrammatic dorsal view of an archaeogastropod (Trochacea) showing the relationship of the gonad to the nephridia. Note that gonad opens into renopericardial canal of right nephridium. *B,* Diagram of the reproductive system of *Littorina,* a mesogastropod. (Both from Fretter, V., and Graham, A., 1962: British Prosobranch Molluscs. Ray Society, London. pp. 283 and 360.)

Species such as *Cerithium* and *Littorina* embed the eggs in jelly masses produced by the jelly gland, but in many of the higher prosobranchs, the eggs are enclosed in a capsule. There is a seminal receptacle or bursa or both for the storage of sperm. Where both are present, the bursa handles the initial reception and short-term storage of sperm (Fig. 11–44*B*). The outgoing eggs pass through the dorsal portion of the oviduct, where membranes are applied. Commonly, there is a groove that conducts soft egg capsules from the female gonopore down to a pedal gland in the foot, where the egg case is molded and attached. The oyster drill *(Urosalpinx),* for example, deposits 7 to 96 such capsules containing 4 to 12 eggs each during a breeding season.

A small number of prosobranchs are hermaphroditic. Protandric hermaphrodites include many patellacean limpets and the slipper shells *(Crepidula).* In the more or less sessile slipper shells, individuals of some species tend to live stacked up on one another (Fig. 11–46*A*). The right shell margins are adjacent, thereby permitting the penis of the upper individual to reach the female gonopore of the individual below. Young specimens of such aggregating species are always males. This initial male phase is followed by a period of transition in which the male reproductive tract degenerates; the animal then develops into a female or another male. The sex of each individual is influenced by the sex ratio of the association, probably

by pheromones. An older male will remain male longer if it is attached to a female. If such a male is removed or isolated, it will develop into a female. The scarcity of females influences certain of the males to become females. Once the individual becomes female, it remains in that state (Hoagland, 1978). The sessile worm shells are dioecious and transfer sperm by rafting spermatophores, which are caught by the mucous nets of other individuals.

Pulmonates are simultaneous hermaphrodites, and opisthobranchs are simultaneous or protandric hermaphrodites. Copulation is the rule. The reproductive systems are very complicated and display endless variations (Figs. 11–31 and 11–45). In land pulmonates there is mutual exchange of sperm, usually in the form of spermatophores, and copulation is commonly preceded by a courtship involving circling, oral and tentacular contact, and intertwining of the bodies. Bizarre sexual behavior occurs in some species, particularly slugs. In the shelled Helicidae *(Helix)* the vagina contains an oval dart sac, which secretes a calcareous spicule. When two snails are intertwined, one drives its spicule dart into the body wall of the other (see Jeppesen, 1976, or Lind, 1976, for detailed description). Copulation follows this rather drastic form of stimulation. A new dart is secreted later. Copulating limacid slugs hang intertwined from a cord of mucus attached to a tree trunk or branch, and in the process of exchanging sperm the penes are unrolled to a length of 10 to 25 cm (sometimes as

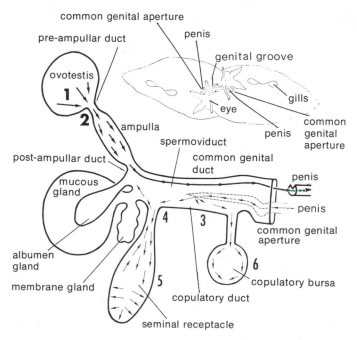

Figure 11–45 Reciprocal sperm transmission in the hermaphroditic sea hare *Phyllaplysia taylori: A,* Dorsal view of copulating pair. *B,* Sperm movements within one individual. Heavy arrows indicate path of sperm emitted; light arrows, the path of sperm received; numbers, the sequential location of sperm before and after emission. (Beeman, R. D., 1970: An autoradiographic study of sperm exchange and storage in a sea hare, *Phyllaplysia taylori,* a hermaphroditic gastropod. J. Exp. Zool., *175*(1):130.)

much as 85 cm) and are twisted together at the tips (Fig. 11–44*C*).

Embryogeny

The eggs of primitive prosobranchs (archaeogastropods) are fertilized externally and are either shed singly or aggregated in gelatinous masses. However, most other gastropods copulate and fertilize eggs internally. Egg deposition in gelatinous strings, ribbons, or masses is characteristic of most mesogastropods and opisthobranchs, but neogastropods and some mesogastropods embed their eggs in an albumin mass surrounded by a capsule or case, which is usually attached to the substratum. The size and shape of the case, the nature of the wall (which may be leathery or gelatinous), and the number of cases attached together are extremely variable and are characteristic of the species (Fig. 11–46*B*).

Aquatic pulmonates deposit their eggs in gelatinous capsules. Terrestrial species produce a relatively small number of eggs, each enclosed with albumin within a separate capsule. The eggs are usually laid in a heap in soil. Among the largest eggs—up to 5 cm in diameter—are laid by a 15 cm South American species of *Strophocheilus* (Fig. 11–46*H*).

In over half of the terrestrial pulmonate families and even in some prosobranch land snails, the capsule wall contains calcite crystals embedded in mucus or consists of a calcareous shell. The shell not only supports and protects the egg contents but serves as a source of calcium for the shell of the developing snail within. A contractile, vesicle-like extension of the embryonic foot, called a podocyst, serves for the utilization of albumin as well as in gas exchange and excretion. The podocyst is thus somewhat analogous to the allantois of reptiles and birds (Cather and Tompa, 1972; Tompa, 1980).

The free-swimming trochophore larva is found only in primitive Archaeogastropoda that shed their eggs directly into the sea water; in all the other gastropods, the trochophore stage is suppressed and is passed before hatching.

More characteristic of marine gastropods is a free-swimming larval type called a veliger. The veliger larva is derived from a trochophore but represents a later, more developed stage (Fig. 11–47). The characteristic feature of the veliger is the swimming organ, called a velum, which consists of two large, semicircular lobes bearing long cilia. The velum forms as an outward extension of the pro-

totroch of the trochophore. The foot, eyes, and tentacles appear; the shell, which first formed in the trochophore, develops spirally in the veliger because of unequal growth. In nudibranchs a shell appears in the veliger and is later cast off, during metamorphosis.

Some gastropods have feeding (planktotrophic) veligers with a larval life that may last as long as three months; others have brief, yolk-laden, nonfeeding (lecithotrophic) veligers. The long cilia of the velum function not only in locomotion but also in suspension feeding. The beating of the long velar cilia brings fine plankton in contact with the shorter cilia of a subvelar food groove. Within the food groove particles become entangled in mucus and are conducted to the mouth (Fig. 11–47*C*).

During the course of the veliger stage, torsion occurs and the shell and visceral mass twist 180 degrees in relation to the head and foot. Torsion may be very rapid (only about 3 minutes in the marine limpet *Acmaea*), or it may be a gradual process.

As development proceeds, the veliger reaches a point at which not only can it swim by means of the velum, but also the foot is sufficiently formed to allow creeping. Settling and metamorphosis occur. The velum is lost and the final features of the adult form are attained. Settling sites are of critical importance in the survival of the larvae, and many species can delay metamorphosis until specific types of substrates can be reached. Some nudibranchs, for example, must contact certain species of hydroids, bryozoans, or ascidians before metamorphosis will occur. Metamorphosis appears to be induced by chemical rather than physical characteristics of the substratum.

Some marine prosobranchs, especially the Neogastropoda, nearly all freshwater prosobranchs, and almost all pulmonates have no free-swimming larvae. At hatching, a tiny snail emerges from the protective shell or case.

The course of development may vary greatly even within a closely related group. For example, among the common intertidal periwinkles (*Littorina*), some release pelagic egg cases from which veliger larvae hatch; some attach their egg cases, with or without larvae; and some brood their eggs, giving birth to larvae or to little snails. The condition is related in part to the particular intertidal level occupied by the species, and the timing of reproductive events may be tied to tidal cycles (Gallagher and Reid, 1974).

Many gastropods reach adult size and sexual maturity at 6 months to 2 years, but slow growth continues and larger species may not reach maxi-

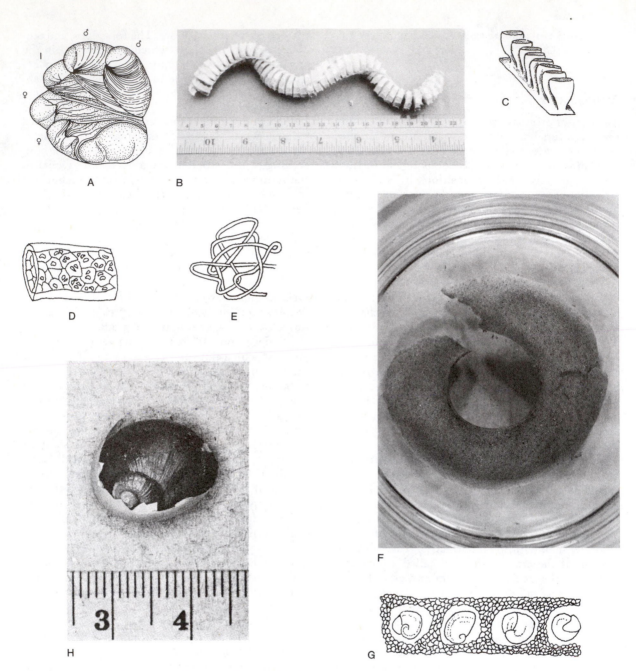

Figure 11–46 *A*, A stack of slipper shells, *Crepidula fornicata*, in life position. (From Hoagland, K. E., 1979: The behavior of three sympatric species of *Crepidula* from the Atlantic, with implications for evolutionary ecology. Nautilus, *94*(4):143–149.) *B*, Parchment egg cases of the whelk, *Busycon carica*. The many cases, each shaped like a pill box, are connected by a cord. Development is direct, and the little whelks emerge from a hole at the margin of each case. Although often washed up on beaches, the cases are produced while the whelk is buried in the sand, so for a time at least the string is anchored in sand. *C*, Parchment egg cases of *Conus*. *D* and *E*, Gelatinous egg string of a sea hare, *Aplysia*; *E*, enlarged. *F*, Egg case, or "sand collar" of the moon shell, *Polinices duplicatus*. *G*, Section through the egg case of a moon shell. (*C–E* after Abbott, R. T., 1954: American Sea Shells. Van Nostrand Co., Princeton, N.J.; *B* and *F* by Betty M. Barnes.) *H*, The very large opened egg (about 16 mm) of the pulmonate *Strophocheilus* showing an enclosed young snail. (From Tompa, A., 1980: A method for the demonstration of pores in calcified eggs of vertebrates and invertebrates. Jour. Microscopy, *118*(4):477–482.)

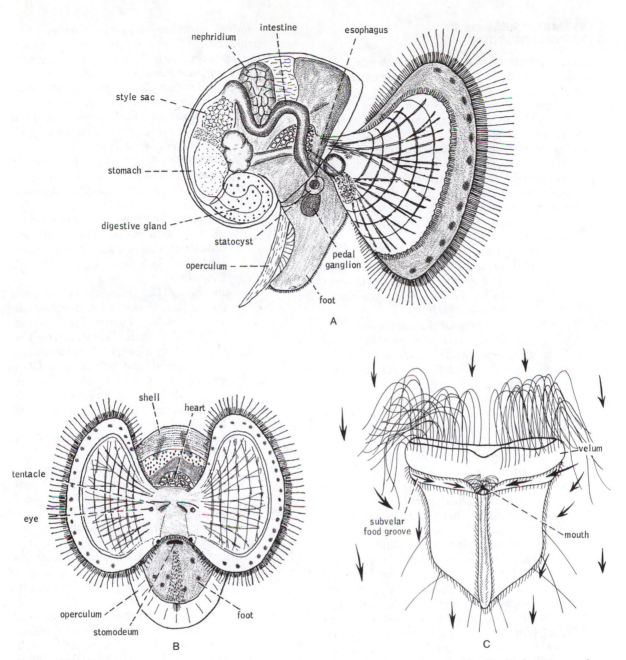

Figure 11-47 Veliger larva of the slipper shell *Crepidula: A*, Lateral view. *B*, Frontal view. (Both after Werner from Raven.) *C*, Suspension feeding in the early veliger of *Archidoris*. Arrows indicate direction of movement of food particles. (After Thompson.)

mum size for many years. The life span is highly variable: 5 to 16 years for the limpet *Patella vulgata*; 4 to 10 years for the periwinkle *Littorina littorea*; 1 to 2 years for many freshwater pulmonates; 5 to 6 years for the land snail *Helix aspersa*; and only 1 year for many nudibranchs. Few juvenile gastropods ever reach sexual maturity. Of the 1000 eggs produced each year by the muricid prosobranch *Thais*, not more than ten ever reach an age of 1 year (Spight, 1975).

SYSTEMATIC RESUMÉ OF CLASS GASTROPODA

Subclass Prosobranchia. Marine, freshwater, and terrestrial forms in which the mantle cavity and contained organs are located anteriorly. Aquatic species possess one or two gills within mantle cavity. Shell and usually an operculum present. Mostly dioecious.

Order Archaeogastropoda (Diotocardia, Aspidobranchia). Primitive forms in which there are usually two bipectinate gills, two auricles, and two nephridia. The right gill may be reduced or absent, but even if only one gill is present, it is bipectinate. Osphradium simple (ridgelike).

Superfamily Pleurotomariacea.* Shell with notch, slit, or row of holes; two gills. The deep water slit shells: *Pleurotomaria, Perotrochus; Scissurella.* The abalones: *Haliotis.*

Superfamily Fissurellacea. Keyhole limpets. Shell with single apical hole. Two gills. *Emarginula, Diodora, Fissurella.*

Superfamily Patellacea (Docoglossa). Limpets without hole or notch in shell. Single auricle; single bipectinate gill, secondary gills, or gills absent. *Acmaea, Collisella, Patelloida, Lottia, Patella, Cellana, Lepeta.*

Superfamily Trochacea. Conical shells with an operculum; single bipectinate gill. Top shells: *Margarites, Calliostoma, Trochus, Tegula, Monodonta, Gibbula.* Turbans: *Turbo.* Star shells: *Astraea.*

Superfamily Neritacea. Globose operculate snails. The presence of a single nephridium and a complex reproductive system excludes these snails from the archaeogastropods in some systems; however, they possess a single bipectinate gill. The common tropical intertidal *Nerita* and *Neritina;* the freshwater *Theodoxus;* the terrestrial Helicinidae.

Order Mesogastropoda (Taenioglossa, or suborder Taenioglossa within the order Monotocardia or Pectinibranchia). Single monopectinate gill, one auricle, and one nephridium; osphradium simple (ridgelike);

complex reproductive system, usually with a penis; radula taenioglossate, i.e., seven teeth in a transverse row. Chiefly marine but with many freshwater and terrestrial genera.

Superfamilies Cyclophoracea and Vivaparacea. Terrestrial, gill-less, operculate Cyclophoridae (absent from the Western Hemisphere); freshwater *Viviparus, Pomacea, Pila.*

Superfamily Littorinacea. The intertidal Littorinidae: *Littorina, Tectarius; Lacuna.* The terrestrial Pomatiasidae: *Pomatias.*

Superfamily Rissoacea. A large assemblage of small marine, freshwater, and terrestrial snails with conical shells. *Hydrobia, Bulimus, Rissoa, Alvania, Cingula.*

Superfamily Cerithiacea. Marine and freshwater snails with high shells. Turret shells (Turritellidae): *Turritella.* Worm shells (Vermetidae): *Vermetus, Serpulorbis, Petaloconchus, Dendropoma,* and (Siliquariidae) *Siliquaria;* the cucumber-like *Caecum;* the freshwater *Goniobasis, Pleurocera;* marine *Cerithrium, Bittium; Batillaria.*

Superfamily Epitoniacea. The pelagic violet snails—*Janthina.*

Superfamily Eulimacea. Predators, commensals and parasites of echinoderms. *Eulima, Stilifer* and the Entoconchidae, endoparasites of sea cucumbers—*Entoconcha, Entocolax* and *Enteroxenos.*

Superfamily Calyptraeacea. Protandric snails with mostly caplike or limpetlike shells. *Capulus; Calyptraea.* Slipper shells: *Crepidula.*

Superfamily Strombacea. Mostly large gastropods having shells with siphonal canals and flaring lip. Carrier shells Xenophoridae; *Struthiolaria; Aporrhais;* the conchs *Strombus, Lambis.*

Superfamily Cypraeacea. Cowries. Spire enclosed within last whorl of shell. *Cypraea, Trivia.*

Superfamily Heteropoda (Atlantacea). Pelagic snails with finlike foot and reduced shell. *Atlanta, Carinaria.*

Superfamily Naticacea. Moon shells. Burrowing snails with globose shells and a drilling apparatus. *Natica, Polinices.*

---

*The Pleurotomariacea and Fissurellacea are sometimes placed together in the superfamily Zeugobranchia, in which all members have two gills.

Superfamily Tonnacea. Marine snails with heavy and often large shells. Helmet shells: *Cassis, Cassidarius.* Bonnets: *Phalium.* Tritons: *Cymatium.* Tuns: *Tonna.*

Order Neogastropoda (Stenoglossa, or suborder Stenoglossa within the order Monotocardia or Pectinibranchia). Members of this order are similar to the Mesogastropoda in having a single monopectinate gill, one auricle, and one nephridium, and in the complexity of the reproductive system. They differ in having a radula with three teeth to a transverse row (rachiglossate) and a complex osphradium (with bipectinate folds). All marine.

Superfamily Muricacea. Heavy conical sculptured shells with a long siphonal canal. The drills (Muricidae): *Murex, Urosalpinx, Eupleura, Purpura, Thais.*

Superfamily Buccinacea. A large assemblage of forms having shells with a short siphonal canal. Whelks (Buccinidae): *Buccinum, Neptunea,* and (Melongenidae) *Busycon* (an exception in having a long siphonal canal). Tulip shells (Fasciolariidae): *Fasciolaria.* Mud snails (Nassariidae): *Nassarius, Ilyanassa.*

Superfamily Volutacea. Shell surface usually smooth and spire conical to low. Olives (Olividae): *Oliva and Olivella.* Miter shells (Mitridae): *Vexillum, Mitra.* Harp shells (Harpidae): *Harpa.* Volutes (Volutidae): *Voluta, Cymbium.*

Superfamily Conacea (Toxoglossa) Predatory species with poison gland and highly modified radula or no radula. Cone shells (Conidae): *Conus; Turris.* The high-spired Terebridae: *Terebra.*

Subclass Opisthobranchia. Have one gill, one auricle, and one nephridium but display detorsion. Reduction and loss of shell and mantle cavity common. Many are secondarily bilaterally symmetrical. Head commonly bears two pairs of cephalic tentacles. Buccal cavity with a pair of jaws. Hermaphroditic. Mostly marine.

Order Cephalaspidea. Bubble shells. Dorsal surface of head shieldlike. Shell generally present but reduced, absent in some. *Acteon, Hydatina, Philine, Scaphander, Bulla.*

Order Pyramidellacea. Ectoparasites of bivalve mollusks and polychaetes. Shell and operculum present. Proboscis contains a stylet instead of a radula. *Pyramidella, Odostomia, Brachystomia.*

Order Acochlidioidea. Small naked species with no gills and having the visceral mass sharply set off from the rest of the body. Some members of this order found in fresh water. *Acochlidium.*

Order Anaspidea, or Aplysiacea. Sea hares. Large opisthobranchs with more or less bilaterally symmetrical external form. Reduced shell buried in mantle; gill and mantle cavity present; foot with lateral parapodia. *Aplysia, Bursatella, Akera.*

Order Notaspidea. Shelled or naked opisthobranchs. Gill present. *Pleurobranchus.*

Order Sacoglossa. Shelled and sluglike opisthobranchs with a radula bearing a single row of teeth adapted for suctorial feeding on algae. *Elysia, Alderia;* the bivalved *Berthelinia.*

Order Thecosomata. Shelled pteropods, or sea butterflies. Shelled pelagic species with large parapodia. *Limacina, Cavolina, Spiratella, Clio, Cymbulia, Gleba.*

Order Gymnosomata. Naked pteropods. Pelagic species with no shell or mantle cavity. Parapodial fins present. *Pneumoderma, Cliopsis.*

Order Nudibranchia. Nudibranchs, or sea slugs. Shell and mantle cavity absent, and body secondarily bilaterally symmetrical. Doridaceans, with secondary gills around anus: *Doris, Chromodoris, Glossodoris, Jorunna, Onchidoris.* Dendronotaceans, with simple to branched cerata: *Tritonia, Dendronotus.* Arminaceans, with platelike gills beneath mantle edge or cerata: *Armina.* Aeolidaceans, with simple cerata: *Aeolidia, Glaucus.*

Subclass Pulmonata. One auricle and nephridium. Gills absent. Mantle cavity, on right side, is converted into a vascularized chamber for gas exchange in air or secondarily in water. Nervous system concentrated and symmetrical. Shell usually present, but operculum lacking. Hermaphroditic.

Order Systellommatophora. Slugs with anus located at posterior end of body, instead of laterally as in other pulmonates. The intertidal Onchidiidae, having a posterior pulmonary sac and the tropical Veronicellidae, which have lost the lung.

Order Basommatophora. Pulmonates with one pair of tentacles; eyes located near tentacle base. Primarily freshwater forms, a few marine. Marine limpets: *Siphonaria, Otina.* Marine *Amphibola,* the only operculate pulmonate. Freshwater snails: *Lymnaea, Planorbis, Helisoma, Bulinus, Physa.* Freshwater limpets: *Ancylus, Ferrissia.*

Order Stylommatophora. Pulmonates with two pairs of tentacles, the upper pair bearing eyes at the tip. Terrestrial. *Partula, Achatina, Retinella, Polygyra, Helix.* Land slugs: *Arion, Limax, Deroceras, Phylomycus, Testacella.*

SUMMARY

1 The evolution of the class Gastropoda involved three important changes: greater cephalization, development of an asymmetrical spiral shell, and torsion.

2 All evidence indicates that shell spiraling preceded torsion and was probably related to the conversion of the shell from a shield to a retreat into which the animal could withdraw. Fossil evidence indicates that the first gastropod shells were planospirals.

3 The evolutionary significance of torsion is uncertain but may have made possible the withdrawal of the head into the shell before the foot, as is true of living gastropods.

4 Living gastropods possess asymmetrical spiral shells, which perhaps have the advantage of compactness over planospiral shells. The shell of modern forms is carried obliquely across the body with the spire apex directed to the right, posteriorly and upward.

5 The asymmetrical shell and its carriage cause some occlusion of the mantle cavity on the right side, which in turn has brought about the reduction or loss of the right gill and associated auricle. The left pedal retractor muscle is reduced or lost.

6 Most gastropods move by waves of muscle contraction that sweep along the length of the broad ventral surface of the foot. Cilia are important in the locomotion of juveniles and forms that live on soft bottoms.

7 The prosobranch archaeogastropods are believed to be the most primitive of living forms. They have retained the bipectinate condition of the gill. The most primitive archaeogastropods are probably the slit shells, keyhole limpets, and abalones, in which shell notches or perforations permit a right angle ventilating current. They are also the only living gastropods to retain the right gill and auricle and the left pedal retractor.

8 The patellacean limpets (archaeogastropods) have a single anterior gill with a posteriorly directed ventilating current, or the original gill has been lost, with or without replacement by laterally paired gills in the mantle groove.

9 The trochaceans and neritaceans exhibit an oblique water current, which enters the mantle cavity on the left side of the head and exits on the right side. This ventilating current has been retained by all higher prosobranchs.

10 The archaeogastropods are largely restricted to firm substrates, where they graze on fine algae and other organisms found on rock surfaces. Their microphagous habit is reflected in the nature of the radula and the presence of a protostyle and sorting region within the stomach.

11 The gametes of archaeogastropods exit by the right nephridium, which continues to serve in excretion. The reproductive circumstances of most archaeogastropods—no copulation, external fertilization, planktonic eggs or eggs that are weakly enveloped when deposited—are correlated with the limitations in gonoduct specialization. A trochophore larva precedes the veliger.

12 The single left gill of higher prosobranchs (mesogastropods and neogastropods) is monopectinate rather than bipectinate. The shift to the monopectinate condition, with the gill axis directly attached to the mantle wall, perhaps had the advantage of reducing surfaces that might be fouled with sediment in the ventilating current. In any event, these prosobranchs are not confined to hard surfaces, and many species live on soft bottoms.

13 The mesogastropods invaded freshwater and land and evolved many species, although the land snails are largely confined to the tropics. Freshwater and terrestrial prosobranchs are operculate.

14 The great diversity of mesogastropods and neogastropods is correlated in large part with their modes of feeding, locomotion, and habitation. Most are macrophagous, which is reflected in the simple, saclike stomach. Feeding habits are also reflected in radula design and in the presence of a proboscis in many species.

15 The right nephridium of mesogastropods and neogastropods has been freed of an excretory function and contributes only to the gonoduct. Another contribution from the floor of the mantle cavity has extended the gonoduct length and potential for complexity of function. Most species copulate, fertilize eggs internally, and deposit the eggs within well-developed envelopes or cases. The eggs hatch into veligers, or development is direct.

16 Pulmonates probably evolved from meso-gastropod ancestors, which lived in some shallow-water marine habitat subjected to frequent, periodic reduction in standing water. The gill was lost and the mantle cavity was converted to a lung for gas exchange in air. Most lower pulmonates invaded fresh water, where they continue to use the lung for gas exchange or have evolved secondary gills. Higher pulmonates are terrestrial and may be readily distinguished from operculate land snails by the absence of an operculum (true of all pulmonates). Higher pulmonates differ from lower pulmonates in having two pairs of tentacles, with the eyes mounted at the top of the second pair.

17 Most pulmonates are herbivores, although there are some carnivorous species. The radula bears a large number of teeth in the transverse row. Higher pulmonates display an array of structural, physiological, and behavioral adaptations for life on land.

18 Pulmonates are hermaphroditic, with copulation and mutual sperm transfer. The eggs are deposited within envelopes and development is direct, except in the few marine species.

19 Opisthobranchs are marine gastropods that exhibit 90-degree detorsion. The gill and mantle cavity are located on the right side. The operculum has been lost except in a single primitive family. Primitively, there is a shell and gill, but in many opisthobranchs reduction or loss of shell, mantle cavity, and gill has occurred. This tendency has developed independently within each of the radiating lines of opisthobranch evolution. The numerous, bilaterally symmetrical, sluglike forms, of which the colorful nudibranchs are the most notable examples, represent the extreme products of this change. Opisthobranchs are very diverse in feeding modes, habitation, and locomotion.

20 Opisthobranchs are simultaneous or protandric hermaphrodites, with copulation, internal fertilization, and egg deposition. The veliger is the hatching stage in indirect development.

Class Monoplacophora

The members of the following two classes, the Monoplacophora and the Polyplacophora, are similar to gastropods in possessing a flat, creeping foot; both are considered primitive groups of mollusks.

In 1952 ten living specimens of *Neopilina*, a group of mollusks previously known only from Cambrian and Devonian fossils, were dredged from a deep ocean trench off the Pacific coast of Costa Rica. Since this discovery, specimens belonging to two genera and eight species have been collected from the shelf edge *(Vema)* and from deep water, 2000 to 7000 meters *(Neopilina)*, in various parts of the world—the South Atlantic, the Gulf of Aden, and a number of places in the eastern Pacific.

Neopilina and *Vema* belong to a group of mollusks known as the Monoplacophora. As the name implies, a monoplacophoran possesses a single, symmetrical shell, which varies in shape from a flattened, shieldlike plate to a short cone. The monoplacophorans had been classified with either the chitons or the gastropods, but all shared a distinctive feature. The undersurface of the shell displayed three to eight pairs of muscle scars.

The living specimens of *Neopilina* are 3 mm to little more than 3 cm long and externally resemble limpets (Fig. 11–48B). The apex of the shell is directed anteriorly. A pallial groove (the mantle cavity) separates the edge of the broad, flat foot from the mantle on each side. The mouth is located in front of the foot, and the anus is located in the pallial groove at the posterior of the body. In front of the mouth is a preoral fold, or velum, which extends laterally on each side as a rather large, ciliated, palplike structure. Another fold lies behind the mouth and projects to either side as a pair of postoral tentacles.

The pallial groove contains five or six pairs of monopectinate gills (Fig. 11–48D). Internally, eight pairs of pedal retractor muscles are present. The digestive system includes a radula and a subradula organ within the buccal cavity. The stomach contains a style sac and style; the intestine is greatly coiled. Stomach contents consist of diatoms, forams, and sponge spicules.

Blood from the gills passes through the nephridia and then into two pairs of auricles, which open into two ventricles, one on each side of the rectum (Fig. 11–48E). The heart is surrounded by a paired, pericardial coelom. Six pairs of nephridia are located on each side of the body, and except for the first pair, each nephridium opens into the coelom. The nephridiopores open into the pallial groove. The nervous system is similar to that of chitons (p. 399).

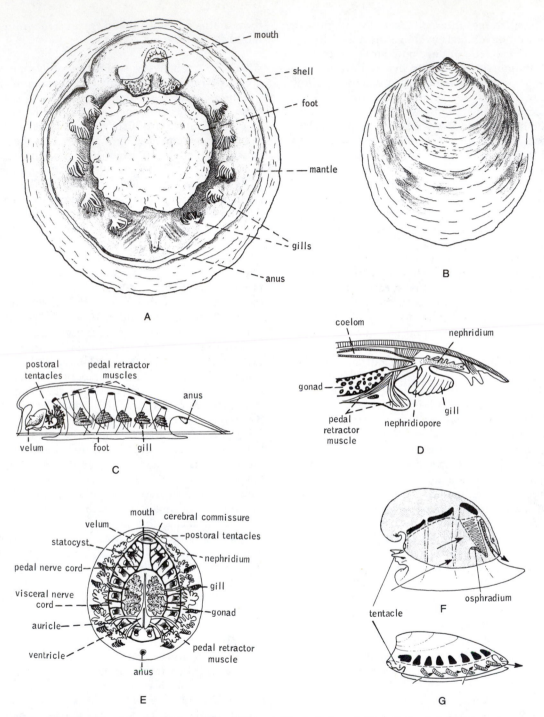

Figure 11–48   The monoplacophoran *Neopilina: A*, Ventral view. *B*, Dorsal view of shell. *C*, Side view. *D*, Transverse section through one half of body. *E*, Internal anatomy. (All adapted from Lemche, H., and Wingstrand, K. G., 1959: The anatomy of *Neopilina galatheae*. Galathea Rep., *3:9–71*.) *F*, Reconstruction of a cyclomyan monoplacophoran. Black bars indicate point of attachment of retractor muscles to shell, and arrows indicate possible ventilating current. *G*, Reconstruction of a tergomyan monoplacophoran. (*F* and *G* from Stasek, C. R., 1972: *In* Florkin, M., and Scheer, B. T. (Eds.): Chemical Zoology. Vol. 7. Academic Press, N.Y. p. 21.)

The sexes are separate, and two pairs of gonads are located in the middle of the body. Each gonad connects through a gonoduct to one of the two pairs of nephridia in the middle of the body. Fertilization must occur externally.

The survival of species of *Neopilina* is undoubtedly correlated with their adaptation for life at great depths, and they are perhaps more specialized than were other members of the class. Fossil species appear to have evolved along two lines. In one group (subclass Cyclomya) there was an increase in the dorsoventral axis of the body, leading to a planospiral shell and a reduction of gills and retractor muscles (Fig. 11–48*F*). Although they disappeared from the fossil record in the Devonian, this group may have been ancestral to the gastropods. The other line (subclass Tergomya) retained a flattened shell with five to eight pairs of retractor muscles (Fig. 11–48*G*). Although this group was thought to have become extinct in the Devonian, when it disappears from the fossil record, *Neopilina* and *Vema* may represent survivors.

SUMMARY

1 The small number of living species of the class Monoplacophora are deep-water relics of a much larger and more widespread group of mollusks that date back to the Cambrian.

2 The repetition of both external and internal structures—gills, retractor muscles, auricles, and nephridia—is a distinctive feature of living monoplacophorans. Fossil species show only multiple muscle scars.

3 Many monoplacophoran features, such as the shield-shaped shell, flat creeping foot, slight cephalization, multiple gills and retractor muscles, radula, and cone-shaped stomach, are thought to be primitive, and the class is believed by many malacologists to be ancestral to the gastropods, bivalves, and cephalopods.

Class Polyplacophora

The class Polyplacophora contains the chitons. Although some features of their structure and embryogeny are primitive, chitons have become highly adapted for adhering to rocks and shells. They are bilaterally symmetrical, with an ovoid body that is greatly flattened dorsoventrally (Fig. 11-49). There are no cephalic eyes or tentacles, and the head is indistinct. The mantle is heavy, and the foot is broad and flat to facilitate adhesion to hard substrata. There are approximately 500 existing species of chitons, of which one fifth are found along the west coast of North America. The fossil record, which dates to the late Cambrian period, is rather sparse. Some 350 fossil chitons are known.

Chitons range in size from 3 mm to 40 cm, the largest being the giant Pacific species *Cryptochiton stelleri* (gumshoe) (Fig. 11–50*A*). However, most species are 3 to 12 cm in length. Chitons are commonly drab shades of red, brown, yellow, and green.

Shell and Mantle

The most distinctive characteristic of a chiton is the shell, which is divided into eight overlapping transverse plates. From the nature of the shell is de-

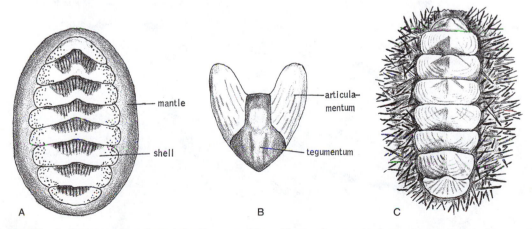

Figure 11–49 *A*, Common American Atlantic coast chiton *Chaetopleura apiculata*, which is only a few centimeters long. (After Pierce.) *B*, Single shell plate of *Katharina*. *C*, *Chiton*. (After Borradaile and others.)

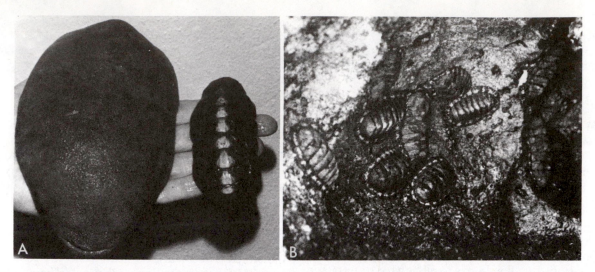

Figure 11–50   *A,* Two species of chitons from the northwest Pacific. The shell plates of the larger species *(Cryptochiton)* are completely covered by the mantle; those of the smaller species *(Katharina)* are partially covered. *B,* The West Indian chiton *Chiton tuberculatus* exposed at low tide. (Both by Betty M. Barnes.)

rived the name of the class, *Polyplacophora*—bearer of many plates (Fig. 11–49). Except for the overlapping posterior edge, the margins of each plate are covered by mantle tissue.

The degree of lateral mantle reflexion varies. In *Lepidochitona, Placiphorella,* and *Chaetopleura* (Fig. 11–49A) most of the plate width is exposed. However, in *Katharina* only the midsection of each plate is uncovered, and in *Cryptochiton* the shell is completely covered by the mantle (Fig. 11–50A).

The shell plates are composed of several layers (Fig. 11–51). The upper layer, called the tegmentum, consists of an organic conchiolin matrix impregnated with calcium carbonate and is sculptured on its exposed surface. The tegmentum is covered by a thin periostracum. Beneath the tegmentum lies a thicker, denser layer, called the hypostracum, which is composed entirely of calcium carbonate. The anterior, articulating projections of each plate are composed only of the hypostracum,

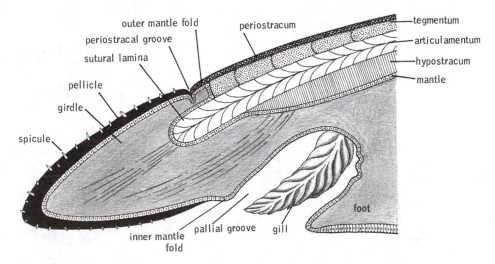

Figure 11–51   Diagrammatic transverse section through girdle, mantle (pallial) groove, and foot on one side of the body of a chiton. (Modified from Mutvei.)

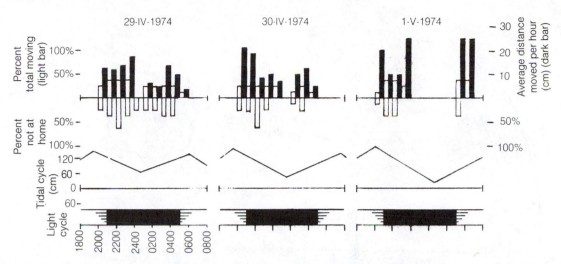

Figure 11–52 Activity of the intertidal homing chiton *Mopalia muscosa* at Pacific Grove, California. Black and white horizontal bars indicate day and night, with tidal cycle above. The vertical white bars below the horizontal line indicate the percentage of the population away from its home. The vertical white bars above the horizontal line indicate the percentage of the population moving at that time. (All limpets away from home may not be moving.) The vertical dark bars above the line indicate the average distance moved in centimeters per hour. (From Smith, S. Y., 1975: Temporal and spatial activity patterns of the intertidal chiton *Mopalia muscosa. Veliger, 18* [Supplement]:58.)

which in these specialized areas of the plate is called the articulamentum.

The peripheral area of the mantle, called the girdle, is heavy and extends a considerable distance beyond the lateral margins of the plates. The girdle surface displays a variety of ornamentation. It may be naked and smooth or covered with scales, bristles, or calcareous spicules (Fig. 11–49C). The bristles or spicules may be so long and dense that the animal has a mossy or shaggy appearance.

Foot and Locomotion

The broad, flat foot occupies most of the ventral surface and functions in adhesion as well as in locomotion (Fig. 11–53A). Chitons creep slowly in the same manner as snails. The division of the shell into transverse plates and their articulation with one another enable them to move over and adhere to a sharply curved surface. Chitons roll up into a ball if dislodged, and although this may serve as a defense mechanism, it also enables the animal to right itself.

Adhesion is effected by both the foot and the girdle. The foot is responsible for ordinary adhesion, but when a chiton is disturbed, the girdle is also employed. The girdle is clamped down especially tightly against the substratum, and the inner margin is then raised. This creates a vacuum that

enables the animal to grip the substratum with great tenacity.

Chitons are common rocky intertidal inhabitants, and like limpets, most species are motionless at low tide. When the rock surface is submerged or splashed, they move about to feed. They are usually negatively phototactic and thus tend to locate themselves under rocks and ledges. They are most active at night if the tide is right. Like limpets, some species exhibit homing (Fig. 11–52) (Smith, 1975; Mook, 1983).

Water Circulation and Gas Exchange

The mantle cavity of chitons consists of a trough, or groove, on each side of the body between the foot and the mantle edge (Fig. 11–53A). This condition is correlated with the extreme dorsoventral flattening of the body, one of the adaptations of chitons for life on rocky surfaces. Six to 88 pairs of bipectinate gills are arranged in a linear series within the two mantle troughs. The number of pairs varies among species and even within a single species, depending on the size of the animal.

The margins of the chiton mantle are held tightly against the substratum, making the pallial groove a closed chamber. On each side of the mantle trough toward the anterior end, the mantle margins are raised to form two inhalant openings (Fig.

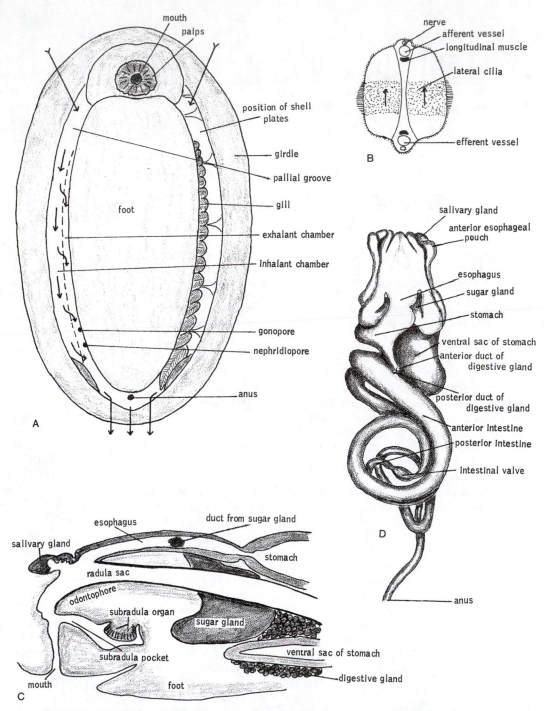

Figure 11–53 The chiton *Lepidochitona cinerea: A*, Mantle groove, showing direction of water currents (ventral view). *B*, Transverse section through gill axis. Arrows indicate direction of water current over filament surface. (*A* and *B* after Yonge, C. M., 1939: On the mantle cavity and its contained organs in the Loricata. Quart. Jour. Microscop. Sci., *81*:367–390.) *C*, Buccal region (lateral view). *D*, Digestive tract (dorsal view). (*C* and *D* after Fretter.)

11–53*A*). Water enters each groove through these anterior openings, runs along the course of the groove, and passes up through the gills. The two exhalant currents in the dorsal part of each groove converge posteriorly and pass to the outside through one or two exhalant openings created by the locally raised mantle.

Nutrition

Most chitons are microphagous and feed on fine algae and other organisms that they scrape from the surface of rocks and shells with the radula. The gut contents of three species from the Maine coast included remains from 14 different algal and animal organisms, and about three quarters of the contents consisted of sediment (Langer, 1983). Some species, however, feed on coarser algae, and one genus, *Placiphorella*, on the west coast of the United States uses its raised and flaring anterior end to catch small crustaceans and other invertebrates. The radula bears 17 teeth in each transverse row, and the lateral teeth are capped with magnetite (iron). The mouth opens into the chitin-lined buccal cavity (Fig. 11–53*C*). A long radula sac projects posteriorly from the back of the buccal cavity, as does a smaller, more ventral evagination called the subradula sac. The latter contains a cushion-shaped sensory structure, the subradula organ, hanging from the roof.

When a chiton feeds, the subradula organ is first protruded and applied against the rock. If food is present, the odontophore and its radula project from the mouth and scrape. Periodically the subradula organ is protruded and tests the substratum again.

From the buccal cavity, mucus-entangled food particles enter the esophagus and are carried along a ciliated food channel toward the stomach (Fig. 11–53*C* and *D*). During this passage, the food particles are mixed with amylase secretions produced by a pair of large pharyngeal glands (or sugar glands), the ducts from which open at the beginning of the esophagus.

The esophagus opens into a stomach, where the food is further mixed with proteolytic secretions from the digestive gland. Digestion is almost entirely extracellular and takes place in the digestive gland, in the stomach, and in the anterior intestine.

The anterior intestine loops and then joins a large, coiled, posterior intestine (Fig. 11–53*D*). Between these two intestinal divisions is a sphincter, forming an intestinal valve. Waste leaving the intestine is compacted in mucus, and the intestinal valve divides it into short fecal pellets. The anus opens at the midline just behind the posterior margin of the foot. The egested fecal pellets are swept out with the exhalant current.

Internal Transport, Excretion, and Nervous System

The pericardial cavity is large and located beneath the last two shell plates. A single pair of auricles collects blood from all of the gills. Each of the large, U-shaped nephridia connects with the pericardial cavity, and the nephridiopore opens into the pallial groove (Fig. 11–53*A*).

The nervous system is primitive (Fig. 11–54*B*). Ganglia are lacking or at least poorly developed. A nerve ring surrounds the esophagus, and nerves issue from it anteriorly to innervate the buccal cavity and subradula organ. Posteriorly the nerve ring gives rise to a large pair of pedal nerve cords, innervating the muscles of the foot, and a large pair of lateral, or palliovisceral, nerve cords, innervating the mantle, visceral organs, and shell esthetes.

The chief sense organs are the subradula organ and the esthetes. Esthetes, which are unique to chitons, are mantle cells lodged within minute vertical canals in the upper tegmentum (Fig. 11–55). The canals and sensory endings terminate beneath a cap on the surface of the shell plates. The density is very great; 1750 terminate on 1 mm² of shell surface in *Lepidochitona cinerea*. Although the structure of esthetes has been studied in detail and the shell plates are involved in the light response, the function of esthetes is still uncertain. Secretory structures are usually more conspicuous than sensory elements, but photoreceptors occur in some (Boyle, 1972, 1974; Fischer, 1978).

In one family of chitons (Chitonidae) distinct eyes are lodged in the shell canals. These ocelli may number in the thousands per individual and are especially concentrated on the anterior shell plates.

Reproduction and Embryogeny

Most chitons are dioecious. A single median gonad is located in front of the pericardial cavity beneath the middle shell plates (Fig. 11–54*A*). The gametes are transported to the outside by two gonoducts, instead of by the nephridia. A gonopore is located in each pallial groove in front of the nephridiopore (Fig. 11–53*A*).

Copulation does not take place. Sperm leave the male in the exhalant currents, and fertilization occurs in the sea or within the mantle cavity of the

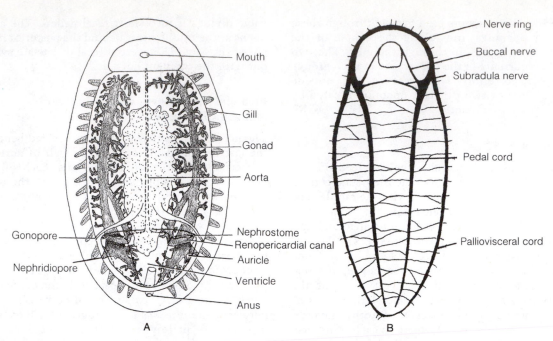

Figure 11–54   *A*, Internal structure of a chiton. (After Lang and Haller.) *B*, Nervous system of a chiton. (After Thiele from Parker and Haswell.)

female. The usual gregariousness of chitons facilitates fertilization. The eggs, which are enclosed within a spiny envelope, are usually shed into the sea either singly or in strings, but in some species the eggs are brooded within the mantle cavity.

There is a free-swimming trochophore, except in some of those forms that brood their eggs (Fig. 11–56*A*), but there is no veliger stage, perhaps a primitive condition. In the metamorphosis of the chiton trochophore, the posttrochal region elongates to form the major part of the body (Fig. 11–56*B*). The prototroch degenerates, and the animal sinks to the bottom as a young chiton. The larval eyes are retained for some time after metamorphosis.

The three orders of chitons are differentiated largely by features of the shell plates, which are too technical to describe here. (See the McGraw-Hill *Synopsis and Classification of Living Organisms*, Vol. I, pp. 953–960.)

SUMMARY

1 Chitons, members of the class Polyplacophora, are adapted for living on hard surfaces, especially in the intertidal zone. The foot and girdle provide for gripping; the low profile reduces water resistance; and the eight articulating shell plates provide protection while permitting folding of the

large body across angles on the substratum.

2 Chitons feed on algae and other organisms that encrust the rocks and shells on which they live. Food is removed by a scraping radula.

3 The poorly developed head, nature of the shell, multiple gills, and lack of a veliger larva suggest that chitons may have diverged early from the main line of molluscan evolution.

Class Aplacophora

The class Aplacophora comprises some 180 species of strange, worm-shaped mollusks, called solenogasters. They are found throughout the oceans of the world to depths of 9000 meters. Some live on the bottom, and others creep on hydroids and corals. Most specimens have been collected by dredging, and the biology of the group is poorly known.

Solenogasters are usually less than 5 cm in length. The head is poorly developed, and the typical molluscan features of shell, mantle, and foot are absent (Fig. 11–57). However, the integument contains layers of embedded calcareous scales or spicules. The body is vermiform because the mantle margins have rolled inward ventrally, and there is a midventral, longitudinal groove containing one or more ridges (Fig. 11–57), which are probably homologous to the foot of other mollusks. The posterior end of the body contains a cavity into which

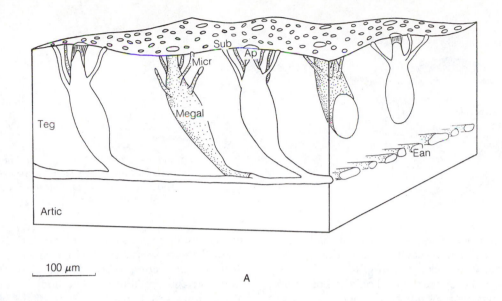

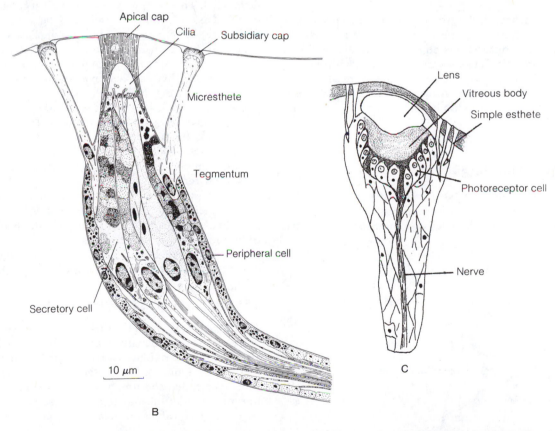

Figure 11–55 Chiton esthetes: *A*, Section of shell showing esthete canals and surface terminations. The larger canal houses a megalesthete; the smaller, a micresthete. On the surface they are covered by an apical or subsidiary cap. *B*, Reconstruction of an esthete of *Lepidochitona*. (*A* and *B* from Boyle, P. R., 1974: The aesthetes of chitons. II. Fine structure in *Lepidochitona cinereus*. Cell Tiss. Res., *153*:383–398.) *C*, Esthete eye of *Acanthopleura*. (After Nowikoff from Parker and Haswell.)

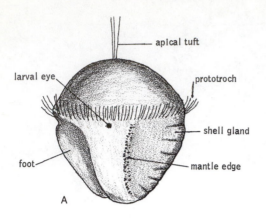

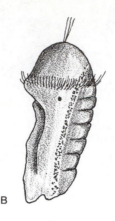

Figure 11–56   Development of chitons: *A*, Trochophore of *Ischnochiton*. *B*, Metamorphosis of *Ischnochiton*. (*A* and *B* after Heath from Dawydoff.)

the anus empties. The posterior cavity is believed to represent a mantle cavity, and in the members of one order it houses a pair of gills.

Solenogasters appear to feed only on cnidarians. A radula may or may not be present, and the gut contains a style sac. Solenogasters are hermaphroditic.

To what extent solenogasters are specialized and to what extent they are primitive mollusks is uncertain, but there is no evidence that a shell was ever present. They share little with chitons, although in the past the two groups were placed together in the class Amphineura. Their position within the phylum has been reviewed by Scheltema (1978).

Class Bivalvia

The class Bivalvia, also called Pelecypoda or Lamellibranchia, comprises mollusks known as bivalves and includes such common forms as clams, oysters, and mussels. Bivalves are laterally compressed and possess a shell with two valves, hinged dorsally, that completely enclose the body. The foot, like the remainder of the body, is also laterally compressed, hence the origin of the name *Pelecypoda*—hatchet foot. The head is very poorly developed. The mantle cavity is the most capacious of any class of mollusks, and the gills are usually very large, having assumed in most species a food-collecting function in addition to that of gas exchange. Most of these characteristics represent modifications that enabled bivalves to become soft bottom burrowers, for which the lateral compression of the body is better suited. Although modern bivalves have invaded other habitats, the original adaptations for burrowing in mud and sand have taken bivalves so far down the road of specialization that they have become largely chained to a sedentary existence.

The class bivalvia contains three subclasses: the Protobranchia, the Lamellibranchia, and the Septibranchia.* The protobranchs are generally believed to be the most primitive of existing bivalves. The septibranchs are highly specialized. The lamellibranchs contain the majority of the bivalve species.

Shell, Mantle, and Foot

A typical bivalve shell consists of two similar, more or less oval, usually convex valves, which are attached and articulate dorsally with each other (Fig. 11–58*A* and *B*). Each valve bears a dorsal protuberance called the umbo, which rises above the line of articulation and is the oldest part of the shell. The two valves are attached by an elastic pro-

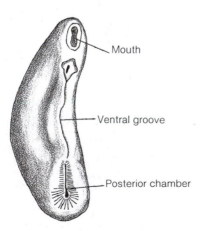

Figure 11–57   *Neomenia*, an aplacophoran. (After Hansen from Parker and Haswell.)

---

*This system is still widely encountered and for the purpose of text discussion has the great advantage of simplicity and ease of reference. The more complex system of modern taxonomists is given in the Systematic Résumé on page 439.

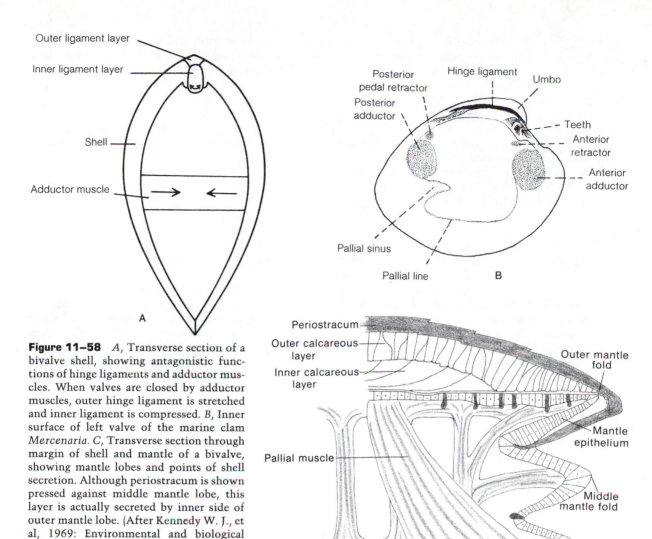

Figure 11–58 *A*, Transverse section of a bivalve shell, showing antagonistic functions of hinge ligaments and adductor muscles. When valves are closed by adductor muscles, outer hinge ligament is stretched and inner ligament is compressed. *B*, Inner surface of left valve of the marine clam *Mercenaria. C*, Transverse section through margin of shell and mantle of a bivalve, showing mantle lobes and points of shell secretion. Although periostracum is shown pressed against middle mantle lobe, this layer is actually secreted by inner side of outer mantle lobe. (After Kennedy W. J., et al, 1969: Environmental and biological controls on bivalve shell minerology. *Biol. Rev.* 44:499–530; copyrighted and reprinted by permission of Cambridge University Press.)

tein band called the hinge ligament, which is covered above with the periostracum. The hinge ligament is so constructed that when the valves are closed, the dorsal or outer part is stretched and the ventral or inner part is compressed (Fig. 11–58*A*). Thus, when the adductor muscles relax, the natural elasticity of the ligament causes the valves to open. To prevent lateral slipping, the two valves in most species are locked together by teeth or ridges and apposing sockets or grooves, located on the hinge line of the shell beneath the ligament (Fig. 11–58*B*). However, a dorsal ridge of mantle tissue still lies between the tooth and socket of the two apposing valves.

The valves of the shell are pulled together by two large dorsal muscles, called adductors, which act antagonistically to the hinge ligament (Fig. 11–58*A* and *B*). The adductors extend transversely between the valves anteriorly and posteriorly; there are scars on the inner surfaces of the valves where these muscles are attached. The adductors of most bivalves contain both smooth and striated fibers, facilitating sustained and rapid closing of the valves.

The bivalve shell is composed of an outer periostracum covering two to four calcareous layers. The periostracum may be very thick, as in many large, freshwater clams, or very thin, as in the edible marine quahog (*Mercenaria*). The primary function of the periostracum is to secrete the shell (see below), but it also protects the underlying calcium carbonate from dissolution and may contribute to a tight seal when the edges of the valves are brought together on closure. The calcareous layers may be entirely aragonite (primitive) or a mixture of aragonite and calcite, and they may be deposited as prisms or as minute laths or tablets arranged in sheets (nacre), lenses, or more complex forms (Fig. 11–59). The basic elements—prisms, tablets, and so forth—are always deposited within an organic framework, which together with the periostracum may account for 12 to 72 per cent of the dry weight of the shell (Price et al, 1974). Although shell structure is not uniform for the class, it is constant and characteristic for different groups of bivalves (Fig. 11–58*C*) (see Taylor, 1973; Carter, 1980).

The shells of bivalves exhibit a great variety of sizes, shapes, surface sculpturing, and colors. Where the rate of shell addition is the same all around the margin, the shell is equilateral; where the rates are different, shell shape changes. As in gastropods, surface sculpturing may contribute to traction, protection, or shell strengthening. The familiar ribbing, or corrugations, of cockle and some scallop shells, for example, increases the shell

strength (Pennington and Currey, 1984). Bivalves range in size from the tiny seed shells of the freshwater family Sphaeriidae, which usually do not exceed 2 mm in length, to the giant clam *Tridacna* (Fig. 11–80*A*) of the South Pacific, which attains a length of over a meter and may weigh over 1100 kg.

Like the shell, the mantle greatly overhangs the body, and it forms a large sheet of tissue lying beneath the valves. The edge of the mantle bears three folds—an inner, a middle, and an outer fold (Fig. 11–58*C*). The innermost fold is the largest and contains radial and circular muscles. The middle fold is sensory in function. The outer fold is related to the secretion of the shell. Shell secretion has been most studied in bivalves. The inner surface of the outer fold lays down the periostracum, and the outer surface secretes the first calcareous layer. The entire mantle surface secretes the remaining calcareous portion. There is a minute, extrapallial space between mantle and shell, except at points of muscle attachment. It is into this space that shell materials are first secreted; from the extrapallial fluid of this space are then deposited both

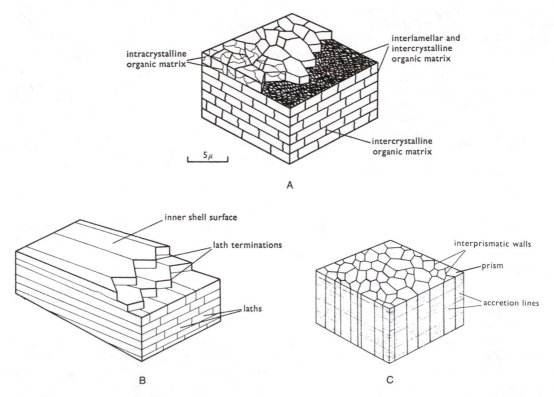

Figure 11–59 Three types of shell ultrastructure: *A*, Nacreous structure. *B*, Foliated structure built of laths. *C*, Prismatic structure. (From Kennedy, W. J., et al, 1969: Biol. Rev., *44*:499–530, copyrighted and reprinted by permission of Cambridge University Press.)

the calcareous elements and the surrounding organic framework (Fig. 11–60). Since the mantle epithelium is in contact with the shell surface only at the periostracal groove, where the periostracum is secreted, the periostracum plays an important role in sealing off the extrapallial space from the external aqueous medium (Fig. 11–58C).

In recent years much has been learned about the various lines and bands produced in shell growth. They provide information about the age of the animal and the environmental conditions under which shell growth took place. Some of the lines are visible on the outside of the shell, but most are only seen when the shell is sectioned radially, i.e., from hinge to ventral margin. The shells of many species that live in intertidal habitats show fine lines between microgrowth increments, reflecting daily tidal cycles as well as the semilunar cycle of spring tides (Fig. 11–61C). The line is believed to be produced when the valves are closed. At that time the production of organic acid from anaerobic respiration causes a slight dissolution of calcium carbonate, leaving a preponderance of organic framework material, which forms the line (Lutz and Rhoads, 1980).

Annual growth increments, which are present in most bivalve shells, can be detected by the seasonal thickness of microgrowth increments or seasonal differences in the density of the shell (Fig. 11–61). In some bivalves annual growth increments are recorded in concentric lines or checks on

the shell surface, resulting from a winter break in the periostracum and outer calcified layer. The previous year's layers may overlap the new like a shingle or be interrupted by a groove (Fig. 11–61B). However, similar lines can be produced by environmental disturbances, such as storms.

The mantle is attached to the shell by the muscle fibers of the inner lobe along a semicircular line a short distance from the shell edge. The line of mantle attachment is impressed on the inner surface of the shell as a scar, called the pallial line (Fig. 11–58B).

Despite the attachment of the mantle, occasionally some foreign object, such as a sand grain or a parasite, lodges between the mantle and the shell. The object then becomes a nucleus around which are laid concentric layers of nacreous shell. In this manner a pearl is formed. If the nucleus is enfolded within the mantle and moved about during secretion, the pearl becomes spherical or ovoid. Commonly, however, the developing pearl adheres to or even becomes completely embedded in the shell.

Pearls can be produced by all shell-bearing mollusks, but only those with shells having an inner nacreous layer produce pearls of commercial value. The finest natural pearls are produced by the pearl oysters, *Pinctada margaritifera* and *Pinctada martensi*, which inhabit most of the warmer Pacific areas. Bead pearls are produced by transplanting a shell bead enclosed within a mantle bag into the gonadial tissue of another pearl oyster. A calcar-

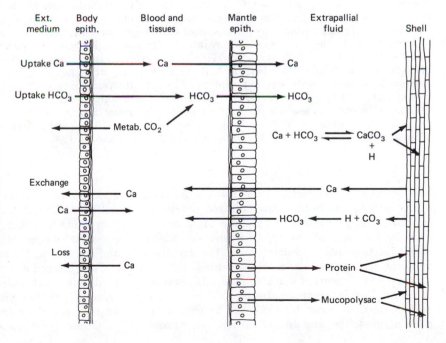

Figure 11–60 Processes of shell secretion in mollusks. Protein and mucopolysaccharides are involved in the production of the organic framework of the shell. Calcium loss occurs during periods of shell dissolution. (After Greenaway, P., 1971: Calcium regulation in the freshwater mollusc *Limnaea stagnalis*. Calcium movements between internal calcium compartments. J. Exp. Biol. *54*:609–620, from Wilbur and Saleuddin, 1983.)

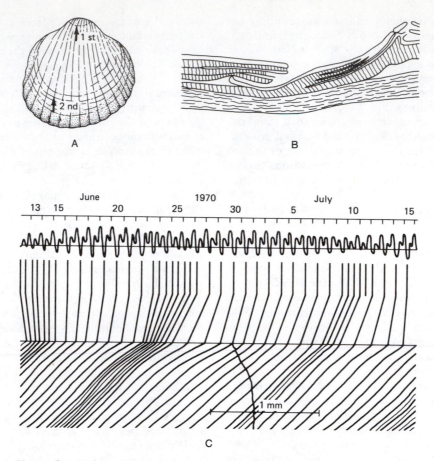

Figure 11–61 *A*, First and second annual growth rings in a European cockle. Ring is produced by winter interruption of shell deposition. Thus, shell produced during second year lies between the faint first annual ring and the more conspicuous second. (From Richardson, C. A., et al, 1980: The use of tidal growth bands in the shell of *Cerastoderma edule* to measure seasonal growth rates under cool temperate and sub-arctic conditions. J. Mar. Biol. Assoc. U.K., *60*:977–989; copyrighted and reprinted by permission of Cambridge University Press.) *B*, Cross section of a winter check line in a freshwater unionacean clam caused by interruption and overlap between two years of growth. (After Coker et al from Tevesz, M. J. S., and Carter, J. G., 1980: Environmental relationships of shell form and structure of unionacean bivalves. *In* Rhoads, D. C., and Lutz, R. A. (Eds.): Skeletal Growth of Aquatic Organisms. Plenum Press, N.Y.) *C*, Microgrowth increments in the shell of the intertidal cockle *(Clinocardium nuttalli)* from Charleston, Oregon. The oblique growth lines of the shell shown at the bottom are correlated with the daily tidal cycles shown at the top. Spring tides are marked by one very low tide per day, neap tides by two equal tides of low amplitude. Cockles will be exposed when tides fall to level of horizontal line. (Adapted from Evans, J. W., 1972: Tidal growth increments in the cockle *Clinocardium nuttalli.* Science, *176*:416–417. Modified from Lutz and Rhoads, 1980.)

eous coating approximately 1 mm thick is then laid down around the bead. Most cultured pearls are started with a microscopic globule of liquid or ground "seeds" of unionid (freshwater clam) shells, placed into a pearl oyster. The resulting year-old seed pearl is then transplanted into another oyster. A pearl of marketable size is obtained three years after transplantation.

The foot of most bivalves has become compressed, bladelike, and directed anteriorly as an ad-

aptation for burrowing. Foot movement, which will be described in detail later in connection with burrowing, is effected by a combination of blood pressure and muscle actions of pedal protractors and retractors. These muscles, which are homologous to the pedal retractors of other mollusks, extend from each side of the foot to the opposite valve, where they are usually attached to the shell near the anterior adductor muscles (Fig. 11–58*B*).

Other specializations of the bivalve shell, man-

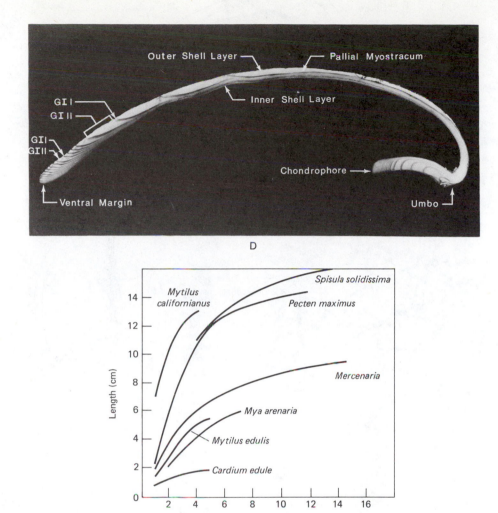

D

E

Figure 11–61 *(continued)* D, Radial cross section of a surf clam shell *(Spisula solidissima)* showing repeating annual growth layers. Each pair of light and dark bands (GL I and GL II) represent one year's growth. Note that most of growth occurs in the early years of the life of the clam. (From Jones, D. S., 1981: Repeating layers in the molluscan shell are not always periodic. Paleontology, *55*(5):1076–1082.) E, Growth curves of shell length for seven bivalves. Approximate age of a specimen can be determined by measuring the shell length and locating the corresponding point on the curve. (Adapted from a number of sources.)

tle, and foot will be described later in connection with different adaptive groups.

Evolution of Bivalve Feeding

It is generally agreed that the early bivalves were shallow burrowers in soft substrata and evolved from a group of now-extinct rostroconch mollusks (Pojeta and Runnegar, 1974). They belonged to the subclass Protobranchia, which is represented by some of the oldest fossil forms (Ordovician, perhaps Cambrian) as well as by some living species— *Nucula, Nuculana, Yoldia, Solemya,* and *Malletia.* Most extant species of protobranchs live in the substratum with the anterior end directed downward and the posterior end directed toward the surface. They possess a single pair of posterolateral bipectinate gills, from which the name *Protobranchia*— first gills—is derived. In most protobranchs, as well as in almost all other bivalves, the ventilation

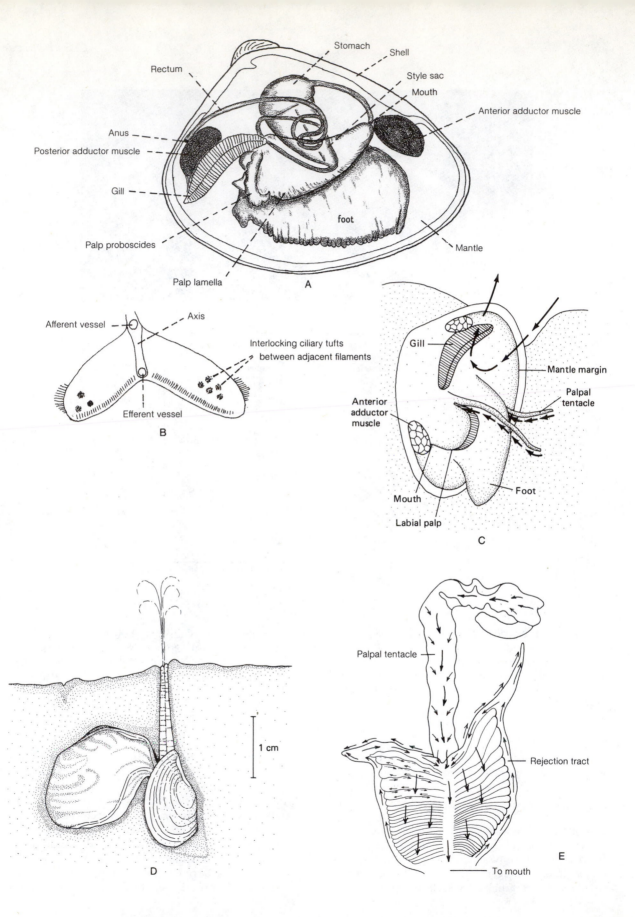

A

Rectum
Stomach
Shell
Style sac
Mouth
Anterior adductor muscle
Anus
Posterior adductor muscle
Gill
Palp proboscides
Palp lamella
foot
Mantle

B

Afferent vessel
Axis
Interlocking ciliary tufts between adjacent filaments
Efferent vessel

C

Gill
Mantle margin
Palpal tentacle
Anterior adductor muscle
Mouth
Labial palp
Foot

D

1 cm

E

Palpal tentacle
Rejection tract
To mouth

◄ **Figure 11–62** Protobranchs: *A*, Body of *Nucula* with right valve and mantle removed (lateral view). *B*, Gill of *Nucula*, showing lateral filaments (transverse section). Note similarity to gill of the primitive gastropod *Haliotis*, shown in Figure 11–1*C*. (*A* and *B* after Yonge, C. M., 1939: The protobranchiate Mollusca, a functional interpretation of their structure and evolution. Phil. Trans. Roy. Soc. London B, *230*:79–147.) *C*, A generalized protobranch, illustrating shallow burrowing and deposit feeding. Small arrows indicate path of food particles along palpal tentacles and labial palps; large arrows show direction of water current. *D*, The protobranch *Yoldia limatula* deposit feeding within chamber excavated below the surface. Siphons come to surface. This species may also extend palpal tentacles to surface and feed on surface particles. (From Bender, K., and Davis, W. R., 1984: The effect of feeding by *Yoldia limatula* on bioturbation. Ophelia, *23*(1):91–100.) *E*, Labial palps and palpal tentacle of *Nuculana minuta*. Apposing surfaces of palps are pulled back and arrows indicate direction of ciliary currents. (After Atkins.)

current enters the mantle cavity through the shell gape posteriorly and ventrally, passes up through the gills, and exits posteriorly and dorsally (Fig. 11–62*C*). Lateral gill cilia create the water current, and frontal cilia remove sediment trapped on the gill surface.

Most living protobranchs are selective deposit feeders, and this is believed by many malacologists to have been the mode of feeding of the early and extinct members of the group. In the ancestral mollusks, the mouth rested against the hard bottom over which the animal crawled. However, when bivalves became adapted for burrowing in sand or mud, the mouth was lifted above the substratum as a result of both the lateral compression of the body and the greatly increased height of the dorsoventral axis. The radula has disappeared. Contact with the substratum is maintained by a pair of tentacles, elongations of the margins of the mouth. Each tentacle is associated with two large, flaplike folds, called labial palps, located to either side of the mouth (Fig. 11–62). During feeding the tentacles are extended into the bottom sediments. Deposit material adheres to the mucus-covered surface of the tentacle and then is transported by cilia back to the palps, which function as sorting devices. The inner apposing surfaces of each pair are ridged and ciliated (similar to Fig. 11–62*E*). Light particles are carried by crest cilia to the mouth; heavy particles are carried by groove cilia to the palp margins, where they are ejected into the mantle cavity. *Solemya*, which is common along the east coast of the United States and in other parts of the world, does not feed in the typical protobranch manner. It lacks palpal tentacles and has greatly reduced labial palps. Members of this genus, including some recently discovered gutless, deep-sea forms, appear

to rely on symbiotic bacteria for their nutrition (Reid and Bernard, 1980).

Although some protobranchs, such as *Nucula*, *Yoldia*, and *Solemya*, live in shallow water, the group is more abundantly represented in deep ocean floors (see review by Knudsen, 1979).

In some group of early protobranch bivalves, filter feeding evolved. An explosive evolution followed this development, and the filter feeders, called lamellibranchs, came to dominate the bivalve fauna. The gills and ventilating current of protobranchs preadapted them for filter feeding. As the lamellibranchs evolved, detrital particles and microorganisms in the ventilating current came to be utilized as a source of food, the gills became filters, and the gill cilia that originally served to keep the gills clean became adapted for the transport of particles trapped in mucus from the filter to the mouth.

The principal modification of the gills for filtering was the lengthening and folding of the gill filaments, which greatly increased their surface area (Fig. 11–63). Many filaments were added to the gills so that they extended anteriorly, reaching the palps. Each gill filament on each side of the axis became folded, or U shaped. The arm of the U that is attached to the axis of the gill is called the descending limb, and the arm next to the mantle or visceral mass, the ascending limb. Since the filaments on both sides of the axis have become folded, the net result has been to transform the original single gill into a pair of gills, or demibranchs; the original outer filaments form one member of the pair, and the original inner filaments form the other (Fig. 11–63). The lengthened, folded filaments and their attachment to one another give the gill a sheetlike form, hence the name of the subclass Lamellibran-

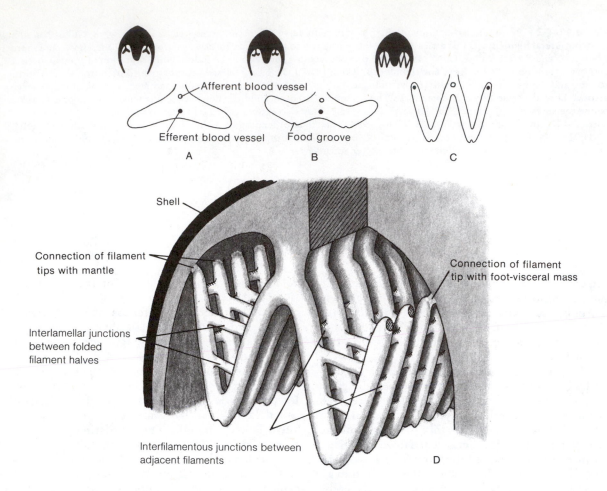

Figure 11–63 Evolution of lamellibranch gills. Arrows indicate direction of water flow. *A*, Primitive protobranch gill (position relative to foot-visceral mass and mantle indicated in cross section. *B*, Development of food groove in hypothetical intermediate condition. *C*, Folding of filaments at food groove to produce the lamellibranch condition. *D*, Tissue connections that provide support for the folded lamellibranch filaments.

chia—sheet gill. Four large, broad, filtering surfaces (lamellae) are present, two on each gill (demibranch). At the angle of flexure, the frontal surface of each filament has developed an indentation, or notch, which, when lined up with the notches of adjacent filaments, forms a food groove that extends the length of the underside of the gill.

These modifications in gill structure have necessitated a change in ciliation. The frontal cilia carry food particles trapped on the gill surface vertically to the food grooves (Figs. 11–65*B* and 11–67). The abfrontal cilia, now inside, are lost. Lateral cilia still produce the water current through the gills. On each side of the filaments, between the lateral and the frontal cilia, is a new ciliary tract composed of laterofrontal cirri. Each cirrus is a

bundle of many fused cilia (Fig.11–64*A*). Opposing cirri form a fine mesh that filters particles from the water entering the gill; the cirri then move them onto the frontal cilia. The pressure of the water stream generated by the lateral cilia is more than sufficient to overcome the resistance offered by the cirri (Fig. 11–64*B*) (Silvester and Sleigh, 1984).

The inhalant, feeding-ventilating current enters the lower part of the mantle cavity at the posterior end of the animal, flows between the filaments, and then moves up between the two lamellae. From the interlamellar spaces, water passes into the exhalant, or suprabranchial, chamber and finally flows out through the posterior exhalant opening.

Support for the long, folded filaments is provided by three kinds of new tissue connections at various points within the gill: (1) cross connec-

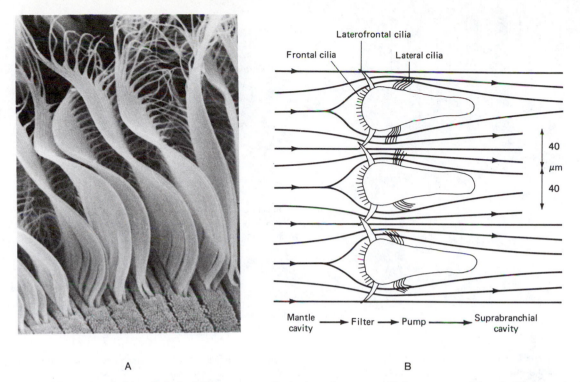

Frontal cilia

Laterofrontal cilia

Lateral cilia

40

µm

40

Mantle
cavity → Filter → Pump → Suprabranchial
cavity

A   B

Figure 11–64 *A,* Laterofrontal cirri of *Venus casina.* Each cirrus is composed of two rows of fused cilia on a single cell. The distal ends of the cilia are unfused. (From Owen, G., 1978: Classification and the bivalve gill. Phil. Trans. Roy. Soc. London B, *284:*377–386.) *B,* Diagrammatic cross section through three gill filaments of *Mytilus.* The frontal cilia beat toward and away from the viewer. The effective stroke of the cirri is to the left, and the effective stroke of the lateral cilia is to the right. Arrows indicate the direction of the feeding-ventilating current produced by the lateral cilia. (From Silvester, N. R., and Sleigh, M. A., 1984: Hydrodynamic aspects of particle capture by *Mytilus.* Jour. Mar. Biol. Assoc. U.K., *64:*859–879; copyrighted and reprinted by permission of Cambridge University Press.)

tions, called interlamellar junctions, between the folded filament halves, or lamellae; (2) connections, called interfilamentous junctions, between adjacent filaments; (3) connections between the tips of the filaments and the mantle or foot (Fig. 11–63D). The extent of these connections varies in different groups of lamellibranchs and accounts for several types of lamellibranch gills.

If the individual filaments are still more or less separate, the gill is known as a filibranch gill (Fig. 11–65A and B). Bars of tissue, the interlamellar junctions, have grown between the two limbs of each U at intervals, and there may be tissue union of adjacent filaments at the bottoms or tops of the lamellae. But throughout most of their length, filaments are attached to adjacent filaments only by specialized ciliary junctions (Reed-Miller and Greenberg, 1982). Filibranch gills are found in such bivalves as ark shells *(Arca),* scallops *(Pecten),* and mussels *(Mytilis).* In the oysters and pen shells, which are said to have a pseudolamellibranch type

of gill, the filaments are bound together with some (though not extensive) interfilamentous tissue junctions.

The most specialized lamellibranch gill is known as a eulamellibranch gill. In this type the union of filaments has developed further, so that the lamellae actually consist of solid sheets of tissue (Fig. 11–65C and D). Furthermore, interlamellar junctions have increased in number and extend the length of the lamellae (dorsoventrally). Thus, the interlamellar space is partitioned into vertical water tubes. Instead of diffusing through the lamellae, the blood is carried in vertical vessels that course within the tissue junctions. The tips of the ascending limbs have become fused with the upper surface of the mantle on the outside and the foot on the inside, morphologically separating the inhalant chamber from the suprabranchial chamber. Ciliation remains the same, for the frontal edges of the filaments are not involved in the interfilamentous fusion. Thus, a frontal section of the eulamel-

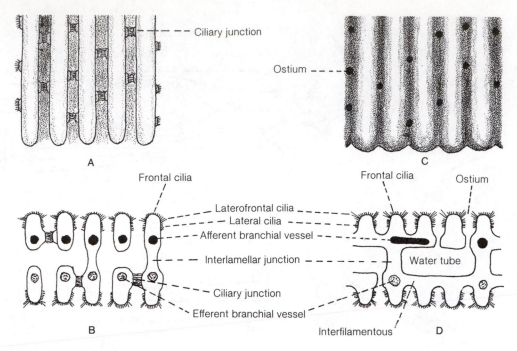

Figure 11–65 *A* and *B*, Filibranch gill: *A*, Five adjacent filaments (surface view). *B*, Frontal section. *C* and *D*, Eulamellibranch gill: *C*, Five fused, adjacent filaments (surface view). *D*, Frontal section.

libranch gill exhibits a lamella with a ridged outer surface, each ridge representing one of the original filaments (Fig. 11–65*C* and *D*).

Water in the inhalant chamber circulates between the ridges and enters the water tubes through numerous pores (ostia) in the lamella. Oxygenation takes place as the water moves dorsally in the water tubes. The water then flows into the suprabranchial cavity and out the exhalant opening.

In the primitive gills of protobranchs, the efferent, or drainage, vessel ran within the axis of the filament beneath the afferent vessel, as in the ancestral molluscan gill. With the elongation and folding of the filament, the old drainage vessel dropped out and a new drainage vessel formed at the junction of the ends of the fused ascending limbs of each filament with the mantle wall or visceral mass (Fig. 11–63*C*). Blood therefore flowed in but one direction through each filament.

In many pseudolamellibranchs and eulamellibranchs, the surface area of the lamellae has been increased further by folding along the length of the gill, and the plicate gill surface of these bivalves presents an undulated appearance (Fig. 11–66*A*).

Most lamellibranchs feed on fine plankton and suspended detritus. Food particles, in some cases as small as 1 μ, are removed from the water currents passing between filaments or entering the ostia. The particles are then passed onto the frontal cilia, where they are entangled in mucus and moved up or down the margin of the filament to a food groove.

The primitive lamellibranch has five food grooves transporting particles anteriorly to the palps. Three of the grooves are located at the top of the gills between and outside the demibranchs; the other two are located ventrally, one along the margin of each gill (Fig. 11–67*G*). The frontal cilia are divided into two tracts of coarse and fine cilia, one carrying particles upward and one downward. Such a two-way vertical tract system with five food grooves is found in oysters and scallops. From such a primitive condition, the great variation in number and location of food grooves and direction of vertical tracts encountered in other lamellibranchs is believed to be derived by deletion. Many of these variations are illustrated in Figure 11–67, in which the orally directed food grooves are indicated by black circles. Note that the dual tract system has disappeared in most lamellibranchs.

Various mechanisms provide for some prepalpal sorting by the gills and for coping with sediment-laden water short of halting ventilation. In

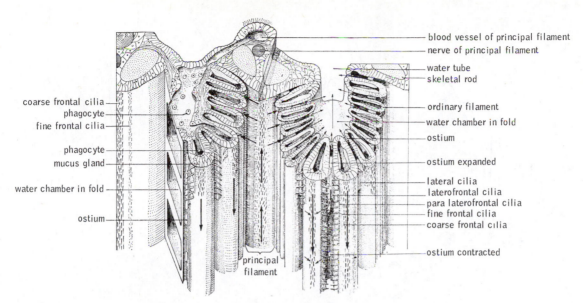

Figure 11–66 Stereodiagram of part of a plicate gill *(Crassostrea)*. (After Nelson from Jorgensen, C. B., 1966: Biology of Suspension Feeding. Pergamon Press, London. p. 71.)

the families Arcidae and Anomiidae the heavy particles that are carried to the ventral margin of the gills are transported *posteriorly* and ejected onto the mantle (Fig. 11–67*F*). In a number of families that have plicate gills, such as the Ostreidae (oysters), Pectinidae (scallops), and Solenidae (razor clams), the filament that lies between the folds, called the principal filament (Fig. 11–66), carries only light particles upward. When the water is relatively clean, the gills are expanded and the upward-moving tracts are largely in operation. When there is a lot of sediment in the water, the gills are stimulated to contract, placing the principal filament deep within the folds (Fig. 11–68*A* and *B*). Coarse cilia are stimulated to activity, driving much of the trapped material to the ventral grooves, which if too heavily laden will drop their loads into the bottom of the mantle cavity.

The lamellibranch palpal lamellae have the same sorting and conveying function as in the Protobranchia (Fig. 11–68*C* and *D*). Particles are sorted by size and weight. Small, light particles are retained for ingestion and are carried up the palpal surface across the crest of the ciliated ridges (Fig. 11–68*D*). Large particles, destined to be rejected, are carried to the edge of the lamellae in the grooves between ridges and fall to the mantle or foot. Rejected material, called pseudofeces, from both the palps and the gills leaves the mantle cavity most commonly by the inhalant aperture. The par-

ticles are carried posteriorly along a ventral ciliated tract of the mantle to accumulate behind the inhalant aperture. When the valves are closed periodically, water is forced out of the inhalant opening, taking these accumulated wastes with it.

The animal can regulate water flow by changing the size of the apertures into the mantle cavity and by gill contraction or expansion, which permits less or more water to pass between the filaments.

From some group of protobranchs evolved still another subclass of bivalves, the Septibranchia (*Poromya* and *Cuspidaria*). The members of this small group have become carnivores or scavengers. The gills have been modified to form a pair of perforated muscular septa, which separate the exhalant, suprabranchial chamber from the inhalant, infrabranchial chamber. By muscular contractions the septum moves up and down, forcing water into the inhalant chamber and out of the exhalant chamber. The force of the pumping septa is sufficient to bring small animals, such as crustaceans and worms, into the mantle cavity. These prey are then seized by the reduced but muscular labial palps and carried to the mouth. Vibrations of small moving crustaceans are detected by siphonal tentacles. The siphon is then shot out in the direction of the prey, which is quickly sucked up by the simultaneous inhalant water current (*Cuspidaria*), or the siphon is everted as a hoodlike trap over the prey (*Poromya*) (Fig. 11–69).

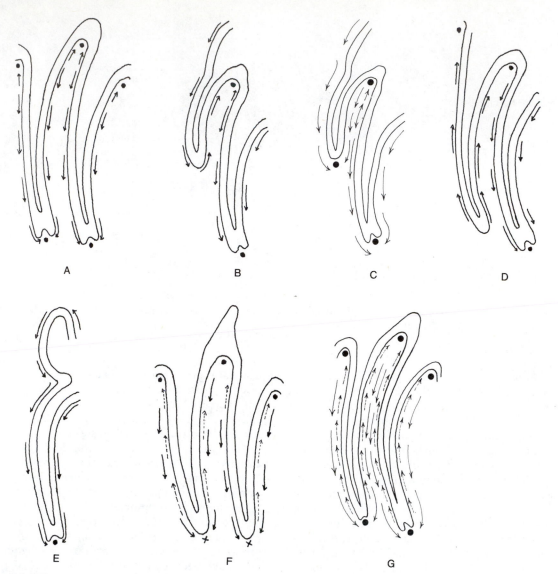

Figure 11–67 Transverse sections of different lamellibranch gills, showing direction of frontal cilia beat and position of anteriorly moving food tracts (black dots): In *F* and *G*, broken arrows indicate fine frontal cilia carrying food particles upward; solid arrows indicate coarse frontal cilia carrying particles ventrally. Crosses represent posteriorly moving channels. Inner demibranch, or gill, is on right in all cases. *A*, Mytilidae and Pinnidae. *B* and *C*, Many eulamellibranchs. *D*, Unionidae (freshwater). *E*, Most Tellinidae and Semelidae. *F*, Arcidae and Anomiidae. *G*, Ostreidae and Pectinidae. (All after Atkins, D., 1937: Q. J. Micr. Sci., 79.)

Digestion

The structure and physiology of the digestive tract of some of the deposit feeding protobranchs have retained a number of primitive features (p. 344) and are similar to those of archaeogastropods (Fig. 11–70). Digestion in most protobranchs is extracellular in the stomach, and absorption occurs in the digestive gland.

The use of finer particles as food in the filter-feeding Lamellibranchia is reflected in a number of stomach modifications. The girdle of chitin, present in the protobranchs, is reduced to a small plate, the gastric shield (Fig. 11–71). A style sac is present, but the mucus has become consolidated into a very compact and often long rod, the crystalline style. In addition to the protein matrix of the style

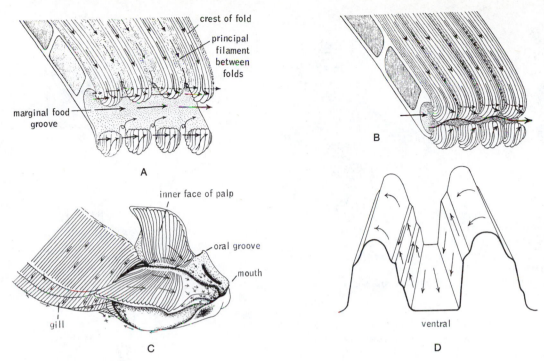

Figure 11–68 *A*, Ventral margin of a section of the plicate gill of *Pinna fragilis* when relaxed and expanded. *B*, Same gill as *A* when contracted. Note that contraction buries the principal filaments that are located between the folds (plicae) and closes the marginal food groove. In members of this family, most frontal ciliary tracts carry particles downward. (See Fig. 11–67*A*.) (From Atkins, D., 1937: Q. J. Micr. Sci., 79:348.) *C*, One pair of palps spread apart and anterior section of gills of *Ostrea edulis*. Arrows indicate ciliary tracts. (After Yonge.) *D*, Section of two ridges of a lamellibranch palp. Mouth lies to left, and ventral edge of palp is toward viewer. Arrows indicate ciliary tracts. (After Stasek.)

itself, the style sac secretes enzymes that are absorbed onto the style in its formation. The projecting anterior end of the style is rotated against the platelike gastric shield by the style sac cilia, and in the process the style end is abraded and dissolved, releasing various carbohydrate-splitting enzymes and lipase, depending on the species (see review by Morton, 1983). Similar enzymes are released from the stomach wall. Thus, starches are digested at least in part extracellularly. Most protein digestion occurs intracellularly within the digestive gland. The rotation of the style also aids in mixing the enzymes with the stomach contents and acts as a windlass to pull food-laden mucous strings from the esophagus into the stomach (Fig. 11–71*A*). The lower pH of the stomach facilitates the dislodgment of particles from the mucous strings.

The length of the style varies, but it is remarkably long considering the size of the animal. In many bivalves, the style is approximately 3 cm long, but a 12-cm *Tagelus* may have a 5-cm style, and Yonge (1975) reported a 36-cm style in a 1-meter *Tridacna*.

The mixing of stomach contents by the crystalline style continually throws partially digested particulate material against a ciliated sorting area. Coarse, heavy particles are segregated and sent to the intestine along a deep intestinal groove. Fine particles and fluid containing digestion products are retained by the cilia of the sorting ridges and directed toward the numerous apertures of the digestive glands. In many lamellibranchs these apertures, along with a typhlosole bordering the intestinal groove, are located in one caecum or, in higher forms, in two caeca of the stomach (Fig. 11–71). Within the main ducts of the digestive diverticula, there is a continuous two-way flow of materials entering for intracellular digestion and absorption and of cell fragments and wastes leaving (Fig. 11–71) (Owen, 1974; Palmer, 1979). The deep intestinal groove by which wastes leave the stomach and pass into the intestine or the actual separation of the style sac from the intestine prevents wastes from becoming incorporated into the style matrix (Fig. 11–71*C*).

In intertidal bivalves, such as *Crassostrea vir-*

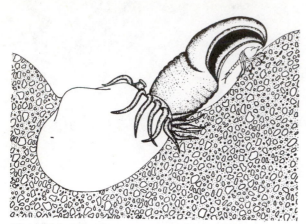

Figure 11–69   The septibranch *Poromya granulata* capturing a crustacean with its hoodlike, inhalant siphon. (From Morton, B., 1981: Prey capture in the carnivorous septibranch *Poromya granulata*. Sarsia, 66:241–256.)

ginica, Lasaea rubra, and *Cardium edule,* which feed only at high tide, the different digestive processes display a tidal rhythm (Fig. 11–72). The crystalline style, which in these forms lies in an intestinal groove, is dissolved at low tide when the animal is not feeding and is re-formed as the tide comes in. A circadian feeding rhythm appears to be present in some species of freshwater clams. It appears that in most shallow-water bivalves, feeding and extracellular digestion in the stomach and extracellular digestion in the digestive diverticula are to varying degrees rhythmic or phasic.

The muscular stomach of the carnivorous septibranchs is lined with chitin and acts as a crushing gizzard. Proteases from the digestive gland initiate extracellular digestion in the stomach. Material digested this way is conveyed into the ducts of the digestive diverticula, where further digestion occurs intracellularly. The style is very reduced and barely projects into the lumen of the stomach.

Adaptive Radiation of Bivalves

The evolution of filter feeding freed lamellibranchs from dependence on deposit material and made possible the colonization of many habitats that were uninhabitable for their protobranch ancestors. The success of this adaptive radiation is reflected in the fact that of the some 20,000 described species and 75 families of bivalves, most are lamellibranchs.

It must be emphasized that the adaptive groups described below do not necessarily constitute

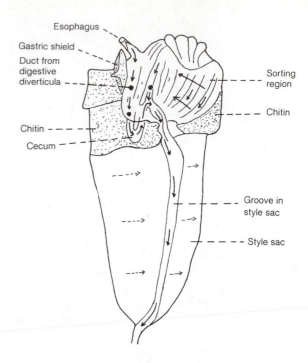

Figure 11–70   Stomach of *Malletia,* opened. Solid arrows indicate food tracts; broken arrows indicate direction of beat of style-rotating cilia. (After Yonge.)

closely related species. Colonization and adaptation for a particular type of habitat have been achieved independently by a number of bivalve families or superfamilies.

SOFT-BOTTOM BURROWERS (INFAUNA)

It should not be thought that all lamellibranchs departed from the ancestral habitat. The majority of species are inhabitants of soft bottoms, exploiting the protection offered by a subterranean life in marine sand and mud while utilizing food suspended in water brought in from above the surface. Some live just beneath the surface, some are adapted for deep burrowing, some move between the surface and lower levels, and some are especially adapted for shallow, rapid burrowing in a shifting environment. In addition to the many marine forms, soft-bottom lamellibranchs also include most of the freshwater clams.

The actual mechanism of burrowing involves the coordinated action of a number of forces (Trueman, 1966 and 1983). The protraction of the foot is initiated by contraction of the pair of pedal protractor muscles (Fig. 11–73). The projecting foot probes and pushes into the surrounding sand. As the foot is protruded, the valves begin closing by

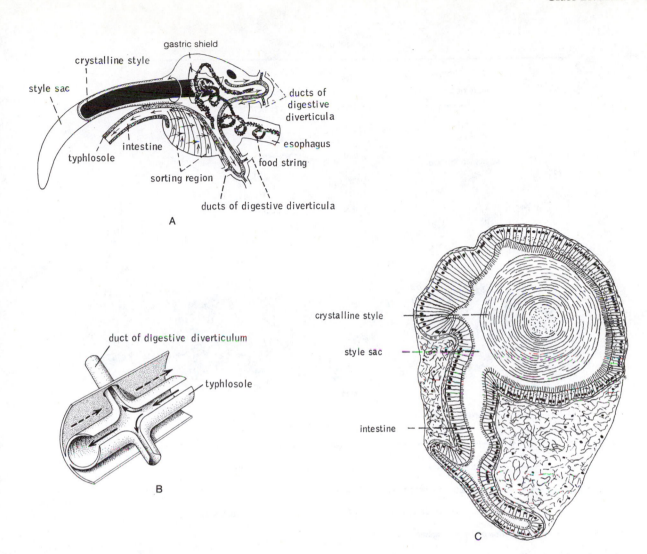

Figure 11–71   *A*, Stomach of a lamellibranch, showing rotation of crystalline style and winding in of mucous food string. Arrows indicate ciliary pathways. (After Morton, J. E., 1967: Molluscs. Hutchinson University Library, London.) *B*, Typhlosole within cecum of stomach, showing extensions into ducts of digestive diverticula. Solid arrows indicate inhalant ciliary currents; broken arrows indicate exhalant currents. *C*, Style sac and intestine of the freshwater clam, *Lampsilis anodontoides* (transverse section). (After Nelson from Yonge.) *D*, Diagram of a section of digestive diverticulum, showing absorption and intracellar digestion of material passed inward from stomach (solid arrows) and outward passage of wastes (broken arrows). (*B* and *D* after Owen, G., 1974: Feeding and digestion in the Bivalvia. Adv. Comp. Physiol. Biochem., *5*:1–35.)

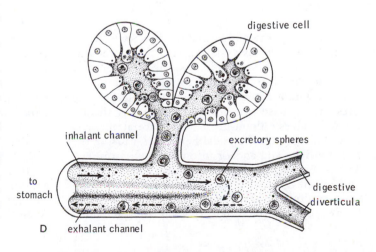

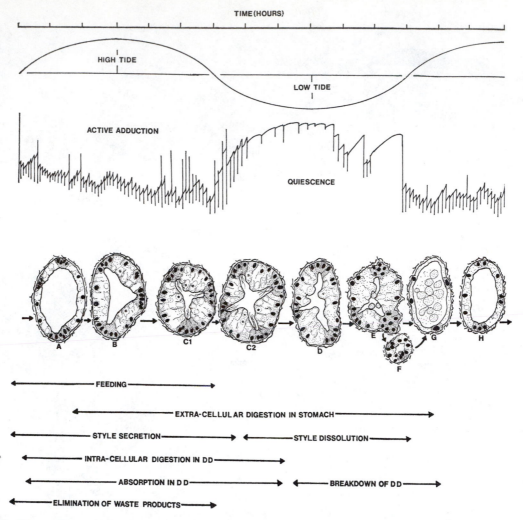

Figure 11–72 Feeding and digestive processes of *Cardium edule* during a period including high and low tides. (From Morton, B., 1970; J. Mar. Biol. Assoc. U.K., *50*:506; copyrighted and reprinted by permission of Cambridge University Press.)

contraction of the adductor muscles. Some water from the mantle cavity is expelled, which loosens the sand or mud and facilitates the movement of the foot. The water remaining in the mantle cavity and the blood act as a hydrostatic skeleton. The pressure of these two fluids is elevated by the adducted valves. Blood from the visceral mass is forced down into the pedal hemocoel, causing the foot to dilate and anchor into the substratum (Fig. 11–74D). With the foot anchored, an anterior pair and a posterior pair of pedal retractor muscles contract, pulling the shell downward (Fig. 11–74B). In many species, retraction by the anterior pedal mus-

cle occurs before that of the posterior muscle. The effect is to rock the shell, which facilitates its movement through the substratum. Ridges or other sculpturing of the shell surface may increase traction. Following pedal retraction, relaxation occurs and the valves gape. To return to the surface or to a higher level within the substratum, most bivalves back out, pushing against the anchored end of the foot.

In the primitive protobranchs, such as *Nucula*, *Solemya*, and *Yoldia*, the foot bears a little flattened sole (Fig. 11–74B). However, the two sides of the sole can be folded together, producing a bladelike

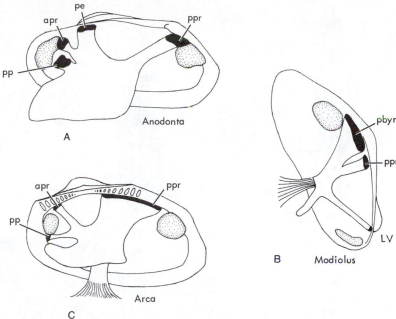

Figure 11–73 Positions of the pedal and adductor muscles in different bivalves: *A,* Freshwater clam *Anodonta. B,* Byssally attached mytilid *Modiolus. C,* Byssally attached *Arca.* Pedal muscles are shown in black; adductors stippled. Abbreviations: apr, anterior pedal retractor; pdm, dorsomedian pedal muscle; pe, pedal elevator; pp, pedal protractor; ppr, posterior pedal retractor. (After Pelseneer from Moore, R. C. (Ed.), 1969: Treatise on Invertebrate Paleontology. (N) Mollusca 6(1), pp. N30 and N33. Geological Society of America and the University of Kansas Press, Lawrence, Kansas.)

edge; in this condition the foot is thrust into the mud or sand. The sole then opens and serves as an anchor.

The common cockles (*Cardium* and allies), which have rounded, convex, ribbed shells, are shallow burrowers (Fig. 11–81). They use the foot not only for conventional burrowing but also for escape. When threatened by a predator, such as a starfish, the cockle leaps by rapidly extending the foot from a folded position against the substratum.

A major problem arising from burrowing in soft bottoms is the sediment brought in by water currents. Blood engorgement within the mantle margin enables the mantle edges to be appressed, even when the valves gape slightly (the muscles of the inner fold provide for retraction), but since circulation of water through the mantle cavity is necessary for gas exchange and, in most species, for feeding, at least the posterior part of the mantle must be opened for entrance of water, with which sediment enters.

There has been a tendency to seal the mantle edges morphologically where openings are not necessary (Fig. 11–75). The most frequent point of fusion of the mantle edges is at the posterior between the inhalant and exhalant openings. This fusion forms a distinct exhalant aperture.

A functional inhalant opening is present since the mantle edges are appressed below the opening,

even though the mantle edges are not morphologically fused. A second point of fusion just below the inhalant opening, forming a distinct inhalant aperture, has evolved in many bivalves.

Still further mantle fusion has occurred in some species, especially deep burrowers, and most of the ventral margin anterior to the inhalant aperture has become sealed. Thus, three apertures remain—the inhalant and exhalant apertures and an anterior pedal aperture through which the foot protrudes. Extensive mantle fusion not only reduces fouling of the mantle cavity but is of primary importance in maintaining the hydraulic pressure within the mantle cavity necessary for burrowing.

Commonly, when structural inhalant and exhalant apertures are present as a result of fusion, the mantle edges surrounding the apertures have become elongated to form tubular siphons of varying lengths (Fig. 11–76). With siphons the animals can be completely buried in the mud and only the siphon tips need project above the bottom. The siphons are extended by blood pressure or by water pressure within the mantle cavity when the valves are closed and are withdrawn by siphon retractor muscles derived from the muscle tissue of the innermost mantle fold. Considerable variation exists in siphon length and degree of fusion between inhalant and exhalant siphons. Where there are separate inhalant and exhalant siphons, siphon for-

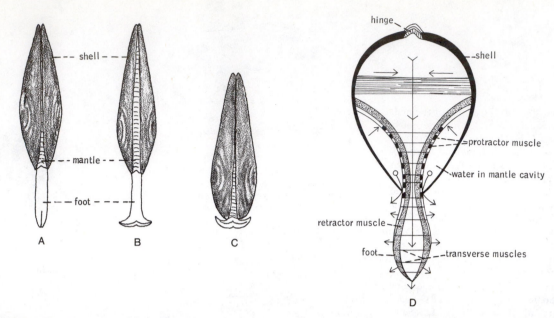

Figure 11–74 Operation of foot of *Yoldia limatula*, a protobranch: *A*, End of foot folded and extended into substratum. *B*, Foot opened and anchored. *C*, Body drawn down into substratum. (All after Yonge.) *D*, Diagrammatic cross section of a bivalve, showing hydrostatic forces that produce dilation of foot. Central vertical arrow indicates flow of blood into foot. (Modified after Trueman, 1966.)

mation involves only the muscular inner fold of the mantle.

Well-developed siphons are indicated by scars on the inner face of the valve. The pallial line impression is recurved sharply inward just below the posterior adductor and represents the point of attachment of the siphon retractor muscles. This bay in the pallial line is called the pallial sinus (Fig. 11–58*B*).

Most modern soft-bottom filter feeders possess siphons. Those that do not are rather sluggish shallow burrowers. The geoducks of the Pacific coast of the United States, *Panopea generosa*, are among the deepest burrowers, going down more than a meter. They have siphons so large that they can no longer be retracted between the shell valves (Fig. 11–77*B*).

Many bivalves that burrow deeply (i.e., to depths greater than the lengths of their bodies) tend to have semipermanent or even permanent burrows. The walls of the burrows become coated with mucus, which also reduces sediment fouling.

The primitive form of lamellibranch soft-bottom burrowers is considered to be one in which the adductor muscles are more or less equal (isomyarian), the mantle is unfused, and the valves are

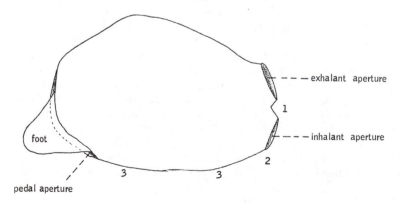

Figure 11–75 Areas of mantle fusion in bivalves: *1*, Between inhalant and exhalant apertures or siphons, the most common point of fusion. *2*, Fusion below inhalant aperture or siphon. *3*, Fusion between inhalant aperture and foot, leaving only a pedal aperture for extension of foot.

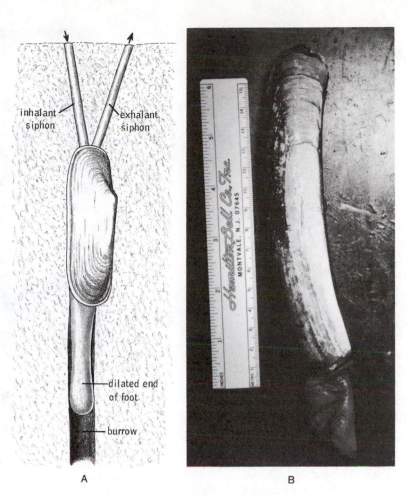

inhalant siphon

exhalant siphon

dilated end of foot

burrow

Figure 11–76 Razor clams from the east coast of the United States. *A, Tagelus plebeius. B, Ensis directus.* The two species belong to different families. Both are common in intertidal sands. *Tagelus* is more common along southern shorelines, where it may occur in enormous numbers. The much shorter siphons of *Ensis* can just be seen at the posterior end of the valves. (*B* by Katherine E. Barnes.)

A

B

equal and circular in outline. We have already described modifications involving mantle fusion. In many groups the valves are also modified, becoming more streamlined for burrowing by being flattened or elongated. Species of *Donax*, which are inhabitants of surf beaches, have shells that are pointed anteriorly and blunt posteriorly. They back out on incoming waves and reburrow with great rapidity as the wave recedes. The margins of the inhalant siphon are fringed with infolded tentacles, which keep out swirling sand grains (Fig. 11–77A).

The razor clams *Ensis* and *Tagelus*, two unrelated but similar groups, have greatly elongated valves and an elongate foot, which enable them to move rapidly within their more or less permanent burrows (Fig. 11–76). *Tagelus* has long siphons, each of which has a separate opening to the surface. *Ensis*, which has short siphons, comes to the surface to feed from the deeper, more protected part of its burrow.

Two further specializations of soft-bottom burrowers should be mentioned. Some tellinaceans have reverted to deposit feeding. *Scrobicularia*, for example, extends its inhalant siphon above the surface at low tide and, like a vacuum cleaner, sucks in deposit material (Fig. 11–78), which is then sorted on the gills. Deposit feeding is generally an addition to, rather than a substitute for, filter feeding.

The lucines, rather small, rounded shells, members of the superfamily Lucinacea (*Linga, Codakia, Lucina, Divaricella, Thyasira*), are remarkable in using the foot not only for burrowing but also for constructing a mucus-lined canal back up to the surface. The feeding-ventilating current enters *anteriorly* by this passageway and leaves by the usual exhalant opening.

ATTACHED SURFACE DWELLERS (EPIFAUNA)
A number of evolutionary lines of lamellibranchs have invaded firm substrates—peat, wood, shell,

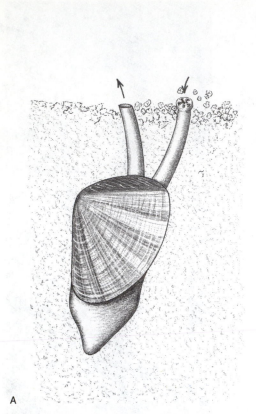

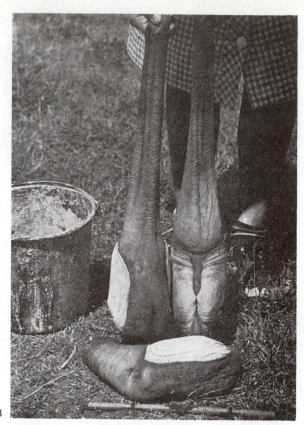

Figure 11—77   *A, Donax variabilis,* a common inhabitant of surf beaches. Rapid burrowing is facilitated by the thin, pointed foot. Opening of inhalant siphon is frilled, preventing entrance of sand grains. *B,* The geoduck, *Panopea generosa,* a giant Californian bivalve in which body and siphon cannot be enclosed within valves. (From Milne and Milne, 1959: *Animal Life.* Prentice-Hall, Englewood Cliffs, N.J.)

coral, rock, or man-made sea walls, jetties, and pilings. Attachment is provided either by byssal threads or by one valve fused to the substratum. Byssal threads are tough protein threads secreted by the foot. In mytilids, which have been most studied, a groove runs from a deep byssus gland to the anterior tip of the foot (Fig. 11–79B and C), and the byssus is formed in the groove by a mixture of secretions produced not only by the byssus gland but also by various glands along the length of the groove (Waite, 1983). After a few minutes, when the fiber has hardened by tanning, the foot is removed, leaving the thread anchored to the substratum by a plaque and to a common stem buried deep within the byssus gland opening (Fig. 11–79A). A byssus retractor muscle may enable the animal to pull against its anchorage.

Among living surface dwellers attached by byssal threads, the widely distributed mussels (Mytil-

idae) are perhaps the most familiar. They live attached to wharf pilings, sea walls, and rocks or among oysters, often in great numbers. The threads, laid down by the little finger-like foot, often radiate outward like guy wires (Fig. 11–79A). Young individuals even use byssal threads to climb walls. Mytilids are widely used as food in many parts of the world and are sometimes "farmed" on ropes suspended from rafts.

Other surface inhabitants attached by byssal secretions include many of the heavy-bodied ark shells (Arcidae) (Fig. 11–79D), which are very common tropical coralline substrates; mangrove oysters *(Isognomon)* (Fig. 11–79F), which hang in clusters from mangrove roots; and winged oysters (pteriids), which live attached to sea fans and other gorgonian corals (Fig. 11–79E).

Some Tridacnidae, which include the giant clams *(Tridacna)* of the Indo-Pacific, are also sur-

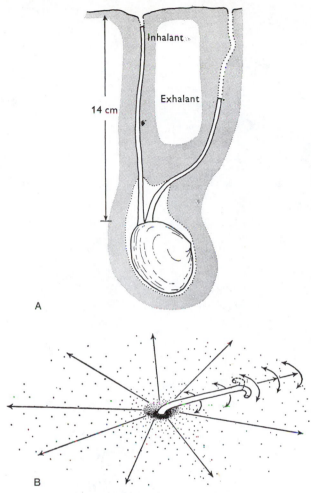

Figure 11–78 Deposit feeding in *Scrobicularia: A*, Animal in burrow with inhalant siphon withdrawn. *B*, Feeding movements of inhalant siphon at low tide. (From Hughes, R. N., 1969: J. Mar. Biol. Assoc. U.K., *49*:807; copyrighted and reprinted by permission of Cambridge University Press.)

face dwellers. The smallest of the six species is only 10 cm long; the largest, *Tridacna gigas*, reaches 1.37 meters. All live vertically oriented with the hinge side down and are initially attached by byssal threads. Some retain the byssus; some lose it and rest on the bottom by the weight of the shell; *Tridacna maxima* bores in coral or coralline rock so that the valve margins are flush with the substrate surface. The gape of all tridacnids is directed upward and with the large mantle surface (actually the siphons) protruded across the fluted shell for maximum exposure to light (Figs. 11–80 and 11–81). Blood sinuses within the mantle contain intercellular, symbiotic zooxanthellae that provide the

clam with an accessory source of nutrition (Trench et al, 1981). The mantle tissue also contains pigments of brilliant green, blue, red, violet, or brown that probably reduce light intensity. Large populations of tridacnids, such as the little boring *Tridacna maxima*, can be a spectacular addition to the beauty of an Indo-Pacific coral reef.

Utilization of byssal threads by adults of some 18 bivalve superfamilies is believed to represent a persistent larval adaptation, for the larvae of many unattached burrowing forms produce a byssus for initial temporary anchorage on settling (Fig. 11–82*A*). Adults of a few living species use byssal threads to anchor rootlike in soft bottoms. The mytilid genus *Modiolus* live partially buried in peat or coarse substrate (Fig. 11–79*H*), and the pen shells *Pinna* and *Atrina* (Fig. 11–79*G*) occupy a similar position in sand, attaching the byssal threads to small stones.

Surface-dwelling bivalves that are attached by cementation lie on one side, fixed to the substratum by either the right or the left valve, depending on the species. Such sessile bivalves include at least eight evolutionary lines, of which the oysters are the most familiar. However, the name oyster is applied to a wide variety of species, some of which attach by byssal threads. In the family Ostreidae, which contains the edible American east coast oyster *Crassostrea virginica* and the European *Ostrea edulis*, the metamorphosing veliger is initially anchored with an organic adhesive produced first by the foot and then by the mantle (see Yonge, 1979, for review). Then the mantle margin attaches the left valve to the substratum in the process of shell secretion (Fig. 11–79*A*).

Shell attachment has led to varying degrees of inequality in the size of the two valves, the lower being larger or smaller (as in the Ostreidae) than the upper one. In *Chama*, the tropical jewel boxes, the upper valve forms a lid over the boxlike lower valve (Fig. 11–83*A*). The extreme condition was reached in the extinct Mesozoic rudists, in which the lower valve was shaped like a tube or horn (Fig. 11–83*B*). The rudists often occurred in reeflike aggregations.

The common jingle or toenail shells, members of the family Anomiidae, possess features of both the byssally attached and the cemented bivalves. They lie on one side but are actually anchored by a large, calcified byssal thread that passes through the attached valve (Fig. 11–83*D*).

Attached bivalves share a number of features. As would be expected, the foot is reduced to varying degrees, and it is completely absent in those

(Text continued on p. 426)

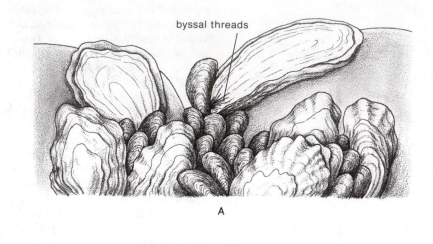

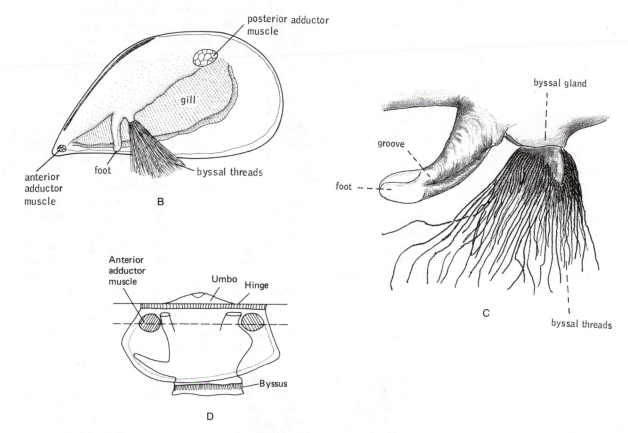

Figure 11–79 Sessile bivalves attached by byssal threads: *A*, Mussels *(Mytilus)* anchored among oysters *(Crassostrea)*. *B*, Lateral view of a mytilid with valve removed. *C*, Foot and byssal gland of a mytilid. *D*, Diagrammatic lateral view of *Arca*. (Modified from Yonge, C. M., 1953: Phil. Trans. R. Soc. London B, 237:365).

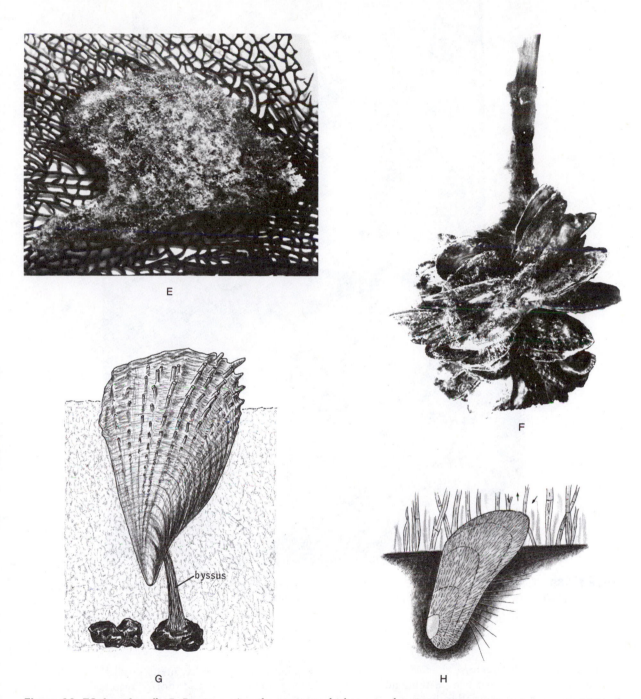

E

F

G

byssus

H

Figure 11–79 *(continued)* *E, Pteria,* a winged oyster, attached to a sea fan. *F, Isognomon,* a mangrove oyster. (*E* and *F* by Betty M. Barnes.) *G,* A pen shell, *Atrina rigida. H, Modiolus* partially buried among intertidal marsh grass. (*H* after Yonge from Stanley, S. M., 1972: J. Paleontol., 46(2):165–212.)

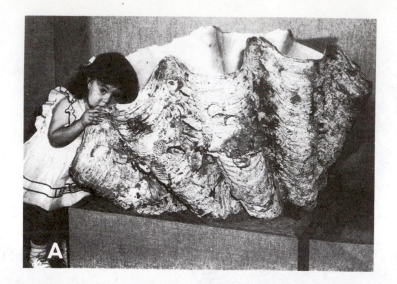

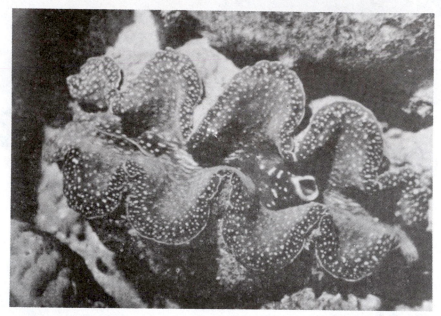

Figure 11–80 The giant clam, *Tridacna: A*, Shell. (Courtesy of Cranbrook Institute of Science.) *B*, Looking down at expanded specimen of *Tridacna derasa*. Mantle extends over shell fluting. Conical aperture is exhalant siphon. (Courtesy of the British Museum.)

bivalves, such as the oysters, that are attached by one valve. There has also been a tendency for the anterior end to become smaller, leading to a reduction of the anterior adductor muscle (anisomyarian) (Fig. 11–79B) or its loss (monomyarian) (Figs. 11–84 and 11–85). It has been suggested that anterior reduction in mussels (Mytilidae) is perhaps an adaptation to elevate the posterior end, thereby reducing the likelihood of osbtruction that the dense aggregations of these bivalves might cre-

ate. Not unexpectedly, many sessile bivalves, such as various scallops and thorny oysters, have a well-developed sensory lobe with tentacles and eyes.

Mantle fusion and siphon formation have not occurred in epifaunal bivalves, since they are above the surface and generally on hard substrata, where sedimentation is less of a problem. However, oysters and mussels, which occur in dense beds, depend on the cleansing action of tidal currents to prevent them from becoming completely buried in

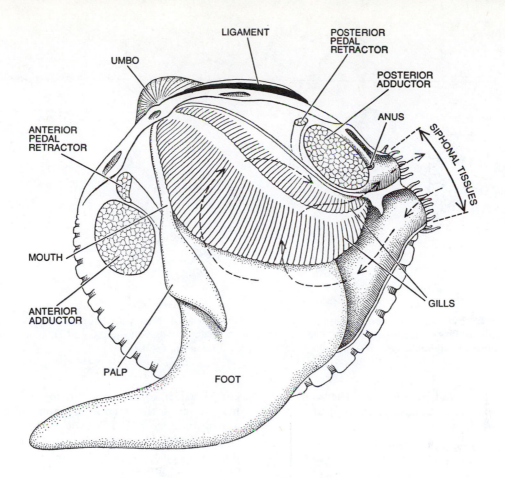

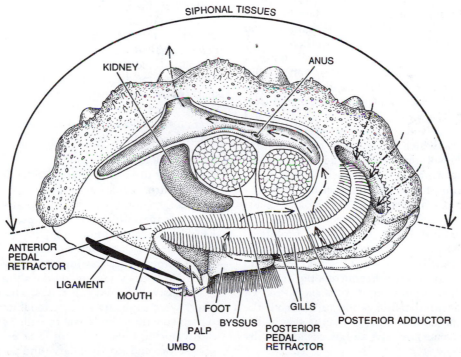

Figure 11–81 Comparison of the structures of a cockle, which is a shallow burrower in soft bottoms, with the related but aberrant giant clam (Tridacnidae), which is byssally attached and oriented with the hinge side down. (From Yonge, C. M., 1975: Giant clams. Sci. Am., *232*(4):96–105.)

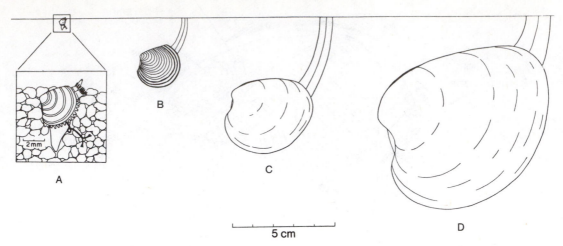

Figure 11–82 Habitation of *Mercenaria mercenaria* at different ages: *A*, Newly settled clam anchored in sand by byssal threads. *B*, First-year individual with heavy, stabilizing, concentric ridge on shell. *C* and *D*, Older individuals, showing changes in siphon length. (From Stanley, S. M., 1972: J. Paleontol., 46(2):165–212.)

their own feces and pseudofeces. The young *Mytilis edulis* actually cleans the exterior of its shell with its long foot.

UNATTACHED SURFACE DWELLERS

Among the small number of bivalves that live free on the surface, some scallops (Pectinidae) and some file shells (Limidae) are the most familiar (Fig. 11–85). There are members of both families that live anchored by byssal threads, but others are unattached or only weakly attached. They lie on their sides, and in many scallops the bottom valve (right) is flattened, the foot is reduced and used in cleaning the mantle cavity, and the anterior adductor muscle has disappeared. Free-living file shells and scallops have evolved the ability to swim by clapping the valves, which forces water from the mantle cavity. The solitary, posterior adductor muscle has shifted to a more central position and is divided into smooth and striated sections (Fig. 11–85). The rapid contraction of the striated fibers provides for swimming, and the sustained contraction of the smooth fibers provides for prolonged closure of the valves. The muscular lobe of the mantle margin, when appressed against the lobe on the opposite mantle surface, controls the direction of the water jet, permitting it (in scallops) to exit on either side of the hinge line or opposite the hinge line. The mantle sensory lobe is also highly developed, bearing tentacles and, in scallops, many small blue eyes.

The swimming ability of scallops and file shells is used primarily to escape predators or other sudden disturbing conditions. For example, if a predatory starfish, or even one tubefoot of such a starfish, touches the mantle margins of a scallop, a swimming response is evoked, and the scallop will be propelled a meter or so away. Some scallops use the water jets to blow out a depression in the sand surface into which they settle. File shells typically nest in crevices beneath stones and swim only when disturbed.

Other free surface dwellers include species of *Placuna*, called windowpane shells, which live free on the surface of mud flats in the Indian Ocean. These flat, translucent shells are commonly seen in shell shops where they are used to make mobiles, lamp shades, and decorative objects. (See Yonge, 1977, for a review of the Anomiacea, a remarkable assemblage of bivalves.)

BORING BIVALVES

The ability to penetrate and live beneath the surface of firm substrates—peat, clay, sandstone, shell, coral, coralline rock, and wood—has evolved in seven superfamilies of lamellibranchs, of which the Pholadacea, containing the piddocks, are among the most conspicuous examples. The ancestors of some of these boring bivalves were soft-bottom inhabitants that evolved the ability to burrow in successively firmer substrates. The ancestors of others were probably surface inhabitants attached to hard substrates by byssal threads.

All boring bivalves begin excavation after the larva settles and slowly enlarge and deepen the burrow with growth. The animal is forever locked within its burrow and only the siphons project to the small surface opening. If a boring bivalve is re-

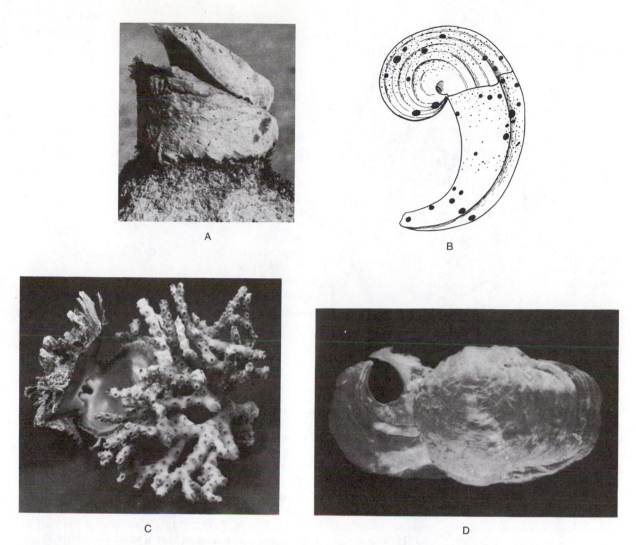

Figure 11–83 *A,* The jewel box *Chama,* a common, tropical, sessile bivalve, whose upper valve forms a lid over the lower valve. *B,* A rudist, a Mesozoic bivalve in which a caplike upper valve covered a hornlike or tubelike lower valve. (From Kauffman, E. G., and Sohl, N. F., 1973: Verh. Nat. Ges.) *C,* A species of thorny oyster, *Spondylus,* attached to a piece of coral *(Oculina).* (*A* and *C* by Betty M. Barnes.) *D,* A jingle shell, or toenail shell (Anomiidae). Attached valve contains large hole for calcareous peduncle, homologous to the byssus of other bivalves. (By Gates, J. B., from Andrews, J., 1971: Sea Shells of the Texas Coast. University of Texas Press, Austin. p. 168.)

moved and placed on the surface, it cannot excavate a new chamber.

Boring bivalves with epifaunal ancestors hold to the side of the burrow by byssal threads. Boring species with burrowing ancestors usually attach by the ventral surface of the foot, which has developed a sucker-like surface. In the great majority of species, drilling is a mechanical process, and the anterior ends of the valves, which are frequently serrated, are the abrading surfaces (Fig. 11–86*A*). The drilling movements are commonly adaptations of the movements found in their nonboring ances-

tors. Depending on the species, cutting force results from (1) an upward movement of the anterior end of valves due to contraction of posterior pedal retractors *(Petricola, Hiatella)*; (2) a downward movement of the anterior end of valves due to contraction of anterior pedal retractors *(Gastrochaena)*; (3) an opening thrust of the anterior ends of the valves due to contraction of the posterior adductor muscle *(Platyodon,* pholads); (4) a back and forth movement due to alternate contraction of the anterior and posterior pedal retractors pulling against two groups of byssus fibers, plus an open-

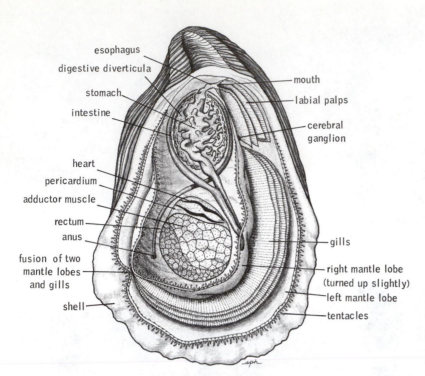

esophagus
digestive diverticula
stomach
intestine
heart
pericardium
adductor muscle
rectum
anus
fusion of two
mantle lobes
and gills
shell

mouth
labial palps
cerebral
ganglion
gills
right mantle lobe
(turned up slightly)
left mantle lobe
tentacles

Figure 11–84 Anatomy of the American oyster *Crassostrea virginica*. (Modified after Galtsoff.)

ing thrust *(Botula)*. Modifications to increase valve articulation and cutting movements include reduction of the hinge ligament and a shift of the anterior adductor muscle to a position *above* the hinge line. This region may then be covered by accessory shell plates (Pholadidae) (Fig. 11–86A).

Drilling rates vary. Over an 18-month period *Penitella penita* and *Chaceia ovoidea* excavated soft shale at rates of 2.6 mm and 11.4 mm per month, respectively (Haderlie, 1981).

Some boring bivalves rotate within the burrow, i.e., change position, and as a result the burrow cross section is round. Others remain attached in one place, and the burrow tube takes the shape of the shell. Such forms may even have two openings to the surface if the ends of the siphons secrete some material between them *(Gastrochaena)* (Fig. 11–86D). Much of the sediment produced in drilling is taken into the mantle cavity and then ejected with the pseudofeces through the inhalant siphon.

Lithophaga, a very common, cigar-shaped, byssally attached borer in shell and coral, excavates chemically. A mucoprotein chelating agent secreted by the mantle margin softens the calcareous substratum, which is then scraped away with the valves.

Many boring bivalves inhabit wood. The wood-boring Pholadidae, such as *Martesia* and *Xylo-*

phaga, are adapted in much the same way as the many rock- and shell-boring members of the same family. Wood panels planted 1830 meters deep on the sea bottom were completely riddled by *Xylophaga* and a related genus when recovered 104 days later (Fig. 11–87) (Turner, 1973). The most specialized wood borers are the shipworms, members of the family Teredinidae. The natural habitats of the some 60 species of this widely distributed family are mangrove roots and timber swept into the sea by rivers, and they play an important ecological role in the reduction of sea-borne wood. They are a serious pest of piers, pilings, and other wooden structures placed by man in the sea, and much expense and research have been devoted to their control. Timbers can become completely riddled with tunnels (Fig. 11–88B).

The body of the shipworm is greatly elongated and cylindrical (Fig. 11–88A). The shell is reduced to two small, anterior valves. Cutting of the wood is effected by opening and rocking motions of the valves while the anterior end of the body is attached to the burrow by the small foot. The mantle, enclosing the greater part of the body behind the valves, produces a calcareous lining within the tunnel. The long, delicate siphons open at the surface of the wood, and the burrow entrance is plugged by calcareous pallets when the siphons are retracted.

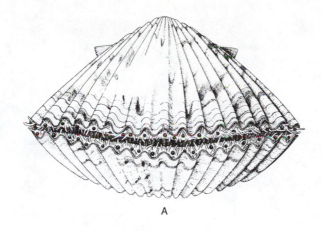

A

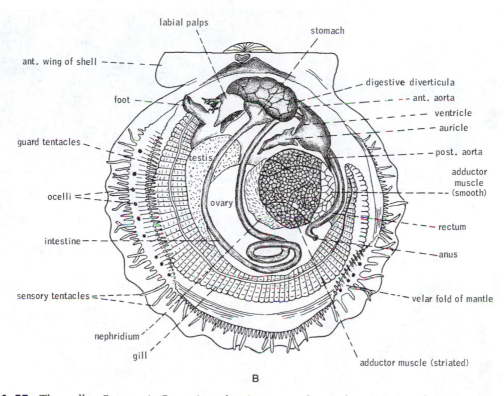

B

Figure 11–85 The scallop *Pecten: A*, Gape view, showing eyes and tentacles on sensory fold of mantle margin. *B*, Internal structure, viewed from left side. Anterior is on viewer's left. (After Pierce, M., 1950. *In* F. A. Brown (Ed.): Selected Invertebrate Types. John Wiley and Sons, N.Y.)

The burrow increases with the growth of the shipworm that fills it, and may reach a length of 18 cm to 2 meters, depending on the species. The life span is one to several years, again depending on the species.

Shipworms use the excavated sawdust for food. The stomach is provided with a caecum for saw-dust storage, and a section of the digestive gland is specialized for handling wood particles. Symbiotic bacteria housed within a special organ that opens into the esophagus not only provide for cellulose digestion but also, by fixing nitrogen, compensate for the low-protein diet (Morton, 1978, 1983; Waterbury et al, 1983).

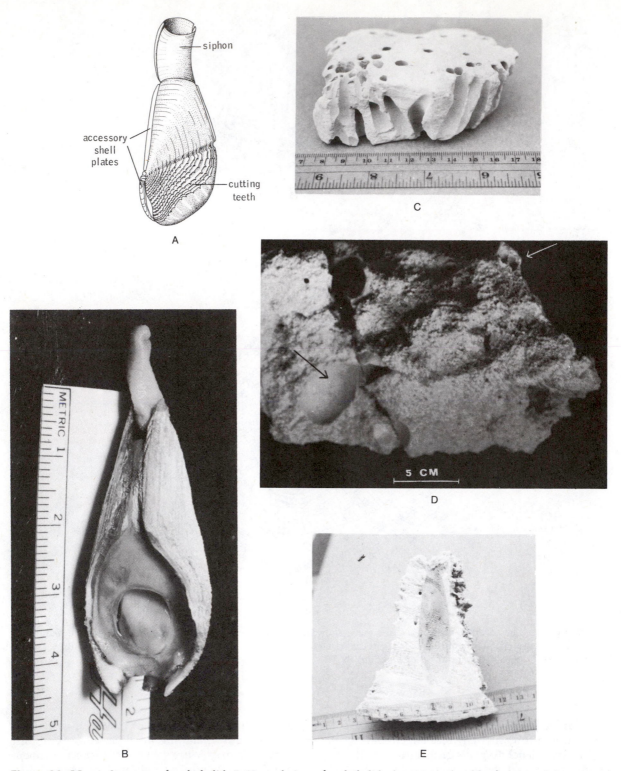

Figure 11–86 *A*, Structure of a pholadid. *B*, Ventral view of a pholadid, showing sucker-like foot in pedal aperture. Siphon is at top. (By Katherine E. Barnes.) *C*, Surface openings and part of burrows of pholadid bivalves in a piece of hard clay. *D*, The separate siphon openings of *Gastrochaena hians*, a common West Indian borer in coralline rock. This bivalve does not turn in its burrow. *E*, Boring of a species of *Lithophaga* in a piece of coral.

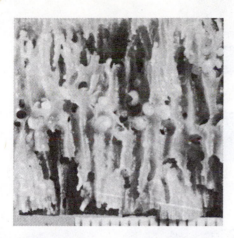

Figure 11–87 Section of a wood panel riddled with burrows of the pholadids *Xylophaga* and *Xyloredo*. The little bivalves are also visible. A millimeter scale is shown at the bottom. The wood panels were planted in the bottom, 1830 meters deep, 180 km south of Woods Hole by the submersible *Alvin*. (From Turner, R. D., 1973: Wood-boring bivalves, opportunistic species in the deep sea. Science, *180*:1379. Copyright 1974 by the American Associaton for the Advancement of Science.)

COMMENSALS AND PARASITES

A small number of bivalves have evolved commensal and parasitic relationships. Most commensals are related to free-living epibenthic forms, such as the little *Kellia* and *Lasaea*, which nestle in crevices. Most attach by byssus threads, but some crawl on the foot like a snail. The hosts are usually burrowing echinoderms, such as heart urchins, brittle stars, sea cucumbers, and burrowing, shrimp-like crustaceans. *Entovalva*, which lives in the gut of sea cucumbers, is the only known parasitic bivalve.

Internal Transport and Gas Exchange

In the majority of bivalves, the ventricle of the heart has become folded around the gut (rectum), so the pericardial cavity encloses not only the heart but also a short section of the digestive tract (Fig. 11–89B). The contractions of the ventricle can be easily observed in a large clam from which one of the valves has been carefully removed. Pulsations are slow, about 20 per minute in *Anodonta*.

An anterior aorta issues from the ventricle, and in the eulamellibranchs there is a posterior aorta as well (Figs. 11–85 and 11–89B).

The typical open molluscan circulatory route—heart, tissue sinuses, nephridia, gills, heart—is exhibited in the bivalves, although some modifications of this circuit have taken place in different species (Fig. 11–90). In all bivalves, there is a more or less well-developed circulatory pathway through the mantle, which is an additional site of gas exchange. Depending on the species, blood may be returned from the mantle or from the kidney directly to the heart.

Gas exchange takes place as water moves dorsally within the gills. The amount of oxygen removed from the water current is low compared with that in other mollusks (2.5 to 6.8 per cent for the scallop compared with 48 to 70 per cent for the abalone, a gastropod). This low oxygen uptake is correlated with the large gill size, which is greater than the respiratory needs of the animal but is required for filter feeding. Moreover, at least in the intertidal clam *Modiolus demissus* (Booth and Mangum, 1978), there is considerable direct passage of oxygen across the body wall, for in highly oxygenated water, blockage of the circulatory system reduces oxygen consumption by less than 15 per cent. When out of water at low tide, the clam utilizes oxygen in air, but deep tissues respire anaerobically.

The blood of most bivalves lacks any respiratory pigment, but in some 21 species, including ark shells (*Noetia, Arca, Anadara*) and *Calyptogena* of deep hydrothermal vents, the blood contains intracellular or extracellular hemoglobin. Hemoglobin and muscle myoglobin may give the mantle and other tissues a bright red color. In *Noetia*, at least, the blood hemoglobin functions both in oxygen transport and in oxygen storage (Deaton and Mangum, 1976).

Recently, hemocyanin has been found in two protobranchs. This is probably additional evidence that hemocyanin is the primitive molluscan respiratory pigment but has been lost in most bivalves (Morse et al, 1986).

Excretion

The two nephridia of bivalves are located beneath the pericardial cavity and above the gills. The nephridium of freshwater clams of the genus *Anodonta*, among the best studied species, is shaped like a hairpin with the two ends close together (Fig. 11–90D). One end opens into the pericardial cavity by way of the nephrostome and the other into the mantle cavity by way of the nephridiopore. The arm of the nephridium associated with the nephrostome has highly folded interior walls. In *Mytilus* the nephridium is a long, branched tube connected to the pericardium by a pericardial canal (Fig. 11–90C). The canal joins the nephridium very near the nephridiopore. The walls of the auricles

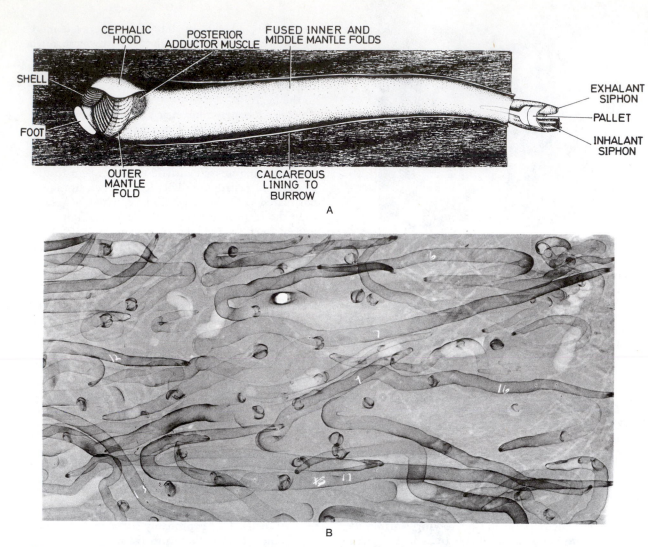

Figure 11–88   *A*, A shipworm, a wood-boring bivalve. (By Brian Morton) *B*, x-ray photograph of a marine timber section showing shipworms. (By C. E. Lane in Scientific American, February 1961.)

and pericardial glands are composed of cells like vertebrate podocytes and are believed to be sites of ultrafiltration. Selective reabsorption and secretion probably occur in the sections of the nephridium with folded walls (Pirie and George, 1979).

Except for random representatives of different marine families, such as the oysters, which can tolerate low salinities and have invaded brackish estuaries and marshes, the freshwater bivalves are members of nine families variously represented in different parts of the world. North American freshwater bivalves are primarily members of the Unionidae and the Sphaeriidae (fingernail clams), although one Asian *Corbicula* has been introduced into many North American drainage systems.

Many species of fingernail clams are adapted for living in temporary bodies of fresh water (McKee and Mackie, 1981, 1983). Like gastropods, freshwater bivalves excrete large amounts of water through the nephridia. The urine is very hyposmotic, and the blood salts are maintained at a very low level. The mantle and gills pick up salts from the circulating water stream.

Nervous System and Sense Organs

The nervous system is bilateral with three pairs of ganglia and two pairs of long nerve cords. On each side of the esophagus is located a cerebropleural ganglion, which is connected to its opposite mem-

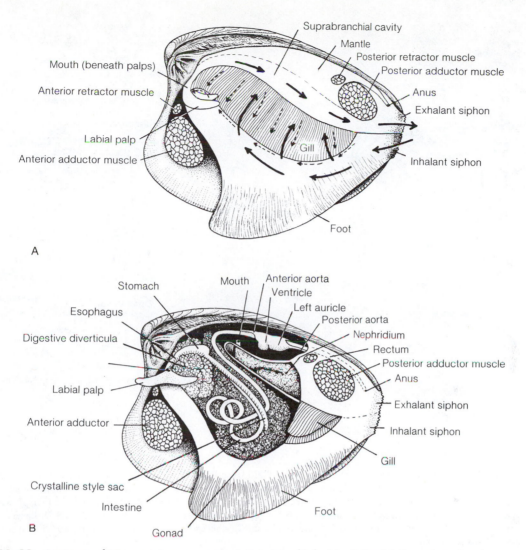

Figure 11–89 Anatomy of *Mercenaria mercenaria. A*, Interior of left side. *B*, Partial dissection, showing some of the internal organs.

ber by a short commissure dorsal to the esophagus. From each cerebropleural ganglion arise two posteriorly directed nerve cords. The upper pair of nerve cords (one from each ganglion) extend directly back through the viscera and terminate in a pair of closely adjacent visceral ganglia on the anteroventral surface of the posterior adductor muscle. The second pair of cords arising from the cerebropleural ganglia extend posteriorly and ventrally into the foot and connect with a pair of pedal ganglia. Foot movement and the anterior adductor muscle are under the control of the pedal and cerebral ganglia, but the visceral ganglia control the posterior adductor muscles and the siphons. Coor-

dination of pedal and valve movements is a function of the cerebral ganglia.

The margin of the mantle, particularly the middle fold, is the principal location of most of the bivalve sense organs. In many species the mantle edge bears pallial tentacles, which contain tactile and chemoreceptor cells. The entire margin may bear tentacles, as in the swimming *Pecten* and *Lima* (Fig. 11–85); more commonly, tentacles are restricted to the inhalant or exhalant apertures or siphons, or they may even fringe the pedal aperture.

A pair of statocysts is usually found in the foot near or within the pedal ganglia, but innervated

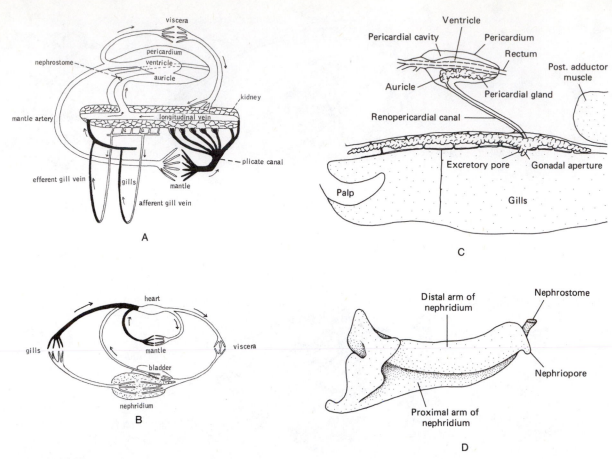

Figure 11–90 *A*, Circulatory system of *Mytilus*. (After Field from Borradaile and others.) *B*, Circulatory system of the freshwater clam *Anadonta*. (After Borradaile and others.) *C*, Lateral view of heart and one nephridium of *Mytilus*. (After Pirie, B. J. S., and George, S. G., 1979: Ultrastructure of the heart and excretory system of *Mytilus edulis*. J. Mar. Biol. Assoc. U.K., *59:*819–829; copyrighted and reprinted by permission of Cambridge University Press.) *D*, Lateral view of nehridium of *Anodonta*. Nephridium is bent like a hairpin and lies between the pericardial cavity above and the mantle cavity below. (Modified after Fernau.)

from the cerebral ganglion. The statocysts of some attached forms, such as oysters, are reduced.

Ocelli may be present along the mantle edge or even on the siphons in some bivalves. They enable the surface-dwelling clam to detect sudden changes in light intensity, like the shadow of a predator. The exposed mantle tissue of giant clams contains several thousand ocelli. In most cases bivalve ocelli are simple pigment-spot or pigment-cup ocelli. However, in ark shells the ocelli are compound, and in scallops and thorny oysters the ocelli are well developed, with a cornea and lens (Fig. 11–85).

Scallop eyes *(Pecten)* are peculiar in having two layers of photoreceptors, one rhabdomeric and one of the ciliary type (see Box 7–2). Behind the retina is a layer of reflecting pigment (tapetum),

which bounces the incoming light back to the ciliary receptor layer, where the image is formed. The function of the rhabdomeric receptor layer is uncertain.

Cephalic eyes present in some filibranch bivalves, including some Mytilidae, persist from the larval stage and are probably homologous to the eyes of gastropods (Rosen et al, 1978). The cephalic eye is located at the anterior end of the gill axis.

Immediately beneath the posterior adductor muscle in the exhalant chamber, there is a patch of sensory epithelium, usually called an osphradium. The osphradium has been considered an organ of chemoreception for monitoring the water passing through the mantle cavity. The position of this sensory tissue in the exhalant chamber makes it doubtful that it is actually homologous with the osphra-dium of a gastropod.

Reproduction

The majority of bivalves are dioecious. The two go-nads encompass the intestinal loops and are usually so close to one another that the paired condition is difficult to detect (Fig. 11–89*B*). The gonoducts are always simple, since there is no copulation. In the protobranchs and the more primitive lamelli-branchs, the short gonoduct opens into the nephri-dium, and sperm and eggs exit by way of the nephridiopores. In most lamellibranchs the gon-oducts are no longer associated with the nephridia and open separately into the mantle cavity, but still very close to the nephridiopore.

The hermaphroditic bivalve species include shipworms, the freshwater Sphaeriidae, a few Unionidae, and some species of cockles, oysters, and scallops. In a hermaphroditic scallop the gonad is divided into a ventral ovary and a dorsal testis, both of which lie on the anterior side of the adduc-tor muscle (Fig. 11–85*B*). The European oyster, *Os-trea edulis*, like other species of *Ostrea*, is a protan-dric hermaphrodite (species of *Crassostrea*, in-cluding the American east coast oyster, *Crassostrea virginica*, are mostly dioecious). *Ostrea edulis* not only shifts from male to female but also changes back from female to male. An individual may ex-hibit active male and female phases each year.

Embryogeny

In most bivalves fertilization occurs in the sur-rounding water; the gametes are shed into the su-prabranchial cavity and then swept out with the ex-halant current. Some bivalves brood their eggs within the suprabranchial cavity, as in some ship-worms, or within the gills, as in *Ostrea edulis* and the freshwater Unionidae and Sphaeriidae. Brood-ed eggs are fertilized with sperm brought in with the ventilating current.

The development of a free-swimming trocho-phore, succeeded by a veliger larva, is typical in marine bivalves (Fig. 11–91). The veliger is sym-metrical and eventually becomes enclosed within the two valves characteristic of bivalves (Fig. 11–92).

Like gastropods, some marine bivalves have long-lived, planktotrophic (feeding) veligers; oth-ers have short-lived, lecithotrophic (nonfeeding) veligers. It is speculated that the larvae of oysters

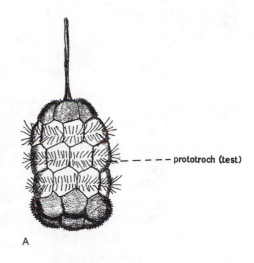

prototroch (test)

A

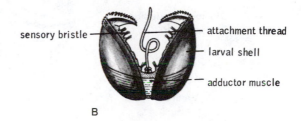

sensory bristle — attachment thread

— larval shell

— adductor muscle

B

Figure 11–91 *A*, Trochophore of *Yoldia limatula*. (After Drew from Dawydoff.) *B*, Glochidium of the fresh-water clam *Anodonta*. (Redrawn from Harms.)

(Ostreidae), for example, are capable of dispersion over a distance as great as 1300 km (Stenzel, 1971). Some bivalves, such as *Ostrea* and species of ship-worms, are larviparous, releasing the veligers fol-lowing an initial period of brooding (eight days in *Ostrea edulis*).

Metamorphosis is characterized by a sudden shedding of the velum. Settling may involve con-siderable testing of the substratum and delayed metamorphosis. *Ostrea edulis* swims upward and attaches to the shaded underside of objects. Ship-worms will settle only on wood.

With the exception of *Dreissena* and *Nausito-ria*, which have free-swimming veliger larvae, freshwater bivalves exhibit modified development. Direct development is characteristic of the fresh-water Sphaeriidae, which brood the eggs in marsu-pial sacs that develop between the gill lamellae. At the completion of development, the young clams are shed from the gills (see Heard, 1977, for review).

The freshwater mussels (Unionacea and Mute-lacea) display an indirect but very specialized de-

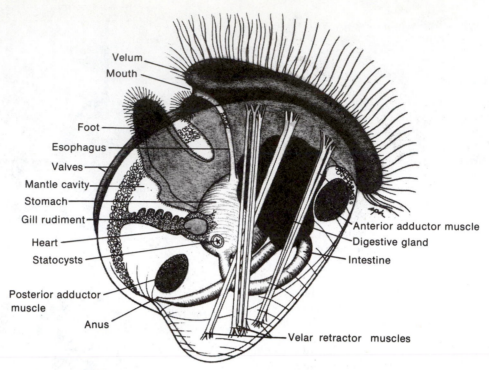

Velum
Mouth
Foot
Esophagus
Valves
Mantle cavity
Stomach
Gill rudiment
Heart
Statocysts
Posterior adductor muscle
Anus
Anterior adductor muscle
Digestive gland
Intestine
Velar retractor muscles

Figure 11–92   A fully developed veliger larva of an oyster. (After Galtsoff.)

velopment. As in the sphaeriids, the eggs are brooded between the gill lamellae, where they develop through the veliger stage. However, in the worldwide Unionacea, the African Mutelidae, and the South American Mycetopodidae the veliger, called a glochidium, haustorium, and lasidium respectively, has become highly modified for a parasitic existence on fish. The fish during this period disperse the rather sedentary freshwater bivalves.

The glochidium larva is enclosed by two valves, each edge of which may bear a hook, as in *Anodonta* (Fig. 11–91B). The shell valves cover a larval mantle, which bears clusters of sensory bristles. A rudimentary foot is present, to which is attached a long adhesive thread. There is neither mouth nor anus, and the digestive tract is rather poorly developed.

When mature, the glochidia range in size from 0.5 mm to 5 mm, depending on the species. In *Unio* and *Anodonta* the glochidia leave the gills through the suprabranchial cavity and exhalant aperture. In *Lampsilis* the glochidia emerge directly from gills through temporary openings, and dispersal into the surrounding water is aided by the movement of special mantle flaps near the exhalant aperture. The way the glochidia are released is re-

lated to the habits of the host. In some species the glochidia are dispersed over the bottom and are picked up by species of fish that have benthic nesting habits. Other clams release their glochidia in the form of colored masses that look like worms. The mass is eaten by the fish host and the glochidia attach to the gills. Most clam species parasitize more than one species of fish, and a species of fish is usually the host for a number of species of clams.

The hooked glochidia of *Anodonta* immediately clamp onto the fins and other parts of the body surface of the fish. A long, sticky thread aids in initial contact and adhesion, and clasping is a response to certain molecules in the fish mucus (Wood, 1974). Hookless glochidia are picked up by the respiratory currents of the fish and attach to the gills. In either case the tissue of the fish in the vicinity of the attached glochidium is stimulated to grow around the parasite and form a cyst. The larval mantle contains phagocytic cells that feed on the tissues of the host and obtain nutrition for the developing clam. During this parasitic period, which lasts 10 to 30 days, many of the larval structures (sensory bristles on the mantle, the adhesive thread, the larval adductor muscle, the larval mantle) disappear, and the adult organs begin developing. Eventually, the immature clam breaks out of the cyst, falls to the bottom, and burrows in the mud. Here the remainder of development is completed, and the adult habit is gradually assumed.

Some of the larger freshwater mussels may produce as many as 3,000,000 glochidia each, and a single fish has been reported to contain 3000 glochidia. Adult fish are apparently not harmed by the parasitic glochidia, but young fry may die from secondary infection.

Growth and Life Span

As in gastropods, the rates of growth and life span of bivalves vary greatly. The common mussel of the California coast *(Mytilus californianus)* may reach a length of 86 mm within one year. In general, most bivalves grow most rapidly during their early years (see Fig. 11–61E). Ages of 20 to 30 years are now known to be common, and for certain species there are records of 150-year-old individuals (Jones, 1983).

The growth stages of species of commercial value are well known. Oysters (Ostreidae), for example, reach marketable size in one to three years depending on the species, latitude, and various environmental conditions. Newly settled oysters, called "spat," are collected on tiles, twigs, or other objects and allowed to grow to a few centimeters in length. These seed oysters are then distributed over a managed bed, where they grow until harvested. In a natural oyster reef, the average life span of individuals that have settled is uncertain. Certainly some live longer than ten years.

Small scallops of the genus *Argopecten* have a life span of only one to two years, but the deep-sea scallops *(Placopecten)* are about ten years old when they reach a maximum size of 15 cm.

SYSTEMATIC RÉSUMÉ OF CLASS BIVALVIA

The relationship of the older subclasses Protobranchia, Lamellibranchia, and Septibranchia to the system below is readily determined. Taxa containing the protobranchs and septibranchs are indicated; all others contain lamellibranchs.

Subclass Palaeotaxodonta (Protobranchia, in part). Shell valves equal and taxodont (row of short teeth along hinge margin; shell structure nacreous or crossed lamellar). Isomyarian. Gills protobranchiate.

> **Order Nuculoida.** Characteristics same as for Palaeotoxodonta. *Nucula, Nuculana, Yoldia, Malletia.*

Subclass Cryptodonta (Protobranchia, in part). Valves thin, equal, somewhat elongate, and without hinge teeth. Gills protobranchiate.

> **Order Solemyoida.** *Solemya.*

Subclass Pteriomorphia. Epibenthic bivalves, most attached by byssus threads or cemented to substratum, but some secondarily free. Unfused mantle margins.

> **Order Arcoida.** Arks. Hinge straight and taxodont. Isomyarian; gills filibranchiate. *Arca, Barbatia, Anadara, Noetia, Glycymeris.*
>
> **Order Mytiloida.** Mussels and pen shells. Two valves equal (except in oysters and some scallops) but inequilateral. Hetero- or monomyarian. Mytilidae (mussels): *Mytilus, Brachydontes, Lithophaga, Modiolus, Botula.* Pinnidae (pen shells): *Pinna, Atrina.* Pteriidae (winged oysters): *Pteria, Pinctada; Isognomon; Malleus.* Pectinidae (scallops): *Pecten, Chlamys, Hinnites, Aequipecten, Argopecten* (Atlantic bay scallop), *Placopecten* (Atlantic deep-sea scallop); *Spondylus* (thorny oysters). Anomiidae (jingle shells or Venus's toenails): *Anomia; Lima* (file shells). Ostreidae (oysters): *Ostrea, Crassostrea, Lopha.*

Subclass Palaeoheterodonta. Equivalve, with a few hinge teeth, in which the elongate lateral teeth, when present, are not separated from the large cardinal teeth.

> **Order Unionoida.** Freshwater bivalves. *Margaritifera.* Unionidae: *Elliptio, Anodonta; Lampsilis.* The African Mutelidae: *Mutela.* Etheriidae (freshwater oysters).
>
> **Order Trigonioida.** *Trigonia.*

Subclass Heterodonta. Equivalve, with a few large cardinal teeth separated by a toothless space from the elongated lateral teeth. Shell without nacreous layer. Siphons usually present; gills eulamellibranchiate.

> **Order Veneroida.** Usually equivalve and isomyarian. Lucinidae: *Lucina, Codakia, Linga, Myrtea, Divaricella, Thyasira.* Chamidae (jewel boxes). Lasaeidae (commensal bivalves): *Erycina, Lasaea* (free living); *Kellia* (free living); *Lepton; Montacuta* (commensal); *Entrovalva* (parasitic). Cardiidae (cockles): *Cardium, Trachycardium, Dinocardium, Laevicardium,* Tridacnidae (giant clams); *Mactra, Spisula.* Solenidae (razor clams): *Solen, Ensis, Siliqua.* The deposit feeders: *Tellina, Macoma, Scrobicularia; Donax* (coquina shells); *Abra.* Solecurtidae (razor clams): *Solecurtus, Tagelus;* the fresh-

water Sphaeriidae; the freshwater and estuarine Dreissenidae. Veneridae (Venus clams): *Venus, Mercenaria, Chione, Callista, Macrocallista, Dosinia, Gemma, Petricola* (rock borer).

Order Myoida. Thin-shelled burrowers with well-developed siphons. One or no cardinal teeth. Shell without nacreous layer. *Mya* (soft-shell clam); *Corbula;* the borers *Hiatella, Gastrochaena; Panopea* (the geoducks). The boring Pholadidae: *Barnea, Pholas, Zirphaea, Martesia* (in wood), *Xylophaga* (in wood). Teredinidae (wood-boring shipworms): *Teredo, Bankia.*

Order Hippuritoida. The fossil rudists.

Subclass Anomalodesmata. Equivalve, with a single hinge tooth or no teeth. Hinge margin thickened or enrolled. Isomyarian. Mantle margins fused. Hermaphroditic. (Members of former subclass Septibranchia included here.)

Order Pholadomyoida. *Lyonsia; Pandora; Clavagella* (watering-pot shells). The septibranch superfamily Poromyacea: *Poromya, Cuspidaria.*

SUMMARY

1 The general characteristics of the class Bivalvia largely reflect adaptations for burrowing in soft substrates, although many species have secondarily colonized other epibenthic habitats. A primary feature, to which many others are related, is lateral compression of the body.

2 Primitive bivalves (protobranchs) possess bipectinate gills and usually a ventilating current that enters and leaves from the posterior. Most are selective deposit feeders, utilizing a pair of palpal tentacles in obtaining and transporting deposit material. The stomach contains a protostyle and sorting region.

3 The great majority of bivalves are filter-feeding lamellibranchs. They probably evolved from some group of protobranchs, which were preadapted for filter feeding. The gills were used as the filter, the ventilating current became the filtering current, and frontal cilia were employed for vertical transport of trapped food particles. Changes necessary for the lamellibranch condition involved lengthening and folding of the gill filament for greater filtering surface and formation of food grooves for horizontal transport.

4 In the more primitive lamellibranch condition (filibranchs) there is relatively little morphological connection between the folded gill filaments. But in the specialized condition (eulamellibranchs) the filaments are fused by a variety of tissue junctions, and the channel for blood leaving the gill runs adjacent to the tips of the ascending limbs of the gill filaments.

5 The stomach of lamellibranchs has retained a number of primitive features associated with a diet of fine particles, but the protostyle has become consolidated into a crystalline style, which liberates enzymes as it is eroded.

6 The evolution of filter feeding led to an explosive evolution of lamellibranchs, for they were no longer chained to deposit material as a food source. The majority of lamellibranchs have remained in soft bottoms, for which varying degrees of mantle fusion and siphons are important adaptations.

7 Many lamellibranchs live on hard substrates, the result of a number of invasions. Anchorage is by byssal threads or by cementation of one valve to the substratum. Reduction or loss of the foot and anterior adductor muscle is common in attached bivalves. Those species that habitually lie on one valve tend to exhibit valve inequality.

8 Ability to drill into hard substrates—rock, shell, coral, and wood—evolved in several groups of lamellibranchs. Most drill mechanically, using the anterior shell margins as cutting tools.

9 A few bivalves live unattached on the surface. Of these, scallops and file shells are capable of escape swimming.

10 The lack of respiratory pigment in most bivalves is correlated with their relatively sluggish habits and their large gill surface.

11 Most bivalves are dioecious. Gametes exit by nephridia in protobranchs and by special gonoducts in other bivalves. Fertilization is usually external and development planktonic, with a trochophore and veliger in marine species.

Class Scaphopoda

The class Scaphopoda contains about 350 species of burrowing marine mollusks that are popularly known as tusk or tooth shells. These names are derived from the shape of the shell, which is an elongated cylindrical tube usually shaped like an elephant's tusk (Fig. 11–93). Both ends of the tube are open. The shells of most scaphopods average 3 to 6 cm in length, but *Cadulus mayori*, found off the Florida coast, does not exceed 4 mm in length. *Dentalium vernedei*, found off Japan, is the largest living species, reaching a length of 15 cm. How-

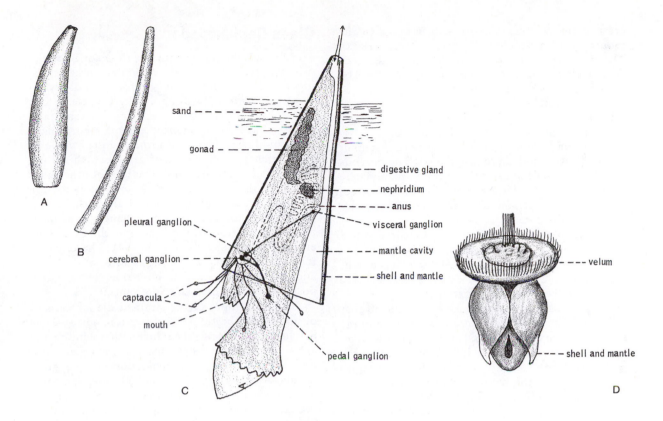

Figure 11–93 Scaphopoda: *A*, Shell of *Cadulus*. *B*, Shell of *Dentalium*. (Both after Abbott, R. T., 1954: American Sea Shells. Van Nostrand Co., Princeton, N.J.) *C*, Structure of *Dentalium*. Arrows indicate water current direction through mantle cavity. (After Naef from Borradaile and others.) *D*, Veliger larva. (After Lacaze Dutheirs from Dawydoff.)

ever, a fossil species of *Dentalium* has a shell 30 cm long with a maximum diameter of well over 3 cm. The shells of most scaphopods are white or yellowish, but an East Indian species of *Dentalium* is a brilliant jade green.

The majority of scaphopods burrow in sand in water depths greater than 6 meters. The living animals are therefore not frequently encountered, but judging from the number of shells washed up on beaches in some areas, scaphopods are not rare benthic animals.

The body of the scaphopod is greatly elongated along the anterior-posterior axis (Fig. 11–93C). The head and foot project from the larger and anterior aperture of the shell. When the shell is slightly curved, the concavity lies over the dorsal surface. Scaphopods live buried in sand, head downward with the body steeply inclined; only the small, posterior aperture projects above the surface.

Adapted to a burrowing habit, the head is reduced to a short, conical projection, or proboscis, bearing the mouth. Scaphopods burrow in a similar manner to bivalves. The cone-shaped foot is projected into the sand.

The scaphopod mantle cavity is large and extends the entire length of the ventral surface. The posterior aperture serves for both inhalant and exhalant water currents. Water slowly enters the mantle cavity as a result of mantle ciliary action and perhaps foot protraction. After 10 to 12 minutes of inhalation, a violent muscular contraction (probably foot retraction) expels the water from the same opening it entered. There are no gills; exchange of gases takes place through the mantle surface.

Scaphopods feed on microscopic organisms, especially forams, in the surrounding sand and water (Bilyard, 1974). Two lobes, located on each side of the head, bear a large number of threadlike tentacles called captacula. Each tentacle has an adhesive knob at the tip and extends into the sand to capture food. Small food particles are conveyed back to the mouth by tentacular cilia; larger particles are grasped by the tentacle tips and brought to the mouth by retraction of the tentacle (Gainey, 1972). The buccal cavity contains a median jaw and a well-developed radula with large, flattened teeth. The radula functions in the ingestion and crushing of the prey seized by the captacula. Digestion is extracellular in the stomach. The intestine empties through the anus into the mantle cavity.

The circulatory system is reduced to a system of blood sinuses, and there is no heart. The nervous system exhibits the typical molluscan plan. There are no eyes, tentacles, or osphradia. A pair of nephridia is present, and the nephridiopores are located near the anus.

Scaphopods are dioecious. The unpaired gonad fills most of the posterior part of the body, and the sperm or eggs reach the outside by way of the right nephridium. The eggs are shed singly and are planktonic. Fertilization is external.

Scaphopod development is very similar to that of the marine bivalves. There is a free-swimming trochopore larva, succeeded by a bilaterally symmetrical veliger (Fig. 11–93D). As in bivalves, the larval mantle and shell in scaphopods are at first bilobed, but then the mantle lobes fuse along their ventral margins (Fig. 11–93D). This fusion thus results in a cylindrical mantle and shell that remain open at each end.

Scaphopods appear to be an offshoot from the ancestral bivalve stock. The symmetry of the body and its orientation within the shell, the reduction of the head, the burrowing habit, the symmetrical veliger, and the embryonic bilobed mantle and shell are strikingly similar to the respective bivalve characteristics.

SUMMARY

1 Members of the class Scaphopoda are burrowing mollusks with cylindrical, tusk-shaped shells open at each end. The animal lives buried in soft bottoms, with the larger, anterior end downward and the small, posterior end, through which the ventilating current enters and leaves, near the surface of the substratum. They burrow like bivalves.

2 Scaphopods feed on interstitial organisms collected by means of small tentacles (captacula) and ingested with a radula.

3 The mantle surface, rather than gills, provides for gas exchange.

4 Scaphopods are dioecious, fertilization is external, and development is planktonic. There is both a trochophore and a veliger larva.

Class Cephalopoda

The class Cephalopoda contains the nautili, cuttlefish, squids, and octopods. Although some cephalopods, such as the octopus, have secondarily assumed a bottom-dwelling habit, the class as a whole is adapted for a swimming existence and contains the most specialized and highly organized of all mollusks. The head projects into a circle of large, prehensile tentacles or arms, which are homologous to the anterior of the foot of other mollusks. In the evolution of the cephalopods, the body became greatly lengthened along the dorsoventral axis, and as a result of a change in the manner of locomotion, this axis became the functional anterior-posterior axis (Fig. 11–94). The circle of tentacles is thus located at the anterior of the body, and the visceral hump is posterior. The original posterior mantle cavity is now ventral.

The cephalopods have attained the largest size of any invertebrate. Although the majority range between about 6 and 70 cm in length, including the arms and tentacles, some species reach giant proportions. The largest cephalopods are the giant squids, *Architeuthis* (Fig. 11–100); one specimen was reported to have measured 16 meters, including the tentacles. The tentacles alone were 6 meters long, and the circumference of the body was 4 meters. Giant octopods with arms 10 to 15 meters long have been observed by divers in the Japan Sea, but no specimens have been taken. The dorsal mantle length of *Octopus dofleini* of the Pacific coast, one of the largest known octopods, does not usually exceed 36 cm, though its rather slender arms may be five times the body length. The record is a specimen with a 9.6-meter radial arm span.

There are only about 600 living species of cephalopods but more than 7500 fossil forms. The class first appeared in the Cambrian period and then, once during the Paleozoic era and once in the Mesozoic era, underwent great periods of evolutionary development with the formation of many species.

Shell

A completely developed shell is found only in the fossil representatives of the class and the four living species of *Nautilus* found in the tropical Indo-western Pacific. In squids and cuttlefish the shell is reduced and internal, and in the octopods it is completely lacking. The shell of *Nautilus* is coiled over the head in a bilaterally symmetrical planospiral (Fig. 11–95). Only the last two whorls are visible,

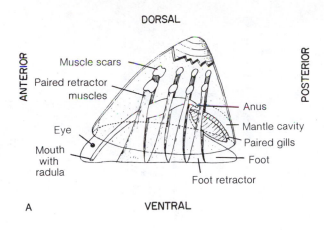

DORSAL

ANTERIOR

POSTERIOR

Muscle scars

Paired retractor
muscles

Anus

Mantle cavity

Eye

Paired gills

Mouth
with
radula

Foot

Foot retractor

A

VENTRAL

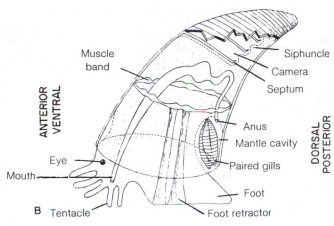

Muscle
band

Siphuncle

Camera

Septum

ANTERIOR
VENTRAL

DORSAL
POSTERIOR

Anus

Mantle cavity

Eye

Paired gills

Mouth

Foot

B Tentacle

Foot retractor

Figure 11–94 Evolution of the cephalopod form: *A*, Reconstruction of a monoplacophoran with a cap-shaped shell. Note that the apex tilts slightly posteriorly. Some high cone monoplacophorans, from which the cephalopods may have evolved, had septate shells but no siphuncle. *B*, Reconstruction of an early cephalopod, such as the Late Cambrian *Plectronoceras*. (From Yochelson, E. L., Flower, R. H., and Webers, G. F., 1973: The bearing of the new Late Cambrian monoplacophoran genus *Knightoconus* upon the origin of the Cephalopoda. Lethaia, 6:275–310.)

since they cover the inner whorls. The shell of *Nautilus*, like all cephalopod shells, is divided by transverse septa into internal chambers, and only the last chamber is occupied by the living animal. As the animal grows, it periodically moves forward, and the posterior part of the mantle secretes a new septum.

Each septum is perforated in the middle, and through the opening a cord of body tissue, called the siphuncle, extends from the visceral mass. The

siphuncle secretes gas into the empty chambers, making the shell buoyant and allowing the animal to swim. The shell is composed of an outer porcelaneous layer, containing prisms of calcium carbonate in an organic matrix, and an inner nacreous layer.

Cephalopods may have evolved from high-cone monoplacophorans (Fig. 11–94), some of which were septate. Such septation, however, was probably a spatial adaptation, and only later did the siphuncle evolve as a true cephalopod innovation (Yochelson et al, 1973). The apical chambers were perhaps first filled with fluid, but with the evolution of the siphuncle, gas production became possible. Initially, such gas-filled chambers may only have aided in keeping the shell upright as the animal moved about over the bottom (Donovan, 1964; Yonge and Thompson, 1976). Swimming and invasion of the pelagic environment were probably later cephalopod developments.

The first cephalopod shells are believed to have been curved cones. From curved shells, straight and coiled shells evolved. The straight shells of some species from the Ordovician period exceeded 5 meters in length, and the aperture was 36 cm in diameter. However, a straight or curved shell is not always an indication of primitiveness, for at different periods in the course of cephalopod evolution, the shells of various groups became secondarily uncoiled and straight again. Most coiled shells were planospirals, but one fossil group (heteromorph ammonites) had asymmetrically or highly irregular coiled shells adapted for planktonic life (Fig. 11–96 *E* and *F*). The largest fossil species with a coiled shell was *Pachydiscus seppenradensis* from the Cretaceous period, which had a shell diameter of 3 meters. However, many fossil cephalopods were small species with shells only 3 cm in diameter.

One of the most important characteristics for the classification of fossil cephalopods is the nature of the internal suture—the junction between the septum and the wall of the shell. As the chambers filled with sediment, the details of the suture pattern, once internal, were beautifully preserved on the outer surface of the fossil. The simplest suture lines were straight or slightly waved, as in *Nautilus* (Fig. 11–95), but one large group, the ammonoids, developed elaborate sutures that were zigzagged, or, more frequently, minutely crinkled (Fig. 11–96*B*). Such sutures indicate a corresponding complexity in the nature of the septal junction and probably represented an adaptation providing for greater strength to compensate for the somewhat thinner ammonoid shell (Ward, 1983).

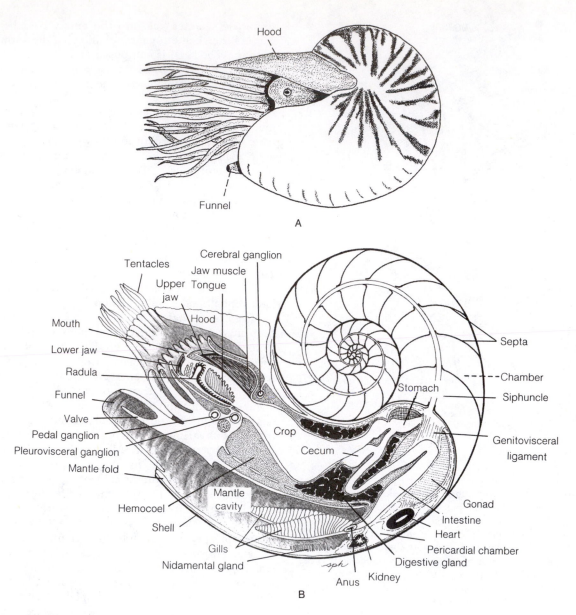

Figure 11–95   *A, Nautilus,* the only living shelled cephalopod, shown swimming. (After Moore and others.) *B,* Sagittal section. (After Stenzel.)

The current system of cephalopod classification separates those forms with complete shells into two subclasses—the Nautiloidea and the Ammonoidea. The Nautiloidea is characterized by straight or coiled shells and simple sutures. The Nautiloidea first appeared in the Cambrian period and is represented today by *Nautilus.* All members of the Ammonoidea were coiled and displayed complex septa and sutures. They appeared in the

Silurian period after the nautiloids and disappeared at the end of the Cretaceous period.

We know little concerning the soft parts of fossil ammonoids and nautiloids and can only assume that they were somewhat similar to those of *Nautilus* (see Ward, 1983).

Except for *Nautilus,* all living cephalopods belong to the subclass Coleoidea, in which the shell is reduced and internal or lacking altogether. These

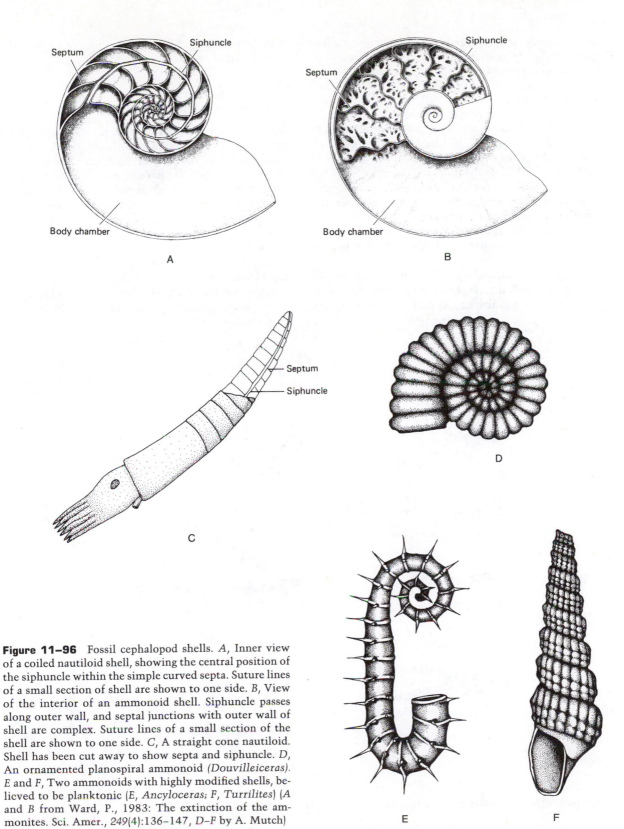

Figure 11–96 Fossil cephalopod shells. *A*, Inner view of a coiled nautiloid shell, showing the central position of the siphuncle within the simple curved septa. Suture lines of a small section of shell are shown to one side. *B*, View of the interior of an ammonoid shell. Siphuncle passes along outer wall, and septal junctions with outer wall of shell are complex. Suture lines of a small section of the shell are shown to one side. *C*, A straight cone nautiloid. Shell has been cut away to show septa and siphuncle. *D*, An ornamented planospiral ammonoid *(Douvilleiceras)*. *E* and *F*, Two ammonoids with highly modified shells, believed to be planktonic (*E*, *Ancyloceras*; *F*, *Turrilites*) (*A* and *B* from Ward, P., 1983: The extinction of the ammonites. Sci. Amer., *249*(4):136–147, *D–F* by A. Mutch)

cephalopods are believed to have evolved from some early straight-shelled nautiloid, whose shell became completely enclosed by the mantle.

From some primitive form the coleoid shell evolved in four directions, all leading to modern forms and all involving a reduction in the weight of the shell. This evolutionary development is illustrated in Figure 11–97. In *Spirula*, a common, worldwide, deep-water cuttlefish, the thickened wall and shelf have disappeared, and the shell has become coiled. In the evolutionary line leading to the squids, the largest group of cephalopods (including *Loligo pealei*, the common squid of the western North Atlantic), the shell is reduced to a long, flattened, chitinous plate, called a pen or gladius.

A third evolutionary line, represented by the European cuttlefish *Sepia*, has retained the septa, but the shelf and thickened parts of the wall have

disappeared (Fig. 11–98). Finally, in *Octopus* the shell has disappeared completely.

Locomotion and Adaptive Diversity in Cephalopods

Most cephalopods swim by jet propulsion, rapidly expelling water from the mantle cavity. The mantle contains both radial and circular muscle fibers. During the inhalant phase of water circulation, the circular fibers relax and the radial muscles contract. This action increases the volume of the mantle cavity, and water rushes in laterally between the anterior margin of the mantle and the posterior end of the head.

During the exhalant phase the contraction of the circular muscles not only increases the water pressure within the cavity but also locks the edges of the mantle tightly around the head. Flap valves

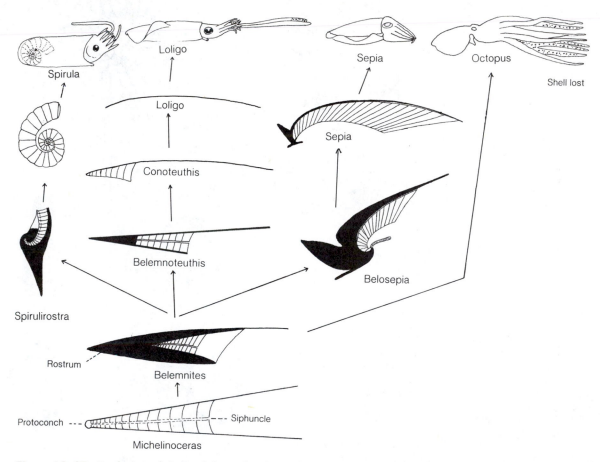

Figure 11–97 Evolution of shells of the Coleoidea. (After Shrock and Twenhofel.)

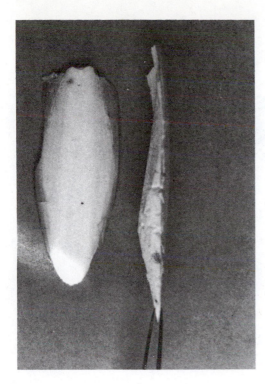

A

B

Figure 11–98 Internal shells of coleoids: *A*, Cuttlefish *Sepia*. Right shell is viewed from the side; the left shell is viewed from the upper surface. *B*, *Spirula*, side and end views. Note siphon at bottom of septum in end view. The shell of *Sepia* is located on the dorsal side of the body; in *Spirula* it is the posterior end of the body. (By Katherine E. Barnes.)

seal the mantle cavity ventrally, and thus water is forced to leave through the ventral tubular funnel. The force of water leaving the funnel propels the animal in the opposite direction. The funnel is highly mobile and can be directed anteriorly or posteriorly, resulting in either forward or backward movement. The fastest movement is achieved in backward escape swimming, when powerful contractions of the mantle eject water from the anteriorly directed funnel.

Squids possess a number of features that increase the efficiency of this jet propulsion system. The powerful contractions of escape swimming are produced by one type of circular muscle fiber; the rhythmic, less powerful contractions of ventilation and slow swimming are produced by another (Bone et al, 1981). The radial muscles function only during escape swimming, when they actually hyperinflate the mantle cavity, i.e., increase its volume beyond the usual ventilating capacity. In slow swimming the mantle cavity is expanded by elastic recoil of stretched collagen fibers extending through the mantle wall, and some additional propulsive power is provided by undulations of the fins (see review of squid swimming by Gosline and DeMont, 1985).

The squids and cuttlefish are best adapted for swimming by water jet. They can hover, perform subtle swimming movements, slowly cruise, or dart rapidly. Squids attain the greatest swimming speeds of any aquatic invertebrate, up to 40 km per hour. The body of a typical squid is long and tapered posteriorly and has a pair of posterior lateral fins, which function as stabilizers, rudders, and even for propulsion at low speeds (Fig. 11–99). The "flying squids" (Onycoteuthidae), which have long, tapered bodies and highly developed fin vanes and funnels, shoot out of the water during escape swimming and glide for some distance. There have been reports of squids accidentally leaping onto the decks of ships 12 feet above the water's surface.

The giant architeuthid squids may inhabit depths of 300 to 600 meters over the continental slopes (Fig. 11–100) and are not rapid swimmers (Roper and Boss, 1982). Two other interesting groups of deep-water squids are the bathypelagic chiroteuthids and the cranchiids. The chiroteuthids have very long, slender bodies and long, whiplike tentacles (Fig. 11–101*A*). The cranchiids are planktonic and are often strangely shaped (Fig. 11–101*C*). Two thirds of the entire body volume of a cranchiid is provided by the enormous, fluid-filled coelom, which functions as a buoyancy

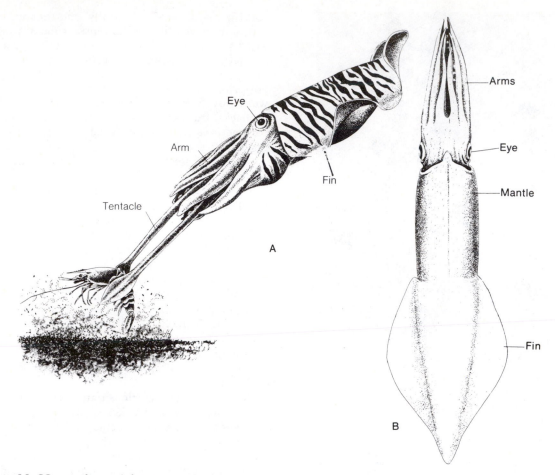

Figure 11-99 *A*, The cuttlefish *Sepia* seizing a shrimp with its tentacles. *B*, Dorsal view of the squid *Loligo* in swimming position. Tentacles and arms are held together, acting as a rudder.

chamber. Most of the seawater cations are replaced in the coelomic fluid with lighter ammonium ions derived from the metabolic wastes. Ammonium is used to achieve neutral buoyancy in about half of the 26 families of oceanic squids, including the chiroteuthids and the giant squids, but is located in special body tissues instead of the coelom (Clark et al, 1979).

The body of a cuttlefish, such as *Sepia* and *Spirula*, tends to be short, broad, and flattened (Fig. 11-99*A*). Cuttlefish are versatile swimmers but not as fast as the more streamlined squids.

The shell of cuttlefish *(Sepia)*, despite its reduction from the ancestral form, still functions in providing buoyancy. Spaces between the thin septa contain fluid and gas—largely nitrogen. By regulating the relative amounts of fluid and gas, the degree of buoyancy can be varied. Light is an important factor controlling the regulating mechanism. During the day the cuttlefish lies buried in the bottom; at night the animal becomes active, swimming and hunting for food. Buoyancy decreases when the animal is exposed to light and increases in the dark.

The deep-water *Spirula*, which lives down to about 1000 meters, swims with the tentacles directed downward. When the animal dies, the little gas-filled shell floats to the surface and is commonly washed ashore (Fig. 11-98*B*). The large number that can often be found on tropical and semitropical beaches attests to the abundance of this cephalopod.

The smallest cuttlefish are species of the genus *Idiosepius*, which are about 15 mm long. They live in tide pools and possess a dorsal disc on the mantle for attaching to algae.

Species of *Nautilus* are mobile epibenthic animals, found at depths as great as 500 meters. When resting, they attach to rubble or the walls of crev-

Figure 11–100 A giant squid of the genus *Architeuthis* stranded at Rahneim, Norway, in 1954. (By Clark, M. R., 1966: Adv. Mar. Biol. 4:103. Copyright by Academic Press, Inc., London.)

ices with their appendages. Whether the animal is swimming or resting, the gas-filled chambers keep the shell upright and provide buoyancy that counteracts the shell and body weight. The work of Denton and Gilpin-Brown (1973) suggests that the regulatory mechanism is essentially the same in *Nautilus*, *Spirula*, and *Sepia*. Gas and fluid are exchanged through the siphuncle to the chambers (Fig. 11–102). Salts are actively pumped in by the living tissue of the siphuncle. Water then passively diffuses out of the chamber and is replaced by gas—chiefly nitrogen—that diffuses into the vacated space. Fluid exchange is restricted to the more anterior chambers. In fact, fluid fills the space between the last septum and the posterior end of the animal when a new septum is being secreted. Fluid is retained in the chamber until the septum is sufficiently strong to withstand pressure changes.

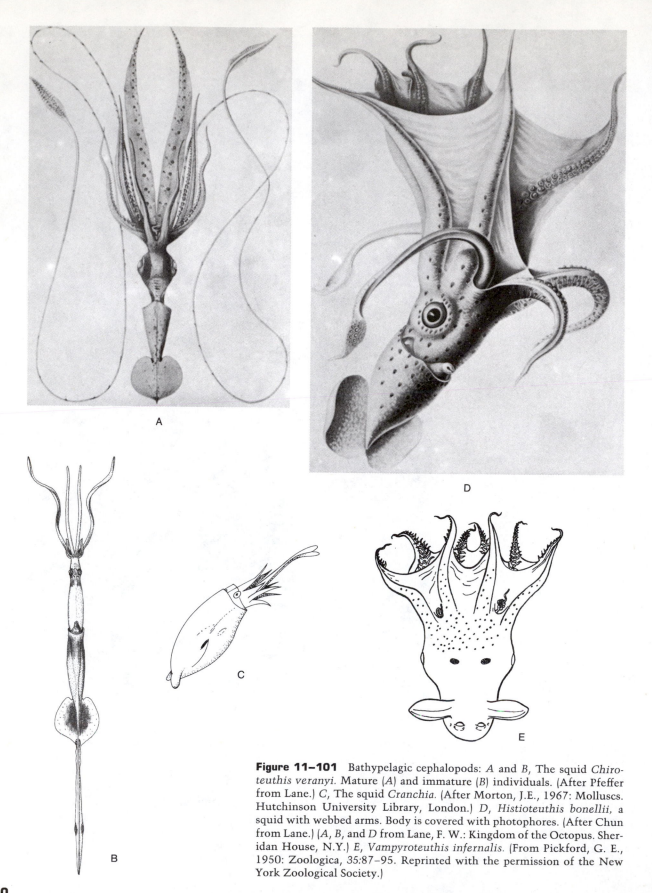

Figure 11–101 Bathypelagic cephalopods: *A* and *B*, The squid *Chiroteuthis veranyi*. Mature (*A*) and immature (*B*) individuals. (After Pfeffer from Lane.) *C*, The squid *Cranchia*. (After Morton, J.E., 1967: Molluscs. Hutchinson University Library, London.) *D*, *Histioteuthis bonellii*, a squid with webbed arms. Body is covered with photophores. (After Chun from Lane.) (*A*, *B*, and *D* from Lane, F. W.: Kingdom of the Octopus. Sheridan House, N.Y.) *E*, *Vampyroteuthis infernalis*. (From Pickford, G. E., 1950: Zoologica, 35:87–95. Reprinted with the permission of the New York Zoological Society.)

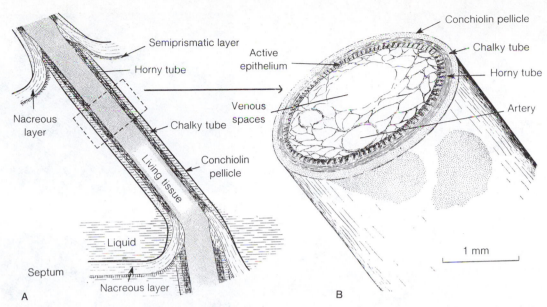

Figure 11–102 Siphuncle of *Nautilus: A,* Longitudinal section between two septa. *B,* Cross section. (From Denton, E. J., 1974: On buoyancy and the lives of modern and fossil cephalopods. Proc. Roy. Soc. London B, *185*:273–299.)

In *Nautilus* the fluid is gradually removed to compensate for the gradual growth of the shell. There is no regulation with short-term changes in depth (Ward et al, 1980).

Except when feeding, *Nautilus* swims backward, at about the same speed as a man doing a slow breast stroke (Haven, 1972). The locomotor mechanism is the same as that of squids, although the ejection of water through the funnel results from the retraction of the body and contraction of the muscles of the funnel rather than by the mantle. The water jet must have been generated in a similar manner in fossil forms, for as in *Nautilus,* enclosure by the shell would have inhibited mantle contraction.

Living cephalopods that still use gas-filled shells to maintain neutral buoyancy—*Nautilus, Spirula,* and *Sepia*—are restricted to depths at which the shell strength can withstand the pressure. Below that depth the shell implodes (caves in). Fossil cephalopods would have been limited in the same manner, and it is therefore reasonable to assume they were all inhabitants of relatively shallow water. During the late Paleozoic and Mesozoic they were the dominant and most highly developed pelagic animals, until they were forced to give way to competition from fish.

The speedier coleoid design perhaps evolved as a result of competition with fish and may account for the many convergent features, such as complex eyes, the two groups share (Packard, 1972). With further modification, the coleoid design permitted invasion of the deep and open ocean, from which the ammonoids and nautiloids were excluded.

The octopods have reverted to more benthic habits. The body is globular and baglike, and there are no fins (Fig. 11–103). The mantle edges are fused dorsally and laterally to the head, resulting in a much more restricted aperture into the mantle cavity. Although octopods are capable of swimming by water jets, with arms trailing, they more frequently crawl about over the rocks around which they live. The arms, which are provided with adhesive suction discs, are used to pull the animal along or anchor to the substratum. Species of *Octopus* usually occupy a den or retreat, from which they make feeding excursions.

A very different mode of existence is exhibited by a number of families of deep-water octopods, some bathypelagic and some abyssalbenthic. These animals have webbed, umbrella-like arms and swim somewhat like jellyfish, by pulsations and by conventional water jet (Fig. 11–104).

Pelagic cephalopods of the open ocean, which includes 46 per cent of the genera, inhabit various levels of the vertical water column. The greatest number live in the upper 100 meters and near the boundary between the mesopelagic and bathypelagic zones, but there are species that live at deeper levels. Many epipelagic and mesopelagic cephalopods exhibit diurnal vertical migration, moving upward during the night and inhabiting lower lev-

Figure 11–103   Lateral view of an octopus.

els during the day (Roper and Young, 1975). Four sperm whales caught off Chile and Peru contained in their stomachs the beaks of 1000 cephalopods, which reflects indirectly the population of pelagic cephalopods and their importance in the pelagic food web (Clarke et al, 1976, 1982).

Gas Exchange

The circulation of water through the mantle not only produces the power for locomotion but, as in other mollusks, also provides oxygen for the gills. The nautiloids have four gills, but the coleoids have the usual two. The surface area of the gill filament has been increased by folding, and cilia have disappeared. The loss of cilia is not surprising. Mantle contractions create the water current passing through the mantle cavity, and the current is strong enough to flush out sediment. In many cephalopods that swim with webbed arms, the gills are vestigial and gas exchange takes place through the general body surface.

Nutrition

Cephalopods are highly adapted for raptorial feeding and a carnivorous diet. Prey is located with the highly developed eyes, and capture is effected by the tentacles or arms. *Nautilus* possesses 38 tentacles, arising from lobes that are arranged in both an inner and an outer circle around the head (Fig. 11–95). The tentacles lack adhesive suckers or discs but have transverse ridges. Above the head and tentacles is a large, leathery, protective hood

A

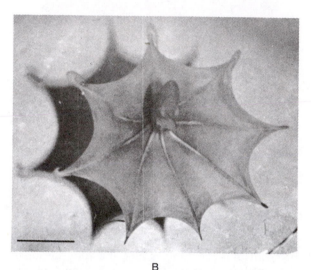

B

Figure 11–104   Deep-sea photographs of a cirrate octopod, *Cirroteuthis*, which lives near the bottom: *A*, Animal perhaps at the beginning of a downward stroke, with arm web closed and fins folded dorsally. *B*, View onto interbrachial web. Cirri on top of arms may have a sensory function. Scale mark equals 30 cm. Photographs taken with a deep-sea camera at 3000 meters in the Virgin Islands Basin. (From Roper, C. E. F., and Brundage, W. L., 1972: Cirrate octopods with associated deep sea organisms: new biological data based on deep benthic photographs. Smithsonian Contrib. Zool., *21*:1–46.)

that acts like an operculum and covers the aperture of the shell when the animal withdraws.

Squids and cuttlefish possess only ten appendages, arranged in five pairs around the head (Figs. 11–99*A* and 11–107). Eight are short and heavy and are called arms; the fourth pair down from the dorsal side are larger and are called tentacles. The inner surface of each arm is flattened and covered

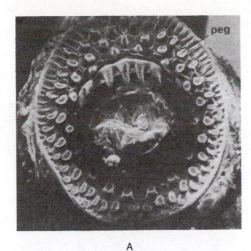

Figure 11–106 Radula teeth of *Octopus briareus*. Only half of a series of seven transverse rows is shown. (From Solem, A., and Roper, C. F. E., 1975: Structures of recent cephalopod radulae. Veliger, *18*(2):127–133.)

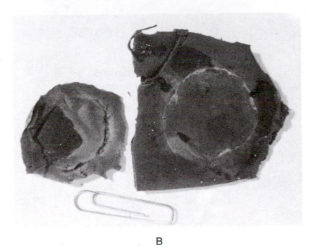

Figure 11–105 *A,* Sucker of an oegopsid squid showing pegged outer ring and toothed inner ring. (From Nixon, M., and Dilly, P. N., 1977: Sucker surfaces and prey capture. *In* Nixon, M., and Messenger, J. B. (Eds.): The Biology of Cephalopods. Symp. Zool. Soc. London, 38, Academic Press, London. p. 449.) *B,* Scars from the suckers of the giant squid *Architeuthis* on the skin of a sperm whale. Sperm whales are the principal predators of the giant squid. (Courtesy of C. Roper.)

with stalked, cup-shaped, adhesive discs that function like suction cups. The rim of the sucker is commonly horny and toothed and the inner wall sometimes has hooks (Fig. 11–105). Suckers are present only on the flattened spatulate ends of the highly mobile tentacles, which are shot out with great rapidity to seize prey. The arms aid in holding the prey after capture (Fig. 11–99*A*). Some squids have curved hooks instead of suckers.

Octopods have only eight arms, which are similar to the arms of squids except that the suckers are stalkless and lack horny rings and hooks.

A radula is present in cephalopods (Fig. 11–106), but more important is the pair of powerful, beaklike jaws in the buccal cavity. The beak can bite and tear off large pieces of tissue, which are then pulled into the buccal cavity by a tongue-like action of the radula and finally swallowed (Fig. 11–107*A*). The location of the buccal mass within a blood sinus permits the animal to turn the entire buccal apparatus and use the jaws with great dexterity. In coleoids two pairs of salivary glands empty into the buccal cavity. The posterior salivary glands of *Sepia* and octopods secrete poison and, at least in *Octopus*, proteolytic enzymes. The poison (a glycoprotein) can enter the tissues of the prey through the wound inflicted by the jaws. The little blue-ringed octopus *Hapalochlaena maculosa*, which feeds on crustaceans in shallow water in the Indo-Pacific, is extremely venomous and its bite has resulted in a few human fatalities.

The diet of cephalopods depends on the habitats in which they live. Pelagic squids, such as *Loligo* and *Alloteuthis*, feed on fish, crustaceans, and other squids (Macy, 1982). *Loligo* will dart into a school of young mackerel, seize a fish with the tentacles, and quickly bite out a chunk behind the head or bite off the head. The fish is devoured with

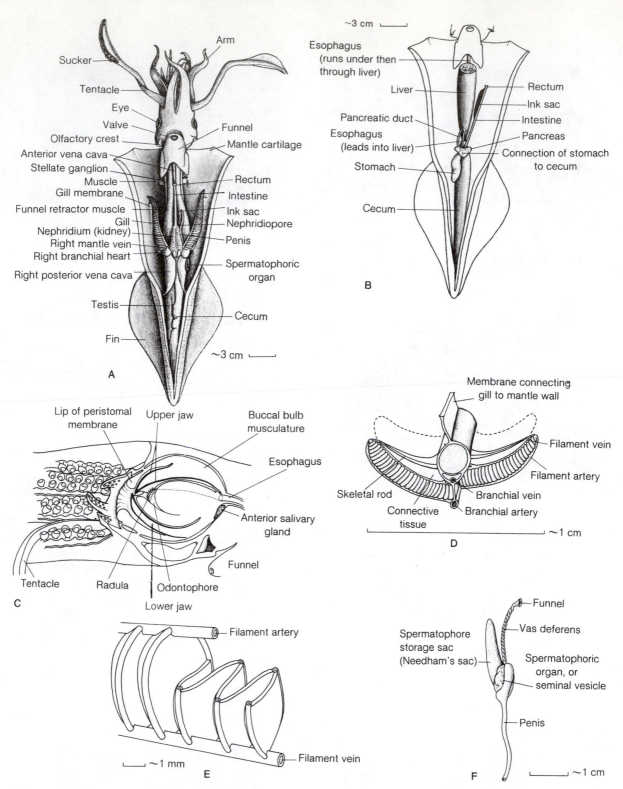

Figure 11–107 Anatomy of the squid *Loligo* (male): *A*, Ventral view, with mantle cut revealing organs within mantle cavity. *B*, Digestive system. The obscuring organs have been removed, the pancreas drawn as semitransparent to show digestive structures on opposite side, and the liver cut to show esophagus. *C*, Longitudinal section through buccal mass, showing base of arms on right side. *D*, Section of gill axis, showing two filaments on either side; only the outlines of the posterior two are shown. *E*, Enlarged view of one filament showing folding. *F*, Male reproductive system. Testis is shown in *A*. (By Mary Ann Nelson.)

small bites of the beak until only the gut and tail remain. These parts are then dropped to the bottom.

Cuttlefish swim over the bottom and feed on surface-inhabiting invertebrates, especially shrimps and crabs. *Sepia* may rest on the bottom, lying in wait for passing prey.

Octopods live in dens located in crevices and holes. They make excursions in search of food or lie in wait near the entrances of their lairs. Clams, snails, crustaceans, and fish are seized and dragged into the lair, where they are eaten. *Octopus dofleini* along the west coast of the United States is active at night; the reef-inhabiting, Indo-Pacific *Octopus cyanea* is active during the day. The latter may make hour-long hunting excursions and cover a distance of 100 meters. Movement is by combined swimming and crawling, and the animal may perch on a rock for a while during the course of a trip. The octopus leaps on motile prey, such as a crab, enveloping it in the outstretched arm web. It also leaps on algal clumps and other objects and then feels beneath the web for a possible catch. The octopus may take several animals, all paralyzed by salivary toxin, home for consumption (Yarnell, 1969; Hartwick et al, 1981).

In contrast to cuttlefish and squids, which tear prey with the jaws, the feeding habit of octopods is rather like that of spiders. The prey is injected with poison, with or without a bite of the jaws, and then, while held, is flooded with enzymes. The partially digested tissues pass into the gut, and the undigestible remains are eventually discarded. To remove gastropods from their shells, octopods drill a hole through the shell with the radula or the toothed salivary papilla and inject poison directly into the occupant.

Nautilus is a scavenger and predator over the bottom, catching decapod crustaceans (Saunders, 1984). When feeding, it swims forward, searching with extended tentacles.

The esophagus is muscular and conducts food by peristaltic action into the stomach or, as in *Nautilus* and *Octopus*, into the crop, which is an expansion of one end of the esophagus.

The stomach is very muscular, and attached to its anterior is a large cecum (Figs. 11–107*A* and *B* and 11–108). The digestive gland in cephalopods is divided into a small spongy diffuse portion, sometimes called the pancreas, and a large solid "liver" (paired in *Sepia*), which is probably homologous to the digestive gland of the other mollusks. In squids the two divisions of the digestive gland are morphologically separated from each other, and the pancreas empties into the liver duct. Digestion is

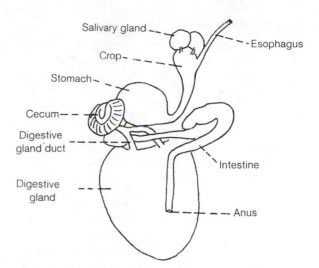

Figure 11–108 Digestive tract of *Octopus vulgaris.* (After Masao.)

entirely extracellular. Enzymes from both digestive gland divisions empty through a common duct into the junction between the stomach and cecum. Secretions from the two glands, which may be released in separate phases, can be sent to the stomach or cecum through regulation of a groove.

Absorption takes place in the cecal walls (*Loligo*) or in the digestive gland (*Sepia, Octopus*). Large, indigestible residues in the stomach can be passed directly into the intestine. Some intestinal absorption takes place. Wastes leave the anus near the funnel and are carried away with the exhalant water jet. The digestive modifications of cephalopods, particularly squids, probably represent an adaptation for rapid digestion correlated with an active pelagic life and a carnivorous diet.

Excretion

There are two nephridia in coleoids and four in *Nautilus*. The conspicuous part of the cephalopod nephridium is a large renal sac (Fig. 11–109). Each sac opens to the mantle cavity through a nephridiopore and communicates with the pericardium by way of a renopericardial canal. The renal sac receives pericardial filtrate via the renopericardial canal and also secretions from large renal appendages, which are evaginations of the wall of the bronchial vein crossing the sac. The wall separating the pericardial coelom and branchial heart contains podocytes, through which a filtrate passes into the pericardial fluid. Selective reabsorption occurs within the coelomic cavity even before the filtrate enters the renal sacs (see review by Martin, 1983).

Internal Transport

The circulatory system of cephalopods is closed. The extensive system of vessels is lined by endothelium, a condition shared with few animals other than vertebrates. Blood within the vessels still follows more or less the same route through the body as in other mollusks (Fig. 11–109). On entering each gill, blood passes through a muscular branchial heart.

The structure and physiology of the cephalopod circulatory system are closely correlated with the higher metabolic rate of these animals compared with other mollusks. The existence of capillaries, some contractile arteries, and the branchial hearts increases the blood pressure and the speed of blood flow. The contraction of the branchial hearts, which receive unoxygenated blood from all parts of the body, boosts the pressure of the blood, sending it through the capillaries of the gills. The two auricles of the heart drain blood from the gills and then pass it into the median ventricle. The ventricle pumps blood out to the body through both an anterior and posterior aorta and eventually through smaller vessels. The blood of cephalopods contains a hemocyanin that loads at the gills and unloads at the tissues at relatively high oxygen pressures.

Nervous System and Sense Organs

The high degree of development of the cephalopod nervous system is unequaled among other invertebrates and is correlated with the locomotor dexter-ity and carnivorous habit of these animals. There is great cephalization. All of the typical molluscan ganglia are concentrated and more or less fused to form a brain that encircles the esophagus and is encased in a cartilaginous cranium (Fig. 11–110A).

In the subesophageal regions of the brain, the pedal ganglia supply nerves to the funnel, and anterior divisions of the pedal ganglia, called brachial ganglia, send nerves to each of the tentacles. The innervations by the pedal and brachial ganglia are evidence that the tentacles and the funnel of cephalopods are homologous to the foot of other mollusks.

A large pair of nerves from the visceral ganglion supplies the mantle. Ordinary swimming and ventilating contractions of the mantle musculature result from impulses conveyed through a system of many small motor neurons radiating from two stellate ganglia, one in each side of the mantle wall. The rapid escape movements of swimming cephalopods, such as squids, result from a highly organized system of giant motor fibers that bring about powerful and synchronous contractions of the circular muscles of the mantle.

The command center of the system is a pair of very large, first-order giant neurons that lie in the median ventral lobe of the fused visceral ganglia (Fig. 11–110B). These neurons are probably fired by a barrage of impulses from the sense organs. They run to another center within the visceral ganglia, where each makes a connection to a second-order giant neuron, which traverses the mantle nerve to the stellate ganglion. Here connections are made with third-order giant neurons supplying the

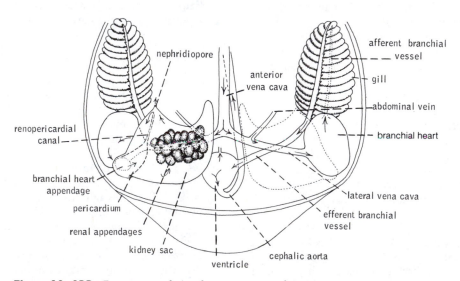

Figure 11–109 Excretory and circulatory systems of *Octopus dofleini.* (After Potts.)

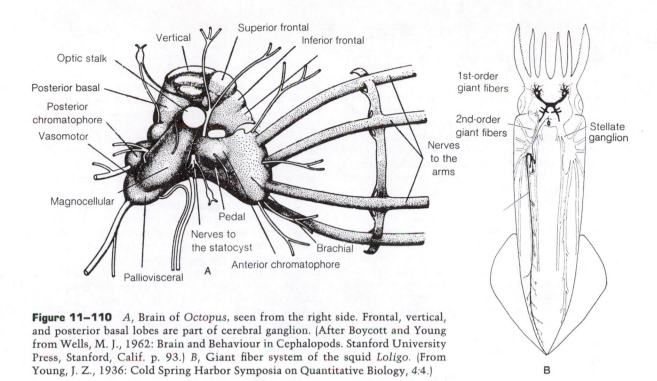

Figure 11–110   *A*, Brain of *Octopus*, seen from the right side. Frontal, vertical, and posterior basal lobes are part of cerebral ganglion. (After Boycott and Young from Wells, M. J., 1962: Brain and Behaviour in Cephalopods. Stanford University Press, Stanford, Calif. p. 93.) *B*, Giant fiber system of the squid *Loligo*. (From Young, J. Z., 1936: Cold Spring Harbor Symposia on Quantitative Biology, 4:4.)

circular muscle fibers. The diameters of the third-order fibers are not uniform. The greater the distance between the stellate ganglion and the muscle terminal, the greater the diameter of the fiber, thus ensuring that all impulses arrive at the muscle fibers simultaneously and produce a powerful, synchronous contraction of the mantle musculature.

The eyes of cephalopods are highly developed and are strikingly similar in structure to those of fishes (Fig. 11–111). A spherical housing containing cartilaginous plates fits into a sort of orbit or socket of cartilages associated with those surrounding the brain. The lens, which is suspended by a ciliary muscle, is a rigid sphere with a fixed focal length, and in front of the lens is an iris diaphragm, which can control the amount of light entering the eye through the slit-shaped pupil. The retina contains closely packed, long, rodlike photoreceptors that are directed toward the source of light. The eye is thus of the direct type, instead of indirect, as in vertebrates. The photoreceptors are connected to retinal cells that send fibers back to an optic ganglion.

The cephalopod eye undoubtedly forms an image, and the optic connections appear to be especially adapted for analyzing vertical and horizontal projections of objects in the visual field. Experimental studies indicate that *Octopus* can dis-

criminate objects as small as 0.5 cm from a distance of 1 meter, which would be a considerable advantage in catching prey (Wells, 1978).

Functioning in an aquatic environment, the cornea of the cephalopod eye contributes little to focusing, because there is almost no light refraction at the corneal surface, as there is in an air-corneal surface (see Box 11–1, p. 458). Accommodation

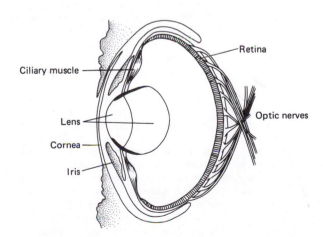

Figure 11–111   Eye of *Octopus*. (After Wells, M. J., 1961: What the octopus makes of it; our world from another point of view. Adv. Sci. Lond., *20*:461–471.)

BOX 11–1 INVERTEBRATE EYES II

The widely distributed pigment-cup eyes described in Box 7–3 are adapted for animals that need to monitor light only for orientation, for detecting the shadow of a possible predator, or for determining photoperiod and setting circadian rhythms. Many animals that move rapidly and search for prey or other objects have evolved eyes capable of some degree of image formation and the neural circuitry to permit object discrimination. In the simplest terms, this requires that each photoreceptor or group of photoreceptors be able to determine the light intensity of some small part of the visual field. When varying intensities across the field are compiled, as are the dots on a television screen, the contrast will stand out as an object. The quality of image formation among invertebrates varies greatly, but in general it is poor. The variation depends on external environmental factors as well as limitations in eye structure. There have been very few studies on the visual physiology of animals, and much of what is thought to be true is based on more readily available anatomical information.

An obvious factor in the ability of an eye to produce a good image is the number of photoreceptors present. This varies enormously. There are about 200 photoreceptors in the pigment-cup eye of *Planaria maculata*, which measures 300 μm across the aperture, compared with 70,000 photoreceptors per square mm in the eye of an octopus. But the large octopus eye is an exception. Most invertebrate eyes have far fewer photoreceptors, in large part because the animals are small. Thus, the small size of most invertebrate eyes limits the size of their visual field and hence their sampling ability.

A lens improves the quality of the image by bending light (focusing) onto the retinal receptors. Light is bent, or refracted, when it passes from one medium into another. In aquatic animals the cornea plays little role in focusing because the index of refraction is not that different from water. The lens plays the principal role in focusing only if it is composed of a material that has a higher index of refraction than that through which light has previously passed. Many invertebrates, such as nereid polychaetes and most prosobranch mollusks, have a "lens" of soft material that fills up much of the interior of the eye. It contributes little to the eye's resolving power. A lens of different material is usually spherical. But a spherical lens produces aberrations because light striking different parts of the curved surface do not all meet at the same point, blurring the image. This problem has been solved in some animals by a gradient of density from the outside of the lens to the center. A good-quality focus also requires space between the back of the lens and the retina. The eyes of very few aquatic invertebrates meet all these conditions—a large number of photoreceptors; a highly developed, dense lens; and a space between the lens and the retina. The notable examples are the pelagic alciopid polychaetes, the heteropod gastropods, and the cephalopods, all of which are raptorial, swimming carnivores.

On land, where light is passing through the air, the cornea plays a primary role in refraction and the lens is secondary. In humans and other terrestrial mammals, for example, the lens functions largely for focal adjustment, the primary task having been accomplished by the cornea. Among terrestrial invertebrates, some spiders with simple eyes and certain crustaceans and insects with compound eyes achieve the highest degree of object discrimination. Their eyes will be discussed later. However, some periwinkles *(Littorina)* and land snails *(Helix)* appear to be able to recognize vertical bars (marsh grass and other vegetation).

Active animals that are nocturnal or live in dimly lighted habitats must capture the maximum amount of light over a given unit of time. Light capture is facilitated by an eye with a large pupil and large photoreceptors. The larger the eye the better, but since the small size of the animal limits the size of the eye, resolving power usually goes down as light-capturing ability goes up. Some animals, such as scallops, wolf spiders and some insects, possess a reflecting layer of pigment (tapetum) behind the retina, which bounces light back to the photoreceptors.

Light waves coming from the sun across the sky oscillate at right angles to the path of the light rays. Sky light is therefore said to be polarized. A few animals, such as certain spiders, amphipods, bees, and sea hares, have the photosensitive pigment in the receptor membrane so oriented that it is sensitive to light only when aligned with the direction of polarization. Such animals can use polarized sky light to determine the sun position even when the sun is hidden. The sun position provides navigational orientation to habitat, food sources, etc.

An excellent account of the comparative structure and function of invertebrate eyes is provided by M. F. Land, 1981: Optics and vision in invertebrates. *In* Autrum, H. (Ed.): Handbook of Sensory Physiology, Vol. III/6B. Springer-Verlag, Berlin. pp. 471–592. A shorter review is provided by T. W. Cronin, 1986: Photoreception in marine invertebrates. Amer. Zool., *26*(2):403–415.

takes place by forward and backward movement of the lens, as in fish.

The cephalopod eye can accommodate itself to light changes both by modifications in the pupil's size and by the migration of pigment in the retina. Squids of the family Histioteuthidae, which live in the mesopelagic zone, have dimorphic eyes. The large left eye is directed upward and responds to faint light from the surface, and the little right eye is directed forward and responds to bioluminescent light (Young, 1975).

The eyes of *Nautilus* are large and are carried at the end of short stalks. Although the eye contains a large number of photoreceptors, strangely it lacks a lens and is open to the external sea water through a small aperture, the pupil. Supposedly, the eye functions like a pinhole camera, but without a lens the resolving power must be very limited. The long evolutionary persistence of such an inefficient eye is puzzling (see Land, 1984, for a discussion of the question).

Statocysts are found in nautiloids but are particularly well developed in coleoids, in which they are large and are embedded in the cartilages located on each side of the brain. They not only provide information about static spatial orientation (i.e., body position in relation to gravitational pull) but are so constructed that, like the semicircular canals of vertebrates, they inform the animal of changing positions in motion, such as turning. Without the statocysts a cephalopod can neither keep the pupil slits of the eyes horizontal nor discriminate between horizontal and vertical surfaces.

Osphradia are present only in *Nautilus*.

The arms, and especially the sucker epithelium, are liberally supplied with tactile cells and chemoreceptor cells, particularly in the benthic hunting octopods. Textural and chemical discrimination of a surface by the arms has been demonstrated in *Octopus*, but these animals cannot determine shape with the tentacles, perhaps because the arms contain no divisions or structures that could function as a unit of measurement, as for example the distance between thumb and forefinger in primates or leg joints in arthropods.

Cephalopod behavior is an area of great interest. Learning, memory, tactile and visual discrimination, and localization of motor and sensory functions within the brain have been studied with rewarding results. Among the many investigators of cephalopod behavior, the British zoologists J. Z. Young, M. J. Wells, J. Wells, and B. B. Boycott are especially prominent. For a more detailed account of the cephalopod nervous system and cephalopod behavior, the reader should refer to the excellent résumés by M. J. Wells (1978, 1979) and J. Z. Young (1972).

Chromatophores, Ink Gland, and Luminescence

The unusual coloration of cephalopods (most other than *Nautilus*) is caused by the presence of chromatophores in the integument. The expansion of these cells results from the action of small muscle cells attached to the periphery of the chromatophore. When the muscles contract, the chromatophore is drawn out to form a large, flat plate; when the muscles relax, the pigment is concentrated and less apparent. Particular species possess chromatophores of several colors—yellow, orange, red, blue, and black—and the chromatophores of a particular color may occur in groups or layers. The chromatophore effect is enhanced by deeper layers of iridocytes, or reflector cells, which differentially reflect light (Fig. 11–112) (see review by Cloney and Brocco, 1983). The coloration of the skin at any particular time is thus a result of the light passing through the chromatophore filters, the particular chromatophores that are expanded, and the iridocyte filters. The chromatophores are controlled by the nervous system and probably by hormones, with vision being the principal initial stimulus. The degree to which these animals can change color and the stimulus for color change vary considerably. The cuttlefish *Sepia officinalis* displays complex color changes and may simulate the background hues of sand, rock, and so forth, but most changes appear to be correlated with behavior. Many species exhibit color change when alarmed. For example, the littoral squid *Loligo vulgaris* is generally very pale and only darkens when disturbed; others lighten (Fig. 11–113). When alarmed, an octopus may flatten its body and present an elaborate "defensive" color display, including color changes flowing over the body and large dark spots around the eyes. Color displays also are associated with courtship in many cephalopods and are described in the next section.

In cephalopods other than *Nautilus* and some deep-water species, a large ink sac opens into the rectum just behind the anus (Fig. 11–107A). The gland secretes a brown or black fluid containing a high concentration of melanin pigment, which is stored in the reservoir. When an animal is alarmed, the ink is released through the anus and the cloud of inky water forms a "dummy," confusing the predator. It is also believed that the alkaloid nature

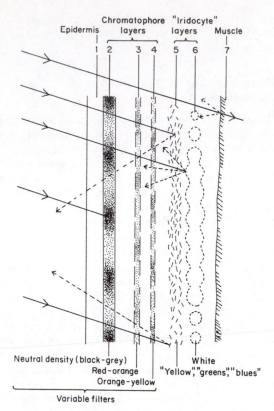

Figure 11–112 Diagram of the layers of chromatophores and iridocytes in the skin of *Octopus* and their interaction. Arrows indicate possible absorption and reflection of light coming from left. See text for further explanation. (From Packard, A., and Hochberg, F. G., 1977: Skin patterning in *Octopus* and other genera. *In* Nixon, M., and Messenger, J. B. (Eds.): The Biology of Cephalopods. Symp. Zool. Soc. London, 38, Academic Press, London. p. 197.)

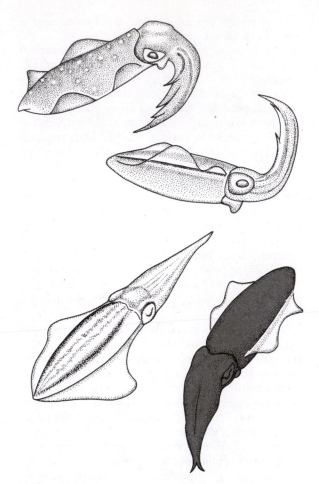

Figure 11–113 Four chromatophoric color patterns of the Caribbean reef squid, *Sepioteuthis sepioidea*. Most of these color patterns, as well as the body and tentacle positions, are probably alarm responses. (From Moynihan, M., and Rodaniche, A. F., 1982: The Behaviour and Natural History of the Caribbean Reef Squid [*Sepioteuthis sepioidea*]. Paul Parey, Publ., Hamburg.)

of the ink may be objectionable to predators, particularly fish, in which it may anesthetize the chemoreceptive senses.

Many of the mid- and deep-water squids are bioluminescent, with luminescent photophores arranged in various patterns over the body, even on the eyeball (Fig. 11–114). In some species, such as *Sepiola*, the luminescence is due to symbiotic bacteria, but in most it is intrinsic (see Box 5–2 on the origin and function of bioluminescence; also the review by Herring, 1977, and Young and Roper, 1976, 1977).

Reproduction

Cephalopods are dioecious, and the single gonad is located at the posterior of the body (Fig. 11–107). In the male the highly coiled vas deferens conducts the sperm from the testis anteriorly to a seminal vesicle, which has ciliated grooved walls. Here in the seminal vesicle the sperm are rolled together and encased in a very elaborate spermatophore. From the seminal vesicle spermatophores pass into a large storage sac (Needham's sac), which opens into the left side of the mantle cavity.

In females the oviduct terminates in an oviductal gland. In octopods and some oceanic squids two oviducts are present.

Fertilization may take place within the mantle cavity or outside, but in either case it involves copulation. One of the arms of the male has become modified as an intromittent organ, called a hectocotylus. The degree of modification varies. In *Sepia*

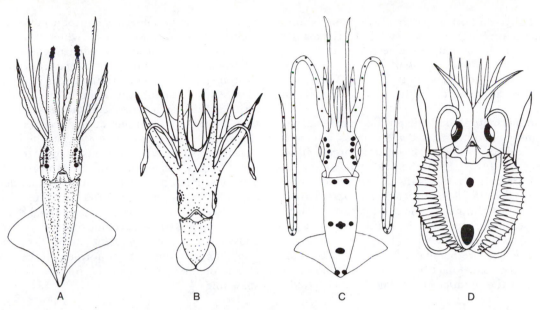

A B C D

Figure 11–114 Distribution of light organs in four squids. Dot on spot indicates size and placement of organ. *A, Abraliopsis.* Small light organs on mantle, head, and arms and large ones on eyes are for counterillumination; arm tip organs are for flashing attention or warning. *B, Histioteuthis.* Small light organs on mantle, head, and arms are for counterillumination; those on arm tips are of unknown function. *C, Nematolampas.* Eye and body light organs serve for counterillumination; those on arms and tentacles are of unknown function. *D, Ctenopteryx.* Light organs on eyes and mantle serve for counterillumination and perhaps flashing. (From Herring, P. J., 1977: Luminescence in cephalopods and fish. *In* Nixon, M., and Messenger, J. B. (Eds.): The Biology of Cephalopods. Symp. Zool. Soc. London, 38, Academic Press, London, pp. 127–159.)

and *Loligo* several rows of suckers are smaller and form an adhesion area for the transport of spermatophores. In *Octopus* the tip of the arm carries a spoonlike depression; in *Argonauta* and others there is actually a cavity or chamber where the sperm are stored.

Before copulation a male cephalopod performs various displays that serve to identify it to the female. The cuttlefish *Sepia* presents a striped color pattern and establishes a temporary bond with a female, swimming above her. The display is also directed toward intruding males, and the weaker male will depart.

In pelagic cephalopods copulation occurs while the animals are swimming, the male seizing the female head-on (Fig. 11–115). During copulation the hectocotylus receives spermatophores from the male's funnel or plucks a mass of spermatophores from the storage sac. The male hectocotylus is then inserted into the mantle cavity of the female and deposits the spermatophores on the mantle wall near the openings from the oviducts or, in *Octopus*, into the genital duct itself. In some octopods the male inserts his copulatory arm into the mantle cavity of the female without grasping her body.

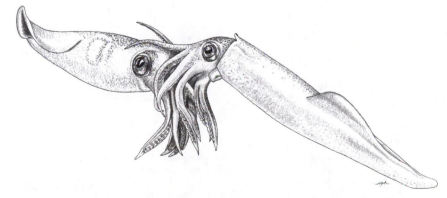

Figure 11–115 Copulation in the squid *Loligo.* (Based on a photograph by Robert F. Sisson.)

In *Loligo* and other genera of the same family, the spermatophores may be received by the mantle cavity as just described, or the hectocotylus may be inserted into a horseshoe-shaped seminal receptacle, located in a fold beneath the mouth. The buccal membrane alone receives the spermatophores in *Sepia*.

The spermatophore is shaped like a baseball bat and consists of an elongated sperm mass, a cement body, a coiled, springlike, ejaculatory organ, and a cap (Fig. 11–116A). With the cap removed as a result of traction in the transfer process or from water uptake, the ejaculatory organ is everted, pulling out the sperm mass. The cement body adheres to the seminal receptacle or mantle wall and the sperm mass disintegrates.

As the eggs are being discharged from the oviduct, each is enveloped by a paired membrane or capsule in the oviduct gland (Fig. 11–117). Addi-tional protective covering is produced by a nidimental gland in most groups, which is located in the mantle wall and opens independently into the mantle cavity. In *Loligo* secretions from the nidimental gland surround the eggs in a gelatinous mass. After leaving the mantle cavity, the egg string is held by the arms, and the eggs may be fertilized by sperm from the seminal receptacle under the mouth. The female then attaches the fertilized eggs to the substratum in a cluster of 10 to 50 elongated strings, each containing as many as 100 eggs. The gelatinous covering of each mass hardens on exposure to sea water, and the individual egg capsules swell to several times the original diameter. Large numbers of *Loligo* come together to copulate and spawn at the same time, and a "community pile" of egg strings may be formed on the bottom. Death of the adult usually follows soon after spawning.

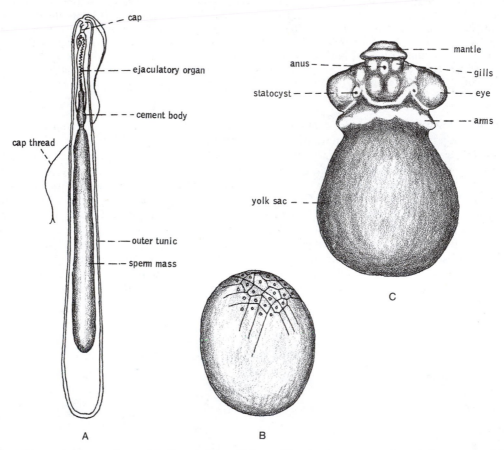

Figure 11–116 *A*, Spermatophore of *Loligo*. *B*, Discoidal meroblastic cleavage in *Loligo*. (After Watase from Dawydoff.) *C*, Embryo of *Loligo*. (After Naef from Dawydoff.) *D*, *Argonauta argo* female in shelly egg case. (After Naef from Kaestner.) *E*, *Argonauta hians*. (After Verrill from Abbott.)

Figure 11–117 Egg cluster of the Caribbean reef squid, *Sepioteuthis sepioidea* One of the young has just hatched, and others can be seen within the eggs. (Photograph by Olga F. Linares in Moynihan, M., and Rodaniche, A. F., 1982: The Behaviour and Natural History of the Caribbean Reef Squid [*Sepioteuthis sepioidea*]. Paul Parey, Publ., Hamburg.)

Sepia deposits its eggs singly but attaches them by a stalk to seaweed or other objects. *Octopus* forms egg clusters that resemble a bunch of grapes and are attached within rocky recesses.

The egg masses of deep-water and some pelagic squids, such as the oegopsids, are commonly free floating rather than attached. Benthic octopods remain to care for the eggs after they are deposited. The females of at least *Octopus vulgaris* die after brooding their eggs.

A remarkable adaptation for egg deposition occurs in the pelagic genus *Argonauta*, commonly known as the paper nautilus. The two dorsal arms of the female secrete a beautiful, calcareous, bivalved case into which the eggs are deposited. The case is carried about and serves as a brood chamber. The posterior of the female usually remains in the case; when disturbed, she withdraws completely into the retreat.

The reproductive system of *Nautilus* is somewhat similar to that of coleoids. There is a finger-like copulatory organ and copulation is head on,

but little is known about spermatophore transfer and fertilization. The eggs are deposited and attached to the substratum singly within rather elaborate capsules.

Cephalopod eggs are much more heavily yolk-laden than other mollusk eggs. The eggs of *Sepia* and *Ozaena* are very large and may reach 15 mm in diameter. The eggs of *Loligo* and *Octopus* contain less yolk and may be small.

Reproduction is under hormonal control, although studies have been largely restricted to octopods. Hormones are produced by a pair of spherical optic glands associated with the optic tracts. The secretions not only regulate the production of eggs and sperm but also, following spawning, cause the female to cease feeding and to brood her eggs. Death follows the reproductive period in both sexes. If the optic glands of brooding females are removed, they stop brooding, resume feeding, and the life span is extended (Wodinsky, 1977).

Development

Cleavage, which is meroblastic, results in the formation of a germinal disc, or cap, of cells at the animal pole (Fig. 11–116B). Here the embryo will form (Fig. 11–116C). The margin of the disc grows down and around the yolk mass and forms a yolk sac; the yolk is gradually absorbed during development.

Although development is direct in that there is no trochophore or veliger larva, the little cephalopods may be planktonic for a while following hatching. This is true of octopods that do not take up a benthic existence until they have reached a larger size (Fig. 11–118). Even in some pelagic squids, the juveniles live at higher levels than do the adults.

Many cephalopods have short lives. Squids (*Loligo*) live only one to three years, depending on the species, and usually die after a single spawning (Holme, 1974). *Octopus vulgaris* also dies after one brood, when the animal is two years old. *Nautilus* may have a life span of 20 years (Saunders, 1983).

SYSTEMATIC RÉSUMÉ OF CLASS CEPHALOPODA

Subclass Nautiloidea. Possess external shells, which may be coiled or straight; sutures not complex. Living forms possess many slender, suckerless tentacles. Two pairs of gills and two pairs of nephridia are present. The class has been in existence since the Cambrian period, but all mem-

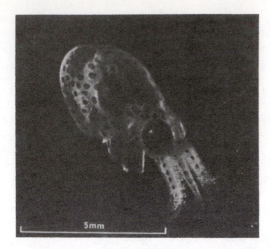

Figure 11–118 Newly hatched planktonic young of the octopus *Eledone cirrosa*. (From Boletzky, S. V., 1977: Post-hatching behaviour and mode of life in cephalopods. *In* Nixon, M., and Messenger, J. B. (Eds.): The Biology of Cephalopods. Symp. Zool. Soc. London, 38, Academic Press, London. p. 562.)

bers are extinct except *Nautilus*. Genera include *Endoceras* and *Nautilus*.

Subclass Ammonoidea. Fossil forms with coiled external shells having complex septa and sutures. Silurian to Cretaceous. Includes *Ceratites*, *Scaphites*, and *Pachydiscus*.

Subclass Coleoidea. Shells are internal, reduced, or absent. The eight or ten appendages bear suckers. One pair of gills and one pair of nephridia. Mississippian to the present.

Order Belemnoidea. Extinct species. Shell internal, chambered but with a posterior solid rostrum and a dorsal, shieldlike extension. *Belemnites, Belemnoteuthis*.

Order Sepioidea. Cuttlefish and sepiolas. Eight arms and two tentacles. Shell with septa, or shell greatly reduced or lost. Body mostly short and broad or saclike. *Spirula, Sepia, Idiosepius, Sepiola, Rossia*.

Order Teuthoidea. Squids. Shell or pen a flattened blade or vane. Body mostly elongate with eight arms and two long tentacles.

Suborder Myopsida. Squids with a transparent corneal membrane over the eye. *Loligo, Lolliguncula, Sepioteuthis*.

Suborder Oegopsida. Squids without a transparent corneal membrane but with eyelids. Pupil circular. Contains most squids, many of which live in deep water. *Architeuthis, Abralia, Abraliopsis, Gonatus,*

Onychoteuthis, Ctenopteryx, Histioteuthis, Bathyteuthis, Illex, Ommastrephes, Chiroteuthis, Cranchia.

Order Vampyromorpha. Vampire squids. Small, deep-water, octopod-like forms with eight arms united by a web, but two small filaments also present. *Vampyroteuthis*.

Order Octopoda. The octopods. Possess eight arms; body globular.

Suborder Cirrata. Finned octopods. Mantle with a pair of fins. Arms, with finger-like cirri, connected by a broad web. Mostly deep-sea species. *Cirrothauma, Opisthoteuthis*.

Suborder Incirrata. Octopods without fins. *Octopus, Ozaena, Eledonella, Vitreledonella, Amphitretus, Argonauta*.

SUMMARY

1 The primary design of the class Cephalopoda is one adapted for a pelagic, raptorial existence.

2 The gas-filled, chambered shells of *Nautilus*, *Sepia*, and *Spirula* provide buoyancy. Regulation occurs by changing the volume of fluid within the chambers via the siphuncle.

3 Cephalopods swim by a water jet produced by expulsion of water from the mantle cavity through the funnel. In *Nautilus*, which is a slow swimmer, force is generated by contraction of the bilobed funnel. This was probably also true of the many fossil cephalopods.

4 Most living cephalopods (coleoids) generate the water jet by contracting the mantle wall, making possible greater force and speed. The shell is reduced and internal, or even lost altogether, freeing the mantle wall for the water pumping function.

5 Octopods have taken up a secondary crawling benthic existence and use jet propulsion only for escape and long-distance swimming.

6 Most cephalopods seize prey with a pair of prehensile tentacles and hold it with the eight arms, all provided with suckers. Octopods lack the tentacles, perhaps because of their crawling habits. *Nautilus* possesses 38 nonsuckered tentacles.

7 Prey is dispatched with a horny, parrot-like beak and a pair of poison glands (modified salivary glands). The radula functions as a tongue, pulling in pieces of tissue torn off by the beak. The digestive system is adapted for rapid digestion.

8 Many cephalopod features are directly or indirectly related to their active life and attendant higher metabolic rate. Such features include:

a. Secondarily folded gills
b. Absence of gill cilia
c. Closed blood-vascular system
d. Accessory branchial hearts
e. Presence of hemocyanin
f. Highly developed eyes
g. Complex nervous system and behavior
h. Chromatophores
i. Ink gland

9 In cephalopod copulation, one of the male arms transfers spermatophores to the female. Before being released, the eggs are fertilized in the oviduct, in the mantle cavity, or from spermatophores deposited by the male in a receptacle around the mouth of the female. The encased eggs are either deposited on the bottom or shed into the sea water. Development is direct in that there is neither trochophore nor veliger larva, although juveniles ("larvae") of some species are planktonic.

The Origin of the Mollusks

The striking similarity between the embryogeny of the mollusks and that of the polychaete annelids has long been recognized. Both kinds of animals exhibit spiral cleavage and virtually identical trochophore larvae. This similarity has been the principal evidence to support the view that the annelids and mollusks arose from some common ancestral stock.

The discovery of living monoplacophorans contributed fresh views not only to a molluscan-annelidan relationship but also to the nature of the ancestral mollusks. The striking feature of *Neopilina* is the linear repetition of structure. *Neopilina* displays eight pairs of retractor muscles, five pairs of gills, and six pairs of nephridia. Some zoologists (Fretter and Graham, 1962; Lemche and Wingstrand, 1959; Wingstrand, 1984) have thought that the repetition displayed by *Neopilina* is evidence that the mollusks constitute a fundamentally metameric phylum. This belief has in turn led to the view that mollusks evolved from the annelids. The molluscan trochophore is thus believed to be additional evidence for an annelidan origin of mollusks.

However, many zoologists do not find the evidence for metamerism very convincing. First, the replicated structures differ in numbers and do not display any particularly striking coincidence in their sequence. Therefore, it is difficult to determine either functionally or structurally what would represent a segment.

The affinity between mollusks and annelids seems probable even without *Neopilina*, but it appears more likely that the two groups diverged from some common ancestor prior to the appearance of either molluscan or annelidan characters, especially metamerism. The nature of such an ancestor is debatable, but some replication of structures along the length of the body probably was present. In annelids this replication developed into metamerism as an adaptation for burrowing. In mollusks the replication tended to become reduced with the assumption of new modes of existence. Vagvolgyi (1967) and Stasek (1972) have suggested the flatworms as being near the common ancestral stock of annelids and mollusks.

Current ideas about molluscan phylogeny have recently been reviewed by Stasek (1972), Runnegar and Pojeta (1974), and Yochelson (1978, 1979). They suggest that the gastropods, bivalves, scaphopods, and cephalopods, all of which appear to be derived from forms possessing a shell of a single piece, evolved from monoplacophorans. The aplacophorans and polyplacophorans may have diverged early in molluscan history, before the formation of the single shell. Thus, the absence of a shell in aplacophorans would be a primitive feature rather than a secondary one.

REFERENCES

The literature included here is restricted to mollusks. The introductory references on page 8 include many general works and field guides that contain sections on mollusks.

More than 15 journals are devoted to studies on mollusks: American Malacological Bull.; Archiv für Molluskenkunde (German); Basteria (Dutch); Boll. Malacologico (Italian); Bull. Malacology Republic of China; Haliotis (French); Jour. of Conchology (British); Jour. of the Malacological Soc. of Australia; Jour. of Molluscan Studies (British); Lavori della Societa Malacologica Italiana; Malacologica (U.S.); Malacological Review (U.S.); Malakologische Abhandlungen (German); Nautilus (U.S.); Veliger (U.S.); Venus (Japanese).

References for Identification

Abbott, R. T., 1974: American Seashells. 2nd Edition. Van Nostrand Reinhold Co., N.Y. 663 pp. (A very complete and excellent guide to the marine mollusks of the Atlantic and Pacific coasts of North America.)

Andrews, J., 1971: Sea Shells of the Texas Coast. University of Texas Press, Austin. 298 pp. (An excellent illustrated systematic account of the mollusks of the Texas coast; useful for much of the Gulf Coast and West Indies.)

Behrens, D. W., 1980: Pacific Coast Nudibranchs, a Guide to the Opisthobranchs of the Northeastern Pacific. Western Marine Enterprises, Ventura, Calif. 112 pp.

Bertsch, H., and Johnson, S., 1981: Hawaiian Nudibranchs. Oriental Publishing Co., Honolulu. 112 pp.

Bouchet, P., 1979: Seashells of Western Europe. Amer. Malacologists, Inc., Melbourne, Fla. 156 pp.

Burch, J. B., 1962: How to Know the Eastern Land Snails. W. C. Brown Co., Dubuque, Iowa. 214 pp.

Burch, J. B., 1973: Freshwater Unionacean Clams of North America. Biota of Freshwater Ecosystems. Ident. Man. No. 11, 176 pp. U.S. Government Printing Office.

Burch, J. B., 1975: Freshwater Unionacean Clams of North America. Rev. Edition. Malacological Publications, Hamburg, Mich. 204 pp.

Burch, J. B., 1982: North American freshwater snails: identification keys, generic synonymy, supplemental notes, glossary, references, index. Walkerana, 4:217–365.

Burch, J. B., and van Devender, A. S., 1980: Identification of eastern North American land snails: the Prosobranchia, Opisthobranchia and Pulmonata (Actophila). Walkerana, 2:33–80.

Burch, J. B., and Tottenham, J. L., 1980: North American freshwater snails: species list, ranges and illustrations. Walkerana, 3:81–215.

Cameron, R. A. D., and Redfern, M., 1976: British Land Snails: Synopses of the British Fauna No. 6. Academic Press, London. 64 pp.

Clarke, A. H., 1981: The Freshwater Molluscs of Canada. National Museum of Natural Science, National Museums of Canada. University of Chicago Press, Chicago. 446 pp.

Ellis, A. E., 1978: British Freshwater Bivalve Mollusks: Synopses of the British Fauna No. 11. Academic Press, London. 109 pp.

Graham, A., 1971: British Prosobranch and Other Operculate Gastropod Molluscs. Synopses of the British Fauna No. 2. Academic Press, N.Y. 112 pp.

Keen, A. M., and McLean, J. H., 1971: Sea Shells of Tropical West America, 2nd Edition. Stanford University Press, Stanford. 1080 pp. (Excellent manual for marine mollusks from San Diego to Peru.)

Lindner, G., 1978: Field Guide to Seashells of the World. Van Nostrand Reinhold Co., N.Y. 271 pp. (Guide to the common shelled mollusks of the world. Translation of a 1975 German work.)

Marcus, E., and Marcus, E., 1967: American Opisthobranch Mollusks. Studies in Tropical Oceanogr. Ser.: No. 6. University of Miami Press, Coral Gables, Fla. 256 pp.

Pilsbry, H. A., 1939–1946: Land Mollusca of North America. Vol. 1, Pts. 1 and 2; Vol. 2, Pts. 1 and 2. Acad. Nat. Sci., Monogr. 3, Philadelphia. (Although old, this is a valuable taxonomic work on the North American land snails.)

Roper, C. F. E., Young, R. E., and Voss, G. L., 1969: An illustrated key to the families of the order Teuthoidea (Cephalopoda). Smithsonian Contributions to Zoology, 13:1–32.

Roper, C. F. E., Sweeney, M. J., and Nauen, C. E., 1984: FAO species catalogue. Vol. 3. Cephalopods of the world. An annotated and illustrated catalogue of species of interest to fisheries. FAO Fisheries Synopsis, (125) Vol. 3. 277 pp.

Thompson, T. E., and Brown, G. H., 1976: British Opisthobranch Molluscs. Synopses of the British Fauna No. 8. Academic Press, London. 204 pp.

Willan, R. C., and Colemann, N., 1984: Nudibranchs of Australasia. Sea Australia Productions; Sidney, Australia.

General

Feder, H. M., 1972: Escape responses in marine invertebrates. Sci. Am., 227:93–100.

Fretter, V. (Ed.), 1968: Studies in the Structure, Physiology and Ecology of Mollusks. Symp. Zool. Soc. London, Academic Press, London. 378 pp.

Giese, A. C., and Pierce, V. S. (Eds.), 1977 and 1979: Reproduction of Marine Invertebrates. Vol. 4 (1977) (Gastropods and Cephalopods). Vol. 5 (1979) (Chitons and Bivalves). Academic Press, N.Y.

Graham, A., 1949: The molluscan stomach. Trans. R. Soc. Edinburgh, Pt. 3, 61(27:)737–778. (A comparative treatment of the molluscan stomach; it deals primarily with gastropods and bivalves.)

Grassé, P. P. (Ed.), 1968: Traité de Zoologie, Vol. V (Pt. II): Introduction to the Mollusks, the Chitons, the Monoplacophorans, and the Pelecypods. (Pt. III) Gastropods and Scaphopods. Masson et Cie, Paris.

Hyman, L. H., 1967: The Invertebrates. Vol. 6. Mollusca. I. McGraw-Hill, N.Y. (This volume, the last of Dr. Hyman's contributions to the Invertebrate Series, covers a part of the phylum Mollusca. Four classes are treated: Aplacophora, Polyplacophora, Monoplacophora, and Gastropoda.)

Joosse, J., and Geraerts, W. P. M., 1983: Endocrinology. In Wilbur, K. M. (Ed.): The Mollusca, Vol. 4, Pt. 1. Academic Press, N.Y. pp. 318–406.

Land, M. F., 1984: Molluscs (Eyes). In Alii, M. A.: Photoreception and Vision in Invertebrates. Plenum Press, N.Y. pp. 699–725.

Lutz, R. A., and Rhoads, D. C., 1980: Growth patterns within the molluscan shell, an overview. In Rhoads, D. C., and Lutz, R. A. (Eds.): Skeletal Growth of Aquatic Organisms. Plenum Press, N.Y.

Mangum, C. P., 1985: Oxygen transport in invertebrates. Amer. Jour. Physiol., 248:505–514.

Martin, A. W., 1983: Excretion. In Wilbur, K. M. (Ed.): The Mollusca, Vol. 5, Pt. 2. Academic Press, N.Y. pp. 353–405.

Moore, R. C. (Ed.), 1957–1971: Treatise on Invertebrate Paleontology. Mollusca, Vol. I–N. Geological Society of America and University of Kansas Press, Lawrence, Kansas. (In addition to descriptions of fossil species, the introductory sections cover various aspects of the biology of mollusks.)

Morton, J. E., 1967: Molluscs. 4th Edition. Hutchinson University Library, London.

Nakazima, M., 1956: On the structure and function of the midgut gland of Mollusca, with a general consideration of the feeding habits and systematic relations. Jap. J. Zool. 2(4):469–566. (A good comparative account of the digestive diverticular of mollusks.)

Potts, W. T. W., 1967: Excretion in molluscs. Biol. Rev., 42(1):1–41.

Price, T. J., Thayer, G. W., Lacroix, M. W., and Montgomery, G. P., 1974: The organic content of shells and soft tissues of selected estuarine gastropods and pelecypods. Proc. Natl. Shellfish Assoc., 65:26–31.

Purchon, R. D., 1977: The Biology of the Mollusca. 2nd Edition. Pergamon Press, N.Y. 596 pp. (Good general account of many aspects of molluscan biology, especially gastropod and bivalve habitation and feeding.)

Raven, C. P., 1958: Morphogenesis: The Analysis of Molluscan Development. Pergamon Press, N.Y. (An account of molluscan embryogeny.)

Runnegar, B., and Pojeta, J., 1974: Molluscan phylogeny: the paleontological viewpoint. Science, 186:311–317.

Salvini-Plawan, L. V., 1972: Zur Morphologie und Phylogenie der Mollusken. Z. wiss. Zool., 184:205–394.

Solem, A., 1974: The Shell Makers: Introducing Mollusks. John Wiley and Sons, N.Y. 289 pp. (This little book is largely concerned with gastropods, especially land snails. A good complement to the volume by Yonge and Thompson.)

Stasek, C. R., 1972: The molluscan framework. In Florkin, M., and Scheer, B. T. (Eds.): Chemical Zoology. Vol. 3. Academic Press, N.Y. pp. 1–44.

Trueman, E. R., 1975: The Locomotion of Soft-Bodied Animals. American Elsevier Publishing Company, N.Y. 200 pp.

Trueman, E. R., 1983: Locomotion in molluscs. In Wilbur, K. M. (Ed.): The Mollusca, Vol. 4, Pt. 1. Academic Press, N.Y. pp. 155–198.

Vagvolgyi, J., 1967: On the origin of mollusks, the coelom, and coelomic segmentation. Syst. Zool., 16:153–168.

Wilbur, K. M., and Saleuddin, A. S. M., 1983: Shell formation. In Wilbur, K. M. (Ed.): The Mollusca, Vol. 4, Pt. 1. Academic Press, N.Y. pp. 236–288.

Wilbur, K. M. (Ed.), 1983–1985: The Mollusca. Academic Press, N.Y. (Ten volumes covering the biochemistry, physiology, development, ecology, and evolution of mollusks.)

Yochelson, E. L., 1978: An alternative approach to the interpretation of the phylogeny of ancient mollusks. Malacologia, 17(2):165–191.

Yochelson, E. L., 1979: Early radiation of Mollusca and mollusc-like groups. In House, M. R. (Ed.): The Origins of Major Invertebrate Groups. Syst. Assoc. Spec. Vol. No. 12. Academic Press, London. pp. 323–358.

Yonge, C. M., and Thompson, T. E., 1976: Living Marine Molluscs. Collins, London. 288 pages. (A brief but excellent general treatment of marine mollusks, with special emphasis on the British fauna. Solem's book provides good coverage of the land snails.)

Gastropods

Boss, K. J., 1974: Oblomovism in the Mollusca. Trans. Am. Micros. Soc., 93:460–481.

Branch, G. M., 1981: The biology of limpets: physical factors, energy flow and ecological interactions. Oceanogr. Mar. Biol., Ann. Rev., 19:235–380.

Brandley, B. K., 1984: Aspects of the ecology and physiology of Elysia cf. furvacauda. Bull. Mar. Sci., 34(2):207–219.

Carriker, M. R., 1981: Shell penetration and feeding by natacean and and muricacean predatory gastropods: a synthesis. Malacologia, 20(2):403–422.

Cather, J. N., and Tompa, A. S., 1972: The podocyst in pulmonate evolution. Malacol. Rev. 5:1–3.

Clark, K. B., Jensen, K. R., Stirts, H. M., and Fermin, C., 1981: Chloroplast symbiosis in a nonelysiid mollusc, Costasiella lilianae Marcus: effects of temperature, light intensity, and starvation on carbon fixation rate. Biol. Bull., 160:43–54.

Conklin, E. J., and Mariscal, R. N., 1977: Feeding behavior, ceras structure, and nematocyst storage in the aeolid nudibranch, Spurilla neapolitana. Bull. Mar. Sci., 27(4):658–667.

Connor, V. M., and Quinn, J. F., 1984: Stimulation of food species growth by limpet mucus. Science, 225:843–844.

Cook, A., 1979: Homing in the Gastropoda. Malacologia, 18:315–318.

Croll, R. P., 1983: Gastropod chemoreception. Biol. Rev., 58:293–319.

Day, R. M., and Harris, L. G., 1978: Selection and turnover of coelenterate nematocysts in some aeolid nudibranchs. Veliger, 21(1):104–109.

Denny, M. W., 1981: A quantitative model for the adhesive locomotion of the terrestrial slug Ariolimax columbianus. J. Exp. Biol., 91:195–218.

Fretter, V., and Graham, A., 1962: British Prosobranch Molluscs. Ray Society, London.

Fretter, V., and Peake, J., 1975 (Eds.): Pulmonates. Vol. 1: Functional Anatomy and Physiology. Academic Press, London. 417 pp.

Gainey, L. F., 1976: Locomotion in the Gastropoda: Functional morphology of the foot in Neritina reclivata and Thais rustica. Malacologia, 15(2):411–431.

Gallagher, S. B., and Reid, G. K., 1974: Reproductive behavior and early development in Littorina scabra angulifera and Littorina irrorata in the Tampa Bay region of Florida. Malacol. Review, 7:105–125.

Garstang, W., 1928: Origin and evolution of larval forms. Report British Assoc., section D, p. 77.

Ghiselin, M. T., 1966: The adaptive significance of gastropod torsion. Evolution, 20:337–348.

Greenwood, P. G., and Mariscal, R. N., 1984: Immature nematocyst incorporation by the aeolid nudibranch Spurilla neapolitana. Mar. Biol., 80:35–38.

Gurin, S., and Carr, W. E., 1971: Chemoreception in Nassarius obsoletus: the role of specific stimulatory proteins. Science, 174:293–295.

Hamilton, P. V., 1977: Daily movements and visual location of plant stems by Littorina irrorata. Mar. Behav. Physiol. 4:293–304.

Haszprunar, G., 1985: The fine morphology of the osphradial sense organs of the Mollusca. I. Gastropoda. Prosobranchia. Phil. Trans. Roy. Soc. London B, 307:457–496.

Hickman, C. S., 1983: Radular patterns, systematics, diversity, and ecology of deep-sea limpets. Veliger, 26(2):73–92.

Hinde, R., 1983: Retention of algal chloroplasts by molluscs. In Goff, L. J. (Ed.): Algal symbiosis. Cambridge University Press, Cambridge. p. 97.

Hoagland, K. E., 1978: Protandry and the evolution of environmentally mediated sex change: a study of the Mollusca. Malacologia, 17(2):365–391.

Hughes, R. N., and Lewis, A. H., 1974: On the spatial distribution, feeding and reproduction of the vermetid gastropod Dendropoma maximum. J. Zool. (London), 172(4):531–548.

James, M. J., 1984: Comparative morphology of radular teeth in Conus: observations with scanning electron microscopy. Jour. Moll. Studies, 46:116–128.

Jensen, K. R., 1980: A review of sacoglossan diets, with comparative notes on radular and buccal anatomy. Malacol. Rev., *13*:55–77.

Jeppesen, L. L., 1976: The control of mating behaviour in *Helix pomatia*. Anim. Behav., *24*:275–290.

Kaelker, H., and Schmekel, L., 1976: Structure and function of the cnidosac of the Aeolidoidea. Zoomorphologie, *86*(1):41–60.

Kandel, E. R., 1979: Behavioral Biology of *Aplysia*. W. H. Freeman and Co., San Franscisco. 463 pp.

Kennedy, D., Selverton, A. I., and Remler, M. P., 1969: Analysis of restricted neural networks. Science, *164*:1488–1496.

Kohn, A. J., 1983: Feeding biology of gastropods. *In* Wilbur, K. M. (Ed.): The Mollusca, Vol. 5, Pt. 2. Academic Press, N.Y. pp. 1–63.

Lind, H., 1976: Causal and functional organization of the mating behaviour sequence in *Helix pomatia*. Behavior, *59*(¾):162–202.

Lindberg, D. R., and Dwyer, K. R., 1982: The topography, formation and role of the home depression of *Collisella scabra*. Veliger, *25*(3):229–233.

Linsley, R. M., 1978: Shell form and the evolution of gastropods. American Scientist, *66*(4):432–441.

Miller, S. L., 1974: Adaptive design of locomotion and foot form in prosobranch gastropods. J. Exp. Mar. Biol. Ecol., *14*:99–156.

Miller, S. L., 1974: The classification, taxonomic distribution, and evolution of locomotor types among prosobranch gastropods. Proc. Malac. Soc. London, *41*:233–272.

Pender, W. F., 1973: The origin and evolution of the Neogastropoda. Malacologia, *12*(2):295–338.

Pennington, J. T., and Chia, F.-S., 1985: Gastropod torsion: A test of Garstang's hypothesis. Biol. Bull., *169*:391–396.

Pratt, D. M., 1974: Attraction to prey and stimulus to attack in the predatory gastropod *Urosalpinx cinerea*. Mar. Biol., *27*(1):37–45.

Price, C. H., 1977: Morphology and histology of the central nervous system and neurosecretory cells in *Melampus bidentatus*. Say. Trans. Am. Microsc. Soc., *96*(3):295–312.

Rollo, C. D., and Wellington, W. G., 1981: Environmental orientation by terrestrial Mollusca with special reference to homing behavior. Can. Jour. Zool., *59*:225–239.

Rudman, W. B., 1981: The anatomy and biology of alcyonarian-feeding aeolid opisthobranch molluscs and their development of symbiosis with zooxanthellae. Zool. Jour. Linn. Soc., *72*:219–262.

Runham, N. W., and Hunter, P. J., 1970: Terrestrial Slugs. Hutchinson University Library, London. 184 pp. (A general biology of pulmonate slugs.)

Smith, C. R., 1977: Chemical recognition of prey by the gastropod *Epitonium tinetum*. Veliger, *19*(3):331–340.

Solem, A., 1978: A theory of land snail biogeographic patterns through time. *In* Pathways in Malacology. 6th European Malacological Congress, Amsterdam.

Solem, A., and Yochelson, E. L., 1979: North American Paleozoic land snails, with a summary of other Paleozoic non-marine snails. U.S. Geol. Surv., Prof. Paper 1072.

Spight, T. M., 1975: On a snail's chances of becoming a year old. Oikos, *26*(1):9–14.

Steneck, R. S., and Watling, L., 1982: Feeding capabilities and limitations of herbivorous molluscs: a functional approach. Mar. Biol., *68*:299–319.

Taylor, J. D., Morris, N. J., and Taylor, C. N., 1980: Food specialization and the evolution of predatory prosobranch gastropods. Paleontology, *23*(2):375–409.

Thiriot-Quievreux, C., 1973: Heteropoda. Oceanography Mar. Biol. Ann. Rev., *11*:237–261.

Thompson, T. E., 1976: Biology of Opisthobranch Mollusca, Vol. I. The Ray Society, London, 206 pp.

Thompson, T. E., and Brown, G. H., 1984: Biology of Opisthobranch Mollusca. Vol. II. The Ray Society Monographs, No. 156. The Ray Society, London. 229 pp.

Todd, C. D., 1981: The ecology of nudibranch molluscs. Oceanogr. Mar. Biol. Ann. Rev., *19*:141–234.

Tompa, A. S., 1980: Studies on the reproductive biology of gastropods: Part III. Calcium provision and the evolution of terrestrial eggs among gastropods. J. Conch., *30*:145–154.

Underwood, A. J., 1979: The ecology of intertidal gastropods. Oceangr. Mar. Biol., Ann. Rev., *16*:111–210.

Willows, A. O. D., 1971: Giant brain cells in mollusks. Sci. Am., *224*:69–76.

Bivalves

Ansell, A. D., and Nair, N. B., 1969: A comparison of bivalve boring mechanisms by mechanical means. Am. Zool, *9*:857–868.

Bayne, B. L., 1976: Marine Mussels: Their Ecology and Physiology. Cambridge University Press, N.Y. 506 pp.

Booth, C. E., and Mangum, C. P., 1978: Oxygen uptake and transport in the lamellibranch mollusc *Modiolus demissus*. Physiol. Zool., *51*(1):17–32.

Carpenter, E. J., and Culliney, J. L., 1975: Nitrogen fixation in marine shipworms. Science, *187*:551–552.

Carter, J. G., 1980: Environmental and biological controls of bivalve shell mineralogy and microstructure. *In* Rhoads, D. C., and Lutz, R. A. (Eds.), Skeletal Growth of Aquatic Organisms. Plenum Press, N.Y. pp. 69–113.

Deaton, L. E., and Mangum, C. P., 1976: The function of hemoglobin in the arcid clam *Noetia ponderosa*. II. Oxygen uptake and storage. Comp. Biochem. Physiol., *53A*:181–186.

Freadman, M. A., and Mangum, C. P., 1976: The function of hemoglobin in the arcid clam *Noetia ponderosa*. I. Oxygenation *in vitro* and *in vivo*. Comp. Biochem. Physiol., *53A*:173–179.

Galtsoff, P. S., 1964: The American Oyster. U.S. Fisheries Bull., *64*:1–480.

Goreau, T. F., Goreau, N. I., and Yonge, C. M., 1973: On the utilization of photosynthetic products from zooxanthellae and of a dissolved amino acid in *Tridacna maxima*. J. Zool., *169*:417–454.

Haderlie, E. C., 1981: Growth rates of *Penitella penita*, *Chaceia ovoidea* and other rock boring marine bivalves in Monterey Bay, California, U.S.A. Veliger, *24*(2):109–114.

Heard, W. H., 1977: Reproduction of fingernail clams (Sphaeriidae: *Sphaerium* and *Musculium*. Malacol., *16*(2):421–456.

Jones, D. S., 1983: Sclerochronology: reading the record of the molluscan shell. American Scientist, *71*:384–391.

Jorgensen, C. B., 1966: Biology of Suspension Feeding. Pergamon Press, N.Y.

Jorgensen, C. B., 1974: On gill function in the mussel *Mytilis edulis*. Ophelia, *13*(½):187–232.

Kat, P. W., 1984: Parasitism and the Unionacea. Biol. Rev., *59*:189–207.

Kennedy, W. J., Taylor, J. D., and Hall, A., 1969: Environmental and biological controls on bivalve shell mineralogy. Biol. Rev., *44*:499–530.

Knudsen, J., 1979: Deep-sea bivalves. *In* van der Spoel, S., et al. Pathways in Malacology. Bohn, Scheltema and Holkema, Utrecht. pp. 195–224.

Kristensen, J. H., 1972: Structure and function of crystalline styles in bivalves. Ophelia, *10*(1):91–108.

Mason, J., 1972: Cultivation of the European mussel, *Mytilus edulis*. Oceanogr. Mar. Biol. Ann. Rev., *10*:437–460.

Mathers, N. F., 1973: Carbohydrate digestion in *Ostrea edulis*. Proc. Malacal. Soc. London, *40*(5):359–367.

McKee, P. M., and Mackie, G. L., 1981: Life history adaptations of the fingernail clams *Sphaerium occidentale* and *Musculium secris* to ephemeral habits. Can. J. Zool., *59*:2219–2229.

McKee, P. M., and Mackie, G. L., 1983: Respiratory adaptions of the fingernail clams *Sphaerium occidentale* and *Musculium securis* to ephemeral habitats. Can. Jour. of Fish. and Aquat. Sci., *40*(6):783–791.

Morse, M. P., Meyhöffer, E., Otto, J. J., and Kuzirian, A. M., 1986: Hemocyanin respiratory pigment in bivalve mollusks. Science, *231*:1302–1304.

Morton, B., 1970: The tidal rhythm and rhythm of feeding and digestion in *Cardium edule*. J. Mar. Biol. Assoc. U.K., *50*:499–512.

Morton, B., 1978: The diurnal rhythm and the processes of feeding and digestion in *Tridacna crocea*. J. Zool. (London), *185*:371–387.

Morton, B., 1978: Feeding and digestion in shipworms. Oceanogr. Mar. Biol. Ann. Rev., *16*:107–144.

Morton, B., 1981: Prey capture in the carnivorous septibranch *Poromya granulata*. Sarsia, *66*:241–256.

Morton, B., 1983: Feeding and digestion in Bivalvia. *In* Wilbur, K. M. (Ed.): The Mollusca, Vol. 5, Pt. 2. Academic Press, N.Y. pp. 65–147.

Newell, N. D., 1969: Classification of Bivalvia. *In* Moore, R. C. (Ed.): Treatise on Invertebrate Paleontology, Vol. N, Pt. 1, Mollusca 6, Bivalvia, pp. N205–N224.

Owen, G., 1974: Feeding and digestion in the Bivalvia. Adv. Comp. Physiol. Biochem., *5*:1–35.

Palmer, R. E., 1979: A histological and histochemical study of digestion in the bivalve *Arctica islandica*. Biol. Bull., *156*:115–129.

Pennington, B. J., and Currey, J. D., 1984: A mathematical model for the mechanical properties of scallop shells. J. Zool., *202*(2):239–264.

Pirie, B. J. S., and George, S. G., 1979: Ultrastructure of the heart and excretory system of *Mytilus edulis*. Jour. Mar. Biol. Assoc., U.K., *59*:819–829.

Pojeta, J. and Runnegar, B., 1974: *Fordilla troyensis* and the early history of pelecypod mollusks. American Scientist, *62*:706–711.

Reed-Miller, C., and Greenberg, M. J., 1982: The ciliary junctions of scallop gills: the effects of cytochalasins and concanavalin A. Biol. Bull., *163*:225–239.

Reid, R. G. B., and Bernard, F. R., 1980: Gutless Bivalves. Science, *208*:609–610.

Reid, R. G. B., and Reid, A. M., 1974: The carnivorous habit of members of the septibranch genus *Cuspidaria*. Sarsia, *56*:47–56.

Rosen, M. D., Stasek, C. R., and Hermans, C. D., 1978: The ultrastructure and evolutionary significance of the cerebral ocelli of *Mytilus edulis*, the bay mussel. Veliger, *21*(1):10–18.

Silvester, N. R., and Sleigh, M. A., 1984: Hydrodynamic aspects of particle capture by *Mytilus*. Jour. Mar. Biol. Assoc. U.K., *64*:859–879.

Stanley, S. M., 1968: Post-Paleozoic adaptive radiation of infaunal bivalve molluscs—a consequence of mantle fusion and siphon formation. J. Paleontol., *42*(1):214–229.

Stanley, S. M., 1970: Relation of Shell Form to Life Habits of the Bivalvia (Mollusca). Geol. Soc. Am., Mem. 125, 296 pp.

Stanley, S. M., 1972: Functional morphology and evolution of byssally attached bivalve mollusks. J. Paleontol., *46*(2):165–212.

Stanley, S. M., 1975: Why clams have the shape they have: an experimental analysis of burrowing. Paleobiology, *1*(1):48.

Stenzel, H. B., 1971: Oysters. *In* Treatise on Invertebrate Paleontology, Part N, Vol. 3. pp. 953–1224.

Taylor, J. D., 1973: The structural evolution of the bivalve shell. Palaeontology, *16*(3):519–534.

Trench, R. K., Wethey, D. S., and Porter, J. W., 1981: Observations on the symbiosis with zooxanthellae among the Tridacnidae. Biol. Bull, *161*:180–198.

Trueman, E. R., 1966: Bivalve mollusks: fluid dynamics of burrowing. Science, *152*:523–525.

Turner, R. D., 1973: Wood-boring bivalves, opportunistic species in the deep sea. Science. *180*:1377–1379.

Waite, J. H., 1983: Adhesion in byssally attached bivalves. Biol. Rev., *58*:209–231.

Waterbury, J. B., Calloway, C. B., and Turner, R. D., 1983: A cellulolytic nitrogen-fixing bacterium cultured from the gland of Deshayes in shipworms. Science, *221*:1401–1403.

Wood, E. M., 1974: Some mechanisms involved in host recognition and attachment of the glochidium larva of *Anodonta cygnea*. J. Zool., *173*:15–30.

Yonge, C. M., 1941: The protobranchiate Mollusca: a functional interpretation of their structure and evolution. Phil. Trans. R. Soc. Lond., B, *230*:79–147.

Yonge, C. M., 1957: Mantle fusion in the Lamellibranchia. Staz. Zool. Napoli, *29*:151–171.

Yonge, C. M., 1975: Giant clams. Sci. Am. *232*(4):96–105.

Yonge, C. M., 1977: Form and evolution in the Anomiacea—*Pododesmus, Anomia, Patro, Enigmonia* (Anomiidae): *Placunanomia, Placuna* (Placunidae Fam. Nov.). Phil. Trans. R. Soc. Lond., B, *276*:453–527.

Yonge, C. M., 1979: Cementation in the bivalves. *In* van de Spoel, S., et al (Eds.): Pathways in Malacology. Bohn, Scheletema and Holkema, Utrecht. pp. 83–106.

Cephalopods

Bidder, A. M., 1950: Digestive mechanisms of European squids. Q. J. Micr. Sci., new series, *91*(1):1–43.

Bone, Q., Pulsford, A., and Chubb, A. D., 1981: Squid mantle muscle. J. Marine Biol. Assoc. U.K., *61*:327–342.

Boyle, P. R. (Ed.), 1983: Cephalopod Life Cycles. Vol. 2. Species Accounts. Academic Press, N.Y. 475 pp.

Clarke, M. R., 1966: A review of the systematics and ecology of oceanic squids. Adv. Mar. Biol, *4*:91–300.

Clarke, M. R., Denton, E. J., and Gilpin-Brown, J. B., 1979: On the use of ammonium for buoyancy in squids. J. Mar. Biol. Assoc. U.K., *59*:259–276.

Clarke, M. R., and Macleod, N., 1982: Cephalopod remains from the stomachs of sperm whales caught in the Tasman Sea. Mem. Nat. Hist. Mus. Victoria, *34*(1):25–42.

Clarke, M. R., Macleod, N., and Paliza, O., 1976: Cephalopod remains from the stomach of sperm whales caught off Peru and Chile. J. Zool. (London), *180*:477–493.

Cloney, R. A., and Brocco, S. L., 1983: Chromatophore organs, reflector cells, iridocytes and leucophores in cephalopods. Amer. Zool., *23*:581–592.

Denton, E. J., 1974: On buoyancy and the lives of modern and fossil cephalopods. Proc. R. Soc. London, B Bio. Sci., *185*(1080):273–299.

Denton, E. J., and Gilpin-Brown, J. B., 1973: Flotation mechanisms in modern and fossil cephalopods. Adv. Mar. Biol., *11*:197–264.

Donovan, D. T., 1964: Cephalopod phylogeny and classification. Biol. Rev., *39*(3):259–287.

Ghiretti-Magaldi, A., Ghiretti, F., and Salvato, B., 1977: The evolution of haemocyanin. *In* Nixon, M., and Messenger, J. B. (Eds.): The Biology of Cephalopods. Symp. Zool. Soc. London, *38*:513–523, Academic Press, London.

Gilpin-Brown, J. B., 1972: Buoyancy mechanisms of cephalopods in relation to pressure. Symp. Soc. Exp. Biol., *26*:251–259. (Good review, with bibliography.)

Gosline, J. M., and DeMont, M. E., 1985: Jet-propelled swimming in squids. Sci. Amer., *252*(1):96–103.

Hartwick, B., Tulloch, L., and MacDonald, S., 1981: Feeding and growth of *Octopus dofleini*. Veliger, *24*(2):129–138.

Haven, N., 1972: The ecology and behavior of *Nautilus pompilius* in the Philippines. Veliger, *15*(2):75–80.

Herring, P. J., 1977: Luminescence in cephalopods and fish. *In* Nixon, M., and Messinger, J. B. (eds): The Biology of Cephalopods. Symp. Zool. Soc. Lond., *38*. Academic Press, London. pp. 127–159.

Holme, N. A., 1974: The biology of *Loligo forbesi* Steenstrup in the Plymouth area. J. Mar. Biol. Assoc. U.K., *54*:481–503.

Lehmann, U., 1981: The Ammonites. Cambridge University Press, Cambridge, 246 pp.

Macy, W. K., 1982: Feeding patterns of the long-finned squid, *Loligo pealei*, in New England waters. Biol. Bull., *162*:28–38.

Moynihan, M., 1985: Communication and Noncommunication by Cephalopods. Indiana University Press, Bloomington, Ind. 160 pp.

Moynihan, M., and Rodaniche, A. F., 1982: The Behavior and Natural History of the Caribbean Reef Squid *(Sepioteuthis sepioidea)*. Paul Parey, Publ. Hamburg.

Nixon, M., and Messenger, J. B. (Eds.), 1977: The Biology of Cephalopods. Symp. Zool. Soc. Lond., *38*. Academic Press, London. 616 pp. (Papers presented at a symposium on cephalopods.)

Packard, A., 1972: Cephalopods and fish: the limits of convergence. Bio. Rev., *47*:241–307.

Roper, C. F. E., and Boss, K. J., 1982: The giant squid. Sci. Amer., *246*(4):96–105.

Roper, C. F. E., Lu, C. C., and Hochberg, F. G. (Eds.), 1983: Proceedings of the workshop on the biology and resource potential of cephalopods, Melbourne, Australia, 1981. Mem. Nat. Mus. Victoria, 44:311.

Roper, C. F. E., and Young, R. E., 1975: Vertical distribution of pelagic cephalopods. Smithsonian Contrib. Zool., *209*:1–51.

Saunders, W. B., 1983: Natural rates of growth and longevity of *Nautilus belauensis*. Paleobiology, *9*(3):280–288.

Saunders, W. B., 1984: The role and status of *Nautilus* in its natural habitat: evidence from deep-water remote camera photosequences. Paleobiology, *10*(4):469–486.

Ward, P., 1983: The extinction of the Ammonites. Sci. Amer., *249*(4):136–147.

Ward, P., Greenwald, L., and Greenwald, O. E., 1980: The buoyancy of the chambered *Nautilus*. Sci. Amer., *243*:190–203.

Wells, M. J., 1978: Octopus: Physiology and Behaviour of an Advanced Invertebrate. Chapman and Hall, London. 417 pp.

Wells, M. J., 1979: The world of a mollusc; brain and behaviour in *Octopus vulgaris*. *In* van der Spoel, S., et al: Pathways in Malacology. Bohn, Scheltema and Holkema, Utrecht. pp. 139–156.

Wodinsky, J., 1977: Hormonal inhibition of feeding and death in *Octopus:* control by optic gland secretion. Science, *198*:948–951.

Yarnell, J. L., 1969: Aspects of the behaviour of *Octopus cyanea* Gray. Animal Behav., *17*(4):747–754.

Yochelson, E. L., Flower, R. H., and Webers, G. F., 1973: The bearing of the new late Cambrian monoplacophoran genus *Knightoconus* upon the origin of the Cephalopoda. Lethaia, *6*:275–310.

Young, J. Z., 1972: The Anatomy of the Nervous System of *Octopus vulgaris*. Oxford University Press, N.Y. 690 pp.

Young, R. E., 1975: Function of the dimorphic eyes in the midwater squid *Histioteuthis dofleini*. Pac. Sci., *29*(2):211–218.

Young, R. E., and Roper, C. F. E., 1976: Bioluminescent countershading in midwater animals: evidence from living squid. Science, *191*:1046–1048.

Young, R. E., and Roper, C. F. E., 1977: Intensity regulation of bioluminescence during countershading in living midwater animals. Fishery Bull. 75(2):239–252.

Aplacophorans, Polyplacophorans, Monoplacophorans, Scaphopods

Bilyard, G. R., 1974: The feeding habits and ecology of *Dentalium entale stimpsoni*. Veliger, *17*(2):126–138.

Boyle, P. R., 1972: The aesthetes of chitons. I. Role in the light response of whole animals. Mar. Behav. Physiol., *1*:171–184.

Boyle, P. R., 1974: The aesthetes of chitons. II. Fine structure in *Lepidochitona cinereus*. Cell Tiss. Res., *153*:383–398.

Boyle, P. R., 1977: The physiology and behavior of chitons. Oceanog. Mar. Biol., Ann. Rev., *15*:461–509.

Boyle, P. R. (Ed.), 1983: Cephalopod Life Cycles. Vol. I. Species Accounts. Academic Press, London. 475 pp.

Emerson, W. K., 1962: A classification of the scaphopod mollusks. J. Paleontol., *36*(3):461–482.

Fischer, v.-F. P., 1978: Photoreceptor cells in chiton esthetes. Spixiana, *1*(3):209–213.

Gainey, L. F., 1972: The use of the foot and captacula in the feeding of *Dentalium*. Veliger, *15*(1):29–34.

Langer, P., 1983: Diet analysis for three subtidal coexisting chitons from the northwestern Atlantic. Veliger, *25*(4):370–377.

Lemche, H., and Wingstrand, K. G., 1959: The anatomy of *Neopilina galatheae*. Galathea Rep., *3*:9–71.

Mook, D., 1983: Homing in the West Indian chiton *Acanthopleura granulata*. Veliger, *26*(2):101–105.

Scheltema, A. H., 1978: Position of the class Aplacophora in the phylum Mollusca. Malacologia, *17*(1):99–109.

Smith, S. Y., 1975: Temporal and spatial activity patterns of the intertidal chiton *Nopalia muscosa*. Veliger, *18*(Supplement):57–62.

Trueman, E. R., 1968: The burrowing process of *Dentalium*. J. Zool. London, *154*:19–27.

Wingstrand, K. G., 1985: On the anatomy and relationships of recent Monoplacophora. Galathea Report, *16*:7–94.

Yonge, C. M., 1939: On the mantle cavity and its contained organs in the Loricata. Q. J. Micr. Sci., *323*:367–390.

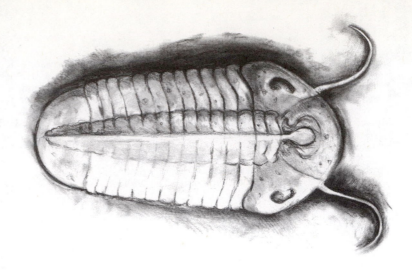

12

Introduction to the Arthropods; the Trilobites

The Arthropods

Arthropods are a vast assemblage of animals. At least three quarters of a million species have been described; this is more than three times the number of all other animal species combined (see figure inside cover). The tremendous adaptive diversity of arthropods has enabled them to survive in virtually every habitat; they are perhaps the most successful of all the invaders of the terrestrial habitat.

Arthropods represent the culmination of evolutionary development in the protostomes. They arose either from primitive stocks of polychaetes or from ancestors common to both, and the relationship between arthropods and annelids is displayed in several ways.

1 Arthropods, like annelids, are metameric (Fig. 12–1A). Metamerism is evident in the embryonic development of all arthropods and is a conspicuous feature of many adults, especially the more primitive species. Within many arthropod groups there has been a tendency for metamerism to become reduced. In such forms as mites, for example, it has almost disappeared. Loss of meta-

merism has occurred in three ways. Segments have become lost; segments have become fused together; and segmental structures, such as appendages, have become structurally and functionally differentiated from their counterparts on other segments. Different structures having the same segmental origin are said to be serially homologous. Thus, the second antennae of a crab are serially homologous to the chelipeds (claws), for both evolved from originally similar segmental appendages.

2 In the primitive condition each arthropod segment bears a pair of appendages (Fig. 12–1B). This same condition is displayed by the polychaetes, in which each metamere bears a pair of parapodia. However, the homology between parapodia and arthropod appendages is uncertain.

3 The nervous systems in both groups are constructed on the same basic plan. In both a dorsal, anterior brain is followed by a ventral nerve cord containing ganglionic swellings in each segment (Fig. 12–1A).

4 The embryonic development of a few arthropods still displays holoblastic determinate cleav-

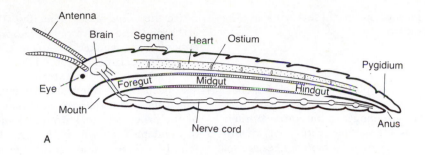

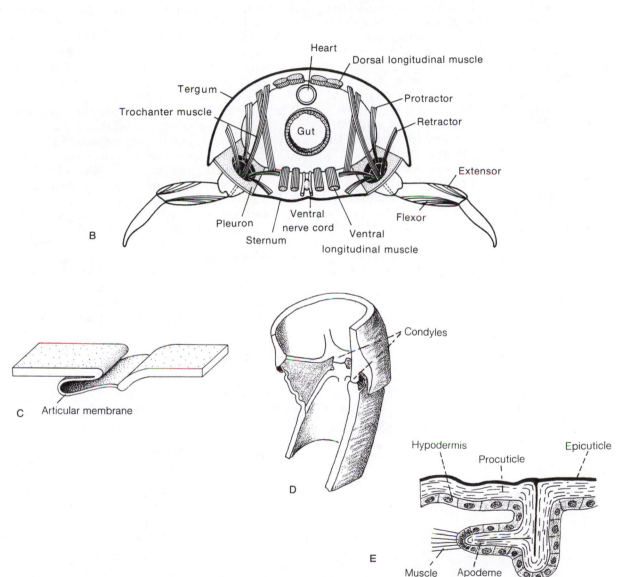

Figure 12–1 *A* and *B*, Structure of a generalized arthropod. *A*, Sagittal section. *B*, Cross section. *C*, Intersegmental articulation. Note articular membrane folded beneath segmental plate. (After Weber from Vandel.) *D*, Appendicular articulation showing condyles. (After Weber from Vandel.) *E*, An apodeme. (After Janet from Vandel.)

age, with the mesoderm in these forms arising from the *4d* blastomere.

Exoskeleton

Although arthropods display these annelidan characteristics, they have undergone a great many profound and distinctive changes in the course of their evolution. The distinguishing feature of arthropods, and one to which many other changes are related, is the chitinous exoskeleton, or cuticle, that covers the entire body (Fig. 12–2). Movement is made possible by the division of the cuticle into separate plates. Primitively, these plates are confined to segments, and the plate of one segment is connected to the plate of the adjoining segment by means of an articular membrane, a region in which the cuticle is very thin and flexible (Fig. 12–1C). Basically, the cuticle of each segment is divided into four primary plates—a dorsal tergum, two lateral pleura, and a ventral sternum (Fig. 12–1B). This pattern has frequently disappeared because of either secondary fusion or subdivision.

The cuticular skeleton of the appendages, like that of the body, has been divided into tubelike segments, or sections, connected to one another by articular membranes, thus creating a joint at each junction. Such joints enable the segments of the appendages, as well as those of the body, to move (hence the name of the phylum, *Arthropoda*—jointed feet). In most arthropods the articular membrane between body segments is folded be-

neath the anterior segment (Fig. 12–1C). In many arthropods the additional development of articular condyles and sockets is suggestive of vertebrate skeletal structures (Fig. 12–1D).

In addition to the external skeleton, there has also been the development of what is sometimes called the endoskeleton. This may be an infolding of the procuticle that produces inner projections, or apodemes, to which the muscles are attached (Fig. 12–1E), or it may involve the sclerotization of internal tissue, forming free plates for muscle attachment within the body.

The arthropod skeleton is secreted by the underlying layer of integumentary epithelial cells known as the hypodermis. It is composed of a thin, outer epicuticle and a much thicker procuticle (Fig. 12–3). The epicuticle is composed of proteins and, in many arthropods, wax. The fully developed procuticle consists of an outer exocuticle and an inner endocuticle. Both layers are composed of chitin and protein bound together to form a complex glycoprotein, but the exocuticle in addition has been tanned; i.e., with the participation of phenols, its molecular structure has been further stabilized by the formation of additional cross linkages. Exocuticle is absent at joints and along lines where the skeleton will rupture during molting. In many arthropods the procuticle is also impregnated with mineral salts. This is particularly true for the Crustacea, in which calcium carbonate and calcium phosphate deposition takes place in the procuticle. Where the exoskeleton lacks a waxy epicuticle and

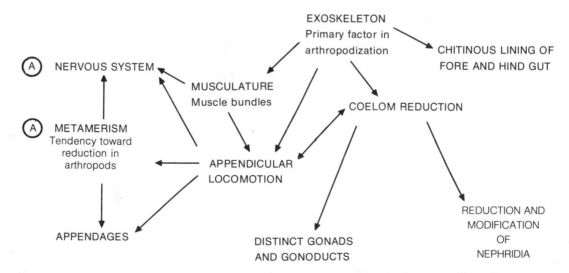

Figure 12–2 Interrelationship of changes resulting from the evolution of the arthropod condition (arthropodization). Ⓐ are annelidan features retained by arthropods.

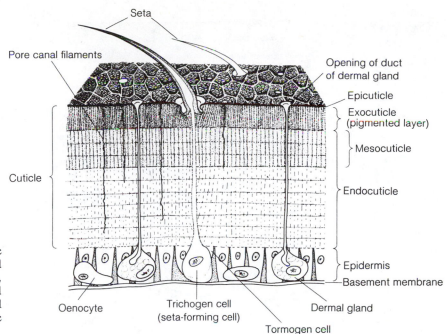

Seta

Pore canal filaments

Opening of duct
of dermal gland

Epicuticle

Exocuticle
(pigmented layer)

Mesocuticle

Cuticle

Endocuticle

Epidermis

Basement membrane

Oenocyte

Trichogen cell
(seta-forming cell)

Dermal gland

Tormogen cell
(socket-forming cell)

Figure 12–3 Diagrammatic section through the arthropod integument. (From Hackman, R. H., 1971: *In* Florkin, M., and Scheer, B. T. (Eds.): Chemical Zoology. Vol. 6. Academic Press, N.Y.)

is thin, it is a relatively permeable covering and allows the passage of gases and water. The cuticle is generally penetrated by fine pore canals, which function as ducts for the passage of secretions of underlying gland cells.

The arthropod cuticle is not restricted entirely to the exterior of the body. The hypodermis develops from the embryonic surface ectoderm, and all infoldings of the original layer, such as the fore- and hindgut, which develop from the stomodeum and the proctodeum, thus are lined with cuticle (Fig. 12–1*A*). Other such ectodermal derivatives include the tracheal (respiratory) tubes of insects, chilopods, diplopods, and some arachnids; the book lungs of scorpions and spiders; and parts of the reproductive systems of some groups. All of these internal cuticular linings are also shed at the time of molting.

The color of arthropods commonly results from the deposition of brown, yellow, orange, and red melanin pigments within the cuticle. However, iridescent greens, purples, and other colors result from fine striations of the epicuticle, which cause light refraction and give the appearance of color. Often, body coloration does not originate directly in the cuticle but instead is produced by subcuticular chromatophores or is caused by blood and tissue pigments, which are visible through a thin, transparent cuticle.

Despite its locomotor and supporting advantages, an external skeleton poses problems for a growing animal. The solution evolved by the arthropods has been the periodic shedding of the skeleton, a process called molting or ecdysis.

Before the old skeleton is shed, the epidermal layer (hypodermis) secretes proenzymes (inactive enzymes) at the base of the skeleton. The hypodermis now detaches from the skeleton, a process referred to as apolysis, and secretes a new epicuticle or at least its outer cuticulin layer (Fig. 12–4*B*). The proenzymes secreted earlier—chitinase and protease—become activated and digest the untanned endocuticle (Fig. 12–4*C*). The products of digestion are reabsorbed through the new cuticulin envelope. With the erosion of the old endocuticle, the hypodermis secretes new procuticle.

At the ultrastructural level there are only three components to the skeleton. The outer cuticulin envelope and chitin fibers, the latter forming most of the arthropod skeleton, are laid down by plasma membrane plaques of the hypodermis (Fig. 12–4*E*). These two skeletal components are separated by the third component, proteins, which are deposited by exocytosis (Locke, 1984).

At this point the animal is encased within both an old and a new skeleton (Fig. 12–4*D*). The old skeleton now splits along certain predetermined lines and the animal pulls out of the old encase-

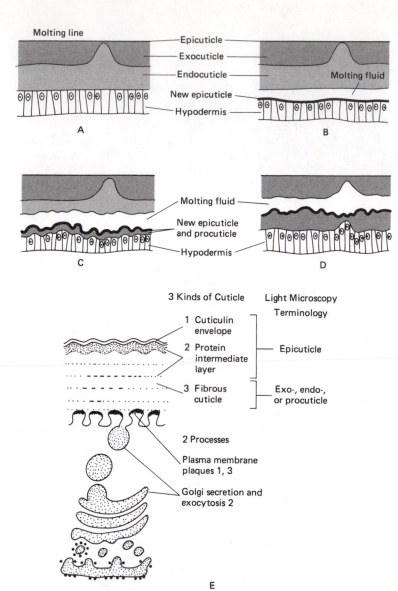

Figure 12–4 Molting in an arthropod. *A,* The fully formed exoskeleton and underlying hypodermis between molts. *B,* Separation of the hypodermis and secretion of molting fluid and the new epicuticle. *C,* Digestion of the old endocuticle and secretion of the new procuticle. *D,* The animal just before molting, encased within both new and old skeleton. *E,* Diagram of the cellular production of the three components of the insect cuticle. See text for explanation. (Modified from Locke, M., 1984: Epidermal cells (Arthropoda). *In* Bereiter-Hahn, J., Matoltsy, A. G., and Richards, K. S. (Eds.): Biology of the Integument. 1. Invertebrates. Springer-Verlag, Berlin. p. 514.

ment. The new skeleton is soft and commonly wrinkled and is stretched to accommodate the increased size of the animal. Stretching is brought about by blood pressure, facilitated by the uptake of water or air by the animal. Hardening of the cuticle results from tanning of the protein and from stretching.

Additional procuticle may be added following ecdysis, and in some arthropods, such as insects, additions are made to the epicuticle by secretions through the pore canals. The final surface of the epicuticle is often formed by a cement layer.

Sensory structures and muscle connections pose special problems for the molting process. Sensory structures, such as hairs, are laid down beneath the old skeleton, usually horizontally against the new skeleton. The dendrite may retain connection with the old hair until broken at ecdysis.

Muscles are attached to the exoskeleton by microtubules in specialized epidermal cells. The microtubules are anchored to an internal fold of the exoskeleton containing a fiber that runs all the way to the epicuticle (Fig. 12–5). The fiber is not digested during the molting process and maintains a connection between the old and new skeletons until severed at ecdysis.

The stages between molts are known as instars, and the length of the instars becomes longer as the

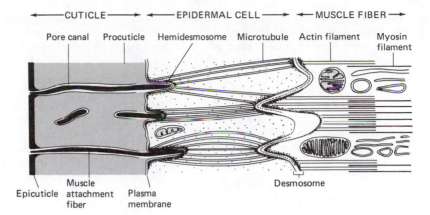

←CUTICLE→ ←EPIDERMAL CELL→ ←MUSCLE FIBER→

Pore canal   Procuticle   Hemidesmosome   Microtubule   Actin filament   Myosin filament

Epicuticle   Muscle attachment fiber   Plasma membrane   Desmosome

Figure 12–5 Diagram of the attachment of an insect muscle to the skeleton. Skeleton, epidermal cells and muscle fibers are not drawn to scale. (After Caveney, 1969, from Chapman, R. F., 1982: The Insects: Structure and Function. 3rd Edition Harvard University Press, Cambridge, Mass. p. 249.)

animal becomes older. Some arthropods, such as lobsters and most crabs, continue to molt throughout their life. Other arthropods, such as insects and spiders, have more or less fixed numbers of instars, the last being attained with sexual maturity.

Although an arthropod is measurably larger and heavier following ecdysis (Fig. 12–6), growth is actually continuous, as in most other animals. Proteins and other organic compounds are synthesized during the intermolt period, replacing fluids taken up following ecdysis.

Molting is under hormonal control. Ecdysone, secreted by certain endocrine glands (for example, the prothoracic glands in insects), is circulated by the bloodstream and acts directly on the epidermal

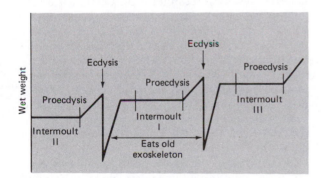

Figure 12–6 Weight changes during sequential stages in the molt cycle of a crab. Weight increases during proecdysis as a result of water uptake but is temporarily lost when old skeleton is shed. Not all crabs eat the old skeleton, but when the new skeleton is completely formed, the weight at intermolt would exceed that of the previous intermolt. In crustaceans much of the endocuticle is secreted after ecdysis. (From Highnam, K. C., and Hill, L., 1977: The Comparative Endocrinology of the Invertebrates. 2nd Edition. Edward Arnold [Publishers], Ltd., London. p. 210.)

cells. The production of ecdysone is in turn regulated by other hormones. Although most studied and best understood in insects and crustaceans, ecdysone controls molting in all arthropods. Molting physiology will be described in some detail for crustaceans (p. 654).

Movement and Musculature

As movement in arthropods has become restricted to flexion between plates and cylinders of the cuticle, a related change has taken place in the nature of the body musculature. In annelids the muscles take the form of longitudinal and circular sheathlike layers of fibers lying beneath the epidermis. Contraction of the two layers exerts force on the coelomic fluid, which then functions as a hydrostatic skeleton. In arthropods, on the other hand, these muscular cylinders have become broken up into striated muscle bundles, which are attached to the inner surface of the skeletal system (Figs. 12–1B and 12–2).

The muscles are attached to the inner side of the exoskeleton by specialized hypodermal cells (Fig. 12–5). Flexion and extension between plates are effected by the contraction of these muscles, with muscles and cuticle acting together as a lever system. This cofunctioning of the muscular system and skeletal system to bring about locomotion is similar to that in vertebrates. Extension, particularly of the appendages, is accomplished, in part or entirely, by an increase in blood pressure.

Arthropods employ as their chief means of locomotion jointed appendages, which act either as paddles in aquatic species or as legs in terrestrial groups. Our knowledge of arthropod locomotion, especially locomotion on land, results largely from the extensive studies of Manton (1978). In contrast to the parapodia of polychaetes, the locomotor appendages of arthropods tend to be more slender, longer, and located more ventrally. Despite the more ventral position of the legs, the body usually sags between the limbs (Fig. 12–7A). In the cycle of movement of a particular leg, the effective step, or stroke, during which the end of the leg is in contact with the substratum, is closer to the body than the recovery stroke, when the leg is lifted and swung forward (Fig. 12–7B to E). Among the several factors determining speed of movement, the length of

stride is of obvious importance, and stride length increases with the length of the leg. The problems of mechanical interference are decreased by a reduction in the number of legs to five, four, or three pairs and by differences in leg length and the relative placement of the leg tip. In arthropods that have retained a large number of legs, such as centipedes, the fields of movement of individual legs overlap those of other legs (Fig. 12–7B). For these animals the difference in proximity of the legs to the body during the effective and recovery strokes prevents mechanical interference.

The arthropodan gait involves a wave of leg movement, in which a posterior leg is put down just before or a little after the anterior leg is lifted. The movements of legs on opposite sides of the body alternate with one another; i.e., one limb of a pair is moving through its effective stroke while its mate is making a recovery stroke. Alternate leg movement tends to induce body undulation. This tendency is counteracted by increased body rigidity, such as the fused leg-bearing segments that form the thorax and cephalothorax of insects, some crustaceans, and arachnids.

An exoskeleton makes a highly efficient locomotor-skeletal system for animals that are only a few centimeters long. It provides protection in addition to support, and there is a large surface area for the attachment of muscles. The tubular construction resists bending. However, the wall will buckle on impact if there is insufficient skeletal material, just as you cannot bend a cylindrical can, but you can buckle one wall by kicking it. Thus, an exoskeleton imposes limits on the maximum size of arthropods. The weight of a large animal and the resulting stress produced when moving would require heavy skeletal walls. But when the arthropod molted, the soft, new skeleton would collapse under the animal's weight before hardening could occur. Significantly, the largest arthropods live in the sea, where the aquatic medium provides much more support than air.

In contrast to the condition in vertebrates, each arthropod muscle contains relatively few fibers and is innervated by only a small number of neurons. Many axon terminals are provided to one muscle fiber (Fig. 12–8), and one neuron may supply more than one muscle. Moreover, several types of motor neurons—phasic (fast) neurons, tonic (slow) neurons, and inhibitory neurons—may supply a single muscle. The terms phasic and tonic, or fast and slow, refer not to the speed of transmission but to the nature of the muscle response. The impulses of phasic motor neurons produce rapid but brief con-

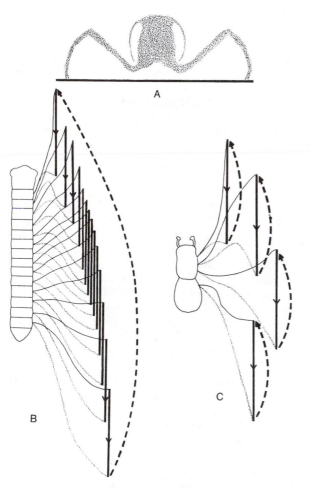

Figure 12–7 Arthropod locomotion: *A*, Stance of a crayfish. *B–C*, Leg movement in the gait of different anthropods. The effective stroke and the stride length are indicated by the heavy bars. Each leg is therefore drawn twice: at the beginning and at the end of the effective stroke. The recovery stroke is shown as a dashed line, and its position is only approximate. *B*, Strokes of the centipede *Scutigera. C*, Strokes of a spider. (All after Manton.)

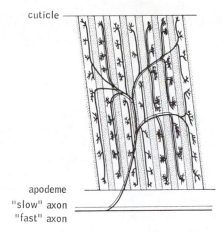

cuticle

apodeme
"slow" axon
"fast" axon

Figure 12–8 Dual innervation of an insect muscle. Fast, or phasic, neuron is depicted with a heavy line. Slow, or tonic, neuron is depicted with a fine line. Note that a slow axon does not innervate every fiber. (After Hoyle.)

tractions, which are often involved in rapid movements. The impulses of tonic motor neurons produce slow, powerful, prolonged contractions, which are involved in postural activities and slow movements. The impulses of inhibitory neurons block contractions.

The neuromuscular system may be further complicated, as in at least the crustaceans, by the differentiation of the muscle fibers into phasic and tonic types, each having a distinctive ultrastructure and physiology. Some muscles are entirely phasic, some are entirely tonic, and some are mixed. Phasic motor neurons innervate only phasic muscle fibers; tonic motor neurons innervate both phasic and tonic fibers or, in some instances, only tonic fibers.

In vertebrates graded responses depend in large part on the number of motor units contracting. Arthropod muscles are not organized as motor units, however, and graded responses depend on the type of muscle fibers contracting, the type of neuron fired, and the interaction of different types of neurons. For example, two different extensor muscles are innervated by the same motor fiber in a crayfish claw, but the two muscles function independently because each is innervated by separate inhibitory neurons.

The organization of arthropod ganglia is like that of annelids and mollusks described on page 321. Giant fiber systems are frequently well developed, and "command" systems have been identified. Arthropod neural networks and neuromuscular systems have been best studied in crustaceans. Interested students should consult Kennedy et al

(1969), Atwood (1973), and Atwood and Sandeman (1982).

Coelom and Blood-Vascular System

The well-developed, metameric coelom characteristic of the annelids has undergone drastic reduction in the arthropods and is represented by only the cavity of the gonads and in certain arthropods by the excretory organs. The change is probably related to the shift from a fluid internal skeleton to a solid external skeleton. The other spaces of the arthropod body do not constitute a true coelom but rather a hemocoel—that is, merely spaces in the tissue filled with blood.

Although derived from the annelids, the arthropod blood-vascular system is an open one. The dorsal vessel of annelids, which is contractile and the chief center for blood propulsion, may be homologous to the arthropod heart. The heart varies in position and length in different arthropodan groups, but in all of them the heart is a muscular tube perforated by pairs of lateral openings called ostia (Fig. 12–1A). Systole (contraction) results from the contraction of heart wall muscles, and diastole (expansion and filling) from suspensory elastic fibers and in some species from the contraction of suspensory muscles. The ostia enable the blood to flow into the heart during diastole from the large, surrounding sinus known as the pericardium. However, in arthropods the pericardium does not derive from the coelom, as in mollusks and vertebrates, but is a part of the hemocoel. After leaving the heart, blood is pumped out to the body tissues through arteries and is eventually dumped into sinuses (collectively the hemocoel) in which it bathes the tissues directly. The blood then returns by various routes to the pericardial sinus.

The blood of arthropods contains several types of cells and in some species the respiratory pigment hemocyanin or, less commonly, hemoglobin. As in mollusks, arthropod hemocyanin is a large molecule dissolved in the plasma; however, the structure of arthropod hemocyanin indicates that it evolved independently from that of mollusks (see reviews by Mangum, 1985; Linzen et al, 1985).

Arthropods possess two types of excretory organs. Malpighian tubules are blind tubules that lie within the hemocoel (blood-filled spaces) and open into the gut. Wastes pass from the blood into the tubules and then into the gut, where they are eliminated through the anus along with fecal material. Malpighian tubules are found in centipedes, millipedes, insects, and arachnids and represent an organ system that evolved independently within these groups or their arthropod ancestors.

The other type of arthropod excretory organ is paired blind saccules that open by ducts to the outside of the body adjacent to an appendage (Fig. 12–9A). The excretory organ takes the name of the appendage with which it is associated—coxal glands, maxillary glands, etc. Since the saccule is derived embryonically from the coelom, the tubule may represent an old metanephridium that originally drained the coelom. Typically, the saccule wall is composed of podocytes (Fig. 12–9B) and is the site of filtration from the surrounding blood. Parts of the tubule may be modified for selective reabsorption and secretion. Although such paired excretory organs may be derived from nephridia, no living arthropod has more than a few such saccules.

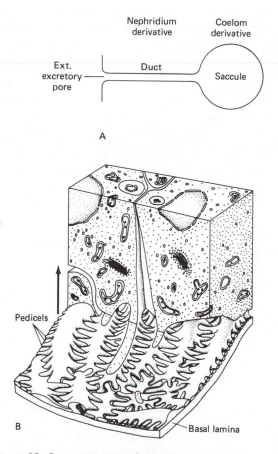

Figure 12–9 *A*, Diagram of an arthropod excretory end sac derived from a nephridium and coelom remnant. *B*, Two podocytes contributing to the saccule wall of a crayfish. (From Kummel, G., 1967: Die Podocyten. Zoolog. Beitr. N. F., *13*:245–263.)

Digestive Tract

The arthropod gut differs from that of most other animals in having large stomodeal and proctodeal regions (Fig. 12–1A). The derivatives of these ectodermal portions are lined with chitin and constitute the foregut and hindgut. The intervening region, derived from endoderm, forms the midgut. The foregut is chiefly concerned with ingestion, trituration, and storage of food; its parts are variously modified for these functions, depending on the diet and mode of feeding. The midgut is the site of enzyme production, digestion, and absorption; however, in some arthropods enzymes are passed forward and digestion begins in the foregut. Very commonly the surface area of the midgut is increased by outpocketings, forming pouches or large digestive glands. The hindgut functions in the absorption of water and the formation of feces.

Brain

There is a high degree of cephalization in arthropods. The increase in brain size is correlated with well-developed sense organs, such as eyes and antennae, and many arthropod groups display complex behavioral patterns. The arthropod brain consists of three major regions—an anterior protocerebrum, a median deutocerebrum, and a posterior tritocerebrum (Fig. 12–10A). The nerves from the eyes enter the protocerebrum, which contains one to three pairs of optic centers (neuropiles). The optic and other neuropiles of the protocerebrum function in integrating photoreception and movement and are probably the centers for the initiation of complex behavior.

The deutocerebrum receives the antennal nerves (first antennae in crustaceans) and contains their association centers. Antennae are lacking in the chelicerates (scorpions, spiders, mites), and in these arthropods there is a corresponding absence of the deutocerebrum (Fig. 12–10B).

The third brain region, the tritocerebrum, gives rise to nerves that innervate the labium (lower lip), the digestive tract (stomatogastric nerves), the chelicerae (claws) of chelicerates, and the second antennae of crustaceans. The commissure of the tritocerebrum is postoral, i.e., behind the foregut.

A debate centers on the extent to which the arthropod head is a segmented structure. Most zoologists agree that the tritocerebrum is a segmental ganglion that has shifted anteriorly. Its postoral commissure alone is good evidence of such an origin. Anderson's review of annelid and arthropod

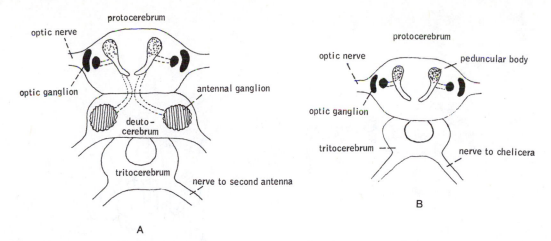

Figure 12–10 Annelid and arthropod brains: *A*, A mandibulate arthropod. *B*, A chelicerate arthropod. (All after Hanstrom from Vandel.)

embryology (1973) supports the belief that the heads of all arthropods contain two or three preoral segments and that the antennae are segmental appendages. A résumé of older ideas on the subject is outlined by Bullock and Horridge (1965).

Sense Organs

The sensory receptors of arthropods are usually associated with some modification of the chitinous exoskeleton, which otherwise would act as a barrier to the detection of external stimuli. A modification of the exoskeleton for the reception of environmental information other than light is called a sensillum. Sensilla have various shapes, depending on the type of signals they are designed to monitor. The most common form is a hair, bristle, or seta, but there are also sensilla in the form of pegs, pits, and slits. Each sensillum is composed of one or more sensory neurons plus a number of cells that produce the housing of the apparatus. Some sensilla contain a single type of receptor neuron; others encompass a number of types. For example, most insect "taste" hairs contain both chemoreceptors and mechanoreceptors. Thus, sensilla cannot always be classified by function. In general, those containing chemoreceptors have perforated walls (Fig. 13–7E). Mechanoreceptors are stimulated by movement of the sensilla, as in the case of hairs (Fig. 13–7D), or by changes in tension on the exoskeleton, as in the case of slit sensilla (Fig. 13–9). The concentrations of sensilla over the arthropod body simply reflect points of most likely contact with signals to be monitored. Arthropods also possess proprioreceptors attached to the inside of the integument or to muscles.

Most arthropods have eyes, but the eyes vary greatly in complexity. Some are simple and have only a few photoreceptors. Others are large, with thousands of retinal cells, and can form a crude image. In all arthropods the skeleton contributes the transparent lens-cornea to the eye.

Insects and many crustaceans, such as crabs and shrimp, have compound eyes composed of many long, cylindrical units, each possessing all the elements for light reception. Each unit, or ommatidium, is covered at its outer end by a translucent cornea derived from the skeletal cuticle (Fig. 12–11). The cornea functions as a lens. The external surface of the cornea, called a facet, is usually hexagonal or sometimes square (as in crayfish and lobsters). Behind the cornea, the ommatidium contains a long, cylindrical or tapered element called the crystalline cone, which functions as a second lens.

The basal end of the ommatidium is formed by the receptor element (the retinula). The center of the retinula is occupied by a translucent cylinder (the rhabdome), around which are arranged elongated photoreceptor, or retinular, cells, commonly seven or eight (Fig. 12–11). The inner photosensitive surfaces of the retinular cells are thrown into tubular folds, or microvilli, oriented at right angles to the axis of the ommatidium (Fig. 12–11E and F). Some of the rhabdomeres may be stacked so that not all are visible at a particular level in the ommatidium. The rhabdome is formed by all the rhabdomeres and the enclosed central space. The rhab-

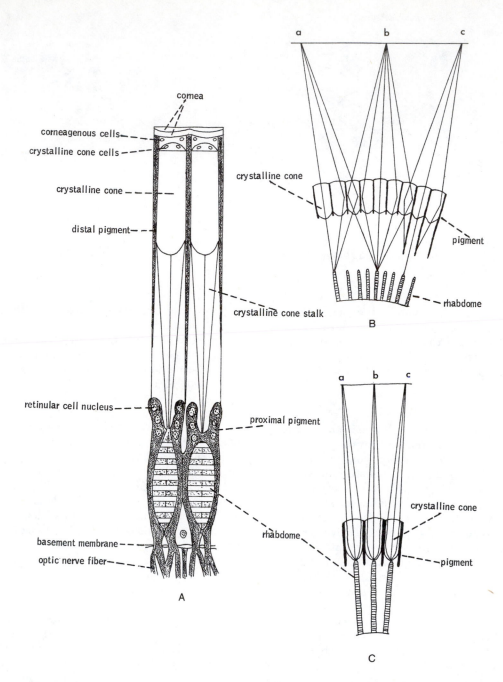

Figure 12–11   *A*, Two ommatidia from the compound eye of the crayfish, *Astacus* (longitudinal view). (After Bernhards from Waterman.) *B*, Compound eye specially adapted for superposition image formation. Light rays from points *a* and *b* are being received as a superposition image. Pigment is retracted, and light rays, initially received by a number of ommatidia, are concentrated on a single ommatidium. Point of light *c* is being received as an apposition image. Pigment is extended, preventing light rays from crossing from one ommatidium to another. *C*, Compound eye especially adapted for apposition image formation. (*B* and *C* after Kühn from Prosser and others.)

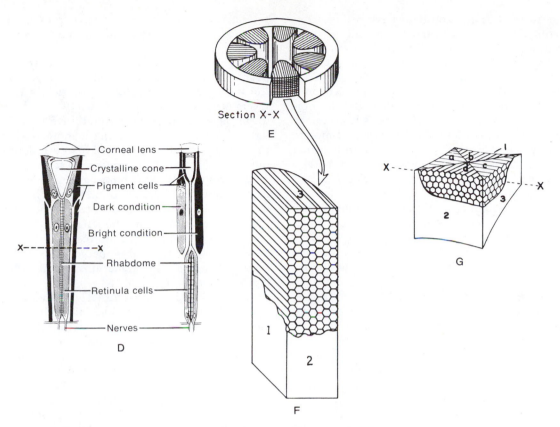

Section X-X

E

Corneal lens
Crystalline cone
Pigment cells
Dark condition
Bright condition
X- - - - - -X
Rhabdome
Retinula cells
Nerves

D

X- - - -
- - -X

G

F

Figure 12–11 (continued) *D*, Insect ommatidia, showing a diurnal type (left) and a nocturnal type (right). In the nocturnal type, pigment is shown in two positions, adapted for very dark conditions on the left side and for relatively bright conditions on the right. *E*, Section through an open type of rhabdome of an insect (level X–X in part *D*). *F*, Microtubules of one of the rhabdomeres in part *E*. *G*, Section through a closed type of rhabdome. (*E–G* from Wolken, J. J., 1971: Invertebrate Photoreceptors. Academic Press, N.Y.)

domeres of an ommatidium function as a single photoreceptor unit and transmit a signal that represents a single light point. From each retinula cell extends an axon, and thus a bundle of seven or eight axons leaves each ommatidium. The axons make connections with second-order neurons in an optic ganglion within the brain or, in some crustaceans, within an eye stalk.

The retinular cells contain black or brown pigment granules, which constitute the proximal retinal pigment. Distally, the ommatidium is surrounded by a number of special pigment cells, forming the distal retinal pigment. The proximal or the distal pigment or both can migrate centrally or distally, depending on the intensity of light (Fig. 12–11*D*).

In bright light most compound eyes produce an apposition image (Fig. 12–11*C*). Light enters an ommatidium either at an angle or perpendicular to the facet. The proximal and distal retinal pigments

are extended and act as a screen to prevent light from passing from one ommatidium to another, and thus light rays are restricted to the axial region of the crystalline cone and rhabdome.

The crystalline cone is essentially a lens cylinder and is formed of concentric lamellae. The outer lamellae have lower refractive indices than the central lamellae and probably function by refraction to eliminate oblique rays and to direct the more axial light rays to the rhabdome. On emerging from the crystalline cone, light rays pass to the rhabdome. Studies by Shaw (1969) have demonstrated that in the apposition eyes of honeybees and locusts, only 0.1 to 1 per cent of the light reaching a rhabdome comes from facets other than its own.

Each ommatidium responds to a patch of light reflected from one part of the visual field with a little overlap between adjacent ommatidia. The retinula cells function more or less as a unit and register the intensity of the light received. Thus, the total image formed by the compound eye results from the compilation of "light spots" from all the stimulated ommatidia. The differences in light intensity they register would provide the contrast from one part of the image to another. The image is therefore analogous to the image produced on a television screen, on which the "picture" is composed of dots of light. The apposition image formed by a compound eye is often called a mosaic image, since the total image results from small pieces put together like a mosaic. Unfortunately, the term *mosaic image* implies a special or different sort of image. Actually, the human eye receives and transmits light stimuli in a somewhat similar manner, a single point of light stimulating a functional unit of approximately seven cones; the principal difference is that the mosaic in the human eye is of a finer grain (composed of smaller and more numerous spots).

The image produced by a compound eye is crude at best, for the small size of the eye limits the number of photoreceptor units (ommatidia). The smallest arc of the visual field subtended by an ommatidium in the honeybee is 1 degree, compared with 1' of arc (1 degree = 60 seconds) in the much larger human eye. Thus, a row of closely spaced vertical bars would appear as a continuous horizontal bar to a bee. The compound eye is therefore not very good for distance vision, compared with the large eyes of birds and mammals, but what matters to most arthropods is the world about 20 cm around them.

An advantage of the compound eye found in many arthropods is its ability to detect movement. A slight shift in a point of light results in a corresponding shift in the ommatidia being stimulated. Many active, diurnal arthropods, such as bees and wasps, have high flicker-fusion rates. Flicker fusion refers to the fusing of a rapid sequence of separate images into one, as in a motion picture. The flicker-fusion rate of the honeybee is three times greater than that of a human. Thus, what we see as one continuous image a bee would see as a succession of separate images. An animal with a high flicker-fusion rate can detect motion more readily than one whose fusion rate is low.

Another advantage of the compound eye is that the total corneal surface is greatly convex, resulting in a wide visual field. This is particularly true for the stalked, compound eyes of crustaceans, in which the cornea may cover an arc of 180 degrees or more (Fig. 14–65A).

In weak light some compound eyes function differently and form what has been called a superposition image (Fig. 12–11B and D). The pigment is retracted, so no screening effect is present. Thus, light can pass from one ommatidium to another, and one rhabdome responds to light rays that originally entered several adjacent facets. In the superposition eyes of the crayfish, at least 50 per cent of the light striking a rhabdome originally entered facets other than its own (Shaw, 1969). This condition appears to be an adaptation for gathering light in semidarkness, making it more likely that the rhabdome will be activated than if it depended only on light received from its own facet.

The compound eyes of most arthropods are able to adapt, to at least some extent, to both bright and weak light, but in general, they tend to be specially modified for functioning under one of these two conditions. Thus, compound eyes can be classified as either apposition or superposition eyes, although there are many gradations between the two types.

Arthropods that are diurnal and live in well-lighted habitats, such as terrestrial and littoral species, usually possess apposition eyes (Fig. 12–11C and D). The screening pigment is well developed. The length of the crystalline cone is approximately equal to its focal length, and the lower end of the cone and the upper end of the rhabdome are contiguous, or nearly so. The retinular cells are quite long, extending from the crystalline cone to the basal membrane of the retina. All of these modifications tend to confine light entering a single ommatidium and to funnel the light down the axis of the ommatidium to the rhabdome.

Superposition eyes are found in nocturnal species or those that live in poorly lighted habitats. However, there are many exceptions, and it is now recognized that not all superposition eyes can be considered nocturnal eyes. The superposition eye is especially modified for collecting and concentrating onto one ommatidium light originally striking a large patch of facets (Fig. 12–11B and D). Screening pigment is usually present but may be reduced or absent in cave-dwelling and bathypelagic species. The crystalline cone tends to be twice as long as its focal length, and there is considerable space between the end of the crystalline cone and the rhabdome, permitting the bent light rays to cross from the crystalline cone of one ommatidium to the rhabdome of another. The retinular cells are much shorter than those in apposition eyes, and they are restricted to the base of the ommatidium. In some insects a reflecting pigment is present around the retinula and may be movable.

Crayfish and lobsters have reflecting superposition eyes, in which "mirrors" at the sides of the ommatidia reflect incoming light across the space between the crystalline cone and rhabdome (Land, 1978 and 1981).

In numerous arthropods with compound eyes, as well as in many animals with other types of eyes, the membrane containing the light-sensitive pigment undergoes rhythmic degeneration and regeneration. Twenty-four-hour cycles are common, and light is a primary controlling factor. The rhythms appear to be a mechanism to bring the eye into optimum condition when the animal is most active (Herman, 1983).

Color vision has been demonstrated in a number of arthropods and has been extensively studied in some insects. Here, the photolabile pigment responds to certain wavelengths of the light spectrum, with different retinula cells exhibiting different sensitivities. For example, within the ommatidium of a bee are two retinula cells that respond to green wavelengths and three that respond to ultraviolet, and some ommatidia contain four retinula cells that respond to blue wavelengths. Variously colored flowers visited by bees would elicit responses from different combinations of retinula cells.

Very important to all the visual responses described here—apposition and superposition imaging, color vision, etc.—is the way incoming signals from the receptor cells are handled by the subsequent levels of neurons. Just as in vertebrates, cross connections by ganglion cells make possible the pooling of signals from a number of retinula cells in very precise ways. There is also the eventual projection of signals to specific areas of the brain. Complex behavioral responses to visual information depend not only on the complexity of the eyes but also on the complexity of the associated neural circuitry.

Good reviews of the structure and functioning of compound eyes are provided by Land (1981); of insects, by Chapman (1982); and of crustaceans, by Cronin (1985).

Reproduction and Development

With few exceptions, arthropods are dioecious. Furthermore, many employ modified appendages during copulation. Fertilization is always internal in terrestrial forms but may be external in aquatic species. The eggs of most arthropods are rich in yolk and are centrolecithal. In the centrolecithal egg the nucleus is surrounded by a small island of nonyolky cytoplasm in the middle of a large mass of yolk (Fig. 12–12A). There is also a peripheral sphere of nonyolky cytoplasm. Most arthropods have a modified type of cleavage that is associated

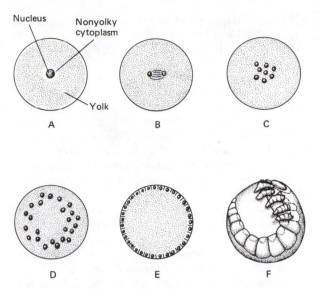

Figure 12–12 *A–E*, Superficial cleavage: *A*, Centrolecithal egg. *B* and *C*, Nuclear division. *D*, Migration of nuclei toward periphery of egg. *E*, Blastula. Cell membranes have developed, separating adjacent nuclei. Embryo of the arachnid *Thelyphonus*. (After Kaestner, A., 1968: Invertebrate Zoology, Vol. II. John Wiley and Sons, N.Y.)

with centrolecithal eggs, called intralecithal or superficial cleavage (Fig. 12–12B to E).

After fertilization the centrally located nucleus undergoes mitotic divisions but without the formation of cell membranes and without any cleavage of the yolk. The result, after several such divisions, is an uncleaved egg containing a large number of syncytial nuclei within the center. As division continues, the nuclei gradually migrate to the periphery, where cell membranes form but do not extend into the yolky interior. This stage of development represents a stereoblastula. Development continues with the formation of a primordial germinal disc located on one side of the embryo. This germinal center proliferates entoderm and mesoderm and eventually forms the embryonic body, which appears to be wrapped around the egg (Fig. 12–12F).

ARTHROPOD CLASSIFICATION

Living arthropods have traditionally been divided into two subphyla. Those that lack antennae have been commonly placed in the subphylum Chelicerata, which takes its name from the chelicerae, the first postoral appendages, which are used in feeding. To this group belong the horseshoe crabs, scorpions, spiders, mites, and ticks. Arthropods with antennae have been placed within the subphylum

Mandibulata, since their first postoral appendages are mandibles. This group includes all the insects as well as shrimp, crabs, millipedes, and centipedes.

Most zoologists today agree, however, that the subphylum Mandibulata is an artificial assemblage of unrelated groups and that there are probably four main lines of arthropod evolution rather than two, as the Mandibulata and Chelicerata suggest.* These lines are believed to be represented by the extinct Trilobita, the Chelicerata, the Crustacea, and the Uniramia. The uniramians include the centipedes, millipedes, and insects. In contrast to the other three arthropod lines, which have marine origins, the uniramians appear to have evolved on land. They are mandibulate and antennate, but the name *uniramian* refers to the fact that the appendages are basically unbranched and have been thought to be derived from an unbranched condition.

There is some evidence from comparative morphology and embryology that at least the uniramians and perhaps even all four of these arthropod groups had a separate origin from different annelidan or near annelidan ancestors and that arthropodization (i.e., the arthropod features of a chitinous exoskeleton and jointed appendages) evolved independently at least twice or maybe four times. If this is true, the Arthropoda should be considered a superphylum, and the Trilobita, Chelicerata, Crustacea, and Uniramia should each be raised to phylum rank. This polyphyletic view of arthropod evolution is still not accepted by all specialists, particularly by many entomologists (Gupta, 1979). The traditional rank of phylum has been retained for the Arthropoda in this edition, but the four lines of arthropod evolution have been recognized as subphyla.

SYNOPSIS OF THE ARTHROPOD CLASSES

Subphylum Trilobita. The fossil trilobites.
Subphylum Chelicerata
    Class Merostomata. The living horseshoe crabs and fossil eurypterids.
    Class Arachnida. The largest of the chelicerate classes; includes the scorpions, harvestmen, spiders, ticks, and mites.

---

*Our changing ideas about the evolution of arthropods are in large part the result of the long and persistent efforts of Manton. Her work and ideas have been reviewed in several publications (see References, p. 490). Anderson's superb survey (1973) of arthropod embryonic development has added weight to her arguments.

    Class Pycnogonida. The sea spiders.
Subphylum Crustacea
Subphylum Uniramia
    Class Insecta. The insects.
    Class Chilopoda. The centipedes.
    Class Diplopoda. The millipedes.
    Class Symphyla. The symphylans.
    Class Pauropoda. The pauropodans.

SUMMARY

1 The arthropods are the largest assemblage of species within the Animal Kingdom. They most probably evolved from annelids or at least from some common ancestral form. The annelidan ancestry of arthropods is reflected in their metamerism, the plan of their nervous system, and their determinate cleavage.

2 In the evolution of the arthropod condition, the chitin-protein exoskeleton was a central development to which many other changes can be related. The division of the skeleton into plates and cylinders makes movement possible, and periodic molting of the exoskeleton permits growth.

3 Arthropods move by jointed segmental appendages rather than by body deformation. The coelom has become vestigial in the adult, and the musculature is organized as bundles attached to the inside of the skeleton. The skeleton and muscles operate together as a lever system.

4 There has been a general tendency among arthropods for metamerism to become reduced through fusion, loss, and differentiation of segments. The number of locomotor appendages has, in general, become reduced as a consequence of differentiation of appendages for other functions and because the small number permits greater maneuverability and speed.

5 Muscle contraction is governed by a system of multiple motor innervations—tonic, phasic, and inhibitory.

6 The circulatory system is open, with a dorsal, primitively tubular heart. Paired lateral ostia permit the passage of blood into the heart from the surrounding pericardial sinus. Hemocyanin is the usual respiratory pigment.

7 Nephridia are probably represented by the paired saccular excretory organs of many arthropods. Malpighian tubules, a second type of excretory organ, are associated with the gut and are a new development in arthropods.

8 The sense organs usually involve some specialization of the exoskeletal barrier, which permits monitoring of environmental stimuli. Hairs or bristles are the most common type of arthropod sensilla. Many crustaceans and most insects have compound eyes, in which each of the units (ommatidia) composing the eye contains all the visual elements.

9 Most arthropods are dioecious. Copulation and internal fertilization are characteristic of the majority of species, with various appendages involved in courtship and sperm transfer.

10 The eggs of most arthropods are centrolecithal, and cleavage is commonly superficial.

The Trilobites

The subphylum Trilobita is the most primitive of all known arthropodan groups and from an evolutionary standpoint represents a good starting point in discussing the arthropod classes. Trilobites are an extinct group of marine arthropods that were once abundant and widely distributed in Paleozoic seas. They reached their height of distribution and abundance during the Cambrian and Ordovician periods and disappeared at the end of the Paleozoic era. Over 3900 species have been described from fossil specimens.

Most trilobites ranged from 3 to 10 cm in length, although some planktonic forms were only 0.5 mm long. The largest trilobites were a little less than a meter in length.

The somewhat oval and flattened trilobite body was divided into three more or less equal sections—a solid, anterior cephalon; an intermediate thorax or trunk region, consisting of a varying number of separate segments; and a posterior pygidium (Fig. 12–13). Each of these body divisions was in turn divided into three regions by a pair of furrows running from anterior to posterior and forming a median axial lobe flanked on each side by a lateral lobe. The name Trilobita refers to this transverse trilobation of the dorsal body surface. The dorsal exoskeleton was heavier than the ven-

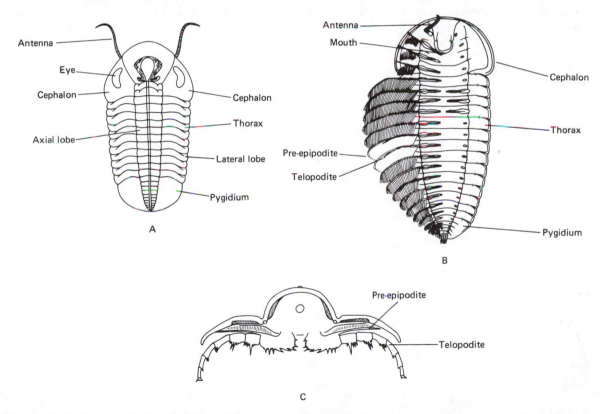

Figure 12–13 *A*, Dorsal view of *Phacops*, believed to have been a crawling-swimming, epibenthic, predacious trilobite with jaws. (From Stürmer, W., and Berström, J., 1973: New discoveries on trilobites by X-rays. Paläont. Z., 47½:104–141.) *B*, Ventral view of *Triarthrus eatoni*. (From Cisne, J. L., 1975: Anatomy of *Triarthrus* and the relationships of the Trilobita. Fossils and Strata, 4:45–63.) *C*, Cross section of *Olenoides serratus*. (From Whittington, H. B., 1975: Trilobites with appendages from the middle Cambrian, Burgess Shale, British Columbia. Fossils and Strata 4:97–136.)

tral surface, which carried the appendages. This difference is the reason that in most cases only the dorsal surface has been preserved in the fossil record.

The anterior body section, the cephalon, was composed of at least five fused segments, counting the antennal segment, in addition to the acron, and was covered by a shieldlike carapace (Bergström, 1973). The cephalic shield not only covered all the dorsal surface but also was folded under at its margins so that the original ventral surface was narrow and restricted. The pleural skeleton was reduced to thin membranes to which the appendages were attached. This condition was true not only for the cephalon but also for the other two divisions of the body. On each side of the middle of the carapace was a pair of eyes that varied considerably in size, depending on the species. X-ray and thin section studies indicate a structure superficially similar to the compound eyes of living arthropods (Fig. 12–14).

The posteriorly directed mouth was located in the middle of the underside of the cephalon just behind and beneath a liplike prominence called the labrum (Fig. 12–13B). On each side of the labrum was a long sensory antenna, believed to be homologous to the first pair of antennae of the crustaceans and the antennae of insects.

Behind the mouth were located four pairs of appendages similar to the appendages of the trunk and pygidium. Each appendage was biramous and consisted of an inner walking leg (telopodite) and an outer filament-bearing branch (pre-epipodite)

(Fig. 12–13). These filaments have been called gills, but in the small number of species in which they have been preserved, the filaments have the form of heavy spines, rakelike teeth, or feather-like barbs, suggesting such functions as digging, filtering, or swimming, depending on the species. The appendages of the cephalon show varying degrees of reduction in different trilobite groups, and in some, such as *Phacops*, certain pairs of coxae (basal segment of the appendage) carried apposing toothed processes that must have functioned as jaws. Moreover, in all trilobites the medial bases of the appendages appear to function in moving food forward to the mouth (Fig. 12–13). The favorable ratio of surface area to volume resulting from the flattened body and the thin exoskeleton over the ventral surface may have reduced the need for specialized organs of gas exchange, at least in smaller species (Stürmer and Bergström, 1973).

The trilobite trunk (thorax) consisted of a varying number of separately articulating segments, each of which bore ventrally a pair of appendages similar to the appendages of the trunk. The pygidium was constructed on the same plan as the thorax except that the segments were fused and formed a solid shield. The appendages of the pygidium usually were successively smaller toward the posterior end.

Differences in size, shape, spination, and eye size and position indicate that trilobites displayed diverse life-styles within a variety of habitats. The majority were bottom dwellers and crawled over sand and mud using the walking legs. The flattened

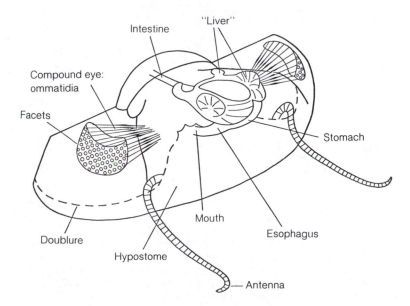

Figure 12–14 Reconstruction of head of *Phacops*. (From Stürmer, W., and Bergström, J., 1973: New Discoveries on trilobites by X-rays. Paläont. Z., 47½:104–141.)

body and dorsal eyes were adaptations for this type of existence. Many could roll up (Fig. 12–15C and D), and in some species projecting spines could have had a defensive function in the rolled position. Some trilobite groups had a shovel-shaped or plow-shaped cephalon adapted for burrowing (Fig. 12–15A). Other trilobites appear to have lived in burrows, keeping the head above or near the surface of the mud or sand, and seized passing prey (Fig. 12–15E).

Some groups of trilobites were apparently not confined to the bottom but took up a swimming ex-

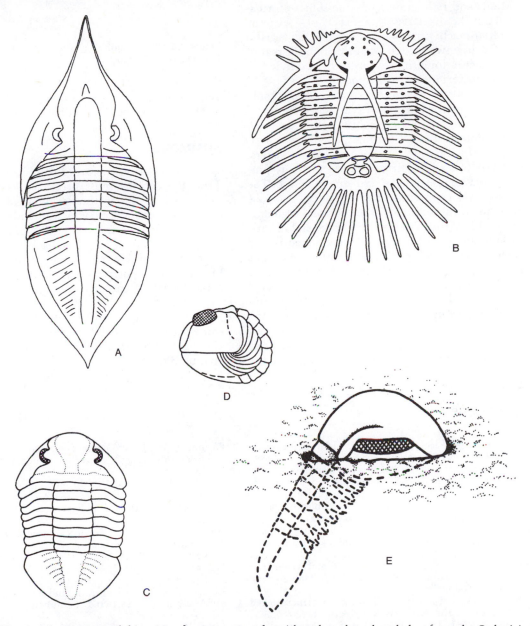

Figure 12–15   *A,* A burrowing trilobite, *Megalaspis acuticauda,* with a plow-shaped cephalon from the Ordovician period. *B,* A planktonic trilobite, *Radiaspis radiata,* from the Devonian period. The long spines may have been flotation devices. (*A* and *B* after Størmer, L., 1949: Classe des Trilobites. *In* Grassé, P.: Traité de Zoologie. Vol. VI. Masson et Cie, Paris.) *C* and *D, Asaphus.* Dorsal view of extended animal (*C*) and side view of animal in enrolled condition (*D*). *E,* Postulated position of trilobite *Panderia* within burrow. (Adapted from Bergström, J., 1973: Organization, life, and systematics of trilobites. Fossils and Strata, 2 Universitetsforlaget, Oslo. 69 pp.)

istence. In these forms the body was narrower, and the eyes were located on the sides of the head. Nothing is known about their appendages. The smallest species of trilobites were planktonic. Some displayed protective spines on the dorsal surface (Fig. 12–15*B*).

Molds and radiographs provide some information about the digestive tract (Fig. 12–14). A pear-shaped stomach surrounded by a digestive gland extended downward to the mouth. From the rear of the stomach a long, narrow intestine extended to near the posterior end of the body.

Fossil material has made possible a remarkably complete understanding of the developmental stages. During their postembryological development, trilobites passed through three larval periods, each of which consisted of a number of instars. The trilobite emerged from the egg as a tiny protaspis larva, measuring 0.5 to 1 mm in length. The protaspis was covered by a single, dorsal carapace consisting of the acron and four postoral segments (Fig. 12–16). After passing through several instars, in which additional segments were added to the carapace, the protaspis passed into a meraspis larva. In this larval stage the pygidium was located behind the cephalon. The thoracic region appeared during succeeding molts as segments were gradually freed from the anterior border of the pygidium.

The final larval stage is known as a holaspis larva, which, though still very small, displayed the

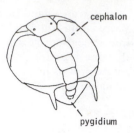

Figure 12–16 Trilobite protaspis larva. (After Moore, Lalicker, and Fischer.)

general adult structure. Additional molts merely involved minor changes and an increase in size.

SUMMARY

1 The marine Paleozoic trilobites (subphylum Trilobita) are the most primitive known arthropods. Each of the postoral segments carried a pair of similar appendages, which included a filament-bearing branch of uncertain function.

2 The body was composed of an anterior cephalon, a middle trunk region of unfused segments, and a posterior pygidium. The cephalon carried a pair of antennae and a pair of dorsolateral eyes.

3 Variations in form indicate that there was some diversity in trilobite life-styles, such as burrowing, epibenthic crawling, planktonic, and swimming.

REFERENCES

Anderson, D. T., 1973: Embryology and Phylogeny in Annelids and Arthropods. Pergamon Press, N.Y. 495 pp. (The closing chapter of this work summarizes the evidence from embryonic development that supports the belief that arthropods are polyphyletic).

Atwood, H. L., 1973: An attempt to account for the diversity of crustacean muscles. Am Zool., 13:357–378.

Atwood, H. L., and Sandeman, D. C. (Eds.), 1982: Biology of the Crustacea, Vol. 3. Neurobiology: Structure and Function. Academic Press, N.Y.

Bergström, J. B., 1973: Organization, life, and systematics of trilobites. Fossils and Strata, No. 2, Universitetsforlaget, Oslo.

Bullock, T. H., and Horridge, G. A., 1965: Structure and Function in the Nervous Systems of Invertebrates, Vol. 2. W.H. Freeman and Co. San Francisco, pp. 801–1719.

Chapman, R. F., 1982: The Insects: Structure and Function. 3rd Edition. Harvard University Press, Cambridge, Mass. 919 pp.

Cisne, J. L., 1974: Trilobites and the origin of arthropods. Science 186:13–18.

Cronin, T. W., 1986: Optical design and evolutionary adaptation in crustacean compound eyes. Jour. Crustacean Biol., 6(1): 1–23.

Gupta, A. P. (Ed.), 1979: Arthropod Phylogeny. Van Nostrand Reinhold, N.Y. 762 pp. (A collection of papers concerned with various aspects of arthropod origins and evolution).

Hackman, R. H., 1971: The integument of arthropods. In Florkin, M., and Scheer, B. T. (Eds.): Chemical Zoology, Vol. 6, Pt. B. Academic Press, N.Y. pp. 1–62.

Herman, K. G. 1983: Rods, rhabdomes and rhythms. BioScience, 33(7):432–438.

Kennedy, D., Selverston, A. I., and Remler, M. P., 1969: Analysis of restricted neural networks. Science, 164:1488–1496.

Land, M. F., 1978: Animal eyes with mirror optics. Sci. Amer., 239(6):126–134.

Land, M. F., 1981: Optics and vision in invertebrates. In Autrum, H. (Ed.): Handbook of Sensory Physiology, vol. VII/6B. Springer-Verlag, Berlin. pp. 471–592.

Levi-Setti, R., 1975: Trilobites: A Photographic Atlas. University of Chicago Press, Chicago 214 pp.

Linzen, B., et al, 1985: The structure of arthropod hemocyanins. Science, *229*:519–524.

Locke, M., 1984: Epidermal cells (Arthropoda). *In* Bereiter-Hahn, J., Matoltsy, A. G., and Richards, K. S.: Biology of the Integument. 1. Invertebrates. Springer-Verlag, Berlin. pp. 502–522.

Mangum, C. P., 1985: Oxygen transport in invertebrates. Amer. Jour. Physiol., *248*:505–514.

Manton, S. M., 1952: The evolution of arthropodan locomotory mechanisms. Pt. 2. General introduction to the locomotory mechanisms of the Arthropoda. J. Linnean Soc. (Zoology), *42*:93–117.

Manton, S. M., 1970: Arthropods: Introduction. *In* Florkin, M., and Scheer, B. T. (Eds.): Chemical Zoology, Vol. 5. Academic Press, N.Y. pp. 1–34.

Manton, S. M., 1973: Arthropod phylogeny—a modern synthesis. J. Zool. *171*:111–130.

Manton, S. M., 1978: The Arthropoda: Habits, Functional Morphology and Evolution. Oxford University Press, London. 527 pp. (An elaboration of the author's lifetime study of the functional morphology of arthropods, especially their limbs.)

Moore, R. C. (Ed.), 1959: Treatise on Invertebrate Paleontology. Part O, Arthropoda 1. Geological Society of America and University of Kansas Press, Lawrence, Kans. (This volume covers trilobites.)

Redmond, J. R., 1971: Blood respiratory pigments—Arthropoda. *In* Florkin, M., and Scheer, B. T. (Eds.): Chemical Zoology, Vol. 6. Academic Press, N.Y. pp. 119–144.

Shaw, S. R., 1969: Optics of arthropod compound eyes. Science, *165*:88–90.

Størmer, L., 1949: Sous-embranchement des Trilobitomorphes. *In* Grassé, P. (Ed.): Traité de Zoologie, Vol. 6. Masson et Cie, Paris. pp. 159–216.

Stürmer, W., and Bergström, J., 1973: New discoveries on trilobites by X-rays. Paläont. Z., *47½*:104–141.

Wolken, J. J., 1971: Invertebrate Photoreceptors. Academic Press, N.Y., 179 pp. (A review of research in some invertebrate photoreceptor systems.)

13

The Chelicerates

All the animals described in this chapter belong to the subphylum Chelicerata, one of the three principal evolutionary lines of living arthropods. The body of a chelicerate is divided into a cephalothorax (or prosoma) and an abdomen (or opisthosoma). There are no antennae, and chelicerates are the only arthropods that lack them. This is the most distinguishing characteristic of the subphylum. The first pair of appendages are feeding structures called chelicerae. The second pair of appendages are called pedipalps and are modified to perform various functions in the different classes.

There are three classes of chelicerates. Two small classes contain marine species, but most chelicerates are terrestrial and belong to the class Arachnida.

Class Merostomata

The Merostomata are aquatic chelicerates characterized by five or six pairs of abdominal appendages modified as gills and by a spikelike telson at the end of the body. The group can be divided into two distinct subclasses—the Xiphosura (horseshoe crabs) and the extinct Eurypterida.

SUBCLASS XIPHOSURA

Although the fossil record of the Xiphosura extends back to the Cambrian period, three genera and four species compose the only living representatives today. One of these is the horseshoe crab, *Limulus polyphemus*, common to the northwestern Atlantic coast and the Gulf of Mexico. All the other members of this group are found along Asian coasts from Japan and Korea south through the East Indies and the Philippines.

Horseshoe crabs live in shallow water on soft bottoms, plowing through the upper surface of the sand. They reach a length of 60 cm and are dark brown. The carapace is smooth, horseshoe shaped, and convex, with the posterior lateral angles prolonged backward to about half the length of the abdomen (Fig. 13–1). Its shape not only facilitates pushing through the sand and mud but also provides a protective covering for the ventral appendages. To the outside of each of two dorsolateral ridges is a large eye, and to each side of a median

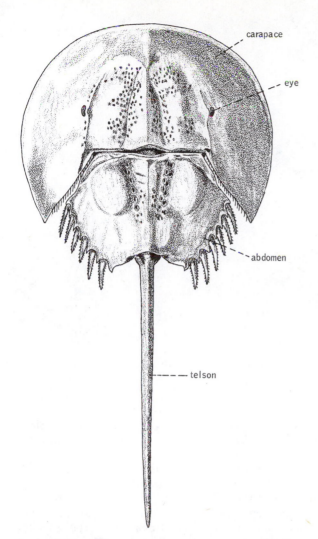

carapace

eye

abdomen

telson

Figure 13–1 A horseshoe crab (dorsal view). (After Van der Hoeven from Fage.)

ridge at the anterior end is one of two small median eyes.

The anterior dorsal surface is reflected ventrally and in the front forms a large, triangular surface that tapers back toward the mouth (Fig. 13–2). A frontal organ and a pair of degenerate eyes are located on the ridge formed by this triangle. A pair of trisegmented chelicerae are attached to each side of the upper lip, or labrum; the last two segments are chelate; i.e., they form a pair of pincers. The mouth is located behind the labrum and is followed by a short, narrow sternum.

Five pairs of walking legs are located posterior to the chelicerae on the underside of the cephalothorax (Fig. 13–2). The first pair is homologous to

the pedipalps of other chelicerates. Except for the last pair, the walking legs are all similar and chelate, and the median side of the coxa (the most prominent segment) of each leg is heavily armed with spines. These spines, called gnathobases, macerate and move food anteriorly. The coxae of the fifth or last pair of legs bear on the median side a short, spatulate process known as a flabellum, which is used for cleaning the gills. Furthermore, the last pair of walking legs are not chelate. Each possesses four leaflike processes attached to the end of the first tarsal segment. This last pair of appendages is used for pushing and for clearing and sweeping away the mud and silt during burrowing.

The last pair of cephalothoracic appendages are known as chilaria and are located behind the sternum between the last pair of walking legs. Each chilarium consists of a single article armed with hairs and spines, as are the gnathobases, and probably functions similarly.

The abdomen (opisthosoma) is unsegmented and fits into the concavity formed by the posterior border of the cephalothorax and its lateral extensions (Figs. 13–1 and 13–2). A long, triangular, spikelike tail or caudal spine (telson) articulates with the posterior of the abdomen. The telson of horseshoe crabs is not actually a true telson, since it does not bear the anal opening. The tail of these animals is highly mobile and may be used for pushing and for righting the body when it is accidentally turned over. It is not used for defense, and a horseshoe crab can be picked up and carried by it.

The abdomen bears six pairs of appendages (Fig. 13–2). The fused first pair forms the genital operculum bearing the two genital pores on the underside.

Posterior to the genital operculum are five pairs of flaplike, membranous appendages modified as gills. The undersurface of each flap is formed into many leaflike folds called lamellae, which provide the actual surface for gas exchange (Fig. 13–2). This arrangement of leaflike lamellae has caused the appendages to be called book gills. The movement of the gills maintains a constant circulation of water over the lamellae, and the gills also function as paddles during swimming.

Horseshoe crabs are scavengers and feed on mollusks, worms, and other organisms, including bottom-dwelling algae. Food material is picked up by the chelate appendages, passed to the gnathobases, where it is broken up and moved anteriorly to the mouth.

The mouth, located just behind the chelicerae, opens into an esophagus; it extends anteriorly through a dilated portion, the crop, and into a

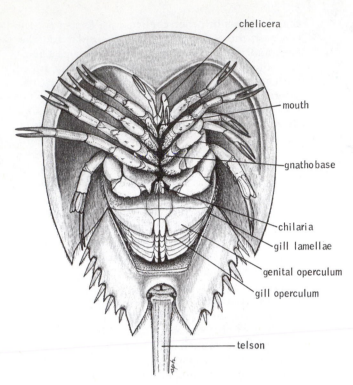

chelicera

mouth

gnathobase

chilaria

gill lamellae

genital operculum

gill operculum

telson

Figure 13–2 *Limulus polyphemus*, showing appendages (ventral view). (From life.)

grinding chamber, the gizzard (Fig. 13–3A). The longitudinal folds of cuticle in the gizzard possess denticles, and the whole structure is provided with strong muscles. After the food is ground in the gizzard, the large, undigestible particles are regurgitated through the esophagus, while the usable food material passes posteriorly through a valve into the enlarged, anterior part of the nonsclerotized midgut known as the stomach. The remainder of the midgut, called the intestine, extends posteriorly into the abdomen. Opening into each side of the stomach is one of two pairs of ducts from two large, glandular, hepatic ceca that ramify throughout the cephalothorax and abdomen. Enzyme production and digestion take place within the midgut region and the hepatic ceca. The hepatic ceca are also the principal areas for absorption of digested food materials.

Wastes are egested through a short sclerotized rectum and out the anus, which is located on the ventral side of the abdomen just in front of the telson.

The circulatory system is well developed. A dorsal tubular heart with eight ostia is located throughout most of the length of the intestine (Fig. 13–3A). From the heart, blood is pumped through a well-developed arterial system. The arteries eventually terminate in tissue sinuses, and the blood collects ventrally in two large, longitudinal sinuses.

From the ventral sinuses the blood flows into the book gills, where it is oxygenated. The movement of the gills not only causes the water to circulate over the lamellae but also pumps blood in and out of these structures. From the gills the blood returns to the pericardium. The blood contains hemocyanin as well as a single type of amebocyte that functions in clotting. Because of the amount of blood that can be obtained from a large horseshoe crab, these animals have been a favorite source for physiologists and biochemists studying hemocyanin.

Excretion takes place through four pairs of coxal glands (Fig. 13–3B), each which open through a common excretory pore at the base of the last pair of walking legs. The coxal glands contribute to osmoregulation by producing a dilute urine when the animal is in brackish water (Towle et al, 1982).

The nervous system displays a large degree of fusion (Fig. 13–3A). The brain forms a collar around the esophagus and includes the ganglia for the remaining first seven segments. Thus, all the appendages anterior to the operculum are directly innervated by the brain. A ventral nerve cord with

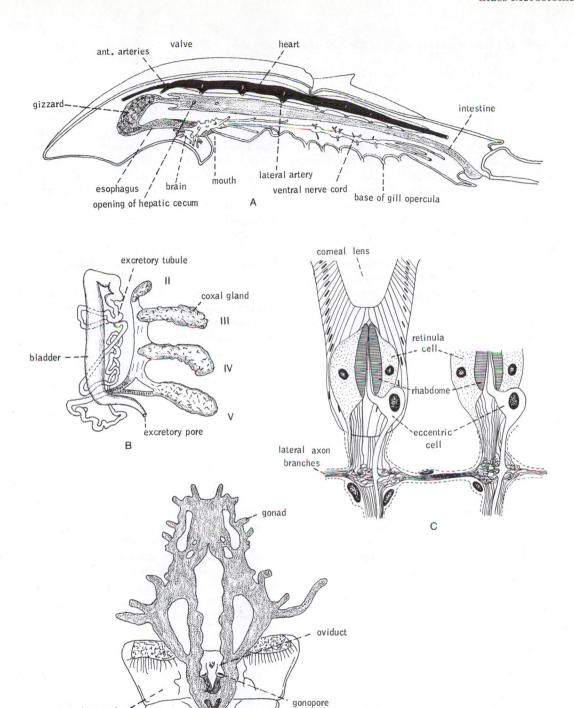

Figure 13–3 *A*, Sagittal section through *Limulus*. (After Patten and Redenbaugh from Kaestner.) *B*, Coxal glands and ducts of *Limulus*. (After Patten and Hazen from Fage.) *C*, Longitudinal section through two ommatidia of the eye of *Limulus*. (After MacNichol.) *D*, Female reproductive system of *Limulus*. (After Owen from Fage.)

five ganglia and lateral nerves extends through the abdomen.

The lateral eyes of horseshoe crabs are peculiar in a number of respects. They are compound eyes made up of units of 8 to 14 retinular cells grouped around a rhabdome (Fig. 13–3C). Each unit has a lens and a cornea and is called an ommatidium. However, in contrast to the ommatidia of the compound eyes of other arthropods, those of horseshoe crabs are not compactly arranged. Although horseshoe crabs may be able to detect movement, there are too few ommatidia in their eyes for image formation. The relationship between light stimulus and axon firing can be studied in a relatively simple state in horseshoe crab eyes, and this has resulted in their being used extensively in neurophysiological research. The median eyes are invaginated cups of retinal cells. The frontal organ is believed to be a chemoreceptor.

Horseshoe crabs are dioecious, and the reproductive system has essentially the same plan in both the male and the female (Fig. 13–3D). The gonad is located subjacent to the intestine, and the sperm or eggs pass to the outside through short ducts that open onto the underside of the genital operculum.

Mating and egg laying in the American *Limulus* take place during the high tides of full and new moons in spring and summer. Males and females migrate into shallow water and congregate in the intertidal zone along the shores of sounds, bays, and estuaries. The smaller male climbs onto the abdominal carapace of the female and maintains its hold with the modified, hooklike first pair of walking legs. Meanwhile, the female, partially buried in the sand, deposits 2000 to 30,000 large eggs. The eggs in each depression are fertilized by the male during their deposition. The mating pair separate, and the eggs are covered and left in the sand (Cohen and Brockmann, 1983).

The eggs are centrolecithal, 2 to 3 mm in diameter, and covered by a thick envelope, or chorion. Cleavage is total. A trilobite larva, so named because of its superficial similarity to trilobites, emerges from the egg. This larva is approximately 1 cm long and actively swims about and burrows in sand. The little caudal spine does not project beyond the abdomen. Only two of the five pairs of book gills are present, although all anterior appendages are present. As successive molts take place, the remaining book gills appear, the caudal spine increases in length, and the young crab assumes the adult form. Juvenile horseshoe crabs, which are common on intertidal sand flats, attain a carapace width of 4 cm after one year (Rudloe, 1981). Sexual maturity is not reached for 9 to 12 years, and the life span may be 19 years.

SUBCLASS EURYPTERIDA

The second group in the class Merostomata is the subclass Eurypterida (or Gigantostraca), the extinct giant arthropods. The eurypterids were aquatic and existed from the Cambrian to the Permian period. The eurypterids probably attained the largest size of any of the arthropods. One species of the genus *Pterygotus* was almost 3 meters long.

Eurypterids were similar to horseshoe crabs in their general body plan (Fig. 13–4), but the cephalothorax was smaller and the abdomen was composed of separate segments. Moreover, the abdomen was divided into a seven-segment preabdomen (mesosoma), bearing six pairs of gills, and a postabdomen (metasoma) of five narrower segments lacking appendages. The telson was attached to the last abdominal segment.

The chelicerae were small, and the last pair of legs was large and paddle-like. Most eurypterids, judging from the appendages, not only crawled over the bottom but also were active swimmers.

The location of fossils indicates that the eurypterids, after a marine origin, gradually invaded both brackish water and fresh water and perhaps even assumed a terrestrial existence.

The aquatic merostomes certainly contain the most primitive known chelicerates. But the origin of the subphylum is still obscure. The chelicerates may have evolved from the trilobites, or they may have had a separate origin from the annelids. However, the account of the eurypterids does not end with their extinction in the Permian period, for it is believed that they may have been the ancestors of the largest and most abundant class of chelicerates, the Arachnida.

SUMMARY

1 The subphylum Chelicerata contains the only nonantennate arthropods. The body is usually divided into an anterior cephalothorax (prosoma) and a posterior abdomen (opisthosoma). The first postoral appendages are a pair of food-handling chelicerae usually followed by a pair of pedipalps and four pairs of legs.

2 The marine origin of chelicerates is evidenced by a long fossil history beginning in the early Paleozoic, but only five species exist today (horseshoe crabs).

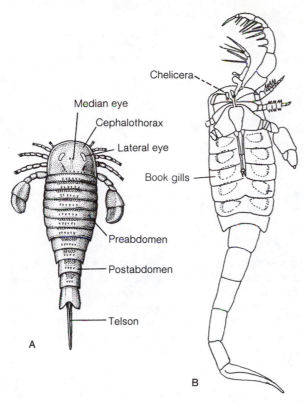

Chelicera

Median eye

Cephalothorax

Lateral eye

Book gills

Preabdomen

Postabdomen

Telson

A

B

Figure 13–4 Eurypterida: *A, Eurypterus remipes* (dorsal view). (After Nieszkowski.) *B, Mixopterus kiaèri* showing appendages of one side (ventral view). (After Størmer from Fage.)

3 Horseshoe crabs (Class Merostomata) are soft-bottom, shallow-water, marine chelicerates. The prosoma is covered by a large, horseshoe-shaped carapace, and the abdominal segments are fused together. The pedipalps are not differentiated from the posterior legs except in the mature male, which uses them to clasp the female. A posterior, spikelike telson is used for pushing and righting. The ventral side of the abdomen carries five pairs of book gills.

4 Fossil evidence indicates that some species of the extinct Paleozoic eurypterids (merostomes) invaded fresh water and may have given rise to the class Arachnida.

Class Arachnida

The arachnids constitute the largest and from a human standpoint the most important of the chelicerate classes; included are many common and familiar forms, such as spiders, scorpions, mites, and ticks. Arachnids also have the dubious honor of being the most unpopular group of arthropods as far as the layman is concerned, a bias largely unwarranted.

The arachnids are an old group. Fossil representatives of all the orders date to the Carboniferous period, and fossil scorpions have been found dating from the Silurian. The early arachnids were undoubtedly aquatic and were contemporaries of the eurypterids, from which they are believed to have evolved. The first terrestrial arachnids appear in the Devonian, well before the first terrestrial scorpions, which appear in the Carboniferous.

Except for the few groups that have adopted a secondary aquatic existence, living arachnids are terrestrial chelicerates. Like other evolutionary conquests of land, this migration from an aquatic to a terrestrial environment required certain fundamental, morphological, and physiological changes. The epicuticle became waxy, reducing water loss. The book gills became modified for use in air, resulting in the development of the arachnid book lungs and tracheae. In addition, the appendages became better adapted for terrestrial locomotion. Once a terrestrial existence was established, a great many unique innovations evolved independently along different lines. The development of silk in spiders, pseudoscorpions, and some mites and of poison glands in scorpions, spiders, and pseudoscorpions are but two examples.

THE GENERAL STRUCTURE AND PHYSIOLOGY OF ARACHNIDS

External Anatomy

Despite the diversity of forms, arachnids exhibit many features in common. The unsegmented prosoma is usually covered dorsally by a solid carapace (Fig. 13–11). The primitive abdomen is segmented and divided into a preabdomen and a postabdomen. In most arachnids other than scorpions these two subdivisions are no longer conspicuous (Fig. 13–41), and a tendency for segments to fuse has developed.

The appendages common to all arachnids are those arising from the prosoma and consist of a pair of chelicerae, a pair of pedipalps, and four pairs of legs (Fig. 13–41). The chelicerae are used in feeding, but the pedipalps serve a number of functions and are variously modified.

Nutrition

The majority of arachnids are carnivorous, and digestion partly takes place outside the body. Prey, usually small arthropods, are captured and killed by the pedipalps and chelicerae. While the prey is held by the chelicerae, enzymes secreted by the midgut are poured out over the torn tissues of the prey. Digestion proceeds rapidly and a partially digested broth is produced. This fluid is then taken into the prebuccal cavity, located in front of the mouth.

The liquid food passes through the mouth and into the sclerotized, pumping pharynx and esophagus of the foregut (Figs. 13–25 and 13–54). The esophagus conveys the food to the midgut or mesenteron, which consists of a central tube with lateral diverticula (Fig. 13–25). The diverticula are located in both the prosoma and the abdomen and become filled with the partially digested liquid. The secretory cells of the midgut wall produce enzymes for external digestion and internal extracellular digestion after the food reaches the mesenteron. The absorptive cells are the sites of intracellular digestion and absorption. The mesenteron extends to the posterior part of the abdomen, where it is connected to the posterior anus. A short sclerotized intestine, forming the hindgut, leads to the anus.

Excretion and Water Balance

The epicuticle of arachnids has a high lipid content, which greatly reduces evaporative water loss. Such a waxy epicuticle also functions as a water repellent (hydrofuge layer), reducing flooding by dew or raindrops. Guanine is the most important nitrogenous waste of arachnids. The excretory organs are coxal glands and malpighian tubules (p. 479). Some groups possess both; some one or the other. Coxal glands are thin-walled, spherical sacs along the sides of the prosoma, which collect wastes from the surrounding blood (Fig. 13–5). Wastes are transported to the outside by a duct that opens onto the coxa of the appendages. Malpighian tubules consist of one or two pairs of slender tubes that arise from the posterior of the mesenteron at its junction with the intestine (Fig. 13–25).

In addition to malpighian tubules and coxal glands, arachnids possess certain large phagocytic cells, called nephrocytes, that are localized in clusters in certain parts of the prosoma and abdomen.

Nervous System

The nervous system is greatly concentrated except in the relatively primitive scorpions (Fig. 13–6A). The brain, composed of the protocerebrum and tri-

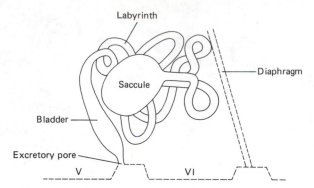

Figure 13–5   Coxal gland of a scorpion. (After Millot, J., and Vachon, M., 1949: Traité de Zoologie, Vol. VI. Masson et Cie, Paris.)

tocerebrum, is an anterior ganglionic mass lying above the esophagus. The protocerebrum contains the optic centers and optic nerves; the tritocerebrum contains the nerves supplying the chelicerae. The remainder of the central nervous system is located below the esophagus. In many orders most or all of the ganglia originally located in the thorax and abdomen have migrated anteriorly and fused with the subesophageal ganglion—the ganglion of the pedipalpal segment (Fig. 13–6B). Thus, the arachnid nervous system commonly resembles a collar or ring surrounding the esophagus. The posterior, ventral half of this collar gives rise on each side to nerves innervating the appendages, and a single posterior nerve bundle extends into the abdomen.

Three types of sense organs are common to most arachnids—sensory hairs, eyes, and slit sense organs. The sensory hairs can be olfactory setae, which are open at the end (Fig. 13–7E); innervated, movable setae located over the body surface (Figs. 13–7C and 13–8); or fine hairs, called trichobothria, which are restricted to the appendages (Fig. 13–7D). The base of a trichobothrium is expanded to form a small ball that fits into a large, structurally complicated socket in the integument. The hair base contains a process from a sensory nerve cell of the hypodermis and is stimulated by very slight vibrations or air currents. For most arachnids the sensory hairs are the primary sense organs.

The eyes of all arachnids are similar (Fig. 13–7A and B). They are always composed of a combined cornea and lens, which is continuous with the cuticle but much thicker. Beneath the lens is a layer of hypodermal cells known as the vitreous body. The retinal layer, containing the photoreceptor cells, lies behind the vitreous body.

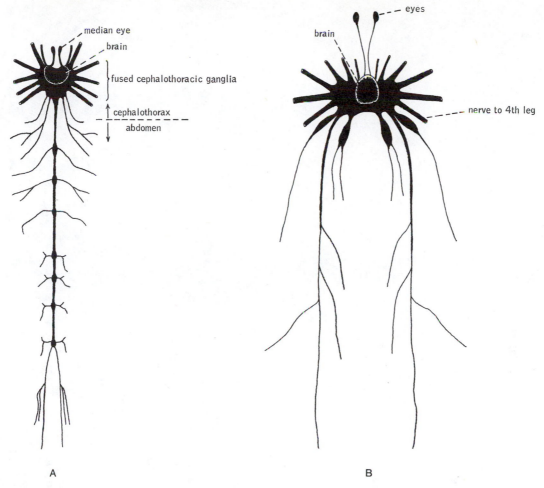

A

B

Figure 13–6 *A*, Scorpion nervous system, in which abdominal ganglia are distinct. *B*, An opilionid nervous system, in which all ventral ganglia have migrated forward and fused. (Both after Millot, J., 1949: Traité de Zoologie, Vol. VI. Masson et Cie, Paris.)

The photoreceptors are oriented either toward the light source (a direct eye) or toward the post-retinal membrane (an indirect eye). In the indirect eyes, the postretinal membrane may function as a reflector, called the tapetum, that reflects the light toward the receptors. Some arachnids possess only direct or only indirect eyes; many, such as spiders, have both.

A slit sense organ consists of a slitlike pit in the cuticle covered by a very thin membrane that bulges inward (Fig. 13–9). The undersurface of the membrane is in contact with a hairlike process, which projects upward from a sensory cell. Slit sense organs may occur in great numbers, either singly or in groups (lyriform organs) on the ap-

pendages and body of most arachnids. For example, the spider *Cupiennius salei* has 3000 slit sense organs; about half of are grouped, and most are on the appendages (Barth and Libera, 1970; Barth, 1978). The grouped, lyriform organs are near joints (Fig. 17–7E), and the simple slits are parallel to the long axis of the leg segment.

Slit sense organs respond to slight changes in the tension of the exoskeleton that result in compression (narrowing of the slit) and bowing in of the covering membrane to which the dendrite is attached. In a lyriform organ the slits are of different lengths, which means that varying numbers will respond to a given amount of tension over that part of the skeleton. Slit sense organs may respond

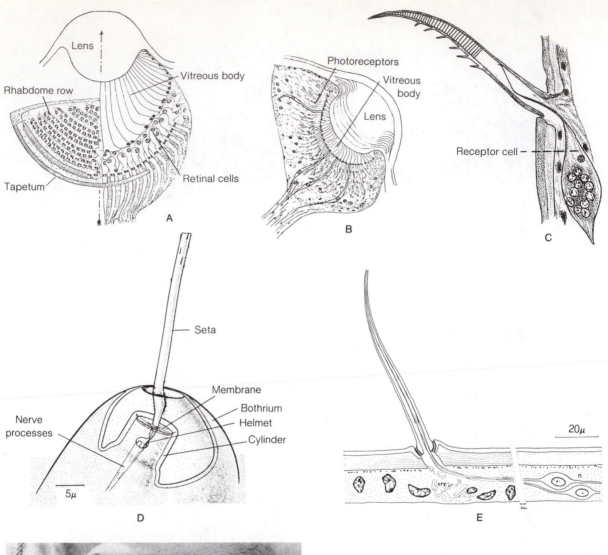

Figure 13–7 Arachnid sensory structures: *A*, Diagram of an indirect eye of a spider having a tapetum. Sagittal view to right of arrow; three-dimensional view to left of arrow. *B*, Sagittal section of a direct eye of a spider. (*A* and *B* from Homann, H., 1971: Z. Morphol. Tiere, *69*(3):201–273.) *C*, Sensory seta of a mite. (After Gossel from Millot.) *D*, Trichobothrium of a spider. Only lower part of hair is shown. (From Gorner, P., 1966; Cold Spring Harbor Symposium Quant. Biol., *30*:69–73.) *E*, Chemosensory hair of a spider (*n*, nerve). (From Foelix, R. F., 1970: J. Morphol., *132*:313–334.) *F*, Surface view of a lyriform sense organ. (Courtesy of D. J. Harris.)

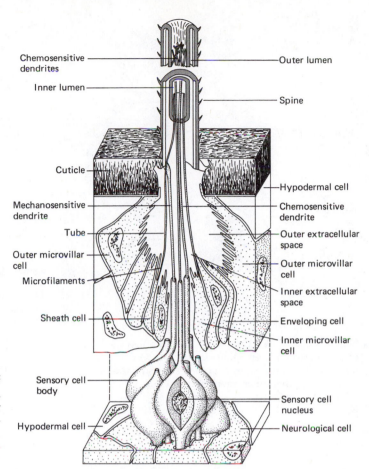

Figure 13–8 Diagrammatic reconstruction of a sensory hair near the tarsal-metatarsal joint on the leg of a spider *(Ciniflo)*. Note that the hair contains both chemoreceptors and mechanoreceptors. (From Harris, D. J., and Mill, P. J., 1973: The ultrastructure of chemoreceptor sensilla in *Ciniflo*. Tissue and Cell, 5(4):679–689.)

to load stresses in locomotion and in this respect are proprioreceptors. They may also respond to gravity, to substrate, and even to air-borne vibrations (see Foelix, 1982, for brief review).

Gas Exchange

Arachnids possess book lungs or tracheae or both. Book lungs, which are always paired, are more primitive and are probably a modification of book gills, an adaptation associated with the migration of the arachnids to a terrestrial environment. Each book lung consists of a sclerotized pocket that represents an invagination of the ventral abdominal wall (Fig. 13–10). The wall on one side of the pocket is folded into leaflike lamellae. The lamellae are held apart by bars that enable the air to circulate freely. Diffusion of gases takes place between blood circulating within the lamellae and the air in the interlamellar spaces (Fig. 13–10). The nonfolded side of the pocket forms an air chamber (atrium) that is continuous with the interlamellar spaces and opens to the outside through a slitlike opening (spiracle). Some ventilation results from the contraction of a muscle attached to the dorsal side of the air chamber. This contraction dilates the chamber and opens the spiracle, but most gas movement is by diffusion.

The tracheal system of arachnids is similar to that of insects in being composed of internal, chitin-lined tubes through which gases diffuse, but the two systems evolved independently. Tracheae tend to be most highly developed in small arachnids, which would be subject to greater water loss with book lungs. Ricinuleids, pseudoscorpions, and some spiders possess sieve tracheae derived from book lungs. In tracheal systems of this type, the spiracle opens into an atrial or tubelike chamber from which arises a great bundle of tracheae. Mites, harvestmen, solifuges, and most spiders possess tube tracheae, in which the tracheae do not arise as a bundle but are simple branched or unbranched tubes (Fig. 13–27B). The spiracles may open into an initial atrium or directly into a tracheal tube. At least in spiders, the trachea terminates in the hemocoel rather than in muscle or

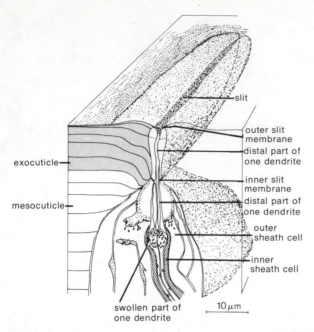

slit

outer slit membrane
distal part of one dendrite

inner slit membrane
distal part of one dendrite

outer sheath cell

inner sheath cell

exocuticle

mesocuticle

swollen part of one dendrite

10 μm

Figure 13–9 Diagrammatic section through a slit sense organ. Two dendrites innervate the slit. One terminates on the outer covering membrane of the slit, and one terminates on the inner membrane. (From Barth, F. G., 1971: Der sensorische Apparat der Spaltsinnesorgane (*Cupiennius salei* Keys.) Z. Zellforsch, *112*:212–246.)

other tissues, as in insects. Blood is still the intermediate gas-transporting agent in these animals. The blood of some arachnids, such as scorpions and many spiders with book lungs, contains the respiratory pigment hemocyanin.

Internal Transport

The heart is almost always located in the anterior half of the abdomen. In its primitive condition the heart bears seven pairs of ostia, each corresponding to a segment. But all degrees of reduction are found in the different arachnid orders (Fig. 13–25).

A large, anterior aorta supplies the prosoma; a small, posterior aorta leads to the posterior half of the abdomen, and from each heart segment emerges a pair of small, abdominal arteries. Small arteries eventually empty the blood into tissue spaces and then into a large ventral sinus that bathes the book lungs. One or more pairs of venous channels conduct blood from the ventral sinus or the book lungs back into the pericardial chamber.

Reproduction

The genital orifice in both sexes is usually found on the ventral side of the second abdominal or eighth body segment (Fig. 13–12). The gonads lie in the abdomen and may be either single or paired.

Indirect sperm transmission with a spermatophore is characteristic of many arachnids (Weygoldt, 1974). It appears to be the original mode of sperm transfer for the class and an adaptation for reproduction on land. The behavior of some living arachnids suggests that, primitively, the spermatophores were probably randomly deposited by the male. If a female encountered a spermatophore, she would be attracted chemotactically and would take it up within her gonopore. However, in most arachnids with spermatophores the male himself has become involved in attracting the female to the spermatophore. There is often a complex "court-

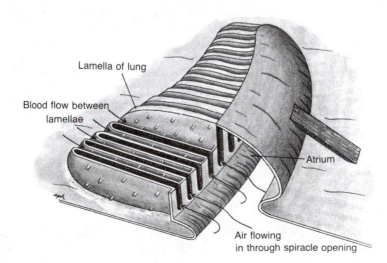

Lamella of lung

Blood flow between lamellae

Atrium

Air flowing in through spiracle opening

Figure 13–10 Diagrammatic section through a book lung.

ship" or behavioral pattern preceding mating. The sexes, especially the female, of different groups respond to chemical, tactile, or visual cues. Such cues provide recognition and elicit the receptivity and posture that indirect sperm transmission demands. This is especially important in highly predatory species, as are many arachnids.

The eggs are yolky and centrolecithal. In some groups cleavage is centrolecithal; in others, such as spiders, pyramids of yolk are associated with the initial cleavage divisions. The blastula, however, typically consists of a uniform blastoderm surrounding a yolky mass.

SUMMARY OF CLASS ARACHNIDA

The class Arachnida includes the following groups:

Order Scorpiones. The scorpions.
Order Palpigradi. The palpigrades.
Order Schizomida. The schizomids.
Order Uropygi. The whip scorpions, vinegarroons.
Order Araneae. The spiders.
Order Amblypygi. The amblypygids.
Order Ricinulei. The recinuleids.
Order Pseudoscorpiones. The false scorpions.
Order Solifugae. The solifuges, or solpugids.
Order Opiliones. Harvestmen, or daddy longlegs.
Acari. An assemblage of orders containing the mites and ticks.

ORDER SCORPIONES

The scorpions are among the oldest known terrestrial arthropods. Their fossil record dates back to the Silurian period. Silurian and Devonian scorpions were aquatic, possessing gills and having no tarsal claws. Terrestrial scorpions appear in the Carboniferous. Although abundant today, the 1500 to 2000 described species of scorpions are most common in tropical and subtropical areas. In North America scorpions are most abundant in the Gulf states and the Southwest.

Scorpions are generally secretive and nocturnal, hiding by day under logs, bark, and stones and in rock crevices or in burrows in the ground. But there are also species associated with vegetation. They are often found near dwellings, and the desert custom of shaking out shoes in the morning is a wise precaution. Scorpions are popularly believed to inhabit desert regions, but although many desert species exist, they are by no means restricted to arid situations. Many scorpions require a humid environment and live in tropical forests.

Scorpions are large arachnids, most ranging from 3 to 9 cm in length. The smallest species is the Middle Eastern *Microbuthus pusillus*, which is only 13 mm long, and the largest is the African *Pandinus*, which reaches 18 cm. However, some Carboniferous scorpions attained lengths of 44 and 86 cm.

External Anatomy

The scorpion body consists of a prosoma covered by a carapace and a long abdomen, ending in a stinging apparatus (Figs. 13–11 and 13–12). In the middle of the carapace is a pair of large, elevated, median eyes. In addition, two to five pairs of small, lateral eyes are present along the anterior, lateral margin of the carapace.

The chelicerae are small and chelate, and they project anteriorly from the front of the carapace (Fig. 13–11). The pedipalps, which are a distinguishing characteristic of scorpions, are greatly enlarged and form a pair of pincers for capturing prey.

The scorpion abdomen is composed of a seven-segment preabdomen and a postabdomen of five narrow segments, so the two regions are clearly differentiated, a primitive feature. A pair of opercular plates hide the genital opening on the midventral side of the first abdominal segment. Posterior to the genital plates and attached to the second abdominal segment is a pair of sensory appendages, known as pectines, that are peculiar to scorpions. Each pectine is a comblike structure that projects to each side from the point of attachment near the ventral midline.

The third through the sixth abdominal segments each bear on the ventral side a pair of transverse spiracles opening into book lungs. The segments of the postabdomen, sometimes called the tail, resemble narrow rings. The last segment bears the anal opening on the posterior ventral side and also bears the stinging apparatus characteristic of scorpions.

Stinging Apparatus and Feeding

The sting is attached to the posterior of the last segment and consists of a bulbous base and a sharp, curved barb that injects the venom (Fig. 13–12). The venom is produced by a pair of oval glands within the base of the apparatus. By a violent contraction of the muscular envelope around the glands, the liquid venom is ejected from the lumen of the glands into a sclerotized common duct that

A   B

Figure 13–11   *A*, The North American desert scorpion *Hadrurus arizonensis* in the alert position, waiting for passing prey. *B*, Successive positions of the abdomen when stinging prey. Total time required for the sequence is about 0.75 second. (From Bub, K., and Bowerman, R. F., 1979: Prey capture by the scorpion *Hadrurus arizonensis.* Jour. Arachnol., 7:243–253.)

leads to the outside through a subterminal opening of the stinging barb. The scorpion raises the post-abdomen over the body, so that it is curved forward, and stabs the prey (Fig. 13–11*B*).

The venom of most scorpions, although sufficiently toxic to kill many invertebrates, is not dangerous to man. At most, the sting is equivalent to that of a hornet. The scorpions of the southeastern United States and Gulf Coast region fall in this category, as well as many of our midwestern and western forms.

However, certain species possess a highly toxic venom that can be fatal to humans. The most notorious are *Androctonus* of North Africa and various species of *Centruroides* in Mexico, Arizona, and New Mexico. Humans stung by *Androctonus australis* may die in 6 to 7 hours. In Mexico species of *Centruroides* have been responsible for deaths, mostly in children.

The neurotoxic venom of scorpions is very painful and may cause paralysis of the respiratory muscles or, in fatal cases, cardiac failure. Antivenoms are available for dangerous species.

Scorpions feed on invertebrates, particularly insects. Many scorpions, perhaps most, sit in an alert position and wait for prey. The pedipalp fingers are open, and the tip of the movable finger and the tip of the pectines touch the ground (Fig. 13–11*A*) (Bub and Bowerman, 1979). Burrowing species will

wait near or at the burrow entrance. The North American desert scorpion *Paruroctonus mesaensis* exploits an area 5 meters in radius about the burrow entrance. The prey is detected by trichobothria on the pedipalps or at least in some scorpions from substrate vibrations through the tarsal hairs and slit sense organs. In a few seconds *Paruroctonus* can locate and dig out a burrowing cockroach 50 cm away (Brownell, 1977; Brownell and Farley, 1979; Brownell, 1984). The prey is caught and held by the large pincers while usually being killed or paralyzed by the sting (Fig. 13–11*B*). However, some species do not use the sting if the prey is readily subdued with the pedipalps, and scorpions with well-developed stings and highly toxic venom (Buthidae) have weaker pedipalps. The prey is transferred to the chelicerae, which slowly crush and tear it, and digestion begins.

Internal Structure and Physiology

Gas exchange is accomplished solely by book lungs. There are two pairs of malpighian tubules and a single pair of coxal glands, which open on the coxae of the third pair of walking legs.

Desert scorpions possess a number of adaptations for life under extremely desiccating conditions. Temperatures lethal for them are high (45°–47°C); evaporative water loss through their almost

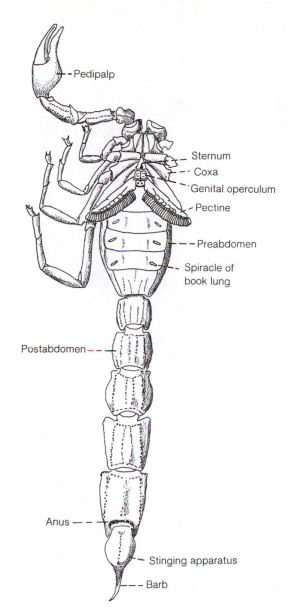

- Pedipalp
- Sternum
- Coxa
- Genital operculum
- Pectine
- Preabdomen
- Spiracle of book lung
- Postabdomen
- Anus
- Stinging apparatus
- Barb

Figure 13–12 *Androctonus australis* (ventral view). (After Lankester from Millot and Vachon.)

impervious exoskeleton is extremely low (0.01 per cent of body weight per hour at 25°C). They can tolerate a water loss of 40 per cent of their body weight (Hadley, 1974). In addition to being nocturnal, many are burrowers. The burrows of some species, such as *Hadrurus arizonensis* of the Sonoran Desert, may extend 90 cm. Some scorpions at times raise their bodies well off the ground. This behavior, called stilting, permits the circulation of air beneath the animal, which tends to prevent excessive elevation of body temperature and desiccation.

The scorpion nervous system, unlike that of other arachnids, retains a distinct nerve cord with seven unfused ganglia (Fig. 13–6*A*). Giant fibers integrate the pedipalpal grasp and the stinging thrust. The precise function of the pectines is still not clear. The ventral side of each tooth of the comb is liberally provided with sensory cells. The distal ends of these cells are contained within little spigot-like projections. During movement of the scorpion, the pectines are held out from the sides of the body in a horizontal position so that the teeth touch the ground. They are sensitive to vibrations and appear to determine some aspect of the substratum surface, perhaps particle size. Their removal prevents spermatophore deposition.

Scorpions exhibit a spectacular fluorescence, unlike other arachnids, and can be easily observed at night with ultraviolet light.

Reproduction and Life History

Male scorpions may have a larger abdomen than females, but the most useful character for distinguishing the sexes in scorpions is the hook present on the opercular plates of the male. The gonads are located between the midgut diverticula in the preabdomen. Prior to emptying into a single genital atrium, each oviduct dilates to form a small, seminal receptacle. In the male each sperm duct near the point of junction with its opposite member is modified for the formation of a spermatophore. In each sex the common genital atrium opens to the outside between the genital opercula on the first abdominal segment.

During the breeding season the male roams about and upon contacting a female initiates an extended courtship (Polis and Farley, 1979a). In some species the male and female face each other; each extends its abdomen high into the air and moves about in circles (Fig. 13–13*B* to *D*); in others the male rocks. The male then seizes the female with his pedipalps, and together they walk backward and forward (promenade). This behavior may last from 10 minutes to hours. Eventually, the male deposits a spermatophore, which is attached to the ground. A winglike lever extends from the spermatophore. The male then maneuvers the female so that her genital area is over the spermatophore. Pressure on the spermatophore lever releases the sperm mass, which is taken up into the female orifice (Fig. 13–13*A*).

All scorpions brood their eggs within the female reproductive tract and give birth to their young. The aplacental viviparous species have large, yolky eggs; development takes place in the lumen of the

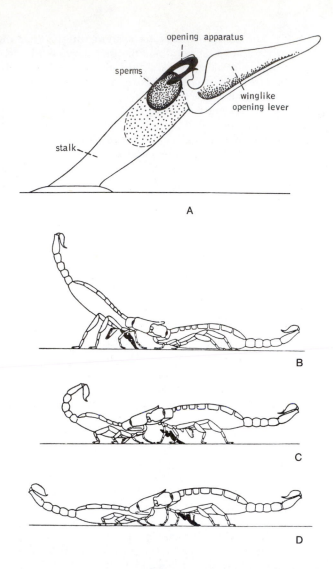

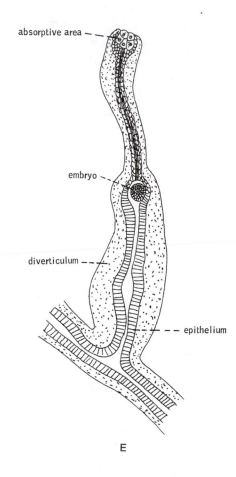

Figure 13–13   *A*, Diagram of a spermatophore of a scorpion. *B–D*, Sperm transfer
in scorpions: *B*, While holding female's pedipalps in his own, male on left deposits
spermatophore on ground. *C*, Female is
pulled over spermatophore. *D*, Spermatophore taken up into female's gonopore. (*A–
D* after Angermann.) *E*, Diverticulum of
the ovary containing a developing embryo
in the tropical Asian scorpion, *Hormurus
australasiae*, a placental viviparous species.
(After Pflugfelder from Dawydoff.) *F*, Female scorpion carrying young. (Courtesy of
H. L. Stahnke.)

ovarian tubules. The eggs of placental viviparous species possess little yolk. This type of development is found in the tropical Asian species, *Hormurus australasiae*. Its eggs develop within the diverticula of the ovary. Each diverticulum in turn develops a distal tubular appendage that contains a cluster of absorbing cells at the end (Fig. 13–13E). These cells rest against the maternal digestive ceca, from which nutritive material is absorbed. The nutritive material passes through the tubule to the embryo at the base.

Development takes several months or even a year or more, and 1 to 95 young are produced, depending on the species. At birth the young are only a few millimeters long, and they immediately crawl upon the mother's back (Fig. 13–13F). The young remain there through the first molt, which occurs in about one week. The young scorpions then gradually leave the mother and become independent. They reach sexual maturity in six months to six years, molting four to seven times (Polis and Farley, 1979b).

ORDERS PALPIGRADI, SCHIZOMIDA, AND UROPYGI

Three small orders of tropical and semitropical arachnids are similar in possessing a terminal, abdominal flagellum composed of numerous, small, articulating pieces (Fig. 13–14A). All live beneath wood and stones or in leaf litter and soil. The some 60 known species of Palpigradi and 80 species of

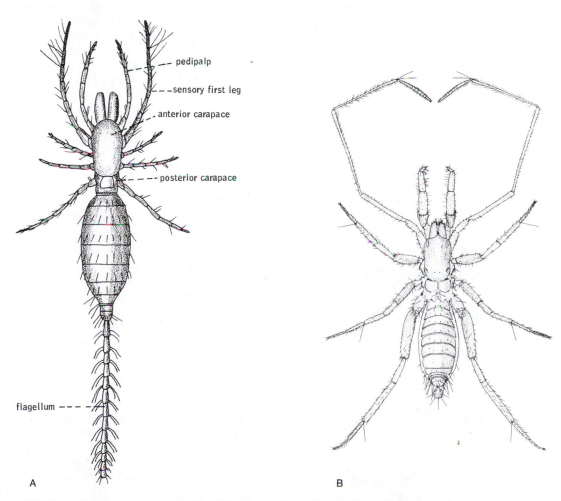

pedipalp

sensory first leg

anterior carapace

posterior carapace

flagellum

A

B

Figure 13–14 *A, Eukoenenia*, a palpigrade. (After Kraepelin and Hansen from Millot.) *B, Schizomus sawadai*, an Asian species of the order Schizomida. The body length is only about 5 mm. (From Sekiguchi, K., and Yamasaki, T., 1972: A redescription of "*Trithyreus sawadai*" from the Bonin Islands. Acta Arachnol., *24*(2):73–81.)

Schizomida are mostly less than 3 mm in length. The palpigrades have the prosomal carapace composed of two plates (Fig. 13–14*A*), and that of the schizomids is composed of three plates (Fig. 13–14*B*).

Members of the order Uropygi, of which there are about 85 described species, are called whip scorpions and include some large arachnids. The American *Mastigoproctus giganteus* reaches 65 mm in length (Fig. 13–15).

The prosoma is covered by a dorsal carapace that carries a pair of anterior median eyes and three or four pairs of lateral eyes. Uropygid chelicerae have two segments, and the distal piece forms a hook or fang that folds against the large basal piece. The pedipalps are stout, heavy, and relatively short; the last two articles of the pedipalps are frequently modified to form a pincer used in seizing prey. The pedipalps and the long, tactile, first pair of legs are held in front of the animal as it moves forward, with the tactile legs frequently touching the ground.

The posterior half of the abdomen contains a pair of large anal glands, which open one on each side of the anus. When irritated, the animal elevates the end of the abdomen and sprays the attacker with fluid secreted from these glands. The secretion in *Mastigoproctus* is 84 per cent acetic acid and 5 per cent caprylic acid, the latter permitting the acetic acid to penetrate the integument of an arthropod predator. The fluid can burn human skin, and the repellent odor has given this animal the common name *vinegarroon*.

During feeding, the prey is seized and torn apart by the pedipalps and then passed to the chelicerae. Two pairs of book lungs are located on the ventral side of the second and third abdominal segments.

Sperm transfer is indirect in all three orders and involves the deposition of a spermatophore. During part of the complex courtship behavior, the male holds the tips of the long, modified sensory legs of the female with his chelicerae (Fig. 13–16*A*). The female picks up the sperm packages with her genital area, and in *Mastigoproctus* and *Thelyphonellus* the male then uses his pedipalps to push them into the gonopore (Fig. 13–16*B*) (Weygoldt, 1978).

The female lays her eggs in a sac attached to her body. She remains in a shelter until they have hatched and undergone several molts.

ORDER ARANEAE

Except for perhaps the Acari, which comprise the mites and ticks, spiders constitute the largest order of arachnids. Approximately 37,000 species have been described, and this probably represents only a portion of the actual number. Also, spider populations are very large. Bristowe (1958) calculated that an acre of undisturbed, grassy meadow in Great Britain contained 2,265,000 spiders.

External Structure

Spiders range from tiny species less than 0.5 mm in length to large, tropical mygalomorphs (called tarantulas, bird spiders, or monkey spiders in different parts of the world) with a body length of 9 cm; leg span can be much greater. The convex carapace usually bears eight eyes anteriorly (Figs. 13–17 and 13–18). A large sternum is present on the ventral surface, and a small, median plate known as the labium is attached directly in front of the sternum (Fig. 13–17*B*).

Each chelicera consists of a fang and a basal piece containing a groove into which the fang folds

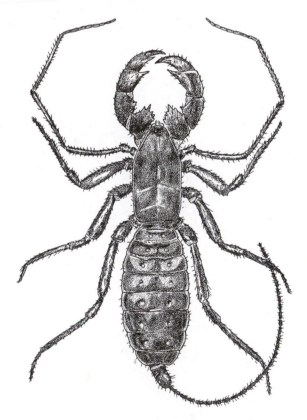

Figure 13–15   American whip scorpion. *Mastigoproctus giganteus.* (After Millot, J., 1949: Traité de Zoologie, Vol. VI. Masson et Cie, Paris.)

A

B

Figure 13–16 Mating in the uropy-gid *Mastigoproctus giganteus: A,* Male grasping the antenniform legs of the female with his chelicerae. *B,* Male inserting the spermatophore into the female gonopore with his chelicerae. (By Weygoldt, P., 1972: Zeitschrift des Kölner Zoo, *15*(3):98 and 99.)

(Fig. 13–18*A*). The female pedipalps are short and leglike, but in the male they have become modified as copulatory organs, with the last segment greatly enlarged and knoblike. The legs vary in length and heaviness, depending on the habits of the species.

The globe-shaped or elongated abdomen is unsegmented except in a few primitive spiders, and it is connected to the prosoma by a short and narrow portion called the pedicel (Fig. 13–17*A*). Anteriorly, on the ventral side of the abdomen is a transverse groove known as the epigastric furrow (Fig. 13–17*B*). The reproductive openings are located in the middle of this furrow, and the spiracles of the book lungs are on each side of it. The end of the abdomen bears a group of modified appendages, the spinning organs, called spinnerets (Fig. 13–17).

Silk

Each spinneret is a short, conical structure bearing many spigots, the openings from the silk glands (Fig. 13–19). The spinnerets are very mobile and can move independently. The silk glands themselves are large and are located within the posterior half of the abdomen.

The silk of spiders is a protein composed largely of glycine, alanine, serine, and tyrosine and is similar to the silk of caterpillars (see Kovoor, 1977). It

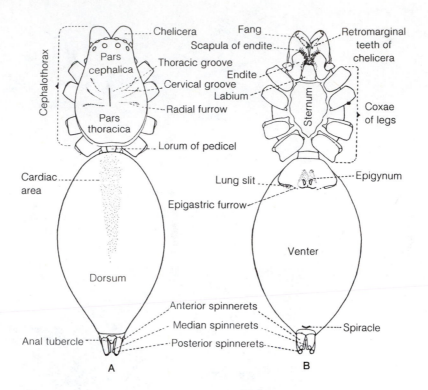

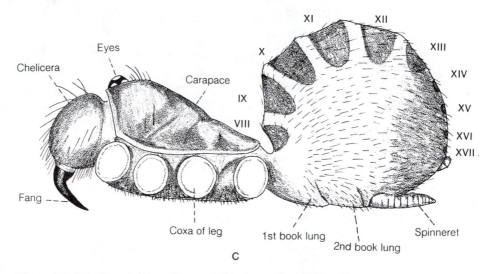

Figure 13–17 Dorsal (*A*) and ventral (*B*) views of a spider, showing external structure. (From Kaston B. J., 1948: Spiders of Connecticut. Bull. 70, State Geol. Nat. Hist. Survey, 13.) *C*, The primitive Asian mygalomorph spider, *Liphistius malayanus*, with legs removed (side view). (After Millot, J., 1949: Traité de Zoologie, Vol. VI. Masson et Cie, Paris.)

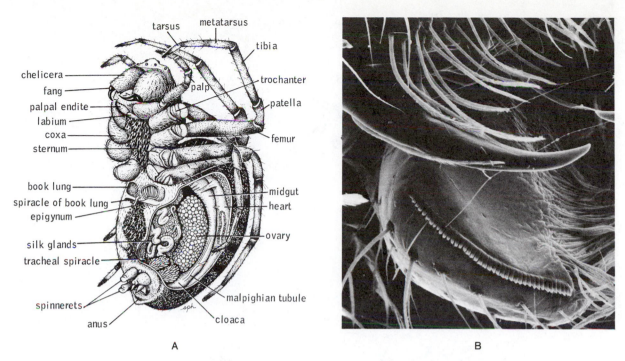

Figure 13–18 *A*, The orb weaver *Araneus diadematus* (ventral view). (After Pfurtscheller from Kaestner.) *B*, Sawlike edge on the palpal endite (coxa) of *Portia*, a jumping spider. Ridge is used for cutting prey. One fang shows above the palpal endite. (Courtesy of R. F. Foelix.)

is emitted as a liquid, and hardening results not from exposure to air but from the drawing-out process itself, which changes the molecular configuration (Work, 1977). Spider silk is about as strong as nylon but more elastic. A single thread is composed of several fibers, each drawn out from liquid silk supplied at a separate spigot. Drawing out occurs as the spider moves away from an anchored thread or pulls the threads with its posterior legs. Spiders produce more than one type of silk, and the various types are secreted from two to six kinds of silk glands.

Silk plays an important role in the life of the spider and is put to a variety of uses, even in the many families that do not build webs to catch prey. One funtion of silk that is common to most spiders is as a dragline (Fig. 13–20). Spiders continually lay out a line of dry silk behind them as they move about. At intervals it is fastened to the substratum with adhesive silk. The dragline acts as a safety line, and the common sight of a spider suspended in midair, after being brushed off some object, results from its continual retention of its dragline. Many spiders build silken nests beneath bark and stones, which they may use as retreats or in which

they may winter. All spiders wrap up their eggs within silken cases, and after hatching, the young of some species are carried on air currents by means of a silken strand. These functions of silk will be more fully described later.

Feeding

Spiders, like most other arachnids, are predatory and feed largely on insects, although small vertebrates may be captured by large species of spiders. The prey either is pounced on by members of the more active cursorial (wandering) groups or is caught in the silken snare of members of web-building families.

Cursorial forms include wolf spiders, fisher spiders, crab spiders, jumping spiders, and many mygalomorph spiders; the mygalomorphs are a more primitive group that contains the so-called tarantulas and the trap-door spiders. Species that stalk their prey typically have heavier legs than web builders. Also, most have a tuft of adhesive hairs (scopula) behind the terminal claws, which aid in adhering to surfaces and in prey capture (Rovner, 1978). Some species roam about catching insects

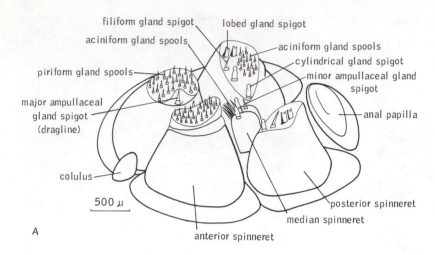

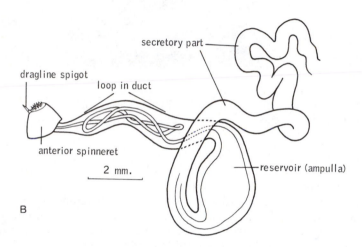

Figure 13–19 Spinnerets of the orb-weaving spider *Araneus diadematus:* *A,* Distribution of spigots for different silk glands on spinnerets. *B,* Anterior spinneret and gland supplying the dragline spigot. (From Wilson, R. S., 1969: Am. Zool., *9*(1):103–111.)

they happen to encounter; others utilize a sit-and-wait strategy. Many crab spiders, for example, sit in flower heads and ambush visiting insects. They can immobilize relatively large prey (Morse, 1979 and 1983). Prey is detected by tactile and visual stimuli, and in some families, such as the wolf spiders and the colorful jumping spiders, the eyes are highly developed (Fig. 13–28*B*) and of primary importance in prey capture. Hunting spiders lay down a dragline, and some ground spiders tie up their prey by running around them. Trap-door spiders construct silk-lined borrows that are closed by a lid covered with moss, soil, and other material (Fig. 13–36*A*). The spider lies in wait beneath the lid, and some species may detect passing prey by means of silk lines that radiate over the ground. If undisturbed, trap-door spiders may occupy the same burrow for many years (Marples and Marples, 1972).

Prey varies with habitat and distribution of the spider. The little wolf spider *Pardosa amentata,* which lives on sunny, moist ground in Europe, depends upon dipterans (flies and gnats) for about 70 per cent of its diet. It catches about one insect per day, and activity is greater in the morning than in the afternoon (Edgar, 1970). The North American green lynx spider, *Peucetia viridans,* which lives in low shrubs and herbaceous vegetation, was found to prey upon 62 species of insects plus a few species of spiders (Turner, 1979).

The dragline may have been the origin of the snare of web-building families. The first web was perhaps the radiating threads of the dragline laid down when the spider emerged from its protective retreat to catch passing prey (Fig. 13–21). Such lines might then have acquired a communicative function, informing the spider of the presence of prey when the line was struck. The retreat would

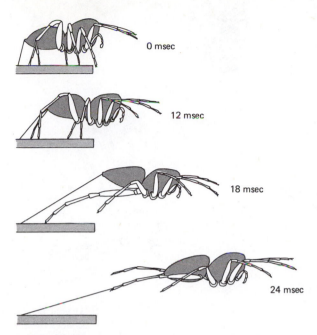

0 msec

12 msec

18 msec

24 msec

Figure 13–20 Tracings from motion picture frames of a jumping spider during a leap. Most of the force is provided by the rapid extension of the fourth pair of legs, which results from sudden elevation of blood pressure. Some jumping spiders jump with the third pair of legs; some with both third and fourth pairs. Note that the dragline is anchored before and retained during the jump. (After photographs by Parry and Brown from Foelix, R. F., 1982: Biology of Spiders. Harvard University Press, Cambridge, Mass. p. 155.)

also have become lined with drag line silk. From such a prototype evolved the highly developed triangles, orbs, funnels, sheets, and meshes characteristic of different spider families and species.

Many webs are composed of both dry and adhesive strands, each produced by different glands. The adhesive lines usually have an outer, unpolymerized, viscous layer of silk that greatly increases the trapping quality of the web; in some spiders (the cribellates) the adhesive is a very fine, entangling, loose "wool" attached to the line.

The construction of an orb web by a spider is a remarkable feat. Web building depends on morphological and physiological factors, such as weight, leg length, silk supply, appetite, and an instinctive behavioral pattern involving the integration of sensory information and locomotor activity. But visual information is not necessary, for a blinded spider can build a perfect web. No aspect is learned. The most complex web can be constructed by members of that species immediately

upon hatching. Considerable knowledge regarding web building has been provided by a pharmacologist, P. Witt, who has found the alterations in a spider's web-building behavior to reflect different drug effects.

Most of the familiar orb webs are produced by members of the family Araneidae. Web construction is begun with a horizontal line, which may be quite long. To place this line, the spider may let out a strand of silk that is carried by air currents. When the free end entangles another surface, it is pulled tight and anchored. Using the original horizontal line or a new one, the spider drops from the center, laying out a vertical line. The vertical line is pulled tight, converting the T to a Y frame, which provides the basic scaffolding of the web (Fig. 13–22). Additional frame lines are placed, and then the radii, or spokes, are attached from outside to center, first on one side and then on the other, using a preexisting spoke as a guide. With the radii in place, a temporary dry spiral of thread is laid from the center to the outside. Using the dry spiral as a scaffold, the permanent adhesive spiral is placed from the outside inward, and the spider removes the scaffold in the process. Since spacing of lines is determined by leg span, the size of the web is partly dependent on the size of the spider and increases with the spider's growth. The entire web, or at least the adhesive spiral, is usually replaced every day or night, for the wet silk loses its stickiness within a few days. The old silk may be eaten, and very quickly much of the protein finds its way back to the silk glands. Webs are replaced in a remarkably short time. *Araneus diadematus* spins a complete web in less than an hour and uses about 20 meters of silk (Foelix, 1982).

Orb webs vary in detail according to the trapping strategies of the species (Levi, 1978). Some have a fine mesh; others have a coarse mesh. Some have a closed hub; others have an open one, which makes it possible for the spider to move quickly from one side of the web to the other (Fig. 13–21). Some species stay in the center of the web; others hide in a curled leaf retreat holding onto a signal thread from the web. There are diurnal orb weavers that add conspicuous zigzag lines of silk in the hub of the web called stabilimenta (Fig. 13–23D). Acting as a warning signal, the stabilimentum reduces the chance that flying birds will destroy the web (Horton, 1981; Eisner and Nowicki, 1983).

The position and location of webs are not random. Some species spin horizontal orbs, catching insects that fly up and out of vegetation. Other species have vertical webs at characteristic elevations

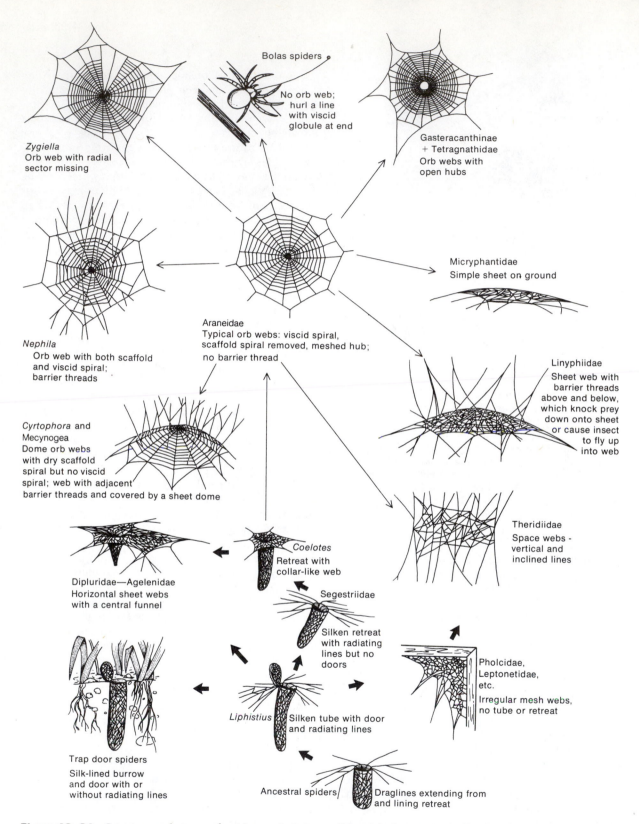

Figure 13–21 Common web types of spiders and their possible evolution. A somewhat parallel evolution of web-building by cribellate families is not included. (Based on Kaston, B. J., 1964: The evolution of spider webs. Am. Zool. 4:191–207; and on Levi, H. W., 1980: Orb webs: primitive or specialized. Proc. 8th Internat. Arach. Congr., Vienna, 367–370.)

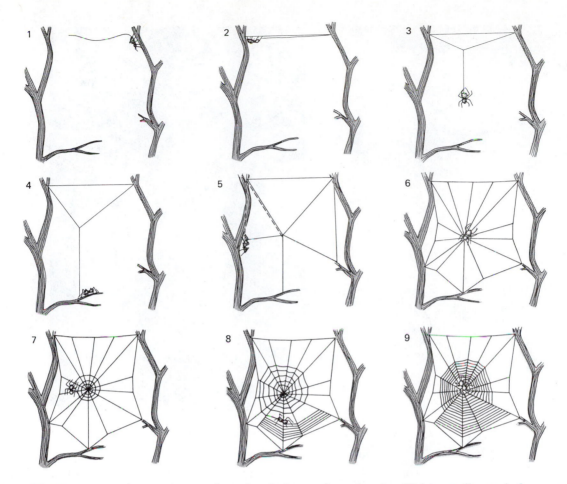

Figure 13–22 Stages in the construction of an orb web. See text for explanation. Web is usually attached to vegetation by additional frame lines, but these have been omitted to save space. (From Levi, H. W., 1978: Orb-weaving spiders and their webs. *American Scientist, 66:*734–742.)

to trap certain species of insects. Thus, the same wooded area may support a number of species of orb weavers, each adapted for trapping a different part of the insect population.

Web-building spiders are aerialists, and they usually have more slender legs than hunting spiders (Fig. 13-23*A* and *B*). Climbing about the web, a spider hooks the lines by a small middle claw that lies between the two large claws found in all spiders, and they are held against two outer setae called accessory claws. Why the spider does not stick to the web and how it can cut silk with its chelicerae are not known. Eyesight is not well developed in web-building spiders, but they are highly sensitive to vibrations. A web builder can determine from thread vibrations the size and location of the trapped prey, and many species will respond to different stimuli with different attack patterns (Fig. 13-24) (Robinson, 1969; Harwood,

1974; Klaerner and Barth, 1982). Orb weavers, tangle-web spiders, and others commonly swathe prey in silk before or after biting it. Swathing aids in immobilizing prey or, when it is used after an immobilizing bite, prevents the prey from falling out of the web or, in the case of hunting spiders, from an elevated position in vegetation. The scales on the wings of moths and butterflies are a defense against the adhesive webs of spiders, and spiders will immediately give a captured moth an immobilizing bite. A few species (kleptoparasites) have abandoned prey trapping and steal the catch of other web builders. Some of the many interesting variations in prey capture are described by Kaestner (1968), Gertsch (1979), and Foelix (1982).

Although most spiders are solitary, some degree of social organization has evolved in a few species within nine families. These social spiders share a communal web and cooperate in the capture of

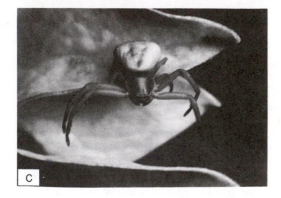

D

Figure 13–23   *A* and *B*, Two common, web-building house spiders: *A*, A species of *Pholcus*, a long-legged spider that builds a web of irregularly placed threads. Newly hatched young can be seen in the web at the top of the photograph. *B*, A comb-footed spider *(Achaearanea)*, which also builds a web of irregularly placed threads. *C*, A crab spider, member of the cursorial family Thomisidae. (*A–C* by Betty M. Barnes.) *D*, The orb weaver *Argiope aurantia* against the stabilimentum in the hub of its web. (From Tolbert, W. W., 1975: Predator avoidance behaviors and web defensive structures in the orb weavers *Argiope aurantia* and *Argiope trifasciata. Psyche,* 82(1):29–52.)

prey (see Kullmann, 1972; Burgess, 1976; Brach, 1977; Lubin and Robinson, 1982).

Spiders bite their prey with the chelicerae, which may also hold and macerate the tissues during digestion. The chelicerae of spiders are unique among arachnids in being provided with poison glands that open near the tip of the fang (Fig. 13–25). The glands themselves are located within the basal segments of the chelicerae and usually extend backward into the head. When the spider bites, the fangs are raised out of the groove in which they lie and are rammed into the prey. Si-

multaneously, muscles around the poison gland contract, and fluid from the gland is discharged from the fang into the body of the prey.

The venom of most spiders is not toxic to man, but a few species have dangerous bites. Among these are species of black widows (*Latrodectus*), which are found in most parts of the world, including the United States and southern Canada (Fig. 13–26*A*). The venom is neurotoxic and, as in most spiders, is composed of a mixture of protein compounds. Although the bite may be unobserved, the symptoms are severe and very painful; they include

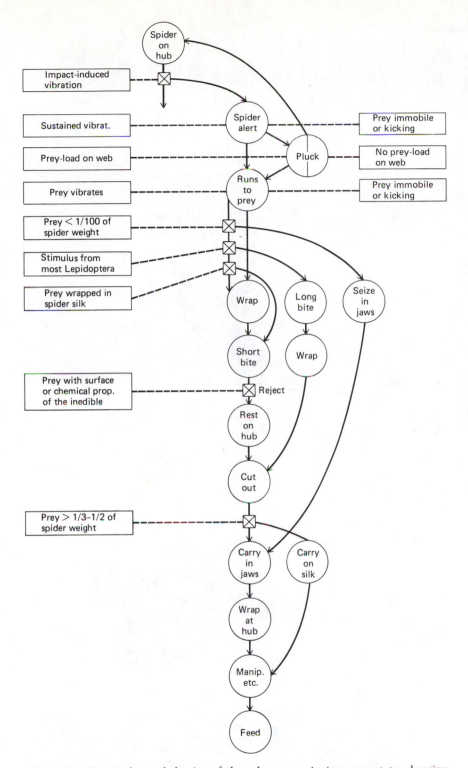

Figure 13–24 Predatory behavior of the orb weaver *Argiope argentata*, showing sequence and relationship to prey size. Circle depicts action of spider; square indicates prey. Sequence begins with spider waiting in center of web (on hub). When insect strikes web, spider becomes alert, following which it may run to prey or pluck web. Vibrations from web-plucking provide information on nature of prey. Large prey are given a long bite and injected with a larger amount of poison than in a short bite. They are then wrapped with silk. Small prey are first wrapped and then given a short bite. (From Robinson, M. H., 1969: Am. Zool. 9:161–173.)

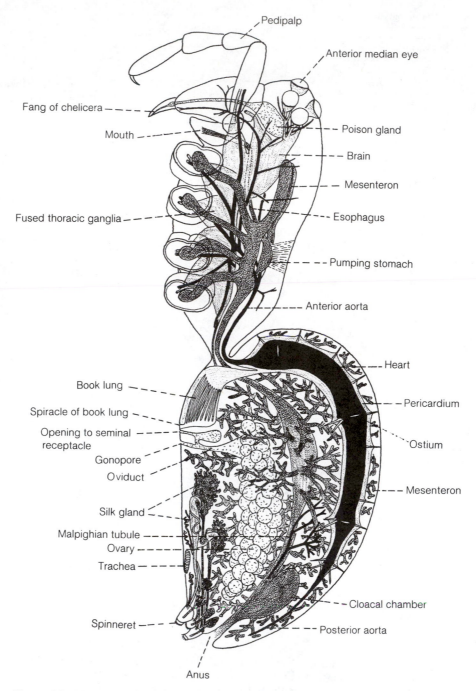

Figure 13–25 Internal anatomy of an araneomorph spider. (After Comstock.)

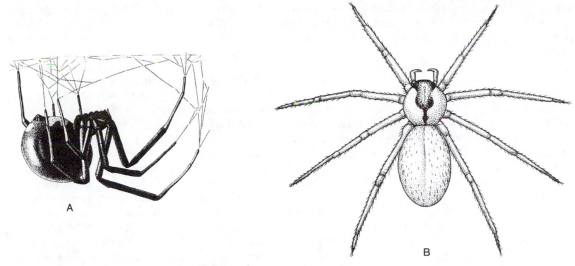

Figure 13–26 Venomous spiders: *A*, Female black widow, *Latrodectus mactans*, hanging in its web. Color is shiny black with red hourglass marking on ventral surface of abdomen. This species of the cosmopolitan genus is found in the southeastern and midwestern United States. (From Kaston, B. J., 1970: Trans. San Diego Soc. Nat. Hist., *16*(3):33–82.) *B*, The brown recluse spider, *Loxosceles reclusa*, whose bite produces a necrotic wound, difficult to heal. The brown recluse spider is found around human dwellings in the midwestern United States but has been introduced into other sections as well.

pain in abdomen and legs, high cerebrospinal fluid pressure, nausea, muscular spasms, and respiratory paralysis. Fatal cases are rare and are more apt to result from incorrect diagnosis and mistreatment (such as removal of the appendix) than from the spider venom. Antivenom and other treatments are available.

The recluse spiders, members of the genus *Loxosceles*, have a hemolytic venom and produce a local necrosis, or ulceration, of tissue that spreads from the bite. The ulcer is slow to heal. The bite of *Loxosceles reclusa*, the brown recluse spider (Fig. 13–26*B*), which is found in the midwestern United States and other localitites, can be dangerous, as can that of the large South American *Loxosceles laeta*. In southeastern Brazil a ctenid, *Phoneutria*, and wolf spider of the genus *Lycosa* have venoms that produce necrosis.

North American mygalomorph tarantulas, despite their size and reputation, are not venomous, but there are mygalomorphs, such as the Australian *Atrax* and the South American *Trechona*, both funnel-web builders, that are dangerous. However, many New World tarantulas (Theraphosidae) have defensive, urticating setae on the abdomen. These setae are barbed and irritating and readily penetrate the skin of a potential predators, such as small mammals, that enter the tarantula's burrow (Cooke et al, 1972). In humans they can cause a rash.

Spiders having chelicerae with teeth (Fig. 13–17*B*), such as many hunting forms, chew their prey, aiding the digestion of tissues by the enzymes poured out from the mouth. The undigestible skeletal remains are discarded as a wad. Spiders with toothless chelicerae do not chew but introduce enzymes through a puncture and suck out the digested tissues. In most spiders each of the palpal coxae (endites), which flank the mouth, has a saw-like edge used in cutting prey tissue and a screen of hairs that act as a filter when juices are sucked into the mouth (Fig. 13–18*B*). Sucking force comes from the pharynx and a posterior enlargement of the esophagus known as the pumping stomach (Fig. 13–25). It is located at about the middle of the prosoma.

The mesenteron fills almost the entire abdomen, as well as much of the prosoma (Fig. 13–25). The posterior part of the midgut becomes enlarged in the back of the abdomen to form a cloacal chamber. Waste is collected here and then discharged from the anus through a short, sclerotized intestine.

As an adaptation for predation and an uncertain food supply, most spiders have relatively low metabolic rates, can tolerate prolonged starvation, and because of the great capacity of the extensive midgut region, can double their body weight in one feeding. For example, the wolf spider *Lycosa lenta*, which has an estimated life span of 305 days, can survive without food for 208 days and reduce its metabolic rate by 30 to 40 per cent during that period (Anderson, 1974; Anderson and Prestwich, 1982).

Although spiders are important insect predators, spiders themselves are preyed upon by some insects and vertebrates. Two groups of wasps hunt spiders as food for their larvae. The wasp paralyzes the spider with a sting and lays an egg on its body. The spider wasps (Pompilidae) put the spider in an underground chamber; the mud daubers (Sphecidae) put a number of paralyzed spiders in a mud tube attached to elevated objects. The pressure of vertebrate predation on spiders is reflected in the fact that densities of spider populations are ten times greater on islands without vertebrate predators (Schoener and Toft, 1983).

Internal Structure and Physiology

Gas exchange organs in spiders are of two forms, book lungs and tracheae. Spiders considered primitive, such as the mygalomorphs, have no tracheae but possess two pairs of book lungs derived from the second and third abdominal segments (Figs. 13–17C and 13–27A). Almost all other spiders have only a single pair of book lungs, the posterior pair having evolved into tracheae. Coinciding with this transformation in most spiders, a posterior mi-

gration and fusion of the tracheal openings have taken place, forming a single spiracle located just in front of the spinnerets (Fig. 13–17B). In several groups of small spiders the anterior book lungs have also been transformed into tracheae, so gas exchange is accomplished entirely by this means (Fig. 13–27B). Although the tracheal system of spiders appears to have developed in most cases from the book lungs, the modification probably evolved independently within a number of spider groups (Levi, 1967).

The spider circulatory system is similar to that of scorpions (Fig. 13–25), but as the book lungs become smaller and the tracheae increase in complexity, there is a corresponding loss in the number of heart chambers, ostia, and arteries, until in the exclusively tracheate spiders, most of which are small, there are often only two pairs of ostia.

Blood pressure in a resting spider is equal to that of humans and may double in an active spider or one ready to molt. Blood pressure also causes the legs to extend in opposition to flexor muscles.

Coxal glands are not as well developed in spiders as in other arachnids. Groups considered primitive have two pairs of coxal glands opening onto the coxae of the first and third pair of walking legs. The coxal glands are highly developed in these forms. In all others only the anterior pair of glands remain, and they display various stages of regression.

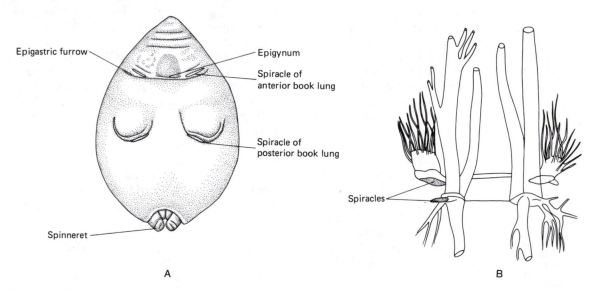

A   B

Figure 13–27 *A*, Abdomen of a mygalomorph spider, showing the two pairs of book lungs (ventral view). (After Comstock.) *B*, Central part of tracheal system of the caponiid spider, *Nops coccineus*, in which both pairs of book lungs have been replaced by tracheae. (After Bertkau from Kaestner.)

Although coxal glands in many spiders have become reduced functionally as excretory organs, it is possible that the silk glands and perhaps even the poison glands evolved from them.

In most spiders the two branched malpighian tubules, which are connected to the cloacal chamber in the posterior part of the abdomen, are more important in excretion than the coxal glands (Fig. 13–25). The excretory wastes are guanine and uric acid. Certain cells located beneath the integument and over the surface of the intestinal diverticula also store guanine crystals and account for the white color of some spiders.

Sense Organs

The eyes of some spiders surpass those of all other arachnids in degree of development. Usually, there are eight eyes arranged in two rows of four each along the anterior dorsal margin of the carapace (Fig. 13–28A), but other arrangements of the eyes are characteristic of certain families. In the wolf spiders and jumping spiders the eyes are situated over the surface of the anterior half of the carapace, resulting in a wide visual field (Fig. 13–28D). Of the two rows of four eyes usually found in spiders, the anterior median eyes are of the direct type, and all the remaining eyes are indirect. The anterior median eyes of jumping spiders are unique in having the photoreceptors arranged in four layers, each of which is believed to respond to different wavelengths. Whether such differentiation makes possible color recognition or greater contrast is still uncertain. The few studies have been largely restricted to the jumping spiders, which have the most highly developed eyes.

Since the indirect eyes are provided with a tapetum, which reflects light rays, the anterior median eyes appear dark and the others often appear pearly white. In some families, notably the hunting wolf spiders, the tapetum has developed to such an extent that these spiders can easily be located at night and captured by using a flashlight to look for the reflection of the eyes. Jumping spiders, on the other hand, have no tapeta at all in the indirect eyes.

The number of receptors is much greater in cursorial (hunting) species than in the sedentary web builders. In the hunting spiders eyes are important for detecting movement and locating objects. The jumping spiders are capable of perceiving a relatively sharp image of considerable size. In this family the number of receptors is large, particularly in the anterior median eyes (about 1000 photoreceptors), in which the tapered ends of the retinal cells are greatly narrowed and compact. In many species of jumping spiders there has been an increase in the depth of the anterior median eyes, resulting in a somewhat tubular telephoto structure (Fig. 13–28C). Muscles attached at the rear can rotate the tubes around the visual axis.

Chemosensitive tubular hairs on tips of appendages, tactile hairs (most of the large body hairs), trichobothria, slit sense organs, and tarsal organs are very important sense organs in all spiders; in most spiders, especially the sedentary, web-building forms, they are more important than the eyes. The funnel-web builder *Agelena*, for example, can determine the position of prey when it is 1 cm distant by means of the trichobothria.

Spiders have the greatest development of slit sense organs of any group of arachnids. Studies on the hunting spider *Cupiennius* have demonstrated the importance of grouped slit sense organs (lyriform organs) on the femur and tibia in kinesthetic orientation (Seyfarth and Barth, 1972). Blinded individuals chased as far as 25 cm from recent captured prey are able to return to their original position, correctly determining the angle and distance of the return path. This ability is greatly reduced if the lyriform organs on the femur and tibia are destroyed.

Slit sense organs in the joint between the tarsus and the metatarsus (the last two long leg segments) of web-building spiders are especially sensitive and enable the spider to discriminate vibration frequencies transmitted through the silk strands of the web or even through the air. In species in which the spiderlings remain in the parental web after hatching, the parent can recognize its young. The spider can also discern the size of the entrapped prey or even the kind of insect if it produces buzzing vibrations. The spiders themselves may produce vibration signals. The male may tweak the strands of the female's web, or the mother may produce vibration signals that are detected by the spiderlings. Tarsal organs, cuplike structures situated near the tips of the legs, are probably olfactory receptors for pheromones (Dumpert, 1978).

Reproduction and Life History

The ovaries of the female consist of two elongated, parallel sacs located in the ventral part of the abdomen (Fig. 13–25). At maturity, the eggs rupture into the lumen of the ovary and pass into the

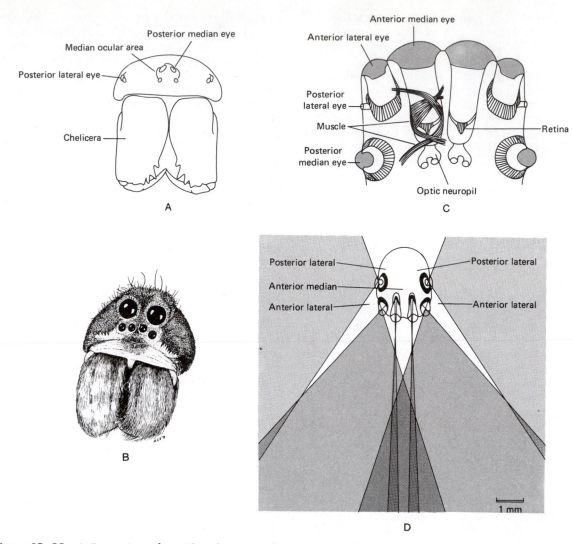

Figure 13–28 *A,* Front view of a spider, showing eight eyes arranged in anterior and posterior rows of four each, a pattern exhibited by many families. (After Kaston.) *B,* Face of a wolf spider. Posterior median eyes are very large; the posterior lateral eyes are located back over the carapace and can just be seen. *C,* Eyes of a jumping spider. Note the tubular, telephoto structure of the anterior median eyes. (*C* modified after Land, M. F., 1969: Structure of the retinae of the principal eyes of jumping spiders in relation to visual optics. Jour. Exp. Biol., *51:*443 [and] Movements of the retinae of jumping spiders in response to visual stimuli. Jour. Exp. Biol., *51:*771.) *D,* Visual fields of the eyes of a jumping spider *(Trite planiceps).* The posterior lateral eyes (PL) function in prey detection; the anterior median eyes (AM) function in tracking the prey after swiveling the body in the prey's direction. (Modified from Forster, L. M., 1979: Visual mechanisms of hunting behavior in *Trite planiceps,* a jumping spider. New Zealand Jour. Zool., *6:*79–93.)

curved oviduct leading from each ovary. Each oviduct converges to form a median tube (Fig. 13–29), which extends ventrally and backward to join with a short, chitinous vagina, opening at the middle of the epigastric furrow.

Associated with the vagina and uterus are two or more seminal receptacles and glands (Fig. 13–29). In most spiders these have separate openings to the outside and are connected internally to the vagina. These external openings are for the reception of the male copulatory organ during mating and are located just in front of the epigastric furrow on a special sclerotized plate called the epigynum (Figs. 13–17*B* and 13–29). Fertilization oc-

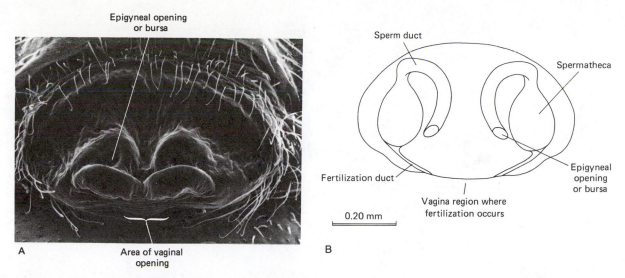

Epigyneal opening
or bursa

Sperm duct

Spermatheca

Fertilization duct

Epigyneal
opening
or bursa

Vagina region where
fertilization occurs

0.20 mm

A

B

Area of vaginal
opening

Figure 13–29 *A*, Epigynum, the chitinous plate bearing the reproductive openings of the female spider of *Philoponella republicana*. *B*, Reproductive ducts beneath the epigynum of the same species. At copulation the embolus of the male palp (see Fig. 13–30*D*) is inserted into the epigyneal opening to the spermatheca. The fertilization duct carries the sperm to the common oviduct (vagina), where fertilization occurs when the eggs are released through the primary genital opening (gonopore), which is hidden in these figures. (From Opell, B. D., 1979: Revision of the genera and tropical American species of the spider family Uloboridae. Bull. Mus. Comp. Zool., *148*(10):443–549.)

curs at the time of egg laying, mating having taken place some time earlier. Sperm are stored in the seminal receptacles, or spermathecae.

The two large, tubular testes lie along each side of the abdomen, and the two convoluted sperm ducts open through a comon genital pore in the middle of the epigastric furrow.

The copulatory organs of the male are not connected to the sperm duct opening but are located at the ends of the pedipalps. The tarsal segment of these appendages has become modified to form a truly remarkable structure for the transmission of sperm. Basically, each palp consists of a bulblike reservoir from which extends an ejaculatory duct (Fig. 13–30*B* and *C*). This leads to a penis-like projection called the embolus. At rest the bulb and embolus fit into a concavity, the alveolus, on one side of the tarsal segment.

During mating (Fig. 13–31*A*), special regions (hematodocha) of the tarsal segment become engorged with blood, causing the bulb and embolus to twist and project out of the alveolus (Fig. 13–30*D*); the embolus is at the same time inserted into one of the female reproductive openings leading to the seminal receptacles.

In primitive spiders the palpal organ consists only of the basic parts previously described (Fig. 13–30*B*). In the majority of families, however, the palp has become much more complicated with the addition of a great many accessory parts, such as the conductor and various processes (Fig. 13–30*D*). These structures, in combination with the sclerotized configurations of the female epigynum, aid in orienting the palp and inserting the embolus into the female openings.

The precise form of the male palp and female epigynum is distinctive for each species, and therefore these organs are the primary structures used by araneologists for classifying and identifying spiders at the species level.

The female and the male reproductive systems and the male palpal organ are not completely formed until the last molt has taken place. The male then fills, or charges, the pedipalps with sperm in the following manner. He first spins a tiny sperm web, or at least a strand of silk, with special silk glands that open onto the anterior ventral surface of the abdomen. Upon this web is ejaculated a globule of semen (Fig. 13–30*A*). Next the palps are dipped into the globule until all of the semen is taken up into the reservoirs. With the palps filled, the male spider then seeks a female with which to mate. Courtship behavior is not dependent on sperm-filled palps, however.

The predatory habits of spiders, as in other arachnids, makes recognition of the sexual partner

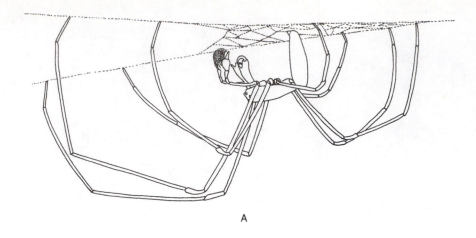

A

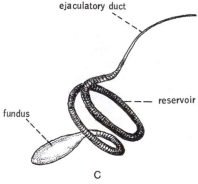

B

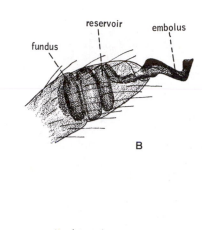

C

Cymbium

Median apophysis spur

Median apophysis bulb

Embolus (inserts into bursa of epigynum)

Femur

Middle hematodocha (here partly expanded or inflated)

Tegular spur (serves as an embolus guide)

D

Figure 13–30 *A*, Male tetragnath spider in sperm web, filling palps from globule of semen. (After Gerhardt from Millot.) *B*, Simple palp of *Filistata hibernalis*, *C*, The semen-containing part of the male palp. (*B–C* after Comstock.) *D*, Left male palp, the copulatory organ, of *Zosis geniculatus*. The severed end of the palp projects at the lower left corner. (From Opell, B. D., 1979: Revision of the genera and tropical American species of the spider family Uloboridae. Bull. Mus. Comp. Zool., *148*(10):443–549.)

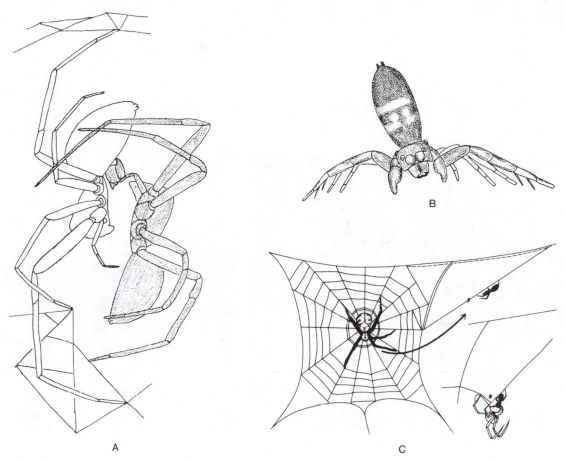

Figure 13–31 *A*, Mating position of *Chiracanthium* (male is shaded). (After Gerhardt from Kaston.) *B*, Courting posture of the male jumping spider, *Gertschia noxiosa*. (After Kaston.) *C*, Courtship behavior of some orb weavers. Male vibrates mating thread he has attached to female's web. Female leaves nub and moves onto mating thread, where copulation occurs (star). The letters *o* and *i* refer to origin and insertion of male mating thread. (From Robinson, M. H., and Robinson, B., 1980: Comparative studies of the courtship and mating behavior of tropical araneid spiders. Pacific Insects Monograph Bishop Museum, Honolulu, *36*:1–218.)

especially important, i.e., the female must identify the male as a potential mate and not as food. As a result, highly complex precopulatory behavior patterns have evolved in many species. In all spiders chemical and tactile cues are of primary importance. On encountering a dragline or web, a male spider can determine whether it was produced by a mature female of the same species, and the pheromone on the silk or body of the female may initiate trail following or the male courting response (Tietjen and Rovner, 1980; Suter and Renkes, 1982). Evidence of such a substance is indicated by the fact that a male will court the severed leg of a female or the evaporated washings of the female's body.

Females respond to a variety of cues produced by the male. In the sedentary web builders, the male often plucks the strands of webbing, producing vibrations that are detected and recognized by the female. The male of orb weavers plucks the radius held by the female or plucks a mating thread that he has attached to the female's web (Fig. 13–31*C*). The message is species specific in the number, frequency, and intensity of plucks (Witt, 1975; Robinson and Robinson, 1980). In some species the male is so small that he clambers ignored over the body of the female (Fig. 13–32*B*).

The cursorial, or hunting, forms display the most unique courtship behavior. The approach may be direct, the male pouncing on the female,

A

B

Figure 13–32 *A,* One of the courtship postures of a male jumping spider. This Australian species is one of a number in various families that mimic ants. Some feed on ants, but most mimic ants as a means of defense. The mimicry includes behavioral as well as structural features. In some species of ant mimics, like this one, the chelicerae are very long and heavy and project forward between the two small pedipalps. (From Jackson, R. R., 1982: The biology of ant-like jumping spiders; intraspecific interactions of *Myrmarachne lupata.* Zool. Jour. Linn. Soc., 76:293–319.) *B,* Male of the giant New Guinea wood spider on the dorsal surface of the female. Abdomen of female is about 3.2 cm. long. (From Robinson, M. H., and Robinson, B., 1973: Ecology and behavior of the giant wood spider *Nephila maculata* in New Guinea. Smithsonian Contrib. Zool., 149.)

palpating her body with his pedipalps and legs and causing her to fall into an immobile state. Some male wolf spiders stridulate with a file and scraper located in the tibiotarsal joint of the palp. During oscillations of the joint, a group of heavy spines at the tip of the palp maintain contact with the substratum, and the vibrations are detected by the female through the substratum rather than as airborne pressure waves (Rovner, 1975; Stratton and Uetz, 1981).

In families with well-developed eyesight, visual cues are also important, and courtship takes the form of dancing and posturing by the male in front of the female (Figs. 13–31*B* and 13–32*A*). This involves various movements and the waving of appendages, which are often brightly colored. Such behavior is most highly developed and has been studied most extensively in the colorful jumping spiders (Jackson, 1982).

Although descriptions of spider courtship behavior emphasize the role of the male, the behavior of each sex depends on a sequence of reciprocal signals that release the next act in one sex or the other. Primitively, according to Platnick (1971), the primary releaser for courtship is body contact. This level is exhibited by many tarantulas, crab spiders, and certain small ground spiders. Body pheromones or chemical or vibration signals of silk lines are more advanced primary releasers for many groups, such as wolf spiders and many web builders. The most highly evolved condition, exhibited by jumping spiders, utilizes visual cues as primary releasers.

Various copulatory positions characterize different families. On attaining the proper position, the male usually scrapes the epignyal surface rapidly with his palp until the proper orienting parts on the palp and epignyal plate connect. Then the palp becomes quickly engorged with blood, driving the embolus into the passageway of the seminal receptacle (Fig. 13–31). Following sperm transmission with one palp, which lasts some seconds to minutes, the other palp is inserted into the opposite seminal receptacle opening, which may require the male to move to the opposite side of the female. The behavioral sequence for the wolf spider *Lycosa rabida* is shown in Figure 13–33. Depending on the species, there may be numerous insertions, and the entire process may last a number of hours.

Adults of some species mate a number of times during their lifetime; others mate only once. The embolus of the male palp breaks off within the female duct in some spiders. There are also many species in which a plug forms, filling the openings into the seminal receptacles and preventing a second mating. However, in very few species does the female eat the male. The peculiar indirect mode of sperm transmission in spiders probably evolved from an ancestral habit in which the male placed a

spermatophore into the genital opening of the female with his pedipalp. As we have already seen, indirect sperm transmission by spermatophores is widespread among other arachnids, and in some groups, such as uropygids and solifugids, the male uses an appendage (the pedipalp in uropygids) to push the spermatophore or sperm mass into the female genital opening.

Some time after copulation, the female lays her eggs. Several to 3000 eggs are laid in one to several cases, depending on the species. Just before the deposition of eggs, the female spins a small basal sheet and then a cuplike wall. The eggs are laid in the cup and fertilized with stored sperm as they leave the female's body. The female now covers the eggs with another sheet of silk, and the entire mass is often given an additional covering of silk so that it assumes a spherical shape, the egg sac or cocoon (Fig. 13–34). The egg sac is attached to webbing (in web-building families), hidden in the spider's retreat, or attached to the spinnerets and dragged about after the mother (as in the wolf spiders and a few others). In many spiders the female dies after completing the egg sac; in others she may produce a series of cases and may remain with the young for some time after hatching.

The spiderlings hatch inside the egg sac and remain there until they undergo the first molt. The spiderlings of many species, as well as the adults of some small forms, use a form of transportation known as ballooning, which aids in the dispersal of the species. The little spider climbs to the top of a twig or blade of grass and releases a strand of silk; as soon as air currents are sufficient to produce a tug on the strand, the spider releases its hold on the plant and is carried away by the currents. Although

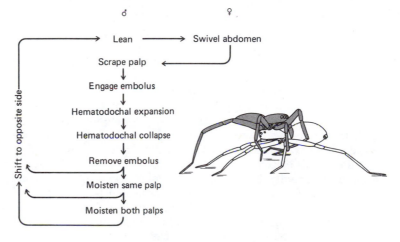

Figure 13–33 Sequence of events during copulation in the wolf spider *Lycosa rabida*. Swivel abdomen refers to the turning of the abdomen of the female to accommodate the position of the male. Figure at right shows pair in copulation, with male (black) inserting right palp into left seminal receptacle of female. (Diagram from Rovner, J. S., 1971: Mechanisms controlling copulatory behavior in wolf spiders. Psyche, 78(1):150–165; figure from Kaston, B. J., 1948: Spiders of Connecticut. Conn. State Geol. Nat. Hist. Survey Bull. 70, p. 2006.)

Figure 13–34 Black widow, *Latrodectus mactans*, with egg case. (By W. Van Riper, Colorado Museum of Natural History.)

ballooning is important in the spread of spider populations into new habitats, desiccation and lack of food would make long-distance transport unlikely (see Platnick, 1976).

Some spiders care for their young following hatching. Among them the female wolf spider carries her young on her back after they hatch (Fig. 13–35A). The mother's dorsal abdominal surface is covered with special, knobbed, spiny hairs that aid in the attachment of the young (Fig. 13–35B). The spiderlings gradually disperse. The mother does not feed while carrying young (Rovner et al, 1973).

Although some of the tarantulas (mygalomorph spiders) have been kept in captivity for as long as 25 years, most spiders live only 1 to 2 years. The number of molts required to reach sexual maturity varies, depending on the size of the species. Large species undergo up to 15 molts before the final instar; tiny species molt only a few times. Males molt fewer times than do females of the same species, and within a species there is always some variation in the number of instars before the final molt.

Almost half of temperate species overwinter as immatures, usually in soil or leaf mold; others overwinter in the egg stage or as adults or in all different stages (Schaefer, 1977). Temperature and photoperiod are the primary controlling factors in determining the pattern of the life cycle.

SYSTEMATIC RÉSUMÉ OF ORDER ARANEAE

Suborder Mesothelae. Original segmentation of abdomen reflected in numerous dorsal plates. Seven or eight spinnerets, the first of which is located at the level of the second pair of book lungs. The tropical Asian Liphistiidae (Fig. 13–17C)

Suborder Orthognatha (Mygalomorphae). Original segmentation of abdomen not apparent externally. Six or fewer spinnerets, located at end of abdomen. Possess two pairs of book lungs. The fang of the chelicera articulates in the same plane as the long axis of the body. This suborder includes the tarantulas (also called bird spiders or monkey spiders in various parts of the world) and trap-door spiders (Fig. 13–36). Although most stalk or lie in wait for their prey, there are some web builders. There are numerous representatives in North America. Many are large.

Suborder Labidognatha (Araneomorphae). This suborder includes most of the spiders. Abdomen not segmented. With few exceptions, only a single pair of book lungs is present, and in all cases the plane of articulation of the chelicerae is at right angles to the long axis of the body.

The araneomorph families can be roughly divided into web-building and cursorial forms. Descriptions of a few of the larger and more common families follows.

> **Family Pholcidae.** Small and often long legged, resembling phalangids or daddy longlegs (Fig. 13–23A). Members spin small webs of tangled threads in sheltered recesses. Several species commonly live in houses and, with other web-building house dwellers, are responsible for cobwebs.
>
> **Family Theridiidae.** A large family of tangled web (cobweb) builders known as comb footed spiders because of the presence of a series of serrated spines on the fourth tarsus. These spines comb out a band of silk used for trussing up the prey. To this family belong the black widow, *Latrodectus* (Fig. 13–26A), and one of the most common house spiders, *Achaearanea tepidariorum* (Fig. 13–23B).
>
> **Family Araneidae.** The orb-weaving spiders (Fig. 13–23D). Members of this family spin circular webs. Many species are of considerable size and brightly colored, such as the black and yellow *Argiope*, or "writing" spider.
>
> **Family Linyphiidae.** The sheet-web spiders. Webs are horizontal silken sheets or bowls

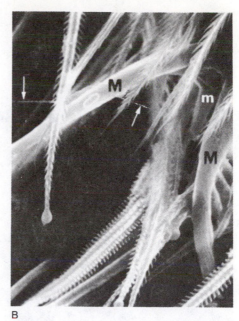

Figure 13–35 *A*, Female wolf spider carrying young on abdomen. (By W. Van Riper, Colorado Museum of Natural History.) *B*, Three tarsal claws (M, m, M) at end of leg of spiderling grasping knobbed, barbed hairs on dorsal surface of abdomen of wolf spider mother. (From Rovner, J. S., Higashi, G. A., and Foelix, R. F., 1973: Maternal behavior in wolf spiders: the role of abdominal hairs. Science, *182*:1153–1155. Copyright 1973 by the American Association for the Advancement of Science.)

(Fig. 13–21). The larger species construct their webs in vegetation. The great number of very tiny species in this family live in fallen leaves and humus.

Family Agelenidae. Funnel-web spiders. Although web builders, they are more closely related to certain cursorial spider groups. The web forms a funnel, the narrowed end acting as a retreat for the spider (Fig. 13–21). The web is constructed in dense vegetation or in crevices of logs or rocks, and it is easily visible, especially in grass covered with dew.

Family Lycosidae. Wolf spiders. These are rapidly moving and rather hairy spiders with dull brown and black coloration (Fig. 13–33). They are most active at night and are common members of the ground fauna.

Family Pisauridae. Fisher spiders. This family is somewhat similar to the wolf spiders in appearance but with longer legs. They are common around the edges of ponds, lakes, and streams.

Family Thomisidae. Crab spiders. This and the following family commonly lie in wait on vegetation. Crab spiders get their name from their crablike movements and the position of their legs (Fig. 13–23*C*).

Family Salticidae. Jumping spiders. Species of this family are heavy bodied and capable of jumping short distances (Fig. 13–20). They are often brightly colored and possess the best eyesight of all spiders. Numerous species live in the temperate and tropical regions of the world.

ORDER AMBLYPYGI

The amblypygids are a tropical and semitropical group of approximately 70 species. There are many common amblypygids, but they are a nocturnal and secretive group, hiding during the day beneath logs, bark, stones, leaves, and similar objects.

Amblypygids range in length from 4 to 45 mm and have a somewhat flattened body, which resembles that of spiders (Fig. 13–37). The carapace bears a pair of median eyes anteriorly and two groups of three eyes each laterally. The chelicerae are also similar to those of spiders, but the pedipalps are heavy and spiny and used to capture insect prey, which are located with the long, antenna-like first pair of legs. The gait of the amblypygid is crablike because of its flattened body and its ability to move laterally. One of the long, tactile legs is always pointed toward the direction of movement; the other may explore areas to either side of the animal. The segmented abdomen is connected to the prosoma by a narrow pedicel, like that of spiders. Two pairs of book lungs are located on the ventral side of the second and third abdominal segments.

A

B

Figure 13–36 Mygalomorph spiders: *A*, Trap-door spiders capturing a beetle. (By W. Van Riper, Colorado Museum of Natural History.) *B*, Tarantulas from the southwestern United States. (From Buchsbaum, R., and Milne, L., 1960: Lower Animals: Living Invertebrates of the World. Doubleday and Co., N.Y.)

In the species of *Charinus*, *Tarantula*, and *Admetus* in which reproductive habits have been observed, the male courts the female with trembling movements of the antenniform legs, alternating with rocking body movements toward the female (Fig. 13–38) (Weygoldt, 1972, 1978). Eventually, a spermatophore is deposited. Using his pedipalps or first legs, the male guides the female over the spermatophore, and she takes up the sperm masses. When egg laying is at hand, the reproductive glands secrete a parchment-like membrane that holds the 6 to 60 large eggs to the underside of the female abdomen. The mother carries the eggs in this manner until hatching and the first molt of the young have taken place. The young climb onto the abdomen of the mother until the next molt.

ORDER RICINULEI

The ricinuleids are a small order of arachnids containing 33 uncommon species, sometimes called tick spiders. They are found in Africa *(Ricinoides)* and in the American hemisphere *(Cryptocellus)* from Brazil to the southern United States, where they have been collected from leaf mold and caves.

Ricinuleids are heavy-bodied animals that measure from 5 to 10 mm in length (Fig. 13–39). The cuticle is very thick and often sculptured. Attached to the anterior margin of the carapace is a curious hoodlike structure that can be raised and lowered. When lowered, the hood covers the mouth and chelicerae. The pedipalps are leglike, and the abdomen is segmented and broadly joined to the carapace.

Sperm transfer is similar to that in spiders but involves the third legs instead of the pedipalps (Fig. 13–40). The first tarsal segment bears a specialized process containing a duct that the male fills with sperm from his genital opening. At copulation the male embraces the female from the back, and the copulatory process is inserted into the female genital atrium by the encircling third legs (Legg, 1977).

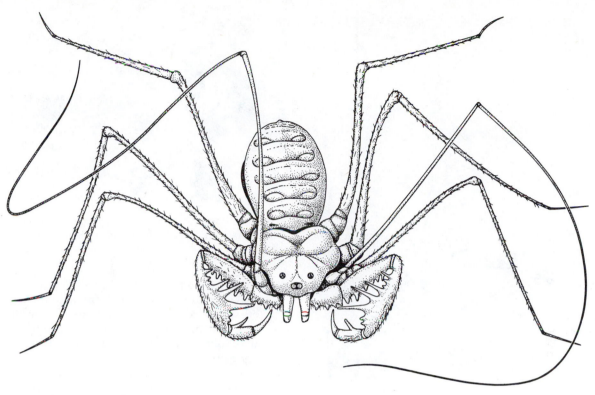

Figure 13–37   The African amblypygid, *Charinus milloti.* (After Millot.)

ORDER PSEUDOSCORPIONES

Pseudoscorpions are tiny arachnids, rarely longer than 8 mm and many only a few millimeters long. They live in leaf mold, in soil, beneath bark and stones, and in moss. Some species of several genera are common inhabitants of algae and drift in the intertidal zone. A cosmopolitan species, *Chelifer cancroides*, is found in houses.

Because of their small size and the nature of their habitat these animals are rarely seen, although they are actually very common. A few handfuls of leaf mold sifted through a Tullgren funnel (Fig. 13–48) usually will yield at least some individuals, and densities of several hundred per square meter have been reported (Johnson and Wellington, 1980). Pseudoscorpions are found throughout the world, and about 2000 species have been described.

Pseudoscorpions superficially resemble the true scorpions but are much smaller and lack the long abdomen and sting (Fig. 13–41). The rectangular carapace bears one or two eyes at each anterior lateral corner, or the eyes may be absent. The rela-

tively wide abdomen forms a broad junction with the prosoma and is rounded posteriorly.

Pseudoscorpion chelicerae are small and chelate and bear several accessory structures, which will be described later in connection with their functions (Fig. 13–42). The pedipalps are similar to those of the scorpions. However, these pedipalps are peculiar in that each usually has a poison gland in one or both fingers or in the hand. A duct issues from the poison gland and opens at the end of a tooth at the tip of the finger. Pseudoscorpions are slow moving and hold their pedipalps, which are supplied with trichobothria, to the front when they walk. If the pseudoscorpion is disturbed, it pulls the pedipalps back over the carapace and becomes immobile.

Pseudoscorpions feed on small arthropods, such as collembolans and mites. The prey is caught and paralyzed or killed by the poison glands in the pedipalps. It is then passed to the chelicerae, which tear open the exoskeleton. Digestion then takes place in typical arachnid fashion. Hairs at the front of the prebuccal chamber strain out the solid par-

Figure 13–38 Male of the amblypygid *Heterophrynus longicornis* tapping the body of a female with his antenniform legs. (By Weygoldt, P., 1972: Zeitschrift des Kölner Zoo, *15*(3):100.)

Figure 13–39 A ricinuleid from the southwestern United States. (Courtesy of Dr. Robert Mitchell.)

ticles. When a large mass of solid particles accumulates, the chelicerae are withdrawn and short, stiff hairs (flagella) on the chelicerae catch the mass of debris and eject it from the prebuccal chamber (Fig. 13–42). After feeding, the buccal pieces are cleaned by comblike structures (serrulae) on the fingers of the chelicerae.

Gas exchange in pseudoscorpions is accomplished by means of a tracheal system that opens through two pairs of spiracles on the ventral side of the third and fourth abdominal segments. Coxal glands that open on the coxae of the third pair of walking legs provide for excretion.

There is little secondary differentiation between sexes. The process of sperm transmission provides some clues as to possible stages in the evolution of indirect sperm transfer by spermatophores in arachnids (Weygoldt, 1969). In some species the male deposits a stalked spermatophore on the substratum in the absence of a female. If a female encounters the spermatophore, she is at-

tracted chemotactically. She takes a stance over the terminal sperm mass, and the sperm are taken into the female orifice. Uptake is facilitated by swelling of the sperm mass, triggered by some substance produced in the female atrium. This is probably the most primitive method of sperm transfer in pseudoscorpions. The course of subsequent evolution appears to have increased the chance of the female's obtaining the spermatophore.

There are certain other species in which the male, on encountering a female, lays down silk signal threads after depositing the spermatophore. When the female comes across the thread, she is guided to the spermatophore.

With pairing behavior, the male himself directs the female to the spermatophore in the more evolved patterns. The male grasps the female with his pedipalps and, following a promenade-like courting behavior, maneuvers the female over the spermatophore until it is in the proper position to be taken up. Finally, in the most specialized pattern, two long, tubelike organs are evaginated from the sexual region of the male abdomen. The female is attracted, and the two sexes promenade without touching each other. The male then attaches a spermatophore to the substratum and backs away. The female follows, and the male then seizes her pedipalpal femurs. When she is in the proper position, he aids in the uptake of the spermatophore by pushing with his forelegs (Fig. 13–43*A* to *C*).

The eggs appear after sperm transmission (a month in *Chelifer cancroides*). Before laying the eggs, the female uses small bits of dead leaves and debris to build a nest and lines it with silk emitted from duct openings on hornlike processes (galeae)

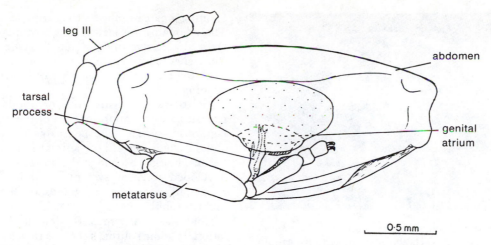

Figure 13–40 Copulatory process of third leg of male *Ricinoides hanseni* inserted into genital atrium of female. (From Legg, G., 1977: Sperm transfer and mating in *Ricinoides hanseni*. J. Zool. (London), *182*(1):51–61.)

on the chelicerae (Fig. 13–42). First the liquid silk is attached, and then the thread is drawn out as the chelicerae are moved to another position.

After the eggs are laid, they remain in a membranous sac attached to the genital opening on the ventral side of the body. Development takes place within the sac (Fig. 13–43*D*). In a later stage of development, the embryo is supplied with a nutritive material secreted by the mother's ovaries. The young undergo one molt before hatching and one during hatching and emerge from the brooding sac during the third instar. Depending on the species, 2 to more than 50 young may be brooded. The young pseudoscorpion molts twice again before becoming an adult. Molting takes place within a nest of silk that is constructed like the nest of the female at the time of egg production. A silk nest may be used for overwintering. Maturity is reached in a

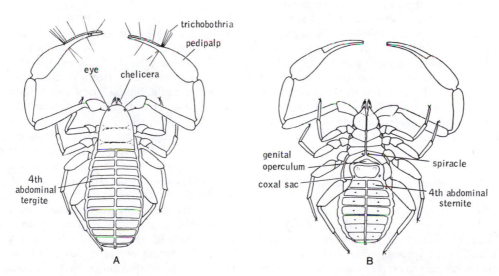

Figure 13–41 Dorsal (*A*) and ventral (*B*) views of the pseudoscorpion *Chelifer cancroides* (male), showing external structure. (After Beier from Weygoldt, P., 1969: Biology of Pseudoscorpions. Harvard University Press, Cambridge, Mass.)

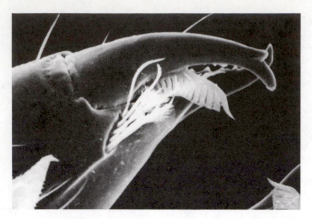

Figure 13–42 Scanning electron micrograph of a pseudoscorpion chelicera. The hornlike projections (galea) at the end of the movable finger contain the openings of the silk glands. The cluster of hairs (flagellum) at the base of the fixed finger is used to remove food particles that accumulate on the screening setae in the prebuccal cavity. The comblike structure (serrula) is used to clean the buccal region after feeding. (By Thomas Pangburn.)

year or less, and individuals may live two to five years. Temperate species may produce several generations a year (Goddard, 1976).

ORDER SOLIFUGAE

The solifuges are a group of about 900 tropical and semitropical arachnids, sometimes called sun spiders because of the diurnal habits displayed by many species, or wind scorpions because of the great speed with which they can run. In the United States a few species have been found in Florida, and more than 100 have been found in the Southwest, some as far north as Colorado. Many solifuges are common in the warm desert regions of the world. They hide under stones and in crevices, and many species burrow.

Solifuges are small to large arachnids, sometimes reaching 7 cm in length. The prosoma is divided into a large, anterior carapace bearing a pair of closely placed eyes on the anterior median border, and a short posterior section (Figs. 13–44 and 13–45). The segmented abdomen is large and broadly joined to the prosoma (Fig. 13–45).

The most striking characteristic of the solifuges is the enormous size exhibited by the chelicerae, which project in front of the prosoma and can be directed upward by the flexing of the subdivided prosoma. Each chelicera is composed of two pieces forming a pair of pincers that articulate vertically. The pedipalps are leglike but terminate in a specialized adhesive organ. The first pair of legs are somewhat reduced in size and are used as tactile organs; the remaining three pairs of legs are used for running.

Solifuges are carnivorous or omnivorous, and termites are an important part of the diet of many American species. The pedipalps locate the prey and the chelicerae kill the animal and tear apart the tissues. For gas exchange the animal uses a highly developed tracheal system; excretion is accomplished by a pair of coxal glands and a pair of malpighian tubules.

Males generally seize the females before copulation. In some solifuges there is then a brief period of stroking and palpation by the male, which throws the female into a passive state. The male turns the female over and opens her genital orifice with his chelicerae. He then emits a globule of semen on the ground, picks it up with his chelicerae, and deposits it into the genital orifice of the female. The entire act takes only a few minutes, and the male then leaps away. In American solifuges sperm transmission is direct, although there is precopulatory behavior and the male inserts the chelicerae into the female orifice before and after sperm transfer. The female deposits 50 to 200 eggs in burrows in the ground.

ORDER OPILIONES

The order Opiliones, or Phalangida, contains the familiar long-legged arachnids known as daddy longlegs or harvestmen (Fig. 13–46).

The more than 4500 described species live in both temperate and tropical climates, and most prefer humid habitats. They are abundant in vegetation, on the forest floor, on tree trunks and fallen logs, in humus, and in caves.

The average body length is from 5 to 10 mm, exclusive of the legs, but some of the tropical giants reach 20 mm and have a leg length of 160 mm. There also are certain tiny, short-legged, mite-like species never larger than 1 mm in length. Most species are shades of brown, but some are red, orange, yellow, green, or spotted. A number of opilionids have bizarre, spiny bodies (Fig. 13–47B).

The prosoma of opilionids is broadly joined to the short segmented abdomen with no constriction between the two divisions (Fig. 13–47A). As a result the body is rather elliptical. In the center of the carapace is a tubercle with an eye located on each side. In some groups, such as the genus *Caddo*, the

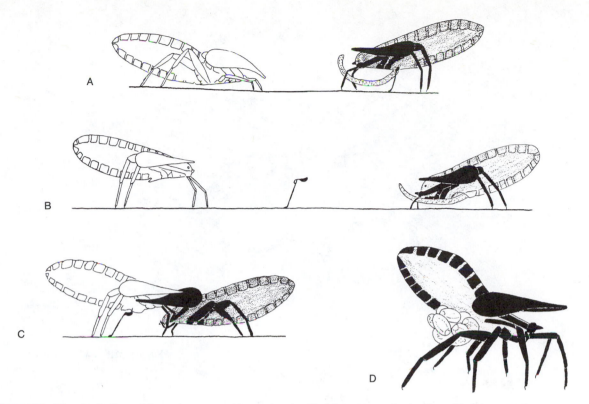

Figure 13–43 *A–C,* Courtship and sperm transmission in *Chelifer cancroides* (male is black). *A,* Male, producing spermatophore, *B,* Spermatophore, attached to substratum. *C,* Male pressing female down onto spermatophore. *D,* Female of *Chelifer cancroides,* carrying embryos. (All after Vachon, M., 1949: Traité de Zoologie, Vol. VI. Masson et Cie, Paris.)

tubercle extends almost the entire width of the carapace, so the eyes appear along the sides of the carapace rather than in the middle.

Along the anterior, lateral margins of the carapace are the openings to a pair of repugnatorial glands (Fig. 13–47A), which produce secretions, often quinones and phenols, having an acrid odor. Some harvestmen spray an intruder; certain species pick up a droplet of secretion mixed with regurgitated gut fluid and thrust it at a would-be predator.

The chelicerae are small, slender, and chelate. The pedipalps are usually short and leglike (Fig. 13–47A), but in one large suborder they are enlarged for capturing prey and possess a sickle-like claw. The legs in many opilionids are extremely long and slender and exceed the body length many times (Fig. 13–46). The tarsus is always multisegmented and flexible. When disturbed, opilionids can run very rapidly and species living in vegetation can climb by wrapping the flexible tarsus around stems or blades of grass. The second pair of legs are usually the longest and have an important sensory function. They are moved about over the

substratum, often in front of the animal. Self-amputation of a leg is an important means of defense against predators, but the legs are not regenerated.

In general, harvestmen are predatory, but scavenging is more important than in other arachnids. North American and European opilionids have been observed feeding on small invertebrates, dead animal matter, and pieces of fruits and vegetables. Predatory species feed on other small arthropods, and some species feed on snails.

The prey or food is seized by the pedipalps and passed to the chelicerae, which hold and crush it. Unlike other arachnids, the ingested food is not limited to liquid material but includes small particles. Thus, a greater part of digestion must take place in the midgut.

A pair of coxal glands provide for excretion. Gas exchange takes place through tracheae, which are probably not homologous with the tracheae of other arachnids. The spiracles are located on each side of the first abdominal segment, and in many active, long-legged harvestmen, there are secondary spiracles on the tibia of the legs.

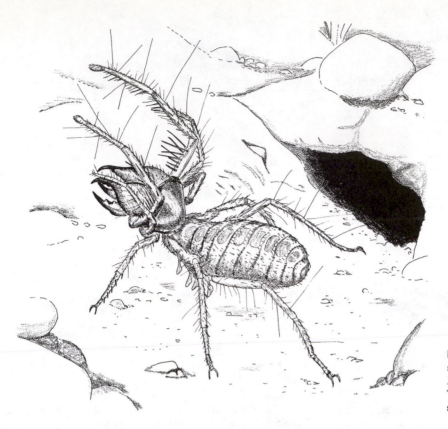

Figure 13–44 North African solifuge, *Galeodes arabs.* (After Millot, J., and Vachon, M., 1949: Traité de Zoologie, Vol. VI. Masson et Cie, Paris.)

The female reproductive system includes an ovipositor, a structure that is absent in other arachnids except certain mites. In opilionids the ovipositor is a tubular, ringed organ lying in a sheath in the midventral part of the abdomen, and at the time of egg laying the ovipositor is telescoped some distance out of the genital orifice.

Mating is not preceded by any elaborate courtship. The male in many species faces the female and projects a tubular penis, which commonly passes between the female's chelicerae before entering the female orifice. Mites are the only other arachnids to possess a penis.

Shortly after mating, the female uses the long ovipositor to deposit her eggs in humus, moss, rotten wood, or snail shells. The number of eggs laid at one time ranges in the hundreds, and several batches are laid during the life of a female. In temperate regions the life span is only one year. An individual may winter over in the egg or as an im-

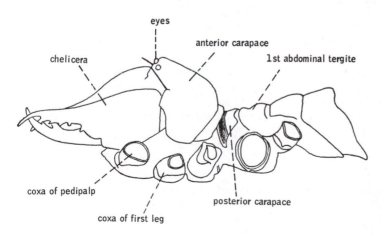

Figure 13–45 Cephalothorax of the solifuge *Galeodes graecus* (lateral view). (After Kaestner from Millot and Vachon.)

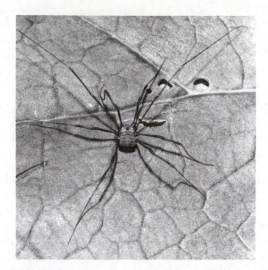

Figure 13–46 Dorsal view of a harvestman on a leaf. (By Betty M. Barnes.)

mature form. The name *harvestmen* refers to certain temperate species that sometimes appear in large numbers in the fall.

SYSTEMATIC RÉSUMÉ OF ORDER OPILIONES

Suborder Cyphophthalmi. Mitelike opilionids living in soil and leaf mold. Pedipalps leglike. All belong to a single family.

Suborder Laniatores. Pedipalps raptorial, with a sickle-like claw. Legs short to long. Most are inhabitants of leaf mold in the tropics and subtropics.

Suborder Palpatores. Pedipalps leglike. Legs short to long. A large and diverse group of opilionids found throughout the world. Includes many common species of temperate regions, such as members of the genera *Phalangium* and *Leiobunum*.

The Acari

The Acari are an enormously diverse arachnid assemblage containing the mites and ticks. Formerly grouped in a single order, the Acarina, they are now placed in three orders, which may not be closely related, for most specialists consider the assemblage to be polyphyletic (van der Hammen, 1977). To simplify the coverage, we will be forced to discuss the assemblage as if it were a relatively uniform group, but exceptions exist for almost every statement.

The Acari are without question the most important arachnids in terms of human economics. Numerous species are parasitic on humans, domesticated animals, and crops; others are destructive to food and other products. Mites rank as one of the most ubiquitous groups in the Animal Kingdom. Free-living, terrestrial species are extremely abun-

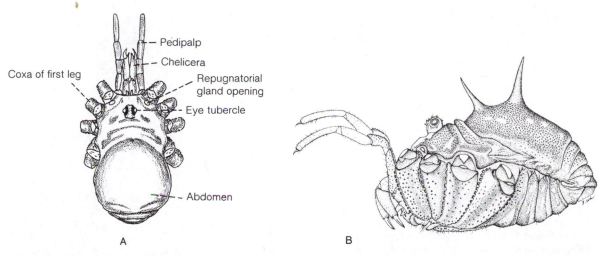

A   B

Figure 13–47 *A*, The opilionid *Leiobunum flavum* with only the bases of legs shown (dorsal view). The abdomen of this species is not as conspicuously segmented as that of many other opilionids. (After Bishop.) *B*, Lateral view of an opilionid whose abdomen bears two large spines. The chelicerae (lower) and pedipalps are shown, but the legs have been removed to reval the body.

dant, particularly in moss, fallen leaves, humus, soil, rotten wood, and detritus, and their success is certainly correlated with their small size and ability to exploit microhabitats. The numbers of individuals are enormous, certainly surpassing all other arachnid orders. A small sample of leaf mold from a forest floor often contains hundreds of individuals belonging to numerous species (Fig. 13–48).

Mites even occur in fresh water and the sea. Despite their abundance, the taxonomy and biology of mites are still not so well known as the other major arachnid orders. To date, 30,000 species have been described (compared with about 32,000 described species of spiders), but some acarologists believe that this number is only a small fraction of the total and that most species of mites will become extinct as rain forests and other habitats disappear before they are ever known.

Because of the economic importance of mites, many zoologists direct their attention entirely to this group of arachnids; as a result, a special field has developed known as acarology, the study of mites.

Much of acarology belongs in the realm of parasitology, but the mites should not be considered an entirely parasitic group. Many species are free living, and others are parasitic for only a brief period during their life cycles.

External Morphology

The majority of adult mite species are 1 mm or less in length, although many are larger. The ticks are the largest members of the order, some species reaching 3 cm in length.

The most striking characteristic is the apparent lack of body divisions (Figs. 13–49 and 13–50A and B). Abdominal segmentation has disappeared in most species, and the abdomen has fused with the prosoma. Thus, the positions of the appendages, the eyes, and the genital orifice are the only landmarks that differentiate the original body regions. Coinciding with this fusion, the entire body has become covered with a single, sclerotized shield, or carapace, in many forms.

Another general feature of the group is the change that has taken place in the head region carrying the mouth parts, this region being called the capitulum or gnathosoma (Fig. 13–51A and B). Ventrally, the large, pedipalpal coxae extend forward to form the floor and sides of the prebuccal chamber. The roof of the chamber is formed by a labrum. These processes, which house the prebuc-

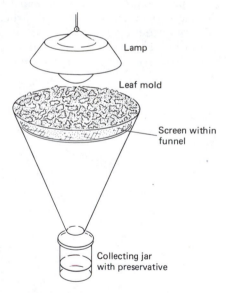

Figure 13–48 *A*, Tullgren funnel method for collecting small arthropods (mites, pseudoscorpions, spiders, insects, and others) from leaf mold and soil. Leaf mold is placed in top of funnel over interior screen. Light, heat, and desiccation drive animals downward. After one to four days many have fallen through screen into collecting jar. Animals can be collected alive if the collecting jar is provided with a damp substratum to prevent desiccation.

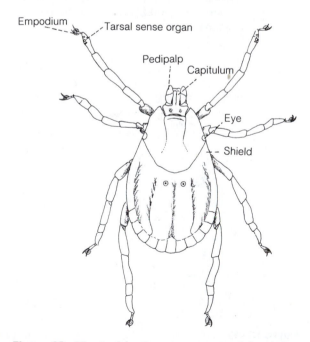

Figure 13–49 A tick, *Dermacentor variabilis.* (After Snodgrass.)

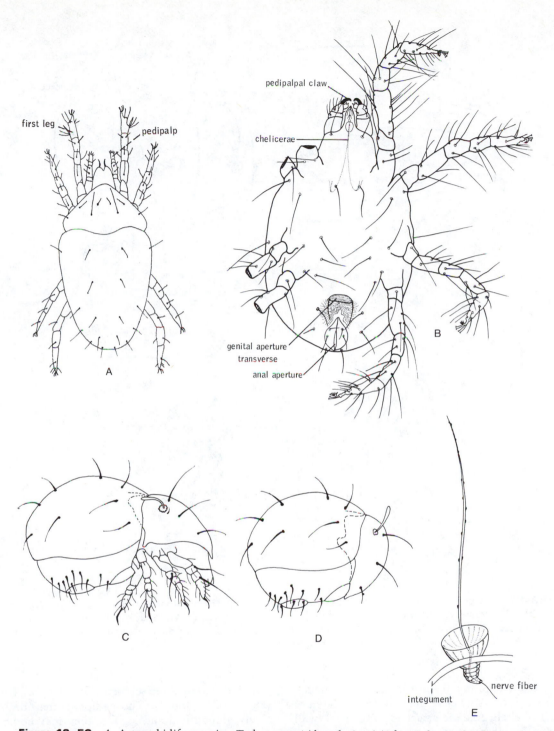

Figure 13–50 *A*, A trombidiform mite, *Tydeus starri* (dorsal view.) (After Baker and Wharton.) *B*, Ventral view of a spider mite, *Tetranychus*. (After Krantz, G. W., 1971: Manual of Acarology. Oregon State University Press, Corvallis, Oreg.) *C* and *D*, Side view of the oribatid mite *Mesoplophora*, (*C*) before and (*D*) after "closing up" the body. The antenna-like structure on the cephalothorax is the pseudostigmatic organ. (After Schaller, F., 1968: Soil Animals. University of Michigan Press, Ann Arbor.) *E*, Pseudostigmatic organ of the oribatid mite *Belba*. (After Beck.)

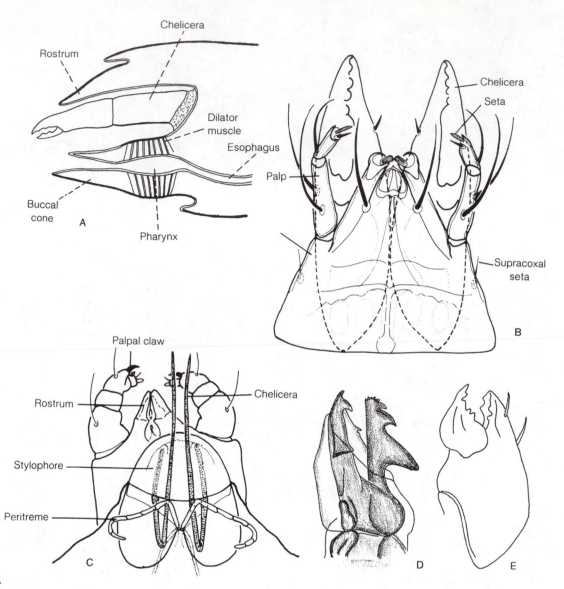

Figure 13–51 *A,* Head of a mite (sagittal section). (After Snodgrass.) *B,* Ventral view of head region of *Glycyphagus.* *C,* Stylet-like chelicerae of a spider mite, *Tetranychus.* (*B* and *C* from Krantz, G. W., 1971: Manual of Acarology. Oregon State University Press, Corvallis, Oreg. pp. 13 and 209.) *D,* Hooked chelicera of the tick, *Ixodes reduvius.* (After Neumann from André.) *E,* Chelate chelicera of *Belba verticillipes.*

cal chamber, together form a buccal cone, which fits into a sort of socket at the anterior of the mite body. The chelicerae are attached to the back wall of the socket above the buccal cone; the pedipalps are attached to both sides of the cone. The attachment of the buccal cone into the anterior socket is such that some species can extend and retract the cone.

The chelicerae and pedipalps vary in structure, depending on their function. They are usually com-

posed of two or three segments and may be chelate (Fig. 13–51*B* and *E*). The pedipalps may be relatively unmodified and leglike; they may be heavy and chelate like an additional pair of chelicerae; or in some parasitic forms they may be vestigial. The four pairs of legs usually have six segments each, and in some groups they have become modified for functions other than walking.

In mites and ticks the ventral side of the body is covered by plates that vary in form and number, depending on the family. The genital plate, located between the last two pairs of legs, bears the genital orifice (Figs. 13–50*B* and 13–52*B*). This location indicates a forward migration of abdominal segments, as in the harvestmen.

Hairs, or setae, cover the mite body; they vary from simple hairs to club-shaped and flattened types. Many are sensory. The nature and position of the setae in mites are extremely important diagnostic characteristics in the classification and identification of species. The symmetrical arrangement of the variously shaped setae and the ornate sculpturing of their cuticle make many mites beautiful and spectacular animals (Fig. 13–52*A*).

Most mites are varying shades of brown, but many display a wide range of hues, such as black, red, orange, green, or combinations of these colors.

Two groups of mites deserve special mention. Oribatida (beetle mites) are the largest and most studied group of free-living mites (145 families). They are very abundant in humus and moss. These little mites are usually globe shaped, with the dorsal surface covered by a convex, highly sclerotized shield, so they resemble tiny beetles (Fig. 13–52*C*). Some oribatids, which are very common in leaf mold, can close up their bodies like armadillos. The cephalothorax is flexed downward and backward so that the legs and head fit into a concavity on the ventral surface of the abdomen. A plate carried on the dorsal surface of the cephalothorax covers over the withdrawn head and legs (Fig. 13–50*C* and *D*).

One group of Acari, the water mites, containing some 2800 species, has returned to an aquatic existence and is found in both fresh and salt water (Fig. 13–52*D*). The marine forms have been found from the intertidal zone to abyssal depths, and shallow-water species are frequently encountered. They do not swim but crawl about over algae, bryozoans, hydroids, and sponges. Most water mites, however, live in fresh water. Some are bright red, and many are active swimmers with long hairs on their legs.

Feeding

Mites exhibit a tremendous diversity and specialization of diets and feeding habits, although, in general, they have retained the arachnid habit of ingesting fluids, and when feeding on solid foods, there is initial external digestion and liquefaction. Carnivores that live in soil and humus feed on nematodes and small arthropods, including eggs, insect larvae, and other mites. Small crustaceans are the principal prey of water mites. The chelicerae of carnivorous mites are variously modified, depend-

ing on the prey. Some mites tear off pieces of prey; others suck out the tissues.

Many herbivorous species, such as spider mites (Tetranychidae), have chelicerae modified as needle-like stylets (Fig. 13–51*C*). The mites pierce plant cells and suck out the contents. A number of spider mites are serious agricultural pests of fruit trees, clover, alfalfa, cotton, and other crops. Spider mites construct protective webs from silk glands that open near the base of the chelicerae. The minute gall mites (Eriophyoidea), which also feed on plant cells, have stylet-shaped chelicerae and include some forms that are agricultural pests (Fig. 13–53*D*). Other herbivores include species that feed on fungi, algae, and mosses.

Many mites are carrion feeders or scavengers. Most soil-inhabiting oribatid mites feed on fungi and on decomposing plant and animal material. A large number of "scavengers" have highly specialized diets. For example, different species of storage mites (Acaridae) and allied families feed on flour, dried fruit, mattress and upholstery stuffing, hay, and cheese. *Dermatophagoides* is commonly associated with house dust. The classification of scavengers should probably also include the feather mites (Fig. 13–52*B*) and some species that live on the fur of animals. They feed on oil, dead skin, and feather fragments and are not actually parasites.

The majority of parasitic mites are ectoparasites of animals, both vertebrates and invertebrates, but other forms of parasitism exist. Some mites have become internal parasites through an invasion of the air passageways of vertebrates and the tracheal systems of arthropods.

Many mites are parasitic only as larvae. For example, the larval stages of freshwater mites are parasitic on aquatic insects and clams. The juvenile stages of the common harvest mites (Trombiculidae) parasitize the skin of vertebrates. Larvae of species of *Trombicula* are the familiar chiggers, or red bugs. The six-legged larva emerges from an egg, which has been deposited on the soil. The larva may attack almost any group of terrestrial vertebrates, biting the host's skin and feeding on the dermal tissue, which is broken down by the external action of proteolytic enzymes (Fig. 13–53*E*). Feeding takes place for up to 10 days or more; then the larva drops off. After a semidormant stage, the larva undergoes a molt and becomes a free-living nymph. A later molt transforms the nymph into an adult. Both nymph and adult are predacious, feeding largely on insect eggs. Although chiggers can cause severe dermatitis, they are of much greater medical importance as vectors for pathogens, such as Asian scrub typhus. The intense itching that re-

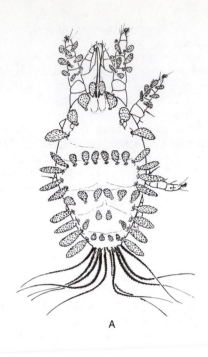

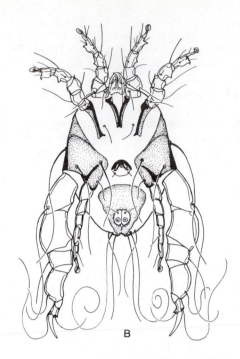

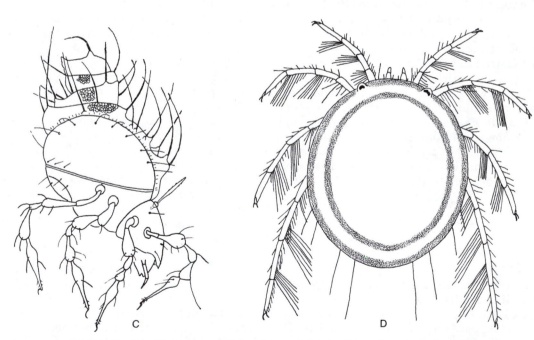

Figure 13–52   *A, Tuckerella,* a phytophagous mite, with symmetrically arranged, clublike setae. *B.* Ventral view of *Analges,* a feather mite. Note greatly enlarged third pair of legs and claws. (*A* and *B* from Krantz, G. W., 1971: Manual of Acarology. Oregon State University Press, pp. 211 and 275.) *C,* An oribatid mite, *Belba jacoti,* carrying five shed nymphal skins. (After Wilson from Baker and Wharton.) *D,* Water mite, *Mideopsis orbicularis.* (After Soar and Williamson from Pennak.)

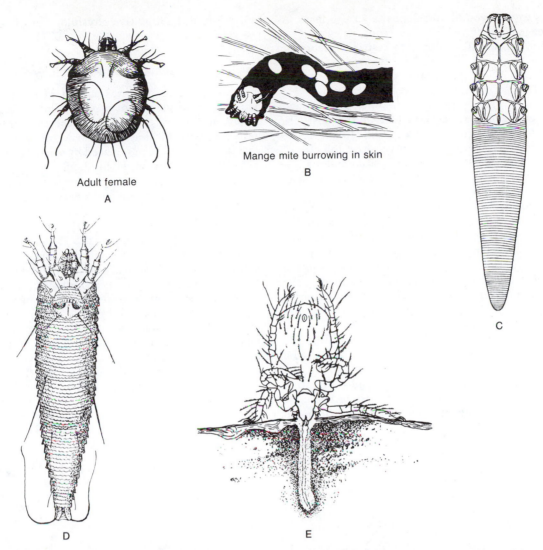

Mange mite burrowing in skin
B

Adult female
A

C

D

E

Figure 13–53 *A* and *B*, The mange mite, *Sarcoptes scabiei: A*, Adult female, *B*, Mite burrowing in skin and depositing eggs in tunnel. (Both after Craig and Faust.) *C*, Ventral view of a sheep hair follicle mite, *Demodex ovis* (0.23 mm). (After Hirst from Kaestner.) *D*, Ventral view of a gall mite, *Eriophyes*, a minute plant parasite (0.2 mm). (After Nalepa from Kaestner.) *E*, A chigger larva of *Neotrombicula*, feeding on skin (0.25 mm). (After Vitzthum from Kaestner. *C–E* from Kaestner, A., 1968: Invertebrate Zoology, Vol. 2. Wiley-Interscience, N.Y.)

sults from the bite of a chigger is caused by the mite's oral secretions and not, as commonly supposed, simply by the presence of the mite. Scratching quickly removes the mite, but the irritation remains for several days.

Many Acari are parasitic during their entire life cycles but are attached to the host only during periods of feeding. The dermanyssid mites of birds and mammals (red chicken and other fowl mites) and the ticks illustrate this type of life cycle. Ticks

penetrate the skin of the host by means of the highly specialized, hooked mouth parts and feed on blood (Fig. 13–51*D*). The body is not highly sclerotized and is capable of great expansion when engorged with blood. This is especially true of the female. With a few exceptions, the tick drops off the host after each feeding and undergoes a molt. Many species can live for long periods, well over a year, between feedings. Copulation occurs while the adults are feeding on the host. The female then drops to the ground and deposits an egg mass. A six-legged "seed" tick hatches from the egg.

Ticks attack all groups of terrestrial vertebrates. In man they are responsible for the transmission of American Rocky Mountain spotted fever, tularemia, Texas cattle fever, and relapsing fever.

Finally, there are parasitic mites that spend their entire life cycles attached to the host. Included in this group are the wormlike follicle mites (Demodicidae), which live in the hair follicles of mammals (Fig. 13–53*C*), and the scab- and mange-producing fur mites (Psoroptidae and Sarcoptidae) of mammals. The human itch mite *(Sarcoptes scabiei)* (Fig. 13–53*A* and *B*), the cause of scabies or seven-year itch, tunnels into the epidermis. The female is less than 0.5 mm and the male less than 0.25 mm in length. Irritation is caused by the mite's secretions. The female deposits eggs in the tunnels for a period of two months, after which she dies. Up to 25 eggs are deposited every two or three days. The eggs hatch in several days and the larvae follow the same existence as the adult. The infection can thus be endless. The mite is transmitted to another host by contact with infected areas of the skin.

Most free-living mites have a typical digestive tract (Fig. 13–54). The different feeding habits of others make it probable that there is considerable variation in the digestive physiology of the group, however.

Internal Structure and Physiology

Excretory organs of mites consist of one to four pairs of coxal glands, or a pair of malpighian tubules, or both (Fig. 13–54). In the trombidiform mites, these typical arachnid excretory organs are lacking and have been replaced by special excretory organs modified from the hindgut.

The circulatory system is reduced and, except in a few groups, consists only of a network of sinuses. Circulation probably results from contraction of body muscles.

Although in some mites the gas exchange organs have completely disappeared, most mites have tracheae. The spiracles vary in number from one to four pairs, located on the anterior half of the body.

Sensory setae are probably the most important of the sense organs. The oribatid mites possess a peculiar form of sensory seta called a pseudostigmatic organ, which is probably similar to a trichobothrium. The seta itself, which has various shapes, arises out of a cupule or pit. Two such pseudostigmatic organs are located on the cephalothorax (Fig. 13–50*C* to *E*). These setae are thought to detect air currents to which the mite re-

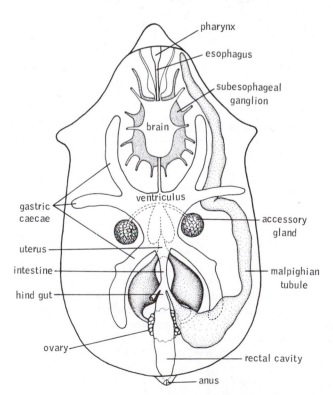

Figure 13–54   Internal anatomy of a mesostigmatid mite, *Caminella.* (After Ainscough from Krantz, G. W., 1971: Manual of Acarology. Oregon State University Press, Corvallis, Oreg.)

sponds by moving down deeper into the leaf mold. These organs would thus represent an adaptation against desiccation.

Although many mites are blind, some trombidiforms and certain other groups possess simple eyes. Innervated pits and slits are common in mites and are perhaps similar to the slit sense organs of other arachnids.

Reproduction and Development

The male reproductive system consists of a pair of lobate testes, located in the midregion of the body. A vas deferens extends from each testis and may join its opposite member ventrally to open through a median gonopore or through a chitinous penis that can project through the genital orifice.

In the female there is usually a single ovary of varying size, which is connected to the genital orifice by an oviduct (Fig. 13–54). A seminal receptacle and accessory glands are also present.

Sperm are transmitted directly in most mites by the male penis, but indirect sperm transmission by spermatophore is known to occur in many species and is probably primitive. The spermatophore may be located and picked up by the female unaided by the male, or it may be transferred to the female orifice by the male chelicerae or, in some water mites, the third pair of legs. Sperm transfer in certain oribatids is remarkably similar to that in some pseudoscorpions. The male deposits stalked cupules, each containing a droplet of semen, on the substratum in large numbers in the absence of any females. When a female encounters a globule of sperm, she apparently is attracted chemotactically. With her open genital orifice, she removes the sperm droplet from its stalked cup.

The number of eggs laid varies in different forms. They are deposited in soil and humus and there are some mites that enclose their eggs within a case. The oribatid *Belba* attaches its large eggs to the bodies of other individuals, which carry them about until hatching (Fig. 13–55). The oribatid mites possess an ovipositor.

After an incubation period of two to six weeks, a six-legged "larva" hatches from the egg. The newly hatched young lacks the fourth pair of legs and differs from the adult in certain other features. The fourth pair of legs is acquired after a molt, and the larva changes into a protonymph. Successive molts transform the protonymph into a deutonymph, a tritonymph, and finally an adult. During these stages adult structures are gradually attained.

The life span of mites varies greatly depending on the species, but it is generally shorter than that

Figure 13–55 An oribatid mite, *Belba*, carrying eggs attached by another individual. (From Schaller, F., 1968: Soil Animals. University of Michigan Press, Ann Arbor.)

of other arachnids. For example, *Amblyseius brazilli*, a tropical, predatory, mesostigmatid mite, reaches adulthood in 7 days, and the female has a life span of 30 days (El-Banhawy, 1975).

SYSTEMATIC RÉSUMÉ OF THE ACARINE ORDERS

Order Opilioacariformes (Notostigmata). Large, leathery, primitive mites having a segmented abdomen. They are brightly colored and resemble harvestmen. Onmivorous or predatory. Live in leaf litter and beneath stones.

Order Parasitiformes. Medium-sized to large mites. Abdomen unsegmented. Tracheal system with ventrolateral spiracles.

Suborder Mesostigmata. A large group of mites with a pair of tracheal spiracles beside the coxae of the third pair of legs. Body covered with plates, which vary in number and position. Parasitic and free-living species. The Dermanyssidae are economically important parasites of birds (red mites) and mammals.

Suborder Holothyrina (Tetrastigmata). Large, predatory mites of Australia, New Zealand, and the American tropics. The body is covered by a convex shield, and there are two pairs of spiracles, one pair by the third coxa and one pair behind the fourth coxa.

Suborder Ixodida (Metastigmata). Species of large, parasitic acarines known as ticks. Mouth with recurved teeth modified for piercing. A tracheal spiracle located behind third or fourth pair of coxae. Hard ticks or wood ticks (Ixodidae)—*Ixodes, Dermacentor;* soft ticks (Argasidae)—*Argas.*

Order Acariformes. Small mites. Chelicerae commonly chelate and toothed, but fingers may be reduced, absent, or needle-like. Tectum covering base of chelicerae is small or absent. Spiracles near mouth parts, or spiracles absent.

Suborder Prostigmata. (Trombidiformes). Mites with a single pair of spiracles located anteriorly near the mouth parts. This large diverse suborder of widely distributed mites contains many plant parasites, including the so-called spider mites (Tetranychidae) and the minute gall mites (Eriophyoidea). Many are parasites on invertebrates and vertebrates. Among the latter are the harvest mites (Trombidioidea). Many of the free-living species, which include the Bdellidae, or snout mites, are predacious. Also included in the suborder are the follicle mites (Demodicidae), marine water mites (Halacaridae), and freshwater mites (Hydrachnoidea, Lebertioidea, and Hygrobatoidea).

Suborder Astigmata (Sarcoptiformes). Tracheal system absent; gas exchange occurs across the weakly sclerotized body wall. Included in this suborder are the storage mites (Acaridae), the scabies, mange, or itch mites (Sarcoptidae and Psoroptidae), and the feather mites.

Suborder Oribatida (Cryptostigmata). Oribatid or beetle mites. A large group of heavily sclerotized mites that are very common in leaf mold and soil. There is usually a tracheal system associated with the bases of the first and third legs, but typical spiracles are absent.

SUMMARY

1 Members of the class Arachnida are terrestrial chelicerates that lack book gills. Those species that are aquatic (some mites) represent a secondary return to fresh water or the sea.

2 Scorpions, the most primitive arachnids, have long, segmented abdomens. The highly specialized mites have lost all external evidence of metamerism, and the cephalothorax and abdomen are broadly joined together.

3 Arachnids are largely predatory chelicerates, other arthropods forming the principal prey. The pedipalps of scorpions and pseudoscorpions are used in seizing and holding prey. Poison is used to immobilize prey in several groups: spiders (with chelicerae), scorpions (with terminal abdominal barb), and pseudoscorpions (with fingers of the pedipalps). In addition, some spiders use silk in prey capture. There is some digestion outside of the body, and fluids and partially digested tissues are sucked into the gut.

4 Harvestmen and many mites are the principal exception to the arachnid predatory habit. The great diversity of feeding habits among mites is in part related to the miniaturization that characterizes the group. Mites may be polyphyletic and have evolved from various arachnid groups, including the opilionids and ricinuleids.

5 Trichobothria and slit sense organs are important sense organs in prey capture. Although most arachnids possess eyes, a relatively small number are capable of object discrimination.

6 Large arachnids (scorpions, some spiders) possess book lungs as gas exchange organs; small forms (pseudoscorpions, some spiders, mites) possess tracheae. The heart is most highly developed in large species with book lungs, and the blood contains hemocyanin.

7 Excretory organs are coxal glands and malpighian tubules. A waxy epicuticle is an important adaptation for water conservation and has certainly contributed to the success of arachnids as terrestrial arthropods.

8 Many are secretive or nocturnal in habit. Leaf mold is the habitat for many small species, especially pseudoscorpions, mites, and spiders.

9 The primitive mode of sperm transfer is indirectly by spermatophores (scorpions, pseudoscorpions, and some mites). The unique indirect sperm transfer of spiders is probably derived from handling of spermatophores by the male with the pedipalps. Sperm transfer is direct in harvestmen and many mites.

Class Pycnogonida

The Pycnogonida, or Pantopoda, are a group of some 1000 described species of marine animals known as sea spiders. The name sea spider is derived from the somewhat spider-like appearance of these animals, which crawl about slowly on long legs. Although largely unknown to the layman, sea spiders are actually common animals. Careful examination of bryozoans and hydroids scraped from a wharf piling or rocks will always yield a few specimens. They live in all oceans from the Arctic and Antarctic to the tropics, and there are many littoral forms, as well as species that live at great depths. The majority of species are found in cold waters.

Pycnogonids are mostly small animals with a body length ranging from 1 to 10 mm, but specimens of the deep sea reach more than 6 cm in body length and have a leg span of almost 75 cm. Although most pycnogonids are drab in color, there are some that are green and some deep-sea forms that are red.

The body is commonly narrow and is composed of a number of distinct segments (Fig. 13–56). The head, or cephalon, bears on its dorsal surface four eyes mounted on a central tubercle and, at its anterior end, a cylindrical proboscis. Posterior to the head is a trunk of four to six cylindrical segments, the first of which is fused with the cephalon. The most striking feature of the trunk is the pair of large processes that projects laterally from each segment. A leg articulates with the end of each process, which often exceeds the length of the segment itself. At the rear of the trunk is a short, dorsal, conical abdomen.

The appendages consist of a pair of chelicerae (also called chelifores), a pair of palps, a pair of ovigerous legs, and four to six pairs of walking legs. The short chelicerae are attached to either side of the proboscis base. In some genera, such as *Pycnogonum*, the chelicerae are absent. The reduction in the chelicerae is correlated with a correspondingly greater development of the proboscis. The palps are leglike but are absent in some species. The ovigerous legs, or ovigers, which are peculiar to pycnogonids, may be used for grooming and, in the male, to carry the eggs. In many species the ovigers are less well developed or are even absent in the female.

The walking legs are attached to the lateral extensions of the trunk segments. There may be four, five, or six pairs of legs, depending on the number of trunk segments. The presence of five or six pairs of legs is not understood, for there are a number of

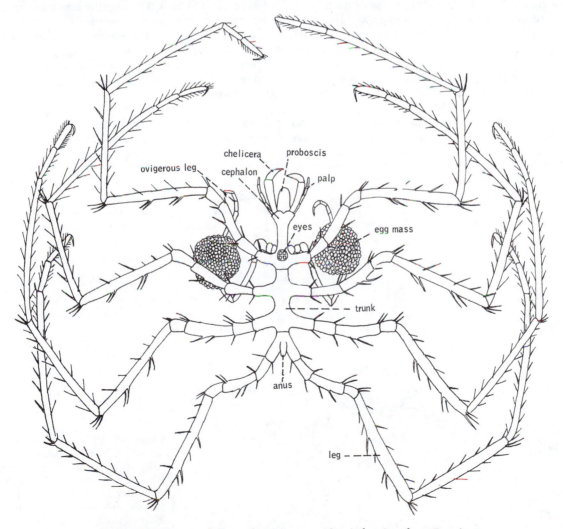

Figure 13–56 *Nymphon rubrum*, a sea spider. (After Sars from Fage.)

genera that contain both eight-legged and ten-legged species.

Most pycnogonids are bottom dwellers and crawl about over hydroids and bryozoans as if on stilts (Figs. 13–57 and 18–7); some are able to swim, flapping the legs vertically.

Many pycnogonids are carnivorous and feed on hydroids, soft corals, anemones, bryozoans, and sponges. Some apply the proboscis directly to the prey and suck up the tissues; others remove pieces of polyps with the chelicerae and pass them to the mouth at the tip of the proboscis. The mouth contains three liplike teeth that not only regulate the size of the opening but also act as a rasp. Some pycnogonids that feed upon bryozoans are able to open the operculum and insert the proboscis into the orifice of the prey. Not all pycnogonids are carnivores. Some feed on algae or microorganisms growing on hydroids and bryozoans or even accumulated detritus. From the mouth a tubular pharynx extends through the proboscis (Fig. 13–58A). The pharynx not only acts as a pump but also masticates the food with bristles projecting into the lumen. The pharynx leads into a short esophagus, which opens through a valve into the intestine.

The intestine constitutes the midgut of the digestive tract and is very extensive. A long, lateral cecum extends into each appendage and in the legs extends almost their entire length. Digestion is intracellular in the walls of the intestine and the ceca. Undigestible waste materials pass into a short rectum and then out through the anus at the tip of the abdomen.

The circulatory system is composed of a heart, or dorsal vessel, and a hemocoel. There are no special organs for gas exchange and excretion.

The brain is located beneath the ocular tubercle and consists of a protocerebrum and tritocerebrum. The brain connects with a subesophageal ganglion, from which extend nerves for the palps and ovigerous legs. A pair of ventral nerve cords extend posteriorly from the subesophageal ganglion and bear a pair of fused ganglia for each trunk segment.

Pycnogonids are dioecious. Females can be distinguished from males by the poorly developed condition of ovigerous legs or by their complete absence. The gonad, either testis or ovary, is single and located in the trunk above the intestine. Branches of the gonad extend far into the legs. In both males and females the reproductive openings are located on the ventral side of the coxae of different pairs of legs. The specific legs and the number of legs possessing gonopores vary in different species and are not necessarily the same in both sexes. On reaching maturity, the eggs migrate into the femurs of the legs containing gonopores (Fig. 13–58B), and in gravid females, the femurs are conspicuously enlarged.

In those pycnogonids in which egg laying has been observed, the male hangs beneath the female so that their ventral surfaces are opposed and their heads are in opposite directions, or the male stands over the female. The eggs are fertilized as they are emitted by the female, and the male then gathers them into his ovigerous legs (Fig. 13–56). Glands on the femurs of the male provide cement for forming as many as 1000 eggs into an adhesive,

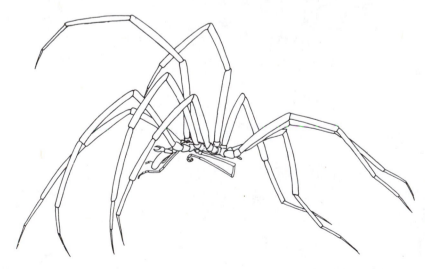

Figure 13–57 Side view of a ten-legged, Antarctic pycnogonid, *Decolopoda australis.* (From Schram, F. R., and Hedgpeth, J. W., 1978: Locomotory mechanisms in Antarctic pycnogonids. Zool. Jour. Linn. Soc., 63:145–169.)

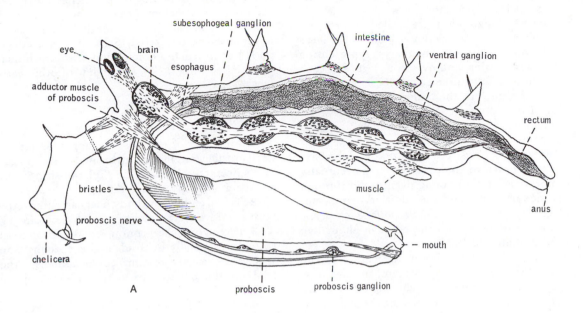

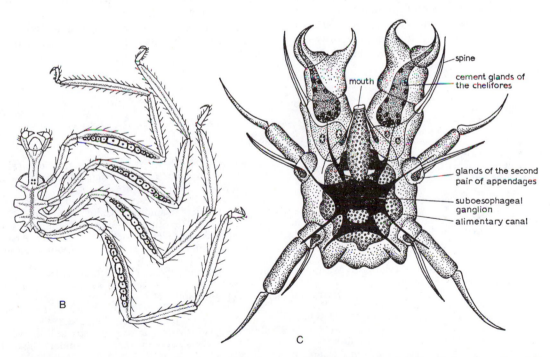

Figure 13–58 *A*, The pycnogonid, *Ascorhynchus castelli* (sagittal section). (After Dohrn from Fage.) *B*, Female of *Pallene brevirostris*, with eggs in femurs. (After Sars from Fage.) *C*, Protonymphon larva of *Achelia echinata*. (From King, P. E., 1973: Pycnogonids. St. Martin's Press, N.Y. 137 pp.)

spherical mass. The egg masses are held around the middle joints of the male's ovigerous legs and brooded until the eggs hatch.

In most pycnogonids a larva called a protonymphon hatches from the egg (Fig. 13–58C). The larva has only three pairs of appendages, representing the chelicerae, palps, and ovigerous legs, and each appendage has only three segments. A short proboscis is present, but the trunk segments are still lacking. Depending on the species, the larva remains on the ovigerous legs of the male, leaves to take up an independent existence, or in many shallow-water species, develops on or within hydroids and corals. In any case, the larva eventually metamorphoses into a young pycnogonid through a sequence of molts and the addition of new appendages. The eastern Pacific *Propallene longiceps* develops from the egg to the adult in about five months (Nakamura, 1981).

The structure of the brain, the nature of the sense organs, the circulatory system, and the presence of chelicerae have been considered by many zoologists as justification for placing the pycnogonids among the chelicerates. However, their exact relationship to arachnids and merostomes is by no means clear, for pycnogonids are aberrant in many respects. The presence of multiple gonopores, ovigerous legs, the segmented trunk, and the additional pairs of walking legs in many species have no counterparts in the other chelicerate classes. Manton (1978) believes that pycnogonids are derived from an early line of marine arachnids that never became terrestrial; Schram and Hedgpeth (1978) think such a connection with arachnids is remote (see King, 1973, for review of other earlier ideas).

SUMMARY

1 Members of the class Pycnogonida (sea spiders) are aberrant marine arthropods that appear to be chelicerates, although their relationship to other chelicerate groups is uncertain.

2 The often narrow body bears an anterior proboscis and four, five, or six pairs of very long legs.

3 Pycnogonids are commonly found on sessile animals, especially hydroids and bryozoans. Some species are carnivores on sessile animals; others feed on the detritus or flora growing on such animals.

4 The absence of gas exchange and excretory organs may be correlated with their large surface and aquatic existence.

5 The eggs are carried by the male with a pair of ovigerous legs located in front of the first walking legs.

REFERENCES

The literature included here is restricted in large part to chelicerates. The introductory references on page 8 include many general works and field guides that contain sections on these animals.

General

Anderson, D. T., 1973: Embryology and Phylogeny in Annelids and Arthropods. Pergamon Press, Oxford. 495 pp. (Includes a detailed survey of chelicerate development.)

Barth, F. G. (Ed.), 1985: Neurobiology of Arachnids. Springer-Verlag, N.Y. 385 pp.

Buskirk, R. E., 1981: Sociality in arachnids. *In* Hermann, H. R. (Ed.): Social Insects. Academic Press, N.Y.

Kaestner, A., 1968: Invertebrate Zoology, Vol. 2 Wiley-Interscience, N.Y. 472 pp. (This volume provides a very good general account of the arachnids.)

Kovoor, J., 1977: Silk and the silk glands of Arachnida. Annee Biol., 16(¾):97–172.

Millot, J., Dawydoff, C., Vachon, M., Berland, L., Andre, M., and Waterlot, G., 1949: Classe des Arachnides. *In* P. Grassé (Ed.), Traité de Zoologie, Vol. 6. Masson et Cie, Paris. pp. 263–905.

Schaller, F., 1968: Soil Animals. University of Michigan Press, Ann Arbor. 144 pp. (An excellent little volume on the biology of animals living in soil and in humus.)

Spiders

Anderson, J. F., 1974: Responses to starvation in the spiders *Lycosa lenta* and *Filistata hibernalis*. Ecology, 55(3):576–585.

Anderson, J. F., and Prestwich, K. N., 1982: Respiratory gas exchange in spiders. Physiol. Zool., 55(1):72–90.

Barth, F. G., 1978: Slit sense organs: strain gauges in the arachnid exoskeleton. *In* Merrett, P. (ed.): Arachnology. Symp. Zool. Soc. London, 42:439–448. Academic Press, London.

Barth, F. G., and Libera, W., 1970: Ein Atlas der Spaltsinnesorgane von *Cupiennius salia*. Z. Morph. Tiere, 68(4):343–369.

Barth, F. G., and Pickelmann, P., 1975: Lyriform slit sense organs. J. Comp. Physiol., 103:39–54.

Brach, V., 1977: *Anelosimus studiosus* and the evolution of quasisociality in theridiid spiders. Evolution, 31:154–161.

Bristowe, W. S., 1958: The World of Spiders. Collins, London.

Buchli, H. H. R., 1969: Hunting behavior in the Ctenizidae. Am. Zool., 9:175–193. (An excellent account of the predatory habits of the ctenizid trap-door spiders.)

Buck, J. B., and Keister, M. L., 1950: Arthropoda: *Argiope aurantia*. *In* Brown, F. A. (Ed.): Selected Invertebrate Types. John Wiley and Sons, N.Y. pp. 382–394. (Directions for a detailed study and dissection of the common garden spider.)

Burgess, J. W., 1976: Social spiders. Sci. Amer. 234(3):101–106.

Cooke, J. A. L., Roth, V., and Miller, F. H., 1972: The urticating hairs of theraphosid spiders. Am. Mus. Novit., 2498:1–43.

Dumpert, K., 1978: Spider odor receptor: electrophysiological proof. Experimentia, 34:754–756.

Edgar, W. D., 1970: Prey and feeding behavior of adult females of the wolf spider Pardosa amentata. Neth. J. Zool., 20(4):487–491.

Eisner, T., and Nowicki, S., 1983: Spider web protection through visual advertisement: role of the stabilimentum. Science, 219:185–187.

Foelix, R. F., and Chu-Wang, I-W, 1973: The morphology of spider sensilla. II. Chemoreceptors. Tissue and Cell, 5(3):461–478.

Foelix, R. F., 1982: Biology of Spiders. Harvard University Press, Cambridge, Mass. 306 pp. (An excellent general account of spiders, with emphasis on adaptive morphology and physiology.)

Forster, L., 1982: Vision and prey-catching strategies in jumping spiders. American Scientist, 70:165–174.

Gertsch, W. J., 1979: American Spiders. 2nd Edition. Reinhold, Van Nostrand, N.Y. 274 pp. (The natural history of spiders.)

Harwood, R. H., 1974: Predatory behavior of Argiope aurantia. Am. Midl. Nat., 91(1):130–138.

Horton, C. C., 1981: A defensive function of the stabilimenta of two orb-weaving spiders. Psyche, 87:13–20.

Jackson, R. R., 1982: The behavior of communicating in jumping spiders. In Witt, P. N., and Rovner, J. S.: Spider Communication: Mechanisms and Ecological Significance. Princeton University Press, Princeton, N.J. 213–247.

Kaston, B. J., 1964: The evolution of spider webs. Am Zool., 4:191–207.

Kaston, B. J., 1970: Comparative biology of American black widow spiders. Trans. San Diego Soc. Nat. Hist., 16(3):33–82.

Kaston, B. J., 1978: How to Know the Spiders. 3rd Edition. W.C. Brown, Dubuque, Iowa.

Kaston, B. J., 1981: Spiders of Connecticut. Rev. Edition. State Biol. Nat. Hist. Surv. Bull., 70:1020. (This work is valuable for any American student interested in the taxonomy of spiders. Keys for most families and genera are provided; almost every species is illustrated. The major part of the Connecticut spider fauna is found throughout the eastern United States.)

Klaerner, D., and Barth, F. G., 1982: Vibratory signals and prey capture in orb-weaving spiders (Zygiella X nota, Nephila clavipes). Jour. Comp. Physiol. and Sensory Neural Behav. Physiol., 148(4):445–456.

Kullmann, E. J., 1972: Evolution of social behavior in spiders. Am. Zool., 12(3):419–426.

Levi, H. W., 1967: Adaptations of respiratory systems of spiders. Evolution, 21(3):571–583.

Levi, H. W., 1978: Orb-weaving spiders and their webs. American Scientist, 66:734–742.

Levi, H. W., and Levi, L. R., 1968: A Guide to Spiders and Their Kin. A Golden Nature Guide, Golden Press, N.Y. (An excellent small, semipopular guide for the identification of common spiders. Many colored figures and much information.)

Lubin, Y. D., and Robinson, M. H., 1982: Dispersal by swarming in a social spider. Science, 216:319–321.

Marples, B. J., and Marples, M. J., 1972: Observations on Cantuaria toddi and other trapdoor spiders in central Otago, New Zealand. J. R. Soc. N. Z., 2(2):179–185.

McCreve, J. D., 1969: Spider venoms: biochemical aspects. Am. Zool., 9:153–156.

Morse, D. H., 1979: Prey capture by the crab spider Misumena calycina. Oecologia, 39:309–319.

Morse, D. H., 1983: Foraging patterns and time budgets of the crab spiders Xysticus emertoni and Misumena vatia on flowers. Jour. Arachnol., 11:87–94.

Platnick, N. I., 1971: The evolution of courtship behavior in spiders. Bull. Brit. Arach. Soc., 2(3):40–47.

Platnick, N. I., 1976: Drifting spiders or continents?: vicariance biogeography of the spider subfamily Laroniinae. Syst. Zool., 25:101–109.

Platnick, N. I., and Gertsch, W. J., 1976: The suborders of spiders: a cladistic analysis. Am. Mus. Novit., 2607:1–15.

Robinson, M. H., 1969: Predatory behavior of Argiope argentata. Am. Zool. 9:161–173.

Robinson, M. H., and Robinson, B., 1980: Comparative Studies of the Courtship and Mating Behavior of Tropical Araneid Spiders. Pacific Insects Monographs, No. 36. Bishop Museum Press, Honolulu. 218 pp.

Roth, V. D., 1982: Handbook for Spider Identification. Published by author, SWRS, Portal, Ariz. 85632. (Keys to families and genera of North American spiders.)

Rovner, J. S., 1971: Mechanisms controlling copulatory behavior in wolf spiders. Psyche, 78(1):150–165.

Rovner, J. S., 1975: Sound production by nearctic wolf spiders: A substratum-coupled stridulatory mechanism. Science, 190:1309–1310.

Rovner, J. S., 1978: Adhesive hairs in spiders: behavioral functions and hydraulically mediated movement. In Merrett, P. (Ed.): Arachnology. Symp. Zool. Soc. Lond., 42:99–108.

Rovner, J. S., Higashi, G. A., and Foelix, R. F., 1973: Maternal behavior in wolf spiders: the role of abdominal hairs. Science, 182:1153–1155.

Schaefer, M., 1977: Winter ecology of spiders. Z. Angew. Entomol., 83(2):113–134.

Schoener, T. W., and Toft, C. A., 1983: Spider populations: extraordinarily high densities on islands without top predators. Science, 219:1353–1355.

Seyferth, E. A., and Barth, F. G., 1972: Compound slit sense organs on the spider leg: mechanoreceptors involved in kinesthetic orientation. J. Comp. Physiol. 78:176–191.

Shear, W. A. (Ed.), 1986: Spiders: Webs, Behavior and Evolution. Stanford University Press, Stanford, Cal.

Stratton, G. E., and Uetz, G. W., 1981: Acoustic communication and reproductive isolation in two species of wolf spiders. Science, 214:575–577.

Suter, R. B., and Renkes, G., 1982: Linyphiid spider courtship: releaser and attractant functions of a contact sex pheromone. Anim. Behav., 30:714–718.

Tietjen, W. J., and Rovner, J. S., 1980: Trail-following behavior in two species of wolf spiders: sensory and etho-ecological concomitants. Anim. Behav., 28:735–741.

Turner, M., 1979: Diet and feeding phenology of the green lynx spider, Peucetia viridans. Jour. Arachnol., 7:149–154.

Wilson, R. S., 1969: Control of drag-line spinning in certain spiders. Am. Zool., 9(1):103–111.

Witt, P. N., 1975: The web as a means of communication. Biosci. Commun., *1*:7–23.

Witt, P. N., Reed, C. F., and Peakall, D. B., 1968: A Spider's Web. Springer-Verlag, N.Y. 107 pp. (An analysis of the regulatory mechanisms in web building.)

Witt, P. N., and Rovner, J. S. (Eds.), 1982: Spider Communication: Mechanisms and Ecological Significance. Princeton University Press, Princeton, N.J. (Papers from a symposium).

Work, R. W., 1977: Mechanisms of major ampullate silk fiber formation by orb-web-spinning spiders. Trans. Amer. Micros. Soc., *96*:170–189.

Other Chelicerates

Baker, E. W., and Wharton, G. W., 1952: An Introduction to Acarology. Macmillan, N.Y.

Balogh, J., 1972: The Oribatid Genera of the World. Akademiai Kiado, Budapest, Hungary. (Keys and descriptions of the 700 described genera of the soil-inhabiting beetle mites.)

Barr, D., 1973: Methods for the collection, presentation and study of water mites. Royal Ontario Museum of Life Science, Misc. Publ., Ontario. 28 pp.

Binns, E. S., 1983: Phoresy as migration—some functional aspects of phoresy in mites. Biol. Rev., *57*:571–620.

Bishop, S. C., 1949: The Phalangida of New York. Proc. Rochester Acad. Sci., *9*(3):159–235. (An old but still useful taxonomic study of New York harvestmen. Keys, figures, and bibliography of taxonomic papers included.)

Bonaventura, J., Bonaventura, C., and Tesh, S. (Eds.), 1982: Physiology and Biology of Horseshoe Crabs: Studies on Normal and Environmentally Stressed Animals. A. R. Liss, N.Y. 334 pp.

Brownell, P. H., 1977: Compressional and surface waves in sand: Used by desert scorpions to locate prey. Science, *197*:479–482.

Brownell, P. H., 1984: Prey detection by the sand scorpion. Sci. Amer., *251*(6):86–97.

Brownell, P., and Farley, R. D., 1979: Dectection of vibrations in sand by tarsal sense organs of the nocturnal scorpion, *Paruroctonus mesaenis*. Jour. Comp. Physiol., *131*:23–30.

Bub, K., and Bowerman, R. F., 1979: Prey capture by the scorpion *Hadrurus arizonensis*. Jour. Arachnol., *7*:243–253.

Carthy, J. D., 1968: The pectines of scorpions. *In* Carthy, J. D., and Newell, G. E. (Eds.): Invertebrate Receptors, Symposia of the Zoological Society of London, No. 23. Academic Press, N.Y. pp. 251–261.

Cohen, J. A., and Brockmann, H. J., 1983: Breeding activity and mate selection in the horseshoe crab, *Limulus polyphemus*. Bull. Mar. Sci., *33*(2):274–281.

Edgar, A. L., 1971: Studies on the biology and ecology of Michigan Phalangida (Opiliones). Misc. Publ. Mus. Zool. Univ. Mich., *144*:1–64.

El-Banhawy, E. M., 1975: Biology and feeding behavior of the predatory mite, *Amblyseius brazilli*. Entomophaga, *20*(4):353–360.

Evans, G. O., Sheals, J. G., and MacFarlane, D., 1961: The Terrestrial Acari of the British Isles. An introduction to their morphology, biology and classification. Vol. I. Introduction and biology. British Museum, London.

Fage, L., 1949: Classe des Merostomaces. *In* P. Grassé (Ed.): Traité de Zoologie, Vol. 6. Masson et Cie, Paris. pp. 219–262.

Goddard, S. J., 1976: Population dynamics, distribution patterns and life cycles of *Neobisium muscorum* and *Chthonius orthodactylus*. J. Zool., *178*:295–304.

Hadley, N. F., 1974: Adaptational biology of desert scorpions. J. Arachology, *2*(1):11–23.

Hedgpeth, J. W., 1948: The Pycnogonida of the Western North Atlantic and the Caribbean. Proc. U. S. Nat. Mus., *97*:157–342.

Hoff, C. C., 1949: The pseudoscorpions of Illinois. Bull. Illinois Nat. Surv., Vol. 24, Art. 4 (A good starting point for students interested in the taxonomy of pseudoscorpions. Keys, figures, and bibliography of taxonomic papers included.)

Johnson, D. L., and Wellington, W. G., 1980: Predation of *Apochthonius minimus* on *Folsomia candida* (Collelbola). I. Predation rate and size-selection. Researches on Population Ecology, *22*(2):339–352.

Griffiths, D. A., and Bowman, C. E. (Eds.) 1983: Acarology VI. Vols. 1 and 2. John Wiley and Sons, N.Y. (Reviews of various aspects of the biology of mites.)

King, P. E., 1973: Pycnogonids. St. Martin's Press, N.Y. 144 pp. (A detailed general account of the pycnogonids.)

King, P. E., 1974: British Sea Spiders. Synopses of the British Fauna (New Series) No. 5. The Linnean Society of London. Academic Press, London. 68 pp. (Guide to the British species of pycnogonids, with a brief introductory general account of the group.)

Krantz, G. W., 1970: A manual of Acarology. Oregon State University Bookstores, Inc., Convallis, Ore. 335 pp. (A very useful introduction to the techniques, the use of keys, and the literature necessary for identifying ticks and mites.)

Legg, G., 1977: Sperm transfer and mating in *Ricinoides hanseni*. J. Zool., *182*(1):51–61.

Lockhead, J. H., 1950: Arthropoda: *Xiphosura polyphemus*. *In* Brown, F. A. (Ed.): Selected Invertebrate Types. John Wiley and Sons, N.Y. pp. 360–381. (Directions for a detailed laboratory study and dissection of the Atlantic horseshoe crab.)

Manton, S. M. 1978: Habits, functional morphology and the evolution of pycnogonids. Zool. Linn. Soc., *63*:1–21.

Martens, J. P., 1978: Opiliones. Die Tierwelt Deutschlands. G. Fischer, Jena.

McCloskey, L. R., 1973: Pycnogonida. Marine flora and fauna of the northeastern U.S. NOAA Technical Reports NMFS Circular 386. U.S. Government Printing Office.

McDaniel, B., 1979: How to Know the Mites and Ticks. W.C. Brown Co., Dubuque, Iowa. 335 pp.

Merrett, P. (Ed.), 1978: Arachnology. Symposia of the Zoological Soc. London, No. 42. Academic Press, London. 530 pp. (Papers from the Seventh International Congress on Arachnology.)

Muma, M. H., 1970: A synoptic review of North American, Central American, and Western Indian Solipugida. Arthropods of Florida. Vol. 5. Contribution No. 154, Bureau of Entomology, Florida Dept. Agriculture and Consumer Services. 62 pp.

Nakamura, K., 1981: Postembryonic development of a pycnogonid *Propallene longiceps*. Jour. Nat. Hist., *15*(1):49–62.

Pasquet, A., 1984: Predatory site selection and adaptation of the trap in four species of orb-weaving spiders. Biol. Behav., 9(1):3–20.

Polis, G. A. (Ed.), in press: The Scorpions. Stanford University Press, Stanford, Cal.

Polis, G. A., and Farley, R. D., 1979a: Behavior and ecology of mating in the cannibalistic scorpion, *Paruroctonus mesaensis.* Jour. Arachnol., 7:33–46.

Polis, G. A., and Farley, R. D., 1979b: Characteristics and environmental determinants of natality, growth and maturity in a natural population of the desert scorpion, *Paruroctonus mesaensis.* Jour. Zool., 187:517–542.

Rudloe, A., 1981: Aspects of the biology of juvenile horseshoe crabs, *Limulus polyphemus.* Bull. Mar. Sci., 31(1):125–133.

Sankey, J. H. P., and Savory, T. H., 1974: British Harvestmen. Synopses of the British Fauna (New Series) No. 4. The Linnean Society of London, Academic Press, London. 76 pp. (Guide to the British species of Opiliones, with an introductory general account of the group.)

Schram, R. F., and Hedgpeth, J. W., 1978: Locomotor mechanisms in Antarctic pycnogonids. J. Linn. Soc. Zool., 63:145–169.

Sharov, A. G., 1966: Basic Arthropodan Stock. Pergamon Press, N.Y.

Størmer, L., 1977: Arthropod invasions of land during late Silurian and Devonian times. Science, 197:1362–1364.

Sturm, H., 1973: On the ethology of *Trithyreus sturmi* (Pedipalpi Schizopeltidia). Z. Tierpsychol., 33(2):113–140.

Towle, D. W., Mangum, C. P., Johnson, B. A., and Mauro, N. A., 1982: The role of the coxal gland in ionic, osmotic, and pH regulation in the horseshoe crab *Limulus polyphemus. In* Bonaventura, J., Bonaventura, C., and Tesh, S. (Eds.), 1982: Physiology and Biology of Horseshoe Crabs: Studies on Normal and Environmentally Stressed Animals. A. R. Liss, N.Y. pp. 147–172.

Van der Hammen, L., 1977: The evolution of the coxa in mites and other groups of Chelicerata. Acarologia, 19(1):12–19.

Weygoldt, P., 1969: The Biology of Pseudoscorpions. Harvard University Press, Cambridge, Mass. 145 pp.

Weygoldt, P., 1972: Geisselskorpione und Geisselspinnen (Uropygi und Amblypygi). Z. des Kölner Zoo, 15(3):95–107. (A good account of the biology of these two arachnid orders.)

Weygoldt, P., 1972: Spermatophorenbau und Samenubertragung bei Uropygen *(Mastigoproctus brasilianus)* und Amblypygen *(Charinus brasilianus)* und *Admetus pumilio.* Z. Morph. Tiere, 71:23–51.

Weygoldt, P., 1974: Indirect sperm transfer in arachnids. Verh. Dtsch. Zool. Ges., 67:308–313.

Weygoldt, P., 1977: Agonistic and mating behaviour, spermatophore morphology and female genitalia in neotropical whip spiders (Amblypygi). Zoomorphologie, 86:271–286.

Weygoldt, P., 1978: Mating behavior and spermatophore morphology in whip scorpions: *Thelyphonellus amazonicus* and *Typopeltis crucifer.* Zoomorphologie, 89:145–156.

Weygoldt, P., and Paulus, H. F., 1979: Untersuchungen zur Morphologie, Taxonomie und Phylogenie der Chelicerata. Z. f. zool. Syst. Evolutionsforsch., 17(2):85–116 and 17(3):177–200.

14

The Crustaceans

The more than 42,000 known species of the subphylum Crustacea include some of the most familiar arthropods, such as crabs, shrimps, lobsters, crayfish, and wood lice. In addition, there are a myriad of tiny crustaceans that live in the seas, lakes, and ponds of the world and occupy an important position in aquatic food chains.

The Crustacea are the only large subphylum of arthropods that is primarily aquatic. Most crustaceans are marine, but there are many freshwater species. In addition, there are some semiterrestrial and terrestrial groups, but in general, the terrestrial crustaceans have never undergone any great adaptive evolution for life on land.

External Anatomy

Structurally, the head is more or less uniform throughout the subphylum. It bears five pairs of appendages (Figs. 14–1A and 14–31A). Anteriormost is a first pair of antennae, or antennules, which are generally considered homologous to the antennae of the other mandibulate classes. The first antennae are followed by the second antennae, often referred to simply as the antennae. The presence of two pairs of antennae is a distinguishing feature of crustaceans. The second antennae are embryologically postoral in origin and are probably homologous to the arachnid chelicerae, both of which are innervated by the tritocerebrum. Flanking and often covering the ventral mouth are the third pair of appendages, the mandibles (Fig. 14–1A). These are usually short and heavy with opposing grinding and biting surfaces. Behind the mandibles are two pairs of accessory feeding appendages, the first and second maxillae.

The trunk is much less uniform than the head. Primitively, the trunk is composed of a series of many distinct and similar segments and a terminal telson bearing the anus at its base (Fig. 14–7). In most crustaceans the trunk segments are characterized by varying degrees of regional specialization, of reduction or restriction in number, of fusion, and of other modifications. In a number of groups the trunk is divided into a thorax and abdomen, (Fig. 14–31A), but the number of segments they contain varies from group to group.

In many common crustaceans the thorax, or anterior trunk segments, is covered by a dorsal carapace (Fig. 14–31A). The carapace generally arises as a posterior fold of the head and may be fused with a varying number of segments behind it (Fig. 14–2). Usually, the lateral margins of the carapace overhang the sides of the body at least to some extent, and in extreme cases the carapace may completely enclose the entire body like the valves of a clam (Fig. 14–12C).

Crustacean appendages are typically biramous (Fig. 14–3). There is a basal protopodite composed of two pieces—a coxopodite (coxa) and a basopodite (basis). To the basopodite is attached an inner branch (the endopodite) and an outer branch (the exopodite), each of which may be composed of one to many segments. There are innumerable variations of this basic plan. Sometimes an appendage has lost one of the branches and has become secondarily uniramous. Often different parts of the appendage bear highly developed processes or extensions that have been given special names.

In primitive crustaceans, the appendages are numerous and similar and together perform a number of functions. But as in other arthropods, the evolutionary tendency in crustaceans has been toward a reduction in the number of appendages and a regional specialization of different appendages for particular functions. Thus, in a family of swimming crabs one pair of appendages is modified for swimming, and others are modified for crawling, prehension, sperm transmission, egg brooding, and food handling.

The cuticle of most large crustaceans, in contrast to that of most other arthropods, is usually calcified. Both the epicuticle and the procuticle contain depositions of calcium salts, and the outer layer of the procuticle is also pigmented and contains tanned proteins.

Locomotion and Nutrition

The ancestral crustaceans were probably small, swimming, epibenthic suspension feeders, and many modern forms have retained this primitive mode of existence. Most larger crustaceans have taken up a benthic habit, and certain appendages have usually become heavier and adapted for crawling and burrowing.

Crustaceans have a great range of diets and feeding mechanisms. Commonly, certain anterior trunk appendages are adapted for suspension feeding, predation, or picking up food, and the maxillae and mandibles function in holding, biting, and directing food into the mouth. The subphylum Crustacea contains the largest number of arthropod filter-suspension feeders. Filter feeding in the arthropods involves setae instead of cilia. Fine

setae on certain appendages function as a filter for the collection of food particles, and the spacing of the setae has been thought to determine the size of the particle collected. However, the process is now known to be more complicated than simple sieving (Box 14–1, pp. 556–557). The necessary water current is produced by the beating of the filtering appendages or, more commonly, by special appendages modified for this purpose. The collected particles are removed from the filter setae by special combing or brushing setae, and these particles are transported to the mouthparts by other appendages or sometimes in a ventral food groove.

Filter feeding undoubtedly evolved independently a number of times within the Crustacea, and virtually every pair of appendages, even the antennae and mandibles, may be modified in one group or another for filter feeding. Filter feeding probably first arose in connection with swimming and is therefore primitively associated with the trunk, the same limbs creating both the swimming and the feeding currents.

The tendency in most groups has been for the filtering apparatus to develop forward, nearer the mouth, and to involve only the anterior trunk appendages or the head appendages.

The mouth is ventral and the digestive tract is almost always straight (Fig. 14–9A). The foregut is commonly enlarged and functions as a triturating stomach, the walls of which bear apposing chitinous ridges, denticles, and calcareous ossicles (Fig. 14–59B). The midgut varies greatly in size, and one to several pairs of ceca are almost always present. One pair of ceca, especially in large crustaceans, has usually become modified to form large, solid digestive glands (the hepatopancreas) composed of ducts and blind secretory tubules. Its secretions are the primary source of digestive enzymes. The action of the digestive fluid takes place in the midgut and in the triturating stomach of the foregut when this chamber is present. Absorption is confined to the midgut walls and the tubules of the hepatopancreas, which also contains cells for glycogen, fat, and calcium storage.

Internal Transport and Gas Exchange

The circulatory system is similar to that of other arthropods, but the dorsal heart varies in form from a long tube to a spherical vesicle. In large crustaceans there is an extensive arterial system (Fig. 14–1B); in very small forms blood vessels are limited or absent.

The blood contains small hyaline and larger granular amebocytes that, in addition to phagocytosis, are also involved in clotting. Under certain irritating conditions, such as amputation, amebocytes called explosive cells disintegrate and liberate a substance that converts plasma fibrinogen to fibrin. Islands of coagulated plasma then appear, to which other islands connect and in which blood cells are trapped, thus forming a clot.

Gills are the usual gas exchange organs of crustaceans and are typically associated with the appendages, but they vary greatly in form, location, and derivation. The water current for ventilation is generally provided by the beating of certain appendages. Within the circulatory medium, oxygen is transported either in simple solution or bound to plasma hemoglobin or hemocyanin. Hemocyanin is found in the large species, but both pigments have a sporadic distribution.

Excretion and Osmoregulation

The excretory organs of crustaceans are similar in structure and origin to the coxal glands of chelicerates. Crustacean excretory organs are paired and composed of an end sac, an excretory canal, and a short exit duct, all located in the head (Fig. 14–4).

The end sac arises from an anterior coelomic compartment adjacent to the antennal or second maxillary segments, and the excretory organs are therefore called antennal or maxillary glands. The excretory pores of the antennal glands open onto the underside of the bases of the second antennae, and those of the maxillary glands open onto or near the bases of the second maxillae. Both antennal and maxillary glands are commonly present in crustacean larvae, but usually only one pair persists in the adult.

The end sac walls are composed of cells similar to the podocytes of vertebrate glomeruli, and it is through the slits between podocyte extensions that filtration from the blood into the end sac occurs. Processing of the filtrate by selective reabsorption or secretion occurs within the excretory canal, and this region may be variously modified, depending on the extent of these processes.

The gills are a principal site for the excretion of ammonia, and therefore in most crustaceans the antennal and maxillary glands must function largely in controlling internal fluid volume. Also, in most crustaceans the antennal and maxillary glands do not play an important role in osmoregulation. Most crustaceans, even many freshwater and terrestrial species, produce a urine that is isosmotic with the blood. This is not true, however, of the freshwater crayfish, in which the antennal

glands do elaborate a hypotonic urine. For most crustaceans, the gills are the chief organs for maintaining salt balance. In freshwater and brackish-water forms the gills actively absorb salts.

Crustaceans, like most other arthropods, possess nephrocytes—cells capable of picking up and accumulating waste particles. Nephrocytes are most commonly located in the gill axes and in the bases of the legs.

Nervous System

In crustaceans, as in most other arthropods, the nervous system displays a tendency toward concentration and fusion of ganglia. In crayfish, for example, medial fusion has taken place, resulting in a single cord with single, rather than obviously paired, ganglia. In most crabs all ventral ganglia have fused with the subesophageal ganglia to form

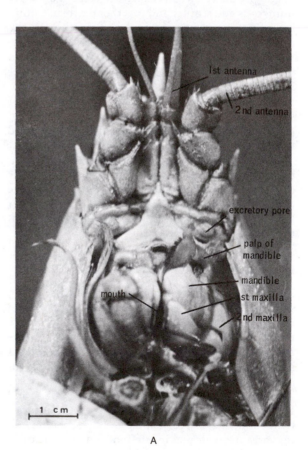

1st antenna
2nd antenna
excretory pore
palp of mandible
mandible
mouth
1st maxilla
2nd maxilla
1 cm

A

Figure 14–1 Crustacean structure: *A,* Head appendages of a lobster. More posterior mouth parts on left side have been removed. (By Betty M. Barnes.)

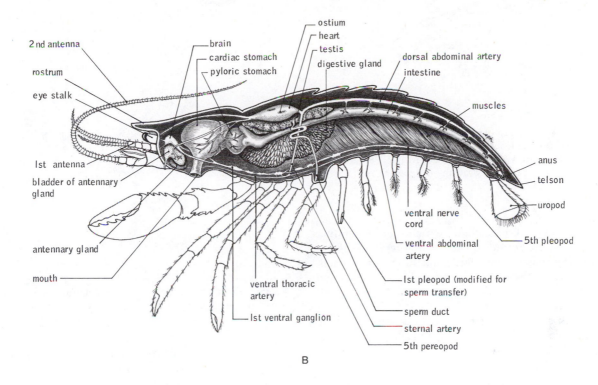

B

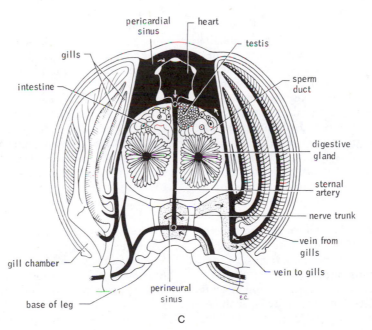

C

Figure 14–1 *(continued)* B, Internal structure of a crayfish (lateral view). C, Cross section of a crayfish just behind third pair of legs. (B and C after Howes.)

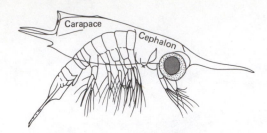

Figure 14–2   Larva of a stomatopod showing the developing of the carapace as a fold of the head (cephalon). (From Newman, W. A., and Knight, M.D., 1984: The carapace and crustacean evolution—a rebuttal. Jour. Crust. Biol., *4*(4):683.)

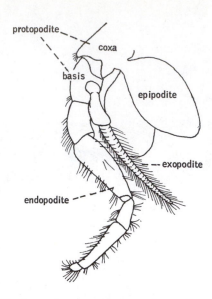

Figure 14–3   Right 5th pereopod, or leg, of the syncarid, *Anaspides tasmaniae,* showing basic structure of a crustacean appendage. (After Waterman and Chace.)

a single mass. Giant fibers, which rapidly conduct impulses, are found in the central nervous system of many crustaceans. The giant fibers are particularly well developed in such forms as shrimps and crayfish, which can dart rapidly backward by a sudden flexion of the tail.

Lack of space does not permit discussion of the numerous studies on crustacean reflexes (see Bliss, 1982, Vol. 4).

Sense Organs

The sense organs of crustaceans include eyes, statocysts, sensory hairs, and proprioceptors. The eyes are of two types, median and compound. A median eye is a characteristic feature of the naupliar larva of crustaceans and is therefore often referred to as the naupliar eye (Figs. 14–6 and 14–13*A*). It may degenerate or persist in the adult. The naupliar eye is composed of three or sometimes four small, pigment-cup ocelli containing a few photoreceptors (Box 7–2, pp. 180–181). The cups are located directly over the protocerebrum, where they either form a compact mass or are somewhat separated. The naupliar eye is probably largely an organ of orientation. As such, the eye would enable the animal to determine the direction of the light source and thus to locate the upper surface of the water or, in burrowing forms, to locate the surface of the substratum.

Two compound (lateral) eyes are found in the adults of most species. The eyes may be at the end of a usually movable stalk (peduncle) (Fig. 14–57*A*), or they may be sessile (Fig. 14–87). The total corneal surface is often greatly convex, resulting in a wide visual field. This is particularly true for stalked compound eyes, in which the cornea may cover an arc of 180 degrees or more (Fig. 14–66*A*). The number of ommatidia in crustacean

compound eyes varies enormously. A single eye in the wood louse, *Armadillidium,* is composed of not more than 25 ommatidia; the eye of the lobster, *Homarus,* may possess as many as 14,000. Those with well-developed compound eyes, such as some shrimps and crabs, show some ability to discriminate form and size (see Cronin, 1986 for review of crustacean compound eyes).

Color discrimination has been demonstrated in a number of crustaceans and may be of wide occurrence in the subphylum. For example, the hermit crab, *Pagurus,* can discriminate between painted yellow and blue snail shells and shells colored different shades of grey. The chromatophores of the shrimp *Crangon* adapt to a background of yellow, orange, or red but not to any shade of grey (the chromatophore changes are mediated through the eyes).

Statocysts are restricted to a few groups of large crustaceans. Only a single pair are present and are located in the base of the antennules or in the base of the abdomen, uropods, or telson. Each statocyst arises as an ectodermal invagination and usually retains an opening to the exterior. The statolith may be secreted, but commonly it is composed of a mass of agglutinated sand grains. The physiology of crustacean statocysts is best known in the decapods (shrimps, crayfish, lobsters, and crabs) and is discussed with these groups.

Various types of sensory hairs are located over the body surface, especially the appendages; among

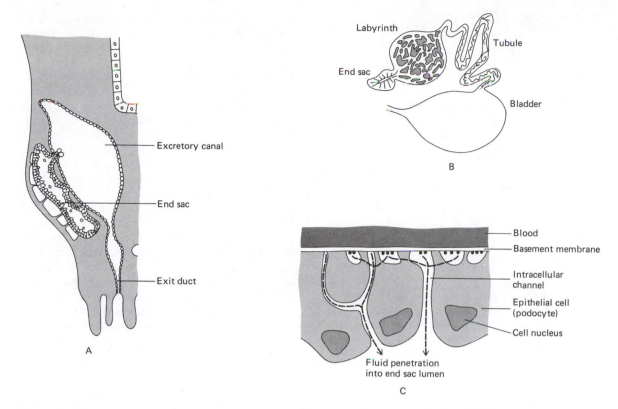

Figure 14—4 *A,* Section through the excretory organ (maxillary gland) of a barnacle. (From White, K. N., and Walker, G., 1981: The barnacle excretory organ. J. Mar. Biol. Assoc. U.K., *61:*529–547. Copyrighted and reprinted by permission of Cambridge University Press.) *B,* The excretory organ of a crayfish. Excretory pore opens onto base of second antenna. *C,* Diagrammatic section through end sac wall, showing site of filtration. (Redrawn from Kummel by A. P. M. Lockwood, 1967: Physiology of Crustacea. W. H. Freeman and Co., San Francisco.)

these are chemoreceptors, called aesthetascs, which are found in the majority of crustaceans. Aesthetascs are sensory hairs that are usually present in rows on the first antennae (Fig. 14–5). In terrestrial crustaceans, the aesthetascs have the form of tiny plates instead of hairs.

Reproduction and Development

Most crustaceans are dioecious. The gonads are typically elongated, paired organs lying in the dorsal portion of the trunk (Fig. 14–9A). The oviducts and sperm ducts are usually paired simple tubules that open either at the base of a pair of trunk appendages or on a sternite. However, the segments that bear the gonopores vary from one group to another.

Copulation is the general rule in crustaceans, the male usually having certain appendages modified for clasping the female. In many crustaceans the sperm lacks a flagellum and is nonmobile. In some crustaceans the sperm are transmitted as

spermatophores (Fig. 14–69). In some groups the sperm ducts open at the end of a penis, or certain appendages may be modified for the transmission of sperm. A seminal receptacle is sometimes present in the female. The seminal receptacle may be located near the base of the oviduct, but frequently it is a separate, pouchlike, ectodermal invagination of the genital segment or a neighboring segment.

Most crustaceans brood their eggs for different lengths of time. The eggs may be attached to certain appendages, contained within a brood chamber in various parts of the body, or retained within a sac formed when the eggs are expelled.

The eggs of many higher crustaceans are centrolecithal, and cleavage is superficial; in lower groups the eggs are small, and holoblastic cleavage is common. In some it is determinate and shows traces of a spiral pattern (barnacles, copepods, cladocerans) (see Anderson, 1973).

A free-swimming, planktonic larva is characteristic of most marine species and even some freshwater forms. The earliest and basic type of crusta-

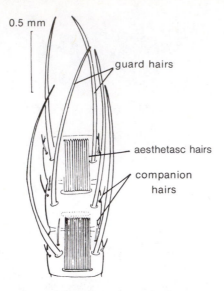

Figure 14–5   Rows of chemosensory hairs, called aesthetascs, on the antenna of the lobster *Panulirus*. Only one of the two rows on each segment has been shown. (From Laverack, M. S., 1968; Oceanogr. Mar. Biol. Ann. Rev., 6:249–324.)

cean larva is a nauplius (Fig. 14–6). Only three pairs of appendages are present—the first antennae, the second antennae and the mandibles. The second antennae and mandibles bear swimming setae. No trunk segmentation is evident, and a single median, or naupliar, eye is borne on the front of the head.

In the course of successive molts, trunk segments and additional appendages are usually gradually acquired. Development proceeds through the proliferation of trunk segments at the anterior

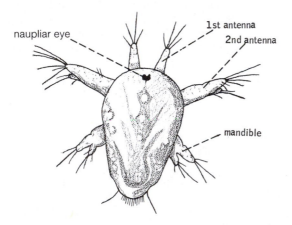

Figure 14–6   First nauplius larva of *Cyclops fuscus*, a copepod. (After Green.)

margin of the telson, which is homologous with the pygidium of annelid larvae. When the first eight pairs of trunk appendages are free of the carapace, the larva in larger crustaceans, such as crabs and shrimps, is called a zoea (Fig. 14–76A to D). In contrast to the naupliar larva, more posterior appendages provide for propulsion in swimming. Since the corresponding stage of development in other crustaceans is not as clearly defined from the earlier stages, the term *zoea* is generally limited to these larger forms. With the acquisition of a full complement of functional appendages, the young crustacean is called a postlarva.

The postlarva may be quite similar to the adult in general appearance or may still be strikingly different in some respects. For example, the postlarva of crabs has a crablike cephalothorax, but the abdomen is still large and extended (Fig. 14–76E). After additional molts all the adult features are attained except for size and sexual maturity.

The basic developmental pattern of nauplius, zoea (or its equivalent), and postlarva is very frequently modified. In many groups there has been a tendency for some or all of the larval stages to be suppressed. In most shrimps and crabs, for example, the naupliar stage is passed in the egg, and the zoea larva is the hatching stage. Both the nauplius and the zoea may be suppressed, and the young hatch as postlarvae. All larval development is suppressed in some crustaceans, such as the crayfish.

Since the postlarval stages of different groups are usually distinctive, these stages have often received special names, such as the megalops of crabs, and the acanthosoma of sergestid prawns. Moreover, the zoea, if particularly distinctive, may be given a special designation, such as the mysis of lobsters. The intermediate stages have also received different names. For example, the later nauplius instars are called metanauplii, and the prezoeal instar is called a protozoea.

Other aspects of crustacean behavior and physiology are discussed at the end of this chapter, following the survey of crustacean groups.

SUMMARY

1   The 42,000 species of the subphylum Crustacea constitute the only major group of aquatic arthropods. Most are marine, but there are many freshwater species, and there have been a number of invasions of the terrestrial environment.

2   Crustaceans are extremely diverse in structure and habit, but they are unique among arthropods in having two pairs of antennae. Other characteristic head appendages are one pair of mandibles and two pairs of maxillae. The trunk specialization varies greatly, but a carapace that covers all or part of the body is common.

3 Crustacean appendages are typically biramous and, depending upon the group, have become adapted for many different functions.

4 Gills, which are absent only in very small species, are typically associated with the appendages, but the location, number, and form vary greatly.

5 Excretory organs are a pair of blind sacs in the head that open onto the bases of the second pair of antennae (antennal glands) or the second pair of maxillae (maxillary glands).

6 The crustacean sense organs include two types of eyes, a pair of compound eyes and a small, median, dorsal naupliar eye, composed of three or four closely placed ocelli. Some groups lack compound eyes, and the naupliar eye, characteristic of the crustacean larva, does not persist in the adult of many groups.

7 Copulation is typical of most crustaceans, and egg brooding is very common. The earliest hatching stage is a naupliar larva, which possesses a median naupliar eye and only the first three pairs of body appendages.

SYSTEMATIC RÉSUMÉ OF CRUSTACEA

The diversity of crustaceans requires the use of a greater hierarchy of taxa in their classification than is usually necessary for other animal groups. The system of classification presented in R. C. Moore's *Treatise on Invertebrate Paleontology* (1969) is preferred by most specialists on the Crustacea. This system makes the Crustacea a superclass or subphylum, or if the phylum Arthropoda is abandoned, a phylum (see page 486). Each of the major crustacean groups is given the rank of class.*

Subphylum Crustacea (44,300+)
  Class Cephalocarida (9)
  Class Branchiopoda (800)
  Class Ostracoda (7000+)
  Class Copepoda (7500)
  Class Mystacocarida (9)
  Class Remipedia (2)
  Class Tantulocarida (4)
  Class Branchiura (130)
  Class Cirripedia (900)

*The name *Entomostraca*, encountered in older literature, included all of the crustacean classes except the Malacostraca.

Class Malacostraca (28,070+)
  Subclass Phyllocarida
    Order Leptostraca (20)
  Subclass Eumalacostraca
    Superorder Syncarida
      Order Anaspidacea (10)
      Order Bathynellacea (25)
    Superorder Hoplocarida
      Order Stomatopoda (300)
    Superorder Eucarida
      Order Euphausiacea (90)
      Order Decapoda (10,000)
    Superorder Peracarida
      Order Spelaeogriphacea (1)
      Order Thermosbaenacea (6)
      Order Mysidacea (780)
      Order Cumacea (800)
      Order Tanaidacea (550)
      Order Isopoda (10,000)
      Order Amphipoda (5500)

Class Cephalocarida

The Cephalocarida† are perhaps the most primitive of the existing classes of crustaceans. Only nine species, belonging to four genera, belong to the class, which was first described in 1955 from specimens collected from the bottom sand and mud of Long Island Sound. At the present time, evidence of cephalocarids has been recorded from many other parts of the world and to depths of 1550 meters.

Cephalocarids are tiny crustaceans less than 4 mm in length (Fig. 14–7). The horseshoe-shaped head is followed by an elongated trunk of 20 segments, of which only the first 8 bear appendages. Both pairs of antennae are short, and eyes are absent, probably as a result of life in the mud-water interface. The trunk appendages are interesting in several respects. Not only are they all similar, but also they are not markedly different from the second pair of maxillae. In addition, the basal section of each appendage bears a large, flattened outer piece (pseudepipodite) that gives the limb a triramous structure, somewhat like that of trilobites (Fig. 14–7B).

The internal anatomy and the physiology and habits of cephalocarids are still not well known. They are selective deposit feeders. The sweeping action of the second antennae and trunk append-

†The root *carid*, so widely used in the names of crustacean taxa, means "shrimp."

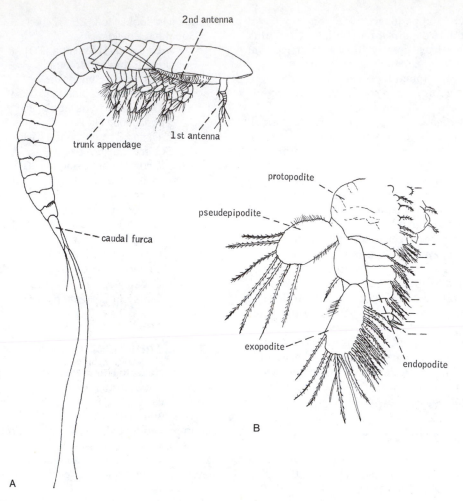

2nd antenna

trunk appendage

1st antenna

caudal furca

protopodite

pseudepipodite

exopodite

endopodite

A

B

Figure 14–7 *A*, The cephalocaridan *Hutchinsoniella macracantha*. (After Waterman and Chace.) *B*, A cephalocaridan trunk appendage. (After Sanders.)

ages throws deposit material into suspension. The suspended detritus is then sucked into the median space between the limbs and carried forward to the mouth by limb spines in a midventral food groove. Unlike most free-living crustaceans, they are hermaphroditic. In *Hutchinsoniella macracantha* a single egg develops within an ovisac carried on the first trunk segment that lacks appendages; the hatching stage is a metanauplius.

Class Branchiopoda

Branchiopods are small crustaceans that are almost entirely restricted to fresh water. Although several structurally diverse groups compose the class, all are characterized by trunk appendages that have a flattened, leaflike structure (Fig. 14–8*B*). The exo-

podite and endopodite each consists of a single, flattened lobe bearing dense setae along the margin. The coxa is provided with a flattened epipodite that serves as a gill, hence the name *Branchiopoda*—"gill feet." In addition to gas exchange, the trunk appendages are usually adapted for suspension feeding and commonly for locomotion. The first antennae and second maxillae are vestigial. The last abdominal, or anal, segment bears a pair of large, terminal processes called cercopods.

The some 800 described species of Branchiopoda belong to four distinct groups. The fairy shrimps (order Anostraca) are characterized by an elongated trunk containing 20 or more segments, of which the anterior 11 to 19 segments bear appendages (Fig. 14–8*A*). There is no carapace, and the compound eyes are stalked. In the tadpole shrimps (order Notostraca) the head and anterior

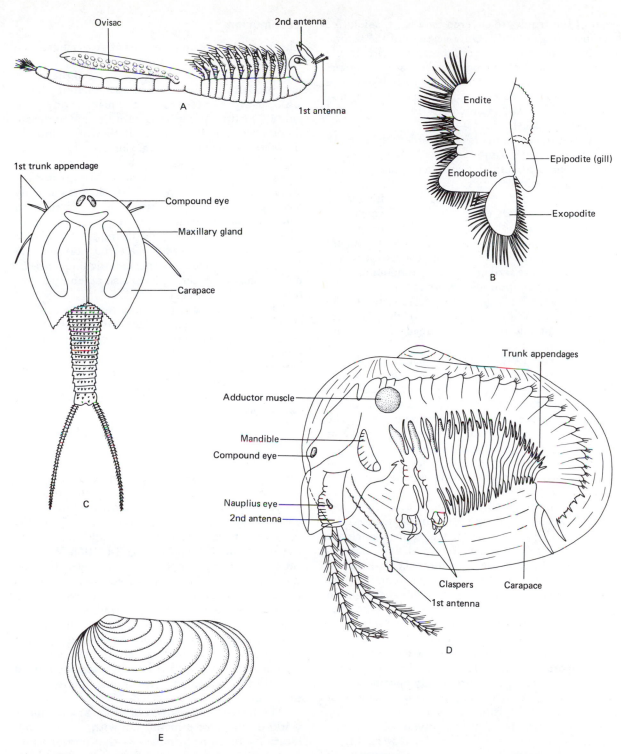

Figure 14—8 Branchiopoda: *A*, Fairy shrimp *Branchinecta*, an anostracan (lateral view). (After Calman.) *B*, Trunk appendage of *Branchinecta paludosa*, showing foliaceous structure. (After Sars from Pennak.) *C*, Tadpole shrimp *Triops*, a notostracan (dorsal view). (After Pennak, R. W., 1978: Fresh-Water Invertebrates of the United States, 2nd Edition. John Wiley and Sons, N.Y.) *D*, Male conchostracan *Cyzicus mexicanus* with left valve removed. (After Mattox from Edmondson, Ward, and Whipple.) *E*, Left shell valve of the conchostracan *Cyzicus*. (After Sars from Calman.)

half of the trunk are covered by a large, shieldlike carapace (Fig. 14–8C). Only vestiges of the second antennae remain. The compound eyes are sessile and are located close together beneath the dorsal shield. There is a long, flexible, ringed abdomen bearing up to 70 pairs of appendages. The telson at the end of the abdomen carries two long, caudal furcae (processes).

Two groups (members of the order Diplostraca), containing clam shrimps and water fleas, are laterally compressed, and the body is at least partially enclosed within a bivalve carapace. In the clam shrimps (suborder Conchostraca) the entire body is nearly or completely enclosed within the carapace, and the animal looks strikingly like a little clam (Fig. 14–8D and E). Ten to 32 trunk segments are present, each bearing a pair of appendages. The second antennae are well developed, biramous, and setose, and the compound eyes are sessile.

The water fleas (suborder Cladocera of the order Diplostraca) constitute half of the branchiopods and include many widespread and common species, such as those belonging to the genus *Daphnia* (Fig. 14–9A). The carapace encloses the trunk but not the head and often terminates posteriorly in an apical spine. The head projects ventrally and somewhat posteriorly as a short beak, so the body has the appearance of a plump bird. The number of trunk appendages is reduced to five or six pairs. The tip end of the trunk, commonly called the postabdomen, is turned ventrally and forward and bears special claws and spines for cleaning the carapace.

The majority of branchiopods are only a few millimeters in length, and some are as small as 0.25 mm. The largest are the fairy shrimps, which are usually more than 1 cm long and may attain a length of 10 cm. Most branchiopods are pale and transparent, but rose or red colors sometimes are found, caused by the presence of hemoglobin within the body.

Branchiopods live almost exclusively in fresh water; only cladocerans inhabit streams, large ponds, and lakes. Most branchiopods are confined to temporary pools, springs, and small ponds. Moreover, most species are highly ephemeral, appearing only briefly during the short existence of pools formed by melting snows or spring rains. The restriction of the larger branchiopods to small bodies of fresh water may be correlated with the absence of fish in such habitats. Some fairy shrimps can tolerate saline pools, and species of *Artemia* (brine shrimps) are found in salt lakes and ponds throughout the world.

Locomotion

The water fleas swim by means of powerful second antennae (Fig. 14–9A). Movement is largely vertical and usually jerky. The downstroke of the antennae propels the animal upward; then it slowly sinks, using the antennae in the manner of a parachute. The little plumose setae at the end of the abdomen act as stabilizers, for if they are removed, the water flea rotates ventral side up.

Other branchiopods use the trunk appendages in swimming, although the clam shrimps use the second antennae as well. Fairy shrimps usually swim upside down; however, when lighted from below in the aquarium, they roll over and swim right side up. Many clam shrimps and tadpole shrimps swim and crawl over the bottom, and some clam shrimps plow through the bottom sediment. Many cladocerans have also taken up a bottom or near-bottom existence.

The naupliar eye is persistent in almost all branchiopods (Fig. 14–9A). The sessile, cladoceran compound eyes are unusual not only in being fused into a single median eye but also in that this single median eye can be rotated by special muscles (Fig. 14–9A). The compound eye, at least in many cladocerans, is used to orient the animal in swimming.

Feeding

Most branchiopods are suspension feeders and collect food particles with fine setae on the trunk appendages. However, particle collection does not involve sieving (see Box 14–1, pp. 566–567). In the fairy shrimps the space between limbs increases as the limbs move forward (Fig. 14–10B). Water is sucked into this space from the midventral line and the lateral setae of the appendages collect particles from the incoming stream. On the backstroke, water is forced out of the interlimb space posteriolaterally and distally (Fig. 14–10A). By several complex mechanisms the collected food particles are transferred to a midventral food groove. Here they become entangled in mucus and then moved forward to the maxillae, which push the food into the mouth.

The feeding mechanism of other branchiopods works on the same principle but with various modifications. In clam shrimps only the anterior trunk appendages are used for collecting food, and the posterior appendages are modified as jaws for grinding large particles. In the cladocerans only certain of the four to six pairs of trunk appendages

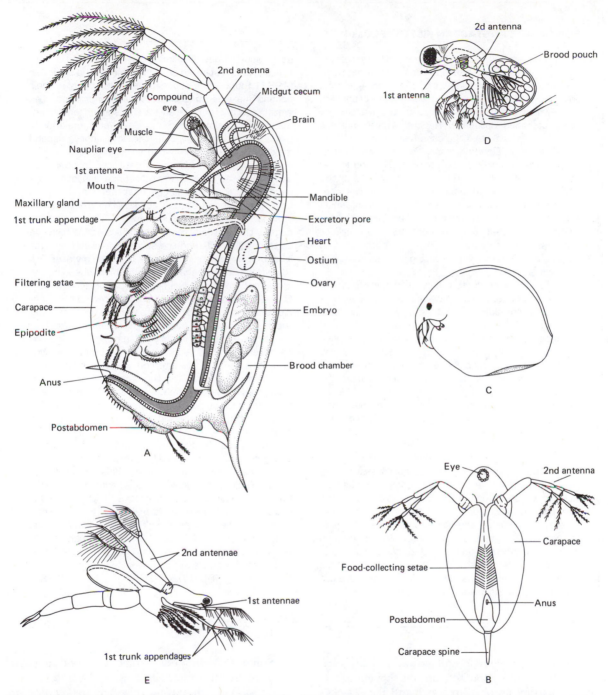

Figure 14—9 *A,* Female of the cladoceran *Daphnia pulex* (lateral view). (After Matthes from Kaestner.) *B,* Ventral view of *Daphnia. C, Chydorus gibbus,* a cladoceran with a more rounded body than *Daphnia pulex.* Appendages are not shown. *D* and *E,* Two aberrant cladocerans: *D, Polyphemus pediculus; E, Leptodora.* (*C* from Pennak, R. W., 1978: Fresh-Water Invertebrates of the United States, 2nd Edition. John Wiley and Sons, N.Y. *D* after Lilljeborg from Pennak. *E* after Sebestyén from Pennak.)

BOX 14–1 CRUSTACEAN FILTER FEEDING

The process by which crustaceans filter particles from a water stream has long been considered to be one of sieving. The mesh size of the filter—i.e., the distance between setae or setules (side branches) *(A)*—would determine the size of the diatoms or other particles collected. Recent studies indicate that this is not always the case (Gerritsen and Porter, 1982). If *Daphnia magna* is fed a mixture of three sizes of polystyrene spheres *(B)*, two larger and one smaller (0.5 μm) than the mesh size of the filter (1 μm between setules), the water flea captures a sizable part of the smallest spheres, which should have passed through the filter. The collecting efficiency is about 60 per cent for the smallest spheres, compared with 100 per cent for the two larger sizes. Clearly, the collecting process is not one of simple sieving.

When a water stream flows over or around an object, a boundary layer develops over the surface because of the adhesive nature of water molecules (viscosity). Water molecules at the surface itself do not move at all, and those near the surface move more slowly than the mainstream current. Away from the surface successive layers of water shear off with greater velocity as the distance from the surface increases. This boundary layer of slowly moving water can be significant when the object over which water is flowing is very small. In *Daphnia*, for example, the boundary layer that develops over one setule of the filter extends beyond the next setule. This means that little water flows between the setules unless subjected to considerable pressure. The smaller and slower a setulated appendage, the less leaky it will be. Thus, in *Daphnia*, particles are captured not by sieving but by attraction of charges between the particles and the filter surface. The proportion of polystyrene spheres collected can be altered in *Daphnia* by changing their surface charges.

In copepods the second maxillae, which are the filtering appendages, have been thought to sieve out particles from a forward moving current set up by other appendages. By filming dye-marked water currents around tethered copepods, Koehl and Strickler (1981) found that these crustaceans move their second maxillae to actively capture particles. The water current generated by the second antennae, mandibular palps, first maxillae, and maxillipeds passes over, not through, the stationary, second maxillae *(C)*. However, when an algal particle in the approaching water stream is detected (probably by sensory setae on the first antennae and other appendages), the appendages beat symmetrically so that the copepod orients toward the particle. The second maxillae are flung open. This movement causes the water stream with the particle to move to the midline *(D)*, and the second maxillae clap over it, squeezing out much of the surrounding water *(E)*. Variations in this behavior can be seen when the animals feed on very small or large particles. The particle is then combed off and transferred to the mandibles and mouth by the endites of the first maxillae. Viscous forces are operative here, as in *Daphnia*. The second maxillae initially act as paddles, and the captured particle is surrounded by a boundary layer of water as it is transferred. As a consequence, Koehl and Strickler liken the particle transfer process from the second maxillae by the first maxillae to removing crumbs from one of your hands with the fingers of the other while both are immersed in molasses.

The viscous forces of water to which these crustaceans are subject would be similar for many other small suspension-feeding animals. It is probable that most collect particles by methods other than sieving. For a brief review, see LaBarbera (1984). *A* and *B* from Gerritsen, J., and Porter, K. G., 1982: The role of surface chemistry in filter feeding by zooplankton. Science, *216*:1225–1227; *C–E* from Koehl, M. A. R., and Strickler, J. R., 1981: Copepod feeding currents: food capture at low Reynolds number. Limnol. Oceanogr., *26*(6):1062–1073.)

---

are adapted for particle collection, and the filter setae are usually arranged on the appendage to form a distinct comb (Box 14–1 and Fig. 14–9*A*). The water current passes from anterior to posterior, and collected particles are transferred into the food groove by special setae (gnathobases) at the basal part of the appendages. The efficiency of the collecting mechanism of *Daphnia* is indicated by the ability of some members of this genus to grow on a diet of bacteria.

Some branchiopods scrape up food material from plant and other surfaces, and some are predacious. The planktonic cladocerans *Leptodora*, *Bythotrephes*, and *Polyphemus* have anterior appendages modified for grasping.

The branchiopod foregut forms a short esophagus, and the midgut is often enlarged to form a stomach. In cladocerans, however, the midgut is more or less tubular and not easily distinguished from other parts of the digestive tract. There are

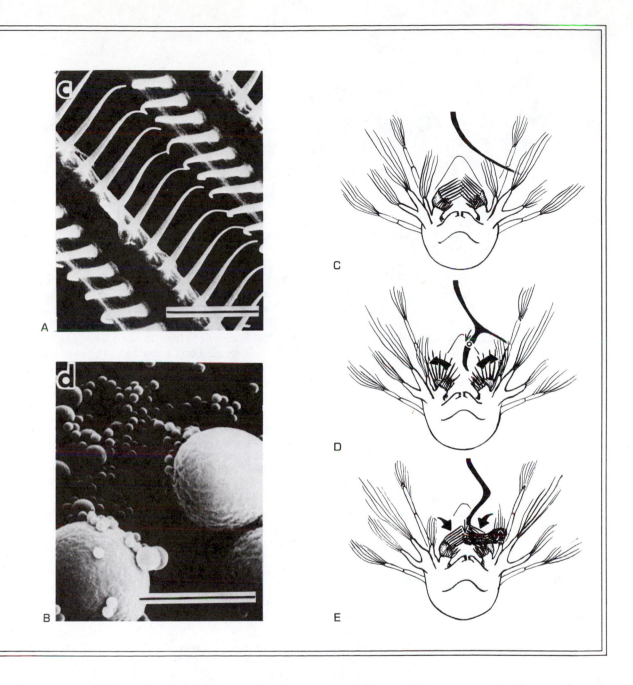

often two small, digestive ceca (Fig. 14–9A). The intestine in some cladocerans is coiled one to several times.

Gas Exchange, Internal Transport, and Osmoregulation

The thin, vesicular or lamellar epipodite on each of the trunk appendages of branchiopods is called a gill, but it is probable that the entire appendage and even the general integumentary surface are also important in gas exchange (Fig. 14–8B).

Hemoglobin has been found in the blood, muscles, nervous tissue, and eggs of many branchiopods. The presence of hemoglobin frequently depends on the amount of oxygen in the water, so the animals are colorless in well-aerated water and pink in stagnant water.

The heart of cladocerans is a small, globular sac with only two ostia, and it lies at the anterior of the

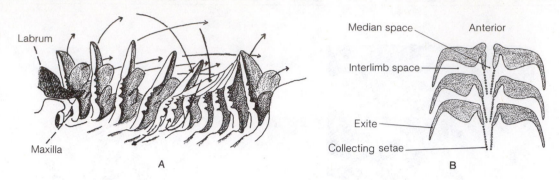

Figure 14–10 *A*, Trunk appendages of the fairy shrimp *Branchinella* showing swimming and feeding currents (lateral view). Setae have been omitted. (After Cannon.) *B, Branchinella*, showing three pairs of trunk appendages (frontal section). (After Cannon from Green.)

trunk (Fig. 14–9*A*). In all other branchiopods the heart is tubular with many ostia. The arterial system is restricted to a short, unbranched anterior aorta.

The excretory organs are maxillary glands, usually called shell glands when the duct can be seen coiled within the carapace wall. Little is known regarding water balance mechanisms except in the brine shrimp *Artemia,* which has been studied extensively. This crustacean can tolerate salinities ranging from 10 percent sea water to the saturation point for sodium chloride. The internal osmotic pressure varies only slightly with external conditions. Ionic regulation is maintained by the absorption or excretion of salts through the gills. Also, *Artemia* is known to be capable of excreting a urine hyperosmotic to its blood. In brine the osmotic pressure of the urine is four times that of the blood.

Reproduction and Development

The united or separate male gonopores of cladocerans open near the anus or the postabdomen, which may be modified in the form of a copulatory organ, and the oviducts open into a dorsal brood chamber located beneath the carapace. In other branchiopod species the gonopores open ventrally on varying segments located more toward the middle of the trunk.

During copulation the male fairy shrimp clasps the dorsal side of the female abdomen with the second antennae, which have become modified for this purpose, and then, twisting the abdomen around, inserts the paired, eversible copulatory processes containing the openings of the vas deferens into the single, median female gonopore

(Wiman, 1981). Similar copulatory behavior is displayed by the water fleas, although the male clasps with the first rather than the second antennae.

In all branchiopods the eggs are brooded for varying lengths of time. In the fairy shrimp a special sac is formed on extrusion of the eggs from the glandular uterine chamber (Fig. 14–8*A*). Both cladocerans and clam shrimps brood their eggs dorsally beneath the carapace (Fig. 14–9*A*).

The eggs are produced in clutches of two to several hundred, and a single female may lay several clutches. Development in most cladocerans is direct, and the young are released from the brood chamber by the ventral flexion of the postabdomen of the female. In the other branchiopods embryonated eggs are released by the female and fall to the bottom after only a brief brooding period, or they may remain attached to the female and reach the bottom when she dies. In any case the eggs typically hatch as nauplii.

Parthenogenesis is common in branchiopods, and in some species males are uncommon or unknown. In cladocerans the reproductive pattern is strikingly like that of many rotifers. Parthenogenetic diploid eggs hatch into females for several generations, and one female can produce a succession of broods. When the young leave the brood chamber beneath the carapace, the skeleton is molted and a new batch of eggs is released into the new brood chamber. The changeover from one brood to the next can take place in five minutes (Frey, 1982). At some point certain factors, such as change in water temperature or the decrease of the food supply as a result of population increase, induce the appearance of males, and fertilized eggs are produced. The fertilized eggs are large, and only two are produced in a single clutch, one from

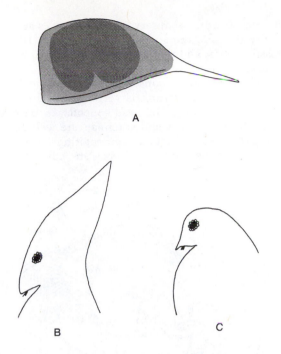

Figure 14–11 *A,* Ephippium of *Daphnia pulex.* (From a photograph by Pennak.) *B* and *C,* Cyclomorphosis in *Daphnia: B,* Summer long-head form. *C,* Spring and fall round-head form. (*B* and *C* after Pennak, R. W., 1978: Fresh-Water Invertebrates of the United States, 2nd Edition. John Wiley and Sons, N.Y.)

each ovary. The walls of the brood chamber are now transformed into a protective, saddle-like capsule (ephippium). This is cast off at the next molt, either separating from or remaining with the rest of the detached exoskeleton (Fig. 14–11*A*). The ephippia float, sink to the bottom, or adhere to objects and can withstand drying and freezing and even passage through the gut of fish and of fish-eating birds and mammals (Mellors, 1975). By means of such protected resting eggs, cladocerans may be dispersed by wind or animals for some distances and can overwinter and survive summer droughts.

Thin-shelled summer eggs and thick-shelled resting (dormant) eggs are also produced by many species of the other branchiopod groups, but both types of eggs may be either parthenogenetic or fertilized. Development in thin-shelled eggs is rapid, and hatching may take place while the eggs are attached to the female. Production of dormant eggs may be stimulated by a variety of external factors, such as population density, temperature, and photoperiod. Dormancy (or diapause) may occur after an initial period of development. Thus, the eggs are prepared for rapid hatching when the precise environmental conditions that are required are present. The controlling factors that break dormancy— oxygen, salinity, temperature, illumination—vary and are related to the type of habitat for which the species is adapted.

Population pulses, or cycles, are common in branchiopods. This is particularly true of many cladoceran species. Some cladocerans, such as species of *Diaphanosoma, Moina,* and *Chydorus,* exhibit a single population rise and fall during the warmer months in some lakes. Others are dicyclic and exhibit both spring and fall population peaks. The phenomenon is by no means always predictable. For example, some species of *Daphnia* may be monocyclic, dicyclic, or even acyclic, depending on the lake or pond in which they live.

In some cladoceran species, such as the lake-inhabiting *Daphnia dubia* and *Daphnia retrocurva,* the head progressively changes from a round to a helmet-like shape between spring and midsummer (Fig. 14–11*B* and *C*). From midsummer to fall the head changes back to the normal, round shape. Cyclomorphosis is still poorly understood. It may, as in rotifers, result from internal factors, or it may result from an interaction between external conditions (temperature) and internal factors (perhaps genetic). In at least some species it appears to reduce predation by producing a size less acceptable to certain predators (Wong, 1981).

A summary of the branchiopods appears on page 591.

Class Ostracoda

Ostracods, sometimes called mussel or seed shrimps, are small crustaceans that are widely distributed in the sea and in all types of freshwater habitats. Over 7000 living species have been described. They superficially resemble clam shrimps in having a body completely enclosed in a bivalve carapace. Perhaps these animals, rather than the Conchostraca, should have the name *clam shrimps,* for the ostracod carapace has evolved along lines that are even more strikingly like those of bivalves. In ostracods the elliptical valves are impregnated with calcium carbonate; there is a distinct, dorsal hinge line formed by a noncalcified strip of cuticle; and the valves are closed by a cluster of transverse adductor muscle fibers that are inserted near the center of each valve. In some ostracods the valves may be locked by hinge teeth and ridges at the hinge line. The surface of the valves may be covered with setae and sculptured with pits, tubercles, or irregular projections (Fig. 14–12*B*). In the order Myodocopa there is a notch in the anterior margin of the valves permitting the protrusion of the antennae when the valves are closed (Fig. 14–12*A*).

The ostracod fossil record is continuous from the Cambrian period and is the most extensive of any group of crustaceans. The small size and the calcification of the valves have undoubtedly been primary factors in preservation. The valves are virtually the sole remains of the more than 10,000 fossil species, and the classification of fossil forms is based entirely on valve morphology.

The head region constitutes much of the ostracod body, for the trunk is very much reduced in size (Fig. 14–12*C*). The head appendages, particularly the antennules and antennae, are well developed. All external trunk segmentation has disappeared, and the trunk appendages are reduced to no

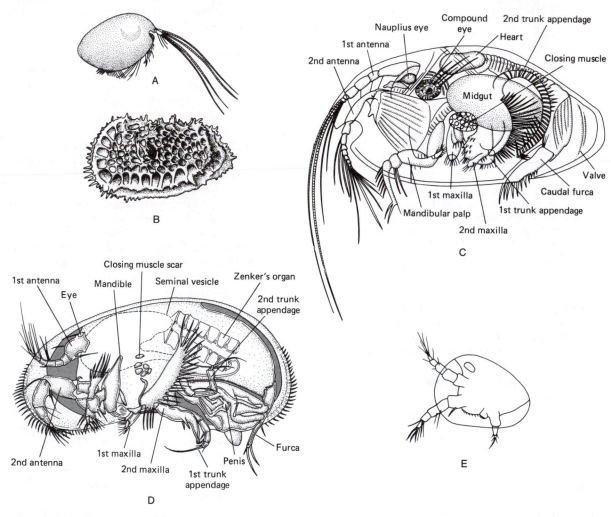

Figure 14–12 *A*, A myodocopid ostracod with antennal notches in valves (lateral view). *B*, Lateral view of the sculptured valve of *Agrenocythere spinosa*, a benthic marine ostracod. *C*, Female marine myodocopid ostracod *Skogsbergia* with left valve and left appendages removed (lateral view). This is a marine scavenging carnivore that is able to swim over the bottom. (After Claus from Calman.) *D*, *Candona suburbana*, a very common freshwater podocopid ostracod, left valve removed. This nonswimming bottom dweller feeds on algae and decomposing vegetation. Zenker's organ, which is part of the male reproductive system, ejects sperm to the penis. (*C* and *D* from Cohen, A. C., 1982: Synopsis and Classification of Living Organisms. Vol. 2. McGraw-Hill Book Co., N.Y. pp. 191 and 192.) *E*, Lateral view of an ostracod naupliar larva. Only one member of each of the three pairs of appendages is shown. (After Schreiber from Pennak, R. W., 1978: Fresh-Water Invertebrates of the United States, 2nd Edition. John Wiley and Sons, N.Y. p. 427.)

more than two pairs. The trunk appendages, including the maxillae, may be more or less leglike or modified for swimming, feeding, clasping, or cleaning debris from within the valves.

Most ostracods are minute, several millimeters or less in length. The giant of the group is the pelagic deep-sea *Gigantocypris mulleri*, which can reach a length of 3 cm. Hues of gray, brown, or green are the most common.

Locomotion and Feeding

Although there are some planktonic ostracods, the majority live near the bottom, where they swim intermittently or crawl over or plough through the upper layer of mud and detritus. There are burrowing and interstitial species and species that live on the surface of algae, water plants, or other submerged objects. Some are commensal on other animals, living, for example, among the leg setae of crayfish. One or both pairs of antennae are the principal locomotor appendages and are variously modified, depending on the habits of the animals. Kicking forward with the caudal furca is a common aid to locomotion in many benthic species. Perhaps the most remarkable ostracods are the New Zealand and South African terrestrial species of *Mesocypris*, which can plough through forest humus.

Ostracods display diverse feeding habits. There are carnivores, herbivores, scavengers, and filter feeders. Algae are a common plant food, and the prey of carnivorous species includes other crustaceans, small snails, and annelids. The large *Gigantocypris* is predacious and is reported to capture other crustaceans and even small fish with its antennae. Detritus particles, often stirred up by the antennae or mandibles, are also a common source of food. In the suspension-feeding species the collecting setae are located on one of the pairs of maxillae.

Internal Structure and Physiology

Gills are lacking and gas exchange is integumentary, the locomotor and feeding currents providing ventilation. A heart and blood vessels are present only in the marine order Myodocopa. Blood circulates between the valve walls in all ostracods.

Some ostracods possess antennal glands, some have maxillary glands, and some are among the few crustaceans that possess both types of excretory organs in the adult. The maxillary glands are large and coiled and lie between the inner and outer walls of the valves.

A naupliar eye is present in all ostracods, but sessile compound eyes appear only in the Myodocopa (Fig. 14–12*C*). The most important sense organs are probably the sensory hairs found not only on the appendages but also on the valves.

Ostracods were the first crustaceans in which luminescence was observed. A cloud of bluish light is produced externally by secretions from a gland in the labrum and occurs as spontaneous flashes lasting only 1 to 2 seconds (see Box 5–1, p. 144).

Reproduction and Development

The female gonopores are located ventrally between the last pair of appendages at the caudal end of the trunk. The paired sperm ducts open between or through one of the two large, sclerotized penes projecting ventrally in front of the caudal furca. The sperm are motile and in some cyprids are of a remarkably large size. *Pontocypris monstrosa*, which is less than 1 mm in length, has sperm as long as 6 mm.

During copulation in some species, the female is clasped dorsally and posteriorly by the second antennae or the first pair of legs of the male, and the penes are inserted between the valves of the female into the gonopores.

Most commonly the eggs are shed freely in the water or are attached singly or in groups to vegetation and other objects on the bottom, but the eggs are brooded in the dorsal part of the shell cavity in some ostracods. Some species, especially those in temporary bodies of fresh water, produce very resistant eggs. The eggs hatch as nauplii, but each nauplius is enclosed in a bivalve carapace like that of the adult. The valve skeleton is shed at each molt in the ostracod life history, along with the rest of the exoskeleton.

Parthenogenesis is common in the freshwater Cyprididae, and there are species in which males are unknown.

SYSTEMATIC RÉSUMÉ OF CLASS OSTRACODA

Order Myodocopa. Marine ostracods having a shell with antennal notches. The second antennae usually adapted for swimming; exopodite larger than endopodite. Two pairs of trunk appendages; the long, second pair is adapted for cleaning interior of valves. *Cypridina, Gigantocypris, Skogsbergia*.

Order Halocyprida. Marine ostracods, with first pair of trunk appendages present or absent and second pair absent or short and leglike. Includes most of the planktonic species.

Order Podocopa. Second antennae with endopodite equal to or longer than expodite. One or two pairs of trunk appendages. This large order includes marine as well as freshwater species. *Cypris, Pontocypris, Candona, Cypridopsis, Mesocypris, Darwinula, Cythere.*

A summary of the ostracods is provided on p. 591.

Class Copepoda

The Copepoda are the largest class of small crustaceans, over 7500 species having been described. Most copepods are marine, but there are many freshwater species and a few that live in moss and soil water films. Also, there are many that are parasitic on various marine and freshwater animals, particularly fish. Marine copepods exist in enormous numbers and are usually the most abundant and conspicuous component of a plankton sample. Since most planktonic species feed on phytoplankton, they are the principal link between phytoplankton and higher trophic levels in many marine food chains. A major part of the diet of many marine animals is composed of copepods.

Most copepods range in length from less than 1 mm to more than 5 mm, although there are larger (17 mm) free-living species. Some parasitic forms are very large, over 32 cm in length (Fig. 14–18). Although most copepods are rather pale and transparent, some species may be brilliant red, orange, purple, blue, or black. Many luminescent species have been reported.

The body of a free-living copepod is commonly tapered from anterior to posterior and is somewhat cylindrical (Fig. 14–13), but there are many exceptions to this generalization. The trunk is composed of a thorax and abdomen. The head is either rounded or pointed. Compound eyes are absent, but the median naupliar eye is a typical and conspicuous feature of most copepods. Also conspicuous are the uniramous first antennae, which are generally long and held outstretched at right angles to the long axis of the body.

The head is fused with the first of the six thoracic segments and sometimes with the second thoracic segment as well. The first pair of thoracic appendages have become modified to form maxillipeds for feeding. The remaining five thoracic appendages, except the last one or two pairs, are all more or less similar and rather symmetrically biramous.

The abdomen is composed of five segments, which are commonly narrower than those of the thorax. There are no abdominal appendages except for the anal segment (telson), which bears two caudal rami. In some planktonic marine species these caudal rami are spectacularly developed. For example, in *Calocalanus pavo*, each ramus is turned laterally and bears four long, feather-like setae.

Of the five free-living orders of copepods, the calanoids are largely planktonic; the harpacticoids, which include over half of the copepod species, are largely benthic; and the cyclopoids contain both planktonic and epibenthic species. Variations in body shape are related to the habitat of the species (Marcotte, 1983). Planktonic forms have rather cylindrical bodies with a narrow abdomen. Those that live high in the water column tend to be more slender and fusiform than those that swim several meters over the bottom. Epibenthic species, which crawl and swim just over the bottom, have somewhat broader bodies. Benthic species that live on algae and sea grasses may be broad and flattened, and interstitial forms are narrow and wormlike (Fig. 14–14).

Feeding

Copepods display a range of feeding habits, depending in part on where they live (see review by Marcotte, 1983). Planktonic copepods are chiefly suspension feeders, and the second maxillae are modified to capture food. However, recent studies indicate that these animals do not really sieve particles from suspension (see Box 14–1). Phytoplankton constitutes the principal part of the diet of most suspension-feeding species, but some rely heavily on detritus particles as well. Using radioactive diatom cultures, Marshall and Orr (1955) reported that *Calanus finmarchicus*, which is about 5 mm long, may collect and ingest from 11,000 to 373,000 diatoms every 24 hours, depending on their size. When feeding on particles of varying size, larger particles will be selected because they can be handled more efficiently, but it is probable that under natural conditions planktonic copepods take whatever food predominates (see review by Marshall, 1973; Frost, 1977).

Not all planktonic copepods are herbivorous suspension feeders; some are omnivorous, and some are strictly predacious. Species of *Anomalocera* and *Pareuchaeta* even capture young fish. Species of *Tisbe*, planktonic epibenthic harpacticoids, are known to swarm over a small fish and eat its fins, thus immobilizing it. They then devour the body as it falls to the bottom (Marcotte, 1977).

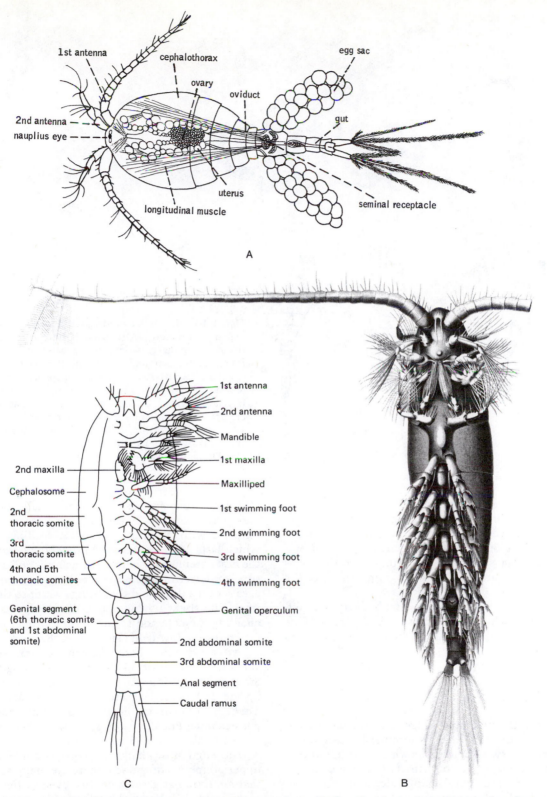

Figure 14–13 *A*, A cyclopoid copepod, *Macrocyclops albidus* (dorsal view). (After Matthes from Kaestner.) *B*, Ventral view of *Calanus*, a typical calanoid copepod, with appendages shown. (After Giesbrecht, W., 1892; Fauna and Flora Golfes Neapel. Monogr., *19*:1–831.) *C*, Diagrammatic ventral view of *Pseudocalanus*, showing appendages. (From Corkett, J., and McLaren, I. A., 1978: The biology of *Pseudocalanus*. Adv. Mar. Biol., *15*:2–231.)

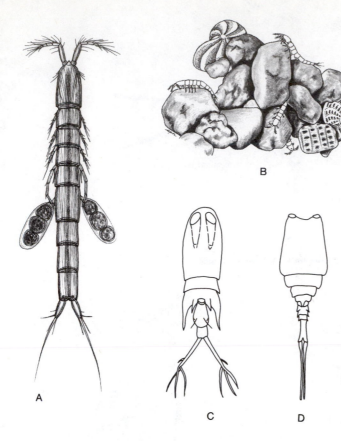

A

B

C

D

Figure 14–14 *A*, A cylindropsyllid harpacticoid copepod (dorsal view). (After Sars). *B*, Interstitial harpacticoid copepods crawling among sand grains. Two foram shells are among sand grains. (Drawn from life.) *C* and *D*, Two marine cyclopoid copepods, *Corycaeus* (*C*) and *Copilia* (*D*), in which the naupliar eye is very large and divided. Copepods of these and related genera are often brilliantly colored. (From Smith, D. L., 1977: A Guide to Marine Coastal Plankton and Marine Invertebrate Larvae. Kendall/Hunt Publishing Co., Dubuque, Iowa.)

Some freshwater Cyclopidae are herbivorous; others are carnivorous. Members of the common freshwater genus, *Cyclops*, are predatory, as are some of the other cyclopoid genera. Most bottom-dwelling harpacticoids feed on microorganisms and detritus attached to sand grains, algae, or sea grasses.

Calanoids store foods in a fat body or in a midgut oil sac, which often gives the body a red or blue color. The oil in many species contributes to buoyancy.

Locomotion

Both the thoracic appendages and the second antennae are used in rapid swimming. The second antennae, the two branches of which beat in a rotary manner, appear to be of primary importance in calanoids, and the thoracic appendages are most important in cyclopoids. The first antennae, which are long and setose in planktonic forms, primarily slow sinking, like parachutes. They may lie back against the body when movement is rapid and then extend laterally again when movement is slowed.

Carnivorous species cruise continually as they seek potential prey. Herbivorous species alternate cruising with feeding. During a feeding bout, which lasts 10 to 30 seconds (*Eucalanus crassus*), the first antennae act both as parachutes and as sensors for detecting algae; the other anterior appendages set up a flow field that brings water to the animal. The slight tendency to sink is important in enabling the copepod to maintain a proper orientation for its flow field (Strickler, 1982).

After a feeding bout the copepod cruises or sinks to a new position for another feeding bout. Swimming positions vary greatly among species—upside down, vertical, etc.—and the caudal rami can be held in varying positions or act as a rudder. For example, *Eucalanus crassus* swims backwards in a vertical position (Fig. 14–15).

Although most planktonic copepods live in the upper 50 meters of the sea, there are many species that are found at greater depths, even in the deep sea (Fig. 14–16). Vertical movement is oriented by light, and many species exhibit daily vertical migrations.

The bottom-dwelling harpacticoids and some cyclopoids crawl over or burrow through the sub-

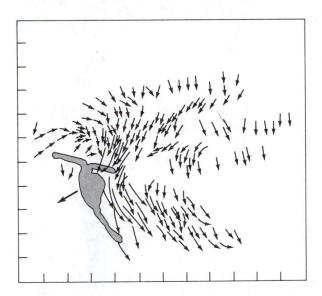

Figure 14–15 Flow field of water over *Eucalanus crassus* during feeding. Axes scale unit is 1 mm; small arrow lengths represent flow distance over an interval of 0.4 second. Note orientation of body and direction of movement (large arrow). (From Strickler, J. R., 1982: Calanoid copepods, feeding currents, and the role of gravity. Science, *218*:158–160. Copyright 1982 by the American Association for the Advancement of Science.)

stratum, and many harpacticoids live between sand grains (Fig. 14–14*B*) (see review by Hicks and Coull, 1983). The thoracic limbs are used in crawling, and this is accompanied in harpacticoids by lateral undulations of the wormlike body.

Internal Structure

There are no gills in free-living copepods, and except in the calanoids and some parasitic species, there is neither heart nor blood vessels. The excretory organs are maxillary glands.

Reproduction and Development

Male copepods are commonly smaller than females and are usually outnumbered by females. Copepods are one of the few small crustaceans that form spermatophores, and the lower end of the sperm duct is modified for this purpose. The male opening is located on the first abdominal segment of most copepods, as are the female gonopores and openings to the seminal receptacles.

During copulation the male clasps the female with one or both first antennae and, in most calanoids with the last pair or modified thoracic appendages as well. The spermatophores are trans-

ferred to the female by the thoracic appendages of the male and adhere to the receptacle openings by means of a special cement.

Most calanoids shed their eggs singly into the water. However, the eggs of other copepods are usually enclosed within an ovisac. The ovisac is produced by oviduct secretions when the eggs are emitted from the oviduct and remain attached to the female genital segment, where they function as brood chambers (Fig. 14–13*A*). One or two sacs are formed, depending on the number of oviducts. Each sac contains a few to 50 or more eggs, and clutches may be produced at frequent intervals. For example, among freshwater copepods, species of diaptomids shift back and forth between a gravid and nongravid condition every four days, and mating is required for each clutch of eggs (Watras and Haney, 1980).

The eggs typically hatch as nauplii. After five or six naupliar instars, the larva passes into the first copepodid stage. The first copepodid larva displays the general adult features, but the abdomen is usually still unsegmented, and there may be only three pairs of thoracic limbs. The adult structure is attained typically after six naupliar and five copepodid stages (Fig. 14–17), and molting then ceases. The entire course of development may take as little as one week or nearly as long as one year. Six months to a little over a year is the maximum life span of most free-living species. Studies on the copepodid stages of 20 planktonic species in the Adriatic indicate that there were from three to six generations a year (Shmeleva and Kovalev, 1974).

Some freshwater calanoids and harpacticoids produce both thin-shelled eggs and thick-shelled dormant eggs, and overwintering resting eggs have now been reported for a number of marine calanoids (Uye and Kasahara, 1978; Grice and Marcus, 1981; Marcus, 1982). In many freshwater copepods the copepodid stages (or even adults) secrete an organic, cystlike covering and become inactive under unfavorable conditions (Sarvala, 1979). Such cysts, buried in mud, are particularly adapted to withstand desiccation and enable the copepod to estivate when its pool or pond dries up. They also provide a means of dispersal when carried on the muddy feet or bodies of birds and other animals.

Parasitic Copepods

The Copepoda contain over 1000 species of parasitic crustaceans. Some copepods are ectoparasites on fish and attach to the gill filaments, the fins, or the general integument (Fig. 14–18). Other copepods are commensal or endoparasitic within poly-

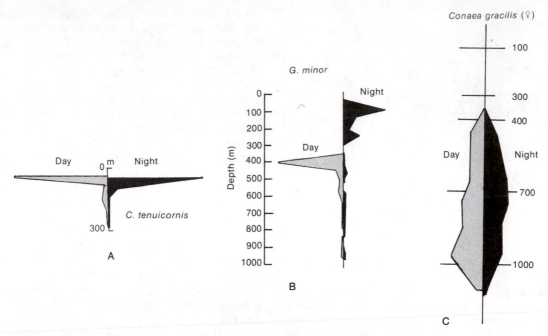

Figure 14–16 Vertical distribution of three species of mid-Atlantic marine planktonic copepods. Black side indicates distribution during night, and gray, during day; width reflects population density. *A* and *B* are calanoids *Calanus tenuicornis* and *Gaetanus minor*. *C* is the cyclopoid *Conaea gracilis*. Note the restricted vertical distribution of *Calanus tenuicornis* as compared with *Conaea gracilis* and the extensive day-night vertical migration of *Gaetanus minor*. (*A* and *B* from Roe, H. S. J., 1972: The vertical distributions and diurnal migrations of calanoid copepods collected on the Sond Cruise, 1965, pt. II. J. Mar. Biol. Assoc. U.K., 52:315–343. *C* from Boxshall, G. A., 1977: The depth distributions and community organization of the planktonic cyclopods of the Cape Verde Islands region. J. Mar. Biol. Assoc. U.K., 57:543–562. All copyrighted and reprinted by permission of Cambridge University Press.)

chaete worms, the intestine of echinoderms (particularly crinoids), and in tunicates and bivalves. Cnidarians, especially anthozoans, are the hosts for many species of copepods. All degrees of modification from the free-living copepod form are exhibited by these parasites. Ancestral forms are usually ectoparasites and resemble free-living species (Fig. 14–19*A*). On the other hand, some ectoparasitic and endoparasitic copepods are so highly modified and bizarre that they no longer have any resemblance to the free-living species (Fig. 14–19*C* to *E*).

In the ectoparasites certain appendages have usually become specialized as holdfast organs, and the mouth parts are adapted for piercing and sucking. In some ectoparasitic copepods a frontal gland produces a button (the bulla), which is attached to the gill filament of the fish. The second maxillae are inserted into the button for permanent attachment (Fig. 14–19*B*).

In most parasitic copepods the adults are adapted for parasitism, and the larval stages are usually typical and free swimming. Contact with the host occurs at various times during the life cycle of the copepod, and modifications appear

with each molt. The salmon gill maggot, *Salmincola salmoneus*, which is parasitic on the gills of the European salmon, has a typical life cycle. When the salmon enters the coastal estuaries on its migration to fresh water, the copepod, in the form of a first copepodid larva, attaches to the gills. The larva attaches by a structure resembling a button and thread (bulla) held by the second maxillae; the larva then undergoes a series of molts. The male matures first, and copulation takes place before the female is mature. The male then dies. The female undergoes a final molt and becomes permanently attached to the host by the second maxillae fused with the bulla, which is embedded in the tissue of the host (Fig. 14–19*B*). Egg sacs are then formed and may be as long as 11 mm. Several clutches are produced by a single female.

SYSTEMATIC RÉSUMÉ OF CLASS COPEPODA

Order Calanoida. Free-living, largely planktonic copepods with very long first antennae (16 to 26 articles). *Calanus, Calocalanus, Diaptomus, Metri-*

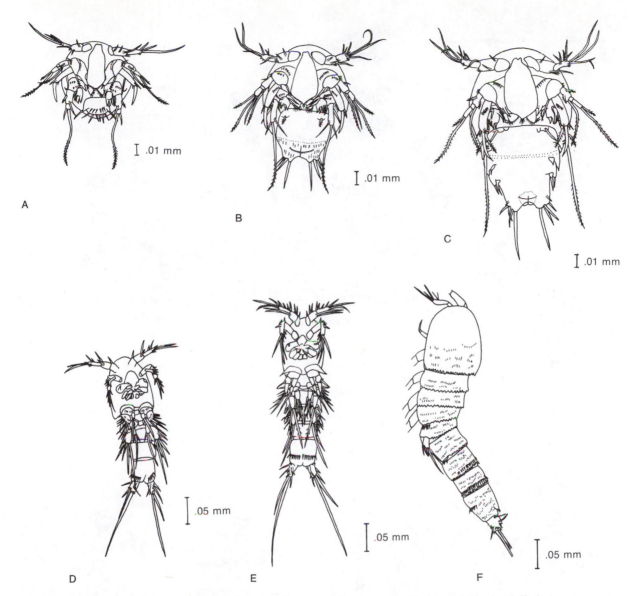

Figure 14–17 Ventral views of the developmental stages of the harpacticoid copepod *Elaphoidella bidens coronata*. The six naupliar and five copepodid stages are each separated by a molt. *A*, First nauplius. *B*, Fourth nauplius. *C*, Sixth nauplius. *D*, First copepodid. *E*, Second copepodid. *F*, Fifth copepodid (dorsal view). (Adapted from Carter, M. E., and Bradford, J. M., 1972: Postembryonic development of three species of harpacticoid copepoda. Smithsonian Contrib. Zool., 119:1–26.)

dia, *Pleuromamma, Centropages, Lucicutia, Acartia.*

Order Misophrioida. Copepods free living on or above the bottom surface. First antennae shorter than those of calanoids (11 to 16 articles). *Misophria.*

Order Harpacticoida. Mostly free-living, marine and freshwater, bottom-dwelling copepods. Some planktonic; many interstitial. First antennae short (fewer than 10 articles). *Harpacticus, Canthocamptus.*

Order Monstrilloida. Marine copepods with larval stages parasitic in polychaetes and gastropods. Nonfeeding planktonic adults lack second antennae and mouthparts. *Monstrilla, Xenocoeloma.*

Order Siphonostomatoida. Freshwater and marine copepods parasitic as adults on fish and invertebrates. *Nemesis, Clavella, Caligus, Lepeophtheirus, Salmincola, Penella.*

Order Cyclopoida. Marine and freshwater, planktonic and benthic copepods. Includes some commensals and parasites. First antennae short (10 to

Figure 14–18 Parasitic copepods, *Penella exocoeti*, on a flying fish. The copepods are in turn carrying the barnacle, *Conchoderma virgatum* (striped body). (Modified after Schmitt.)

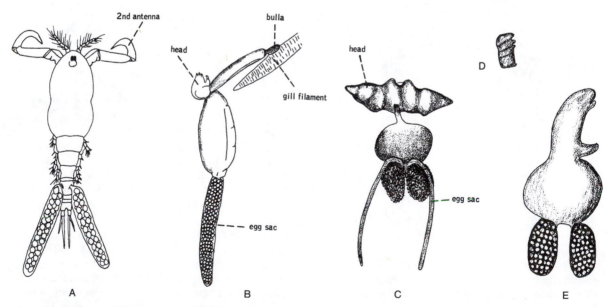

Figure 14–19 Parasitic copepods: *A, Ergasilus versicolor*, a cyclopoid parasite. It lives on gills of freshwater fish. Only adult female is parasitic, hooking to fish with clasping antennae. (After Wilson from Pennak.) *B, Salmincola salmonea*, mature female attached to gill of European salmon. (After Friend.) *C, Spyrion*. Head is embedded in skin of fish, remainder of body hangs free. (From Parker and Haswell.) *D* and *E*, Male (*D*) and female (*E*) of *Brachiella obesa* live on gills of red gurnard. (After Green, J., 1961: A Biology of Crustacea. Quadrangle Books, Chicago, p. 113.)

16 articles) and second antennae uniramous. *Cyclops, Sapphirina, Oncaea, Lernaea, Doropygus.*
Order Poecilostomatoida. Marine copepods parasitic as adults on invertebrates and fish. *Ergasilus, Chondracanthus.*

A summary of the copepods is provided on p. 591.

Classes Mystacocarida, Branchiura, Tantulocarida, and Remipedia

The Mystacocarida and the Branchiura are two small classes that are related to the copepods and barnacles. The Mystacocarida were first described in 1943 by Pennak and Zinn from specimens collected off Massachusetts, but nine species have since been reported from many other coasts, principally around the North and South Atlantic oceans. The majority of these little crustaceans are only approximately 0.5 mm in length and are adapted for living between intertidal sand grains (Fig. 14–20*A*). The body is long and cylindrical but divided like that of copepods. Both antennae are long and prominent, and the mouth appendages are provided with setae, probably for collecting detritus and other particulate matter. Only the naupliar eye is present. The sexes are separate, and a naupliar larva is the hatching stage.

The class Branchiura contains approximately 130 species of ectoparasites on the skin or in the gill cavities of freshwater and marine fish. The most striking differences between branchiurans and copepods are the presence of a pair of sessile compound eyes and a large, shieldlike carapace covering the head and thorax (Fig. 14–20*B*).

The abdomen is small, bilobed, and unsegmented. Both pairs of antennae are very small, and the first pair is provided with a large claw for attachment to the host. Also important in attach-

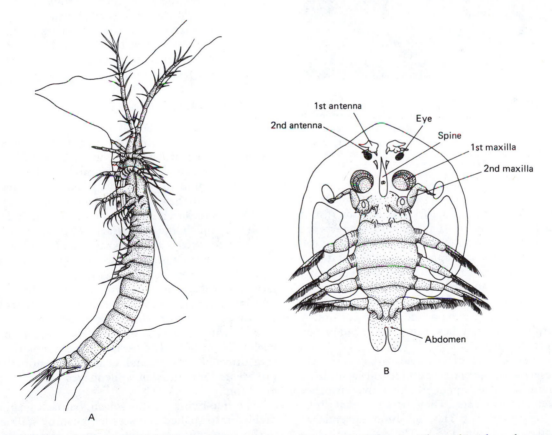

1st antenna
2nd antenna
Eye
Spine
1st maxilla
2nd maxilla
Abdomen

A

B

Figure 14–20 *A*, The interstitial *Derocheilocaris typica* crawling among sand grains. The pushing force for movement is produced by the second antennae and mandibles, the branches of each appendage operating against the substratum both above and below the head of the animal. (From Lombardi, J. and Ruppert, E. E., 1982: Functional morphology of locomotion in *Derocheilocaris typica.* Zoomorph., *100*:1–10.) *B*, *Argulus foliaceus*, a branchiuran fish parasite. (After Wagler from Kaestner.)

Figure 14–21 *Lasionectes entrichomas,* a member of the newly discovered, very primitive class Remipedia from caves in the Turks and Caicos islands. Note the long, wormlike body with numerous pairs of similar appendages. (By Dennis W. Williams.)

ment (except in *Dolops*) are two large suckers modified from the bases of the first maxillae, the rest of the appendage being vestigial. The mouthparts are adapted for feeding on the mucus and blood of the host. The four thoracic appendages are well developed; branchiurans can detach and swim or crawl from one host to another.

Copulation occurs while the parasites are on the host, but the eggs are deposited on the bottom. A postnaupliar stage hatches from the egg and is parasitic like the adult.

Two new classes of crustaceans have recently been described. The class Tantulocarida contains a small number of ectoparasites that live on other deep-water crustaceans. They are somewhat similar to copepods but lack all hind appendages (Boxshall and Lincoln, 1983). The class Remipedia is represented by eight species, all from island caves that connect to the sea. They have been reported from the Turks and Caicos, Bahamas, and Canary islands. The body is very long, resembling

that of a polychaete, with numerous biramous appendages (Fig. 14–21) (Yager, 1981). The remipedians will probably prove to be the most primitive crustaceans, more so than cephalocarids.

Class Cirripedia

The Cirripedia include the familiar marine animals known as barnacles. Cirripedes are the only sessile group of crustaceans, aside from the parasitic forms, and as a result they are one of the most aberrant groups of Crustacea. In fact, it was not until 1830, when the larval stages were first discovered, that the relationship between the barnacles and other crustaceans was fully recognized, and the barnacles were removed from the phylum Mollusca.

Barnacles are exclusively marine. Approximately two thirds of the nearly 900 described species are free living, attaching to rocks, shells, coral, floating timber, and other objects. Some barnacles are commensal on whales, turtles, fish, and other animals, and a large number of barnacles are parasitic. The wholly parasitic Rhizocephala are so highly specialized that all traces of arthropod structure have disappeared in the adult.

External Structure

Louis Agassiz described a barnacle as "nothing more than a little shrimplike animal, standing on its head in a limestone house and kicking food into its mouth." A further analogy might be drawn if one can imagine an ostracod turned upside down and attached to the substratum by the anterior end.

The cypris larva of a barnacle, which indeed looks very much like the ostracod *Cypris* for which it is named, settles to the bottom and attaches to the substratum by means of cement glands located in the base of the first antenna (Fig. 14–29B). The larval carapace, which encloses the entire body, as in ostracods, persists as the enveloping carapace or mantle of adult barnacles. It becomes covered externally with calcareous plates in the ordinary barnacles (Thoracica). The carapace opening is therefore directed upward and enables the animal to project its long, thoracic appendages for scooping plankton.

The free-living, or thoracican, barnacles can be divided into stalked and sessile (meaning stalkless in this case) types. In stalked barnacles, sometimes called goose barnacles, there is a muscular, flexible stalk (peduncle) that is attached to the substratum at one end and bears the major part of the body (capitulum) at the other (Fig. 14–22A). The peduncle represents the preoral end of the animal and contains the vestiges of the larval first antennae and the cement glands that opened on them. The capitulum contains all but the preoral part of the body and includes the surrounding carapace (mantle). The mantle surface is covered by two pairs of calcareous plates (scuta and terga) (Fig. 14–23). The carapace margin can be pulled together for protection or opened for the extension of the appendages. A large adductor muscle runs transversely between two of the plates (scuta) (Fig. 14–22A).

There is no peduncle in sessile barnacles (Fig. 14–24). The attached undersurface of the barnacle is called the basis and is either membranous or calcareous. This is the preoral region of the animal and contains the cement glands. A vertical wall of plates completely rings the animal (Fig. 14–24B and C), and within the wall the top of the animal is covered by an operculum, formed by the paired movable terga and scuta. The plates composing the wall overlap one another and may be held together by living tissue only or by interlocking teeth or may actually be fused to some extent.

Within the mantle the body of either a stalked or a sessile barnacle is flexed backward so that the appendages are directed upward toward the mantle aperture rather than toward the side (Figs. 14–22A and 14–24A). The major part of the body consists

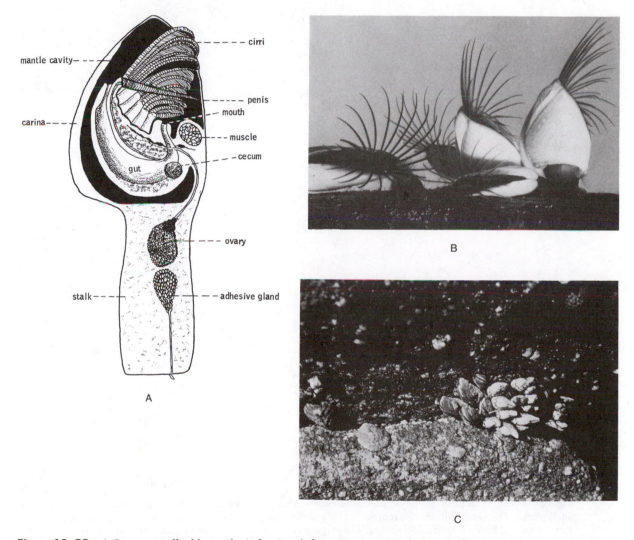

Figure 14–22 *A, Lepas*, a stalked barnacle. (After Broch from Kaestner.) *B*, A species of *Lepas* with extended cirri. *C, Polycipes polymerus*, a stalked, scalpellid barnacle that occurs in large numbers on intertidal rocks along the west coast of the United States. Several mytilid bivalves are to the left of the cluster of barnacles. (©Jen and Des Bartlett, The National Audobon Society Collection.)

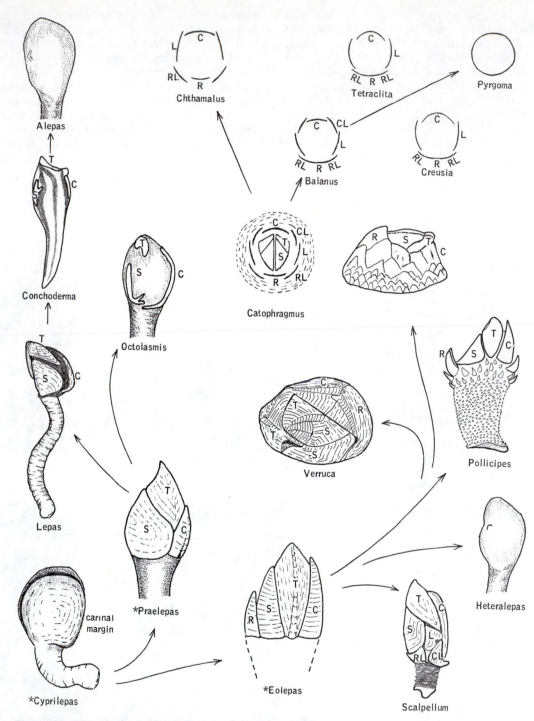

Figure 14–23 The probable phylogeny of the barnacles, showing the tendency toward reduction of shell plates. The genera illustrated represent types within the principal phylogenetic lines. *Cyprilepas*, *Praelepas*, and *Eolepas* are extinct genera. The tergum and scutum are omitted from plans of the sessile barnacles except for *Verruca* (Verrucomorpha) and *Catophragmus*. Plates designated include carina (C), carinolateral (CL), lateral (L), rostrum (R), rostrolateral (RL), scutum (S), and tergum (T).

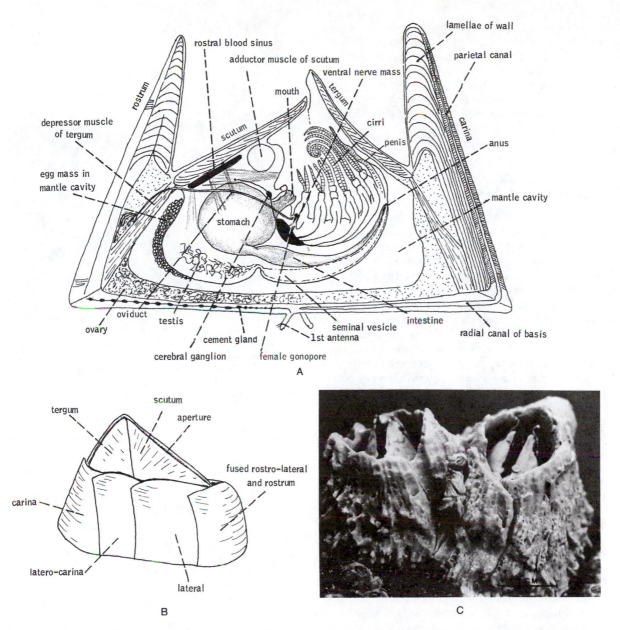

Figure 14–24   *A, Balanus,* a sessile barnacle (vertical section). (After Gruvel from Calman.) *B, Balanus* showing number and position of shell plates (diagrammatic lateral view). (After Broch from Kaestner.) *C, Tetraclita,* a sessile barnacle, showing the circular wall and the projecting movable tergal plates. (By Betty M. Barnes.)

of a cephalic region and an anterior trunk (thoracic) region. External segmentation is very indistinct.

The first antennae are vestigial except for the cement glands, and the second antennae are present only in the larva. The oral appendages are variously modified. There are typically six pairs of long, biramous thoracic appendages used in sus-

pension feeding. They are called cirri, from which the name *Cirripedia* is derived, and each branch is provided with many long setae.

When not encrusted with other sessile organisms, barnacles are usually white, pink, or purple. Stalked barnacles, including the peduncle, range from a few millimeters to 75 cm in length. The majority of sessile species are a few centimeters in diameter, but some are considerably larger. *Balanus psittacus* from the west coast of South America reaches a height of 23 cm and a diameter of 8 cm and is a popular local seafood.

The calcareous plates of barnacles have provided an extensive fossil record that dates back to the Silurian.

Evolution and Ecological Distribution of Barnacles

Of the two types of common free-living (thoracican) barnacles, the stalked barnacles are considered more primitive. On the basis of fossil forms, the cypris larva common to all cirripedes (Fig 14–29B), and other evidence, the ancestral barnacle was probably a cypris-like, bivalved crustacean attached to the substratum by its first antennae (Fig. 14–23). Initially, each valve was covered by a single plate, but perhaps facilitating the opening of the valves and extension of the cirri without exposing the entire body, the plate became subdivided. The ventral aperture became guarded by the paired terga and scuta, and these were supported by a posterior dorsal plate, the carina. From such an ancestral stalked barnacle, two principal lines are believed to have evolved: one leading to the existing stalked barnacles known as the lepadids, and another leading to the living stalked barnacles known as the scalpellids. The scalpellids, in turn, gave rise to the sessile barnacles. Not surprisingly, a commensal habit—attachment on other animals—has evolved in all three lines. Wherever this has occurred, there has been a tendency for the calcareous plates to become reduced or even lost, since the host offers some protection to the barnacle.

In the lepadids, which attach to floating objects, wood, coconuts, bottles, tar, or to other animals, the peduncle, or stalk, has remained naked, and the capitulum is covered with no more than the five original, basic plates—one carina, two terga, and two scuta (Fig. 14–23). This form is well illustrated by the many common and widely distributed species of *Lepas*. Commensals include *Conchoderma*, which lives on whales, turtles, and ships' bottoms, and *Alepas*, which occurs only on jellyfish (Figs. 14–23 and 14–25B).

A related family of small, stalked barnacles (Poecilasmatidae) contains many common commensals of crustaceans, such as a species of *Octolasmis* that lives on the gills of lobsters and crabs.

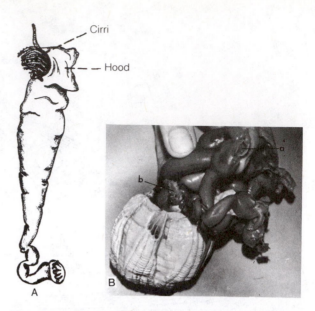

Figure 14–25 *A, Xenobalanus,* a sessile barnacle that lives commensally on fins of cetaceans. Note loss of opercular plates. (After Gruvel from Calman.) *B,* A cluster of the stalked barnacle, *Conchoderma,* attached to *Coronula,* a sessile whale barnacle. Opening into mantle cavity of *Conchoderma* (a) and *Coronula* (b). (By V. B. Scheffer, U.S. Fish and Wildlife Service.)

The scalpellids are bottom dwellers, and although some, such as the common Californian *Pollicipes polymerus,* live on intertidal rocks, most are found in deep water. The peduncle is covered with calcaerous plates or scales, which generally increase in size toward the capitulum. In many genera the base of the capitulum is surrounded by several whorls of accessory plates (Fig. 14–23). The extreme in plate reduction is seen in the naked *Heterolepas,* which is commensal on other crustaceans, such as lobsters. One group of scalpellids has become adapted for boring in coralline rock (Fig. 14–26C).

The sessile barnacles (balanomorphs) are thought to have arisen during the Cretaceous from the stalked scalpellids by a shortening and disappearance of the peduncle. The large attachment surface and the low, heavy, circular wall make sessile barnacles highly adapted for life on current-swept and wave-pounded intertidal rocks. The wall (mural plates) is believed to have formed from the lateral plates covering the capitulum of stalked barnacles. In the primitive genus *Catophragmus* the wall is composed of many whorls of imbricated plates, with the interior and largest whorl consist-

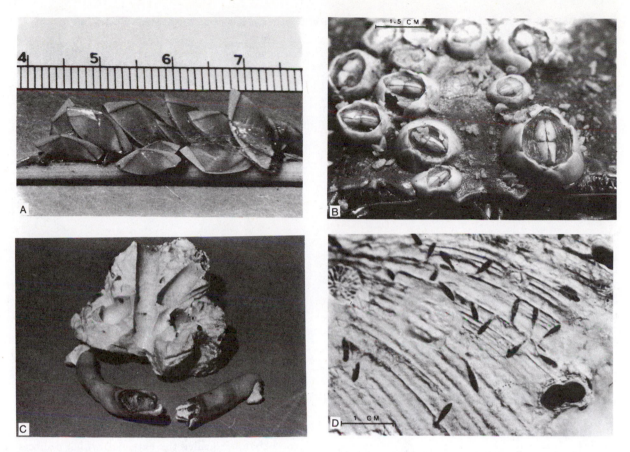

Figure 14–26 *A*, A species of *Poecilasma* on a sea urchin spine. The members of this family of stalked barnacles (Poecilasmatidae) are related to the lepadids but are attached to bottom objects, including sea urchins and crustaceans. *Octolasmis* (shown in Fig. 14–23), which is found on crab gills, is also a member of this family. *B*, *Chelonebia* on the carapace of a blue crab. *C*, *Lithotrya*, a scalpellid barnacle that bores into coralline rock. *D*, Slitlike openings of the burrows of the boring acrothoracican barnacle *Kochlorine* in an old clam shell. The dumbbell-shaped openings are the burrows of a boring clam. (All by Betty M. Barnes.)

ing of eight plates—three paired mural plates, the carina, and the rostrum (Fig. 14–23). In most other sessile barnacles only this inner whorl of eight plates remains and composes the wall; these plates are usually reduced in number through loss or fusion (Fig. 14–23), perhaps as an evolutionary response to predation by snails, which attack sessile barnacles at plate junctions (Palmer, 1982).

Although there are some deep-water balanomorphs, the group as a whole is intertidal or just subtidal, and species are typically restricted to particular zones. Some are limited to the low tide mark, and some live in the midintertidal zone. A few are adapted for life in the spray zone at the high tide mark on wave-lashed rocks. Intertidal

barnacles commonly occur in enormous numbers, as indicated by the figures in Table 14–1, which give the location and aggregation of two species of barnacles on the English coast. Barnacles are less conspicuous on tropical rocky shores, perhaps because of prolonged high temperature that must be tolerated at the upper intertidal levels (Foster, 1974).

Many sessile barnacles have become adapted for life on other surfaces besides rocks; a number of species have colonized intertidal grasses and mangroves. They are commensal with a wide range of hosts—sponges, hydrozoans, octocorals, scleractinian corals, crabs (Fig. 14–26B), sea snakes, sea turtles, manatees, porpoises, and whales.

TABLE 14–1   **Vertical Distribution of Two Species of Sessile Barnacles on the Plymouth Coast of England***

| Height above Mean Tide | Numbers of Individuals per Square Meter | | | | | |
| | *Balanus balanoides* | | | *Chthamalus stellatus* | | |
	Adults	Young	Total	Adults	Young	Total
+3.4 m.	0	0	0	0	0	0
+2.7 m.	0	0	0	15,200	9,200	24,400
+1.8 m.	0	400	400	54,000	38,000	92,000
+0.8 m.	4,000	12,400	16,400	55,600	35,200	90,800
−0.2 m.	40,400	20,400	60,800	400	4,800	5,200
−2.1 m.	0	0	0	0	0	0

*(Moore, 1936.)

Barnacles are among the most serious fouling problems on ship bottoms, buoys, and pilings, and many species have been transported all over the world by shipping. Sessile barnacles are acquired by ships in coastal waters; larvae from floating lepadids settle on ships at sea. The speed of a badly fouled ship may be reduced by 30 per cent. Much effort and money have been expended toward the development of special paints and other antifouling measures.

Most peculiar are the boxlike, mostly deep-water species of the suborder Verrucomorpha, in which one tergum and scutum form the lid, and the other tergum and scutum are part of the wall (Fig. 14–23).

The barnacles described thus far and illustrated in Figure 14–23 are all members of the order Thoracica. In the early evolution of barnacles there appeared another line, represented today by some 30 species of the order Acrothoracica, which are believed to have stemmed from the scalpellids (Newman, 1974). Acrothoracicans are the smallest cirripedes, usually only a few millimeters in length. They bore into coral or old mollusk shells using chitinous teeth on the mantle as well as chemical dissolution (Fig. 14–26D). When feeding, the naked, sac-shaped animal projects its cirri through the slitlike opening of its burrow, sometimes holding the appendages open like a fan and turning them several times before retracting them into the burrow.

Parasitic Barnacles

Considering the common occurrence of commensal barnacles, it is not surprising that parasitism has evolved within the class. There are a few parasitic Thoracica, and two other orders are exclusively parasitic. The most important of these is the order Rhizocephala, which is largely parasitic on decapod crustaceans. The body is saccular, and the mantle is devoid of calcareous plates. There is also a complete absence of appendages and segmentation.

A cypris larva of *Lernaeodiscus porcellanae*, one destined to be a female, enters the gill chamber of the host crab and attaches to the gill. Following metamorphosis of the cypris, a perforation in the host's integument is made, permitting the entrance of a mass of dedifferentiated cells of the parasite (Fig. 14–27A). Within the host growth of the parasite takes place through the ramification of a nutrient rootlike system (called the interna). Sexual development involves the formation of an external brood chamber (called the externa), which projects mushroom-like through a new opening produced in the crab's integument near the base of the abdomen. A male cypris can then attach itself to the opening of the brood chamber and extrude dedifferentiated cells, which migrate into a special chamber in the female, in which they redifferentiate into a testis. The female is now a hermaphrodite, and fertilization and development to the naupliar stage occur within the brood chamber. The effects of this parasitism on the crab are numerous and severe. Two of the most striking effects are the inhibition of molting and parasitic castration. The development of the gonads is retarded, or the gonads atrophy.

Members of the order Ascothoracica are ectoparasites on sea lilies and serpent stars, or endoparasites in octocorallian corals and in the coelom of sea stars and echinoids (Fig. 14–27B).

The ascothoracicans are now believed to be the most primitive of the Cirripedia, despite their adaptations for parasitism. A chitinous, basically bivalved carapace encloses the body, which includes a limbless abdomen. There are no second antennae, but all species possess prehensile first antennae.

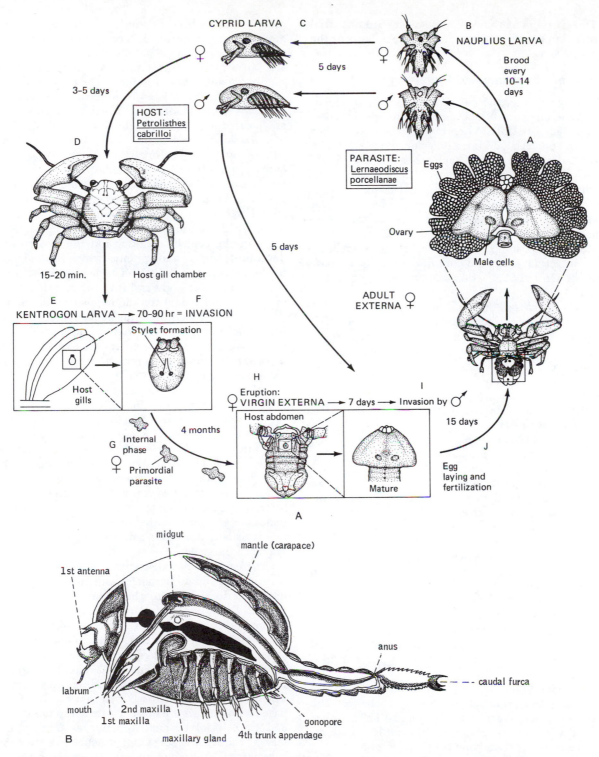

CYPRID LARVA C

NAUPLIUS LARVA B

Brood every 10–14 days

5 days

3–5 days

HOST: Petrolisthes cabrilloi

D

PARASITE: Lernaeodiscus porcellanae

Eggs

Ovary

Male cells

5 days

15–20 min.

Host gill chamber

ADULT EXTERNA ♀

E KENTROGON LARVA → 70–90 hr = INVASION F

Stylet formation

Host gills

H Eruption: VIRGIN EXTERNA → 7 days → Invasion by ♂ I

Host abdomen

15 days

Mature

G Internal phase

♀ Primordial parasite

4 months

J Egg laying and fertilization

A

B

1st antenna

midgut

mantle (carapace)

anus

caudal furca

labrum

mouth

2nd maxilla

1st maxilla

maxillary gland

4th trunk appendage

gonopore

Figure 14–27 *A*, Life cycle of a rhizocephalian barnacle, *Lernaeodiscus porcellanae*, parasitic on the porcelain crab *Petrolisthes cabrilloi* on the California coast. See text for additional explanation. (From Ritchie, L. E., and Hoeg, J. T., 1981: The life history of *Lernaeodiscus porcellanae* and coevolution with its porcellanid host. Jour. Crust. Biol., *1*(3):334–347. *B*, *Ascothorax ophioctenis*, a parasite in the bursae of brittle stars. (After Wagin.)

The six pairs of thoracic appendages are not cirriform but are like the swimming appendages of the cypris larva.

Internal Anatomy and Physiology

During feeding, the paired scuta and terga open, and the cirri unroll and extend through the aperture (Fig. 14–22B). When outstretched, a number of cirri on each side form one side of a basket. The two sides sweep toward each other and downward, each half acting as a scoop net. The action is somewhat analogous to the opening and closing of your two fists simultaneously when the bases of the palms are placed together. In currents some barnacles may hold the cirri outstretched for a period of time, and some, such as *Pollicipes polymerus*, the intertidal scalpellid common on the west coast of the United States, do not rhythmically beat the cirri.

On the closing stroke, food particles suspended in the water are trapped by the setae, and the first one to three pairs of cirri are used to scrape these particles off and transfer them to the mouthparts. The size of plankton used for food varies. Some species of *Lepas*, *Pollicipes*, and *Tetraclita* capture copepods, isopods, amphipods, and other relatively large organisms and could therefore be classified as predacious rather than suspension feeders.

Many barnacles feed on small plankton, but in at least some of these barnacles, such as *Balanus improvisus* and *Balanus balanoides*, the last four pairs of cirri perform the usual scooping motion and capture particles over several microns in diameter. The water current produced by scooping is directed over the first two pairs of cirri, which remain within the mantle cavity close to the mouth. These anterior cirri have the setae closely placed, and the cirri collect fine phytoplankton from the water current.

A heart and arteries are absent, but circulation of the blood is facilitated by a blood pump, consisting of a large sinus located between the esophagus and the adductor muscle (Fig. 14–24A). Blood circulates through the mantle walls and also into the peduncle.

Gills are lacking, and the mantle and cirri are probably the principal sites of gas exchange. The excretory organs are maxillary glands. The naupliar eye divides and is retained in the adult as two lateral components and one median component.

Reproduction and Development

Most parasitic barnacles and the boring acrothoracicans are dioecious. Thoracican barnacles are mostly hermaphroditic and are the only large group of hermaphroditic crustaceans. However, cross fertilization is generally the rule, for a suitable substratum almost always contains a large number of adjacent individuals. The ovaries of sessile barnacles (Fig. 14–24A) lie in the basis and in the walls of the mantle, and in stalked forms they are located in the peduncle (Fig. 14–22A). The paired oviducts open at or near the basis of the first pair of cirri. Just before reaching the gonopore, each oviduct dilates as an oviducal gland that secretes a thin, elastic ovisac at the time of egg deposition. As the ovisac receives eggs, it swells and stretches, emerging from the gonopore and coming to lie within the mantle cavity. The testes are located in the cephalic region, and the two sperm ducts unite within a long penis, which lies in front of the anus (Fig. 14–22A). The penis can be protruded out of the body and into the mantle cavity of another individual for the deposition of sperm (Fig. 14–28). Functional males can recognize a functional female, which may be inseminated by more than one male. The sperm are deposited as a mass near the first cirri and must penetrate the ovisac to reach the eggs.

Dwarf males that attach themselves to female or hermaphroditic individuals are found in some of the pedunculate genera, such as *Scalpellum* and *Ibla*, in a few species of *Balanus*, and in the boring Acrothoracica (Fig. 14–28B–D). In addition to being very small, these males show all degrees of modification through degeneracy or loss of structures. When attached to a hermaphrodite, the males are called complemental males. When the males are greatly modified (Fig. 14–28D), the species is usually dioecious, and the larger, host individuals are females.

The eggs are brooded within the ovisac in the mantle cavity in all barnacles, but the ovisac gradually deteriorates during the incubation period. In *Balanus balanoides*, for example, eggs are fertilized in the autumn and brooded until March. A nauplius represents the hatching stage in most species and can be easily recognized by the triangular, shield-shaped carapace (Fig. 14–29A). Six naupliar instars are succeeded by a nonfeeding cypris larva. The entire body is enclosed within a bivalve carapace and possesses a pair of sessile compound eyes and six pairs of thoracic appendages. The cypris larva is the settling stage and attaches to a suitable substratum, at first temporarily, with the secretions of discs located on the first antennae, and then permanently, with the antennal cement glands (Fig. 14–29B) (Walker and Yule, 1984).

A number of factors appear to increase the like-

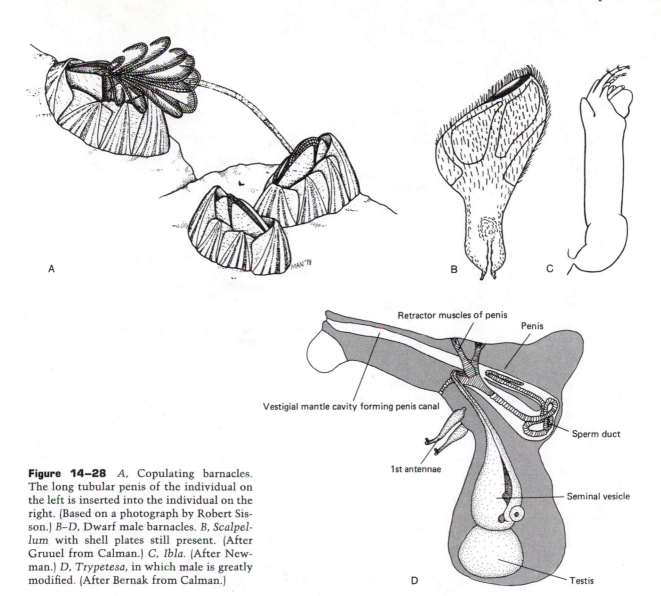

Figure 14–28 *A*, Copulating barnacles. The long tubular penis of the individual on the left is inserted into the individual on the right. (Based on a photograph by Robert Sisson.) *B–D*, Dwarf male barnacles. *B*, *Scalpellum* with shell plates still present. (After Gruuel from Calman.) *C*, *Ibla*. (After Newman.) *D*, *Trypetesa*, in which male is greatly modified. (After Bernak from Calman.)

Labels in figure D: Retractor muscles of penis; Penis; Vestigial mantle cavity forming penis canal; Sperm duct; 1st antennae; Seminal vesicle; Testis

lihood of dense settling, upon which reproduction in the adult depends. Simultaneous shedding of the nauplius by aggregations of individuals has been observed in some species. A protein in the exoskeleton of older attached individuals has been demonstrated to attract settling larvae. Illumination, roughness, position, depth, and bacterial film of the substrate may also be important in selection, and attachment may not occur at the first surface contact (see Stubbings, 1975; review by Lewis, 1978; Moyse and Hui, 1981). Following attachment metamorphosis takes place: the cirri elongate, the body undergoes flexion, and the primordial plates appear on the next exoskeleton lying beneath the old cypris valves.

Shell growth is more or less continuous and independent of body growth and ecdysis. In *Balanus improvisus*, the first 20 molts take place two to three days apart, on average. The cuticle, or exoskeleton, lining the interior of the mantle cavity and covering the appendages is molted periodically, as in other arthropods. The calcareous plates are secreted by the underlying mantle and are not shed at ecdysis. Growth of the plates takes place by the continual addition of material to their margins and interior surfaces, thus increasing their thickness and diameter. Microscopic growth bands are produced, as in the calcareous shells and skeletons of other animals (Bourget and Crisp, 1975). In sessile barnacles a wedge of mantle tissue lying be-

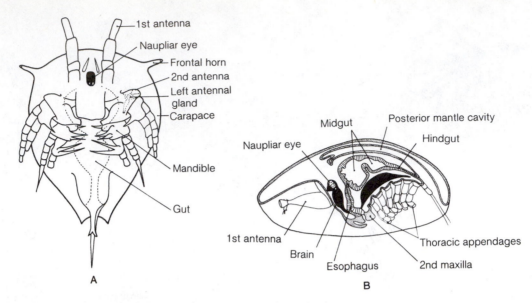

Figure 14–29   *A*, Naupliar larva of *Balanus*. Setae omitted. *B*, Free-swimming cypris larva of *Balanus*. (All from Walley, L. J., 1970: Phil. Trans. Roy. Soc. London, B, Biol. Sci., 256(807):237–280).

tween the junction of the basis and the mural plates adds material to the periphery of the basis and to the lower margin of the mural plates (Fig. 14–30). This mantle tissue thus permits a continual outward and upward growth of the mural plates to accommodate the increase in diameter and height. The young barnacle is about 3 mm in diameter at the end of a month.

Cement is elaborated throughout the life of an individual, and repair of partial detachment is possible. However, detachment is most likely rare, for

the protein cement provides a very strong bond that can withstand enormous pressures.

Most barnacles that survive the heavy mortality in the period immediately following settling probably live for one to ten years. In *Balanus glandula* on the Pacific coast of the United States, where the reproductive biology is well known, two to six broods of 1,000 to 30,000 nauplii each are produced in the winter and spring. The maximum basal diameter of the adult is 22 mm, of which 7 to 12 mm is reached by the end of the first year and

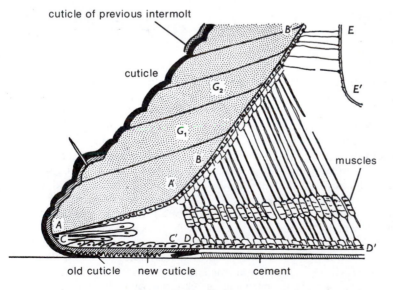

Figure 14–30 Diagrammatic section through junction of wall and basis of *Balanus*, showing growth zones. A–A' is region of maximum calcium carbonate secretion; B–B' region of minimum secretion; C–C' cuticle-secreting region; D–D' region in basis of cement ducts and no secretion of calcium carbonate; E–E' inner wall of mantle, in which exoskeleton is shed at each molt. G_1 and G_2 are growth bands of calcium carbonate. (From Bourget, E., and Crisp, D. J., 1975: Analysis of the growth bands and ridges of barnacle shell plates. J. Mar. Biol. Assoc. U.K., 55(2):439–462. Copyrighted and reprinted by permission of Cambridge University Press.)

14 to 17 mm at the end of the third. The life span is eight to ten years. However, longevity and growth rates are greatly influenced by environmental conditions. Smothering by other individuals and other sessile organisms is a common cause of death.

SYSTEMATIC RÉSUMÉ OF CLASS CIRRIPEDIA

Order Thoracica. Free-living and commensal barnacles with six pairs of well-developed cirri. Mantle usually covered with calcareous plates.

    Suborder Lepadomorpha. Pedunculate barnacles. *Lepas, Scalpellum, Pollicipes, Conchoderma, Heteralepas.*

    Suborder Verrucomorpha. Asymmetrical, sessile barnacles. Wall composed of carina and rostrum, one tergum, and one scutum; the other tergum and scutum form the operculum. *Verruca.*

    Suborder Balanomorpha. Symmetrical, sessile barnacles with a wall surmounted by the paired movable terga and scuta. *Balanus, Chthamalus, Catophragmus, Octomeris, Tetraclita, Pyrgoma, Coronula, Xenobalanus.*

Order Acrothoracica. Naked, boring barnacles with a chitinous attachment disc and four to six pairs of cirri (one at mouth, remainder terminal). They live in any calcareous substratum, especially shells and corals. *Trypetesa, Kochlorine, Berndtia.*

Order Ascothoracica. Naked barnacles that parasitize echinoderms and corals. Bivalve or saccular mantle. Prehensile first antennae and abdomen are usually present. *Dendrogaster, Ascothorax.*

Order Rhizocephala. Naked barnacles. Parasitic primarily on decapod crustaceans; a few parasitic on tunicates. Appendages and digestive tract are absent; peduncle forms footlike absorptive processes. *Peltogasterella, Lernaeodiscus, Sacculina.*

SUMMARY (CLASSES THROUGH CIRRIPEDIA)

1 The nine known species of the class Cephalocarida are of special interest because all the trunk appendages are similar to each other and to the second maxillae. In other respects these tiny crustaceans show specializations for living in the mud-water interface.

2 The three major classes of small crustaceans are the Branchiopoda, the Ostracoda, and the Copepoda. Branchiopods are distinguished by their foliaceous appendages, which in many species are adapted for suspension feeding. In other respects branchiopods are diverse. Water fleas (cladocerans)

have a carapace that encloses the trunk but not the head; clam shrimps (conchostracans) have a bivalve carapace that encloses the entire body; fairy shrimps and brine shrimps (anostracans) lack a carapace.

3 Branchiopods are largely inhabitants of fresh water, especially temporary pools and ditches. Only cladocerans are also found in the sea and in lakes.

4 Ostracods have a bivalve carapace impregnated with $CaCO_3$ that encloses the entire body. Most ostracods are less than 2 mm long. They are mostly marine benthic animals, but some species live in fresh water. Ostracods have an extensive fossil record.

5 Copepods possess more or less cylindrical, tapered bodies. Long, laterally projecting first antennae and a persistent naupliar eye are distinctive features of many species. The trunk appendages are markedly biramous. Most copepods are less than 2 mm long.

6 Most copepods are marine, but some species are common in freshwater lakes and pools. There are planktonic, epibenthic, and interstitial species. There are also many parasitic forms.

7 There is a diversity of feeding habits among these classes of small crustaceans. Suspension feeding is especially characteristic of most branchiopods and most planktonic copepods. Suspension-feeding, planktonic copepods are of great importance in marine food chains.

8 Many freshwater forms, especially branchiopods, undergo parthenogenesis, produce both dormant and rapidly hatching eggs, and encyst.

9 Barnacles, members of the class Cirripedia, are unique among crustaceans, indeed most arthropods, in being sessile. A number of peculiarities of barnacles, such as a carapace covered with calcareous plates, suspension-feeding cirri, hermaphroditism, long tubular penis, and dwarf males, can be correlated with their sessility.

10 Living barnacles are attached by a stalk, or peduncle, or are stalkless (sessile). The peduncle represents the preoral part of the body and contains the cement glands. The oldest known barnacles are pedunculate. Lepadids, which are living pedunculate barnacles, attach to floating, inanimate objects, such as wood, or to pelagic animals. Scalpellids attach to rocks and, in addition to the five large principal plates, have many small plates covering the peduncle and capitulum.

11 The sessile barnacles, which are believed to have evolved from the scalpellids, are stalkless, the preoral region being represented by the basis, which contains the cement glands. Only the paired terga and scuta are movable. The other capitular plates form a circular wall around the barnacle. Sessile barnacles are especially adapted for intertidal life on hard substrates that are subjected to waves, surge, and currents.

12 Commensalism, which has evolved in all three major lines of barnacles, has resulted in reduction of the protective calcareous plates. Commensalism has undoubtedly been the avenue to parasitism, which is characteristic of one third of the species of cirripedes.

Class Malacostraca

The class Malacostraca contains almost three quarters of all the known species of crustaceans, as well as most of the larger forms, such as crabs, lobsters, and shrimps.

The trunk of a malacostracan is typically composed of 14 segments, plus the telson, of which the first 8 segments form the thorax and the last 6 the abdomen (Fig. 14–31A). The thorax may or may not be covered by a carapace. All the segments bear appendages. The first antennae are often biramous. The exopodite of the second antenna is often in the form of a flattened scale. The mandible usually bears a palp.

In the primitive condition the thoracic appendages, or legs, are similar, and the endopodite is the more highly developed of the two branches of these appendages, being used for crawling or prehension (Fig. 14–31B). In most malacostracans the first one, two, or three pairs of thoracic appendages have turned forward and become modified to form maxillipeds.

The anterior (usually the first five pairs) abdominal appendages, called pleopods, are similar and biramous. The pleopods may be used for swimming, burrowing, ventilating, carrying eggs in the female, or sometimes for gas exchange. In the male the first one or two pairs of pleopods are usually

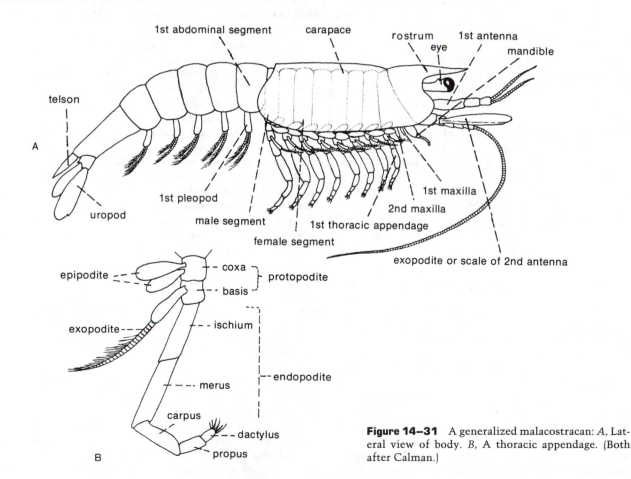

Figure 14–31 A generalized malacostracan: *A*, Lateral view of body. *B*, A thoracic appendage. (Both after Calman.)

modified, forming copulatory organs. Usually, each ramus of the sixth abdominal appendages (uropods) is composed of a large flattened piece and, together with the usually flattened telson, forms a tail fan, which is most frequently used in escape swimming.

In most malacostracans the foregut is modified as a two-chambered stomach bearing triturating teeth and comblike, filtering setae. Within the stomach food is chewed and digestion begins. The fine particulate products of this action are filtered out and passed to the midgut outpocketings, called the digestive gland or hepatopancreas.

The female gonopores are always located on the sixth thoracic segment, and the male gonopores are on the eighth. The naupliar larva is usually passed within the egg.

Because of the size and diversity of the Malacostraca, each of the orders is considered separately.

SUBCLASS PHYLLOCARIDA; ORDER LEPTOSTRACA

This cosmopolitan order is composed of about 20 species of small marine crustaceans that differ from the basic malacostracan plan in that they have eight abdominal segments instead of six. Morphologically, the phyllocaridans are believed to represent the most primitive existing malacostracans.

The fossil record bears this out, for the earliest known malacostracans were phyllocaridans and appeared in the Cambrian period.

Most are shallow-water suspension feeders. *Nebalia bipes*, which reaches about 12 mm in length, lives in bottom mud and in seaweed along the Atlantic coast, as well as in many other parts of the world (Fig. 14–32).

SUBCLASS EUMALACOSTRACA; SUPERORDER SYNCARIDA; ORDERS ANASPIDACEA AND BATHYNELLACEA

The syncarida are represented by two small orders of primitive freshwater crustaceans, the Anaspidacea and the Bathynellacea, in which the thoracic appendages are biramous and similar. A carapace is absent. Although living syncarids are found only in fresh water, there are many fossil marine species. Members of the Bathynellacea are long, blind, interstital or groundwater species and include some of the smallest malacostracans (Schminke, 1981). *Brasilibathynella florianopolis* from Brazil is just over 0.5 mm long, and *Parabathynella neotropica* is only 1.2 mm in length (Fig. 14–33). Bathynellaceans have been reported from most parts of the world, including North America.

The anaspidaceans, which may reach a length of 5 cm, inhabit lakes, pools, and groundwater in Australia, New Zealand, and South America (Fig.

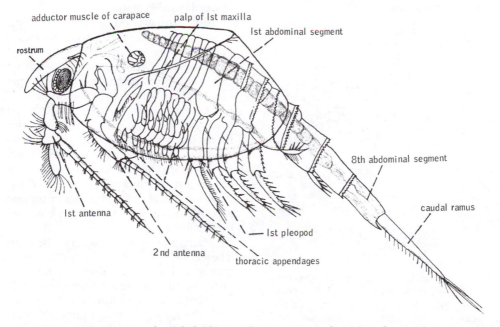

Figure 14–32 Female *Nebalia bipes*, a leptostracan. (After Claus from Calman.)

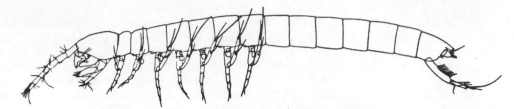

Figure 14–33 Lateral view of *Parabathynella neotropica*, a bathynellacean. Total length is only 1.2 mm. (From Noodt, W., 1965: Natürliches System und Biogeographie der Syncarida. Gewässer und Abwässer, 37–38:77–186.)

14–34). The larger species swim and crawl over the bottom or on aquatic vegetation. They are suspension feeders or omnivores.

SUPERORDER HOPLOCARIDA AND ORDER STOMATOPODA

The 300 marine crustaceans called mantis shrimps constitute the Stomatopoda, which is the only order of hoplocaridans. Mantis shrimps are highly specialized predators of fish, crabs, shrimps, and mollusks, and many of their distinctive features are related to their predatory behavior. The body is dorsoventrally flattened with a small, shieldlike carapace and a large, broad, distinctly segmented abdomen (Fig. 14–35).

The well-developed compound eyes are large and stalked, and between them is a naupliar eye. A distinctive feature of stomatopods is the structure of the thoracic appendages. The first five pairs are uniramous and subchelate, and the second pair is enormously developed for raptorial feeding. The inner edge of the movable finger is provided with long spines or is shaped like the blade of a knife. The finger folds back into a deep groove in the heavy penult segment, which has a sharp and finely toothed margin. *Hoplocarida* means "armed shrimp." The pleopods are well developed and bear filamentous gills (Fig. 14–35*B*). The uropods and telson are very large.

Mantis shrimps range in size from small species approximately 5 cm long to giant forms greater than 36 cm in length. Most mantis shrimps are tropical, but *Squilla empusa*, which is about 18 cm in length, is a common species inhabiting the North American Atlantic coast and is frequently caught in shrimp trawls. Many stomatopods are brilliantly colored. Green, blue, and red with deep mottling are common, and some species are striped or display other patterns.

Most stomatopods live in rock or coral crevices or in burrows excavated in the bottom. *Squilla empusa* lives in U-shaped burrows but may build single-opening vertical burrows extending to 4 meters for winter quarters. Some stomatopods use burrows excavated by other animals. One coral-inhabiting species, *Echinosquilla guerini*, has radiating spines that ornament the entire surface of its telson. This armored telson is used to plug the entrance to the burrow when the mantis shrimp is inside. From the exterior, the telson mimics a small sea urchin attached to the coral surface. *Gonodactylus bredini* closes its burrow entrance with debris at night and blocks it during the day with the raptorial appendages.

Many mantis shrimps leave the burrow to feed. They crawl about over the bottom or swim by the powerful oarlike beating of the pleopods. The large, antennal scales and uropods serve as rudders.

Mantis shrimps either spear or smash their prey with the large, second, thoracic raptorial appendage. Some forage for prey; others lie in wait at the mouths of their burrows. Spearers feed on soft-

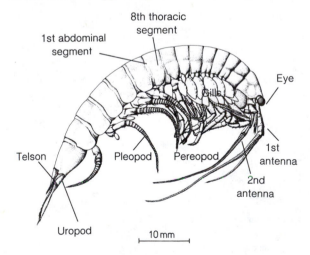

Figure 14–34 *Anaspides tasmaniae*, a syncaridan. (From Schminke, Von H. K., 1978: Die phylogenetische Stellung der Stygocarididae (Crustacea, Syncarida)— unter besonderer Berücksichtigung morphologischer Ähnlichkeiten mit Larvenformen der Eucarida. Z. zool. Syst. Evolut.-forsch., *16*:225–229.)

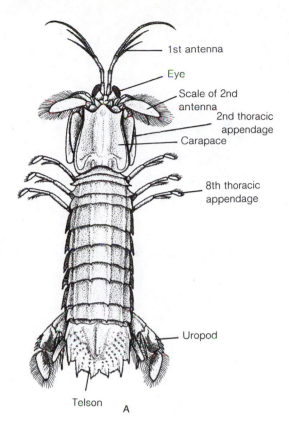

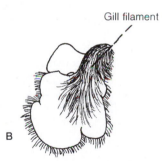

Figure 14–35 *Squilla empusa*, a stomatopod: *A*, Dorsal view. Pleopods hidden (see Fig. 14–41). *B*, Second pleopod and gill. (*B* after Calman.)

bodied invertebrates, such as shrimps and fish. The prey is speared by an extremely rapid extension and retraction of the movable finger of the second thoracic appendage, which is provided with barbed spines (Fig. 14–36) (Dingle and Caldwell, 1978). Smashers, especially species of the Gonodactylidae, which live in holes and crevices on rocky or coralline bottoms, stalk their prey, mostly snails, clams, and crabs. Their hard-bodied victims are smashed with the heavy heel of the unfolded rap-

torial appendage. A crab is disabled by a blow from behind or by having its claws broken immediately. Its legs and carapace are then smashed, and the carcass is dragged back to the hole, where it is consumed. Snails and clams are broken within the burrow, and the shells are deposited outside. The blows of the raptorial appendage are so powerful that captured specimens have cracked the glass walls of aquaria (Caldwell and Dingle, 1976).

Many stomatopods defend their burrow territories against other stomatopods, and those species that live in holes and crevices within coralline rock and rubble have especially complex social behavior (Caldwell, 1979; Reaka and Manning, 1981). The raptorial appendages are the offensive weapons and are used as clubs even in those species that spear their prey. A gonodactylid defends itself against opponents' clubs with its large telson, which is heavily armored and thrust forward by the abdomen while the animal lies on its back (Fig. 14–37). Some gonodactylids have large, colored depressions on the basal piece (merus) of the raptorial appendage, which functions in threat display (Steger and Caldwell, 1983).

Mantis shrimps have the most highly developed compound eyes among crustaceans. Not only do the eyes detect moving objects but, judging from the proximity and position in which the eyes can be held and the accuracy with which the animal can swim to prey, depth perception must be possible. The antennae are important sites of chemoreception and are also used in prey detection when the range is short enough.

The eggs are agglutinated to form a globular mass by means of an adhesive secretion. The egg mass, which may be the size of a walnut with as many as 50,000 eggs *(Squilla)*, is carried by the smaller, subchelate appendages and is constantly turned and cleaned (Fig. 14–38). The female does not feed while she is brooding. Some species keep the egg mass inside the burrow. For example, a Bahamian species of *Gonodactylus*, which lies curled up in a coral crevice, holds the egg mass over the back of the carapace (Fig. 14–38).

A zoea larva is the hatching stage and bears a much larger carapace than the adult (Fig. 14–39). Planktonic larval life may last for three months.

SUPERORDER EUCARIDA

The Eucarida contain most of the larger malacostracans. The basic form is shrimplike (caridoid) (Fig. 14–31). The carapace is highly developed and fused with all the thoracic segments. The eyes are stalked. The eggs are usually carried beneath the

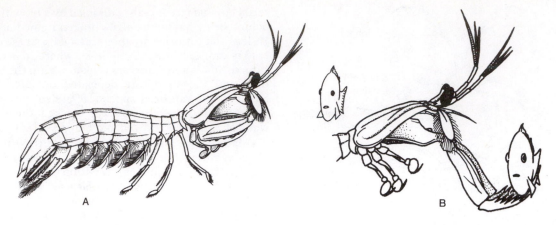

Figure 14–36 Prey capture by a mantis shrimp, which spears its prey. The second thoracic appendate rapidly unfolds, and the barbs on the terminal finger are driven into the body of the prey.

abdomen, and development is generally indirect, with a zoea larva. The Eucarida consist of only the small order Euphausiacea, containing the krill, and the very large order Decapoda, containing the shrimps, lobsters, and crabs.

ORDER EUPHAUSIACEA

Euphausiaceans, known as krill, are pelagic, shrimplike crustaceans approximately 3 cm in length. They are all marine. The sides of the carapace do not tightly enclose the gills, as is true of many decapods (Fig. 14–40A). The thoracic appendages are biramous, and none are specialized as maxillipeds.

Euphausiaceans swim with the large, setose pleopods, and most are suspension feeders.

Each of the 6 to 8 leglike thoracic appendages bears a long fringe of setae on one side, and together with the other limbs of that side they form one half of a funnel-shaped net or basket (Fig. 14–40A). In at least those species, such as members of the genus *Euphausia*, that feed on phytoplankton, the filter-feeding process occurs in bouts in response to chemical cues, at which time water is pumped out of the basket (Hamner et al, 1983; Hamner, 1984). Many consume zooplankton, and a few are predacious. Of the 28 species in the central Pacific, 22 depend largely on animal food and only 4 on phytoplankton (Roger, 1975).

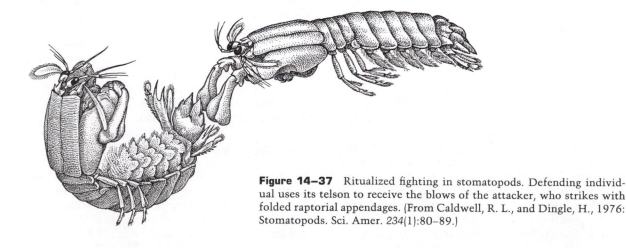

Figure 14–37 Ritualized fighting in stomatopods. Defending individual uses its telson to receive the blows of the attacker, who strikes with folded raptorial appendages. (From Caldwell, R. L., and Dingle, H., 1976: Stomatopods. Sci. Amer. 234(1):80–89.)

Figure 14–38 Two species of stomatopods caring for eggs. *Squilla mantis (left)* carrying egg mass; *Gonodactylus oerstedi (right)* with egg mass in burrow. (Modified after Schmitt.)

Large, exposed, filamentous gills are present on the thoracic appendages (Fig. 14–40*B*), and the ventilating current is produced by the thoracic exopodites.

Many euphausiaceans are luminescent. The light-producing material is not secreted, as in ostracods and other entomostracans, but is intracellular, located within special light-producing organs (photophores) (Fig. 14–40*B*) (Box 5–1, p. 144) (Herring and Locket, 1978). The luminescence is probably an adaptation for swarming and reproduction.

The sperm are transferred to the female as spermatophores, and the eggs may be liberated into the sea water or brooded briefly. A nonfeeding, naupliar larva is the hatching stage.

Many species of euphausiaceans are important components of the pelagic fauna (Fig. 14–41). Such species as the Antarctic *Euphausia superba*, which is about 5 cm in length and transparent, are surface forms; others live at deeper levels or undergo vertical migrations. Many Antarctic species, such as *Euphausia superba*, live in great swarms and constitute the chief food of many other animals (Ross and Quetin, 1986). Blue whales may consume a ton of euphausiaceans at one feeding and make up to four such feedings a day. A euphausiacean swarm may cover an area equivalent to several city blocks,

and one seen from the air looks like a giant ameba slowly moving and changing shape. Although the swarm may occupy a layer 5 or more meters thick, the several meters of surface water contain the greatest concentration and may reach densities of 60,000 individuals per cubic meter, all of which are adults or near adults. Swarming *Euphausia superba* molt very rapidly—within a second—and if alarmed, many will literally jump out of their skins. The shed molt remains behind and perhaps functions as a decoy (Hamner et al., 1983, Hamner, 1984).

Krill, especially *Euphausia superba*, is receiving increasing attention as a potential human food source, and the Russians and Japanese have initiated krill fisheries. Russian trawlers crush and press the animals to obtain juices from which the protein is extracted by heat coagulation. The product has largely been used for enrichment of other foods.

ORDER DECAPODA

The order Decapoda contains the familiar shrimps, crayfish, lobsters, and crabs and is the largest order of crustaceans. The approximately 10,000 described decapods represent almost one quarter of

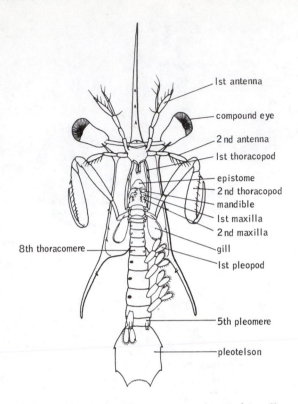

Figure 14–39 labels:
- 1st antenna
- compound eye
- 2nd antenna
- 1st thoracopod
- epistome
- 2nd thoracopod
- mandible
- 1st maxilla
- 2nd maxilla
- gill
- 1st pleopod
- 8th thoracomere
- 5th pleomere
- pleotelson

Figure 14–39   First pelagic, antizoea larva of *Squilla*, a mantis shrimp. (After Brooks, W. K., 1878: The larval stages of *Squilla empusa*. Chesapeake Zoological Laboratory: Scientific Reports, pp. 143–170.)

the known species of crustaceans. Most of the decapods are marine but the crayfish and some shrimps and crabs have invaded fresh water. There are also some terrestrial crabs.

Decapods are distinguished from the euphausiaceans, as well as from other malacostracans, in that their first three pairs of thoracic appendages are modified as maxillipeds. The remaining five pairs of thoracic appendages are legs, from which the name Decapoda is derived (Fig. 14–47*B*). The first pair, or sometimes the second, is frequently much enlarged and chelate, and when so constructed, the limb is called a cheliped (Fig. 14–47*A* and 14–48). The legs usually lack exopodites; i.e., they are not biramous. The sides of the overhanging carapace enclose the gills within well-defined, lateral branchial chambers (Fig. 14–60*A* and *E*).

Among the smallest decapods are species of the brachyuran crab, *Dissodactylus*, which are commensal on sand dollars (Fig. 14–50*B*). The cephalothorax of this crab is only a few millimeters wide. *Macrocheira kaempferi*, a Japanese spider crab, has the greatest leg span of any living arthropod (Fig. 14–50*A*). The cephalothorax may attain a length of 45 cm, and the chelipeds, a span of 4 meters. However, some lobsters have longer bodies.

Decapod Diversity—Locomotion and Habitation

The diversity of forms encountered among decapods can more easily be appreciated if examined from the standpoint of adaptations for locomotion and habitation.

There are three separate, unrelated groups of shrimps, or prawns, one of which—the Penaeidea—contains the most primitive decapods (see p. 629). The bodies of shrimps tend to be cylindrical or laterally compressed with well-developed abdomens (Fig 14–42), and the cephalothorax often bears a keel-shaped, serrated rostrum. The legs are usually slender, and chelipeds may be present or absent. The exoskeleton is commonly thin and flexible.

The only pelagic decapods are shrimps, mostly members of three families (Penaeidae, Sergestidae,

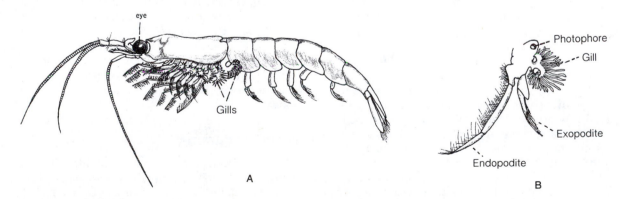

Figure 14–40 labels:
- eye
- Gills
- Photophore
- Gill
- Exopodite
- Endopodite
- A
- B

Figure 14–40   *A, Meganyctiphanes*, a euphausiacean. *B*, Seventh thoracic appendage. (All after Calman.)

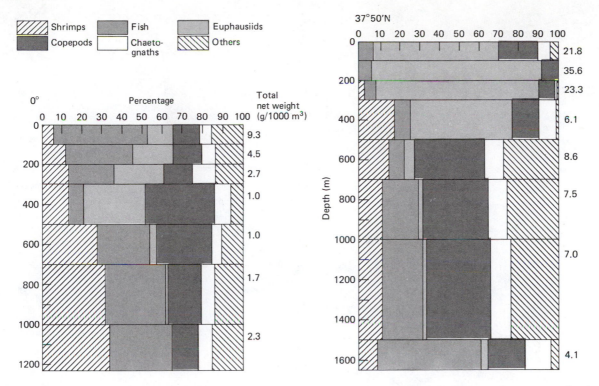

Figure 14–41 Percentage composition of the major groups of planktonic and small nectonic animals at various depths at night at tropical and temperate stations in the Pacific along 150°E. (From Omori, M., 1974: The biology of pelagic shrimps in the ocean. Adv. Mar. Biol. (Academic Press, London), *12*:233–324.)

and Oplophoridae) (Fig. 14–43*B*). Although pelagic shrimps occur at all depths, the majority are found in the epipelagic and mesopelagic zones (the upper 1000 meters) and typically migrate from 100 to over 800 meters upward at night (Fig. 14–44).

The pleopods, which are large and fringed, are the principal swimming organs, although rapid ventral flexion of the abdomen with the tail fan is used for quick backward darts. Abdominal flexion also provides for vertical steering.

Species living in the upper 500 meters (epipelagic and upper mesopelagic) are transparent or semitransparent; those pelagic forms living below about 500 meters during the day are red. Many of the latter group also possess luminescent organs, or photophores, located internally or anywhere on the body surface. Among the best known luminescent species are those belonging to the genera *Sergestes* and *Sergia* (see Box 5–1, p. 144).

Most shrimps are not pelagic. They are bottom dwellers, using the legs for crawling, and swim intermittently with the pleopods. They live among algae and sea grasses, beneath stones and shells, and within holes and crevices in coral and rock.

Some, including many penaeids and the sand shrimps *(Crangon)*, are shallow burrowers in soft bottoms and use the beating pleopods for excavating. Their activity patterns may be controlled by light or tides. The commercially important *Penaeus duorarum* on the east cost of the United States are quiescent on the surface or buried during the day and are active at night. The European *Crangon crangon* lies buried near the low water mark at low tides on sandy beaches but emerges at high tide to swim over the lower half of the beach.

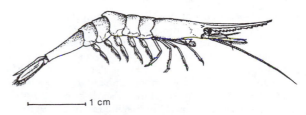

Figure 14–42 *Tozeuma carolineuse* (Caridea), a common shrimp of eel grass beds along the east coast of the United States. Green in life.

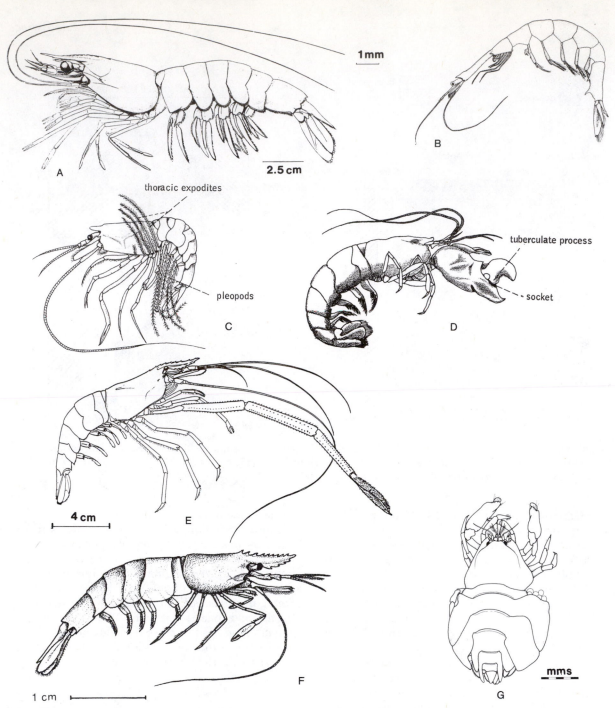

Figure 14–43 Shrimps: *A, Penaeus setiferus* (Penaeidea), an important species for the shrimping industry along the east coast of the United States. *B, Lucifer faxoni* (Penaeidea), a widely distributed pelagic species. *C, Psathyrocaris fragilis* (Caridea) with well-developed thoracic exopodites. *D, Alpheus* (Caridea), a pistol, or snapping, shrimp. *E, Macrobrachium acanthurus* (Caridea), a river shrimp of eastern United States that ranges from brackish water to as much as 100 miles above the river's mouth. *F, Palaemonetes* (Caridea), a large genus of small, transparent, estuarine or freshwater shrimps; many are common inhabitants of eel grass and other submerged vegetation. *G, Paratypton siebenrocki* (Caridea), an almost spherical Indo-Pacific, pontoniid shrimp that lives in cystlike chambers induced in the scleractinian coral *Acropora*. (*A,B,* and *E* from Williams, A. B., 1965: Marine decapod crustaceans of the Carolinas. U.S. Fishery Bull., 65(1):1–298. *C* after Alcock from Calman. *D* after Schmitt from Bruce, A. J., 1976: Shrimps and prawns of coral reefs, with special reference to commensalism. *In* Jones, O. A., and Endean, R. (Eds.): Biology and Geology of Coral Reefs. Vol. III: Biology 2. Academic Press, New York. p. 49.)

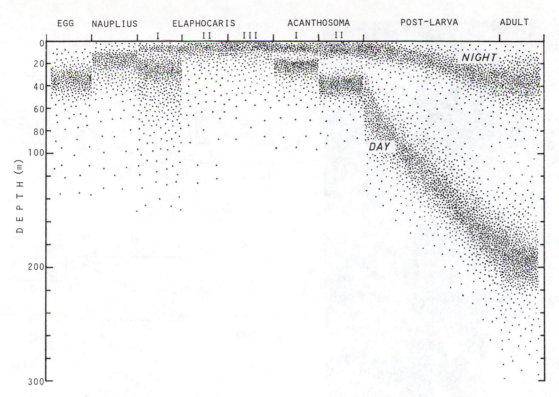

Figure 14–44 Vertical distribution of the pelagic shrimp *Sergia lucens* from egg to adult. (After Omori, M., 1974: The biology of pelagic shrimps in the ocean. Adv. Mar. Biol. (Academic Press, London), *12*:233–324.)

The pistol, or snapping, shrimps (Alphaeidae) are a common and widely distributed family. These little shrimps, which are 3 to 6 cm long, have one of the chelipeds greatly enlarged (Fig. 14–43D). The base of the movable finger contains a large, tuberculate process that fits into a socket on the immovable finger. The movable finger is locked, or cocked, when contact is made between two specialized discs, one at the base of the elevated finger and one on the hand. The adhesive force between the discs prevents closing of the finger until the contracting muscle has generated a counteracting pull. Then the finger closes with great rapidity and force, producing a snapping or popping noise and a little water jet. Although some species have been reported to use the snapping mechanism in predation and defense, it also functions in threat (agonistic) displays between individuals and probably ensures spacing of a population (Nolan and Salmon, 1970).

Snapping shrimps live in holes and crevices beneath shells, rocks and coral rubble, or they construct retreats or burrows. There are a number of snapping shrimps, as well as other species of shrimp, that live in sponges, tunicates, bivalves, and corals or with mollusks, sea urchins, and sea anemones (see Bruce, 1976). A large sponge may become a veritable apartment house of shrimps.

A number of unrelated groups of shrimps, called cleaning shrimps, remove ectoparasites and other unwanted materials from the surfaces of certain reef fish. The shrimp may climb over the fish and even insert its chelipeds into the fish's gill region (Fig. 14–45A). Some species of *Periclimenes* use the tentacles of sea anemones as their homes, and cleaning stations to which the fish must come are located nearby. The shrimp signals the fish with a "dance" of antennae waving and body rocking, and the fish responds with a distinct stationary pose (Fig. 14–45B). The shrimp then strokes the fish with its antennae and boards it to clean. The shrimp are protected from the nematocysts of the sea anemone in much the same manner as are clown fish that live in sea anemones (p. 126) (Levine and Blanchard, 1980).

Species of *Penaeus* and related groups are the most important commercial shrimps throughout the world (Fig. 14–43A). In the United States the

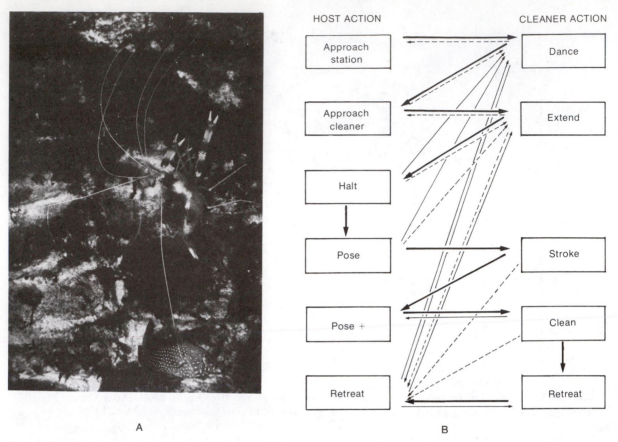

HOST ACTION		CLEANER ACTION

A   B

Figure 14–45 *A*, A cleaning shrimp. (Courtesy of Photo Researchers, N.Y.) *B*, Diagram of the sequence of behavior in the interaction between a host fish and the cleaning shrimp. *Periclimenes anthophilus.* Arrow from approach station to dance, for example, means that the approach of the fish to the cleaning station initiated dancing behavior in the shrimp. The heaviest arrows indicate the most frequently observed patterns. (Modified from Sargent, R. C., and Wagenbach, G. E., 1975: Cleaning behavior of the shrimp, *Periclimenes anthophilus.* Bull. Mar. Sci., 25(4):466–472.)

shrimp fisheries are centered primarily along the southeastern Atlantic coast and the Gulf of Mexico. In the United States and other parts of the world, marine shrimps are caught by trawling, in which a very long, V-shaped net is towed behind the shrimp boat. Each arm of the V is connected by ropes to the boat, and at the apex the net forms a bag. The nets are towed along the bottom at slow speeds for half an hour to an hour, depending on the catch. Shrimp farming is carried out in various parts of Asia but has been most highly developed in Japan, where shrimps are reared from eggs to marketable size (Wickins, 1976; Bliss, 1982). Shrimp farming is not yet economically feasible in the United States, except perhaps in Hawaii.

All the remaining decapod groups are benthic animals that have become more highly adapted for crawling than most shrimps. The body tends to be dorsoventrally flattened at least to some degree. The legs are usually heavier than those of shrimps,

and the first pair of legs usually are powerful chelipeds. The pleopods are never adapted for swimming.

In many groups the more primitive, long-bodied form of the shrimp has been greatly shortened through folding of the abdomen ventrally beneath the anterior part of the body (Fig. 14–46). This evolution of the "crab" form occurred independently a number of times in decapod evolution as an adaptation for locomotion or habitation.

Lobsters, crayfish, and burrowing shrimps still possess a large, extended abdomen bearing the full complement of appendages, and the carapace is always longer than it is broad (Fig. 14–47). Lobsters and crayfish crawl with the legs but can move rapidly backward to escape by flexing the abdomen ventrally.

Lobsters are heavy-bodied decapods and are generally inhabitants of holes and crevices of rocky and coralline bottoms. The Nephropidae have large

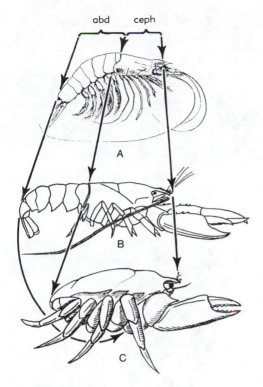

abd   ceph

A

B

C

Figure 14–46 Comparison of cephalothorax and abdomen in a shrimp, lobster, and crab and the origin of the short body form. (From Glaessner, M. F., 1969; Decapoda. *In* Moore, R. C.: Treatise on Invertebrate Paleontology, Part R, Arthropoda 4, Vol. 2. Geol. Soc. America and University of Kansas Press, Lawrence, Kan. p. 401.)

chelipeds and are similar in form to crayfish. The American lobster *Homarus americanus* may reach a length of 60 cm and a weight of 22 kg. They are caught commercially in pots, or traps, which the animal enters seeking shelter or bait. The lobster population along the New England and Canadian coasts has been badly overfished, and the largest catches are now taken offshore along the continental shelf and shelf edge. The European lobster *Homarus gammarus* is similar but smaller (Fig. 14–47B and C). The frozen lobster tails sold in foodmarkets are mostly species of spiny and slipper lobsters shipped from various tropical and subtropical parts of the world. Neither of these two groups of lobsters has chelipeds (Fig. 14–47D and E).

Members of the infraorder Brachyura, the true crabs, have the most highly specialized short body form and, judging from the number of species (over 4500), are probably the most successful decapods. The abdomen is greatly reduced and fits tightly beneath the cephalothorax (Fig. 14–48). Uropods

have disappeared in all but a few primitive forms. In the female, pleopods are retained for brooding eggs (Fig. 14–48C); in the male, only the anterior two pairs of copulatory pleopods are present (Fig. 14–48D). The carapace is very broad, often as wide as it is long and commonly much wider, increasing the flattened appearance of the body.

The evolution of abdominal reduction and flexion in brachyurans was probably a locomotor adaptation, shifting the center of gravity foward to a point beneath the locomotor appendages. Crabs can crawl forward slowly, but they commonly move sideways, especially when crawling rapidly. In such a gait, the leading legs pull by flexing, and the trailing legs push by extending. Chelipeds are not used in crawling.

Brachyuran crabs are found in all types of habitats and to great depths. Many deep-sea crabs have long, slender legs for crawling about over soft bottoms. At the other extreme is the common tropical *Grapsus grapsus*, which lives just above the water on wave-washed rocks, climbing and clinging to vertical surfaces with great agility. Terrestrial and freshwater crabs will be described later (p. 619).

Most crabs cannot swim, but members of the family Portunidae, which includes the common edible blue crab *Callinectes sapidus* of the Atlantic coast, are the most powerful and agile swimmers of all crustaceans. The last pair of legs in members of this family terminate in broad, flattened paddles (Fig. 14–48B) and during swimming describe figure eights in their movement. The action is essentially like that of a propeller, and the counterbeating fourth pair of legs act as stabilizers. Portunids can swim sideways, backward, and sometimes forward with great rapidity. However, they are benthic animals, like other crabs, and only swim intermittently.

Although the chelipeds are important in defense, other protective devices and habits have evolved in many brachyurans. Some species carry sea anemones with their chelipeds. Some spider crabs, which have triangular convex bodies and slender legs, are covered with hooked setae to which foreign objects become attached. The decorating habit has been highly developed in some species, and the body becomes completely overgrown with algae, sponges, and other sessile organisms (Fig. 14–49A). The decorator crab remains relatively immobile and camouflaged during the day, when predators are active (Wicksten, 1980).

The Dromiidae and Dorippidae use the small, dorsally directed fourth and fifth pairs of legs to hold objects over the body. *Hypoconcha* and *Dorippe* usually cover themselves with half of a bi-

Figure 14—47 Lobsters, crayfish, and mud shrimps: *A, Cambarus coosae* (bottom), a crayfish from Georgia, beside a specimen of the giant crayfish *Astacopsis gouldi* from Tasmania. The rule is 15 cm. *B, Homarus americanus,* the lobster of commercial importance along the northeastern coast of the United States. *C, Nephrops norvegicus* (Norway lobster). This species and *Homarus gammarus* are small lobsters taken commercially along European coasts. *D, Panuliris argus* (spiny lobster) from the West Indies. Members of this genus have large antennae but lack large chelipeds.

valve shell, although *Dorippe* may use other objects, including an old fish head. Species of *Dromia* and related genera cut out a cap of sponge with the chelipeds and fit the cap upon the back like an oversized beret (McLay, 1983).

Most of the commensal crabs belong to the family Pinnotheridae, called pea crabs because of the small size of many species. In addition to inhabitants of polychaete tubes and burrows, there are species that live in the mantle cavities of bivalves and snails, on sand dollars (Fig. 14—50B), in tunicates, and in other animals. Often the body has become considerably modified for commensal existence. For example, the female of the oyster crab, *Pinnotheres ostreum,* has a soft exoskeleton. The male of this species, which leaves its host to find a female, has the usual chitinization. Pea crabs re-

spond positively to substances produced by the host and can detect and move up a water stream that has passed over a host.

The family Hapalocarcinidae contains the coral gall crabs. These little tropical crabs cause certain species of corals to form gall-like chambers, within which the female lives (Fig. 14—50C). Small openings in the coral allow the entrance of the tiny male, as well as the plankton on which the crab feeds.

Many species of crabs are caught and eaten by humans throughout the world. The portunid swimming crab, *Callinectes sapidus,* or blue crab, is the commercially important crab occurring along the east and Gulf coasts of the United States, especially in the Chesapeake Bay. It is caught in shallow water with a trap or line, but commercial fish-

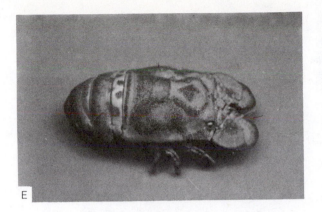

Figure 14–47 *(continued)* E, *Scyllarides aequinoctialis* (Spanish or shovel-nosed lobster) from the West Indies. Members of this family have short, flat antennae and lack large chelipeds. F, A species of burrowing shrimp *(Upogebia)* from the west coast of the United States. (By Betty M. Barnes.) G, Mold of burrow of *Upogebia affinis* from the coast of Georgia. (By Frey, R. W., and Howard, J. D., 1969: Trans. Gulf Coast Assoc. Geol. Soc., *19*:427–444.)

ermen also catch them in large numbers by trawling when fishing for shrimp, or in winter by means of a crab dredge, which dislodges crabs that have buried themselves in the sea bottom. Soft-shell crabs are simply newly molted blue crabs. On the west coast of the United States and in Europe, species of *Cancer*, a nonswimming crab, are used as food and are caught by trapping. *Cancer magister*, the dungeness or market crab, is the most important species off the coast of California.

The remaining decapods are a diverse assemblage of forms composing the infraorder Anomura. Some are crab-like. However, the abdomen is never as reduced as in brachyurans and uropods are usually present. All anomurans are similar in having the fifth pair of legs small and either located beneath the sides of the carapace or directed dorsally (Fig. 14–56A). The Anomura is probably an artificial grouping, and fifth leg reduction has clearly occurred independently within the different anomuran lines and has varying adaptive significance.

Burrowing shrimps or mud shrimps (*Callianassa*, *Upogebia*, and *Thalassina*) are shallow-water or intertidal decapods that live in long, deep burrows excavated in sand or mud (Fig. 14–47F). There are many common species along both temperate and tropical coastlines. The fifth pair of legs is not greatly reduced, and these decapods are not always placed within the Anomura. The exoskeleton is soft and flexible and the coloration is typically pale.

Decapods of the anomuran superfamily Paguroidea have evolved the habit of housing the abdomen within gastropod shells and are called hermit crabs. The hermit crab condition probably had its origin in forms that utilized crevices and holes as protective retreats. Indeed, among the primitive, symmetrical Pomatochelidae and even the more

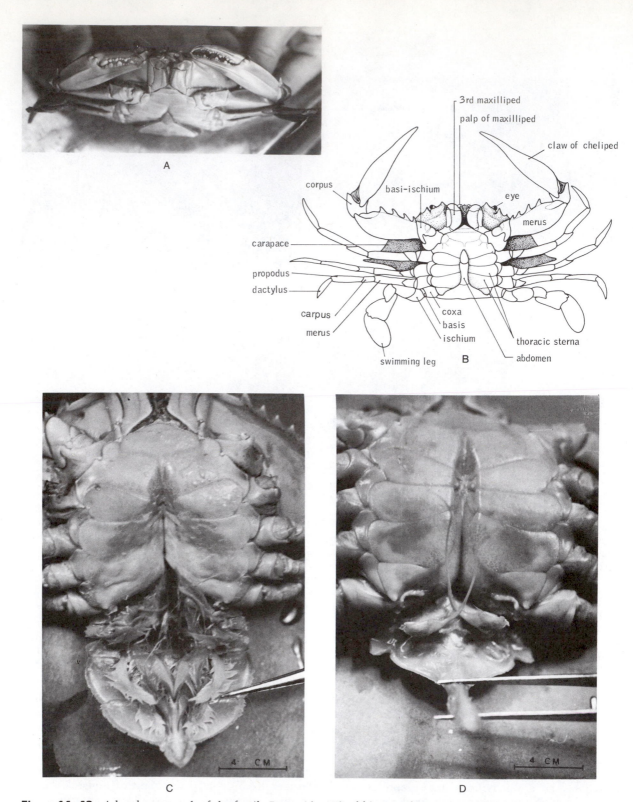

Figure 14–48 A brachyuran crab of the family Portunidae. The fifth pair of legs is adapted for swimming *A,* Frontal view, showing the dimorphic claws. The crushing claw is on the crab's right, and the cutter claw is on the left. Between the claw tips are the large, third maxillipeds covering the other mouthparts. *B,* Ventral view. (After Schmitt from Rathbun.) *C,* Abdomen of a female. Pleopods are used for carrying eggs. *D,* Abdomen of a male. Note that it is much narrower than that of the female. Only the anterior two pairs of pleopods, which are used as copulatory organs, are present. (*C* and *D* by Betty M. Barnes.)

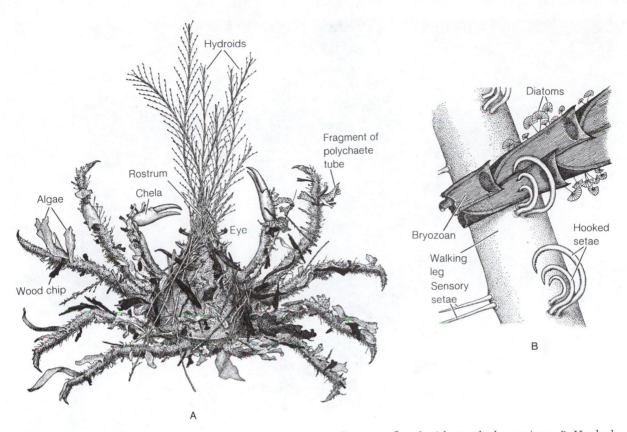

Figure 14–49 *A*, decorator crab (spider crab), *Oregonia gracilis*, camouflaged with attached organisms. *B*, Hooked setae on the leg of the decorator crab *Podochela hemphilli*. (From Wicksten, M. K., 1980: Decorator crabs. Sci. Amer., 242(2):149.)

specialized Paguridae there are species that still utilize such retreats; others live in bamboo or hollow mangrove roots (Fig. 14–51F and 14–53).

In most hermit crabs the abdomen is not flexed beneath the cephalothorax but is modified to fit within the spiral chamber of gastropod shells (Fig. 14–51). The abdomen is asymmetrically developed, with a thin, soft, nonsegmented cuticle, and the pleopods on the short side have been lost. Those on the long side are retained in the females to carry eggs. The twist of the abdomen is adapted for right-handed spirals, although left-handed shells may be used.

The shell is held in several ways. The uropods are modified, and the larger left one is used for hooking to the columella of the shell. Contraction of the longitudinal abdominal muscles presses the surface of the abdomen against the inner walls of the shell, and the last two pairs of legs are also pushed against the wall of the shell opening. The contact surfaces on both the legs and the uropods are covered with tiny tubercles that provide traction in gripping the shell (Fig. 14–51D). One or

both chelipeds may be adapted for blocking the aperture of the shell when the crab is withdrawn. It is the shell-jamming function that probably accounts for modification of the fifth pair of legs in the paguroids, and significantly the fifth legs are not greatly modified in the primitive symmetrical pomatochelids.

Hermit crabs always use empty shells and never kill the original occupant. Most species inhabit the shells of a number of different gastropods, depending on what is available, and there can be a relatively high turnover of the shell supply (Wilber and Herrnkind, 1982).

When the crab becomes too large for its shell, it seeks another but does not leave the old shell until a suitable new one has been found. Hermit crabs locate a prospective shell with the eyes and then inspect it with the chelae, inserting one into the interior. They can even determine that it is calcareous (Mesce, 1982). If the shell appears suitable, the crab will leave its old shell and try out the new one. If the fit, weight, or movability is bad, the crab returns to the old shell (Fig. 14–52) (Reese, 1963).

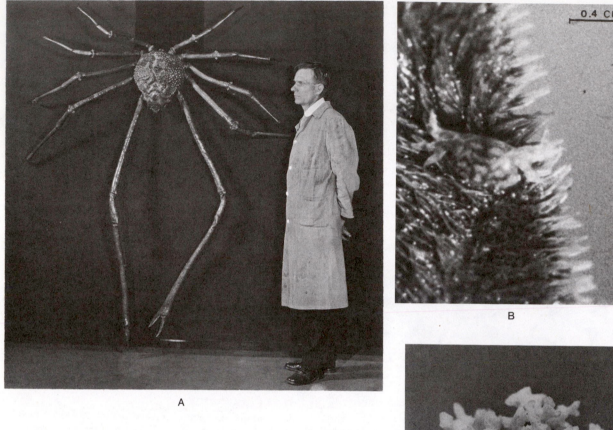

Figure 14–50   Brachyurans. *A*, The Japanese spider crab, *Macrocheira kaempferi*, the largest living arthropod. (Neg./Trans. No. 312007. Courtesy Department Library Services, American Museum of Natural History.) *B*, *Dissodactylus*, a commensal pea crab that lives on sand dollars. This is one of the smallest decapods. *D*, Three coral galls formed by crabs of the family Hapalocarcinidae. (*C* by Betty M. Barnes.)

Two hermit crabs will fight for possession of an empty shell or the shell inhabited by one of the combatants. Crabs carrying visually larger or heavier shells more frequently win shell fights than do those carrying small shells (Hazlett, 1970). The importance of shells to hermit crabs as a housing resource is reflected by the fact that in some areas hermit crab growth and reproduction are limited by a short shell supply (Bertness, 1981).

Coral, stones, tooth shells, wood, and other structures have been adapted as houses by certain species (Fig. 14–53). There are also species of hermit crabs that live commensally with species of the sea anemone *Calliactis*. By tapping and massaging

the sea anemone, the hermit crab is able to bring about the sea anemone's release and transfer to its own shell. *Calliactis parasitica*, however, will transfer to the shell of the hermit crab without the aid of the crab.

From hermit ancestors evolved two groups of anomurans, the coconut crabs (*Birgus*) and the lithodid crabs, that have short body forms and no longer house the abdomen within shells (Fig. 14–53). The abdomen, although folded crablike beneath the cephalothorax, has asymmetrical appendages (Fig. 14–55*B*).

Coconut crabs are large, terrestrial crabs inhabiting tropical Pacific Islands (Fig. 14–54) (p.

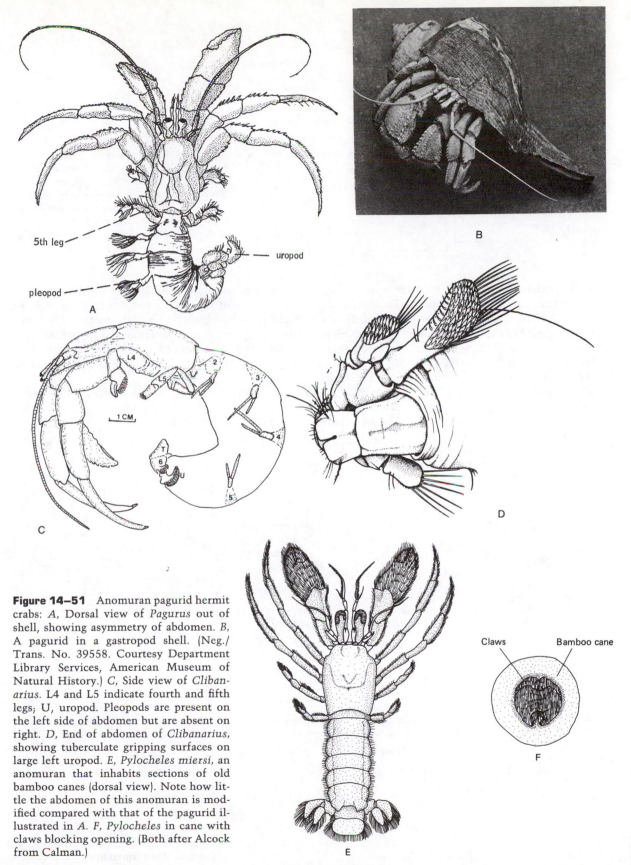

5th leg

uropod

pleopod

A

B

L4
L5
2
3
1 CM
4
T
6
U
5

C

D

Figure 14–51 Anomuran pagurid hermit crabs: *A*, Dorsal view of *Pagurus* out of shell, showing asymmetry of abdomen. *B*, A pagurid in a gastropod shell. (Neg./ Trans. No. 39558. Courtesy Department Library Services, American Museum of Natural History.) *C*, Side view of *Clibanarius*. L4 and L5 indicate fourth and fifth legs; U, uropod. Pleopods are present on the left side of abdomen but are absent on right. *D*, End of abdomen of *Clibanarius*, showing tuberculate gripping surfaces on large left uropod. *E*, *Pylocheles miersi*, an anomuran that inhabits sections of old bamboo canes (dorsal view). Note how little the abdomen of this anomuran is modified compared with that of the pagurid illustrated in *A*. *F*, *Pylocheles* in cane with claws blocking opening. (Both after Alcock from Calman.)

Claws   Bamboo cane

F

E

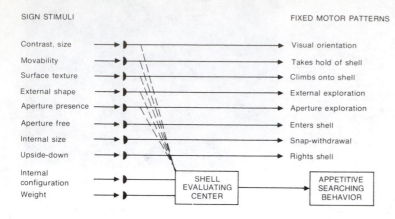

SIGN STIMULI

Contrast, size — Visual orientation
Movability — Takes hold of shell
Surface texture — Climbs onto shell
External shape — External exploration
Aperture presence — Aperture exploration
Aperture free — Enters shell
Internal size — Snap-withdrawal
Upside-down — Rights shell
Internal configuration
Weight

SHELL EVALUATING CENTER

FIXED MOTOR PATTERNS

APPETITIVE SEARCHING BEHAVIOR

Figure 14–52 Sequential behavioral responses of a shell-seeking hermit crab to various shell stimuli. The searching behavior is reduced with increasingly positive responses to shell conditions. (From Reese, E. S., 1963: The behavioral mechanisms underlying shell selection by hermit crabs. *Behaviour, 21:*78–126.)

620).The lithodid crabs, which have very heavy and sometimes sculptured carapaces, are mostly inhabitants of cold oceans (Fig. 14–55*A* and *B*). To this group belongs the large Alaskan king crab, *Paralithodes camtschatica,* which is trapped commercially in the north Pacific (Fig. 14–55*C*).

The anomuran Galatheoidea include the little porcelain crabs. This animal has a flexed, crablike abdomen, but the abdomen is symmetrical and not greatly reduced, as in brachyurans. The porcelain crabs are common, shallow-water decapods in many parts of the world and often occur in large numbers—hundreds per square meter. They live beneath stones and, in their general appearance and locomotion, look very much like brachyuran crabs (Fig. 14–56). Perhaps the abdominal flexion is an adaptation for motility.

The superfamily Hippoidea contains the sand crabs, or mole crabs. Abdominal flexion in this group appears to be an adaptation for burrowing

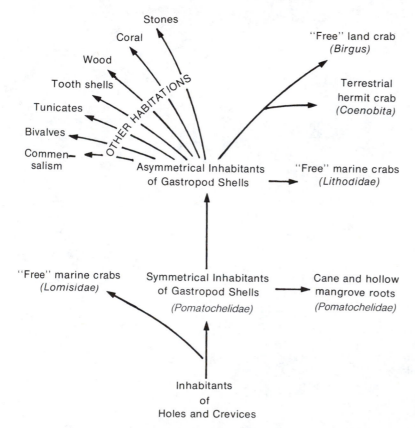

Figure 14–53 Adaptive evolution of the anomuran superfamily Paguroidea.

Figure 14—54 *A*, The South Pacific robber, or coconut, crab, *Birgus latro*. (By Betty M. Barnes.) *B*, Juvenile individual of *Birgus*, in which the abdomen is housed within a gastropod shell in typical hermit crab manner. (From Reese, E. S., 1968: Science, *161*:385–386. Copyright 1968 by the Association for the Advancement of Science.)

backward in sand. The mole crab body is somewhat cylindrical, and there are no chelipeds (Fig. 14–57). Species of *Emerita* live on open beaches and dig with the uropods and fourth pair of legs. In surf mole crabs are usually washed out with each wave; while the wave is receding, they rapidly burrow backward in the soft sand until only the antennae are visible.

Nutrition

The ventral mouth of a decapod is flanked by the feeding appendages, which lie on top of one another (Fig. 14–1*A*). The third maxillipeds are the outermost appendages and cover the other appendages. In a brachyuran crab the third maxillipeds are rectangular plates and completely fill the usually square buccal frame, covering the inner mouth appendages like double doors (Fig. 14–48*A*). Food is

caught or picked up with the chelipeds and then passed to the third maxillipeds, which push it between the other mouthparts. While a portion is bitten or held by the mandible, the remainder is torn away by the maxillae and maxillipeds. The severed piece is then directed into the mouth.

Decapods exhibit a wide range of feeding habits and diets, but the greatest number couple predacious feeding with the scavenging. The relative importance of the two habits varies with the species and also with the available food resource. Large invertebrates are common prey. For example, echinoderms and bivalves are the main food of the Alaskan king crab, and polychaetes, crustaceans, and bivalves are prey for snow crabs in the same area.

Herbivores include most freshwater and terrestrial decapods and some marine species. Herbivorous and scavenging habits grade into detritus feeding. Scavenging-detritus feeding is characteristic of many decapods, including most hermit crabs and many shrimps. *Palaemonetes pugio*, the abundant grass shrimp of tidal marshes along the east cost of the United States, plucks out and consumes small bits of cellulose matrix from large detrital fragments and thereby plays an important role in the breakdown of algae and grasses of tidal marsh ecosystems (Welsh, 1975).

The chelipeds of crabs and other decapods often reflect feeding habits. Species that scrape algae from rocks or feed on detritus from the surface of sand and mud commonly have chelipeds with spoon-shaped fingers. The many species that include mollusks in their diet have dimorphic chelipeds. The heavier right claw bears blunt proximal teeth in the fingers and is adapted for crushing; the more slender left claw is adapted for cutting (Figs. 14–48 and 14–58).

Detritus feeding grades into filter feeding. For example, fiddler crabs scoop up mud and detritus with the cheliped. The doorlike third maxillipeds open, and the collected material is dumped into the buccal frame. Using water pumped in from the branchial chamber, the material is worked over and shifted by brush and straining setae on the second and first maxillipeds. After organic material has been removed, the mineral residue is spit out as round pellets, which may eventually surround the burrows or cover the surface of the beach in forms that feed on detritus left by the receding tide (Fig. 14–65*A*).

Decapods that filter feed from a water current include burrowing shrimps (*Callianassa, Upogebia*), the commensal pea crabs, most porcelain crabs, and the mole crabs.

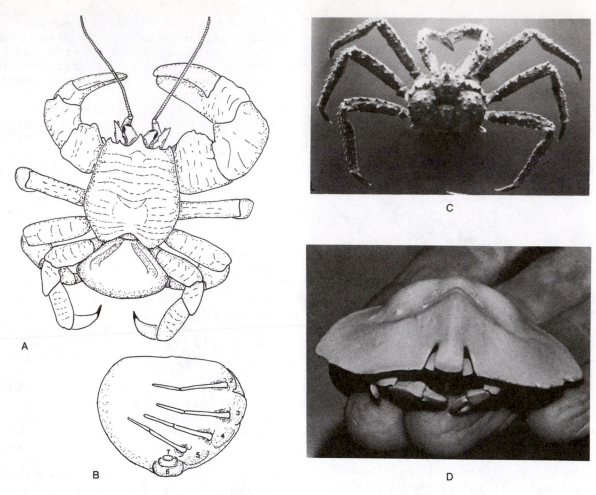

Figure 14–55 Paguroid lithodid crabs: *A*, Dorsal view of *Dermaturus mandteii* from Alaska. Fifth pair of legs is hidden by carapace. *B*, Ventral view of abdomen, showing asymmetrical pleopods. Numbers indicate abdominal segments, and T stands for telson. *C*, *Paralithodes camtschatica* (king crab). This large north Pacific species is a lithode crab, although it looks somewhat like a brachyuran spider crab. The leg span of the specimen in the photograph is over 1 meter. *D*, *Cryptolithodes sitehansis*, a little lithodid crab with a helmet-like carapace. (*C* and *D* by Betty M. Barnes.)

The maxillipeds are used as filtering appendages in many anomurans and in the commensal pea crabs, although some pea crabs living in oysters are known to feed on the mucous food strings collected by the host. The anomuran porcelain crabs filter feed with the third pair of maxillipeds as well as scraping detritus from rocks with their chelipeds (Caine, 1975).

Among the most interesting filter feeders are the mole crabs, *Emerita* (Fig. 14–57). The long, densely fringed, second antennae project above the sand surface after the crabs are buried. In species that inhabit open beaches, such as *Emerita talpoida*, the second antenna filter plankton and de-

tritus from the receding wave current. The crab buries itself seaward and its outstretched second antennae form a characteristic V on the sand surface during the backwash of each wave; in species that live in quieter water, the projecting antennae are moved about.

The typical decapod digestive tract consists of a short esophagus leading into a capacious cardiac chamber (Figs. 14–1*B* and 14–59) and a smaller, posterior, pyloric chamber separated from the cardiac portion by a valve.

The esophagus and the cardiac and pyloric chambers are lined with chitin, which is variously thickened to form a number of ossicles in the walls

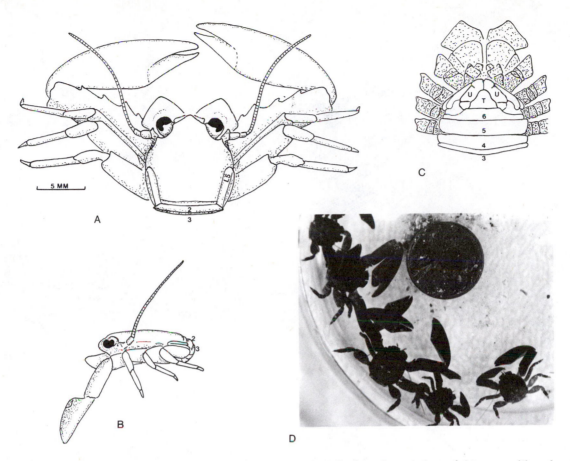

Figure 14–56 The anomuran superfamily Galatheoidea: *A–C, Petrolisthes,* The members of this genus, like others in the family Porcellanidae (porcelain crabs), are often very common beneath stones in shallow water along rocky coasts. They look remarkably like brachyuran crabs, but note the reduced and folded fifth pair of legs and the retention of the uropods and telson on the folded abdomen. *A,* Dorsal view. *B,* Lateral view. *C,* Ventral view of abdomen and overlying cephalothorax. Numbers indicate abdominal segments or legs (L): U and T refer to uropods and telson. *D,* Intertidal porcelain crabs collected from beneath stones in the Gulf of California.

of the cardiac and pyloric chambers. The ossicles provide support and sites for muscle attachment externally. Certain ossicles give rise internally to a median dorsal tooth and to two lateral teeth, one on either side of the median tooth. These three teeth, which sit internally at the posterior portion of the cardiac chamber, form the so-called gastric mill, where food is broken down mechanically. The triturating action of the gastric mill and the movement of the stomach walls are controlled by a complex series of external muscles. The pyloric chamber is divided into a dorsal portion, which leads directly to the intestine, and a ventral, bilobed gland filter (ampulla), which leads to the hepatopancreas via two large ducts, one from each

lobe of the gland filter. The dorsal portion of the pyloric chamber is separated from the ventral by a row of paired denticles, which prevent large particles from entering the gland filter.

The hepatopancreas, or digestive gland, is a large, bilobed organ consisting of numerous blind tubules. Each tubule is composed of cells serving several functions: enzyme secretion, absorption and storage of food, and packaging and removal of digestive waste products through the vacuoles (see review by Gibson and Barker, 1979). Digestive secretions produced by the hepatopancreas move into both the cardiac and the pyloric chambers.

Material that is too large to enter the gland filter and hence the hepatopancreas is passed from the

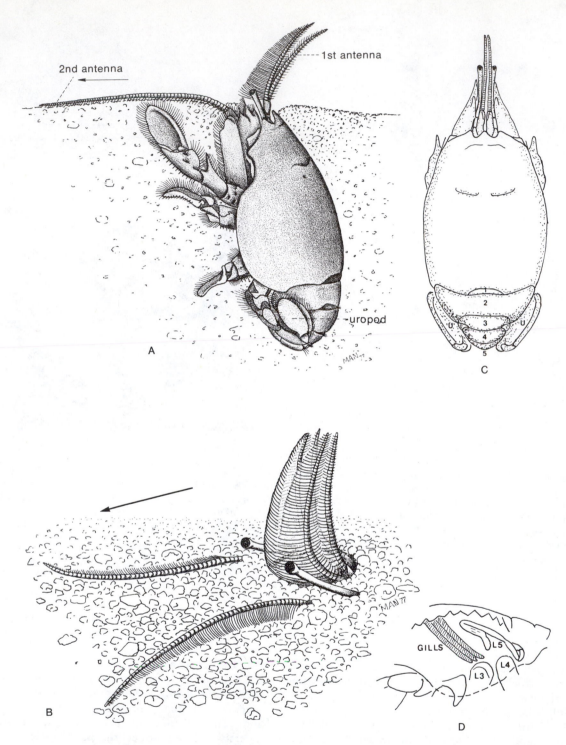

Figure 14—57   The anomuran superfamily Hippoidea, mole crabs. *Emerita talpoida*, a common mole crab on surf beaches along the east coast of the United States. *A*, Lateral view of animal buried in sand with second antennae in filtering position. *B*, Surface view of buried animal. First antennae form siphon for ventilating current. Arrow indicates direction of receding wave. *C*, Dorsal view. *D*, Lateral view of one side of body showing tiny fifth pair of legs folded beneath carapace, which has been cut away.

Figure 14–58 The crab *Calappa flammea* opening the shell of the marine snail *Fasciolaria*. The right crushing claw contains large tubercles at the base of the movable finger. (From Shoup, J. B., 1968: Shell opening by crabs of the genus *Calappa*. Science, *160*:887–888. Copyright 1968 by the American Association for the Advancement of Science.)

dorsal portion of the pyloric chamber into the intestine. Here the midgut epithelium at the anterior end of the intestine secretes a clear, membranous tube, the peritrophic membrane, which encloses the material to be voided as fecal pellets.

Gas Exchange

Primitively, there are four gills on each side of every thoracic segment. The gills arise from the body wall on or near the point of attachment of the appendage and have been given different names depending on their position (Fig. 14–60E). If all the gill series were present on all segments, there would be a total of 32 gills on each side of the body, but no decapod has retained this maximum number. The penaeid shrimp *Benthesicymus* has the greatest number, 24, but reduction to far less than this is the general rule. For example, the lobster *Homarus* has a complement of 20 gills on each side of the body, distributed as follows: one gill on the second and three gills on the third maxillipeds, and four gills on each of the legs except the first, which bears three, and the last, which bears only one. A similar pattern is found in crayfish. Nine pairs of gills are common in marine crabs, but the little pea crab *Pinnotheres* has only three gills to a side.

The gill is composed of a central axis along which are arranged lateral extensions or branches. The structure of the gill branches varies among decapods; the types are illustrated in Figure 14–61. In the axis of each gill runs an afferent and an efferent

branchial channel (Fig. 14–60A). From the afferent channel blood flows into each filament or lamella and then back into the efferent channel.

The blood of decapods contains hemocyanin dissolved in the blood plasma, and in large, active forms, such as the swimming crabs, the hemocyanin transports about 90 per cent of the blood's oxygen (Mangum and Weiland, 1975).

The ventilating current is produced by the beating of a paddle-like scaphognathite, or gill bailer, a projection of the second maxilla (Fig. 14–60D). Water is pulled forward, and the exhalant current flows out anteriorly in front of the head. In the shrimps the ventral margins of the carapace fit loosely against the sides of the body, and water can enter the branchial chamber at any point along the posterior and ventral edges of the carapace (Fig. 14–60A). In other decapods the carapace fits somewhat more tightly, and the entrance of water is limited to the posterior carapace margins and around the bases of the legs (Fig. 14–60B). The point of entrance of the ventilating stream is most restricted in the brachyuran crabs, in which the inhalant opening is located around the bases of the chelipeds (Figs. 14–60C and 14–62A). The forward position of the inhalant openings in brachyurans results in water taking a U shaped course through the gill chambers. On entering the inhalant opening, the water passes posteriorly into the hypobranchial part of the chamber and then moves dorsally, passing between the gill lamellae. The exhalant current flows anteriorly in the upper part of the gill cham-

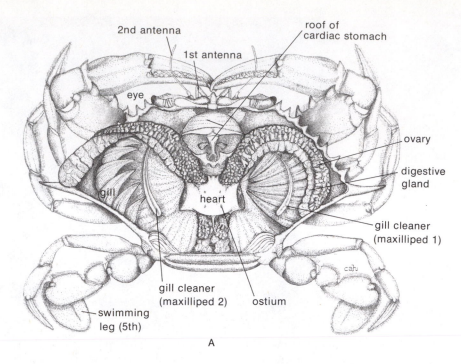

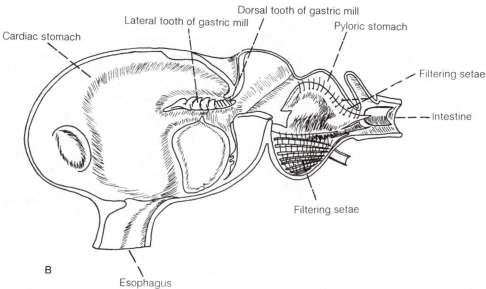

Figure 14–59 *A*, The brachyuran blue crab *Callinectes sapidus* (dorsal dissection). (By Carolyn Herbert.) *B*, Stomach of the crayfish *Astacus* (lateral view). (After Kaestner, A., 1970: Invertebrate Zoology. Vol. 3. Wiley-Interscience, N.Y.)

ber and issues from paired openings in the upper lateral corners of the buccal frame (Fig. 14–62*A*).

Since the majority of decapods are bottom dwellers and include many burrowers, a variety of mechanisms have evolved to prevent clogging the gills with silt and debris. The bases of the chelipeds in crabs and the coxae of the legs of crayfish and lobsters bear setae that filter the incurrent stream. The gills are cleaned in some shrimps with the first and second pairs of legs and in some anomurans with the reduced last pair of legs (see review by Bauer, 1981). The gills are cleaned in crabs by the

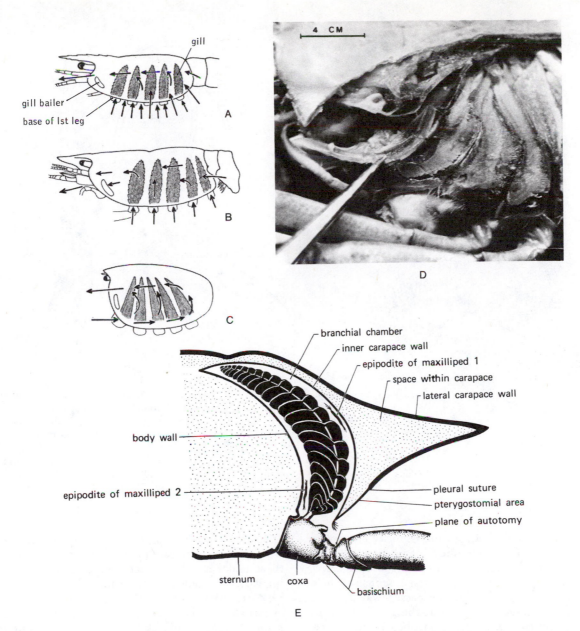

Figure 14–60 *A–C*, Paths of water circulation through gill chamber of three decapods, showing progressive restriction of openings into chamber. *A*, Shrimp. Water enters along entire ventral and posterior margin of carapace. *B*, Crayfish. Water enters at bases of legs and at posterior carapace margin. *C*, Crab. Water enters only at base of cheliped. *D*, Gill bailer (held by forceps) of a crayfish, showing its position within branchial chamber. (By Betty M. Barnes.) *E*, Cross section through the gill chamber of a crab. (From Kaestner, A., 1970: Invertebrate Zoology. Vol. 3. Wiley-Interscience, N.Y.)

fringed epipodites of the three pairs of maxillipeds. These processes are elongated, especially that of the first maxilliped, and they sweep up and down the surface of the gills, removing detritus (Fig. 14–60*E*). As a further aid to the cleaning of the gills and branchial chamber, the gill bailer of many de-

capods periodically reverses its beat and thus reverses the direction of flow through the chamber.

In some burrowing species, when the ventral parts of the body are covered by mud and sand, a reversed current is used for ventilating purposes; when the body is free, a forward current is used. A

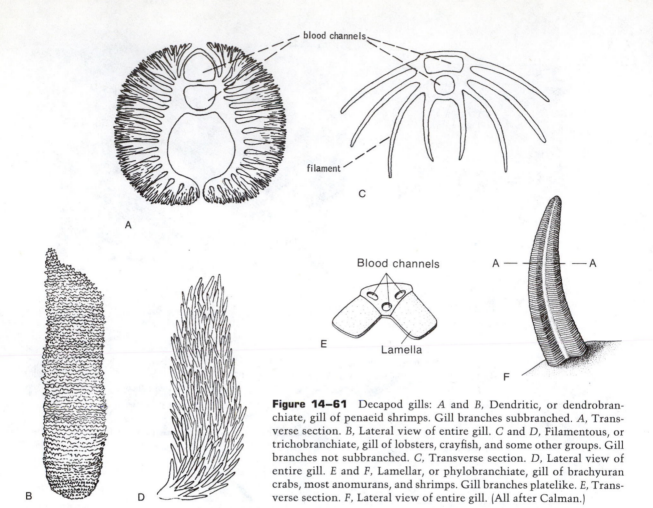

Figure 14–61 Decapod gills: *A* and *B*, Dendritic, or dendrobranchiate, gill of penaeid shrimps. Gill branches subbranched. *A*, Transverse section. *B*, Lateral view of entire gill. *C* and *D*, Filamentous, or trichobranchiate, gill of lobsters, crayfish, and some other groups. Gill branches not subbranched. *C*, Transverse section. *D*, Lateral view of entire gill. *E* and *F*, Lamellar, or phylobranchiate, gill of brachyuran crabs, most anomurans, and shrimps. Gill branches platelike. *E*, Transverse section. *F*, Lateral view of entire gill. (All after Calman.)

further ventilating adaptation in many burrowing decapods is the development of inhalant siphons. For example, in some species of brachyuran crabs, mole crabs, and shrimps, the first antennae are held together, forming an inhalant passageway between them. The large box crabs of the genera *Calappa* and *Hepatus*, which burrow just below the surface, have greatly flattened chelipeds, which when folded fit tightly against the face but leave an inhalant channel between the inner side of the cheliped and the carapace (Fig. 14–62*B*).

Internal Transport

The heart is a box-shaped vesicle located in the thorax and is provided with three pairs of ostia (Fig. 14–59*A*). Five arteries leave the heart anteriorly (Fig. 14–1*B*). A median abdominal artery leaves the heart posteriorly, and a sternal artery arises either from the underside of the heart or from the base of the abdominal artery. Each of these major arteries branches extensively, supplying various organs and structures.

After passing into the tissue sinuses, blood eventually drains into a large, median, sternal sinus prior to passing through the gills and returning to the heart. Complete circulation has been estimated to take from 40 to 60 seconds in large decapods.

Excretion and Salt Balance

Antennal (or green) glands are the excretory organs of decapods and reach their highest degree of development in this group. The end sac, which lies in front of and to both sides of the esophagus, is typically divided into a saccule, where fluid collects by filtration, and a labyrinth. The labyrinth walls are greatly folded and glandular, converting the original vesicle into a spongy mass, and appear to be an

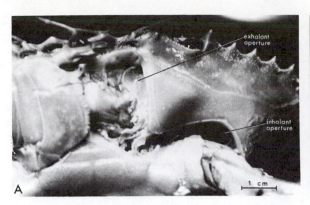
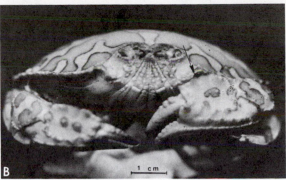

Figure 14–62 *A*, Anterior view of a blue crab, *Callinectes*. The left maxillipeds have been removed, and the gill bailer can be seen within the exhalant aperture. *B*, The box crab *Hepatus*. Arrow indicates opening of left siphon formed by the opposing carapace and cheliped. The siphonal groove on the carapace can be seen on the right side. (Both by Betty M. Barnes.)

important site of reabsorption (Peterson and Loizze, 1974). The labyrinth leads into a bladder of varying complexity by way of an excretory tubule (Fig. 14–4*B*). From the bladder a short duct extends into the basal segment of the second antenna, where it opens to the outside on the summit of a little papilla (Fig. 14–1*A*). In brachyurans the excretory pore is covered by a little movable operculum.

Although the antennal glands are called excretory organs, most nitrogenous waste (NH^{4+}) is diffused from the body surface where the exoskeleton is thin, as on the gills. The antennal glands appear to function in controlling internal fluid pressure and, to some extent, ion content, such as Mg^{++}. Uptake of water by the blood increases the filtration pressure within the antennal gland and the passage of fluid into the saccule. A copious amount of urine may be excreted, depending on internal and external conditions. In undiluted seawater (3.4 percent) the European crab, *Carcinus maenas*, produces a daily amount of urine equivalent to 3.6 per cent of the body weight; in brackish water (1.4 per cent) daily urine production is equal to one third of the body weight. The maintenance of a constant body volume is indicated by the fact that the size increase of a newly molted blue crab (*Callinectes*) is the same regardless of the salinity of the environment. The antennal glands apparently function to maintain this constancy.

Although the antennal glands may function to regulate specific ions, such as magnesium (Cornell, 1979), they do not play a major role in osmotic regulation. In most decapods the urine is isosmotic with the blood even when the animal is in brackish water. Within the higher salinity ranges of sea water, most decapods act as osmoconformers; i.e., the blood salt content is the same as that of the external medium. Those species that can tolerate lower salinities become osmoregulators, maintaining a minimum blood salt concentration that is hyperosmotic to the surrounding medium. As in fish, the gills pick up salts from the ventilating current, replacing ions lost in the urine and elsewhere. The crayfish antennal gland contains a long tubule between the labyrinth and the bladder. Reabsorption of salts by the tubule enables these crustaceans to produce a dilute urine that aids in osmoregulation.

Freshwater and Terrestrial Decapods

There have been numerous invasions of fresh water and land by decapods, but except for the crayfish and the true freshwater crabs (superfamily Potamoidea), the record has not been especially successful.

Freshwater shrimps belong to the families Atyidae and Palaemonidae. The Atyidae are scraping-filter feeders inhabiting streams, pools, and lakes in tropical and subtropical parts of the world. The Palaemonidae include many marine and brackish shallow-water species, such as *Palaemonetes*, in both tropical and temperate regions. Freshwater species include some river, stream, and pool inhabitants, such as the large, pantropical *Macrobrachium*, which is trapped or farmed in many parts of the world. The Palaemonidae osmoregulate by means of the gills, and this is probably also the case with Atyidae.

The crayfish are the most successful freshwater decapods. The more than 500 species are found throughout the world in streams, ponds, lakes, and caves. Some live beneath stones or within debris. Many species excavate burrows, which they use as retreats and for overwintering. Burrowing crayfish commonly inhabit bottom lands where the water table is not too far below the surface, and they may be semiterrestrial. A chimney of excavated mud typically rises above the burrow opening. Most crayfish are about 10 cm long, but some Australian species reach the size of lobsters (Fig. 14–47*A*). Crayfish are omnivores. The antennal glands of crayfish, unlike those of other decapods, can excrete a hypoosmotic urine and play an important role in salt balance.

The anomurans contain one group of terrestrial decapods, the tropical land hermit crabs (*Coenobita,*) which are often sold in pet shops, and the closely related coconut crab (*Birgus*) (Fig. 14–54*A*). Both are found in Indo-Pacific regions and *Coenobita* also occurs in the West Indies. The land hermits live close to the shore and use either burrow water or sea water to wet the body and interior of the shell. Adult coconut crabs have abandoned the hermit crab habit and have acquired a crablike form with a flexed abdomen. The adults live in burrows farther back from the sea but are still a coastal species. Coconut crabs feed on carrion and both decaying and fresh vegetation, and they can husk and open fallen coconuts. They obtain water by drinking. Both *Coenobita* and *Birgus* have reduced gills and have converted the moist branchial chamber into lungs with highly vascularized areas for gas exchange.

There are many brachyuran crabs that can tolerate either very brackish or fresh water but must return to salt water to breed. In this category belongs the Chinese mitten crab, *Eriocheir* (Fig. 14–63*A*), which is found in the rivers and rice paddies of Asia and has invaded the rivers of Europe, probably by ship ballast, beginning about 1912. Three specimens were collected from Lake Erie in 1973 (Nepszy and Leach, 1973). Another is the brachyuran, *Rhithropanopeus*, of the American coast, which is found in salt water as well as in freshwater ditches. *Rhithropanopeus* has been introduced into northern Europe and San Francisco Bay, possibly in the same way as *Eriocheir*. The blue crab, *Callinectes sapidus*, is quite euryhaline and ranges far up into brackish estuaries and bays. Some very successful but poorly known freshwater decapods are the so-called river crabs (superfamily Potamoidea). These occur in all tropical regions from the mouths of rivers to mountain elevations of 5000 meters. They complete their entire life cycle in fresh water.

All of these crabs, both temporary and permanent residents of brackish and fresh water, excrete urine that is isosmotic with the blood and regulate their salts by means of ion absorption through the gills.

Except for the land hermits and coconut crabs, most of the decapods that have invaded land are brachyuran crabs. This is not surprising, considering the motility of crabs and their tightly enclosed branchial chambers. Although the living terrestrial crabs are derived from a number of different land invasions, most display similar adaptations.

Gills continue to be utilized for gas exchange, but there is also a tendency for the branchial cham-

Figure 14–63   *A*, The Chinese mitten crab, *Eriocheir sinensis*, in a rice paddy. This amphibious freshwater brachyuran is native to southern Asia but has been introduced into the Rhine and Elbe rivers of Europe. (After Schmitt.) *B*, A river crab, *Potamon anomalus* (Potamonidae). (By Betty M. Barnes.)

ber to become rather like a lung, with some surface other than the gills given over to gas exchange (Díaz and Rodriquez, 1977). The gill bailer continues to provide for ventilation, but it moves air rather than water.

Water loss by evaporation is much greater in land crabs than in other terrestrial arthropods (arachnids, insects), and most of this is through the carapace. Water is replaced, for the most part, by drinking, and many can maintain gill moisture by taking up water from soil, dew, or rain (Wolcott, 1984). No land crab can excrete a hyperosmotic urine. Such a capability would aid in conservation of water. However, land crabs save water by excreting very little urine.

All land crabs live in burrows, or at least beneath stones, and are usually active only at night. They are primarily vegetarians and scavengers. All can run very rapidly. Eyes are generally well developed.

The name *land crab* usually refers to the members of the family Gecarcinidae, all of which are terrestrial. They are found in tropical and subtropical America, West Africa, and the Indo-Pacific area. The family includes *Cardisoma*, which lives in fields and woods in southern Florida and the West Indies, and *Gecarcinus*, which inhabits grasslands and forests along the coasts of Florida and tropical America and in the West Indies (Fig. 14–64).

Although terrestrial, the gecarcinids are nevertheless coastal in distribution, for females must return to the sea to release their spawn. The terrestrial potamoids are found the greatest distance from the coast, often hundreds of miles. However, these crabs have direct development and do not need to return to the sea to breed.

The family Ocypodidae contains some of the most familiar amphibious and terrestrial crabs, such as the fiddler crabs and ghost crabs. The known species of semiterrestrial fiddler crabs (*Uca*) number 62. They live on protected sand and mud beaches of bays and estuaries, in brackish marshes, and in mangroves. Although most species are tropical and semitropical, fiddler crabs are found on both the east and the west coasts of North America.

The burrows of fiddler crabs are located in the intertidal zone, and at low tide the crabs come out to feed and court (Fig. 14–65A); enormous numbers may cover a beach. Tropical species tend to be active only during diurnal low tides, but temperate species emerge at night as well, perhaps compensating for the winter period, when they must remain dormant within their burrows. The simple burrows are usually L shaped; those of North American species are generally no deeper than about 36 cm (Fig. 14–65B). The excavated sand is carried to the surface as small balls cradled in the legs of one side. As the tide comes in, the crabs return to their burrows and plug the entrance with sand or mud. The burrows of some tropical species that inhabit upper zones of mangrove are flooded only during spring tides.

Some fiddler crabs feed within 20 to 50 cm of the burrow, making spokelike movements from the burrow mouth; others may feed up to 50 meters away. When disturbed, fiddler crabs quickly flee into their burrows. Many crabs do not find their original burrows, however, and they either occupy a vacant hole or dig a new one.

Ghost crabs (*Ocypode*) are common in many parts of the world, including the east coast of the United States (Fig. 14–66). They are several times larger than fiddler crabs and never occur in large aggregations. Their widely separated burrows are found mostly above the high tide on the upper beach or in dunes. Ghost crabs are largely nocturnal and many go down to the lower beach to prey on clams and mole crabs or to scavenge for food.

The Indo-Pacific soldier crabs of the family Mictyridae are similar in habit to *Uca* (Fig. 14–67). When disturbed, a soldier crab quickly burrows sideways like a corkscrew.

The family Grapsidae contains perhaps the most ecologically diverse assemblage of crabs. There are marine, brackish-water, freshwater, amphibious, and terrestrial species. Amphibious grapsids include *Sesarma*, which lives beneath drift and stones; the agile, rock-inhabiting *Grapsus*; and numerous mangrove inhabitants, such as *Aratus*.

Figure 14–64 The gecarcinid land crabs *Cardisoma guanhumi* of the West Indies and Florida. (By Betty M. Barnes.)

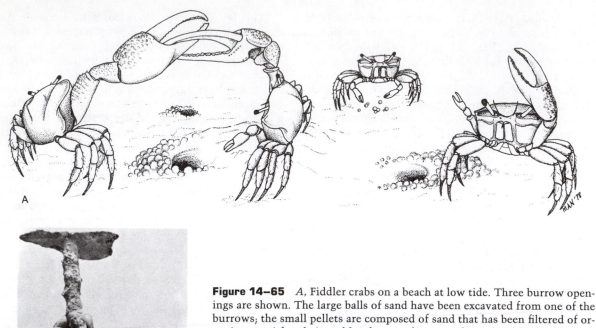

Figure 14–65   *A*, Fiddler crabs on a beach at low tide. Three burrow openings are shown. The large balls of sand have been excavated from one of the burrows; the small pellets are composed of sand that has been filtered of organic material and ejected by the mouth parts. The crab in the background with the two small claws is a female. The two males on the left are engaging in ritualized combat. The male on the right is waving the large claw in courtship display. *B*, Mold of burrow of the fiddler crab *Uca pugilator*. (By Frey, R. W., and Howard, J. D., 1969: Trans. Gulf Coast Assoc. Geol. Soc., *19*:427–444.)

Nervous System and Sense Organs

The ventral nerve cord of shrimps and lobsters, which have large extended abdomens, contain separate ganglia for all but the first two thoracic segments. The ventral ganglia in other groups show varying degrees of additional fusion, culminating in the brachyurans, in which all of the abdominal ganglia have migrated anteriorly to fuse with the thoracic ganglia, forming a single ventral mass.

As in other crustaceans, the legs and antennae are important sites for the reception of environmental information, and aesthetascs, the chemosensory hairs on the first pair of antennae, are commonly well developed (Fig. 14–5) (Fontaine et al, 1982; Derby, 1982). The aesthetascs aid not only in locating food but also in recognizing other individuals and their sexual state. The frequent grooming of the antennae by many decapods prevents fouling of the receptor sites (Bauer, 1981).

Although there are some blind decapods, particularly among deep-sea forms and cave-dwelling crayfish, compound eyes are usually highly developed, and the eye stalk is more mobile than in most other crustaceans.

A pair of statocysts is present in nearly all decapods and is located in the basal segment of the first antennae. The sac is always open to the exterior, although in crabs and in some others the opening is reduced to a slit and is functionally closed. The statolith may be composed of fine sand grains bound together by secretions from the statocyst wall (Fig. 14–68). Since the statocyst is an ectodermal invagination, its lining along with the statolith is shed at each molt. The sand grain statoliths are replaced when the head is buried in sand or the animal actually inserts sand grains into the sac. Along the floor of the sac are a number of rows of sensory hairs with which the statolith is directly connected or is in intermittent contact. The hairs arise from receptor cells in the sac wall, and the receptor cells are innervated by a branch of the antennular nerve.

When the animal is in a normal horizontal position, the floor of the statocyst is inclined about 30 degrees, which results in a medial gravitational pull on the statolith, but one statocyst counterbalances the other (Fig. 14–68B). When the floors of both statocysts are tilted, the one in which the receptor cells are most stimulated dominates the

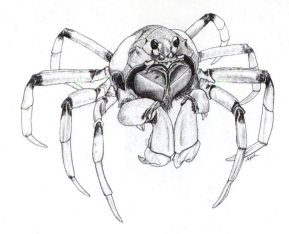

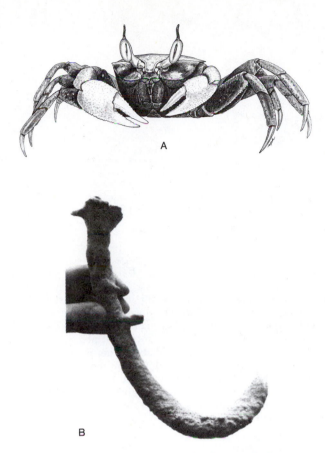

Figure 14–67 The Indo-Pacific soldier crab *Mictyris longicarpus*. (Based on a photograph by Healy and Yaldwyn.)

Figure 14–66 *A*, The Indo-Pacific ghost crab *Ocypode ceratophthalma*. (Based on a photograph by Healy and Yaldwyn.) *B* Cast of burrow of the ghost crab *Ocypode quadrata* from the coast of Georgia. (From Frey, R. W., 1971: Decapod burrows in holocene barrier island beaches and washover fans, Georgia. Senckenbergiana Maritima, 3:53–77.)

other (Fig. 14–68E). Crabs and some other decapods have complex statocysts, which indicate not only the gravitational position of the animal but also its position in regard to movement. Such statocysts are functionally similar to the semicircular canals in vertebrates.

Reproduction and Development

The decapod sperm is tack or star shaped and lacks a middle piece and flagellum (Fig. 14–69D). Most decapods transmit sperm in spermatophores, which are delivered to the female by the anterior two pairs of copulatory pleopods of the male. The paired but connected testes lie in the thorax but may extend into the abdomen. The sperm duct is glandular and modified, depending on the degree of

spermatophore formation, and the terminal end of the sperm duct is a muscular ejaculatory duct, which opens onto or near the coxae of the last pair of legs. At copulation the spermatophores pass from the gonopore to the copulatory pleopods for transmission. For example, in brachyuran crabs the first pleopod is in the form of a cylinder into which the piston-like second pleopod fits. Soft spermatophores are delivered to the pleopods by a nipple-like penis associated with the gonopore opening on each side and pumped through the anterior conducting pleopod of each pair (Fig. 14–70).

The ovaries are similar in structure and location to the testes (Fig. 14–59A). The oviducts are usually unmodified and open onto the coxae or near the coxae of the third pair of legs. In most brachyuran crabs the terminal end of each oviduct is modified as a seminal receptacle and vagina for the reception of the male pleopod. A median seminal receptacle occurs in most other decapods and is a depression created by processes of the last one or two thoracic sternites.

Mating in most aquatic decapods occurs shortly after molting, and the sexes are attracted to each other by pheromones before or after molting, depending on the group. Some sort of precopulatory courtship is typical. For example, the male hermit crab holds the female with one cheliped and taps and strokes her with the other, or pulls her back and forth.

In some brachyuran families, such as the Cancridae (*Cancer*) and Portunidae (*Callinectes*, *Portunus*, and *Carcinus*), there is a premolt attendance of the female by the male in which the male carries

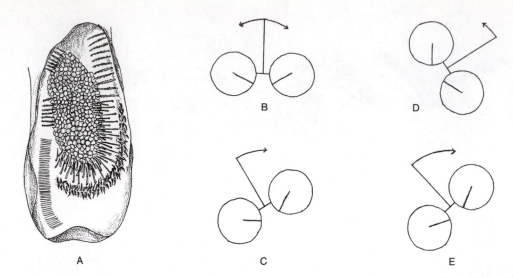

Figure 14–68 *A*, Statocyst of the American lobster, *Homarus americanus* (with dorsal wall removed). Note the four crescent-shaped rows of sensory hairs; the inner three are in contact with the statolith. The row of long, delicate hairs (lower left) are thread hairs. The opening of the statocyst to the exterior is at upper left beyond the edge of the illustration. (After Cohen from Cohen and Dijkgraaf.) *B–E*, Statocyst function. Circle represents statocyst chamber with floor inclined 30 degrees (bar). *B*, Counterbalancing pull exhibited by two statocysts when animal is in normal horizontal position. *C*, Animal rotated 30 degrees to right, placing floor of right statocyst in horizontal position and inclining left statocyst floor 60 degrees. Impulses initiated by left statocyst cause animal to rotate back to left (arrow). *D*, Animal rotated to left so floor of right statocyst is inclined 90 degrees, initiating impulses that cause animal to rotate back to the right. *E*, Animal rotated to right. The more inclined position of the left statocyst floor causes it to dominate the right statocyst, and the animal rotates back to the left. (*B–E* greatly modified from Schöne.)

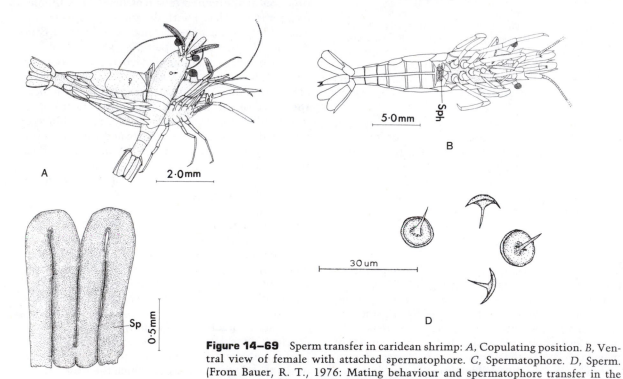

Figure 14–69 Sperm transfer in caridean shrimp: *A*, Copulating position. *B*, Ventral view of female with attached spermatophore. *C*, Spermatophore. *D*, Sperm. (From Bauer, R. T., 1976: Mating behaviour and spermatophore transfer in the shrimp *Heptacarpus pictus*. J. Nat. Hist., *10*:415–440.)

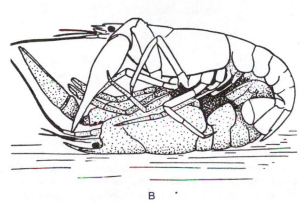

Figure 14–70 Copulatory structures of the brachyuran crab *Ranina*. The nipple-like structure in front of the end of the dissecting needle is the penis. Just above the penis is the base of the second copulatory pleopod, which projects into the somewhat cylindrical first pleopod. The two pleopods are directed forward and lie in the sternal depression covered by the abdomen (By E. Ryan). *B*, Copulating crayfish. Female is stippled. (From Andrews in Kaestner, A., 1970: Invertebrate Zoology. Vol. 3. Wiley-Interscience, N.Y.)

the female about beneath him, her carapace beneath his sternum. He releases her so that she can molt. Copulation occurs shortly afterwards. In contrast to these brachyuran crabs, some male mole crabs are very small and neotenous and attach to the body of the female (Fig. 14–73).

Visual and acoustical signals are of special importance for attraction in terrestrial forms. The semiterrestrial fiddler crabs *(Uca)* go through an elaborate courtship behavior, which has been studied in considerable detail (Crane, 1975). Sexually active males entice females into their burrows, where mating occurs, and the females remain there until their eggs hatch. Males use the greatly enlarged claw to attract females and to defend their burrow territory against other males (Fig. 14–65*A*).

In male-male encounters there is a highly ritualized combat, in which the large claw is held like a shield. Combat movements involve variations of pushing and extension and have been best described in *Uca pugilator* and *Uca pugnax* of the east coast of North America (Hyatt and Salmon, 1978).

The male fiddler crab atracts a female by waving the large claw in a semophore fashion. Each species has a characteristic pattern of movement (Fig. 14–71). The male also attracts the female with acoustical signals produced by rapping the elbow (propodus) of the claw against the substrate or by rapid flexion of the ambulatory legs. The number of raps (pulses) in a series and the interval between series is also characteristic of the species. The sounds are picked up largely through the substrate and have been studied with special amplifying microphones. Females can detect the signals 50 to 100 cm away from the male by means of special myochordotonal organs in the legs.

The courting of tropical species is restricted to daylight hours and, depending on the species, reaches a peak each month whenever the low tides occur within certain time periods. Temperate species court during both diurnal and nocturnal low tides, but activity is concentrated during two periods within the tidal month (Fig. 14–72) (Christy, 1978 and 1982b; Greenspan, 1982). An *Uca pugilator* male attracts females during the day by waving the claw. If a female approaches, the male will also rap his cheliped against the substrate and eventually back into the burrow, where rapping alone entices the female to join him (Fig. 14–71). At night the male attracts the female by rapping only at the mouth of his burrow.

Male ghost crabs do not wave the cheliped to attract females but use acoustical signals like fiddler crabs. Some male ghost crabs, such as *Ocypode ceratophthalmus* along the Red Sea, construct sand pyramids that attract females and also provide orientation.

In some brachyuran families such as the Cancridae (*Cancer*) and Portunidae (*Callinectes, Portunus,* and *Carcinus*), there is a premolt attendance of the female by the male, in which the male carries the female about beneath him, her carapace be-

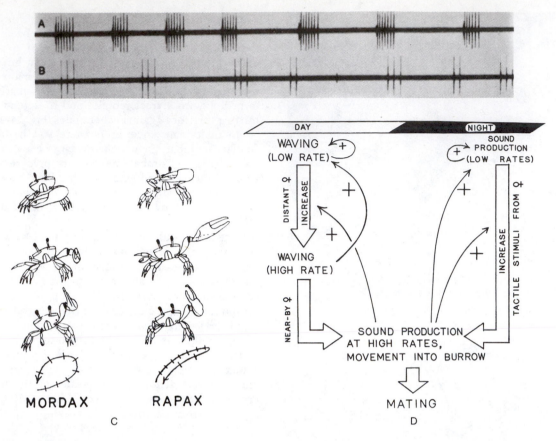

Figure 14–71 Courtship behavior in fiddler crab: *A* and *B*, Acoustical signals produced by cheliped rapping in *Uca pugilator* (*A*) and *Uca speciosa* (*B*). Each vertical line on the oscillograph recording is one rap (pulse). *C*, Cheliped waving in two species of *Uca*. The starting position is shown at the top of each of the figures and the path of the cheliped is described at the bottom by an arrow. The cheliped is commonly moved in a series of jerks, which are indicated by the cross bar on the arrow path. *D*, Diagram of sequence of behavior in courtship ending in mating. See text for explanation. (*A–C* from Salmon, M., 1967: Coastal distribution, display and sound production by Florida fiddler crabs (genus *Uca*). Animal Behavior, *15*:449–459. *D* from Salmon, M., and Horch, K. W., 1972: Acoustic signalling and detection by semi-terrestrial crabs of the family Ocypodidae. *In* Winn, H. E., and Olla, B. L. (Eds.): Behavior of Marine Animals. Vol. I. Plenum Press, N.Y. pp. 60–96.)

neath his sternum. He releases her so that she can molt. Copulation occurs shortly afterwards. In contrast to these brachyuran crabs, some male mole crabs are very small and neotenous and attach to the body of the female (Fig. 14–73).

In shrimps the copulating pair are commonly oriented at right angles to each other with the genital regions opposing each other (Fig. 14–69*A*). The somewhat modified first and second pairs of pleopods are used to transfer a spermatophore to the median receptacle between the thoracic legs of the female. In crayfish and lobsters the male turns the female over and pins back her chelipeds with his own (Fig. 14–70*B*). The first two pairs of pleopods

are then lowered to a 45-degree angle and held in this position by one of the last pair of legs (folded beneath the pleopods).

In crayfish possessing a seminal receptacle, the tips of the first pleopods are inserted into the chamber, and spermatophores flow along grooves in the pleopods. When the seminal receptacle is absent in lobsters the spermatophores are attached to the body of the female, particularly at the bases of the last two pairs of legs. These are frequently seen as "tar spots" on the females of the spiny lobsters, *Panulirus*.

Hermit crabs partially emerge from their shells to mate. The ventral surfaces are appressed, and

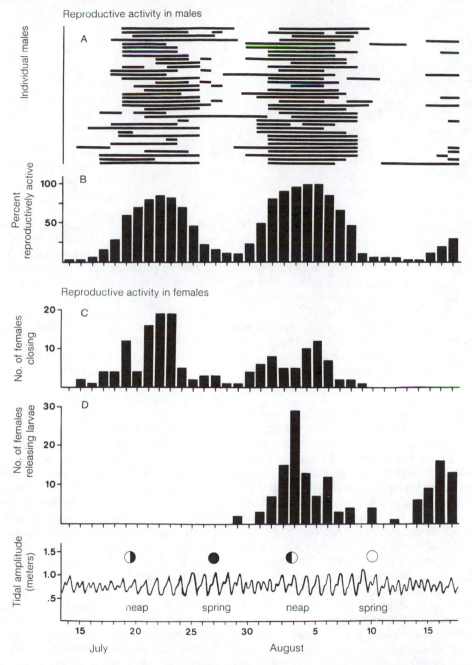

Figure 14–72 Relationship of the reproductive activity of the fiddler crab *Uca pugilator* to the tidal cycle on the west coast of Florida. Number of females closing refers to the number of females entering the male's burrow, where copulation occurs. The male closes the burrow and waits with the female until she emerges to release larvae. (From Christy, J. H., 1978: Adaptive significance of reproductive cycles in the fiddler crab *Uca pugilator:* a hypothesis. Science, *199*:453–455. Copyright 1978 by American Association for the Advancement of Science.)

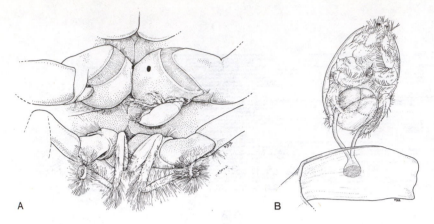

Figure 14–73 Dwarf male of the mole crab *Emerita rathbunae*. *A,* Male (2.5 mm) attached to base of third leg of female. *B,* Enlarged view of male showing attachment by pair of large spermatophores. (From Efford, I. E., 1967: Neoteny in sand crabs of the genus *Emerita*. Crustaceana, *13*:81–83)

A   B

spermatophores and eggs are released simultaneously.

During copulation in crabs, the female lies beneath the male or in the reverse position, with ventral surfaces opposing each other. The first pair of pleopods, which conduct the sperm, are inserted into the openings of the female (Fig. 14–74).

Egg laying takes place shortly after copulation in forms with no seminal receptacles, but when sperm are deposited in seminal receptacles, which are commonly sealed and plugged by the semen (as in brachyuran crabs), egg laying may not take place until some time later.

Penaeid and related shrimps shed their eggs directly into the sea water. In all other decapods the eggs on extrusion are typically attached to the pleopods with cementing material. The cementing material is associated with the egg membrane.

Fertilization is internal in brachyurans, but in most decapods the eggs are probably fertilized at the moment of egg laying. In crayfish the female lies on her back and curls the abdomen far forward, creating a chamber into which the eggs are driven by a water current produced with the beating pleopods. In crabs the usually tightly flexed abdomen is lifted to a considerable degree to permit brooding, and the egg mass, which often becomes orange in color, is sometimes called a sponge.

The hatching stage varies greatly. In all dendrobrachiate shrimps (penaeids and sergestids) the eggs are shed directly into the water and hatch as naupliar or metanaupliar larvae (Fig. 14–75). In all other decapods the eggs are carried on the pleopods of the female. In marine species hatching takes place at the protozoea and zoea stage. The special names given the zoea and postlarval stage of the different decapod groups are listed in Table 14–2.

The zoea larval stages of most crabs are easily recognized by the very long, rostral spine and sometimes by a pair of lateral spines from the posterior margin of the carapace (Fig. 14–76*A*). The postlarval stage, called a megalops, has a large or flexed abdomen and the full complement of appendages (Fig. 14–76*E*).

Many decapods with planktonic larvae inhabit shallow coastal waters, estuaries, or salt marshes as adults or during juvenile stages. For example, estuaries and salt marshes are the nursing grounds for species of penaeid shrimps. The larvae of such decapods display various locomotor responses to different environmental cues, such as salinity and tidal changes, which help to keep larvae from being dispersed out to sea (Cronin and Forward, 1979).

As is true of many other invertebrates, there is a tendency for larval life to be shortened in decapods that inhabit cold oceans or abyssal depths. Larval stages are usually absent in the strictly freshwater decapods (see p. 619). Brackish-water forms, such as blue crabs (*Callinectes*), and river immigrants, such as *Eriocheir*, return to more saline water for breeding.

Likewise, development in terrestrial anomurans and brachyurans takes place in the sea. The female migrates to the shore and into the water, at which time the zoeae hatch and are liberated. The reproductive events in the common American land crab *Gecarcinus lateralis* are facilitated by zonation of the population from the shore landward (Klaassen, 1975; Bliss et al, 1978, Wolcott and Wolcott, 1982). During the reproductive period, which usually coincides with the rainy season, the zone nearest the shore is occupied by courting males, females near ovulation, and females carrying eggs, and it is here that reproduction occurs. The female migrates into the water to liberate the larvae, vigorously shaking herself as she is washed

Figure 14–74   Copulating land crabs *(Gecarcinus lateralis)*. Female is uppermost. From Bliss, D. E., van Montfrans, J., van Montfrans, M., and Boyer, J. R., 1978: Behavior and growth of the land crab *Gecarcinus lateralis* (Fréminville) in southern Florida. Bull. Am. Mus. Nat. Hist., *160*(2):137.)

by waves. A female may produce three broods during the reproductive period. The composition of the various population zones changes over the years, and individuals will migrate shoreward to establish themselves within the reproductive zone.

The potamonid crabs, which are terrestrial or amphibious, return to fresh water to breed, although there are no free larval stages.

SYSTEMATIC RÉSUMÉ OF ORDER DECAPODA

Suborder Dendrobranchiata. Shrimps. Gills dendrobranchiate; body laterally compressed; first three pairs of legs chelate but not with enlarged chelipeds. Eggs planktonic (not carried by female on pleopods), and a nauplius is the first larval stage.

Infraorder Penaeidea.* Rostrum well developed. Pleura of second abdominal segment not over-

lapping those on the first. *Penaeus, Sicyonia,* and the pelagic shrimps *Sergestes, Acetes,* and *Lucifer.*

Suborder Pleocyemata. Decapods with phyllobranchiate and trichobranchiate gills; eggs carried by female on pleopods and hatch as zoeae.

Infraorder Stenopodidea.* Shrimps. Cephalothorax more or less cylindrical; first three pairs of legs chelate, and one member of the third pair is enlarged; pleura of second abdominal segment not overlapping those of the first. Gills trichobranchiate. *Stenopus.*

Infraorder Caridea.* Shrimps. Cephalothorax more or less cylindrical; first two pairs of legs chelate or subchelate, and either first or second pair commonly heavier or longer than the others, third pair chelate; pleura of second abdominal segment overlapping those of the first and third. Gills phyllobranchiate. The Caridea include the greatest number of species of shrimps. Sand shrimps, *Crangon;* the snapping shrimps *Alpheus* and *Synalpheus;* the marine, brackish-water, and freshwater Palaemoni-

*Those groups marked with an asterisk were formerly included in old suborder Natantia. All others were placed in the older suborder Reptantia.

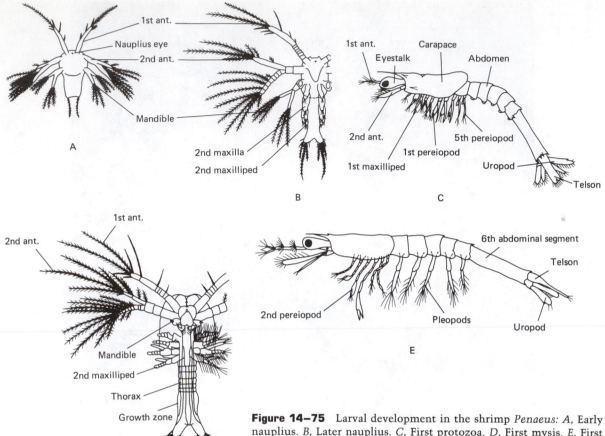

Figure 14–75 Larval development in the shrimp *Penaeus: A*, Early nauplius. *B*, Later nauplius. *C*, First protozoa. *D*, First mysis. *E*, First postlarva. (After Dobkin from Anderson, D. T., 1973: Embryology and Phylogeny in Annelids and Arthropods. Pergamon Press, Oxford. p. 495.)

dae—*Palaemonetes, Macrobrachium, Leander,* and *Periclimenes* (cleaning shrimps); the freshwater Atyidae—*Atya;* the pelagic Olophoridae; *Hippolysmata* (cleaning shrimps), *Hippolyte,* and *Tozeuma.*

Infraorder Astacidea. Crayfish and lobsters with large claws. Cephalothorax more or less cylindrical; abdomen well developed and somewhat flattened dorsoventrally. First three pairs of legs chelate and first pair greatly enlarged. The lobsters *Nephrops* and *Homarus;* the freshwater crayfish *Astacus, Parastacus,* and *Cambarus.*

Infraorder Palinura. Cephalothorax more or less cylindrical; abdomen well developed and somewhat flattened dorsoventrally. Legs may be chelate or subchelate, but first pair in most species not enlarged. The relic *Neoglyphea;* spiny lobsters, *Panulirus,* with long, heavy an-

tennae; Spanish, locust slipper, or shovel-nosed lobsters, *Scyllarus,* with flattened carapace and large, short, flattened antennae; the deep-water *Polycheles,* with broad, depressed carapace and first four pairs of legs chelate.

Infraorder Anomura. Carapace usually depressed; third legs never chelate; fifth legs reduced or turned upward. Abdomen variable. In crablike forms, eyes medial to antennae.

Superfamily Thalassinoidea. Compressed carapace. First pair of legs in form of chelipeds but asymmetrical; third pair not chelate. Abdomen well developed, extended, and flattened. Marine burrowing shrimps. *Thalassina, Axius, Calianassa,* and *Upogebia.*

Superfamily Paguroidea. Carapace oval. Abdomen usually asymmetrical and housed within a gastropod shell or other

TABLE 14–2 Types of Postembryonic Development and Larvae in Decapods

Group	Postembryonic Development	Larvae
Suborder Dendrobranchiata		
Family Penaeidae	Slightly metamorphic	Nauplius→protozoea→mysis→mastigopus (zoea) (postlarva)
Family Sergestidae	Metamorphic	Nauplius→elaphocaris→acanthosoma→mastigopus (protozoea) (zoea) (postlarva)
Suborder Pleocyemata		
Infraorder Caridea	Metamorphic	Protozoea→zoea→parva (postlarva)
Infraorder Stenopodidea	Metamorphic	Protozoea→zoea→parva
Infraorder Palinura	Metamorphic	Phyllosoma→puerulus, nisto, or pseudibaccus (zoea) (postlarva)
Infraorder Astacidea	Slightly metamorphic	Mysis→postlarva (zoea)
Infraorder Anomura	Metamorphic	Zoea→glaucothoë in pagurids, grimothea (postlarva)
Infraorder Brachyura	Metamorphic	Zoea→megalopa (postlarva)

(Modified from Waterman and Chace, *In* T. H. Waterman (Ed.): Physiology of Crustacea. Vol. 1.)

objects (hermit crabs) or folded beneath the carapace. First legs in form of chelipeds. The hermit crabs, *Pomatocheles, Petrochirus, Clibanarius, Coenobita* (land hermit crab), *Pagurus,* and *Pylopagurus.* This group also contains a number of crablike forms that are probably secondarily derived from the hermit crabs. The abdomen is flexed beneath the thorax as in the true crabs and well chitinized, but like that in hermit crabs, the abdomen is not symmetrical, and females have pleopods on left side. Includes the coconut crab—*Birgus*—and the lithodid (stone) crabs—*Lithodes* and *Paralithodes* (the commercial king crab of the North Pacific).

Superfamily Galatheoidea. Mostly crablike forms with symmetrical abdomen curled or flexed beneath thorax. Well-developed tail fan. Rostrum often well developed; carapace not fused with epistome. First legs in form of chelipeds; fifth legs very small and folded along side of carapace. The lobster-like *Galathea* and *Munida* with abdomen only curled beneath cephalothorax; the porcelain crabs, *Petrolisthes, Pachycheles, Porcellana,* and *Polyonyx;* the South American freshwater *Aegla.*

Superfamily Hippoidea. Symmetrical abdomen flexed beneath thorax. Rostrum reduced or absent. Cephalothorax flattened or more or less cylindrical. First legs chelate or subchelate and never in form of chelipeds; fifth legs greatly reduced and folded, often located beneath carapace. Posterior end of abdomen folded ventrally and forward. Commonly in sand in surf zone. The sand, or mole, crabs—*Hippa, Emerita, Blepharipoda, Lepidopa,* and *Albunea.*

Infraorder Brachyura. Carapace broad and fused with epistome. First legs in form of heavy chelipeds; third legs never chelate. Eyes usually lateral to second antennae. Symmetrical abdomen reduced and tightly flexed beneath cephalothorax.

Section Dromiacea. Primitive brachyurans; marine. Last pair of legs often dorsal in position and modified for holding objects over the crab. Carapace usually not broader than long. Uropods present but much reduced. Genital apertures on coxae. *Homolodromia, Dromia, Hypoconcha,* and *Homola.*

Section Oxystomata. Last pair of legs normal or modified as in Dromiacea. Mouth frame triangular rather than quadrate. Uropods absent. Female reproductive openings on sternum. Marine. *Dorippe;* the box crabs, *Calappa* and *Hepatus; Persephona* and *Ebalia.*

Section Oxyrhyncha. Carapace narrowed anteriorly into a rostrum. Body shape roughly triangular. Mouth frame quadrate. Marine. The decorator and spider crabs—*Maja, Inachus, Macrocheira, Hyas, Libinia, Pelia, Par-*

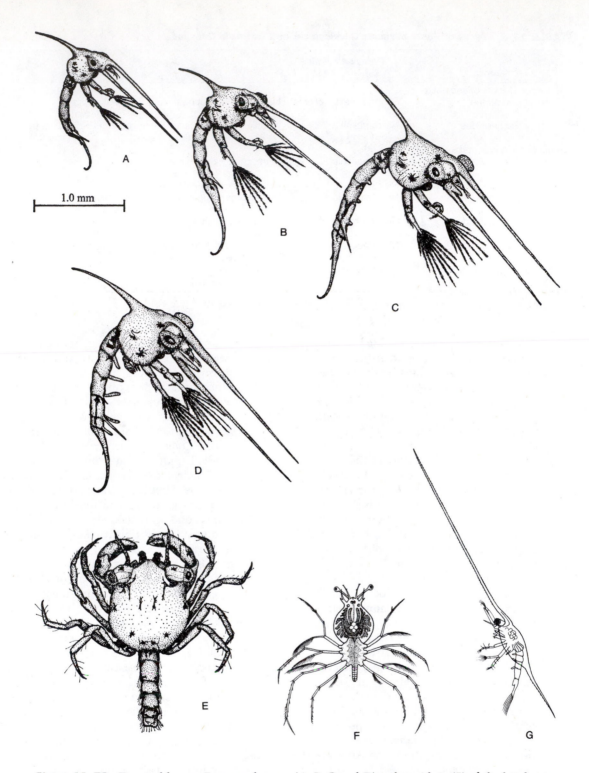

Figure 14—76   Decapod larvae: Four zoeal stages (*A, B, C,* and *D*) and megalops (*E*) of the brachyuran crab *Rhithropanopeus harrisii.* (From Costlow, J. D., and Bookhout, C. G., 1971: Fourth European Marine Biology Symposium, Cambridge University Press, London. p. 214. Copyrighted and reprinted by permission of the publisher.) *F,* Phyllosoma (zoea) larva of the spiny lobster. *Palinurus.* (Modified after Claus.) *G,* Metazoea of the anomuran crab *Porcellana.* (Adapted from various sources.)

thenope, *Pugettia*, *Loxorhynchus*, and *Stenorhynchus*.

Section Cancridea. Carapace elongated or transversely oval or hexagonal; teeth on front margin. Orbits of eyes with two fissures above. First antennae folded longitudinally or obliquely. Mouth frame quadrate. Female openings on sternum. Corystidae—*Corystes*. The cancer crabs, Cancridae—*Cancer*.

Section Brachyrhyncha. Carapace not narrowed anteriorly. Body shape round, transversely oval, or square. Orbits well developed and complete. Mouth frame quadrate. Female openings on sternum. This section contains the majority of crabs. The swimming crabs, Portunidae—*Portunus*, *Callinectes*, *Carcinus*, *Arenaeis*, and *Ovalipes*. The freshwater crabs, Superfamily Potamoidea—*Potamon* and *Pseudothelphusa*. The mud crabs, Xanthidae—*Xantho*, *Menippe* (stone crab), *Pilumnus*, *Rhithropanopeus*, *Panopeus*, *Neopanopeus*, and *Eurypanopeus*. The commensal pea crabs, Pinnotheridae—*Pinnotheres*, *Pinnixa*, and *Dissodactylus*. The amphibious crabs, Ocypodidae—*Ocypode* (ghost crabs), *Uca* (fiddler crabs), *Dotilla*, and *Ucides*; Mictyridae (soldier crabs). Marine, freshwater, amphibious, and terrestrial Grapsidae—*Geograpsus*, *Grapsus*, *Pachygrapsus*, *Planes*, *Sesarma*, *Eriocheir* (Chinese mitten crab), *Aratus*, *Metaplax*, and *Metopaulias*. The land crabs, Gecarcinidae—*Cardisoma*, *Epigrapsus*, *Gecarcoidea*, and *Gecarcinus*. The coral gall crabs, Hapalocarcinidae—*Hapalocarcinus*, *Cryptochirus*, and *Troglocarcinus*.

SUMMARY (THROUGH DECAPODA)

1 The class Malcostraca contains the largest number of crustacean species and most of the larger forms. The trunk is composed of a thoracic region of eight segments and an abdominal region of six segments. All trunk segments bear appendages. Usually, one or more thoracic appendages have turned forward and function in food handling. The other thoracic appendages (legs, or periopods) are used in prehension or crawling. The abdominal appendages are pleopods and one or more pairs of terminal uropods.

2 The superorder Syncarida contains a small number of primitive, mostly freshwater, malacos-

tracans. There is no carapace, and the thoracic appendages are similar and biramous.

3 The superorder Hoplocarida, order Stomatopoda (mantis shrimps), is a group of relatively large, raptorial malacostracans in which the large, subchelate, second thoracic appendage is adapted for spearing or smashing prey. They live in burrows or in natural retreats and tend to be territorial.

4 The superorder Eucarida is distinguished by stalked, compound eyes and a carapace that is fused with all the thoracic tergites.

5 The small eucarid order Euphausiacea contains marine planktonic species. They are shrimplike in appearance, but none of the anterior thoracic appendages are modified as maxillipeds. Some of the filter-feeding species occur in enormous numbers in waters of high primary productivity and are important in marine food chains.

6 The eucarid order Decapoda contains almost one quarter of the species of crustaceans. The first three pairs of thoracic appendages are modified as maxillipeds, leaving five pairs of legs, hence the name of the order. The first pair of legs is commonly modified as a large claw, or cheliped. Most of the larger species of crustaceans are decapods.

7 Shrimps, lobsters, and crayfish have well-developed abdomens (long-bodied forms), but many decapods have the abdomen reduced and folded beneath the thorax. The short body form evolved independently a number of times, and all such species, called crabs, are not necessarily related.

8 The three groups of shrimps are adapted for swimming. The body is laterally compressed, and the pleopods, which are the swimming appendages, are larged and fringed. The legs are long and slender, chelipeds may or may not be present, and the exoskeleton is relatively flexible. Most shrimps are bottom dwellers and swim intermittently.

9 Most other decapod groups (crayfish, lobsters, and crabs) are benthic decapods and adapted for crawling. The legs are heavier than those of shrimps and the pleopods are never used in swimming. The pleopods are retained only for reproductive functions. The body is somewhat dorsoventrally flattened, and the exoskeleton is relatively rigid. Crayfish and lobsters (infraorders Astacidea and Palinura) have retained a well-developed abdomen.

10 The infraorder Brachyura, or "true" crabs, are the most successful of the decapods with short bodies, for this group includes more than 4500 species (one half of the decapods) and they are diverse and widely distributed. In brachyurans, folding of the abdomen beneath the thorax, which throws the center of gravity forward beneath the legs, appears to have enhanced motility, for these crabs move sideways.

11 The infraorder Anomura is characterized by reduction or dorsal position of the fifth pair of legs, but the modification appears to have evolved independently in the different anomuran groups.

12 Mole crabs (anomuran superfamily Hippoidea) are adapted for suspension feeding and burrowing backward in sand. Some inhabit surf beaches. Porcelain crabs (anomuran superfamily Galatheoidea) live beneath stones and move very rapidly. The short body form of this group appears to be an adaptation for motility.

13 The central group of the anomuran superfamily Paguroidea is the hermit crabs, which are adapted for housing the abdomen within gastropod shells. The ancestors of this group probably were decapods that backed into holes or other retreats, and the primitive modern forms have symmetrical abdomens. Various species have become secondarily adapted for utilizing other objects for housing, and the coconut crabs (*Birgus*) and the lithodid crabs (*Lithodidae*) have abandoned the hermit habit and evolved the short body form. However, their hermit ancestry is revealed in the unpaired pleopods of the female.

14 The greatest number of decapods couple predation with scavenging, but there are species that are filter feeders, detritus feeders, and herbivores. Enzymes produced by the midgut hepatopancreas pass forward into the large foregut cardiac stomach, where digestion begins. The cardiac stomach is adapted for trituration and the pyloric stomach for filtering material into the hepatopancreas, where absorption occurs.

15 The gills of decapods are dorsal evaginations of the body wall near the junction of the thoracic appendages and the trunk. They are protected by the overhanging carapace, which forms a gill chamber. The ventilation current is produced by the gill bailer, a process of the second maxilla. It generally pulls water through the gill chamber and therefore is at the point of exit of the ventilation current. The intake point is determined by how closely the carapace fits against the trunk.

16 The majority of decapods are marine osmoconformers, but numerous species have invaded fresh water and land. The gills are a major site for excretion of ammonia, however, and except for crayfish and a few shrimps, the urine of the antennal glands is isosmotic with the blood. For all freshwater decapods, the gills are an important site of ion absorption to balance loss through the urine.

17 All terrestrial decapods are burrowers and nocturnal. None can produce a hyperosmotic urine. Except for the Potamoidea, land crabs must return to the sea to breed, and development is indirect.

18 Most male decapods have the anterior pleopods adapted as copulatory organs. Most female decapods brood their eggs attached to the pleopods. A zoea larva is the typical hatching stage. Development in many freshwater species is direct.

SUPERORDER PERACARIDA

The superorder Peracarida* is a large assemblage of seven orders, which together with the decapods make up the majority of malacostracans, indeed the majority of crustaceans. Approximately 40 percent of crustaceans are peracaridans. Although most are less than 2 cm in length and therefore not as conspicuous as decapods, peracaridans, especially isopods and amphipods, are abundant and widespread in the sea, in fresh water and on land.

The distinctive characteristic of the group is the presence of a ventral brood pouch, or marsupium, in the female. The marsupium is formed by large, platelike processes (oostegites) on certain thoracic coxae. The oostegites project inward horizontally and overlap with one another to form the floor of the marsupium (Fig. 14–88). The thoracic sternites form the roof. The marsupial oostegites may make their appearance as small projections during juvenile instars, but their development is under hormonal control, and they are not completely formed until the reproductive instar. Development is direct, and on release from the brood chamber the young, called postlarvae, have most of the adult features.

A carapace may be present or absent. The naupliar eye never persists in the adult. Of the eight pairs of thoracic appendages, the first is usually a pair of maxillipeds and the remaining seven are legs.

Primitively, peracaridans are maxillary suspension feeders, as are many mysids, cumaceans, and tanaidaceans, but the tendency in most higher peracaridans has been toward other modes of feeding.

*Five orders are described here. Two other orders, the Thermosbaenacea, containing eight species (which are exceptional among the peracaridans in that they carry eggs in a dorsal brood pouch) and the Spelaeogriphacea, containing a single species, are omitted.

ORDER MYSIDACEA

Mysidaceans look much like little shrimps, and since they possess a ventral marsupium, they are sometimes called opossum shrimps. The majority are from 2 to 30 mm in length, but some, such as the bathypelagic *Gnathophausia*, may attain a length of 35 cm (Fig. 14–77A). Some species live in fresh water, but most are marine and are found at all depths.

The thorax is covered by a carapace, but unlike true shrimps, the carapace is not united with the last four thoracic segments. Anteriorly, the carapace often extends forward as a rostrum, below which project stalked, compound eyes. The first thoracic appendages and sometimes the second pair as well (in the Mysidae) are modified as maxillipeds. The remaining six or seven thoracic appendages are more or less similar, and the exopodites are filamentous and sometimes bear swimming setae. The pleopods are commonly reduced. Many mysidaceans (suborder Mysida) have a statocyst in the uropod endopite. Its visibility through the translucent exoskeleton is a convenient way to distinguish mysidaceans from other, similar planktonic crustaceans.

When the pleopods are reduced, as in *Mysis*, the thoracic exopodites are used for swimming. Benthic forms crawl over the bottom or plough through the surface sand or mud.

Thoracic gills are present in some species, but in the Mysidae, the inner surface of the carapace acts as a gill.

Some mysidaceans feed on small particles, which they apparently collect by grooming the body surface. Others, including the freshwater *Mysis relicta* and *Neomysis mercedes*, are predacious on other planktonic animals, especially copepods (Grossnickle, 1982). Many bathypelagic forms are scavengers.

About 780 mysidaceans have been described (see review by Mauchline, 1980). Marine species often live in large swarms and form an important part of the diet of such fish as shad and flounder, but many marine forms are found in algae and tidal grass. There are some 42 freshwater species, including those which live in ground water. *Mysis relicta* is confined to cold, deep-water lakes of the northern United States, Canada, and Europe (Fig. 14–77B). Lake trout feed extensively on this species.

The stalked, compound eyes, well-developed carapace, thoracic gills, long, tubular heart, and the

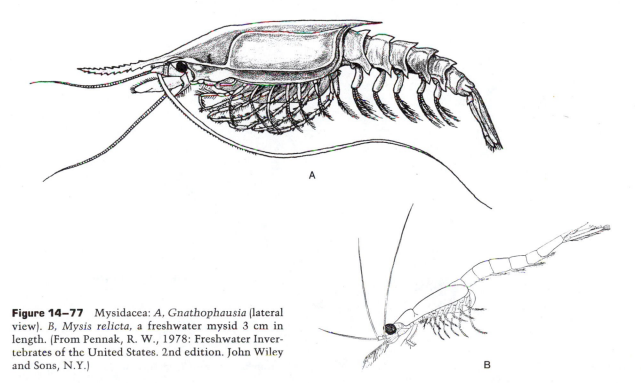

Figure 14–77 Mysidacea: *A, Gnathophausia* (lateral view). *B, Mysis relicta*, a freshwater mysid 3 cm in length. (From Pennak, R. W., 1978: Freshwater Invertebrates of the United States. 2nd edition. John Wiley and Sons, N.Y.)

A

B

presence of both antennal and maxillary glands are believed by many specialists to be primitive features and suggest that the mysidaceans stem from near the base of the peracaridan line.

ORDER CUMACEA

Cumaceans are marine peracaridans that live buried in sand and mud. The body is several centimeters long and distinctively shaped (Fig. 14–78). The head and thorax are greatly enlarged; the abdomen is very narrow and terminates in slender, elongated uropods. When eyes are present, they are located on a common median prominence situated above the base of the rostrum. The antennae are vestigial in the female; in the male they are extremely long and are borne folded back along the sides of the body, sometimes in a groove. The animal burrows backward, using the more posterior legs, and ends up in an inclined position with its head projecting above the surface (Fig. 14–78).

A series of filamentous gills is located on each first maxilliped, which also function as a ventilating pump. In *Diastylis* the inhalant ventilating current, while passing over the mouthparts, is filtered for food particles by setae on the second maxillae. Not all cumaceans are filter feeders. Some scrape organic matter from sand grains.

Swarming at the time of mating is characteristic of many cumaceans. Large numbers leave their burrows and swim to the surface. In the European cumacean *Diastylis rathkei*, swarming, which occurs at night, is associated with molting, and successive pelagic phases over the late summer involve successively older juveniles, with the first adults appearing in the fall (Valentin and Anger, 1977).

More than 800 species of cumaceans have been described. Most live at depths of less than 200 meters; where bottom conditions are favorable, some species may attain population densities of hundreds per square meters.

ORDER TANAIDACEA

The Tanaidacea display similarities to both the Cumacea and the Isopoda, with which they were formerly classified. These peracaridans, of which some 550 species have been described, are almost exclusively marine and generally small, most of them being only 2 to 7 mm in length (Fig. 14–79). The majority are bottom inhabitants of the littoral zone, where they live buried in mud, construct tubes, or live in small holes and crevices in rocks, but there a number of species that inhabit the deep sea. *Leptochelia dubia* is a widespread, tubicolous inhabitant of shallow water in many parts of the

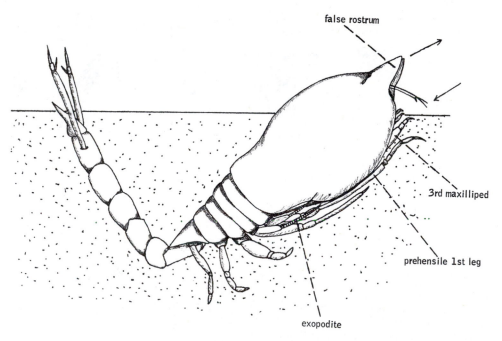

Figure 14–78 *Diastylis*, a cumacean, buried in sand. Arrows indicate direction of feeding-ventilating current. (Modified from several authors.)

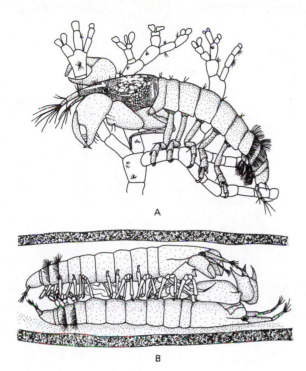

A

B

Figure 14–79 *Tanais cavolinii*, an intertidal tanaidacean from the Norwegian coast. This species lives within tubes constructed on calcareous algae. *A,* The male leaves its tube to find a female. (From Johnson, S. B., and Attramadal, Y. G., 1982: A functional morphological model of *Tanais cavolinii* adapted to a tubicolous life-strategy. Sarsia, 67:29–42.) *B,* Copulating pair of *Tanais cavolinii* within tube of female. The male genital cones penetrate slits in the ovisac to release sperm. (From Johnson, S. B., and Attramdal, Y. G., 1982: Reproductive behaviour and larval development of *Tanais cavolinii.* Mar. Biol., 71:11–16.)

world. In the soft bottoms of Tomales Bay, California, this tanaidacean reaches densities as great as 30,000 per square meter (Mendoza, 1982).

A small carapace covers the anterior part of the body and is fused to the first and second thoracic segments. The inner surface of the carapace functions as a gill. Many species lack eyes, but when eyes are present, they are located laterally on immovable processes. The first pair of thoracic appendages are maxillipeds, and the second pair (gnathopods) are large and chelate, a distinctive feature of the tanaidaceans. The third pair of thoracic appendages are adapted for burrowing. The inner surface of the carapace functions as a gill.

Some tanaidaceans are suspension feeders using the second maxillae or maxillipeds, but *Leptochelia dubia* and *Tanais cavolinii* collect diatoms,

algae, and other material from around the mouths of their burrows with their chelipeds (Johnson and Attramadal, 1982). There are also raptorial species, and deep-sea tanaidaceans are believed to feed on organic detritus.

Some tanaidaceans are hermaphroditic, but the eggs are brooded as in other peracaridans.

ORDER ISOPODA

The order Isopoda is one of the largest orders of crustaceans. Most of the 10,000 described species live in the sea, where they are widely distributed and occupy all types of habitats. One group (suborder Paraselloidea) forms one of the most abundant components of the deep-sea benthic fauna. There are a considerable number of freshwater isopods, and the pill bugs, or wood lice, are the largest group of truly terrestrial crustaceans. The order also includes many parasitic forms.

The most striking characteristic of isopods is the dorsoventrally flattened body (Figs. 14–80 and 14–81B). The head is usually shield shaped, and the terga of the thoracic and the abdominal segments tend to project laterally. A carapace is absent, although the first one or two thoracic segments are fused with the head. The abdominal segments may be distinct or fused to varying degrees. The last abdominal segment is almost always fused with the telson; in the Asellota, which includes the greatest number of freshwater species, all but the first or second abdominal segments are fused with the telson to form a large abdominal plate (Fig. 14–80A). The abdomen is usually the same width as the thorax, so that the two regions may not be clearly demarcated dorsally.

The first antennae are short and uniramous, and in terrestrial isopods they are vestigial. The compound eyes are always sessile.

The first pair of thoracic appendages are modified to form maxillipeds; the remaining seven pairs of thoracic appendages are legs usually adapted for crawling. In some groups the more anterior pairs are modified as prehensile gnathopods (Fig. 14–80A). Unlike those in most other crustaceans, some of the isopod pleopods are used for gas exchange.

Most isopods are 5 to 15 mm in length. The giant of the group is the deep-sea *Bathynomus giganteus*, which reaches a length of 42 cm and a width of 15 cm. The coloration in isopods is usually drab, shades of gray being the most common. Chromatophores adapt the body coloration to the background in many species.

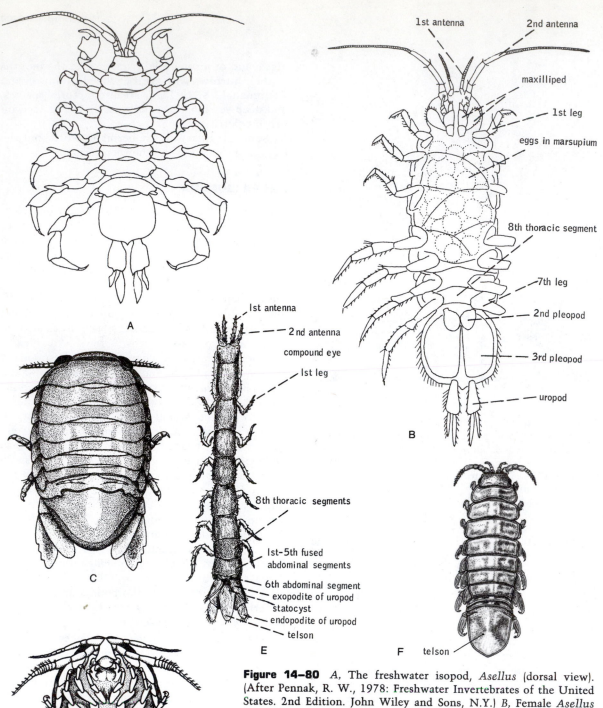

Figure 14–80 *A*, The freshwater isopod, *Asellus* (dorsal view). (After Pennak, R. W., 1978: Freshwater Invertebrates of the United States. 2nd Edition. John Wiley and Sons, N.Y.) *B*, Female *Asellus* (ventral view). Appendages complete only on one side. (After Van Name, W. G., 1936: The American land and freshwater isopod Crustacea. Bull. Amer. Mus. Nat. Hist., 71:7.) *C* and *D*, *Sphaeroma quadridentatum*, a common, shallow-water marine isopod found on algae and pilings. All but one abdominal tergite are fused with the telson. The members of this genus are capable of rolling into a ball, as are many terrestrial isopods. *C*, Dorsal view. *D*, Ventral view. *E*, *Cyathura*, a genus of marine isopods adapted for living in burrows in mud and sand bottoms. The many other species of this large family have a similar habit. (After Gruner from Kaestner.) *F*, *Idotea pelagica* (dorsal view). This and other members of the large marine family Idoteidae are common inhabitants of algae in shallow water. (After Sars.)

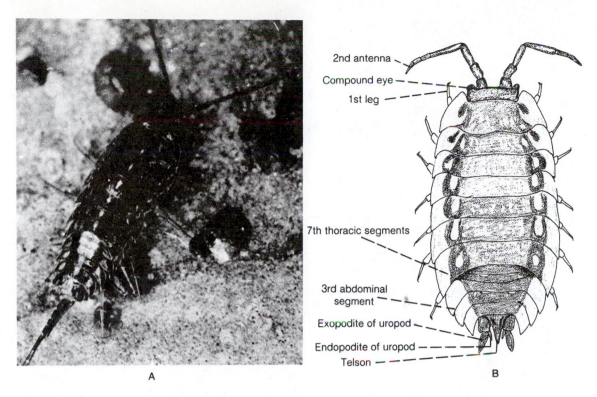

2nd antenna

Compound eye

1st leg

7th thoracic segments

3rd abdominal segment

Exopodite of uropod

Endopodite of uropod

Telson

A

B

Figure 14–81 Terrestrial oniscoidean isopods: *A*, *Ligia* on a sea wall at low tide. The barnacles are little less than 1 cm in diameter. The members of this widely distributed genus are fast-moving species that live at the edge of the sea just above the water's edge on rocks, pilings, and sea walls. (By Betty M. Barnes.) *B*, *Oniscus asellus*, a European pill bug common in gardens, hot houses, and human habitations. It has been introduced into North America. (After Paulmier from Van Name.)

Locomotion

Isopods are benthic animals, and most are adapted for crawling. *Ligia*, one of the most common isopods along coastlines, can run very rapidly over exposed wharf pilings and rocks, even upside down. Many aquatic isopods burrow, and some construct tunnels through the substratum, packing excavated material against the walls. Some species of *Limnoria* tunnel through wood and can cause extensive damage to docks and pilings (Fig. 14–82); others bore into the holdfasts of kelps. *Sphaeroma tenebrans* bores into the prop roots of mangroves.

Aquatic isopods can usually swim as well as crawl. Most commonly the pleopods are used for swimming, and in the families Sphaeromatidae and Serolidae the first three pairs of pleopods are especially adapted for swimming; gas exchange is restricted to the more posterior pleopods.

The ability to roll up in a ball has evolved in many terrestrial Oniscoidea. Many marine sphaeromids also roll up, with the sharp and spiny tips of the uropods and telson exposed.

Feeding

Most isopods are scavengers and omnivores, although some tend toward a herbivorous diet. Deposit feeding is common. Wood lice feed on algae, fungi, moss, bark, and any decaying vegetable or animal matter. A few wood lice are carnivorous, as are some marine species, such as the intertidal *Cirolana* and the large, pelagic *Bathynomus*.

The maxillipeds form a protective operculum over the mouthparts. During feeding, food is usually held up by the anterior legs, which may be subchelate.

Wood-boring marine isopods feed on wood, and their hepatopancreatic secretions include cellulase. At settling, the wood-boring species of *Limnoria* are attracted to fungi in the wood (Geyer and Becker, 1980). The fungi add nitrogen to their largely cellulose diet. In wood lice cellulose digestion results from bacteria, and the hindgut plays a major role in the digestive process (Hassall and Jennings, 1975).

There are several groups of parasitic isopods.

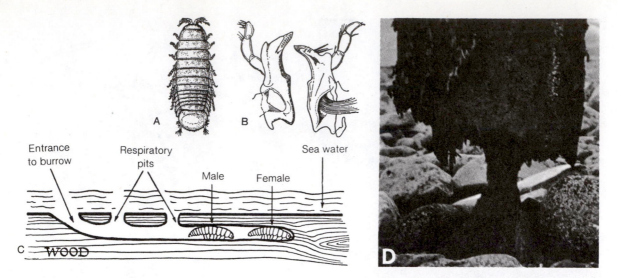

Figure 14–82   *A*, The wood-boring isopod *Limnoria lignorum*. *B*, Mandibles of *Limnoria lignorum*. *C*, Diagrammatic section of burrow of *Limnoria lignorum*. *D*, Jetty piling nearly eaten through at base by *Limnoria lignorum*. (*A* after Sars from Yonge, *B* after Hoek from Yonge, *C* from Yonge, C. M., 1949: The Seashore. Collins, London.)

The Gnathiidae in the larval stages and the adult Cymothoidae are ectoparasitic on the skin of fish and have mandibles adapted for piercing (Fig. 14–83*C*). Piercing mouth parts are also present in the parasites composing the suborder Epicaridea, all of which are bloodsuckers, parasitic on many crustacean groups. Many epicaridans are highly modified and show little resemblance to free-living forms (Fig. 14–83*D* to *I*).

Gas Exchange and Excretion

The pleopods of isopods provide for gas exchange. In primitive forms, each pleopod ramus is modified as a large, flat lamella, and both rami of each pleopod function in both gas exchange and swimming (Fig. 14–80*B*). However, there is usually some modification of this arrangement. In some isopods gas exchange and swimming are divided among the pleopods, the anterior ones being fringed and coupled for swimming and the posterior pleopods being for gas exchange.

The pleopods typically lie flat against the underside of the abdomen and are often protected by a covering (the operculum) formed by the first pair of pleopods or by the exopodites of one or more pairs. In the marine Valvifera the uropods are greatly elongated and meet at the midline ventrally to form a gill covering resembling two doors. If gas exchange through the gill is blocked in the terrestrial *Ligia* and *Oniscus*, oxygen consumption is reduced by only about 50 percent, indicating that the general integumentary surface is equally important as a site of gas exchange. The blood of isopods contains hemocyanin.

The excretory organs are maxillary glands.

Adaptations for Life on Land

The terrestrial isopods—the wood lice or pill bugs—are members of the suborder Oniscoidea. They are believed to have invaded land directly from the sea, rather than by way of fresh water, and have come to occupy a wide range of habitats and to display varying degrees of toleration to desiccating conditions (Fig. 14–84). Some live at the edge of the sea and some in marshes. Many species live beneath stones, in bark, and in leaf mold in both temperate and tropical regions, but there are also species that live in deserts.

Shore-inhabiting forms include the widespread *Ligia*, which lives on pilings, jetties, and rocks at the water's edge, and *Tylos*, which lives beneath beach drift or sand at the high tide mark. *Tylos* emerges at low tide to feed on algae and other debris and orients to the high tide mark by means of beach slope, sand moisture, or horizon height. Some species can also use the sun's angle for orientation (see Herrnkind, 1972, and Farr, 1978, for *Lygia exotica*).

Most terrestrial isopods possess some adaptations to reduce water loss, but the group as a whole

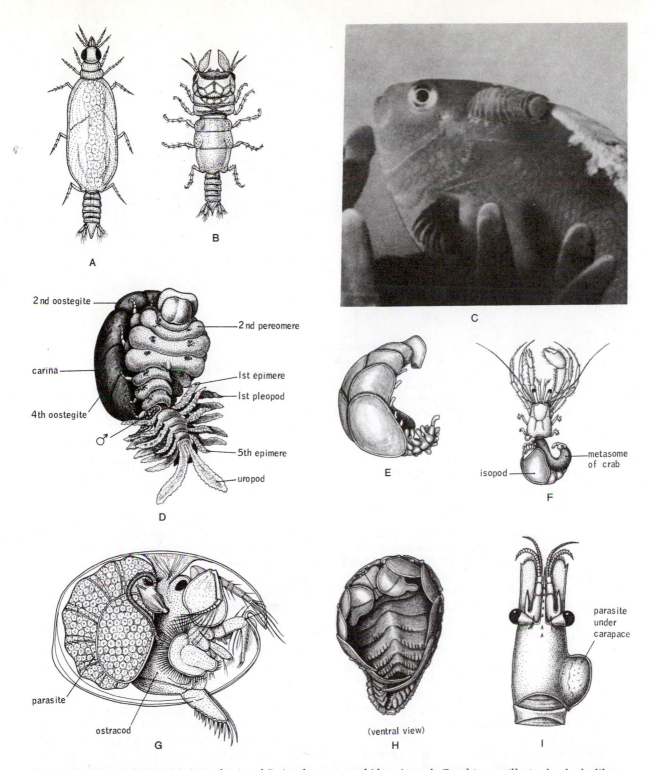

2nd oostegite

2nd pereomere

carina

1st epimere

1st pleopod

4th oostegite

♂

5th epimere

uropod

D

E

isopod

metasome of crab

F

parasite

ostracod

G

(ventral view)

H

parasite under carapace

I

Figure 14-83 *A-C,* Parasitic isopods: *A* and *B,* An aberrant gnathidean isopod, *Gnathia maxillaris,* that looks like an insect. The larval stage (*B*) is parasitic on fish and possesses crowded sucking mouthparts. Both larva and adult (*C*) are less than 3 mm long. (Based on living specimens and figures by Sars.) *C,* An isopod fish louse, *Rocinela,* on a fish living with a sea anemone. (By Mariscal, R. N., 1969: Crustaceana, 14:7–104.) *D-I,* Epicaridean parasitic isopods: *D,* The bopyrid *Cancricepon elegans,* parasitic in the gill chambers of certain crabs. *E* and *F, Athelges tenuicaudis* on abdomen of hermit crab. *G,* Isopod *Cyproniscus* in ostracod *Cypridina. H* and *I,* Ventral view of *Bopyrus squillarum* and location in branchial cavity of shrimp. (*D-I* after Sars.)

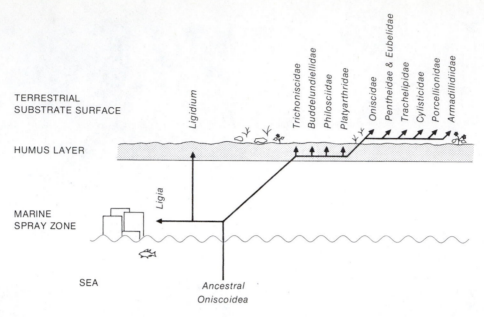

Figure 14–84 Stages in the isopod invasion of land. Adaptations to reduce water loss, such as pseudotracheae and rolling into a ball, are best developed in the surface-dwelling families. (Adapted from Schmalfuss, H., 1975: Morphologie, Funktion und Evolution der Tergithöcker bei Landisopoden. Z. Morph. Tiere, *80*:287–316.)

is considerably less well adapted in this regard than other terrestrial arthropods. Wood lice tend to be nocturnal and live beneath stones and in other places where the environment is humid. They have never evolved a waxy epicuticle of the type responsible for reducing integumental evaporation in insects and spiders. The thin, ventral exoskeleton is a primary site of evaporation, and the ability to roll up into a ball is probably, for some species, an adaptation to cut down on water loss. In general, wood lice are photonegative and strongly thigmotactic and can discriminate between relatively slight differences in humidity, all of which tends to keep them beneath protective retreats during the day. The commonly observed aggregation of wood lice under certain stones or wood may result in part from their being attracted by the body odor of other individuals of their own species and from a common response to the same environmental conditions.

The eyes of wood lice are poorly developed, probably because of their nocturnal, secretive behavior and their diet of decaying vegetation. Repugnatorial glands are used in defense against such predators as spiders and ants. Tubercles and tergal plates, which in some species may be large and even spinelike, serve in protection, especially in forms that roll up into a ball (Fig. 14–85). In addi-

tion to protection, the tergal tubercles may function in digging, strengthening of the tergal plates, and reducing evaporative water loss (Schmalfuss, 1975).

Wood lice continue to use the gills for gas exchange, but in the more terrestrial species the operculum contains a lunglike cavity (in the Oniscidae) or tubelike invaginations, or pseudotracheae (as in the Porcellionidae and Armadillidiidae) (Fig. 14–86B). Wood lice with pseudotracheae can tolerate much drier air.

The replacement of water lost from integumental evaporation usually comes from moist food and drinking, but some desert species replace lost water by ingesting moist burrow sand and by cutaneous absorption of water from moist burrow air (Coenen-Strasse, 1981). The gills must retain a covering film of moisture. The fact that they lie within a depression of the covering exopodite facilitates water retention, but water must be replaced periodically (Fig. 14–86A). In some wood lice the two endopodites of the uropods are held together like a tube and dipped into a droplet of dew or rain. Water taken up is distributed to the gills. Other wood lice possess a system of surface channels that carry any water that comes in contact with the animal's back to the ventral surface and then back to the gills.

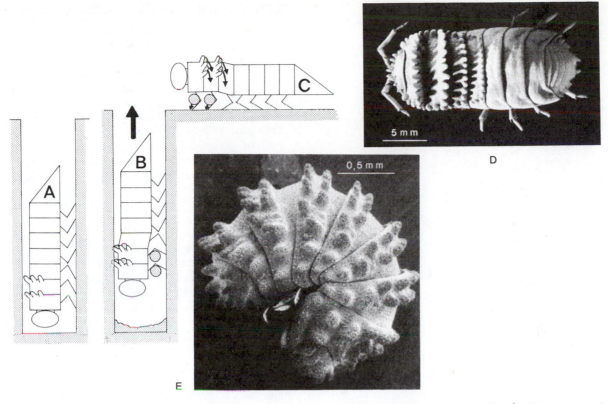

Figure 14–85 Adaptations of tergal tubercles in terrestrial isopods: *A–C*, A desert species, *Hemilepistus reaumuri*, that uses the anterior tergal tubercles for digging. The soil is transported with the anterior legs. *D, Hemilepistus aphganicus*, a related species in which the anterior tergal tubercles are used for both digging and transporting soil. *E, Armadillo tuberculatus* with large projecting protective tubercles. (From Schmalfuss, H., 1975: Morphologie, Funktion und Evolution der Tergithöcker bei Landisopoden. Z. Morph. Tiere, *80*:287–316.)

The maxillary glands of wood lice are poorly developed, and these isopods release nitrogenous wastes as gaseous ammonia.

Reproduction and Development

The gonads are paired and separate. The male sperm ducts open onto the sternum of the genital segment by way of either separate or united papillae. The first pleopod of the male bends the papilla back to the endopodites of the second pleopods, and the cavity in the second pleopods is filled with sperm. The female gonopores are also median, paired, sternal openings, and each oviduct is enlarged basally, forming a seminal receptacle. During copulation the male presses the ventral side against one side of the female and injects sperm into one of her gonopores with the second, copulatory pleopod. The male then moves to the other side of the female's body, where the process is repeated. The eggs are fertilized in the oviduct. In many species copulation occurs during or just after the female's molt, and there may be a long precopulatory attendance by the male. In the freshwater asellids mating can occur only during the short period between the molting of the posterior half and the anterior half of the female's body.

In most of the terrestrial oniscoids (wood lice) copulation occurs during the intermolt, and the blind, saclike seminal receptacles do not make connection with the oviducts until the next molt takes place. The first two pairs of pleopods are copulatory organs in most terrestrial isopods.

The eggs are usually brooded in the marsupium (Fig. 14–80B). The marsupium of a wood louse is kept filled with fluid, so development of the young is essentially aquatic despite the terrestrial habit of the adults.

Few to several hundred eggs are usually brooded, and the hatching stage is a postlarva

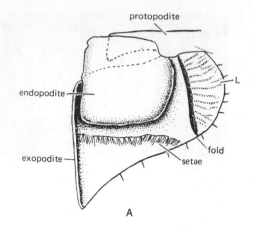

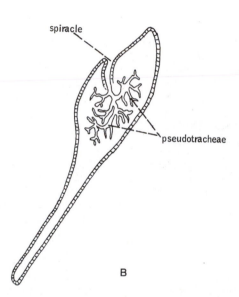

Figure 14–86   *A,* Fifth pleopod of the wood louse *Oniscus asellus,* showing depression in expodite in which respiratory endopodite lies. Setae keep out foreign material in water entering depression. *B,* Section through pleopod exopodite of the wood louse *Porcellio scaber,* showing spiracle and pseudotracheae. (Both after Unwin from Kaestner, A., 1970: Invertebrate Zoology. Vol. 3, Wiley-Interscience, N.Y.)

(manca stage), with the last pair of legs incompletely developed. The young usually do not remain with the female after they leave the marsupium, but in *Arcturus,* the female carries the young about attached to her long antennae. Most isopods in temperate regions—marine, freshwater, or terrestrial—produce one to two broods each summer and live for two to three years.

SYSTEMATIC RÉSUMÉ OF ORDER ISOPODA

Suborder Gnathiidea. Thoracic segments 1 and 7 reduced, so that only five large segments are visible dorsally. Eighth thoracic appendages are absent. Abdomen small and much narrower than thorax. Manca stage is ectoparasitic on marine fish; adult nonfeeding. *Gnathia.*

Suborder Anthuridea. Body long and cylindrical. First pair of legs are heavy and subchelate; first pair of pleopods form an operculum covering other pairs. Most are marine burrowers except for a few freshwater species. *Anthura, Cyathura, Cruregens.*

Suborder Microcerberidea. A small group of minute, blind, interstitial and freshwater cave species.

Suborder Flabellifera. Body is more or less flattened; some abdominal segments may be fused together; the last segment is fused with the telson. The uropods are fan shaped and form a tail fan together with the telson. This suborder contains the most common shallow-water marine species. A few are found in fresh water. The wood borer, *Limnoria; Cirolana; Bathynomus; Serolis; Sphaeroma;* the ectoparasites of fish, the Cymothoidae.

Suborder Valvifera. Abdominal segments 3 to 6 always fused with telson, and in some species segments 2 and 1 as well. Uropods form an operculum over gills. Mostly marine, and many species are common in seaweed. *Astacilla, Arcturus, Idotea.*

Suborder Asellota. Abdominal segments 3 to 6 always fused with telson, and in some species segments 2 and 1 as well. Uropods are commonly styliform. Pleopods 3 to 5 form gills and covered by anterior two opercular pleopods. A large suborder that includes many common marine species as well as some freshwater forms. *Asellus, Lirceus, Munnopsis, Jaera.*

Suborder Phreatoicoidea. Abdominal segments are not fused with each other. Elongate body is laterally compressed. Uropods are styliform. Includes certain freshwater isopods of Australia, New Zealand, and South Africa. *Phreatoicus.*

Suborder Epicaridea. Parasites on crustaceans. Females are often greatly modified, some without segmentation or appendages. *Bopyrus, Entoniscus, Portunion, Liriopsis.*

Suborder Oniscoidea. Five abdominal segments are usually distinct dorsally. First antennae are vestigial. Amphibious and terrestrial members (wood lice). *Ligia, Oniscus, Porcellio, Armadillidium, Tylos.*

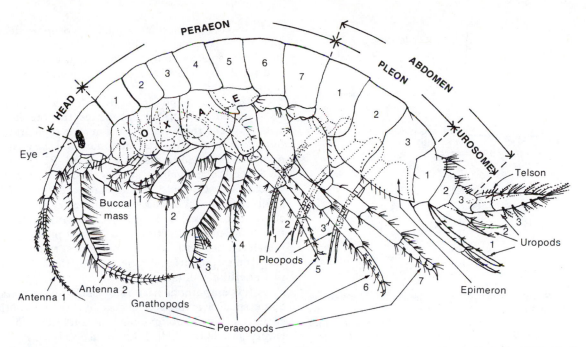

Figure 14–87 External structure of a gammaridean amphipod in lateral view. (From Bousfield, E. L., 1973: Shallow-Water Gammaridean Amphipoda of New England. Cornell University Press, Ithaca, N.Y. 344 pp.)

ORDER AMPHIPODA

The more than 5500 species of amphipods constitute the second of the two major groups of peracaridans. The order contains a great diversity of species, which are placed within over 100 families. Most are marine, but there are many freshwater species and a family of terrestrial forms. The structure of amphipods displays some convergence with that of isopods (Fig. 14–87). The eyes are sessile. There is no carapace, although the first thoracic segment and sometimes the second are fused with the head. Also, the abdomen is usually not distinctly demarcated from the thorax in either size or shape. In contrast to isopods, however, the amphipod body tends to be laterally compressed, giving the animal a somewhat shrimplike appearance (Fig. 14–87).

Sessile compound eyes are typical in most amphipods. The first and second antennae are usually well developed, and the first pair of thoracic appendages are modified to form maxillipeds. The coxae of the seven pairs of legs are usually long, flattened plates that increase the appearance of lateral body compression (Fig. 14–88). The second and third thoracic appendages (pereopods, or legs) are usually enlarged and subchelate for prehension and are called gnathopods. The anterior three pairs

are pleopods and are used in swimming and for ventilating. The posterior three pairs are uropods and are directed backward.

Exceptions to all of these typical amphipod features can be found, and some groups have diverged widely from the general plan.

Amphipods are about the same size as isopods. The giant of the order was thought to be the marine *Alicella gigantea*, which may reach 14 cm in length, but in 1968, an undescribed 28-cm benthic lysianassid amphipod was photographed from 5300 meters in the Pacific. The smallest forms, mostly blind interstitial types, are less than 1 mm long. Most amphipods are translucent, brown, or gray, but some species are red, green, or blue-green.

Locomotion and Habitation

Hyperiidean amphipods are pelagic, and for at least part of their life cycle probably all live commensally with such gelatinous planktonic animals as salps, siphonophores, jellyfish, and ctenophores (Harbison et al, 1977; review by Laval, 1980). Members of the family Phronimidae live in "barrels" constructed from salps and other zooplankton (Fig. 14–89D). The strange swollen shape of the head and thorax of some species (*Hyperia, Mimonectes,* and *Rhabdosoma*) may be related to

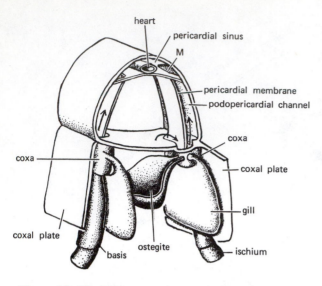

Figure 14–88 Diagrammatic cross section through the body of an amphipod. Only one marsupial plate, or oostegite, is shown. (After Klövekorn and others, from Kaestner, A., 1970: Invertebrate Zoology. Vol. 3. Wiley-Interscience, N.Y.)

their commensal habits or contribute to flotation. Many have very large eyes (Fig. 14–89).

Gammarids include many of the common amphipods. They are essentially bottom dwellers, but most can swim, even if infrequently. Propulsion for swimming is provided by the pleopods and in many groups by the uropods. Usually, they swim intermittently between crawling and burrowing; in leaving the substratum, initial thrust is commonly gained by a backward flip of the abdomen. Walking is effected by the legs, but in rapid movements over the bottom, both legs and pleopods are used and the animal often leans far over to one side.

Some amphipods, especially the caprellids, are adapted for climbing. Caprellids, called skeleton shrimps, have long slender bodies with greatly reduced abdomens (Fig. 14–90B). The tips of the legs are provided with grasping claws for clinging to hydroids, bryozoans, and algae over which many species crawl like inchworms (Caine, 1978). Some caprellids live on sea stars or even on spider crabs (Vader, 1978).

Many amphipods are accomplished burrowers, and some construct tubes of mud or of secreted material (Fig. 14–91). The burrows may be horizontal, vertical, or U shaped with two openings. Sometimes the walls are simply packed, but often they are reinforced with secreted material.

Tube construction is peculiar to corophioidean and ampeliscoidean amphipods, and a variety of materials, such as mud, clay, sand grains, and shell and plant fragments may be used. The material is usually bound together with a cementing secretion produced from glands in the fourth and fifth thoracic appendages. *Haploops* mixes the secretion with clay, and the resulting stiff mass is then drawn out between the end of the abdomen and the gnathopods and applied to the walls (Fig. 14–91A).

Several tube dwellers, such as species of *Siphonoecetes* and *Cerapus*, build unattached tubes of shell fragments and sand grains and carry the tubes about with them (Fig. 14–91C). *Cyrtophium* makes its tubes out of a section of a hollow plant stem, which it lines with secreted materials; the animal can even swim with the tube, beating the antennae in an oarlike fashion.

A few amphipods have rather unusual retreats. Some species of *Siphonoecetes* live in old tooth shells, and *Photis conchicola* builds its tubes in an empty gastropod shell. Like the isopod *Limnoria*, the amphipod *Chelura terebrans* bores in wood. The Stenothoidae are commensals of sponges and ascidians, and *Dulichia rhabdoplastis* lives on sea urchin spines (Fig. 14–92E).

Members of the suborder Ingolfiellidea, some less than 1 mm long, are adapted for an interstitial existence in both marine and freshwater habitats (Fig. 14–92F).

Freshwater amphipods, mostly gammarids, are common benthic animals in algae and other vegetation of streams, ponds, and lakes and in subterranean waters and are sometimes found in great numbers. Pennak (1978) reported population densities for *Gammarus lacustris* of 10,000 per square meter in certain springs.

Lake Baikal in Siberia contains a remarkable endemic gammarid fauna of nearly 300 species, some brightly colored red and blue and some reaching a size of over 6 cm. These are both pelagic and benthic species.

In contrast to isopods, amphipods have been less successful in adapting to life on land. The terrestrial species are members of the family Talitridae. The beach fleas live beneath drift or stones or burrow in sand near the high tide mark (Fig. 14–93), but the leaf litter hoppers of the Southern Hemisphere are found in moist humus and soil away from the shore.

Beach fleas scull rapidly over the sand, gaining additional power with pushing strokes from the abdomen. They can jump, using a sudden back-

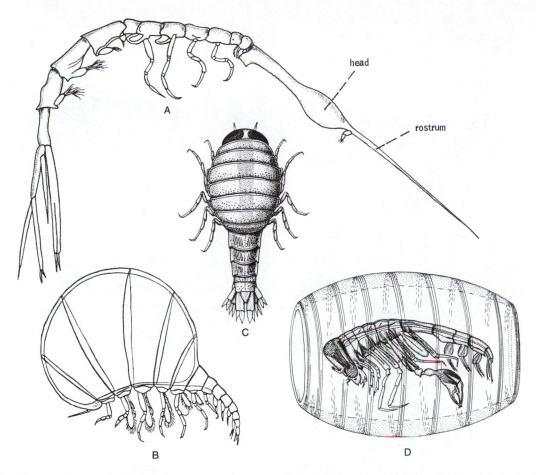

head

rostrum

A

B

C

D

Figure 14–89 Pelagic hyperiidean amphipods: *A, Rhabdosoma,* an amphipod with elongated head and needle-like rostrum. (After Stebbing from Calman.) *B, Mimonectes,* 2.5 cm long. (After Schellenberg from Kaestner, A. 1970: Invertebrate Zoology.Vol. 3. Wiley-Interscience N.Y.) *C, Hyperia,* 2 cm long. Species of this genus are commonly found attached to pelagic coelenterates and ctenophores. (After Sars.) *D, Phronima sedentaria,* 3 cm long, which lives within salps. (After Claus.)

ward extension of the abdomen and telson. *Talorchestia,* 2 cm in length, can leap forward 1 meter. Most beach fleas also burrow. In *Talorchestia* the body is braced with the second and third pairs of legs, while the first pair of gnathopods sweep the material back to the uropods and telson, which flip it away.

Like the isopod *Tylos,* the beach flea *Talitrus* has been shown to use the eyes to obtain astronomical clues for locating the high tide zone, in which these amphipods normally live. If displaced either above or below the high tide mark, the animals migrate accurately back to their normal zone. The angle of the sun is used as a compass in conjunction with a map sense of the east-west orientation of the particular beach they inhabit.

The angle of the sun is a primary clue for orientation, since the direction of movement of the animal can be controlled in experiments that reflect the rays of the sun from different angles. An internal clock mechanism provides interpolation for the changing angle of the sun during the course of the day. This aspect of the mechanisms was proved by transporting *Talitrus* in the dark to a beach at different longitude. On liberation, the immigrants operated on the same time as that of the original location. In the absence of direct sunlight, these amphipods are reported to use sky polarization in the same manner.

Other factors beside the celestial clues are utilized in orientation—horizon level, beach slope, sand moisture, and grain size—but visual clues are

Figure 14–90 Skeleton shrimps (Caprellidea). *A*, Diagram of a typical caprellidean skeleton shrimp. Abdomen is stippled; marsupial plates and bat-shaped gills in center of thorax. (From Laubitz, D. R., 1970: Studies on the Caprellidea of the American North Pacific. Nat. Mus. Can. Publ. Biol. Oceanogr., *1*:1–89.) *B*, *Caprella equilibra* clinging to seaweed. (By D. P. Wilson.)

most important (see Herrnkind, 1972; Hardwick, 1976).

Feeding

Most amphipods are detritus feeders or scavengers. Mud or animal and plant remains are picked up with the gnathopods, or detritus is raked from the bottom with the antennae, particularly the second pair. Some burrowing forms scrape detritus and diatoms from sand grains. Daily consumption of

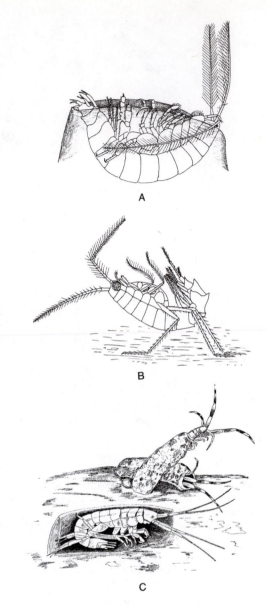

Figure 14–91 *A*, *Haploops tubicola*, hanging on edge of tube in feeding position. Margin of tube is shaded but shown as transparent. *B*, *Melphidippella macra* in feeding stance. (*A* and *B* after Enequist.) *C*, *Cerapus*, a tubicolous amphipod. The tube, which the animal can carry about, is composed of an inner, secreted layer covered with fragments of foreign material. (Modified after Schmitt.)

detritus in gammarids may be as great as 100 per cent of the body weight for juveniles and 60 per cent for adults (Bek, 1972).

A number of amphipods are suspension feeders, but different appendages are adapted as collectors.

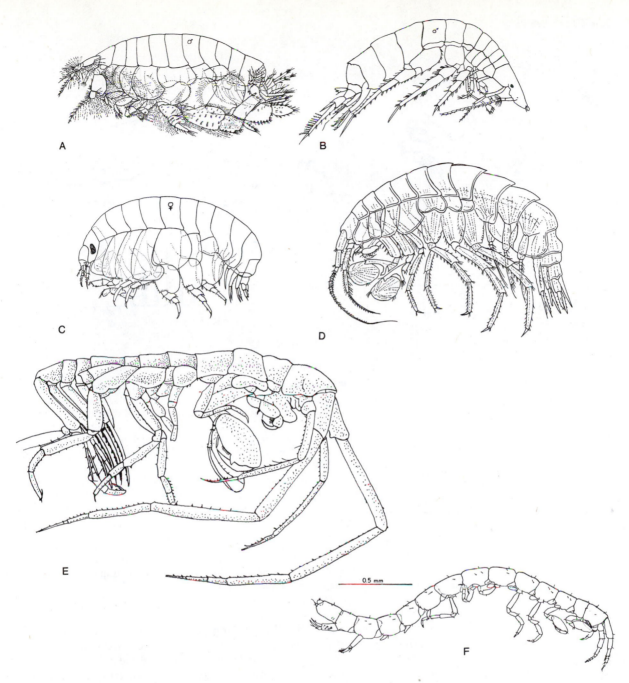

Figure 14–92 Diversity in amphipods: *A*, Burrowing *Haustorius canadensis* (Haustoriidae), with powerful digging legs and antennae and with filter-feeding mouthparts. *B*, Burrowing *Platyischnopus herdmani* (Platyischnopidae), with slender body and "shark snout." *C*, *Orchomenella minuta* (Lysianassidae), a flesh-scavenger, *D*, *Eusirus cuspidatus* (Eusiridae), a predatory and free-swimming species. (*B* from Bernard, J. L., 1969: The families and genera of marine gammaridean Amphiphoda. U.S. Nat. Mus. Bull., 27:1–535; *D* after Sars; others from Bousfield, E. L., 1973: Shallow-Water Gammaridean Amphipoda of New England. Cornell University Press, Ithaca, N.Y.) *E*, *Dulichia rhabdoplastis*, commensal on sea urchins along the north Pacific coast of the United States. The amphipod lives on detritus strands stretched between the sea urchin spine tips and constructed from its own fecal pellets. The amphipod feeds on diatoms growing on the strands or filters detritus and plankton from the surrounding water with outstretched antennae. (From McCloskey, L. R., 1970: A new species of *Dulichia* commensal with a sea urchin. Pacific Science, *24*:90–98. By permission of the University of Hawaii Press.) *F*, *Ingolfiella putealis* (Ingofiellidea), an interstitial amphipod, with vermiform body and no eyes. (From Stock, J. H., 1976: A new member of the crustacean suborder Ingolfiellidea from Bonaire, with a review of the entire suborder. Vitg. Naturwet. Studiekring Suriname Ned. Antillen, 86:57–75.)

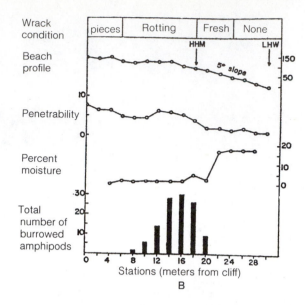

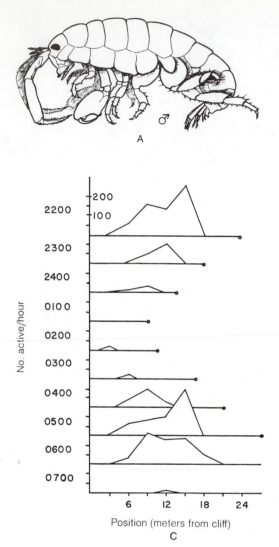

Figure 14–93 Distribution of the beach flea *Orchestoidea corniculata* on a California beach. Distance across beach is measured from base of a cliff toward water. *A*, Lateral view of male. *B*, Distribution of population burrowed in sand and environmental parameters of beach. HHM and LHW refer to highest high water mark and lowest high water. *C*, Distribution of emergent adults at night (caught in pitfall trap). Dot at end of each horizontal line indicates location of mean water level (tide level), with waves washing beach 2 to 4 meters above that point. (*A* from Smith, R. I., and Carlton, J. T. (Eds.), 1975: Light's Manual: Intertidal Invertebrates of the Central California Coast. 3rd Edition. University of California Press, Berkeley, Cal. p. 356; *B* from Craig, P. C., 1973: Behaviour and distribution of the sand-beach amphipod *Orchestoidea corniculata*. Mar. Biol., 23:101–109; *C* from Craig, P. C., 1973: Orientation of the sand-beach amphipod *Orchestoidea corniculata*. Anim. Behav., 21:699–706.)

Aora, Corophium, and other tube builders strain fine detritus through filter setae on the gnathopods, the feeding current being provided by the pleopods. Many gammarids, including the tube-dwelling *Haploops,* extend the antennae into the natural water current as a net. *Haploops* hangs upside down in the mouth of the tube, clinging to the rim with its specialized legs (Fig. 14–91*A*). *Melphidippella,* which sits upside down on the bottom, uses not only the second antennae but also the third and fourth pairs of thoracic appendages as a filter. All three limbs are held outstretched and catch falling detritus (Fig. 14–91*B*). At intervals the limbs are scraped by the gnathopods. The beach burrower *Haustorius arenarius* uses a maxillary filter.

Although many amphipods supplement their diet by catching small animals, strictly predacious feeding is not common. The most notable examples of predacious feeders are the pelagic hyperiids, two families of gammarid (eusirids and pardaliscids), and some caprellids (skeleton shrimps). A raptorial caprellid attaches itself to a hydroid or bryozoan stem with the last pair of thoracic appendages and projects itself motionless and outstretched, waiting to seize passing prey, such as copepods, with the gnathopods. However, many caprellids feed on diatoms and detritus that collects on the surfaces of the organisms on which they live, or filter feed with the antennae (Caine, 1974).

Parasitism is much less prevalent among amphipods than among isopods. There are a few ectoparasites of fish, with suctorial mouth parts. The

Figure 14–94 *Cyamus boopis*, an amphipod commensal on whales. (After Sars.)

cyamids, called whale lice, have legs adapted for clinging to the host, but they probably feed on diatoms and debris that accumulates on the whale's skin (Fig. 14–94).

Gas Exchange, Internal Transport, and Excretion

The gills, which are thoracic in contrast to those of isopods, are usually lamellae or vesicles attached to the inner face of the coxae of the legs (pereopods 2 to 7). Typically, the posteriorly flowing ventilating current is produced by the pleopods. A ventilating current is particularly important in those amphipods that dwell in burrows or tubes. Oxygen is transported in the blood by hemocyanin.

Although they live out of water, the Talitridae have retained the gills. The air must be humid, and talitrids are thus restricted to living in moist sand beneath drift or in damp forest leaf litter away from the sea. They feed at night, when there is less danger of desiccation.

The amphipod heart is tubular, with one to three pairs of ostia, and lies in the thorax above the coxal gills. The arterial system is not as greatly developed as in the isopods. The excretory organs are antennal glands.

Reproduction

The gonads are paired and tubular. The male gonopores open at the end of a pair of long penis papillae on the sternum of the last thoracic segment, and the female oviducts open on the sixth thoracic

coxae. In most species, pleopods do not function as copulatory organs. By means of sensory aesthetascs on the first antennae, males are attracted by female pheromones. In species of *Gammarus* and some other genera, the male carries the female around beneath him for days, clasping the thoracic region on the coxal plates of the female with his gnathopods. The animals seperate briefly to permit the final, preadult molt of the female. Actual sperm transfer is accomplished quickly. The male twists his abdomen around so that his uropods touch the female marsupium; when sperm is emitted, they are swept into the marsupium by the ventilating current of the female. The pair separate, and the eggs are soon released into the brood chamber, where fertilization takes place. Development is direct. The ventilating current also provides for the ventilation of the eggs in the marsupium. The marsupia of most gammarideans bear interlocking marginal setae, which aid in preventing the eggs from falling out. One annual brood of 15 to 50 eggs is common in most temperate freshwater species. In marine forms there may be 2 to 750 eggs in a clutch, and more than one brood per year is common, but the life span is generally about a year. For example, the intertidal estuarine *Gammarus palustris* along the east coast of the United States overwinter as adults; the first eggs appear in February (Maryland). The overwintering population dies out during the spring and early summer, and the new generation produces broods during late summer and early fall (Rees, 1975).

SYSTEMATIC RÉSUMÉ OF ORDER AMPHIPODA

Suborder Gammaridea. Head not fused with second thoracic segment. Maxilliped with palp. Thoracic legs with well-developed coxal plates. Eyes normally present, pigmented, lateral. Abdomen strong; pleopods and uropods well developed. This is the largest suborder of amphipods with more than 4700 described species. Marine forms include the free-burrowing Phoxocephalidae, Haustoriidae, Platyischnopidae, and Oedicerotidae; the flesh-scavenging Lysianassidae; the free-swimming and predatory Eusiridae and Pardaliscidae; the tube-building Ampeliscidae and Corophioidea; the sponge and ascidian inhabiting Leucothoidae and Colomastigidae; the parasitic and predatory Stenothoidae and Acanthonotozomatidae; the fish parasites Lafystiidae and Laphystiopsidae; and a host of free-swimming, clinging, crawling, and nestling species of Melitidae, Pontogeneiidae, Pleustidae, and Atylidae, among others.

The principal freshwater families include (in the Northern Hemisphere) the mainly epigean Gammaridae and Pontoporeiidae; the mainly hypogean Crangonyctidae, Niphargidae, and Hadziidae; (mainly in the Southern Hemisphere) the essentially epigean Hyalellidae, Paramelitidae, Neoniphargidae, and Calliopiidae; and in tropic and warm-temperate regions, the hypogean Bogidiellidae.

Terrestrial and semiterrestrial species are contained in the single family Talitridae.

Suborder Hyperiidea. Head not fused with second thoracic segment. Maxilliped lacks a palp. Thoracic coxae often small or fused with the body. Abdomen and pleopods powerfully developed, last two abdominal segments fused. Eyes often very large, covering greater part of head. Body generally transparent. The more than 350 known species are entirely marine and may all be commensal on gelatinous pelagic animals. The infraorder Physosomata contains species with inflated bodies: *Scina, Mimonectes, Lanceola*. The infraorder Physocephalata contains species with inflated heads: *Vibilia, Parathemisto, Hyperia, Primno, Phronima, Rhabdosoma, Platyscelus*.

Suborder Caprellidea. Head partly fused with second thoracic segment. Maxilliped with palp. Thoracic coxae vestigal or lacking. Abdominal segments very reduced, with vestigial appendages. Eyes small. Body elongate, cylindrical, or short and flattened. The 250 known species are entirely marine and include the skeleton shrimps, Caprellidae and others—*Cercops, Aeginella, Deutella*, and *Caprella*; the whale lice, Cyamidae—*Cyamus*; and the monotypic family Caprogammaridae, intermediate between gammarids and caprellids.

Suborder Ingolfiellidea. Head fused or not with second thoracic segment. Maxillipeds with palp. Thoracic coxae very small. Body elongate, cylindrical. Abdominal segments not fused; uropods independent, pleopods very reduced. Pigmented sessile eyes lacking, but ocular lobes present. The some 30 species of this primitive group are mostly hypogean or intertidal in fresh and brackish water. There are a few deep-sea forms.

SUMMARY (PERACARIDA)

1 The superorder Peracarida contains approximately 40 per cent of the described species of crustaceans. Although most are smaller in size than the average decapod, peracaridans are diverse and widespread. They have invaded all of the major habitats, including land. Peracaridans are distinguished by the presence of a brood chamber, or marsupium, beneath the thorax formed by shelflike coxal plates.

2 The shrimplike Mysidacea have stalked compound eyes and a thorax covered by a carapace. Species that occur in large swarms are an important food source for certain fish and other animals.

3 The Amphipoda and Isopoda, the two largest orders, comprise most of the peracaridans. In addition to a somewhat similar size range (most are between 0.5 and 1.5 cm), these two groups share a number of characteristics: no carapace, sessile compound eyes, one pair of maxillipeds and seven pairs of legs, and no sharp demarcation between thorax and abdomen.

4 Amphipods, however, are usually laterally compressed; most isopods are dorsoventrally flattened. Moreover, amphipods possess thoracic gills, and isopods have abdominal pleopods modified as gills.

5 Most amphipods and isopods are benthic crustaceans, although intermittent swimming is common, especially among amphipods. There are few truly pelagic species, but hyperiidean amphipods live with planktonic gelatinous animals, such as jellyfish. There are many burrowing members of both groups, and there are numerous tube-dwelling amphipod species.

6 Many isopods and amphipods are scavengers and detritus feeders, but there are some carnivorous members of both groups. The amphipods also include some suspension feeders. There are many parasitic isopods, which, like parasitic copepods, are called fish lice.

7 Although most amphipods and isopods are marine, there are species of both groups that live in fresh water and on land. The terrestrial amphipods are members of a single family, the Talitridae (beach fleas and leaf litter hoppers), and are common inhabitants of drift along the shore throughout the world and of forest litter in New Zealand. Terrestrial isopods (wood lice) are widely distributed throughout the world, including deserts, and are the largest and most successful group of terrestrial crustaceans.

8 The eggs of peracaridans are deposited, fertilized, and brooded within the marsupium, and development is direct.

Aspects of Crustacean Physiology
Hormones

The hormones of crustaceans and insects are the best known of any invertebrates. Crustacean hormones are either neurosecretions or secretions of

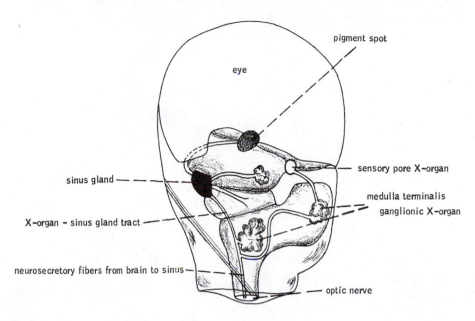

Figure 14–95 Eye and eyestalk of the shrimp *Palaemon*, showing neurosecretory centers and tracts. (After Carlisle, D. B., and Knowles, F., 1959: Endocrine Control in Crustaceans. Cambridge University Press, London. Copyrighted and reprinted by permission of the publisher.)

one of three endocrine tissues—the Y-organ, the androgenic gland, or the ovary. Located between the two basal optic ganglia is a body called the sinus "gland," which is a center for hormone release (Fig. 14–95). The sinus gland is composed of the swollen endings of nerve fibers that have their origin in neurosecretory cell bodies located in several different places within the eye stalk ganglia. The cell body clusters are called X-organs with various designations to indicate the specific site, such as sensory pore X-organ. The neurosecretory material synthesized within the cell bodies, probably a different hormone in each cluster, migrates along the axons to the swollen endings composing the sinus gland. Here it is released into the hemolymph. In addition to the sinus glands, neurosecretions are released into the blood in the tritocerebrum (postcommissural organs) and in the region of the heart (pericardial organs).

Hormones are known to regulate many functions in crustaceans, including various aspects of general metabolism (see Bliss, 1982, Vol. 4).

Regulation of Reproduction

Although sex in crustaceans is determined genetically, the development and function of the gonads and the development of secondary sexual characteristics is under hormonal control (Fig. 14–96). If the Y-organ, located on either side in the anterior part of the cephalothorax, is removed prior to sexual maturity, gonadal development is seriously impaired; if the animal is adult, the gonads are unaffected.

In female malacostracans there exists a hormonal interrelationship between the ovary and the X-organ–sinus gland complex that is somewhat similar to the pituitary–ovary relationship of vertebrates. The sinus gland produces a hormone that inhibits the development of eggs during the nonbreeding periods of the year. During the breeding season a gonad-stimulating hormone is secreted, probably by the central nervous system (Fig. 14–96). The blood level of gonad-inhibiting hormone declines, egg development begins, and the ovary elaborates a hormone, initiating structural changes preparatory for egg brooding, such as the development of ovigerous setae on the pleopods or the development of oostegites (marsupium) in peracaridans. These characteristics appear at the next molt.

The development of the testes and male sexual characteristics in malacostracans is controlled by hormones produced in a small mass of secretory tissue, the androgenic gland. This gland is located at the end of the vas deferens (except in isopods, in which it appears to be located in the testis itself). Removal of the androgenic gland is followed by a loss of male characteristics and conversion of the testes into ovarian tissue. If an androgenic gland is

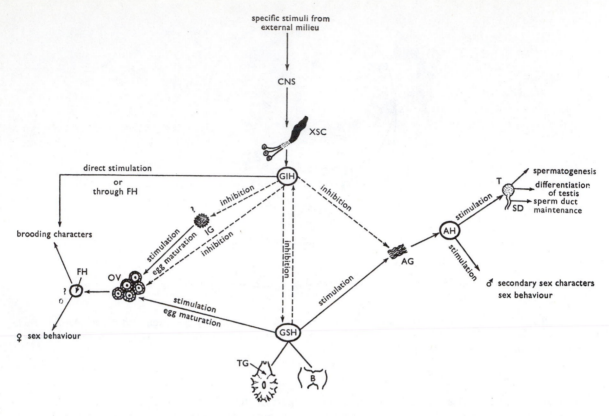

Figure 14–96 Diagram illustrating possible hormonal control of reproduction in a decapod crustacean. Key to lettering: AG, androgenic gland; AH, androgenic hormone; B, brain; CNS, central nervous system; FH, female hormone; GIH, gonad inhibiting hormone; GSH, gonad stimulatory hormone; IG, hypothetical intermediate gland; OV, ovaries; SD, sperm duct; T, testes; TG, thoracic ganglion; XSC, X-organ and sinus-gland complex. (From Adiyodi, K. G., and Adiyodi, R. G., 1970: *Biol. Rev.*, *45:*121–165. Copyrighted and reprinted by permission of Cambridge University Press.)

transplanted into a female, the ovaries become testes and male characteristics appear. (See Bliss, 1982, for a good review of hormonal regulation of crustacean reproductive development.)

Molting and Growth

In many crustaceans, including barnacles, crayfish, and the lobster *Homarus*, molting and growth continue throughout the life of the individual, although molts become spaced further and further apart. Such crustaceans may live to be quite old, and some may become very large. In others, such as some crabs, molting and growth cease with the attainment of sexual maturity or of a certain size or after a certain number of instars. The process of molting has probably been investigated in more detail in crustaceans, especially decapods, than in any other arthropods.

Molting is virtually a continuous process in the life of a crustacean; 90 per cent or more of the period between actual molts may be involved with concluding and preparatory processes associated with the preceding and the future molts. This is especially true in species that molt all year round. In species that molt seasonally, such as species of the crayfish *Cambarus* and the fiddler crab, *Uca*, there is a rather definite intervening rest period during the intermolt. But even during the rest period food reserves are accumulated for the next molt.

Physiologists generally recognize four stages in the molt cycle: proecdysis, ecdysis, postecdysis, and intermolt (see Fig. 12–6). The preparatory phase, or proecdysis or premolt, is marked by a continuing accumulation of food reserves and a rise in blood calcium probably resulting from the activity of the hepatopancreas and resorption of calcium from the cuticle. In some crustaceans, such as crayfish and land crabs, the stomach epithelium secretes calcareous concretions, called gastroliths, that function as calcium storage centers. Eventually, the membranous layers and part of the calcified layers of the old exoskeleton are digested away (Fig. 14–97). Resorption of calcium and digestion of the calcified layer are especially great where splitting later occurs or where the old skeleton must be stretched or broken to permit extraction of a large terminal part of an appendage, such as a claw. After the separation of the old cuticle from the epidermis and the secretion of the new epicuticle and exocuticle, the animal is prepared for the actual brief process of ecdysis and usually seeks some protected retreat or remains in its burrow. The body swells from the uptake of water through the gills or midgut and quickly emerges from the old skeleton, which is commonly eaten later for its calcium salts. The precise mechanism of water absorption is uncertain and may be under neurosecretory control. The amount of water absorbed may equal almost one half of the premolt body weight (Mykles, 1978). The gastroliths, if present, are dissolved by digestive secretions, and the calcium salts are absorbed back into the blood.

During postecdysis, or metecdysis, the endocuticle is secreted, and calcification and hardening of the skeleton take place (Fig. 14–97). The animal re-

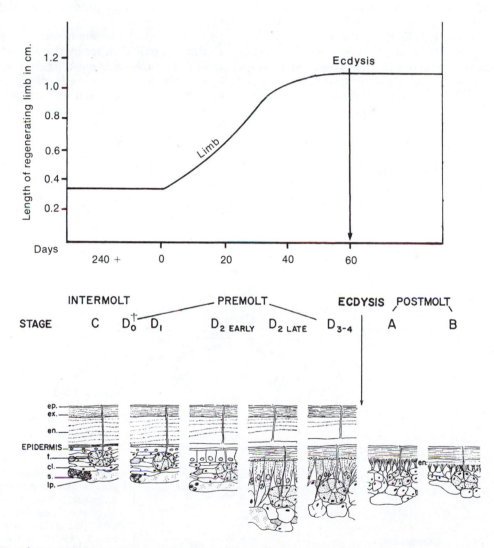

Figure 14–97 Changes in the exoskeleton and associated tissues during the molt cycle of the land crab *Gecarcinus*. Levels of the molting hormone, ecdysone, and growth of regenerating limbs at corresponding states are plotted above. (Adapted from Skinner, D. M., 1962: The structure and metabolism of a crustacean integumentary tissue during a molt cycle. Biol. Bull., *123*(3):635–647.)

mains in its retreat and does not feed during the first part of this phase. (Cameron, 1985, provides a good review of the physiology of molting in the blue crab, *Callinectes*.)

The intermolt may be long or short depending on whether or not the animal molts seasonally. Although the exoskeleton is completely formed, food reserves are accumulated for the next molt.

The physiological processes involved in molting are regulated by hormonal interactions that are essentially like those of insects (Fig. 14–98). The Y-organs, which are analogous to the prothoracic glands of insects, produce ecdysone, which acts on the hypodermal cells and the hepatopancreas to initiate proecdysis. The production of ecdysone by the Y-organs is inhibited by a hormone released from the sinus gland. Thus, the removal of the Y-organs prevents molting; removal of the eyestalks initiates premature proecdysis. The inhibitory action of a sinus gland hormone on the production of ecdysone by the Y-organs is an important difference from control in insects, where the corpora cardiaca stimulate rather than inhibit the prothoracic glands to produce ecdysone (Fig. 14–98).

The regulation of molting hormones, and thereby the regulation of the actual molt cycle, depends on different stimuli operating upon the central nervous system. In crayfish, which molt seasonally, day length is the controlling factor (Aiken, 1969); in the crab *Carcinus*, tissue growth is the controlling factor (Adelung, 1971).

Limb loss is a common event in the life of many crustaceans, and in decapods limbs may even be self-amputated. Severance always takes place at a preformed breakage plane, which runs across the basi-ischium, a proximal leg joint. Internally, there is a corresponding double membranous fold, which is perforated by a nerve and blood vessels. When the limb is cast off, the plane of severance passes between the two membranes, leaving one membrane attached to the basal stub. The membrane constricts around the perforations, so there is very little bleeding.

In some species severance can take place only if the limb is pulled either by the animal itself or by an outside force; in its most highly developed state, however, as in most decapods (except shrimps), autonomy is a unisegmental reflex. If a leg is caught or damaged by a predator, a reflex is set up, and an automizer muscle (one of the locomotor muscles), is stimulated to undergo extreme contraction, fracturing the limb along the breakage plane.

Following severance and scab formation, a small papilla representing the new limb bud grows out from the stub. Growth then halts until the pre-

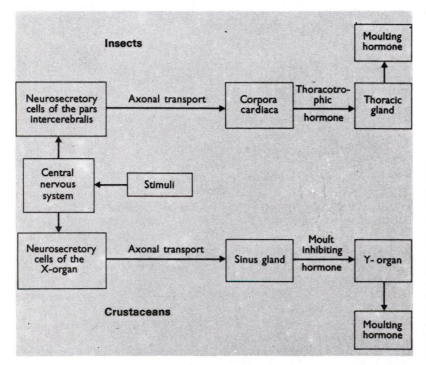

Figure 14–98 Comparison of the hormonal control of molting in crustaceans with that in insects. The insect thoracic glands and the crustacean Y-organs are analogous, producing ecdysone in both. But in insects the hormone from the corpora cardiaca stimulates secretion of ecdysone, while in crustaceans the sinus gland hormone inhibits secretion. (From Highnam, K. C., and Hill, L., 1977: The Comparative Endocrinology of the Invertebrates. 2nd Edition. Edward Arnold [Publishers], Ltd., London. p. 215.)

molt period, when rapid growth and regeneration are completed. The new limb unfolds from a sac at the time of molting. Multiple amputation of limbs induces molting, but the premolt regeneration phase must first be accomplished. If partially regenerated limbs are removed, the molting cycle is delayed until new limb buds are formed (McCarthy and Skinner, 1977).

Chromatophores

A characteristic and striking feature of the integument of many malacostracans is the presence of chromatophores. The crustacean chromatophore is a pigment-bearing cell with branched, noncontractile processes. Each cell is tightly bound with a number of similar chromatophores to form a multicellular organ with radiating processes (McNamara, 1981) (Fig. 14–99A). Pigment granules flow into the processes in the dispersed, or stellate, state and are confined to the center of the cell in the concentrated, or punctate, state. The chromatophores are located in the subhypodermal connective tissue, and where present, the overlying exoskeleton is sufficiently thin or transparent to make them visible.

White, red, yellow, blue, brown, and black pigments may be present. The red, yellow, and blue pigments are carotene derivatives obtained from the diet. The red compound so conspicuous in boiled crabs, lobsters, and shrimps is astaxanthin (carotenalbumin). In the exoskeleton of the living animal, this pigment is conjugated with a protein and is blue or some other color characteristic of the conjugated state. Curiously, a single chromato-

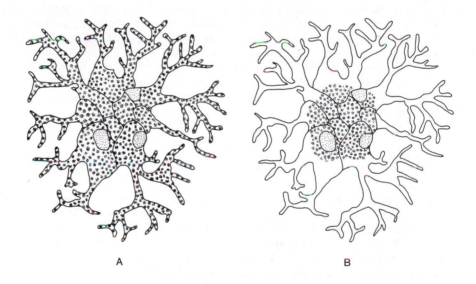

A   B

C

Figure 14–99 *A* and *B*, Crustacean chromatophores in which pigment is dispersed (*A*) and concentrated (*B*). Each body is actually a cluster of cells, each a chromatophore. Processes radiate in all directions, not just in the plane shown above. (Based on a photograph by McNamara, 1981.) *C*, Fiddler crabs in pale (nighttime) and dark (daytime) phases. (From Prosser, A. L., 1973: Comparative Animal Physiology. W. B. Saunders Co., Philadelphia. p. 924.)

phore may possess one, two, three, or even four color pigments, any one of which can move independently of another. In general, polychromatic chromatophores are found only in shrimp.

The most common type of rapid (physiological) color change is a simple blanching (or lightening) and darkening. This response is typical of many crabs, such as the fiddler crab, *Uca.* Many crustaceans, however, especially shrimps, can adapt to a wide range of colors. The little shrimp *Palaemonetes*, for example, possesses trichromatic chromatophores with red, yellow, and blue pigments; through the independent movement of these three primary colors it can adapt to any background color, even black. Other species have similar abilities.

The movement of chromatophore pigments is controlled by hormones elaborated by the neurosecretory system in the eyestalk (or below the eye, when the eyes are sessile) and in other parts of the central nervous system. In shrimps (aside from *Crangon*) that have red, yellow, blue, and white pigments, removal of the eyestalks results in a darkening of the body through dispersion of the red and yellow pigments. If these animals are then injected with sinus gland extract, the white pigment disperses, and the body color rapidly blanches. The opposite occurs in some crabs that have black, red, yellow, and white pigments. Removal of the eyestalks causes blanching of the body color, resulting from a concentration of the black pigments and dispersal of the white. Injections of sinus gland extract cause a darkening.

The current understanding of hormonal control of chromatophores, although still incomplete, provides an explanation for the opposite effects of eyestalk removal in shrimps and crabs. There appear to be separate, antagonistic pairs of chromatophorotropins (hormones) for each pigment; one brings about dispersion and one concentrates. Moreover, some at least appear not to be species specific but are found in a large number of decapods. These hormones are released not only by the sinus gland in the eyestalk but also by the commissural gland in the tritocerebrum. Whether blanching of the body occurs when the eyestalks are removed depends on whether the sinus gland is the releasing site for a number of those chromatophorotropins that have a dispersing effect on the pigment granules of the chromatophores.

Eyestalk hormones control not only the functioning of the chromatophores, but other pigment changes as well. In different crustaceans any one or a combination of the three retinal pigments—dis-tal, proximal, and reflecting pigments—may migrate distally or proximally in adapting the eye to bright or weak light. Experimental evidence indicates that at least the movement of the distal retinal pigment and the reflecting pigment is under the control of sinus gland hormones.

Physiological Rhythms

Pigment movement and other physiological processes often display a rhythmic activity in crustaceans. Such physiological rhythms, or physiological "clock mechanisms," have been studied extensively in a number of species, especially the green crab, *Carcinus*, and the fiddler crab, *Uca* (Palmer, 1975 and 1976). Both crabs live on sand flats and in the intertidal zone of protected beaches, but *Carcinus* is active at high tide and *Uca* at low tide. Through dispersion and concentration of chromatophore pigments, both species are pale at night and dark during the day (Fig. 14–100). The rhythm persists when the crabs are kept in constant light or darkness, and fiddler crabs flown from Woods Hole, Massachusetts, to California continued to blanch and darken on Woods Hole time. Removal of the eyestalks, which secrete the pigment-controlling hormone, disrupts the rhythm in *Carcinus* and reduces its amplitude in *Uca.*

Locomotor activity in both crabs follows the tide and is governed by a lunar rhythm or a biological clock set on lunar time (Fig. 14–100). After about a week under constant conditions, the rhythm is lost. It can be restored by simply immersing *Uca* in sea water and cooling *Carcinus.* The clocks of both species can be set to a new tidal rhythm by subjecting the crabs to periods of cooling or high pressure that correspond to the desired tidal rhythm.

Similar rhythms have been recorded for other crustaceans, such as the blue crab, *Callinectes sapidus*, and the intertidal isopod *Ligia exotica.* Species of crayfish display a diurnal rhythm of locomotor activity, but such a rhythm is never present in cave species.

A discussion of the possible mechanisms underlying biological clocks in living organisms is beyond the scope of this text but is provided by Palmer (1976) and DeCoursey (1983).

SUMMARY (ASPECTS OF PHYSIOLOGY)

1 Hormones are known to control a number of functions in crustaceans, of which reproduction, molting and growth, and chromatophore changes

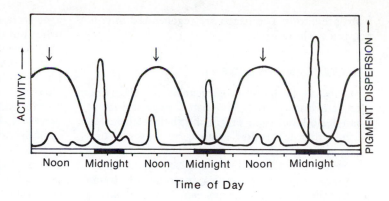

Figure 14–100 Rhythmic motor activity and chromatophore changes in the crab *Carcinus*. Motor activity exhibits a tidal rhythm and is greatest at high tides. The chromatophores exhibit a diurnal rhythm, with pigment dispersed (animals dark) during the day (indicated by arrows). (Modified from Palmer, J. D., 1976: An Introduction to Biological Rhythms. Academic Press, N.Y. p. 99; and 1975: Biological clocks in the tidal zone. Sci. Am., 232(2):74.)

have been most studied (in decapods). There are several centers of hormone secretion, and the sinus gland in the eyestalk of decapods is a principal center of hormone release.

2 Some crustaceans molt throughout their lives; others cease on reaching sexual maturity. Many important aspects of molting physiology take place during the long preparatory phase (calcium resorption), during the concluding phase (calcium deposition), and during the intermolt (accumulation of food reserves).

3 The integument of many malacostracans contains branched chromatophores, within which pigment granules of one or more colors may become dispersed or concentrated, changing the coloration of the animal. Adaptation to background is a common provision of chromatophores.

4 Chromatophore changes and other functions of crustaceans may display rhythmic activity that coincides with tidal or diurnal rhythms.

5 Many malacostracans are capable of self-amputation of limbs (autotomy), aiding in escape

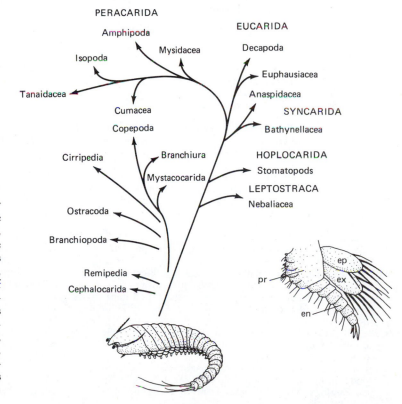

Figure 14–101 A possible phylogeny of the Crustacea. The figures illustrate the hypothetical ancestral crustacean, dorsal view of entire animal, and one trunk appendage. All trunk appendages were similar, including the two maxillae, and every trunk segment carried a pair of appendages. In the illustration of the entire animal, the second pair of antennae is hidden beneath the head. Label abbreviations refer to protopodite, endopodite, exopodite and epipodite. (From Hessler, R. R., and Newman, W. A., 1975: A trilobite origin for the Crustacea. Fossils and Strata, 4:437–459.)

from predators. The limbs are regenerated in connection with molting.

CRUSTACEAN PHYLOGENY

Crustaceans, both entomostracans and malacostracans, are known to have existed since the Cambrian period, but their origin and their relationship to the other arthropod groups are very obscure. The patchy fossil record is not very informative (see reviews by Cisne, 1982, and Schram, 1982).

Given the comparative anatomy of living forms, the ancestral crustacean was probably a small, swimming, epibenthic animal possessing a head and a trunk of numerous similar segments. The head bore two pairs of antennae, a pair of mandibles, two pairs of maxillae, a pair of stalked compound eyes, and a naupliar eye (Fig. 14–101). The mouth was directed backward. The trunk append-

ages were numerous and similar, not only to each other but probably also to the maxillae, and probably served in locomotion, gas exchange, and feeding. Certainly among living crustaceans, the cephalocarids and the one known remipedian most closely resemble such a hypothetical ancestral crustacean. In addition, Hessler and Newman (1975) believe that the appendages were primitively triramous, not biramous (Fig. 14–101).

During the evolution of the existing crustacean groups, the ancestral stock probably divided early into three principal lines—one leading to the Branchiopoda, one to the Branchiura, Cirripedia, and Copepoda, and one to the malacostracans. The Cephalocarida and Remipedia are considered by many to be sufficiently generalized to stand near the stem line of all three (Fig. 14–101). The malacostracan line culminates in two major groups, the peracaridans and eucaridans.

REFERENCES

The literature included here is restricted in large part to crustaceans. The introductory references on page 8 include many general works and field guides that contain sections on these animals.

A. Kaestner's (1970) and F. R. Schram's (1986) volumes provide detailed general accounts of the Crustacea. The *Journal Crustaceana* and the *Journal of Crustacean Biology* are devoted to the publication of research on crustaceans.

General

Anderson, D. T., 1973; Embryology and Phylogeny in Annelids and Arthropods. Pergamon Press, Oxford, 495 pp. (Includes a detailed survey of crustacean embryology.)

Bliss, D. E. (Ed.), 1982–1985: Biology of the Crustacea. 10 vols. Academic Press, N.Y. (This series covers many aspects of the ecology, behavior, and physiology of crustaceans.)

Cisne, J. L., 1982: Origin of the Crustacea. *In* Bliss, D. E., (Ed.): Biology of the Crustacea. Vol. 1, pp. 65–92.

Cronin, T. W., 1986: Optical design and evolutionary adaptation in crustacean compound eyes. Jour. Crustacean Biology, 6(1): 1–23.

DeCoursey, P. J., 1983: Biological timing. *In* Bliss, D. E. (Ed.): Physiology of the Crustacea. Vol. 7 pp. 107–162.

Fitzpatrick, J. F., 1983: How to Know the Freshwater Crustacea. W. C. Brown Co., Dubuque, Iowa.

Gibson, R., and Barker, P. L., 1979: The decapod hepatopancreas. Oceanogr. Mar. Biol. Ann. Rev., 17:285–346.

Green, J., 1961: A Biology of Crustacea. Quadrangle Books, Chicago. (A brief general account of crustaceans.)

Gregoire, C., 1971: Hemolymph coagulation in arthropods. *In* Florkin, M., and Scheer, B. T. (Eds.): Chemical Zoology. Vol. 6. Academic Press, N.Y. pp. 145–189.

Hessler, R. R., and Newman, W. A., 1975: A trilobite origin for the Crustacea. Fossils and Strata, 4:437–459.

Kaestner, A., 1970: Invertebrate Zoology. Vol. 3. Crustacea. Wiley-Interscience, N.Y. 523 pp. (An excellent general account of the crustaceans; systematically arranged and covering all aspects of the biology.)

LaBarbera, M., 1984: Feeding currents and particle capture mechanisms in suspension feeding animals. Amer. Zool., 24:71–84.

Lockwood, A. P. M., 1967: Physiology of Crustacea. W. H. Freeman and Co., San Francisco. 328 pp.

McLaughlin, P., 1980: Comparative Morphology of Recent Crustacea. W. H. Freeman and Co., San Francisco. 159 pp. (Detailed anatomical descriptions of representatives of the higher taxa of crustaceans.)

McNamara, J. C., 1981: Morphological organization of crustacean pigmentary effectors. Biol. Bull., 161:270–280.

Moore, R. C. (Ed.), 1979: Treatise on Invertebrate Paleontology, Pt. R., Arthropoda 4. Vols. 1 and 2. Geological Society of America and University of Kansas Press, Lawrence, Kan. (These two volumes cover the Crustacea except for ostracods, which are treated in Part Q, Arthropoda 3, 1961.)

Palmer, J. D., 1976: An Introduction to Biological Rhythms. Academic Press. N.Y.

Rebach, S., and Dunham, D. W. (Eds.), 1983: Studies in Adaptation: The Behavior of Higher Crustacea. John Wiley and Sons, N.Y. 282 pp.

Riegel, J. A., 1971: Excretion—Arthropods. *In* Florkin, M., and Scheer, B. T. (Eds.): Chemical Zoology. Vol 6, Academic Press, N.Y. pp. 249–277.

Schmitt, W. L., 1975: Crustaceans. University of Michigan Press, Ann Arbor. (An interesting popular treatment of crustaceans.)

Schram, F. R., 1982: The fossil record and evolution of Crustacea. *In* Bliss, D. E. (Ed.): Biology of the Crustacea. Vol. 1, pp. 93–147.

Schram, F. R. (Ed.), 1983: Crustacean Phylogeny. A. A. Balkema, Rotterdam. 372 pp. (A collection of papers from a symposium on crustacean phylogeny.)

Schram, F. R., 1986: Crustacea. Oxford University Press, N.Y. 700 pp. (A detailed account of the crustacean groups.)

Van Weel. P. B., 1979: Digestion in Crustacea. *In* Florkin, M., and Scheer, B. T. (Eds.): Chemical Zoology, Vol. 5, Pt. A. Academic Press, N.Y. pp. 97–115.

Whittington, H. B., and Rolfe, D. W. I. (Eds.), 1963: Phylogeny and Evolution of Crustacea. Spec. Pub. Mus. Comp. Zool., Harvard University Press. Cambridge, Mass. (A collection of papers devoted to various aspects of crustacean evolution.)

Branchiopods, Ostracods, Copepods, Barnacles

Barnes, H., Barnes, M., and Klepal, W., 1977: Studies on the reproduction of cirripedes: I. Introduction: copulation, release of oocytes, and formation of the egg lamellae. Jour. Exp. Mar. Biol. Ecol., 27(3):195–218.

Bate, R. H., Robinson, E., and Sheppard. L. M. (Eds.), 1982: Fossil and Recent Ostracods. Horwood, Chichester, England. 494 pp. A collection of papers on ostracods.

Benson, R. H., 1966: Recent marine podocopid ostracods. Oceanogr. Mar. Biol. Ann. Rev., 4:213–232.

Bourget, E., and Crisp, D. J., 1975: Factors affecting deposition of the shell in *Balanus balanoides*. J. Mar. Biol. Ass. U.K., 55:231–249.

Boxshall, G. A., and Lincoln, R. J., 1983: Tantulocarida, a new class of Crustacea ectoparasitic on other crustaceans. Jour. Crust. Biol., 3(1):1–16.

Corkett, C. J., and McLaren, I. A., 1978: The biology of *Pseudocalanus*. Adv. Mar. Biol., 15:1–231.

Coull, B. C., 1977: Copepoda: Harpacticoida. Marine flora and fauna of the northeastern U.S. NOAA Technical Report NMFS circular 399. U.S. Government Printing Office, Washington, D.C.

Darwin, C., 1851–1854: A Monograph on the Subclass Cirripedia. 2 Vols. Ray Society, London. (A classic and still valuable account of the barnacles.)

Foster, B. A., 1971: Desiccation as a factor in the intertidal zonation of barnacles. Mar. Biol., 8:29.

Foster, B. A., 1974: The barnacles of Fiji, with observations on the ecology of barnacles on tropical shores. Pac. Sci., 28(1):35–56.

Frey, D. G., 1982: Contrasting strategies of gametogenesis in northern and southern populations of Cladocera. Ecology, 63(1):223–241.

Frost, B. W., 1977: Feeding behavior of *Calanus pacificus* in mixtures of food particles. Limnol. Oceanogr., 22(3):472–491.

Gerritsen, J., and Porter, K. G., 1982: The role of surface chemistry in filter feeding by zooplankton. Science, 216:1225–1227.

Gooto, R. V., 1979: The association of copepods with marine invertebrates. Adv. Mar. Biol., 16:1–109.

Grice, G. D., and Marcus, N. H., 1981: Dormant eggs of marine copepods. Ann. Rev. Oceanogr. Mar. Biol., 19:125–140.

Griffiths, A. M., and Frost, B. W., 1976: Chemical communication in the marine planktonic copepods *Calanus pacificus* and *pseudocalanus* sp. Crustaceana, 30(1):1–8.

Hartmann, G., 1975: Ostracoda. 4th Part *In* Bronn's Klassen and Ordnungen des Tierreichs. Fuenster Band:

Arthropoda. I. Abteilung: Crustacea. 2. Buch IV. Reil: Ostracoda, 4 Lieferung. Gustav Fisher Verlag, Jena.

Hicks, G. F., and Coull, B. C., 1983: The ecology of marine meiobenthic harpacticoid copepods. Oceanogr. Mar. Biol., Ann. Rev., 21:67–175.

Ho, J. S., 1977: Copepoda: Lernaeopodidae and Sphyriidae. Marine flora and fauna of the northeastern U.S. NOAA Technical Reports NMFS Circular 406. U. S. Gov. Printing Office, Washington D.C. 13 pp.

Ho, J. S., 1978: Copepoda: Cyclopoids parasitic on fishes. Marine flora and fauna of the northeastern U.S. NOAA Technical Reports NMFS Circular 409. U.S. Gov. Printing Office, Washington, D.C. 11 pp.

Knox, G. A., and Fenwick, G. D., 1977: *Chiltoniella elongata* n. gen. and sp. (Crustacea: Cephalocarida) from New Zealand. J. R. Soc. N. Z., 7(4):425–432.

Koehl, M. A. R., and Strickler, J. R., 1981: Copepod feeding currents: food capture at low Reynolds number. Limnol. Oceanogr., 26(6):1062–1073.

Lewis, C. A., 1978: A review of substratum selection in free-living and symbiotic cirripeds. *In* Chia, F.-S., and Rice, M. E. (Eds.): Settlement and Metamorphosis of Marine Invertebrate Larvae. Elsevier, North Holland, N.Y. pp. 207–218.

Marcotte, B. M., 1977: An introduction to the architecture and kinematics of harpacticoid feeding: *Tisbe furcata*. Mikrofauna Meeresboden, 61:183–196.

Marcotte, B. M., 1983: The imperatives of copepod diversity: perception, cognition, competion and predation. *In* Schram, F. R. (Ed.): Crustacean Phylogeny. A. A. Balkema, Rotterdam. pp. 47–72.

Marcus, N. H., 1982: Photoperiodic and temperature regulation of diapause in *Labidocera aestiva*. Biol. Bull., 162:45–52.

Marshall, S. M., 1973: Respiration and feeding in copepods. Adv. Mar. Biol., 11:57–120.

Marshall, S. M., and Orr, A. P., 1955: On the biology of *Calanus finmarchicus*. VIII. Food uptake, assimilation and excretion in adult and stage V *Calanus*. J. Mar. Biol. Assoc. U.K., 35:495–529.

Mellors, W. K., 1975: Selective predation of ephippial *Daphnia* and the resistance of ephippial eggs to digestion. Ecology, 56:975–980.

Moyse, J., and Hui, E., 1981: Avoidance by *Balanus balanoides* cyprids of settlement on conspecific adults. J. Mar. Biol. Assoc. U.K., 61:449–469.

Newman, W. A., 1974: Two new deep-sea Cirripedia (Ascothoracica and Acrothoracica) from the Atlantic. J. Mar. Biol. Assoc. U.K., 54:437–456.

Newman, W. A., and Ross, A., 1976: Revision of the balanomorph barnacles; including a catalog of the species. San Diego Soc. Nat. Hist. Memoir. 9:1–108.

Palmer, A. R., 1982: Predation and parallel evolution: recurrent parietal plate reduction in balanomorph barnacles. Paleobiology, 8(1):31–44.

Pennak, R. W., and Zinn, D. J., 1943: Mystacocarida, a new order of Crustacea from intertidal beaches in Massachusetts and Connecticut. Smithsonian Misc. Collect., 103:1–11.

Pilsbry, H. A., 1907: The barnacles contained in the collections of the U.S. Natural Museum. Smithsonian Inst. Bull. 60:1–114. (An old but still valuable taxonomic treatment of the American stalked barnacles. The following paper covers the sessile barnacles.)

Pilsbry, H. A., 1916: The sessile barnacles contained in the collection of the U.S. National Museum, including a monograph of the American species. Smithsonian Inst. Bull., 93:1–357.

Poulet, S. A., 1976: Feeding of *Pseudocalanus minutus* on living and nonliving particles. Mar. Biol., 34(2):117–125.

Sanders, H. L., 1957: The Cephalocarida and crustacean phylogeny. Syst. Zool., 6:112–128.

Sanders. H. L., 1963: The Cephalocarida. Memoirs Conn. Acad. Arts and Sci., 15:1–180.

Sarvala, J., 1979: Benthic resting periods of pelagic cyclopoids in an oligotrophic lake. Holarctic Ecology. 2:88–100.

Shmeleva, A. A., and Kovalev, A. V., 1974: Biologic cycles of copepods of the Adriatic Sea. Boll. Pesca Piscic. Hydrobiol., 29(1):49–70.

Southward, A. J., 1955: Feeding of barnacles. Nature, 175:1124–1125.

Strickler, J. R., 1982: Calanoid copeods, feeding currents, and the role of gravity. Science, 218:158–160.

Stubbings, H. G., 1975: *Balanus balanoides.* Liverpool Mar. Biol. Comm. Mem. No. 37 (A detailed account of one of the most common sessile intertidal barnacles of north temperate Atlantic and Pacific waters.)

Tomlinson, J. T., 1969: The burrowing barnacles (Cirripedia: Order Acrothoracica). Bull. U.S. Nat. Mus., 269:1–162.

Uye, S., and Kasahara, S., 1978: Life history of marine planktonic copepods in neritic region with special reference to the role of resting eggs. Bull. Plankton Soc. Japan, 25(2):109–122.

Walker, G., and Yule, A. B., 1984: Temporary adhesion of the barnacle cyprid: the existence of an antennular adhesive secretion. J. Mar. Biol. Assoc. U.K., 64:679–686.

Watras, C. J., 1983: Mate location by diaptomid copepods. Jour. Plankton Research, 5(3):417–423.

Watras, C. J., and Haney, J. F., 1980: Oscillations in the reproductive condition of *Diaptomus leptopus* and their relation to rates of egg-clutch production. Oecologia, 45:94–103.

Wiman, F. H., 1981: Mating behavior in the *Streptocephalus* fairy shrimps. The Southwestern Naturalist, 25(4):541–546.

Wong, C. K., 1981: Cyclomorphosis in *Bosmina* and copepod predation. Can. Jour. Zool., 59:2049–2052.

Yager, J., 1981: Remipedia, a new class of Crustacea from a marine cave in the Bahamas. Jour. Crust. Biol., 1(3):328–333.

Decapods

Adelung, D., 1971: Studies on the moulting physiology of decapod crustaceans as exemplified by the shore crab *Carcinus maenas.* Helgolander Wiss. Mceresunters. 22(1):66–119.

Adiyodi, K. G., and Adiyodi, R. G., 1970: Endocrine control of reproduction in decapod Crustacea. Biol. Rev., 45:121–165.

Aiken, D. E., 1969: Photoperiod, endocrinology, and the crustacean molt cycle. Science, 164:149–155.

Allen, J. A., 1972: Recent studies on the rhythms of postlarval decapod crustacea. Ann. Rev. Oceanography and Mar. Biol., 10:415–436.

Barker, P. L., and Gibson, R., 1977: Observations on the feeding mechanism, structure of the gut, and diges-

tive physiology of the European lobster *Homarus gammarus.* J. Exp. Mar. Biol. Ecol., 26(3):297–324.

Bauer, R. T., 1977: Antifouling adaptations of marine shrimps: Functional morphology and adaptive significance of antennular preening by the third maxillipeds. Mar. Biol., 40:261–276.

Bauer, R. T., 1981: Grooming behavior and morphology in the decapod Crustacea. Jour. Crust. Biol., 1(2):153–173.

Bertness, M. D., 1981: Pattern and plasticity in tropical hermit crab growth and reproduction. Amer. Nat., 117(5):754–773.

Binns, R., 1969: Physiology of the antennal gland of *Carcinus.* J. Exp. Biol., 51(1):1–10.

Bliss, D. E., 1982: Shrimps, Lobsters and Crabs. New Century Publishers, Piscataway, N.J. 242 pp. (A semipopular account of the decapods.)

Bliss, D. E., van Montfrans, J., van Montfrans, M., and Boyer, J., 1978: Behavior and growth of the land crab *Cecarcinus lateralis* (Fréminville) in southern Florida. Bull. Am. Mus. Nat. Hist., 160(2):113–151.

Bruce, A. J., 1976: Shrimps and prawns of coral reefs, with special reference to commensalism. *In* Jones, O. A., and Endean, R. (Eds.): Biology and Geology of Coral Reefs. Vol. III: Biol. 2. Academic Press, N.Y. pp. 37–94.

Caine, E. A., 1975: Feeding and masticatory structures of selected Anomura. J. Exp. Mar. Biol. Ecol., 18:277–301.

Cameron, J. N., 1985: Molting in the blue crab. Sci. Amer., 252(5):105–109.

Chace, F. A., and Hobbs, H. H., 1969: The freshwater and terrestrial decapod crustaceans of the West Indies with special reference to Dominica. Smithsonian Institution, Bull. U.S. Nat. Mus., 292:1–258.

Christy, J. H., 1978: Adaptive significance of reproductive cycles in the fiddler crab *Uca pugilator:* A hypothesis. Science, 199:453–455.

Christy, J. H., 1982a: Burrow structure and use in the sand fiddler crab, *Uca pugilator.* Animal Behaviour, 30:687–694.

Christy, J. H., 1982b: Adaptive significance of semilunar cycles of larval release in fiddler crabs: test of an hypothesis. Biol. Bull., 163:251–263.

Cobb, J. S., and Phillips, B. F. (Eds.), 1980: The Biology and Management of Lobsters. Vol. 1, Physiology and Behavior, 462 pp. Vol. 2, Ecology and Management, 390 pp. Academic Press, N.Y.

Cornell, J. C., 1979: Salt and water balance in two marine spider crabs, *Libinia emarginata* and *Pugettia producta.* I. Urine production and magnesium regulation. Biol. Bull., 157(2):221–233.

Crane, J., 1975: Fiddler Crabs of the World. Princeton University Press, Princeton, N.J. 660 pp. (An exhaustive treatment of these familiar intertidal crabs. First part covers taxonomy; second part, biology.)

Cronin, T. W., and Forward, R. B., 1979: Tidal vertical migration: an endogenous rhythm in estuarine crab larvae. Science, 205:1020–1022.

Derby, C. D., 1982: Structure and function of cuticular sensilla of the lobster *Homarus americanus.* Jour. Crust. Biol., 2(1):1–21.

Díaz, H., and Rodriguez, G., 1977: The branchial chamber in terrestrial crabs: a comparative study. Biol. Bull., 153:485–504.

Fontaine, M. T., Passelecq-gerin, E., Bauchau, A. G., 1982: Structures chemoreceptrices des antennules du crabe *Carcinus maenas*. Crustaceana 43(3):271–283.

Gibson, R., and Barker, P. L., 1979: The decapod hepatopancreas. Oceanogr. Mar. Biol., Ann. Rev., 1979:285–346.

Greenspan, B. N., 1982: Semi-monthly reproductive cycles in male and female fiddler crabs, *Uca pugnax*. Sci. Amer., 30:1084–1092.

Hart, J. F. L., 1982: Crabs and Their Relatives of British Columbia. British Columbia Provincial Museum, Victoria. 267 pp.

Hartnoll, R. G., 1979: Mating in Brachyura. Crustaceana, 16:161–181.

Hazlett, B. A., 1970: The effect of shell size and weight on the agonistic behavior of a hermit crab. Z. Tierpsychol., 27(3):369–374.

Hazlett, B. A., 1982: Chemical induction of visual orientation in the hermit crab *Clibanarius vittatus*. Animal Behaviour, 30(4):1259–1260.

Herreid, C., 1963: Observations on the feeding behavior of *Cardisoma guanhumi* in southern Florida. Crustaceana, 5:176–180.

Herring, P. J., 1976: Bioluminescence in decapod Crustacea. J. Mar. Biol. Assoc. U.K., 56:1029–1047.

Hobbs, H. H., 1972: Crayfishes (Astacidae) of North and Middle America. Biota of Freshwater Ecosystems. Identification Manual No. 9. EPA. U.S. Government Printing Office, Washington, D.C. 173 pp.

Hopkins, S. P., and Nott, J. A., 1980: Studies on the digestive cycle of the shore crab *Carcinus maenas* with special reference to the B cells in the hepatopancreas. J. Mar. Biol. Assoc. U.K., 60:891–907.

Horch, K. W., and Salmon, M., 1969: Production, perception and reception of acoustic stimuli by semiterrestrial crabs. Forma et Functio, 1:1–25.

Hyatt, G. W., and Salmon, M., 1978: Combat in the fiddler crabs *Uca pugilator* and *U. pugnax*: a quantitative descriptive analysis. Behaviour, 65(¾):182–211.

Ingle, R. W., 1983: Shallow-Water Crabs. Synopses of the British Fauna No. 25. Cambridge University Press, London. 206 pp.

Ivanov, B. G., 1970: On the biology of the Antarctic krill *Euphausia superba*. Mar. Biol., 7:340.

Johnson, P. T., 1980: The Histology of the Blue Crab, *Callinectes sapidus*. Praeger Publ., N.Y. 440 pp.

Klaassen, F., 1975: Ecological and ethological studies on the reproductive biology in *Gecarcinus lateralis*. Forma et Functio, 8(2):101–174.

Levine, D. M., and Blanchard, O. J., 1980: Acclimation of two shrimps of the genus *Periclimenes* to sea anemones. Bull. Mar. Sci., 30(2):460–466.

Mangum, C. P., and Weiland, A. L., 1975: The function of hemocyanin in respiration of the blue crab *Callinectes sapidus*. J. Exp. Zool., 193(3):257–263.

Mauchline, J., and Fischer, L. R., 1969: The biology of euphausiids. Adv. Mar. Biol., Vol. 7.

McCarthy, J. F., and Skinner, D. M., 1977: Proecdysial changes in serum ecdysone titers, gastrolith formation and limb regeneration following molt induction by limb autotomy and/or eyestalk removal in the land crab *Gecarcinus lateralis*. Gen. Comp. Endocrin. 33:278–292.

McLaughlin, P. A., 1974: The hermit crabs of northwestern North America. Zool. Verh., Leiden, 130:1–396.

McLay, C. L., 1983: Dispersal and use of sponges and ascidians as camouflage by *Cryptodromia hilgendorfi*. Mar. Biol., 76:17–32.

Mesce, K. A., 1982: Calcium-bearing objects elicit shell selection behavior in a hermit crab. Science, 215:993–995.

Miller, D. C., 1961: The feeding mechanisms of fiddler crabs, with ecological considerations of feeding adaptations. Zoologica, 46(8):89–101.

Molenock, J., 1975: Evolutionary aspects in the courtship behavior of anomuran crabs *(Petrolisthes)*. Behaviour, 53(½):1–30.

Mykles, D. L., 1978: The mechanism of water uptake at ecdysis in the lobster *(Homarus americanus* and *H. gammarus)* and dungeness crab *(Cancer magister)*. Sea Grant 8(11,12):18–19.

Nepszy, S. J., and Leach J. H., 1973: First records of the Chinese mitten crab, *Eriocheir sinensis* from North America. J. Fish Res. Board Can., 30:1909–1910.

Nolan, B. A., and Salmon, M., 1970: The behavior and ecology of snapping shrimp. Forma et Functio, 4:289–335.

Omori, M., 1974: The biology of pelagic shrimps in the ocean. Adv. Mar. Biol., 12:233–324.

Palmer, J. D., 1975: Biological clocks of the tidal zone. Sci. Amer., 232(2):70–79.

Peterson, D. R., and Loizzi, R. F., 1974: Ultrastructure of the crayfish kidney—coelomosac, labyrinth, nephridial canal. J. Morph., 142(3):241–263.

Reese, E. S., 1963: The behavior mechanisms underlying shell selection by hermit crabs. Behaviour, 21:78–126.

Roger, C., 1975: Feeding rhythms and trophic organization of a population of pelagic crustaceans. Mar. Biol., 32(4):365–378.

Salmon, M., 1967: Coastal distribution, display and sound production by Florida fiddler crabs. Animal Behavior, 15:449–459.

Salmon, M., 1971: Signal characteristics and acoustic detection by fiddler crabs, *Uca rapax* and *Uca pugilator*. Physiol. Zool., 44:210–224.

Salmon, M., and Horch, K. W., 1972: Acoustical signalling and detection by semiterrestrial crabs of the family Ocypodidae. *In* Winn, H. E., and Olla, B. L. (Eds.): Behavior of Marine Animals. Vol. 1. Invertebrates. Plenum Press, N.Y. pp. 60–96.

Sargent, R. C., and Wagenbach, G. E., 1975: Cleaning behavior of the shrimp, *Periclimenes anthophilus*. Bull. Mar. Sci., 25(4):466–472.

Schembri, P. J., 1982: Feeding behaviour of fifteen species of hermit crabs from the Otago region, southeastern New Zealand. Jour. Nat. Hist., 16:859–878.

Schöne, H., 1968: Agonistic and sexual display in aquatic and semiterrestrial brachyuran crabs. Am. Zool., 8:641–654.

Spight, T. M., 1977: Availability and use of shells by intertidal hermit crabs. Biol. Bull., 152:120–133.

Stevcic, Z., 1971: The main features of brachyuran evolution. Syst. Zool., 20(3):331–340.

Warner, G. F., 1977: The Biology of Crabs. Van Nostrand Reinhold Co., N.Y. 202 pp.

Warner, J. A., Latz, M. I., and Case, J. F., 1979: Cryptic bioluminescence in a midwater shrimp. Science, 203:1109–1110.

Welsh, B. L., 1975: The role of grass shrimp, *Palaemonetes pugio*, in a tidal marsh ecosystem. Ecology, *56*(3):513–530.

Wickins, J. F., 1976: Prawn biology and culture. Oceanogr. Mar. Biol. Ann. Rev., *14*:435–507.

Wicksten, M. K., 1980: Decorator crabs. Sci. Amer., *242*(2):146–154.

Wilber, T. P., and Herrnkind, W., 1982: Rate of new shell acquisition by hermit crabs in a salt marsh habitat. Jour. Crustacean Biol., *2*(4):588–592.

Williams, A. B., 1974: Crustacea: Decapoda. Marine flora and fauna of the northeastern U.S. NOAA Technical Reports NMFS Circular 389. U.S. Government Printing Office, Washington, D. C. 49 pp.

Williams, A. B., 1984: Shrimps, Lobsters, and Crabs of the Atlantic Coast of the Eastern United States, Maine to Florida. Smithsonian Institution Press, Washington, D. C.

Wolcott, T. G., 1984: Uptake of interstitial water from soil: mechanisms and ecological significance in the ghost crab *Ocypode quadrata* and two gecarcinid land crabs. Physiol. Zool., *57*(1):161–184.

Wolcott, T. G., and Wolcott, D. L., 1982: Larval loss and spawning behavior in the land crab *Gecarcinus lateralis*. Jour. Crust. Biol., *2*(4):477–485.

Other Malacostracans

Barnard, J. L., 1969: The families and genera of marine gammaridean Amphipoda. U.S. Nat. Mus. Bull., *271*:1–535.

Barnard, J. L., and Barnard, C. M., 1983: Freshwater Amphipoda of the World: I, Evolutionary patterns. II, Handbook and bibiliography. Hayfield Associates, Alexandria, Va., 830 pp.

Bek, T. A., 1972: Feeding of intertidal gammarids. Vestn. Mosk, Univ. Ser. G.. Biol. Pochvoved, *27*(1):106–107.

Berkes, F., 1975: Some aspects of feeding mechanisms of euphausiid. Crustaceana. *29*(3):266–270.

Bousfield, E. L., 1973: Shallow-water Gammaridean Amphipoda of New England. Cornell University Press, Ithaca, N.Y. 344 pp.

Bousfield, E. L., 1977: A new look at the systematics of Gammaroidean amphipods of the world. Crustaceana (Supplement), *4*:282–316.

Bousfield, E. L., 1978: A revised classification and phylogeny of amphipod crustaceans. Trans. R. Soc. Can., (Ser. IV) *16*:343–390.

Bowman, T. E., and Gruner, H. E., 1973: The families and genera of Hyperidea (Amphipoda). Smiths. Contr. Zool. No., *146*:1–64.

Brooks, H. K., 1962: On the fossil Anaspidacea, with a revision of the classification of the Syncarida. Crustaceana *4*(3):229–242.

Burrows, M., 1969: The mechanics and neural control of the prey capture and strike in the mantid shrimps *Squilla* and *Hemisquilla*. Z. Vergl. Physiol., *62*(4):361–381.

Caine, E. A., 1974: Comparative functional morphology of feeding in three species of caprellids from the northwestern Florida Gulf Coast. J. Exp. Mar. Biol. Ecol., *15*:81–96.

Caine, E. A., 1978: Habitat adaptations of North American caprellid Amphipoda. Biol. Bull., *155*:288–296.

Caldwell, R. L., 1979: Cavity occupation and defensive behaviour in the stomatopod *Gonodactylus festai*: evidence for chemically mediated individual recognition. Animal Behaviour, *27*:194–201.

Caldwell, R. L., and Dingle, H. 1976: Stomatopods. Sci. Amer. *234*(1):80–89.

Coenen-Strasse, D., 1981: Some aspects of water balance of two desert woodlice, *Hemilepistus aphganicus* and *Hemilepistus reaumuri*. Comp. Biochem. Physiol. A Comp. Physiol., *70*(3):405–420.

Dingle, H., and Caldwell, R. L., 1978: Ecology and morphology of feeding and agonistic behavior in mudflat stomatopods. Biol. Bull., *155*(1):134–149.

Enequist, P., 1949: Studies on the soft-bottom amphipods of the Skagerrak. Zool. Bidrag. Från Uppsala, *18*:297–492.

Farr, J. A., 1978: Orientation and social behavior in the supralittoral isopod *Ligia exotica*. Bull. Mar. Sci., *28*(4):659–666.

Fox, R. S., and Bynum, D. H., 1975: The amphipod crustaceans of North Carolina estuarine waters. Chesapeake Sci., *16*(4):223–237.

Geyer, H., and Becker, G., 1980: Attractive effects of several marine fungi on *Limnoria tripunctata*. Mater Org., *15*(1):53–78.

Grossnickle, N. E., 1982: Feeding habits of *Mysis relicta*: an overview. Hydrobiologia, *93*(½):101–108.

Hamner, W. M., 1984: Aspects of schooling in *Euphausia superba*. Jour. Crust. Biol., *4*(spec. no. 1):67–74.

Hamner, W. M. Hamner, P. P., Strand, S. W., and Gilmer, R. W., 1983: Behavior of Antarctic krill, *Euphausia superba*: chemoreception, feeding, schooling, and molting. Science, *220*:433–435.

Hamner, W. M., Smith M., and Mulford, E. K., 1969: The behavior and life history of a sand-beach isopod, *Tylos punctatus*. Ecology, *50*(3):442–453.

Harbison, G. R., Biggs. D. C., and Madin, L. P., 1977: The associations of Amphipoda Hyperiidea with gelatinous zooplankton: II. Associations with Cnidaria, Ctenophora and Radiolaria. Deep-Sea Res., *24*(5):465–488.

Hartwick, R. F., 1976: Beach orientation in talitrid amphipods: Capacities and strategies. Behav. Ecol. Sociobiol., *1*(4):447–458.

Hassall, M., and Jennings, J. B., 1975: Adaptive features of gut structure and digestive physiology in the terrestrial isopod *Philoscia muscorum*. Biol. Bull., *149*:348–364.

Herring, P. J., and Locket, N. A., 1978: The luminescence and photophores of euphausiid crustaceans. Jour. Zool., *186*:431–462.

Herrnkind, H. F., 1972: Orientation in shore living arthropods, especially the sand fiddler crab. In Winn, H. E., and Olla, B. L. (Eds.): Behavior of Marine Animals. Vol. 1: Invertebrates. Plenum Press, N. Y. pp. 1–59.

Hessler, R. R., and Sanders, H., 1966: *Derocheilocaris typicas* Pennak and Zinn (Mystacocarida) revisited. Crustaceana, *11*(2):141–155.

Holdich, D. M., and Jones, J. A., 1983: Tanaids. Synopses of the British Fauna No. 27. Cambridge University Press, London. 98 pp.

Holsinger, J. R., 1972: The freshwater amphipod crustaceans (Gammaridae) of North America. Biota of Freshwater Ecoystems, Identification Manual 5, U.S. Environmental Protection Agency, Washington, D.C.

Johnson, S. B., and Attramadal, Y. G., 1982: Reproductive behaviour and larval development of *Tanais cavolinii*. Mar. Biol., *71*:11016.

Johnson, S. B., and Attramadal, Y. G., 1982: A functional-morphological model of *Tanais cavolinii* adapted to a tubicolous life-strategy. Sarsia, 67:29–42.

Jones, N. S., 1976: British Cumaceans. Synopses of the British Fauna No. 7. Linn. Soc. London. Academic Press, London. 62 pp.

Laval, P., 1980: Hyperiid amphipods as crustacean parasitoids associated with gelatinous zooplankton. Oceanogr. Mar. Biol. Ann. Rev., 18:11–56.

Madin, L. P., and Harbison, G. R., 1977: The associations of Amphipoda Hyperiidea with gelatinous zooplankton. I. Associations with Salpidae. Deep-Sea Res. 24:449–463.

Manning, R. B., 1969: Stomatopod Crustacea of the Western Atlantic, University of Miami Press, Miami. 380 pp.

Manning, R. B., 1974: Crustacea: Stomatopoda. Marine flora and fauna of the northeastern United States. NOAA Technical Report NMFS Circular 386. U.S. Government Printing Office, Washington, D.C. 6 pp.

Mauchline, J., 1980: The biology of mysids and euphausids. Adv. Mar. Biol., 18:3–677.

McCain, J. C., 1968: The Caprellidae (Crustacea: Amphipoda) of the western North Atlantic. Bull. U.S. Nat. Mus., 278:1–147.

Mendoza, J. A., 1982: Some aspects of the autecology of *Leptochelia dubia*. Crustaceana 43(3):225–240.

Naylor, E., 1972: British Marine Isopods. Synopses of the British Fauna No. 3. Linn. Soc. London. Academic Press, London. 86 pp.

Noodt, W., 1970: Zur Eidonomie der Stygocaridacea, einer Gruppe interstitieller Syncarida. Crustaceana, 19:242–244.

Reaka, M. L., and Manning, R. B., 1981: The behavior of stomatopod Crustacea, and its relationship to rates of evolution. Jour. Crust. Biol., 1(3):309–327.

Rees, C. P., 1975: Life cycle of the amphipod *Gammarus palustris*. Estuarine Coastal Mar. Sci., 3(4):413–419.

Roger, C., 1975: Feeding Rythms and Trophic Organization of a Population of Pelagic Crustaceans. *Mar. Biol.*, 32(4): 365–378.

Ross, R. M., and Quetin, L. B., 1986: How productive are Antarctic krill? BioScience, 36(4):264–269.

Schmalfuss, H., 1975: Morphologie, Funktion und Evolution der Tergithöcker bei Landisopoden Z. Morph. Tiere, 80:287–316.

Schminke, H. K., 1974: Mesozoic intercontinental relationships as evidenced by Bathynellid Crustacea (Syncarida: Malacostraca). Syst. Zool., 23(2):157–164.

Schminke, H. K., 1978: Die phylogenetische Stellung der Stygocarididae unter besonderer Berücksichtigung morphologischer Ähnlichkeiten mit Larvenformen der Eucarida. Z. f. zool. Syst. Evolutionsforsch., 16(3):225–239.

Schminke, H. K., 1981: Adaptation of Bathynellacea to life in the interstitial ("Zoea Theory"). Int. Revue ges. Hydrobiol., 66(4):575–637.

Schmitz, E. H., and Schultz, T. W., 1969: Digestive anatomy of terrestrial Isopoda: *Armadillidium vulgare* and *Armadillidium nasutum*. Amer. Midl. Nat., 82(1):163–181.

Schultz, G. A., 1969: How to Know the Marine Isopod Crustaceans. W. C. Brown Co., Dubuque, Iowa. 359 pp.

Steger, R., and Caldwell, R. L., 1983: Intraspecific deception by bluffing: a defense strategy of newly molted stomatopods. Science, 221:558–560.

Stock, J. H., 1976: A new genus and two new species of the crustacean order Thermosbaenacea from the West Indies. Bidr. Tot. Dierkunde, 46(1):47–70.

Sutton, S. L., 1972: Woodlice. Ginn & Co., London. 144 pp. (An informative general account of terrestrial isopods plus a guide to British species.)

Sutton, S. L., and Holdich, D. M. (Eds.), 1984: The Biology of Terrestrial Isopods. Oxford Univ. Press, Oxford. 518 pp. (Papers from a symposium.)

Vader, W., 1978: Associations between amphipods and echinoderms. Astarte, 11:123–134.

Valentin, C., and Anger, K., 1977: *In situ* studies on the life cycle of *Diastylis rathkei*. Mar. Biol., 39(1):71–76.

Van Name, W. G., 1936: The American land and freshwater isopod crustaceans. Amer. Mus. Bull., 71:1–535. (A very complete taxonomic treatment of the American wood lice and freshwater isopods.)

Williams. W. D., 1972: Freshwater Isopods (Asellidae) of North America. Biota of Freshwater Ecosystems, Identification Manual No. 7. EPA U.S. Government Printing Office, Washington, D.C. 45 pp.

15

The Myriapods

Five groups of related arthropods, including centipedes, millipedes, and insects, are believed to have had a common origin separate from that of other arthropods. They were formerly linked with crustaceans as members of the subphylum Mandibulata, but such shared features as mandibles and compound eyes are believed by many zoologists to represent convergence.

These arthropods are called uniramians because of the apparent unbranched nature of the appendages. The appendages of crustaceans and those of primitive chelicerates, on the other hand, are composed of several divisions (e.g., the biramous, abdominal appendages of shrimps and crabs). Moreover, in crustaceans the mandibles are multiple, jointed appendages in which the basal piece (gnathobase) performs the grinding, biting, or cutting function. Grinding appears to have been the primitive function of the crustacean mandibles, with food moving forward into a posteriorly directed mouth. This is also the primitive condition in chelicerates, such as horseshoe crabs. In uniramians, on the other hand, the mandibles are supposedly unjointed whole limbs without palps and are used to handle food from below, not from behind.

The most distinctive characteristic of uniramians is the presence of a single pair of antennae, which are believed to be appendages of the second head somite, thus corresponding to the first antennae of crustaceans. The gas exchange organs are tracheae, and the excretory organs are malpighian tubules, but both structures originated independently from those of arachnids.

Manton has been the principal proponent of the uniramian concept, and her extensive work on the functional anatomy of centipedes and millipedes has done much to persuade other zoologists. Uniramians are terrestrial animals, and Manton has argued that the arthropods evolved on land independently from the marine chelicerates and crustaceans.

Not all zoologists, especially many entomologists, have been convinced by Manton's arguments, and some recent evidence has raised further doubts. Fossils of some giant Paleozoic insects have been described as having a branch (exite) at the base of their legs (Kukalová-Peck, 1983); they therefore could not be uniramians, that is, have unbranched appendages. Moreover, the molecular structure of the hemocyanin of certain centipedes can be most easily explained as having a common origin with that of chelicerates and crustaceans (Mangum et al, 1985). The oldest uniramians are terrestrial Devonian forms similar to centipedes and millipedes. The small number of marine millipedes and insects and the more numerous freshwater insects are all clearly secondary invaders of the aquatic environment. The freshwater larvae of some insects also represent a specialization, for primitive insects lack larvae.

The most likely ancestors of uniramian arthropods are believed by many zoologists to be the onychophorans (p. 730). These small, terrestrial, caterpillar-like animals are known from the beginning of the Paleozoic and are represented today by a small number of tropical and south temperate species.

Myriapodous Arthropods

Four groups of uniramians comprising some 10,500 species—the centipedes, the millipedes, the pauropods, and the symphylans—have a body composed of a head and an enlongated trunk with many leg-bearing segments. This common feature was formerly considered sufficient reason for uniting all four groups within a single class, the Myriapoda. Although these arthropods are perhaps more closely related to each other than to the Insecta, they exhibit fundamental differences. Most zoologists have therefore abandoned the Myriapoda except as a convenient collective name, and each of the four groups is now considered a separate class.

Most myriapods require a humid environment, for the relatively permeable epicuticle usually lacks the high lipid content found in spiders and insects. The lipids present are probably more important in repelling water (functioning as a hydrofuge) than in reducing water loss. Myriapods live beneath stones and wood and in soil and humus and are widely distributed in both temperate and tropical regions.

The head bears a pair of antennae and usually ocelli, but except in certain centipedes, true compound eyes are never present. The mouthparts lie on the ventral side of the head and are directed forward. An epistome and labrum form the upper lip and the roof of a preoral cavity. The lower lip is formed by either a first or second pair of maxillae, and enclosed within the preoral cavity are the pair of mandibles and a hypopharynx.

Gas exchange is typically by a tracheal system in which the spiracles cannot be closed (another path of water loss). Excretion takes place through malpighian tubules. The heart is a dorsal tube extending through the length of the trunk, with a pair of ostia in each segment. A branched system of arteries is rarely present. The nervous system is not concentrated, and the ventral nerve cord contains a ganglion in each segment. Indirect sperm transfer by spermatophore is highly developed, and the myriapods parallel the arachnids in many aspects of this process.

CLASS CHILOPODA

The members of the class Chilopoda, known as centipedes, are perhaps the most familiar of the myriapodous arthropods. They are distributed throughout the world in both temperate and tropical regions, where they live in soil and humus and beneath stones, bark, and logs. The approximately 2500 described species are distributed within four principal orders (Fig. 15–1). The order Geophilomorpha is composed of long, threadlike centipedes that are adapted for living in soil. The orders Scolopendromorpha and Lithobiomorpha both contain heavy-bodied, flat centipedes that live in crevices beneath stones, bark, and logs and in soil. The Scutigeromorpha are long-legged forms, some of which live in and around human habitations. *Scutigera coleoptrata*, which is found in both Europe and North America, is frequently found trapped in bathtubs and wash basins.

The largest centipede is the tropical American *Scolopendra gigantea*, which may reach almost 30 cm in length. Many other tropical forms, particularly scolopendrids, are over 20 cm in length, but most North American and European species are only 3 to 6 cm long. Temperate zone centipedes are most commonly reddish brown, but many tropical forms, especially the scolopendromorphs, are red, green, yellow, blue, or combinations of colors.

The head is convex in scutigeromorphs but flattened in other centipedes, with the antennae on the front margin (Figs. 15–1A and 15–2A). The mandible, which bears teeth and a thick fringe of setae, lies beneath the ventrolateral surface of the head. Beneath the mandibles are two pairs of maxillae. Covering the mouth appendages are a large pair of forcipules, commonly called poison claws, which are actually the appendages of the first trunk segment (Fig. 15–2A and C). Each appendage is curved toward the midventral line and bears a terminal, pointed fang, which is the outlet for the duct of a poison gland. The large coxae of the forcipules and the associated sternite of that segment form a large plate that covers the underside of the head.

Posterior to the first trunk segment, which carries the forcipules, are 15 or more leg-bearing seg-

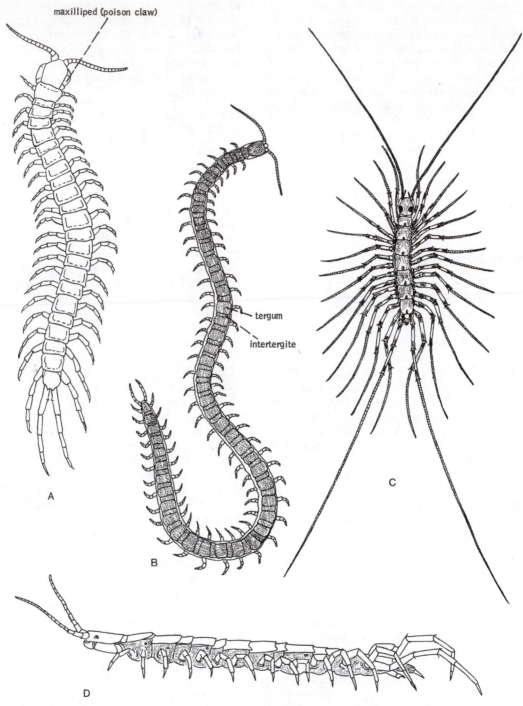

Figure 15–1 Chilopoda: *A, Otocryptops sexspinnosa,* a scolopendromorph centipede. *B,* A geophilomorph centipede. *C, Scutigera coleoptrata,* the common house centipede, a scutigeromorph. *D, Lithobius,* a lithobiomorph centipede. (All after Snodgrass.)

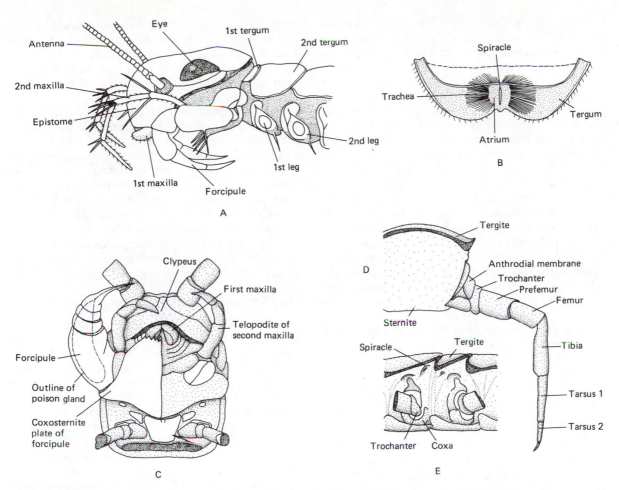

Figure 15-2 *A,* The head of *Scutigera coleoptrata* (lateral view). *B,* Posterior of *Scutigera* tergum showing pair of tracheal lungs. (*A* and *B* after Snodgrass.) *C,* Ventral view of head of *Lithobius forficatus.* Only one forcipule (poison claw) is shown. *D,* Cross section of *Lithobius forficatus. E,* Lateral view of two segments of *Lithobius forficatus.* Only basal leg segments are shown. (*C* after Rilling, 1968. *D* and *E* after Manton, 1965. All from Lewis, J. G. E., 1981: The Biology of Centipedes. Cambridge University Press, London. pp. 9, 11. Copyrighted and reprinted by permission of the publisher.)

ments. The last pair of legs are sensory or defensive and not used in locomotion. The tergal plates vary considerably in size and number, and the differences are correlated with locomotor habits.

Protection

The hiding places of centipedes afford some protection not only from possible predators but also from desiccation. At night they emerge to hunt for food or new living quarters. Scolopendromorphs construct a burrow system, with a chamber into which the animal retreats, in soil or beneath stones and logs.

Although centipedes are equipped with poison claws, there are other adaptations for defense. The last pair of legs in centipedes is the longest, and in some scolopendromorphs they can be used in defense by pinching. Many scolopendromorphs and geophilomorphs possess repugnatorial glands on the ventral side of each segment, and some lithobiomorphs bear large numbers of unicellular repugnatorial glands on the last four pairs of legs, which they will kick in the direction of an enemy, throwing out adhesive droplets.

Locomotion

Two orders of centipedes are adapted for running. The scolopendromorphs have long legs, all approximately the same length, and correspondingly long strides. The scutigeromorphs can run three times faster, however, and many of their structural peculiarities are associated with the evolution of a rapid gait (Manton, 1952). In scutigeromorphs the effective leg stroke is faster than the recovery stroke, unlike the scolopendromorphs, whose strokes are equal in duration. Moreover, scutigeromorphs have a marked progressive increase in leg length from anterior to posterior, which enables the posterior legs to move to the outside of the anterior legs, thus reducing interference. For example, in *Scutigera* the posterior legs are twice as long as the first pair (Fig. 15–1*C*).

To overcome the tendency to undulate, the trunk is strengthened by more or less alternately long and short tergal plates in the lithobiomorphs, and by a reduced number of large, overlapping, tergal plates in the scutigeromorphs. Finally, the annulated distal leg segments of the scutigeromorphs enable the animal to place a considerable section of the end of the leg against the substratum, very much like a foot, to decrease slippage.

In contrast to the other centipedes, the wormlike geophilomorphs are adapted for burrowing through loose soil or humus. The pushing force is provided not by the legs, as in millipedes, but by extension and contraction of the trunk, as in earthworms. A British species of *Stigmogaster*, for example, can increase its body length by as much as 68 per cent. Powerfully developed longitudinal muscles of the body wall, an elastic pleural wall, an increased number of segments, and a small intersternite and intertergite between each of the larger, main sternal and tergal plates—all these facilitate great extension and contraction of the trunk in burrowing (Fig. 15–1*B*). The legs are short and in burrows anchor the body like the setae of an earthworm. Geophilomorphs walk with the legs, but there is little overlap of leg movement.

Nutrition

Most centipedes are believed to be predacious. Small arthropods form the major part of the diet, but there are numerous reports of large scolopendrids feeding on frogs, toads, snakes, birds, and even mice. Some centipedes, especially geophilomorphs, feed on earthworms, snails and nematodes. Prey is detected and located with the antennae, or with the legs in *Scutigera*, and then is captured and killed or stunned with the forcipules. *Scolopendra* will attack glass beads (smeared with a food extract) when it first touches them with its antennae; *Lithobius* will not feed if deprived of its antennae. Large tropical centipedes are often held in dread, but the neurotoxic venom of most forms,

although painful, is not sufficiently toxic to be lethal to man, even to small children. Reports of fatalities from older literature are difficult to authenticate. The effect of the bite of such forms is generally similar to that of a very severe yellow jacket or hornet sting.

Following capture, the prey is held by the second maxillae and the forcipules while the mandibles and first maxillae perform the manipulative action required for ingestion. Geophilomorphs, which possess weakly armed and less mobile mandibles, may partially digest their prey before ingestion. The digestive tract is a straight tube, with the foregut occupying from one seventh to two thirds of the length, depending on the species (Fig. 15–3). The hindgut is short. Salivary secretions are provided by glands associated with the feeding appendages and buccal region.

Gas Exchange and Excretion

Except in the scutigeromorphs, the spiracles of the tracheal system lie in the membranous pleural region above and just behind the coxae (Fig. 15–2*E*). There is basically one pair of spiracles per segment, but some segments lack them, and the pattern of distribution varies in different groups. The unclosable spiracle opens into an atrium lined with cuticular hairs (trichomes), which may reduce desiccation or prevent the entrance of dust particles (Fig. 15–4). The tracheal tubes open at the base of the atrium.

Perhaps associated with their more active habits, and thus with a higher metabolic rate, the tracheal system of the scutigeromorphs is lunglike and probably evolved independently from that of the other centipedes. The spiracles are located middorsally near the posterior margin of the tergal plates covering the leg bearing segments (Fig. 15–2*B*). Each spiracle opens into an atrium from which extend two large fans of short tracheal tubes. The tracheae are bathed in the blood of the pericardial cavity. The blood of scutigeromorphs contains the respiratory pigment hemocyanin, which is absent from other uniramians (Mangum et al, 1985).

There is usually a single pair of malpighian tubules (Fig. 15–3), but much of the nitrogenous waste is excreted as ammonia rather than as uric acid.

Sense Organs

All geophilomorphs and some scolopendromorphs lack eyes. Other centipedes possess few to many ocelli. In the scutigeromorphs the ocelli are so clustered and organized that they form compound

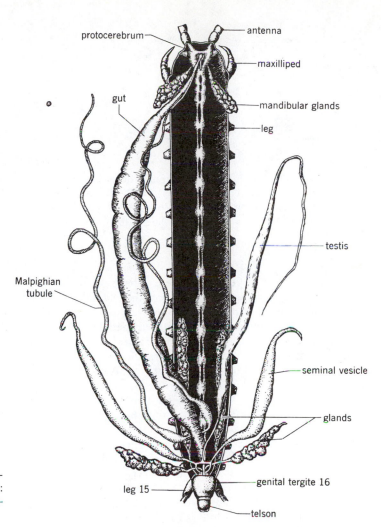

protocerebrum

antenna

maxilliped

gut

mandibular glands

leg

testis

Malpighian tubule

seminal vesicle

glands

genital tergite 16

leg 15

telson

Figure 15–3 Internal structure of the centipede *Lithobius.* (From Kaestner, A., 1968: Invertebrate Zoology. Vol. II. Wiley-Interscience, N.Y.)

eyes. The optical units, of which there up to 200, form a compact group on each side of the head and tend to be elongated with converging optic rods. In *Scutigera* the combined corneal surface is greatly convex, as are the compound eyes of insects and crustaceans, and each unit is remarkably similar to an ommatidium (Fig. 15–2*A*). However, there is no evidence that the compound eyes of *Scutigera*, nor the eyes of any other centipedes, function in more than the simple detection of light and dark. Many centipedes are negatively phototactic.

A pair of organs of Tömösvary are present on the head at the base of the antennae in lithobiomorphs and scutigeromorphs. Each sense organ consists of a disc with a central pore into which the endings of subcuticular sensory cells converge. The few studies on the function of organs of Tömösvary are conflicting. There is evidence for vibration detection and monitoring of humidity.

Centipedes possess various types of sensory hairs and other cuticular sensory structures on the antennae and other parts of the body. The long, last pair of legs of many centipedes have a sensory function, especially in lithobiomorphs and scutigeromorphs; they are modified to form a pair of posteriorly directed, antennae-like appendages.

Reproduction and Development

The ovary is a single, tubular organ located above the gut, and the oviduct opens through a median atrium and aperture onto the ventral surface of the posterior, legless genital segment. A pair of seminal receptacles also open into the genital atrium. In the male one to many testes are located above the midgut. The testes are connected to a single pair of sperm ducts, which open through a median gonopore on the ventral side of the posterior genital seg-

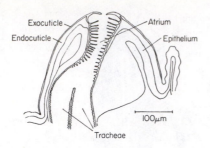

Figure 15–4 Longitudinal section through the spiracle of *Lithobius forficatus*. (From Curry, A., 1974: The spiracle structure and resistance to desiccation of centipedes. *In* Blower, J. G., (Ed.): Myriapoda. Academic Press, London. p. 368.)

ment (Fig. 15–3). The genital segment of both sexes carries small appendages, called gonopods, which function in reproduction.

Sperm transmission is indirect in centipedes, as in other myriapods. Except in scutigeromorphs, the male constructs a little web of silk strands secreted by a spinneret located in the genital atrium. A spermatophore as large as several millimeters is emit-, ted and placed on the webbing. The female picks up the spermatophore and takes it into her reproductive opening. The gonopods of each sex aid in handling the spermatophore. In centipedes the male usually does not produce a spermatophore until a female is encountered. Moreover, there is often initial courtship behavior that varies in detail from species to species. The two sexes may palpate one another's posterior end with their antennae while moving in a circle. This behavior may last as long as one hour before the male spins a spermatophore web and deposits a spermatophore. Following spermatophore deposition, the male "signals" the female in various ways. For example, in species of *Lithobius* the male keeps his posterior pair of legs to either side of the spermatophore and webbing while turning the anterior part of his body and stroking the antennae of the female. She responds by crawling across the posterior end of his body, picking up the spermatophore.

Both the scolopendromorphs and the geophilomorphs brood their eggs in clusters of 15 or more. These centipedes locate themselves in cavities hollowed out in a piece of decayed wood or soil and then wind themselves about the egg mass. The female guards the eggs in this manner through the hatching period and the dispersal of the young (Fig. 15–5*C*). Female lithobiomorphs and scutigeromorphs carry the eggs about for a short time be-

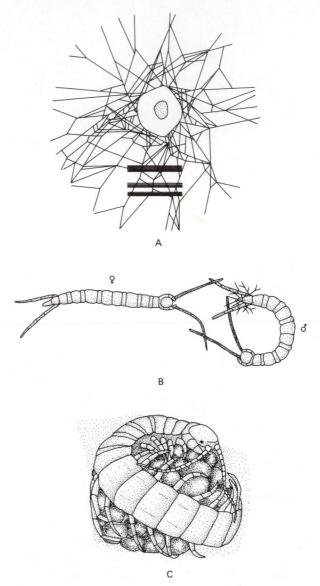

Figure 15–5 *A*, Web and spermatophore of *Lithobius forficatus*. *B*, Male *Lithobius forficatus* with posterior and over web signaling female to pick up spermatophore. (Both after Klingel, 1960, from Lewis, J. G. E., 1981: The Biology of Centipedes. Cambridge University Press, London. p. 281.) *C*, A female of *Scolopendra* brooding her eggs. (After Brehm from Lewis, J. G. E., 1981: The Biology of Centipedes. Cambridge University Press, London. p. 272.)

tween the gonopods and then deposit them singly in the soil.

In the brooding orders, Scolopendromorpha and Geophilomorpha, development is epimorphic; i.e., the young display the full complement of segments when they hatch. Development in the other two orders is anamorphic; i.e., the young on hatching have only a part of the adult complement of segments. For example, newly hatched young in *Scutigera* have 4 pairs of legs and in the subsequent six molts pass through stages with 5, 7, 9, 11, and 13 pairs of legs. The life-span of many centipedes is from four to six years or more (Albert, 1983).

SYSTEMATIC RÉSUMÉ OF CLASS CHILOPODA

Subclass Epimorpha. Eggs brooded; young possess all segments on hatching. Adult has 21 or more pairs of legs.

> Order Geophilomorpha. Slender, burrowing centipedes, with 31 to 170 pairs of legs. Intercalary tergal plates located between tergal plates of more or less equal length. Eyes absent. Widely distributed in both temperate and tropical regions throughout the world. *Geophilus, Strigamia, Mecistocephalus.*

> Order Scolopendromorpha. Most species have 21 pairs of legs, but some possess 23 pairs. Tergal plates not alternating in size. With or without eyes. Many species distributed throughout the world, especially in the tropics. *Scolopendra, Theatops, Otocryptops.*

Subclass Anamorpha. Brooding absent; young do not possess full complement of segments on hatching. Adult has 15 pairs of legs.

> Order Lithobiomorpha. Alternating large and small tergal plates; spiracles paired and lateral. Worldwide in distribution, but most genera and species are found in temperate and subtropical zones. *Lithobius, Bothropolys.*

> Order Scutigeromorpha. Legs and antennae very long. Eyes large and compound. Spiracles unpaired and located middorsally on tergal plates. A single family distributed throughout the world, especially in the tropics. *Scutigera.*

CLASS SYMPHYLA

The Symphyla are a small class containing approximately 120 known species that live in soil and leaf mold in most parts of the world. They have evoked considerable interest among some zoologists as being myriapods that display a number of characteristics like those of insects.

Symphylans are between 2 and 10 mm long and superficially resemble lithobiomorph centipedes (Fig. 15–6A). The trunk contains 12 leg-bearing segments, which are covered by 15 to 24 tergal plates. The last (13th) segment carries a pair of spinnerets, or cerci, and a pair of long, sensory hairs (trichobothria). The trunk terminates in a tiny oval telson.

The head projects in front of the laterally placed antennae (Fig. 15–6B). The mandibles are covered ventrally by a pair of long, first maxillae. The second pair of maxillae are fused, forming a labium (Fig. 15–6C). The apparent similarity of symphylan mouthparts to those of insects has often been cited as evidence for the supposed affinity of the two groups. However, Manton (1977) and others argue that the mouthparts are only superficially similar and are functionally very different. They reject the idea that insects are most closely related to symphylans.

The trunk structure, especially the presence of the additional tergal plates, which increases dorsoventral flexibility, is undoubtedly correlated with the locomotor habits of these animals. Most symphylans can run very rapidly and can twist, turn, and loop their bodies when crawling through the crevices within humus. This ability is probably an adaptation to escape predators, for symphylans feed on living and decayed vegetation. Scutigerellids attack plant roots and can be a serious pest of vegetable and flower crops, especially in greenhouses. A single pair of spiracles open onto the sides of the head, and the tracheae supply only the first three trunk segments. Attached to the body wall beneath the base of each leg are an eversible coxal sac and a small appendage (the stylus), structures also present in primitive insects. The coxal sacs take up moisture. The function of the stylus is unknown, although it is probably of a sensory nature. There are no eyes, but two organs of Tömösvary are well developed (Fig. 15–6B).

The genital openings are located on the ventral side of the fourth trunk segment. The copulatory behavior of *Scutigerella* is known and is most unusual. The male deposits 150 to 450 spermatophores, each at the end of a stalk. The female, on encountering a spermatophore, eats it, but instead of swallowing the sperm, stores the contents of the spermatophore in special buccal pouches. After removing the eggs with her mouth from the single gonopore, she attaches them to the substratum and then works them over with her mouth, smearing each egg with sperm and fertilizing it (Fig. 15–6D and E).

The eggs are laid in clusters of about 8 to 12 and are attached to the walls of crevices or to moss or lichen. Parthenogenesis is common. The role of the

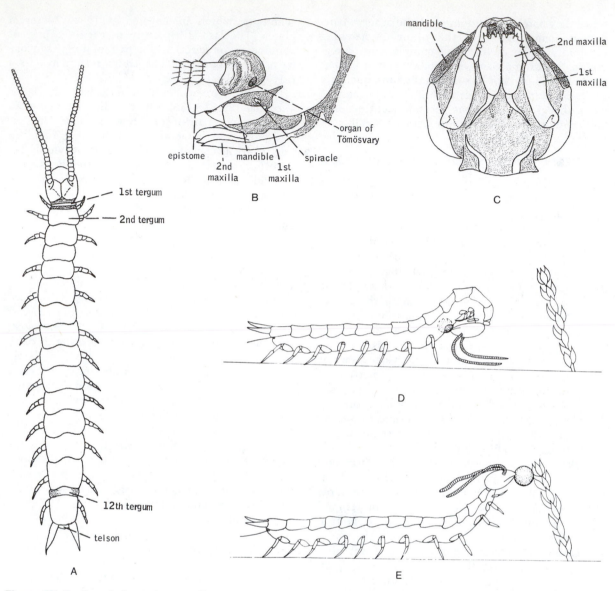

Figure 15–6   Symphyla: *A, Scutigerella immaculata* (dorsal view). *B,* The head of *Hanseniella* (lateral view). *C,* The head of *Scutigerella immaculata* (ventral view). (All after Snodgrass.) *D* and *E,* Female of *Scutigerella* removing egg from gonopore with her mouthparts and attaching it to moss. When carried by the mouthparts, the egg is smeared with semen stored in buccal pouches. (After Juberthie-Jupeau.)

spinning organs in reproduction is unknown. Development is anamorphic; the young on hatching have six or seven pairs of legs. *Scutigerella immaculata* lives as long as four years and molts throughout its lifetime.

CLASS DIPLOPODA

The Diplopoda are commonly known as millipedes or thousand-leggers. They are secretive and largely shun light, living beneath leaves, stones, bark, and logs and in soil. Quite a number of millipedes are cave inhabitants. The more than 7500 described species constitute the greatest number of myriapodous arthropods. They live throughout the world, especially in the tropics, but the best known faunas are those of North America and Europe.

A distinguishing feature of the class is the presence of doubled trunk segments, or diplosegments, derived from the fusion of two originally separate somites. Each diplosegment bears two pairs of legs, from which the name of the class is derived (Figs. 15–7 and 15–8*B* and *C*). The diplosegmented condition is also evident internally, for there are two pairs of ventral ganglia and two pairs of heart ostia within each segment.

The diplopod head tends to be convex dorsally and flattened ventrally (Fig. 15–9). The sides of the head are covered by the convex bases of the very large mandibles. The biting edge of the mandible bears teeth and a rasping surface. The floor of the preoral chamber is formed by the maxilla, often called the gnathochilarium (Figs. 15–8*A* and 15–9). It is a broad, flattened plate attached to the posterior, ventral surface of the head and bearing distally four sensory palps. The head of a diplopod does not contain a second maxillary segment.

The trunk may appear to be dorsoventrally flattened in the so-called flat-backed millipedes because of lateral, shelflike projections of the terga (Fig. 15–7*C*). In the familiar millipedes of the order Julida and a number of large tropical orders, the trunk is essentially cylindrical (Fig. 15–7*F*). Such species are said to be juliform. In primitive diplopods the tergal, sternal, and pleural sclerites composing a segment may be separate and distinct, but coalescence of varying degrees has usually taken place. In flat-backed and juliform millipedes all the sclerites are fused, and in the latter group they form a nearly cylindrical ring.

The extreme anterior segments differ from the others in that the first (the collum) is legless and forms a large collar behind the head (Fig. 15–9), and the second, third and fourth segments carry only a single pair of legs. In some millipedes, such as the flat-backed species, the last one to five segments are also legless. The body terminates in the telson, on which the anus opens ventrally.

The integument is hard, particularly the tergites, and like the crustacean integument, it is impregnated with calcium salts. The surface is often smooth, but in some groups the terga bear ridges, tubercles, spines, or isolated bristles and in the little, soft-bodied pselaphognaths, the body is covered with tufts and rows of scalelike spines (Fig. 15–7*B*).

Diplopods vary greatly in size. The Penicillata contains minute forms, some species of *Polyxenus* being only 2 mm long. The largest millipedes are tropical species of the family Spirostreptidae, which may be 28 cm long.

Most diplopods are black and shades of brown; some species are red or orange, and mottled or spotted patterns are not uncommon. Some southern California millipedes are luminescent.

Locomotion

In general, most species of diplopods crawl slowly about over the ground. According to the extensive studies of Manton (1954) on arthropod locomotion, the gaits of diplopods, although slow, are adapted for exerting a powerful pushing force, enabling the animal to push its way through humus, leaves, and loose soil. The force is exerted entirely by the legs, and it is with the evolution of such a gait that the diplosegmented structure is probably associated. The backward, pushing stroke is activated in waves along the length of the body and is of longer duration than the forestroke. Thus, at any moment, more legs are in contact with the substratum than are raised. The number of legs involved in a single wave is proportional to the amount of force required for pushing. Thus, while the animal is running, as few as 12 legs or less may compose a wave, but when pushing, a single wave may involve as many as 52 legs in some juliform millipedes.

The head-on pushing habit has been most highly developed in the juliform species, which burrow into relatively compact leaf mold and soil. This habit is reflected in the smooth, fused, rigid cylindrical segments, the rounded head, and the placement of the legs close to the midline of the body. The flat-backed millipedes, which are the most powerful, open up cracks and crevices by pushing with the whole dorsal surface of their bodies. The lateral keels in these millipedes provide a protected working space for the more laterally placed legs. Ability to climb is particularly striking in some colobognaths and lysiopetalids, which inhabit rocky situations. These millipedes can climb up smooth surfaces by gripping with opposite legs. These rock dwellers also include the swiftest of the millipedes. Speed is correlated with their predatory and scavenging feeding habits and the need to cover great distances in finding food.

Protection

To compensate for the lack of speed in fleeing from predators, a number of protective mechanisms have evolved in millipedes. The calcareous exoskeleton offers some protection to the upper and lateral sides of the body. The long, many-segmented millipedes, such as the colobognaths and the juliform groups, protect the more vulnerable ventral surface by coiling the trunk spirals when at rest or disturbed. Members of the order Glomerida (superorder Oniscomorpha), called pill millipedes, as well as some others, can roll up into

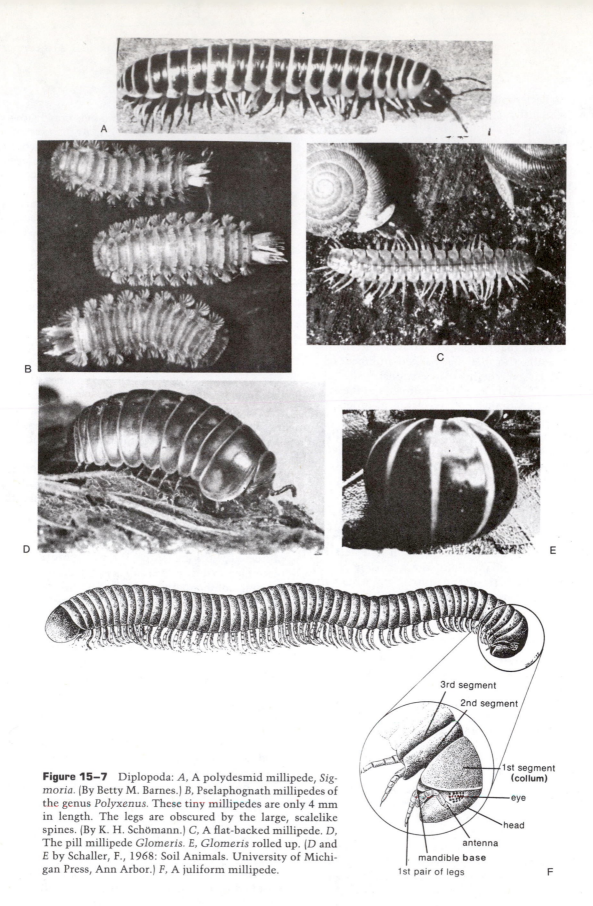

Figure 15–7 Diplopoda: *A*, A polydesmid millipede, *Sigmoria*. (By Betty M. Barnes.) *B*, Pselaphognath millipedes of the genus *Polyxenus*. These tiny millipedes are only 4 mm in length. The legs are obscured by the large, scalelike spines. (By K. H. Schömann.) *C*, A flat-backed millipede. *D*, The pill millipede *Glomeris*. *E*, *Glomeris* rolled up. (*D* and *E* by Schaller, F., 1968: Soil Animals. University of Michigan Press, Ann Arbor.) *F*, A juliform millipede.

3rd segment
2nd segment
1st segment (collum)
eye
head
antenna
mandible base
1st pair of legs

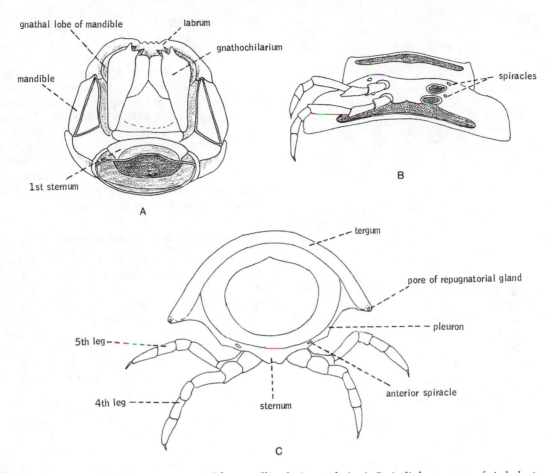

Figure 15–8 *A*, Head of *Habrostrepus*, a juliform millipede (ventral view). *B*, A diplosegment of *Apheloria*, a flat-backed millipede (ventral view). *C*, A diplosegment of *Apheloria* (transverse section). (All after Snodgrass.)

a ball (Fig. 15–7*D* and *E*). When rolled up, some tropical species are larger than golf balls.

Repugnatorial glands are present in many millipedes, including the flat-backed and juliform groups. There is usually only one pair of glands per segment, and the openings are located on the sides of the tergal plates or (in the flat-backed millipedes) on the margins of the tergal lobes (Fig. 15–8*C*). Each gland consists of a large secretory sac, which empties into a duct and out through an external pore. The principal component of the secretion varies in different species. Aldehydes, quinones, phenols, and hydrogen cyanide have been identified. The hydrogen cyanide is liberated when a precursor and an enzyme are mixed from a double-chambered gland. The secretion is toxic or repellent to other small animals; the secretion of some large tropical species is reportedly caustic to human skin. The fluid is usually exuded slowly, but large, tropical, juliform spirobolids can discharge it as a spray or jet for 20 to 30 cm.

Nutrition

Most millipedes are herbivorous, feeding mostly on decomposing vegetation. Food is usually moistened by secretions and chewed or scraped by the mandibles. However, in the tropical Siphonophoridae, the labrum and gnathochilarium are modified to form a long, piercing beak for feeding on plant juices.

A carnivorous or omnivorous diet has been adopted by the rock-inhabiting lysiopetalids and some other millipedes. It has been reported that prey includes phalangids, insects, centipedes, and earthworms. Like earthworms, some millipedes ingest soil from which organic matter is digested.

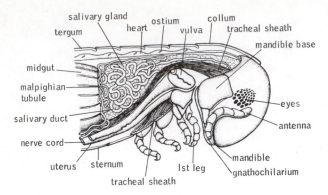

Figure 15–9 Head and anterior trunk segments of the juliform millipede *Narceus* (lateral view). (After Buck and Keister.)

The digestive tract is typically a straight tube with a long midgut. Salivary glands open into the preoral cavity. The midgut produces a peritrophic membrane that surrounds the food, as in insects (see p. 691).

Gas Exchange, Internal Transport, and Excretion

There are four spiracles per diplosegment, located on the sterna, an unusual position among arthropods (Fig. 15–8*B* and *C*). Each spiracle opens into an internal tracheal pouch from which arise numerous trachae.

The heart ends blindly at the posterior end of the trunk, but anteriorly a short aorta continues into the head (Fig. 15–9). There are two pairs of lateral ostia for each segment. Two malpighian tubules arise from each side of the midgut-hindgut junction and are often long and looped. Like centipedes, millipedes excrete more ammonia than uric acid. Also like centipedes, millipedes possess nephridia-like maxillary glands, but their function is still unknown.

Although most millipedes cannot tolerate desiccating conditions, some colobognaths and lysiopetalids live in arid habitats. These species possess coxal sacs, which supposedly take up water, such as dew drops. The ability of many millipedes to coil the body or roll up into a ball may contribute to reduction of water loss when they are inactive. The appearance of large numbers of ground millipedes on tree trunks, rocks, or walls is probably related to humidity, the animals tending to move upward when the air is more saturated with water.

Sense Organs

Eyes may be totally lacking, as in the flat-backed millipedes, or there may be 2 to 80 ocelli arranged about the antennae (Fig. 15–9). Most millipedes are negatively phototactic, and even those without eyes have integumental photoreceptors. The antennae contain tactile hairs and peglike and conelike projections richly supplied with what are probably chemoreceptors. The animal continually taps the substratum with the antennae as it moves along.

As in centipedes, organs of Tömösvary are present in many millipedes and may have an olfactory function or monitor water vapor.

Reproduction and Development

A pair of long, fused, tubular ovaries lie between the midgut and the ventral nerve cord. Two oviducts extend anteriorly to the third, or genital, segment, where each opens into a protractable, pouchlike atrium, or vulva behind the coxae of the second pair of legs (Fig. 15–9 and 10*C* and *D*). When retracted, a vulva is covered externally by a sclerotized, hoodlike piece. At the bottom of the vulva, a groove leads into a seminal receptable.

The testes occupy positions corresponding to those of the ovary but are paired tubes with transverse connections. Anteriorly, near the region of the genital segment (the third), each testis passes into a sperm duct, which either opens through a pair of penes on or near the coxae of the second pair of legs or opens through a single median penis into a median groovelike depression between the coxal bases.

Sperm transfer in millipedes is indirect. The actual copulatory organs are usually modified trunk appendages (gonopods) (Fig. 15–10*A* and *B*), and these are critical structures in identification of species. In most millipedes one or both pairs of legs of the seventh diplosegment serve as gonopods. In the flat-backed millipedes, *Apheloria*, for example, in which the first pair of legs on the seventh segment are gonopods, the male charges the gonopods with sperm by bending the anterior part of the body ventrally and posteriorly. In this position sperm are deposited from the two coxal penes on the third segment into the reservoir of the gonopods. In the juliform order Spirobolida, both pairs of legs of the seventh segment are gonopods, but the first pair forms a protective shield over the second pair, which contains sperm reservoirs and canals.

Males communicate to females their identity and intent in a variety of ways (Haacker, 1974).

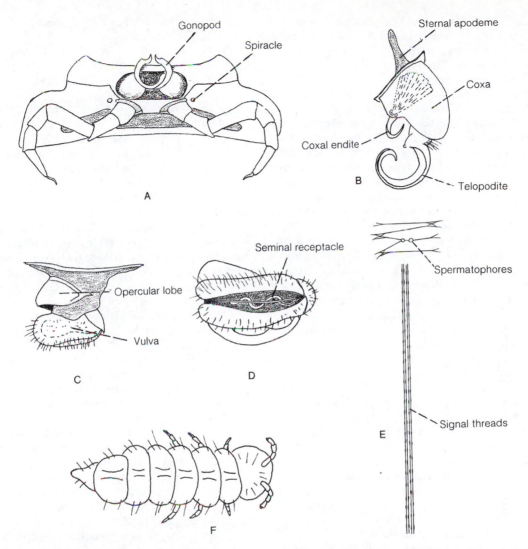

Figure 15–10 *A*, Seventh diplosegment of a male *Apheloria*, showing gonopods and legs (ventral view). *B*, Left gonopod of *Apheloria*. *C*, Right vulva of third segment of *Apheloria* (lateral view). *D*, Vulva (ventral view). (*A* to *D* after Snodgrass.) *E*, Signal threads leading to spermatophore web of *Polyxenus*. (After Schömann.) *F*, A newly hatched millipede. (After Cloudsley-Thompson, J. L., 1958: Spiders, Scorpions, Centipedes and Mites. Pergamon Press, N.Y.)

The signal is tactile in most juliform millipedes, when the male climbs on the back of the female, clinging by special leg pads. Antennal tapping, head drumming, and stridulating are utilized by various other millipedes. Many millipedes produce pheromones, which either initiate mating behavior or continue the sequence of steps initiated by other signals.

During copulation the body of the male is twisted about or stretched out against that of the female so that the gonopods are opposite the vul-vae, and the body of the female is held by the legs of the male. The gonopods are protracted, and sperm are transferred through the ringlike tip of the gonopod (telopodite) into the vulva.

The diplopod eggs are fertilized at the moment of laying, and anywhere from 10 to 300 eggs are produced at one time, depending on the species. Some deposit their eggs in clusters in soil or humus; others, such as *Narceus*, regurgitate a material that is molded into a cup with the head and anterior legs. A single egg is laid in the cup, which is then sealed and polished. The capsule is deposited in humus and crevices, and it is eaten by the young millipede on hatching. The European pill millipede *Glomeris* has similar habits but forms the capsule with excrement.

Many millipedes construct a nest for the deposition of the eggs. Some flat-backed species and colobognaths construct the nest from excrement, building a thin-walled, domed chamber topped by a chimney. The vulvae are applied against the chimney opening, and the eggs fall into the chamber as they are laid. The opening is then sealed, and the chamber is covered with grass and other debris. The female, and in some species the male, may remain coiled about the nest for several weeks.

Some flat-backed and juliform millipedes construct the nest in soil, reinforcing the walls from the inside with excrement. The flat-backed *Strongylosoma pallipes* builds several such nests, each containing 40 to 50 eggs, and closes them from the outside. The Chordeumatida enclose their eggs within silk cocoons, and *Polyxenus* covers its egg clusters with shed tail setae.

Development is anamorphic. The eggs of most species hatch in several weeks, and the newly hatched young usually have only the first three pairs of legs and not more than seven trunk rings (Fig. 15–10*F*). With each molt, additional segments and legs are added to the trunk. Many millipedes undergo ecdysis within specially constructed molting chambers similar to the egg nests, and it is within the molting chamber that many tropical species survive the dry season. The shed exoskeleton is generally eaten, perhaps to aid in calcium replacement. Millipedes live from one to ten or more years, depending on the species.

Parthenogenesis is common in the pselaphognaths, and males are rare.

SYSTEMATIC RÉSUMÉ OF CLASS DIPLOPODA

Subclass Pencillata (Pselaphognatha). Minute millipedes with soft integument bearing tufts and rows of serrated scalelike setae. Trunk bears 13 to 17 pairs of legs. No gonopods. No repugnatorial glands. Fewer than 100 species, but the group is represented throughout the world. *Polyxenus, Lophoproctus.*

Subclass Pentazonia. Tergal plates arched. Last two pairs of legs modified for clasping.

  Order Glomeridesmida (Limacomorpha). Small, eyeless, tropical millipedes. Trunk is composed of 22 arched segments. Cannot roll

into a ball. No repugnatorial glands. *Glomeridesmus.*

  Order Sphaerotheriida. Giant pill millipedes. Trunk with 13 tergites. Mostly tropical; South Africa, Asia, Australia, and New Zealand.

  Order Glomerida (Oniscomorpha). Pill millipedes. Trunk is covered with 11 to 12 arched tergites; the second and last are enlarged, enabling body to roll into a tight ball that conceals head and legs. No repugnatorial glands. Largely Palearctic. *Glomeris* is common in Europe.

Subclass Helminthomorpha. Segments more or less cylindrical or somewhat flattened in cross section. At least one pair of legs (gonopods) of the seventh segment in the male is modified for sperm transfer. This subclass contains the greatest number of millipede species. Only half of the 11 orders are mentioned here.

  Order Polyzoniida. Gonopods leglike and preceded by eight pairs of ordinary legs. *Polyzonium.*

  Order Spirobolida. Cylindrical body composed of 35 to 60 ringlike segments. Gonopods modified. Chiefly tropical millipedes, somewhat similar to the temperate juliform species. *Narceus, Rhinocricus.*

  Order Spirostreptida. Segments cylindrical. First pair of legs modified in male. Anterior pair of gonopods of male modified for sperm transfer; posterior pair commonly reduced. This tropical order contains the largest species of millipedes. *Orthoporus.*

  Order Julida. Trunk composed of 30 to 90 cylindrical segments. Sternites fused with pleurotergal arch. In general, smaller than millipedes of the previous two orders. Both pairs of legs of seventh segment modified as gonopods in male. Mostly temperate species. Widespread. *Julus, Blaniulus, Nemasoma, Cylindroiulus.*

  Order Polydesmida. Flat-backed millipedes. No eyes. Trunk usually composed of 20 rings with prominent, lateral, tergal keels. Sternites fused with pleurotergal arch. Widely distributed: many species. *Polydesmus, Oxidus, Apheloria.*

  Order Chordeumatida. Usually with eyes. Trunk composed of 26 to 32 rings, which are either cylindrical or with lateral tergal keels. End of trunk bears no spinnerets. Sternites not fused with pleurotergal arch. No repug-

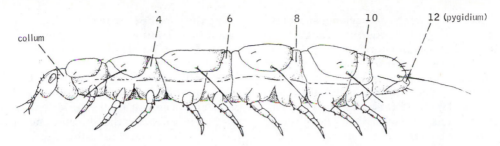

Figure 15–11 The pauropod *Pauropus silvaticus* (lateral view). (After Tiegs from Snodgrass.)

natorial glands. These species have longer legs and move more rapidly than most other millipedes. Widely distributed. *Cleidogona*, *Chordeuma*.

CLASS PAUROPODA

The pauropods constitute a small class of soft-bodied, rather grublike animals that inhabit leaf mold and soil (Fig. 15–11). All are minute, ranging from 0.5 to 2 mm in length. Although once considered rare, pauropods have now been found to be frequently abundant in forest litter. There are approximately 380 described species, which are widespread in both temperate and tropical regions.

Pauropods are similar to millipedes in a number of ways. The trunk usually contains 11 segments, 9 of which each bear a pair of legs. The first segment (collum) and the 11th segment and telson are legless. Certain of the dorsal tergal plates are very large and overlap adjacent segments. Five of the terga carry a pair of long, laterally placed setae. Unlike the collum of a diplopod, that of the pauropod is very inconspicuous dorsally and expanded ventrally.

On each side of the head there is a peculiar, disclike sensory organ that is perhaps homologous to the organ of Tömösvary of other myriapods. The antennae are biramous. One division terminates in a single flagellum; the other in two flagella and a peculiar club-shaped sensory structure. The mandibles are adapted for grinding or piercing. The lower lip is probably homologous to the gnathochilarium of diplopods, for it apparently represents the first maxillae.

Most pauropods feed on fungi or decomposing plant tissue, but some are predatory. There is neither heart nor (except in some primitive species) trachea, their absence probably being associated with the small size of these animals.

As in diplopods, the third trunk segment is the genital segment. Sperm are transferred via a spermatophore, which is deposited by the male along with two signal threads in the female's absence. The eggs are laid in humus, either singly or in clusters. Development is anamorphic, and as in diplopods, the young hatch with only three pairs of legs. In *Pauropus sylvaticus*, development to sexual maturity takes about 14 weeks.

Summary

1 Members of the subphylum Uniramia include the centipedes, millipedes, and insects: terrestrial arthropods with appendages that are primitively unbranched. They possess one pair of antennae, and the mouthparts include a pair of mandibles. The mandible is an unsegmented whole limb, and in the primitive condition food is not brought forward from behind, as in other arthropods, but is picked up directly beneath the mouth.

2 The uniramians are believed to have evolved from terrestrial ancestors, which may have been early members of the phylum Onychophora.

3 The myriapodous arthropods include the centipedes (class Chilopoda) and millipedes (class Diplopoda), plus two other small classes (Symphyla and Pauropoda). All have long trunks with many segments and appendages. Tracheae provide for gas exchange and malpighian tubules for excretion.

4 Myriapods live in leaf litter and beneath stones, logs, and bark. Many of their structural features are adaptations for locomotion.

5 Centipedes possess one pair of legs per segment. In many groups the trunk has been strengthened for a running gait by overlapping tergites or tergites of unequal size, the larger extending onto adjacent segments.

6 Centipedes are largely predacious, and prey, mostly other small arthropods, are caught and killed with a pair of anterior forcipules.

7 Millipedes possess two pairs of legs per segment, a condition derived from the fusion of two original segments. The millipede diplosegments appear to be an adaptation for a pushing gait. The trunk is strengthened to withstand the pushing force generated by a large number of legs.

8 Most millipedes feed on decomposing vegetation. Depending on the group, protection is gained from repugnatorial glands, coiling and rolling up.

9 Both centipedes and millipedes transfer sperm indirectly by spermatophores. The gonopores are located at the posterior end of the trunk in centipedes and at the anterior end of the trunk (third trunk segment) in millipedes.

REFERENCES

The literature included here is restricted in large part to myriapods. The introductory references on page 8 include many general works and field guides that contain sections on these animals.

Albert, A. M., 1983: Life cycle of Lithobiidae, with a discussion of the r- and K-selection theory. Oceologia, 56:272–279.

Anderson, D. T., 1973: Embryology and Phylogeny in Annelids and Arthropods. Pergamon Press, N.Y. 495 pp. (Detailed accounts of the embryology of onychophorans and uniramian arthropods and the phylogenetic implications of the embryonic patterns.)

Attems, G., 1926–1940: *In* W. Kukenthal and T. Krumbach (Eds.): Handbuch der Zoologie, Vol. 4, Progoneata, Chilopoda, 1926; Vol. 52, Myriapodia, Geophilomorpha, 1929; Vol. 54, Chilopoda, Scolopendromorpha, 1930; Vols. 68–70, Diplopoda, Polydesmoidea, 1937–1940. W. de Gruyter, Berlin and Leipzig. (This and the works of Verhoeff (see below) contain the most extensive and detailed accounts of the myriapodous classes.)

Blower, J. G., (Ed.), 1974: Myriapoda. Academic Press, London. (Papers presented at the Second International Congress of Myriapodology at the University of Manchester in 1972.)

Buck, J. B. and Keister, M. L., 1950: *Spirobolus marginatus. In* F. A. Brown (Ed.): Selected Invertebrate Types. John Wiley and Sons, N.Y., pp. 462–475.

Camatini, M. (Ed.), 1979: Myriapod Biology. Academic Press, London. 456 pp. (A collection of papers from an international symposium on myriapods.)

Cloudsley-Thompson, J. L., 1948: *Hydroschendyla submarina* in Yorkshire: with an historical review of the marine Myriapoda. Naturalist, 827:149–152.

Cloudsley-Thompson, J. L., 1958: Spiders, Scorpions, Centipedes, and Mites. Pergamon Press, N.Y. (A discussion of the natural history and ecology of the myriapodous arthropods is presented in Chapters 2, 3 and 4.)

Haacker, U., 1974: Patterns of communication in courtship and mating behavior of millipedes. (See Blower, J. G., pp. 317–328.)

Hoffman, R. L., 1979: Classification of the Diplopoda. Museum d'Histoire Naturelle, Geneve. 237 pp.

Hoffman, R. L., and Payne, J. A., 1969: Diplopods as carnivores. Ecology, 50(6):1096–1098.

Kaestner, A., 1968: Invertebrate Zoology. Vol. 2. Wiley-Interscience, N.Y. 472 pp.

Klingel, H., 1960: Die Paarung des *Lithobius forficatus.* Verh. dt. zool. Ges., 23:326–332.

Kukalová-Peck, J., 1983: Origin of the insect wing and wing articulation from the insect leg. Canadian Jour. Zool., 61(7):1618–1669.

Lewis, J. G. E., 1981: The Biology of Centipedes. Cambridge University Press, London. 476 pp.

Mangum, C. P., et al, 1985: Centipedal hemocyanin: Its structure and its implications for arthropod phylogeny. Proc. Natl. Acad. Sci., 82:3721–3725.

Manton, S. M., 1952–1961: The evolution of arthropodan locomotory mechanisms. J. Linn. Soc. (Zool.), 1952, Pt. 3, The locomotion of the Chilopoda and Pauropoda, 42:118–166; 1954, Pt. 4, The structure, habits, and evolution of the Diplopoda, 42:229–368; 1956, Pt. 5, The structure, habits, and evolution of the Pselaphognatha (Diplopoda), 43:153–187; 1958, Pt. 6, Habits and evolution of the Lysiopetaloidea (Diplopoda), some principles of the leg design in Diplopoda and Chilopoda, and limb structure in Diplopoda, 43:487–556; 1961, Pt. 7, Functional requirements and body design in Colobognatha (Diplopoda), together with a comparative account of diplopod burrowing techniques, trunk musculature, and segmentation, 44:383–461.

Manton, S. M., 1964: Mandibular mechanisms and the evolution of Arthropods. Phil. Trans. R. Soc., London, B, 247:1–183.

Manton, S. M., 1965: The evolution of arthropod locomotory mechanisms. Pt. 8, Functional requirements and body design in Chilopoda, together with a comparative account of their skeletomuscular systems and an appendix on the comparison between burrowing forces of annelids and chilopods and its bearing upon the evolution of the arthropodan haemocoel. Jour. Linn. Soc. (Zool.), 46:251–483.

Manton, S. M., 1973: Arthropod phylogeny—a modern synthesis. J. Zool., 171:111–130

Manton, S. M., 1973: The evolution of arthropodan locomotory mechanisms. Pt. 2, Habits, morphology and evolution of the Uniramia (Onychophora, Myriapoda, Hexapoda) and comparisons with the Arachnida, together with a functional review of uniramian musculature. Zool. J. Linn., 53:257–375.

Manton, S. M., 1977: The Arthropoda: Habits, Functional Morphology, and Evolution. Clarendon Press, Oxford. 527 pp. (A synthesis of the author's lifelong study of the functional morphology of arthropod limbs and its implications for arthropod evolution.)

Moore, R. C. (Ed.), 1969: Treatise on Invertebrate Paleontology. Pt. R., Arthropoda 4, Vol. 2. Geological Society of America and University of Kansas Press, Lawrence, Kansas. (This volume covers the myriapods.)

Rilling, G., 1968: *Lithobius forficatus. In* Grosses Zoologisches Praktikum, Pt. 13b. Stuttgart: Fischer.

Sakwa, W. N., 1974: A consideration of the chemical basis of food preference in millipedes (see Blower, J. G., pp. 329–346).

Schaller, F., 1968: Soil Animals. University of Michigan Press, Ann Arbor. 144 pp.

Sharov, A. G., 1966: Basic Arthropodan Stock. Pergamon Press, N.Y.

Verhoeff, K. W., 1926–1934: *In* H. G. Bronn (Ed.): Klassen und Ordnungen des Tierreichs. Chilopoda, Bd. 5, II (1); Diplopoda, Bd. 5, II (2); Symphyla and Pauropoda, Bd. 5, II (3).

16

The Insects

The class Insecta, or Hexapoda, containing more than 750,000 described species, is the largest group of animals; in fact, it is larger than all the other animal groups combined. Only a brief, rather superficial treatment of insects is possible here. For more extensive accounts, especially those dealing with the details of morphology of the insect orders, the student must refer to textbooks of entomology.*

Insects are distinguished from other arthropods by having three pairs of legs and usually two pairs of wings carried on the middle, or thoracic, region of the body. In addition, the head typically bears a single pair of antennae and a pair of compound eyes. A tracheal system provides for gas exchange, and the gonoducts open at the posterior end of the abdomen.

The success of insects is evidenced by the tremendous number of species and individuals and by their great adaptive radiation. Although they are

*This chapter has been designed to meet the requirements of those invertebrate zoology courses that include a brief coverage of insects. This account is not intended to be equivalent to the extensive treatment accorded to the other invertebrate groups.

essentially terrestrial animals and have occupied virtually every environmental niche on land, insects have also invaded the aquatic habitats and are absent only from the deeper waters of the sea. This success of insects can be attributed to a number of factors, but certainly the evolution of flight endowed these animals with a distinct advantage over other terrestrial invertebrates. Dispersal, escape from predators, and access to food or optimum environmental conditions were all greatly enhanced. The powers of flight also evolved in reptiles, birds, and mammals, but the first flying animals were insects.

Insects are of great ecological significance in the terrestrial environment. Two thirds of all flowering plants depend on insects for pollination. Insects are also of enormous importance for humans. Mosquitoes, lice, fleas, bedbugs, and a host of flies can contribute directly to human misery. More importantly, these and others affect us indirectly as vectors of human diseases or of diseases of domesticated animals: mosquitoes (malaria, elephantiasis, and yellow fever); tsetse fly (sleeping sickness); lice (typhus and relapsing fever); fleas (bubonic plague); and the housefly (typhoid fever and dysentery). Our domesticated plants are dependent on some insects for pollination but are destroyed by others. Vast sums are expended to control insect pests, which can greatly reduce the highagricultural yields neccessary to support large human populations. But the overzealous use of pesticides can in turn be hazardous to the environment.

External Morphology

Although the exoskeleton of an arthropod segment is composed of a tergum, a sternum, and two pleura, some parts are more highly sclerotized or are more conspicuous than others. Such thickened areas of cuticle, called sclerites, are prominent features of the insect body surface. They are often separated by sutures, which reflect ridges on the inner side of the skeleton. The sclerites and sutures have received detailed attention from insect morphologists, and an extensive anatomical nomenclature has been developed. Only those general features of insect anatomy that provide a basis for comparison with other arthropods will be discussed in the following paragraphs.

Body Regions

The heads of most insects are oriented so that the mouthparts are directed downward (hypognathus; Fig. 16–1*B*). A more specialized, anteriorly directed position (prognathus) is found in some predacious species, such as the carabid and tiger beetles; a posteriorly directed position (opisthognathus) is present in the hemipterans and homopterans, which have sucking beaks. The head skeleton of most insects forms a complete, external capsule surrounding the soft, inner tissues. The more lateral and dorsal surfaces of the head bear one pair of compound eyes and one pair of antennae. Between the eyes and the antennae are usually three ocelli (Fig. 16–1*A*). Three pairs of appendages contribute to the mouthparts (Fig. 16–1*A* and *B*). One pair of mandibles is located anteriorly, followed by a pair of maxillae and then by the labium. Although single, the labium actually represents a fused pair of second maxillae. Anteriorly, the mandible is covered by a shelflike extension of the head, forming an upper lip, or labrum. From the floor of the prebuccal cavity projects a median lobelike process, called the hypopharynx. The hypopharynx arises behind the maxillae near the base of the labium. The modifications of the mouthparts associated with different feeding habits will be discussed later.

The thorax, which forms the middle region of the insect body, is composed of three segments—a prothorax, a mesothorax, and a metathorax. On each of the three segments a pair of legs articulates with the pleura (Fig. 16–1*C*). The thoracic terga of insects are called nota, and it is with the notal and pleural processes of the mesothorax and the metathorax that the two pairs of wings articulate. The basal section of the leg articulating with sclerites in the pleural area is the coxa, which is followed by a short trochanter (Fig. 16–1*D*). The remaining sections consist of a femur, a tibia, a tarsus, and a pretarsus. The tarsus is composed of one to five segments. The pretarsus is represented chiefly by a pair of claws. In other arthropods, such as arachnids, the pretarsus is not usually considered separate from the tarsus. The legs of insects are generally adapted for walking or running, and during their effective strokes, the forelegs pull while the middle and hind legs push. The middle legs step outside of the other two pairs, which reduces interference. One or more pairs may be modified for such functions as grasping prey, jumping, swimming, and digging.

The abdomen is composed of 9 to 11 segments, but the 11th segment is complete only in the primitive proturans and in embryos (Fig. 16–1*E*). The telson is vestigial. The only abdominal appendages in the adult are a terminal pair of sensory cerci borne on the 11th segment. The reproductive structures are thought by some entomologists to represent segmental appendages, but except for the female ovipositors, this is by no means certain. A variety of abdominal appendages serving different functions are present in many insect larvae.

Insect Wings and Flight

Wings are characteristic features of insects, but a wingless condition occurs in a number of groups. In some the absence of wings is obviously secondary. For example, ants and termites have wings only at certain periods of the life cycle; workers always lack wings. Some parasitic insect orders, such as the lice and fleas, have lost the wings completely. On the other hand, it is fairly certain that the insects in the orders Protura, Thysanura, Collembola, and a few others arose from ancestral wingless insects, independently from the stock in which wings evolved. Those groups in which the wingless condition is considered primary are clas-

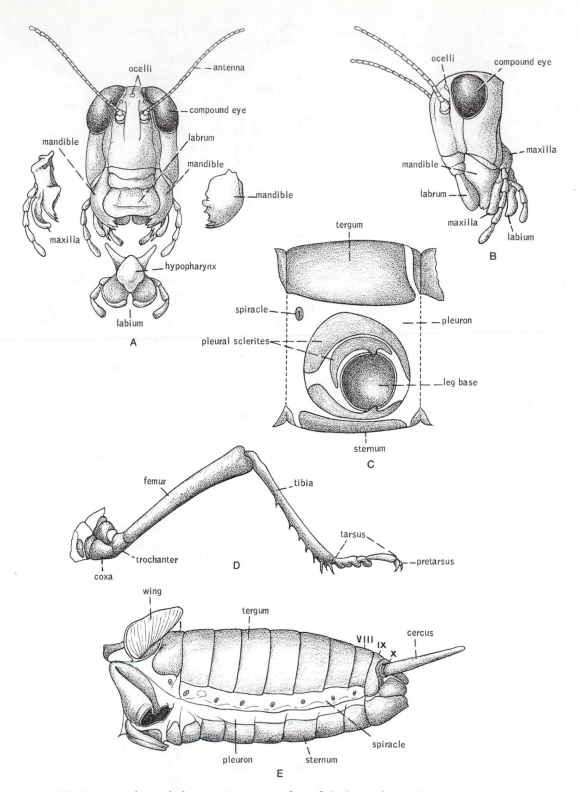

Figure 16–1   External morphology: *A*, Anterior surface of the head of a grasshopper. *B*, Lateral view of the head of a grasshopper. (After Snodgrass from Ross.) *C*, Lateral view of a wingless thoracic segment. *D*, Leg of a grasshopper. *E*, Lateral view of the abdomen of a male cricket. (*C–E* from Snodgrass.)

sified as the subclass Apterygota and are believed to represent the most primitive members of the class. Winged insects and those that are secondarily wingless form the subclass Pterygota. Many contemporary entomologists prefer to limit the class to the pterygotes. They consider the wingless apterygotes to be related to insects but not insects themselves.

The sequence of events in the evolutionary development of insect wings is unknown. The earliest known fossil insects are from the Devonian period and include both wingless and a primitive, winged species. No intermediate types have yet been discovered. The most widely accepted theory of wing origin is that wings were originally flat, lateral flanges of the notum, which enabled the insect to alight right side up when jumping. These flanges then gradually enlarged into winglike structures, making gliding possible. The last step was the development of hinges, enabling wings to move. The earliest functional wing was thought to be a fan-shaped membrane with trusslike supporting veins.

Since wings are evaginations, or folds, of the integument, they are composed of two sheets of cuticle. At a vein the two cuticular membranes are thickened and separated by a trachea, resulting in a tubular lumen surrounded by heavy cuticle; the veins thus form an effective supporting skeletal rod for the wing. Wing veins open into the body and contain circulating blood. The lumina of the main veins contain, in addition to blood, tracheoles and sensory nerve branches.

The wings of the more primitive insects are net-like, but there has been a general tendency in the evolution of wings toward reduction to a few longitudinal veins and cross veins, thus giving a stronger support system to the wing. The arrangement of veins in a wing is very specific in certain genera and families of insects and provides a useful tool for systematists; the principal veins and their branches are all named and numbered.

Primitive wings are held outstretched, as in dragonflies. The evolution of sclerites in the wing base, which permits many insects to fold the wings over the abdomen and thus keep them out of the way when at rest, was an important event in the evolution of the class. The ancestors of many modern orders were then able to radiate into microhabitats, beneath bark and stones and in soil, dung, and wood, where outstretched wings would have been a serious impediment. Wing folding was probably accompanied by reduction in the body size of many groups.

There is great variation in the wings of insects. Many of these variations represent modifications that accommodate the demands of flight characteristic to the particular group of insects. Primitively, as in damselflies, roaches, and termites, the two pairs of wings beat independently of each other, but this requires that the hind wings operate in the air turbulence created by the forewings. Thus, in many insects the two wings on each side are coupled by interlocking devices or by simple overlapping so that the wings operate together. Only the second pair of wings is used for flight in beetles; the front pair has been adapted as hard protective plates, called elytra (Fig. 16–13).

Each wing articulates with the edge of the tergum, but its inner end rests on a dorsal pleural process, which acts as a fulcrum (Fig. 16–2). The wing is thus somewhat analogous to a seesaw off center. Upward movement of the wing results indirectly from the contraction of vertical muscles within the thorax, depressing the tergum. Downward movement of the wings is produced directly, by contraction of muscles attached to the wing base (dragonflies and roaches), indirectly by the contraction of transverse horizontal muscles raising the tergum (bees, wasps, and flies), or by both direct and indirect muscles (grasshoppers and beetles).

Up and down movement alone is not sufficient for flight. The wings must at the same time be moved forward and backward. A complete cycle of a single wing beat describes an ellipse (grasshoppers) or a figure eight (bees and flies), during which the wings are held at different angles to provide both lift and forward thrust (Fig. 16–2C and D).

The raising or lowering of the wings resulting from the contraction of one set of flight muscles stretches the antagonistic muscles, which then also contract. Insect wing beat thus involves the alternate contraction of the antagonistic elastic systems. The beat frequency varies greatly—4 to 20 beats per second in butterflies and grasshoppers; 190 beats per second in the honeybee and housefly; and 1000 beats per second in certain midges. At low frequencies (30 beats per second or less) there is usually one nerve impulse to one muscle contraction. At higher frequencies, however, the contraction is myogenic, originating from the stretching caused by the contraction of the antagonistic muscles, and there are a number of beats, or oscillations, associated with each nerve impulse.

Rapid contraction is facilitated by the nature of the muscle insertion. A very slight decrease in muscle length during contraction can bring about a large movement of the wing (as a seesaw with the fulcrum near one end). The elastic nature of the thoracic skeleton and the joints of wing articulation also contribute to the beat motion. If, for ex-

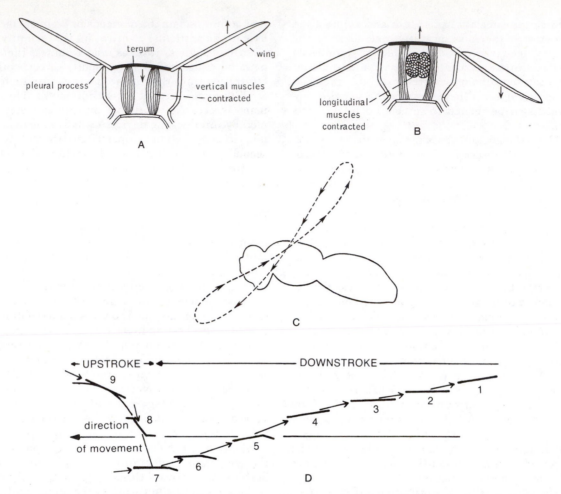

Figure 16–2 Diagrams showing relationship of wings to tergum and pleura and the mechanism of the basic wing strokes in an insect: *A*, Upstroke resulting from the depression of the tergum through the contraction of vertical muscles. *B*, Downstroke resulting from the arching of the tergum through the contraction of longitudinal muscles. *C*, An insect in flight, showing the figure eight described by the wing during an upstroke and a downstroke. *D*, Changes in the position of the forewing of a grasshopper during the course of a single beat. Short arrows indicate direction of wind flowing over wing, and numbers indicate consecutive wing positions. (*A* and *B* after Ross, H. H., 1965: A Textbook of Entomology. 3rd Edition. John Wiley and Sons, N.Y. *C* after Magnan from Fox and Fox. *D* after Jensen from Chapman, 1971: The Insects: Structure and Function. University of London Press Ltd., London. Courtesy of the American Elsevier Publishing Co., Inc.)

ample, the stable position is down, as in grasshoppers, elastic forces to return to the down position are set up when the wings are raised, much like stretching a rubber band. Some flies and beetles have both stable down and stable up positions, and at some point in the movement of the wing from one position to the other, the opposite elastic forces take over. The arrangement is called a click mechanism.

Flying ability varies greatly. Many butterflies and damselflies have a relatively slow wing beat

and limited maneuverability. At the other extreme, some flies, bees, and moths can hover and dart. The fastest flying insects are hummingbird moths and botflies, which have been clocked at 25 mph. Honeybees can cruise at 15 mph. Gliding, an important form of flight in birds, occurs in only a few large insects.

There is no flight control center in the insect nervous system, but the eyes and sensory receptors on the antennae, head, wings and other parts of the body provide continuous feedback information for flight control. Horizontal stability is maintained in part by a dorsal light reaction: the insect keeps the dorsal ommatidia of the eyes under maximum illumination from above. Deviation because of rolling is corrected by slight changes in wing position to bring the dorsal part of the eyes back to maximum illumination.

Members of the order Diptera (flies, gnats, and mosquitoes) have the second pair of wings reduced to knobs, called halteres (Fig. 16–17R). The halteres beat with the same frequency as the forewings and function as gyroscopes to control flight instability. Receptors on the haltere base detect deviating forces, such as tendencies to pitch, roll, and yaw, and from this information corrections in wing position are made.

Flight speed is probably determined by air flow over receptors on the antennae and movement of objects from front to back across the eyes. Flight is inhibited by contact of the tarsi with the ground.

Insect flight muscles are very powerful. The fibrils are relatively large and the mitochondria are huge (about half the size of a human red blood cell), reflecting the high respiratory rate of these cells. Insects are the only poikilothermic fliers, and a low body temperature and a correspondingly low metabolic rate impose limitations on mobility. On a cool morning many flying insects literally warm up before flight. They remain stationary on a tree trunk or some other location and move the wings up and down until sufficient internal heat is generated to permit the stroke rate necessary for flight, or more commonly, contract the flight muscles while in a decoupled state or "neutral gear."

Internal Anatomy and Physiology

Nutrition

Insects have adapted to all types of diets. The mouthparts may be highly modified, but the modifications are associated less with diet than with the method by which food is obtained. Primitive mouthparts are adapted for chewing, and it is mouthparts of this type that are described and illustrated in the beginning section on external structure (Fig. 16–1A and B). The mandibles are heavy and capable of cutting, tearing, and crushing, and the maxillae and labium function in food handling. The hypopharynx aids in swallowing. Insects with chewing mouthparts include the primitive apterygotes, dragonflies, crickets, grasshoppers, beetles, and many others. The larvae of such insects as

moths and butterflies have chewing mouthparts, although the adult's mouthparts are highly modified. The diets of chewing insects may be herbivorous or carnivorous, and some diets are very restrictive.

The specialization of insect mouthparts has been primarily in modifications for piercing and sucking. However, adaptations for the same feeding habits are not uniform, since a sucking or piercing feeding habit evolved independently in different insect orders. Moreover, the mouthparts may be adapted for more than one function—chewing and sucking, cutting and sucking, or piercing and sucking.

The mouthparts of moths and butterflies are adapted for sucking liquid food, such as nectar, from flowers. A part (the galea) of each of the two greatly modified maxillae forms a long tube through which food is sucked (Fig. 16–3A). When the insect is not feeding, the tube is coiled. The other mouthparts are absent or are vestigial.

Piercing mouthparts are characteristic of herbivorous insects, such as aphids and leafhoppers, which feed on plant juices. Predacious insects, such as assassin bugs and mosquitoes, which utilize the body fluids of other animals as food, also have piercing mouthparts. In all these insects the mouthparts are elongated and are organized in various ways to form a beak. For example, the beak of the plant-feeding and predacious bugs (Hemiptera and Homoptera) consists of a stylet composed of the mandibles and the maxillae that lie in a groove on the heavier labium. The stylet contains one lumen for the outward passage of salivary secretions and another for sucking in fluids (Fig. 16–3B to F). Other parts of the beak do not penetrate.

Bees and wasps have mouthparts adapted for both chewing and sucking. A bee, for example, gathers nectar by the elongated maxillae and the labium. Pollen and wax are handled by the labrum and mandibles, which have retained the chewing form.

In biting flies, such as horseflies, the knifelike mandibles produce a wound. Blood is collected from the wound by a spongelike labium and is conveyed to the mouth by a tube formed from the hypopharynx and epipharynx (the inner side of the labrum) (Fig. 16–4A). Some predatory flies and hemipterans inject salivary secretions into the prey and suck up the digested tissues. Certain nonbiting flies, such as houseflies, use the spongelike labium alone for obtaining food, the mandibles and maxillae being reduced. Such insects are not restricted to liquid foods. Saliva can be exuded through the

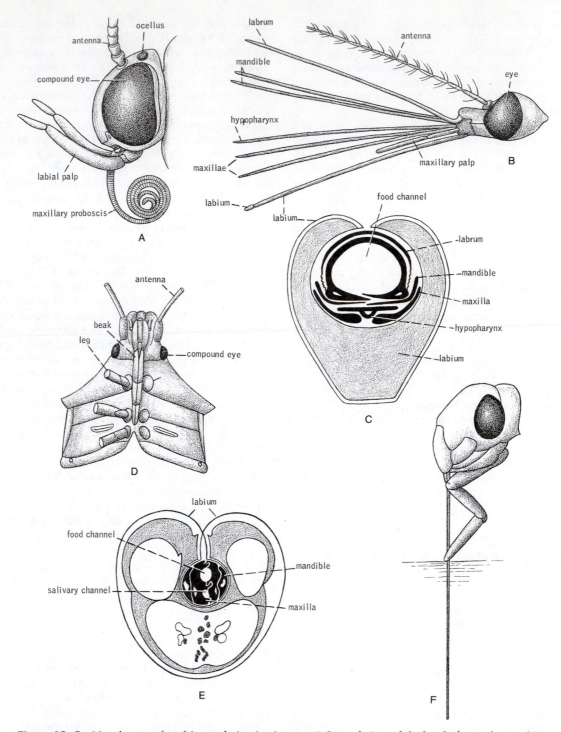

Figure 16–3   Mouthparts of sucking and piercing insects: *A*, Lateral view of the head of a moth, a sucking insect (After Snodgrass.) *B*, Lateral view of the head of a mosquito, showing separated mouthparts. *C*, Cross section of mouthparts of a mosquito in their normal functional position. (*B* and *C* after Waldbauer from Ross.) *D*, Ventral view of anterior half of a hemipteran, showing beak. (After Hickmann.) *E*, Cross section through a hemipteran beak, showing the food and the salivary channels enclosed within the stylet-like maxillae and the mandibles. (After Poisson.) *F*, A hemipteran penetrating plant tissue with its stylets. (After Kullenberg.)

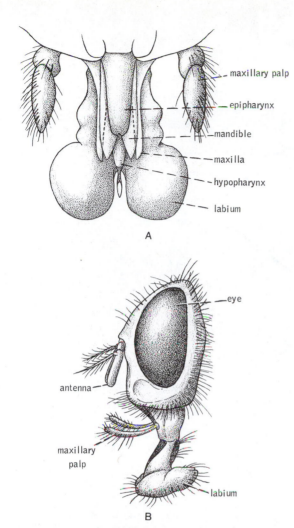

maxillary palp

epipharynx

mandible

maxilla

hypopharynx

labium

A

eye

antenna

maxillary palp

labium

B

Figure 16–4 *A*, Mouthparts of false blackfly adapted for cutting and sponging. (After Ross, H. H., 1965: A Textbook of Entomology. 3rd Edition. John Wiley and Sons, N.Y.) *B*, Lateral view of the head and sponging mouthparts of a housefly. (After Snodgrass.)

labium onto the solid material, and the fluid then can be sucked back into the mouth (Fig. 16–4*B*).

Food taken into the mouth passes into the foregut, which is commonly subdivided into an anterior pharynx, an esophagus, a crop, and a narrower proventriculus (Fig. 16–5*A*). The pharynx is highly modified as a pump in sucking insects. The crop, when present, is a storage chamber. The proventriculus is variable in structure and function. In insects that eat solid food, the proventriculus is usually modified as a gizzard and bears teeth or hard protuberances for triturating food (Fig. 16–5*C*). In sucking insects, on the other hand, the proventri-

culus consists only of a simple valve opening into the midgut. Between these two extremes are some beetles and honeybees, in which the proventriculus acts as a regulatory valve permitting fluids but not solid food to enter the midgut. This function is particularly important in the separation of pollen from nectar in bees.

Most insects possess a pair of salivary, or labial, glands, that lie below the midgut and have a common duct opening into the buccal cavity (Fig. 16–5*B*). Mandibular glands, in addition to salivary glands, are functional in the apterygotes and in a few groups of pterygotes. The function of salivary glands varies and has not been determined in all insects. The glands usually secrete saliva, which moistens the mouthparts and may be a solvent for the food. The salivary glands may also produce digestive enzymes, which are mixed into the food mass before it is swallowed. In some lepidopterans (moths) and hymenopterans (bees and wasps) the glands secrete silk used to make the pupal cells. Other special secretions of the salivary glands in various insects include mucoid materials, a pectinase that hydrolyzes the pectin of cell walls, venomous spreading agents, anticoagulants and agglutins, and an antigen that produces the typical mosquito-bite reaction in man.

A stomodeal valve separates the foregut from the midgut. The insect midgut, which is also called the ventriculus or stomach, is usually tubular and, as in other arthropods, is the principal site of enzyme production, digestion, and absorption. A characteristic feature of the midgut of many insects is the presence of a peritrophic membrane. This membrane, composed of a very thin layer of protein and chitin, is periodically delaminated by the midgut lining (grasshoppers) or is continuously secreted by the epithelial cells near the valve at the end of the foregut (flies). The membrane forms a covering around the food mass moving through the midgut (Fig. 16–6). Supposedly, the peritrophic membrane protects the delicate midgut walls from abrasion by the food mass and perhaps more importantly conserves enzymes by dividing the gut lumen into two compartments. The membrane is permeable to some enzymes and the products of digestion. The initial products of digestion pass through the membrane, where they are attacked by a second order of enzymes that are restricted to the space between the gut wall and peritrophic membrane. Some insects that live on a liquid diet do not secrete a peritrophic membrane.

Most insects possess outpocketings of the midgut called gastric ceca, commonly located at the anterior end of the midgut (Fig. 16–5). The gastric

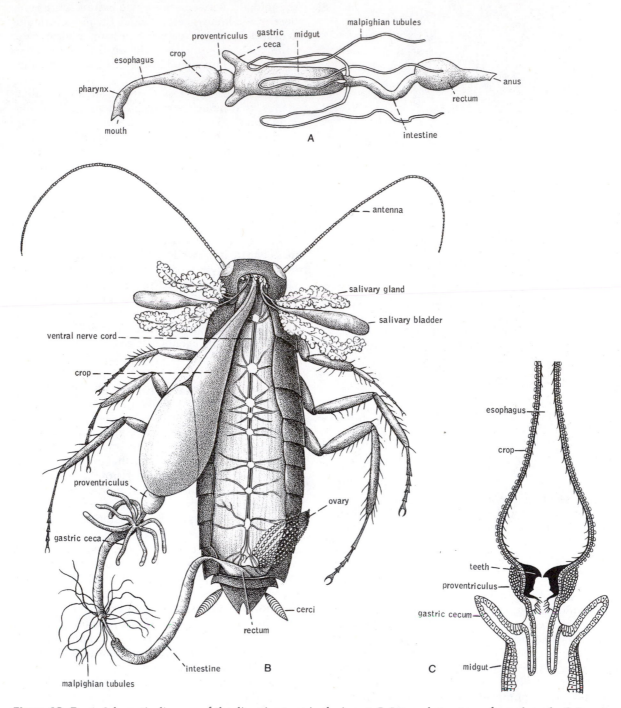

Figure 16–5   *A,* Schematic diagram of the digestive tract in the insect. *B,* Internal structure of a cockroach. *C,* Longitudinal section through the foregut and anterior part of the midgut of a cockroach. (*A* and *C* after Snodgrass; *B* after Rolleston.)

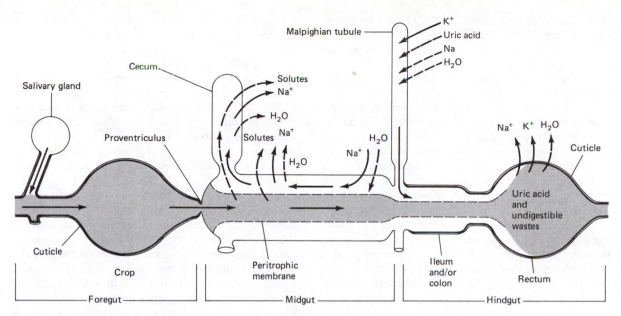

Figure 16–6 Diagram of the digestive tract of an insect showing passage of food (small circles) through the gut, absorption of food products in the ceca, and secretion of wastes in the malpighian tubule. Active transport of salts (solid arrows) leads to passive diffusion of water and other substances (dashed arrows). Modified from Berridge, M. J., 1970: A structural analysis of intestinal absorption. *In* Neville, A. C. (Ed.): Insect Ultrastructure. Sympos. Roy. Ent. Soc., 5:135–151; and from Evans, H. E., 1984: Insect Biology: A Textbook of Entomology. Addison-Wesley Publishing Co., Reading, Mass. p. 85.)

ceca and anterior end of the gut are the principal sites of food absorption. Water is also absorbed here, although some water enters at the posterior end of the midgut (Fig. 16–6).

The hindgut, or proctodeum, consists of an anterior intestine and a posterior rectum, both of which are lined by cuticle (Fig. 16–6). The hindgut functions in the egestion of waste and in water and salt balance. In most insects rectal pads, or glands, occur in the epithelium. These organs are the principal sites of water reabsorption. Digestion of cellulose by termites and certain wood-eating roaches is made possible by the action of enzymes produced by protozoa that inhabit the hindgut (p. 20). Acetic acid formed by the breakdown of wood is actively absorbed by the hindgut epithelium in these insects.

Fat bodies are present in various places within the hemocoel, depending on the species, and function somewhat like the chlorogogen tissue in annelids and the liver of vertebrates. This tissue is a site of synthesis and long- and short-term storage. Glycogen reserves in the fat body can be rapidly mobilized by hormones to release sugar into the blood. Many insects that do not feed as adults rely on the fats, proteins, and glycogen stored in the fat body during immaturity.

Internal Transport

The heart of the insect lies within a dorsal pericardial sinus that is separated by a perforated, dorsal diaphragm from the perivisceral sinus surrounding the gut. The heart is tubular and in most species extends through the first nine abdominal segments. In each segment a pair of alary muscles extends laterally from the heart to the body wall within the double-layered dorsal diaphragm. Contractions of the alary muscles cause the heart to expand and blood to pass through the heart ostia. This filling phase is followed by contractions of the heart wall muscle, and blood is driven forward. The heart is closed posteriorly but anteriorly is continuous with an aorta that runs into the head. Blood normally flows from posterior to anterior in the heart and from anterior to posterior within the perivisceral sinus. Blood flow may be augmented by accessory pulsating structures in the head, thorax, legs or wings and by contractions of the dorsal diaphragm. In many rapid-flying insects there is an additional thoracic "heart," which draws blood through the wings and discharges it into the aorta. Blood flow is also facilitated by various body movements, such as the ventilating abdominal contractions. In addition to bringing about blood transport, localized elevations of blood pressure may serve a variety of functions, such as the removal of wings from termites, the unrolling of the proboscis in Lepidoptera, the eversion of various organs, the egestion of fecal pellets, and the swelling of the body during molting and hatching.

The blood of insects is usually colorless or green with a number of types of amebocytes. Some insects possess clotting agents in the blood, but most species close wounds with a plug of cells, largely phagocytes. Since tissue gas exchange is handled directly by the tracheal system, the blood plays a very minor role in gas transport. Most animals rely on inorganic ions, such as sodium and chloride ions, as osmotic regulators of the body fluid. In insects organic molecules, especially free amino acids, are more important in this function. Hemolymph also contains high concentrations of dissolved uric acid, organic phosphates, and a nonreducing sugar, trehalose.

Gas Exchange

Gas exchange in insects occurs through a system of tracheae, which has been more extensively studied than that of any other arthropods. A pair of spiracles is usually located above the second and third pairs of legs or only above the last pair. The first seven or eight abdominal segments possess a spiracle on each lateral surface (Figs. 16–1C and E and 16–7B). Thus, there are a maximum of ten pairs of spiracles. Tracheal spiracles in their simplest form are merely holes in the integument, as in some of the Apterygota. However, in most insects the spiracles open into a pit, or atrium, from which the tracheae arise. The spiracle is generally provided with a closing mechanism, and in most terrestrial insects the atrium contains filtering devices such as sieve plates and felt pads (Fig. 16–7A). The closing mechanism of the spiracle reduces water loss, and the filtering structures prevent the entrance of dust and parasites as well as reduce water loss. The opening and closing of the spiracles are controlled in part by direct innervation, and the frequency and duration of opening or closing is apparently related to the oxygen–carbon dioxide tension of the blood. More spiracles are therefore open during flight than when the insect is at rest. Since an insect must balance oxygen need against the danger of water loss, the number and duration of open spiracles are generally held to the lowest possible level.

The pattern of the internal tracheal system is variable, but a pair of longitudinal trunks with cross connections form the ground plan of most species (Fig. 16–7B). The larger tracheae are supported by thickened spiral rings of cuticle, the taenidia (Fig. 16–7A); the smaller ones by annular rings. The rings resist compression (i.e., prevent collapse) but permit stretching of the tube. The epicuticle of the tracheae lacks the waxy component typical of the external skeleton. The tracheae themselves are seldom uniform in size but widen in various places, forming internal air sacs (Fig. 16–7D). The air sacs have no taenidia and can be compressed.

The smallest subdivisions of the tracheae, the tracheoles, are generally less than 1 μ in diameter. These fine, cuticular tubes are often given off in clusters from the tracheae and then further branch into a fine network over the tissue cells (Fig. 16–7C). A number of tracheoles may be formed by a single tracheole cell. In the flight muscles of some insects, the tracheoles even push into the fibrils. The tracheole cuticle is not shed during molting, as is that of the tracheae, and after molting new tracheae are joined to old tracheoles.

Within the tracheal system gas transport takes place by diffusion along a concentration gradient, by a ventilating mass flow of air down a pressure gradient, or by a combination of both. Ventilating pressure gradients result from body movements, largely abdominal, which bring about compression of the air sacs and of certain elastic tracheae. Ventilation is facilitated by the sequence in which certain spiracles are opened and closed. Diffusion along a concentration gradient can supply enough oxygen for small insects, but forms that weigh more than about 1 g or are highly active require some degree of ventilation. At the tissue-tracheole level gases are exchanged by diffusion across a concentration gradient. Tracheoles are permeable to liquids, and in most insects their tips are filled with fluid. This fluid is believed to be involved in the final transport of oxygen. The fluid has been shown to rise and fall, depending on the osmotic pressure in the surrounding tissue.

Some very small insects that live in moist surroundings, such as collembolans and proturans, lack tracheae, and gas exchange occurs over the general body surface. Some aquatic immature insects also lack tracheae, especially during the early stages. However, tracheae are usually present in aquatic immatures and always in adult insects that live in water. The adults merely utilize air from air bubbles or films held against the body surface by special "unwettable" (hydrofuge) hairs, but the nymphs and larvae of certain groups may possess

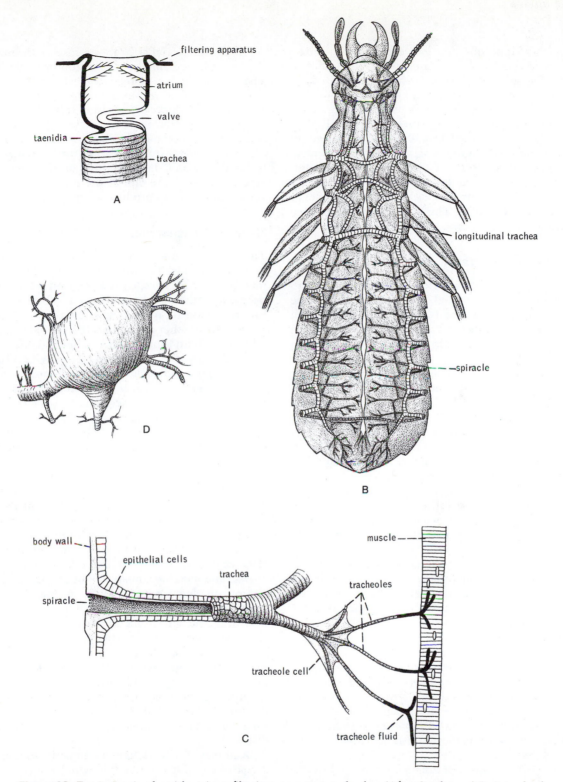

Figure 16–7 *A*, A spiracle with atrium, filtering apparatus, and valve. (After Snodgrass.) *B*, A tracheal system of an insect. (After Ross.) *C*, Diagram showing relationship of spiracle and tracheoles to tracheae. (*B* and *C* after Ross, H. H., 1965: A Textbook of Entomology, 3rd Edition. John Wiley and Sons, N.Y.) *D*, An air sac. (After Snodgrass.)

special adaptations for gas exchange in water. Damselfly and mayfly nymphs possess abdominal gills. The gills are provided with tracheae, and gas exchange occurs across the gill surface between the water and tracheae. Dragonfly nymphs pump water in and out of the rectum, which contains gills supplied with tracheae. Usually, gas exchange in aquatic immature insects occurs across the general integument between the tracheae and the water. Some larvae, such as those of mosquitoes, have a few functional spiracles associated with one or more breathing tubes. The larva rises to the surface periodically and obtains air through the tube.

Excretion and Water Balance

The chief organs of excretion in insects are the malpighian tubules, although they are absent in some forms (collembolans and aphids). The malpighian tubules lie more or less free in the hemocoel, with the proximal end usually attached at the junction of the midgut and the hindgut (Figs. 16–5 and 16–6). The number of tubules varies from 2 (coccids) to about 250 (grasshopper *Schistocerca*). The tubule lumen is lined by large, cuboidal, epithelial cells. The outer layer of the tubule wall, which is in contact with the hemolymph, is composed of elastic connective tissue and muscle fibers. The tubules are capable of peristalsis and can undergo some movement within the hemocoel.

The contents of the malpighian tubules are derived from secretion rather than from filtration. Water enters passively as a consequence of the active transport of ions, particularly potassium, into the lumen (Fig. 16–6). Uric acid, formed in the tissues and passed into the hemolymph, is then taken up by the malpighian tubule cells along with amino acids and salts. Together these substances are secreted into the lumen of the tubule. The lower pH that develops in the contents toward the junction with the hindgut causes the uric acid to precipitate. Reabsorption of water, some of the salts, and other nutritive substances occurs in the course of elimination. Some reabsorption of water and inorganic ions may take place in the proximal parts of the tubules themselves, these substances being returned to the hemolymph, but in most insects, the rectal epithelium transfers these substances back into the hemolymph.

Not all the waste products are removed by the malpighian tubules. Pericardial cells, or nephrocytes, typically located on or near the heart, pick up particulate or complex waste for intracellular degradation.

Of the terrestrial arthropods, insects are among the best adapted for the prevention of water loss. The epicuticle is impregnated with waxy compounds, which reduce surface evaporation, and spiracle closure reduces evaporation from the tracheal system. The excretion of uric acid reduces loss of water due to protein metabolism. And the reabsorption of water by the rectum further conserves water that would be lost through excretion and egestion. Insects are one of the few groups of invertebrates that can produce a hyperosmotic urine.

Nervous System and Sense Organs

The insect nervous system is basically like that of other arthropods. The brain is composed of a protocerebrum with eyes, deutocerebrum with antennae, and a tritocerebrum. The ventral nerve cord forms a chain of median segmental ganglia (Fig. 16–5*B*). As in other arthropods the ventral segmental ganglia, both thoracic and abdominal, are often fused. The greatest number of free ganglia is three in the thorax and eight in the abdomen (apterygotes). The subesophageal ganglion is always composed of three pairs of fused ganglia, which control the mouthparts, the salivary glands, and some of the cervical muscles. Giant fibers are of general occurrence and have been well studied in the cockroach.

Associated with a hypocerebral ganglion lying over the foregut and just beneath the brain are two pairs of glandular bodies, the corpora cardiaca and the corpora allata. These two bodies, together with the prothoracic glands and certain neurosecretory cells in the protocerebrum, are the principal endocrine centers of insects. Their role in controlling growth and metamorphosis will be described in the section on development. Other endocrine functions include regulation of water reabsorption, heartbeat, and certain metabolic processes.

Sensilla, sense organs other than eyes and ocelli, are scattered over the body but are especially numerous on the appendages. Most are believed to be derived from simple setae, but many have become modified as bristles, pegs, scales, domes, plates, and so on (Figs. 16–8 and 16–9). They exhibit a range of receptor functions, as described on page 481.

Auditory receptors, called tympanic organs, are a type of chordotonal organ in which the receptors are stretched between two points on the inner side of the integument and respond to tension changes. Tympanic organs are found in grasshoppers, crickets, and cicadas, which also have sound-producing

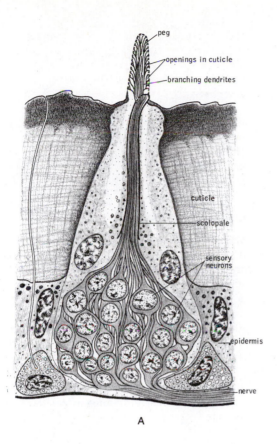

peg

openings in cuticle

branching dendrites

cuticle

scolopale

sensory neurons

epidermis

nerve

A

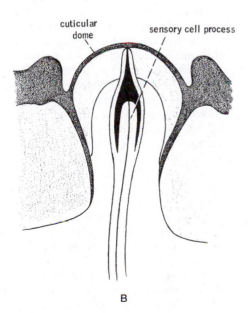

cuticular dome

sensory cell process

B

Figure 16–8 *A*, A chemosensory peg organ from the antenna of a grasshopper. (After Slifer et al.) *B*, A campaniform sense organ. (After Snodgrass.)

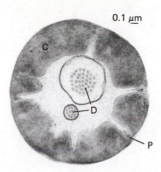

0.1 μm

C

D

P

Figure 16–9 Cross section of an olfactory hair from the antenna of a male moth. The thick cuticular wall (C) is perforated (P) and contains two dendrites (D). (From Steinbrecht, R. A., 1984: Arthropoda: Chemo-, hygro-, and thermoreceptors. *In* Bereiter-Hahn, J., Matoltsy, A. G., and Richards, K. S.: Biology of the Integument. 1. Invertebrates. Springer-Verlag, Berlin. p. 528.)

organs. Tympanic organs develop from the fusion of parts of a tracheal dilation and the body wall. The sensory receptors are attached to the tympanum. An air sac beneath the tympanum permits vibrations that excite the attached receptors.

The photoreceptors are the ocelli and the compound eyes. Ocelli are absent in many adult insects, but when present, there are usually three, found on the anterior dorsal surface of the head (Fig. 16–1*A*). The photoreceptor cells on each ocellus are organized somewhat like those of a single ommatidium of a compound eye. Ocelli can detect changes in light intensity and may be very sensitive to low intensities. They function in orientation and in some way appear to have a general stimulatory effect on sensitivity, enhancing the reception of stimuli by other sensory structures. The number of facets in the compound eyes is greatest in flying insects that depend on vision for feeding, and the facets are larger in nocturnal than in diurnal insects.

Reproduction and Development

Reproduction

The typical female reproductive system consists of two ovaries, one on each side, and two lateral oviducts (Fig. 16–10). Each ovary consists of a group of tubules, the ovarioles. The paired oviducts usually unite to form a common oviduct, which leads into a vagina. The vagina in turn opens onto the ventral surface behind the seventh, eighth, or ninth

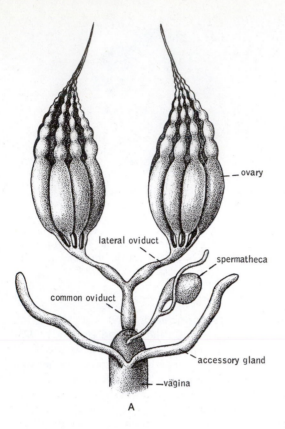

A

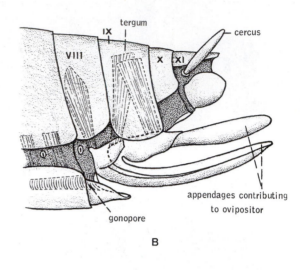

B

Figure 16–10 *A*, Reproductive system in a female insect. *B*, Lateral view of the posterior end of the abdomen, showing reproductive opening and appendages, forming ovipositor. (Both after Snodgrass.)

segment. Diverticula of the common oviduct or vagina include a spermatheca (seminal receptacle), a bursa copulatrix, and paired accessory glands.

The male reproductive system includes a pair of testes, a pair of lateral ducts, and a median duct opening through a ventral penis, called the aedeagus, associated with the eighth segment (Fig. 16–11*A* and *B*). Each testis consists of a group of sperm tubes containing spermatozoa in various stages of development. These tubes empty into a lateral duct, the vas deferens, which joins the duct from the other side to form a common ejaculatory duct. A section of each vas deferens is usually enlarged into a seminal vesicle, where sperm are stored. Accessory glands, which secrete seminal fluid, are commonly present as pouches from the upper end of the ejaculatory duct.

In copulation the often extensible, or eversible, penis of the male is inserted into the female vagina or bursa copulatrix. The males of many orders—dragonflies, true flies, butterflies, and moths, to name a few—possess clasping organs to hold the female abdomen (Fig. 16–11*C*). The clasping organs are derived from parts of the terminal segments and vary greatly in structure.

Although some insects, such as honeybees, do not have spermatophores, most insects use spermatophores to transfer sperm. Among a few primitive, wingless insects, such as thysanurans and collembolans, the spermatophore is deposited on the ground and then taken up by the female. However, in most insects the spermatophores are deposited directly into the female bursa copulatrix at copulation. Following deposition, sperm are released from the spermatophores and are soon found within the spermatheca, where they are stored until the eggs are laid. Fertilization occurs as the eggs pass through the oviduct at the time of egg deposition. At each mating a large number of sperm are transferred to the spermatheca of the female. This number is sufficient for fertilization of more than one batch of eggs. Many insects mate only once in their lifetime, and none mate more than a few times.

When the eggs reach the oviduct, they are already surrounded by a shell-like membrane (chorion) secreted by ovarian follicle cells. This shell may be up to seven layers thick and is perforated by one or more micropyles. It is through these minute openings that the sperm enter. The evolution

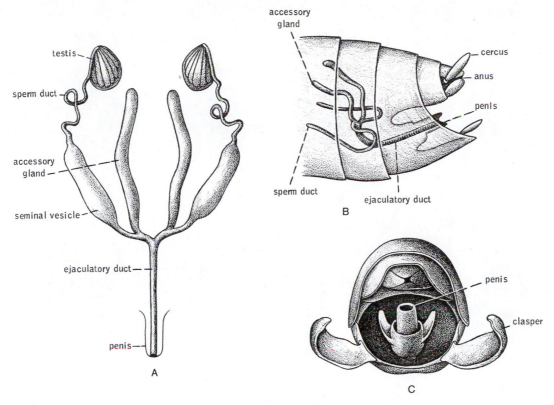

Figure 16–11 Reproductive system in a male insect: *A*, General plan of system. *B*, Lateral view of the posterior end of the abdomen, showing reproductive opening and other structures. *C*, Posterior view of the abdomen, showing penis and claspers. (All after Snodgrass.)

of a protective encasement for the development of the egg in a desiccating terrestrial environment is another factor that has contributed to the success of insects. Egg deposition in the majority of insects is through an ovipositor, derived from parts of the eighth or ninth segment (Fig. 16–10*B*). The site of egg laying varies tremendously and in large part depends on the habitat and life-style of the adult. Adhesive materials for attaching the eggs to the substratum or to each other are produced by the accessory glands.

An interesting type of egg deposition is that associated with gall formation. The females of gall wasps (hymenopterans) and gall gnats (dipterans) deposit their eggs in plant tissues. The plant tissue surrounding the eggs is induced to undergo abnormal growth and forms a gall, which has a shape characteristic of the insect producing it. The gall forms a protective chamber for the developing eggs, larvae, and often even the pupae. The larvae feed on the gall tissues. The gall-inducing agent is a substance secreted by the female when she deposits the egg and by another substance produced later by the larva. In North America alone there are about 2000 species of gall-inducing insects.

Development

Superficial cleavage is characteristic of most insects. Insect young vary in the degree of development at hatching. Young apterygotes are like the adults except in size and sexual maturity. Newly hatched grasshoppers, cockroaches, stoneflies, leaf hoppers, and bugs resemble the adults, except that the wings and reproductive organs are undeveloped. The wings of early nymphs are merely external pads, which begin to look like wings only at the preadult molt (Fig. 16–12). The adult form is reached gradually with successive molts. This type of development is called gradual, or incomplete, metamorphosis (hemimetabolous development);

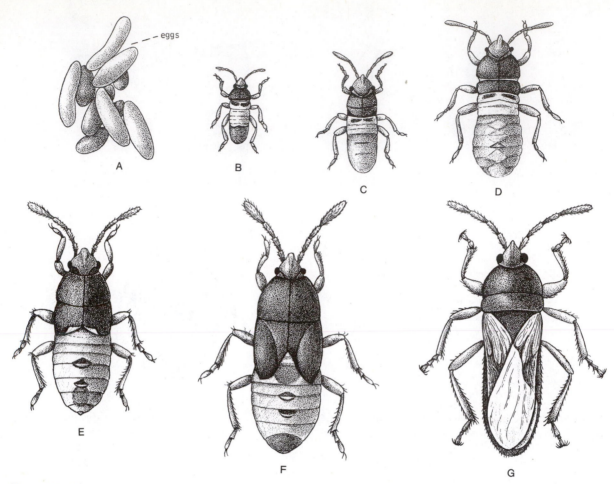

Figure 16–12   Stages in the gradual metamorphosis of a chinch bug. (After Ross, H. H., 1965: A Textbook of Entomology, 3rd Edition. John Wiley and Sons, N.Y.)

all the immature stages from hatching to the adult are termed nymphs or, when aquatic, naiads.

In many insects, including bees, wasps, flies, and beetles, the wing rudiments develop internally; the wings seem to appear suddenly in the adults. This type of development is a complete metamorphosis (holometabolous development) and consists of three distinct stages (Fig. 16–13). The newly hatched larval stage, which has no wings, is the caterpillar of butterflies, the maggot of flies, and the grub of beetles. This is an active feeding stage, although the food is usually quite different from that of the adult. In some species the larvae and the adults have different kinds of mouthparts. For example, caterpillar larvae have chewing mouthparts, whereas the adults have sucking mouthparts. Some parasitic groups may have two or more different

larval habits and structures (hypermetamorphosis). It is important to keep in mind that insect larvae represent a specialized, not a primitive, condition and evolved through suppression and delay of the development of adult features.

At the end of the larval period the young become nonfeeding and quiescent. This stage is called a pupa and is usually passed in protective locations, such as the ground, a cocoon, or plant tissues. During pupation adult structures are developed from embryonic rudiments. Few larval structures are carried over to the adult stage. The number of molts required to reach the adult stage ranges from as few as three to more than 30 and depends in large part on the type of development.

Holometabolous development was of great adaptive significance in the evolution of the higher

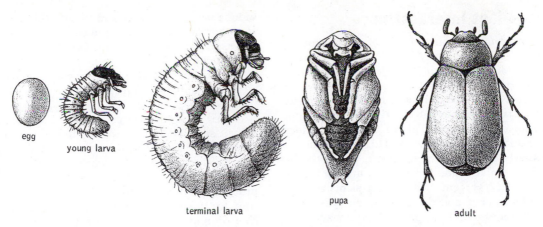

egg

young larva

terminal larva

pupa

adult

Figure 16–13 Stages in the complete metamorphosis of a beetle. (After Ross, H. H., 1965: A Textbook of Entomology, 3rd Edition. John Wiley and Sons.)

orders of insects, for the larvae can utilize different food sources, habitats, and life-styles than those of adults. Of the 26 orders of living insects, 9 have holometabolous development, but these 9 include about 80 percent of the species.

The transformation of immature insects into reproducing adults is known to be under endocrine control. A hormonal secretion from the brain stimulates a gland in the prothorax, the prothoracic gland, which produces ecdysone, a hormone that stimulates growth and molting. During the larval stages another hormone, the juvenile hormone, is secreted by the corpora allata of the brain. This hormone is responsible for the maintenance of larval structures and inhibits metamorphosis. The juvenile hormone can exert its effect only after the molting process has been initiated. It thus must act in conjunction with the prothoracic hormone. When a relatively high level of juvenile hormone is present in the blood, the result is a larva-to-larva molt. When the level of the juvenile hormone is lower, the molt is larva-to-pupa, and in the absence of the juvenile hormone, there is a pupa-to-adult molt. The hormonal control of molting in insects and in crustaceans is compared in Figure 14–98.

There are many insects whose life cycle includes a period of arrested development, called diapause. Diapause, which is under hormonal control, may be passed in any stage of development, depending on the species, and enables the insect to survive adverse environmental conditions, such as long dry or cold periods. Diapause is typically passed in the ground, a cocoon, or some other protected place.

There are several groups of insects in which multiple generations live in different habitats and have different methods of reproduction. A life cycle of this type is exemplified by the aphids, which have remarkable powers of reproduction. Eggs laid in the autumn hatch in the spring and develop into wingless, parthenogenetic females. The females, sometimes called stem mothers, are viviparous and may give birth to any number of broods of wingless, parthenogenetic females like themselves. One of these generations eventually gives birth to winged forms, which are still parthenogenetic and viviparous females. This generation usually moves to other plants but continues to produce either winged or wingless parthenogenetic females. As fall approaches, a sexual generation of both males and females is produced. These mate, eggs are laid, and the cycle is repeated. The reproductive abilities and life cycles of aphids appear to be adaptations for rapid population increase from low initial levels and are similar to those of water fleas, rotifers, and other inhabitants of temporary bodies of fresh water.

Polyembryony, in which the initial mass of embryonic cells gives rise to more than one embryo, is highly developed in some parasitic hymenopterans and is a classic example of this developmental phenomenon. Two to many larvae are formed from a simple egg deposited in the body of another insect host. The extreme occurs in the tiny chalcid *Litomastix*. The female deposits a few eggs into the body of a large caterpillar. From these eggs several thousand chalcid larvae may develop, completely devouring the caterpillar host.

Insect-Plant Interactions

Plants are a major resource for thousands of species of insects. Virtually every part of a plant is a food source for some adult or juvenile insect. Not surprisingly, the evolution of plants and insects has been closely connected: insects have been a major factor in the selection of certain features in the evolution of plants, and plants have determined various adaptations in insects. This coevolution is particularly striking in pollination, where the interaction of plants and insects is mutually beneficial. About 67 percent of flowering plants are pollinated by insects, and the great diversity of floral structure reflects, in large part, adaptations for pollination. Bees, wasps, butterflies, moths, and flies are the principal pollinators. Colors, odors, and nectaries are insect attractants, and floral form often determines landing sites and orientation. Concealment of nectaries deep in the flower forces the pollinator to touch reproductive structures. There has been a general tendency in the evolution of flowers for the floral design to restrict possible pollinators to a relatively small number of insect species, thus increasing the likelihood that the visiting insects will bring pollen from the same plant species.

Insects can also damage plants, and in recent years there has been much interest in plant defenses against insects and insect counteroffenses. The prinicipal plant defense against being eaten is chemical. Plants have evolved certain toxic compounds deposited in leaves or other parts, or the tissues contain high concentrations of indigestible compounds, such as tannins, resins, and silica. The plant defenses have led to various evolutionary strategies among insect herbivores. Some feed on a wide range of plant species (polyphagous insects) but must be selective or utilize less desirable parts because of toxic compounds. Other insects are restricted to one or a few plant species (monophagous insects) and have evolved enzyme systems that can detoxify young, desirable tissues. The life cycles of insects may also be adapted to coincide with the availability of the plant food source.

Parasitism

There are many parasitic insects, and the condition has evolved many times within the class.

Insect parasitism represents an adaptation to meet the habitat-nutrition needs of different stages in the life cycle. For some insects parasitism provides a new food source and habitat for the adults; for others it offers a new food source for the larvae. For example, adult fleas and lice (Fig. 16–16J) are bloodsucking ectoparasites on the skin of birds and mammals (there is one group of chewing lice). The eggs and immature stages of fleas develop off the host in its nest or den.

Many species of wasps and flies illustrate larval parasitism. The screwworm fly, a species of blowfly and a pest of domestic animals, lays its eggs in the wounds of mammals, and the larvae feed on living tissue. The parasitic condition was probably preceded by the deposition of eggs in carrion, for this is the habit of many nonparasitic species of blowflies. These few paragraphs cannot begin to provide an appreciation of our present knowledge of the many parasitic insect species. The interested reader should consult Askew (1971).

Communication

Both social and nonsocial insects utilize chemical, tactile, visual, and auditory signals to communicate. Chemical communication by pheromones has been studied more extensively in insects than in any other group of animals, and many examples are now known (see Evans, 1984). Many species utilize pheromones to attract one sex to another, and the much-studied moth *Bombyx mori* is a classic example. Pheromones also mark trails or territories in some species. For example, substances deposited on the ground by ants returning from a foraging trip serve as a trail marker for other ants. This type of communication is especially important in the complex movements of tropical army ants. Substances produced by the death and decomposition of the body of an ant within the colony stimulate other workers to remove the body. If a live ant is painted with an extract from a decomposing body, the painted ant will be carried live and struggling from the nest. There are also pheromones that produce alarm responses in other individuals of an aggregation or colony.

Among the more unusual visual signals are the luminescent flashings of fireflies, which function in sexual attraction. In species of *Photinus*, for example, flying males flash at definite intervals. Females located on vegetation will flash in response if the male is sufficiently close. The male will then redirect his flight toward her and further flashing will occur.

Sound production is especially notable in grasshoppers, crickets, and cicadas. The chirping sounds of the first two are produced by rasping. The front margin of the forewing (crickets) or the hind leg (katydids) acts as a scraper and is rubbed over a file formed by a vein of the forewing. Where scraper and file are both located on the forewings, as in crickets, the wings cross over, and one forewing functions as a scraper and the other as a file. Each species of cricket produces a number of songs that differ from the songs of other species. Cricket songs function in sexual attraction and aggression. The static-like sounds of cicadas, which serve to aggregate individuals, are produced by vibrations of special chitinous, abdominal membranes.

The remarkable mechanisms of communication in bees are described in the next section.

Social Insects

Colonial organization has evolved in a number of animal phyla, but only among a few spiders and some insects and vertebrates are individuals functionally interdependent yet morphologically separate. The condition is therefore usually described as a social organization.

Social organizations have evolved in two orders of insects, the Isoptera, which comprise the termites, and the Hymenoptera, which include the ants, bees, and wasps. In all social insects no individual can exist outside of the colony nor can it be a member of any colony but the one in which it developed. There is a cooperative brood care and an overlap of generations. All social insects exhibit some degree of polymorphism, and the different types of individuals of a colony are termed castes. The principal castes are male, female (or queen), and worker. Males function for the insemination of the queen, which produces new individuals for the colony. The workers provide for the support and maintenance of the colony. Caste determination is a developmental phenomenon regulated by the presence or absence of certain substances provided in the immature stages by other members of the colony.

Termites live in a nest usually constructed in soil, and in many species the nest may be huge and structurally complex. Termites differ from social hymenopterans in that workers are sterile individuals of both sexes and since termites are hemimetabolous, workers may be juveniles or adults. The reproductive male is a permanent member of the colony. The colony is built and maintained by

workers and may include a soldier caste (Fig. 16–14). The soldiers have large heads and mandibles and defend the colony. Workers and soldiers are wingless; wings are present in the males and queens only during a brief nuptial flight, during which pairing and dispersion occur.

Except for the fungus-growing species, termites depend on cellulose as a food source and on symbiotic flagellates for cellulose digestion. The symbiosis was probably an important factor in the evolution of social behavior in termites, which are generally agreed to be closely related to cockroaches (Wilson, 1971).

Ant colonies resemble those of termites and are usually housed within a gallery system in soil or wood or beneath stones. There may be a soldier caste in addition to workers. Some species of ants raid the nests of other species and carry away their larvae and pupae to be raised as "slaves." As in bees and wasps, ant soldiers and workers are always sterile females. Wings are present only during the nuptial period of reproductive males and females. Copulation occurs at this time, and the male never becomes a functional part of the colony.

Polymorphism is less highly developed in wasps and bees. There is no soldier caste and workers are winged, but many of these insects exhibit remarkable adaptations for a social organization.

The honeybee, *Apis mellifera*, is the best known social insect. This species is believed to have originated in Africa and to be a recent invader of temperate regions, for unlike other social bees and wasps of temperate regions, the honeybee colony survives the winter. Multiplication occurs by the division of the colony, a process called swarming. Stimulated at least in part by the crowding of workers (20,000 to 80,000 in a single colony), the mother queen leaves the hive along with part of the workers (a swarm) to found a new colony. The old colony is left with developing queens. On hatching, a new queen takes several nuptial flights during which copulation with males (drones) occurs, and she accumulates enough sperm to last her lifetime. The male dies following copulation, when his reproductive organs are literally exploded into the female. A new queen may also depart with some of the workers as an afterswarm, leaving the remaining workers to yet another developing queen. Eventually, the old colony will consist of about one third of the original number of workers and their new queen.

Honeybee colonies are large. The workers' lifespan is not long, and a queen may lay 1000 eggs per day. The diet provided these larvae by the nursing

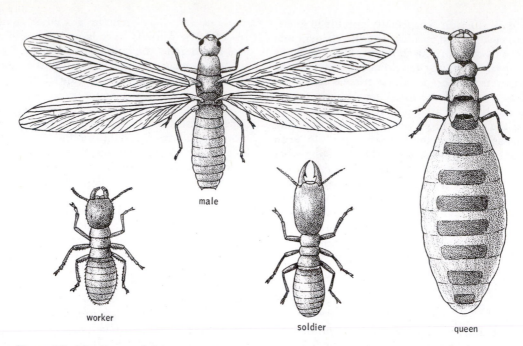

male

worker

soldier

queen

Figure 16–14   Castes of the common North American termite *Termes flavipes*. (After Lutz.)

workers results in their developing into sterile females, i.e., additional workers. The nursing behavior of the workers is a response to a pheromone
("queen substance") produced by the queen's mandibular glands. At the advent of swarming, or when
the vitality of the queen diminishes, the production of this pheromone declines. In the absence of
the inhibiting effect of the pheromone, the nursing
workers construct royal cells, into which eggs,
royal jelly, and a greater amount of food are placed.
Royal jelly is a secretion from the hypopharyngeal
glands of the worker nurses, and those larvae fed
upon it develop into queens in about 16 days. At
the same time that queens are being produced, unfertilized eggs are deposited into cells similar to
those for workers. These haploid eggs develop into
drones.

A remarkable feature of honeybee social organization is the temporal division of the workers. As
can be seen from the graph in Figure 16–15*A*, the
first activities of the worker are maintenance tasks
within the hive. During this period there is secretion by wax, mandibular, and other glands involved in comb construction, food storage, and larval care. After about three weeks, such glandular
activity declines, and the bee begins a period of foraging outside of the hive, its final service to the colony. The many functions performed in the lifetime
of a worker are not strictly sequential; rather, a

worker shifts from one task to another (Fig.
16–15*A*). A large amount of time is spent by the
older worker bees in resting and patrolling. Patrolling, or "determination" of hive needs, plus the
ability of workers to change tasks, enables a colony
to adjust to changing environmental conditions
and is believed to have contributed to the success
of the species.

Communication between members of a honeybee colony is highly evolved; some aspects, such as
the tail-wagging dance, set the honeybees apart
from all other social insects. A successful foraging
scout returns to the hive and communicates to
other workers the nature, direction, and distance
of a food source. The nectar and pollen on the
scout's body provide the information about the
kind of food that has been found. The scout bee
also executes an excited dance that is a ritualization
of the flight path. The dancing bee circles to the
right and to the left, with a straight-line run between the two semicircles (Fig. 16–15*B*). During
the straight-line run, the bee wags her abdomen
and emits audible pulsations. Von Frisch, a pioneer
in the study of communication in bees, discovered
that the orientation of the circular movements
shows the direction of the food and that the frequency of the tail-wagging runs indicates the distance. The closer the food source is to the hive, the
greater the frequency of tail-wagging runs. The

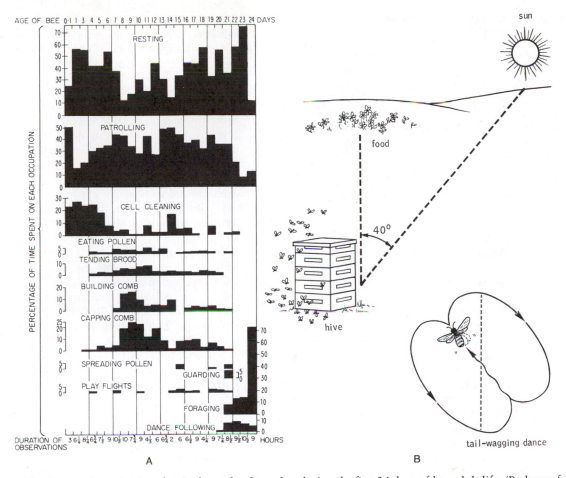

Figure 16-15 *A,* The activities of a single worker honeybee during the first 24 days of her adult life. (Redrawn from Ribbands, 1953; based on data of Lindauer, 1952.) *B,* Diagram illustrating the inclination of the straight tail-wagging run by a scout bee to indicate the location of a food source by reference to the sun. The food source is located at an angle 40 degrees to the left of the sun. The tail-wagging run of the scout bee is therefore upward (indicating that food is toward the sun) and inclined 40 degrees to the left (indicating the angle of the food source to the sun). (After Von Frisch).

sound pulsations of the dancer apparently also indicate the distance of the food source from the hive. The average number of vibrations is proportional to the distance of the food from the hive.

· Bees use the angle of the sun and light polarization as a means of orientation, and the dance of the scout bee indicates the location of the food in reference to the sun's position. If the tail-wagging run is directed upward, the food is located toward the sun; if the tail-wagging run is directed downward, the food is located away from the sun. The inclination of the run to the right or to the left of vertical indicates the angle of the sun to the right or to the left of the food source. An internal "clock" compensates for the passage of time be-

tween discovery of the food and the start of the dance, so the information is correct even though the sun has moved during the interval. On cloudy days the polarization of the light rays and ultraviolet light act as indirect references in the absence of the sun. If the food source is closer than 80 meters, the clues provided by chemoreception are sufficient for finding the food, and the tail-wagging dance is not performed by the scout bee.

Although the tail-wagging dance has been decoded, the sensory modality by which it is transmitted to other bees is still uncertain. The hive is dark so that the dance cannot be easily detected visually. The surrounding bees must receive the dancer's vibrations through their antennae or legs.

SYSTEMATIC RÉSUMÉ OF CLASS INSECTA, OR HEXAPODA

Living insects may be placed in 26 orders, comprising nearly 1000 families and many thousands of genera. More than 84,000 insects from all 26 orders have been described from North America, but it is estimated that nearly 25,000 species remain to be discovered on that continent alone. And there may be as many as 1,000,000 to 10,000,000 more undescribed insect species in the world.

Subclass Apterygota. Insects whose wingless condition is primary. The members of the following three orders are believed to represent the most primitive living insects.

Order Protura. Small eyeless insects with a cone-shaped head. They live in damp humus and soil and feed on decayed organic matter. Proturans are sometimes considered a separate subclass, the Myrientomata (Fig. 16–16A).

Order Thysanura. Silverfish and bristletails. Fast-running insects with two or three styliform appendages on the abdomen. They live in dead leaves and wood and around stones. Some species are found in houses where they eat books and clothing (Fig. 16–16B).

Order Collembola. Springtails. Small insects with an abdominal jumping organ, well-developed legs, and antennae; eyes either absent or represented by isolated ommatidia. Abundant in moist leaf mold, soil, and rotten wood. They are sometimes considered a separate subclass, the Oligoentomata (Fig. 16–16C).

Subclass Pterygota. Winged insects, or if lacking wings, the wingless condition is secondary.

Order Ephemeroptera. Mayflies. Elongate insects with net-veined wings, of which the first is larger than the second; wings held vertically at rest. Two or three caudal, filiform appendages. Antennae small and mouthparts of short-lived adults vestigial. Gradual metamorphosis with aquatic nymphs (Fig. 16–17A).

Order Odonata. Dragonflies and damselflies. Predacious insects with long, narrow, net-veined wings, large eyes and chewing mouthparts. Gradual metamorphosis. Nymphs aquatic. Dragonflies are stout bodied; damselflies are slender and delicate, weak fliers. Wings are held vertically at rest (Fig. 16–17B).

Order Orthoptera. Grasshoppers, katydids, crickets, roaches, mantids, and walking-sticks. Large-headed insects with strong chewing mouthparts and compound eyes. Femur of hind leg enlarged for jumping in many species. Winged and wingless species. Most winged forms have membranous hindwings folded fanlike beneath leathery forewings. Largely herbivorous, sometimes causing vast crop damage. Gradual metamorphosis. Some classifications place the roaches, mantids, and walkingsticks each in separate orders (Fig. 16–17C).

Order Isoptera. Termites. Social insects. Winged and wingless individuals composing the colony. Soft-bodied, pale, with abdomen broadly joined to thorax. Forewings and hindwings of equal size and held horizontally over abdomen (unlike flying ants, which are darker, with elbowed antennae, narrow waists, hindwings smaller than forewings, and wings held vertically over abdomen). Gradual metamorphosis (Fig. 16–14.)

Order Plecoptera. Stone flies. Adults have long antennae, chewing mouthparts, and two pairs of well-developed, membranous wings, or vestigial wings. Abdomen with two multisegmented, caudal cerci of varying length. Gradual metamorphosis. Nymphs aquatic. Adults emerge during winter in certain groups (Fig. 16–17D).

Order Dermaptera. Earwigs. Elongate insects resembling beetles, with chewing mouthparts, compound eyes, and large, forceps-like cerci. Most species have fan-shaped wings and elytra. Nocturnal, with omnivorous food habits. Gradual metamorphosis (Fig. 16–17E).

Order Embioptera. Web spinners. Small slender, soft-bodied insects with large heads and eyes. They feed on plants and live in silken tunnels, which they weave ahead of themselves to create routes. Silk glands and spinning hairs located on front tarsi. They are gregarious, and many individuals may live together. Mostly tropical. Gradual metamorphosis (Fig. 16–17F).

Order Psocoptera, or Corrodentia. Book lice, bark lice, and psocids. Small, fragile, pale insects with chewing mouthparts. Wings membranous and front pair a little larger than hind pair, or wingless. Gradual metamorphosis. They live in a wide variety of habi-

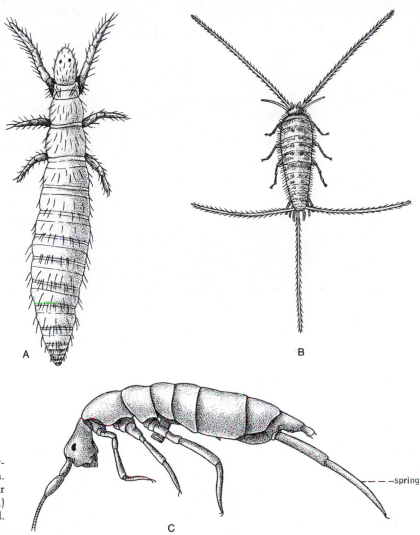

Figure 16–16 Subclass Apterygota: *A*, Order Protura: a proturan. (After Ewing from Ross.) *B*, Order Thysanura: a silverfish. (After Lutz.) *C*, Order Collembola: a springtail. (After Willem.)

spring

tats—bark, foliage, under stones. Some species infest buildings and are found in books (Fig. 16–17*G*).

Order Zoraptera. Small, pale, soft-bodied insects resembling tiny termites with chewing mouthparts. Both winged and wingless forms. Gradual metamorphosis. These are rare insects, living in colonies under dead wood in warm climates. Only some 20 known species and one genus (Fig. 16–17*H*).

Order Mallophaga. Chewing lice and bird lice. Wingless, flattened insects that live as ectoparasites on birds; two families parasitic on mammals. The eyes are reduced or absent; legs are short; and thorax is small. Gradual metamorphosis. Feed on scales, feathers, hair, skin, and sometimes dried blood around wounds. Many species infest domestic birds and livestock and cause considerable damage by skin irritation (Fig. 16–17*I*).

Order Anoplura. Sucking lice. Similar to the chewing lice but mouthparts adapted for sucking. Ectoparasites of birds and mammals. A number of species are parasitic on domestic animals, and the head louse and crab louse are parasites of man. More serious than the irritation produced by these parasites is their role as vectors of disease, such as typhus fever (Fig. 16–17*J*).

(Text continued on p. 713)

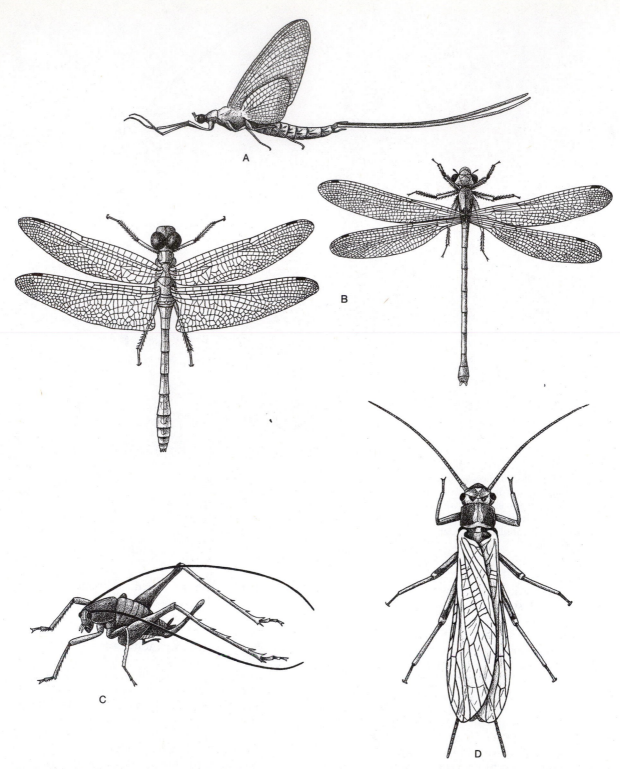

Figure 16–17   Subclass Pterygota: *A*, Order Ephemeroptera: a mayfly. (After Ross.) *B*, Order Odonata: a dragonfly (left) and a damselfly (right). (After Kennedy from Ross.) *C*, Order Orthoptera: a camel cricket. (After Lutz.) *D*, Order Plecoptera: a stone fly. (After Ross from Illinois Nat. Hist. Survey.)

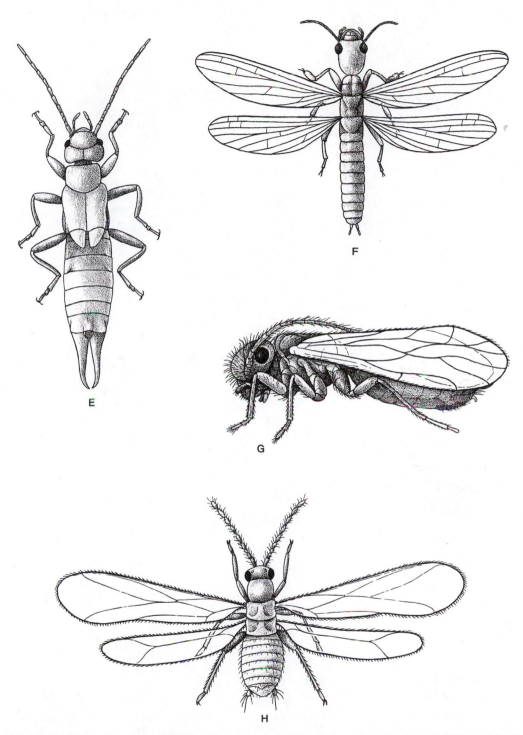

Figure 16–17 *(continued)* *E*, Order Dermaptera: an earwig. (After Fulton from Borror and DeLong.) *F*, Order Embioptera: a web spinner. (After Enderlein from Comstock.) *G*, Order Psocoptera: a psocid. (After Sommerman from Ross.) *H*, Order Zoraptera: *Zorotypus*. (After Caudell from Comstock.)

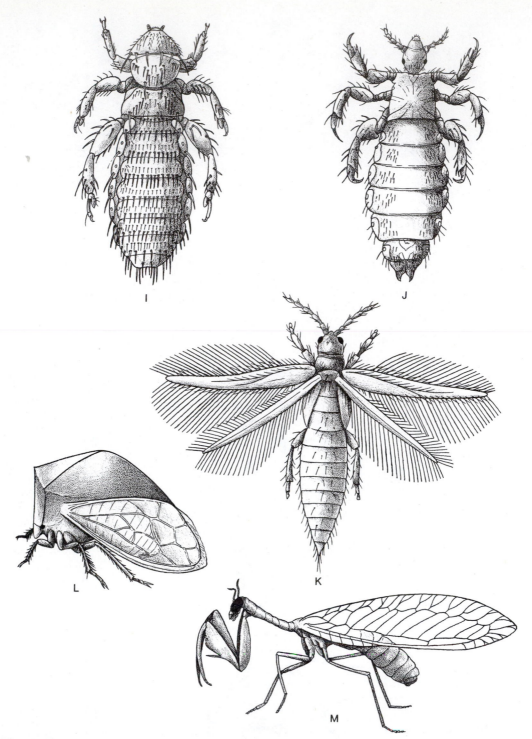

Figure 16–17 *(continued)*   *I*, Order Mallophaga: a guinea pig louse. (From Grassé, P. (Ed.): Traité de Zoologie.) *J*, Order Anoplura: a body louse of man. (From P. Grassé (Ed.): Traité de Zoologie.) *K*, Order Thysanoptera: a thrip. (After Moulton from Grassé, P. (Ed.): Traité de Zoologie.) *L*, Order Homoptera: buffalo treehopper. (After Irving from H. Curran, 1954: Golden Playbook of Insect Stamps. Simon & Schuster, New York.) *M*, Order Neuroptera: mantispid. (After Banks from Borror & DeLong.)

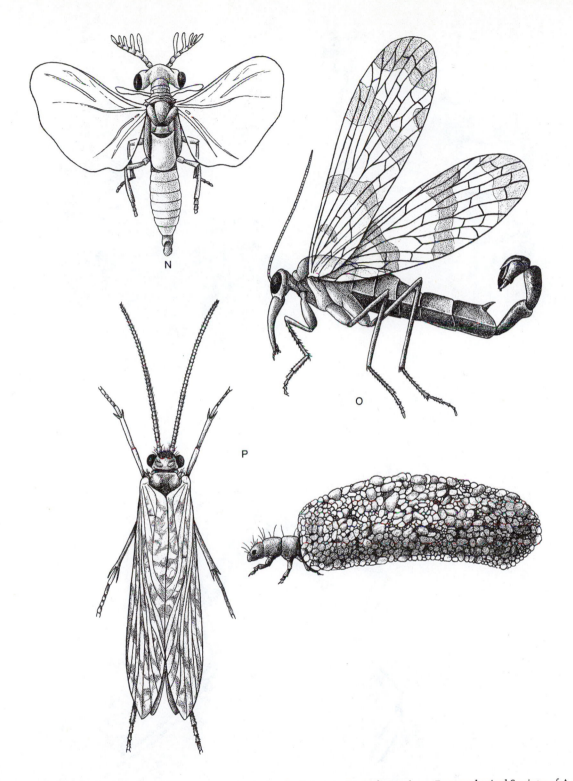

Figure 16–17 (continued)   N, Order Strepsiptera: twisted wing parasite. (After Bohart; Entomological Society of America from Borror & DeLong.) O, Order Mecoptera: a scorpion fly. (After Taft from Borror & DeLong.) P, Order Trichoptera: an adult caddis fly and larva in case. (After Mohr; Illinois Nat. Hist. Survey, from Ross.)

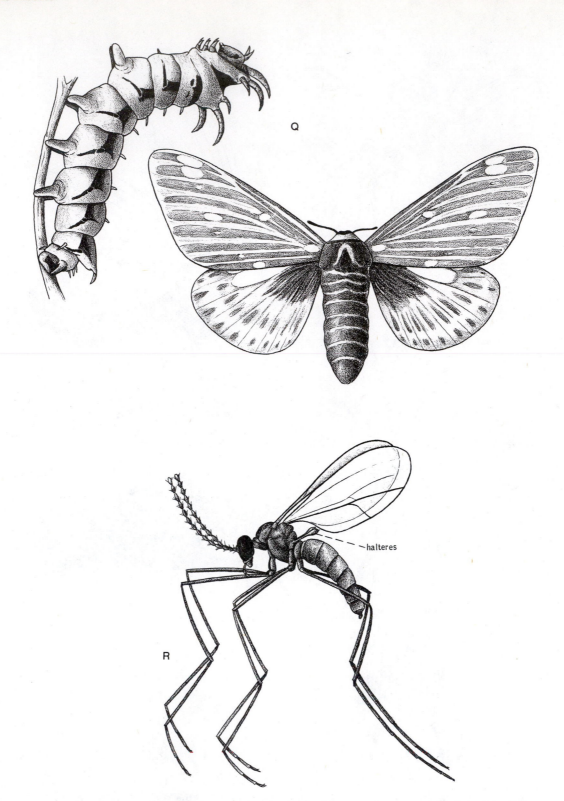

Figure 16–17 *(continued)* Q, Order Lepidoptera: the royal walnut moth and its caterpillar, the hickory-horned devil. (After Lutz.) R, Order Diptera: a gall gnat. (After Usda from Borror & DeLong.)

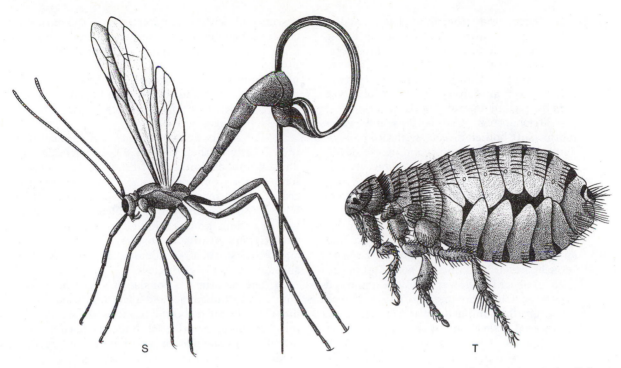

Figure 16–17 *(continued)* *S*, Order Hymenoptera: a parasitic female ichneumon fly. (After Lutz.) *T*, Order Siphonaptera: a flea. (After Bouché from Borror & DeLong.)

Order Thysanoptera. Thrips. Small, slender insects with mouthparts adapted for rasping and sucking. Winged and wingless. Wings are narrow, with few veins, and fringed with hairs. Development peculiar in that there are nymphlike early instars but a preadult pupa stage. A large number of species feed on sap in flowers, but some thysanopterans are predacious and feed on mites and smaller insects (Fig. 16–17*K*).

Order Hemiptera. True bugs. Piercing and sucking mouthparts, in which the beak arises from the front of the head and extends ventrally and posteriorly. Forewings with a thickened basal and distal membranous section. The membranous portions of the forewings overlap when at rest over the abdomen. Hindwings entirely membranous. Gradual metamorphosis. Herbivorous and predacious (Fig. 16–12).

Order Homoptera. Cicadas, leafhoppers, and aphids. Herbivorous insects related to the hemipterans, but beak arises from back of head. Forewings typically membranous. Wings are commonly held in a tentlike position over the body. Gradual metamorphosis (Fig. 16–17*L*).

Order Neuroptera. Lacewings, ant lions, mantispids, snake flies, and dobsonflies. Adults have chewing mouthparts and long antennae. The two pairs of similar membranous wings with many veins are held tentlike over abdomen. Complete metamorphosis. Larvae are predacious and usually terrestrial (Fig. 16–17*M*).

Order Coleoptera. Beetles, weevils. Largest order of insects (over 300,000 species), with hard bodies and chewing mouthparts. Adults usually have two pairs of wings, of which the front pair are modified as heavy protective covers (elytra); the hind pair are membranous. Complete metamorphosis. The majority are plant feeders but there are predatory families. Some aquatic species (Fig. 16–13).

Order Strepsiptera. Minute, beetle-like insects, mostly parasitic on other insects. Only males possess wings; in females forewings reduced to club-shaped appendages. Complete metamorphosis (Fig. 16–17*N*).

Order Mecoptera. Scorpion flies. Slender-bodied insects, often vividly colored. Biting mouthparts prolonged as a beak. Most species with long narrow wings, which have many veins. Complete metamorphosis. Adults are omnivorous, and grublike larvae feed on organic matter (Fig. 16–17O).

Order Trichoptera. Caddis flies and water moths. Soft-bodied insects, with two pairs of hairy, membranous wings and poorly developed, chewing mouthparts. Complete metamorphosis. Larvae are aquatic and build portable cases of various materials (Fig. 16–17P).

Order Lepidoptera. Butterflies and moths. Soft-bodied insects with wings, body, and appendages covered with pigmented scales. Mouthparts are modified as coiled proboscis used for sucking flower nectar. Compound eyes large. Complete metamorphosis. Larvae are caterpillars and usually are plant feeders; adults feed little or not at all (Fig. 16–17Q).

Order Diptera. True flies. Large order, all of which have functional front wings and reduced, knoblike hind wings (halteres). Mouthparts are variable, as is body form. Complete metamorphosis. Group includes mosquitoes, horseflies, midges, and gnats. Adults often vectors of diseases; larvae frequently damaging to vegetables and domestic animals (Fig. 16–17R).

Order Hymenoptera. Ants, bees, wasps, and sawflies. A large and varied order, all with chewing mouthparts but also modified for lapping or sucking in some forms. Winged and wingless species. Wings are transparent with only a few veins. Complete metamorphosis. Larvae are caterpillars or are grublike with chewing mouthparts (Fig. 16–17S).

Order Siphonaptera. Fleas. Small, wingless insects with laterally flattened bodies. Legs are long with large coxae and are adapted for jumping. These insects have piercing and sucking mouthparts and feed on the blood of mammals and birds. They are vectors of bubonic plague. Complete metamorphosis (Fig. 16–17T).

Summary

1 The class Insecta contains the largest number of species of any group of animals. At least three quarters of a million species have been described. Insects rank with vertebrates as being the most successful inhabitants of the terrestrial environment.

2 Insects are distinguished from other uniramian arthropods in having the body divided into a head, thorax, and abdomen. The head bears one pair of antennae and the feeding appendages; the thorax carries three pairs of legs; the abdomen lacks appendages.

3 The ability of most insects to fly has contributed greatly to their success. Flight has enhanced distribution, exploitation of food sources and habitats, escape from predators, and reproductive processes. Most insects have two pairs of thoracic wings, although one pair is reduced, modified, or lost in various groups. Flight evolved early in the evolutionary history of insects, but some groups (apterygotes), such as collembolans and thysanurans, are primitively wingless.

4 The mouthparts consist of a pair of mandibles, a pair of maxillae, and a labium (fused second maxillae). Primitively, the mouthparts are adapted for chewing plant material, but they have become modified for a wide range of diets and feeding modes, including piercing and sucking.

5 A tracheal system provides for gas exchange. Spiracles are located along the sides of the thorax and abdomen but vary in number depending on the species.

6 The nitrogenous waste of insects is uric acid, which is excreted through malpighian tubules. Insects are capable of producing a hyperosmotic urine, which together with the waxy epicuticle, is an important adaptation for reducing water loss and has contributed to the success of insects as terrestrial animals.

7 The tubular heart is located in the dorsal part of the abdomen and propels blood anteriorly through a short aorta. The remainder of the blood-vascular system is open.

8 Most insects possess a pair of large, lateral, compound eyes, three ocelli on the top or front of the head, and a great variety of types of sensilla located over the body surface, especially on the antenna and legs.

9 Most insects transfer sperm in spermatophores. Primitively, transfer is indirect, as in many other terrestrial arthropods, but in the majority of insects the male deposits the spermatophores directly within the female reproductive system. The female deposits eggs encased in protective coverings. Cleavage is typically superficial.

10 In primitive insects the juvenile stages are similar to the adult. In higher orders the juvenile gradually acquires certain structures, such as the wings, during the course of development. Development with larval stages and complete metamorphosis is a specialization of the orders containing beetles, flies, bees, and wasps. Development of this type enables juveniles and adults to exploit different habitats and food sources.

11 Parasitism has evolved a number of times in the evolution of insects. Juveniles, adults, or both may be parasitic.

12 Highly developed social (colonial) organization has evolved within two orders, the Isoptera (termites) and the Hymenoptera (ants, bees, and wasps). Only some hymenopterans are social, and there is a great range in the complexity of social organization.

REFERENCES

The literature here is restricted to insects. The introductory references on page 8 include many general works and field guides that contain sections on these animals. The literature on insects is voluminous. The following introductory entomology texts provide good general accounts of insects. They vary in their approach and emphasis, but all contain references to more specialized topics. An introductory entomology text is also a useful starting point for the student interested in taxonomy.

Borror, D. J., De Long, D. M., and Triplehorn, C. A., 1981: An Introduction to Insect Biology and Diversity, Saunders College Publishing, Philadelphia. 728 pp.

Daly, H. V., Doyen, J. T., and Ehrlich, P. R., 1978: An Introduction to Insect Biology and Diversity. McGraw-Hill Book Co., N.Y. 538 pp.

Elzinga, R. J., 1981: Fundamentals of Entomology. 2nd Edition. Prentice-Hall, Inc., Englewood Cliffs, N.J. 464 pp.

Evans, H. E., 1984: Insect Biology: A Textbook of Entomology. Addison-Wesley Publishing Co., Reading, Mass. 436 pp. (This text emphasizes the relationships of insects to the world around them.)

Gillot, C., 1980: Entomology. Plenum Publishing, N.Y. 747 pp.

Little, V. A., 1972: General and Applied Entomology, 3rd Edition, Harper and Row, N.Y. 527 pp.

Richards, O. W., and Davies, R. G., 1978: Imm's Outline of Entomology, 6th Edition. Chapman and Hall, London. 254 pp.

Romoser, W. S., 1981: The Science of Entomology, 2nd Edition. Macmillan Publishing Co., N.Y. 544 pp.

Ross, H. H., Ross, J. R. P., and Ross, C. A., 1982: A Textbook of Entomology, 4th Edition. John Wiley and Sons, N.Y. 696 pp.

Other General References

Arnett, R. H., 1985: American Insects: A Handbook of the Insects of North America. Van Nostrand Reinhold N.Y. 850 pp.

Arnett, R. H., Downie, N. M., and Jaques, H. E., 1980: How to Know the Beetles, 2nd Edition. W. C. Brown, Dubuque, Iowa, 416 pp.

Askew, R. R., 1971: Parasitic Insects. Elsevier-North Holland Publishing Co., N.Y. (See also the general parasitology texts listed on p. 204.)

Bland, R. G., and Jaques, H. E., 1978: How to Know the Insects. 3rd Edition. W. C. Brown, Dubuque, Iowa. 409 pp.

Blum, M. S. (Ed.), 1985: Fundamentals of Insect Physiology. Wiley-Interscience, N.Y. 598 pp.

Borror, D. J., and White, R. E., 1974: A Field Guide to the Insects of America North of Mexico. Peterson Field Guide Series. Houghton Mifflin Co., Boston.

Brian, M. V., 1983: Social Insects: Ecology and Behavioural Biology. Chapman and Hall, London. 377 pp.

Chapman, R. F., 1982: The Insects: Structure and Function. 3rd Edition. Harvard University Press, Cambridge, Mass. 919 pp.

Chu, H. F., 1949: How to Know the Immature Insects. W. C. Brown, Dubuque, Iowa. 234 pp.

Covell, C. V., 1984: A Field Guide to the Moths of Eastern North America. Peterson Field Guide Series. Houghton Mifflin Co. Boston. 496 pp.

Ehrlich, P. R., and Ehrlich, A. H., 1961: How to Know the Butterflies. W. C. Brown, Dubuque, Iowa. 262 pp.

Merritt, R. W., and Cummins, K. W., 1978: An Introduction to the Aquatic Insects of North America. Kendall Hunt Publishing Co., Dubuque, Iowa. 464 pp.

Metcalf, C. L., and Flint, W. P., and Metcalf, R. L., 1962: Destructive and Useful Insects. 4th Edition. McGraw-Hill Book Co., N.Y. 1087 pp.

Price, P. W., 1984: Insect Ecology. 2nd Edition. John Wiley and Sons, N.Y. 607 pp.

Rockstein, M. (Ed.), 1973–1974: The Physiology of Insecta. 2nd Edition. Vols I–VI. Academic Press, N.Y. (An extensive treatment of the physiology and behavior of insects. Well organized by topics contributed by many physiologists and entomologists.)

von Frisch, K., 1971: Bees: Their Vision, Chemical Senses and Language. 2nd Edition. Cornell University Press, Ithaca, N.Y. 176 pp.

Wilson, E. O., 1971: The Insect Societies. Harvard University Press, Cambridge, Mass. 548 pp. (An excellent detailed account of social insects.)

17

Some Lesser Protostomes

This chapter treats six small phyla of coelomates, which probably stemmed from various points along the protostome line.

Phylum Pogonophora

The Pogonophora were unknown prior to the 20th century. Their discovery awaited the development of oceanographic vessels capable of sampling deep-water habitats. The first specimen was dredged from Indonesian waters in 1900. Since that time more than 80 species have been described and more are being discovered. The northwest Pacific has yielded the richest pogonophoran fauna, but this discovery perhaps reflects the concentrated work of oceanographic vessels in this area. It is now becoming apparent that the phylum is widespread in the world's seas, especially along the continental slopes. In the North Atlantic pogonophorans have been collected from European waters, from Nova Scotia to Florida, and from the Gulf of Mexico.

Pogonophorans are almost exclusively deep-water animals (more than 100 meters). They are

sessile, living in secreted, stiff, chitinous tubes that are usually fixed upright in bottom ooze, and they often occur in dense aggregations, as many as 200 per square meter. There are a few species that construct their tubes in decaying wood or other debris. The long, wormlike body of most species ranges in length from 10 to 85 cm, but in 1977 the research submarine *Alvin*, while investigating the fauna around the warm-water vents on the 9000-meter floor of the Galapagos rift, discovered clusters of large pogonophorans 1.5 meters long and almost 4 cm in diameter (Fig. 17–2*B*).

The body is composed of a short forepart, a long trunk, and a short opisthosoma (Fig. 17–1). The forepart consists of an anterior cephalic lobe and a posterior glandular region that provides secretions for the formation of the tube. Beneath the cephalic lobe, the forepart bears long tentacles, which are the distinguishing feature of the phylum (*Pogonophora* means "beard bearer"). Depending on the species and age of the individual, the tentacles range in number from a single spiral tentacle to more than 250 (Fig. 17–1*A* and *C*). *Riftia*, found in the Galapagos hydrothermal vents, has many ten-

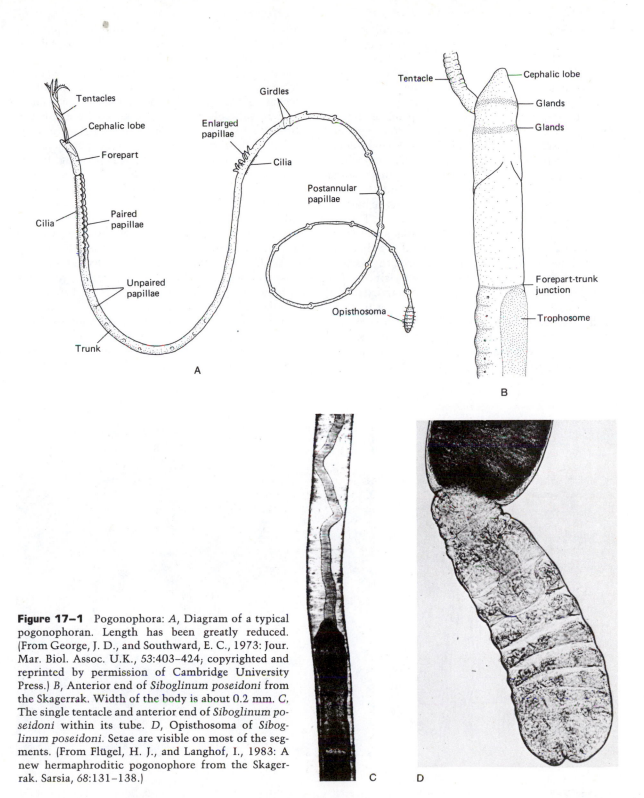

Figure 17–1 Pogonophora: *A*, Diagram of a typical pogonophoran. Length has been greatly reduced. (From George, J. D., and Southward, E. C., 1973: Jour. Mar. Biol. Assoc. U.K., 53:403–424; copyrighted and reprinted by permission of Cambridge University Press.) *B*, Anterior end of *Siboglinum poseidoni* from the Skagerrak. Width of the body is about 0.2 mm. *C*, The single tentacle and anterior end of *Siboglinum poseidoni* within its tube. *D*, Opisthosoma of *Siboglinum poseidoni*. Setae are visible on most of the segments. (From Flügel, H. J., and Langhof, I., 1983: A new hermaphroditic pogonophore from the Skagerrak. Sarsia, 68:131–138.)

tacles fused at the base as sheets and collectively forming a large plume that emerges from the tube (Fig. 17–2*B*). The tentacles are peculiar in usually bearing minute pinnules that are extensions of single epithelial cells. Tracts of cilia run the length of each tentacle (Fig. 17–2*A*).

The long trunk typically bears papillae and, in the midregion, two girdles of short setae. The terminal opisthosoma (Fig. 17–1*A* and *D*) is composed of 5 to 23 short segments with setae, which are similar to those of annelids. The opisthosoma and setae probably function in anchoring the body within the tube and aid in burrowing in the bottom ooze below the open lower end of the tube. The opisthosoma is fragile and easily breaks away from the rest of the body when the animal is pulled. As a result, most specimens dredged from the bottom lack this region, and it has been only since 1964 that the opisthosoma has been known to exist.

Internally, there is a coelomic compartment in each of the body divisions, and there are extensions of the coelom into the tentacles. In the opisthosoma there are septa between the segmental coelomic compartments. The nature of the pogono-

phoran body cavity is uncertain, for Gupta and Little (1975) were unable to find a peritoneal lining. A well-developed, closed blood-vascular system is present, and each tentacle is supplied with two vessels (Fig. 17–2*A*). There is a nerve plexus at the base of the epithelium and a ventral nerve cord.

A remarkable feature of pogonophorans is the complete absence of a mouth and a digestive tract. In fact, zoologists who examined the first specimen thought that part of the body was missing. In the absence of a digestive tract, the mode of nutrition in these animals was puzzling. The large size of the pogonophorans from the Galapagos rift made possible the discovery that in the trunk of the worm is a mass of tissue, called the trophosome, that is packed with symbiotic bacteria (Jones, 1981; Cavanaugh et al, 1981). The bacteria fix carbon chemosynthetically and share the organic products with their pogonophoran host (Gnassle, 1985). The vascular hemoglobin, which accounts for the red color of these large worms, is important in delivering the large amounts of oxygen required by the bacteria. Similar trophosomal tissue was subsequently found in small pogonophorans living in

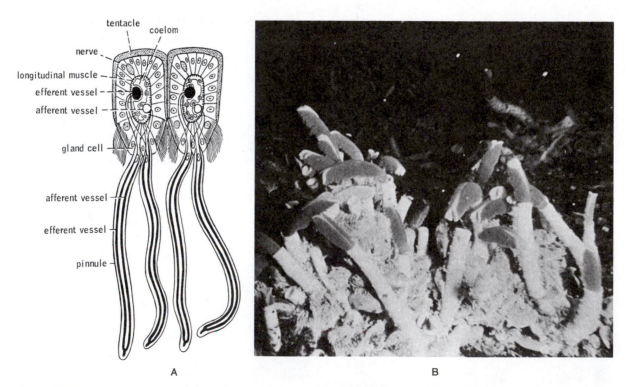

A   B

Figure 17–2   *A*, Cross section of two adjacent tentacles of *Lamellisabella*. (After Ivanov.) *B*, Living specimens of the giant pogonophoran *Riftia pachyptila* at the Galapagos rift. The red plume of tentacles projects from the opening of the white tubes. (Courtesy of Meredith Jones, National Museum of Natural History.)

BOX 17–1   UTILIZATION BY INVERTEBRATES OF DISSOLVED ORGANIC NUTRIENTS IN SEA WATER

It has long been recognized that sea water contains considerable amounts of dissolved organic material (DOM), including free amines. Might gutless animals such as pogonophorans and certain bivalves augment the organic compounds provided by their bacterial symbionts with dissolved organic materials from the surrounding sea water?

With radioactive tracers, uptake of DOM across body surfaces has been demonstrated in a number of invertebrate groups. Filter feeders, such as bivalves, which move large volumes of water over the gills or other surfaces, have been most commonly investigated. The uptake involves active transport, since the dissolved substances must be moved across a steep concentration gradient; however, there is also some leakage of amines and other subtances in the outward direction. So is there any net influx? A net influx has been demonstrated in mussels (*Modiolus* and *Mytilis*), the polychaete *Nereis diversicolor*, and the sand dollar *Dendraster excentricus*. Most at-

tention has been given to the movement of free amines. Less is known about sugars and very little about lipids.

Assuming that a net influx occurs in most marine invertebrates, how important is the influx in the animal's nutrition? Absorbed DOM coupled with organic materials supplied by algal symbionts appears significant in the total energy budget of one soft coral studied (*Heteroxenia fuscescens*), but we are still a long way from understanding the physiological or ecological significance of DOM uptake for most invertebrates. Good reviews of this topic are provided by M. G. Stewart (1979: Absorption of dissolved organic nutrients by marine invertebrates. Oceanogr. Mar. Biol Ann. Rev., 17:163–192) and G. C. Stevens et al (1982: The role of uptake of organic solutes in the nutrition of marine organisms. Amer. Zool., 22(3):611–733 [ten papers from a symposium]).

water around Scandinavia and in the Bay of Biscay (Southward et al, 1981). However, there may be varying degrees of supplementary nutrition by absorption of organic compounds dissolved in the surrounding sea water (Box 17–1) (Southward and Southward, 1980). Dependence on symbiotic bacteria for nutrition is not unique to pogonophorans. This mode of nutrition is found in certain gutless clams (p. 409) and some other groups. Most pogonophorans are dioecious, and two cylindrical gonads are located one on each side in the trunk coelom. The two male gonopores are located at the anterior part of the trunk, and the terminal part of the sperm duct packages the sperm into spermatophores. In the female the two oviducts open through separate gonopores farther back on the trunk.

Sperm transfer, fertilization, and egg deposition have not been observed, but one specimen of *Siboglinum* was seen to move spermatophores to the mouth of the tube with its tentacles. The spermatophores perhaps reach the tubes of neighboring females by floating.

Since the eggs of many species are brooded within the tube, we have some knowledge of pogonophoran development. Cleavage follows a bilateral pattern, which could perhaps be derived from either a spiral or a radial type. Descriptions of mesoderm and coelom formation are contradictory (see Southward in Giese and Pearse, 1975).

Late embryos collected from tubes are wormlike and yolk laden and have ciliary girdles. Whether they are dispersed by currents or quickly sink to the bottom is uncertain.

The segmented setiferous terminal part of the body, the similarity of the setae to those of annelids, and the segmentation of the mesoderm indicate a protostome position of the Pogonophora and suggest that they are closely related to the Annelida. Although this view is held by a number of zoologists, there are others who place pogonophorans between deuterostomes and protostomes or on a separate line of evolution (Cutler, 1975; Siewing, 1975).

The giant pogonophorans of the Galapagos rift are so different from other members of the phylum that they and a number of related species from other hydrothermal areas and elsewhere have been thought to justify a new grouping, the Vestimentifera. Jones (1985) has proposed that they be placed within a separate phylum.

Phylum Sipuncula

The sipunculans are a group of approximately 320 species of marine animals, sometimes called peanut worms. Sipunculans are rather drab-colored worms that range in length from 2 mm to more than 72 cm, although most are less than 10 cm long. All are bottom dwellers, the majority in shallow water. Some live in sand and mud where, like *Sipunculus*, they are active burrowers; some live in mucus-lined excavations. Others live in coral crevices, in empty mollusk shells *(Phascolion)* or annelid tubes, and in other sorts of protective retreats (Fig. 17–3B). There are a number of species that bore in coralline rock, the animal directing its anterior end toward the opening of its burrow (Fig. 17–4A). Densities as great as 700 individuals per square meter of coralline rock have been reported from Hawaii. The mechanism of boring is not completely understood but appears to involve both chemical and mechanical processes (Williams and Margolis, 1974).

External Structure

The cylindrical body of sipunculans is divided into an anterior narrowed section, called the introvert, and a larger posterior trunk (Fig. 17–3A and 17–5A). The introvert represents the head and the anterior part of the body, but it can be retracted into the anterior end of the trunk. The anterior end of the introvert contains the mouth, which is located below or is surrounded, at least in part, by a scalloped fringe, lobes, or tentacles. Most of these projections are ciliated, and each bears a deep ciliated groove on its inner side. Behind the anterior end, the surface of the introvert is typically covered with spines, tubercles, and other ornamentations.

The trunk is cylindrical, and in some rock-boring sipunculans, the surface of the trunk is thickened at the anterior end to form a dorsal or a collar-like shield, which is used to block the opening of the retreat when the introvert is invaginated (Fig. 17–4A).

Internal Structure and Physiology

The body wall is constructed like that of the annelids, and a large coelom extends the length of the body. Elevated coelomic fluid pressure through contraction of the body wall brings about the protrusion of the introvert, and retractor muscles withdraw it. Separate contractile sacs receive fluid from, or supply fluid to, the hollow tentacles when they are contracted or expanded (Fig. 17–5B).

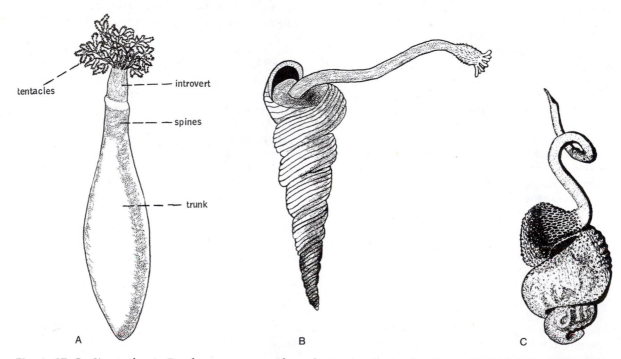

tentacles

introvert

spines

trunk

A   B   C

Figure 17–3 Sipuncula: *A, Dendrostomum pyroides,* a burrowing sipunculan (⅔ natural size). (After Fisher from Tétry.) *B, Phascolion* in a snail shell. Introvert is extended out of aperture. (After Cuénot from Tétry.) *C, Aspidosiphon spinalis,* from the southeastern coast of the United States, removed from the snail shell that it inhabits. (From Cutler, E. B., 1973: Sipuncula of the western North Atlantic. Bull. Am. Mus. Nat. Hist., *152*:103–204.)

B

A

Figure 17–4 *A*, Rock-boring sipunculans. From left to right: *Phascolosoma antillarum, Phascolosoma dentigerum, Paraspidosiphon steenstrupi.* All from the Caribbean. The introvert of *Paraspidosiphon steenstrupi* is displaced ventrally by the anterior dorsal shield. Scale is 5 mm. (From Rice, M. E., 1969: Am. Zool., 9:803–812.) *B*, Specimens of the West Indian *Sipunculus* dug from sand in shallow water. (By Betty M. Barnes.)

Sipunculans are mostly nonselective deposit feeders, drawing in sediment with the extended introvert and tentacles (Walter, 1973). In burrowing forms, such as *Sipunculus*, the worm ingests the sand and silt through which it burrows. Some rock borers, such as *Phascolosoma antillarum*, hold their tentacular crown open at the mouth of the burrow and perhaps are suspension feeders. Other hard-bottom forms extend the introvert out over the rock surface. The expanded tentacles are placed against the substratum, and surface detritus is trapped in mucus and driven into the mouth by the beating cilia. After the introvert collects material, it then invaginates and the food is ingested.

The digestive tract is U shaped and coiled (Fig. 17–5B). The anus is located middorsally at the anterior end of the trunk except in a single genus, *Onchnesoma*, in which it opens on the introvert.

There is no blood-vascular system, but the coelomic fluid functions in circulation and contains abundant corpuscles bearing hemerythrin (see p.

295). The tentacles are probably an important site of gas exchange.

Sipunculans possess a single pair of large, sac-like metanephridia, and the nephridiopores open anteriorly and ventrally at about the same level as the anus (Fig. 17–5B). Associated with excretion in sipunculans are peculiar, vase-shaped clusters of peritoneal cells called urns. Free urns are fixed urns that have become detached from the peritoneum and move about in the coelomic fluid, gathering particulate waste material that is eventually dumped in various places within the coelom or removed by the nephridia.

The nervous system is essentially similar to that of the annelids, but the single ventral nerve cord, which runs the length of the body, lacks metameric ganglionic swellings (Fig. 17–5B).

Sensory cells are particularly abundant on the end of the introvert that is used to probe the surrounding environment. Many species of *Golfingia* have a pair of ciliated pits (nuchal organs) at the

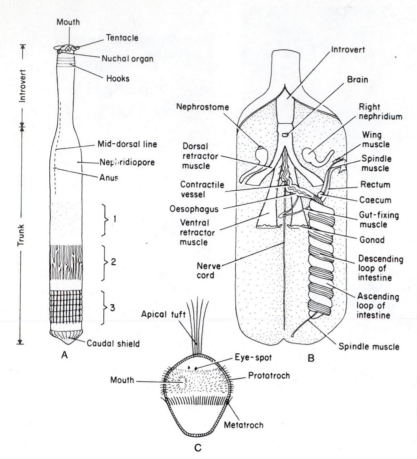

Figure 17–5 Diagrammatic representation of the structure of sipunculans: *A*, External structure. Numbers refer to the arrangement of the body wall muscles. In 1 the circular and longitudinal muscles are arranged in continuous sheets; in 2 the circular muscles are arranged as a continuous sheet and the longitudinal muscles are arranged in bands; and in 3 both are arranged in bands. *B*, Internal structure, showing the body wall opened. *C*, Trochophore larva of *Golfingia vulgaris*. (From Gibbs, P. E., 1977: British Sipunculans. Synopses of the British Fauna No 12. Linnean Society of London, Academic Press, London. p. 3.)

end of the introvert and a pair of pigment-cup ocelli embedded in the brain.

Reproduction

Sipunculans are mostly dioecious, and the immature gametes, which arise from the peritoneum, are shed into the coelom, where maturation is completed; ripe eggs or sperm leave the body by way of the nephridia. Fertilization is external, and the fertilized eggs undergo spiral cleavage. Later development may be direct or lead to a trochophore larva. In *Golfingia* the trochophore is typical (Fig. 17–5C), but in *Sipunculus* it is elongated. Of the some 20 species whose life histories are known, 17 have larvae, and of these over half are remarkable in having a long larval life-span (three to eight months) (Rice, 1981). Such a long larval life involves the development of a later larval stage, called a pelagosphera, following the trochophore. The pelagosphera swims with a large metatroch.

Although sipunculans are not metameric animals, they are probably related to the annelids. The construction of the body wall, the nature of the nervous system, and the embryology are annelidan in character. Sipunculans perhaps diverged from the line leading to the annelids at some point before the development of metamerism.

Phylum Echiura

Echiurans are marine worms that are somewhat similar to sipunculans in size and general habit. Many species, such as *Echiurus, Urechis,* and *Ikeda,* live in burrows in sand and mud; others live in rock and coral crevices (Fig. 17–6D). *Thalassema mellita,* which lives off the southeastern coast of the United States, inhabits the test of dead sand dollars. When the worm is very small, it enters the test and later becomes too large to leave. The majority of echiurans live in shallow water,

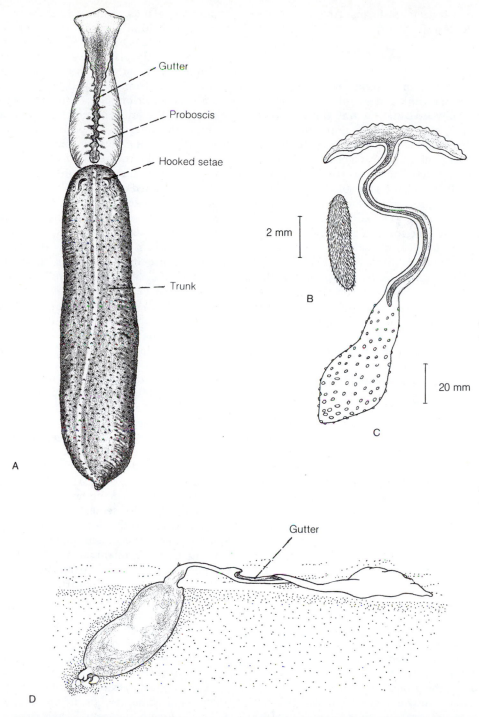

Figure 17–6 Echiura: *A, Echiurus* species from life (ventral view). *B* and *C,* Dwarf male (*B*) and female (*C*) of *Bonellia viridis,* showing the extreme sexual dimorphism in this species. (After MacGinitie, G. E., and MacGinitie, N., 1949: Natural History of Marine Animals. McGraw-Hill Book Co., N.Y.) *D, Tatjanellia grandis* in feeding position with trunk buried and proboscis extended over surface of bottom. (After Zenkevitch from Dawydoff.)

but there are also deep-sea forms. About 140 species have been described.

External Structure

The body of an echiuran is composed of a sausage-shaped, cylindrical trunk and an anterior proboscis (Fig. 17–6A). They are usually a drab gray or brown color, but some, such as *Bonellia*, are green, and others are red or rose. A few are transparent. The proboscis is a large, flattened projection of the head and cannot be retracted into the trunk. The proboscis is actually a cephalic lobe and, since it contains the brain, is probably homologous with the prostomium of annelids. The edges are rolled ventrally, so the underside forms a gutter. The mouth opens at the base of the proboscis.

The length of the proboscis varies considerably. In the Japanese echiuran, *Ikeda*, a specimen with a trunk 40 cm long may have a proboscis 1.5 meters in length. However, in the common *Echiurus*, which lives on both sides of the North Atlantic, the proboscis is much shorter than the trunk. The proboscis is used in obtaining food and is very extensible in some echiurans. For example, a specimen of *Bonellia* 8 cm long can extend its proboscis 2 meters. Extension takes place by ciliary creeping, like a flatworm.

On the underside of the cylindrical trunk, just back of the anterior end, is a pair of large, closely placed, chitinous setae, which are curved or hooked (Fig. 17–6A). In addition to anterior setae, *Urechis* and *Echiurus* possess one or two circlets of setae around the posterior end of the trunk.

Echiurans move by peristaltic contractions within their burrows or lodgment, and some can squeeze through very confined spaces (Schembri and Jaccarini, 1977). Setae provide for traction or gripping burrow walls.

Internal Structure and Physiology

Most echiurans, such as *Echiurus*, are detritus feeders. The proboscis is projected from the burrow or retreat, and the ventral face is stretched out over the substratum (Fig. 17–6D). Detritus adheres to the mucus on the proboscis surface, from which it is then driven into a median ciliated groove. Here the particles are conducted back into the mouth (Jaccarini and Schembri, 1977).

Urechis caupo, which lives along the California coast in U-shaped burrows, has a mode of feeding that is somewhat similar to the polychaete, *Chaetopterus*. The proboscis is very short, and a circlet of mucous glands girdles the anterior end of the trunk just behind the setae. During feeding the glandular girdle is brought in contact with the burrow wall, and the glands begin secreting mucus. As the mucus is spun out, the worm backs up, and eventually a funnel-shaped mucous collar surrounds the anterior end of the body. After the formation of the mucous collar, water is pumped through the burrow by peristaltic action of the trunk. Because the mucous "net" extends from the wall of the tube to the body of the worm, all sea water must pass through it, and even the finest particles are strained out. When loaded with food material, the net is detached from the body. The worm then moves backward, seizes the accumulated food mass with its short proboscis, and swallows the mass. This worm is sometimes called the innkeeper worm because of the various commensal guests (crabs, gobies, clams) that share its burrow.

The digestive tract is extremely long and greatly coiled (Fig. 17–7), with the mouth located at the base of the proboscis and the anus opening at the posterior of the body.

A closed blood-vascular system is present, except in *Urechis*, and is constructed on the same general plan as that of the annelids (Fig. 17–7). The blood is colorless, but some coelomocytes contain hemoglobin. The proboscis is an important surface for gas exchange.

The excretory organs are metanephridia, which may number one (in *Bonellia*), two (in *Echiurus*), or up to hundreds of pairs (in *Ikeda*). When there are only one or two pairs of nephridia, they are located anteriorly, and the nephridiopores open just behind the anterior setae.

In addition to the nephridia, echiurans possess a peculiar pair of accessory organs called anal sacs. These are simple, or sometimes branched, contractile diverticula that arise from each side of the rectum. Distributed over the surface of the diverticula and often arranged in tufts are numerous ciliated funnels similar to nephrostomes. The funnels open into the coelom and lead into the lumen of the diverticula. The anal sacs function to eliminate waste from the coelomic fluid by way of the anus (Harris and Jaccarini, 1981).

The nervous system is similar to that of sipunculans and annelids but is not metameric (Fig. 17–7). There are no special sense organs.

Reproduction

The sexes are separate, and the gametes that arise from the peritoneum are freed into the coelom,

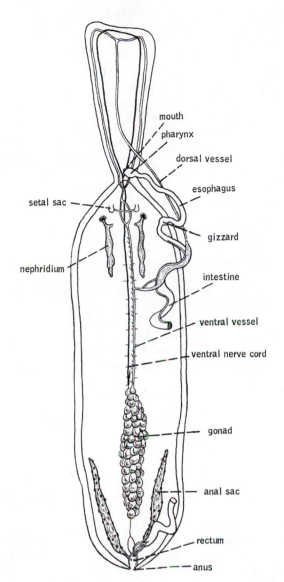

mouth

pharynx

dorsal vessel

esophagus

setal sac

gizzard

nephridium

intestine

ventral vessel

ventral nerve cord

gonad

anal sac

rectum

anus

Figure 17–7   Internal structure of an echiuran. (After Delage and Hérourard from Dawydoff.)

reduced. The entire body surface is ciliated, and there is a system of genital ducts. Sex is determined at the larval stage. Any larva that comes in contact with a female and enters her body is induced by a female hormone to develop into a dwarf male. First contact is made with the female's proboscis, which may extend as much as 1 meter from the rocky retreat hiding the trunk. It is the proboscis that secretes the hormone (Jaccarini et al, 1983). After a few days on the proboscis, the now presumptive male passes into the female's esophagus and eventually into the nephridia or a special male tube, where it is sexually mature in one to two weeks. A single female will house some 20 males. Larvae that do not come in contact with females develop into females, and over a year is required to reach sexual maturity.

Spiral cleavage and free-swimming trochophore larvae characterize the embryology of echiurans. During the course of development, the mesodermal bands in some echiurans, such as *Echiurus*, are segmented and develop ten pairs of rudimentary coelomic pouches, which coincide with a similar transitory metamerism of the nerve cord.

The circulatory system, the setae, the presence of multiple nephridia in some species, and the transitory metamerism that appears during the course of embryonic development, in addition to the annelidan character of the body wall and nervous system, all indicate a close phylogenetic relationship between the echiurans and the annelids, closer it would seem than between the sipunculans and the annelids.

It appears probable that the echiurans represent a group of worms that stemmed from the early ancestral polychaetes or at least from the line leading to the polychaetes. Some degree of metamerism had certainly been attained in the ancestral group from which the echiurans arose.

Phylum Tardigrada

The tardigrades are a group of very tiny but highly specialized animals called water bears. Some reach 1.2 mm, but the majority are no longer than 0.3 to 0.5 mm. Although not uncommon, tardigrades are seldom encountered unless looked for because of their minute size and rather specialized habitats.

There are a few marine interstitial tardigrades that have been collected from both shallow and deep water. Also, there are some freshwater species that live in bottom detritus or on aquatic algae and mosses, and there are even species that live in the

where maturation is completed; the mature eggs or sperm are released through the nephridia. Fertilization is usually external in the sea water.

Species of *Bonellia* are peculiar in displaying an extreme sexual dimorphism. In *Bonellia viridis*, the female trunk may be 8 cm long, whereas the minute male is only 1 to 3 mm in length (Fig. 17–6B) and lives within the female's esophagus or nephridia. Besides being smaller, the male is structurally modified in many ways. There is no proboscis or circulatory system, and the digestive tract is water films of soil and forest litter. Many tardigrades, however, live in the water films surrounding the leaves of terrestrial mosses and lichens, especially those that grow on stones and trees. This latter habitat is shared primarily with some bdelloid rotifers, and the two groups present many parallel adaptations. About 375 species of tardigrades have been described, many of which are cosmopolitan. A single fossil tardigrade has been described from Cretaceous amber.

External Structure

The bodies of tardigrades are short, plump, and cylindrical, and there are four pairs of ventral, stubby legs (Fig. 17–8). Each leg terminates in four to eight claws or discs (interstitial species) (Fig. 17–8B). The body is covered by either a smooth or an ornamented cuticle, which in some tardigrades, such as *Echiniscus*, is divided into symmetrically arranged plates (Fig. 17–8A).

The cuticle contains chitin, mucopolysaccharides, proteins, and lipids, composing several layers (see review by Greven, 1984). Periodically, the old cuticle is shed, and the epidermis secretes a new one. During molting, the body contracts, pulling away from the old cuticle, which is then slipped off and left behind as a relatively intact casing.

Internal Structure and Physiology

The musculature consists of separate muscle bands, each composed of single, smooth muscle cell, extending from one subcuticular point of attachment to another (Fig. 17–8C). Tardigrades move about slowly, crawling on their legs and using the hooks at the ends of the legs to grasp the substratum.

The coelom is confined to the gonadial cavity, but a fluid-filled hemocoel extends between the muscle bands and the other internal organs. Correlated with their size is the lack of gas exchange organs and internal transport system.

The majority of tardigrades feed on the contents of plant cells, which are pierced with a stylet apparatus resembling that of herbivorous nematodes and rotifers (Figs. 17–8C and 17–9B). Soil tardigrades feed on algae and probably detritus, and some are predators on nematodes and other minute soil animals. The anterior mouth opens into a chitin-lined buccal tube and bulbous muscular pharynx. The pharynx is myoepithelial; i.e., there is no separate epithelial lining, another feature shared with gastrotrichs. The pointed ends of two stylets project into the anterior end of the buccal tube.

During feeding, the mouth is placed against the plant cells, and the stylets are projected to puncture the cell wall. The contents of the cell are then sucked out by the pharyngeal bulb.

The pharyngeal bulb passes into the tubular esophagus, which opens in turn into a large midgut, or intestine (Fig. 17–8C). There digestion and absorption take place. The end of the intestine leads into a short hindgut (rectum), which opens to the outside through the terminal anus.

At the junction of the intestine and rectum of some tardigrades are three large glands, sometimes called malpighian tubules, that are thought to be excretory in function (Fig. 17–8C).

The nervous system is distinctly metameric (Fig. 17–8C). The lobed brain is connected to a subpharyngeal ganglion by a pair of commissures surrounding the buccal tube. From the subpharyngeal ganglion, a double ventral nerve cord extends posteriorly, connecting a chain of four ganglia. From each ganglion issue several pairs of lateral nerves, one of which terminates in pedal ganglia in each pair of legs. Some of the bristles and spines are sensory, particularly those in the head region. Most tardigrades also possess a pair of simple eye spots, each containing a single red or black pigmented cell.

Reproduction

Tardigrades are all dioecious and have a single saccular gonad, testis or ovary, located above the intestine. In the male two sperm ducts open through a single median gonopore just in front of the anus. In the female the single oviduct opens either above the anus or into the ventral side of the rectum, which then serves as a cloaca (Fig. 17–8C). In the latter case a small, adjacent seminal receptacle also opens into the rectum.

Females are often more numerous than males, and in some genera, such as *Echiniscus*, males are unknown. Mating and egg laying occur at the time of a molt. In some aquatic species sperm are deposited into an old female skin containing eggs; in most terrestrial forms sperm are deposited into the female tract before the cuticle is completely shed, and fertilization occurs in the ovary.

One to 30 eggs are laid at a time, depending on the species. Aquatic tardigrades either deposit them in the old cuticle or attach them singly or in groups to various objects. Like rotifers and gastrotrichs, some aquatic tardigrades are reported to

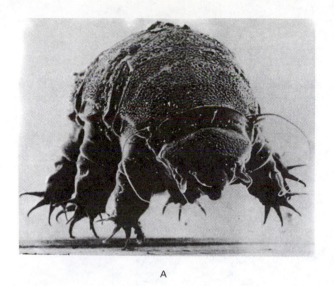

A

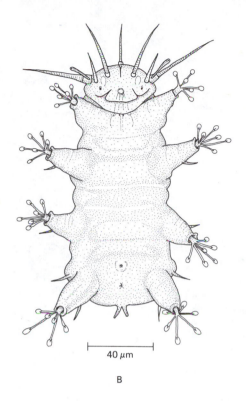

40 µm

B

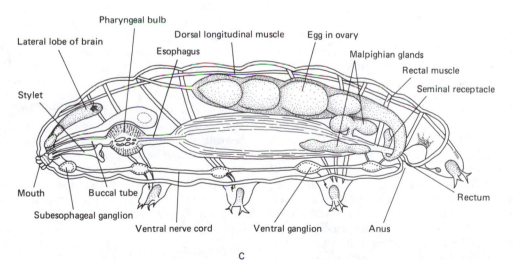

Pharyngeal bulb

Lateral lobe of brain

Esophagus

Dorsal longitudinal muscle

Egg in ovary

Malpighian glands

Rectal muscle

Seminal receptacle

Stylet

Mouth

Buccal tube

Subesophageal ganglion

Ventral nerve cord

Ventral ganglion

Anus

Rectum

C

Figure 17–8 Tardigrada: *A*, Anterior view of *Echiniscus arctomys*. Mouth is located at end of snoutlike, anterior projection. (By Crowe, J. H., and Cooper, A. F., 1971: Sci. Amer., 225:30–36. Copyright © 1971 by Scientific American, Inc. All right reserved.) *B*, *Batillipes noerrevangi*, a marine interstitial tardigrade. (From Kristensen, R. M., 1981: Sense organs of two marine arthrotardigrades. Acta Zool., 62(1):27–41.) *C*, Internal structure of *Macrobiotus hufelandi* (lateral view). (After Cuénot, L., 1949: Les Tardigrades. *In* Grassé, P. (Ed.) Traité de Zoologie. Vol. VI. Masson ct Cie, Paris.)

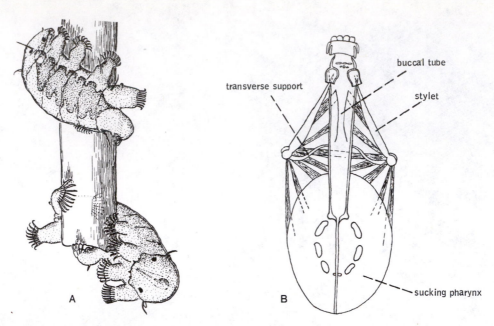

transverse support

buccal tube

stylet

sucking pharynx

A

B

Figure 17–9 *A*, Tardigrades climbing on an algal filament. (From Marcus, E., 1929: *In* Bronn, H. G. (Ed.), Klassen und Ordnungen des Tierreichs. Vol. 5, Pt. 4. Akademische Verlagsgesellschaft, Frankfurt. p. 156.) *B*, Pharynx and buccal apparatus of a tardigrade (dorsal view). (After Pennak, R. W., 1978: Freshwater Invertebrates of North America, 2nd Edition. John Wiley and Sons, N.Y.)

produce thin-shelled eggs when environmental conditions are favorable and thick-shelled eggs when environmental conditions are adverse. The eggs of terrestrial species typically possess a thick, sculptured shell that can resist the frequent periods of desiccation to which mosses are subjected (Fig. 17–10*A*). Parthenogenesis is common.

Development is direct and rapid (see Nelson, 1982a, for review). Cleavage is holoblastic but does not follow a typical spiral or radial pattern. The early report by Marcus (1929) on coelom formation by enterocoelic pouching is difficult to interpret and has never been confirmed. Development is completed within 14 days or less, and the little tardigrades hatch by breaking the shells with their stylets. Further growth is attained by increase in the size of cells rather than by the addition of cells. As many as 12 molts may take place in the life of a tardigrade, which has been estimated to last 3 to 30 months.

Anabiosis

Tardigrades exhibit remarkable ability to withstand extreme desiccation and low temperatures. During dry weather, when the moss becomes desiccated, the tardigrade inhabitants pull in their legs, lose water, become contracted and shriveled,

and pass into an anabiotic (or cryptobiotic) state. An animal in such a condition is called a tun (Fig. 17–10*B*). In this state metabolism proceeds at a very low rate, and the animal can withstand abnormal environmental conditions. For example, specimens have recovered after immersion in liquid helium (−272°C), brine, ether, absolute alcohol, and other substances.

When water is again present, the animal swells and becomes active within a few hours. There are records of tardigrades emerging from a state of anabiosis lasting seven years. Under natural conditions the life of moss-inhabiting tardigrades is undoubtedly frequently interrupted by anabiosis, which may well lengthen the life-span of these animals to 60 to 70 years.

Phylogeny

The phylogenetic position of tardigrades is uncertain. They show some similarities to arthropods, but they also show many similarities to the aschelminths, especially gastrotrichs, and indeed they may turn out to be aschelminths. Study to confirm or refute the existence of an embryonic coelom is needed. Many of the tardigrade structural peculiarities are undoubtedly specializations for living in their restricted habitats.

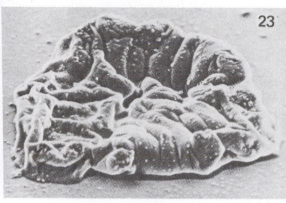

Figure 17–10   *A*, Attached egg of the moss-inhabiting *Macrobiotus tonollii*. (From Nelson, D. R., 1982: Developmental biology of the Tardigrada. *In* Harrison, F. W., and Cowden, R. R. (Eds.): Developmental Biology of Freshwater Invertebrates. Allan R. Liss, N.Y. p. 382.) *B*, Tun of *Microbiotus hufelandi* on a glass slide. (From Greven, H., 1971: On the morphology of tardigrades: A stereoscan study of *Macrobiotus hufelandi* and *Echiniscus testudo*. Forma et Functio, 4:283–302.)

Phylum Pentastomida, or Linguatulida

Pentastomids constitute a little phylum of approximately 90 species related to the arthropods. All members of the phylum are parasitic and live within the lungs or nasal passageways of vertebrates. The hosts of 90 per cent of pentastomids are reptiles, such as snakes and crocodiles, but some species parasitize mammals and birds. Although largely tropical, pentastomids have been reported in North America, Europe, Australia, and even in Arctic birds.

The wormlike body is 2 to 13 cm long and bears five short, anterior protuberances, from which the inappropriate name *pentastomida* ("five mouths") is derived (Fig. 17–11*A*). Four of these projections, two on each side of the body, are leglike, bearing claws. The fifth projection is an anterior, median, snoutlike process bearing the mouth. Not infrequently the legs are reduced to nothing more than the claws, which are used for clinging to the tissues of the host.

The body is covered by a chitinous cuticle that is ringed over the abdomen and molted periodically during larval development. The body wall muscles

are striated but arranged in circular and longitudinal layers. There is also a hemocoel, as in arthropods. The digestive tract is a relatively simple, straight tube with the anterior end modified to pump in the blood of the host, on which these parasites subsist (Fig. 17–11*B*). The nervous system is similar to that of arthropods, with paired, metamerically arranged ganglia located along the ventral nerve cord. There are no gas exchange, circulatory, or excretory organs.

The sexes are separate, with a well-developed genital system (Fig. 17–11*B*) (Riley, 1983). Fertilization is internal, and the embryonated eggs are passed into the digestive system of the host and then to the outside in the feces of the host. In most pentastomids the life cycle requires an intermediate host, which may be fish (pentastomids in crocodiles) or herbivorous or omnivorous mammals, such as rodents, rabbits (intermediate host for *Linguatula* in dogs), or small ungulates. Actually, larval pentastomids have been reported in almost every class of vertebrates. Larval development takes place within the intermediate host and involves a number of molts. The larvae possess four to six leglike appendages (Fig. 17–11*C*). When the intermediate host is eaten, the parasite is trans-

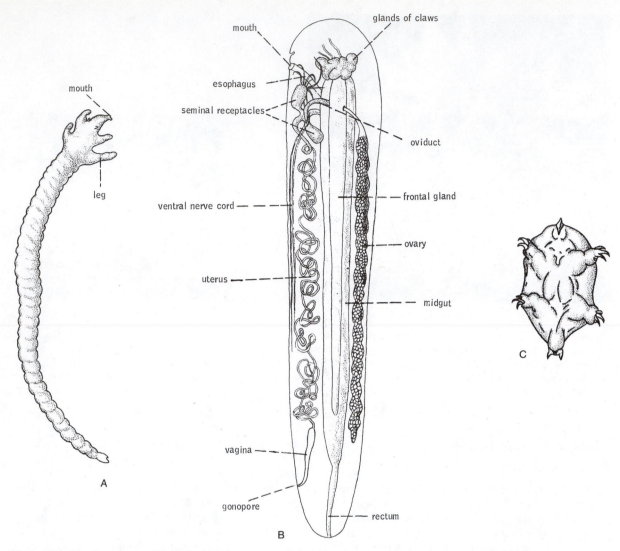

Figure 17–11 Pentastomida: *A*, *Cephalobaena tetrapoda* from the lung of a snake. (After Heymons from Cuénot.) *B*, Internal structure of a female *Waddycephalus teretiusculus*, parasitic in certain Australian snakes. (After Spencer from Cuénot.) *C*, Larva of *Porocephalus crotali*. (After Penn.)

ferred to the stomach of the primary host and reaches the lungs and nasal passageways through the esophagus. Pentastomid larvae have been reported in humans but are killed by calcareous encapsulation.

The taxonomic status of the pentastomids is uncertain, and their being ranked here as a phylum is undoubtedly arbitrary. There is little question that the group is closely related to the arthropods, and they probably should be considered members of that phylum. But there is little agreement among those who consider them arthropods as to where they should be placed within the phylum. Some zoologists argue that the pentastomids had an acarine

origin, but the evidence appears rather superficial. A myriapod relationship has also been suggested. Recently, a crustacean origin for pentastomids has been postulated, specifically from branchiurans, which are parasites of fish (see Riley et al, 1978, who review previous ideas of pentastomid relationships).

Phylum Onychophora

There are only about 70 existing species of onychophorans, but the phylum is an ancient one and does not appear to have changed greatly since the Cambrian period, from which the only certain fossil specimen has been taken. The geographical distribution of existing forms is relatively restricted. All onychophorans live in tropical regions (the East Indies, the Himalayas, the Congo, the West Indies, and northern South America) or south temperate regions (Australia, New Zealand, South Africa, and the Andes). No species have been found north of the Tropic of Cancer.

Most onychophorans are confined to humid habitats, such as in tropical rain forests, beneath logs, stones, and leaves, or along stream banks. During winter snows and low temperatures or dur-ing dry periods, they become inactive and remain in protective burrows or other retreats.

External Structure

Onychophorans have been described as a missing link between annelids and arthropods because of their many similarities to both groups. Onychophorans look very much like slugs with legs (Fig. 17–12A); in fact, they were thought to be mollusks when first discovered by Guilding in 1825. The body is more or less cylindrical and ranges from 1.4 cm to 15 cm in length. The anterior carries a pair

A

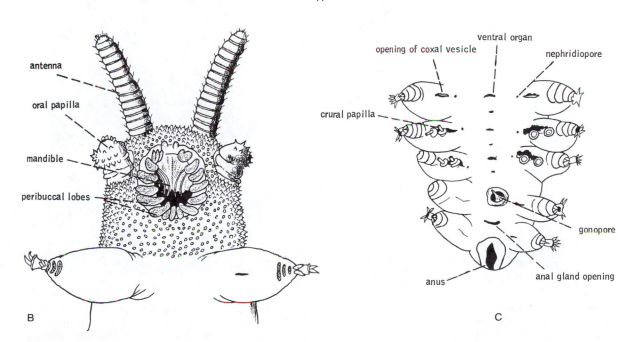

Figure 17–12   Onychophora: *A, Peripatus.* (By H. Sturm.) *B*, Anterior of *Peripatopsis capensis* (ventral view). (After Cuénot, L., 1949: Les Onychophores. *In* Grassé, P. (Ed.): Traité de Zoologie. Vol. 6. Masson et Cie, Paris. pp. 3–75.) *C*, Posterior of a male *Peripatus corradoi* (ventral view). (After Bouvier from Cuénot.)

of large, annulated antennae and a ventral mouth, which is flanked by a pair of clawlike mandibles and by a short, conical, oral papilla. The mandibles represent modified segmental appendages, as in arthropods. The legs vary in number from 14 to 43 pairs, depending on the species and the sex. Each leg is a large, conical, unjointed protuberance bearing a pair of terminal claws (Fig. 17–12*B*). The entire surface of the body is covered by large and small tubercles, which are arranged in rings or bands encircling the legs and trunk. The tubercles are covered by minute scales. Onychophorans are blue, green, orange, or black, and the papillae and scales give the body surface a velvety and iridescent appearance.

Internal Structure and Physiology

The body surface is covered by a chitinous cuticle that is composed of the same layers as in arthropods and is molted, but unlike the exoskeleton of arthropods, the cuticle of onychophorans is thin, flexible, and very permeable, and it is not divided

into articulating plates (Fig. 17–13*A*). The absence of a rigid exoskeleton enables onychophorans to squeeze their bodies into confining places. Beneath the cuticle are a single layer of epidermis and three layers of muscle fibers—circular, diagonal, and longitudinal. The body wall is thus constructed on the typical annelidan plan. However, the coelom is reduced to the gonadial cavities and to small sacs associated with the nephridia.

Onychophorans crawl slowly by means of the legs and by extension and contraction of the body, which is held off the ground. When a segment is extended, the legs are lifted from the ground and moved forward. As in arthropods, the legs are located more ventrally than are the parapodia of annelids.

Most species are predacious and feed on small invertebrates, such as snails, insects, and worms. For prey catching and defense, onychophorans secrete an adhesive material from glands opening at the ends of the oral papillae (Fig. 17–12*B*). The secretion is discharged as two streams from a distance as great as 15 cm; it hardens almost imme-

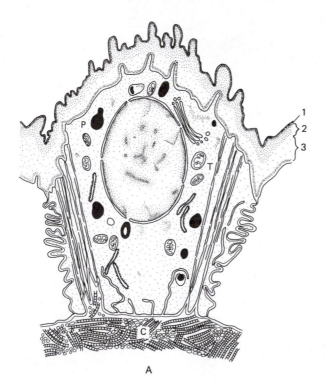

A

B

Figure 17–13 *A*, An epidermal cell of *Peripatus acacioi*, showing the three layers of the surface cuticle: epicuticle (1), exocuticle (2) and endocuticle (3). Tonofibrils anchoring the muscle layer to the cuticle are labeled T. (After Lavallard from Storch, V., 1984: Onychophora. *In* Bereiter-Hahn, J., Matoltsy, A. G., and Richards, K. S.: Biology of the Integument. 1. Invertebrates. Springer-Verlag, Berlin. p. 704.) *B*, SEM of the body surface of *Peripatopsis moseleyi*, showing a papilla with terminal sensory bristle. (Storch, V., and Ruhberg, H., 1977: Fine structure of the sensilla of *Peripatopsis moseleyi*. Cell Tiss. Res., *177*:539–553.)

diately, entangling the prey in a net of adhesive threads.

The mouth is located at the base of the prebuccal depression (Fig. 17–12*B*). Within the depression lie the lateral clawlike mandibles, which are used for grasping and cutting prey. Salivary secretions are passed into the body of the prey, and the partially digested tissues are then sucked into the mouth. The prebuccal depression opens into the chitin-lined foregut, composed of a pharynx and an esophagus. A large, straight intestine is immediately posterior to the esophagus and is the site of the remaining digestion and of absorption. The tubular hindgut (rectum) opens through the anus on the ventral side at the end of the body.

The circulatory system is similar to that of the arthropods. A tubular heart, open at each end and provided with a pair of lateral ostia in each segment, lies within the pericardial sinus and propels blood forward into the general hemocoel. The partitions between sinuses are perforated by openings that facilitate blood circulation. The blood is colorless and contains phagocytic amebocytes.

Each segment contains a single pair of nephridia located in the ventrolateral sinuses. The ciliated funnel and nephrostome lie within an end sac, which represents a vestige of the coelom. The nephridiopore is located on the inner base of each leg (Fig. 17–12*C*).

The gas exchange organs are tracheae. The spiracles are minute openings and are present in large numbers all over the surface of the body between bands of tubercles. Each spiracle opens into a very short atrium, at the end of which arises a tuft of minute tracheae. Each trachea is a simple, straight tube and extends directly to the tissue that it is supplying.

The nervous system is composed of a large, bilobed brain lying over the pharynx and a pair of ventral nerve cords connected by commissures. The brain supplies nerves to the tentacles, the eyes, and the mouth region. In each segment the ventral nerve cords contain a ganglionic swelling and give rise to a number of paired nerves supplying the legs and body wall.

There is a small eye at the base of each antenna. Onychophorans avoid light and are largely nocturnal. Sensory bristles are located on the antennae and on the tubercles scattered over the body surface (Fig. 17–13*B*).

Reproduction

The sexes are always separate. The ovaries are a pair of fused, elongated organs, located in the posterior part of the body. Each ovary is connected to an elaborate genital tract that in some species contains a seminal receptacle and a uterus. The ends of each uterus join together and open to the exterior through a common genital pore, situated ventrally near the posterior of the body.

The male system contains two elongated, nonfused testes and relatively complex, paired genital tracts. Prior to reaching the exterior, the two tracts join to form a single tube in which the sperm are formed into spermatophores. The male gonopore is ventral and posterior, like that of the female (Fig. 17–12*C*).

In the mating of the South African *Peripatopsis*, which lacks seminal receptacles, the male crawls over the body of the female and deposits a spermatophore at random on her side or back, and over a period of time a female may accumulate many spermatophores. The spermatophore stimulates blood amebocytes to bring about dissolution of the underlying integument. Sperm then pass from the spermatophore into the female hemolymph. They eventually reach the ovaries, where fertilization of the eggs takes place.

Sperm transfer in onychophorans with seminal receptacles is not understood.

Onychophorans are oviparous or viviparous. Oviparous forms lay large, yolky eggs in moist situations, each egg enclosed in a chitinous shell. As in most arthropods, cleavage is superficial.

All other onychophorans are viviparous, with either yolk (aplacental) or the mother (placental) providing for embryonic nutrition, and the eggs develop within the uterus. The eggs of those possessing placental viviparity are small and have little yolk, and cleavage either is superficial or has become holoblastic. Uterine secretions provide for the nutrition of the embryo, and the nutritive material is obtained by the embryo through a special embryonic membrane or through a "placental" connection to the uterine wall.

During the life-span of some six years (*Peripatopsis*), molting occurs frequently, as often as every two weeks.

Geographical Distribution

The geographical distribution of onychophorans is peculiar in a number of respects. The phylum consists of two families. Each has a wide, discontinuous distribution around the world, but neither is found in the same area with species of the other family. The Peripatidae are more or less equatorial in distribution, extending no further north than the Himalayas, Caribbean Sea, and central Mexico;

the Peripatopsidae are limited to the Southern Hemisphere. Thus, the South African species are more closely related to those in Chile than to those in equatorial Africa.

Considering the antiquity of the onychophorans and the improbability of their being spread by other animals, the distribution of the phylum can be accounted for by the past geological connection of the American and African land masses.

Phylogenetic Relationships

The structure of the body wall, the nephridia, the thin, flexible cuticle, and the nonjointed appendages are certainly annelidan in character. But in other respects onychophorans are more like arthropods. The coelom is reduced, and the cuticle is chitinous and is molted. A pair of appendages are modified for feeding, the tracheae are gas exchange organs, and there is an open circulatory system with a dorsal, tubular heart containing ostia. Although onychophorans reflect an annelidan ancestry, their origins are obscure. We know nothing about their initial invasion of land, but they clearly have never become adapted for withstanding desiccating conditions.

Manton (1977) included onychophorans within the uniramian arthropods. Such a classification is based largely on the belief that they are the ancestors of uniramians. The soft body and nonjointed appendages, among other annelidan characteristics, which are certainly primary, present serious problems, and to consider onychophorans arthropods demands a fundamental redefinition of the Arthropoda. It is questionable whether such a redefinition is justified or necessary even if they did give rise to the uniramian groups.

Summary

1 The Pogonophora are a phylum of 80 or more species of wormlike, marine animals that live in chitinous tubes fixed upright in the bottom at depths of 100 meters or more.

2 The elongate body of pogonophorans is composed of three sections. The anterior section bears 1 to 250 tentacles, depending on the species, and the posterior section is segmented and bears setae. There is no gut, and the mode of feeding is uncertain.

3 The Sipuncula are a phylum of about 320 species of nonsegmented marine worms, which

burrow in sand and mud or live in coral excavations or crevices or in old mollusk shells.

4 The cylindrical trunk of sipunculans bears an anterior, retractile introvert. At the end of the introvert are the mouth and a partial or complete circlet of tentacles or lobes, which are used in deposit feeding.

5 Cleavage is spiral, and some species have trochophore larvae.

6 The phylum Echiura contains some 140 species of marine worms. Like sipunculans, they burrow in sand and mud or live in rock or coral crevices. The cylindrical body bears an anterior, flattened, nonretractile, cephalic lobe called a proboscis. A single pair of large, lateral setae are located near the anterior end of the trunk.

7 Echiurans feed on detritus material that is collected by the proboscis and conveyed by cilia to the mouth.

8 Spiral cleavage, a trochophore larva, a transitory embryonic metamerism, and setae indicate that echiurans are probably related to annelids.

9 The phylum Tardigrada is composed of approximately 375 species of microscopic animals, called water bears, that live in the sea, fresh water, soil, and water films of mosses. Most species are moss inhabitants.

10 Tardigrades crawl on four pairs of short, stubby legs. Moss-inhabiting species feed on plant cell contents, utilizing a pair of stylets and a sucking pharynx, but some species are predatory, and some are detritus feeders.

11 Soil and moss-inhabiting tardigrades are able to tolerate extreme desiccation, passing into an anabiotic or cryptobiotic state of dormancy.

12 The phylum Pentastomida is a group of some 90 parasites of the lungs and respiratory tracts of vertebrates, mostly reptiles. The wormlike body bears at the anterior end two pairs of hooked protuberances and a protuberance housing the mouth.

13 Pentastomids possess a chitinous cuticle that is molted, a hemocoel, and a metameric nerve cord. They will probably be found to be most closely related to some class of arthropods.

14 The 70 species of onychophorans are terrestrial, caterpillar-like animals of the tropics and Southern Hemisphere. The soft body, which is covered by a thin, flexible, chitinous cuticle, is adapted for squeezing beneath stones, logs, and other objects.

15 Onychophorans possess a pair of antennae, a pair of clawlike mandibles, and many pairs of unjointed, peglike legs. Internally, there is a combination of arthropod and annelidan features: body wall of circular and longitudinal muscles, segmental nephridia, reduced coelom, open blood-vascular system, and tracheae. The chitinous cuticle is periodically molted.

16 In some species sperm are transferred as spermatophores. Some onychophorans are oviparous, but many brood their eggs internally and give birth to their young.

REFERENCES

The literature included here is restricted in large part to the lesser protostome phyla. The introductory references on page 8 include many general works and field guides that contain sections on these animals.

Anderson, D. T., 1973: Embryology and Phylogeny in Annelids and Arthropods. Pergamon Press, N.Y. 495 pp. (Detailed accounts of the embryology of onychophorans and uniramian arthropods and the phylogenetic implications of the embryonic patterns.)

Arp, A. J., and Childress, J. J., 1983: Sulfide binding by the blood of the hydrothermal vent tube worm *Riftia pachyptila*. Science, *219*:295–297.

Bussers, J. C., and Jeuniaux, C., 1973: Structure et composition de la cuticle de *Macrobiotus* et de *Milnesium tardigradum*. Ann. Soc. R. Zool. Belg., *103*(2/3):271–279.

Cavanaugh, C. M., Gardiner, S. L., Jones, M. L., Jannasch, H. W., and Waterbury, J. B., 1981: Prokaryotic cells in the hydrothermal vent tube worm *Riftia pachyptila* Jones: Possible chemoautotropic symbionts. Science, *213*:340–342.

Clark, R. B., 1969: Systematics and phylogeny; Annelida, Echiura, Sipuncula. *In* Florkin, M., and Scheer, B. T. (Eds.): Chemical Zoology. Vol. 4. Academic Press, N.Y. pp. 1–68.

Crowe, J. H., and Cooper, A. F., Jr., 1971: Cryptobiosis. Sci. Amer., *225*(12):30–36.

Cuénot, L., 1949: Les Onychophores, Les Tardigrades, et Les Pentastomides. *In* Grassé, P. (Ed.): Traité de Zoologie. Vol. 6. Masson et Cie, Paris. pp. 3–75.

Cutler, E. B., 1973: Sipuncula of the western North Atlantic. Bull. Am. Mus. Nat. Hist., *152*:103–204. (Key and detailed descriptions of sipunculans on the western side of the North Atlantic.)

Cutler, E. B., 1975: The phylogeny and systematic position of the Pogonophora. Syst. Zool., *24*(4):512–513. (A short review of a symposium held in Copenhagen in 1973.)

Cutler, E. B., 1977: Sipuncula. Marine flora and fauna of the northeastern U.S. NOAA Technical Report NMFS Circular 403. U.S. Government Printing Office, Washington, D.C.

Dawydoff, C., 1959: Classes des Echiuriens et Priapuliens. *In* Grassé, P. (Ed.): Traité de Zoologie. Vol. 5, Pt. 1. Masson et Cie, Paris. pp. 855–926.

George, J. D., and Southward, E. C., 1973: A comparative study of the setae of Pogonophora and polychaetous Annelida. J. Mar. Biol. Assoc. U.K., *53*(2):403–414.

Gibbs, P. E., 1977: British Sipunculans. Synopses of the British Fauna No. 12. Academic Press, London.

Giese, A. C., and Pearse, J. S. (Eds.), 1975: Reproduction of Marine Invertebrates. Vol. II. Entoprocts and lesser Coelomates. Vol. III. Annelids and Echiurans. Aca-

demic Press, N.Y. (Volume II includes tardigrades, priapulids, sipunculids, and pogonophorans.)

Grassle, J. F., 1985: Hydrothermal vent animals: distribution and biology. Science, *229*:713–717.

Greven, H., 1984: Tardigrada. *In* Bereiter-Hahn, J., Matoltsy, A. G., and Richards, K. S.: Biology of the Integument. 1. Invertebrates. Springer-Verlag, Berlin. pp. 714–727.

Gupta, B. L., and Little, C., 1975: Ultrastructure, phylogeny and Pogonophora. pp. 45–63. (See Nørrevang, 1975.)

Harris, R. R., and Jaccarini, V., 1981: Structure and function of the anal sacs of *Bonellia viridis*. Jour. Mar. Biol. Assoc. U.K., *61*:413–430.

Heymons, R., 1935: Pentastomida. *In* Bronn, H. G. (Ed.): Klassen und Ordnungen des Tierreichs. Bd. 5, Abt. IV. Akademische Verlagsgesellschaft, Frankfurt.

Hyman, L. H., 1951 and 1959: The Invertebrates. Vol. 3. Acanthocephala, Aschelminthes, and Entoprocta. (Covers entoprocts, pp. 521–554.) Vol. 5. The Smaller Coelomate Groups. (Covers sipunculans, pp. 610–696.) McGraw-Hill, N.Y.

Ivanov, A. V., 1963: Pogonophora. Consultants Bureau, N.Y. 479 pp.

Ivanov, A. V., 1975: Embryonalentwicklung der Pogonophora und ihre systematische Stellung. pp. 10–44. (See Nørrevang, 1975.)

Jaccarini, V., and Schembri, P. J., 1977: Feeding and particle selection in the echiuran worm *Bonellia viridis*. Jour. Exp. Mar. Biol. Ecol., *28*:163–181.

Jaccarini, V., Agius, L., Schembri, P. J., and Rizzo, M., 1983: Sex determination and larvae sexual interaction in *Bonellia viridis*. Jour. Exp. Mar. Biol. Ecol., *66*:25–40.

Jones, M. L., 1981: *Riftia pachyptila* Jones: Observations on the vestimentiferan worm from the Galapagos rift. Science, *213*:333–336.

Jones, M. L., 1985: On the Vestimentifera, a new phylum: six new species, and other taxa, from hydrothermal vents and elsewhere. *In* Jones, M. L. (Ed.): The Hydrothermal Vents of the Eastern Pacific: An Overview. Bull. Biol. Soc. Wash., *6*:117–158.

Kohn, A. J., and Rice, M. E., 1971: Biology of Sipuncula and Echiura. Bioscience, *21*:583–584. (A brief review of a symposium on the biology of these two phyla.)

Manton, S. M., 1977: The Arthropoda: Habits, Functional Morphology, and Evolution. Clarendon Press, Oxford. 527 pp.

Marcus, E., 1929: Tardigrada. *In* Bronn, H. G. (Ed.): Klassen und Ordnungen des Tierreichs. Bd. 5, Abt. IV. Akademische Verlagsgesellschaft, Frankfurt.

Morgan, C. I., and King, P. E., 1976: British Tardigrades. Synopses of the British Fauna No. 9. Academic Press, London. 133 pp.

Nelson, D. R., 1982a: Developmental biology of the Tardigrada. *In* Harrison, F. W., and Cowden, R. R.: Developmental Biology of Freshwater Invertebrates. Alan R. Liss, N.Y. pp. 363–398.

Nelson, D. R. (Ed.), 1982b: Proceedings of the Third International Symposium on the Tardigrada. East Tennessee State University Press, Johnson City, Tenn.

Nørrevang, A., 1970: The position of Pogonophora in the phylogenetic system. Z. Zool. Syst. Evolutionsforsch., 8, H. 3, 161–172.

Nørrevang, A., 1970: On the embryology of *Siboglinum* and its implications for the systematic position of the Pogonophora. Sarsia, 42:7–16.

Nørrevang, A. (Ed.), 1975: The Phylogeny and Systematic Position of Pogonophora. Z. Zool. Syst. Evolutionsforsch. Sonderheft, 1975. (Proceedings of a symposium held at the University of Copenhagen in 1973.)

Pollock, L. W., 1976: Tardigrada. Marine flora and fauna of the northeastern U.S. NOAA Technical Reports NMFS Circular 394 U.S. Government Printing Office, Washington, D.C. 25 pp.

Rice, M. E., 1969: Possible boring structures of sipunculids. Am. Zool., 9:803–812.

Rice, M. E., 1970: Asexual reproduction in a sipunculan worm. Science, 167:1618–1620.

Rice, M. E., 1986: Larvae adrift: patterns and problems in life histories of sipunculans. Amer. Zool., 21:605–619.

Rice, M. E., and Todororic, M. (Eds.), 1970: Proceedings of the International Symposium on the Biology of the Sipuncula and Echiura. Vols. I. (355 pp.) and II (254 pp.). Published by the Institute for Biological Research, Yugoslavia and Smithsonian Institution, Washington, D.C.

Riggin, G. T., Jr., 1962: Tardigrada of southwest Virginia, with the addition of a description of a new marine species from Florida. Virginia Agr. Exp. Sta. Tech. Bull., 152:1–145. (This paper is a good source for taxonomic work on North American tardigrades.)

Riley, J., 1983: Recent advances in our understanding of pentastomid reproductive biology. Parasitology, 86:59–83.

Riley, J., Banaja, A. A., and James, J. L., 1978: The phylogenetic relationships of the Pentastomida: the case for their inclusion within the Crustacea. Int. J. Parasit., 8:245–254.

Schembri, P. J., and Jaccarini, V., 1977: Locomotory and other movements of the trunk of *Bonellia viridis*. Jour. Zool., 182:477–494.

Southward, A. J., 1975: On the evolutionary significance of the mode of feeding of Pogonophora. pp. 77–85. (See Nørrevang, 1975.)

Southward, E. C., 1971: Pogonophora of the Northwest Atlantic: Nova Scotia to Florida. Smithsonian Contrib. Zool., 88. 29 pp.

Southward, E. C., 1971: Recent researches on the Pogonophora. Oceangr. Mar. Biol. Ann. Rev., 9:193–220.

Southward, A. J., and Southward, E. C., 1980: The significance of dissolved organic compounds in the nutrition of *Siboglinum ekmani* and other small species of Pogonophora. Jour. Mar. Biol. Assoc. U.K., 60:1005–1034.

Southward, A. J., Southward, E. C., Dando, P. R., Rau, G. H., Felbeck, H., and Flügel, H., 1981: Bacterial symbionts and low $^{13}C/^{12}C$ ratios in tissues of Pogonophora indicate unusual nutrition and metabolism. Nature, 293(5834):616–620.

Stephen, A. C., and Edmonds, S. J., 1972: The Phyla Sipuncula and Echiura. British Museum, London. 528 pp.

Walter, M. D., 1973: Feeding and studies on the gut content in sipunculids. Helgol. Wiss. Meeresunters, 25(4):486–494.

Williams, J. A., and Margolis, S. U., 1974: Sipunculid burrows in coral reefs: Evidence for chemical and mechanical excavation. Pac. Sci., 28(4):357–359.

18

The Lophophorates

Four phyla, the Bryozoa, the Entoprocta, the Phoronida, and the Brachiopoda, are similar in possessing a food-catching tentacular organ called a lophophore and are often grouped together as the lophophorate coelomates. However, they are probably not closely related. Bryozoans and entoprocts appear to be closer to the protostome line of evolution, and the phoronids and brachiopods appear to be closer to the deuterostome line (Nielson, 1985).

The lophophore is a circular or horseshoe-shaped fold of the body wall that encircles the mouth and bears numerous ciliated tentacles (Fig. 18–1A). The tentacles are hollow outgrowths of the body wall. Each contains an extension of the coelom. The ciliary tracts on the tentacles drive a current of water through the lophophore, and plankton is collected in the process.

In addition to possessing a lophophore, nearly every member of these phyla is sessile, has a reduced head, secretes a protective covering, and except for some of the brachiopods, possesses a U-shaped digestive tract (Fig. 18–1A). But all these characteristics are correlated with a sessile existence and, in part at least, represent evolutionary convergence.

Phylum Bryozoa

The phylum Bryozoa, or Polyzoa or Ectoprocta, is the largest and the most common of the lophophorate phyla and contains approximately 4000 living species. The group constitutes one of the major animal phyla, even though it is much less familiar than many other invertebrate groups.

Bryozoans are colonial and sessile animals, and the individuals composing the colonies are usually about 0.5 mm in length. In most species each individual is encased in a protective covering that contains an opening for the protrusion of the lophophore. The interior of the body is occupied largely by the rather spacious coelom and the U-shaped digestive tract. There are no gas exchange, circulatory, or excretory organs, probably because of the small size of these animals. Most of the pe-

culiarities of bryozoans are associated with miniaturization, the evolution of colonial organization,
and a skeletal covering.

The phylum is divided into three classes—the
Phylactolaemata, the Gymnolaemata, and the Stenolaemata. The class Stenolaemata contains some
living marine species and over 500 fossil genera.
The class Gymnolaemata is almost entirely marine
and includes the great majority of living bryozoans,
as well as many fossil species. The class Phylacto-

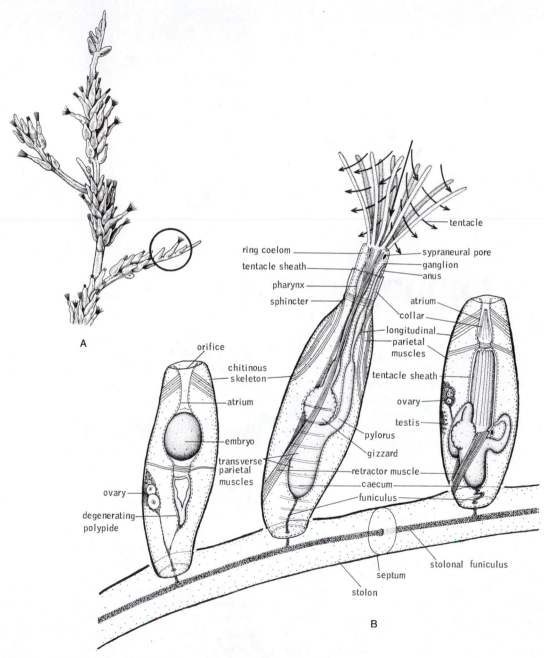

Figure 18–1 *Bowerbankia*, a stoloniferous gymnolaemate bryozoan: *A*, Section of colony. (Modified from Gay in Pren-
ant and Bobin.) *B*, Stolon, brooding individual, and two autozooids of *Bowerbankia*, from circled portion of *A*. One
autozooid has lophophore protruded and the other has it retracted. (Modified from Ryland and others.)

laemata is restricted to fresh water and, although it is widely distributed, contains only about 50 species.

Structure of the Zooid

The bryozoan individual is boxlike, oval, or tubular (Figs. 18–1 and 18–5). In gymnolaemates, on which the following description is based, the covering (zoecium) consists of a protein-chitin cuticle or of cuticle overlying calcium carbonate (Fig. 18–19). Thus, many bryozoans possess a heavy, rigid exoskeleton, although in others, such as the common *Bugula*, calcification is light. Moreover, some impregnation of the chitinous layer with calcium carbonate may be present, even when a calcareous layer is absent.

An opening (the orifice) enables the lophophore to protrude, and in many marine bryozoans the orifice is provided with a lid (operculum), which closes when the lophophore withdraws (Fig. 18–5A). Just within the orifice is a chamber, called the atrium, the bottom of which may be provided with a sphincter or, in a few forms, a collar (Fig. 18–1).

The outer covering is secreted by the epidermis of the body wall (Figs. 18–1 and 18–19A). As would be expected, there are no muscle layers composing the body wall underlying the rigid exoskeleton of gymnolaemates, and the epidermis overlies a thin, delicate peritoneum.

Within the body wall is a large coelom surrounding the digestive tract. From the underside of the sphincter, a sheath of body wall, called the tentacular sheath, extends downward and is attached to the base of the lophophore (Fig. 18–1). The tentacular sheath encloses the tentacles of the lophophore when the animal is retracted. When the animal is feeding, the lophophore protrudes through the sphincter, atrium, and orifice, and the tentacular sheath everts, forming the covering of the extended, necklike anterior of the trunk.

In gymnolaemates the lophophore is circular and consists of a simple ridge bearing 8 to 30 or more tentacles (Fig. 18–1). When retracted, the tentacles are bunched together within the sheath; when protruded, the tentacles fan out, forming a funnel with the mouth at the base.

The lateral surface and the inner median surface of each tentacle bear a longitudinal tract of multiciliated cells. Within the tentacle are longitudinal muscle fibers and nerves, and through the center runs a collagen tube and an extension of the coelom.

The mouth at the center of the lophophore opens into a U-shaped digestive tract (Figs. 18–1 and 18–3). The anus opens through the dorsal side of the tentacular sheath outside the lophophore, hence the name *Ectoprocta* ("outside anus"), which is sometimes used for the phylum. The coelom is partially divided by a septum into an anterior portion occupying the lophophore with extensions into the tentacles, and a larger posterior or trunk coelom. The two divisions are connected by a pore. The trunk coelom is crossed by muscle fibers and by a single or branching tube of peritoneal tissue, which constitutes the funiculus. In a ctenostome like *Bowerbankia*, the funiculus extends between the posterior end of the stomach and the back body wall (Fig. 18–1).

The nervous system is composed of a nerve ring around the pharynx with a ganglionic mass on its dorsal side (Fig. 18–1). The ganglion and ring give rise to nerves extending to each of the tentacles and to other parts of the body. There are no specialized sense organs in bryozoans.

The terms *cystid* and *polypide* will be encountered by students who deal with the literature on bryozoans. The term *cystid* refers to the exoskeleton and body wall. The term *polypide* refers to the contents of the zooid within the body wall, i.e., the lophophore, gut, muscles, and other structures.

Zooids of the freshwater Phylactolaemata differ from those of gymnolaemates in many ways. The body wall contains circular and longitudinal muscle layers, and the epidermis is covered by a cuticle or, as in *Pectinatella*, by a gelatinous layer. The lophophore of freshwater phylactolaemates, with the exception of the circular lophophore of *Fredericella*, is horseshoe shaped and is composed of two ridges bearing a total of 16 to 106 tentacles (Fig. 18–3). Such a shape provides a greater food-collecting surface for these zooids, which tend to be larger than the marine gymnolaemates. As in phoronids, one ridge passes above and one below the mouth at the bend of the horseshoe. A dorsal hollow lip (epistome) overhangs the mouth.

Colony Organization

Gymnolaemates are very common and abundant marine animals. Although some species have been recorded from depths as great as 8200 meters, most species are found in coastal waters and attach to rocks, pilings, shells, algae, and other animals. Members of such genera as *Bowerbankia* and *Amathia* form stoloniferous colonies, which possibly

(Text continued on p. 742)

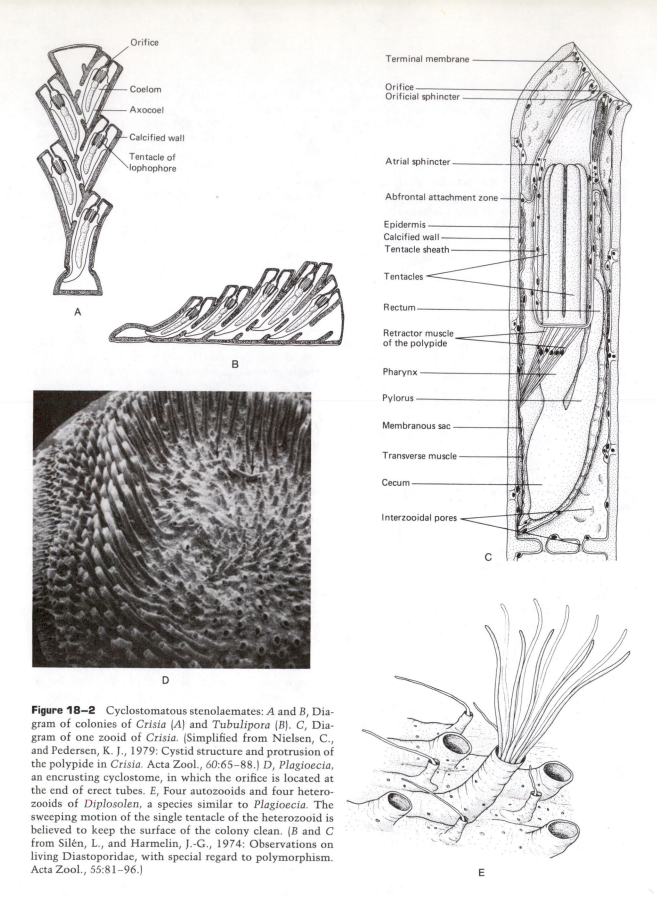

Figure 18–2 Cyclostomatous stenolaemates: *A* and *B*, Diagram of colonies of *Crisia* (*A*) and *Tubulipora* (*B*). *C*, Diagram of one zooid of *Crisia*. (Simplified from Nielsen, C., and Pedersen, K. J., 1979: Cystid structure and protrusion of the polypide in *Crisia*. Acta Zool., 60:65–88.) *D*, *Plagioecia*, an encrusting cyclostome, in which the orifice is located at the end of erect tubes. *E*, Four autozooids and four heterozooids of *Diplosolen*, a species similar to *Plagioecia*. The sweeping motion of the single tentacle of the heterozooid is believed to keep the surface of the colony clean. (*B* and *C* from Silén, L., and Harmelin, J.-G., 1974: Observations on living Diastoporidae, with special regard to polymorphism. Acta Zool., 55:81–96.)

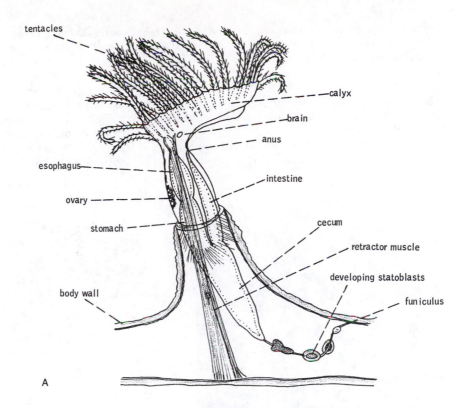

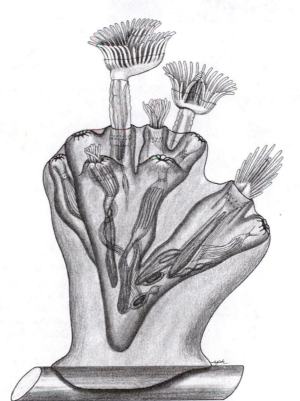

Figure 18–3 Phylactolaemate (freshwater) bryozoans: *A*, Structure of *Plumatella*. (After Pennak, R. W., 1978: Freshwater Invertebrates of the United States. 2nd Edition. John Wiley and Sons, N.Y.) *B*, Small colony of *Lophopus crystallinus* attached to a water plant. External covering is gelatinous. (After Allman.)

represent the more primitive condition (Fig. 18–1). The erect or creeping stolons are composed of modified zooids that give the stolon a jointed appearance. Those zooids that are unmodified are attached by the posterior end to the stolon and are often completely separate from one another. The exoskeleton of stoloniferous bryozoans usually lacks calcium carbonate.

The vast majority of marine bryozoans are not stoloniferous, and the colony is formed by the more direct attachment and fusion of adjacent zooids. Moreover, the orientation of the body to the substratum is different. The dorsal surface is attached to the substratum, or to other zooids, and the ventral surface becomes the exposed surface, now called the frontal surface (Fig. 18–5A).

The growth patterns of nonstoloniferous bryozoan colonies vary greatly (Fig. 18–4). Many slightly calcified species, such as the common Atlantic *Bugula*, form erect, branching colonies that

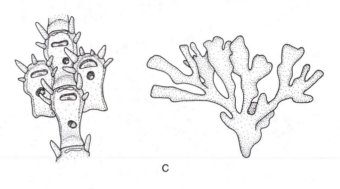

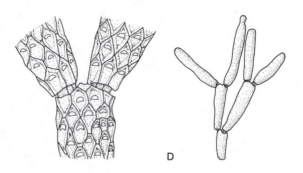

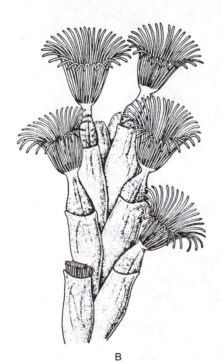

Figure 18–4 Erect marine bryozoans: *A*, Colony of *Bugula neritina*, a common and widespread species, in which biserially arranged zooids form arborescent colonies. *B*, A section of a colony of *Bugula neritina* showing some zooids with protruded lophophores. *C*, *Flustra foliacea*, a common European species forming bushy leaflike colonies. Outline of one "leaf" formed by two layers of zooids arranged back to back. *D*, *Cellaria sinuosa*, a European species, in which the zooids are arranged to form jointed clubs composing bushy colonies. (*C* and *D* from Ryland, J. S., and Hayward, P. J., 1977: British Anascan Bryozoans. Synopses of the British Fauna No. 10. Academic Press, London. pp. 77 and 120.)

look like seaweed. In *Bugula*, for example, such a plantlike growth form is attained commonly through a biserial attachment of zooids (Fig. 18–4*A*). The most common type of colony is the encrusting form, in which the zooids are organized as a sheet attached to rocks and shells. The exoskeleton is usually calcareous, and since the lateral and end walls of the zooids are fused, the orifice has typically migrated toward the exposed ventral, or frontal, surface. *Membranipora, Microporella,* and *Schizoporella* are very common encrusting genera (Figs. 18–5, 18–6, and 18–7), and a large colony may be composed of as many as 2,000,000 members.

There are erect foliaceous colonies composed of one sheet of zooids or two sheets attached back to back (Fig. 18–4*C*). Other colonies are tuftlike, and in some the zooids are radially arranged.

The zooids of a colony communicate by pores through the transverse end walls or the lateral walls, or both, depending on the growth pattern of the colony (Figs. 18–1, 18–5*A*, and 18–8).

In two classes (Phylactolaemata and Stenolaemata) at least some of the pores are open, and coelomic fluid may flow between the member zooids of the colony. The pores of gymnolaemates differ in being closed by a plug of cells, which are arranged in an ordered and polarized manner. Some of these cells are associated with cells of the funiculus, the peritoneal cord that extends back from the underside of the stomach (Fig. 18–8). The funiculus is an important means of communication between zooids, and there is clear histochemical evidence for a movement of lipid along the funiculus and out of actively feeding zooids (Bobin, 1977).

Electrophysiological studies (Thorpe et al, 1975) confirmed that a colonial nervous system exists in at least some nonstoloniferous bryozoans, such as *Electra* and *Membranipora*. Certain nerves, which encircle the zooid wall, enter the pores and make connections of some sort with similar nerves of adjacent zooids. This colonial nervous system contributes to the integration of lophophore activity and position within the colony.

The colonies of most gymnolaemates are polymorphic. The zooid is a typical feeding individual, called an autozooid, and makes up the bulk of the colony. Reduced or modified zooids that serve other functions are known as heterozooids. A common type of heterozooid is one modified to form stolons, attachment discs, rootlike structures, and other such vegetative parts of the colony. These individuals are so reduced that they consist of little more than the body wall and the strands of funiculus tissue passing through their interior (Fig. 18–1).

Two defensive types of heterozooids, called avicularia and vibracula, are found in many cheilostomes. An avicularium is usually smaller than an autozooid, and the internal structure is greatly reduced (Fig. 18–5*C*). However, the operculum is typically highly developed and modified, forming a movable jaw. Avicularia may be sessile or stalked; when stalked, they are capable of rapid bending or "pecking" movements. Such stalked avicularia are found in *Bugula* (but not in the common *Bugula neritina*) and resemble little bird heads attached to the colony. Avicularia defend the colony against small organisms, including perhaps settling larvae of other animals. However, the avicularia of *Bugula* appear to be more important in defending against larger crawling animals (0.5 to 4 mm), such as tube-building amphipods and polychaetes, whose appendages are seized by the avicularium jaws.

In a vibraculum the operculum is modified to form a long bristle, sometimes called a seta, which can be moved in one plane and is apparently used to sweep away detritus and settling larvae (Fig. 18–5*D*). Modification of zooids for reproductive purposes is very common and will be discussed later.

Although the zooids are microscopic, the colonies themselves are one to several centimeters in diameter or height and may attain a much greater size. Some erect species reach a height of more than 24 cm, and some encrusting colonies may attain a diameter of more than 50 cm. White or pale tints are typical of most colonies, but darker colors may be found. The taxonomy of marine bryozoans is based almost exclusively on the structure of the exoskeleton and the colonial organization.

The colonies of freshwater phylactolaemates are of two types. In forms such as *Lophopus, Cristatella,* and *Pectinatella,* the zooids project from one side of the soft colony sac, resembling the fingers of a glove (Fig. 18–3*B*). *Pectinatella* colonies secrete a gelatinous base, sometimes several feet in diameter, to which surface the zooids adhere. The other type of colony, of which *Plumatella, Fredericella,* and *Stolella* are examples, has a more or less plantlike growth form, in which there are either erect or creeping branches composed of a succession of zooids. The colonies of freshwater bryozoans are attached to vegetation, submerged wood, rocks, and other objects. *Cristatella,* in which the colony is a flattened gelatinuous ribbon,

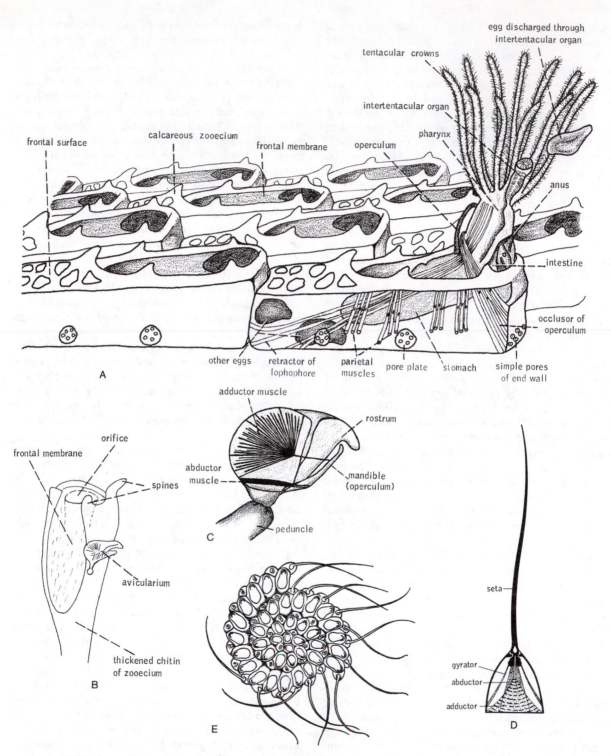

Figure 18–5 *A*, Part of a colony of the encrusting cheilostome *Electra*. (Modified from Marcus.) *B*, Exoskeleton of *Bugula*. (After Hyman, L. H., 1959: The Invertebrates. Vol. 5. Smaller Coelomate Groups. McGraw-Hill Book Co., N.Y.) *C*, Avicularium of *Bugula*. (After Maturo.) *D*, Vibraculum. (Based on Marcus and Ryland.) *E*, Part of colony of *Heliodoma*, showing vibracula. (From Moore, R. C. (Ed.): Treatise on Invertebrate Paleontology. Geological Society of America and University of Kansas Press, Lawrence, Kans.)

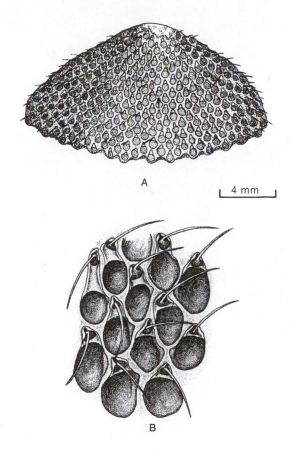

A

⊢ 4 mm ⊣

B

Figure 18–6 *A, Cupuladria.* Cap-shaped colonies resting on soft bottoms. *B,* An enlarged section of *Cupuladria.* Note the long vibraculum at the end of each individual.

is not fixed and creeps over the substratum up to a centimeter a day.

Physiology

During feeding the lophophore is pushed outward through the diaphragm and vestibule, causing the tentacular sheath to evert. The tentacles then expand, forming a funnel. Protraction of the lophophore is effected in all cases by the elevation of the coelomic fluid pressure, although the mechanism accomplishing this varies. In the freshwater phylactolaemates the coelomic fluid pressure is elevated by contraction of the body wall musculature. In marine species with flexible, chitinous body coverings, the contraction of transverse parietal muscle bands, which are attached to the walls (Fig. 18–1), causes the flexible wall to bow inward, elevating the coelomic fluid pressure.

In calcareous species and even some forms with rigid chitinous exoskeletons, the exposed frontal surface contains a window covered by a thin chitinous membrane (the frontal membrane) (Figs. 18–5A and B and 18–12). The parietal muscles are inserted on the inner side of the membrane. On contraction, the flexible frontal membrane is bowed inward, increasing the coelomic pressure.

Retraction of the lophophore through the orifice is effected by the contraction of the lophophoral retractor muscles.

When the lophophore is protruded, the lateral ciliated tracts on the tentacles create a current that sweeps downward into the funnel and passes outward between the tentacles (Fig. 18–9). Regardless of the tentacle length and number, the distance between adjacent tentacle tips is generally about 110 μ. Small phytophanktonic organisms, which are probably the principal food of bryozoans, are driven into the funnel with the water current and, on touching the lateral cilia, effect a local reversal of beat, which bounces the particle back onto the upstream side of the tentacle and down toward the mouth (Fig. 18–9B). This is thus an upstream ciliary collecting system.

A survey of many different species by Winston (1978) indicates that bryozoans are more diverse in feeding habit than previously thought and that the range of particle size being utilized as a food source probably varies with lophophore size and behavioral differences. Tentacle flicking is a common accessory feeding mechanism in many species (even primary in a few). A particle is batted or rolled down the funnel by a rapid or slow lashing of one tentacle. *Bugula neritina* captures zooplankton by closing the tips of the tentacles to form a cage around the prey. Many species scan for particles by rotating the lophophore (Fig. 18–10).

Ingestion is facilitated by the ciliation of the pharyngeal lining, as well as by rapid dilations of the lower part of the pharynx. Particles may be rejected by mouth closure, tentacle flicking, or funnel closure or by passage between the tentacles.

Adjacent expanded funnels may form an extensive filtering surface. In encrusting forms the large volume of water passing below the filters exists through "chimneys," blank areas formed by lophophores tilted away from each other or from a space occupied by one or more modified zooids (Banta et al, 1974; Winston, 1978).

From the mouth and pharynx, food particles pass into the large stomach, which composes much of the U-shaped gut (Fig. 18–11). The anterior part of the stomach, called the cardia, is separated from the pharynx by a valve. A valvular constriction sep-

(*Text continued on p. 748*)

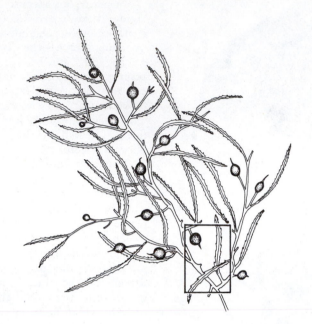

A. *Litiopa* (gastropod mollusk)
B. *Amphithoë* (amphipod crustacean)
C. *Luconacia* (caprellid amphipod)
D. *Anemonia sargassensis* (sea anemone)
E. *Platynereis* (nereid polychaete)
F. *Sertularia* (hydroid cnidarian)
G. *Clytia* (hydroid cnidarian)
H. *Scyllaea* (nudibranch gastropod)
I. *Spirorbis* (serpulid polychaete)
J. Copepod crustacean
K. *Zanclea* (hydroid cnidarian)
L. *Ceramium* (an epiphytic red alga)
M. *Membranipora* (bryozoan)
N. *Doto* (nudibranch gastropod)
O. *Gnescioceros* (polyclad flatworm)
P. *Obelia* (hydroid cnidarian)
Q. *Anoplodactylus* (pycnogonid) feeding on a nudibranch
R. *Fiona* (nudibranch gastropod)

Figure 18–7 Invertebrate animals that live on floating *Sargassum* (Gulf weed), a brown algae, in the Sargasso Sea. Although probably derived from West Indian ancestors, many of these species have become especially adapted for an epiphytic life on floating *Sargassum*. The figure at top shows a large piece of *Sargassum* from which the figure on the opposite page has been taken.

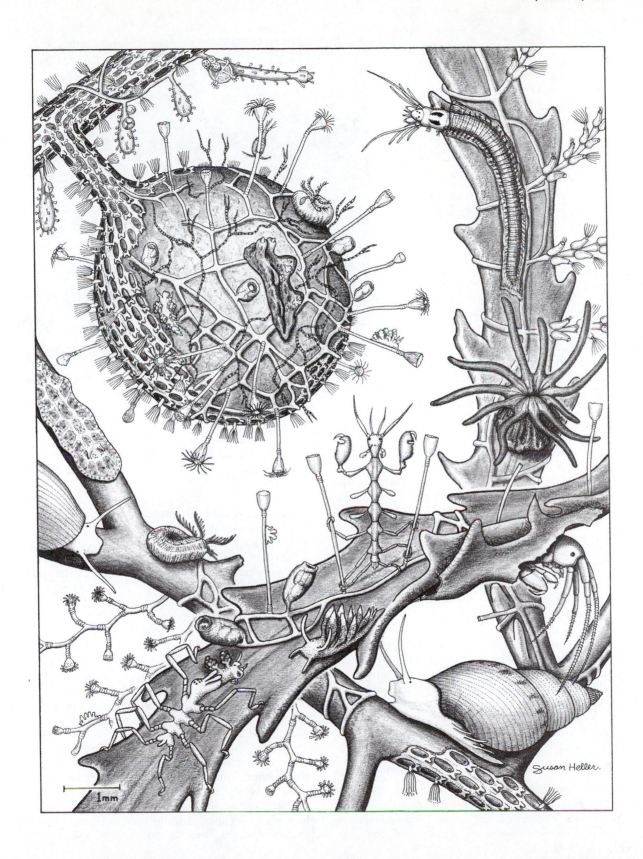

1mm

Susan Heller.

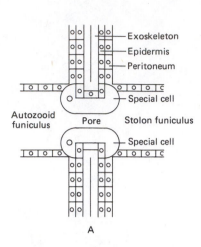

Figure 18—8 *A*, Diagram of the junction of the funiculus of an autozooid with the funiculus of a heterozooid (stolon) of *Zoobotryon verticillatum*. The pore is bounded by a rosette of special cells, but the small pore opening connects the lumens of the two funiculi. *B*, Cross section through a pore and the rosette of special cells. (From Carle, K. J., and Ruppert, E. E., 1983: Comparative ultrastructure of the bryozoan funiculus: a blood vessel homologue. Z. f. zool. Systematik u. Evolutionsforschung, *21*:181–193.)

arates the posterior stomach region, or pylorus, from the rectum. A large cecum projects backward from the central stomach.

Digestion is both extracellular and intracellular within the stomach, the cecum being the principal site of intracellular digestion. Food passes through the stomach by peristaltic contractions, but the pylorus functions to rotate and compact waste materials, which then pass into the rectum.

In some bryozoans, such as *Bowerbankia* and *Amathia*, the cardia is modified as a gizzard. There is a well-developed, circular muscle layer, and the lining epithelial cells bear teeth.

Gas exchange occurs across the exposed body surface, and internal transport of gases, some food, and wastes is provided by the coelomic fluid. The coelomic fluid contains coelomocytes, which engulf and store waste materials. The funiculus provides for at least some nutrient transport and is the main system for the colony-wide dispersal of metabolites in gymnolaemates.

Evolution and Distribution of Bryozoans

Of the three classes of living bryozoans, the freshwater phylactolaemates are considered the most primitive. The cylindrical zooids, the anterior orifice, the horseshoe-shaped lophophore, the presence of an epistome, and the nonpolymorphic colonies are all considered primitive features. In many other ways, however, phylactolaemates are specialized.

Although not usually common, freshwater bryozoans are widely distributed in lakes and streams that do not contain excessive mud or silt. Many species, such as *Fredericella sultana* and *Plumatella repens*, are cosmopolitan.

Unfortunately, there are no fossil phylactolaemates, and thus we know little about their relationship with the marine classes. The first known marine bryozoan is a questionable fossil species from the late Cambrian. But beginning with the Ordovician there is a rich fossil record, and thousands of fossil species have been described. Stenolaemates, of which there are three distinct orders, dominate the Paleozoic fauna, although there were Paleozoic ctenostome gymnolaemates. Cheilostomes, the dominant marine forms today, made their appearance in the late Jurassic.

Marine bryozoans are highly successful animals, exploiting all types of hard surfaces—rock, shells, coral, and wood (e.g., mangrove roots)—and capable of utilizing very restricted spaces. A few species even bore in calcareous substrates. The stolons of boring species are located in tunnels within the substratum, and the autozooids open to the

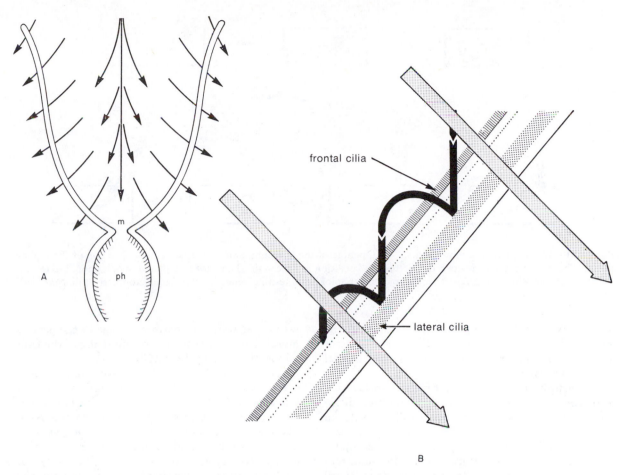

Figure 18—9 Filter feeding by the bryozoan lophophore: *A*, Path of water stream into lophophore funnel. *B*, Diagrammatic lateral view of one tentacle. Large gray arrows indicate direction of water current. Black arrow depicts path of a food particle that has been thrown back onto upstream side of tentacle by localized reversal of the beat of the lateral cilia. In this manner the particle bounces down tentacle to mouth at base of lophophore. (From Ryland, J. S., 1976: Physiology and ecology of marine bryozoans. Adv. Mar. Biol., *14*:285–433, Academic Press, London.)

surface. A few bryozoans also live on soft bottoms. The circular, shield-shaped colony of *Cupuladria* reaches the size of a small coin. It rests free on the bottom so that the convex surface bearing the frontal surface of the zooids is directed upward (Fig. 18–6).

From an economic standpoint, marine bryozoans are one of the most important groups of fouling organisms. About 120 species, of which different species of *Bugula* are among the most abundant, have been taken from ship bottoms.

Along with hydroids, bryozoans rank among the most abundant marine epiphytic animals. Large brown algae are colonized by many species, which display distinct preferences for certain types of algae. Evidence indicates that at the time of settling, the larvae of epiphytic species are attracted to the algal substrate, perhaps by some substance produced by the algae. The widely distributed, encrusting *Membranipora*, a species of which is abundant on floating *Sargassum* (Fig. 18–7*M*), displays a number of adaptations for an epiphytic life. The frontal surface is uncalcified and the lateral walls are jointed (i.e., broken), permitting some flexibility of the sheet of zooids attached to a bending algal thallus. Although the colony is encrusting on kelp weed, growth in the direction of the stalk of the thallus predominates and is controlled by the movement of water over the algal surface.

Stoloniferous colonies with cylindrical, upright, more or less separate autozooids probably represent the more primitive organization, but as we have seen, the great majority of bryozoans have contiguous autozooids with frontal surfaces ex-

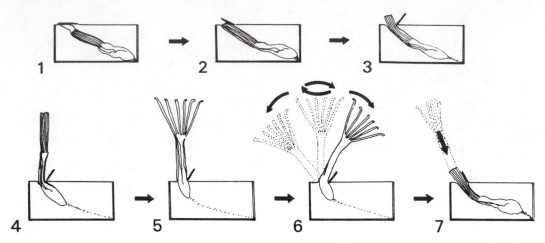

Figure 18–10   Lophophore movements: *1–5*, Opening of the operculum and protrusion and expansion of the lopho-phore. There is commonly a pause at the "testing" position in 3. Scanning behavior is depicted in *6*; lophophore retrac-tion in 7. (From Winston, J. E., 1978: Polypide morphology and feeding behavior in marine ectoprocts. Bull. Mar. Sci., *28*(1):1–31).

posed. Of various trends that are evident in the evolution of marine bryozoans, adaptations asso-ciated with protection of the vulnerable frontal surface are especially striking.

In primitive bryozoans the frontal surface be-hind the orifice is covered only by the frontal mem-brane, which must be thin and flexible in order to bow inward in the process of elevating the coe-lomic fluid pressure and protruding the lopho-phore. Such a frontal surface is found in *Membran-ipora*, in which the contrasting calcified, lateral walls are especially conspicuous (Figs. 18–7*M* and 18–12*A* and *B*).

A number of groups have reduced the vulnera-bility of the frontal membrane by reducing its area to an oval. Protection has also been afforded by the development of spines on the frontal margin of the lateral walls (Fig. 18–12*C*) or, in a few species, by a single long spine that curves back over the frontal membrane.

Elaboration of the protective spines led to the cribrimorphs, which include many fossil and some living species. In these forms overarching spines form a protective cover but permit access of water to the frontal membrane. In some the spines merely touch at their tips, but in others the spines also fuse laterally except for porelike openings (Fig. 18–12*E* and *F*).

A different kind of adaptation, illustrated by the genera *Cellaria*, *Micropora*, and others, is the protection of the zooid by the development of a wall (called a cryptocyst) beneath the frontal mem-brane, resulting in a double-chambered zooid. The

wall is perforated to permit passage of the parietal muscle bands responsible for inbowing of the fron-tal membrane (Fig. 18–12*G*).

On the basis of the numbers of species, how-ever, the most successful adaptations for the pro-tection of the zooid are found in the assemblage of forms at present grouped together as the Asco-phora (ascus bearers). In these the frontal wall has become calcified, and in many species a sac, or ascus, develops as an invagination of the body wall near the orifice, projecting backward into the coe-lom (Fig. 18–13*A*). The parietal muscles are at-tached to the underside of the sac; when they re-tract, the sac dilates, water enters, coelomic fluid pressure rises, and lophophore is protruded. The operculum, which covers not only the orifice but also the entrance to the ascus, is pivotally hinged so that when the back of the operculum goes down, opening the ascus, the front goes up and opens the orifice (Figs. 18–13*A* and 18–14).

Two principal predators of bryozoans, aside from nonselective scrapers, such as sea urchins, are sea spiders (pycnogonids) and nudibranchs. Anas-can species, with exposed frontal membranes, have more such predators than ascophorans, despite the large number of the latter group, perhaps attesting to the protective value of a calcified frontal wall (Ryland, 1976).

Modifications described thus far developed in the Gymnolaemata. The Stenolaemata, of which the cyclostomes are the only living representatives, evolved a method of lophophore protrusion that, although it depended on elevation of coelomic

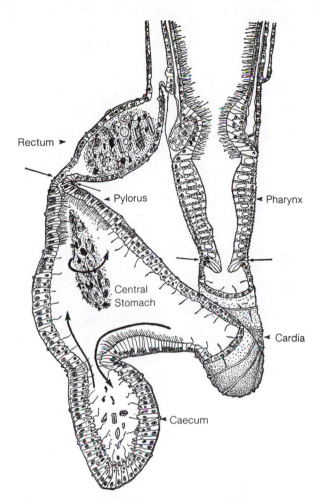

Rectum ➤

Pylorus

Pharynx

Central
Stomach

Cardia

Caecum

Figure 18–11 Digestive system of *Cryptosula palla-siana*. Outer arrows indicate valves, or sphincters; inner arrows show passage of food into and out of the cecum and the rotation of wastes. (From Gordon, D. P., 1975: Ultrastructure and function of the gut of a marine bry-ozoan. Cah. Biol. Mar., *16*:367–382.)

fluid pressure, did not involve deformation of the body wall. Moreover, there is no evidence that stenolaemates, which are represented by an enor-mous number of extinct Paleozoic forms, evolved from forms that possessed a frontal membrane. The zooids are tubular and calcified, and the ori-fice, which lacks an operculum, is located at the an-terior, or distal, end (Fig. 18–2). A space (exocoel) lies between the epidermis, which secretes the skel-eton, and the peritoneal mesoderm, which forms a membranous sac (Fig. 18–2*C*). The sac wall con-tains transverse muscle bands, and the space be-tween the epidermis and the membranous sac per-

mits the sequential contraction of these muscle bands free of the body wall. The pressure of the coelomic fluid is elevated and the lophophore is protruded. This description, based on a recent study of *Crisia* by Nielsen and Pedersen (1979), is probably applicable to all stenolaemates.

Reproduction

All freshwater bryozoans and most marine species are hermaphroditic. Simultaneous production of eggs and sperm may take place, but a tendency to-ward protandry is not uncommon. A colony may thus be composed of male and female zooids. (Fig. 18–15). The one or two ovaries and the one to many testes are masses of developing gametes cov-ered by peritoneum; the masses bulge into the coe-lom. The ovaries are located in the distal end of the animal and the testes in the basal end (Fig. 18–1). There are no genital ducts, and the eggs and sperm rupture into the coelom.

Some marine species (*Electra* and *Membrani-pora*) shed small eggs directly into the sea water. The vast majority of bryozoans brood their eggs, which are almost always large, yolky, and few in number. A variety of brooding mechanisms are em-ployed. A few species brood their eggs within the coelom, but most brood them externally. The digestive tract and lophophore often degenerate to provide space for the egg. The cavity of the tenta-cular sheath or invaginations of the atrial wall are common sites for brooding (Fig. 18–1). Many chei-lostomes, including the common *Bugula*, brood their eggs in a special external chamber called an ovicell (Figs. 18–13*A* and 18–16). The body wall at the distal end of the zooid grows outward, forming a large hood (the ovicell). A second, smaller evagin-ation, which is directly connected to the coelom, bulges into the space formed by the ovicell. A sin-gle egg is brooded in the space between the two evaginations. The developing embryo may derive its nutrition entirely from contained yolk, but in many species, including *Bugula*, placenta-like con-nections to the ovicell provide food material from the maternal zooid (Fig. 18–16*B*). An ovicell is thought to represent a modified individual and thus contributes to the polymorphic nature of the colony.

When the eggs are shed into the sea water or have been brooded internally, they escape from the coelom by way of a special opening in the region of the lophophore. This coelomopore may be a simple opening or may be mounted at the end of a projec-tion called an intertentacular organ (Fig. 18–5*A*).

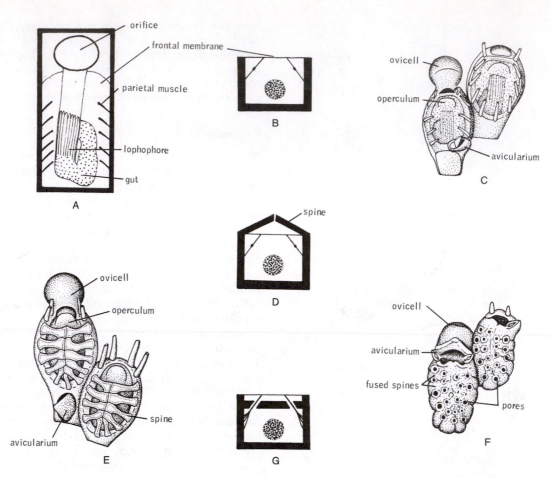

Figure 18–12 Modifications for protection of frontal surface: *A*, Frontal view of primitive condition, in which frontal membrane occupies most of frontal surface. *B*, Cross section through *A*. *C*, Frontal view of *Callopora*, in which oval frontal membrane is protected by spines. *D*, Cross section of cribrimorph condition, in which frontal membrane is covered by overarching spines. *E*, A species of *Callopora* in which opposing spines contact each other distally. *F*, *Cribrillina*, with pores remaining between fused spines. *G*, Cross section through a zooid with calcified wall beneath frontal membrane. (All after Ryland, J. S., 1970: Bryozoans. Hutchinson and Co., London.)

In those species with ovicells, the egg is extruded though the coelomopore as a "stream" and then rounds up within the ovicell cavity.

In all those species that have now been studied (*Electra, Membranipora, Schizoporella, Bugula,* and others) the sperm are shed though terminal pores of two or more tentacles of the lophophore and are disseminated into the sea water. The liberated sperm, when caught in the feeding currents of other individuals, adhere to the tentacular surfaces of the lophophore or enter the intertentacular organ. In the former case, the eggs are fertilized as they leave the intertentacular organ. In those species in which the sperm enter the intertentacular organ, this structure is the site of fertilization. The

coelomopore may be the means by which sperm enter the coelom in species that brood their eggs internally. Fertilization between zooids of the same colony is probably common, but sufficient cross fertilization between colonies occurs to ensure outbreeding.

Cleavage in marine species is bilateral and leads to a larva, which escapes from the brood chamber. The stenolaemates are remarkable in exhibiting a polyembryony, in which the primary embryo produces numerous secondary embryos, which may then produce tertiary embryos. The larvae of bryozoans vary considerably in form (Fig. 18–17), but all possess a locomotor ciliated girdle or corona, an anterior tuft of long cilia, and a posterior adhesive

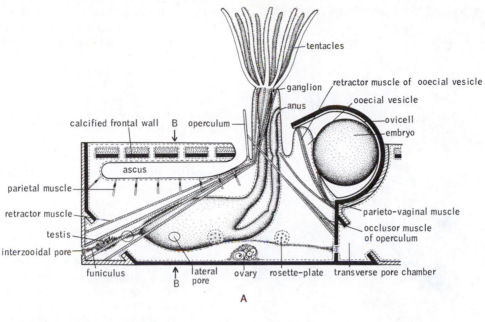

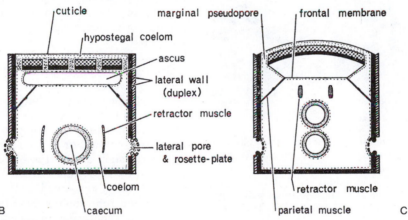

Figure 18-13 *A*, Diagrammatic sagittal section through an ascophoran cheilostome. The lateral pivoting points of the operculum are not visible in this view. (Modified from Ryland.) *B*, Transverse sections through same individual. *C*, Type of ascophoran in which ascus is a shallow space beneath an overarching vault. (*B–C* from Ryland, J. S., 1970: Bryozoans. Hutchinson and Co., London.)

sac. The larvae of some species of nonbrooding gymnolaemates, such as *Electra* and *Membranipora*, called cyphonautes larvae, are triangular and greatly compressed, and each lateral surface may be covered by a chitinous valve. Only the larvae of nonbrooding bryozoans possess a functional digestive tract and feed during larval existence. Feeding larvae may have a larval life of several months, but the larvae of brooding species, which are nonfeeding, have a very brief larval existence prior to settling.

During settling the adhesive sac everts and fastens by means of secretions to the substratum. The larval structures of the attached larva undergo retraction and histolysis, followed by development into an adult. The first zooid is called an ancestrula. By budding, the ancestrula gives rise to a series of other zooids that often show changes in size and shape. These zooids, in turn, bud off new individuals, and thus by subsequent asexual reproduction the colony gradually increases in size (Fig. 18-18). Budding involves the cutting off of a part of the

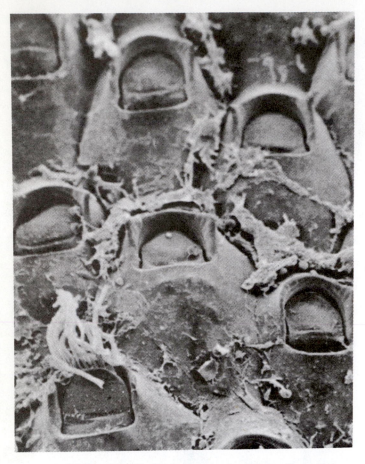

Figure 18–14 SEM photograph of the surface of an encrusting ascophoran bryozoan. Orifice is covered by a conspicuous operculum, and lophophore tentacles can be seen protruding in lower middle and left individuals. Considerable debris fouls the exposed frontal wall. (From Sieburth, J. M., 1975: Microbial Seascapes. University Park Press, Baltimore.)

parent zooid by the formation of a body wall partition. The new chamber evaginates by the mitotic activity of the distal cells either before or after its formation, and the skeleton is "stretched" by the addition of new material (Fig. 18–19). New internal structures develop from the ectoderm and peritoneum of the body wall. The exact pattern of budding (i.e., the number and location of buds) determines the growth pattern of the colony. In the erect, dendritic *Bugula*, growth occurs at the tips of each branch. Encrusting species have a peripheral growing edge (Fig. 18–19*B*).

The life-span of bryozoan colonies varies greatly. Some live only a single year, with growth taking place with increase of water temperatures in temperate regions; liberation of larvae at some point in the summer marks the end of the life of the colony. Annual life spans are especially characteristic of bryozoans that live on algae. Some epiphytic species may pass through several generations in one season. Many species live for two or more years, as long as 12 years in the European *Flustra foliacea*, with growth slowing down or halting during the winter. The colonies of some, such as *Bugula*, may die back to the stolons or holdfast and re-form the following season. Sexual reproduction may occur during a restricted period, or, commonly in perennial species, it may occur throughout the growing period.

Regardless of the longevity of the colony, the lophophore and gut of a zooid degenerate after a few weeks. Some components are phagocytized, but a large residual mass of necrotic cells is lodged in the coelom as a conspicuous dark ball, called a brown body. Regression is followed by regeneration of a new lophophore and gut. In some species the old brown body is a permanent resident of the coelom; in others it is incorporated within the regenerating stomach and expelled at the first defecation.

Alternating phases of degeneration and regeneration are especially common in bryozoan colonies that live a number of years, and even in some annual species. A large colony may exhibit zones of individuals in various stages of development (Fig. 18–20). For example, at the outer perimeter of an encrusting colony, or at the tips of erect branches in a form such as *Bugula*, budding and development of new individuals occur. Further inward is a zone of fully developed individuals, which are feeding and reproducing. Still further inward is a zone of degenerating members containing dark brown bodies. To the inner side of the degenerate zone may be a zone of regenerated feeding individuals.

Development in freshwater phylactolaemates takes place within an embryo sac that bulges into the coelom. Development leads to the formation of a cystid sac, which then proceeds to bud off one to several zooids. This young ciliated colony or "larva" is released from the parent colony. It swims about for a short time prior to settling. After attachment, retraction and degeneration of the once-ciliated outer cystid wall ensues. Meanwhile, the young colony that had been developing inside the confines of the cystid wall continues to bud successive zooids until an adult colony is formed. The parent zooid dies after producing a number of daughter individuals. Thus, in branching colonies, only the tips of branches contain living zooids; in

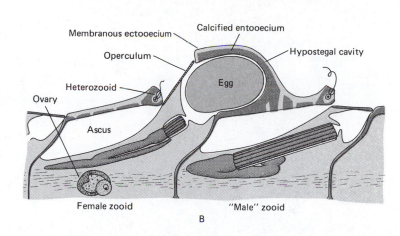

A
1 mm

Ovary

Heterozooid

Operculum

Membranous ectooecium   Calcified entooecium

Hypostegal cavity

Egg

Ascus

Female zooid

"Male" zooid

B

Figure 18–15 *A,* Surface view of part of a colony of *Hippodiplosia insculpta* from the coast of California. In this species each female zooid (asterisk) brooding an egg in the domelike ovicell is surrounded by six male zooids (dot). *B,* Diagrammatic longitudinal section of a male and female zooid of *Hippodiplosia.* At the time of reproduction, the female gut and lophophore degenerate and are replaced by a smaller, nonfeeding gut and lophophore. (From Nielsen, C. 1981: On morphology and reproduction of "Hippodiplosia" insculpta and Fenestrulina malusii. Ophelia, *20*(1):91–125.)

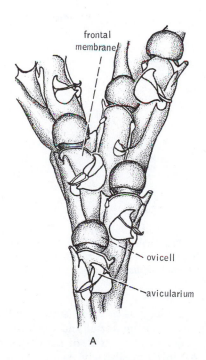

frontal membrane

ovicell

avicularium

A

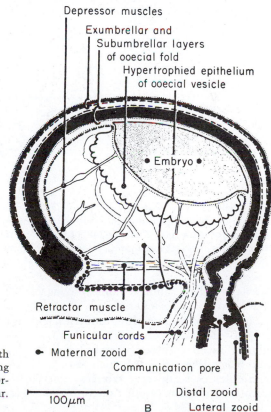

Depressor muscles

Exumbrellar and Subumbrellar layers of ooecial fold

Hypertrophied epithelium of ooecial vesicle

Embryo

Retractor muscle

Funicular cords

Maternal zooid

Communication pore

Distal zooid

Lateral zooid

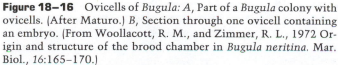
100 μm

B

Figure 18–16 Ovicells of *Bugula: A,* Part of a *Bugula* colony with ovicells. (After Maturo.) *B,* Section through one ovicell containing an embryo. (From Woollacott, R. M., and Zimmer, R. L., 1972 Origin and structure of the brood chamber in *Bugula neritina.* Mar. Biol., *16*:165–170.)

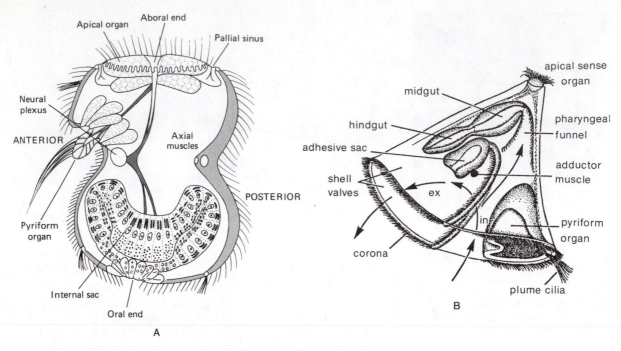

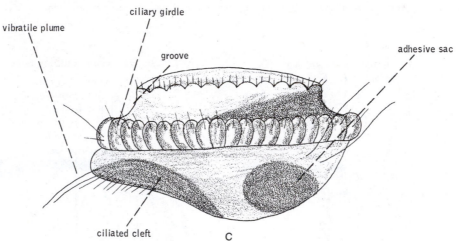

Figure 18–17 Bryozoan larvae: *A, Bugula.* (Based in part on Reed, C. G., and Woollacott, R. M., 1982: Mechanisms of rapid morphogenetic movements in the metamorphosis of the bryozoan *Bugula neritina.* I. Attachment to the substratum. Jour. Morph., 172:335–348.) *B,* Cyphonautes larva of *Membranipora.* (From Ryland, J. S., 1970: Bryozoans. Hutchinson and Co., London.) *C, Alcyonidium mytili.* (After Barrois from Hyman.)

the flattened gelatinous colonies, living zooids are restricted to the periphery.

In addition to sexual reproduction and to budding, freshwater bryozoans also reproduce asexually by means of special resistant bodies called statoblasts (Fig. 18–21), which are similar to the gemmules of freshwater sponges. One to several statoblasts develop on the funiculus and bulge into

the coelom as masses of peritoneal cells that contain stored food material and epidermal cells that have migrated to the site of statoblast formation. After organizing cellularly, each mass secretes both an upper and a lower chitinous valve that form a protective covering for the internal cells. Since the rims of the valves often project peripherally to a considerable extent, the statoblasts are usually

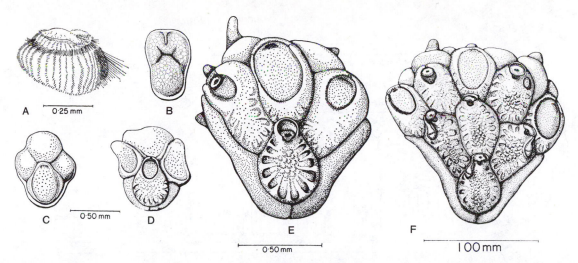

Figure 18–18 Ancestrula and early colony formation in *Metrarabdotos unquiculatum*, an encrusting, shallow-water bryozoan found on stones on the west coast of Africa: *A*, Larva. *B–C*, Formation of ancestrula, which is one of an initial tetrad of zooids. *B* is 2 hours after settlement; *C* is 28 hours after settlement. *F*, Formation of additional members of colony at 140 hours after settlement. (Cook, P., 1973: Settlement and early colony development in some cheilostomata. *In* Larwood, G. P. (Ed.): Living and Fossil Bryozoa. Academic Press, London, pp. 65–71.)

somewhat disc-shaped. Statoblasts are continuously formed during the summer and fall. Some types of statoblasts adhere to the parent colony or fall to the bottom; others contain air spaces and float. These floating statoblasts are sometimes armed with hooks around the margins.

Statoblasts remain dormant for a variable length of time. During this period they may be spread considerable distances by animals, floating vegetation, or other agents and are able to withstand desiccation and freezing. When environmental conditions become favorable, as in the spring, germination takes place, the two valves separate, and a zooid develops from the internal mass of cells. The number of statoblasts produced by a freshwater bryozoan is enormous. Brown (1933) reported drifts of statoblast valves 1 to 4 feet wide along the shores of Douglas Lake in Michigan and estimated that *Plumatella repens* colonies in 1 square meter of littoral lake vegetation produce 800,000 statoblasts.

SYSTEMATIC RÉSUMÉ OF PHYLUM BRYOZOA

Class Phylactolaemata. Freshwater bryozoans in which the cylindrical zooid possesses a horseshoe-shaped lophophore (except in *Fredericella*), an epistome, a body wall musculature, and a noncalcified covering. Coelom continuous between individuals. Colonies nonpolymorphic. *Frederi-*

cella, Plumatella, Pectinatella, Lophopus, Cristatella.

Class Stenolaemata. Marine bryozoans. Zooids tubular, with calcified walls that are fused with adjacent zooids. Orifices circular and terminal. Lophophore protrusion not dependent on deformation of body wall. Order Cyclostomata contains some living and many fossil species: *Crisia, Lichenopora, Stomatopora, Tubulipora.* Orders Cystoporata, Trepostomata, and Cryptostomata all became extinct at the end of the Paleozoic.

Class Gymnolaemata. Primarily marine bryozoans with polymorphic colonies. Zooid cylindrical or flattened; lophophore circular; an epistome and body wall intrinsic musculature are lacking. Protrusion of the circular lophophore depends on body wall deformation.

Order Ctenostomata. Stoloniferous or compact colonies in which the uncalcified exoskeleton is membranous, chitinous, or gelatinous. The usually terminal orifice lacks an operculum. *Amathia, Alcyonidium, Aeverrillia, Bowerbankia,* the freshwater *Paludicella.*

Order Cheilostomata. Colonies composed of boxlike zooids that are adjacent but have separate calcareous walls. Orifice is provided with an operculum (except in *Bugula*). Avicularia, vibracula, or both may be

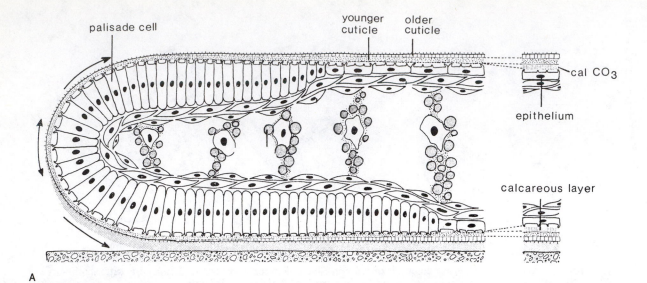

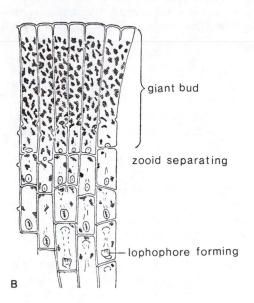

Figure 18–19 Colony development: *A*, Wall formation and development at the growing margin of a cheilostome colony. Completely developed wall is diagrammed at extreme right. (From Tavener-Smith, R., and Williams, A., 1972: The secretion and structure of the skeleton of living and fossil Bryozoa. Phil. Trans. R. Soc. B, *264*:101.) *B*, Growing edge of the encrusting *Membranipora*. New zooids are cut off at the rear of the giant buds. (From Lutaud, G., 1961: Annales de la Société Royale Zoologique de Belgique, *91*:157–300.)

present. Eggs are commonly brooded in ovicells.

Suborder Anasca. Frontal membrane membranous. *Aetea, Callopora, Electra, Flustra, Membranipora, Tendra, Bugula, Scrupocellaria, Cupuladria, Thalamoporella, Cellaria.*

Suborder Cribrimorpha. Frontal membrane covered by a vault of overarching spines, which may be partially fused. *Cribrilina.*

Suborder Ascophora. Frontal wall calcified. Lophophore protrusion involving the dilation of an invaginated sac, or ascus. *Microporella, Schizoporella, Smittina, Watersipora.*

Phylum Entoprocta

The Entoprocta are a small phylum of some 150 species of mostly sessile animals that are very similar to bryozoans. They were formerly included in the phylum Bryozoa but were removed because of a number of differing features, including the uncertain nature of the body cavity. The body cavity has been considered a pseudocoel by some zoologists. However, if entoprocts are indeed pseudocoelomates, the relationship with the other pseudocoelomate phyla is very obscure. The entoprocts are therefore discussed at this point in order that they may be more easily compared with bryozoans.

Except for the single freshwater genus *Urnatella* (Fig. 18–22A), all entoprocts are marine and live attached to rocks, shells, or pilings or are commensal on sponges, polychaetes, bryozoans, and other marine animals. Members of the commensal family Loxosomatidae are solitary; the other two families are colonial. All are very small and never exceed 5 mm in length. The body (Figs. 18–23 and 18–24) consists of a somewhat ovoid or boat-shaped structure called the calyx, which contains the internal organs, and a stalk by which the calyx is attached to the substratum (Fig. 18–22B). The attached underside of the calyx was the original dorsal surface. The upper margin of the calyx bears an encircling crown of 8 to 30 tentacles, which represent extensions of the body wall.

The area enclosed by the tentacles, the vestibule or atrium, contains the mouth at one end and the anus at the other. Both mouth and anus, however, are located within the tentacular crown, hence the name *Entoprocta*—"inside anus." The mouth and anus mark the anterior and posterior of the animal. The bases of the tentacles are connected by a membrane that is pulled over the crown when the tentacles contract and fold inward over the vestibule (Fig. 18–24B).

There may be a single stalk, as in the solitary *Loxosoma* (Fig. 18–24), several stalks from a common attachment disc, or, as in *Pedicellina*, numerous stalks arising from a horizontal, creeping stolon or upright, branching stems (Fig. 18–22). The stalk is separated from the calyx by a septum-like fold of the body wall and is commonly partitioned into short cylinders, or segments. In many species

Figure 18–20 Cycle of generation and degeneration in *Chartella papyracea*, a bryozoan with colonies organized like that of *Flustra* (Fig. 18–4C). Colony lives over the winter, but growth is limited to the summer. During the first two thirds of the growing season, sperm-producing zooids (androzooids) are formed. Egg-producing zooids (gymnozooids) are formed during the last third. The age of a zooid increases with distance from the growing edge of the colony, where new zooids are formed. (From Dyrynda, P., 1981: A preliminary study of patterns of polypide generation-degeneration in marine cheilostome Bryozoa. *In* Larwood, G. P., and Nielsen, C. (Eds.): Recent and Fossil Bryozoa. Olsen and Olsen, Fredensborg, Denmark. p. 79.)

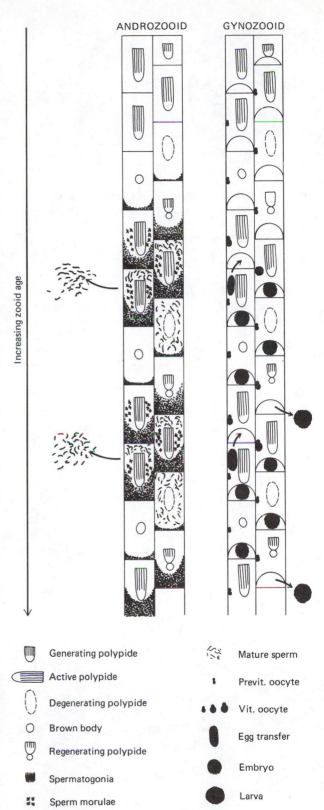

Increasing zooid age

ANDROZOOID   GYNOZOOID

Generating polypide

Active polypide

Degenerating polypide

Brown body

Regenerating polypide

Spermatogonia

Sperm morulae

Mature sperm

Previt. oocyte

Vit. oocyte

Egg transfer

Embryo

Larva

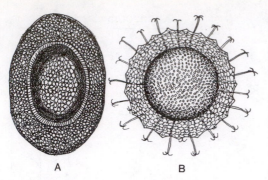

Figure 18–21 Statoblasts of freshwater bryozoans. *A*, A floating statoblast of *Hyalinella punctata.* (After Rogick from Pennak.) *B*, Statoblast with hooks from *Cristatella mucedo.* (After Allman.)

certain segments are swollen with longitudinal muscle fibers that, on sudden contraction, produce a curious nodding motion in members of the colony. The solitary *Loxosoma* can move about on its stalk, which is provided with a sucker (Fig. 18–24).

The body wall consists of a cuticle and underlying epithelium. The muscle layer is limited to longitudinal fibers along the inner wall of the tentacles, in the tentacular membrane, and in certain areas of the calyx. The body cavity, which also extends into the tentacles, is filled with a gelatinous "mesenchyme" containing both free and fixed cells.

Entoprocts are filter feeders, consuming organic particles and small plankton. The beating of cilia on the sides of the tentacles causes a water current

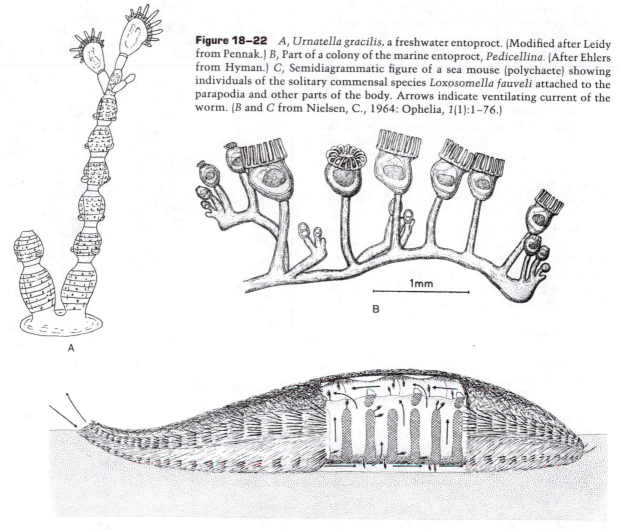

Figure 18–22 *A, Urnatella gracilis,* a freshwater entoproct. (Modified after Leidy from Pennak.) *B,* Part of a colony of the marine entoproct, *Pedicellina.* (After Ehlers from Hyman.) *C,* Semidiagrammatic figure of a sea mouse (polychaete) showing individuals of the solitary commensal species *Loxosomella fauveli* attached to the parapodia and other parts of the body. Arrows indicate ventilating current of the worm. (*B* and *C* from Nielsen, C., 1964: *Ophelia*, 1(1):1–76.)

1mm

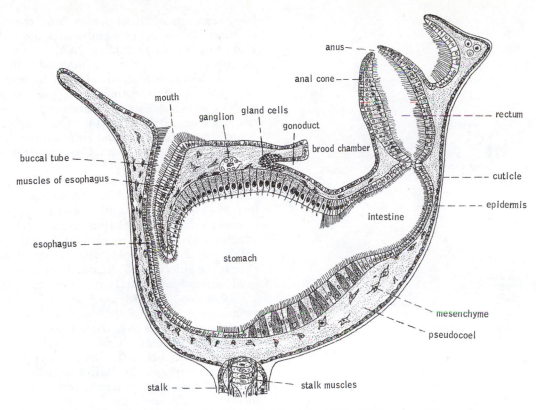

Figure 18–23   *Pedicellina* (median sagittal section.) (After Becker from Hyman.)

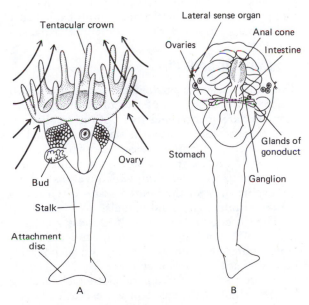

Figure 18–24   Solitary marine entoproct, *Loxosoma: A,* Expanded. Arrows show water current. *B,* Retracted. (Both after Atkins, *A* greatly modified.)

to pass into the vestibule between the tentacles and then to pass upward and out (Fig. 18–24). When suspended food particles pass between the tentacles, they become trapped by the lateral cilia (a downstream collecting system, instead of the upstream system of bryozoans). They are then transported by the frontal cilia, which are located on the inner tentacular face and beat downward, carrying the food particles to the base of the tentacle. Here the food particles are carried in ciliated vestibular grooves that run along the inner base of the tentacular crown on both sides toward the mouth.

The digestive tract is U shaped with a large, bulbous stomach making up the major part (Fig. 18–23). There are two protonephridia, which open through a common nephridiopore just behind the mouth. The nervous system consists of a single, large, median ganglion situated between the stomach and the vestibule, with nerves to the tentacles, stalk and calyx (Fig. 18–23).

Asexual reproduction by budding is common in all entoprocts, and it is by this means that extensive colonies are formed. In most species the buds arise

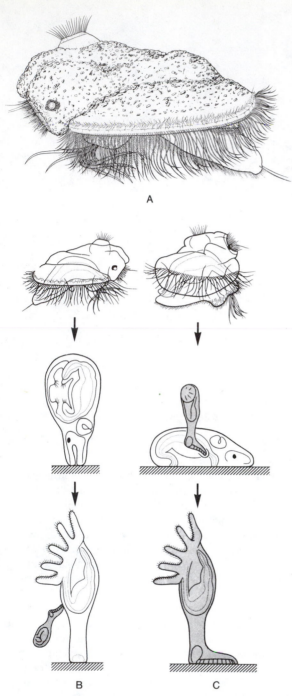

Figure 18–25 *A*, Lateral view of larva of *Loxosomella harmeri*, showing apical tuft at the top, frontal organ to the left, circular prototroch, and foot below. *B*, Attachment and metamorphosis of *Loxosomella harmeri*, in which adult individual is derived directly from larva. Gray, pendant structure attached to adult is a bud. *C*, Development in *Loxosomella leptoclini*, in which larva produces bud that develops into adult. (All from Nielsen, C., 1971: Ophelia, 9:209–341.)

from segments of the stolon or from the upright branches. In the solitary entoprocts the buds develop from the calyx (Fig. 18–24*A*), separate from the parent, and then attach as new individuals.

The phylum is probably entirely hermaphroditic, with some protandric species. The one or two pairs of gonads are located between the vestibule and the stomach (Fig. 18–24). The simple gonoducts become confluent and empty through a single, median gonopore located just posterior to the nephridiopore.

The eggs are believed to be fertilized in the ovaries and are brooded externally in the vestibule (Fig. 18–23). Cleavage is spiral, and a ciliated, free-swimming larva hatches from the egg. The larva, which is a trochophore more or less like that of annelids and mollusks, possesses an apical tuft of cilia at the anterior end, a frontal organ, a ciliated girdle around the ventral margin of the body, and a ciliated foot (Fig. 18–25*A*). After a short free-swimming existence, the larva settles to the bottom where it creeps over the surface with the ciliated foot and eventually attaches with the frontal organ. In some species the larva undergoes a complex metamorphosis, in which the future calyx rotates 180 degrees to attain the inverted condition of the adult. In some the larva does not develop into an adult but produces buds from which the adults are derived (Fig. 18–25*C*).

On the basis of their development, Nielsen (1985) believes that entoprocts are related to other groups with trochophore larvae and probably had a common ancestry with the coelomate bryozoans.

Phylum Phoronida

The phoronids are a small group consisting of two genera and ten species of wormlike animals. All members are marine and live within a chitinous tube that is either buried in sand or attached to rocks, shells, and other objects in shallow water (Fig. 18–26). A few species bore into mollusk shells or calcareous rock. *Phoronis vancouverensis* along the Pacific coast of the United States often forms masses of intertwined individuals.

The cylindrical body, which in most species is less than 20 cm long, bears no appendages or regional differentiation except the conspicuous lophophore at the anterior end (Fig. 18–27). The lophophore consists of two parallel, tentacle-bearing ridges curved in the shape of a crescent or horseshoe. The bend of the crescent is located ventrally, with one ridge passing above and the other ridge passing below the mouth. The horns of the ridges are directed dorsally and may be rolled up as a spi-

Figure 18–26 *Phoronis hippocrepia* from the coasts of Europe, South Africa, and Brazil. Aggregations of tubes encrust rocks, shells, and coral. (From Emig, C. C., 1974: The systematics and evolution of the phylum Phoronida. Z. f. zool. Systematik u. Evolutionsforsch., *12*(2):128–151.)

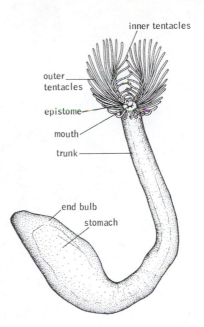

Figure 18–27 *Phoronis architecta* removed from tube. (After Wilson.)

ral. The anus opens above the mouth but outside the upper lophophoral ridge.

The body wall is composed of an outer epithelium, under which lie a thin layer of circular muscle fibers and a thick longitudinal muscle layer. The coelom is divided into a posterior region, which occupies the trunk region, and an anterior region, which extends into the lophophore and each of the tentacles.

Like all lophophorates, phoronids are filter feeders. The tentacular cilia beat downward, creating a water current from which plankton and suspended detritus are collected and entangled in mucus on contact with the tentacles. Cilia in the groove between the two ridges of the lophophore convey the food particles toward the mouth. The digestive tract is U shaped, and digestion occurs extracellularly within the esophagus and stomach (Fig. 18–28).

Phoronids possess a blood-vascular system with corpuscles that contain hemoglobin, a pair of metanephridia with nephridiopores to either side of the anus, and an epidermal nerve ring at the base of the lophophore from which nerves arise supplying the tentacles and body wall muscles (Fig. 18–28).

Phoronids reproduce sexually and asexually, by budding and by transverse fission. The majority are

hermaphroditic. The sex cells arise from the peritoneum associated with the ventral blood vessel, sperm on one side and eggs on the other. The gametes are shed into the coelom of the trunk and escape to the outside by way of the nephridia. The sperm are transferred in spermatophores produced by a pair of lophophoral organs. The eggs appear to be fertilized internally and either are planktonic or are brooded in the concavity formed by the two arms of the lophophore.

Cleavage in most species is biradial, and the mesoderm and coelom are somewhat enterocoelous in origin (Emig, 1977). The gastrula develops into an elongated, tentacular, ciliated larva called an actinotroch (Fig. 18–28C). A telotroch, which is probably the principal locomotor organ, rings the posterior of the trunk. After a long or short free-swimming and feeding planktonic existence, the actinotroch undergoes a rapid metamorphosis and sinks to the bottom, where it secretes a tube and takes up an adult existence.

The phylogenetic relationships of the phoronids are discussed at the end of the next section on brachiopods.

Phylum Brachiopoda

The last of the three lophophorate phyla is the phylum Brachiopoda, commonly known as lamp shells. These animals resemble bivalve mollusks in

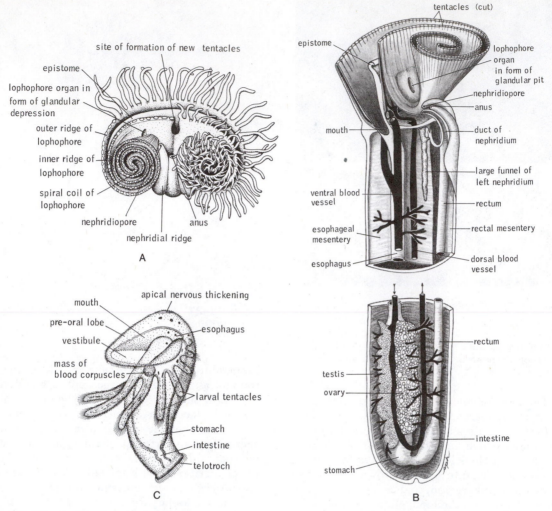

Figure 18–28   *A*, Anterior view of *Phoronis australis*. Tentacles on right side have been cut away to show spiral arrangement of lophophore. (After Shipley.) *B*, Internal structure of anterior and posterior of *Phoronis australis*. (After Benham.) *C*, Actinotroch larva of *Phoronis*. (After Wilson.)

possessing a mantle and a calcareous shell of two valves that approximates that of small mollusks in size. In fact, the phylum was not separated from the mollusks until the middle of the 19th century. However, the resemblance to mollusks is superficial, for in brachiopods the two valves enclose the body dorsally and ventrally instead of laterally, and the ventral valve is typically larger than the dorsal (Fig. 18–29*A* and *C*).

All brachiopods are marine, and are found from the intertidal zone to the deep sea. Most species live attached to rocks or other firm substrata, but some forms, such as *Lingula*, live in vertical burrows in sand and mud bottoms (Fig. 18–30*A*). Al-

though fossil species were widely distributed and abundant on reefs, modern forms are largely inhabitants of cold waters and are of rather spotty occurrence.

The approximately 325 species of living brachiopods are but a fraction of the 12,000 described fossil species that flourished in the seas of the Paleozoic and Mesozoic eras. The phylum made its appearance in the Cambrian period and reached its peak of evolutionary development during the Ordovician period. Brachiopods were especially hard hit by the great Permian extinction, and their present, more limited numbers and rather restricted distribution date from that event (Gould and Cal-

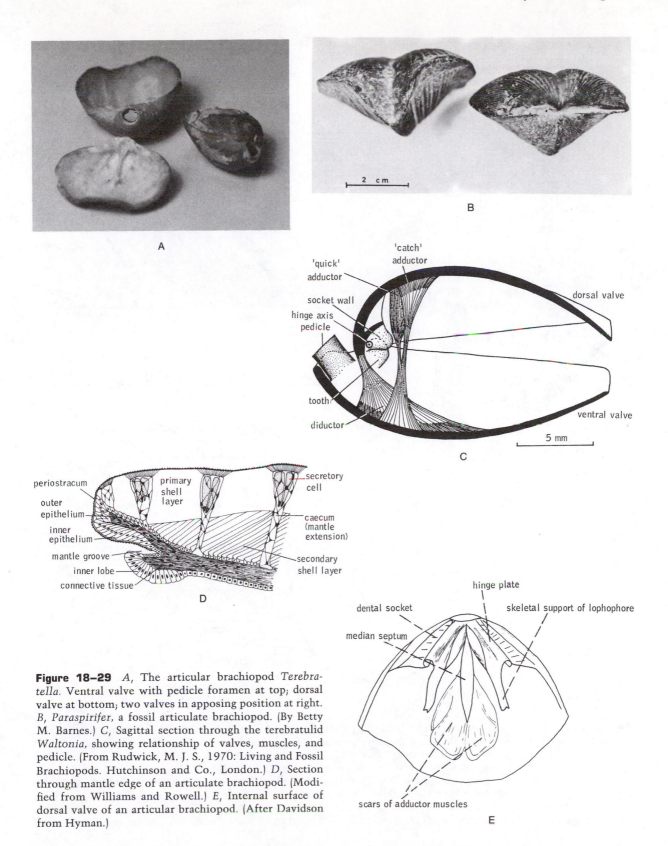

A

B

'quick'
adductor

'catch'
adductor

socket wall

hinge axis
pedicle

dorsal valve

tooth

diductor

ventral valve

5 mm

C

periostracum

outer
epithelium

inner
epithelium

mantle groove

inner lobe

connective tissue

primary
shell
layer

secretory
cell

caecum
(mantle
extension)

secondary
shell layer

D

hinge plate

dental socket

skeletal support of lophophore

median septum

scars of adductor muscles

E

Figure 18–29 *A*, The articular brachiopod *Terebra-tella*. Ventral valve with pedicle foramen at top; dorsal valve at bottom; two valves in apposing position at right. *B*, *Paraspirifer*, a fossil articulate brachiopod. (By Betty M. Barnes.) *C*, Sagittal section through the terebratulid *Waltonia*, showing relationship of valves, muscles, and pedicle. (From Rudwick, M. J. S., 1970: Living and Fossil Brachiopods. Hutchinson and Co., London.) *D*, Section through mantle edge of an articulate brachiopod. (Modified from Williams and Rowell.) *E*, Internal surface of dorsal valve of an articular brachiopod. (After Davidson from Hyman.)

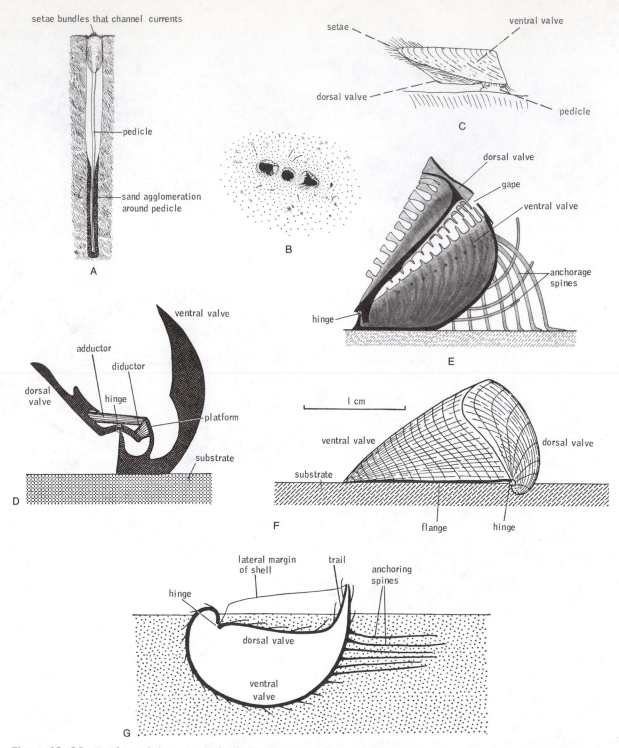

Figure 18–30 Brachiopod diversity and relationship to substratum: *A*, The inarticulate *Lingula* in feeding position within burrow. (Modified from Francois.) *B*, Burrow opening of *Lingula* when feeding. Setae surround middle exhalant and lateral inhalant apertures. (Modified from Rudwick.) *C*, Living articulate *Discinisca*, attached by pedicle. (After Morse from Hyman.) *D*, Diagrammatic lateral view of living articulate *Lacazella*, which lives attached directly to substratum by ventral valve. *E*, Sagittal section through Permian articulate *Chonosteges*, which cemented ventral valve directly to substratum. (Modified from Rudwick.) *F*, Devonian articulate *Spyringospira*, which is believed to have lived unattached on soft bottoms. *G*, Permian articulate *Waagenoconchia*, which is believed to have lived partially buried in soft bottoms. (*D*, *F*, and *G* from Rudwick, M. J. S., 1970: Living and Fossil Brachiopods. Hutchinson and Co., London.)

loway, 1980). The genus *Lingula* (but none of the present living species) dates back to the Ordovician period.

External Anatomy

Each of the two valves is bilaterally symmetrical and is usually convex. The smaller dorsal shell fits over the larger ventral shell, the apex of which in some groups contains a hole like a Roman lamp—hence the name *lamp shell* (Fig. 18–29A and C). In the burrowing lingulids the valves are flattened and more equal in size. The valves may be ornamented with concentric growth lines and a fluted, ridged, or even spiny surface. The shells of most living brachiopods are dull yellow or gray, but some species have orange or red shells.

The two valves articulate with one another along the posterior line of contact, called the hinge line (Fig. 18–29A and C), and the nature of the articulation is the basis for the division of the phylum into two classes, the Inarticulata and the Articulata. In inarticulate brachiopods, such as *Lingula* and *Glottidia*, the two valves are held together only by muscles. Valve opening occurs by backward retraction of the soft body, which produces outward pressure through the coelomic fluid on the dorsal and ventral valves (Gutmann et al, 1978). In articulate brachiopods, on the other hand, the ventral valve bears a pair of hinge teeth that fit into opposing sockets on the underside of the hinge line of the dorsal valve (Fig. 18–29E). This articulating mechanism locks the valves securely together and allows only a slight anterior gape of approximately 10 degrees. A pair of adductor muscles closes the valves. Another pair of muscles (diductors) opens the valves.

Although inarticulates are considered more primitive than articulate forms, both groups are present from the beginning of the fossil record.

The shell is secreted by the underlying dorsal and ventral mantle lobes (Fig. 18–29D). The shell of inarticulates is usually composed of a mixture of calcium phosphate and chitin (chitinophosphate), and the shell of articulates is composed of calcium carbonate in the form of calcite. The chitinophosphate shell appears to be the more primitive, for the Cambrian shells were of this type; calcite shells did not appear until the Ordovician. The outer surface of the shell is covered by a thin organic periostracum. As in most mollusks, the periostracum and the outer layers of mineral deposition are secreted by the mantle edge, and the inner layer of shell is secreted by the entire outer mantle surface.

In addition to secreting the shell, the mantle edge in most species bears long, chitinous setae that are believed to have a protective and perhaps sensory function (Fig. 18–30C).

The body proper of brachiopods occupies only the posterior part of the chamber formed by the two valves (Fig. 18–32). Anteriorly, the space between the mantle lobes (the mantle cavity) is filled by the lophophore.

Most brachiopods are attached to the substratum by a cylindrical extension of the ventral body wall, called the pedicle. The pedicle of the inarticulate lingulids (Lingula and Glottidia) is long and provided with muscles; it emerges at the posterior of the animal between the two valves (Figs. 18–29C and 18–30C). The lingulids live in U-shaped burrows excavated in sand and mud. The anterior ends of the valves are directed toward the burrow opening, and the pedicle extends downward toward the bottom of the burrow and is encased in sand. When the animal is feeding, the gaping valves are near the burrow opening, and the long mantle setae appear to function in preventing fouling by sediment (Fig. 18–30A). When the animal is disturbed, the pedicle contracts and pulls the animal downward into the burrow.

The pedicle of an articulate brachiopod may be short and lacking muscles, or it may be a flexible, muscular tether (Figs. 18–30C and 18–33). Moreover, the pedicle emerges either from a notch at the hinge line of the ventral valve or through a hole at the upturned apex (Fig. 18–29A). This means, of course, that the pedicle emerges from the dorsal side of the ventral valve, which extends posteriorly considerably beyond the dorsal valve. An articulate brachiopod is attached to the substratum upside down with the valves held in a horizontal position (Fig. 18–30C) or with the hinge end down and the gape directed upward. Muscles within the valves, which are inserted on the pedicle base, permit erection, flexion, and even rotation of the animal on the pedicle. The end of the pedicle adheres to the substratum by means of rootlike extensions or short papillae.

The pedicle has been lost completely in a few brachiopods of both classes, such as *Crania* (Inarticulata) and *Lacazella* (Articulata). Such species are cemented directly to the substratum by the ventral valve and are thus oriented in a normal manner with the dorsal side up (Fig. 18–30D). The more posterior part of the ventral valve being the point of attachment, the anterior margin is directed somewhat upward and clear of the substratum. Some species of fossil brachiopods that attached by

Figure 18–31   A rock wall on the coast of New Zealand supporting three species of brachiopods. The white, branching organism is a stylasterine hydrocoral. Photograph covers about 1 square meter. (By P. J. Hill, DSIR, New Zealand.)

cementation were important contributors to Paleozoic reefs.

The shell form of a number of fossil groups suggests that they were adapted for living free on the surface of soft bottoms. Spines, long shell "wings," and flattened ventral surfaces appear to have been devices to prevent sinking (Fig. 18–30*F* and *G*). Among living forms the New Zealand species *Neothyris lenticularis* lives free on gravel or coarse sand bottoms, and *Terebratella sanguinea* lives either attached to rock or free on sand and mud (Richardson, 1981) (Fig. 18–31).

Lophophore and Feeding

As in other lophophorates, the brachiopod lophophore is basically a crown of tentacles surrounding the mouth, but here, resulting in increased surface area, the lophophore projects anteriorly as two arms, or brachia, from which the name *brachiopod* is derived. In its simplest form the lophophore is horseshoe shaped, each arm, or brachium, projecting anteriorly. The arms may be further looped or spiraled in complicated ways, greatly increasing the collecting surface area of the lophophore (Fig. 18–32). Each brachium bears a row of tentacles, and the tentacle-bearing ridge is flanked by a brachial groove at its base. The lophophore is supported by a cartilaginous axis and a fluid-filled canal within each brachium. Also, in many brachiopods the dorsal valve bears complicated supporting processes, and the processes and inner valve surface may be grooved and ridged for the reception of the lophophore (Fig. 18–29*E*).

When the brachiopod is feeding, water enters and leaves the valve gape through distinct inhalant and exhalant apertures and chambers created by the lophophore (Fig. 18–32). In its circuit over the lophophore, water passes between the tentacles, driven by lateral tentacular cilia. Particles, especially fine phytoplankton, are screened by the lateral cilia of two adjacent filaments and then transported down the tentacles to the brachial groove in the same manner as in bryozoans (see Fig. 18–9). The brachial groove conducts food to the mouth. Rejected particles are carried away in the median, outward-flowing current.

The mouth leads into an esophagus that extends dorsally and a short distance forward prior to turning posteriorly and joining a dilated stomach (Fig. 18–33). The stomach is surrounded by a digestive gland that opens through the stomach wall by means of one to three ducts on each side. At least in *Lingula* digestion is largely intracellular within the digestive gland. In an inarticulate an intestine from the stomach opens to the outside through a rectum and posterior anus located between the valves. In an articulate the intestine is blind.

Internal Anatomy and Physiology

The coelom of a brachiopod extends into the mantle, where it is partitioned into channels. The coelomic fluid contains coelomocytes of several sorts, one of which contains hemerythrin. There are no specialized gas exchange organs. The lophophore and mantle lobes are probably the principal sites of gas exchange. Oxygen transport is probably provided by the coelomic fluid, for there is a definite circulation of coelomic fluid through the mantle channels, and oxygen is carried, at least in part, by hemerythrin in the coelomocytes.

Brachiopods possess an open circulatory system separate from the coelomic channels. There is a contractile vesicle (heart) located over the stomach in the dorsal mesentery, and from the heart extend an anterior and a posterior channel supplying various parts of the body. The blood is colorless, and the relatively few formed elements are all coelomocytes. The exact role of the blood-vascular system in the physiology of brachiopods is not definitely known. The circulation of food materials is perhaps the primary function of the blood-vascular system.

One or two pairs of metanephridia are present in brachiopods. The nephrostomes open into the metacoel on each side of the posterior end of the stomach, and the tubules then extend anteriorly to open into the mantle cavity through a nephridio-

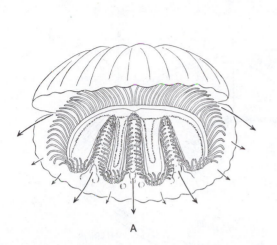

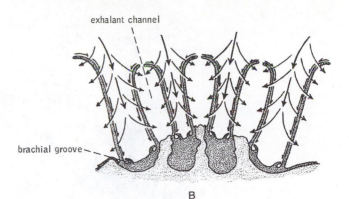

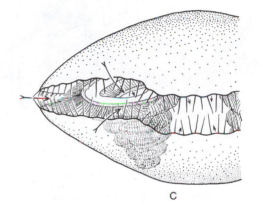

Figure 18–32 *A,* View into anterior gape of *Megathyris,* showing lophophore and exhalant water currents. *B,* Diagrammatic cross section through lophophore of *Megathyris,* showing water currents (large arrows) passing between filaments into exhalant channels, and the direction of ciliary tracts (small arrows) carrying food particles to the brachial grooves. (*B* and *C* after Atkins.) *C,* Front view of gape of *Notosaria,* showing setal grille and spiral lophophore. (Modified from Rudwick.)

pore situated posteriorly and to each side of the mouth (Fig. 18–33).

An esophageal nerve ring with a small ganglion on its dorsal side and a larger ganglion on the ventral side forms the nerve center of brachiopods. From the ganglia and their connectives, nerves extend anteriorly and posteriorly to innervate the lophophore, the mantle lobes, and the valve muscles.

As in bivalve mollusks, the mantle margin of brachiopods is probably the most important site of sensory reception. The mantle setae, although not directly associated with sensory neurons, probably do transmit tactile stimuli to receptors in adjacent mantle epidermis. In some brachiopods the setae are long and form a protective "sensory grille" over the gaping valves (Fig. 18–32*C*).

Reproduction and Embryogeny

With a few exceptions, brachiopods are dioecious, and the gonads, usually four in number, are masses of developing gametes in the peritoneum of the mantle coelom (articulates) or visceral mesenteries (inarticulates) (Fig. 18–33). When ripe, the gametes pass into the coelom and are discharged to the exterior by way of the nephridia.

Except for a few brooding species, the eggs are shed into the sea water and fertilized at the time of spawning. Cleavage is radial and nearly equal and leads to a coeloblastula that usually undergoes gastrulation by invagination. In contrast to the typical method of mesoderm formation in protostomes, the mesoderm in brachiopods appears to be enterocoelic, as in deuterostomes—that is, it arises from the archenteron.

The embryo eventually develops into a free-swimming and feeding larva. The inarticulate larva resembles a minute brachiopod (Fig. 18–34*A*). The pair of mantle lobes and the larval valves enclose the body and the ciliated lophophore, which acts as the larval locomotor organ. The pedicle, which in this group is derived from the mantle, is coiled in the back of the mantle cavity. As additional shell is laid down, the larva becomes heavier and sinks to the bottom. There is no metamorphosis in *Lingula;* the pedicle attaches to the substratum, and the young brachiopod takes up an adult existence. The articulate larva differs in having a ciliated anterior lobe representing the body and lophophore, a posterior lobe that forms the pedicle, and a mantle lobe that is directed backward (Fig. 18–34*B*). In *Terebratulina* the larva settles after a short free-

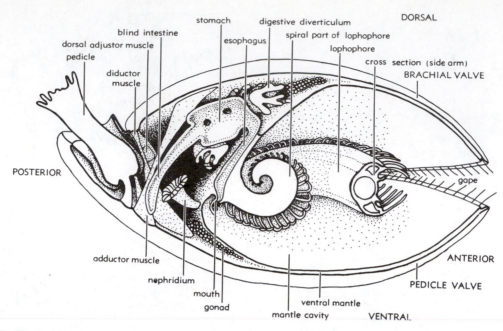

Figure 18–33   Section through an articulate brachiopod. (From Williams, A., and Rowell, A. J., 1965: *In* Moore, R. C. (Ed.): Treatise on Invertebrate Paleontology. Geological Society of America and University of Kansas, Lawrence, Kans.)

swimming existence of approximately 24 to 30 hours, and then it undergoes metamorphosis. The mantle lobe reverses position and begins the secretion of the valves, and the adult structures develop from their larval precursors.

The phoronids are often suggested as the ancestors of the brachiopods. The embryology of these two lophophorate phyla is somewhat similar, and both have monociliated tentacular cells with an upstream ciliary collecting system. Nielsen (1985) believes they are deuterostomes and unrelated to the bryozoans, which have multiciliated tentacular cells. The brachiopod and bryozoan lophophores would thus be convergent structures.

Summary

1   Lophophorates are a group of four phyla that possess an anterior, ciliated, tentacular structure, called a lophophore. Each tentacle contains an extension of the coelom. The lophophore is used in suspension feeding.

2   The phylum Bryozoa contains minute sessile animals that form colonies of a centimeter or more in length or diameter. The majority of the more than 4000 species are marine, but there are some freshwater species.

3   Many of the peculiarities of the bryozoans are related to miniaturization, colonial organization, and sessility.

4   The body of most species (marine forms) is covered by a chitinous cuticle or a cuticle overlying calcium carbonate. The outer covering accounts for the great fossil record of bryozoans.

5   The lophophore can be retracted within the body encasement. It is protruded by elevation of coelomic fluid pressure, usually produced by compression of some area of the body. The skeletal covering, as well as the colonial organization, has restricted the area of possible compression to the frontal surface of most bryozoans.

6   Particles collected by the lophophore are passed into the large stomach, which makes up the greater part of the U-shaped gut. Digestion is both extracellular and intracellular.

7   Bryozoans' lack of special systems for gas exchange, internal transport, and excretion is correlated with their minute size.

8   Some bryozoan colonies are stoloniferous, but most species have erect or encrusting colonies of contiguous individuals.

9   Polymorphism is common, and there is physiological exchange between zooids of a colony via wall pores and a mesenchymal cord, called the funiculus.

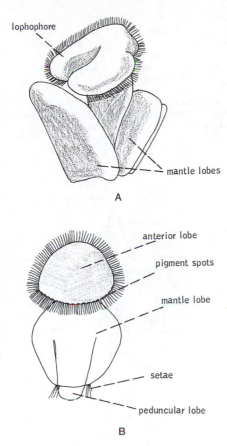

Figure 18–34 *A*, Larva of the inarticulate *Lingula.* (After Yatsu from Hyman.) *B*, Larva of the articulate *Terebratella inconspicua.* (After Percival from Hyman.)

10 Bryozoans are hermaphroditic but without special gonoducts. Gametes break into the coelom and exit by way of tentacular pores (sperm) or by an elevated intertentacular organ (eggs).

11 Some bryozoans are oviparous, but most brood their eggs. A larva is typically present.

12 The phylum Entoprocta contains approximately 150 species of minute, tentaculate, sessile animals that were formerly included in the Bryozoa. However, in contrast to bryozoans, the anus of entoprocts is located within the circle of tentacles, the body cavity may not be a coelom, and there is no rigid body covering. Entoprocts are almost entirely marine and may be solitary or colonial.

13 The phylum Phoronida contains ten species of elongate marine lophophorates that live in chitinous tubes fixed in sand or attached to rocks. When the animal feeds, the lophophore is projected from the end of the tube.

14 The phylum Brachiopoda contains about 325 marine lophophorates with two calcareous valves, or shells, enclosing the body. The calcareous valves account for the long fossil history of brachiopods. In contrast to other lophophorates, most brachiopods are relatively large (1 to 6 cm).

15 Although brachiopods look like bivalve mollusks, the shell valves are oriented dorsoventrally, and in most species the ventral valve is larger than the dorsal one.

16 Like other lophophorates, brachiopods are sessile, and the body is usually anchored to the substratum by a flexible pedicle, which emerges through the articulating end of the ventral valve.

17 The lophophore is horseshoe shaped, with the arms often looped or spiraled.

18 Brachiopods possess an open blood-vascular system and one or two pairs of metanephridia. The sexes are separate. Gametes mature in the coelom and exit through the nephridia.

19 Fertilization is external, cleavage is radial, and there is a free-swimming larva.

REFERENCES

The literature included here is restricted to books and papers on lophophorates alone. The introductory references on page 8 lists many general works and field guides that contain sections on these animals.

Banta, W. C., McKinney, F. K., and Zimmer, R. L., 1974: Bryozoan monticules: excurrent water outlets? Science, *185*:783–784.

Bobin, G., 1977: Interzooecial communications and the funicular system. See Woollacott and Zimmer.

Brown, C. J. D., 1933: A limnological study of certain freshwater Polyzoa with special reference to their statoblasts. Trans. Am. Microsc. Soc., 52:271–316.

Chuang, S. H., 1959: Structure and function of the alimentary canal in *Lingula unguis*. Proc. Zool. Soc. London, *132*:293–311.

Cook, P. L., 1977: Colony water currents in living Bryozoa. Cah. Biol. Mar., *18*:31–47.

Cooper, G. A., 1977: Brachiopods from the Caribbean Sea and adjacent waters. Studies in Tropical Oceanography No. 14. University of Miami Press, Coral Gables. 211 pp.

Emig, C. C., 1974: The systematics and evolution of the phylum Phoronida. Z. Zool. Syst. Evolutionsforsch., *12*:(2):128–151.

Emig, C. C., 1977: Embryology of Phoronida. Am. Zool., *17*:21–37.

Emig, C. C., 1979: British and other Phoronids. Synopses of the British Fauna No. 13. Academic Press, London. 58 pp.

Emig, C. C., 1981: Observations on the ecology of *Lingula reevei.* Jour. Exp. Mar. Biol. Ecol., *52*(1):47–62.

Emig, C. C., 1982: The biology of Phoronida. Adv. Mar. Biol., *19*:1–89.

Farmer, J. D., Valentine, J. W., and Cowen, R., 1973: Adaptive strategies leading to the ectoproct groundplan. Syst. Zool., *22*(3):233–239.

Gordon, D. P. 1975: Ultrastructure and function of the gut of a marine bryozoan. Cah. Biol. Mar., *16*:367–382.

Gould, S. J., and Calloway, C. B., 1980: Clams and brachiopods—ships that pass in the night. Paleobiology, *6*(4):383–396.

Gutmann, W. F., Vogel, K., and Zorn, H., 1978: Brachiopods: Biochemical interdependences governing their origin and phylogeny. Science, *199*:890–893.

Hayward, P. J., and Ryland, J. S., 1979: British Ascophoran Bryozoans. Synopsis of the British Fauna No. 10. Academic Press, London. 312 pp.

Hyman, L. H., 1951: The Invertebrates. Vol. 3. Acanthocephala, Aschelminthes and Entoprocta. McGraw-Hill, N.Y.

Hyman, L. H., 1959: The Invertebrates. Vol. 5. Smaller Coelomate Groups. McGraw-Hill, N.Y. pp. 228–609.

Larwood, G. P. (Ed.), 1973: Living and Fossil Bryozoa: Recent Advances in Research. Academic Press, London. 652 pp.

Larwood, G. P., and Abbott, M. B. (Eds.), 1979: Advances in Bryozoology. Syst. Assoc. Spec. Vol. 13. Academic Press, London. 638 pp.

Larwood, G. P., and Nielsen, C. (Eds.), 1981: Recent and Fossil Bryozoa. Olsen and Olsen, Fredensborg, Denmark. 334 pp. (Papers from a symposium.)

Moore, R. C. (Ed.), 1953: Treatise on Invertebrate Paleontology. Bryozoa, Pt. G. 253 pp. 1965: Brachiopoda, Pt. H. Vols. 1 and 2. 926 pp. Geological Society of America and University of Kansas Press, Lawrence, Kan.

Mundy, S. P., 1980: Stereoscan studies of phylactolaemate bryozoan statoblasts including a key to the statoblasts of the British and European Phylactolaemata. Jour. Zool., *192*:511–530.

Nielson, C., 1964: Studies on Danish Entoprocta. Ophelia, *1*(1):1–76.

Nielson, C., 1971: Entoproct life cycles and the entoproct/ectoproct relationship. Ophelia, *9*(2):209–341.

Nielson, C., 1977: The relationship of Entoprocta, Ectoprocta, and Phoronida. Am. Zool., *17*:149–150.

Nielson, C., 1985: Animal phylogeny in the light of the trochaea theory. Biol. Jour. Linn. Soc., *25*:243–299.

Nielsen, C., and Pedersen, K. J., 1979: Cystid structure and protrusion of the polypide in *Crisia.* Acta Zool., *60*:65–88.

Nielsen, C., and Rostgaard, J., 1976: Structure and function of an entoproct tentacle with a discussion of ciliary feeding types. Ophelia, *15*:115–140.

Richardson, J. R., 1981: Brachiopods in mud: resolution of a dilemma. Science, *211*:1161–1162.

Rider, J., and Cowen, R., 1977: Adaptive architectural trends in encrusting ectoprocts. Lethaia, *10*:29–41.

Rudwick. M. J. S., 1970: Living and Fossil Brachiopods. Hutchinson University Library, Hutchinson and Co., London. 199 pp. (An excellent general account of living and fossil brachiopods.)

Ryland, J. S., 1970: Bryozoans. Hutchinson University Library, Hutchinson and Co., London. (An excellent general account of living and fossil bryozoans.)

Ryland, J. S., 1976: Physiology and ecology and marine bryozoans. Adv. Mar. Biol., *14*:285–443.

Ryland, J. S., and Hayward, P. J., 1977: British Anascan Bryozoans. Synopses of the British Fauna No. 10. Academic Press, London. 188 pp.

Silen, L., 1972: Fertilization in the Bryozoa. Ophelia, *10*(1):27–34.

Steele-Petrovic, H. M., 1976: Brachiopod food and feeding processes. Paleontology (Lond.)*19*(13):417–436.

Strathmann, R., 1973: Function of lateral cilia in suspension feeding of lophophorates. Mar. Biol., *23*(2):129–136.

Suchanek, T. H., and Levinton, J., 1974: Articulate brachiopod food. J. Paleontol., *48*(1):1–5.

Thorpe, J. P., Shelton, G. A., and Laverack, M. S., 1975: Colonial nervous control of lophophore retraction in cheilostome Bryozoa. Science, *189*:60–61.

Vandermeulen, J. H., 1970: Functional morphology of the digestive tract epithelium in *Phoronis vancouverensis:* an ultrastructural and histochemical study. Jour. Morphol., *130*(3):271–286.

Winston, J. E., 1978: Polypide morphology and feeding behavior in marine ectoprocts. Bull. Mar. Sci., *28*(1):1–31.

Woollacott, R. M., and Zimmer, R. L. (Eds.), 1977: Biology of Bryozoans. Academic Press, N.Y. 566 pp. (A collection of reviews of various aspects of bryozoan biology.)

19

The Echinoderms

Members of the phylum Echinodermata are among the most familiar marine invertebrates, and such forms as the sea stars have become virtually a symbol of sea life. The phylum contains some 6000 known species and constitutes the only major group of deuterostome invertebrates.

Echinoderms are exclusively marine and are largely bottom dwellers. All are relatively large animals, most being at least several centimeters in diameter. The most striking characteristic of the group is their pentamerous radial symmetry—that is, the body can usually be divided into five parts arranged around a central axis. This radial symmetry, however, has been secondarily derived from a bilateral ancestral form, and the echinoderms are not closely related to the other radiate phyla.

Characteristic of all echinoderms is the presence of an internal skeleton. The skeleton is composed of calcareous ossicles that may articulate with one another, as in sea stars and brittle stars, or may be sutured together to form a rigid skeletal test, as in sea urchins and sand dollars. Commonly, the skeleton bears projecting spines or tubercles that give the body surface a warty or spiny appearance, hence the name *echinoderm*—"spiny skin."

The most distinctive feature of echinoderms is the presence of a unique system of coelomic canals and surface appendages composing the water-vascular system. Primitively, the water-vascular system probably functioned in collecting and transporting food, but in many echinoderms it has assumed a locomotor function.

Echinoderms possess a spacious coelom in which is suspended a well-developed digestive tract. There are no excretory organs. Gas exchange structures vary from one group to another and appear to have arisen independently within the different classes. Most members of the phylum are dioecious. The reproductive tracts are very simple, for there is no copulation and fertilization is usually external in sea water.

Echinoderm Development

A brief introduction to echinoderm development will aid in understanding the secondary nature of their symmetry as well as some of the structural features that will be described in the following section. The eggs are typically homolecithal, and the early embryogeny is relatively uniform throughout

773

the phylum and displays radial and indeterminate cleavage, the basic features of deuterostome development (p. 62).

The blastula contains a large blastocoel, and gastrulation takes place primarily by invagination, forming a narrow, tubular archenteron (primitive gut). The archenteron grows forward and eventually connects with the anterior stomodeum, which will form the mouth. The blastopore remains as the larval anus.

Before the mouth forms, the advancing distal end of the archenteron, primitively at least, gives rise to two lateral pockets or pouches that eventually separate from the archenteron (Fig. 19–41A and 3–8). The cavities of the pouches represent the future coelomic cavity, and the cells composing the pouch wall become the mesoderm. The two original pouches, one on each side, each give rise by evagination or subdivision to coelomic vesicles arranged one behind the other and called respectively the axocoel, the hydrocoel, and the somatocoel (Fig. 19–1A). These coelomic vesicles correspond to the paired protocoel, mesocoel, and metacoel of other deuterostome coelomates (hemichordates)

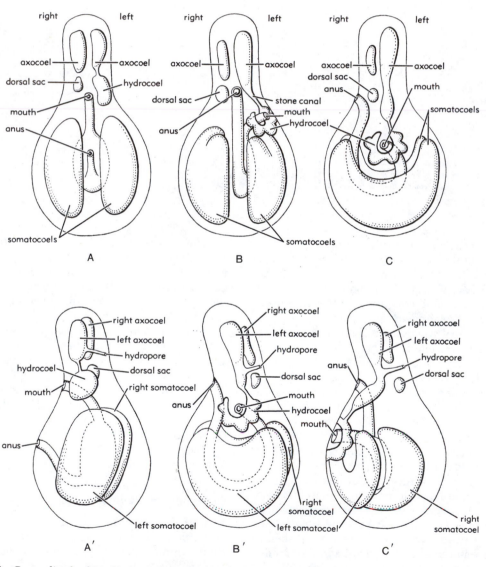

Figure 19–1 Generalized echinoderm metamorphosis. Top row, ventral view; bottom row, side view. (From Ubaghs, G., 1967: *In* Moore, R. C. (Ed.): Treatise on Invertebrate Paleontology. Pt. S, Vol. 1. Courtesy of the Geological Society of America and the University of Kansas, Lawrence, Kan.)

with a tripartite (oligomerous) body. The two somatocoels meet above and below the gut to form the gut mesenteries. The left axocoel opens dorsally through a pore called the hydropore. This somewhat generalized plan of the coelomic vesicles is considerably modified in the development of existing echinoderms.

The gastrula rapidly develops into a free-swimming larva (Fig. 19–2). The most striking feature of the echinoderm larva is its bilateral symmetry, which is in marked contrast to the radial symmetry of the adult. Wound over the surface of the larva are a varying number of ciliated locomotor bands.

There is a complete, functional digestive tract, and food particles are obtained by the ciliated bands and stomodeal cilia. Later larval development in many echinoderms involves the formation of short or long slender projections (arms) from the body wall, and the nature and position of these arms, or the lack of them, distinguishes the larvae of the different echinoderm classes. The arms disappear in later development and are not equivalent to the arms of certain adult echinoderms, such as sea stars and brittle stars.

After a free-swimming planktonic existence, the bilateral larva undergoes a metamorphosis in

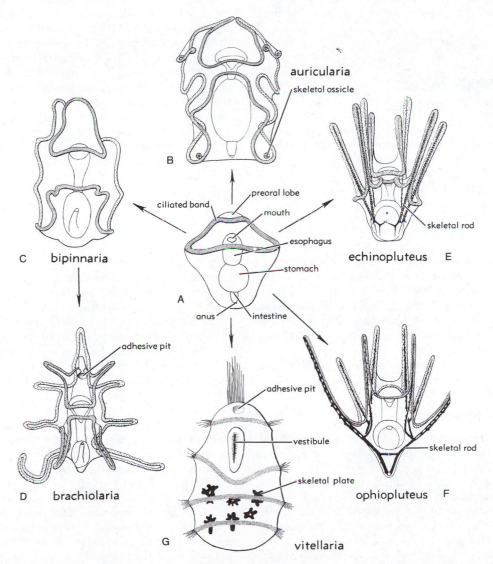

Figure 19–2 Comparison of different types of echinoderm larvae showing relationship to a hypothetical dipleurula-type ancestral larva. (From Ubaghs, G., 1967: *In* Moore, R. C. (Ed.): Treatise on Invertebrate Paleontology. Pt. S, Vol. 1. Courtesy of the Geological Society of America and the University of Kansas, Lawrence, Kan. pp. S3–S60.)

which the larval arms and, in some echinoderms, the mouth and parts of the gut degenerate. The left side of the body becomes the oral surface and the right side the aboral surface. Only the left axohydrocoel forms the water-vascular system (Fig. 19–1*C*). Such a planktotrophic larval development is characteristic of only 40 to 50 per cent of the phylum. The remainder have lecithotrophic larvae or direct development.

Although the crinoids (sea lilies) are the most primitive class of living echinoderms, the more familiar asteroids (sea stars) are treated first in the discussion of the echinoderm classes and serve to introduce the basic features of echinoderm structure.

Class Stelleroidea, Subclass Asteroidea

The class Stelleroidea contains those star-shaped, free-moving echinoderms in which the body is composed of rays, or arms, projecting from a central disc. All living species are members of one of two subclasses, the Asteroidea (containing the sea stars, or starfish) and the Ophiuroidea (containing the brittle stars).

The 1500 described species of sea stars are common and familiar animals that crawl about over rocks and shells or live on sandy or muddy bottoms. They are found throughout the world, largely in coastal waters, but the northeast Pacific, particularly from Puget Sound to the Aleutians, possesses the greatest concentration of asteroid species. Seventy species are endemic to the Vancouver Island area alone.

Sea stars are commonly red, orange, blue, purple, or green or exhibit combinations of colors.

External Structure

Sea stars are typically pentamerous, with most species possessing five arms. However, the sun stars possess 7 to 40 or more arms (Fig. 19–3*B*). Most asteroids range from 12 to 24 cm in diameter, but there are some that are less than 2 cm in diameter, and the many-rayed star *Pycnopodia* of the northwest coast of the United States may measure almost 1 meter across.

The arms of asteroids are not sharply set off from the central disc—i.e., the arm usually grades into the disc, and some species have very short arms. In the cushion stars *Plinthaster* and *Goniaster* each arm has the shape of an isosceles triangle, and in *Culcita* (Fig. 19–3*C*) the arms are so short that the body appears pentagonal.

The mouth is located in the center of the underside of the disc, and the entire undersurface of the disc and arms is called the oral surface. The opposite, or upper, side of the body is the aboral surface. From the mouth a wide furrow extends radially into each arm (Fig. 19–9*B*). Each furrow (ambulacral groove) contains two or four rows of small, tubular projections, called tube feet or podia.

The margins of the ambulacral grooves are guarded by movable spines that are capable of closing over the groove. The tip of each arm bears one or more small, tentacle-like sensory tube feet and a red pigment spot. The aboral surface bears both the inconspicuous anus, when present, in the center of the disc, and a large, button-like structure (the madreporite) toward one side of the disc between two of the arms. The general body surface may appear smooth or be covered with spines, tubercles, or ridges. In some species the arms and disc are bordered by large, conspicuous plates (Fig. 19–3*A*).

Body Wall

The outer surface of the body is covered by an epidermis composed of monociliated and nonciliated epithelial cells, mucous cells, and ciliated sensory cells (Fig. 19–4). Detritus that falls on the body is trapped in the mucus and then swept away by the epidermal cilia. At the base of the epidermis is a layer of nerve cells forming a subepidermal plexus.

Below the integument a thick layer of body wall connective tissues (the dermis) houses the skeletal system. The asteroid skeleton is composed of separate ossicles in the shape of rods, crosses, or plates. The ossicles are so arranged that they form a lattice network and are bound together by connective tissue (Fig. 19–5*A* and *B*).

The calcareous ossicles that compose the echinoderm skeleton display many interesting features. As in vertebrate bone, there is a considerable turnover of the mineral constituents. The ossicles are irregularly perforated (fenestrated), perhaps representing an adaptation for reduction of weight and increase in strength. The most remarkable characteristic is that each ossicle of the echinoderm skeleton represents a single crystal of magnesium-rich calcite, $6(Ca,Mg)CO_3$. The crystal is formed within a cell of the dermis, but as the crystal increases in size, it becomes surrounded by a large

Figure 19–3   *A, Astropecten irregularis,* a burrowing sea star (aboral view). *B,* The sun star *Crossaster papposus* (aboral view). (Both by D. P. Wilson.) *C,* Oral view of *Culcita,* a sea star with very short arms. (Courtesy of the National Museum of Natural History.) *D,* The Pacific coral-eating *Acanthaster planci.* (By Betty M. Barnes.)

number of cells, all of which are daughter cells of the original cell that initiated the formation of the crystal.

Spines and tubercles are also part of the skeleton, and each either consists of separate pieces resting on the deeper dermal ossicles or represents an extension of the dermal ossicles that projects to the outer surface. In the paxillosid and valvatid sea stars the aboral surface bears special ossicles, and the central portion of each is raised above the body

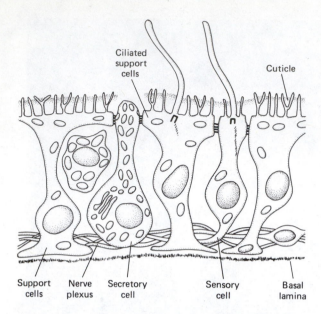

Ciliated support cells

Cuticle

Support cells

Nerve plexus

Secretory cell

Sensory cell

Basal lamina

Figure 19–4 Diagram of the echinoderm integument. (From Holland, N. D., 1984: Epidermal cells (Echinodermata). *In* Bereiter-Hahn, J., Matoltsy, A. G., and Richards, K. S. (Eds.): Biology of the Integument. 1. Invertebrates. Springer-Verlag, Berlin. p. 757.)

surface and even extended out like a parasol. The raised part of the ossicle is crowned with small, movable spines. Such an ossicle and its associated spines are called a paxilla; it is an adaptation for the burrowing existence of many sea stars belonging to this group. Adjacent paxillae create a protective space above the aboral integument. Through this space flow respiratory and feeding currents. The surrounding sediment is held back by the paxillae (Fig. 19–5C and D). Paxillae account for the smooth appearance of the aboral surface of many sea stars.

Beneath the dermis is a muscle layer composed of outer circular and inner longitudinal smooth fibers, which are involved in the bending of the arms. The inner surface of the muscle layer borders the coelom and is covered with peritoneum.

In two orders of sea stars the body surface bears small, specialized jawlike appendages (pedicellariae) that are used for protection, especially against small animals or larvae that might settle on the body surface of the sea star. The pedicellariae are of two types—stalked and sessile. The stalked pedicellariae are characteristic of the order Forcipulata, which includes *Asterias, Pycnopodia,* and *Pisaster.* Each pedicellaria consists of a short, fleshy stalk surmounted by a jawlike apparatus composed

of three small, movable ossicles that are arranged to form forceps or scissors (Fig. 19–6B). Stalked pedicellariae may be scattered over the body surface or situated on the spines, or commonly form a wreath around the base of the spines (Figs. 19–7 and 19–10).

Sessile pedicellariae are largely limited to the order Valvatida and are composed of two or more short, movable spines on the same or adjacent ossicles. The spines, which in some species are shaped like the valves of a clam, oppose each other and articulate against one another to act as pincers (Fig. 19–6A).

The papulae, which are numerous, small evaginations of the body wall scattered over the body surface, and the podia are discussed in connection with gas exchange and the water-vascular system.

Water-Vascular System

The water-vascular system, unique to echinoderms, consists of canals and appendages of the body wall. Since the entire system is derived from the coelom, the canals are lined with a ciliated epithelium and filled with fluid. The water-vascular system is well developed in asteroids and functions as a means of locomotion.

The internal canals of the water-vascular system connect to the outside through the button-shaped madreporite on the aboral surface (Fig. 19–8A). The surface of the madreporite is creased with many furrows covered by the ciliated epithelium of the body surface. The bottom of each furrow contains many pores that open into pore canals passing downward through the madreporite. The pore canals eventually lead into a vertical stone canal that descends to the oral side of the disc (Fig. 19–8A). The stone canal is so named because of the calcareous deposits in its walls. On reaching the oral side of the disc, the stone canal joins a circular canal (the water ring) just to the inner side of the ossicles that ring the mouth.

The inner side of the water ring gives rise to four or, more usually, five pairs of greatly folded pouches called Tiedemann's bodies (Fig. 19–8A). Each pair of these pouches has an interradial position. Also attached interradially to the inner side of the water ring in many asteroids, although not in *Asterias,* are one to five elongated, muscular sacs, which are suspended in the coelom. These sacs are known as polian vesicles.

From the water ring, a long, ciliated, radial canal extends into each arm (Fig. 19–8A). The radial canal runs on the oral side of the ossicles that

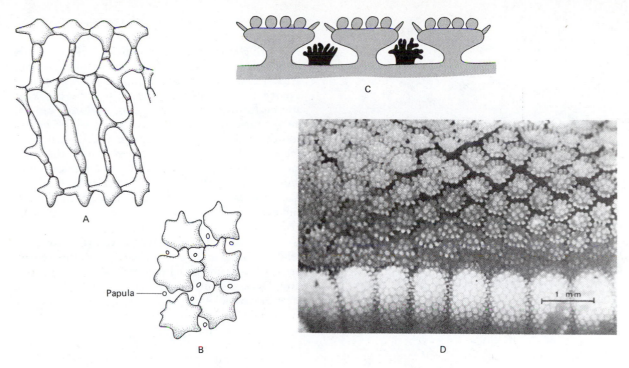

Figure 19–5 *A,* Lattice-like arrangement of skeletal ossicles in the arm of an asteriid. (After Fisher from Hyman.) *B,* Small section of endoskeletal system of a paxillosid sea star. (After Hyman, L. H., 1955: The Invertebrates. Vol. IV. McGraw-Hill Book Co., N.Y.) *C,* Diagrammatic cross section through several paxillae of *Luidia.* The raised, table-shaped ossicles bear small rounded spines on the surface and flat movable spines along the edge. Dendritic papulae (black) are located in the spaces between the projecting edges of the paxillae and associated spines. *D,* Surface of paxillae of *Astropecten* (compare with Fig. 19–3A). (By Betty M. Barnes.)

form the center of the ambulacral groove (ambulacral ossicles) (Fig. 19–19), and it ends in a small, external tentacle at the tip of the arm. Lateral canals arise alternately from each side of the radial canal along its entire length. These lateral canals pass between the ambulacral ossicles on each side of the groove and enter the coelom (Fig. 19–8B).

Each lateral canal is provided with a valve and terminates in a bulb and a tube foot (Fig. 19–8). The bulb, or ampulla, is a small, muscular sac that bulges into the aboral side of the coelom. The ampulla opens directly into a canal that passes downward between the ambulacral ossicles and leads into the tube foot, or podium.

The podium is a short, tubular, external projection of the body wall located in the ambulacral groove. Commonly, the tip of the podium is flattened, forming a sucker. Like the body wall, the podium is covered on the outside with a ciliated epithelium and internally with peritoneum. Between these two layers lie connective tissue and longitudinal muscle fibers. Contraction of the muscles on

one side of the podium brings about bending of the appendage.

The podia are arranged in two or four rows along the length of the ambulacral groove, and the ampullae occupy a corresponding position on the coelomic side of the ambulacral ossicles. The difference in the number of rows results from the length of the lateral canals. For example, in asteroids such as *Asterias* that possess four rows of podia, the lateral canals along each side are alternately long and short, giving the appearance of four rows of podia instead of two.

The entire water-vascular system is filled with fluid that is similar to sea water except that it contains coelomocytes, a little protein, and a high potassium ion content. The system operates during locomotion as a hydraulic system. When the ampulla contracts, the valve in the lateral canal closes, and water is forced into the podium, which elongates. When the podium comes in contact with the substratum, the sucker adheres. Adhesion is largely chemical, the podium secreting a substance that

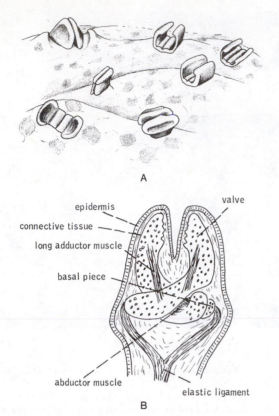

A

epidermis

connective tissue

long adductor muscle

basal piece

valve

abductor muscle

elastic ligament

B

Figure 19–6 *A*, Sessile valvate pedicellariae on the surface of *Anthenea* (drawn from a dried specimen). *B*, Distal ends of a scissors-type pedicellaria from *Asterias*. (After Hyman, L. H., 1955: The Invertebrates. Vol. IV. McGraw-Hill Book Co., N.Y.)

bonds with surface films. Another secretion breaks the bonds and brings about release (Thomas and Hermans, 1985). Note the similarity to the duogland systems of flatworms and gastrotrichs.

After the sucker adheres to the substratum, the longitudinal muscles of the podium contract, shortening the podium and forcing fluid back into the ampulla. It has been generally thought that other parts of the water-vascular system—the madreporite, the stone canal, the water ring, the muscular polian vesicles, and the radial canal—perhaps function in maintaining the proper water pressure necessary for the operation of the ampullae and podia, for there is some leakage across the podial wall during fluid pressure elevation. Using radioactive tracers, Furguson (1984) demonstrated that sea water does enter the madreporite, and some appears to pass through the Tiedemann's bodies into the perivisceral coelom. The Tiede-

mann's bodies may thus function in the production of coelomic fluid.

During movement each podium performs a sort of stepping motion. The podium swings forward, grips the substratum, and then moves backward. In a particular section of an arm most of the tube feet are performing the same step, and the animal moves forward. The action of the podia is highly coordinated. During progression one or two arms act as leading arms, and the podia in all the arms move in the same direction, but not necessarily in unison (Fig. 19–9). The combined action of the podial suckers exerts a powerful force for adhesion and enables sea stars to climb vertically over rocks or up the side of an aquarium.

If a sea star is turned over, it can right itself by folding. The distal end of one or two arms twists, bringing the tube feet in contact with the substratum. Once the substratum has been gripped, these arms move back beneath the animal so that the rest of the body is folded over. The sea star may also right itself by arching the body and rising on the tips of the arms. It then rolls over onto its oral surface. In general, sea stars move rather slowly and tend to remain within a more or less restricted area.

Podia of sea stars that live on soft bottoms, such as *Astropecten* and *Luidia*, lack suckers. Rather, the tip is pointed to facilitate thrusting of the podium into the sand (Fig. 19–8*C*). Associated with this adaptation of the podia is the presence of doubled (bilobed) ampullae, which provide increased force for driving the podia into the substratum. In addition to enabling a sea star to move over a soft bottom, podia of this type may also be used to burrow and even to plaster the walls of the burrow with mucus.

Nutrition

The digestive system is radial, extending between the oral and aboral sides of the disc (Fig. 19–10). The mouth is located in the middle of a tough, circular, peristomial membrane that is muscular and provided with a sphincter. The mouth leads into a short esophagus, which opens in turn into a large stomach that fills most of the interior of the disc and is divided by a horizontal constriction into a large, oral chamber (the cardiac stomach) and a smaller, flattened, aboral chamber (the pyloric stomach). The walls of the cardiac stomach are pouched and connected to the ambulacral ossicles of each arm by ten pairs of triangular mesenteries called gastric ligaments.

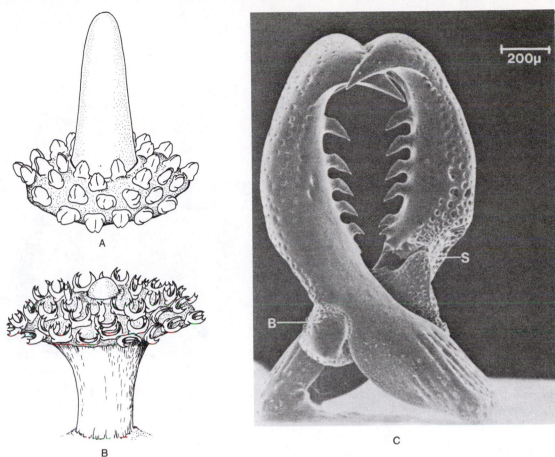

Figure 19–7 Fish-catching pedicellariae of *Stylasterias: A*, Wreath of pedicellariae at rest around spine. *B*, Wreath raised when stimulated by potential prey. *C*, SEM of ossicle. (*B*, basal piece; *S*, muscle attachment scar.) (From Chia, F. S., and Amerongen, H., 1975: On the prey-catching pedicellariae of a starfish, *Stylasterias forreri*. Can. J. Zool., 53:748–755. Reproduced by permission of the National Research Council of Canada.)

The smaller, aboral, pyloric stomach is often star shaped because of the entrance of the ducts from the pyloric ceca (Fig. 19–10). There are two pyloric ceca, or digestive glands, in each arm, each of which is composed of an elongated mass of glandular cells suspended in the coelom of the arm by a dorsal mesentery (Fig. 19–8B).

A short, tubular intestine extends from the aboral side of the pyloric stomach to open through the anus in the middle of the aboral surface of the disc. The intestine commonly bears a number of small outpocketings called rectal ceca (Fig. 19–10). The entire digestive tract is lined with a ciliated epithelium, and in the ducts of the pyloric ceca the cilia are so arranged as to create fluid currents, both incoming and outgoing. Gland cells are particularly abundant in the cardiac stomach lining.

Most asteroids are scavengers and carnivores and feed upon all sorts of invertebrates, especially snails, bivalves, crustaceans, polychaetes, other echinoderms, and even fish (see review by Sloan, 1980, and Jangoux, 1982). Some have very restricted diets; others utilize a wide range of prey but may exhibit preferences, depending on availability. For example, the Chilean *Meyenaster* feeds on 40 types of echinoderm and molluscan prey found within its habitat. Most asteroids detect and locate prey by prey substances released into the water (Sloan and Campbell, 1982), and many prey species have evolved escape responses to the slow-moving asteroids. Some soft-bottom sea stars, including species of *Luidia* and *Astropecten*, can locate buried prey and then dig down into the substratum to reach it. *Stylasterias forreri* and

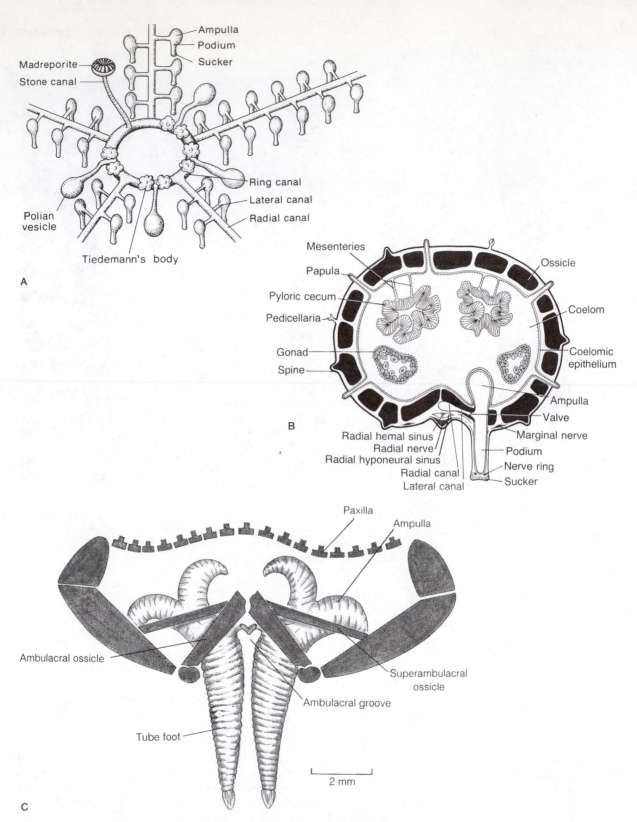

Figure 19–8 *A*, Diagram of the asteroid water-vascular system. *B*, Diagrammatic cross section through the arm of a sea star. *C*, Transverse section through arm of *Astropecten*, a soft-bottom sea star. (From Heddle, D., 1967: Echinoderm Biology. Academic Press, N.Y.)

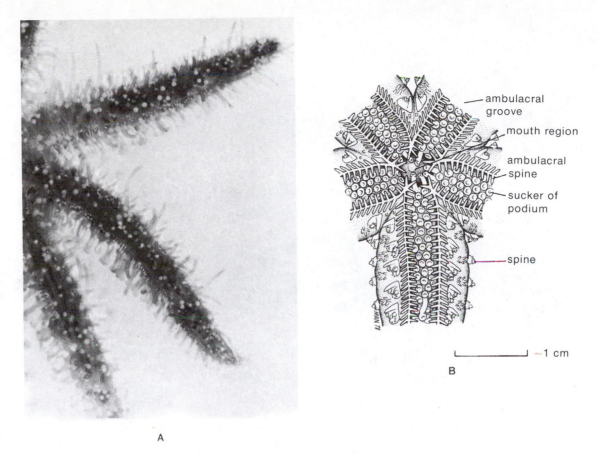

A

Figure 19–9 *A*, Oral view of three arms of *Coscinasterias*, showing tube feet. The round, white discs are the suckers. Sea star is moving upward; note that podia are swinging at right angles to the long axis of the upper arm but parallel to the long axis of the lower arm. (By Betty M. Barnes.) *B*, Oral surface of *Asterias* (disc and part of one arm).

Astrometis sertulifera along the west coast and *Leptasterias tenera* of the east coast of the United States catch small fish, amphipods, and crabs with the pedicellariae when the prey comes to rest against the aboral surface of the sea star.

There are some asteroids that feed on sponges, sea anemones, and the polyps of hydroids and corals. The tropical Pacific *Acanthaster planci* (crown-of-thorns sea star) has attained considerable notoriety as a result of its consumption of coral polyps. High population levels of this sea star, as great as 15 adults per square meter, have temporarily devastated large numbers of reef corals in some areas. Branching and plate corals are preferred over massive and encrusting types, and with its everted stomach one sea star will consume in one day an area as great as its own disc. Whether or not the high population levels of this sea star have resulted from human modification of the environment has been debated. Current evidence suggests that the

larval periods of high population levels coincided with plankton blooms. The blooms appear to have resulted from nutrient runoff following heavy rains, which in turn followed a dry period of several years (Birkeland, 1982).

Some sea stars are suspension feeders. Plankton and detritus *(Porania, Henricia)* or mud *(Ctenodiscus)* that comes in contact with the body surface is trapped in mucus and then swept toward the oral surface by the epidermal cilia. On reaching the ambulacral grooves, the food-laden mucous strands are carried by ciliary currents to the mouth. Some sea stars, such as *Astropecten* and *Luidia*, which are largely carnivorous, utilize ciliary feeding as an auxiliary method of obtaining food.

In primitive groups of sea stars, including *Astropecten* and *Luidia*, which cannot evert their stomachs and have suckerless tube feet, the prey is swallowed whole and digested within the stomach, although the stomach wall must be in contact with

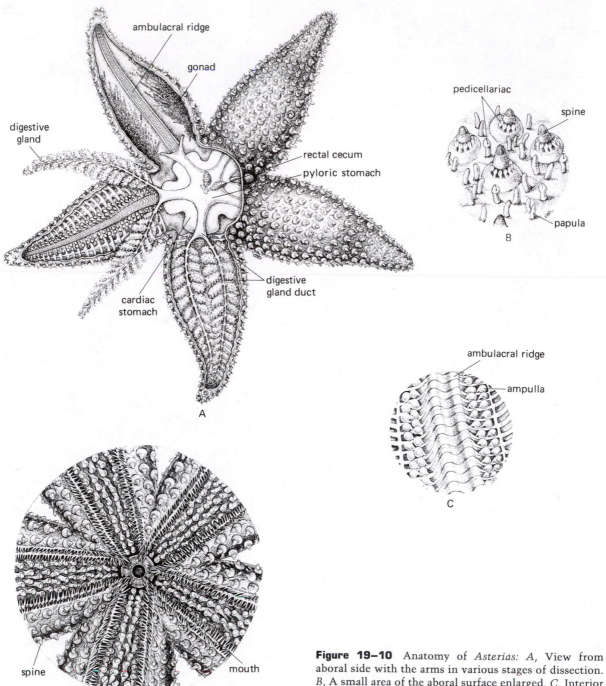

ambulacral ridge

gonad

digestive gland

rectal cecum

pyloric stomach

digestive gland duct

cardiac stomach

A

pedicellariae

spine

papula

B

ambulacral ridge

ampulla

C

spine

podia

mouth

D

Figure 19–10 Anatomy of *Asterias: A,* View from aboral side with the arms in various stages of dissection. *B,* A small area of the aboral surface enlarged. *C,* Interior view of a part of one of the ambulacral ridges, showing ampullae to either side. *D,* Oral surface of disc.

the tissues being digested. Shells and other indigestible material are then cast out of the mouth. Other asteroids (Valvatida, Spinulosida, Forcipulata) feed extraorally. Through the contraction of the body wall muscles, the coelomic fluid exerts pressure on the cardiac stomach, causing it to be everted through the mouth. The everted stomach, which is anchored by the gastric ligaments, engulfs the prey. The prey then may be brought into the stomach by retraction, or digestion may begin outside the body. The soft parts of the victim are reduced to a thick broth, which is then passed into the body in ciliated gutters. When digestion is completed, the stomach muscles contract, retracting the stomach into the interior of the disc.

Many sea stars feed almost exclusively on bivalves and are notorious predators of oyster beds. During feeding, such a sea star extends itself over a clam, holding the gape of the clam upward against its mouth and applying its arms against the sides of the clam valves. The sea star inserts the everted stomach through minute openings between the imperfectly sealed edges of the valves, or the pull exerted by the sea star produces a very slight gape in the clam. The gape is produced quite rapidly and not by causing the clam to fatigue over a long period. The everted cardiac stomach of some sea stars can squeeze through a space as slight as 0.1 mm. The gape increases as digestion ensues, and the clam's adductor muscles are attacked. However, according to Feder (1955), an increased gape is not necessary for continued digestion, at least not in the Pacific *Pisaster ochraceus*. This species can consume mussels that are bound with wire. A Japanese species of *Asterias* requires 2.5 to 8 hours to consume a bivalve, depending on the species of bivalve being attacked.

Asteroids are of considerable economic importance as predators of oysters, and they are sometimes removed from commercial oyster beds by a large, moplike apparatus dragged over the bottom. The sea stars grasp or become entangled in the mop threads with their pedicellariae and are brought to the surface and destroyed.

The everted stomach is also an effective feeding organ for many omnivores and nonpredacious sea stars. The American west coast bat star, *Patiria miniata*, spreads its stomach over the bottom, digesting all types of organic matter encountered. In the same manner the tropical cushion sea star *Culcita* and the oreasterids, which inhabit reef flats, feed on sponges, algal felt, and epibenthic film (Thomassin, 1976; Scheibling, 1980).

Digestion in asteroids appears to be primarily extracellular, and a complex of enzymes is produced by the stomach wall and pyloric ceca. Products of stomach digestion are carried by ciliary tracts up the stomach wall and through the pyloric ducts into the pyloric ceca, where digestion, both intracellular and extracellular, and absorption occur. Products of digestion may be stored in the cells of the pyloric ceca or passed through the ceca into the coelom for distribution. The pyloric ducts also convey wastes out to the rectum, where the rectal ceca, when present, act as pumps for expulsion through the anus (see Jangoux, 1982, for review).

Internal Transport, Gas Exchange, and Excretion

The large, fluid-filled coelom surrounding the internal organs within the disc and the arms provides the principal means for internal transport. The coelomic peritoneum is ciliated, and there is continuous circulation of the coelomic fluid. The body fluids of all asteroids, as well as those of other echinoderms, are isosmotic with sea water. Their inability to osmoregulate prevents most species from inhabiting estuarine waters. The coelomic fluid contains phagocytic coelomocytes that are produced by the coelomic peritoneum and can form a clot in response to tissue damage.

The hemal, or blood-vascular, system found in asteroids consists of small, fluid-filled sinus channels that lack a distinct lining. The channels are surrounded by special separate extensions of the coelom called perihemal spaces or sinuses. The principal hemal channels include oral and aboral hemal rings and a radial hemal sinus extending into each arm, as shown in Figure 19–11. From the oral hemal ring a channel ascends through a dark, elongated mass of spongy tissue (the axial gland). The hemal system appears to function in the distribution of food materials, especially to the tube feet and gonads (Ferguson, 1984).

Removal of nitrogenous wastes (NH_4^+) is accomplished by general diffusion through thin areas of the body surface, such as the tube feet and the papulae. Studies involving the injection of dyes into the coelom indicate that coelomocytes engulf waste, and when laden, some migrate to the papulae, where they collect at the distal end. The tip of the papula then constricts and pinches off, discharging the coelomocytes to the outside. Other coelomocytes may pass to the outside through the epithelium of the suckers of the podia or at other sites.

The papulae and tube feet provide the principal gas exchange surfaces for asteroids. The ciliated

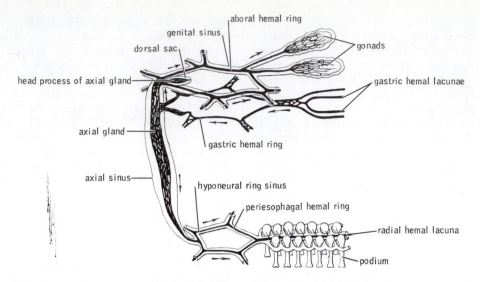

Figure 19–11 Asteroid axial complex and hyponeural and hemal systems. (From Ubaghs, G., 1967: *In* Moore, R. C. (Ed.): Treatise on Invertebrate Paleontology. Pt. S, Vol. 1. Courtesy of the Geological Society of America and the University of Kansas, Lawrence, Kan.)

peritoneum that forms the internal lining of the papulae and tube feet produces an internal current of coelomic fluid; the outer, ciliated epidermal investment produces a current of sea water flowing over the papulae (Fig. 19–12). In burrowing species the branched papulae are protected by the paxillae, and the ventilating current flows through the channel-like spaces beneath these spines (Fig. 19–5C).

Nervous System

The asteroid nervous system, like that of some other echinoderms, is not conspicuously ganglionated, and the greater part of the system is intimately associated with the epidermis. The nervous center is a somewhat pentagonal, circumoral nerve ring that lies just beneath the peristomial epider-

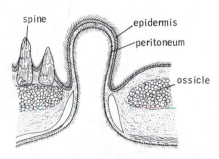

Figure 19–12 Section through an asteroid papula. (After Cuénot, L, 1948: Echinoderms. *In* Grassé, P.: Traité de Zoologie, Vol. XI. Masson et Cie, Paris.)

mis. From each angle of the ring a large, radial nerve extends into each of the arms, forming a large, subepidermal, V-shaped mass along the midline of the oral surface of the ambulacral groove (Fig. 19–8B). The radial nerve supplies fibers to the podia and ampullae and is continuous with the general subepidermal nerve plexus.

At the margins of the ambulacral groove, the subepidermal nervous layer is thickened to form a pair of marginal nerve cords that extend the length of the arm (Fig. 19–8B). The radial and superficial system thus far described is mainly sensory, and the radial cords provide through-conducting pathways. The muscles of the podia, ampullae, and body wall are supplied by a deeper, predominantly motor system. This deep system consists of fibers running within but adjacent to the superficial radial nerves and motor centers located in the vicinity of the podia and the podia-ampullae junctions.

There are many experimental studies on the role of the nervous system in the movement of sea stars. The integrity of both the radial nerves and the circumoral nerve ring is essential in the coordination of the podia. Although the podia may not all "step" in unison, they are coordinated to the extent of their involvement in stepping—that is, to step or not to step—and they are coordinated in that they step in the proper direction, depending on which arm is leading.

Each arm has a nerve center, probably at the junction of the radial nerves and nerve ring. A leading arm exerts a temporary dominance over the nerve centers of the other arms. In the majority of sea stars, including *Asterias*, any arm can act as a dominant arm, and such dominance is determined by reaction to external stimuli. In a few species one arm is permanently dominant. Of all the reactions to external stimuli, contact of the podia with the substratum appears to be dominant and probably accounts for the righting reaction.

With the exception of the eye spots at the tips of the arms, there are no specialized sense organs in the asteroids. The dispersed sensory cells contained within the epidermis are the primary sensory receptors and probably function for the reception of light, contact, and chemical stimuli. This is true of other echinoderms as well. These epidermal sensory cells are particularly prevalent on the suckers of the tube feet, on the terminal, tentacle-like sensory tube feet, and along the margins of the ambulacral groove, where 70,000 sensory cells per square millimeter have been reported.

The eye spot at the end of each arm lies beneath the tentacle on the oral side of the arm tip and is composed of a mass of 80 to 200 pigment-cup ocelli that form an optic cushion. The importance of the optic cushions in reactions to light stimuli varies in different species, but most asteroids are positively phototactic.

Regeneration and Reproduction

Asteroids exhibit considerable powers of regeneration. Any part of the arm can be regenerated, and destroyed sections of the central disc are replaced. Studies on *Asterias vulgaris* have shown that if there is at least one fifth of the central disc attached to an arm, an entire starfish will be regenerated. If the remaining section of the disc contains the madreporite, even less of the disc is required. Regeneration is typically slow and may require as long as one year for complete re-formation to take place.

A number of asteroids normally reproduce asexually (see review by Emson and Wilkie, 1980). Commonly, this involves a division of the central disc so that the animal breaks into two parts. Each half then regenerates the missing portion of the disc and arms, although extra arms are commonly produced (Fig. 19–13*A*). Species of *Linckia*, a genus of common sea stars in the Pacific and other parts of the world, are remarkable in being able to cast off their arms near the base of the disc. Unlike those of other asteroids, the severed arm regenerates a new disc and rays. Such regenerating specimens with small regenerating arms at the base of the original arm are popularly called comets (Fig. 19–13*B*).

A

Figure 19–13 *A*, Regenerating arms in a specimen of *Coscinasterias*, which reproduces asexually by division of the disc. (By Betty M. Barnes.) *B*, Comet of *Linckia*. Regeneration of body at base of detached arm. (After Richters from Hyman.)

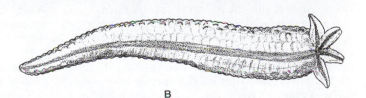

B

With few exceptions asteroids are dioecious, and there are ten gonads, two in each arm (Fig. 19–10). The gonads are double walled and tuftlike or resemble a cluster of grapes. Normally, they occupy only a small area at the base of the arm. When filled with eggs or sperm, however, the gonads almost completely fill each of the arms. There is a gonopore or gonopore cluster for each gonad, usually located between the bases of the arms. In a number of astropectinids, as well as in some other groups, each arm contains many gonads, which are arranged in rows along the length of the arm. In such species the gonopores open on the oral surface. There are a few hermaphroditic asteroid species, such as the common European sea star, *Asterina gibbosa*, which is protandric.

In the majority of asteroids the eggs and sperm are shed freely into the sea water, where fertilization takes place. There is usually only one breeding season per year, and a single female may shed as many as 2,500,000 eggs.

In most asteroids the liberated eggs and individuals in the later developmental stages are planktonic. However, some sea stars, especially Arctic and Antarctic species, brood large, yolky eggs beneath the disc in depressions on the aboral surface of the disc or in brooding baskets formed by spines between the bases of the arms. *Leptasterias groenlandica*, a circumpolar Arctic species, broods its eggs in the cardiac stomach. In all the brooding species development is direct. Although not a brooding species, *Asterina gibbosa* is unusual in that it attaches its eggs to stones and other objects.

Embryogeny

The early stages of development conform to the pattern described in the introduction of this chapter. In most species the coelom arises from the tip of the advancing archenteron as two lateral pouches. The tricoelomate condition is attained by a complex subdivision, but the axocoel and hydrocoel never completely separate (Fig. 19–14A and 19–15). The left hydrocoel connects with the dorsal surface to form the hydropore.

The asteroid embryo becomes free swimming at some point between the blastula and gastrula stages. At first the entire larval surface is covered with cilia, but as development proceeds, the surface ciliation becomes confined to a definite locomotor band (Fig. 19–14A). This locomotor band consists of two lateral, longitudinal bands that connect anteriorly in front of the mouth and posteriorly in front of the anus. The preoral loop later separates or in some cases arises separately from the rest of the locomotor band to form an anterior ciliated ring around the body. After the formation of the locomotor bands, projections, called arms, arise from the body surface. The locomotor bands extend along the arms. This larval stage is then known as a bipinnaria larva.

The ciliated bands function in both locomotion and feeding, and the larval arms increase the surface area of the bands (Fig. 19–14B) (Strathmann, 1975). Phytoplankton and other fine suspended particles that constitute the food of echinoderm larvae are kept on the upstream side of the ciliary bands by localized reversal of ciliary beat. In this manner they are bounced down the ciliated bands to the stomodeum. The bipinnaria larva becomes a brachiolaria larva with the appearance of three additional arms at the anterior end (Fig. 19–14B). These arms are short, ventral in position, and covered with adhesive cells at the tip. Between the bases of the three arms is a glandular, adhesive area that forms a sucker. The three arms and the sucker represent an attachment device, and the brachiolaria then settles to the bottom. The tips of the arms provide a temporary attachment to some object until the sucker itself is attached. There are some species, such as *Luidia* and *Astropecten* and the common European sea star *Asterina*, in which a brachiolaria never forms.

Metamorphosis then takes place. The anterior end of the larva degenerates and forms only an attachment stalk, and the adult body develops from the rounded, posterior end of the larva (Fig. 19–15E and F). The left side becomes the oral surface, and the right side becomes the aboral surface. The adult arms appear as extensions of the body.

Internally, the mouth, the esophagus, part of the intestine, and the anus degenerate. All these parts are formed anew and in a position coinciding with the adult radial symmetry. The somatocoel forms the major part of the coelom. The left axohydrocoel forms the water-vascular system, and from the hydrocoel develop five pairs of projections, two in each of the developing arms. These projections represent the cavity and coelomic lining of the first pair of podia in each arm. As soon as additional podia are formed, they begin to grip the substratum and soon free the body from the substratum. At about this time the skeletal system appears. As in other echinoderms, the first adult ossicles to be formed are those around the aboral pole, which contains the anus. New ossicles are then added peripherally to the initial skeleton. The detached baby starfish is less than 1 mm in diameter, with very short, stubby arms.

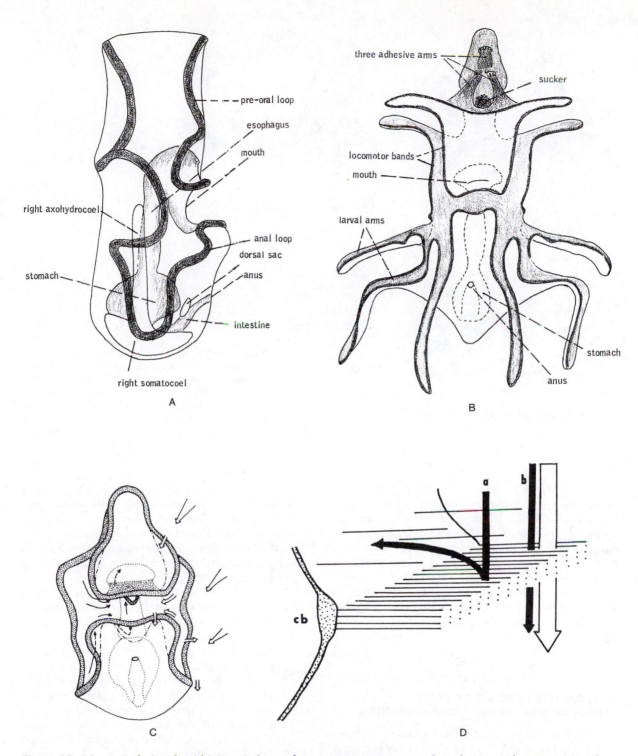

Figure 19–14 *A,* Early (14 days) bipinnaria larva of *Astropecten auranciacus,* lateral view. (After Hörstadius from Hyman.) *B,* Brachiolaria larva of *Asterias,* ventral view. (After Agassiz from Cuénot.) *C,* Water currents (white arrows) and paths of food particles (black arrows) produced by the ciliated band of a bipinnaria larva. *D,* Diagrammatic view of a section of the ciliated band (cb) of an echinoderm larva, showing passage of water current (white arrow), uncollected particle (black arrow b), and collected particle (black arrow a) resulting from a localized reversal of ciliary beat. (From Strathmann, R. R., 1972 and 1975: Biol. Bull., *142*:505–519, and Am. Zool., *15*:717–730.)

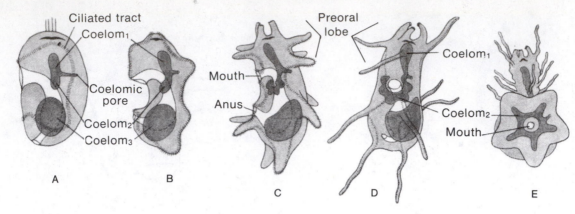

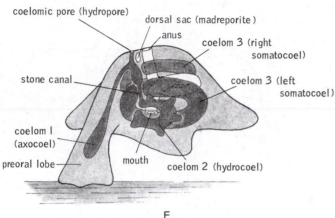

Figure 19–15 *A–F*, Diagrammatic lateral views of larval development and metamorphosis of a sea star, showing development of coelom and water-vascular system. *A* and *B*, Early bipinnaria larva; *C*, brachiolaria larva; *D*, attached metamorphosing larva; *E* and *F*, young starfish developing from posterior part of old larva. (Adapted from various sources.)

Growth rates and life spans are variable, as illustrated by two intertidal species from the Pacific Coast of the United States. *Leptasterias hexactis* broods a small number of yolky eggs during the winter, and the young mature sexually in two years, when they weigh about 2 g. The average life span is ten years. *Pisaster ochraceus* releases a large number of eggs each spring, and development is planktonic. Sexual maturity is reached in five years, at which time the animal weighs 70 to 90 g. Individuals may live 34 years, reproducing annually (Menge, 1975).

SYSTEMATIC RÉSUMÉ OF CLASS STELLEROIDEA, SUBCLASS ASTEROIDEA

Unattached echinoderms having a body composed of flattened central disc and radially arranged arms, or rays.

Subclass Somasteroidea. Mostly fossil Paleozoic sea stars in which the skeletal structure of the arms displays a number of primitive and some extinct features, which are intermediate between asteroids and ophiuroids and show homologies with crinoids. One living species, *Platasterias*, from the Pacific coast of Central America.

Subclass Asteroidea. Stelleroid echinoderms with open ambulacral grooves and a large coelomic cavity in the relatively wide arms.

  Order Platyasterida. Feet without suckers. Primitive, mostly extinct sea stars, and the common soft-bottom *Luidia*.

  Orders Paxillosida and Valvatida. These two orders were formerly united within the order Phanerozonia. The Paxillosida lack suckers on the tube feet; the Valvatida possess suckers. Sea stars with marginal plates and usually with paxillae on the aboral surface. Pedicellariae of the sessile type. *Astropecten, Ctenodiscus, Culcita, Goniaster, Oreaster, Linckia, Porania.*

  Order Spinulosida. The members of this order are not always distinct from those of the two orders above, but in general, conspicuous marginal plates are absent, and the tube feet are suckered. The aboral surface is covered with low spines, which give the order its name. There are no pedicellariae. *Asterina, Patiria, Echinaster, Henricia, Acanthaster, Crossaster, Pteraster.*

Order Forcipulata. Sea stars with pedicellariae composed of a short stalk and three skeletal ossicles. Tube feet with suckers. *Heliaster, Pycnopodia, Asterias, Leptasterias, Pisaster, Brisinga, Zoroaster.*

The systematic résumé of the class Stelleroidea is continued on page 799 following the discussion of the other subclass, the Ophiuroidea.

SUMMARY

1 The Phylum Echinodermata is composed of marine animals that are distinguished by a pentamerous radial symmetry, an endoskeleton of calcareous osssicles, spiny ossicles on the body surface, and a water-vascular system of coelomic canals and body appendages, or podia, that is used in feeding or locomotion.

2 In general, the sexes are separate, fertilization is external, and development is planktonic. There is commonly a bilateral larva that swims and feeds by means of ciliated bands wound over the body.

3 The class Stelleroidea contains echinoderms in which the body is composed of a central disc and radiating arms. In the subclass Asteroidea the arms are not sharply set off from the central disc.

4 Asteroids move by means of podia, which are located within ambulacral grooves. Podia are extended by hydraulic pressure generated by the contraction of a bulblike ampulla. In many species suckers at the ends of the podia permit attachment to the substratum.

5 The arms can bend and twist, permitting the sea star to move over irregular surfaces, grasp prey, and right itself. Arm movement is made possible by a flexible, lattice-like arrangement of ossicles within the dermis and by circular and longitudinal muscle layers in the body wall.

6 The large coelom provides for internal transport, and evaginations of the body wall (papulae) are the sites of excretion and gas exchange. However, the thin walls of the podia are a significant additional exchange surface.

7 Feeding behavior is related not only to diet but also to arm length. Predatory species with short arms swallow the prey entire. Those with long arms evert the stomach and partially digest the prey outside the body. Those sea stars that prey on bivalve mollusks slide the stomach between the valves of the mollusk. Some species use the everted stomach like a mop to remove organic material from various surfaces.

8 Sea stars inhabiting soft bottoms generally possess pointed tube feet and double ampullae; paxillae keep the papulae clear of sediment.

9 Pedicellariae, which are restricted to certain groups of sea stars, probably function to clear the body surface of settling organisms.

10 There are usually two gonads in each arm, and the gametes exit by interradial gonopores. Development leads to a bipinnaria larva, in which ciliated bands are located on long larval arms. With the formation of attachment structures, the larva is called a brachiolaria and is prepared for settling. Following settlement and attachment, the larva undergoes metamorphosis, in which the larval arms degenerate, the left side becomes the oral surface, and the adult body is derived from the posterior part of the larval body.

Class Stelleroidea, Subclass Ophiuroidea

The subclass Ophiuroidea contains those echinoderms known as basket stars and serpent stars, or brittle stars (Fig. 19–16). The 2000 described species make this the largest of the major groups of echinoderms. They are found in all types of marine habitats, and they are often abundant on soft bottoms in the deep sea (see review by Tyler, 1980).

Ophiuroids resemble asteroids in that they also possess arms. However, in other respects the two classes are quite different. The extremely long arms of ophiuroids are more sharply set off from the central disc. There is no ambulacral groove, and the podia play no role in locomotion. Moreover, the arms have a relatively solid construction compared with those of the asteroids.

External Structure

Ophiuroids are relatively small echinoderms. The disc in most species ranges from 1 to 3 cm in diameter, although the arms may be quite long. The basket stars are the largest members of the class, and the disc in some species of this group may attain a diameter of almost 12 cm. A great variety of colors are found in the ophiuroids, and mottled and banded patterns are common.

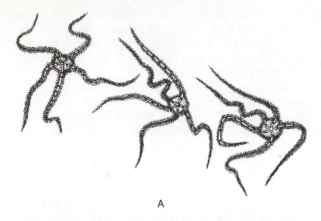

A

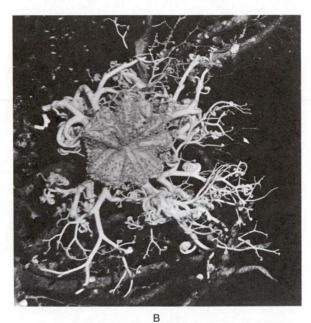

B

Figure 19–16 A Caribbean brittle star, shown in a drawing based on repetitive flash photographs, pulling itself along with its two anterior arms and shoving with the other three. Ophiuroids are far more agile and flexible than are sea stars. (Based on a photograph by Fitz Goro.) *B*, A basket star. (Courtesy of Photo Researchers, N.Y.)

The central disc is flattened, displaying a rounded or somewhat pentagonal circumference (Fig. 19–17). The aboral surface varies from smooth to granular and may bear small calcareous plates, called shields, and small tubercles or spines (Fig. 19–17*B*). There are typically only five arms; however, in basket stars the arms branch at either the base or more distally, and the subdivisions repeatedly branch to produce a great mass of coils that resemble tentacles (Fig. 19–16*B*).

The arms of ophiuroids appear jointed because of the presence of four longitudinal rows of shields (calcareous plates) (Figs. 19–17*A* and *B* and 19–18). There are two rows of lateral shields, one row of aboral shields, and one row of oral shields. A single set—that is, one aboral, one oral, and two lateral shields—completely surrounds the arm and corresponds in position to an internal skeletal ossicle, which is described later. Not infrequently, the oral and aboral shields are reduced by the large size of the lateral shields, which may even meet on the oral and aboral surfaces (Fig. 19–17*A*). Each lateral shield usually bears 2 to 15 large spines arranged in a vertical row (Fig. 19–18). These spines vary considerably in size and shape, depending on the species.

In contrast to the asteroids, there is no ambulacral groove (open ambulacrum) on the oral surface of the arms. The ambulacral ossicles have sunk inward and enlarged to form the vertebrae, and the radial water canal lies below the ossicles (Fig. 19–19). The ambulacrum is said to be closed. The podia are small, tentacle-like papillate appendages that extend between the oral and the lateral shields, and there is typically one pair of podia per joint (Fig. 19–18). Neither papulae nor pedicellariae are present in ophiuroids.

The center of the oral surface of the disc is occupied by a complex series of large plates that frame the mouth area and also form a chewing apparatus with five triangular, interradial jaws (Fig. 19–17*A*).

In most ophiuroids one oral shield is modified, forming a madreporite. Thus, the madreporite is located on the oral surface, in contrast to its aboral position in asteroids. The arms extend inward to the jaws on the oral surface of the disc, leaving five large, somewhat triangular, interradial areas having essentially the same surface structure as the aboral side of the disc.

Body Wall and Skeleton

The epidermis lacks conspicuous surface cilia except in certain areas. The dermis contains the more superficial skeletal shields as well as the large, deeper ossicles of the arms. Each ossicle is a large, bilaterally symmetrical, skeletal piece (a vertebra) that almost fills the entire interior of the arm and greatly reduces the coelom (Figs. 19–18, 19–19 and 19–20). The vertebral ossicles are arranged linearly from one end of the arm to the other, and each ossicle is covered by the four superficial arm shields.

The two end surfaces of a vertebral ossicle bear nodes and sockets, which articulate with corre-

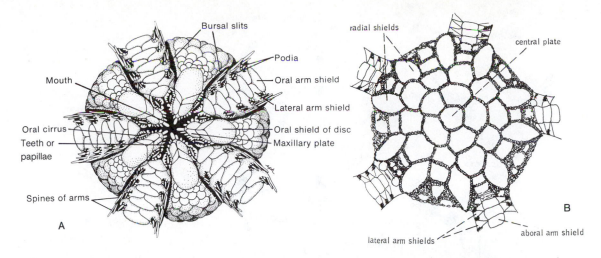

Figure 19–17 *A*, The disc of *Ophiura sarsi* (oral view). (After Strelkov.) *B*, The disc of *Ophiolepis* (aboral view). (After Hyman, L. H., 1955: The Invertebrates. Vol. IV. McGraw-Hill Book Co., N.Y.)

sponding surfaces on adjacent vertebrae, and pits for the insertion of large intervertebral muscles, which move the arm (Fig. 19–18). In many brittle stars this articulation allows great lateral mobility of the arm but little vertical movement. However, in the basket stars and some brittle stars, which lack aboral arm shields, the arms can bend and coil in any direction.

Locomotion and Habitation

The ophiuroids are the most mobile echinoderms. During movement the disc is held above the substratum, with one or two arms extended forward and one or two arms trailing behind. The remaining two lateral arms perform a rapid rowing movement against the substratum that propels the ani-

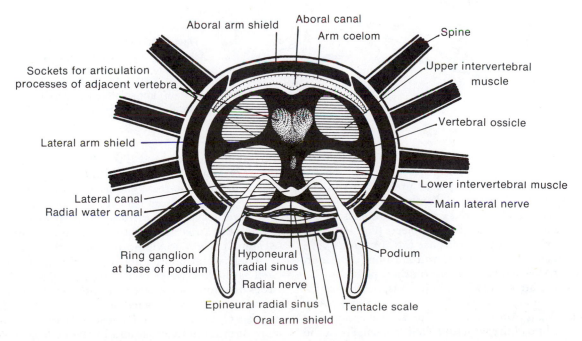

Figure 19–18 Diagrammatic section through the arm of a brittle star.

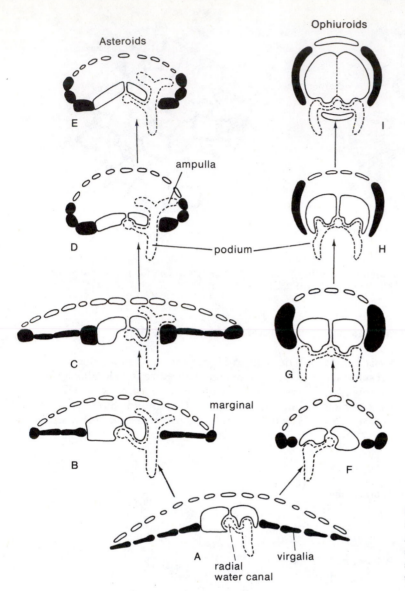

Figure 19–19 Evolution of the arm skeleton in asteroids and ophiuroids. Arms are shown in cross section with ossicles white or black and water-vascular system dotted. *A* is an ancestral somasteroid. (Adapted from Nichols, 1969.)

mal forward in leaps or jerks. The spines provide traction.

Brittle stars show no arm preference and can move in any direction. In clambering over rocks or in seaweed and hydroid colonies, the ends of the supple arms often coil about objects (Fig. 19–21C). Although most ophiuroids use the arms to move, a few, such as *Ophionereis annulata* along the American west coast, creep about on their podia.

There are burrowing ophiuroids, notably members of the family Amphiuridae (*Amphiura, Amphiodia* and *Amphipholis*). The animal excavates a mucus-lined burrow with tubelike channels to the surface in mud or sand (Fig. 19–21D), using undulatory waves of the arms and digging movements of the tube feet. The latter swing laterally, moving substrate from beneath the body. The animal never leaves the burrows unless dislodged, although the arms may be projected to the surface. Arm undulation provides for burrow ventilation.

Underwater photographs from the deep sea as well as shallow water reveal high population densities of ophiuroids where conditions are favorable. Some species live in dense aggregations. The European *Ophiothrix fragilis* responds to the presence of other individuals and in favorable currents

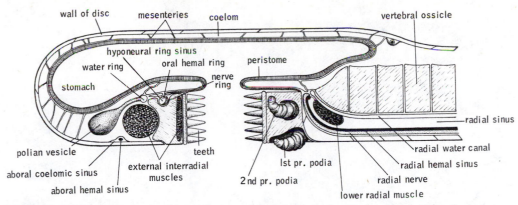

Figure 19–20 The disc and base of an arm of a brittle star (vertical section). (After Ludwig.)

on muddy gravel bottoms attains densities as great as 1000 to 2000 per square meter (Broom, 1975).

Some echinologists consider the ophiuroids the most successful group of living echinoderms; they attribute this success in part to their motility, small size, and ability to utilize the protective cover of crevices, holes, spaces beneath stones, and other natural retreats (Fig. 19–21*A*). Although they are inconspicuous, the number of ophiuroids, both species and individuals, on many tropical reefs greatly exceeds that of other echinoderms.

The only commensal echinoderms are ophiuroids. Large sponges may contain great numbers of ophiuroids living in the water canals (Fig. 19–21*B*). Other ophiuroids inhabit arborescent octocorallians and scleractinian corals. Species of *Ophiomaza*, an Indo-Pacific brittle star, live on the oral surfaces of feather stars, clutching the calyx with the arms. There are also several tiny Indo-Pacific brittle stars that live on the undersurfaces of sand dollars.

Water-Vascular System

The oral shield that forms the madreporite in ophiuroids usually bears but a single pore and canal, and the stone canal ascends to the water ring, which is located in a groove on the aboral surface of the jaws. The water ring bears four polian vesicles and also gives rise to the radial canals, which penetrate the lower side of the vertebral ossicles of the arms (Fig. 19–18). In each ossicle the radial canal gives rise to a pair of lateral canals, which lead to the podia. The paired lateral canals of ophiuroids contrast with the staggered arrangement of other echinoderms. Ampullae are absent, probably correlated with the reduction of the arm coelom, but a valve is present between the podium

and the lateral canals. Fluid pressure for protraction is generated by a dilated, ampulla-like section of the podial canal and in some forms by localized contraction of the radial water canal. As in asteroids, the radial canal terminates in a small external tentacle at the tip of the arm.

Nutrition

Ophiuroids are carnivores, scavengers, deposit feeders, or filter feeders (review by Warner, 1982). Most use several feeding modes, but one is generally predominant. The Atlantic brittle star *Ophiocomina nigra*, which is primarily a suspension feeder, can be used to illustrate all four feeding habits.

In filter feeding the arms of *Ophiocomina* are lifted from the bottom and waved about in the water. Plankton and detritus adhere to mucous strands strung between the adjacent arm spines. The trapped particles may be swept downward toward the tentacular scale by ciliary currents or collected from the spines by the tube feet, which extend upward for this purpose (Fig. 19–22). A tentacular scale is a reduced spine. The tube feet are then scraped across the tentacular scales, depositing collected particles in front of the scale (Fig. 19–22). This is also where the ciliary tracts deposit their material. On each side the food particles are picked up by adjacent podia, compacted into a bolus, and passed along the midoral line of the arm toward the mouth. The food balls are moved by the podia until they reach the proximal parts of the arm, where movement toward the mouth is facilitated by cilia.

Deposit feeding on intermediate particles is performed by the podia. The podia collect the par-

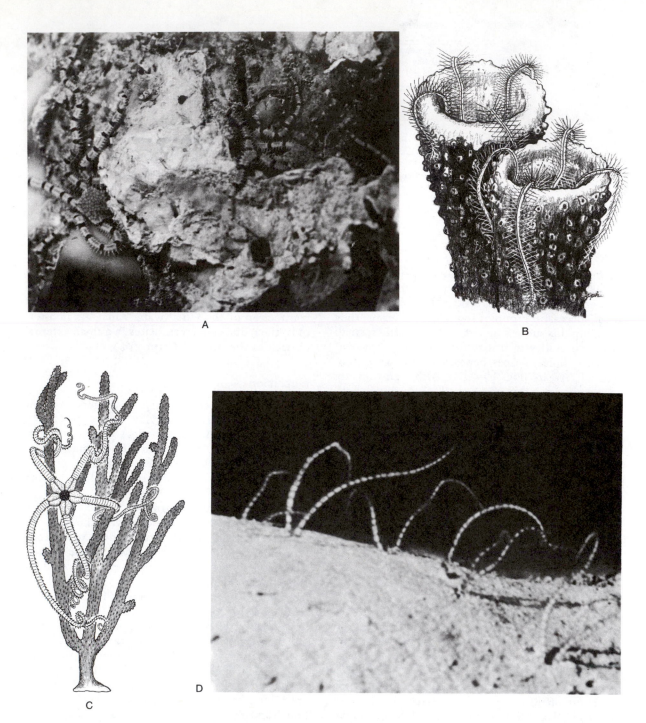

Figure 19–21 *A*, Two specimens of a West Indian brittle star *(Ophionereis)* lodged in crevices on the underside of a coral head. (By Betty M. Barnes.) *B*, Two brittle stars in a sponge. *C*, A euryalous brittle star climbing on a gorgonian coral. These brittle stars are related to the basket stars and are capable of coiling their arms vertically. (Modified from Hyman, L. H., 1955: The Invertebrates. Vol. IV. McGraw Hill Book Co., N.Y.) *D*, Specimens of *Amphioplus* projecting two arms from the tubelike burrows and trapping suspended particles from the passing water current. (From Fricke, H. W., 1970: Helgolander wiss. Meeresunters, *21*:124–133.)

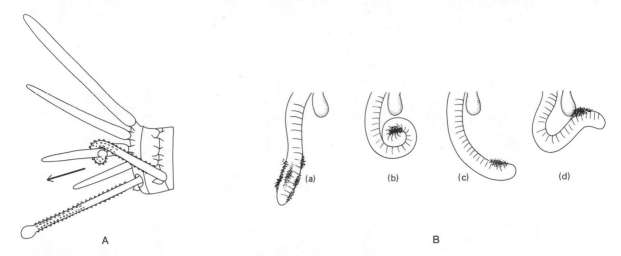

Figure 19–22 Feeding activity of podia in brittle stars: *A*, Spine wiping in *Ophiocoma wendtii*, a West Indian brittle star. Note that the podium of one side wipes the spines on the opposite side of the arm. (From Sides, E. M., and Woodley, J. D., 1985: Niche separation in three species of *Ophiocoma* in Jamaica, West Indies. Bull. Mar. Sci., 36(3):701–715.) *B*, Particle consolidation and transfer in the suspension-feeding brittle star *Ophionereis fasciata*. *a*, Particles collected by podium from spines. *b*, Particles consolidated by podium into one mass. *c* to *d*, Mass transferred from podium to tentacle scale. (From Pentreath, R. J., 1970: J. Zool., 161:395–429.)

ticles from the substratum, compact them into food balls, and move them toward the mouth.

Large food material, such as dead animal matter, is swept into the mouth by the looping motion of an arm. Browsing over algae or carrion, the animal utilizes its teeth or oral tube feet.

Ophiothrix fragilis uses its papillate podia for filter feeding. The feeding arms are elevated and so twisted that the oral surface is directed toward the current. The podia are extended well beyond the spines, forming comblike filtering series on either side of the arm (Fig. 19–23*A*). Collected particles are periodically removed and transported as a growing bolus by a wave action of the podia that travels down the arm toward the mouth (Fig. 19–23*B* and *C*) (Warner and Woodley, 1975).

Such mechanisms of deposit and filter feeding have the advantage of permitting the animal to extend only two or three feeding arms from its protective retreat as well as to utilize a variety of food sources (Fig. 19–21).

Brittle stars that are predominantly carnivores feed largely on polychaetes, mollusks, and small crustaceans. Food is usually captured and brought to the mouth by arm looping.

Basket stars, of which there are only about 100 species, are suspension feeders but capture zooplankton of relatively large size (10 to 30 mm—crustaceans, polychaetes, and others). Perched above the bottom, the basket star extends its arms in a parabolic filtration fan with the concave and aboral sides directed toward the prevailing water current (Fig. 19–23*D*) (Fricke, 1968; Meyer and Lane, 1976; Hendler, 1982). Prey is seized with the ends of the many arm branches, which coil about the catch, and minute hooks on the arm surface prevent escape. Periodically, the basket star removes the collected plankton from the arms by passing them through comblike oral papillae. The tube feet do not play the role of food transport, as in brittle stars.

The digestive tract is extremely simple (Fig. 19–20). The jaws frame a shallow, prebuccal cavity, which is roofed aborally by the peristomial membrane containing the mouth.

The esophagus connects the mouth with a large, saclike stomach. The stomach fills most of the interior of the disc, and in most ophiuroids the margins are infolded to form ten pouches. There is no intestine or anus, and the digestive tract does not extend into the arms. Extracellular and intracellular digestion and absorption occur largely within the stomach pouches, but this restriction seems less likely for carnivorous species.

Gas Exchange, Excretion, and Internal Transport

Gas exchange in ophiuroids takes place by means of ten internal sacs (bursae) that represent invaginations of the oral surface of the disc. The bursae are

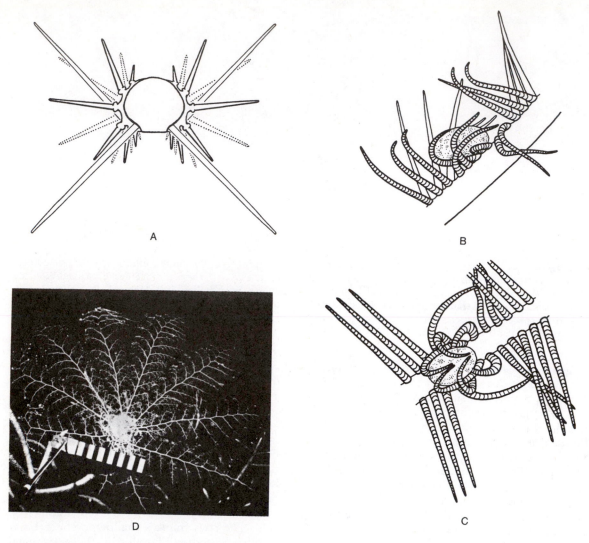

A

B

D

C

Figure 19–23 Filter feeding in ophiuroids: *A*, End view of arm section of the brittle star *Ophiothrix fragilis*, showing spines and position of tube feet. Alternate tube feet are directed orally and aborally. The tube feet may also be extended laterally, forming a single filtering series on either side of the arm. *B* and *C*, Particle collection and transport in *Ophiothrix fragilis* seen in side view (*B*) and orally (*C*). Particles are added to a bolus, which is about 1 mm in diameter on reaching the mouth. (From Warner, G. F., and Woodley, J. D., 1975: Suspension-feeding in the brittle star *Ophiothrix fragilis*. J. Mar. Biol. Assoc. U.K., 55:199–210. Copyrighted and reprinted by permission of Cambridge University Press.) *D*, Feeding position of the basket star *Astrophyton*. The arms form a parabolic fan with the tips directed toward the current, which in this figure is moving away from the viewer. (Courtesy of Meyer, D. L., and Lane, N. G., 1976: The feeding behavior of some Paleozoic crinoids and recent basket stars. J. Paleontol., 50(3):472–480.)

connected to the outside by slits that run along the margins of the arms on the oral surface of the disc (Fig. 19–17*A*). The bursae may be lined with ciliated epithelium, especially the slits. The beating cilia create a current of water that enters the peripheral end of the slit, passes through the bursae, and flows out the oral end of the slit. Many species also pump water into and out of the bursae by rais-

ing and lowering the oral or aboral disc wall or by contracting certain disc muscles associated with the bursae.

The thin-walled respiratory bursae may well be the principal center for removing wastes, including waste-laden coelomocytes.

The coelom in ophiuroids is much reduced compared with that of other echinoderms. The vertebral ossicles restrict the coelom to the aboral part of the arms (Fig. 19–18); the stomach, bursae, and gonads leave only small coelomic spaces in the disc. The hemal system is essentially like that of asteroids.

Nervous System

The nervous system is composed of a circumoral nerve ring and radial nerves, as in asteroids (Figs. 19–18 and 19–20). There are no specialized sense organs; dispersed epithelial sensory cells compose the sensory system. Most ophiuroids are negatively phototropic, as are many asteroids, and are also able to detect food without contact.

Regeneration and Reproduction

Many ophiuroids can cast off, or autotomize, one or more arms if disturbed or seized by a predator. A break can occur at any point beyond the disc; the lost portion is then regenerated. There are some ophiuroids, notably species of *Ophiactis*, in which asexual reproduction takes place by division of the disc into two pieces, each piece with three arms. Fission can take place along any plane; the missing half is then regenerated (see review by Emson and Wilkie, 1980).

The majority of ophiuroids are dioecious. The gonads are small sacs attached to the coelomic side of the bursae near the bursal slit. There may be one, two, or numerous gonads per bursa with various positions of attachment. Hermaphroditic species are not uncommon. Some bear separate testes and ovaries; others are protandric.

When the gonads are ripe, they discharge into the bursae, probably by rupture, and the sex cells are carried out of the body in the ventilating water current. Fertilization and development take place in the sea water in many species, but brooding is common. The bursae are commonly used as brood chambers, but the female of some species broods her eggs in the ovary or coelom. Development

takes place within the mother until the juvenile stage is reached. In most species only a few young are brooded in each bursa.

Embryogeny

In nonbrooding, oviparous ophiuroids early development is similar to that in the asteroids. The larva of many species, called an ophiopluteus, displays four pairs of elongate arms supported by calcareous rods and bearing ciliated bands. The shape of the larva is distinctive and different from that of the brachiolaria of asteroids (Fig. 19–24). Metamorphosis takes place while the larva is still free swimming, and there is no attachment stage. The tiny brittle star sinks to the bottom and takes up an adult existence. For the some six species for which information is available, development to this point takes 14 to 40 days. To reach the same stage in the single brooding species studied takes three to seven months. Of the 2000 described species of ophiuroids, larvae are known for 71, and 55 are known to brood. The developmental patterns of the remainder are still unknown (Hendler, 1975).

SYSTEMATIC RÉSUMÉ OF CLASS STELLEROIDEA, SUBCLASS OPHIUROIDEA

Subclass Ophiuroidea

Order Oegophiurida. A largely fossil group with a single living species (from Indonesia). No dorsal and ventral arm shields or bursae. Madreporites at edge of disc.

Order Phrynophiurida. Ophiuroids in which dorsal arm shields are absent.

Suborder Ophiomyxina. Primitive brittle stars in which disc and arm plates are covered by a thick soft skin. *Ophiocanops*, *Ophiomyxa*.

Suborder Euryalina. Arms simple or branched (basket stars) but capable of coiling vertically. *Asteronyx*, *Gorgonocephalus*.

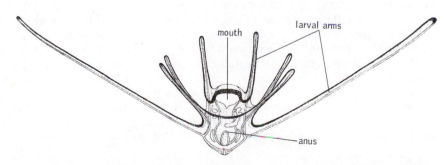

Figure 19–24 Ophiopluteus larva of *Ophiomaza* (oral view). (After Mortensen.)

Order Ophiurida. Mostly small ophiuroids, usually with five arms. Arms capable of transverse movement only. Dorsal arm shields present. This order contains most of the brittle stars, or serpent stars. *Amphiura, Amphipholis, Ophiopholis, Ophiactis, Ophiothrix, Ophioderma, Ophiocoma, Ophiolepis, Ophiomusium, Ophiomaza, Ophionereis.*

SUMMARY

1 In the stelleroid subclass Ophiuroidea the long, narrow arms are sharply set off from the central disc.

2 Ophiuroids are considered the most successful group of echinoderms. Their success is probably correlated with their motility, diversity of feeding habits, and small size, which have enabled them to exploit habitats unavailable to most other echinoderms.

3 Ophiuroids move rapidly by pushing and pulling with the flexible arms. Lateral arm spines provide traction. The arm is occupied by large ossicles (vertebrae) that articulate with each other in a horizontal column. Intervertebral muscles provide movement. Podia are not used in locomotion.

4 The vertebrae are covered by flat, superficial ossicles, called shields, with which the spines are associated. The vertebral ossicles restrict the coelom to a small, dorsal chamber.

5 Perhaps because of coelom reduction, the water-vascular system lacks ampullae. The lateral canal or radial canals assume the function of ampullae. The madreporite is located on one of the oral shields.

6 The reduced arm coelom restricts much of gas exchange to five pairs of pouchlike invaginations (bursae) on the oral side of the disc.

7 The feeding of ophiuroids includes one or all of the following mechanisms for a given species: scavenging by arm raking, deposit feeding by the podia, and suspension feeding by podia and mucous strands slung between spines. These methods enable many species to feed without leaving protective retreats. The principal functions of ophiuroid podia are food collection and transport. Basket stars use their arms to form a parabolic fan at right angles to the water current and catch zooplankton with the tips of the arm branches.

8 The gonads of ophiuroids are associated with the coelomic side of the bursae, which provide the exit for the gametes and the site of development in brooding species. In nonbrooding species development leads to an ophiopluteus larva, which undergoes metamorphosis prior to settling.

Class Echinoidea

The echinoids are free-moving echinoderms commonly known as sea urchins, heart urchins, and sand dollars. About 950 species have been described. The name *Echinoidea*, which means "like a hedgehog," refers to the movable spines that cover the bodies of these animals. The echinoid body does not possess arms. Rather, the shape is circular or oval, and the body is spherical or greatly flattened along the oral-aboral axis. The class is particularly interesting from the standpoint of symmetry, for although the sea urchins are radially symmetrical, many members of the class that live in soft bottoms display various stages in the attainment of a secondary bilateral symmetry. A third distinctive feature of echinoid structure is the flattening and suturing of the skeletal ossicles into a solid case (the test).

External Structure

REGULAR ECHINOIDS

The radial, or regular, members of the class are known as sea urchins. In these forms the body is more or less spherical and armed with relatively long, movable spines (Fig. 19–25). Sea urchins are brown, black, purple, green, white, and red, and some are multicolored. Most are 6 to 12 cm in diameter, but some Indo-Pacific species may attain a diameter of nearly 36 cm.

The sea urchin body can be divided into an aboral and an oral hemisphere, with the parts arranged radially around the polar axis. The oral pole bears the mouth and is directed against the substratum. The mouth is surrounded by a peristomial membrane that bears a number of different structures arranged in a radial manner. There are five pairs of short, heavy, modified podia, called buccal podia, and five pairs of bushy projections, called gills (Fig. 19–26A). In addition to the buccal podia and the gills, the area around the peristome bears small spines and pedicellariae.

The aboral pole contains the anal region, known as the periproct (Figs. 19–26B and 19–27A). The periproct is a small, circular membrane containing the anus, usually in the center, and a varying number of embedded plates (Fig. 19–26B and 19–27). The globose body surface can be divided into ten radial sections, which converge at the oral and

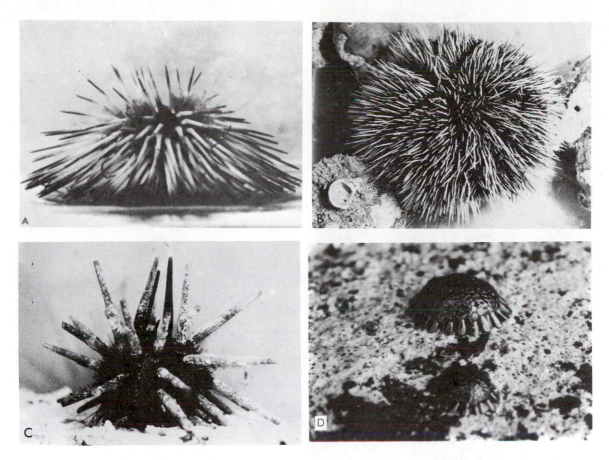

Figure 19–25   Regular urchins: *A*, Side view of the common Atlantic sea urchin *Arbacia punctulata* showing long spines and podia. *B*, A West Indian species of *Tripneustes*, viewed from above. *C*, A species of *Eucidaris* with very small secondary spines around the base of the heavy primary spines. *D*, *Colobocentrotus*, a Pacific sea urchin with blunt aboral spines that fit together to form a smooth surface. Such spines are perhaps an adaptation for living on intertidal rocks. (All by Betty M. Barnes.)

aboral poles (Fig. 19–26*A*). Five sections contain tube feet and are called ambulacral areas. The ambulacral areas alternate with sections devoid of podia, known as interambulacral areas.

The skeletal plates are arranged in rows running from the oral pole to the aboral pole. Each ambulacral area is composed of two rows of ambulacral plates, and each interambulacral area is composed of two rows of interambulacral plates. There are thus 20 rows of plates—10 ambulacral and 10 interambulacral (Fig. 19–26*B*). The ambulacral plates are pierced by holes forming canals which connect the internal ampullae and external podia (Fig. 19–26*B*). In sea stars and brittle stars the canals pass between ossicles.

Around the periproct is a series of plates. These consist of five large, genital plates, one of which is porous and serves as the madreporite, and five smaller ocular plates (Figs. 19–26*B* and 19–27*A*). The genital plates, each of which bears a gonopore, line up with the interambulacral areas and alternate with the ocular plates, which coincide with the ambulacral areas.

The movable spines, which are so characteristic of sea urchins, are arranged more or less symmetrically in the ambulacral and interambulacral areas. The spines are longest around the equator and shortest at the poles. Most sea urchins possess long (primary) and short (secondary) spines, the two types being more or less equally distributed over the body surface. However, *Arbacia punctulata*, the common sea urchin along the Atlantic coast of North America, possesses only the long type (Fig. 19–25*A*).

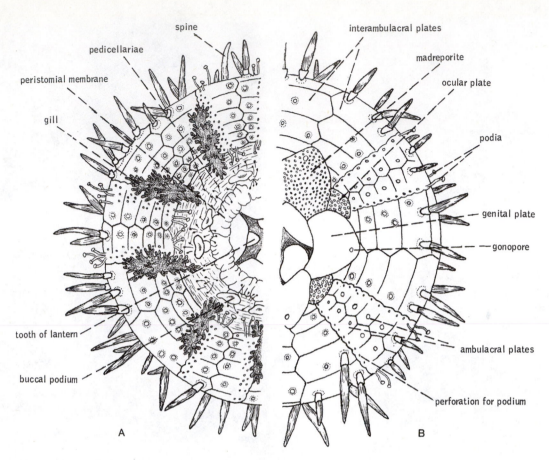

Figure 19–26   The regular urchin *Arbacia punctulata: A,* Oral view. *B,* Aboral view. (After Ried, W. M. *In* Brown, F. A.: Selected Invertebrate Types. John Wiley and Sons, N.Y.)

Each spine contains a concave socket at the base that fits over a corresponding tubercle on the test (Fig. 19–28A). Two sheaths of fibers that extend between the spine base and the test encircle the ball-and-socket joint. Contractions of the outer muscular sheath serve to incline the spines in one direction or another. The inner sheath of collagen fibers (catch fibers) can rapidly shift from a soft to a hard condition on stimulation, thereby causing the spine to be rigidly erect (Motokawa, 1984).

The spines are usually cylindrical and taper to a point, but many species depart from this generalization. Species of *Diadema,* which are common on tropical reefs, have very long, needle-like spines, which can be tilted and waved in the direction of intruders (Fig. 19–29A). The spines, which can be regenerated, are hollow, brittle, and provided with an irritant, and the outer surface is covered with circlets of small barbs directed toward the spine tip

(Fig. 19–29B). This urchin can inflict serious, painful wounds if stepped upon. The primary spines of the slate-pencil urchins, species of *Heterocentrotus,* and some species of cidarids are heavy and blunt (Fig. 19–25C). The aboral spines of the intertidal Indo-Pacific genus *Colobocentrotus* are short and heavy, and the blunt tips are polygonal in cross section (Fig. 19–25D). These spines fit together like tiles, providing an effective wave-resistant surface and protection against desiccation. The deep-water sea urchin of the family Echinothuridae bears special poison spines on the aboral surface (Fig. 19–28B).

Pedicellariae, which are characteristic of all echinoids, are located over the general body surface as well as on the peristome. The echinoid pedicellaria is composed of a long stalk surmounted by jaws. The stalk may contain a supporting skeletal rod, and there are usually three opposing jaws (Fig.

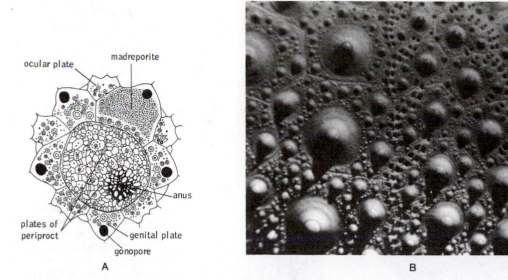

A

B

Figure 19–27 *A*, Periproct and surrounding plates of the regular urchin *Strongylocentrotus*. (After Lovén from Ludwig.) *B*, Surface view of a section of a sea urchin test, showing tubercles on which spines are located, paired perforations for tube feet, and junction line (groove) of sutured ossicles. Compare with Figure 19–35. (By Betty M. Barnes.)

19–35). Muscles at the base of the stalk provide for elevation and direction of the pedicellariae in response to certain stimuli.

Members of the widespread subtropical and tropical family Toxopneustidae *(Lytechinus, Tripneustes)* possess several types of pedicellariae, one of which contains poison glands (Fig. 19–35).

The outer side of each jaw is surrounded by one or two large poison sacs that open by one or two ducts just below the terminal tooth of the jaw. The poison has a rapid paralyzing effect on small animals and drives larger enemies away. The spines frequently incline away from the poison pedicellariae so that these pedicellariae are more exposed. The

Figure 19–28 Spines of regular urchins. *A*, Section through the base of a *Cidaris* spine, showing muscular and collagen sheaths. (After Cuénot, L., 1948: Echinoderms. *In* Grassé, P.: Traité de Zoologie. Vol. XI. Masson et Cie, Paris.) *B*, Poison spine of *Asthenosoma varium*, an Indo-Pacific species. After Sarasins from Hyman.)

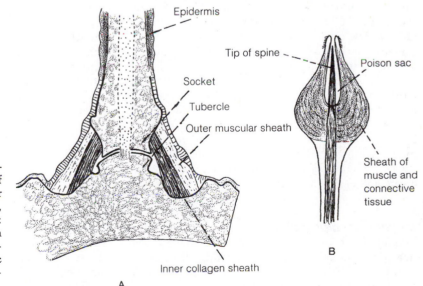

Epidermis

Tip of spine

Poison sac

Socket

Tubercle

Outer muscular sheath

Sheath of muscle and connective tissue

Inner collagen sheath

A

B

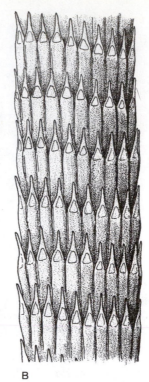

Figure 19–29 *A, Diadema.* Species of this genus, which occurs in both the Caribbean and the Indo-Pacific, possess long, hollow, needle-like spines, which can inflict painful punctures when handled or stepped upon. This West Indian species is common on reefs, where it lives in sheltered or protected recesses on exposed or sand bottoms. (By C. Gebelein.) *B,* Section of a spine of *Diadema.* Circlets of barbs are directed toward spine tip.

poison pedicellariae of the Indo-Pacific *Toxopneustes* look like little parasols when open and can produce a painful reaction in humans.

Other types of pedicellariae are used for defense or for cleaning the body surface, biting and breaking up small particles of debris , which are then removed by the surface cilia. When the pedicellariae are touched on the outside, they snap open; when touched on the inside, they snap shut. Pedicellariae also respond to chemical stimuli.

In most sea urchins the ambulacral areas bear hard, stalked, spherical or ovoid bodies (spheridia), each of which contains a statocyst vesicle with statoliths. The spheridia may be limited in number and located only on the oral side, or there may be many throughout the length of the ambulacrum. In *Arbacia* there is only one spheridium per ambulacrum, and each is located near the peristome.

IRREGULAR ECHINOIDS

The bilateral, or irregular, echinoids include the heart urchins, cake urchins, and sand dollars. Most of their peculiarities are adaptations for burrowing in sand. In contrast to sea urchins, the test is clothed with many small spines, which are used in locomotion and keeping sediment off the body surface.

The heart urchins (spatangoids) are more or less oval, the long axis representing the anteroposterior axis of the body (Fig 19–30). The oral surface is flattened, and the aboral surface is convex. The entire center of the oral surface, containing the mouth and peristome, has migrated anteriorly. The center of the aboral surface usually remains in the center of the upper or dorsal surface, but the periproct and anus have migrated to the posterior end in what now becomes the posterior interambulacrum.

Podia are degenerate or absent around the circumference of the body, so functional podia are confined to the oral and aboral surfaces. The conspicuous aboral ambulacral areas are each shaped like a petal radiating from the center and are known as petaloids. The podia of the petaloids are modified for gas exchange. The oral ambulacral areas (phyllodes) contain specialized podia for obtaining food particles. The small spines form a dense covering over the body surface but have the same basic structure as those of sea urchins.

The cake urchins or sand dollars (clypeasteroids) differ from the heart urchins in a number of

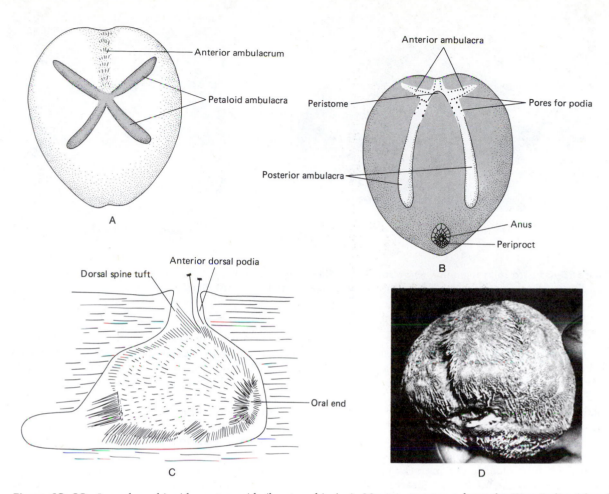

Figure 19–30 Irregular echinoids, spatangoids (heart urchins): *A, Meoma ventricosa* from the West Indies (aboral view). *B, Meoma* (oral view). (Hyman, L. H., 1955: The Invertebrates. Vol. IV. McGraw-Hill Book Co., N.Y.) *C, Echinocardium flavescens* in its sand burrow (lateral view). (Modified after Gandolfi-Hornyold.) *D,* Anterior end of *Moira atropos,* a heart urchin from the Atlantic coast of the southeastern United States. (By Betty M. Barnes.)

respects. A few species, such as the sea biscuits, *Clypeaster* (Fig. 19–31*A*), are shaped somewhat like heart urchins, but the typical sand dollar has a greatly flattened body displaying a circular circumference (Fig. 19–32). Burrowing and covering are thus facilitated. The aboral center and the oral center, which contains the mouth, are both centrally located. The periproct, however, is ventral and, like that of the heart urchins, is located in the posterior interambulacrum. The aboral surface bears conspicuous petaloids, and the oral surface contains distinct radiating grooves. Spheridia are located near the peristome and are buried in the test.

The bodies of some common sand dollars, called keyhole sand dollars (*Mellita*), contain large, elongated notches or openings known as lunules (Fig. 19–32*A* to *C*). Lunules vary in number from two to many and are symmetrically arranged. In most cases the lunules arise from indentations that form along the circumference of the animal and then become enclosed in the process of growth.

Body Wall

The body wall of echinoids is composed of the same layers as in asteroids. A ciliated epidermis covers the outer surface, including the spines. Beneath the epidermis lie a nervous layer and then a connective tissue dermis that contains the flattened and sutured skeletal plates. A muscle layer is absent, since the ossicles are immovable, and the inner surface of the test is covered by the peritoneum, composed of columnar epithelium.

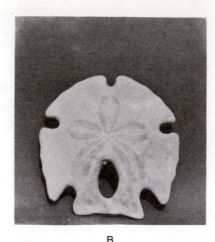

A   B

Figure 19–31 Irregular echinoids, clypeasteroids. *A,* Side view of the test of sea biscuit, *Clypeaster. B,* Aboral surface of the test of the arrowhead sand dollar, a species of *Encope.* (Both by Betty M. Barnes.)

Locomotion

Sea urchins are adapted for life on both hard and soft bottoms, and spines and podia are used in movement. The tube feet function in the same manner as those of the sea stars, and spines may be used for pushing and raising the oral surface off the substratum. Sea urchins can move in any direction, and any one of the ambulacral areas can act as the leading section. If overturned, these animals right themselves by attaching the more aboral podia of one of the ambulacral areas. Attachment of the podia progresses in an oral direction, gradually turning the animal over onto the oral side. Righting may also involve specialized movements of the spines.

Movement of sea urchins is closely related to feeding activity. For example, *Strongylocentrotus franciscanus* in kelp beds off the California coast exhibits mean movements of 7.5 cm per day, but where food supplies are lower, movement may be as great as 50 cm per day.

Some sea urchins tend to seek rocky depressions, and some species are actually capable of increasing the depth of such depressions or even of excavating burrows in rock and other firm material. Boring is performed largely by the scraping action of the chewing apparatus.

Boring behavior appears to be an adaptation to counteract excessive wave action, and these species are largely found in habitats that are exposed to rough water. One of the most notable boring sea urchins is *Paracentrotus lividus,* which lives along the coast of Europe. This sea urchin riddles rock walls with burrows. When the burrows are shal-low, the animal leaves to feed, but it remains permanently within deeper burrows, which often have entrances too small to permit exit. Echinometrids are common boring species on tropical reefs. The urchins can usually be seen within their shallow, irregular excavations but are very difficult to remove without breaking the surrounding rock. The West Indian *Echinometra* honeycombs coralline rock in surge areas (Fig. 19–33), but nonburrowing populations of this species are sometimes encountered. *Strongylocentrotus purpuratus* is a surge-loving sea urchin found along the Pacific coast of North America that commonly burrows in soft rock.

The irregular echinoids are adapted for a life of burrowing in sand. The animal burrows with its anterior end forward, and movement results from action of the spines, the podia being modified for other functions. A heart urchin burrows into the sand by inclining the anterior end downward and moving sand with specially modified, paddle-shaped spines on the anterior sides of the body. Some heart urchins tend to remain buried in one spot below the surface of the sand. Contact with the surface in those species that bury themselves more than several centimeters is maintained by a funnel-like opening in the sand over the aboral side of the animal. Maintenance of the opening and of the subterranean chamber wall is carried out by specialized podia and apical spines (Fig. 19–30C). A short, blind tunnel extends posteriorly behind the anus. Tracts of ciliated spines (fascioles) pump water into the surface opening and out the posterior tunnel. Some intertidal species, such as the

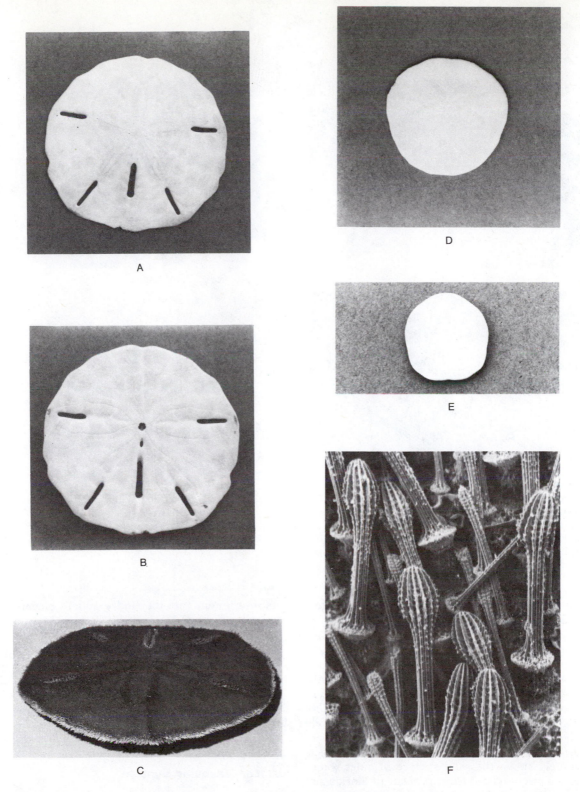

Figure 19–32 Irregular echinoids, clypeasteroids. *A–C, Mellita quinquiesperforata*, the five-slotted sand dollar from the Atlantic coast of the United States. *A*, Oral view of test. *B*, Aboral view of test. *C*, Anterior view of a specimen with spines intact. *D*, Aboral view of test of *Dendraster excentricus* from the Pacific coast of the United States. Aboral center is closest to posterior edge of test. *E*, Aboral view of test of species of *Laganum* from the Indo-Pacific. *F*, SEM of the aboral spines of *Mellita*. (Courtesy of J. Ghiold.)

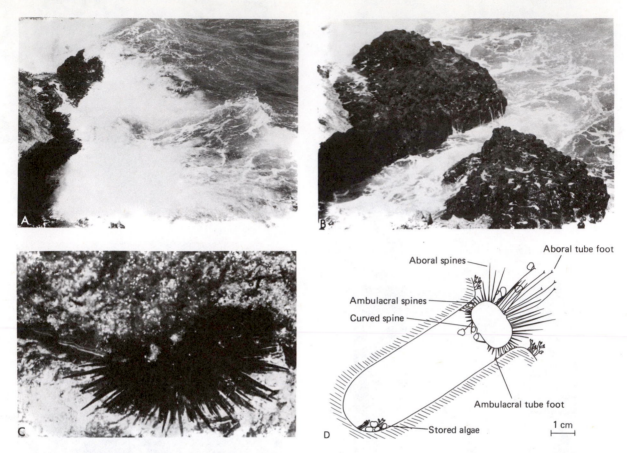

Figure 19–33 The West Indian sea urchin *Echinometra lacunter: A* and *B,* Rock honeycombed with *Echinometra* burrows, shown covered (*A*) and uncovered (*B*) during wave surges. *C,* Urchin in burrow. (All by Betty M. Barnes.) *D,* The Indo-Pacific *Echinostrephus molaris* feeding on pieces of algae collected at the entrance to its burrow. (From De Ridder, C., and Lawrence, J. M., 1982: Food and feeding mechanisms: Echinoidea. *In* Jangoux, M., and Lawrence, J. M. (Eds.): Echinoderm nutrition. A. A. Balkema, Rotterdam. p. 90.)

Indo-Pacific *Lovenia,* come to the surface at low tide and rebury themselves when the tide comes in. *Lovenia* can bury itself in 1 minute, but other species may take as long as 50 minutes (Ferber and Lawrence, 1976).

Sand dollars burrow just beneath the sand surface. Some species, such as the common sand dollars of the east coast of the United States, *Mellita quinquiesperforata,* cover themselves completely with sand; in others, such as the Pacific coast *Dendraster excentricus,* the posterior end projects obliquely above the sand surface in quiet water. The sea biscuit *Clypeaster rosaceus* of Florida and the West Indies does not burrow but sits on the surface of sandy bottoms.

Some sand dollars can right themselves if turned over. In righting itself, the animal burrows its anterior end into the sand, gradually elevates its posterior end, and eventually flips its body over. The Atlantic five-lunuled species partially elevates its body and then apparently depends on water currents to be turned back over onto the oral surface. The flattened shape of the sand dollars lying just beneath the surface subjects them to lift and dislodgment by water currents. The slots or lunules characteristic of many species appear to be an adaptation to reduce lift (Telford, 1983).

Water-Vascular System

The water-vascular system of echinoids is essentially like that of the sea stars (Fig. 19–34). One of the genital plates around the periproct contains pores and pore canals and functions as the madre-

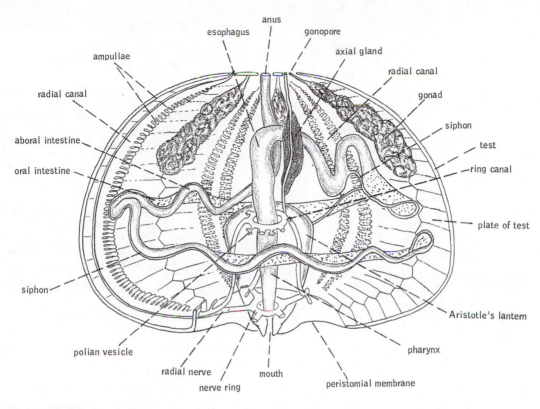

Figure 19–34   Internal structure of the regular urchin *Arbacia* (side view). (Modified after Petrunkevitch from Reid.)

porite (Fig. 19–27*A*). A stone canal descends orally to the water ring, which lies above the peristome in heart urchins or just above the chewing apparatus in regular urchins and sand dollars. The radial canals extend from the water ring and run along the underside of the ambulacral areas of the test.

Each radial canal terminates in a small protrusion called a terminal tentacle, which penetrates the most apical ambulacral plate. The lateral canals of one side of the radial canal alternate with those of the other side. The canals connecting the ampullae and podia, unlike those in other echinoderms, penetrate the ambulacral ossicles rather than pass between them. These canals are also peculiar in being doubled—that is, from each ampulla two canals pierce the ambulacral plate and become confluent on the outer surface to enter a single podium (Fig. 19–35). The suckers of the podia of sea urchins are highly developed and have a system of muscles and supporting ossicles.

In irregular urchins, which use the spines to move, the podia are modified for a number of functions that will be described in the following sec-

tions. Those on the oral surfaces of sand dollars are widely dispersed, penetrate the test via a single canal, and are served by elaborately branching canals.

Nutrition

Sea urchins feed with a highly developed scraping apparatus called Aristotle's lantern. The apparatus is composed of five large, calcareous plates called pyramids, each of which is shaped somewhat like an arrowhead with the point projected toward the mouth (Fig. 19–36). The pyramids are arranged radially, with each side connected to that of the adjacent pyramid by means of transverse muscle fibers. Passing down the midline along the inner side of each pyramid is a long, calcareous band. The curled, upper end of the band is enclosed within a dental sac and is the area of new tooth formation. In *Paracentrotus lividus* new tooth material is produced at a rate of about 1 to 1.5 mm per week. The oral end of the band projects beyond the tip of the pyramid as an extremely hard, pointed tooth. Since

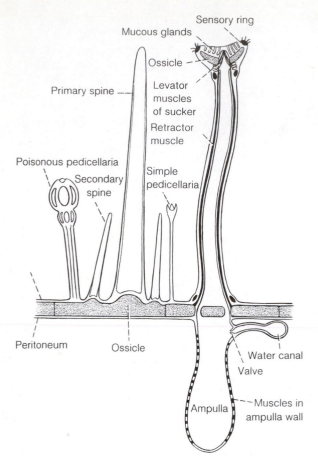

Figure 19–35 Diagrammatic section through the body wall of a sea urchin, showing one ambulacral and one interambulacral ossicle and associated structures. (After Nichols, in part.)

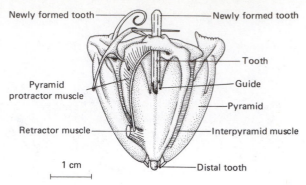

Figure 19–36 Lateral view of Aristotle's lantern of *Sphaerechinus granularis*. (From De Ridder, C., and Lawrence, J. M., 1982: Food and feeding mechanisms: Echinoidea. *In* Jangoux, M., and Lawrence, J. M. (Eds.): Echinoderm Nutrition. A. A. Balkema, Rotterdam. p. 81.)

The majority of sea urchins are grazers, scraping the substrate surface on which they live with their teeth. Although algae are usually the most important food, most sea urchins are generalists and include a wide range of plant and animal material in their diets (Lawrence, 1975). Moreover, the diet of a particular species varies from area to area, depending on what is available. *Lytechinus variegatus*, an inhabitant of turtle grass beds, consumes about 1 g of grass per week (Greenway, 1976).

Sea urchins living on soft bottoms in the deep sea are probably detritus feeders, consuming minute organic particles in the bottom ooze. Some, such as *Hygrosoma*, feed exclusively on terrestrial or shallow-water plant material carried down to the deep-sea floor.

Boring sea urchins feed on encrusting and endolithic algae on the walls of their burrows as well as algal fragments and other organic debris that is washed in. *Echinostrephus molaris* of the Indo-Pacific sits at the mouth of its burrow, and when debris touches the spines or long podia, the spine tips close and grasp it.

The interior of Aristotle's lantern contains both a buccal cavity and a pharynx that ascends through the apparatus and passes into an esophagus (Fig. 19–36). The esophagus descends along the outer side of the lantern and joins a tubular stomach (Fig. 19–34). At the junction of the esophagus and stomach, a blind pouch or cecum is usually present. The stomach makes a complete turn around the inner side of the test wall, to which it is suspended. It then passes into the thinner-walled aboral intestine, which makes a complete turn in the opposite

there is one such tooth band for each pyramid, there are five teeth projecting from the oral end of the lantern.

In addition to the teeth and pyramids, the Aristotle's lantern is composed of a number of smaller, rodlike pieces at the aboral end. By means of special muscles the lantern can be partially protruded and retracted through the mouth. Other muscles control the teeth, which can be opened and closed. In the more primitive cidaroids the scraping action of the lantern is largely restricted to opening and closing the teeth, but in higher groups the lantern can also be swung laterally. The ability of the lantern to be protruded and retracted makes possible pulling and tearing in addition to scraping. Aristotle's lantern is absent from heart urchins but functional in sand dollars.

direction. The intestine then ascends to join the rectum, which empties through the anus within the periproct.

In most echinoids a narrow tube, called a siphon, parallels the stomach for about half its length. The ends of the siphon open into the lumen of the intestine. Extracellular digestion begins in the stomach and is completed in the intestine, where absorption also occurs. The siphon functions to remove excess water from the food.

Irregular urchins are selective deposit feeders. All feed on organic material in the sand in which they burrow. The heart urchins obtain food by means of modified podia on the oral surface. During feeding these podia grope about the sand surface of the chamber, picking up food particles (see review by De Ridder and Lawrence, 1982).

Sand dollars had been thought to feed on small particles that fell between the aboral spines and were carried by cilia to the oral food groove. Recent work by Ellers and Telford (1984) and Telford, Mooi, and Ellers (1985) has demonstrated that most such particles fall off the edge of the test. Particles used as food are picked up by podia from the substrate beneath the oral surface of the animal. These particles are passed from podia to podia to the food grooves and then down the grooves to the mouth (Fig 19–37).

For *Dendraster excentricus* on the west coast of the United States, suspended particles are a major source of food because in quiet water this sand dollar lives with the posterior half of the body projecting above the sand surface. Food includes not only particles that pass between the spines but also diatoms and algal fragments collected by the tube feet and small crustacean prey caught by the pedicellariae (Timko, 1976).

The alimentary canal of irregular echinoids is more or less like that described for sea urchins, although the rectum extends posteriorly to the anus. Most spatangoids, including *Echinocardium*, build one or two drains at the back of the burrow to collect feces (Fig 19–30C). The drain is constructed by specialized posterior podia around the anus.

Predators of sea urchins include sea otters, fish, certain gastropods, and sea stars. Sea urchin eggs are eaten by humans in various parts of the world, especially in Japan. The Japanese support a large sea urchin fishery and import large quantities of sea urchin roe from other parts of the world. Two million pounds of *Strongylocentrotus* were harvested from the west coast of the United States and Canada in 1972.

Internal Transport, Gas Exchange, and Excretion

Coelomic fluid is the principal circulatory medium, and coelomocytes are abundant. The hemal system has the same basic plan of structure as in the asteroids.

In regular echinoids the five pairs of peristomial gills are probably the chief centers of gas exchange (Fig. 19–26A). Each gill is a highly branched outpocketing of the body wall and is therefore lined within and without by a ciliated epithelium. Coelomic fluid is pumped into and out of the gills by a system of muscles and ossicles associated with Aristotle's lantern.

As in other echinoderms, all the podia contribute to gas exchange. In most sea urchins the more aboral podia are modified in various ways for this function (Fenner, 1973). The podia and ampullae are commonly septate with a two-way circulation of fluid through these structures, one current (on the side nearest the radial canal) flowing toward the base of the podium and the opposite current flowing toward the podial tip (Fig. 19–38). The aboral podia may lack suckers and on *Arbacia* are flattened (Fig. 19–38B).

There are no peristomial gills in heart urchins and sand dollars. In these animals the modified podia of the petaloids, which are short and flat-

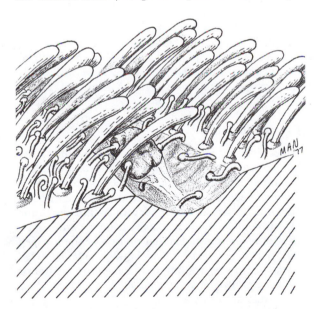

Figure 19–37 Section through food groove on oral surface of a sand dollar, showing food bolus being moved by podia. Many club-shaped spines rise above the podia.

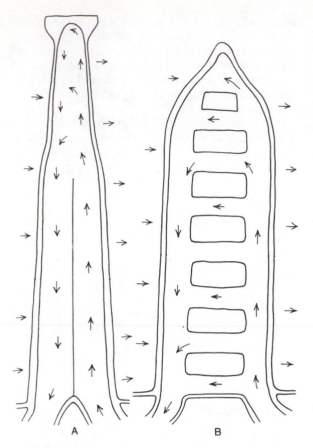

Figure 19–38 Respiratory adaptations of sea urchin podia: *A*, Oral-aboral view of a podium of *Strongylocentrotus*. All are suckered and partially septate, dividing the interior into two channels with a circulatory flow of fluid through podium, test, and ampulla. *B*, Flattened, suckerless, respiratory podium of *Arbacia*. Rippled wall divides interior into two channels with circulating fluid. Arrows indicate internal and external fluid flow. (From Fenner, D. H., 1973: The respiratory adaptations of the podia and ampullae of echinoids. Biol. Bull., *145*:323–339.)

tened, act as gas exchange structures (Fig. 19–39). The water current produced by external cilia flows in the opposite direction of the current within the podium, the countercurrent system ensuring a gradient favoring uptake of oxygen by the coelomic fluid. A similar countercurrent exists between the fluid within the ampulla and the surrounding fluid of the somatic coelom.

The coelomocytes are active in the removal of particulate waste and carry these accumulations to the gills, podia, and axial gland for disposal.

Nervous System

The nervous system is basically like that of the asteroids (Fig. 19–34). The circumoral ring encircles the pharynx inside the lantern, and the radial nerves pass between the pyramids of the lantern and run along the underside of the test, lying just beneath the radial canals of the water-vascular system.

The numerous sensory cells in the epithelium, particularly on the spines, pedicellariae, and podia, compose the major part of the echinoid sensory system. The buccal podia of sea urchins, the podia around the circumference of heart urchins, and the podia of the oral surface of sand dollars are all important in sensory reception. The spheridia are statocysts that function in orienting the animal to gravitational pull. It has been found, for example, that if the spheridia of a sand dollar are removed, the righting reaction is greatly delayed.

Echinoids are in general negatively phototropic and tend to seek the shade of crevices in rocks and shells. Some species of sea urchins, such as *Tripneustes, Lytechinus, Strongylocentrotus,* and the sea biscuit *Clypeaster,* cover themselves with shell fragments and other objects, using the tube feet (Fig 19–40*A*). The significance is uncertain but clearly seems to be a light response in some, for *Tripneustes* covers more in the summer than in the winter, and the related *Lytechinus* drops its cover at night and will cover only an experimental band of light crossing the body (Millott, 1975) (Fig. 19–40*B*).

A few sea urchins, notably species of *Diadema,* possess chromatophores, permitting some diurnal color change. However, this function is still poorly understood (see Binyon, 1972; Millott, 1975).

Reproduction

All echinoids are dioecious. A regular echinoid has five gonads suspended along the interambulacra on the inner side of the test (Fig. 19–34), but in most irregular echinoids the gonad of the posterior interambulacrum has disappeared. A short gonoduct extends aborally from each gonad and opens through a gonopore located on one of the five genital plates (Figs. 19–27*A*, 19–34, and 19–39*A*). Some burrowing sand dollars have long genital papillae, permitting release of eggs or sperm above the sand surface.

Sperm and eggs are shed into the sea water, where fertilization takes place. Brooding is dis-

A

Figure 19–39 *A*, Surface of the aboral center of the test of the sand dollar *Mellita.* Holes opposite the points of the central star are the gonopores. There is no gonad in the posterior interambulacrum. In the petaloid included in the lower half of the photograph, the grooves in which the modified branchial podia are located are conspicuous (compare with diagram in *B*). At the ends of the grooves can be seen the perforations for the podial canals penetrating the test. The inner of the two canals is very conspicuous; in this photograph, the outer canals can be seen only in the areas opposite the gonopores. (By Betty M. Barnes.) *B*, Diagram of branchial podia across one petaloid of a sand dollar. Arrows show direction of flow of sea water (external) and fluid of water-vascular system. (Adapted from Fenner.)

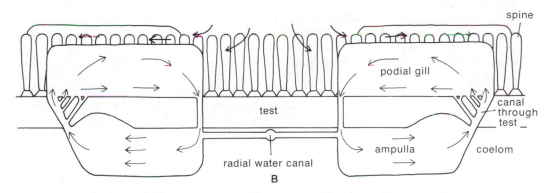

B

played by some cold-water sea urchins and heart urchins, and there is a brooding species of sand dollar. Brooding sea urchins retain their eggs on the peristome or around the periproct and use the spines to hold the eggs in position. The irregular forms brood their eggs in deep concavities on the petaloids.

Embryogeny

Cleavage is equal, up to the eight-cell stage, after which the blastomeres at the vegetal pole proliferate a number of small micromeres. A typical blastula ensues and becomes ciliated and free swimming within 12 hours after fertilization.

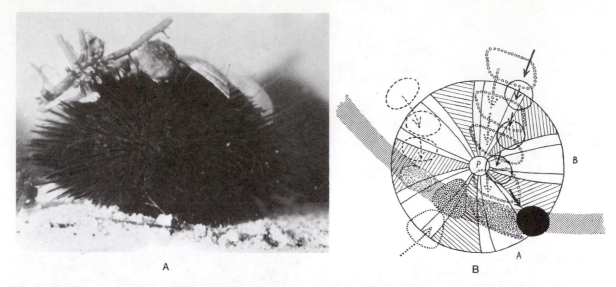

Figure 19–40   *A, Lytechinus variegatus,* a sea urchin that covers the aboral surface with shells, stones, and algae. (By Betty M. Barnes.) *B,* Specimen of *Lytechinus* is subject to a narrow band of light, which cuts across one side of the test. In response, the sea urchin picks up small stones and passes them from three different directions onto the illuminated region of the test. (From Millott, N., 1956: The covering reaction of sea-urchins. Jour. Exp. Biol., 33:508–523.)

Gastrulation is typical but is preceded by an interior proliferation of cells by the micromeres, which form the mesenchyme. The coelom is formed by the separation of the free end of the archenteron. This separated portion then divides into right and left pouches, or lateral divisions may appear before the end separates from the main portion of the archenteron (Fig. 19–41*A*). The gastrula becomes somewhat cone shaped and gradually develops into a planktonic larva, the echinopluteus, which bears six pairs of long larval arms and is very similar to the ophiopluteus of ophiuroids (Fig. 19–41*B*). The close similarity represents convergence, for the larval arms appear to be adaptive for increasing the surface area of the feeding-locomotor bands. The echinopluteus swims and feeds as described for asteroid larvae. Its complete development may take as long as several months. During later larval life the adult skeleton begins to form, first the five genital plates, then the ocular. The echinopluteus gradually sinks to the bottom; however, there is no attachment as in asteroids, and metamorphosis is extremely rapid, taking place in about an hour. Young urchins are no larger than 1 mm.

The larvae of *Dendraster excentricus,* the common sand dollar along the Pacific coast of the United States, has been found to settle and metamorphose in response to a substance released by adults located in the same site. Such preferential settlement would explain why this species, like many other sand dollars, occurs in sand beds with high population densities. The life-span of *Dendraster* is about eight years, and the sand dollar beds persist for decades (Highsmith, 1982; Burke, 1984).

Growth rates are known for only a few echinoids. Two sand dollars from the Gulf of California, *Encope grandis* and *Mellita grantii,* which reach diameters of 74 mm and 38 mm, respectively, require five years to attain 95 per cent of their maximum size. The annual mortality rate is 18 per cent for *Encope grandis* and 58 percent for *Mellita grantii* (Ebert and Dexter, 1975). *Strongylocentrotus purpuratus* off the California coast, one of the best known sea urchins, reaches sexual maturity during its second year when only 25 mm in diameter but may live 30 years or more.

SYSTEMATIC RÉSUMÉ OF CLASS ECHINOIDEA

Subclass Perischoechinoidea. Largely primitive fossil urchins of the Paleozoic seas, which made their first appearance in the Ordovician period with *Bothriocidaris.*

   Order Cidaroida. Of the four orders of the subclass Perischoechinoidea, this is the only one

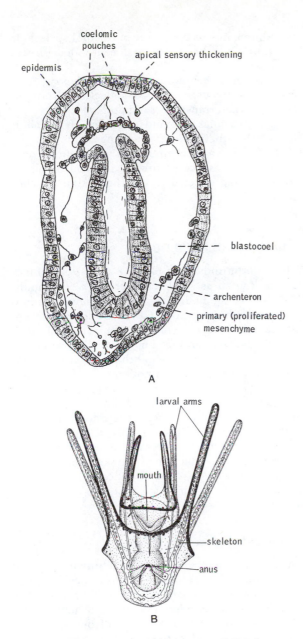

Figure 19–41 *A*, Gastrula of the sea urchin *Echinus esculentus.* (After McBride from Hyman.) *B*, Echinopluteus larva of the sand dollar *Fibularia craniola.* (After Mortensen.)

with two rows of plates for each ambulacrum and interambulacrum. It is also the only one that survived the Paleozoic era and became the ancestor of the remaining echinoids, most of which belong to the next subclass. The existing members of the Cidaroida are characterized by widely separated pri-

mary spines and small secondary spines. Gills are absent. *Eucidaris, Cidaris, Notocidaris.*

Subclass Euechinoidea. This subclass contains the majority of living species of echinoids.

Superorder Diadematacea. Sea urchins with perforated tubercles. Gills usually present.

Order Pedinoida. Rigid test with solid spines. Ten buccal plates on peristomial membrane. *Caenopedina* is the only living genus.

Order Diadematoida. Rigid or flexible test with hollow spines. Ten buccal plates on peristomial membrane. *Diadema, Plesiodiadema.*

Order Echinothuroida. Flexible test with poisonous secondary spines. Simple ambulacral plates on peristomial membrane. Gills inconspicuous or lost. *Asthenosoma.*

Superorder Echinacea. Sea urchins with rigid test and solid spines. Gills present. Peristomial membrane with ten buccal plates.

Order Arbacioida. Periproct with four or five plates. *Arbacia.*

Order Salenoida. Anus located eccentrically within periproct because of the presence of a large plate (suranal plate). *Acrosalenia.*

Order Temnopleuroida. Test sculptured in some. Camarodont lantern (large epiphyses are fused across top of each pyramid). *Tripneustes, Toxopneustes, Lytechinus.*

Order Phymosomatoida. Like the previous order but primary tubercles imperforate. *Glyptocidaris.*

Order Echinoida. Camarodont lantern and nonsculptured test with imperforate tubercles. *Echinus, Psammechinus, Paracentrotus, Echinometra, Echinostrephus, Colobocentrotus, Heterocentrotus, Strongylocentrotus.*

Superorder Gnathostomata. Irregular urchins. Mouth is in center of oral surface but anus has shifted out of apical center. Lantern present.

Order Holectypoida. No petaloids. Many fossil members were essentially regular in shape. The two living genera, *Echinoneus* and *Micropetalon,* are oval.

Order Clypeasteroida. True sand dollars. Petaloids present. Test greatly flattened. No phyllodes. *Clypeaster, Fibularia, Mellita, Encope, Rotula.*

Superorder Atelostomata. Irregular urchins. No lantern.

Order Holasteroida. Oval or bottle-shaped echinoids with thin delicate test. Petaloids and phyllodes not developed. Deep-water species. *Pourtalesia.*

Order Spatangoida. Heart urchins. Oval and elongated echinoids. Oral center shifted anteriorly, and anus shifted out of aboral apical center. Petaloids present but may be sunk into grooves. Phyllodes present. *Spatangus, Echinocardium, Moira, Meoma, Lovenia.*

Order Cassiduloida. Mostly extinct echinoids with round to oval test and a central or slightly anterior apical center. Phyllodes with intervening smaller areas (bourrelets). Poorly developed petaloids. The few existing species are tropical burrowers and somewhat similar to heart urchins. *Echinolampas.*

SUMMARY

1 In the class Echinoidea the spherical or flattened body is not drawn out into arms. The surface is covered with movable spines, which articulate on a test of sutured ossicles. Ambulacral areas containing the podia alternate with interambulacral areas arranged in meridians around the body. The plates of the test are perforated for the exit of gametes and for the canals connecting podia and ampullae. One genital plate functions as the madreporite. Correlated with the presence of a rigid skeletal test, the body wall lacks an internal muscle layer. Stalked, tridentate pedicellaria provide protection against settling organisms.

2 The regular echinoids, or sea urchins, are in general adapted for living on firm substrates. The radial globose body with long spines is believed to be primitive for the class. Sea urchins move by using the podia and pushing with the spines.

3 Most sea urchins feed by scraping algae, encrusting organisms, and detritus from hard surfaces. The scraping apparatus is a complex organ composed of numerous ossicles, five of which function as teeth.

4 Five pairs of oral evaginations (gills) function in gas exchange.

5 The irregular echinoids are adapted for burrowing in soft bottoms. The body is covered with a great number of minute spines. The spines serve not only for locomotion and burrowing but also to keep sediment off the body surface. The greatly flattened form of sand dollars is probably an adaptation for shallow burrowing.

6 Because of the burrowing habit, the same ambulacrum is always directed anteriorly, and varying degrees of secondary bilaterality have developed. In all irregular echinoids the anus has moved out of the aboral center to the posterior margin or posterior lunule. In sea biscuits and sand dollars the mouth remains in the center of the oral surface; in heart urchins the entire oral center has shifted forward.

7 Irregular echinoids are largely deposit feeders. Podia are used in food collection (heart urchins) or food transport (sand dollars).

8 In irregular echinoids modified aboral podia (petaloids) function in gas exchange.

9 The larva of echinoids is an echinopluteus. Metamorphosis occurs toward the end of planktonic life and at settling, but there is no attached stage.

Class Holothuroidea

The holothuroids are a class of some 900 echinoderms known as sea cucumbers. Like echinoids, the body of the holothuroid is not drawn out into arms, and the mouth and anus are located at opposite poles. Also, there are ambulacral and interambulacral areas arranged meridianally around the polar axis. However, holothuroids are distinguished from other echinoderms in having the polar axis greatly lengthened, which results in the elongated cucumber shape (Fig. 19–42). This shape forces the animal to lie with the side of the body, rather than the oral pole, against the substratum. The class is further distinguished from other echinoderms by the reduction of the skeleton to microscopic ossicles and by the modification of the buccal podia into a circle of tentacles around the mouth.

External Anatomy

Most sea cucumbers are black, brown, or olive green, but other colors and patterns are encountered. There is considerable range in size. The smallest species are less than 3 cm in length (oral to aboral end), whereas *Stichopus* from the Philippines may attain a length of 1 meter and a diameter of 24 cm. Most of the common North American and European species, such as *Cucumaria, Holothuria, Thyone,* and *Leptosynapta,* range from 10 to 30 cm in length.

The body shape varies from almost spherical to long and wormlike, as in the synaptid sea cucumbers (Fig. 19–43). Not infrequently, the mouth and anus are turned dorsally at the ends of the long axis of the body, and some sea cucumbers are even U shaped. Although there are a few genera, such as *Psolus* and *Ceto,* that have a protective armor of calcareous plates (modified surface ossicles), the

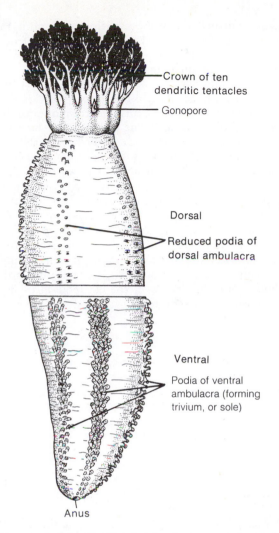

Figure 19–42 The North Atlantic sea cucumber *Cucumaria frondosa.*.

Figure 19–43 *Leptosynapta*, a member of order Apodida, wormlike burrowing holothuroids that lack podia, from the southeastern coast of the United States. Tentacles retracted.

body surface of the majority of sea cucumbers is leathery.

Holothuroids lie with one side of the body against the substratum, and this ventral surface is composed of three ambulacral areas (the trivium), commonly called the sole (Fig. 19–42). The dorsal surface consists of two ambulacral areas. As might be expected, the dorsal podia are usually reduced to warts or tubercles or are absent altogether, thus producing a secondary bilateral symmetry. Note that the bilateral symmetry of holothuroids has evolved in an entirely different manner from that of the irregular echinoids.

The podia of some sea cucumbers, whether reduced or not, have lost their radial distribution and become more or less randomly scattered over the body surface. The primitive radial configuration is seen in *Cucumaria*, where the podia are more or less restricted to the five ambulacral areas; in *Thyone*, on the other hand, podia are scattered over the entire body surface. Members of orders Apodida (*Synapta*, *Leptosynapta*, and *Euapta*), which are elongated and wormlike, and Molpadiida completely lack podia (Fig. 19–43).

The mouth is always surrounded by 10 to 30 tentacles, which represent modified buccal podia and are thus part of the water-vascular system. The tentacles are highly retractile, and the animal can completely retract both mouth and tentacles by pulling the adjacent body wall over them. The form and branching of the tentacles vary considerably.

Body Wall

The epidermis is nonciliated and covered externally by a thin cuticle. The thick dermal layer contains microscopic ossicles (called sclerites), which display a great variety of shapes (Fig. 19–44). These different shapes are important in the taxonomy of holothuroids. Beneath the dermis is a layer of circular muscle that overlies five single or double bands of longitudinal fibers located in the ambulacral areas. Recent studies have disclosed that the mechanical properties of the dermis can change under chemical stimulation. Thus, the body wall can be sufficiently flexible to permit the sea cucumber to squeeze through restricted passages or become so rigid that the animal cannot be dislodged. The mechanism by which the change occurs is still unknown (Motokawa, 1982).

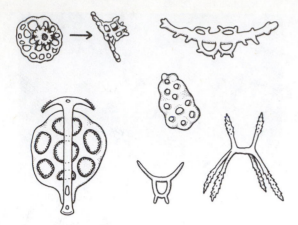

Figure 19—44   Microscopic ossicles of sea cucumbers. (After Bell.)

The body wall of sea cucumbers is a culinary delicacy in the Orient. Large species of sea cucumbers are collected and boiled, which causes the bodies to contract and thicken and also brings about evisceration of the internal organs. The body wall is then dried and sold, mostly to Chinese, as trepang or bêche-de-mer. Trepang imparts a distinctive flavor to food.

Locomotion

Sea cucumbers are relatively sluggish animals and live on the bottom surface or burrow in sand and mud. Forms with podia may creep along on the sole, with the podia functioning as in asteroids. Righting is accomplished by twisting the oral end around until the podia touch the substratum. Many hard-bottom forms live beneath stones, in rock and coral crevices, and among large algal holdfasts. Some, such as species of *Holothuria*, *Cucumaria*, and *Psolus*, are so sedentary that the podia are used more for attachment than for locomotion. Others, such as species of the large, tough *Stichopus*, crawl exposed on the surface (Fig. 19–45*A*). There are also a few species that live on algae.

Burrowing species include the Apodida and the order Molpadiida, which lack podia, as well as some pedate holothuroids, such as the common *Thyone* (Fig. 19–46). Burrowing is accomplished by alternate contraction of longitudinal and circular muscle layers of the body wall in the manner employed by earthworms. The tentacles aid by pushing away the sand. Some Apodida burrow completely beneath the surface. Other burrowers, the Molpadiida and species of *Thyone* and *Cucumaria*, are relatively sedentary and excavate U-

shaped burrows or burrows with one opening to the surface (Fig. 19–46). The tentacles extend from one opening of the burrow, and a pulsating, anal ventilating current is maintained through the opposite opening. Sedentary burrowers move very little once they have attained the proper position.

The Elasipodida are a curious group of entirely deep-sea holothuroids, some of which are benthic and some pelagic. The podia may be greatly enlarged and used for walking, for most deep-sea holothuroids live on the surface of the sea floor. The sole is quite flattened, and the tentacular crown is turned ventrally, giving them a markedly bilateral symmetry (Fig. 19–47). Pelagic species have papillae webbed together in various ways to form fins or sails. The transparent bathypelagic *Peniagone diaphana* lives up to 70 meters off the bottom, holding the body in a vertical position, tentacles upward. About half of the species of the Elasipodida are believed capable of swimming.

Nutrition

Sea cucumbers are chiefly deposit or suspension feeders. They stretch out their branched tentacles and either sweep them over the bottom or hold them out in the sea water. In either case, particulate material is trapped on adhesive papillae on the tentacular surfaces; the structure of the surface is related to the particle sizes normally selected for food (Smith, 1983). One at a time the tentacles are then stuffed into the pharynx, and the adhering food particles are wiped off as the tentacles are pulled out of the mouth.

Many sedentary species that live on hard surfaces beneath stones, such as members of the genus *Cucumaria*, are suspension feeders. More mobile epibenthic forms, such as the large *Stichopus* and the deep-sea elasipods, are deposit feeders, grazing on the bottom with their tentacles. They may be selective or not. Nonselective forms may literally shovel sand into the mouth with the tentacles, and the sand or mud castings of such species are conspicuous. *Parastichopus parvimensis*, a shallow-water, epibenthic deposit feeder from the west coast of North America, has been found to be about 22 per cent efficient in its utilization of organic deposits, of which the plant components are largely undigested (Yingst, 1976).

The epibenthic and infaunal Apodida and Molpadiida are also deposit feeders. The column of ingested sand in some of these wormlike species is easily visible through the transparent body wall. The European *Leptosynapta tenuis*, which lives in U-shaped burrows, ingests sand from the bottom of

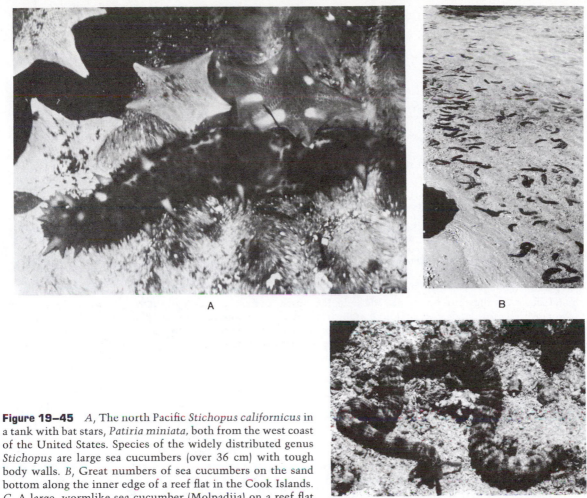

Figure 19–45 *A*, The north Pacific *Stichopus californicus* in a tank with bat stars, *Patiria miniata*, both from the west coast of the United States. Species of the widely distributed genus *Stichopus* are large sea cucumbers (over 36 cm) with tough body walls. *B*, Great numbers of sea cucumbers on the sand bottom along the inner edge of a reef flat in the Cook Islands. *C*, A large, wormlike sea cucumber (Molpadiia) on a reef flat in Fiji. (All by Betty M. Barnes.)

a funnel-like opening to the surface. Species of *Molpadia* live upside down in the sediment and produce a pile of castings where the anus comes to the surface.

The mouth is located in the middle of a buccal membrane at the base of the tentacular crown. The mouth opens into a pharynx, usually muscular, which is surrounded anteriorly by a calcareous ring of ossicles (Fig. 19–48). The calcareous ring provides support for the pharynx and the water ring, and it serves as the site for the anterior insertion of the longitudinal muscles of the body wall and the retractor muscles of the tentacles and mouth region. The tentacles and mouth can be pulled completely within the anterior end of the body when the animal is disturbed. Protraction is brought about through elevation of the coelomic fluid pressure.

The pharynx opens into an esophagus and then into a stomach (Fig. 19–48). The stomach is absent in many holothuroids, but when present, it functions as a gizzard. A long, looped intestine composes most of the digestive tract and is the site of digestion and absorption. The digestive process is still poorly understood. In many holothuroids, the intestine terminates in a cloaca prior to opening to the outside through the anus.

Gas Exchange, Excretion, and Internal Transport

The coelom is large, and the peritoneal cilia produce a current of coelomic fluid that contributes to the general circulation of materials within the body. Of the several types of coelomocytes, one flattened discoidal type, called a hemocyte, con-

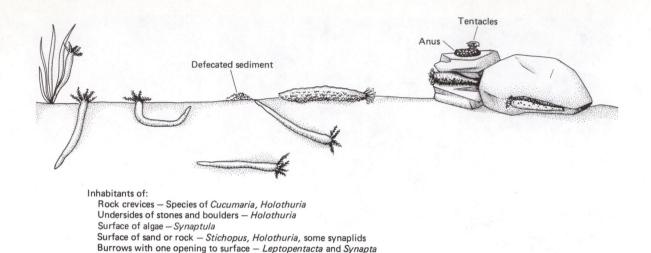

Inhabitants of:
    Rock crevices — Species of *Cucumaria, Holothuria*
    Undersides of stones and boulders — *Holothuria*
    Surface of algae — *Synaptula*
    Surface of sand or rock — *Stichopus, Holothuria,* some synaplids
    Burrows with one opening to surface — *Leptopentacta* and *Synapta*
    U-shaped burrows — *Thyone, Cucumaria, Echinocucumis*
    Subsurface burrows — *Leptosynapta*

Figure 19–46 Life-styles of shallow-water sea cucumbers. Animals not drawn to scale. The genera named contain common species that illustrate a particular life-style. Inhabitants of rock crevices (*Cucumaria, Holothuria*); inhabitants of the undersurfaces of large stones (*Holothuria*); rock and sediment surface dwellers (*Stichopus, Holothuria,* some synaptids); inhabitants of marine plant surfaces (synaptids); burrowers with two openings to sediment surface (*Thyone, Cucumaria, Echinocucumis*); burrowers with tentacles projecting to surface (*Synapta*); burrowers with anus projecting to surface (*Leptopentacta, Caudina*); burrowers with no opening to the surface (*Leptosynapta*).

tains hemoglobin. When present in large numbers, hemocytes give a red color to the coelomic fluid and hemal fluid (*Thyone, Cucumaria, Molpadia,* and others).

Except for the pelagic Elasipodida and burrowing Apodida, which obtain oxygen through the general body surface, gas exchange in holothuroids is accomplished by a remarkable system of tubules called respiratory trees. The respiratory trees are located in the coelom on the right and left sides of the digestive tract (Fig. 19–48). Each tree consists of a main trunk with many branches, each of which ends in a tiny vesicle. The trunks of the two trees emerge from the upper end of the cloaca either separately or by way of a common trunk.

Water circulates through the tubules by means of the pumping action of the cloaca and the respiratory trees. The cloaca dilates, filling with sea water. The anal sphincter then closes, the cloaca contracts, and water is forced into the respiratory trees. Water leaves the system because of the contraction of the tubules and the reverse action of the cloaca. Pumping is slow; *Holothuria* requires six to ten cloacal dilations and contractions to fill the trees, each contraction taking a minute or more. All the water is expelled in one action.

Gases are exchanged with the coelomic fluid and indirectly with the hemal system described later.

An interesting commensal relationship exists between the slender tropical pearlfish and sea cucumbers. This little fish, which is about 15 cm long, makes its home in the trunk of a respiratory tree of certain sea cucumbers. The fish leaves the host at night while it searches for food; after such excursions the fish forces its way into the anus and cloaca and back to the shelter of the respiratory tree.

Most ammonia probably exits by diffusion through the respiratory trees. Particulate waste, as well as nitrogenous material in crystalline form, is carried by coelomocytes from various parts of the body to the gonadal tubules, the respiratory tree, and the intestine. Waste then leaves the body through these organs.

Holothuroids, especially the Aspidochirotida (*Holothuria* and *Stichopus*), also possess the most highly developed hemal system of any of the echinoderm classes (Fig. 19–49). The general organization of the system is essentially like that of other echinoderms. A hemal ring and radial hemal sinuses parallel the water ring and the radial canals of the water-vascular system. The most conspicuous features of the system, at least in larger species, are a dorsal and a ventral vessel that accompany the intestine. These main intestinal sinuses supply the intestinal wall with a large number of smaller channels. In the region of the descending small intes-

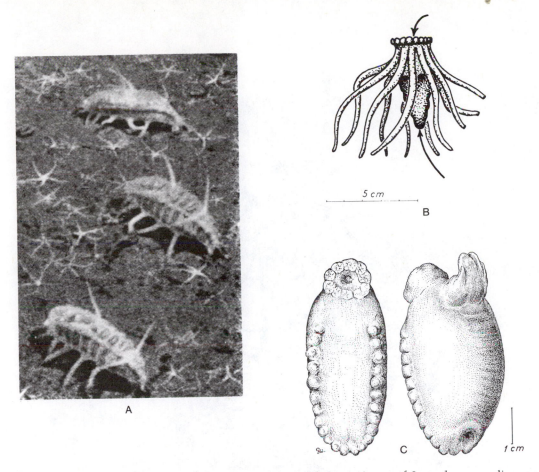

Figure 19–47 Deep-sea holothuroids of the order Elasipodida: *A*, Three specimens of *Scotoplanes* crawling over the bottom in the San Diego trough (1060 meters). The leglike structures are podia. (By R. F. Dill through courtesy of Barham, E. G. *In* Hansen, B., 1972: Deep Sea Research, 19:461–462.) *B, Pelagothuria*, a pelagic species having a circlet of long, webbed papillae behind the mouth and tentacles. (After Nichols, D., 1969: Echinoderms. Hutchinson and Co., London.) *C*, Benthic species *Ellipinion solium* with fused anterior papillae. (From Hansen, B., 1975: Galathea Report. Vol. 13. Systematics and Biology of the Deep-Sea Holothurians. pp. 111 and 164.)

tine, at least in *Isostichopus badionotus*, 120 to 150 single-chambered hearts pump blood from the dorsal vessel through a system of intestinal lamellae that project into the intestinal lumen. The left respiratory tree is intimately associated with the hemal system of the ascending small intestine.

Most of the hemal vessels are well developed, containing a muscle and a connective tissue layer, and covered on the outside with ciliated peritoneum and on the inside with endothelium (Herreid et al, 1976). The blood is essentially like the coelomic fluid; in fact, coelomocytes are formed in the walls of certain vessels. Peristaltic contractions of the dorsal vessel are of primary importance in propelling blood, but although circulation is largely unidirectional, it is not rapid, and there is some ebb

and flow. The holothuroid hemal system is clearly involved in some gas transport and appears to play some role in absorption or food transport.

Water-Vascular System

Although the water-vascular system of holothuroids is basically like that of other echinoderms, the madreporite in most species is peculiar in having lost connection with the body surface and in being unattached in the coelom (Fig. 19–48). Coelomic fluid rather than sea water enters and leaves the system. The madreporite hangs just beneath the base of the pharynx and is connected to the water ring by a short stone canal.

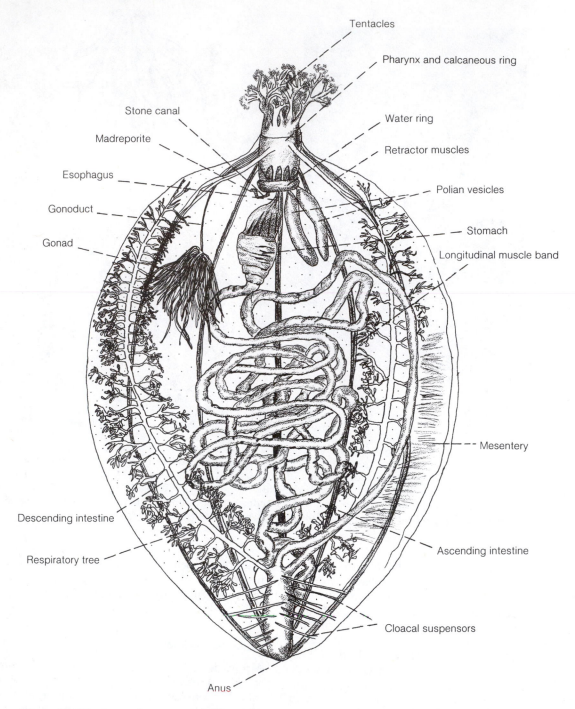

Tentacles

Pharynx and calcaneous ring

Stone canal

Water ring

Madreporite

Retractor muscles

Esophagus

Polian vesicles

Gonoduct

Stomach

Gonad

Longitudinal muscle band

Mesentery

Descending intestine

Respiratory tree

Ascending intestine

Cloacal suspensors

Anus

Figure 19–48 Internal structure of *Thyone briaereus*, a common sea cucumber that inhabits North Atlantic coastal waters. (After Coe from Hyman.)

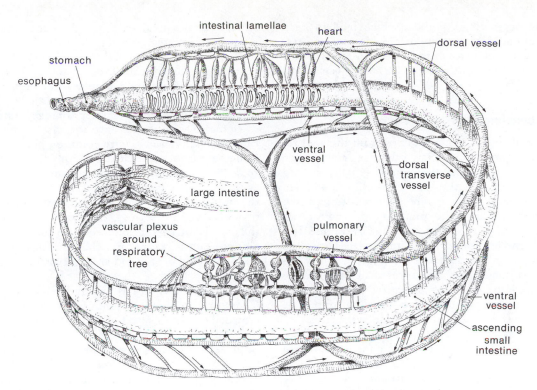

Figure 19–49 Hemal system of *Isostichopus badionotus*. Arrows indicate direction of blood flow. (After Herreid, C. F., LaRussa, V. F., and DeFesi, C. R., 1976: Blood vascular system of the sea cucumber *Stichopus moebii*. J. Morph., *150*(2):423–451.)

The water ring encircles the base of the pharynx and gives rise to polian vesicles, which hang into the coelom (Fig. 19–48). The vesicles are believed to function as expansion chambers in maintaining pressure within the water-vascular system. From the water ring, five radial canals pass upward to the inner side of the calcareous ring and then outward through a notch at the end of each radial plate (Fig. 19–50). Just before leaving the calcareous ring, each radial canal gives off canals to the tentacles. On leaving the ring, the radial canals then pass posteriorly within the body wall along the length of the ambulacra. Here lateral canals supply the podia. Ampullae are present for both podia and tentacles, although when the podia are reduced, there is a corresponding reduction in the ampullae.

In the Apodida, which lack tube feet, the water-vascular system is limited to the oral water ring, the polian vesicles, and the buccal podia (tentacles).

Nervous System

The circumoral nerve ring lies in the buccal membrane near the base of the tentacles (Fig. 19–50). The ring supplies nerves to the tentacles and also to the pharynx. The five radial nerves, on leaving the ring, pass through the notch in the radial plates of the calcareous ring and run the length of the ambulacra in the coelomic side of the dermis.

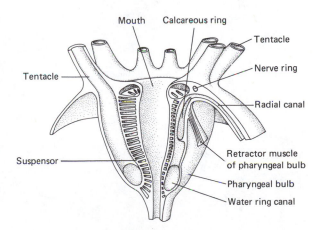

Figure 19–50 Section through anterior end of *Ocnus planci*. (After Herouard from Feral, J.-P., and Massin, C., 1982: Digestive systems: Holothuroids. *In* Jangoux, M., and Lawrence, J. M. (Eds.): Echinoderm Nutrition. A. A. Balkema, Rotterdam. p. 199.)

The burrowing Apodida, which tend to keep the oral end directed downward, possess one statocyst adjacent to each radial nerve, located near the point at which the nerve leaves the calcareous ring.

Evisceration and Regeneration

The expulsion of sticky tubules from the anal region is commonly associated with sea cucumbers, but this defensive phenomenon is actually limited to some species of the genera *Holothuria* and *Actinopyga*. Such sea cucumbers possess from a few to a large mass of white, pink, or red blind tubules (tubules of Cuvier) attached to the base of one (frequently the left) or both respiratory trees or to the common trunk of the two trees (Fig 19–51). When these sea cucumbers are irritated or attacked by some predator, the anus is directed toward the intruder, the body wall contracts, and by rupture of the cloaca the tubules are shot out of the anus.

The Cuvierian tubules of some species are not adhesive but liberate a toxic substance, holothurin (a saponin). Holothurin is also found in the body wall of some species. South Pacific islanders have long used the macerated bodies of certain sea cucumbers to catch tide pool fish.

During the process of expulsion each tubule is greatly elongated by water forced into its lumen, and the tubules break free from their attachment to the respiratory tree. In *Holothuria* the detached tubules are sticky and entangle the intruder in a mesh of adhesive threads. Small crabs and lobsters may be rendered completely helpless and left to die slowly, while the sea cucumber crawls away. After discharge the tubules of Cuvier are regenerated.

Sometimes confused with the discharge of the tubules of Cuvier is a more common phenomenon (called evisceration) that occurs in many holothuroids. Evisceration in the case of some genera, such as *Holothuria*, *Stichopus*, and *Actinopyga*, involves the rupture of the cloaca and the expulsion of one or both respiratory trees, the digestive tract, and the gonads. In *Thyone* and other sea cucumbers the anterior end ruptures, and the tentacles, pharynx and associated organs and at least part of the intestine are expelled.

The phenomenon has largely been observed in laboratory specimens that were subjected to crowded conditions, to foul water, to the injection of chemicals into the coelom, or to other abnormal conditions. But eviscerated specimens or individuals in the process of regeneration have been reported from natural habitats during certain times of the year, and it is a normal seasonal phenomenon in some species, perhaps eliminating a waste-laden gut at the end of a long feeding period (Byrne, 1985). Evisceration is followed by regeneration of

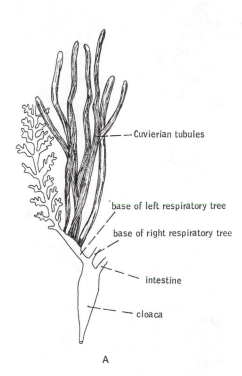

A

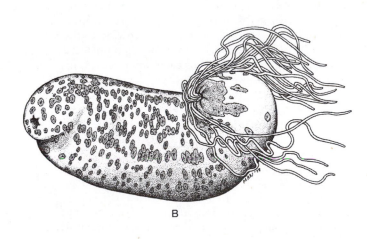

B

Figure 19–51 *A*, Base of a respiratory tree of *Holothuria impatiens*, showing Cuvierian tubules. (After Russo from Hyman.) *B*, Specimen of *Holothuria* releasing tubules of Cuvier. (Adapted from a photograph by Isobel Bennett.)

the lost part; the remaining stubs of the eviscerated organs or the associated mesenteries are the sites of the initial regenerative growth.

Reproduction

Holothuroids differ from all other living echinoderms in possessing a single gonad. Most cucumbers are dioecious, and the gonad is located anteriorly in the coelom beneath the middorsal interambulacrum (Fig. 19–48). The gonopore is located middorsally between the bases of two tentacles or just behind the tentacular collar (Fig. 19–42).

Some 30 brooding species are known, over half of which are cold-water forms, largely Antarctic. During spawning the eggs are caught by the tentacles and transferred to the sole or to the dorsal body surface for incubation. Even more remarkable is coelomic incubation, which takes place in the Californian *Thyone rubra,* in *Leptosynapta* from the North Sea, and in a few species from other parts of the world. The eggs pass from the gonads into the coelom and are fertilized in an undiscovered manner. Development takes place within the coelom, and the young leave the body of the mother through rupture in the anal region.

Embryogeny

Except in brooding species, development takes place externally in the sea water, and the embryo is planktonic. Development through gastrulation is like that of asteroids. The anterior half of the archenteron separates to develop as the coelom, leaving a shorter, posterior portion to become the gut. The right axohydrocoel never forms.

By the third day of development a larval stage called an auricularia has been reached (Fig. 19–52*A*). The auricularia is very similar to the bipinnaria of the asteroids and possesses a ciliated locomotor band that conforms to the same development as the locomotor band of the bipinnaria. Further development leads to a barrel-shaped larva, called a doliolaria, in which the original ciliated band has become broken up into three to five ciliated girdles (Fig. 19–52*B*).

There are many species of holothuroids (Dendrochirotida) that possess a nonfeeding, barrel-shaped vitellaria (Fig. 19–2). This type of larva, which is found in crinoids and a few ophiuroids, possesses cilated bands but no arms and is probably a specialized condition. Gradual metamorphosis during the latter part of planktonic existence results in a young sea cucumber with little loss of larval features. The tentacles, which are equivalent to buccal podia, appear prior to the appearance of the functional podia. At this stage the metamorphosing animal is sometimes called a pentactula larva. Eventually, the young sea cucumber settles to the bottom and assumes the adult mode of existence. The life-span of many sea cucumbers is between five and ten years.

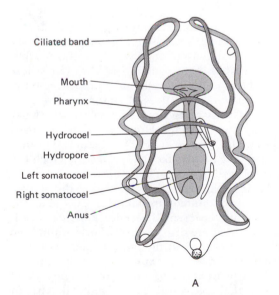

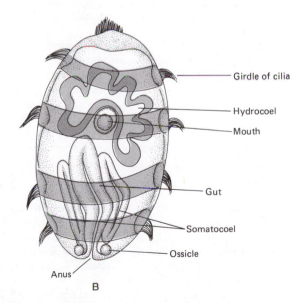

Figure 19–52 *A,* An auricularia larva (oral view). (After Mortensen from Hyman.) *B,* The doliolaria larva of *Leptosynapta inhaerens;* a common North Atlantic holothuroid (oral view). (After Runnström from Cuénot.)

SYSTEMATIC RÉSUMÉ OF CLASS HOLOTHUROIDEA

Order Dactylochirotida. Primitive sea cucumbers. Tentacles are simple, and the body is enclosed within a flexible test. Body U shaped. *Sphaerothuria*, *Echinocucumis*.

Order Dendrochirotida. Buccal podia, or tentacles, are dendritic and not provided with ampullae. Podia occurring on the sole, on all the ambulacra, or over the entire surface. *Cucumaria*, *Thyone*, *Psolus*.

Order Aspidochirotida. Tentacles peltate, or shield-like. Podia present, sometimes forming a well-developed sole. *Holothuria*, *Actinopyga*, *Stichopus*.

Order Elasipodida. Aberrant sea cucumbers with large, conical papillae and other appendages. Tentacles peltate. Almost all are deep-sea species. *Pelagothuria*, *Peniagone*.

Order Molpadiida. Posterior end of body narrowed to a tail. Fifteen digitate tentacles, but regular podia absent. *Molpadia*, *Caudina*.

Order Apodida. Wormlike sea cucumbers with only buccal podia, or tentacles, present. Tentacles digitate or pinnate. *Leptosynapta*, *Synapta*, *Euapta*.

SUMMARY

1 Members of the class Holothuroidea are distinguished by cylindrical bodies, in which the oral-aboral axis is greatly elongated, by reduction of the skeleton to microscopic ossicles, and by tentacular oral podia.

2 As a consequence of the elongated oral-aboral axis, holothuroids lie on their side. Since most species lie on the same three ambulacra (sole), this posture has led to some secondary bilateral symmetry. The ventral ambulacra in bilateral forms have well-developed podia; the dorsal ambulacra have reduced podia.

3 Some holothuroids are bottom surface dwellers, some live beneath stones or lodge in crevices, some burrow, and a few (mostly deep-sea forms) are pelagic. The podia are used for crawling and gripping the substratum. Two groups of wormlike burrowers has lost the locomotor podia and moves by peristaltic contractions.

4 Holothuroids are deposit feeders and suspension feeders. The mucus-covered tentacular surface traps particles when swept across the bottom or held in the water. Collected material is removed by the sucking action of the pharynx when the tentacles are stuffed in the mouth.

5 The water-vascular system is peculiar only in having the madreporite in the coelom. Branched internal evaginations of the posterior gut wall (respiratory trees) are gas exchange organs. The coelomic fluid contributes to internal transport, but many holothuroids have in addition a well-developed hemal system.

6 Gametes from the single gonad exit through an intertentacular gonopore. Development leads to a barrel-shaped doliolaria larva. Metamorphosis occurs prior to settling.

Class Crinoidea

The crinoids are the most ancient and in some respects the most primitive of the living classes of echinoderms. Attached, stalked crinoids, called sea lilies, flourished during the Paleozoic era, and some 80 species still exist today. However, modern sea lilies live at depths of 100 meters or more and are therefore not commonly encountered. The majority of living crinoids belong to a more modern branch of the class, the order Comatulida. The comatulids, or feather stars, are free-living crinoids that live from the intertidal zone to great depths, and some occur in large numbers on coral reefs. There are approximately 550 species found primarily in Indo-Pacific and polar waters.

External Structure

The body of existing crinoids is composed of a basal attachment stalk and a pentamerous body proper, called the crown (Fig. 19–53*A*). A well-developed stalk is present in sea lilies but is lost during the postlarval development of the free-moving feather stars (Fig. 19–53*B*). In the sessile sea lilies the stalk may reach almost 1 meter in length but is usually much shorter. However, there are fossil species with 20-meter stalks. The basal end bears a flattened disc or rootlike extensions by which the animal is fixed to hard or soft substrata.

The internal skeletal ossicles give the stalk a characteristic jointed appearance. The stalk of many crinoids bears small, slender, jointed appendages (cirri) that are displayed in whorls around the stalk (Fig. 19–54*A*). Although the stalk is lost in comatulids, the most proximal cirri of the stalk remain and spring as one or more circles from around the base of the crown (Fig. 19–53*B*). The cirri of comatulids are used for grasping the substratum when the animal comes to rest. They are long and

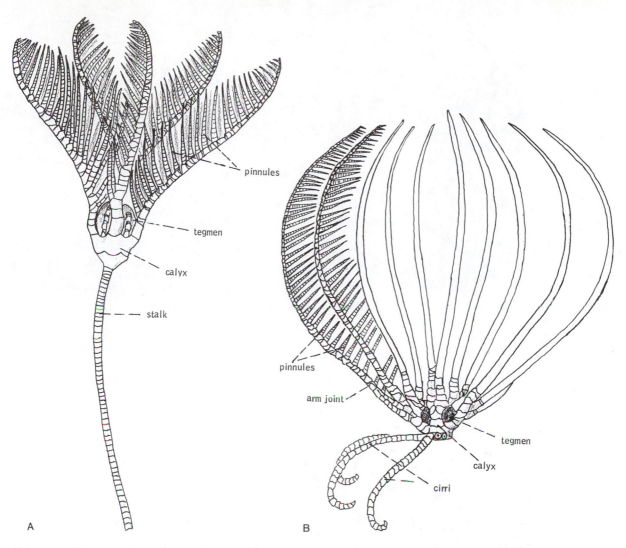

Figure 19–53 *A, Ptilocrinus pinnatus*, a stalked crinoid (or sea lily) with five arms. *B*, A Philippine 30-armed comatulid (or feather star), *Neometra acanthaster*. Not all arms shown. (Both after Clark from Hyman.)

slender in forms that rest on soft bottoms and stout and curved in species that grasp rocks, seaweed, and other objects.

The pentamerous body (the crown) is equivalent to the body of other echinoderms and, like those of the asteroids and the ophiuroids, is drawn out into arms. The crown is attached to the stalk by its aboral side; thus, in contrast to other living echinoderms, the oral surface is directed upward. The skeletal ossicles are best developed in the aboral body wall, usually called the calyx, which is thus somewhat cuplike. The oral wall (tegmen) forms a more or less membranous covering for the calyx cup (Figs. 19–53A and 19–54B). The mouth

is located in or near the center of the oral surface; five ambulacral grooves extend peripherally from the mouth to the arms. The anus opens onto the oral surface and is usually located in one of the interambulacral areas at the top of a prominence called the anal cone (Fig. 19–54B).

The arms issue from the periphery of the crown and have a jointed appearance like the stalk. Although there are some primitive species that possess five arms (Fig. 19–53A), in most crinoids each arm forks immediately upon leaving the crown, forming a total of ten arms. Further branching results in additional arms, and some comatulids possess 80 to 200 arms. The arms are usually less than

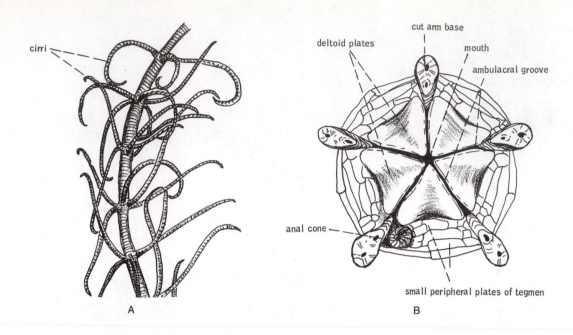

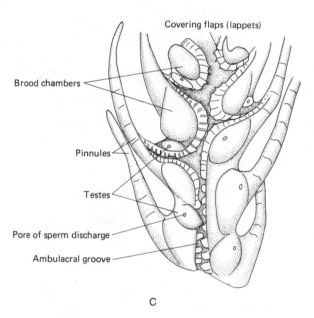

Figure 19–54 *A*, Part of stalk of the West Indian crinoid, *Cenocrinus asteria*, showing whorls of cirri. *B*, Tegmen of *Hyocrinus*, a stalked crinoid (oral view). (*A* and *B* after Carpenter from Hyman.) *C*, An arm section from *Notocrinus virile*, a comatulid (oral view). Podia not shown. (Modified after Hyman, L. H., 1955: The Invertebrates. Vol. IV. McGraw Hill Book Co., N.Y.)

10 cm in length but may reach almost 35 cm in some species.

On each side of the arm is a row of jointed appendages called pinnules, from which the name *feather star* is derived (Figs. 19–53 and 19–54*C*). The ambulacral grooves on the oral surface extend along the length of both the arms and the pinnules. The margins of the grooves are bordered by movable flaps, called lappets, which can expose or cover the groove. On the inner side of each lappet are three podia united at the base (Figs. 19–54*C* and 19–56*B*). Both podia and lappets also extend onto the pinnules.

Cold-water crinoids and those of the eastern Pacific are usually brown, but littoral species from tropical waters, especially comatulids, display a variety of colors in brilliant solid and variegated patterns.

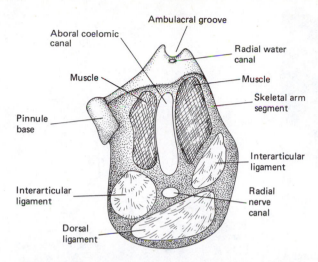

Aboral coelomic canal

Ambulacral groove

Radial water canal

Muscle

Muscle

Skeletal arm segment

Pinnule base

Interarticular ligament

Interarticular ligament

Radial nerve canal

Dorsal ligament

Figure 19–55 Diagrammatic cross section of the arm of the feather star *Florometra serratissima*. The muscles and ligaments are attached to the face of the large arm ossicle, which fills most of the section. (Based on a photograph by Meyer, D. L., 1971: The collagenous nature of problematical ligaments in crinoids. Mar. Biol., 9(3):238.)

Body Wall

The epidermis is unciliated except in the ambulacral grooves, and most of the dermis is occupied by the skeletal ossicles. The stalk, cirri, arm, and pinnules are of a solid construction, being composed almost entirely of a series of thick, disc-shaped ossicles; this accounts for the jointed appearance of these appendages (Figs. 19–53 and 19–55).

The surfaces of the arm ossicles articulate to permit at least some movement, similar to the ossicles composing the ophiuroid arms. The ossicles of the stalk are more securely interlocked than those of the arms, but even here some bending is possible. The ossicles of the stalk, cirri, and arms are bound together by collagen fiber bands, called ligaments, which penetrate into the porous skeletal material. As in other echinoderms, these fibers can rapidly change from a soft to a rigid state, permitting the animal to roll up the arms by the flexor muscles or lock them in an extended position (Meyer, 1971; Motokawa, 1984).

Locomotion

The sessile sea lilies are limited to bending movements of the stalk and flexion and extension of the arms. The stalkless comatulids, however, are free moving and are capable of both swimming and crawling. The oral surface is always directed upward, and the cirri are strongly thigmotactic and appear to control the righting reflex.

A sea lily swims by raising and lowering one set of arms alternately with certain others. In the ten-arm species every other arm sweeps downward while the alternate set moves upward. In species with more than ten arms, the arms still move in sets of five, but sequentially. For example, in a 40-arm comatulid there are eight sets of arms acting in sequence. To crawl, the animals lifts the body from the substratum and moves about on the arms. The arms and pinnules, which have minute terminal hooks, are often used to grasp and pull the animal over irregular and vertical surfaces.

Feather stars swim and crawl only for short distances, and swimming is largely an escape response. They cling to the bottom for long periods by means of the grasping cirri. Many shallow-water species are nocturnal. Three Red Sea species of *Lamprometra*, *Capillaster*, and *Comissia* inhabiting coral reefs hide during the day in crevices and deep within branching corals, keeping their arms tightly rolled (Fig. 19–57*A*). Stimulated by the lowered light intensities at sunset, they crawl upward out of their hiding places to exposed positions, where the arms are extended to feed. Species inhabiting deeper water may be stationary. Fishelson (1974) reports 200 specimens of two species of *Decametra* and *Oligometra* clinging to a gorgonian coral at about 30 meters for several months in the same position. They roll their arms in response to daytime illumination.

Nutrition

Crinoids are suspension feeders. During feeding the arms and pinnules are held outstretched and the podia are erect. The podia are shaped like small tentacles and bear mucus-secreting papillae along their length (Fig. 19–55). The three podia forming the triplets along the pinnules have different functions in the feeding process (Byrne and Fontaine, 1981; Meyer, 1982; Lahaye and Jangoux, 1985). The primary podium is the long, conspicuous, outstretched member. On touching a food particle, the primary podium whips into the ambulacral groove with the particle adhered to its surface (Fig. 19–56). Depending on the species, the animal removes the particle by wiping the podium against the ciliary current of the ambulacral groove or against the teriary podium (Fig. 19–56) or by scrap-

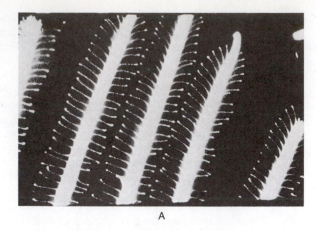

A

Figure 19–56 Function of crinoid podia in feeding: *A*, Extended position of the primary podia along the pinnules of the feather star *Florometra serratissima*. (From Byrne, M., and Fontaine, A. R., 1980: The feeding behavior of *Florometra serratissima*. Can. Jour. Zool., 59:11–18.) *B*, View of the ambulacral groove of a section of pinnule of *Antedon bifida*. One podium is wiping against ciliary current of ambulacral groove and against tertiary podium. (From Lahaye, M. C., and Jangoux, M., 1985: Functional morphology of the podia and ambulacral grooves of the comatulid crinoid *Antedon bifida*. Mar. Biol., 86:307–318.)

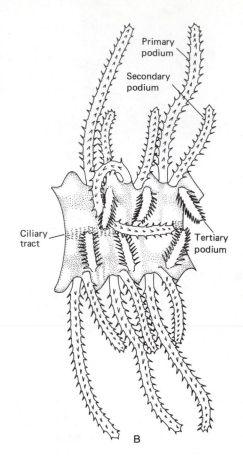

B

ing it between adjacent lappets. The secondary podium functions much like the primary podium. Transport of particles down the groove is by ciliary action. The lappets function largely as side walls for the ambulacral groove.

There is still uncertainty as to the precise nature of crinoid food. The gut contents of Red Sea crinoids were largely zooplankton (Rutman and Fishelson, 1969), but La Touche and West (1980) found that *Antedon bifida* fed largely on resuspended detritus and that detrital bacteria may be the principal food source.

The feeding position of the arms varies with the environment. Many crinoids form a planar, vertical filtration fan that is oriented more or less at a right angle to the prevailing current or a circular fan that is tilted into the current. Sea lilies achieve the tilt by bending the stalk. The arm tips may be turned back toward the current so that the animal has the shape of an umbrella (Fig. 19–57C). In the fan position the adjacent pinnules and podia form a relatively tight mesh and an efficient filter. Shal-

low-water comatulids may employ a vertical fan held at a right angle to currents, or they may remain in more protective positions within crevices or coral beds and extend the arms in several directions (Fig. 19–57B) (La Touche, 1978; Meyer et al, 1984). Only some deep-water crinoids hold the arms in the form of an upward-directed collecting funnel (Macurda and Meyer, 1974, 1976; Meyer, 1982).

There is a correlation between podial spacing and habitat. Those species that live on reefs in spaces between corals have longer and more widely spaced podia than those that live exposed and form more regular filtration fans (Meyer, 1982). There is also some correlation between the number and length of the arms and the food supply of the habitat. Crinoids living at great depths or in cold water, where detritus or plankton is rich, usually have a small number of arms (ten or fewer); the reverse is true of littoral, warm-water species. The total length of food-trapping ambulacral surface may be enormous. The Japanese stalked crinoid

Figure 19–57 Arm positions of crinoids: *A*, Diurnal, nonfeeding, rolled arm position of the feather star *Heterometra. B*, The Red Sea feather star *Commisia*, which inhabits crevices in coral reefs, projecting its arms to feed. (*A* and *B* from Fishelson, L., 1974: Ecology of the northern Red Sea crinoids and their epi- and endozoic fauna. Mar. Biol., 26:183–192.) *C*, Feeding crinoids in the Straits of Florida at 600 to 700 meters. A feather star atop a sponge is holding its arms in the form of a vertical fan. Three inclined sea lilies *(Diplocrinus)* form parabolic fans with the arm tips directed toward the current. (By C. Neumann, from Macurda, D. B., and Meyer, D. L., 1976: The identification and interpretation of stalked crinoids from deep-water photographs. Bull. Mar. Sci., 26(2): 205–215.)

C

Metacrinus rotundus, with 56 arms 24 cm in length, possesses a total ambulacral groove length of 80 meters.

Crinoids are believed to display the primitive method of feeding used by the echinoderms and also to illustrate the original function of the water-vascular system—that is, the water-vascular system originally evolved as a means of capturing food and secondarily assumed a locomotor function in those groups that have become free moving and inverted.

The mouth leads into the short esophagus, which then opens into an intestine (Fig. 19–58). The intestine descends and makes a complete turn around the inner side of the calyx wall. The terminal portion then passes upward into the short rectum, which opens through the anus at the tip of the anal cone.

The details of digestion are still unknown. Wastes are egested as large, compact, mucus-cemented balls that fall from the anal cone onto the surface of the disc and then drop off the body.

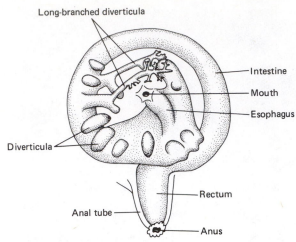

Figure 19–58 Digestive tract of the comatulid *Antedon.* (After Chadwick from Hyman.)

Internal Transport, Gas Exchange, and Excretion

The coelom is reduced to a network of communicating spaces as a result of invasion by connective tissue. In the oral side of the arms the coelom extends as five parallel canals, and in the stalk five coelomic canals pass through a central perforation in the ossicles and give rise to one canal into each cirrus (Fig. 19–55). The hemal system is a network of spaces and sinuses within the connective tissue strands invading the coelom.

The podia are undoubtedly the principal sites of gas exchange, and the great surface area presented by the branching arms makes unnecessary any special respiratory surfaces, such as are found in most other echinoderms.

Wastes gathered by coelomocytes are believed to be deposited in little saccules located in rows along the sides of the ambulacral grooves. Supposedly, the saccules periodically discharge to the exterior.

Water-Vascular System

There is no madreporite in crinoids. The water ring encircles the mouth and gives off at each interradius a large number of short stone canals, which open into the coelom (the feather star *Antedon* possesses about 50 canals at each interradius). At each radius of the ring canal a radial canal extends into each arm just beneath the ambulacral groove and forks into all the branches and into the pin-

nules (Fig. 19–55). From the radial canals extend lateral canals supplying the podia. There are no ampullae, and one lateral canal supplies the cluster of three podia except in the buccal region. Hydraulic pressure for extension of the podia is generated by contraction of the radial water canal, which is provided with cross-muscle fibers.

Nervous System

The crinoid nervous system is composed of three interconnecting divisions. The chief motor system is an aboral (or entoneural) system located as a cup-shaped mass in the apex of the calyx. The aboral system provides nerves to the cirri and five brachial nerves to the arms and pinnules (Fig. 19–55). The oral (ectoneural) nerve system, which is sensory, is homologous to the principal system of other echinoderms and consists of a subepidermal radial nerve that runs just beneath the ambulacral groove in the arms and a nerve ring around the mouth. Just below the oral system is a deeper, hyponeural sensory system having a central ring from which arises a pair of lateral brachial nerves to each arm. These nerves innervate the pinnule and podia.

Regeneration and Reproduction

Crinoids possess considerable powers of regeneration and in this respect are similar to the asteroids and the ophiuroids. Part or all of an arm can be cast off if seized or if subjected to unfavorable environmental conditions. The lost arm is then regenerated. The visceral mass within the calyx can be regenerated in several weeks; such regeneration may be important in surviving fish predation.

Crinoids are all dioecious, and there are no distinct gonads. The gametes develop from germinal epithelium within an expanded extension of the coelom (the genital canal) located within the pinnules, as in *Antedon*, or within the arms (Fig. 19–55). Not all the pinnules are involved in the formation of sex cells, but only those along the proximal half of the arm length.

When the eggs or sperm are mature, spawning takes place by rupture of the pinnule walls, and the eggs and sperm are shed into the sea water. In *Antedon* and others the eggs are cemented to the outer surface of the pinnules by the secretion of epidermal gland cells. Hatching takes place at the larval stage. Brooding by cold-water crinoids (many Antarctic forms) is displayed as in other echinoderms. The brood chambers are saclike invaginations of the arm or the pinnule walls adjacent to the genital canals, and the eggs probably enter the brood chamber by rupture (Fig. 19–54C).

Embryogeny

Development through the early gastrula stage is essentially like that in asteroids and holothuroids. During the formation of the coelomic sacs the embryo elongates, and development proceeds toward a free-swimming larval stage. The crinoid larva, a nonfeeding vitellaria, is essentially like the vitellaria of holothuroids, being somewhat barrel shaped with an anterior apical tuft and a number of transverse, ciliated bands (Fig. 19–59A).

After a free-swimming existence the vitellaria settles to the bottom and attaches, employing a glandular midventral depression (the adhesive pit) located near the apical tuft. There ensues an extended metamorphosis resulting in the formation of a minute, stalked, sessile crinoid. In the comatulids, metamorphosis also results in a stalked sessile stage (the pentacrinoid larva) that resembles a minute sea lily (Fig. 19–59B).

The pentacrinoid of *Antedon* is a little over 3 mm long when the arms appear, and it requires about six weeks from the time of attachment of the vitellaria to attain this stage. After up to several months as a pentacrinoid, during which time the cirri are formed, the crown breaks free from the stalk, and the young animal assumes the adult, free-swimming existence.

SYSTEMATIC RÉSUMÉ OF CLASS CRINOIDEA

Subclasses Inadunata, Flexibilia, and Camerata. Stalked Paleozoic crinoids with or without cirri, some without pinnules. Organization of calyx ossicles important in distinguishing these fossil groups.

Subclass Articulata. Extinct as well as all living crinoids belong to this subclass. The latter are placed within five orders.

Order Millericrinida. Sea lilies without cirri. *Hyocrinus, Calamocrinus.*

Order Cyrtocrinida. This order contains two living species from the Caribbean and mid-Atlantic in which the aboral end of the crown is attached directly to the substratum. *Holopus.*

Order Bourgueticrinida. Mostly small sea lilies. The slender stalk lacks cirri. *Rhizocrinus, Bathycrinus.*

Order Isocrinida. Sea lilies with cirri. *Metacrinus, Cenocrinus.*

Order Comatulida. Feather stars. Stalkless, unattached crinoids.

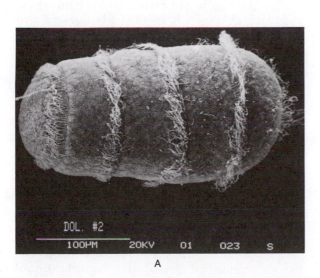

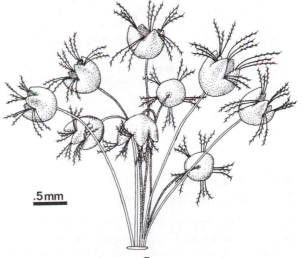

Figure 19–59 *A,* Scanning electron micrograph of the doliolaria larva of the feather star *Florometra serratissima.* The cilated bands are numbered and the arrow points to the apical tuft of cila. *B,* A cluster of pentacrinoids of *Florometra serratissima* one month after settlement. (From Mladenov, P. V., and Chia, F. S., 1983: Development, settling behavior, metamorphosis and pentacrinoid feeding and growth of the feather star *Florometra serratissima.* Mar. Biol., 73:309–323.)

SUMMARY

1 Members of the class Crinoidea, which include the stalked and attached sea lilies and the stalkless and free feather stars, are the only living echinoderms in which the oral surface is directed upward. This condition is also true of most Paleozoic echinoderms.

2 The crown of both stalked and stalkless crinoids is composed of multiple arms around a heavy, central calyx, which is covered by a membranous oral wall, or tegmen. The tegmen contains the mouth in the center and the anus to one side.

3 The multiplicity of arms results from basal branching of an originally pentamerous arrangement. The arms bear numerous small, lateral branches (pinnules), and the oral surface of all branches, including the pinnules, contains a ciliated, ambulacral groove.

4 Heavy ossicles compose much of the relatively solid stalk, cirri, arms and pinnules. Heavy ossicles are also located within the calyx wall.

5 The attached sea lilies can bend the stalk and unroll the arms when feeding. Feather stars perch with their cirri and crawl and swim with the arms.

6 Crinoids are suspension feeders, and the podia, on touching zooplankton or other suspended particles, undergo flicking action, driving the particles into the ambulacral groove. Ambulacral cilia carry the mucus-entrapped particles down the arms into the mouth. The arms are held as a funnel or, when in a current, as a planar or circular fan. The multiple arms and pinnules provide the necessary surface area for this mode of feeding.

7 Gametes are produced in the arms, which are also the site of brooding, when it occurs. Development leads to a barrel-shaped vitellaria larva. Metamorphosis occurs following settling and attachment. Feather stars pass through a stalked stage (pentacrinoid) before the crown becomes free.

Fossil Echinoderms and Echinoderm Phylogeny

The echinoderms rank with mollusks, brachiopods, and arthropods in having one of the richest and oldest fossil records of any group in the Animal Kingdom. Echinoderms first appeared in the early Cambrian period and were extremely abundant during the later periods of the Paleozoic era, when a number of fossil classes reached the peak of their evolutionary development.

The crinoids are the only living echinoderms that are attached, but an attached condition was characteristic of the majority of Paleozoic forms, including a number of extinct classes. These groups are believed to display many primitive features of the phylum.

Like living crinoids, most fossil forms were probably suspension feeders in which the podia and upward-directed ambulacral grooves were food-catching and food-conducting structures. In most attached fossil species the grooves branched onto a slender, pinnule-like projection (the brachiole) (Fig. 19–62B). The term *brachiole* is used to distinguish these body extensions from the heavier arms of crinoids and stelleroids. Brachioles varied in number from a few to hundreds and, like the arms and pinnules of crinoids, represented an adaptation that increased the surface area for food collection. Since these fossil forms were sessile, and since the oral surface was always directed upwards, the original function of the podia could not have been locomotor.

The skeletal system of the extinct attached echinoderms was somewhat like that of the crinoids, with the exception that the ossicles (plates) of the crown were not limited to the aboral surface (calyx) but extended orally to the mouth, thus enclosing the internal organs within a test (or theca). Protective cover plates, which could fold over the ambulacral grooves and mouth, were generally present. The anus was usually located eccentrically in one of the interradii, as in the crinoids.

The oldest group of extinct echinoderms is the Eocrinoidea, known from the early Cambrian to the Ordovician. Eocrinoids were stalked or stalkless echinoderms with an enclosed theca (Fig. 19–60C). At the upper, or oral, end were five ambulacra and five to many brachioles. The pentamerous and oral position of ambulacra and brachioles gives these animals a superficial resemblance to crinoids (Fig. 19–61A), although they are actually more similar to cystoids, the following group.

The cystoids (classes Rhombifera and Diplorita) were attached species that ranged from the middle Ordovician period to the Permian. The theca of cystoids was more or less oval, with the oral end directed upward and the aboral end attached to the substratum directly or sometimes by a stem (Fig. 19–62). A characteristic feature of cystoids was a system of pores that perforated the theca. The pores were either dispersed over the body surface (diplopores) or confined to special areas (rhombopores). In either case the pores were probably part of a system for gas exchange permitting either an outward flow of coelomic fluid or an inward flow of sea water. The three or, more commonly, five ambulacra were radially arranged and extended

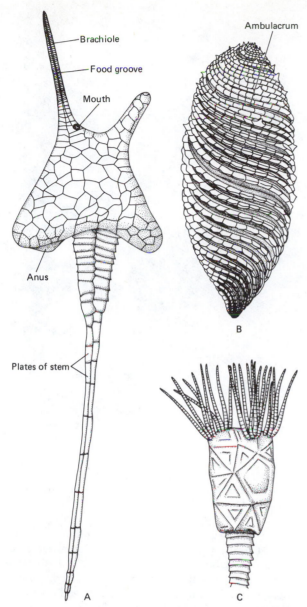

Figure 19–60 *A, Dendrocystites,* a later carpoid possessing one brachiole. (Modified from Bather.) *B,* A hypothetical restoration of a partially contracted helicoplacoid, *Helicoplacus.* (After Durham and Caster.) *C, Mimocystites,* an eocrinoid. (After Ubaghs, G., 1967: Treatise on Invertebrate Paleontology, Pt S, Vol. 1. Courtesy of Geological Society of America and the University of Kansas, Lawrence, Kan.)

outward and downward to varying degrees over the sides of the theca, in some species to the aboral pole. Small brachioles were either located around the mouth or mounted on the plates along each side of the ambulacral grooves (Fig. 19–62). Branches from the groove extended up each brachiole.

The class Blastoidea is another group of extinct, attached echinoderms that flourished in the Paleozoic seas (Fig. 19–63). The blastoids first appeared in the Ordovician period, reached their peak in the Mississippian, and became extinct during the Permian. *Pentremites* is a very common blastoid fossil found in Mississippian limestone (Fig. 19–63*B*).

Like cystoids, blastoids possessed an oval theca that was attached aborally by a short stalk or attached directly to the substratum. Five ambulacra extended from the mouth at the oral pole down the sides of the theca and frequently were elevated as broad ridges that alternated with the depressed interambulacral areas. The margins of the ambulacra were bordered by a single row of slender, closely placed brachioles, into which extended branches from the ambulacral groove. In oral view the brachioles gave the appearance of a thick fringe along the border of a five-pointed star.

Peculiar to blastoids was a system of folds, or folds and pores, called hydrospires, located at each side of the ambulacra. The hydrospires are thought to represent a gas exchange mechanism by which water circulated inwardly through the thecal plates, allowing greater surface area for the exchange of gases between the coelom and the sea water. The hydrospires were thus somewhat like reversed asteroid papulae, the folds projecting into the coelom instead of externally from the coelom (Fig. 19–64).

The Edrioasteroidea is a class of extinct attached echinoderms that first appeared in the early Cambrian period and ranged into the Pennsylvanian period. The theca was oval or discoid and composed of imbricated, abutting, or fused plates. The presence of tubercles on the plates indicates that some species possessed spines. Some edrioasteroids had stalks (Fig. 19–65*A*), but the discoid species were stalkless and lived attached to shells or other objects (Fig. 19–65*B*). There were no brachioles, and the five ambulacra, which extended over the theca, were straight or curved. Some of these animals thus looked like a brittle star wrapped around a ball. Of particular interest was the presence of pores between the ambulacral plates through which the podia extended. Movable cover plates protected the podia and food grooves (Fig. 19–65*C*).

The crinoids possess the richest fossil record of all the echinoderm classes and are the only attached species that still exist today. Typical crinoids first appeared in the Ordovician, and they left a fossil record in every succeeding period. The class reached its climax in the Mississippian, although

Figure 19–61 Reconstructions of the marine fauna of Paleozoic seas: *A*, Crinoids; a coiled cephalopod is on right. *B*, A long stalked crinoid in front of a cluster of rugose corals. To the left of the rugose corals is a brachiopod, and in the front left is a gastropod. *C*, Straight cone nautiloid cephalopods, with rugose corals in lower right corner. *D*, A large coiled cephalopod resting on bottom. Brachiopods in lower left corner and rugose corals in background. (Photographs by Betty M. Barnes. Models are in the National Museum of Natural History, Smithsonian Institution.)

there was a second somewhat lesser evolutionary development during the Permian. During these periods certain shallow seas supported enormous faunas. The Burlington (Mississippian) limestone of Iowa, Illinois, and Missouri is composed almost entirely of crinoids, with 15,000 crinoids per cubic meter of limestone (Macurda and Meyer, 1983). Paleozoic crinoids were all stalked, and the earlier species lacked pinnules. Modern crinoids, which contain the stalkless comatulids and belong to different orders than the Paleozoic species, did not appear until the Mesozoic era.

An early, small group of extinct echinoderms was the carpoids, known from the Cambrian to the Devonian. Formerly placed in the class Carpoidea, they are now separated into four classes (p. 837). These animals are of especial interest because they were asymmetrical (Fig. 19–60A). Some were stalkless and rested directly on the bottom; others were stalked, but the body was apparently bent over so that the crown was oriented horizontally to the

substratum—i.e., one side of the oral-aboral axis faced toward the substratum and the other side faced away from the substratum. The oral surface carried openings that have been interpreted as anus and mouth. Two of the classes possessed a single, armlike projection bearing an ambulacral groove.

Another nonradial class of fossil echinoderms is the Helicoplacoidea (Durham and Caster, 1963). The helicoplacoids were spindle-shaped animals with the mouth at one end of the body (Fig. 19–60B). There was a single, branched ambulacrum. The body wall was covered with pleated plates, which allowed expansion and contraction. This flexibility perhaps permitted the animal to retract its body, which may have been situated vertically in the sand. The expanded anterior end projected above the surface of the sand when the animal was feeding.

As would be expected from the nature of the skeleton, the holothuroids have the poorest fossil record and the echinoids have the best fossil record

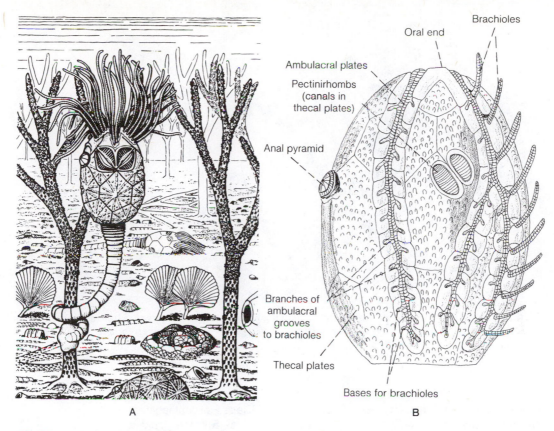

Figure 19–62 *A*, Reconstruction of an Ordovician cystoid, *Lepadocytis,* attached to an erect bryozoan. An edrioasteroid is attached to the bottom in front of two brachiopods. (From Kesling, R. V., 1967: *In* Moore, R. C. (Ed.): Treatise on Invertebrate Paleontology. Pt. S, Vol. 1. Courtesy of the Geological Society of America and the University of Kansas. pp. S85–S286.) *B*, Lateral view of the cystoid *Callocystites.* (After Hyman, L. H., 1955: The Invertebrates. Vol. IV. McGraw-Hill Book Co., N.Y.)

of the living classes of unattached echinoderms, but compared with attached forms, the fossil record of unattached species is sparse. Although fossil asteroids and echinoids first appeared in the Ordovician period and ophiuroids in the Mississippian, Paleozoic species are relatively rare; these classes are much better represented in Mesozoic and Cenozoic rocks. The earlier echinoids, including a number of extinct orders, were all regular forms, and the first fossil heart urchins and sand dollars appeared in the Triassic period.

The classes of echinoderms, both living and extinct, were formerly placed into two subphyla, the Pelmatozoa and the Eleutherozoa. The Pelmatozoa contained attached species having the oral surface directed upward; the Eleutherozoa contained unattached echinoderms having the oral surface directed downward. It is now recognized that such a division is an artificial one from the standpoint of phylogenetic relationships, and these subphyla have been abandoned. The following arrangement of echinoderm classes is used in the *Treatise on In-*

vertebrate Paleontology (see Moore). It has been adopted by most echinologists.

Subphylum Homalozoa. Carpoids. Paleozoic echinoderms lacking any evidence of radial symmetry.
  Class Homostelea.
  Class Homoiostelea.
  Class Stylophora.
  Class Ctenocystoidea.
Subphylum Crinozoa. Radially symmetrical echinoderms having a globoid or cup-shaped theca and brachioles or arms. Attached, with oral surface directed upward.
  Class Eocrinoidea.
  Class Paracrinoidea.
  Class Crinoidea.
  Class Cystoidea (now divided into classes Rhombifera and Diploporita).
  Class Blastoidea.
  Class Parablastoidea.
Subphylum Asterozoa. Unattached, radially symmetrical, star-shaped echinoderms.

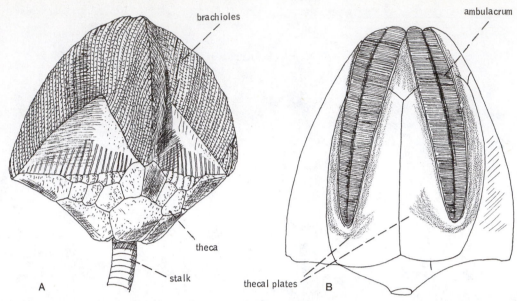

Figure 19–63 Class Blastoidea: *A, Blastoidocrinus,* showing brachioles (lateral view). (After Jaekel from Hyman.) *B, Pentremites* (lateral view, brachioles not shown). (After Hyman, L. H., 1955: The Invertebrates. Vol. IV. McGraw-Hill Book Co., N.Y.)

Class Stelleroidea.
   Subclass Somasteroidea.
   Subclass Asteroidea.
   Subclass Ophiuroidea.
Subphylum Echinozoa. Radially symmetrical globoid or discoid echinoderms without arms or brachioles. Mostly unattached.

Class Cyclocystoidea.
Class Helicoplacoidea.
Class Edrioasteroidea.
Class Holothuroidea.
Class Echinoidea.
Class Ophiocistioidea.

The origin of the echinoderms and the phylogenetic relationships of the subphyla continue to be unsolved questions and subjects of much speculation. Despite the extensive fossil record, paleontological evidence is still insufficient in many important areas. Before reviewing some of the older ideas that are still current, as well as some more recent views, it may be helpful to enumerate a few reasonable assumptions that would be accepted by many zoologists.

1 Echinoderms evolved from motile, bilaterally symmetrical ancestors that possessed a tripartite coelom.

2 As is true of other animals, the echinoderm skeleton and radial symmetry probably represent adaptations to a sessile existence, at least initially.

3 The original function of the water-vascular system was feeding and not locomotion.

A classical theory, and one that still finds many supporters, holds that from a bilateral, tricoelomate, free-moving ancestor, some group became attached to the bottom and assumed a sessile mode

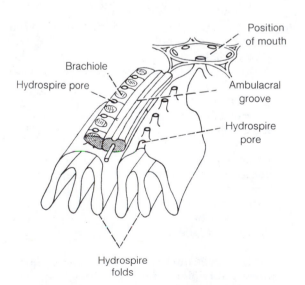

Figure 19–64 Hypothetical gas exchange system of a blastoid. (After Nichols, D., 1969: The Echinoderms. Hutchinson and Co., London.)

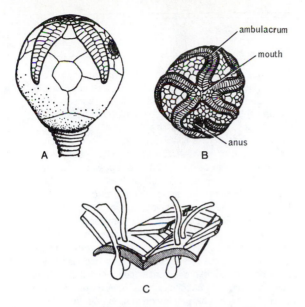

Figure 19–65 Edrioasteroids: *A, Steganoblastus,* a stalked species with straight ambulacra. *B, Edrioaster,* an unattached species with curved ambulacra. *C,* Ambulacral groove, showing ampullae, podia, and cover plates. (All after Nichols, D., 1969: Echinoderms. Hutchinson and Co., London.)

of life. As in many other animal groups, such a sessile existence resulted in a shift to a more adaptive radial symmetry. Attachment apparently took place at the anterior end of the animal. Based on the metamorphosis of living echinoderms, the change in symmetry involved a clockwise, 90 degree rotation of the animal so that the left side became the upper (oral) side and the right side became the lower (aboral) side. Simultaneously, the mouth moved around to the original left side, now the upper side. The two right, anterior, coelomic sacs (axohydrocoel) were reduced, and the two left sacs became the water-vascular system, which functioned in suspension feeding.

During the course of these changes, the echinoderm skeleton probably evolved as a supportive and protective structure for the sessile animal. Nichols (1967) suggests that the pentamerous form of radial symmetry arose in conjunction with the skeleton. The suture planes—the junction between two skeletal plates—represent a weak point in the body wall from a structural standpoint, and it would have been advantageous to the animal not to have had two such suture planes opposite one another. This advantage could be attained only by an odd number of ossicles forming the circumference of the body wall. The smallest number would be five if the animal were to be truly radial (Fig.

19–66). At this point we have a suspension-feeding echinoderm that was both sessile and radial. This stage in the evolution of echinoderm symmetry is illustrated by the extinct and living Crinozoa.

After attaining a radial symmetry, some of these sessile echinoderms became detached and reassumed a free-moving existence. The radial symmetry was retained, but the oral surface, which was directed upward in sessile forms, was placed against the substratum; the aboral surface became the functional upper side of the animal, and the water-vascular system was utilized in locomotion. Sea stars, brittle stars, and sea urchins all illustrate such a free-moving, radial existence.

This theory is supported by the extinct and living crinozoan fauna and by embryological evidence, such as the attached metamorphosis of asteroids. However, there are also some objections to this theory. The oldest echinoderms, all from the early Cambrian, are three very diverse groups: the eocrinoids, the edrioasteroids, and the carpoids. The upward-directed oral surface of the eocrinoids and the edrioasteroids is compatible with the idea that the first echinoderms were radially symmetrical and attached animals, but the irregular carpoids are not. Either the carpoid asymmetry is secondary, in which case the earliest echinoderms are Precambrian, or the first echinoderms were not pentamerous, radially symmetrical animals (Box 19–1).

A second principal problem with this classical theory of echinoderm evolution is the lack of any forms that bridge the great gap between those echinoderms in which the oral surface is directed upward and those, like sea stars and sea urchins, in which it is directed downward. The echinozoan edrioasteroids have frequently been suggested as a

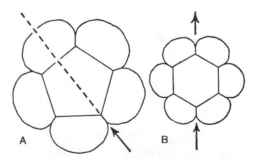

Figure 19–66 *A,* Diagram of a pentamerous radial symmetry involving five ossicles. The suture plane between any two ossicles is never located directly across from a suture plane on the opposite wall. *B,* Diagram of a hexamerous radial symmetry involving six ossicles. Note that every suture is on the opposite wall. (Based on the views of Nichols, 1969.)

BOX 19–1   WHAT DOES THE FOSSIL RECORD TELL US?

A rich fossil record of metazoan animals has been uncovered since its systematic exploration began over 200 years ago. Taxonomic and anatomical descriptions of fossils have occupied much of this investigation, and more recent techniques using thin sections and x-rays have yielded considerable information. Contemporary paleontologists are also searching for clues about the life-styles of fossil species and the environments in which they lived.

The fossil record provides a fairly good acccount of the animals that have inhabited our planet and the sequence of their appearance and disappearance in geological history. It also provides some information about the evolution of lower taxa. However, the fossil record tells us almost nothing about the evolutionary origins and relationships of phyla and classes. *Knightoconus*, an apparent intermediate stage between high cone monoplacophorans and cephalopods, is one of the very few animal "missing links" at the level of class and phylum.

A number of factors account for the lack of such links. The fossil record is uneven, for the likelihood of preservation depends on the age of the entrapping sediments, the kind of environment in which the animal lived, and most important, the type of skeleton the animal possessed. Thus, soft-bodied animals covered by very ancient sediments will be rare specimens in the fossil record. Yet such were most of the ancestral forms that gave rise to the known phyla.

Almost all of the animal phyla and many of the classes had made their appearance by the end of the great Cambrian radiation (550 million years ago) at the beginning of the Paleozoic, which is the starting point for most of the animal fossil record. The best of the Cambrian record is in the Burgess shales of British Columbia. This remarkable preservation contains 120 fossil species, including cnidarians, annelids, priapulids, mollusks, arthropods, onychophorans, echinoderms, and chordates, plus ten species that fit into no existing phylum. Without the Burgess shales our knowledge of Cambrian fossils would be very limited, for fossil-bearing Cambrian strata are limited and often covered by younger sedimentary rocks. Still older pre-Cambrian rocks have an even more restricted distribution. The few fossil-bearing outcrops that have been found in various parts of the world contain the oldest known animals. The most ancient animal fauna, called the Edicarian fauna, includes jellyfish, certain corals, and worms—mostly soft-bodied animals.

There is nothing in these early or later fossil records that reveals the evolutionary origins of major animal groups. Our speculations about the evolutionary origins and relationships of animal phyla and classes, therefore, continue to be based largely on comparative morphology and development.

bridge group, for they were directed upward and some were unattached. However, it is difficult to visualize how the oral surface could have come to be directed downward with sufficient intermediate stages to permit the necessary functional adjustments.

Fell (1965) postulates that the free habit, not the attached one, was the most primitive mode of existence in echinoderms, and that the sessile condition arose secondarily, a number of times independently, within the phylum. The living echinoids and most of the holothuroids are believed to be descended from forms that were always free moving. The helicoplacoids are regarded as being closest to the ancestral echinoderms, and the earliest echinoderms were perhaps burrowing animals. Fell's views have the advantage of reconciling the two conflicting types of body orientation within the

phylum, but they do not explain the original adaptive significance of the echinoderm skeleton, which is a basic feature of the phylum, nor do they explain the pentamerous radial symmetry of such forms as the echinoids.

There has also been much speculation about the nature of the ancestral preechinoderms. Was the water-vascular system a new development of the coelom in the evolution of the phylum or was there a functional precursor? A number of zoologists have noted the considerable similarity between the coelom associated with a lophophore and that which is part of the echinoderm water-vascular system. Both are derived from the mesocoel, both have a water ring (base of lophophore) from which extend radial canals (lophophore tentacles). Could the ancestors of echinoderms have possessed a lophophore-like structure from which the water-vascular system was derived? Perhaps one of the lophophorate phyla and echinoderms had a common ancestor.

SUMMARY

1 The Paleozoic echinoderms, which include a number of extinct classes in addition to the crinoids, were largely attached and had the oral surface directed upward.

2 Assuming a feeding mechanism like that of living crinoids, the food-collecting surface area in Paleozoic forms was provided either by brachioles (eocrinoids, blastoids, and cystoids) or by arms and pinnules (crinoids).

3 A bilateral larva bearing ciliated bands is a characteristic developmental feature of some mem-

bers of all echinoderm classes and indicates that the pentamerous radial symmetry of echinoderms is secondary. The ancestors of echinoderms were probably bilateral, motile tricoclomates.

4 The evolution of a pentamerous radial symmetry and an endoskeleton of calcareous ossicles was perhaps correlated with an early assumption of a sessile existence.

5 The evolution of the water-vascular system may have originally been an adaptation for suspension feeding.

6 The relationship of most modern classes of echinoderms, in which the oral surface is directed downward, to the Paleozoic forms, in which the oral surface is directed upward, is obscure.

REFERENCES

The literature included here is restricted in large part to echinoderms. The introductory references on page 8 include many general works and field guides that contain sections on these animals.

Anderson, J. M., 1960: Histological studies on the digestive system of a starfish, *Henricia*, with notes on Tiedemann's pouches in starfishes. Biol. Bull., 119(3):371–398.

Anderson, J. M., 1978: Studies on functional morphology in the digestive system of *Oreaster reticulatus*. Biol. Bull., 154:1–14.

Arshavskii, Y. I., Kashin, S. M., Litvinova, N. M., Orlovskii, G. N., and Feldman, A. G., 1976: Coordination of arm movements during locomotion of *Ophiura*. Neirofiziologiya, 8(5):529–537.

Bakus, G. J., 1973: The biology and ecology of tropical holothurians. *In* Jones, O. A., and Endean, R. (Eds.): Biology and Geology of Coral Reefs. Vol. II: Biology 1. Academic Press, N.Y., pp. 326–368.

Benson, A. A., Patton, J. S., and Field, C. E., 1975: Wax digestion in a crown-of-thorns starfish. Comp. Biochem. Physiol. B., 52(2):339–340.

Binyon, J., 1964: On the mode of functioning of the water vascular system of *Asterias rubens*. J. Mar. Biol. Assoc. U.K., 44:577–588.

Binyon, J., 1972: Physiology of Echinoderms. Pergamon Press, Oxford. (An excellent general physiology of echinoderms.)

Binyon, J., 1976: The permeability of the asteroid podial wall to water and potassium ion. J. Mar. Biol. Assoc. U.K., 56:639–647.

Birkeland, C., 1982: Terrestrial runoff as a cause of outbreaks of *Acanthaster planci*. Mar. Biol., 69:175–185.

Boolootian, R. A. (Ed.), 1966: Physiology of Echinodermata. John Wiley and Sons, N.Y. (A review of echinoderm ecology, behavior, and physiology.)

Brehm, P., and Morin, J. G., 1977: Localization and characterization of luminescent cells in *Ophiopsila cali-*

fornica and *Amphipholis squamata*. Biol. Bull., 152:12–25.

Broom, D. M., 1975: Aggregation behavior of the brittlestar *Ophiothrix fragilis*. J. Mar. Biol. Assoc. U.K., 55:191–197.

Burke, R. D., 1984: Pheromonal control of metamorphosis in the Pacific sand dollar, *Dendraster excentricus*. Science, 225:442–443.

Burnett, A. L., 1960: The mechanism employed by the starfish *Asterias forbesi* to gain access to the interior of the bivalve *Venus mercenaria*. Ecology, 41:583–584.

Byrne, M., 1985: Evisceration behaviour and the seasonal incidence of evisceration in the holothurian *Eupentacta quinquesemita*. Ophelia, 24(2):75–90.

Byrne, M., and Fontaine, A. R., 1981: The feeding behaviour of *Florometra serratissima*. Can. Jour. Zool., 59:11–18.

Chia, F. S. 1969: Some observations of the locomotion and feeding of the sand dollar, *Dendraster excentricus*. J. Exp. Mar. Biol. Ecol., 3(2):162–170.

Chia, F. S., and Amerongen, H., 1975: On the prey-catching pedicellariae of a starfish, *Stylasterias forreri*. Can. J. Zool., 53:748–755.

Christensen, A. M., 1970: Feeding biology of the sea star *Astropecten irregularis*. Ophelia, 8(1):1–134.

Clark, A. H., 1915, 1921, 1931, 1941, 1947, 1950: A monograph of the existing crinoids. Vol. 1, Pts. 1, 2, 3, 4a, 4b, 4c. Bull. U.S. Nat. Mus. 82. (A monumental work on living crinoids.)

Clark, A. M., 1962: Starfishes and Their Relations. British Museum, London.

Cuénot, L., 1948: Anatomie, Éthologie, et Systématique des Échinodermes. *In* Grassé, P. (Ed.): Traité de Zoologie. Vol. II. Échinodermes, Stomocordes, Procordes. Masson et Cie, Paris, pp. 1–363.

De Ridder, C., and Lawrence, J. M., 1982: Food and feeding mechanisms: Echinoidea. pp. 57–115. See Jangoux and Lawrence, 1982.

Donnay, G., and Pawson, D. L., 1969: X-ray diffraction studies of echinoderm plates. Science, *166*:1147–1150.

Downey, M. E., 1973: Starfishes from the Caribbean and the Gulf of Mexico. Smithsonian Contributions to Zoology 126. 158 pp.

Durham, J. W., and Caster, K. E., 1963: Helicoplacoidea, a new class of echinoderms. Science, *140*:820–822.

Ebert, T. A., and Dexter, D. M., 1975: A natural history study of *Encope grandis* and *Mellita grantii*, two sand dollars in the northern Gulf of California. Mar. Biol., *32*(4):397–407.

Ellers, O., and Telford, M., 1984: Collection of food by oral surface podia in the sand dollar, *Echinarachnius parma*. Biol. Bull., *166*:574–582.

Emson, R. H., and Wilkie, I. C., 1980: Fission and autotomy in echinoderms. Oceanogr. Mar. Biol. Ann. Rev., *18*:155–250.

Endean, R., 1977: *Acanthaster planei* infestations of reefs of the Great Barrier Reef. Proceedings of the 3rd International Coral Reef Symposium, University of Miami, *I*(Biology):185–191.

Fankbonner, P. V., 1981: A re-examination of mucus feeding by the sea cucumber *Leptopentacta (Cucumaria) elongata*. Jour. Mar. Biol. Assoc. U.K., *61*:679–683.

Feder, H. M., 1955: On the methods used by the starfish *Pisaster ochraceus* in opening three types of bivalved mollusks. Ecology, *36*:764–767.

Fell, H. B., 1965: The early evolution of the Echinozoa. Breviora, *219*:1–19.

Fell, H. B., 1966: Ancient echinoderms in modern seas. Oceanogr. Mar. Biol. Ann. Rev., *4*:233–245.

Fenner, D. H., 1973: The respiratory adaptations of the podia and ampullae of echinoids. Biol. Bull., *145*:323–339.

Ferber, I., and Lawrence, J. M., 1976: Distribution, substratum preference, and burrowing behavior of *Lovenia elongata* in the Gulf of Elat and Red Sea. J. Exp. Mar. Biol. Ecol., *22*:207–225.

Furguson, J. C., 1984: Translocative functions of the enigmatic organs of starfish—the axial organ, hemal vessels, Tiedemann's bodies, and rectal caeca: an autoradiographic study. Biol. Bull., *166*:140–155.

Fish, J. D., 1967: Biology of *Cucumaria elongata*. J. Mar. Biol. Assoc. U.K., *47*:129–143.

Fishelson, L., 1974: Ecology of the northern Red Sea crinoids and their epi- and endozoic fauna. Mar. Biol., *26*:183–192

Fontaine, A. R., 1965: Feeding mechanisms of the ophiuroid *Ophiocomina nigra*. J. Mar. Biol. Assoc. U.K., *45*:373–385.

Frankel, E., 1977: Previous *Acanthaster* aggregations in the great Barrier Reef. Proceedings of the 3rd International Coral Reef Symposium, University of Miami, *I*(Biology):201–208.

Fricke, H. W., 1968: Beiträge zur Biologie der Gorgonenhäupter *Astrophyton muricatum* (Lamarck) und *Astroboa nuda* (Lyman). Ernst-Reuter Gesellschaft, Berlin. 197 pp.

Ghiold, J., 1979: Spine morphology and its significance in feeding and burrowing in the sand dollar, *Mellita quinquiesperforata*. Bull. Mar. Sci., *29*(4):481–490.

Greenway, M., 1976: The grazing of *Thalassia testudinum* in Kingston Harbor, Jamaica. Aquat. Bot., *2*(2):117–126.

Hamann, A., 1885: Beiträge zur Histologie der Echinodermen. Heft 2, Die Asteriden.

Hendler, G., 1975: Adaptational significance of the patterns of ophiuroid development. Am. Zool., *15*:691–715.

Hendler, G., 1982: Slow flicks show star tricks: elapsed-time analysis of basketstar (*Astrophyton muricatum*) feeding behavior. Bull. Mar. Sci., *32*:909–918.

Herreid, C. F., La Russa, V. F., and DeFesi, C. R., 1976: Blood vascular system of the sea cucumber, *Stichopus moebii*. J. Morphol., *150*(2):423–451.

Highsmith, R. C., 1982: Induced settlement and metamorphosis of sand dollar (*Dendraster excentricus*) larvae in predator-free sites: adult sand dollar beds. Ecology, *63*(2):329–337.

Hyman, L. H., 1955: The Invertebrates. Vol. 4. Echinodermata. McGraw-Hill, N.Y. (The chapter Retrospect in Vol. 5 [1959] of this series summarizes the literature on echinoderms from 1955 to 1959.)

Jangoux, M. (Ed.), 1980: Echinoderms: Present and Past. Proceedings of the European Colloquium on Echinoderms, Brussels, 1979. A. A. Balkema, Rotterdam.

Jangoux, M., 1982: Food and feeding mechanisms: Asteroidea; Digestive systems: Asteroidea and Ophiuroidea. See Jangoux and Lawrence, 1982.

Jangoux, M., and Lawrence, J. M. (Eds.), 1982: Echinoderm Nutrition. A. A. Balkema, Rotterdam. 654 pp.

Jangoux, M., and Lawrence, J. M. (Eds.), 1983: Echinoderm Studies. Vol. I. A. A. Balkema, Rotterdam. 203 pp. (A collection of reviews on various aspects of echinoderm biology.)

Kanatani, H., and Shirai, H., 1970: Mechanism of starfish spawning. Develop. Growth Differ. *12*(2):119–140.

Khripounoff, A. and Sibuet, M., 1980: La nutrition d'echinodermes abyssaux. I. Alimentation des holothuries. Mar. Biol., *60*:17–26.

LaHaye, M. C., and Jangoux, M., 1985: Functional morphology of the podia and ambulacral grooves of the comatulid crinoid *Antedon bifida*. Mar. Biol., *86*:307–318.

Langeloh, H., 1937: Uber die Bewegungen von *Antedon*. Zool. Jahrb. Abt. Allg. Zool., *57*(3):235–279.

La Touche, R. W., 1978: The feeding behavior of the featherstar *Antedon bifida*. J. Mar. Biol. Assoc. U.K., *58*:877–890.

La Touche, R. W., and West, A. B., 1980: Observations on the food of *Antedon bifida*. Mar. Biol., *61*(1):39–46.

Lawrence, J. M., 1975: On the relationship between marine plants and sea urchins. Oceanogr. Mar. Biol. Ann. Rev., *13*:213–286.

Laxton, J. H., 1974: Aspects of the ecology of the coral-eating starfish, *Acanthaster planci*. Biol. J. Linn. Soc., *6*(1):19–45.

Lewis, J. B., 1968: The function of the sphaeridia of sea urchins. Can. J. Zool., *46*:1135–1138.

Macurda, D. B., and Meyer, D. L., 1974: Feeding posture of modern stalked crinoids. Nature, *247*:394–396.

Macurda, D. B., and Meyer, D. L., 1976: The identification and interpretation of stalked crinoids from deep-water photographs. Bull. Mar. Sci., *26*(2):205–215.

Macurda, D. B., and Meyer, D. L., 1983: Sea lilies and feather stars. Amer. Scientist, *71*:354–364.

Mauzey, K. P., Birkeland, C., and Dayton, P. K., 1968: Feeding behavior of asteroids and escape responses of their prey in the Puget Sound region. Ecology, 49(4):603–619.

Menge, B., 1975: Brood or broadcast? The adaptive significance of different reproductive strategies in the two intertidal sea stars *Leptasterias hexactis* and *Pisaster ochraceus*. Mar. Biol., 31(1):87–100.

Meyer, D. L., 1971: The collagenous nature of problematical ligaments in crinoids. Mar. Biol., 9(3):235–241.

Meyer, D. L., 1982: Food and feeding mechanisms: Crinozoa. pp. 25–45. See Jangoux and Lawrence, 1982.

Meyer, D. L., and Lane, N. G., 1976: The feeding behavior of some Paleozoic crinoids and recent basket stars. J. Paleontology, 50(3):472–480.

Meyer, D. L., LaHaye, C. A., Holland, N. D., Areson, A. C., and Strickler, J. R., 1984: Time-lapse cinematography of feather stars on the Great Barrier Reef, Australia: demonstrations of posture changes, locomotion, spawning and possible predation by fish. Mar. Biol., 78:179–184.

Millott, N. (Ed.), 1967: Echinoderm Biology. Symp. Zool. Soc. London, No. 20. Academic Press, N.Y. 240 pp. (A collection of papers presented at a symposium in 1966.)

Millott, N., 1975: The photo sensitivity of echinoids. Adv. Mar. Biol., 13:1–52.

Mladenov, P. V., and Chia, F. S., 1983: Development, settling behaviour, metamorphosis and pentacrinoid feeding and growth of the feather star *Florometra serratissima*. Mar. Biol., 73:309–323.

Moore, R. C. (Ed.) 1966 to 1978: Treatise on Invertebrate Paleontology, Echinodermata. Pts. S to U. Geological Society of America and University of Kansas Press, Lawrence, Kan.

Motokawa, T., 1982: Factors regulating the mechanical properties of holothurian dermis. Jour. Exp. Biol., 99:29–41.

Motokawa, T., 1984: Connective tissue catch in echinoderms. Biol. Rev., 59:255–270.

Nichols, D., 1960: The histology and activities of the tube-feet of *Antedon bifida*. Q. J. Micr. Sci., 101(2):105–117.

Nichols, D., 1967: The origin of echinoderms. *In* Millott, N. (Ed.): Echinoderm Biology. Academic Press, N.Y. pp. 209–229.

Nichols, D., 1969: Echinoderms. 4th Edition. Hutchinson University Library, London.

Ogden, J. C., Brown, R. A. and Salesky, N., 1973: Grazing by the echinoid *Diadema antillarum* Philippi: Formation of halos around West Indian patch reefs. Science, 182:715–717.

Pawson, D. L., 1977: Echinodermata: Holothuroidea. Marine flora and fauna of the northeastern U.S. NOAA Technical Reports NMFS Circular 405. U.S. Government Printing Office. 13 pp.

Rutman, J., and Fishelson, L., 1969: Food composition and feeding behavior of shallow water crinoids at Eilat (Red Sea). Mar. Biol., 3(1):46–57.

Scheibling, R. E., 1980: The microphagous feeding behavior of *Oreaster reticulatus*. Mar. Behav. Physiol., 7(3):225–232.

Shick, J. M., Edwards, K. C., and Dearborn, J. H., 1981: Physiological ecology of the deposit-feeding sea star *Ctenodiscus crispatus*: ciliated surfaces and animal-sediment interactions. Mar. Ecol. Prog. Ser., 5:165–184.

Sloan, N. A., 1980: Aspects of the feeding biology of asteroids. Oceanogr. Mar. Biol. Ann. Rev., 18:57–124.

Sloan, N. A., and Campbell, A. C., 1982: Perception of food. See Jangoux and Lawrence, 1982.

Sloan, N. A., and Northway, S. M., 1982: Chemoreception by the asteroid *Crossaster papposus*. Jour. Exp. Mar. Biol. Ecol., 61:85–98.

Smith A., 1984: Echinoid Paleobiology. Allen and Unwin, London. 190 pp.

Smith, T. B., 1983: Tentacular ultrastructure and feeding behavior of *Neopentadactyla mixta*. Jour. Mar. Biol. Assoc. U.K., 63:301–311.

Strathmann, R. R., 1975: Larval feeding in echinoderms. Am. Zool., 15:717–730.

Telford, M., 1983: An experimental analysis of lunule function in the sand dollar *Mellita quinquiesperforata*. Mar. Biol., 76:125–134.

Telford, M., Mooi, R., and Ellers, O., 1985: A new model of podial deposit feeding in the sand dollar, *Mellita quinquiesperforata*: The sieve hypothesis challenged. Biol. Bull., 69:431–448.

Thomas, L. A., and Hermans, C. O., 1985: Adhesive interactions between the tube feet of a starfish, *Leptasterias hexactis*, and substrata. Biol. Bull., 169:675–688.

Thomassin, B. A., 1976: Feeding behavior of the felt-, sponge-, and coral-feeding sea stars, mainly *Culcita schmideliana*. Helgol. Wiss. Meeresunters., 28(1):51–65.

Timko, P., 1976: Sand dollars as suspension feeders: A new description of feeding in *Dendraster excentricus*. Biol. Bull., 151(1):247–259.

Tyler, P. A., 1980: Deep-sea ophiuroids. Oceanogr. Mar. Biol. Ann. Rev., 18:125–153.

Ubaghs, G., 1969: General characteristics of the Echinodermata. *In* Florkin, M., and Scheer, B. T. (Eds.): Chemical Zoology. Vol. 3. Echinodermata. Academic Press, N.Y. pp. 3–46.

Vine, P. J., 1973: Crown-of-thorns plagues: The natural causes theory. Atoll Res. Bull., 166:1–10.

Warner, G. F., 1982: Food and feeding mechanisms: Ophiuroidea. pp. 161–181. See Jangoux and Lawrence.

Warner, G. F., and Woodley, J. D., 1975: Suspension-feeding in the brittle star *Ophiothrix fragilis*. J. Mar. Biol. Assoc. U.K., 55:199–210.

Woodley, J. D., 1975: The behavior of some amphiurid brittle stars. J. Exp. Mar. Biol. Ecol., 18:29–46.

Yingst, J. Y., 1976: The utilization of organic matter in shallow marine sediments by an epibenthic deposit-feeding holothurian. J. Exp. Mar. Biol. Ecol., 23:55–69.

Yoshida, M., Takasu, N., and Tamotsu, S., 1984: Photoreception in echinoderms. *In* Ali, M. A.: Photoreception and Vision in Invertebrates. Plenum Press, N.Y. pp. 743–772.

20
The Lesser Deuterostomes

In addition to the echinoderms, there are four other small groups of invertebrate deuterostomes—the Hemichordata, the Chaetognatha, the Urochordata, and the Cephalochordata. The last two groups are subphyla of the phylum Chordata. Although all are deuterostomes, they do not represent a close phylogenetic unit but stem from different points along the deuterostome line. Many are widely distributed and common invertebrate animals, and all are highly specialized.

Phylum Hemichordata

The hemichordates are wormlike marine animals that once were considered a subphylum of the chordates. They are composed of two classes—the Enteropneusta (acorn worms) and the Pterobranchia. The acorn worms are the most common and best known hemichordates. The pterobranchs consist of three genera of small, tube-dwelling animals that are not frequently encountered.

CLASS ENTEROPNEUSTA

The 70 species of Enteropneusta, or acorn worms, are inhabitants of shallow water. Some live under stones and shells, but many common species, including *Saccoglossus*, burrow in mud and sand. Exposed tidal flats are frequently dotted with the coiled, ropelike castings of these animals.

Body Structure

Acorn worms are relatively large animals, the majority ranging from 9 to 45 cm in length. A Brazilian species, *Balanoglossus gigas*, may exceed 2.5 meters in length and constructs burrows 3 meters long. The cylindrical and rather flaccid body is composed of an anterior proboscis, a collar, and a long trunk (Fig. 20–1A). These regions correspond to the typical deuterostome body divisions—protosome, mesosome, and metasome.

The proboscis is usually short and conical, from which the name acorn worm is derived, and is connected to the collar by a stalk. The collar is a short

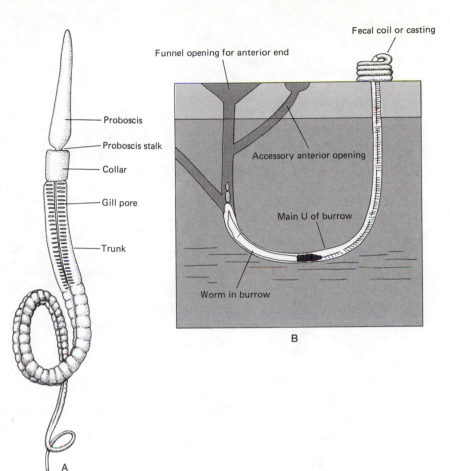

Figure 20–1 *A, Saccoglossus kowalevskii,* an acorn worm common to the Atlantic Coast of North America and Europe. *B,* Burrow system of the Mediterranean *Balanoglossus clavigerus.* (After Stiasny from Hyman.)

cylinder that anteriorly overlaps the proboscis stalk and contains the mouth on the ventral side. The trunk constitutes the major part of the body. Behind the collar the trunk bears a longitudinal row of gill pores at each side of a middorsal ridge. More laterally, the anterior half of the trunk contains the gonads, and in some hemichordates, such as *Balanoglossus,* the lateral body wall in this region is drawn out on each side, forming winglike plates. Hemichordates display a tricoelomate structure as in echinoderm larvae. A single coelomic cavity occupies the proboscis, a pair of cavities are present in the collar, and a pair of cavities are present in the trunk. However, connective tissue and muscle fibers fill much of the original cavity, and a distinct peritoneal lining is lacking. Moreover, this coelomic musculature in large part replaces the typical body wall musculature. The proboscis and collar coelomic cavities open to the exterior by dorsal pores (Fig. 20–2).

Locomotion, Habitation, and Nutrition

Acorn worms have limited locomotor powers and are rather sluggish animals. Many burrowing species construct mucus-lined excavations in mud and sand. The burrows of species of *Balanoglossus, Saccoglossus,* and other genera may be U shaped, with two openings to the surface; one or both ends of the worm may protrude from the openings (Fig. 20–1B). Burrowing or movement within the burrow is accomplished largely by the proboscis, which is lengthened and anchored by peristaltic contractions. Other acorn worms live in masses of seaweed, under rocks and stones, or buried in sand and mud, and they move about relatively little.

Many borrowing enteropneusts consume sand and mud, from which organic matter is digested. The quantity of substrate ingested is indicated by the great piles of castings that accumulate at the posterior opening of the burrow (Fig. 20–3B).

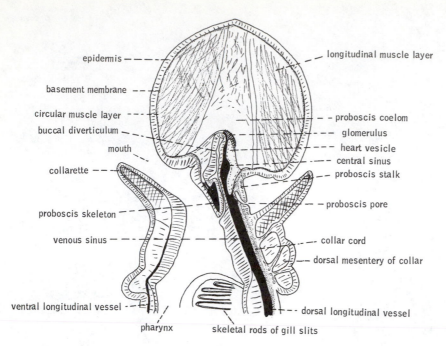

Figure 20–2 Anterior of *Glossobalanus minutus* (sagittal section). (After Spengel from Hyman.)

Suspension feeding is an important method of obtaining food for some species. Detritus and plankton that come in contact with the proboscis surface are trapped in mucus and carried posteriorly by strong ciliary currents. This mode of feeding is utilized by nonburrowing and even many burrowing species. Some species that inhabit burrows project the proboscis from the mouth of the burrow, moving it about. At the base of the proboscis the cilia beat ventrally toward the mouth

(Fig. 20–3*A*). Some of the food particles carried ventrally by these cilia pass into a groove, forming the preoral ciliary organ. Within the groove the particles are conducted ventrally. Besides conduction, the function of the preoral ciliary organ is not certain.

Although particles are carried into the mouth by cilia, their passage is probably facilitated by the water current flowing into the mouth and out of the pharyngeal clefts. The animal may stop feeding

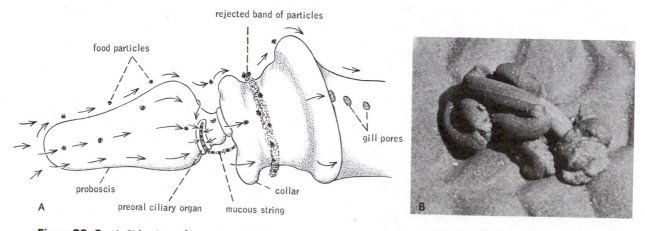

Figure 20–3 *A*, Side view of anterior end of *Protoglossus kohleri*, showing the direction of food and rejected particles carried by cilia of the proboscis and collar. (Burdon-Jones, C., 1956: Observations on the enteropneust, *Protoglossus kohleri*. Proc. Zool. Soc. London, *127*:35.) *B*, Photograph of the anal end of *Balanoglossus aurantiacus* depositing casting on the surface of an exposed sand flat at low tide.

or reject large particles by covering its mouth with the edge of the collar. Particles then pass over the collar instead of into the mouth.

The digestive tract is a straight tube that is histologically differentiated into a number of regions. The large mouth, located between the ventral anterior margin of the collar and the dorsal proboscis stalk, leads into a buccal tube within the collar (Fig. 20–2). Dorsally, a long, narrow diverticulum, once thought to be a notochord, extends from the buccal tube and projects forward into the proboscis.

The buccal tube passes posteriorly into the pharynx, which occupies the branchial region of the trunk and is laterally perforated by the gill slits. The gill slits are usually limited to the dorsal half of the pharynx, and the alimentary portion of the tract occupies the ventral half of the pharynx.

Behind the pharynx the gut continues as an esophagus, where food particles and mucus are molded into a cord. In some families excess water is eliminated from the esophagus through openings to the dorsal body surface. The intestine constitutes the remainder of the gut, and it is here that digestion and absorption occur.

Some enteropneusts, such as *Balanoglossus aurantiacus* and *Balanoglossus biminiensis*, smell strongly of iodine, but the odor is due to the presence of 2,6-dibromophenol. The biological significance of this compound is still unknown. There are, however, some species of enteropneusts that do accumulate iodine.

Internal Transport, Excretion, and Gas Exchange

Enteropneusts possess an open blood-vascular system composed of a system of sinus channels, a pulsating vesicle in the proboscis, and two main contractile vessels, one middorsal, carrying the colorless blood forward, and one midventral, carrying blood posteriorly (Fig. 20–2).

Little is known about excretion in acorn worms, but a special system of blood sinuses, collectively called the glomerulus, at the base of the proboscis is thought to have an excretory function.

The pharyngeal gill slits in the anterior trunk region are assumed to be gas exchange organs. The number of slits can range from a few to 100 or more pairs, since new slits are continually being formed during the life of the worm. Each slit opens through the side of the pharyngeal wall as a U-shaped cleft (Fig. 20–4B). The pharyngeal wall between clefts (the septum) and that part of the wall projecting downward between the arms of the U

(the tongue bar) are supported by skeletal thickenings of the basement membrane of the pharyngeal epithelium.

Each cleft opens into a branchial sac, which in turn opens to the exterior by a dorsolateral gill pore. All the pores of one side are often located in a longitudinal groove. The walls around the pharyngeal clefts are ciliated and contain a plexus of blood sinuses, which are involved in gas exchange. The beating cilia produce a stream of water passing into the mouth and out through the gill slits.

The gill slits of the hemichordates and the chordates probably evolved originally as a feeding mechanism, in which small particles were strained out of the water current passing through the pharyngeal clefts. This is still the method of feeding in the tunicates and the cephalochordates. A gas exchange function has been secondarily assumed by the gill slits.

Nervous System and Sense Organs

The nervous system is relatively primitive (Fig. 20–4A). In different regions of the body the nerve plexus beneath the surface epithelium has become thickened to form epidermal nerve cords, in which the nerve fibers are arranged longitudinally. The principal nerve cords are the midventral and the middorsal of the proboscis and trunk. The ventral trunk cord terminates at the collar, but the dorsal cord continues into the collar as the collar cord.

The collar cord actually becomes internal—i.e., it is separated from the epidermis above and is continuous with the general epidermal nerve plexus only at the two ends. In some acorn worms the collar cord is hollow and is perhaps homologous to the dorsal hollow nerve cord of chordates. The collar cord possesses giant nerve cells and is apparently a conduction path. Neurosensory cells scattered throughout the surface epithelium constitute the sensory system of hemichordates.

Reproduction

Acorn worms are fragile animals, and the larger species are especially difficult to collect intact. Most species probably can regenerate at least missing parts of the trunk. Asexual reproduction has been reported for several species, including members of the genera *Glossobalanus* and *Balanoglossus*.

Enteropneusts are all dioecious. The saclike, paired gonads are located in the trunk coelom, beginning in the branchial region or just behind the

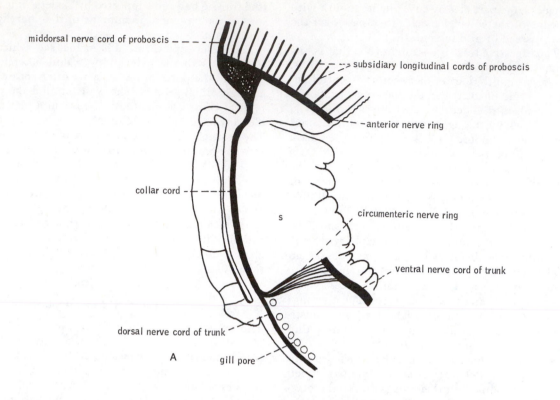

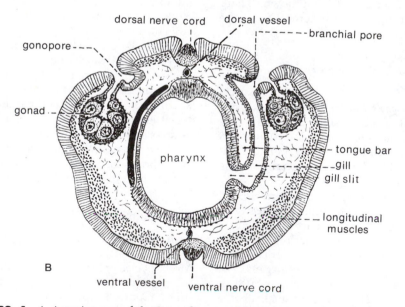

Figure 20–4 *A,* Anterior part of the *Saccoglossus cambrensis* nervous system (longitudinal view). Nerve networks not shown. (After Knight-Jones from Hyman.) *B,* Transverse section through the trunk of *Saccoglossus.* (After Dawydoff, C., 1948. *In* Grassé, P. (Ed.): Traité de Zoologie. Vol. 11. Masson et Cie, Paris. p. 400.)

gill slits. Each gonad opens to the exterior through a pore that is often located in the same groove as the gill pores (Fig. 20–4B).

Masses of eggs embedded in mucus are shed from the burrow and are fertilized externally by sperm emitted from nearby males that are apparently stimulated by the presence of the released eggs. The mucous masses are soon broken up by tidal currents and the eggs are dispersed.

Early development is strikingly like that of the echinoderms. Equal holoblastic cleavage leads to a coeloblastula, which then undergoes invagination to form a narrow archenteron. The blastopore, which marks the future posterior end of the embryo, closes, and the embryo lengthens along the anteroposterior axis and becomes ciliated. At this stage, hatching takes place, and the embryo assumes a planktonic existence.

The anterior tip of the archenteron separates to form a coelomic vesicle (the protocoel), which will form the proboscis coelom. Later the protocoel gives rise to two posterior extensions that form the collar and the trunk coelom. In some species the two posterior coelomic divisions arise as evaginations of the archenteron independent of the formation of the protocoel.

Development from this point may be either direct or indirect. In indirect development the embryo develops into a free-swimming tornaria larva. The ciliation becomes restricted to a distinct band, which at first is very similar to the ciliated band in the bipinnaria larva of sea stars (Fig. 20–5A). Gradually, the band becomes more winding, and there develops a separate, posterior girdle of cilia that forms the principal locomotor organ of the larva (Fig. 20–5B). The anterior, winding band is involved only in feeding, which is like that of echinoderm larvae (Strathmann and Bonar, 1976). After a planktonic feeding existence of several days to several weeks, the larva becomes girdled by a constriction initiating the division between proboscis and collar (Fig. 20–5). The larva elongates, sinks to the bottom, and assumes an adult existence. A small number of giant tornaria larvae (0.7 to 2.2 cm) have been collected from tropical and semitropical oceanic waters of the Atlantic since 1932. Although assigned to the species *Planctosphaera pelagica*, the adults are unknown.

Development is direct in a number of enteropneusts, including the Atlantic acorn worm, *Saccoglossus kowalevskii*. A ciliated gastrula may hatch from the egg, or hatching may take place at a later stage, but in any case a tonaria larva never forms and development proceeds directly, terminating in the young worm. In *Saccoglossus kowalevskii* the eggs hatch as young worms.

CLASS PTEROBRANCHIA

The Pterobranchia consist of about 15 species belonging to three genera. They are not commonly encountered. Most are inhabitants of relatively deep water, and the greatest number of species are found in the Southern Hemisphere. However, *Rhabdopleura* has been dredged up off European coasts, and species of both *Rhabdopleura* and *Ce-*

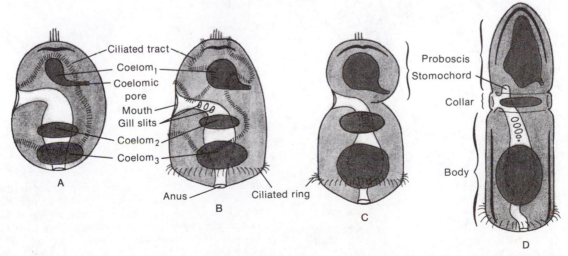

Figure 20–5 Development of *Balanoglossus clavigerus: A–D*, Diagrammatic side views of the larval development of a hemichordate. Compare with Figure 19–2. *A*, Early larva. *B*, Later larva. *C* and *D*, Metamorphosis.

phalodiscus have recently been found at wading depths off Bermuda. With the exception of *Atubaria*, pterobranchs live in secreted tubes, which are organized in aggregations *(Cephalodiscus)* or colonies *(Rhabdopleura)* on the bottom (Fig. 20–6). Species of both genera are attached by a stalk, but individuals forming the colonies *Rhabdopleura* are connected by a stolon.

The proboscis is shield shaped and functions in creeping up inside the tube as well as in secreting the tube. The most striking features of these worms are the arms and tentacles carried on the dorsal side of the collar. In *Rhabdopleura* there are two recurved arms, and in *Cephalodiscus* there are five to nine pairs of arms (Fig. 20–6). The arms bear

numerous small tentacles that are heavily ciliated. The tentacles supposedly capture minute organisms, which are then driven by the cilia to the mouth; however, the entire ciliated body surface of *Cephalodiscus* may collect suspended particles. Both the arms and the tentacles are hollow, each containing an extension of the mesocoel, and in this respect they are thus somewhat similar to a lophophore. There are no gill clefts in the genus *Rhabdopleura*, and there is only one pair in *Cephalodiscus*. And as is true of many sessile, tube-dwelling animals, the gut is U shaped, the anus opening anteriorly on the dorsal side of the collar. The sexes are separate, and a ciliated larva, unlike a tornaria, is known for *Cephalodiscus* and *Rhab-*

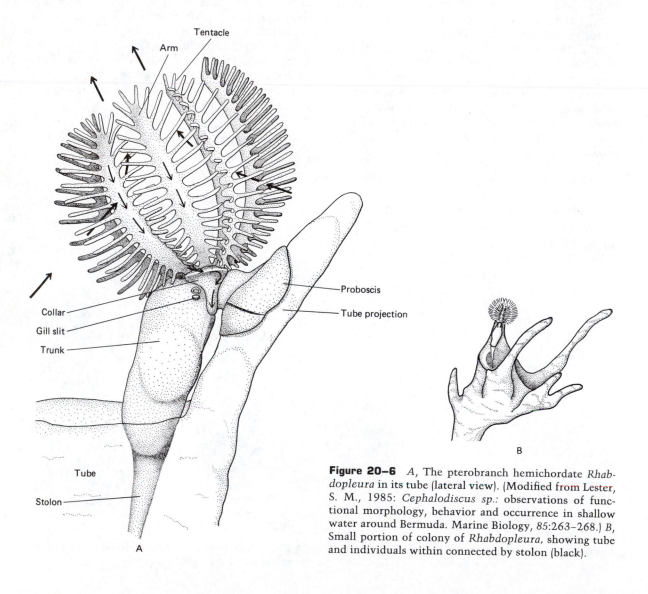

Figure 20–6 *A,* The pterobranch hemichordate *Rhabdopleura* in its tube (lateral view). (Modified from Lester, S. M., 1985: *Cephalodiscus sp.:* observations of functional morphology, behavior and occurrence in shallow water around Bermuda. Marine Biology, 85:263–268.) *B,* Small portion of colony of *Rhabdopleura,* showing tube and individuals within connected by stolon (black).

dopleura. From the individual that develops from the larva, the colony, or aggregation, is formed by budding of the stalk or stolon.

A fossil group, called graptolites, which have long puzzled paleontologists, are now believed to be pterobranchs or closely related to them. The chemistry of the tubes of the two groups is very similar (Armstrong et al, 1984).

HEMICHORDATE PHYLOGENY

The evidence for a close phylogenetic relationship between hemichordates and both the echinoderms and the chordates is convincing. Although the adults are different, the early embryogeny of the hemichordates is remarkably like that of the echinoderms. The formation of the gastrula and the coelom is very similar to these stages in echinoderms, and the early tornaria larva is virtually identical to the bipinnaria of the asteroids. Of the two major classes of hemichordates, the pterobranchs are considered the more primitive, and the lophophore-like arms and tentacles are thought to represent a primitive feature of the phylum that has been lost in the enteropneusts.

Some zoologists, such as Hyman (1959), have speculated that the pterobranchs may be similar to the common ancestor of both the echinoderms and the hemichordates, and it may have been from such arms and tentacles that the echinoderm water-vascular system, which is believed to have been originally a food-catching device, arose. In fact, Nielsen (1985) considers pterobranchs more closely related to the deuterostome lophophorates (phoronids and brachiopods) than to the enteropneusts.

An affinity with the chordates is also indicated, although not as close a one as that with the echinoderms. Only in the hemichordates and the chordates are pharyngeal clefts found. Also, the dorsal collar nerve cord of hemichordates, which is sometimes hollow, is somewhat similar to the dorsal hollow nerve cord of chordates, and perhaps the two structures are homologous. However, the lack of a notochord and the difference in the general body structure exclude hemichordates from the phylum Chordata.

SUMMARY

1 The phylum Hemichordata contains a small number of marine species called acorn worms. They have long bodies, and most are burrowers in sand or mud or live in algae or beneath stones.

2 The body is divided into an anterior proboscis region, a short collar region, and a long trunk. The mouth opens just in front of the collar.

3 Paired lateral gill clefts, or pores, which open through the pharyngeal and body walls in the anterior part of the trunk, are a distinctive feature of hemichordates and indicate a phylogenetic relationship with the chordates.

4 Hemichordates move by peristaltic contractions of the proboscis. They are deposit feeders or suspension feeders, using the proboscis as a trapping surface for suspended particles, which are then driven by cilia to the mouth.

5 The sexes are separate, fertilization is external, and development may be indirect or direct. When indirect, there is a tornaria larva that is strikingly like the larva of echinoderms.

6 The hemichordate class Pterobranchia contains a small number of species that are mostly tubicolous and bear a pair of tentaculate arms on the collar that are utilized in suspension feeding.

Phylum Chordata, Subphylum Urochordata

The chordates are the largest phylum of deuterostomes, but most chordates are vertebrates and fall outside the scope of this book. Two subphyla, the Urochordata and Cephalochordata, lack a backbone but possess the three distinguishing chordate characteristics—at some time in the life cycle, there can be found a notochord, a dorsal hollow nerve cord, and pharyngeal clefts. Since cephalochordates (amphioxus) initiate most courses in comparative vertebrate anatomy, they are not treated here. The urochordates, however, deserve some attention, for members of this group are less familiar, even though they are very common marine animals.

Adult urochordates, commonly known as tunicates, little resemble other chordates. Most are sessile, and the body is covered by a complex secreted tunic. There is a highly developed, perforated pharynx, but the notochord and nerve cord are absent in the adult. Only the larval stage, which looks like a microscopic tadpole, possesses distinct chordate characteristics. The tunicates consist of three classes—the Ascidiacea, the Thaliacea, and the Larvacea. The ascidians contain the majority of species and the most common tunicates. The other two classes are specialized for planktonic existence.

About 1250 species of urochordates have been described.

CLASS ASCIDIACEA

External Structure

Ascidians, often called sea squirts, are sessile tunicates and are common marine invertebrates throughout the world. The majority are found in shallow waters; they attach to rocks, shells, pilings, and ship bottoms or are sometimes fixed in mud and sand by filaments or a stalk. Piling surfaces in temperate waters are often covered with large species. A great diversity of species inhabit tropical reefs, and many minute colonial forms live in crevices and beneath old coral heads. Others form large, conspicuous clusters on gorgonian corals. Over 119 species have been taken from depths greater than 2000 meters (Monniot and Monniot, 1975), and there are even a few interstitial ascidians.

The bodies of solitary species range from spherical to cylindrical in shape (Fig. 20–7). One end is attached to the substratum, and the opposite end contains two openings, the buccal and atrial siphons. Although gray and green colors are common, all shades are found in ascidians, and some colonial species are very beautiful. The body ranges in size from that of a seed (a millimeter or so in diameter) to that of a large potato, which some species closely resemble. *Halocynthia pyriformis*, which is found on the Atlantic coast north of Cape Cod, is called the sea peach because of its similarity to the fruit in size, shape, and color. Some large, irregular ascidians are commonly covered by other smaller sessile organisms, which make the ascidians even more inconspicuous.

In contrast to shallow-water ascidians, 95 per cent of which live attached to rigid surfaces, most deep-sea species inhabit soft bottoms. The greatest number are tiny, spherical species anchored by fibrils, but there are also some large, stalked forms and some strange, transparent, irregular, raglike species that float over a small attachment point (Fig. 20–8*D*) (Monniot and Monniot, 1975).

Larval Origins of Adult Body Form

The attached condition of adult ascidians has led to such specialization that it is necessary to examine the larva and its metamorphosis to fully appreciate the chordate affinities and the origins of the adult

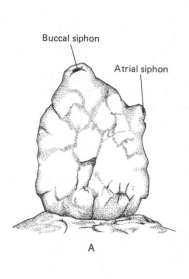

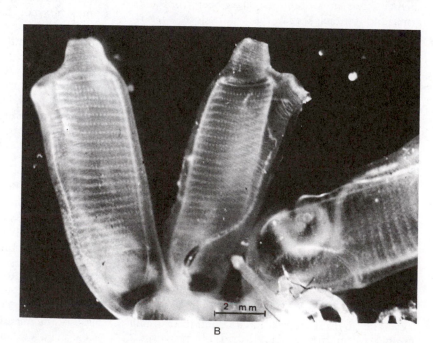

Figure 20–7   *A, Polycarpa pomaria,* a European, shallow-water, solitary ascidian with an irregular tough tunic. *B,* A transparent ascidian. The pharyngeal basket can be seen through the tunic. (By Betty M. Barnes.)

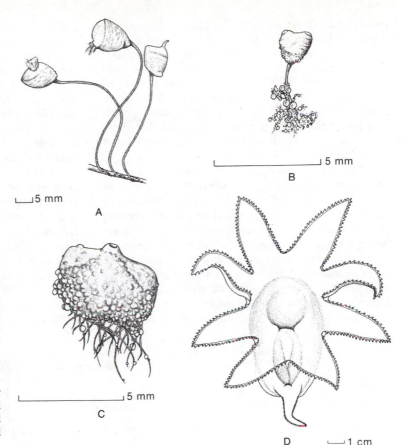

Figure 20–8 Deep-sea tunicates. Most live on soft bottoms, in contrast to the predominately hard-bottom forms of shallow water. Note the long slender stalks or rootlike processes by which fixation in soft bottoms is made possible. *A, Coleolus suhmi. B, Polycarpa delta. C, Bolteniopsis sessilis. D, Octacnemus ingolfi.* This last species, unlike most other tunicates, is predacious. Prey is trapped by means of the large siphon lobes. (All drawn from photographs by Monniot and Monniot, 1975.)

form. The tadpole larva swims with a long, posterior tail, which contains a notochord and neural tube (Fig. 20–15*A*). Dorsally, the anterior half of the larva* contains a pigmented cup, a statocyst, and the dilated end of the neural tube, which becomes the cerebral ganglion. The mouth, which will become the buccal siphon, is located anteriorly but may not be open during larval life. The mouth leads into the pharynx, which in turn is followed by a twisted digestive loop with a dorsally directed intestine. The pharynx is perforated by a few clefts that open into a surrounding pocket, the atrium. The atrium exits dorsally through what will become the atrial siphon.

At the end of the free-swimming stage, the larva settles to the bottom, attaching by three anterior

*This description is based on aplousobranch larvae. The larvae of many tunicates are less well developed, and differentiation of structures occurs only during metamorphosis.

fixation papillae (Fig. 20–16). A radical metamorphosis now ensues, in which the tail with the notochord and neural tube are resorbed. As a result of rapid growth of the area between the adhesive papillae and the mouth, the entire body is rotated 180 degrees. The mouth, or buccal siphon, is shifted backward to open at the end opposite that of attachment. The atrium expands to capture the anus, and the number of pharyngeal clefts rapidly increases. The siphons become functional, and the metamorphosed larva has become a young ascidian (Fig. 20–15*D*).

Body Wall, Atrium, and Pharynx

The body of an ascidian is covered by a single layer of epithelial cells, but this epithelial covering does not form the external surface. Instead, the entire body is invested with a special covering, the tunic, which is characteristic of most members of the subphylum and from which the name *tunicate* is derived (Figs. 20–7*A* and 20–9*A*). The tunic is usually quite thick but varies from a soft, delicate consistency to one that is tough and similar to cartilage. The tunic of *Amaroucium stellatum,* called sea pork, has both the appearance and the texture of salted pork. The tunic may be colored and commonly looks and feels like marble or glass. Not infrequently the tunic is translucent, and the colors of the internal organs account for the coloration of the animal.

The tunic is composed of water and varying amounts of proteins and carbohydrates (see review by Welsch, 1984). Curiously, a principal fibrous constituent is a type of cellulose, called tunicin, but there are some species in which it occurs in relatively small amounts or may even be absent. In some species calcium salts are precipitated in the form of distinct spicules. Another unusual feature of the tunic is the presence of ameboid cells and blood cells that have migrated from the body mesenchyme. Moreover, in some species, such as *As-*

cidia, the tunic is supplied with blood vessels. It is by means of the tunic that ascidians adhere to the substratum, and the tunic is often roughened or papillose in this region. Often, rootlike extensions called stolons ramify from the base of the body, and these too are covered by the tunic (Fig. 20–10*A*) The tunic functions as an external supportive and protective skeleton, but it is a living, not a dead, skeletal envelope and is a little like connective tissue.

Within the tunic the body of an ascidian can be conveniently divided into three regions—an anterior or distal pharyngeal region containing the pharynx, an abdominal region containing the digestive tract and other internal organs, and a postabdomen (Fig. 20–10*A*). The postabdomen, which may be very long, is the most basal part of the body and contains the heart and reproductive organs (Fig. 20–10*B*). The three regions are distinct only in primitive species. Most ascidians lack a postabdomen, and many species (*Ascidia, Styela, Mol-*

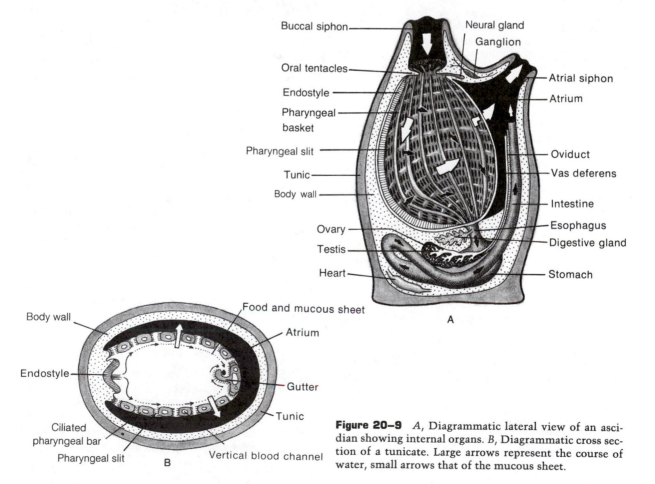

Figure 20–9 *A,* Diagrammatic lateral view of an ascidian showing internal organs. *B,* Diagrammatic cross section of a tunicate. Large arrows represent the course of water, small arrows that of the mucous sheet.

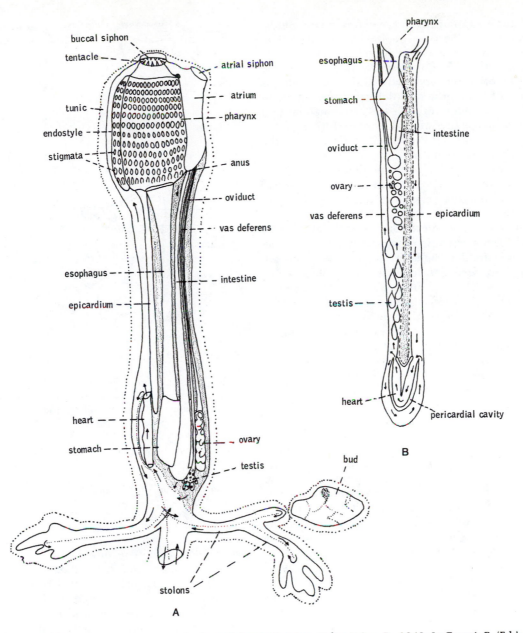

Figure 20–10 *A, Clavelina,* an ascidian with a long abdominal region. (After Brien, P., 1948: *In* Grassé, P. (Ed.): Traité de Zoologie. Vol. 11. Masson et Cie, Paris.) *B,* Abdominal and postabdominal region of *Sydnium.* (After Berril from Brien.) Arrows indicate paths of circulating blood.

gula) even lack an abdomen, the visceral organs being located to one side of the pharyngeal basket.

The anterior buccal siphon opens internally into a large pharyngeal chamber, and a circlet of tentacles projecting from within the siphon prevents large objects from coming in with the water current (Fig. 20–9A). The walls of the pharynx are perforated with small slits, permitting water to pass from the pharyngeal cavity into the surrounding atrium and then out by way of the atrial siphon.

The pharynx is completely surrounded by the atrium except along the midventral line, where the pharynx is attached to the body wall. In addition, the atrium is crossed by cordlike strands of tissue that apparently limit the expansion of the cavity during the flow of water through the body. Dorsally, the atrium opens to the exterior through the atrial siphon. The atrial region just inside the siphon is sometimes called the cloaca because the anus and the gonoducts empty here. All of the atrium, both the pharyngeal and the outer sides, is lined with an epithelium derived from ectoderm and is continuous through the atrial siphon with the external epidermis. The inner lining of the pharynx is thus derived from entoderm and the outer covering from ectoderm; between these two layers lies mesenchymal tissue. The body wall in the atrial region consists of an inner and an outer layer of ectodermal epithelium and mesenchyme in between.

The body wall mesenchyme contains striated muscle bands extending longitudinally toward the siphons. Circular bands are also present and are particularly well developed in the siphon walls, where they act as sphincters. The body wall muscles can cause a limited degree of general body contraction, depending on the thickness and rigidity of the tunic. When an animal is exposed at low tide or taken out of the water, contractions of the body and siphons cause the water in the pharynx and atrium to be forced from the siphons as jets—hence the name *sea squirt*.

Nutrition

Tunicates are filter feeders and remove plankton from the current of water that passes through the pharynx. The water current is produced by the beating of lateral cilia on the margins of the stigmata, and an enormous quantity of water is strained for food. A specimen of *Phallusia* only a few centimeters long can pass 173 liters of water through its body in 24 hours.

On the ventral side of the pharynx—the side opposite the atrial siphon—a deep groove, the endostyle, extends the length of the pharyngeal wall (Fig. 20–9*A* and *B*). The endostyle, which is the forerunner of the vertebrate thyroid gland, is the principal center for the elaboration of mucus—a complex mucoprotein containing iodine bound to the amino acid tyrosine. The long flagella at the bottom of the gutter deflect mucus to the side, and the keyhole shape of the organ perhaps ensures proper mixing from different zones of secretion before the final product is driven onto the transverse bars between rows of pharyngeal slits, called stigmata (singular, stigma) (Figs. 20–9*A*, 20–11 and 20–12). Frontal cilia on the bars carry the mucus in the form of a continuous sheet across the pharynx toward the opposite side (Fig. 20–9*B*). During the course of transport across the pharynx, plankton

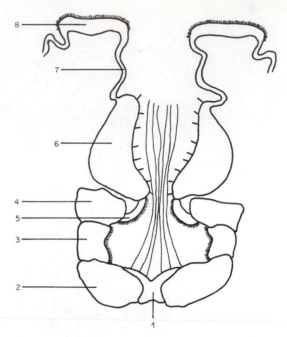

Figure 20–11 Diagrammatic section through the endostyle showing functional zones. Production of the complex iodinated protein filtering sheet begins with secretions from zones 2 and 4, which are mixed in zone 3 and passed to zone 5. More protein is added by zone 6 and iodinated compounds by zone 7. The flagella of zone 1 probably contribute to the outward movement of material from the endostyle. (From Thorpe, A., Thorndyke, M. C., and Barrington, E. J. W., 1972: Ultrastructural and histochemical features of the endostyle of the ascidian *Ciona intestinalis* with special reference to Figure 20, 19:559–571.)

suspended in the stream of water passing through the stigmata becomes trapped on the mucous film, which has a meshlike ultrastructure. Particles as small as 0.1 μm can be filtered out (Monniot, 1979).

The stigmata are arranged in transverse rows with transverse bars separating successive rows and vertical (longitudinal) bars separating adjacent stigmata within each row. The overall structure is therefore much like a grid (Figs. 20–9*A* and 20–12*A*). In the primitive condition the branchial sac has only six rows of stigmata, but most groups have additional rows (because of subdivision) and more stigmata per row. In some groups, including most stolidobranch ascidians, folding of the internal wall has increased the surface area of the branchial sac for filtering food from the passing water (Fig. 20–12*E*). The tendency for the surface area to increase reaches its highest development in the molgulids, in which the orderly, gridlike pattern

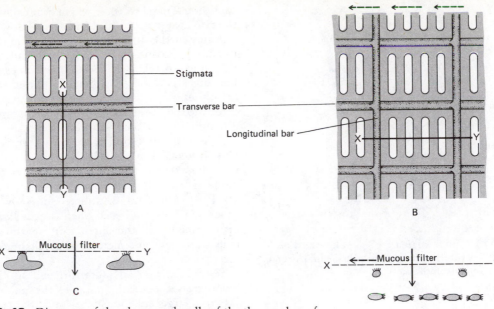

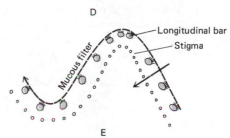

Figure 20–12 Diagrams of the pharyngeal walls of the three orders of ascidians: *A*, Inner surface of the pharyngeal wall of an aplousobranch, in which the mucous filter is carried by raised transverse bars. *B*, Section taken between X and Y. *C*, Inner surface of the pharyngeal wall of a phlebobranch, in which the mucous filter is carried by longitudinal bars. *D*, Section between X and Y. *E*, Section through pharyngeal wall of a stolidobranch, in which the mucous filter is carried on longitudinal bars as in the phlebobranchs but the pharyngeal basket is folded. *F*, Cross section of a stolidobranch, showing folding pharyngeal basket. In all figures, solid arrows indicate direction of water current and dashed arrows direction of mucous filter.

has been replaced by spiralled stigmata (Fig. 20–13). Although the shape of the stigma varies, the distance across the opening remains about twice the length of the lateral cilia. Thus, the cilia on either side of a stigma more or less fill the opening.

In the simplest condition (aplousobranchs) the bars of the branchial grid bear lateral cilia for propelling water and frontal cilia for moving the mucous film. However, in many ascidians (phlebobranchs and stolidobranchs) longitudinal bars with frontal cilia are elevated toward the inside of the pharynx, lifting the mucous sheet off the branchial basket (Fig. 20–12*B*).

The two moving mucous sheets, one on each side of the pharynx, converge onto the dorsal side of the pharynx—the side opposite the endostyle. Here there is a large projecting ridge, the dorsal lamina, or a row of tonguelike processes (languets) that run posteriorly to the esophagus. The margin of the ridge curves to the right and forms a gutter, where the converging, food-laden sheets are rolled

up into cords. The cords are carried posteriorly toward the esophageal opening at the bottom of the pharyngeal basket. Ascidians can halt feeding by closing the buccal siphon and by stopping the ciliary beat or the flow of mucus from the endostyle.

There are a few soft-bottom ascidians that feed on deposit material from surrounding sediment. There are also some deep-water species that feed on small animals such as nematodes and small, epibenthic crustaceans caught with lobes around the buccal siphon (Fig. 20–8*D*). The pharynx is small and glandular with only a few perforations.

The base of the pharynx on the dorsal side contains the esophageal opening, toward which the dorsal lamina is directed. The postpharyngeal part of the digestive tract is located in the abdomen and is arranged in a U-shaped loop (Figs. 20–9*A* and 20–10*A* and *B*).

The esophagus forms the descending arm. The stomach is an enlargement of the digestive loop at the turn of the U. It is lined with secretory cells and

Figure 20–13 Spiral stigmata of *Corella intestinalis.* Heavy black lines are elevated longitudinal bars. (After de Selys Longchamps from Brien.)

is the site of extracellular digestion. The ascending arm of the digestive tract is formed by the intestine, whose terminal end is modified as a rectum and opens through the anus into the cloacal region of the atrium.

In all tunicates there is a network of vesicles and tubules (pyloric glands) on the outer walls of the intestine. By way of one or many collecting canals, this network drains into the proximal end of the intestine. Although they are found in the same region as the excretory vesicles, the pyloric glands appear to be distinct from the excretory vesicles. The products of the pyloric glands are secretory in nature, but whether they are enzymatic, are involved in pH regulation, or have some other function is uncertain.

Some members of the colonial family Didemnidae contain symbiotic algae within the tunic or cloacal region. The alga, called *Prochloron,* is found only in these ascidians and has characteristics of both the blue-green and the green algal divisions (Pardy et al, 1983). Excess photosynthate produced by the algae is assumed to be utilized by the tunicate as an accessory food source.

Epicardium, Internal Transport, and Excretion

Associated with the digestive loop of many tunicates is a structure called the epicardium, which plays a role in development. The epicardium is usually a simple tube that parallels the digestive loop (Fig. 20–10*A* and *B*). It arises as a double, endodermal evagination of the base of the pharynx on

each side. Distal to their origin from the pharynx, the two evaginations fuse and extend downward to one side of the digestive loop. In *Ciona* the distal, unpaired portion has disappeared, but the basal right and left tubes are present, one on each side of the digestive loop. Moreover, these tubes are greatly enlarged, and they surround the stomach, intestine, and other organs in the same manner as a coelom, which they may represent.

The tunicate blood-vascular system is remarkable in many respects. The heart is a short curved or U-shaped tube lying in a pericardial cavity at the base of the digestive loop (Figs. 20–9*A* and 20–10*A*). One end is directed dorsally and the other ventrally, and each opens into a large vessel or channel. All the circulatory pathways lack true walls and are merely sinus channels in the mesenchyme.

From the ventral (anterior) end of the heart blood passes beneath (outside) the endostyle by way of a large, subendostylar sinus that runs along the ventral side of the pharynx supplying branches to the bars of the pharyngeal basket. From the branchial channels blood is delivered to channels supplying the digestive loop and other visceral organs. These channels eventually drain into the dorsal end of the heart. In those ascidians, such as *Ciona,* in which the tunic is provided with blood vessels, a vessel to the tunic usually arises from each of the main channels at the opposite ends of the heart.

An unusual feature of the tunicate circulatory system is the periodic reversal of blood flow. For 2 to 3 minutes the heart pumps blood out of its dorsal end into the dorsal abdominal sinus, and blood enters the heart from the subendostylar sinus. After a slight period of rest, the direction of contraction is reversed, and blood is pumped in the opposite direction. There are no heart valves, but pericardial fluid pressure appears to form a functional valve at the intake end of the heart. Contraction is myogenic, and there is an excitation center at each end of the heart that is responsible for initiating contractile waves over the heart from that point, each alternating in dominance over the other. Rising back pressure may generate the necessary stimulus for beat reversal, but why reversal should be necessary is unknown.

The blood of tunicates contains several kinds of cells that can be grouped into four categories: lymphocytes, which are formed in the mesenchyme and give rise to all other types of cells; nutritive phagocytic amebocytes; morula cells; and storage cells (Goodbody, 1974).

The morula cells, so called because of their berry-like appearance under certain conditions, are involved in tunic formation. In the Ascidiidae and Perophoridae the morula cells (vanadocytes) are green and contain vanadium that appears to be loosely bound with sulfuric acid and protein. Considering the low concentration of vanadium in sea water (0.3 to 3 parts per million), the ability of these ascidians to take up and concentrate vanadium to high levels (3700 parts per million) is remarkable. The element is removed from the sea water by cells of the pharyngeal basket. The morula cells of the Pyuridae contain iron instead of vanadium. Morula cells disintegrate in the tunic, and the vanadium and iron compounds function as reducing agents in the deposition of tunicin microfibrils. Vanadium is found in some other ascidians, either in other cells or at lower concentrations, but its function in these species is still unclear.

The storage cells, or nephrocytes, are filled with crystals or pigment of various colors. After accumulating waste in the course of circulation, the laden cells in most tunicates tend to become fixed in certain regions of the body, such as the mantle, digestive loop, and gonads.

Nervous System

The nervous system is relatively simple and consists of a cylindrical to spherical cerebral ganglion, or "brain," located in the body wall between the two siphons (Figs. 20–9*A*). The nerves arising from the anterior end of the ganglion supply the buccal siphon and mantle musculature; those issuing from the posterior end innervate the greater part of the body—the atrial siphon, the mantle musculature, the pharyngeal basket, and the visceral organs.

Beneath the cerebral ganglion lies a glandular body called the neural gland (Fig. 20–9*A*). The lumen of the gland extends anteriorly as a duct opening into the pharynx by way of a large, ciliated funnel. The gland exhibits cyclical histological activity and is exocrine, not endocrine, but its function is unknown.

There are no special sense organs in ascidians, but sensory cells are abundant on the internal and external surfaces of the siphons, on the buccal tentacles, and in the atrium, and very likely play a role in controlling the current of water passing through the pharynx.

Colonial Organization

Most of the larger ascidians, such as *Styela*, *Ascidia*, and *Molgula*, are solitary forms and are often called simple ascidians. There are, however, many colonial, or compound, species. Colonial organization has arisen independently a number of times

within the class, and a number of types of colonies occur. In general the individuals composing a colony are very small, although the colony itself may reach a considerable size.

In the simplest colonies the individuals are separated but are united by stolons. For example, in *Perophora* the colony is like a vine with a long, trailing, branching stolon to which are attached the globular individuals (Fig. 20–14*A*). In others of these compound forms, such as species of *Clavelina*, the stolons are short, and the individuals form tuftlike groups (Fig. 20–7*B*).

A more intimate association is seen in some ascidians, in which not only the stolons but also the basal parts of the body tunics are joined to other individuals.

In the most specialized colonial families, all the individuals composing the colony are completely embedded in a common tunic. Usually, there is a very regular arrangement of individuals within the test. *Botryllus* and other species form flat, encrusting colonies in which the members are organized in a star-shaped pattern (Fig. 20–14*B* and *C*). The buccal siphons of each member open separately to the exterior, but the atria open into a common cloacal chamber, which has one aperture in the middle of the colony. The individuals of *Botryllus* are only a few millimeters in diameter, but since a single tunic may contain a number of star-shaped clusters, the entire colony may be 12 to 15 cm in diameter.

Asexual Reproduction

Regeneration and asexual reproduction are highly developed in colonial ascidians but absent in many families. Asexual reproduction takes place by means of budding but is complex and exceedingly variable, perhaps more so than in any group of metazoans.

A tunicate bud is called a blastozooid and originates in different parts of the body in different groups of ascidians (see review by Nakauchi, 1982). With the variation of the site of bud formation, there is a corresponding variation in the germinal tissues included within the bud. The most primitive type of budding appears in *Clavelina* and other forms, in which the bud arises from the stolon (Fig. 20–10*A*). In other families the buds may arise in the abdomen or postabdomen or even precociously in the larval stage. The individuals produced by such primary buds are eventually freed from the budding parent. The star-shaped pattern of many adult colonies results from secondary budding from the body wall.

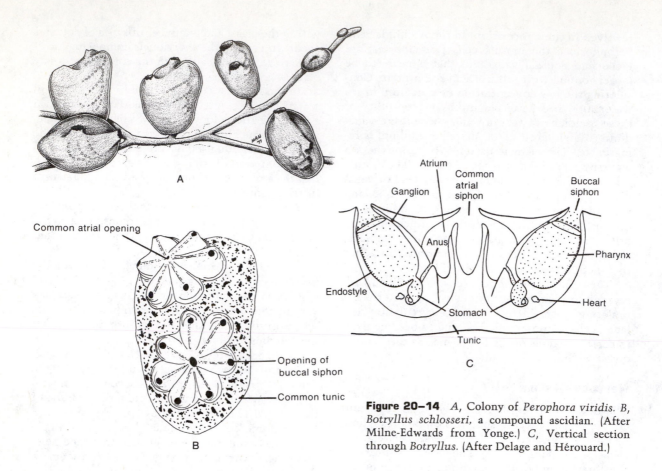

Figure 20–14 *A*, Colony of *Perophora viridis*. *B*, *Botryllus schlosseri*, a compound ascidian. (After Milne-Edwards from Yonge.) *C*, Vertical section through *Botryllus*. (After Delage and Hérouard.)

Sexual Reproduction

With few exceptions tunicates are hermaphroditic. There are usually a single testis and a single ovary that lie in close association with the digestive loop (Figs. 20–9*A* and 20–10*A* and *B*). The oviduct and sperm duct run parallel to the intestine and open into the cloaca in front of the anus. In the order Stolidobranchia (*Molgula* and *Styela*) one to many gonads are usually located in the body wall and not in the intestinal loop.

Solitary ascidians generally have small eggs with little yolk. The eggs are shed from the atrial siphon, and fertilization takes place in the sea water. The eggs of such oviparous species are frequently surrounded by special membranes that act as flotation devices. The eggs of colonial species are typically richer in yolk material and may be brooded in the atrium, which sometimes contains special incubating pockets. Hatching usually takes place at the larval stage, and the larva then leaves the parent; but the entire course of development may take place within the atrial cavity. In general,

development in brooding species with considerable yolk material is more rapid and condensed than in nonbrooding forms.

Cleavage is complete and leads to a coeloblastula. Gastrulation is by epiboly and invagination, and the large archenteron obliterates the old blastocoel. The blastopore marks the posterior end of the embryo and closes while the embryo elongates along the anteroposterior axis. Along the middorsal line, the archenteron gives rise to a supporting rod—the notochord. Laterally, the archenteron proliferates mesodermal cells that form a cord of cells along each side of the body. In this respect, development departs from that shown in amphioxus and other deuterostomes, because there is no pouching of the archenteron. A coelomic cavity never appears, nor is there any segmentaion. The ectoderm along the middorsal line differentiates as a neural plate, sinks inward, and rolls up as an internal neural tube.

Continued development leads to the tadpole larva described at the outset of the section (Figs.

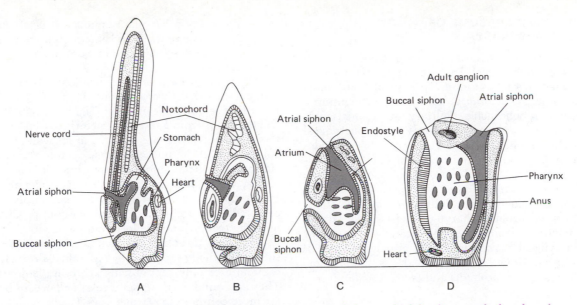

Figure 20–15 *A*, Diagrammatic lateral view of a urochordate tadpole larva, which has just attached to the substratum by the anterior end. *B* and *C*, Metamorphosis. *D*, A young individual just after metamorphosis. (Modified from Seeliger.)

20–15 and 20–16). The only additional features that should be mentioned here are that the larva is covered by a tunic secreted by the surface ectoderm and that in the tail region the tunic is extended dorsally and ventrally to form a fin. The pharynx contains a ventral endostyle. The tadpole larva has a free-swimming period of only 36 hours or less, and in *Botryllus* the larva may settle in a few minutes.

There has been a tendency toward a shorter free-swimming larval stage among tunicates that inhabit sand and mud bottoms. This reduction seems associated with the fact that in such species reaching a suitable substratum is no problem. For species requiring certain kinds of firm substrata for attachment, a free-swimming larva is indispensable.

Most ascidians have a life-span of one to three years, although colonies may have a longer life.

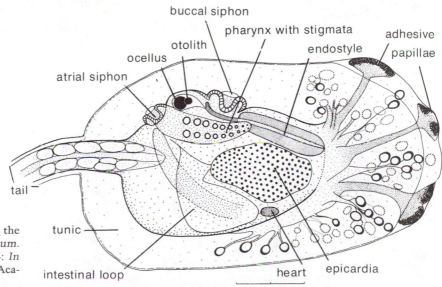

Figure 20–16 Structure of the tadpole larva of *Amaroucium.* (From Godeaux, J. E. A., 1974: *In* Chemical Zoology, Vol. 8. Academic Press, N.Y. pp. 4–60.)

SYSTEMATIC RÉSUMÉ OF CLASS ASCIDIACEA

Order Aplousobranchia. Colonial tunicates with simple pharyngeal baskets. Epicardia present. Gonads within gut loop. Some species with postabdomen. *Clavelina, Amaroucium, Didemnum.*

Order Phlebobranchia. Solitary or colonial tunicates with longitudinal vessels within pharyngeal basket. Gonads within gut loop. No postabdomen. Gut loop behind or to one side of pharyngeal basket. *Diazona, Perophora, Ascidia, Ciona, Phallusia, Corella.*

Order Stolidobranchia. Solitary or colonial tunicates with folded pharyngeal baskets containing longitudinal bars. Gut loop located to one side of pharyngeal basket. Epicardia absent. *Styela, Polycarpa, Molgula, Botryllus;* the macrophagous, deep-sea *Hexacrobylus* and *Gasterascidia.*

CLASS THALIACEA

The other two classes of tunicates, Thaliacea and the Larvacea, are both specialized for a free-swimming, planktonic existence. Thaliaceans, often called salps, differ from ascidians in having the buccal and atrial siphons at opposite ends of the body. The water current is thus utilized not only for gas exchange and feeding but also for locomotion. Like many other planktonic animals, thaliaceans are transparent. The class contains only six genera, and most species live in tropical and semitropical waters.

The tropical *Pyrosoma* are brilliantly luminescent colonial thaliaceans, having the form of a cylinder that is closed at one end (Fig. 20–17*A*). The length ranges from a few centimeters to over 3 meters. The individuals are oriented in the wall of the colony so that the buccal siphons open to the outside and the atrial siphons empty into the central cavity, which acts as a common cloacal chamber. *Pyrosoma* is thus organized in a manner similar to some colonial ascidians. Pharyngeal cilia produce both the feeding and the locomotor currents.

Salps and doliolids are thaliaceans in which a solitary, sexually reproducing individual alternates with an asexually reproducing aggregate (colony). They are found in the upper levels of all oceans but are more common in warmer seas. The body of the solitary zooid in both is somewhat barrel shaped, particularly in doliolids (Fig. 20–17*C*). Circular muscle bands, complete in doliolids and incomplete in salps, produce contractions of the body wall that drive water through the atrial cavity. Doliolids have retained the ciliary feeding current, but salps use the current generated by muscle contractions for the feeding current as well. Salps are peculiar in having only two gill clefts, which are so

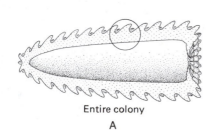

Entire colony

A

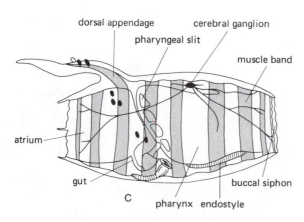

dorsal appendage

pharyngeal slit

cerebral ganglion

muscle band

atrium

gut

buccal siphon

C

pharynx endostyle

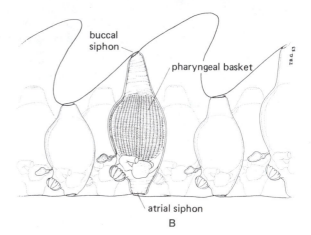

buccal siphon

pharyngeal basket

atrial siphon

B

Figure 20–17 *A,* A colony of the thaliacean *Pyrosoma* (longitudinal section). *B,* Section through part of the colony wall (circled area of *A*) showing three individuals. *C,* Solitary phase of the thaliacean *Doliolum.* Buds develop on the ventral stolon and migrate to the dorsal process, where they attach and grow. (Modified after Uljanin and Barrois from Borradaile and others.)

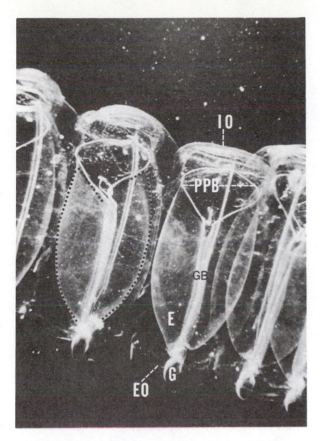

Figure 20–18 Aggregated salps *(Pegea confoederata)* filter feeding. Dotted line outlines mucous net extending back to esophagus (E) from peripharyngeal band (PPB). Rows of cilia on gill bar (GB) appear as bars. Water enters the inhalant opening (IO) and leaves through the exhalant opening (EO) in the region of the gut (G). Exhalant opening is very faint in photograph. Each individual is 6 cm long. (From Madin, L. P., 1974: Field observations on the feeding behavior of salps. Mar. Biol., 25:143–147.)

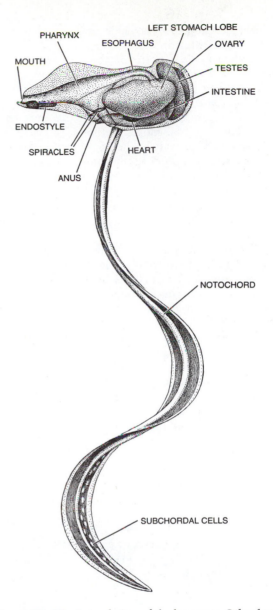

Figure 20–19 Lateral view of the larvacean *Oikopleura albicans*. (From Alldredge, A., 1976: Sci. Am., 235(1)95–102.)

enormous that there are virtually no side walls of the pharynx remaining. They feed with a mucous film, as do ascidians; the film, secreted by the endostyle, projects back into the pharynx from the peripharyngeal bands like a plankton net. As the net moves posteriorly, it is rolled up and passed into the esophagus by esophageal cilia and cilia on the gill bar (Fig. 20–18) (Madin, 1974; Alldredge and Madin, 1982).

CLASS LARVACEA

The Larvacea, or Appendicularia, contain some 70 species of small, transparent animals that are the most specialized of all tunicates. They are found in surface marine plankton throughout the world. The Larvacea are so named because the adults, most of which are about 5 mm long, are neotenic and have retained some larval characteristics (Fig. 20–19). A tail is present, and the body looks somewhat like a typical ascidian tadpole larva bent at right angles or in the shape of a U.

The mouth is located at the anterior of the body, and the intestine opens directly to the outside on the ventral side. There are only two pharyngeal clefts, one on each side, and each opens directly to the exterior.

A remarkable feature of the Larvacea is the "house" in which the body is enclosed or to which it is attached (Fig. 20–20). There is no cellulose tunic in larvaceans, but the surface epithelium secretes a delicate gelatinous material that in several genera completely encloses the body. In *Fritillaria* the animal lies outside of the house but is attached beneath it. In *Oikopleura* the somewhat spherical, gelatinous enclosure is about the size of a walnut and thus much larger than the body of the animal. But some giant species with houses 100 cm in diameter have recently been reported (Barham, 1979).

Of the three families of appendicularians, the common Oikopleuridae is the best known. The interior of the oikopleurid house in which the animal is suspended contains a number of interconnecting passageways and both an incurrent and an excurrent orifice (Fig. 20–20). By beating its tail, the animal creates a water current that passes through the house. The orifice through which water enters is covered by a grid or screen of fine fibers that keep out all but the finest plankton. During its passage through the house, the water is strained a second time through two very fine filters. The collected plankton is pumped into the mouth through a long mucous tube and upon entering the body is collected in the mucous film produced by the endostyle. Appendicularians feed on very fine phytoplankton (nannoplankton) that is unavailable to many other planktonic filter feeders. They exhibit very efficient clearing rates; one animal can remove 250,000 phytoplankton cells from the 300 ml of water filtered through the house each day (Paffenhöfer, 1973). The house is continually shed and replaced, and a single house is kept no longer than 4 hours in *Oikopleura*. House construction involves two stages. It is first secreted by trunk epithelium and carried as a compact surface layer in preparation for later use. When the old house is discarded, the new one is expanded in a few seconds by movements of the animal's body and filling with water. One animal can produce 4 to 16 houses per day, depending on temperature and food availability, and up to 46 houses may be secreted during the life span. House production ceases at the time of spawning (Fennaux, 1985). Appendicularians are common planktonic animals and can be very numerous. A bloom of *Oikopleura dioica* off the coast of British Columbia in 1968 reached population densities of 10,000 per cubic meter and colored the water pink (Seki, 1973).

Only sexual reproduction occurs in the Larvacea. Development leads to a free-swimming tadpole larva that undergoes metamorphosis without settling.

UROCHORDATE PHYLOGENY

The traditional view of chordate origins holds that the phylum evolved from attached ancestors related to the sessile echinoderms and pterobranchs

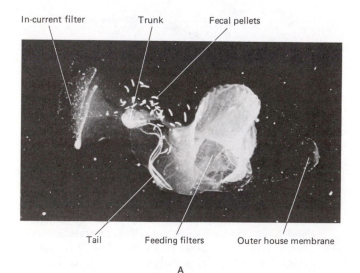

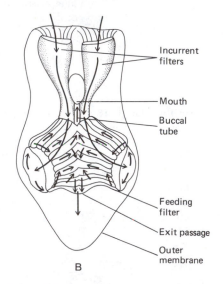

Figure 20–20 *A,* The larvacean *Megalocercus* within its mucous house. (By James M. King.) *B,* Diagram of the house of *Megalocercus.* Large arrows indicate path of water current; small arrows path of filtered particles. Oval represents trunk of animal. (After Alldredge, A., 1976: Appendicularians. Sci. Am., *235*(1):95–102.)

and that the motile cephalochordates and vertebrates arose by neoteny from a larva similar to that of ascidians. There is little evidence to support such an evolution. The so-called calcichordates, an enigmatic fossil group on which this scheme is partly based, are of very questionable interpretation. The ancestral chordates may well have been motile, ciliary feeders (Gans and Northcutt, 1983; Nielsen, 1985). If so, the tunicates undoubtedly departed early from the chordate line of evolution. Most of their adult peculiarities are associated with a sessile mode of existence, but they differ in two basic respects from other chordates. There is no evidence of segmentation, and the coelom is either lost or reduced, in the latter case being represented by the epicardia and renal vesicles. Regarding metamerism, the complete lack of any segmentation, even in embryonic development, seems to indicate that the ancestral chordate was not a metameric animal and that metamerism evolved in the line leading to the cephalochordates and vertebrates *after* the departure of the tunicates.

SUMMARY

1 The great majority (class Ascidiacea) of the some 1300 species of the chordate subphylum Urochordata are adapted for a sessile existence. The tunic, pharyngeal basket, and hermaphroditism can all be correlated with sessility.

2 The external tunic of urochordates is unique in containing cellulose and in housing amebocytes and blood vessels (in many), although it lies external to the epidermis.

3 The pharynx has become highly specialized for filter feeding. A mucous film (the filter) is produced in the endostyle and carried across the inner surface of the pharynx by frontal cilia.

4 Ascidians possess a blood-vascular system that supplies not only the internal organs and the pharyngeal basket but in some species the tunic as well. The system is unique in the periodic reversal of flow through the circuit. Some ascidians possess a type of blood cell (vanadocyte) within which vanadium has been concentrated from the trace amounts in sea water. The cell passes into the tunic where the vanadium compound functions in the deposition of cellulose.

5 Most ascidians are simultaneous hermaphrodites. Fertilization occurs externally or within the atrium. Brooding within the atrium is common.

6 Development leads to a tadpole larva, which possesses all of the chordate characters. Following a free-swimming existence of varying duration, the larva settles and attaches by the anterior end. Metamorphosis involves degeneration of the tail, containing the notochord and dorsal nerve cord. Differential growth results in rotation of the siphons to the end opposite the point of attachment.

7 There are two small classes of pelagic urochordates, the Thaliacea and the Larvacea. The solitary thaliaceans and colonial thaliaceans have the buccal and atrial siphons at the opposite ends of the body. They swim by using the water currents passing through the pharynx and atrium. The current is generated by contraction of the body wall. The larvaceans are neotenous urochordates that live within unique, secreted mucous houses. Plankton is filtered from the water stream that flows through the house.

Phylum Chaetognatha

The chaetognaths, known as arrowworms, are common animals found in marine plankton. The entire phylum of some 70 species is marine, and except for the benthic genus *Spadella*, all arrowworms are adapted for a planktonic existence. Most are found in tropical waters and in the upper 900 meters, but the phylum is represented in the plankton of all oceans. The adults possess none of the features common to the other deuterostome phyla, and they are like aschelminths in many respects. Only the embryogeny of arrowworms would suggest a deuterostome position for these animals.

External Structure

The body of an arrowworm is shaped like a torpedo or feathered dart and ranges in length from 6 to 70 mm (Fig. 20–21). The body is divided into a head, a trunk, and a postanal tail region; a narrowed neck separates the head and trunk. On the underside of the rounded head is a large chamber (the vestibule) that leads into the mouth (Fig. 20–22).

Hanging down from each side of the head and flanking the vestibule are 4 to 14 large, curved, chitinous spines that are used in seizing prey. Several rows of much shorter spines (anterior and posterior teeth) that are curved around the front of the head also assist in capturing prey. A pair of eyes is located posteriorly on the dorsal surface. In the neck region is a peculiar fold of the body wall, the hood, that can be pulled forward to enclose the entire head. The hood is thought to perhaps protect the spines when they are not in use and reduce water resistance during swimming.

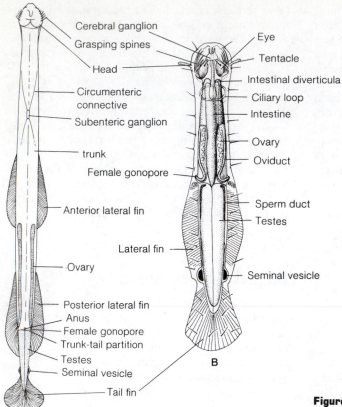

Cerebral ganglion
Grasping spines
Head
Circumenteric connective
Subenteric ganglion
trunk
Female gonopore
Anterior lateral fin
Ovary
Posterior lateral fin
Anus
Female gonopore
Trunk-tail partition
Testes
Seminal vesicle
Tail fin
A

Eye
Tentacle
Intestinal diverticula
Ciliary loop
Intestine
Ovary
Oviduct
Sperm duct
Testes
Lateral fin
Seminal vesicle
B

Figure 20–21 Phylum Chaetognatha: *A, Sagitta elegans* (ventral view). (After Ritter-Zahony). *B, Spadella* (dorsal view). (After Hertwig).

The remainder of the body is composed of the elongated trunk and the tail. A characteristic feature of the chaetognaths is the lateral fins that border these regions of the body. In some arrowworms, such as *Sagitta*, there are two pairs of lateral fins, but in most species there is a single pair (Fig. 20–21). Posteriorly, a large, spatula-like caudal fin encompasses the end of the tail.

Internal Structure and Physiology

The epidermis, which is covered on the outer surface with a thin cuticle, is multilayered except in the head region and contains large, vacuolated cells around the neck (Ahnelt, 1984). A basement membrane lies beneath the epithelium and is thickened to form the supporting rays between the two epithelial layers of the fins. The longitudinal muscles of the body walls are arranged in two dorsolateral and two ventrolateral bands. In the head are special muscles for operating the hood, the teeth, the grasping spines, and other structures.

The compartmented coelom is lined with a thin peritoneum, which around the gut is myoepithelial (Welsch and Storch, 1982). The head contains a single coelomic space that is separated by a septum from the paired trunk coelomic spaces. One or two coelomic compartments occupy the tail.

Chaetognaths alternately swim and float, with the fins acting as flotation devices. When the body begins to sink, the longitudinal trunk muscles contract rapidly, and the animal darts swiftly forward. This forward motion is then followed by an interval of gliding and floating. The benthic *Spadella* adheres to bottom objects by means of special adhesive papillae, but it can swim short distances.

Arrowworms are all carnivorous and feed on other planktonic animals, particularly copepods, which they detect from vibrations. *Sagitta* has reportedly consumed young fish and other arrowworms as large as itself. Arrowworms are voracious feeders. *Sagitta nagae*, for example, consumes 37 per cent of its own weight each day in prey (Nagasawa and Marumo, 1972). In captur-

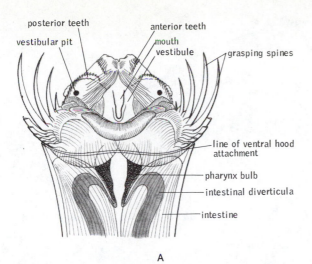

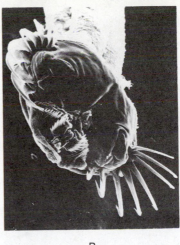

A

B

Figure 20–22 *A*, Head of *Sagitta elegans* (ventral view). (After Ritter-Zahony.) *B*, Anterior-ventral SEM view of head of *Sagitta.* Courtesy of H. Spero, D. Hagan, and A. Vastano.)

ing prey, chaetognaths dart forward or laterally, the hood is withdrawn, and the grasping spines are spread. The prey is seized with the grasping spines and its body is pierced with the teeth, the chaetognath perhaps injecting a toxin produced by the vestibular pits.

The digestive tract is simple (Figs. 20–21*B* and 20–22). The mouth leads into the bulbous, muscular pharynx that penetrates the head-trunk septum to join a straight intestine. The intestine extends through the length of the trunk, and at its anterior gives rise to a pair of lateral diverticula.

After capture the prey is pushed into the mouth, where it is lubricated by pharyngeal secretions and then passed to the posterior of the intestine. Here the food is rotated and often moved back and forth until it is broken down. Digestion is probably extracellular.

There are no gas exchange or excretory organs, and the coelomic fluid acts as a circulatory medium.

The nervous center of chaetognaths is a nerve collar surrounding the pharynx. The ring contains dorsally a large cerebral ganglion, and a number of lateral ganglia, from which extend nerves to more distal parts of the body.

Sense organs include the eyes and sensory hairs (Figs. 20–21*B* and 20–23). The eyes, at least in *Sagitta,* are of the pigment-cup type, rather like those of flatworms (Gotto and Yoshida, 1984). The sensory hairs are arranged in fanlike arrays in longitudinal rows along the length of the trunk and function in detecting water-borne vibrations, as does the lateral line system of fish (Feigenbaum, 1978).

Reproduction

Chaetognaths are hermaphroditic; a pair of elongated ovaries are located in the trunk coelom in front of the trunk-tail septum, and a pair of testes are behind the septum (Fig. 20–21). From each tes-

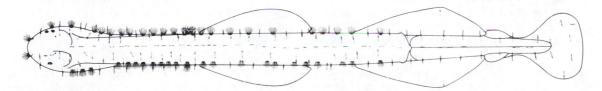

Figure 20–23 Distribution of sensory hair-fans in *Sagitta hispida* (dorsal view). (From Feigenbaum, D. L., 1978: Hair-fan patterns in the Chaetognatha. Can. J. Zool., 56(4):436–546. Reproduced by permission of the National Research Council of Canada.)

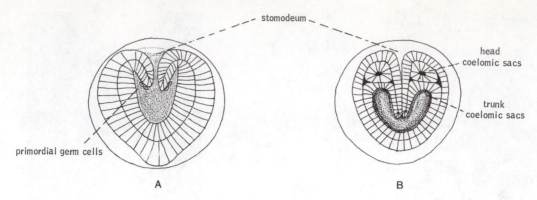

Figure 20–24 Coelom formation in *Sagitta: A,* Initial folding of archenteron walls. *B,* Separation of head coelomic sacs. (Both after Burfield from Hyman.)

tis a sperm duct passes posteriorly and laterally to terminate in a seminal vesicle embedded in the lateral body wall. Sperm leave the testis as spermatogonia, and spermatogenesis is completed in the tail coelom. When mature, the sperm pass into the ciliated funnel of the sperm duct and from there into a seminal vesicle in which the sperm are formed into a single spermatophore. The seminal vesicle ruptures, enabling the spermatophore to escape.

An oviduct runs along the lateral side of each ovary and opens to the exterior through two gonopores, one on each side of the body just in front of the trunk-tail septum. The eggs do not begin to mature until after spermatogenesis has commenced in the tail coelom.

The process of spermatophore transfer is known in detail only for species of *Spadella.* In *Spadella schizoptera* one individual acts as a donor and one as a receiver, and the spermatophore is transferred very rapidly to the recipient's female gonopore after a brief, preliminary signaling behavior (Gotto and Yoshida, 1985).

In *Sagitta* the eggs are fertilized when they pass into the oviduct and are then emitted through the gonopore. The eggs are planktonic and are surrounded by a coat of jelly. In other arrowworms the eggs may be attached to the body surface of the parent and carried about for some time. *Spadella* deposits its eggs in small clusters on algae or other objects.

Cleavage is radial, complete, and equal and leads to coeloblastula. Gastrulation is accomplished by invagination, obliterating the blastocoel. The anterior end wall of the archenteron invaginates, folding backward on each side and cutting off two pairs of lateral, coelomic sacs (Fig. 20–24). The coelom is thus enterocoelic in origin.

Further development is direct, and although the young are called larvae when they hatch, they are similar to the adult, and no metamorphosis occurs.

Despite the similarities of adult chaetognaths to aschelminths, the embryogeny of the phylum appears to be deuterostome in nature. There are, however, some peculiarities. For example, the coelom is enterocoelic in origin but does not arise by a direct outpocketing of the archenteron, and only two pairs of coelomic pockets are formed instead of three. Moreover, there is no larval stage comparable to that of the echinoderms and the hemichordates. Thus, the chaetognaths cannot be allied with any specific deuterostome phylum. If chaetognaths are really deuterostomes, the phylum must have departed very early from the base of the deuterostome line and is only remotely related to the other deuterostome groups. Nielsen (1985), who places chaetognaths with the aschelminths, reviews embryonic and adult features of the group.

SUMMARY

1 Members of the phylum Chaetognatha, called arrowworms, are common, marine planktonic animals. The small, translucent, torpedo-shaped body bears lateral and caudal fins, which enable the animal to glide and float following rapid, dartlike swimming.

2 Chaetognaths are predacious on other planktonic animals, which are seized with grasping spines located on either side of the head. There are no special organs for gas exchange, excretion, or internal transport.

3 Chaetognaths are hermaphroditic with internal fertilization via spermatophores. Development is planktonic and direct.

REFERENCES

The literature included here is restricted to books and papers on urochordates, hemichordates, and chaetognaths. The introductory references on page 8 list many general works and field guides that contain sections on these groups.

Ahnelt, P., 1984: Chaetognatha. *In* Bereiter-Hahn, J., Matoltsy, A. G., and Richards, K. S. (Eds.): Biology of the Integument. 1. Invertebrates. Springer-Verlag, Berlin. pp. 746–755.

Alldredge, A. L., 1976: Appendicularians. Sci. Am., *235*(1):94–102. (An excellent brief account of the Larvacea.)

Alldredge, A. L., 1977: House morphology and mechanisms of feeding in the Oikopleuridae (Tunicata, Appendicularia). J. Zool., London, *181*:175–188.

Alldredge, A. L., and Madin, L. P., 1982: Pelagic tunicates: unique herbivores in the marine plankton. BioSci., *32*(8):655–663.

Alvarino, A., 1965: Chaetognaths. Oceanogr. Mar. Biol. Ann. Rev., *3*:115–194.

Armstrong, W. G., Dilly, P. N., and Urbanek, A., 1984: Collagen in the pterobranch coenecium and the problem of graptolite affinities. Lethaia, *17*(2):145–152.

Barham, E. G., 1979: Giant larvacean houses: observations from deep submersibles. Science, *205*:1129–1131.

Barnes, R. D., 1977: New record of a pterobranch hemichordate from the Western Hemisphere. Bull. Mar. Sci., *27*(2):340–343.

Barrington, E., 1965: Observations of feeding and digestion in *Glossobalanus*. Q. J. Micr. Sci., *82*:227–260.

Barrington, E., 1965: The Biology of Hemichordata and Protochordata. W. H. Freeman, San Francisco.

Berrill, N. J., 1975: Chordata: Tunicata. *In* Geise, A. C., and Pearse, J. S. (Eds.): Reproduction of Marine Invertebrates. Vol. II. Academic Press, N.Y. pp. 241–282.

Brien, P., 1948: Embranchement des Tuniciers. *In* Grassé, P. (Ed.): Traité de Zoologie. Vol. 11. Echinodermes, Stomocordes, Procordes. Masson et Cie, Paris. pp. 553–930.

Bullough, W. S., 1958: Practical Invertebrate Anatomy. Macmillan, N.Y. pp. 446–464. (Descriptions of representative tunicates, including Thaliacea and the Larvacea.)

Cloney, R. A., 1982: Ascidian larvae and the events of metamorphosis. Amer. Zool., *22*:817–826.

Dawydoff, C., 1948: Embranchement des Stomocordes. *In* Grassé, P. (Ed.): Traité de Zoologie. Vol. 11. Echinodermes, Stomocordes, Procordes. Masson et Cie, Paris. pp. 367–551.

Feigenbaum, D. L., 1978: Hair-fan patterns in the Chaetognatha. Can. J. Zool., *56*(4):536–546.

Fennaux, R., 1985: Rhythm of secretion of oikopleurid's houses. Bull. Mar. Science, *37*(2):498–503.

Gans, C., and Northcutt, R. G., 1983: Neural crest and the origin of vertebrates: a new head. Science, *220*:268–274.

Ghirardelli, E., 1968: Some aspects of the biology of the chaetognaths. Adv. Mar. Biol., *6*:271–375.

Goodbody, I., 1974: The physiology of ascidians. Adv. Mar. Biol., *12*:1–149. (An excellent review.)

Gotto, T., and Yoshida, M., 1984: Photoreception in Chaetognatha. *In* Ali, M. A.: Photoreception and Vision in Invertebrates. Plenum Press, N.Y. pp. 727–742.

Gotto, T., and Yoshida, M., 1985: The mating sequence of the benthic arrowworm *Spadella schizoptera*. Biol. Bull., *169*:328–333.

Hadfield, M. G., 1975: Hemichordata. *In* Geise, A. C., and Pearse, J. S. (Eds.): Reproduction of Marine Invertebrates. Vol. II. Academic Press, N.Y., pp. 185–240.

Harbison, G. R., and McAlister, V. L., 1979: The filter-feeding rates and particle retention efficiencies of three species of *Cyclosalpa*. Limnol. Oceanogr. *24*(5):875–892.

Hyman, L. H., 1959: The Invertebrates. Vol. 5. Smaller Coelomate Groups. McGraw-Hill, N.Y. pp. 1–71. (A general account of the chaetognaths.)

Jones, J. C., 1971: On the heart of the orange tunicate, *Ecteinascidia turbinata*. Biol. Bull., *141*:130–145.

Katz, M. J., 1983: Comparative anatomy of the tunicate tadpole, *Ciona intestinalis*. Biol. Bull., *164*:1–27.

Knight-Jones, E. W., 1952: On the nervous system of *Saccoglossus cambrensis*. Phil. Trans. R. Soc. London B, *236*:315–354.

Madin, L. P., 1974: Field observations on the feeding behavior of salps. Mar. Biol., *25*:143–147.

Millar, R. H., 1970: British Ascidians. Synopses of the British Fauna No. 1. Academic Press, London. (Keys and notes for the identification of British species.)

Millar, R. H., 1971: The biology of ascidians. Adv. Mar. Biol., *9*:1–100 (A good review of ascidian biology.)

Monniot, F., 1979: Microfiltres et ciliatures branchiales des ascidies littorales en microscopie electronique. Bull. Mus. Hist. Nat., Paris, Ser. 4, Vol. 1, Sec. A (4):843–859.

Monniot, C., and Monniot, F., 1975: Abyssal tunicates: an ecological paradox. Ann. Inst. Oceanogr., *51*(1):99–129.

Monniot, C., and Monniot, F., 1978: Recent work on the deep-sea tunicates. Oceanogr. Mar. Biol. Ann. Rev., *16*:181–228.

Nagasawa, S., and Marumo, R., 1972: Feeding of a pelagic chaetognath, *Sagitta nagae*, in Suruga Bay, central Japan. J. Oceanogr. Soc. Japan, *28*(5):181–186.

Nakauchi, M., 1982: Asexual development of ascidians: its biological significance, diversity, and morphogenesis. Amer. Zool., *22*:753–763.

Nielsen, C., 1985: Animal phylogeny in the light of the trochaea theory. Biol. Jour. Linn. Soc., *25*:243–299.

Paffenhöfer, G. A., 1973: The conservation of an appendicularian through numerous generations. Mar. Biol., *22*(2):183–185.

Pardy, R. L., Lewin, R. A., and Lee, K., 1983: The *Prochloron* symbiosis. *In* Goff, L. J. (Ed.): Algal Symbiosis. Cambridge University Press, Cambridge, p. 91.

Pennachetti, C. A., 1984: Functional morphology of the branchial basket of *Ascidia paratropa*. Zoomorphology, *104*:216–222.

Plough, H. H., 1978: Sea Squirts of the Atlantic Continental Shelf from Maine to Texas. Johns Hopkins University Press, Baltimore. 118 pp. (A systematic account of western Atlantic species along with much information on the biology and evolution of tunicates.)

Reeve, M. R., and Lester, B., 1974: The process of egg-laying in the chaetognath *Sagitta hispida*. Biol. Bull., *147*(1):247–256.

Seki, H., 1973: Red tide of *Oikopleura* in Saawich Inlet. Mer (Tokyo), *11*(3):153–158.

Stebbing, A. R. D., 1970: The status and ecology of *Rhabdopleura compacta* from Plymouth. J. Mar. Biol. Assoc. U.K., *50*:209–221.

Stebbing, A. R. D., and Dilly, P. N., 1972: Some observations on living *Rhabdopleura compacta*. J. Mar. Biol. Assoc. U.K., *52*:443–448.

Strathmann, R., and Bonar, D., 1976: Ciliary feeding of tornaria larvae of *Ptychodera flava*. Mar. Biol., *34*:317–324.

Welsch, U., 1984: Urochordata. *In* Bereiter-Hahn, J., Matoltsy, A. G., and Richards, K. S. (Eds.): Biology of the Integument. 1. Invertebrates. Springer-Verlag, Berlin. pp. 800–816.

Welsch, U., and Storch, V., 1982: Fine structure of the coelomic epithelium of *Sagitta elegans*. Zoomorph. *100*:217–222.

Index

Page numbers in *italics* indicate illustrations.